AF513045

Fundamentals of Engineering Thermodynamics

—An Innovative Approach

(Revised 8th Edition)

Fundamentals of Engineering Thermodynamics

—An Innovative Approach

(Revised 8th Edition)

Dr. R. Yadav

Ph.D., FIE, M.I.S.T.E

Former Director General, M.I.T., Moradabad (U.P.)

Former Dean., C.E.T., SHIATS, Allahabad

Former Principal, R.I.T., (Now NIT) Jamshedpur

Former Prof & Head, Mech. Engg. Deptt., M.N.R.E.C.

(Now M.N.N.I.T.) Allahabad

Emeritus Professor of Mech. Engg., M.N.N.I.T., Allahabad

and

Dr. Sanjay

BE., ME., Ph.D.

Professor of Mechanical Engineering

N.I.T., Jamshedpur

and

Er. Rajay

B. Tech. (IT BHU, now IIT BHU)

Manager, B.H.E.L., Haridwar

I.K. International Pvt. Ltd.

New Delhi

Published by
I.K. International Pvt. Ltd.
4435-36/7, Ansari Road, Daryaganj
New Delhi-110 002 (India)
E-mail: info@ikinternational.com
Website: www.techsarworld.com

ISBN: 978-93-90620-87-6

© 2023 I.K. International Pvt. Ltd.

All rights reserved. No part of this publication may be reproduced, stored in a retrieval system, or transmitted in any form or any means: electronic, mechanical, photocopying, recording, or otherwise, without the prior written permission from the publisher.

Published/Printed by Krishan Makhijani for I.K. International Pvt. Ltd., 4435-36/7, Ansari Road, Daryaganj, New Delhi-110 002.

Dedicated to

My Late Parents,

Elder Brother and Sister-in-Law

Preface

The present edition of the book is the outgrowth of long teaching and research experience. The purpose is to make a broad based book keeping the following objective in view to:

- present a thorough treatment of Engineering Thermodynamics from classical point of view;
- provide a sound basis for subsequent courses in Thermal Engineering;
- prepare students to use Thermodynamics to Engineering practice;
- give emphasis on fundamentals and real-life applications;
- provide brief design experiences for Thermal Systems;
- develop creativity and engineering judgement among students;

The book contains 21 chapters.

Salient features of the book:

- A large number of typical and real-life solved problems in each chapter with diagrams have been included.
- For an idea, some model solutions of some examples have been given in some chapters which include: **Schematic, Data given, To find, Assumptions, Analysis and Comments on results**.
- Emphasis on physical explanation of equations.
- Emphasis on fundamentals.
- The unsolved problems, include **Things to Think About, Short and Long Theoretical Questions, Numerical Questions. Software Based Questions, Design and Open Ended Questions and Multiple-choice Answer Type Questions.**

The questions given in **Things to Think About** enthuse the students to think about the physics involved and this may enhance the reasoning capability. **Short and Long Theoretical Questions** help in developing the core knowledge of the subject while **Numerical Questions** will enhance the analytical capability in the subject. **Software Based Questions** will enhance the capability of students to develop software for engineering problems by developing models, going through parametric study and finally designing. These questions may take the form of projects. The **Design and Open Ended Questions** provide brief design experience that afford students opportunities to develop their creativity and engineering, judgement, formulate design task statements, apply realistic constraints, and consider alternatives. These questions are such that some of them may lead the students up to high research level. **Multiple-choice Answers Type Questions** put some check on the understanding of the subject.

The depth, richness, emphasis on fundamentals, creativity, innovative approach and judgement enhancement capabilities are the beauties and strength of the book.

Finally, the authors thank Meera Yadav, wife of the main author and mother of last two co-authors who has been a consistent source of encouragement and pillar of strength for the authors and endured certain hardships due to preoccupation of this venture.

R. Yadav
Sanjay
Rajay

Acknowledgements

Sincere acknowledgements and thanks are due to various authors of books and research papers listed in References and elsewhere which were consulted while revising the book.

R. Yadav
Sanjay
Rajay

Contents

Nomenclature

A	area, availability
a	sound velocity; specific availability
B	constant
C	velocity
c	constant
c_p	specific heat at constant pressure
c_v	specific heat at constant volume
D	diameter
E	energy (kJ)
e	specific energy (kJ/kg)
E_x	exergy (kJ/kg)
e_x	specific exergy (kJ/kg)
F	force
G	Gibb's function
H	Helmholtz free energy, total enthalpy, kJ
h	specific enthalpy, kJ/kg; heat transfer coefficient
HCV	higher calorific value of fuel
$h^{\circ}{}_f$	enthalpy of formation
J	mechanical equivalent of heat
K	constant
k	thermal conductivity
LCV	lower calorific value of fuel
LMTD	log mean temperature difference
$\overline{m}$	mass
$\dot{m}$	mass flow rate
N	speed
n	index
P	power
Q	total heat, kJ
q	specific heat, kJ/kg; heat transfer rate
R	gas constant
r	radius
S	total entropy, kJ/K
S_{gen}	entropy generated
s	specific entropy, kJ/kg K
T	absolute temperature, K
t	temperature, °C
U	total internal energy, kJ
u	specific internal energy, kJ/kg
V	volume, m^3
v	specific volume, m^3/kg
W	total work, kJ
w	specific work, kJ/kg
x	dryness fraction
η	efficiency
COP	coefficient of performance
$\in$	effectiveness
η_{ex}	exergetic efficiency
Δh_{R}	enthalpy of reaction
σ	Stefan-Boltzmann constant

CHAPTER

1

Introduction: Basic Concepts, Definitions and Applications

OVERVIEW

Thermodynamics is the science that deals with energy interactions between material systems. In other words, *it is the science which deals with the transformation of energy of all kinds from one form to another*. In fact, one very excellent definition of thermodynamics is that *it is the science of* "*3Es*" namely, **energy, entropy and equilibrium.** Generally speaking, thermodynamics is primarily concerned with two forms of energy, **heat and work**. Carnot, Joule. Kelvin and Clausius were the main scientists who developed this branch of science in the past century. In 1876, J.W. Gibbs advanced classical thermodynamics to such an extent that in its scope it could be applied to almost any physical or chemical phenomena.

The science of thermodynamics is based on the ***four laws of thermodynamics*** known as **Zeroth**, **First, Second and Third laws**. Although, the formulation of these laws is simple but their applications are remarkably extensive. And surprisingly, there is no mathematical proof for these laws. These laws were deduced from experimental observations and are based on logical reasonings. ***The Zeroth law gives the concept of Temperature, the First law gives the concept of Internal Energy while the Second law gives the concepts of Entropy***.

The application of thermodynamics is extremely wide. Its principles are used in the designing of energy converting devices. The credit for proving that mechanical work can be converted to heat energy goes to James Joule and for converting heat into mechanical work to James Watt, the inventor of steam engine.

When we consider matter from the ***macroscopic point of view***, the subject is called ***classical thermodynamics*** and when from the ***microscopic point of view***, it is called ***statistical thermodynamics***. In this book, we will deal mainly with the *classical thermodynamics*.

In this chapter, some basic concepts and definitions of various terminologies related with the study of thermodynamics are presented.

1.1 AREAS OF APPLICATIONS OF ENGINEERING THERMODYNAMICS

Engineers use principles drawn from thermodynamics and other engineering sciences, such as fluid mechanics and heat and mass transfer, to analyse and design products (devices, appliances, machines, systems, etc.) intended to meet human needs. The following are some of the areas of applications of knowledge of engineering thermodynamics.

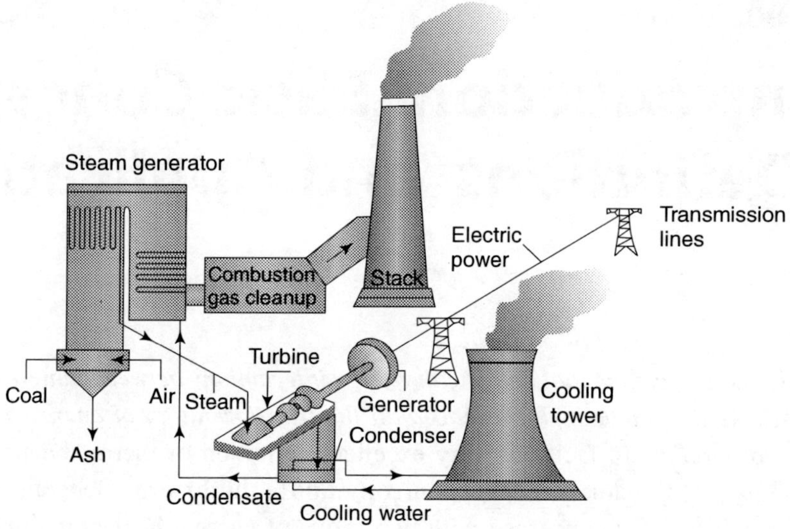

(a) Electrical Power Plant

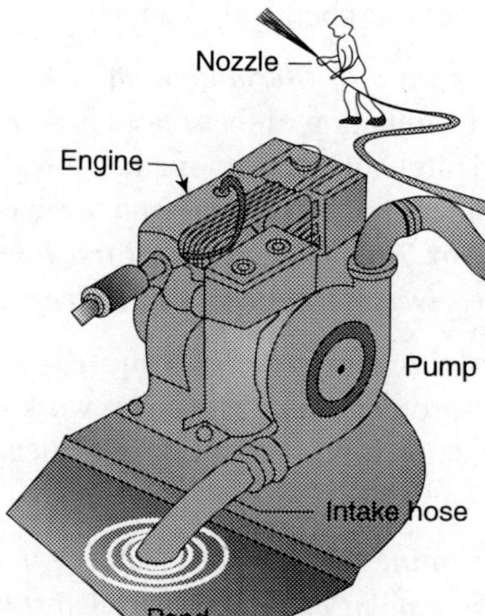

(b) Diesel Engine Pumping Set

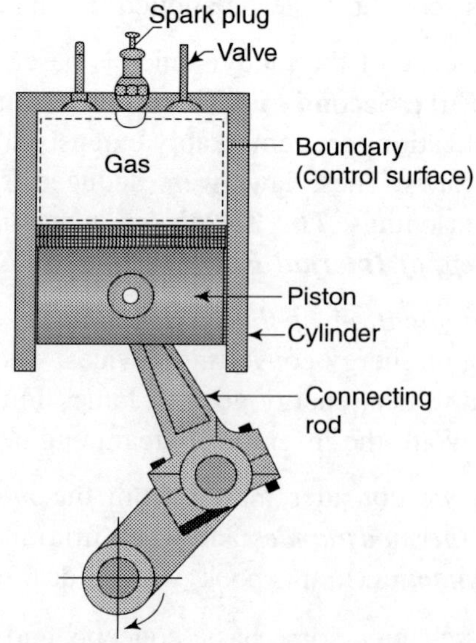

(c) S.I. Combustion Engine Cylinder

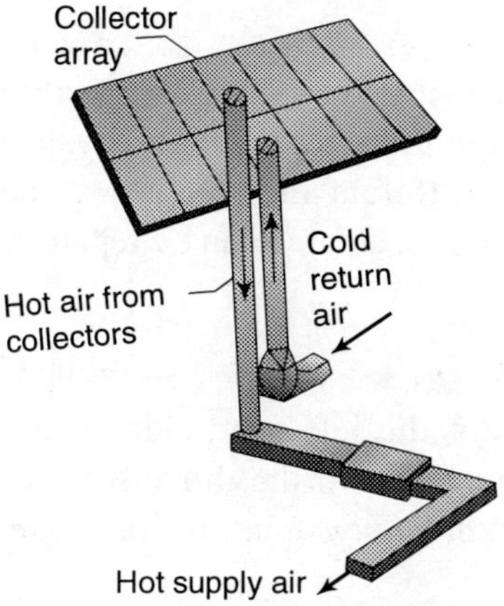

(d) Solar Heating

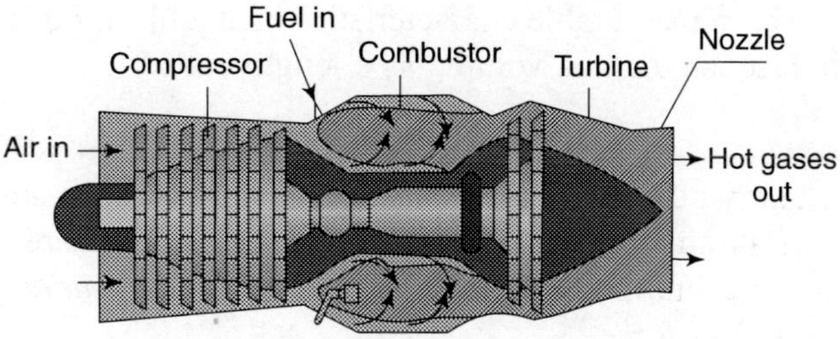

(e) Turbojet Engine

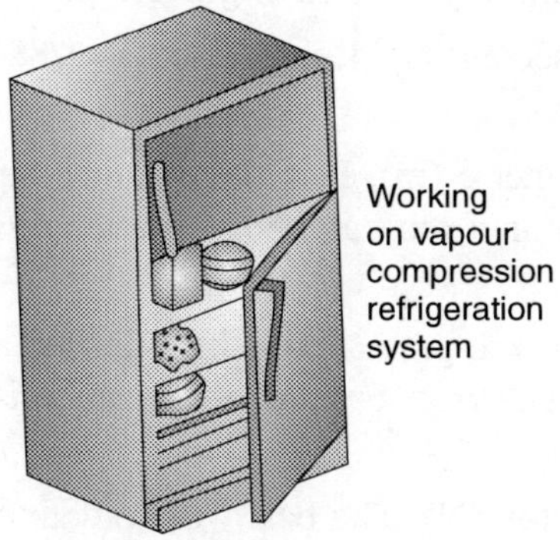

(f) Refrigerator

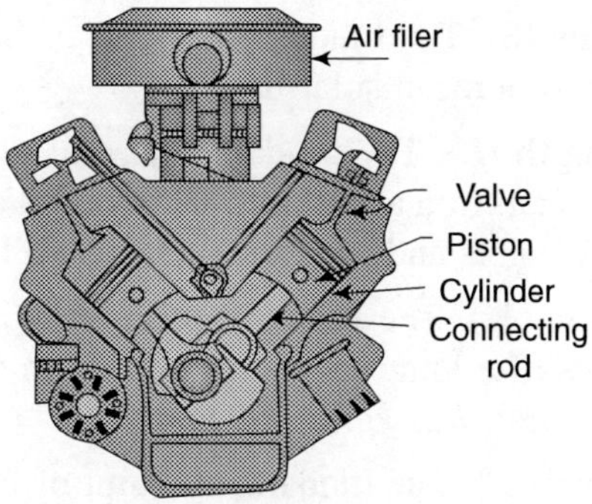

(g) Automobile Engine

Fig. 1.1 Devices where knowledge of thermodynamics is applicable

(i) Automobile engines, (ii) Steam and gas turbines, (iii) Compressors, (iv) Fossil and nuclear fuelled power plants, (v) Propulsion systems for aircraft and rockets, (vi) Combustion systems, (vii) Heating, ventilating and air-conditioning systems, such as (a) vapour compression refrigeration, (b) vapour absorption refrigeration and (c) heat pumps, (viii) Alternative (non-conventional) energy systems such as (a) Fuel cells, (b) Thermoelectric and thermionic devices, (c) Magnetohydrodynamic (MHD) converters, (d) Solar-activated heating, cooling and power generations, (e) Geothermal systems, (f) Ocean thermal wave and tidal power generation, (g) Wind power, (ix) Biomedical applications such as (a) Life support systems, (b) Artificial organs, etc.

Figure 1.1 shows some of the devices where knowledge of thermodynamics is used for analysis and design.

1.2 DIMENSIONS AND UNITS

Dimensions and units are the main tools of engineering. The dimensions of a physical quantity may be defined as the properties in terms of quality not of magnitude, by which a physical quantity may be described. Length, area and volume are all different dimensions which describe certain measurable characteristics of an object. For example, an area can be regarded as a length squared and a volume as a length cubed.

$$(A) = (L^2) \quad \text{and} \quad (V) = (L^3) \qquad ...(1.1)$$

The right hand side of these equations shows dimensions of the left hand side. These equations state that dimensions of an area and volume are equivalent to the dimensions of length squared and length cubed respectively. *The fundamental dimensions are the time, length, mass and force.*

A unit is a definite standard by which a dimension is to be measured, for instance, in this book, we will follow the S.I. units.

The following are the dimensions and the units commonly used in engincering:

(i) Time (θ). The fundamental unit of time is the second (s). It is defined as 1/886, 400 part of a mean solar day.

(ii) Length (*L*). The fundamental unit of length is metre (m). It is the distance between two marks on a platinum-iridium rod at 0°C, kept in a vault at the International Bureau of Weights and Measures at Sevres, France.

(iii) Mass (*M*). The fundamental unit of mass is the kilogram mass (kg), defined as the mass of a lump of platinum iridium also kept at Sevres, France. *Kilogram mass will be denoted as kg in this book.*

(iv) Force (*F*). The fundamental unit of force is newton (N). The best way to define force is the Newton's Second Law of Motion. *According to this law, force is proportional to the product of mass and acceleration.* Thus,

$$F \propto (m)\,(a) \qquad ...(1.2)$$

$$F = \frac{1}{g_c}(m)\,(a) \qquad ...(1.3)$$

where F = force, m = mass, a = acceleration, and $\frac{1}{g_c}$ = a proportionality constant

Let us find the dimension and unit of g_c.

(a) In the S.I. system, the unit of force is newton (N) defined as *that force which will cause a mass of I kg to accelerate* 1 m/s^2. So, by substituting these values in Eq. (1.3), we have

$$1 \text{ newton} = \frac{1}{g_c}(1 \text{ kg})\left\{1\frac{m}{s^2}\right\} \quad \text{or } g_c = \left\{\frac{\text{kg}\cdot\text{m}}{\text{newton}\cdot\text{s}}\right\} \qquad ...(1.4)$$

Writing; dimensions, we have $[g_c] = \left[\dfrac{ML}{F\,\theta^2}\right]$

So in the book, g_c will not be written since $g_c = 1$ in S.I. unit

Therefore, the form of force is expressed as $F = ma$

For S.I. units, see the Appendix.

(b) In M.K.S. system, the unit of force is kg_f. *This is defined as that force which will cause a mass of 1 kg to accelerate* 9.80665 m/s^2, so that

$$1\ kg_f = \frac{1}{g_c}(1\ \text{kg})\left(9.80665\,\frac{\text{m}}{\text{s}^2}\right)$$

or

$$g_c = \left\{9.80665\,\frac{\text{kg}\cdot\text{m}}{\text{kg}_f\cdot\text{s}^2} \approx 9.81\,\frac{\text{kg}\cdot\text{m}}{\text{kg}_f\ \text{s}^2}\right\} \qquad ...(1.5)$$

(c) In the C.G.S. system, the unit of force is dyne. *It is defined as that force which will cause a mass of 1 gram to accelerate* 1 cm/s^2, so that

$$1\ \text{Dyne} = \frac{1}{g_c}(1\ \text{gm})\left(1\,\frac{\text{cm}}{\text{s}^2}\right)$$

or

$$g_c = 1\left(\frac{\text{gm}\cdot\text{cm}}{\text{dyne}\cdot\text{s}^2}\right) \qquad ...(1.6)$$

Writing dimensions, we have $[g_c] = \left(\dfrac{ML}{F\,\theta^2}\right)$

(d) In F.P.S. system, the unit of force is the pound force (lb_f). *It is defined as that force which will cause a mass of 1 pound to accelerate* 32.1739 ft/s^2, so that

$$1\ lb_f = \frac{1}{g_c}\ (l/b_m)\ (32.1739\ \text{ft/s}^2)$$

or

$$g_c = \left(32.179\,\frac{lb_m\ \text{ft}}{lb_f\cdot\text{s}^2}\right) \qquad ...(1.7)$$

Also, $\quad g_c = 1.0$ slug ft/lb_f . s^2

Writing dimensions, we have $[g_c] = \left[\dfrac{ML}{F\theta^2}\right]$

Introduction of g_c reveals that force is a fundamental dimension. **g_c is called the dimensional constant which includes four basic dimensions – mass, length, force and time**.

(e) A complete dimensional system can be built up with only three basic dimensions. For example, force can be expressed in terms for only 3 fundamental dimesnions, [θ], [*L*] and [*M*]. According to this, Newton's Second Law becomes

$$F = (m)\ (a)$$

$$[F] = \left[\frac{ML}{\theta^2}\right]$$

Writing units, we have 1 newton = (1 kg) $\left(1\frac{\text{m}}{\text{s}^2}\right)$

$$1 \text{ newton} = 1\frac{\text{kg}\cdot\text{m}}{\text{s}^2} \qquad ...(1.8)$$

(f) g_c as a Unit Conversion Factor As already discussed, 1 kilogram force will accelerate a kilogram mass at the rate of 9.80665 m/s^2 and 1 newton will accelerate 1 kilogram mass at the rate of 1 m/s^2. So, by implication,

$$1\ \text{kg}_f = 9.80665 \text{ newton.} \approx 9.81 \text{ newton.}$$

Since kg$_f$ and newton are both units of force, we have

$$[F] = [F]$$

$$9.80665\frac{\text{Newton}}{\text{kg}_f} = 1 \qquad ...(1.9)$$

From Eq. (1.8), we have, $9.80665\ \dfrac{\text{kg}\cdot\text{m}}{\text{kg}_f\cdot\text{s}^2} = 1\dfrac{\text{kg}\cdot\text{m}}{\text{N}\cdot\text{s}^2} = 1$ ...(1.10)

In Eq. (1.10), the left hand side is equal to g_c. And since the absolute value of g_c is unity as is obvious from the right hand side of the Eq. (1.10), g_c can be considered as ***a unit conversion factor*** because by unity we can multiply or divide any expression without changing the value. The main point to be noted is that the multiplication or division must be done by the composite factor 9.80665 kg-m/kg$_f$ s or 1 kg . m/N . s^2 and not by numerical value of g_c. To check the unit and dimension, whether, balanced or not, g_c plays an important role and Newton's *Second Law of Motion is not dimensionally balanced unless g_c is put in the denominator.*

1.3 MASS AND WEIGHT

Mass is the measure of quantity of matter as well as the property called inertia. It is true that mass is related to weight but the two terms are different. Weight signifies the sense of force and it refers to the force exerted by gravity on the mass.

Weight is proportional to mass at a particular location but the proportionality factor changes from place to place. The weight of an object varies from place to place but its mass remains (if the velocity is constant) constant. As we know, the value of acceleration due to gravity decreases as we go above or below the surface of the earth, so it follows that the weight of an object on the moon will be 1/6 that of the earth because the value of g on the moon is 1/6 that of the earth.

On the earth (i.e., sea level) that value of mass and weight is the same but with a different sense. According to Newton's law, force due to gravity is

i.e., $$\text{Weight} = \frac{\text{Mass} \times \text{Acceleration due to gravity}}{g_c}$$

$$w = \frac{m \times g}{g_c} \quad ...(1.11)$$

Now, $$F(N) = \frac{\text{kg} \times (\text{m/s}^2)}{\text{kg} \cdot \text{m/Ns}^2}$$

From Eq. (1.11), we have $w/g = m/g_c$

In S.I. unit, $$w = mg \quad ...(1.12)$$

Note that $F = (m)\,(a)$ is only valid in S.I. (unit of force is newton) and in C.G.S. unit (i.e., unit of force is dyne) because g_c is equal to unity in these two cases. It is obvious from the Eq. (1.12) that 1 kg will produce 9.81 N in gravitational field.

Note : Conversion Table from M.K.S. to S.I. units is given at the end of the book in Appendix.

S.I. units of various parameters used in thermodynamics are tabulated as follows in Table 1.1.

Table 1.1 S.I. units of various parameters

Parameters	Unit	Relations
Length	metre (m)	1 m = 100 cm
Mass	kilogram (kg)	1 kg = 1000 g
Force	newton (N)	1 kN = 10^3 N
Time	second (s)	
Volume	m^3	
Pressure	newton/metre^2 (N/m^2) or pascal (Pa)	Standard atmospheric pressure = 76 cm of Hg = 1.01325 barometric = 1.01325 barometric (bar) 1 bar = 10^5 N/m^2 = 10^5 Pa 1 kN/m^2 = 10^3 N/m^2 1 MN/m^2 = 10^6 N/m^2
Velocity	m/s	
Specific volume	m^3/kg	
Density	kg/m^3	

Contd.

Table Contd. 1.1

Heat energy	joule (J) or newton-metre (Nm)	1 kJ = 1000 J
Power	J/s or watt (W)	1 kW = 10^3 W 1 MW = 10^6 W 1 J/s = 1 W
Temperature	kelvin (K)	T = t°C + 273 K

Example 1.1 Calculate the force exerted in M.K.S. and S.I. units and by 10 kg mass because of gravity at a place where the acceleration due to gravity is 9.80665 m/s^2.

Solution We know that

$$F = \frac{(m)(a)}{g_c} = \frac{(m)(a)}{g_c} = \frac{10 \times 9.80665}{9.80665} = 10 \text{ kg}_f \qquad \textbf{Ans.}$$

or

$$F = \frac{10 \times 9.80665}{1} = 98.0665 \text{ N} \qquad \textbf{Ans.}$$

Example 1.2 A certain mass is weighed at the poles where acceleration due to gravity g = 9.85 m/s^2 by a spring balance which was calibrated at sea level where g = 9.80665 m/s^2. The spring balance reads 500 newton. Find the mass of the body.

Solution The spring scale registers newton, hence the body is under the influence of gravitational force of 500 Newton is the weight of the mass at sea level. From Newton's Law

$$F = \frac{mg}{g_c} = mg = 500$$

or

$$m = \frac{500 \times 1}{9.85} = 50.761 \text{ kg} \qquad \textbf{Ans.}$$

1.4 THERMODYNAMIC SYSTEM, SURROUNDING AND UNIVERSE

1. System *The term system is defined as a prescribed region or space of finite quantity of matter surrounded by an envelope called the boundary.* The boundary may be a real physical surface, such as the walls of a vessel, I.C. engine cylinder, etc., or it may be an *imaginary* surface enclosing some matter such as steam, gas, vapour, etc. The boundary may be fixed or it may be moving, as when a system containing a gas is compressed or expanded. Thus, *we can say that the system may be a quantity of steam, mixture of vapour and gas or an I.C. engine cylinder or its contents.*

In a broad sense, the system is defined as a specified region wherein changes due to transfer of mass or energy or both are to be studied. In a system, it is not necessary that the volume or shape should remain fixed.

2. Surrounding *The space and matter external to the thermodynamic system and outside boundary is called the surrounding.*

3. Universe When system and surrounding are put together it is called universe.

In order to understand these, consider a pistone cylinder arrangement shown in Fig. 1.2 in which the system, boundary and surrounding are described. In this case, there are *three fixed*

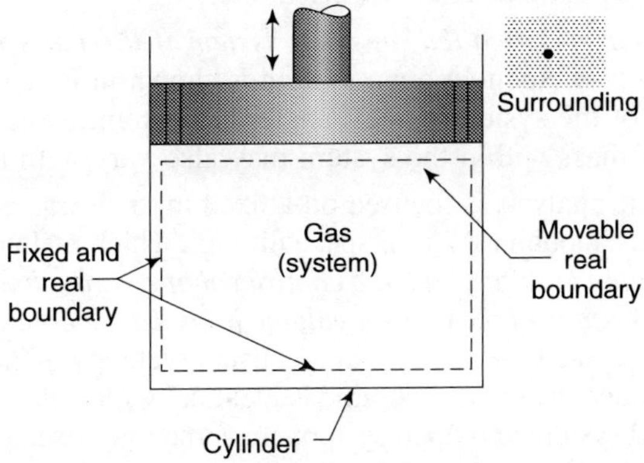

Fig. 1.2 Representation of system, boundary and surrounding

and one movable boundary formed by the bottom surface of the piston. The gas contained in the cylinder and enclosed by the boundaries is the system. All the matter and space external to these four boundaries are known as the surrounding. It is to be noted that all transfers of mass and energy between the system and surrounding are **evaluated at the boundaries**.

1.4.1 Types of System

There are three type of systems : (i) **closed**, (ii) **open** and (iii) **isolated** systems used in thermodynamics.

1. Closed System *A system is called a closed system if the mass within the boundary of the system remains constant and only the energy (heat and work) may transfer across its boundary.* Such a closed system is shown in Fig. 1.3 in which the gas is confined between the piston and the cylinder. The mass of the gas is constant within the system even though

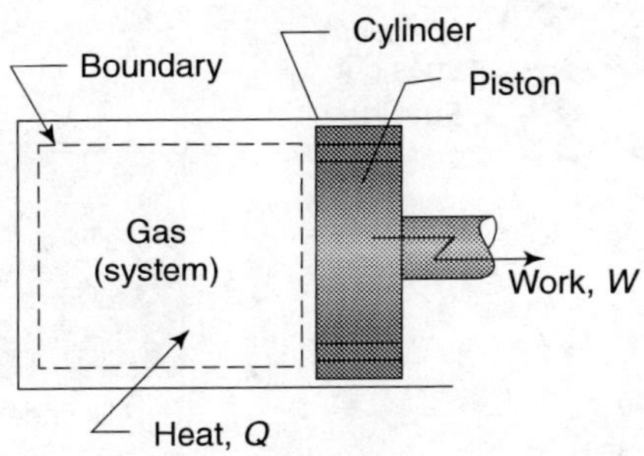

Fig. 1.3 Closed system

the gas may be expanded or compressed because it is closed from all sides, A variation in the volume of the gas may occur. But the transfer of heat and work energy are taking place in the closed system. Other examples of closed system include a *free-body and a point mass as used in mechanics, pressure cooker before whistle, gas contained in piston-cylinder arrangement with non-flow process when both valves are closed, etc.*

2. Open System *A system is called an open system if the mass as well as the energy transfer across its boundaries.* Such an open system is shown in Fig. 1.4 in which the gas is entering and coming out of the system while transfer of heat and work energy is also taking place. The net amount of mass within the system may also vary with time.

In the closed system, the analysis is focused on a fixed mass of matter, whereas in the open system, the analysis centres about a region in space through which matter flows. *It is frequently useful to think of a fixed region in space, called a control volume, through which mass, momentum and energy may flow. The surface of the control volume is called the control surface.* The control volume may be stationary or may be moving at a constant velocity. If no mass transfer across the control surface takes place the control volume is identical with a closed system. It should be noted that the terms closed system and open system are sometimes used as the equivalent of the terms system (fixed mass) and control volume (involving a flow of mass). Control volume is a very useful concept for the study of fluid flows. *Examples of the open systems are compressors, turbines, nozzles, diffusers, steam engines, I.C. engines*, flow through pipes, etc.

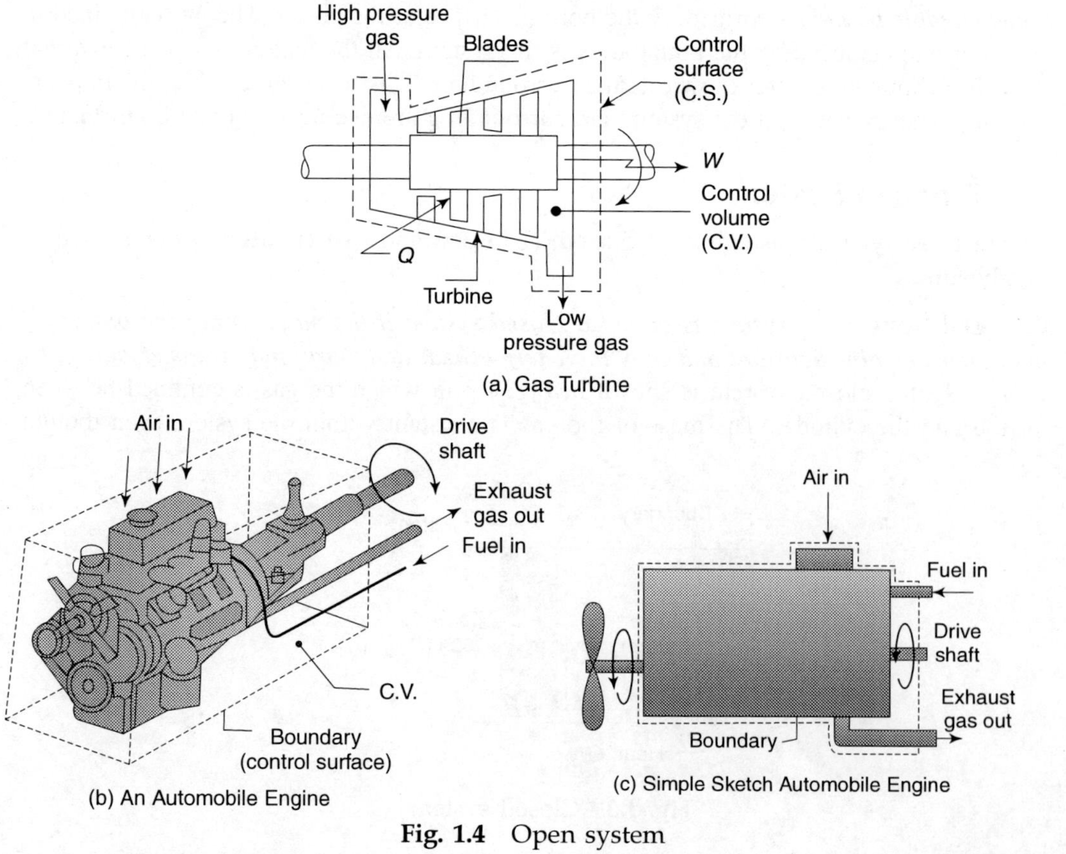

Fig. 1.4 Open system

3. Isolated System *A system is called an isolated system if neither mass nor energy transfers across its boundaries.* Example of isolated system is fluid enclosed in a perfectly insulated closed vessel (thermos flask) as shown in Fig. 1.5.

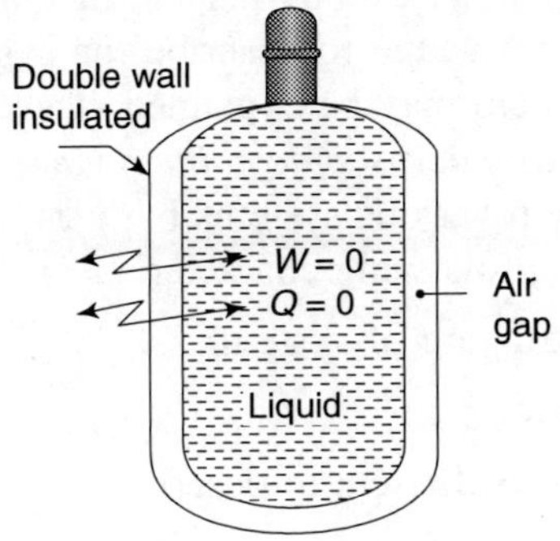

Fig. 1.5 Isolated system (thermos flask)

1.5 PHASE

If a quantity of matter is homogeneous or uniform throughout in physical structure and chemical composition, it is termed as a phase. Homogeneity of physical structure means that the material is either all solid or all liquid or all gas. Homogeneity of chemical composition includes one or more homogeneous chemical species. Thus, a piece of iron, the gas contained in a vessel and a liquid contained in a pot will be called, one phase. A mixture of gases is *one phase*; a liquid and gas is *two phases* (i.e., a liquid phase and a gas phase) and a mixture of solid, liquid and gas is three phases. A system consisting of two miscible liquids is said to be in a *single liquid phase* and system consisting of two immiscible liquids forms *two liquid phases*.

On the basis of phases, the system may be classified as (i) homogeneous system and (ii) heterogeneous system.

(i) Homogeneous System *If the system consists of single phase, it is called homogeneous system.* Examples of homogeneous system are mixture of air and water vapour, solution of NH_3 in water, water and nitric acid, octane and heptane, etc.

(ii) Heterogeneous System If a system consists of more than single phase, it is called heterogeneous system. Examples of heterogeneous system are water and steam, ice and water, water and oil, etc.

1.6 MACROSCOPIC AND MICROSCOPIC POINT OF VIEW

If the analysis of a thermodynamic system is explained by the measurable *properties such as pressure, volume, etc., of the system, it is termed as macroscopic point of view. The microscopic point of view focuses on the statistical behaviour of mass consisting of numerous individual molecules and correlates macroscopic properties of the matter with molecular configuration and with intermolecular forces*. Thus, the microscopic point of view deals with the time average behaviour of the thermodynamic system. The difference between two approaches may be explained as follows.

Consider a gas confined in a cylinder. At any instant, the gas will have a certain pressure, which can be measured by fitting a gauge to describe the system. Similarly, by measuring volume, temperature, etc., the system may be explained. Thus, *explaining the system by the properties of the system is termed as macroscopic point of view*. Further, different macroscopic co-ordinates are required to explain various systems but it is important to remember that all macroscopic co-ordinates have the following characteristics:

(i) They do not explain the structure of system.

(ii) They are readily measurable.

(iii) Only a few co-ordinates are sufficient to describe the system.

According to the *microscopic point of view*, gas is composed of a large number of molecules, say n, each *molecule* having, at a given instant, certain characteristics, such as, velocity, momentum and position. Each molecule may be described by three co-ordinates x, y and z and three components of velocity u, v, and w. This means that **six variables** are required to explain **one molecule** and the system requires ***6n* variables**. Similarly, the pressure exerted by a gas at a given instant depends on the momentary behaviour of the molecules in the neighbourhood of the particular point. It is clear that pressure does fluctuate with time owing to the random motion of the molecules. This concept employs probability, which predicts that the average behaviour of the molecule remains uniform although the behaviour of the molecule does not. Thus, *the microscopic point of view deals with the structures of the system and time average behaviour.*

1.7 CONCEPT OF CONTINUUM

The concept of continuum is very useful in dealing with classical *thermodynamics. From the continuum point of view, the matter is seen as being distributed through space and not, as in the particle view, localised.* Matter exists in "big chunks" having mass, energy and momentum, as do particles, but it also possesses some additional continuum characteristics, such as volume, density and temperature. *In other words, continuum idealisation treats the substance as being continuous disregarding the action of individual molecules.*

From the continuum stand point of view we can speak of properties at a point. For example, let us consider the density of a fluid at a point P which is surrounded by a small size volume $\Delta V'$ and let the mass contained by this volume be Δm. It is obvious from the graph as shown in Fig. 1.6 that if the volume $\Delta V'$ becomes smaller and smaller, the density, i.e., ratio $\Delta m/\Delta V$ deviates very much from its asymptotic value at $\Delta V'$ and the value of $\Delta m/\Delta V'$ is either very large or very small. Thus, the volume $\Delta V'$ of the point P is the limit below which there is a large variation in density, but at $\Delta V'$ and slightly above this, the density may be referred to as **density at a point**. Mathematically, density at a point may be expressed as

$$\rho = \lim_{\Delta V \to \Delta V'} \frac{\Delta m}{\Delta V} \qquad \text{...(1.13)}$$

where $\Delta V'$ is the smallest volume about point P in which the fluid can be called in continuum.

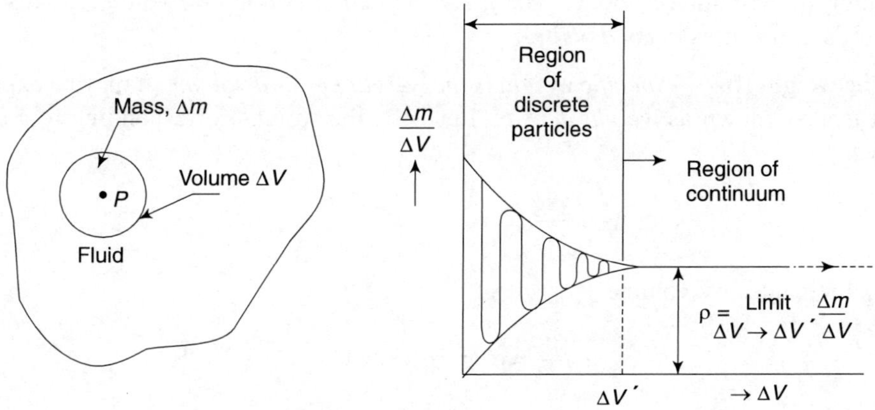

Fig. 1.6 Density at a point

The continuum idealisation is therefore not useful for volumes which are too small. If the spatial variations in $\Delta m/\Delta V$ are too great, a limit may not exist and the continuum method would be invalid.

Other continuum concepts include, mass flow rate, electric current, mass flux and the pressure exerted by a fluid on the walls of a container. In the continuum analysis, calculus is employed. For example, the mass of a body can be expressed as

$$m = \int_{\text{vol.}} \rho \, dV .$$

The above discussion leads into say that *continuum volume is extremely small compared to the dimension of the installation*. But this is *not true in the case of ballistic missiles, supersonic* rockets and earth satellites. At about 2000 km altitude (high vacuum), the mean molecular path of a molecule may be as large as 3.5 m, i.e., approximately equal to the vehicle itself. In *such circumstances (high vacuum), the fluid cannot be treated as continuum.* The microscopic point of view can be applied for the study.

Except for a few exceptions, we can conclude that thermodynamics is a continuum science.

1.8 DENSITY AND SPECIFIC WEIGHT

1. Density (ρ) *The average density of a system is the ratio of its total mass to its total volume.* The density of a continuum of a particle is

$$\rho = \lim_{\Delta V \to \Delta V'} \frac{\Delta m}{\Delta V}; \text{kg/m}^3 \qquad (1.14)$$

Here, the volume ΔV must contain a large number of molecules and yet it must be small compared to the dimension of the system, so that volume $\Delta V'$ is the smallest volume capable of maintaining the continuum model. Here, $\Delta V'$ is called continuum volume as discussed earlier. It is also known as ***mass density***.

2. Specific weight (ρw) *Specific weight is the weight per unit volume.* It may be expressed in N/m^3. It is also known as *weight density*. Let w be the weight corresponding to a certain mass m, then

$$w = \frac{mg}{g_c}$$

Dividing both sides by volume

$$\rho_w = \frac{\rho}{g_c} = \rho \cdot g \cdot \text{as } g_c = 1 \qquad ...(1.15)$$

where ρ_w is the weight denisty and ρ is mass density.

Numerically, specific weight is the same as mass density if g/g_c is unity. On the earth g/g_c is unity in S.I. and C.G.S. unit.

1.9 SPECIFIC VOLUME (v)

The specific volume (v) is defined as the volume per unit mass and may be expressed in m^3/kg. Thus, it is reciprocal of the density.

or
$$v = \frac{1}{\rho}; \text{ m}^3\text{/kg} \qquad ...(1.16)$$

1.10 PRESSURE

The pressure exerted by a system is defined as the force exerted normal to unit area of the boundary. When a fluid is contained within a vessel, the pressure exerted on the vessel is equal to the mean change of momentum of the molecules exerted perpendicular to the confining boundary per unit area, and per unit time. Hence, pressure of fluids is analogous to normal stress in solids.

From the continuum point of view, the pressure at a point is the force per unit area in the limit; where the area tends to be very small, i.e., approaching zero. Thus,

$$p = \lim_{\Delta A \to \Delta A'} \frac{\Delta F}{\Delta A} \text{ N/m}^2 \qquad ...(1.17)$$

where ΔF in N is the force normal to the area ΔA and $\Delta A'$ (m^2) is the smallest possible area capable of maintaining the continuum model.

A fluid is defined as a substance in which a fluid is at rest, only normal stresses exist but when a fluid is in motion shear and tangential stresses exist in addition to compressive stresses. When a fluid is in equilibrium, the pressure becomes a scalar point function, i.e., it becomes *independent of direction, this is called static pressure*. When a fluid is not in equilibrium, the pressure may vary according to direction. Examples of this are *ideal* or *inviscid fluid*.

In the case of an ideal fluid, shear stresses are absent even if there is relative motion within the fluid. Under such a condition, pressure is independent of direction whether the ideal fluid is in motion or at rest.

The following terms are generally associated with the measurement of pressure.

(a) Atmospheric Pressure (p_{atm}) *This is the pressure exerted by the atmospheric air*. The standard atmospheric pressure is defined as the pressure produced by a column of mercury 760 mm high, the mercury density being 13.5951 kg/m^3 and the acceleration due to gravity being 9.80665 m/s^2, the standard atmospheric pressure is 101.325 kN/m^2 or 1.01325 bar or 76 cm of Hg or 1.01325 × 10^2 Pa.

(b) Gauge Pressure (p_{gauge}) *The pressure measured from the gauge and instrument such as Bourdon and manometers is called the gauge pressure*. The gauge actually measures the difference between the fluid pressure and the pressure of atmosphere surrounding the gauge.

(c) Absolute Pressure (p_{abs}) *The pressure measured from the level of absolute zero pressure is called absolute pressure*. Absolute zero pressure will only occur when the molecular momentum of the fluid is zero. Such a situation can only arise if there is a perfect vacuum. A state of perfect vacuum is treated as reference point, hence

$$p_{abs} = p_{atm} + p_{gauge}; \text{ N/m}^2 \quad \text{...(1.18)}$$

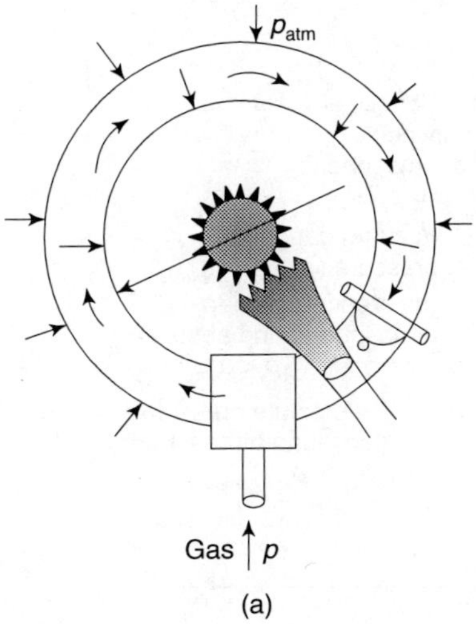

(a)

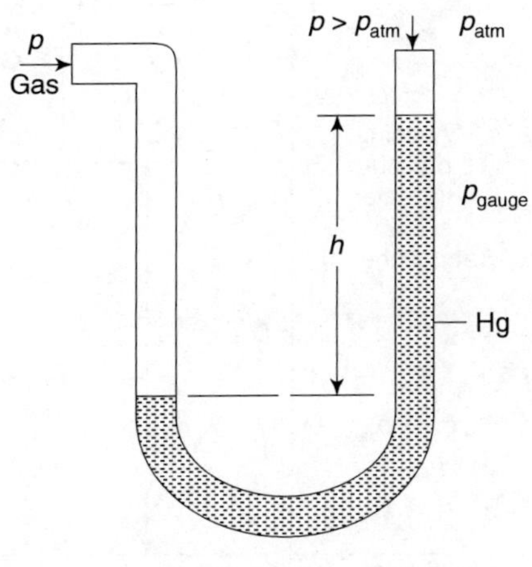

(b)

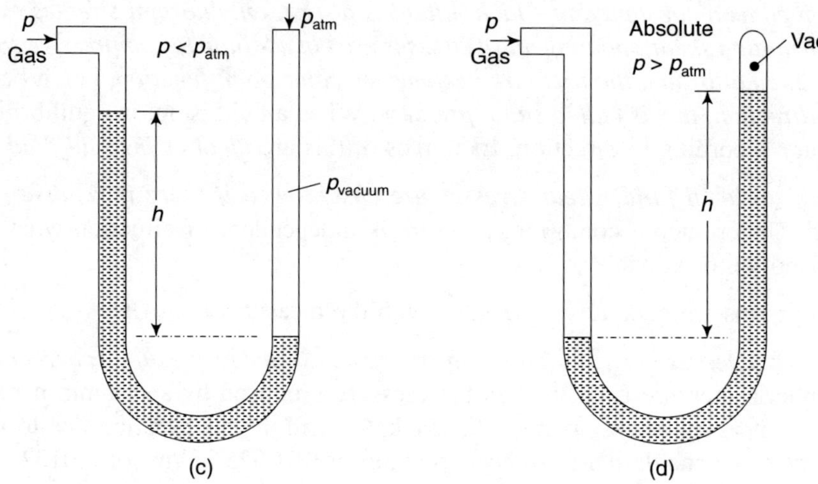

Fig. 1.7 Pressure gauges, (a) Bourdon gauge, (b) Open U-tube indicating gauge pressure, (c) Open U-tube indicating vacuum, (d) Closed U-tube indicating absolute pressure

(d) Vcuum Pressure (p_{vac}) *If the pressure of fluid to be measured is less than atmospheric pressure, the gauge will read on the negative side of the atmospheric pressure which is known as vacuum, rarefaction or negative pressure*, hence

$$p_{abs,\ vacuum} = p_{atm} - p_{vac,\ gauge};\ N/m^2 \qquad ...(1.19)$$

The various pressure gauges are shown in Fig. 1.7.

The relationship of pressure is shown in Fig. 1.8.

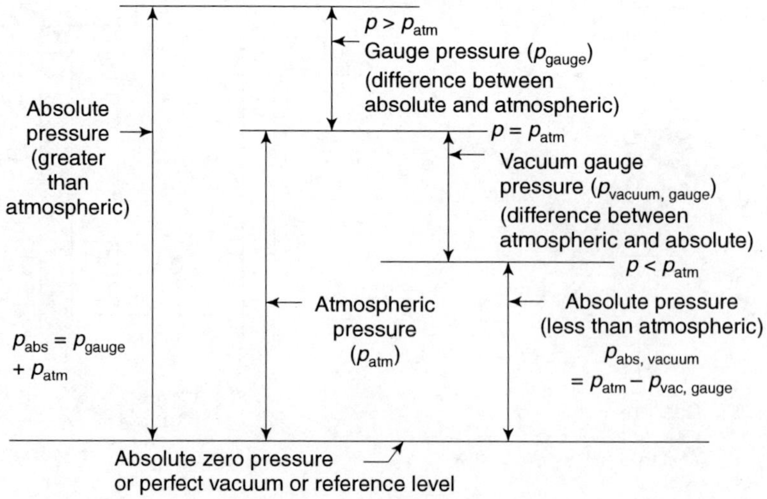

Fig. 1.8 Relationship of pressure

In problems connected with the flow, various types of pressures such as *static, impact* and *velocity* are commonly used.

(i) *Static Pressure* *Static pressure is defined as the pressure sensed by a measuring device (such as a manometer, pitot tube, etc.) moving with the same velocity as the fluid stream.*

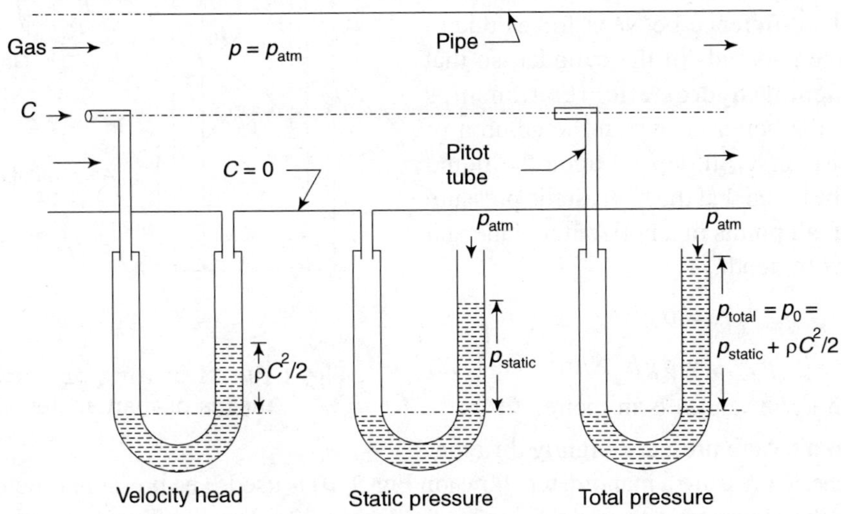

Fig. 1.9 Types of pressure

(ii) *Impact or Stagnation or Total Pressure* *Stagnation pressure is defined as the force per unit area perpendicular to the direction of the force when the fluid is brought to rest reversibly.* For a constant density fluid, the total pressure is given by

$$p_0 = p_{total} = p_{stattic} + \frac{\rho C^2}{2}; \text{ N/m}^2 \qquad ...(1.20)$$

where C is the fluid velocity (m/s) and ρ is the density of fluid (kg/m^3). *The difference between the stagnation and static pressure is the velocity pressure.* The stagnation pressure is measured by an instrument called *pitot tube*.

(iii) *Velocity (Dynamic) Pressure* The velocity or dynamic head is termed as velocity pressure. Thus, velocity pressure

$$p_d = \frac{\rho C^2}{2}; \text{ N/m}^2 \qquad ...(1.21)$$

Figure 1.9 shows the measurement of different types of pressure.

In fluid flow through a duct, the pressure measured at the wall is static pressure because the velocity at the wall is zero. If the fluid particles move parallel to the centre line of the duct, static pressure is uniform across any section of the duct. The pressure measured at the centre of the duct is the stagnation pressure.

1.10.1 Pressure as the Result of a Depth of Fluid

Consider the case of a homogeneous fluid of density ρ in static equilibrium. There exists a pressure difference between two points that are separated by distance h in the vertical direction. Hence, the mass of cylinder of fluid can be equated to the difference between forces due to pressure at the two ends of the cylinder, so that the fundamental hydrostatic relationships governed by the height to which the column of liquid will rise and will depend upon the liquid used. It is to be noted that the hydrostatic pressure is the same at all points in a horizontal plane and varies only with depth.

$$p_{gauge} A = \rho g h A$$

or $$p_{gauge} = \rho g h; \text{ N/m}^2 \quad ...(1.22)$$

where ρ is in kg/m^3 and h is in metre.

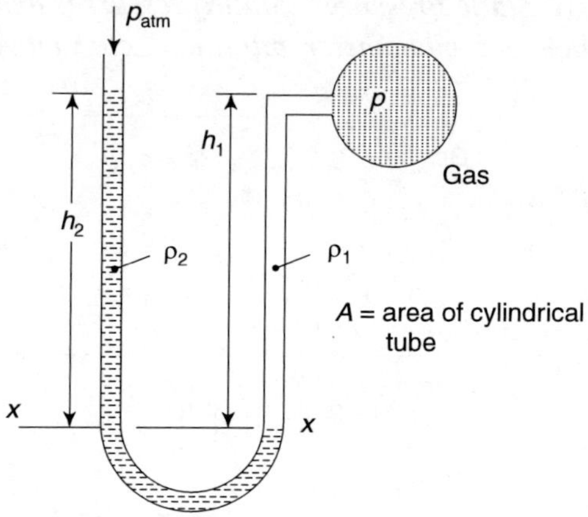

Fig. 1.10 Measuring pressure by means of manometer

In order to measure pressure slightly different from atmospheric pressure a manometer shown in Fig. 1.10 is used. The pressure is determined according to the hydrostatic formula given by the Eq. (1.22). The liquid used in manometer may be mercury, water, or alcohol, etc. Since the manometer fluid is in equilibrium, the pressure along a horizontal line x–x is the same of either branch of manometer. Thus,

$$p + \rho_1 g h_1 = p_a + \rho_2 g h_2 \quad ...(1.23)$$

In general,

$$p_1 + \rho_1 g h_1 = p_2 + \rho_2 g h_2 \quad ...(1.23a)$$

where p is the absolute pressure in the bulb, p_a is the atmospheric pressure exerted on liquid free surface and ρ_1 and ρ_2 are the densities of the fluid in the bulb and the manometer respectively.

If $\rho_1 << \rho_2$, then

$$p - p_a = \rho_2 g h_2 \quad ...(1.24)$$

Example 1.3 A pressure gauge reads 72 cm of Hg, find the gauge pressure in N/m^2.

Solution 76 cm Hg = 101.3 kN/m^2

$$\therefore \quad 72 \text{ cm of Hg} = \frac{101.3}{76} \times 72 = 95.96 \text{ kN/m}^2$$ **Ans.**

Example 1.4 A pressure gauge fixed in a boiler reads 1200 kN/m^2. The barometer reads atmospheric pressure as 735.5 mm of Hg. Determine the corresponding absolute pressure.

Solution From the standard value 76 cm Hg = 101.3 kN/m^2

$$\therefore \qquad 735.5 \text{ mm of Hg} = \frac{101.3}{76} \times 73.55 = 98.03 \text{ kN/m}^2$$

Hence, barometer reads atmospheric pressure equal to 98.03 kN/m^2

Absolute pressure = $p_{gauge} + p_{atm}$ = 1200 + 98.03 = 1298.03 kN/m^2 **Ans.**

Example 1.5 A vacuum gauge fixed in steam condenser reads 71 cm of Hg. The barometric reading is 76 cm of Hg. Calculate the absolute pressure in the condenser.

Solution

$$p_{abs} = p_{atm} - p_{vac} = 76 - 71 = 5 \text{ cm of Hg}$$

$$= \frac{5 \times 101.3}{76} = 6.66 \text{ kN/m}^2 \qquad \textbf{Ans.}$$

Example 1.6 A forced fan supplies air to the furnace of a boiler at a head of 40 mm of water column. Determine the absolute pressure of air supply if the barometric reading is 760 mm of Hg.

Solution Since it is forced fan, hence the absolute pressure will be above atmospheric. We know that $p_{gauge} = \rho_1 g h_1 = \rho_2 g h_2$

13.6 mm of water column = 1 mm of Hg

$$\therefore \qquad 40 \text{ mm of water column} = \frac{1 \times 40}{13.6} = 2.94 \text{ mm of Hg}$$

Hence, $p_{abs} = 760 + 2.94 = 762.94 \text{ mm of Hg} = \dfrac{762.94}{760} \times 101.3 = 101.69 \text{ kN/m}^2$ **Ans.**

Example 1.7 Air having density 1 kg/m^3 is flowing at speed of 60 m/s through a duct. The static pressure is 400 kN/m^2. Find the stagnation or impact pressure.

Solution

$$p_{stagnation} = p_0 = p_{static} + \frac{\rho C^2}{2} = 400 + \frac{1 \times 60^2}{2 \times 1000} = 401.8 \text{ kN/m}^2$$

Example 1.8 Convert the following:

(a) 400 kN/m^2 abs to kN/m^2 gauge.

(b) 50 cm Hg vacuum to cm Hg abs and kN/m^2 abs.

(c) 60 kN/m^2 to cm Hg vacuum.

(d) 30 cm Hg gauge to cm Hg abs and to bar. Assume barometer reads 76 cm of Hg.

Solution Refer to Fig. 1.7. 76 cm of Hg = 101.3 kN/m^2

(a) $p_{gauge} = p_{abs} - p_{atm} = 400 - 101.3 = 298.7 \text{ kN/m}^2$ gauge

(b) $p_{abs} = p_{atm} - p_{vac} = 76 - 50 = 26 \text{ cm of Hg abs} = \dfrac{26 \times 101.3}{76} = 34.655 \text{ kN/m}^2$ abs

(c) $p_{vac} = p_{atm} - p_{abs} = 101.3 - 60 = 41.3 \text{ kN/m}^2 = \dfrac{76}{101.3} \times 41.3 = 30.98$ cm Hg vacuum

(d) $p_{abs} = p_{atm} + p_{gauge} = 76 + 30 = 106$ cm of Hg $= \dfrac{106 \times 101.3}{76} = 141.286$ kN/m^2 abs.

Example 1.9 A vessel is shown in Fig. 1.11 in which there are two compartments at different pressures. The pressure gauge *A* reads 4 bar gauge and *B* reads 2 bar gauge. The barometer reads 760 mm of Hg, calculate the reading of gauge *C*.

Solution We know that the pressure gauge reads the gauge pressure, i.e., it reads the difference in pressures between the inside and outside region in which the dial is located. Let *X* and *Y* be the absolute pressures in bar in both compartments:

$$760 \text{ mm of Hg} = 1.013 \text{ bar}$$

$$\therefore \quad Y - 1.013 = \text{reading on gauge } A = 4 \text{ bar gauge} \quad ...(1)$$

$$Y - X = \text{reading on gauge } B = 2 \text{ bar gauge} \quad ...(2)$$

$$X - 1.013 = \text{reading on gauge } C \quad ...(3)$$

From Eq. (1), $Y = 5.013$ bar abs

Putting the value of *Y* in Eq. (2), we have, $5.013 - X = 2$

or $X = 3.013$ bar abs

Putting the value of *X* in Eq. (3); the reading of the pressure gauge *C*

$$= 3.103 - 1.013 = 2.0 \text{ bar gauge}$$ **Ans.**

The absolute pressure for reading *C* is given by

$$= 2.0 + 1.013 = 3.130 \text{ bar}$$ **Ans.**

Example 1.10 Generally, the atmospheric pressure air is measured by mercury barometer. It is calibrated for 0°C. The density of mercury and length of the scale change with temperature. For actual pressure of air, the values read on barometric scale should be corrected for 0°C. The correction formula is given by

$$B_0 = B\,(1 - 0.000172\,t)$$

B = corrected barometer pressure at 0°C.

where, t = temperature in °C

For this case, calculate the absolute pressure if the vacuum gauge on the condenser fitted in steam turbine condenser reads 65 cm of Hg. The barometer reads 75 cm of Hg at a temperature of 20°C in the boiler house.

Suddenly, if the barometric pressure drops to 66 cm of Hg and pressure in condenser is constant, find the vacuum gauge pressure.

Solution We have

$$B_0 = B\,(1 - 0.000172\,t) = 75\,(1 - 0.000172 \times 20) = 74.742 \text{ cm of Hg}$$

$$\therefore \quad p_{abs} = p_{atm} - p_{vac}$$

$$= 74.742 - 65 = 9.742 \text{ cm} = \frac{9.742}{76} \times 1.013 = 0.12985 \text{ bar}$$ **Ans.**

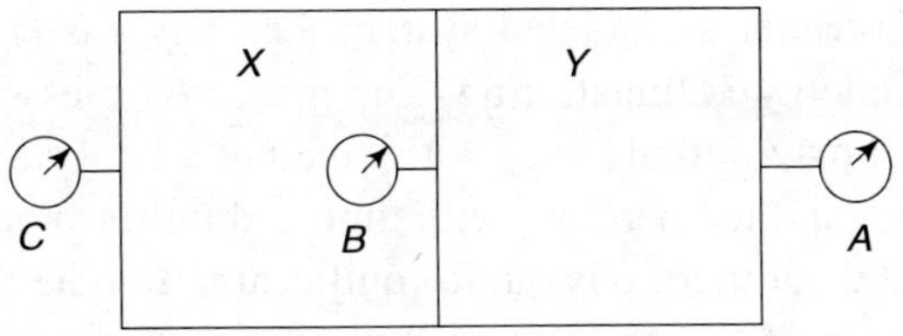

Fig. 1.11 Example 1.9

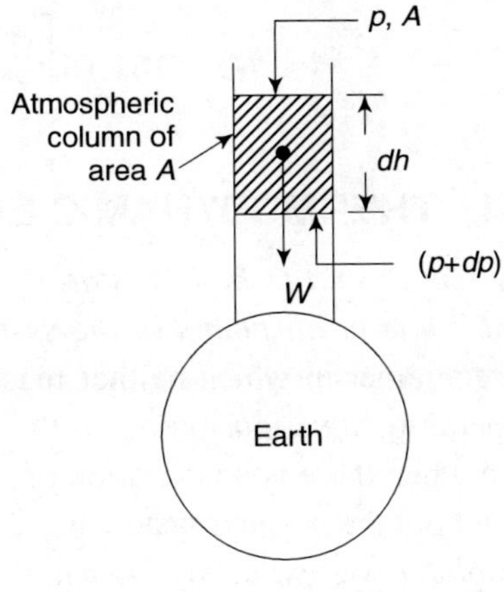

Fig. 1.12 Example 1.11

Let $B_0' =$ new barometric reading reduced to 0°C

$\therefore$ $$B_0' = 66\ (1 - 0.000172 \times 20) = 65.77 \text{ cm of Hg}$$

Hence, the new vacuum gauge reading

$$p_{\text{vac}} = 65.77 - 9.742 = 56.028 \text{ cm of Hg} \qquad \textbf{Ans.}$$

Example 1.11 The pressure and specific volume of the atmospheric air are related according to the equation $pv^{1.4} = 0.25 \times 10^5$ where p is in N/m^2 and v is in m^3/kg. Calculate the depth of atmosphere required to produce a pressure of 1.032 bar at the earth's surface.

Solution Refer to Fig. 1.12. Let H be the depth of atmosphere required to produce a pressure of 1.033 bar at the earth's surface. Considering an element of length dh, then by the force balance, we have

$$A\ (p + dp) = mg + pA = \rho\ A\ dh\ g + pA$$

or $$pA + A\ dp = \rho\ A\ dh\ g + pA$$

Integrating $$\int_1^2 dp = \int_1^2 \rho\, g\, dh = \int \frac{1}{v} g\, dh$$

or $$\int_0^H dh = \frac{1}{g} \int_0^{1.032 \times 10^5} v\, dp = \frac{1}{g} \int_0^{1.032 \times 10^5} \left(\frac{0.25 \times 10^5}{p} \right)^{\frac{1}{1.4}} dp$$

or $$H - 0 = \frac{1}{9.81} \times (0.25 \times 10^5)^{1/1.4} \left[\frac{p^{0.286}}{0.286} \right]_0^{1.032 \times 10^5}$$

or $$H = 141.16\left[\frac{(1.032\times10^5)^{0.286}}{0.286}\right] = 13408.364 \text{ m} = 13.408 \text{ km}$$ **Ans.**

1.11 THERMODYNAMIC EQUILIBRIUM

A system is said to be in a state of thermodynamic equilibrium if the value of the property is the same at all points in the system. Consider an isolated system, i.e., any gas enclosed in a container in which neither mass nor energy is transferring. The measured pressure and temperature at various points of the system may initially vary with time but a final stage will come when there is no variation of pressure and temperature with time and uniform property throughout the isolated system is called a state of thermodynamic equilibrium. In other words, *an isolated system always reaches in course of time a state of thermodynamic equilibrium and can never depart from it spontaneously.*

It is to be noted that the thermodynamic equilibrium is a complete equilibrium and it includes the following equilibria.

(a) Mechanical equilibrium (b) Chemical equilibrium

(c) Thermal equilibrium (d) Electrical equilibrium

(a) Mechanical Equilibrium *A system is said to be in a state of mechanical equilibrium if there exists no unbalanced force either in the interior of the system or between the system and the surroundings.* This is possible only if pressure is the same throughout the system which, in turn, is also equal to that of surroundings. For example, if the pressure is not uniform throughout the system, internal changes in the state of the system will take place till the mechanical equilibrium is restored. Similarly, if the weight on the piston from piston–cylinder arrangement as shown in Fig. 1.2 is removed, system boundary will expand until the mechanical equilibrium is again restored.

(b) Chemical Equilibrium *A system is said to be in a state of chemical equilibrium if there exists no chemical reaction or transfer of matter from one part of the system to another.* It is to be noted that even though the system may be in mechanical equilibrium yet the system may undergo spontaneous change of internal structure due to chemical potential. For example, a mixture of oxygen and gasoline is not in chemical equilibrium once the mixture is ignited.

(c) Thermal Equilibrium *When a system existing in mechanical and chemical equilibria is separated from its surrounding by a diathermic wall (diathermic means heat flow is allowed) and if there is no spontaneous change in the property of the system, the system is said to exist in thermal equilibrium.* This definition never means that for thermal equilibrium mechanical and chemical equilibrium must always exist. This type of definition has been used to show that changes in the state can be attributed to thermal potential alone. *In other words, a system is said to be in thermal equilibrium if there exists an uniformity of temperature throughout the system or between the system and the surroundings.*

(d) Electrical Equilibrium *A system is said to be in a state of electrical equilibrium if there exists an uniformity of electrical potential throughout the system.*

When the conditions for any one of the four types of equilibrium are not satisfied system is said to be in a non-equilibrium state.

To determine whether the system is in the state of thermodynamic equilibrium or not, we can formulate the following test.

(i) *A system is said to be in thermodynamic equilibrium only, if the system is incapable of undergoing a spontaneous change in any macroscopic property when it is isolated.* For example, metal bar conducting heat when isolated by stopping the heat flow is not under thermodynamic equilibrium because the temperature level in the bar will increase. But a gas being compressed very slowly in a cylinder piston arrangement is in thermodynamic equilibrium because no further change in state takes place if the system is isolated by stopping the piston.

1.12 PROPERTY

A property of a system is its measurable characteristics describing the system. The measurable characteristics are mass, physical compositions, temperature, pressure, volume, surface area, velocity, elevation, electrical potential, etc. The above-mentioned characteristics are also called thermodynamic properties or thermodynamic co-ordinates. Properties are classified into two categories; (i) ***intensive*** and, (ii) ***extensive properties***.

(i) Intensive Properties *If the value of the property does not depend upon the mass of the system, it is called an intensive property.* Examples of intensive properties are pressure, temperature, density, velocity, height, and viscosity.

(ii) Extensive Properties *If the value of the property depends upon the mass of the system, it is called an extensive property.* Examples of extensive properties are volume, surface area, internal energy, potential energy, kinetic energy, electric charges, and magnetisation.

In order to find whether a given property is intensive or extensive, consider the system uniform throughout and divide it into segments (subsystems) by drawing lines across the system in any manner and see whether the property changes or remains constant as one goes from one subsystem to another. If it does not change then it is an intensive property otherwise an extensive property.

The property depends only on the state of the system and not upon the manner in which the state was reached. This means that for measuring the pressure of a system, the previous change in pressure is not essential. *If the property is represented by ϕ, then $d\phi$ should be an* exact differential. That is,

$$\int_1^2 d\phi = \phi_2 - \phi_1 \qquad \text{...(1.25)}$$

In this context, it will be better to distinguish between internal, i.e., thermostatic property and external, i.e., mechanical property. Internal property is defined as a characteristic of the matter within the equilibrium system. It is measured by an observer at rest relative to the system. *Mass, composition, internal energy, etc., are examples of internal property. External property is defined as a characteristic of either motion or the position of the system in a gravitational field. It is measured relative to an external datum. Kinetic energy, potential, etc., are examples of mechanical properties.*

The number of properties necessary to define a system depends on the complexity of the system. In a simple system, the intensive state has two degrees of freedom. If such a system is in equilibrium, the intensive state is specified by two independent intensive properties. System which contains more than one component or one phase substance requires more than two independent properties to specify its state.

On the basis of the above discussion we can define "*independent*" and "*dependent*" properties. *Independent properties are those properties which define the state of the system, while dependent properties are those properties which become fixed when the state of the system is defined by independent properties.*

It is to be noted that macroscopic properties are manifestations of the microscopic behaviour of the particles of a system. The microscopic properties continuously fluctuates about a time-average value, so that the macroscopic property represents the time average value of microscopic properties.

Generally speaking, properties are expressed as functional relationship known as the *equation of state of the system.* For example, an equation of state relating three to properties of a system such as x, y, z, will be

$$f(x, y, z) = 0 \qquad ...(1.26)$$

Assuming that it is possible to solve the above equation explicitly for each of the variables

$$x = x\,(y, z),\ y = y\,(z, x),\ z = z\,(x, y) \qquad ...(1.27)$$

Now, if we suppose that in the last equation z is a dependent variable then x and y are independent variables. The differential of the dependent property may be written as in terms of partial derivatives, and the differential of independent properties will be

$$dz = \left(\frac{\partial z}{\partial x}\right)_y dx + \left(\frac{\partial z}{\partial y}\right) dy$$

If

$$M = \left(\frac{\partial z}{\partial x}\right)_y \text{ and } N = \left(\frac{\partial z}{\partial y}\right)_x$$

then

$$dz = M\,dx + N\,dy \qquad ...(1.28)$$

The subscripts on the partial derivatives indicate the independent variables. If Eq. (1.28) has a solution, then dz is called an *exact or perfect differential*, and the equation is integrable. Otherwise, it is an inexact differential which cannot be integrated and has to be written as ∂z to distinguish from dz.

In case z is adequately defined by the thermodynamic coordinates x, y, its value is the same regardless of the order taken in approaching it. Suppose $z = z\,(x, y)$ where partial derivatives are continuous, then the second derivative of x, with respect to x and y, is independent of the order of successive differentiation.

and $$\frac{\partial}{\partial y}\left(\frac{\partial z}{\partial x}\right) = \frac{\partial}{\partial x}\left(\frac{\partial z}{\partial y}\right) \quad \text{...(1.29)}$$

or $$\frac{\partial M}{\partial y} = \frac{\partial N}{\partial x} \quad \text{...(1.30)}$$

Equation (1.30) is a necessary condition for the existence of a function of x and y satisfying Eq. (1.28). Equation (1.30) is also a sufficient condition when this equation is integrated twice, the expression of z is obtained. *A function which satisfies Eq. (1.30) is called a point function or a property of the system.*

It is possible to show that the sum, or the difference, of two point functions is also a point function, provided that functions and their derivatives are continuous. Similarly, the product of two point functions is a point function.

Example 1.12 Discuss whether the following quantities can be used as properties or not?

(a) $\int pdV + \int Vdp$, (b) $\int pdV$, (c) $\int Vdp$

Solution If the differentials of the above terms are exact differential, then it will be called a property. All differentials are exact ones if integration between the two end states can be carried out without requiring any information on the variation of variables.

(a) $$\int pdV + \int Vdp = \int (pdV + Vdp) = \int d(pV) = pV$$

This is an exact differential, therefore, it is a property.

(b) $\int pdV$.

p and V are interrelated to each other and they are connected by a line on the p-V diagram. Thus, the integral $\int pdV$ is not an exact differential, hence it is not a property.

(c) Similarly, $\int Vdp$ is not an exact differential and so it is not a property.

Example 1.13 $dq = f(T)\, dT + \frac{RT}{v} dv$ is property or not?

Solution $$\frac{\partial f(T)}{\partial v} = 0 \quad \text{and} \quad \frac{\partial (RT/v)}{\partial T} = \frac{R}{v}.$$

Since $\frac{R}{v} \neq 0$, hence dq is not an exact differential. Therefore, no state function exists that can have a differential equal to dq, hence, dq is not a property.

1.13 STATE

The state is a unique condition of the system. Broadly speaking, state is the *condition of a system at an instant of time described by its properties such as pressure, temperature, density, etc.* This is shown on pressure–volume diagram (Fig. 1.13) as state 1 and 2.

The above definition of state clearly indicates that each property has a single value at each state, i.e., all properties are state or point functions. The following test determines whether a given variable is a property or not.

(i) A variable is a property, and only if it has a single value at each equilibrium state.

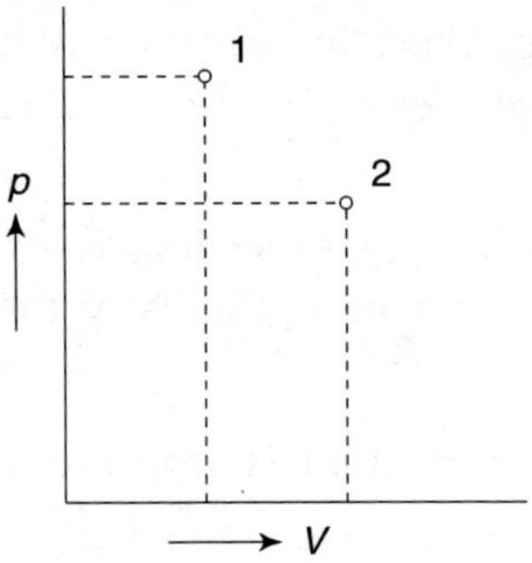

Fig. 1.13 Representation of state

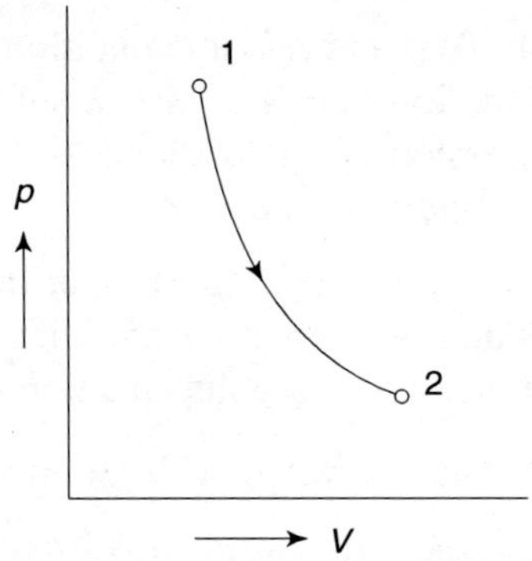

Fig. 1.14 Representation of a process

(ii) A variable is a property, and only if the change in its value between any two equilibrium states is a single valued.

These conditions show that *a variable is a property if its changes are fixed by the end states.*

1.14 PATH

A thermodynamic system passing through a series of states constitutes a path.

1.15 A PROCESS

A process is defined as a transition in which a system changes from one initial state to a final state. Not only this, energy transfer to and from a system at steady state also constitutes a process. For example, consider a case shown in Fig. 1.14. State *A* is the initial state of the system. Due to expansion of the system the final state is *B*. The line *AB* is the process that has taken place.

1.16 CYCLIC PROCESS

If any system undergoes through a series of processes from one state and ends with the same initial state by forming a complete cycle, so that the properties of the system at the end of the cycle are the same as at the beginning, the system is said to undergo a cyclic process.

A cyclic process may be represented on *p-v* diagram as shown in Fig. 1.15 if the various states and paths are known.

1.17 QUASI-STATIC (QUASI-EQUILIBRIUM) PROCESS

A quasi-static process is one in which the system deviates from one equilibrium state by only infinitesimal amounts throughout the entire process. In other words, *a process closely approximating to a succession of equilibrium states is known as quasi-static process.* For example, consider the pressure p_1 of a gas on one face of a frictionless piston and a pressure p_2 on the opposite face. If the two pressures are equal, the piston is in equilibrium. If p_1 is infinitesimally larger than p_2, the gas on both the sides still remains essentially in equilibrium. This is known as the quasi-static process. This quasi-static process only can be represented on a thermodynamic plane. As shown in Fig. 1.16, let *AB* represent a quasi-static process on a thermodynamic plane, say *p*-*V* plane. Then at the successive states such as 1, 2, 3, etc.; the system is nearly in thermodynamic equilibrium.

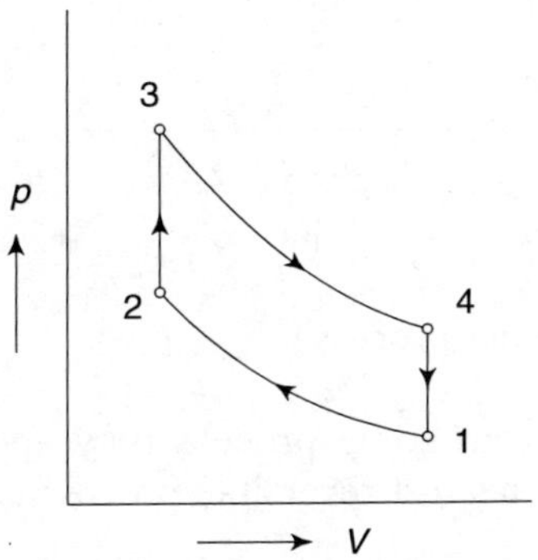

Fig. 1.15 Cyclic process

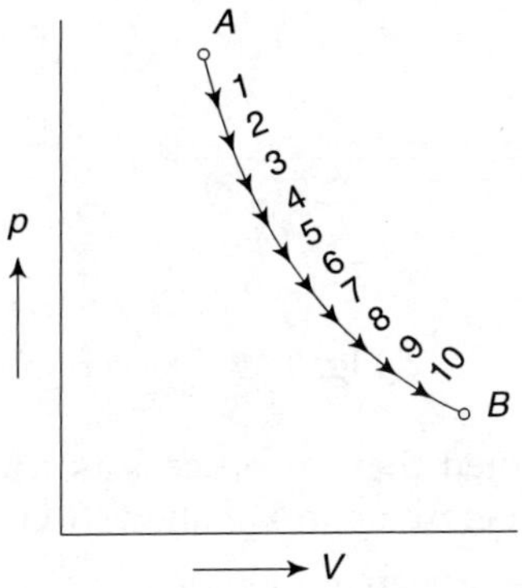

Fig. 1.16 Quasi-static process

1.18 REVERSIBLE PROCESS

A process is reversible if the system passes through a continuous series of equilibrium states. Hence, these states can be located on a pressure and volume diagram and a line representing the path of the process can be drawn through all the points as shown in Fig. 1.17(a). *In other words, a process is called reversible if the initial state together with all energies transferred or transformed during the process can be completely restored in the system and environment.* Thus, if a process is really reversible there is no after effects or changes are evident in the system or in the environment when the process occurs in the forward direction and then in the backward direction as is obvious from Fig. 1.17(a).

In a reversible process, any energy transferred would also have to be transferred reversibly. Hence, the processes involving fluid flow with friction and convective heat transfer, turbulent flow cannot be reversible. Consider a piston moving in a cylinder in which gas expands from high to low pressure, some energy is expended in overcoming the friction between the piston and the cylinder; this friction produces heat. If the path is restored, i.e., if the gas is compressed back to its original volume, more heat is produced, and the energy thus expended in overcoming friction is not recovered. Hence, the expansion process of the fluid is not reversible. The process will be reversible only when there are *ideal processes or reversible processes*. A quasi-static process is very near to reversible process and may be called a reversible process because if such a process is reversed it restores the original state.

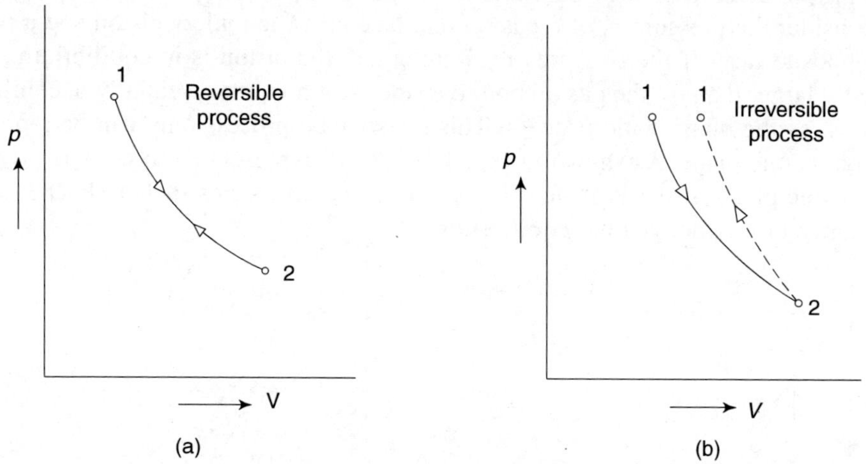

Fig. 1.17 Reversible and irreversible process

It is to be noted that no process is truly reversible, but some process may approach reversibility to a close approximation. Some examples of nearest reversible process are:

(i) Frictionless relative motion.

(ii) Frictionless adiabatic expansion or compression of fluid.

(iii) Extension and compression of a fluid.

(iv) Electrolysis.

(v) Polytropic expansion or compression of a fluid.

(vi) Isothermal expansion or compression.

The conditions for a process to be reversible are:

(i) There should be no friction present in the process.

(ii) There should be no heat transfer across finite temperature difference.

(iii) The process should be carried out with absolute slowness to be quasi-static throughout.

(iv) After the process, both the system and surroundings must be restored to the original state, i.e., there should not be any past event either in the system or surroundings after the process.

1.19 IRREVERSIBLE PROCESS

A process is irreversible if a system passes through a sequence of non-equilibrium states. These states cannot be located on any property diagram because each property does not have a unique value in the entire system. However, assuming that the system is in a state of thermodynamic equilibrium before and after the process, the process may be represented on the property diagram, say *p-V* diagram as shown in Fig. 1.14(b). Now, if this irreversible process is made to reverse path, it will not restore the original state 1 as is obvious from Fig. 1.17(b). Thus, we can say that if the process is not reversible, it is called an irreversible process.

We have already seen that expansion of gas from high to low pressure or vice versa is not reversible because of the presence of friction. Consider another example of heat transfer between two systems. For heat transfer to occur, there must be a temperature difference between the two systems. If heat transfer takes place from a system A to B during a process, then system A must be at a higher temperature than B. For the reverse process to occur (i.e., heat transfer from B to A), heat would have to flow in the opposite direction, i.e., from lower to higher temperature. This is not possible.

Thus, if a system after undergoing an irreversible process is restored to its initial state, it leaves a history of the events in the surroundings. Some examples of irreversible processes are:

(i) Fluid flow with friction. (ii) Combustion of air and fuel,

(iii) Diffusion of gases. (iv) Throttling.

(v) Free expansion. (vi) Heat transfer.

(vii) Plastic deformation. (viii) Electricity flow through a resistance.

(ix) Adiabatic mixing.

Hence, we can say that *all real or natural processes are irreversible processes* since they take place with finite differences of potential between parts of the system, and between the system and the surroundings. *For theoretical analysis, processes are assumed as reversible because reversibility indicates the limit of possible change in a real system. Any system going through a reversible processes will result in maximum efficiency.*

Note The reasons for irreversibilities have been discussed in chapter 7. By considering the efficiency or effectiveness or irreversibiiity the performance of real processes may be evaluated.

1.20 ENERGY

Energy is defined as the capacity to exert a force through a distance, i.e., capacity to do work. The concept of energy was first introduced in mechanics by Newton when he hypothesized about kinetic and potential energies. Energy is a scalar quantity that cannot be observed but can be recorded and evaluated by indirect measurements. The dimension of energy is

$$[E] = [L]\,[F]$$

where L = length and F = force. [= MLT^{-2}]

The unit of energy is joule (J).

Energy may be broadly classified as **stored energy** *and* **energy in transition.**

(a) *Stored energy is contained within the system boundaries.* Examples of this type energy are (i) Potential energy, (ii) Kinetic energy, (iii) Internal energy, etc.

(b) *Energy in transition is one which crosses the system boundaries.* Examples of this type of energy are (i) Heat, (ii) Work, (iii) Electricity, etc.

1.21 POTENTIAL ENERGY (PE)

Potential energy is the energy due to position and the elevation of mass with respect in the earth. A force F is exerted on mass m due to gravitational attraction, is given by

$$F = mg$$

Potential energy of a mass kept at an elevation z_1 is $= F \cdot z_1 = mg\, z_1$

If we wish to calculate the change in potential energy due to change in the elevation of mass from z_1 to z_2 then

$$\Delta\text{PE} = (\text{PE})_2 - (\text{PE})_1 = F\,(z_2 - z_1) = mg\,(z_2 - z_1) \qquad ...(1.31)$$

Normally, potential energy is expressed in the forms of some datum where the potential energy is zero. Suppose z represents the elevation of mass m above the datum of zero potential (say the earth's surface). By this condition in Eq. (1.31), we have

$$(\text{PE})_{z=0} - (\text{PE})_z = mg\,(0 - z)$$

Since $$(\text{PE})_{z=0} = 0$$

Therefore, $$(\text{PE})_z = mgz \qquad ...(1.32)$$

1.22 KINETIC ENERGY (KE)

Kinetic energy is defined as the energy created from the motion of a mass. In other words, it is the energy that could be recovered when the system is brought to the stagnation state in the absence of friction. We know that when a mass undergoes a change in velocity, a force is exerted. Thus, we have,

$$F = ma = m\,\frac{dC}{d\theta}$$

If the force is exerted over the period θ_1 to θ_2 while the mass covers the distance from L_1 and L_2, the change in energy is

$$\Delta\text{KE} = \int_{L_1}^{L_2} F dL = \int_{\theta_1}^{\theta_2} FC \cdot d\theta = m \int_{C_1}^{C_2} = CdC$$

or $$\Delta\text{KE} = (\text{KE})_{C=C_2} - (\text{KE})_{C=C_1} = m\,\frac{1}{2}(C_2^2 - C_1^2) \qquad ...(1.33)$$

Usually, the kinetic energy is expressed in terms of some datum where kinetic energy of mass is zero when it is at rest with respect to the earth. By putting this condition in Eq. (1.33), we have

$$(\text{KE})_{C=0} - (\text{KE})_C = \frac{1}{2} m (0 - C^2)$$

Since $$(\text{KE})_{C=0} = 0$$

Therefore, $$(\text{KE})_C = \frac{m C^2}{2} \qquad \text{...(1.34)}$$

It is to be noted that the *potential and kinetic energies are external forms of energy.*

1.23 INTERNAL ENERGY (U)

Internal energy of a system is defined as the energy associated with the configuration and motion of its molecules, atoms and subatomic particles relative to its centre of mass. In other words, internal energy is a property consisting of the *combined molecular kinetic and potential energy*, and is determined by intensive properties such as pressure and temperature. Generally, the change in internal energy is established rather than the absolute value of it.

Internal energy is invisible energy but its effects can be sensed. Internal energy is a property, (see the first law of thermodynamics Art. 5.2 for the proof).

It is worth to discuss all various configurations and motion of molecules (Fig. 1.18) and associated various energies which contribute towards internal energy, They are as follows:

(i) *Translational, rotational and vibrational energy* The molecules of the system move in space in all directions with all possible velocities. As a result of such motions, the molecules possess kinetic energy. When the motion of molecules does not change (i.e., the motion is translational) then the acquired kinetic energy is termed *translational energy.*

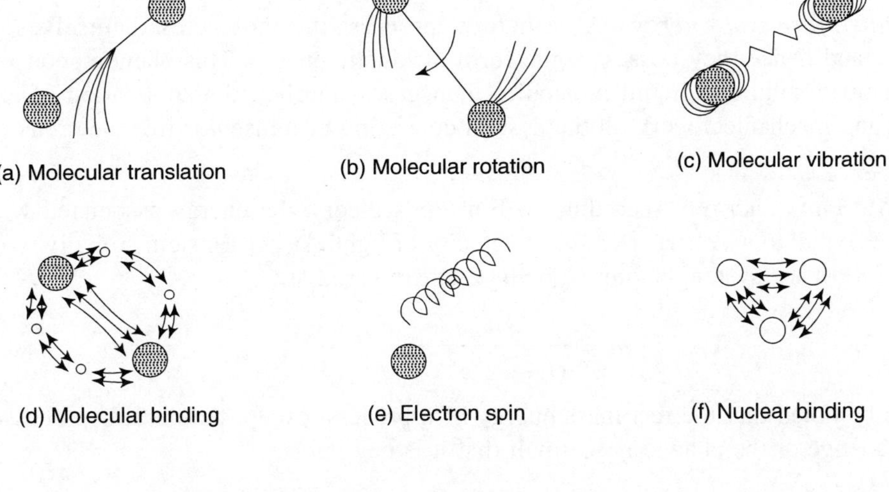

(a) Molecular translation (b) Molecular rotation (c) Molecular vibration

(d) Molecular binding (e) Electron spin (f) Nuclear binding

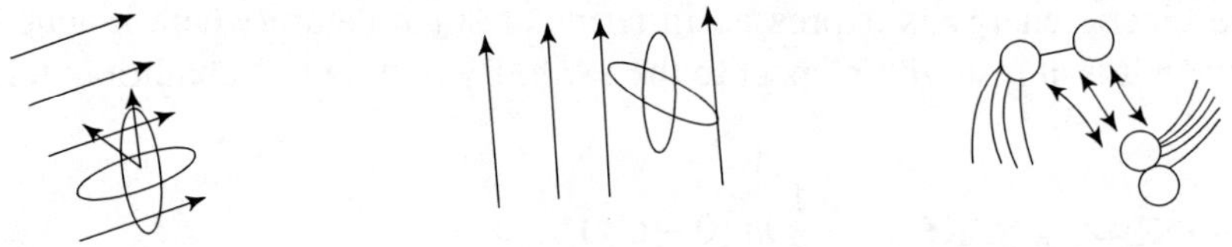

(g) Magnetic dipole moment (h) Electric dipole moment (I) Intermolecular (weak)

Fig. 1.18 Various components of internal energy

If the molecules rotate about their axis during movements as it always happens in the case of diatomic and polyatomic gases, the kinetic energy acquired is termed *rotational energy*. At high temperature, the molecules start vibrating and hence they *acquire vibrational energy*.

(ii) *Spin energy* The electrons orbit around the nucleus and their spins may be either clockwise or anticlockwise. The energy associated with the spin of these electrons is termed spin energy.

(iii) *Potential energy* The molecules of a gas experience intermolecular forces (conservative force) which are associated with potential energy. These forces include coulomb and gravitational forces which are the main parameters to keep molecules together.

(iv) *Molecular binding energy* The nucleus of any atom consists of protons, neutrons, mesones and many other particles. To keep them together there exists a molecular binding forces such as force of attraction between nucleus and electrons, nucleus forces, etc., and consequently possess the molecular binding energy.

(v) *Dipole energy* A thermodynamic system may have magnetic or electric dipoles. A magnetic dipole formed as a result of attraction between nucleus and electrons, which are electric current carriers. An electric dipole is produced when in an electrically neutral atom the centre of gravity of positive and negative charges does not coincide. These dipoles twist in the presence of external field resulting in work done or change in the energy of the system. These energies are more pronounced in dielectric and paramagnetic substances.

(vi) *Other forms of energy* As a natural process, the molecules themselves attract each other and hence they possess some form of kinetic energy. This phenomenon is more predominant in solid and liquid molecules than gases. Further, the application of external electrical and mechanical work on the system consisting of ionised or free electrons change their energy.

(vii) *Rest mass energy* According to Einstein's theory, the energy associated with rest mass, $m_{0,}$ is equal to $m_0 c^2$ where c is the velocity of light. As per Einstein's relativity theory, the mass (m) of the system having velocity C is expressed as

$$m = \frac{m_0}{(1 - C^2/c^2)^{1/2}}$$

It is to be noted that the rest mass energy of a process, except the nuclear process either does not change or the change is so small that it is negligible.

1.24 TOTAL ENERGY (E)

Total energy possessed by a mass is the sum of all forms of energy internal as well as external. It is expressed in joule.

Hence, $$E = U + \text{KE} + \text{PE} = U + \frac{mC^2}{2} + mgz \quad ...(1.35)$$

Various forms of energies with its manifestation are given below in Table 1.2.

Table 1.2 Forms of energy and manifestation

Forms of energy	Created by
(A) *External*	
(a) Potential	Elevation
(b) Kinetic	Velocity
(B) *Internal*	
(a) Molecular	
(i) Kinetic	Intermolecular or atomic velocities
(ii) Potential	Intermolecular or interatomic forces
(b) Chemical	Changes in molecular composition
(c) Nuclear	Changes in atomic composition

Example 1.14 A satellite of mass 600 kg is moving towards the moon. Find the kinetic and potential energies for the following cases:

(i) relative to the earth when 50 km from launching and moving at 2500 km/h and when the acceleration of the earth's gravitational field is 7.9 m/s^2.

(ii) relative to the moon when moving at the same velocity and 50 km from its destination where 1 kg mass has a weight of 4 N.

Solution (i) Given, $m = 600$ kg, $g = 7.9$ m/s^2, $z = 50$ km $= 50 \times 1000$ m.

$$\text{PE} = mgz = 600 \times 7.9 \times 50000 = 23.7 \times 10^7 \text{ J} \quad \textbf{Ans.}$$

$$\text{KE} = \frac{mC^2}{2} = \frac{600}{2} \times \left(\frac{2500 \times 10^3}{3600}\right)^2 = 1.446 \times 10^8 \text{ J} \quad \textbf{Ans.}$$

(ii) Since 1 kg mass has weight 4N, hence 600 kg will weigh relative to the moon as $600 \times 4 = 2400$ N

$$\text{PE} = mgz = wz = 2400 \times 50 \times 10^3 = 12 \times 10^7 \text{ J} \quad \textbf{Ans.}$$

$$\text{KE} = \frac{mC^2}{2} = \frac{600}{2} \times \left(\frac{2500 \times 10^3}{3600}\right)^2 = 1.446 \times 10^8 \text{ J} \quad \textbf{Ans.}$$

Example 1.15 A mass of 5 kg possesses 13000 J of kinetic energy at a place where the acceleration due to gravity (g) is 9.7 m/s^2. Calculate the maximum height to which this will be displaced if all the kinetic energy is converted into potential energy.

Solution $$\text{PE} = mgz = \text{KE} = \frac{mC^2}{2} = 13000 \text{ J}$$

$$z = \frac{13000}{mg} = \frac{13000}{5 \times 9.7} = 268.4 \text{ m} \qquad \textbf{Ans.}$$

Example 1.16 A satellite in the earth's atmosphere is moving with a velocity of 1000 m/s relative to the earth. Find the kinetic energy possessed by it for the following cases:

(a) Where acceleration due to gravity is 9.6 m/s^2 and the gravitational force is 5000 N.

(b) Where acceleration due to gravity is 8.5 m/s^2 and the gravitational force is 4300 N.

Solution (a) We have force = mg or 5000 = $m \times 9.6$

or $$m = \frac{5000}{9.6} = 520.833 \text{ kg}$$

$$\text{KE} = \frac{mC^2}{2} = \frac{520.833 \times 1000^2}{2} = 26.04 \times 10^7 \text{ J} \qquad \textbf{Ans.}$$

(b) $$\text{KE} = \frac{mC^2}{2} = \frac{w}{2g}C^2 = \frac{4300 \times 1000^2}{2} = 23.5 \times 10^8 \text{ J} \qquad \textbf{Ans.}$$

1.25 THERMODYNAMIC MEDIA (WORKING FLUID)

In thermodynamics, the energy transfer to and from a system/device machines/equipment occurs through a medium called **working fluid**. *It is the changes in the properties of working fluid that gives the desired effect.* The working fluid or media may be liquid, vapour, or a gas. The liquid may be in a different states ranging from sub-cooled to boiling, vapour or superheated states. The following are the commonly used working fluid:

- **Water** in liquid state is used in boilers to generate steam.
- **Vapour** (a mixture of water and steam), i.e., wet steam gets condensed in the condenser so that it may be fed back to the boiler. Other types of vapour say mercury, KOH may be used.
- **Air** (either cold, hot or humid) is used in air compressor and air-conditioning plants, etc.
- **Fuel-air mixture in gaseous form** is used in internal combustion engines and gas turbines.
- **Refrigerants** such as Freon-12, Freon-22, R-134a, NH_3, CO_2, etc. are used in various forms ranging from sub-cooled state to superheated gaseous state in refrigerators, air-conditioners and refrigerating plants.
- **Steam** in superheated state is used in steam turbines, steam engines (now obsolete) and process heating.
- **Gases** such as air, CO_2, N_2, etc., may be used in compressors.
- **Cooling and heating fluids** are used in heat exchangers.

1.26 THERMODYNAMIC DEVICES

The following are some of the thermodynamic devices which are used in engineering practices (Fig. 1.19).

(i) *Piston-cylinder arrangement* This is employed in internal combustion engines, reciprocating compressor, steam engines, pumps, etc. As engines, they produce power and as compressors, they absorb power to raise the pressure of working fluid. They are open system.

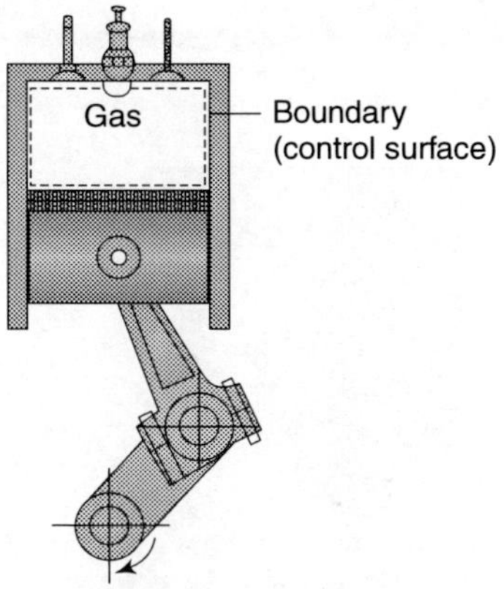

(a) Piston-Cylinder

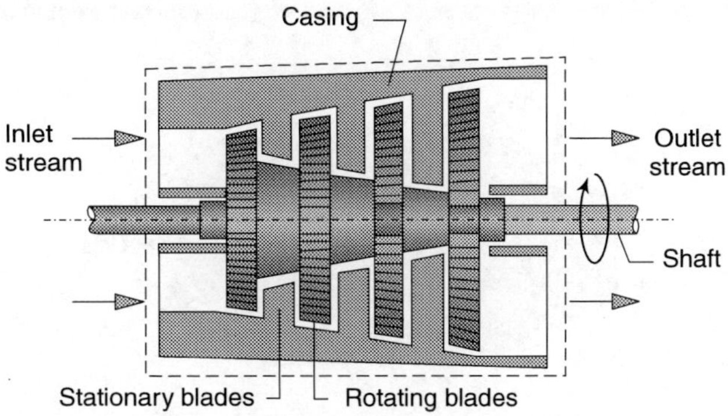

(b) Turbine

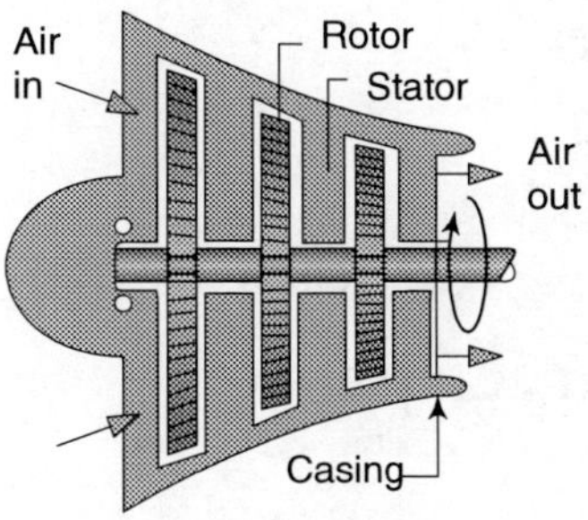

(c) Axial flow compressor

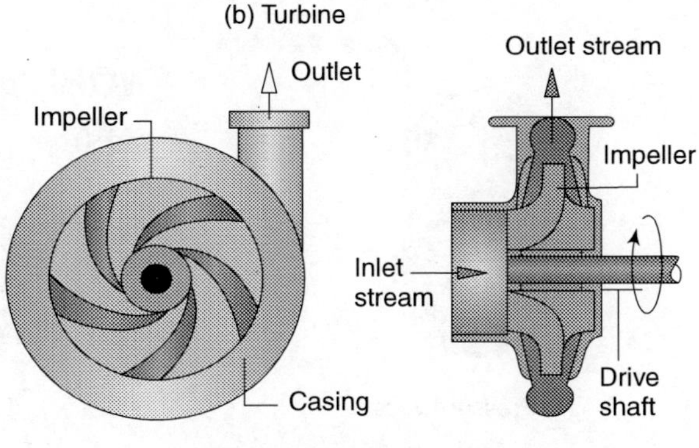

(d) Centrifugal compressor

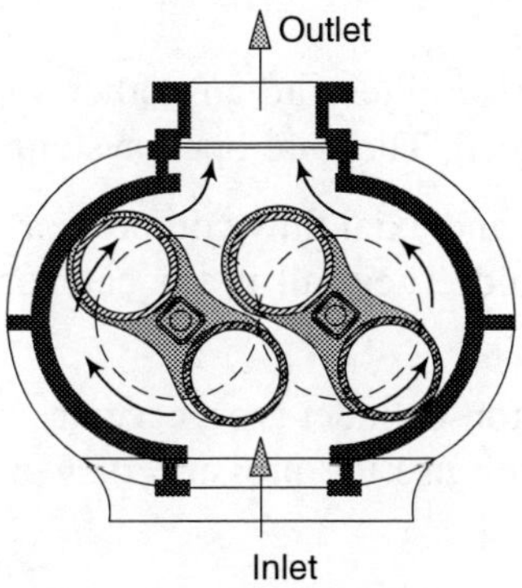

(e) Roots type compressor

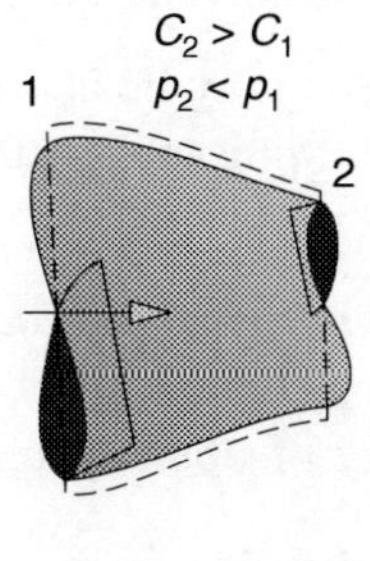

(f) Nozzle (Subsonic)

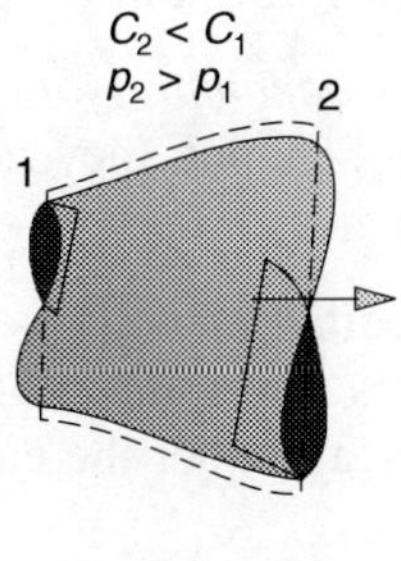

(g) Diffuser (Subsonic)

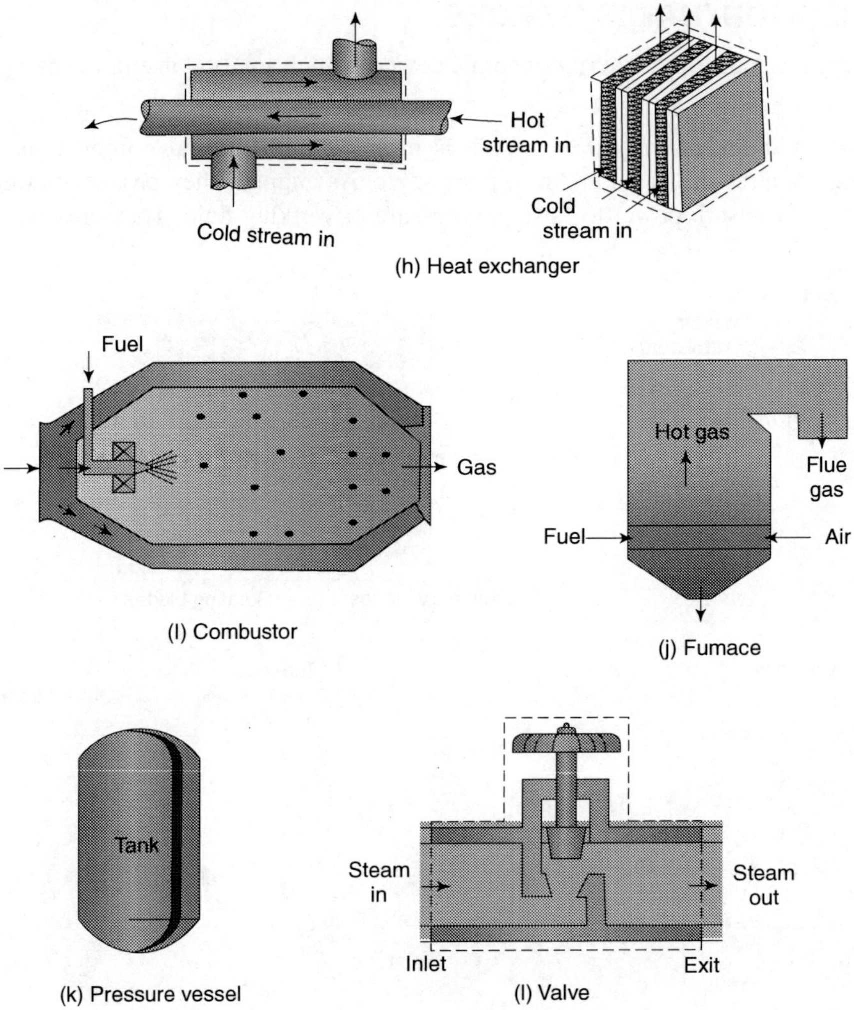

Fig. 1.19 Some thermodynamic devices

(ii) *Turbines* These are employed as steam turbines, gas turbines and any other vapour turbine. Working fluid expands in turbine and work is produced. They are open system.

(iii) *Turbocompressors* These may be centrifugal, mixed and axial flow compressors. By taking work input, they raise the pressure of the working fluid. They fall in the category of open system.

(iv) *Positive displacement compressors* These include piston-cylinder type, eccentric vane type, roots types and gear types. By taking work input, they raise the pressure of working fluid and fall under the category of open system.

(v) *Nozzles* These are used in turbines, jet engines, etc. The function of nozzle is to convert the thermal (pressure energy) into kinetic energy. They are open systems.

(vi) *Diffusers* These are used in turbocompressors, jet engines, etc. The function of a diffuser is to convert the kinetic energy into pressure energy. They are open systems.

(vii) *Heat exchangers* These include boilers, condensers, intercoolers, aftercoolers. evaporators, etc. The heat energy of one fluid is exchanged into other fluid. They are open system.

(viii) *Combustor* These are employed in gas turbine plants, jet engines, rocket, internal combustion engines, etc. As a result of combustion of fuel, a large amount of energy is released. It is an open system.

(ix) *Furnaces* It may be a boiller or other furnaces. As a result of combustion, a large amount of energy is released. They are open systems.

(x) *Pressure vessel* High pressure fluids are stored in pressure vessels and are used as per need. They include boiler drum, tankers, receivers in compressor system, etc. They also fall under the categories of open system.

(xi) *Pipings* Fluid at high pressure and temperature flows through piping. These are available in various sizes.

(xii) *Valves* Valves are used to allow and stop the fluid flow in pipe lines. Throttling action takes place in valve. They are open systems.

(xiii) *Pumps* It may be reciprocating, rotary and turbopump types to pump fluids from low pressure to high pressure by taking work input. They fall under the category of open system. Centrifugal pump is similar to centrifugal compressor.

1.27 ENGINEERING DESIGN AND ANALYSIS

One of the most important engineering function is to design and analyse things intended to meet human needs.

1.27.1 Design

Engineering design is a decision making process involving the principles drawn from engineering and other fields such as economics and mathematics which are applied usually iteratively, to design a system, a system component or process.

Fundamental elements of design consist of establishment of objective, synthesis, analysis, construction, testing and evaluation. It is worth to note that designs typically are subject to a variety of *constraints* related to economics, safety, environmental impact, etc. The initial steps in engineering design include pinning down quantitative performance specification and identifying workable designs including conceptual design and drawings that meet the specifications. Workable designs may be one or more that are *best* according to some criteria, namely, lowest cost, highest efficiency, smallest size, lightest weight, elegant and beautiful shape, etc. There are other important factors in the selection of a final design that include reliability, manufactureability, maintainability and market place construction. The above dimensions suggest that a compromise must be sought among competing criteria and there may be alternative design solutions that are very similar.

1.27.2 Analysis

Synthesis (i.e., selecting and putting together components to form a co-ordinated whole) is the main requirement of a design. The thermodynamic analysis is based on the following aspects

- Mass balance (i.e., the conservation of mass principle)
- Energy balance (i.e., the conservation of energy principle, first law)
- Exergy balance (i.e., the second law of thermodynamics)
- Momentum balance (i.e., Newton's second law of motion, if fluid dynamics is involved).

The first step in thermodynamic analysis is the *engineering model. By modeling in engineering problems, we mean developing and using the appropriate combination of assumptions and equations that permit critical features of the process to be analyzed.*

1.28 SUMMARY OF FORMULAE

1. $\rho = \lim_{\Delta V \to \Delta V'} \frac{\Delta m}{\Delta V}$

2. $\rho_w = \rho\, g$

3. $v = \frac{1}{\rho}$

4. $p = \lim_{\Delta A \to \Delta A'} \frac{\Delta F}{\Delta A}$

5. $p_{\text{abs}} = p_{\text{atm}} + p_{\text{gauge}}$

6. $p_{\text{abs, vac}} = p_{\text{atm}} + p_{\text{vac, gauge}}$

7. $p_0 = p_{\text{total}} = p_{\text{stag}} = p_{\text{static}} + \rho C^2/2$

8. $p_{\text{gauge}} = \rho\, gh$

9. $p_1 + \rho_1\, gh_1 = p_{\text{atm}} + \rho_2\, gh_2$ and $p_1 + \rho_1\, gh_1 = p_2 + \rho_2\, gh_2$

10. $F = mg$

11. $\text{KE} = mC^2/2$ and $\text{PE} = mgz$

12. $E = U + \text{KE} + \text{PE} = U + mC^2/2 + mgz$

PROBLEMS

Things to Think About

1.1 List the possible boundaries for studying each of the following. Also give a sketch.

(a) a bicycle tyre inflating

(b) a cup of water being heated in a microwave oven

(c) a household refrigerator in operation

(d) a jet engine in flight

(e) cooling a desktop computer

(f) a residential gas furnace in operation

(g) a rocket launching

1.2. Air at normal temperature and pressure contained in a closed tank adheres to the continuum hypothesis. But when sufficient air has been withdrawn from the tank, the continuum hypothesis no longer applies to the remaining air. Explain this fact.

1.3. Can we think of the value of an intensive property uniform with position throughout of a system? Be constant with time? Both?

1.4. What is the reason that we commonly ignore the pressure variation with elevation for a gas inside a storage tank?

1.5. It has been observed that when buildings have large exhaust fans, exterior doors can be difficult to open due to pressure difference between the inside and outside. Is it possible for you to open a door of 1 m × 2.5 m if the inside pressure is 2.5 cm of water (vacuum).

Short and Long Theoretical Questions

1.6. Define the term thermodynamics.

1.7. Define dimensions and units.

1.8. Define the units for time, length and mass.

1.9. Define force in general and in terms of different system of units.

1.10. Give the dimensions and units in S.I. system of the following:

(a) Velocity (b) Acceleration (c) Work (d) Energy

(e) Power (f) Momentum (g) Pressure (h) Discharge

(i) Temperature (j) Density.

1.11. Differentiate between mass and weight.

1.12. Explain thermodynamic system, surrounding and universe. Distinguish between closed, open, isolated, homogeneous and heterogeneous systems. Illustrate with examples.

1.13. What do you mean by phase of a system?

1.14. Discuss the macroscopic and microscopic point of view of thermodynamics.

1.15. Discuss the concept of continuum in thermodynamics.

1.16. Define the following terms—

(a) density, (b) specific weight, (c) specific volume, (d) pressure,

(e) atmospheric pressure, (f) absolute pressure, (g) gauge pressure,

(h) vacuum pressure, (i) static pressure, (j) stagnation pressure, and

(k) velocity pressure.

1.17. What do you mean by thermodynamic equilibrium?

1.18. What do you mean by "Property"? Distinguish between intensive and extensive properties.

1.19. State whether the following properties of a system are intensive or extensive—mass, weight, volume, pressure, temperature, density, surface area, velocity, magnetisation and potential energy.

1.20. Explain the terms—state, path, process and cyclic process.

1.21. What are state functions and how do they differ from path functions? Give examples of each. How will you test whether any physical variable is a state function or not?

1.22. Determine whether in each of the following example, the system shown in italics is open or closed, homogeneous or heterogeneous, adiabatic or non-adiabatic.

(a) Small metal *sphere* is rotating within a rigid nonconducting hemispherical *bowl* and eventually comes to rest at the bottom.

(b) A *block* descends on inclined plane against the force of friction.

(c) An *iron ball* is held at the top by a magnetic field, the ball falls to the bottom of the field.

(d) A *spring mass* oscillating within an evacuated rigid envelope.

1.23. What do you mean by reversible and irreversible processes? Give examples.

1.24. What do you understand by cyclic and quasi-static process?

1.25. Explain the term "Energy". How will you find the presence of energy? What are the different forms of energies? Explain and give some examples. What is total energy?

1.26. What do you mean by N.T.P. and S.T.P.?

1.27. What do you mean by stagnation and static pressure?

1.28. Prove that the weight of a body at an elevation z above sea level is given by $w = mg\ [r/(r + z)]^2$ when r is the radius of earth at equator.

1.29. Choose whether the following is open, or closed system?

(a) Water pump, (b) Pressure cooker, (c) Water wheel, (d) Thermometer surrounded by a medium at high temperature, (e)Automobile engine, (f) Boiler including all piping and radiators, (g) Air compressor, (h) A dash pot consisting of a cylinder, piston and containing liquid.

1.30. Indicate the number of phases in the following—(a) Pure water, (b) Water alcohol mixture, (c) Coffee, (d) Water oil mixture, (e) Mild steel at room temperature, (f) Liquid oxygen at room temperature.

1.31. If z is a function of three properties x, y and z such that $dU = Mdx + Ndy + Pdz$ where M, N and P are functions of properties x, y and z. Prove the following conditions that are essential for U to be an exact differential.

$$\frac{\delta P}{\delta y} = \frac{\delta N}{\delta M}\frac{\delta M}{\delta z} = \frac{\delta P}{\delta N}\frac{\delta N}{\delta x} = \frac{\delta M}{\delta y}$$

1.32. Draw the diagrammatic sketches of the following systems, in each case, emphasizing the system, boundary, surrounding and the direction of heat and work. Write whether the system is open or closed and indicate the energy conversions:

(a) Mercury in glass thermometer surrounded by a medium at high temperature.

(b) A perfectly insulated constant pressure chamber containing a mixture of ice and water constantly agitated by a motor driven stirrer.

(c) A non-insulated nozzle receiving gas at a high pressure with negligible velocity and expanding it down to a lower pressure. Assume no loss due to friction or turbulence during the process.

(d) Inflation of a flat tyre by forcing air into the tyre with tyre walls as the system boundary, the tyre walls being non-conducting.

1.33. Distinguish clearly between the following, giving examples wherever necessary—

(a) Closed system and open system.

(b) Point functions and path functions,

(c) Enthalpy and internal energy .

Numerical Questions

1.34. On the surface of the moon, the gravitational effect is 1/6.2 of that on the earth's surface. Calculate the weight of the mass of a solid. [*Ans.* m/6.2 where m is the mass]

1.35. A man weighs 760 N at a location where acceleration due to gravity g = 9.8 m/s^2. What will be the weight of the man on a planet where local g = 2.45 m/s^2? [*Ans.* 190 N]

1.36. A ball with an initial kinetic energy of 700 J is shot vertically from a cannon and the ball rises to an altitude of 250 m. Calculate the mass of the ball. [*Ans.* 2.854 kg]

1.37. A force of 390 N is applied on a mass of 2 kg along an inclined surface at a location where g = 9.081 m/s^2. If the force and the inclined surface make a 30° angle with the vertical, determine the acceleration of the mass if the force is acting (a) upwards, (b) downwards.

1.38. The pressure of steam generated in a boiler as recorded by the Bourdon pressure gauge was 9.2 bar. Calculate the absolute pressure of steam if the barometer reads 774 mm of Hg.

1.39. The pressure of the oil delivered by an oil pump as recorded by the Bourdon pressure gauge was 9 bar gauge. What must be the height of an oil column which corresponds to this pressure? The density of oil is 0.9 gm/m^3.

1.40. The gas used in a gas engine trial was tested. The pressure of gas supply was 7 cm of water column. Find the absolute pressure of the gas if the barometric pressure is 760 mm of Hg.

1.41. A barometer reads 76 cm of Hg. What would be the absolute pressure if (a) a pressure gauge connected to a turbine inlet reads 28 bar and (b) a vacuum gauge connected in the exhaust line of the same turbine reads 70 cm of Hg?

1.42. While measuring the air pressure in an air chamber, a liquid of specific gravity 0.8 shows height of 4.25 cm 80 cm. To what pressure is this equivalent in cm of water? What is the absolute pressure in Pa?

1.43. A vertical composite liquid column with its upper end exposed to atmosphere comprises 54 cm of Hg. (sp. gravity of Hg = 13.6), 65 cm of water (sp. gravity of water is 1) and 80 cm of oil (sp. gravity of oil is 0.8). Calculate the pressure (a) at the bottom of the column, (b) at the centre surface of oil, and water and (c) at the inter surface of water and Hg.

1.44. The level of the water in an enclosed water tank is 50 m above the ground. The pressure in the air space above the water in the tank is 140 kPa. The average density of water is 1000 kg /m^3. Calculate the pressure of water at the ground level.

1.45. A diver descends 100 m to a sunken ship. A container is found with a pressure gauge reading 100 kPa (gauge). Atmospheric pressure is 100 kPa. What is the absolute pressure of the gauge in the container expressed in kPa? The density of water is 100 kg/m^3. [*Ans.* 1181 kPa]

146. A certain engine is supplied with 1.59 kg of fuel per minute. The potential energy stored per kg of fuel is equivalent to 24,000 kJ/kg. If 20 per cent of the energy supplied to engine is converted into mechanical energy, find out the rate at which the engine is developing power.

1.47. Calculate the height of a column of water equivalent to atmospheric pressure of 1 bar if the water is at 15°C. What is the height if the water is replaced by mercury?

1.48. If all the atmosphere is liquified what will be the equivalent height of this liquid air in metre of water at 15°C if the earth's surface is assumed flat?

1.49. Find the mass of air in kg of the atmosphere surrounding the earth if the pressure is 1 bar everywhere on the surface. Assume that the earth is a sphere of 12850 km.

1.50. A special manometric fluid on specific gravity of 2.98 is used to measure a pressure of 2.2 atm at a location where the barometric pressure is 750 mm Hg. What height will the manometer fluid register?

1.51. (a) With reference to Fig. 1.20, find $(p_a - p_b)$ in N/m^2.

(b) A vacuum gauge indicates that the pressure of air in a closed chamber is 0.2 bar (vacuum). The pressure of the surrounding atmosphere is equivalent to 750 mm column of mercury. The density of mercury is 13.59 g/cm^3; and the acceleration of gravity is 9.81 m/s^2. Calculate the absolute pressure within the chamber, in bar. [*Ans.* 0.8 bar]

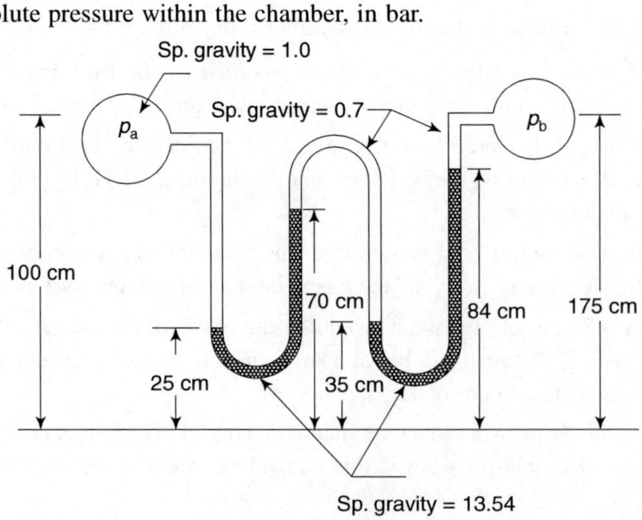

Fig. 1.20 Example 1.51

1.52. Calculate the total force, in kN, on the bottom of 120 × 60 m swimming pool. The depth of the pool varies linearly along its length from 1 m to 5 m. Also, calculate the pressure on the floor at the centre of the pool, in kPa. The atmospheric pressure is 0.99 bar, the density of the water is 999 kg/m^3 and the local acceleration of gravity is 9.81 m/s^2.

Software Based Questions

1.53. Develop a software to know the pressure exerted on a submarine moving beneath the sea at different depths of water.

1.54. 1500 kg of natural gas at 100 bar and 260 K is stored in a tank. If the pressure (p), temperature (T) and volume (v) of the gas is related by the following equations

$$p = [(5.18 \times 10^{-5})\ T/(v - 0.002669)] - (8.91 \times 10^{-3})/v^2$$

where v is in m^3/kg, T is in K and p is in bar. Calculate the volume of the tank in m^3. Also, plot pressure versus specific volume for the isotherms T = 250 K, 500 K, 1000 K and 1500 K.

1.55. Plot the variation of atmospheric pressure in atm, versus elevation z above sea level, in km, ranging from 0 to 10 km if specific volume $v = 72.435/p$ when v is m^3/kg and p is in kPa.

Design and Open Ended Questions

1.56. Prepare a report on the subject of global warming as the issue is receiving considerable importance worldwide. The report should start with the meaning of global warming and the scientific evidence of its occurrence.

1.57. Draw a schematic diagram of an automatic data acquisition system for sampling pressure data inside a cylinder of petrol engine. Choose a suitable type of pressure transducer for this purpose. Search or develop an appropriate computer software for running the system. Prepare a complete report covering all aspects.

1.58. The American National Standards Institute (ANSI) and the American Society of Heating, Refrigerating and Air-conditioning Engineers (ASHRAE) have adopted standards for pressure and temperature measurement. A facility is to be developed for testing valves in nuclear power plants. The pressure and temperature of flowing gases must be accurately measured as per test program. Using these standards, prepare a memorandum discussing what standards must be met in the design of the facility and the requirements of standards.

1.59. There is outcry of sustainable development by economists and others as a means for meeting present human needs without compromising the ability of future generations to meet their own needs. Prepare a paper concerning this with its underlying principles and objective.

1.60. Prepare a paper using manufacturer's data on thermocouple and thermistor temperature sensors, for measuring temperature of hot combustion gases from a furnace. Describe operating principles sensitivity, accuracy calibration, cost, advantages and disadvantages of each.

Multiple-Choice Answer Type Questions

1.61. The units of temperature, mass, time, length and energy in SI system of units are given by

(a) Celcius, kilogram, second, metre and watt

(b) Kelvin, kilogram, second, metre and joule

(c) Kelvin, gram, hour, metre and joule

(d) Kelvin, kilogram, second, centimetre and joule

1.62. Time, length, mass and temperature are

(a) Primary dimension (b) Secondary dimensions

(c) Both (a) and (b) (d) None of these

1.63. The difference between the pressure of fluid and the pressure of atmosphere is called

(a) Barometric pressure (b) Absolute pressure

(c) Gauge pressure (d) None of these

1.64. Which of the following sets has all intensive properties?

(a) Pressure, volume, energy, heat, specific volume

(b) Pressure, energy, volume, heat, density

(c) Pressure, volume, temperature, density, specific volume

(d) Pressure, volume, heat, density, specific volume

1.65. Which of the following sets has all open systems?

(a) Boiler, gas turbine, compressor, condenser

(b) Pump, thermoflask, refrigerator, petrol engine

(c) Window air conditioner, scooter engine, thermometer, diesel engine

(d) Jet engine, gas engine, pressure cooker, steam turbine

1.66. Which of the following sets has all properties as point functions?

(a) Entropy, enthalpy, work (b) Pressure, temperature, heat,

(c) Heat, work enthalpy (d) Temperature, enthalpy, internal energy

1.67. Which of the following is not the property of thermodynamic system?

(a) Work (b) Temperature (c) Internal energy (d) Pressure

Answers to Multiple-Choice Questions

1.61 (b) 1.62 (a) 1.63 (c) 1.64 (c) 1.65 (a) 1.66 (d)

1.67 (a)

CHAPTER

2

Work and Heat—Energy Interaction

OVERVIEW

A closed system (non-flow process) and its surroundings can interact in two ways, namely, by work transfer and by heat transfer. Such processes are called ***energy interaction*** and they bring about changes in the property of the system. It is essential for the students of thermodynamics to understand clearly the definition of both work and heat because the correct analysis of many thermodynamic problems depends on distinguishing between them. This chapter deals with the definition, distinction and similarities between work and heat.

2.1 WORK

In mechanics, mechanical work is defined as the product of an external force acting at the boundaries of a system and the distance through which the force moves along its line of action. Thus, mathematically

$$W = \int F dL \qquad ...(2.1)$$

The unit of work is joule (J) or newton metre (Nm) and 1 Nm = 1 J.

However, this definition of work is not adequate for the study of thermodynamics, because of the fact that thermodynamics deals with energy transfers between a system and surroundings and, the energy transfers are of two kinds—*heat and work*. In thermodynamics, there is a difference between heat and work. The definition of work used in mechanics fails to distinguish between them. For example, consider a hot body in contact with water resulting in boiling, and producing steam. This steam may drive a steam engine, which in turn, may lift a weight through a distance. In thermodynamics, ***work is defined as energy in transition,*** i.e., **a form of energy interaction between two systems**. According to *Obert*:

Work is defined as the energy transferred, without transfer of mass across the boundary of a system because of an intensive property difference other than temperature that exists between system and surroundings.

The above definition by Obert clearly presents a distinction between heat and work. Work may also be defined as follows:

Work is an interaction between system and surroundings, and is done by a system if the sole effect external to the boundaries of the system could have been the raising of a weight.

The above definitions give a broader interpretation of work than the purely mechanical one and also distinguish between heat and work. Consider the following examples to understand the meaning of work.

Assume a storage battery as a system in which the external terminals of the system are connected with a resistance by means of a switch as shown in Fig. 2.1 (a). When the switch is closed the current will flow through the battery and consequently the resistance will get warmer. The system has, therefore, interacted with the surroundings. Can this interaction be classified as work? According to the definition of work used in mechanics, the answer is "**no**" because no force has moved through a distance. But according to the definition of thermodynamics, the answer is '**yes**' as explained below.

Assume the resistance is replaced by an electric motor with a pulley on its shaft to wind up the rope on which weight is suspended as shown in Fig. 2.1 (b). If the current flow is the same as when the resistance was there, then there is absolutely no difference between the two systems. Assuming negligible friction in the motor bearing pulley, *the sole effect external to the system boundary has been the rise of weight. Therefore, the system does work.*

The words "***sole effect***" used in the definition of work differentiate between heat and work. In the previous example, on a hot body in contact with water produced steam and drove a steam engine which carried weight is not the sole effect only. But there are always some other effects either the steam does not return to its original state after passing through the engine or the surrounding becomes warmer. Thus, *heat is not the same as work.*

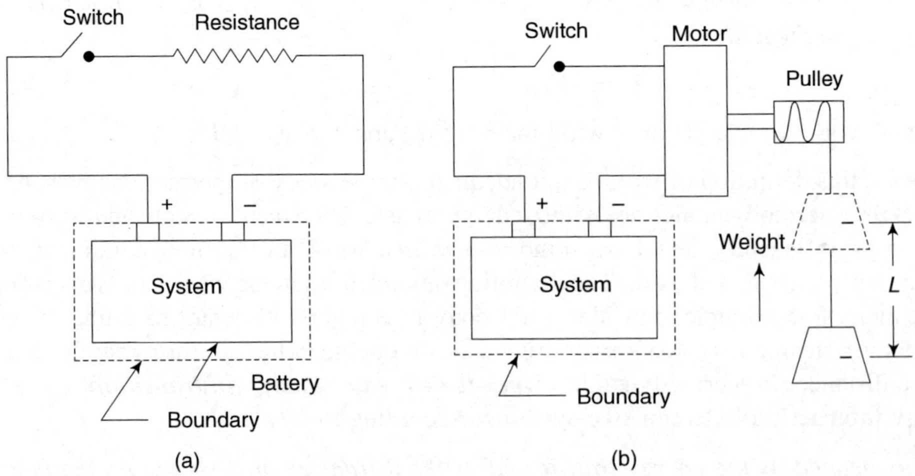

Fig. 2.1 Concept of work

The word '***external***' used in the definition of work has a distinct meaning. This word "*external*" emphasizes that work is defined only with reference to a system boundary, i.e., *work is an interaction across the boundary. It is to be noted that work like heat, is a transient quantity and cannot be stored in a system.* Only during a transfer of energy work occurs, i.e., *it is an inexact differential* ***represented as δW instead of dW***.

Work done by a system on its surrounding is regarded as ***positive work*** while work done *on a system by surrounding is regarded* as ***negative work***.

2.2 POWER

Power may be defined as rate of doing work, i.e.,

$$\text{Power} = P = \frac{\text{Work done}}{\text{Time}} = \frac{W}{\tau}\text{; J/s or Watt} \qquad \text{...(2.2)}$$

The unit of power is joule/s or watt and 1 watt = 1 joule/s.

Example 2.1 Figure 2.2 shows a storage battery, which drives a motor and, in turn, a fan. Apply the definition of work.

Solution The motor drives the fan. Thus, the effect of the work transfer from the system boundary is the driving of the fan. To satisfy the definition of work used in thermodynamics, this effect should be reduced to only raising of a weight. Now, if the fan is replaced by a pulley and weight arrangement as shown in Fig. 2.3 such that this system is identical with the previous one with the flow of same amount of current, the turning of motor will raise the weight and the sole effect external to the boundary will be raising of weight only. Thus, this system does work.

Example 2.2 A system having a horizontal force of 500 N pushes a block kept on a rough surface 10 m horizontally as the system changes the shape. Find the work done form the concept of mechanics and thermodynamics. The block and the rough surface may be assumed as surrounding.

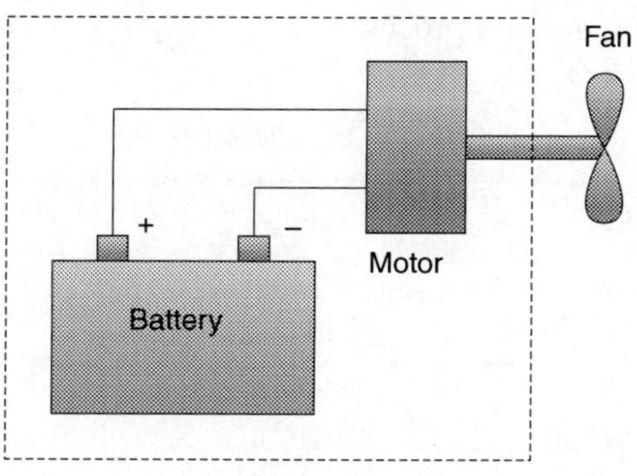

Fig. 2.2 Example 2.1

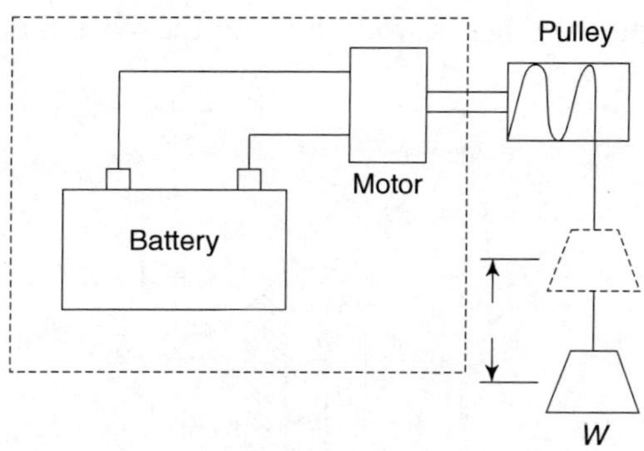

Fig. 2.3 Example 2.1

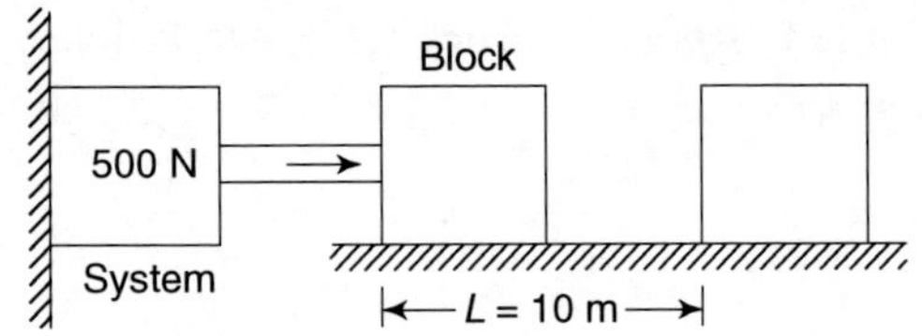

Fig. 2.4 Example 2.2

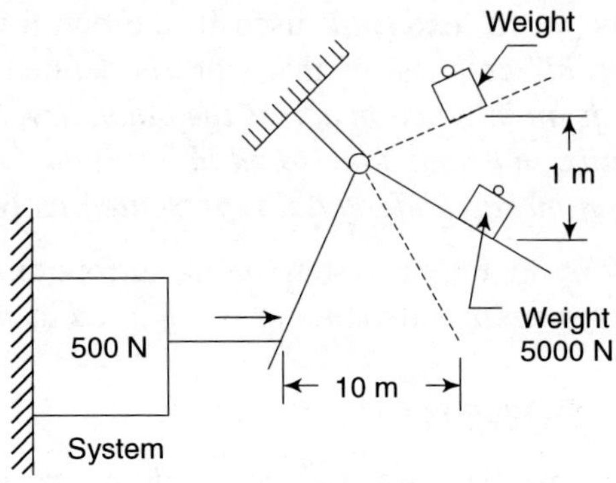

Fig. 2.5 Example 2.2

Solution Refer to Fig. 2.4. Accorcing to the definition of work in mechanics

$$\text{Work done} = \text{F.L.} = 500\ (\text{N}) \times (10\ \text{m}) = 5000\ \text{J} \qquad \textbf{Ans.}$$

The above arrangement shown in Fig. 2.4 may be replaced by an arrangement as shown in Fig. 2.5 by changing a fictitious surrounding to suit thermodynamical definition. The sole effect external to the system is raising of weight of 5000 N through a distance of 1 metre. Hence, the work done = 5000 J.

2.3 FORMS OF WORKS DURING QUASISTATIC OR REVERSIBLE PROCESS

2.3.1 Expansion or Compression Work of Fluid or *pdV* Work or Diplacement Work

For the expansion or compression work of a fluid during a quasistatic process, let us consider a piston-cylinder arrangement shown in Fig. 2.6. In a piston-cylinder arrangement type machines, two types of works are observed, one is the work done by the system due to *expansion* process and the other is work done on the system due to *compression* process.

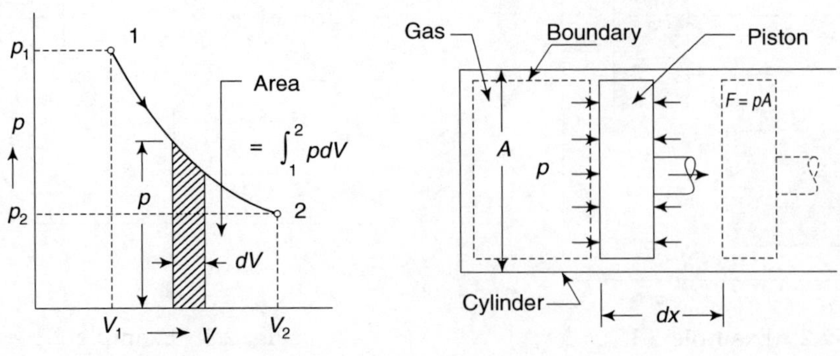

Fig. 2.6 *pdV* work done in quasistatic process

Let the system be the fluid enclosed between the piston and the cylinder. The system will exert a pressure p on the piston face of area A. If the piston is allowed to move a distance dx, the force F acting on the piston may vary but at any instant, it is given by

$$F = p\,.\,A\;;\mathrm{N} \qquad \text{...(2.3)}$$

$$\text{Work} = \text{Force} \times \text{Distance}$$

$$\delta W = pAdx$$

But

$$Adx = dV = \text{Change in volume}$$

$$\delta W = pdV \quad \text{Nm or J} \qquad \text{...(2.4)}$$

The total work done at the moving system boundaries can be calculated by integrating Eq. (2.4). To integrate this equation, the relationship between p and V must be known. This relationship may be known from the p-V diagram of a quasi-static expansion process, in which the volume variation is from V_1 to V_2 with a infinitesimal departure from thermodynamic equilibrium. Thus,

$$W_{1-2} = \int_1^2 pdV \qquad \text{...(2.5)}$$

or

$$W_{1-2} = p\int_1^2 dV = p\,(V_2 - V_1);\ \mathrm{J} \qquad \text{...(2.6)}$$

Similarly, the work done during compression process, i.e., from 2 to 1, is given by $\delta W = -\,pdV$.

The minus sign is used to indicate that during compression, work is done on the system. Expansion work is assumed *positive*, i.e., work done by the system while the compression is assumed negative. *The area under p–V diagram is the work done.*

Equation (2.4) *applies to any pressure as long as pressure* (*p*) *is the pressure at the moving boundary.* The pressure on the face of the moving piston is the same as the pressure throughout the system only if the process goes slowly, proceeding through a series of equilibrium states that is quasistatic process. **This *quasistatic process* may be taken as a *reversible process*.** A process involving rapid expansion or compression is usually accompanied by fluid friction or mechanical friction; in such non-quasistatic processes, the fluid pressure at the boundary differs from that at the interior of the system. Thus, $W \neq \int pdV$ in a non-quasistatic process i.e., irreversible process.

Equation (2.4) *is valid under the following conditions.*

(a) That the process is frictionless.

(b) That the system is closed.

(c) That there is no viscous effects within the system.

(d) That the process is quasistatic, i.e., reversible process. In other words, the pressure as well as other properties have the same value on all boundaries of the system.

(e) That the effects due to gravity, electricity of magnetism are negligible.

The expansion and compression processes may be achieved by various thermodynamic processes (see chapter 4).

2.3.2 Work Done in Moving an Electric Charge through an Electric Field

In the presence of an electric field, an electric charge, Q, experiences a force. Work is done when the charge moves with respect to the field. The potential difference (V) across any given portion of the electric field is defined in terms of the work necessary to move a unit charge across the potential difference if the differential charge is dq and work is expressed as

$$\delta W = V dq \quad \text{...(2.7)}$$

The equation is valid when the process *is reversible.*

2.3.3 Work of Polarization and Magnetization

According to the theory of electromagnetic fields, work is done on a substance contained within an electric or magnetic field when the field is altered. For a dielectric material which lies within an electric field, the work per unit volume supplied externally to increase polarization of the dielectric is given by

$$\delta W = - E dP \quad \text{...(2.8)}$$

where E is the electric field strength and P is the polarization.

Similarly, work done per unit volume in increasing the magnetization of a substance due to a change in the magnetic field is given by

$$\delta W = - \mu_0 H . dM \quad \text{...(2.9)}$$

where H = magnetic field strength

M = magnetization per unit volume

μ_0 = permeability of free space.

It is to be noted that the *system must be reversible.*

2.3.4 Work of Chemical Cell

A galvanic cell is a device for converting chemical energy into electric energy. The electrical work delivered by the cell for the passage of a differential quantity of charge dq is found as follows. For a potential difference, i.e., electromotive force E, a current I does the work at the rate of $\delta W d\tau = E\, I$.

But $$I = -\frac{dq}{d\tau}$$

where $d\tau$ = time interval

Hence, $$\delta W = - E dq = - E I d\tau \quad \text{...(2.10)}$$

Here the negative sign is added to make the work done positive during discharge because dq is negative during discharge. *The above equation is valid when the process is reversible.*

2.3.5 Work in Stretching a Liquid Surface

Consider a liquid sheet suspended between two plates as shown in Fig. 2.7. The molecules near the liquid surface are attracted more strongly by internal liquid molecules than by the very distant gas molecules, and this gives rise to a macroscopically measurable force as in

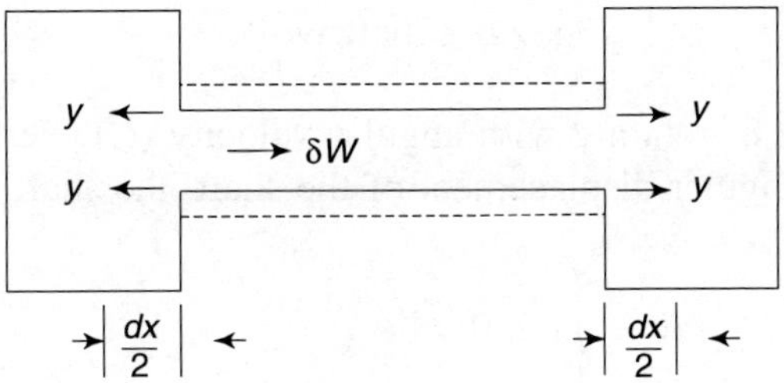

Fig. 2.7 Stretching a liquid

the liquid-gas surface. *The interfacial tension* (y) *is defined as force per unit length acting to a line in the surface*. Hence, the total interfacial tension force exerted on each plate is 2 yb where b is the depth of the liquid sheet normal to the plane of paper. The work done from the plates to the liquid when the plate separation is increased by an amount dx is given by

$$\delta W = -4\, y\, b \frac{dx}{2} = -2\, y\, b\, dx$$

But $$2 b\, dx = dA$$

$$\delta W = -\, y\, dA \qquad \text{...(2.11)}$$

The negative sign is put because work is done on the system to increase the area.

2.3.6 Charging a Capacitor

The energy required to charge a capacitor can be evaluated by noting that the work needed to move Q coulomb of charge from the upper to the lower plate is equal to the potential. That is,

$$\delta W = -\, V \delta Q \qquad \text{...(2.12)}$$

But $$V = \frac{Q}{C} = \text{voltage}$$

where Q is the charge in coulomb and C is capacity. Hence,

$$\delta W = -\frac{Q}{C} dQ$$

Integrating to determine the work required to increase the change from 0 to Q, we have

$$W_{1-2} = -\text{ PE} = \frac{1}{2}\frac{Q^2}{C} = \frac{1}{2} C V^2 \qquad \text{...(2.13)}$$

2.3.7 Shaft Work

A rotating shaft is commonly encountered in machine element.

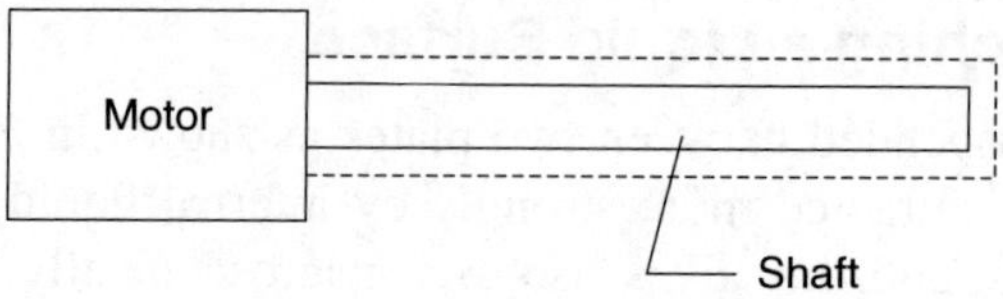

Fig. 2.8 Shaft work

Consider a shaft (Fig. 2.8) rotating with angular velocity (ω) exerting a torque (T) on its surrounding. If $d\theta$ is the angular displacement of the shaft, the shaft work is

$$W = \int_1^2 T\, d\theta \qquad \text{...(2.14)}$$

The shaft power, $$P_{shaft} = \frac{W}{d\tau} = \int_1^2 T \frac{d\theta}{d\tau} = T\,\omega \qquad \text{...(2.15)}$$

Here, the torque has been considered constant. The work is done on the surrounding, so it is a positive work.

2.3.8 Paddle-Wheel Work or Stirring Work

Figure 2.9 shows the arrangement for paddle-wheel or stirring work. As the weight is lowered, and the paddle wheel turns, there is a work transfer into the fluid system which gets stirred. Since the volume of the system remains constant, so $\int p\, dV = 0$. If m is the mass of the weight lowered through a distance dz and T is the torque transmitted by the shaft through an angle $d\theta$, the differential work transfer to the fluid is expressed as

$$\delta W = mg\, dz = T\, d\theta \qquad \text{...(2.16)}$$

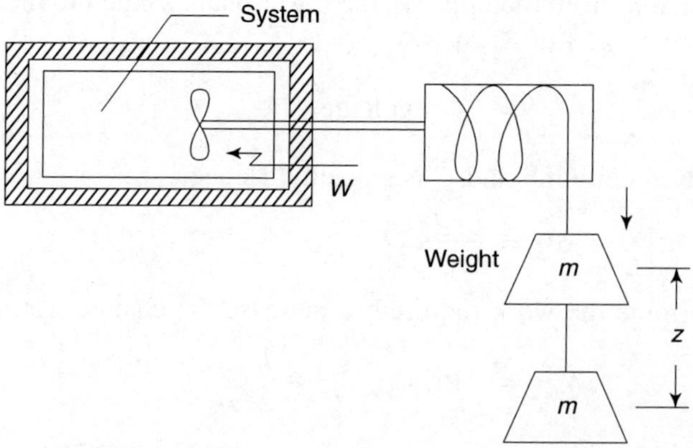

Fig. 2.9 Paddle-wheel work

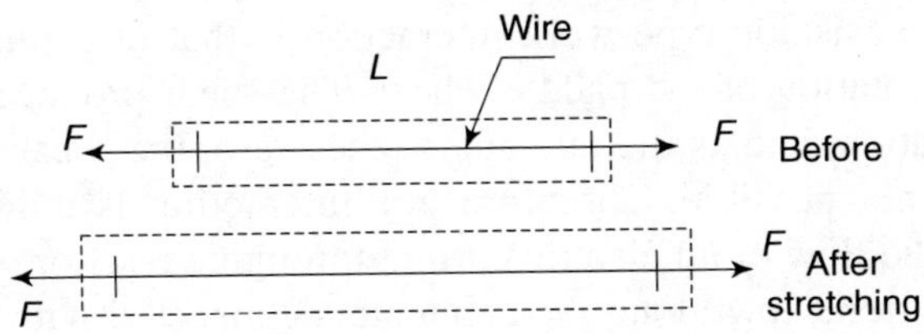

Fig. 2.10 Stretching a wire

or $$W_{1-2} = \int_1^2 mg\,dz = \int_1^2 T\,d\theta \qquad ...(2.17)$$

2.3.9 Work Done in Stretching a Wire

Consider a wire system as shown in Fig. 2.10. The wire is under the tension due to force applied at both ends and as a result the length of the wire changes from L to $L + dL$. The infinitesimal amount of work that is done is equal to

$$\delta W = -F\,dL$$

The minus sign shows the work done on the system.

Thus, $$W_{1-2} = W = -\int_1^2 F\,dL \qquad ...(2.18)$$

Since $F = \sigma A$ where σ is the normal stress developed and A is the cross-sectional area of the wire, thus

$$W = -\int \sigma A\,dL \qquad ...(2.19)$$

If $d\epsilon = \dfrac{dL}{L}$ is the longitudinal strain, and $\dfrac{\sigma}{\epsilon} = E$ = bulk modulus of elasticity, then

$$\delta W = -A\,\epsilon\,E\,L\,d\epsilon$$

or $$W_{1-2} = W = -A\,E\,L\int_1^2 \epsilon\,d\epsilon = -\frac{AE\,L}{2}(\epsilon_2^2 - \epsilon_1^2) \qquad ...(2.20)$$

2.4 NON-QUASISTATIC FORMS OF WORKS

There are other important work interactions which are not quasistatic types and thus called *non-quasistatic works*. The non-quasistatic forms of work have a number of characteristics which differentiate them from quasistatic interactions. These characteristics are:

(i) The force F depends upon the rate of change of state.

(ii) The work interaction is unidirectional, and the process is irreversible.

(iii) The force F approaches zero as the rate of change of state approaches zero resulting in the work approaching zero for a differential change of state.

An example of non-quasistatic type work interaction is that of a fluid undergoing shear deformation due to the rotation of the paddle wheel. The shear produced is proportional to the rate of fluid deformation, and as the rate approaches zero, the shear force does likewise. The reverse process is not possible. Therefore, the interaction is unidirectional. Another example of this type is the flow of an electric current through a resistor which lies within the system boundaries. A current flow along the conductors connected with resistor resulting in a rise of temperature within the system. The reverse is not possible.

2.5 WORK AS A PATH FUNCTION

Consider several reversible processes such as A, B and C from state 1 to state 2 or from 2–1 as shown in Fig. 2.11. The work interaction in each case is represented by the area underneath the corresponding path. From the figure, it is clear that the work is different in each case because the path depends on the nature of the process. *This suggests that work is not a property or a state function rather it is a path function and an infinitesimal increment of work is an inexact differential.* Hence, a system does not possess work but work is a mode of transfer of energy and this transfer of energy occurs only at the boundaries of the system during a change of state of the system. Hence, work is expressed by

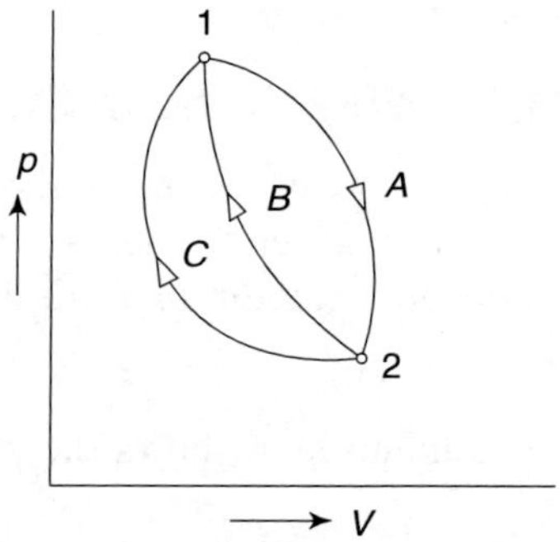

Fig. 2.11 Work depends on path

$$W_{1-2} = \int_1^2 \delta W \text{ and not by } W_2 - W_1 \quad ...(2.21)$$

2.6 NET WORK DONE BY A SYSTEM

In many engineering situations, different forms of work transfer occur simultaneously during a process executed by a system. Under such circumstances, the total or net work done by the system would be equal in the algebraic sum of these given as follows.

$$W_{total} = W_{displacement} + W_{shear} + W_{stirring} + W_{electrical} + W_{magnetic} + ... \quad ...(2.22)$$

Example 2.3 Calculate the work done when the volume changes from 4 m^3 to 8 m^3 through a non-flow quasistatic process in which the pressure p is given by

$$p = (4V - 5) \text{ bar}$$

Solution

$$W_{1-2} = \int_{v_1 = 4}^{v_2 = 8} p\, dV$$

But

$$p = (4V - 5) \times 10^5 \text{ N/m}^2$$

$$W_{1-2} = \int_{v_1 = 4}^{v_2 = 8} (4V - 5) \times 10^5 \, dV = \left[\frac{4V^2}{2} - 5V\right]_4^8 \times 10^5$$

$$= \left[\left(\frac{4\times 8^2}{2} - 5\times 8\right) - \left(\frac{4\times 4^2}{2} - 5\times 4\right)\right]\times 10^5 = 76\times 10^5 \text{ J}$$ **Ans.**

Example 2.4 Calculate the work done in a piston-cylinder arrangement during an expansion process where the process is given by the equation

$$p = (V^2 + 6V) \text{ bar}$$

Assume that during the expansion process the volume changes from 1 m^3 to 4 m^3.

Solution

$$p = (V^2 + 6V) \times 10^5 \text{ N/m}^2$$

$$W_{1-2} = \int_{v_1=1}^{v_2=4} p\,dV = \int_{v_1=1}^{v_2=4} (V^2 + 6V)\times 10^5\, dV = \left[\frac{V^3}{3} + \frac{6V^2}{2}\right]_1^4 \times 10^5$$

$$= \left[\left(\frac{4^3}{3} + \frac{6\times 4^2}{2}\right) - \left(\frac{1^3}{3} + \frac{6\times 1^2}{2}\right)\right]\times 10^5 = 66\times 10^5 \text{ J}$$ **Ans.**

Example 2.5 In a piston-cylinder arrangement, the non-flow reversible process is given by $V = 200/p$ in m^3 where p is in bar. Find the work done when the pressure increases from 1 bar to 10 bar. Indicate whether the process is compression or expansion.

Solution Given

$$V = \frac{200}{p} \text{ m}^3$$

$\therefore$

$$V_1 = \frac{200}{1} = 200 \text{ m}^3 \text{ and } V_2 = \frac{200}{10} = 20 \text{ m}^3$$

Also

$$p = \frac{200}{V}\times 10^5 \frac{N}{m^2}$$

$$W_{1-2} = \int_{v_1=200}^{v_2=20} \left(\frac{200}{V}\right)\times 10^5\, dV = 200\times 10^5 [\ln V]_{200}^{20} = -4.6066\times 10^7 \text{ J}$$

The negative sign shows that the process occurring is the compression process. This is also obvious from the decrease in volume or increase in pressure during the process.

Example 2.6 A capacitor has a capacitance of 1 microfarad. How much work is required to change it to the level of 1000 volts?

Solution

$$W_{1-2} = -\frac{1}{2}C\cdot V^2 = -\frac{1}{2}\times 10^{-6}(1000)^2 = -0.5 \text{ joules}$$ **Ans.**

2.7 HEAT (Q)

Heat is an energy interaction between the system and surroundings. Assume that two closed systems at different temperatures are brought into contact with each other so that they have a common boundary. Due to the temperature difference, energy in the form of heat is transferred from the high temperature system to the low temperature system, this process is continued until thermal equilibrium is attained. Here, heat is not observed but instead heat transfer is inferred through temperature changes or through some similar effects. This heat transfer may be regarded as energy interaction between two closed systems in which there is no work interaction. In the words of **Obert**:

Heat is energy transferred, without transfer of mass, across the boundary of system, because of temperature difference between the system and surroundings.

The unit of heat is joule (J) in S.I. system.

An adiabatic process is defined as one in which no heat interaction occurs. This adiabatic condition exists if the system is surrounded by an enclosure impermeable to heat.

A wall which is impermeable to the flow of heat is an ***adiabatic wall*** *whereas a wall which permits the flow of heat is a* ***diathermal wall****. Heat transfer modes include conduction, radiation and convection.*

2.8 COMPARISON BETWEEN HEAT AND WORK

Following are the some similarities and dissimilarities between heat and work.

Similarities

(a) Heat and work transfer are *energy interactions*.

(b) Heat and work are both *transient phenomena*. They are not possessed by the system but either or both cross the system boundary when a system undergoes a change.

(c) Both heat and work are *boundary phenomena* and are observed only at the boundary of the system. Both represent energy crossing the boundary of the system. Once the heat or work transfer is over, no heat or work is present, only, the energy, i.e., the result of heat or work transfer.

(d) Heat and work are not the property of the system.

(e) Both heat and work are path functions and inexact differentials. So instead of dQ and dW, these are expressed as δQ **and** δW, respectively.

Dissimilarities

(a) Heat transfer is the energy interaction due to temperature difference only. All other energy interactions are termed work transfer.

(b) Heat is a low grade energy while work is a high grade energy.

The following example will clearly demonstrate the difference between heat and work.

In order to understand the difference between heat and work consider a gas contained in a rigid vessel (Fig. 2.12) on which the resistance coils are wound. When current flows through the resistance coils, the temperature of the gas increases. Now question arises whether heat or work cross the boundary of the system.

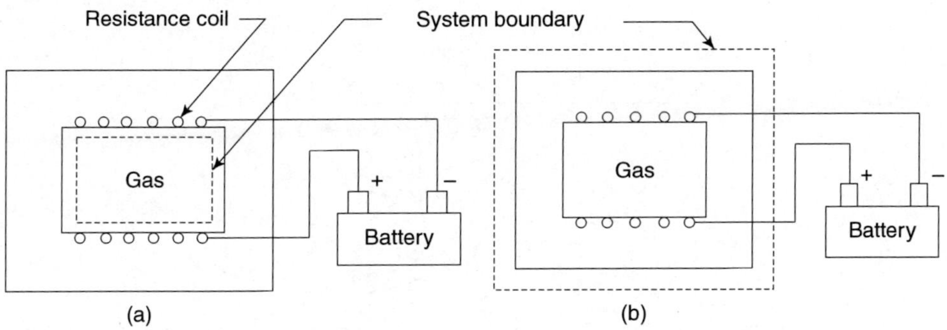

Fig. 2.12 Difference between heat and work

In the first case (Fig. 2.12 a), we consider the gas as the system. Here, the energy crosses the boundary of the system because the temperature of the walls is higher than the temperature of the gas. Thererfore, this energy transfer is ragarded as the heat crossing the boundary of the system.

In the second case (Fig. 2.12 b), we consider the vessel and the resistance heater as the system. Electrical energy crosses the boundary of the system. It can be shown similar to the cases discussed about Fig. 2.1 that the electrical energy crossing the boundary of the system is the work.

2.9 SIGN CONVENTIONS OF VARIOUS ENERGIES IN THEIR ALGEBRAIC SENSE

Heat received by system : Q is positive

Heat rejected from system : Q is negative

Heat neither received nor rejected : $Q = 0$

Internal energy increases : ΔU is positive

Internal energy decreases : ΔU is negative

Internal energy neither increases nor decreases : $\Delta U = 0$

Work done by system : W is positive

Work done on system : W is negative

No work done on or by system : $W = 0$

2.10 VARIOUS THERMODYNAMIC PROCESSES

Various thermodynamic processes used are given below (Fig. 2.13). These processes may undergo either *compression* or *expansion* process.

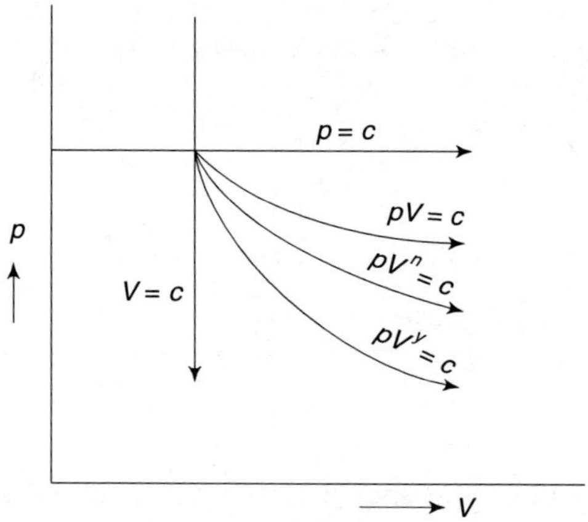

Fig. 2.13 Various thermodynamic processes

In compression process, volume decreases while pressure and temperature increase. In expansion process, volume increases while pressure and temperature decrease.

(i) Constant pressure process ($p = c$)

(ii) Constant volume process ($V = c$)

(iii) Isothermal process ($pV = c$, or $T = c$)

(iv) Adiabatic process ($pV^\gamma = c$)

(v) Polytropic process ($pV^n = c$)

By giving the proper value of index n in the polytropic process, the other above-mentioned processes may be described. For example, $n = 0$, gives $p = c$, $n = \pm\infty$, gives $V = c$, $n = 1$ gives $pV = c$ and $n = \gamma$ gives $pV^\gamma = c$.

(vi) Free expansion process ($U = c$)

(vii) Throttling process ($H = c$)

Applications of these processes are given in chapter 4.

Example 2.7 A pump delivers 1200 litres of water per hour against a constant head of 100 metres. Calculate the work done per hour in heat units.

Solution The mass of 1 litre of water is 1 kg. Therefore, the quantity of water raises per hour = 1200 × 1 = 1200 kg. Work done by pump/hr = Force per hr × distance.

$$= m \times g \times h = 1200 \times 9.81 \times 100 = 1177200 \text{ J} \quad \textbf{Ans.}$$

Example 2.8 Compressed helium gas kept in a bottle is used to inflate an inelastic flexible balloon, originally folded completely flat to a volume of 0.6 m^3. If the barometer reads 76 cm Hg, calculate the amount of work done upon the atmosphere by the balloon. Sketch the system showing the boundary before and after the process.

Solution Refer to Fig. 2.14. The firm and dotted lines show the boundary of the system before and after the process. The displacement work is given by

$$W_{disp} = \underset{\text{Balloon}}{\int_1^2} pdV + \underset{\text{Bottle}}{\int_1^2} pdV = p\Delta V + 0$$

$$= 1.01325 \times 10^5 \times 0.6 = 0.60795 \times 10^5 \text{ J} = 60.795 \text{ kJ} \qquad \textbf{Ans.}$$

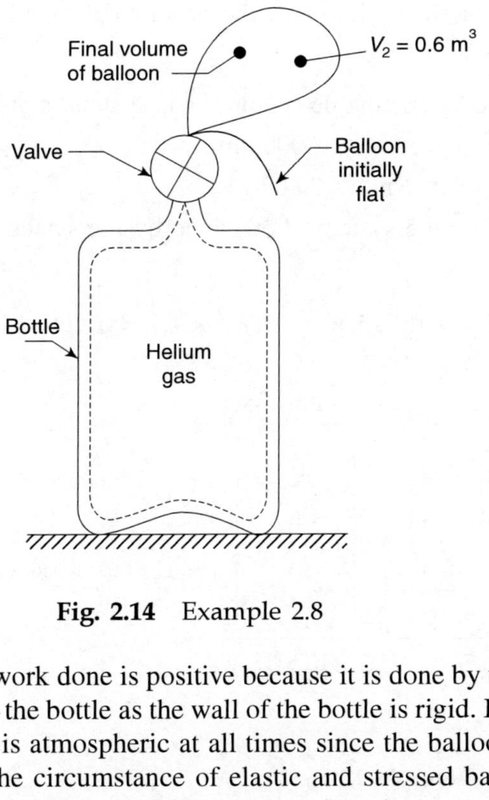

Fig. 2.14 Example 2.8

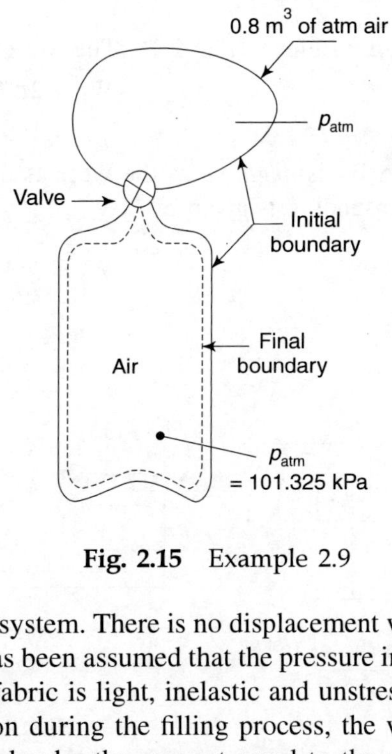

Fig. 2.15 Example 2.9

The work done is positive because it is done by the system. There is no displacement work done by the bottle as the wall of the bottle is rigid. It has been assumed that the pressure in the balloon is atmospheric at all times since the balloon fabric is light, inelastic and unstressed. Under the circumstance of elastic and stressed balloon during the filling process, the work done by the gas would be greater than the present value by the amount equal to the work done in stretching the balloon. But it is to be noted that if the system includes both bottle and gas as the system, the displacement work would be as calculated above.

Example 2.9 When the valve of an evacuated bottle is opened atmospheric air rushes into it. The atmospheric air is at 101.325 kPa and 0.8 m^3 of air enters the bottle at atmospheric condition. Draw the system and calculate the work done.

Solution Refer to Fig. 2.15.

The displacement work done by the air is given by

$$W_{disp} = \int_{\text{Bottle}} pdV + \int_{\text{Free air boundary}} pdV = 0 + p\Delta V = 101.325 \times 0.8 = 81.06 \text{ kJ}$$

$$= 101.325 \times 0.8 = 81.06 \text{ kJ}$$ **Ans.**

Since the free-air boundary is contracting, i.e., ΔV is negative, so the work done by the system is negative and the surroundings do positive work upon the system.

Example 2.10 A piston cylinder arrangement contains a fluid system and is equipped with stirring device driven by lowering the weight. The piston is frictionless and is held down against the fluid due to the atmospheric pressure of 101.325 kPa. The stirring devices makes 15,000 revolutions with an average torque against the fluid 1.9 mN. As a result of stirring work, the piston having 0.8 m diameter moves out 1.2 m. Draw the system and calculate the net work transfer.

Solution Refer to Fig. 2.16. The work done by stirring device upon the system is given by

$$W_1 = 2\pi\, TN = 2\pi \times 1.9 \times 15000 \text{ Nm}$$

$$= 179.07 \times 10^3 \text{ Nm} = 179.07 \text{ kJ}$$

Here W_1 is *negative* as the work is done on the system. Work done by the system upon the surrounding is given by

$$W_2 = (pA)\,.\,L = 101.325 \times \frac{\pi}{4}(0.8)^2 \times 1.2 = 61.117 \text{ kJ}$$

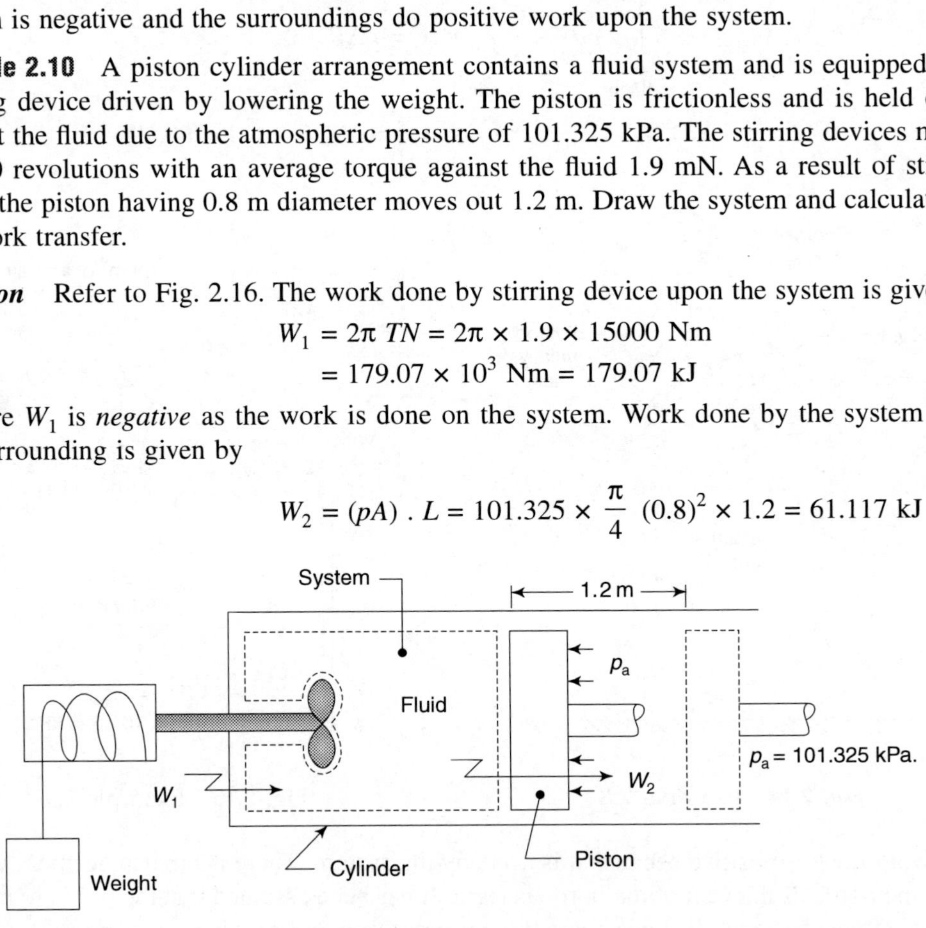

Fig. 2.16 Example 2.10

Here, W_2 is the work done on the system, hence it is positive.

Thus, $$W = W_1 + W_2 = -179.07 + 61.117 = -117.953 \text{ kJ}$$ **Ans.**

Example 2.11 In a cylindrical furnace, 10 tonnes/hr of iron is to be melted from a charge at 15°C to a molten metal at 1660°C. The melting point is 1540°C and the latent heat is 270 kJ/kg. The specific heats in solid and liquid states are 0.503 and 0.534 kJ/kg K. Assuming the electric furnace efficiency is 70%, calculate the rating of the furnace in kW.

If the density in molten state is 6900 kg/m^3 and bath volume is three times the hourly melting rate, calculate the diameter of the cylindrical furnace by assuming the length to diameter ratio is 2.

Solution Heat required to melt 1 kg of iron at 15°C to a molten metal temperature of 1660°C

= Heat required to raise the temperature of metal from 15°C to 1540°C + Latent heat + Heat required to raise the temperature from 1540°C to 1660°C

$$= c_{pm}\,(t_{mp} - t_{initial}) + \text{Latent heat} + c_{pm}\,(t_{final} - t_{mp})$$

$$= 0.503\,(1540 - 15) + 270 + 0.534\,(1660 - 1540)$$

$$= 1101.15 \text{ kJ/kg}$$

The rate of heat supplied required

$$= \text{Melting rate} \times 1101.15 = 10 \times 10^3 \times 1101.15$$

$$= 1101.15 \times 10^4 \text{ kJ/h}$$

$$\text{The rate of furnace} = \frac{\text{Rate of heat supplied per second}}{\text{Furnace efficiency}}$$

$$= \frac{1101.15 \times 10^4}{3600 \times 0.7} = 4.369 \times 10^3 \text{ kW} \qquad \textbf{Ans.}$$

$$\text{Volume of furnace needed} = V = \frac{3 \times 10 \times 10^3}{6900} = 4.347 \text{ m}^3$$

If d and l be the diameter and length of furnace, then

$$V = \frac{\pi}{4} d^2 \times l = 4.369 = \frac{\pi}{4} d^2 \times 2d = \frac{\pi\, d^3}{2}$$

or

$$d = \left(\frac{4.369 \times 2}{\pi}\right)^{1/3} = 1.405 \text{ m}$$

and

$$l = 2\,d = 1.405 \times 2 = 2.81 \text{ m} \qquad \textbf{Ans.}$$

Example 2.12 Calculate the non-flow work done for a gas which expands from initial pressure 5 bar and volume 4 m^3 to final volume 20 m^3 under the following reversible processes. Show the process on p-V diagram.

(a) $p = c$, (b) $V = c$, (c) $pV = c$, (d) $pV^{\gamma} = c$ where $\gamma = 1.4$ and (e) $pV^n = c$

where $n = 1.25$. Comment on the results.

Solution Refer to Fig. 2.17.

(a) When $p = c$, i.e., $p_1 = p_2$ (Isobaric process)

$$W_{1-2} = \int_1^2 p\,dV = p\,(V_2 - V_1) = 5 \times 10^2\,(20 - 4) = 8 \times 10^3 \text{ kJ} = 8 \text{ MJ} \qquad \textbf{Ans.}$$

(b) When $V = c$, i.e., $dV = 0$ (Isobaric process)

$$W_{1-2} = \int_1^2 p\,dV = 0 \qquad \textbf{Ans.}$$

(c) When $pV = c$, i.e., $p_1 V_1 = p_2 V_2$ (Isothermal process)

$$W_{1-2} = \int_1^2 p\,dV = \int_1^1 \frac{C}{V}\,dV = p_1 V_1 \int_1^2 \frac{dV}{V} = p_1 v_1 \ln \frac{V_2}{V_1}$$

$$= 5 \times 10^2 \times 4 \ln \frac{20}{4} = 3218.8 \text{ kJ} = 3.2188 \text{ MJ} \qquad \textbf{Ans.}$$

Here, $$p_2 = p_1 \frac{V_1}{V_2} = 5 \times \frac{4}{20} = 1 \text{ bar}$$

(d) When $pV^\gamma = c$ (Adiabatic process)

$$W_{1-2} = \int_1^2 p\,dV = \int_1^2 \frac{c\,dV}{V^\gamma} = \left[\frac{V^{-\gamma+1}}{1-\gamma}\right]_1^2$$

or $$W_{1-2} = \frac{p_2 V_2^\gamma V^{-\gamma+1} - p_1 V_1^\gamma V_1^{-\gamma+1}}{1-\gamma} = \frac{p_1 V_1 - p_2 V_2}{\gamma - 1}$$

Here, $$p_2 = p_1 \left(\frac{V_1}{V_2}\right)^{1.4} = 5\left(\frac{4}{20}\right)^{1.4} = 0.525 \text{ bar}$$

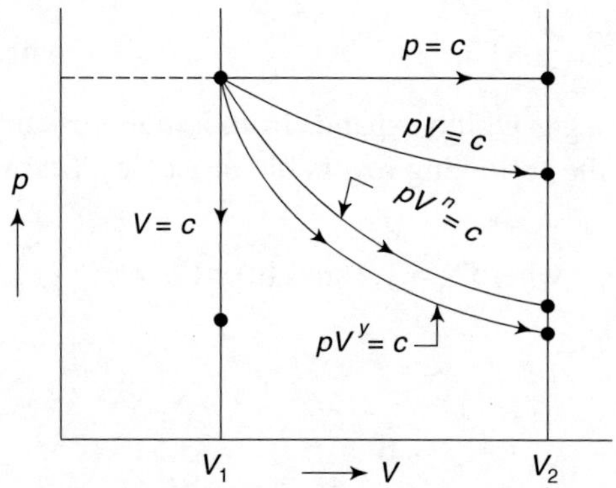

Fig. 2.17 Example 2.12

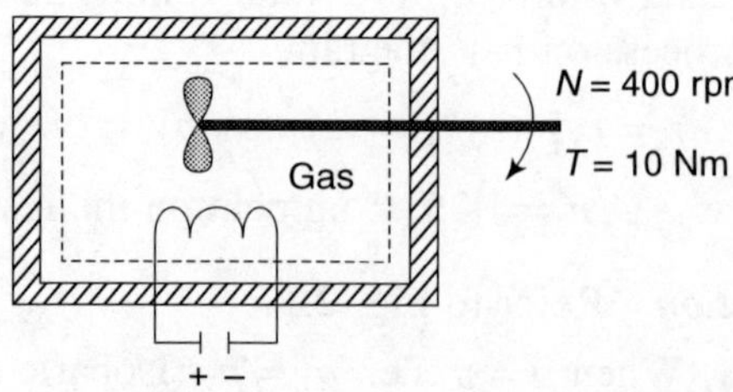

Fig. 2.18 Example 2.13

$$\therefore \quad W_{1-2} = \frac{5\times10^2 \times 4 - 0.525\times10^2 \times 20}{0.4} = 2375 \text{ kJ} = 2.375 \text{ MJ} \quad \textbf{Ans.}$$

(e) When $pV^n = n$ when $n = 1.25$ (Polytropic process)

$$W_{1-2} = \int_1^2 p\,dV = \int_1^2 \frac{c\,dV}{V^n} = c\left[\frac{V^{-n+1}}{1-n}\right]_1^2 = \frac{p_1 V_1 - p_2 V_1}{n-1}$$

Here
$$p_2 = p_1\left(\frac{V_1}{V_2}\right)^{1.25} = 0.668 \text{ bar}$$

$$\therefore \quad W_{1-2} = \frac{5\times10^2 \times 4 - 0.668\times10^2 \times 20}{0.25} = 2656 \text{ kJ} = 2.656 \text{ MJ} \quad \textbf{Ans.}$$

The work done is the maximum when $p = c$ but practically, it is not possible. The work done is zero in the case of $V = c$. The work done processes are $pV^\gamma = c$ as and $pV^n = c$.

Example 2.13 An insulated tank contains a gas, a paddle wheel connected to the shaft, and an electric resistor connected to a battery. The shaft produces a torque of 10 Nm at the speed of 5400 rpm. At the same time, an electric current of 8 ampere is supplied to the resistor from a battery with a potential difference of 9 V. Draw the system and calculate the net power in watts and the network on the system if the process continues for 1 minute.

Solution Refer to Fig. 2.18. The system is the gas contained inside the tank within the system boundary.

$$\text{Net work} = W_{net} = W_{paddle} + W_{elect} \text{ or } P_{net} = P_{paddle} + P_{elect}$$

or
$$P_{net} = 2\,\pi\,NT + VI = 2\,\pi \times \frac{400}{60} \times 10 + 9 \times 8 = 490.879 \text{ W} \quad \textbf{Ans.}$$

$$W_{net} = \int_{\tau_1}^{\tau_2} P_{net}\,d\tau = P_{net}\,(\tau_2 - \tau_1) = 490.879 \times 60 = 29452.74 \text{ J} \quad \textbf{Ans.}$$

2.11 SUMMARY OF FORMULAE

1. $W = \int F dL$

2. $P = \dfrac{W}{\tau}$

3. $W_{1-2} = \int_1^2 pdV = p\,(V_2 - V_1)$

4. $(\delta w)_{\text{electric charge}} = -\,Vdq$

5. $(\delta w)_{\text{polar}} = -\,E\,dP$

6. $(\delta W)_{magn} = -\mu_0 H\, dM$
7. $(\delta W)_{cell} = -EI\, d\tau$
8. $(\delta W)_{lig.\ surf} = -y\, dA$
9. $(\delta W)_{capacitor} = -V\delta Q$
10. $(\delta W)_{shaft} = T\, d\theta$ and $P_{shaft} = T\,\omega$
11. $(\delta W)_{stirring} = T\, d\theta$
12. $(W)_{1-2})_{wire} = -\dfrac{AFL}{2}(\varepsilon_2^2 - \varepsilon_1^2)$
13. $W_{total} = W_{disp} + W_{shear} + W_{stirring} + W_{elect} + W_{magnetic} + \ldots$

PROBLEMS

Things to Think About

2.1 Why is it incorrect to say that a system contains heat?

2.2. When a falling object impacts the earth and comes to rest what is the state of its kinetic and potential energies?

2.3. When a cup of coffee is stirred, what happens to the energy transferred to the coffee by work?

2.4. For a moving automobile, mention the various energy transfer by work and heat.

2.5. What form does the energy balance take for an isolated system? Interpret the expression you derive.

2.6. If the change in energy of a closed system is known for a process between two end states, identify if the energy change was due to work, to heat transfer, or to some combination of work and heat transfer.

2.7. How will you define the appropriate efficiency of a gear box?

Short and Long Theoretical Questions

2.8. Define work. What are the different forms of work? Explain.

2.9. Show work is a path function and not a state function.

2.10. Define the term 'heat' in detail. Justify the statement that work and heat are not properties.

2.11. Differentiate between heat and work by citing similarities and dissimilarities and by taking an example.

2.12. Derive the expressions of work done for various quasi-static or reversible processes.

2.13. Show that work is a path function and not the property of system.

2.14. What are positive or negative work interactions?

2.15. Give the sketches of various thermodynamic devices with their major functions and applications.

2.16. Derive the expression for work done for the following processes, (a) $p = c$, (b) $V = c$, (c) $pV = c$, (d) $pV^{\gamma} = c$, (e) $pV^n = c$

Numerical Questions

2.17. A piston-cylinder arrangement is containing a fluid at 10 bar, the initial volume being 0.05 m^3. Find the work done by the fluid when it expands reversible for the following cases :

(a) at constant pressure to a final volume of 0.2 m^3.

(b) according to the linear law to a final volume of 0.2 m^3 and a final pressure of 2 bar.

(c) according to law pV = constant to a final volume of 0.06 m^3.

(d) according to law pV^3 = constant to a final volume of 0.06 m^3.

(e) according to law = $(a/V^3) - (B/V)$ to final volume of 0.1 m^3 and a final pressure of 1 bar. Where A and B are constant. Sketch all process on diagram p-V.

[*Ans.* 150 kJ, 90 kJ, 34.7 kJ, 7.64 kJ, 17.67 kJ]

2.18. A cylinder-piston arrangement is containing a fluid at a pressure of 3 bar and with specific volume of 0.18 m^3/kg allows the fluid to expand reversible to a pressure of 0.6 bar according to the law $p = c/v^2$ where c is a constant. Determine the work done by the fluid on the piston.

[*Ans.* 29.84 kJ]

2.19. An escalator raises a 100 kg bucket of sand 10 m in 1 minute. Calculate the total amount of work done and the instantaneous rate of work during the process.

2.20. A linear spring whose force characteristics is $F = K_s\,(x - x_0)$, with spring constant K_s = 500 N/m, is stretched until it is 100 mm longer than original. Calculate the required force and work input. [*Ans.* 50 N; 2.5 J]

2.21. In a two-part process with an expansion from 0.1 to 0.2 m^3 at a constant pressure of 150 kPa is followed by an expansion, from 0.2 to 0.4 m^3 with a linearly rising pressure from 150 kPa ending at 300 kPa. By showing the process on p-V diagram calculate the boundary work.

[*Ans.* 60 kJ]

2.22. The gas space above the water in a closed storage tank contains nitrogen gas at 30°C, 100 kPa. The total tank volume is 5 m^3, and there is 600 kg of water at 30°C. An additional 600 kg of water is forced into the tank. Assuming constant temperature throughout, calculate the final pressure of the nitrogen and the work done on the nitrogen in this process.

2.23. Air at 300 kPa, 30°C is contained in a cylinder-piston arrangement with initial volume of 0.1 m^3. The inside pressure balances ambient pressure of 100 kPa plus an externally imposed force that is proportional to $V^{0.5}$. Now heat is transferred to the system to a final pressure of 400 kPa. Calculate the final temperature and the work done in the process.

2.24. A spherical elastic balloon initially containing 6 kg of ammonia as saturated vapour at 20°C is connected by a valve to a 3 m^3 evacuated tank. The balloon is made such that the pressure inside is proportional to diameter. The valve is now opened, allowing ammonia to flow into the tank until the pressure in the balloon has dropped to 600 kPa at which point the valve is closed. The final temperature in both the balloon and the tank is 20°C. Calculate (a) the final pressure in the tank, and (b) the work done by the ammonia in the process.

2.25. A 0.5 m long steel rod with 1 cm diameter is stretched in a tensile test. Calculate the work required to obtain a relative strain of 0.1%. The modulus of elasticity of steel is 2 × 108 kPa.

[*Ans.* 3.93 J]

2.26. A battery is well insulated while being charged by 12.3V at a current of 6A. Assuming the battery as a control mass, calculate the instantaneous rate of work and the total work done over 3 hours.

2.27. 680 kg of fish at 5°C are to be frozen and stored at – 12°C. The specific heat of fish above freezing point is 3.182 kJ/kgK and below freezing point is 1.717 kJ/kgK. The freezing point of fish is – 2°C, and the latent heat of fusion is 234.5 kJ/kg. Calculate the heat removed to cool the fish and its percentage of latent heat. [*Ans.* 186.27 MJ, 85.65%]

2.28. The drag force, F_D, imposed by the surrounding air on a vehicle moving with velocity V is given by $F_D = c_d\, A(1/2)\, \rho C^2$.

where c_d is the coefficient of drag, A is the projected frontal area of the vehicle and ρ is the air density. Calculate the power, in kW, required to overcome aerodynamic drag for a truck moving at 110 km/h. Take $c_d = 0.65$, $A = 10\ \text{m}^2$, $\rho = 1.1\ \text{kg/m}^3$.

2.29. At a speed of 50 km/h, the resistance to motion of a car is 900 N. Neglecting losses, calculate the power of the engine of the car at this speed. Determine also the engine developing power.

2.30. Two objects having different masses fall freely under the influence of gravity from rest and the same initial elevation. By neglecting the effect of air resistance, show that the magnitude of the velocities of the objects are equal at the moment just before the strike the earth.

2.31. The drag force, imposed by the surroundings air on a vehicle moving with velocity C in expressed as

$$F_d = c_d A\ (1/2)\ \rho\ C^2$$

where c_d is the drag coefficient, A is the projected frontal area of the vehicle and ρ is the air density. Calculate the power required to overcome the aerodynamic drag for a bus moving at 100 km/h. Take $c_d = 0.65$, $A = 10\ \text{m}^2$, and $\rho = 1.1\ \text{kg/m}^3$.

2.32. A major force opposing the motion of a vehicle is the rolling resistance of the tires, F_r which is expressed as

$$F_r = f\ .\ W$$

where f is a constant called the rolling resistance coefficient and W is the weight of the vehicle. Calculate the power, in kW, required to overcome rolling resistance of the tires, for a truck weighing 350 kN which is moving at a speed of 110 km/h. Take $f = 0.0069$.

2.33. 2 kg of a gas contained with the piston-cylinder assembly undergoes a constant-pressure process beginning at $v_1 = 0.75\ \text{m}^3/\text{kg}$. For a gas as the system, the work is – 90 kJ. Calculate the final volume of the gas, in m^3.

Software Based Questions

2.34. Develop a software for finding out work done by considering various thermodynamic processes and plot the results by taking real life data using any graphic package.

2.35. Develop a software to calculate the power required to overcome rolling resistance and aerodynamic drag by various types of vehicles. Take real data from manufacturers and plot the results by varying the independent variables using any graphic package. Suggest the exterior shape of vehicle to reduce aerodynamic drag.

2.36. Develop a software to calculate the forms of works during quasi-static or reversible processes.

Design and Open Ended Questions

2.37. Methane-laden gas generated by the decomposition of landfill trash is more commonly flared than exploited for some useful purpose. Write a report on the possible uses of landfill gas through the research literature on this topic.

2.38. The concept of specially designed flywheels are being used by electric utilities to store electricity. Automotive applications of flywheel energy storage also have been proposed. Prepare a report discussing promising uses of flywheels for energy storage with consideration of flywheel materials, their properties and costs.

Multiple-Choice Answer Type Questions

2.39. The elastic work (δW) per unit volume required for stretching a wire of length is given by the expression

(a) $-\dfrac{\sigma}{d\varepsilon}$ (b) $-\sigma\, d\varepsilon$ (c) $-\sigma - d\varepsilon$ (d) none of the above

where σ is stress and $d\epsilon$ is infinitesimal strain.

2.40. The work of polarization and magnetization is expressed as

(a) $-\mu_0 HdM$ (b) $-\mu_0 \dfrac{H}{dM}$ (c) $-\mu_0 \dfrac{dM}{H}$ (d) none of the above

2.41. Which does not show the similarities between work and heat

(a) Energy interaction (b) Transient phenomena

(c) Boundary phenomena (d) All energy transfer

2.42. Which does not show the similarities between work and heat

(a) Property of system (b) Path function

(c) Inexact differentials (d) Exact differentials

Answers to Multiple-Choice Questions

2.39. (b) 2.40. (a) 2.41. (d), 2.42. (d)

CHAPTER

3

Temperature and Zeroth Law of Thermodynamics

OVERVIEW

The credit for formulating zeroth law of thermodynamics goes to Ralph H. Fowler. By the time scientist Fowler felt the need of zeroth law of thermodynamics, the first and second laws had become so popular, so its name was given zeroth law as it gives the concept of temperature.

This chapter deals with the concept of temperature, zeroth law of thermodynamics and basic principle of various thermometers used in the measurement of temperature.

3.1 CONCEPT OF TEMPERATURE

The concept of temperature arises from the sensory perception of hotness of coldness. The housewife thinks of temperature as how hot she must set the oven to cook the roast. The computer engineer thinks of temperature as how cool he must keep the silicon chips and integrated circuit to make them work reliably. The plasma physicist views temperature as a measure of kinetic energy of the molecules or electrons. And the astronomer views temperature as a measure of the radiant energy emission from the stars. Truly speaking, such physiological sensation is insufficient for precise knowledge of temperature. But all these diverging concepts have one thing in common; they all relate to energy or energy transfer, hence clearly mark *temperature as a thermodynamic property.*

Actually, the word temperature is derived from the Latin word "*Temperature*" meaning proper mixing or tempering implying attainment of thermal equilibrium. *According to Maxwell "Temperature of body is its thermal state considered with reference to its ability to communicate heat to other bodies.* In a precise way, *temperature is that property of a system which determines whether or not it is in thermal equilibrium with the other systems. Temperature is an intensive property of the system* because it is independent of mass. It is *not a function of path but a function of state of the system.*

From the microscopic of view, temperature is a manifestation of activity of the molecules. An increase in temperature is accomplished by simultaneous increase in kinetic energy of the molecules. When two systems are in thermal equilibrium, the average kinetic energy of the molecules is the same for two systems.

It is important to have a clear distinction in one's mind about the **concepts of heat, temperature and internal energy**. *Internal energy is the energy possessed by molecules that are "hidden" from direct macroscopic view by the disorganized character of the microscopic state.* That energy may or may not have been put in the matter through transfer of energy as heat. *Heat is energy transfer* that is not accounted for in macroscopic evaluation of energy transfer as work; heat is microscopic work that is hidden from direct macroscopic view by the disorganized nature of the energy transfer process. Temperature is a property of matter, if the temperature of one body is greater than temperature of the other, then any energy transfer as heat will take place from the hotter to the colder body. *The internal energy of a substance will depend in part, but not wholly on the temperature. The temperature, therefore, is not in general a complete measure of the internal energy.* Temperature can only be defined when a body is in equilibrium; the body has energy regardless of whether it is in equilibrium or not. A body does not have any heat, it may have received some energy as heat, but now that energy appears as internal, and one cannot tell from the body's state whether that energy entered the body as heat or as work. *It is to be noted thai temperature and internal energy are properties of matter, heat is not a property.* Temperature and internal energy are also inherently different kinds of properties. The temperature of a small piece of a large body is the same as the temperature of the body as a whole; the energy of the small piece is but a small fraction of energy of the entire body.

The concept of temperature will be more clear when we define the equality of temperature and Zeroth Law, as it forms the concept of temperature.

3.2 EQUALITY OF TEMPERATURE

In order to understand equality of temperature, i.e., a method of defining and measuring temperature, consider two systems *A* and *B*, perfectly insulated from the surroundings. Both of them are brought in good contact, the energy in the form of heat will transfer from the body at a higher temperature to the body at the lower temperature. After sufficient time is elapsed, the bodies *A* and *B* will approach a state at which no further change is observed. The two bodies are then in a state of thermal *equilibrium*. At this point, the two systems are said to be at the same temperature. *This is called equality of temperature*. This forms the basis of concept of temperature.

3.3 ZEROTH LAW OF THERMODYNAMICS

Consider three systems *A*, *B* and *C* as shown in Fig. 3.1 perfectly insulated from surrounding and apply the equality of temperature. If *A* and *C* are brought into good contact, energy in the form of heat will transfer from the body at a higher temperature to the body at a lower temperature. After a certain time they will be in thermal equilibrium. If *B* and *C* are brought into good contact, after some time, these two will be in thermal equilibrium. Now if *A* and *B* are brought in good contact they will be also in thermal equilibrium. Hence, this suggests that *A* and *B* are in thermal equilibrium with each other. From this experiment we come to the conclusion that:

If two systems are each in thermal equilibrium with a third system, they are also in thermal equilibrium with each other. This statement is known as the Zeroth Law of Thermodynamics *and is the basis of concept of temperature as well as the basis of all temperature measurement. The third body C is called the thermometer.*

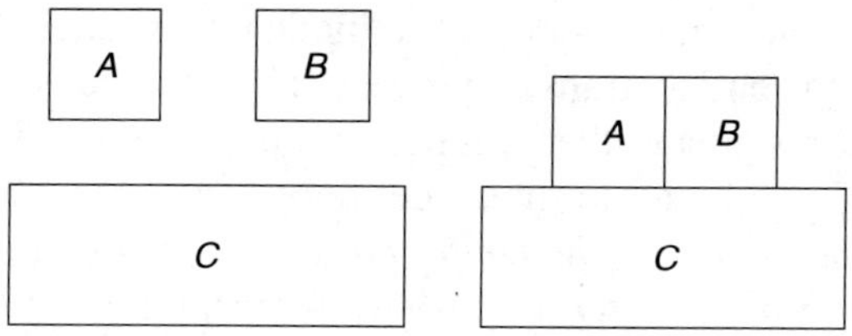

Fig. 3.1 Concept of Zeroth law

The Zeroth law permits us to test the equality of temperature without actually bringing the system in thermal contact. In order to understand this, consider two beakers of water at the same temperature. If we insert a thermometer in first beaker and keep it till the mercury column to a stand still and then insert the same thermometer in the second beaker and if there is no change in the mercury column, we conclude that the two beakers are at the same temperature. Such an instrument will indicate equality or inequality of temperature without determining a numerical value of temperature.

3.4 MEASUREMENT OF TEMPERATURE (THERMOMETRY)

Thermometry may be defined as the art or device of measuring temperature with scientific precision. The basic principle of thermometry is that the two bodies in contact and in thermal equilibrium have the same temperature. To measure temperature we use indirect methods. Generally, heat is transferred to a measuring instrument and the change due to temperature in some property is recorded. *A property which changes in value as a function of temperature is called the* ***thermometric property****. The substance which have the thermometric property is is called the* ***thermometric substance***. In practice, thermometry involves:

(a) *The Construction of an Instrument known as Thermometer*. The thermometer is based on the certain physical properties of matter, *like expansion, electric resistance, etc.* which are readily affected by the thermal state of the matter in a fairly regular manner; the choice of thermometric substance which defines both the matter and its state chosen, i.e., solid, liquid, or gas. The thermometric substance is selected on the basis of range of temperature to be measured as well as the accuracy of the result desired. Various types of thermometers with their thermometric properties are shown in Table 3.1.

(b) *The Calibration of the Instrument*. In order to calibrate the thermometer, two fixed points are selected, one is the ice point corresponding to the temperature at which pure ice melts and the other, the steam point which is boiling point of pure water. After this, interval between these two fixed points is divided in equal parts which are assigned arbitrary numerical values of temperature, known as **thermometric scale**, e.g., the centigrade, Kelvin. **Now, one fixed point is in common practice**.

Table 3.1 Thermometers and thermometric properties

Thermometers	Thermometric properties
1. Liquid thermometer	Length (L)
2. Constant volume gas thermometer	Pressure (p)
3. Constant pressure gas thermometer	Volume (V)
4. Electric resistance thermometer	Resistance (R)
5. Thermocouple	Thermal e.m.f. (ε)
6. Radiation pyrometer	Thermal radiation (E)
7. Optical pyrometer	Monochromatic radiation (E_λ)
8. Pyrometric cone	Fusion

(c) *The Sensitiveness of the Instrument*. A thermometer is considered sensitive when it (i) quickly takes up the temperature of the body with which it is put in contact and (ii) reflects even small changes of temperature.

The use of any particular kind of thermometer depends on the range of temperature to be measured and the accuracy needed. For example, the mercury thermometer is a simple, convenient device for measuring ordinary temperature with a fair degree of accuracy. A well constructed gas thermometer can measure much wider ranges of temperature with a much greater sensitivity; the resistance and thermoelectric thermometers are used to measure fairly high and low temperature; the radiation pyrometers are best suited for inaccessible sources and for much higher temperatures than those covered by all the others; optical pyrometers and pyrometric cones are also suitable for high temperatures.

3.5 TEMPERATURE SCALE

In order to measure the temperature of a system by an instrument such as a thermometer, a **temperature scale** (i.e., assigning a numerical value to the thermometer) must be provided on the thermometer. Since temperature is a property, it has a unique value at each equilibrium state of the system, hence in order to define or establish a temperature scale, the first step is to ***select some thermometric property***.

After selecting the suitable thermometric property, the next step is to select the form of variation of thermometric property as a function of temperature and to select reference point or fixed point. It is the form of thermometric function which determines the temperature scale. Let us consider for simplicity that the temperature t is a linear function or thermometric property (say x in general and in particular length L of mercury column).

It may be any one of the following forms:

$$t = ax + b = aL + b \quad ...(3.1)$$

and
$$t = ax = aL \quad ...(3.2)$$

where a and b are arbitrary constants.

To calculate the constants, fixed point have to be considered.

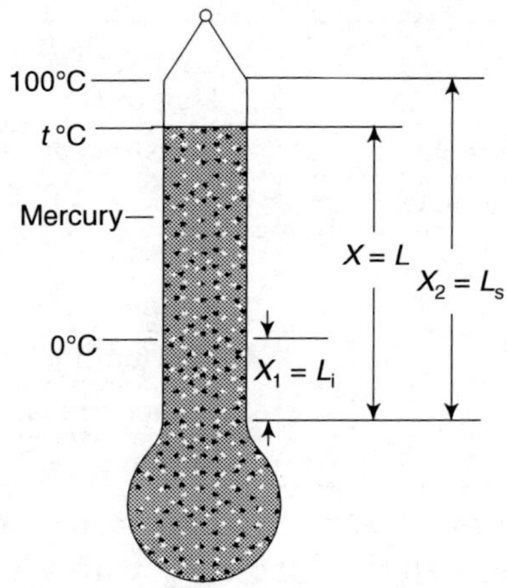

Fig. 3.2 Mercury thermometer

(i) Method in Use before 1954 Before 1954, it was usual to take ice point (corresponds to the equilibrium temperature of ice and air saturated pure water) and the steam point (corresponds to the equilibrium temperature of pure water vapour) at one standard atmospheric pressure as the two fixed points for fixing the scale. Let these temperature be represented by t_1 and t_2 on this scale and the corresponding thermometric property be x_1 (or L_1) and x_2 (or L_s) where 1 or i corresponds to the ice point and 2 or s corresponds to the steam point (Fig. 3.2).

Using Eq. (3.1), we get

$$t_1 = ax_1 + b \text{ and } t_2 = ax_2 + b$$

or
$$t = t_1 + \frac{(t_1 - t_2)}{(x_1 - x_2)}(x - x_1)$$

The fixed points assigned to the ice and steam point in Centigrade (or Celsius) scale are 0°C and 100°C and in Fahrenheit scales, these values are 32°F and 212°F respectively. Thus, in Centigrade scale, $t_1 - t_2 = 100$°C and $t_1 = 0$°C hence the above equation takes the form of

$$t^\circ\ C = 100\left(\frac{x - x_1}{x_2 - x_1}\right) = 100\left(\frac{L - L_i}{L_s - L_i}\right) \quad ...(3.3)$$

Note that Eq. (3.3) defines the centigrade temperature on the scale of this particular thermometer. A second thermometer is constructed from different kind of glass and using a different liquid and is calibrated in the above manner. Let L'_s and L'_i be the reading of their fixed points. The new thermometer is placed in contact with the same body for which the first thermometer, gave the temperature t. Let t' be the temperature of same body by second thermometer, then t' will be given by

$$t' = 100 \frac{(L' - L'_i)}{(L'_s - L'_i)} \qquad ...(3.4)$$

Generally, it is found that t' is not equal to t even though these are temperatures of the same body. This is because of the fact that the two liquids do not expand in the same manner. In addition, different variety of glasses have different expansion coefficients and the methods of construction also affects the readings. *This drawback is applicable to all thermometers which are based upon single thermometric property.*

The use of two fixed points was found unsatisfactory and later abandoned because of (a) the difficulty of achieving equilibrium between pure ice and ice and air saturated water and (b) extreme sensitiveness of the steam point to the change in pressure.

(ii) Method in Use after 1954 Since 1954 only one fixed point has been in use, i.e., *the triple point of water. It is the state at which ice, liquid water and vapour coexist in equilibrium.* The temperature at which this state exists is arbitrarily assigned a value of 273.16 degrees Kelvin or 273.16 K. Let x, be the thermometric property and the body is placed in contact with the triple point of water whose temperature be t_t then as per Eq. (3.2).

$$t = ax \quad \text{and} \quad t_t = ax_t$$

or

$$a = \frac{t_t}{x_t} = \frac{273.16}{x_t}$$

$$t = 273.16 \frac{x}{x_t} \qquad ...(3.5)$$

Comparison of Thermometers Applying the method in use after 1954 to the following thermometers, the temperature recorded by them will be given by

(a) Liquid thermometer: $t\,(L) = 273.16 \dfrac{L}{L_t}$ ...(3.6)

(b) Thermocouples: $t\,(\varepsilon) = 273.16 \dfrac{\varepsilon}{\varepsilon_t}$ ...(3.7)

(c) Electric resistance thermometers: $t\,(R) = 273.16 \dfrac{R}{R_t}$ ...(3.8)

(d) Constant pressure gas thermometers: $t\,(V) = 273.16 \dfrac{V}{V_t}$ ...(3.9)

(e) Constant volume gas thermometers: $t\,(p) = 273.16\,\dfrac{p}{p_t}$ ...(3.10)

If above thermometers are employed to measure the temperature of a system simultaneously, it will be observed that there is a considerable difference among the readings due to the reasons stated above. *The minimum variation is found in the case of gas thermometers and that is why gas is chosen as* **standard thermometric substance**.

3.6 LIQUID THERMOMETERS (MERCURY)

Out of the many liquid thermometers, the mercury thermometers is the most commonly used in scientific work for ordinary temperature. This is because of the fact that mercury has many advantages as the thermometric fluid over other fluids. It consists of a cylindrical glass bulb (Fig. 3.2) containing mercury to which a graduated capillary be of uniform bore is attached. The upper point (steam point) and the lower point (ice point) are set by using hypsometer and ice respectively.

Since ice and steam point are fixed, hence

$$\frac{C}{100} = \frac{F-32}{180} = \frac{\text{Re}}{80}$$

$$\frac{C}{5} = \frac{F-32}{9} = \frac{\text{Re}}{4} \qquad \text{...(3.11)}$$

For ordinary thermometers, "Vender" glass is in use while for high temperature measurement "Juna 16" and Boro Silicate are used.

As pointed out, thermometers are generally used in the rough work but when they are employed for precision work, many errors have to be taken into consideration and these have to be corrected.

Limitation of Liquid Thermometers The main drawback of liquid thermometers is that, they use different liquids such as mercury and alcohol and are calibrated as stated above, they do not agree at any temperature except 0° and 100°. This is due to the fact that expansion of the different liquids is neither regular nor similar. Thus, such thermometers give an arbitrary scale of temperature. *In addition, the mercury thermometer needs many corrections, it is hardly suited for precision work because its nearest accuracy is 0.05°C.*

Considering the freezing point (– 38.87°C and boiling point 357°C) of mercury, however, it is mostly used in the range of 0°C to 300°C.

3.7 GAS THERMOMETERS (STANDARD OR ABSOLUTE SCALE OF TEMPERATURE)

The Advantages of Gas as Thermometric Substance Over Liquids. We have already seen that the thermometers which are based on a single property are not entirely satisfactory for temperature scale. The gas thermometer fills this gap. The advantages possessed by gas as a thermometric substance over liquids are:

(i) For the same increase of temperature gases expand much more than liquids. For 1 degree rise in temperature, a permanent gas expands twenty times as much as mercury. Gas thermometers, will lead to a negligible correction.

(ii) The expansion of the gases is more regular over a much wider range of temperature compared to liquids. For example, a permanent gas when not subjected to high pressure, maintains its state and behaves in the same manner at very high or low temperature while mercury is limited by its freezing and boiling point.

(iii) Gases have much lower thermal capacity than liquids, it is thus more sensitive to gain or loss of heat.

(iv) The gases possess very low specific gravity and as obtainable in the same purity all over the world. A perfect dry state of gas is essential for this purpose.

(v) The scale furnished by different gases are almost identical. This is because of the fact that the volume and pressure coefficients of all permanent gases nearly equal. This means that the temperature recorded is independent of thermometric substance. That is why *gas thermometers are taken as primary standards and all the other thermometers are compared with them for calibration and hence called secondary thermometers.*

Principle of Gas Thermometers The use of gas as a thermometric substance is based on the expansion or contraction of the gas which is a function of temperature. In other words, it is based on perfect gas law, i.e., the combination of Boyle's and Charles' laws which is given as

$$\frac{p_1 v_1}{T_1} = \frac{p_2 v_2}{T_2} = \text{constant}$$

or $$\left(\frac{pv}{T}\right) = \text{constant} \qquad \text{...(3.12)}$$

It is obvious from the Eq. (3.12) that *pv product of all gases is a function of temperature alone and is independent of the thermometric property.* Thus, gas may be used as a thermometric property to measure temperature and thermometers using gas as thermometric property will give a *standard scale of temperature or an absolute scale of temperature.* Equation (3.12) indicates two ways of measuring temperature with gas thermometer—

(i) Keeping the pressure constant and observing the change in volume – *constant pressure thermometer.*

(ii) Keeping the volume constant and noting the change in pressure – *constant volume thermometer.*

Two thermometers filled with a perfect gas obeying strictly the gas equation will agree throughout the scale in their indications of temperature measured by change of volume under constant pressure by one and by change of pressure at constant volume by the other. The temperature registered by the two thermometers is called the absolute temperature. *But pressure measurement at constant volume is easier than measurement of volume at constant pressure.* Hence, **the constant volume gas thermometer is more commonly used. Generally hydrogen gas is used for this purpose.**

Constant Volume Gas Thermometer A constant volume gas (say hydrogen) thermometer as shown in Fig. 3.3 consists of a platinum-iridium bulb *B* which contains the gas connected with U-tube filled with mercury through a platinum-iridium capillary tube. A flexible hose pipe connects the two limbs *D* and *D* of the U-tube. The volume of gas in the gas bulb *B* is kept constant by adjusting the level of mercury in the limb *D* at the tip of the capillary marked *A*. This is achieved by raising or lowering the limb *D*.

The difference in level *h* between the two mercury columns gives gauge pressure of the gas in the bulb *B*.

When the bulb is at the triple point of water, the pressure p_{tp} is equivalent to say 1000 mm of Hg absolutely. The bulbs is then brought to an unknown temperature *T* and the corresponding pressure measured, say *p*. Let this ratio p/p_{tp} be determined and called $(p/p_{tp})_{1000}$. Some gas is then removed such that at the triple point the pressure is equivalent to 900 mm of Hg abs. In the same manner, the bulb is brought to an unknown temperature and $(p/p_{tp})_{900}$ is determined. The above process is repeated to collect more and more values of (p/p_{tp}).

It is noted that the values of (p/p_{tp}) plotted against p_{tp} are straight lines for all gases, such as, hydrogen, helium, oxygen, etc., shown in Fig. 3.4. On extrapolation, these lines give (p/p_{tp}) at $p/p_{tp} = 0$ which will meet at a point. Thus, the thermodynamic temperature *T* can then be defined as

$$\frac{T}{T_{tp}} = \lim_{p_{tp} \to 0} \frac{p}{p_{tp}} \quad \text{or} \quad T = 273.16 \lim_{p_{tp} \to 0} \frac{p}{p_{tp}} \qquad \text{...(3.13)}$$

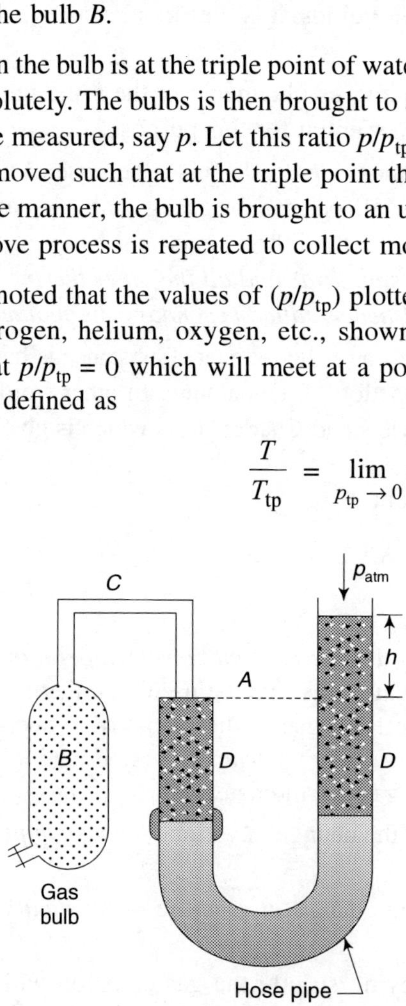

Fig. 3.3 Constant volume gas thermometer

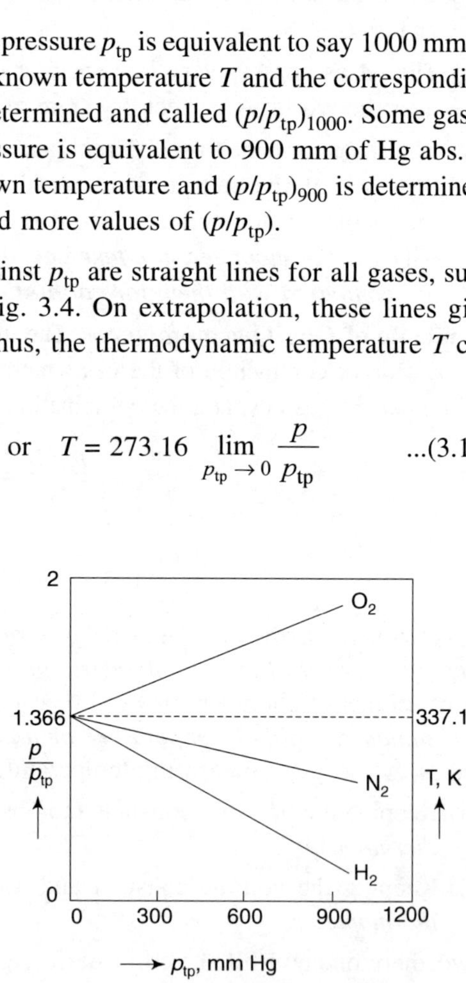

Fig. 3.4 p/p_{tp} versus p_{tp}

The temperature *T* is called the *absolute temperature*. Since the value of T_{tp} is fixed, other temperature can be found out.

Two absolute scale exist the Kelvin and Rankine scale. At the Tenth Conference of Weights and Measures in 1954 the Kelvin scale assigned a value of 273.16 to the triple point of water. The Kelvin scale has the same numerical intervals as the Celsius scale (Centigrade scale), but its value at the ice point corresponds to 0.01°C rather to zero. Thus, the conversion between the Kelvin and Celsius temperature scales are accomplished through the following manner:

$$T°K = t°C + 273.15 \qquad ...(3.14)$$

The unit of temperature in the Celsius is degree Celsius (i.e., °C) and the Kelvin scale is degree Kelvin (i.e., K). The Kelvin scale is also called the absolute Centigrade scale. *Instead of °K only K is used.*

In the Rankine scale, the triple point of water is 491.69°R whereas the ice pointis 491.69°R. The unit denoted by °R is called the absolute Fahrenheit degree thus, the Kelvin and Rankine scale is related as

$$\frac{T°R}{T°K} = \frac{491.69}{273.16} = \frac{180}{100} = \frac{9}{5} \qquad ...(3.15)$$

The Rankine scale has the same unit temperature intervals as the Fahrenheit scale, but the zero values is 491.67 – 32 = 459.67°F below the zero of the Fahrenheit scale. Hence, the relationship between the rankine and the Fahrenheit scale is

$$T° R = t° F + 459.67 \qquad ...(3.16)$$

In the Eq. (3.11), if an unknown temperature is the steam point, then

$$\frac{T_s}{T_{tp}} = \lim_{p_{tp} \to 0} \frac{p_s}{p_{tp}}$$

The triple point may be taken as ice point because there is negligible difference between ice point and triple point (i.e., 273.16 – 273.15 = 0.01°C and 491.69 – 491.67 = 0.02°F).

Thus,
$$\frac{T_s}{T_i} = \lim_{p_{tp} \to 0} \frac{p_s}{p_i}$$

The above process for drawing the graph between p_s/p_i and p_i is carried out them on extrapolation of lines on the graph result in p_s/p_i = 1.366 at p_i = 0 (since at $p \to 0$, the gas behaves at ideal gas), hence,

$$\frac{T_s}{T_i} = \lim_{p_{tp} \to 0} \frac{p_s}{p_i} = 1.366 \qquad ...(3.17)$$

Also
$$T_s - T_i = 100 \qquad ...(3.18)$$

These equations define Kelvin or absolute centigrade scale and are independent of the properties of the individual substance. **Hence, they give a standard scale of temperature.**

The constant volume hydrogen thermometer can be used over a wide range of temperature ranging from – 200°C to 500°C. Above this temperature, hydrogen diffuses through platinum. But using glazed porcelain bulb temperature up to 1100°C may be measured. For temperature upto 1500°C, nitrogen gas is used. For temperature below 200°C, helium is used.

For correct measurement corrections have to be applied for the change in volume of the bulb due to variations of temperature and internal pressure and capillary, etc.

The gas thermometers are chiefly used for standardising other thermometers, such as resistance thermometer, thermocouple, etc. because they are very cumbersome to handle.

Note For simplicity the value 273.15 is taken as 273 and 459.67 is taken as 460 to get absolute value of temperature.

Table 3.2 compares the several temperature scales and Fig. 3.5 shows the comparison of thermometers.

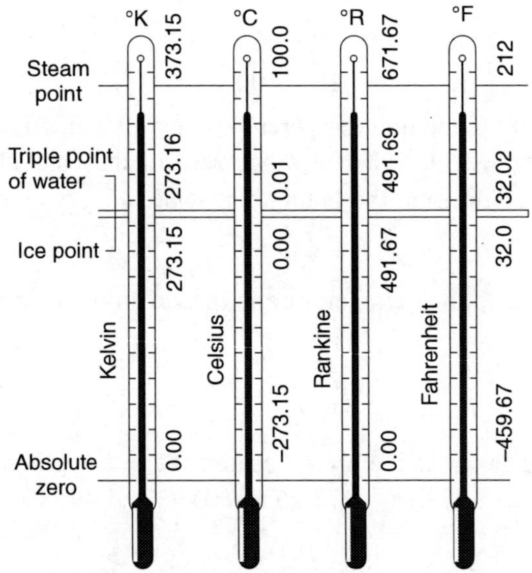

Fig. 3.5 Comparison of thermometers

Table 3.2 Comparison of different scales

Fixed point	Celcius	Farhrenheit	Kelvin	Rankine
Absolute zero	273.15	- 459.67	0.00	0.00
Ice point	0.00	32.00	273.15	459.67
Triple point of water	0.00	32.02	273.16	459.69
Steam point	100.00	212.00	373.15	671.67

Example 3.1 Define a new temperature scale, say °M. At ice and steam point the temperature are 80°M and 300°M respectively. Correlate this with the centigrade scale. The °N reading

on this scale is a certain number of degrees on a corresponding absolute temperature scale. Find this absolute temperature at N.

Solution Let

$$t = aL + b$$

$$300 = aL_s + b$$

and

$$80 = aL_i + b$$

or

$$a = \frac{220}{L_s - L_i} \quad \text{and} \quad b = 80 - \frac{220\,L_i}{L_s - L_i}$$

Hence,

$$t°\text{M} = \frac{220\,L}{L_s - L_i} + 80 - \frac{220\,L_i}{L_s - L_i}$$

or

$$t°\ \text{M} = \frac{220\,(L - L_i)}{L_s - L_i} + 80$$

Also, we have,

$$t°\text{C} = 100\frac{(L - L_i)}{L_s - L_i}$$

Combining above equations, we have, $t°\ \text{M} = 2.2\ t°\text{C} + 80$ **Ans.**

We have

$$\frac{T_s}{T_i} = \lim_{p_i \to 0} \frac{p_s}{p_i} = 1.366$$

∴

$$T_s - T_i = 300 - 80 = 220$$

$$1.366\ T_i - T_i = 220$$

or

$$T_i = \frac{220}{0.366} = 601.1\ °\text{N abs.}$$

∴

$$T_s = 601.1 + 300 = 901.1\ °\text{N abs.}$$

$$T°\text{N abs} = (t°\text{N} - 80) + 601.1 = t°\ \text{N} + 521.1$$

$$T°\text{N abs} = 0°\text{N} + 521.1°\text{N abs.}$$ **Ans.**

Example 3.2 Define a new temperature scale, °F in which the ice and steam points are 32°F and 212°F respectively. Find the absolute temperature at 0°F and 200°F. Solve from the first principle.

Solution We know

$$\frac{T_s}{T_i} = \lim_{p_i \to 0} \frac{p_s}{p_i} = 1.366$$

or

$$T_s - T_i = 212 - 32 = 180$$

$$1.366\ T_i - T_i = 180$$

or

$$T_i = \frac{180}{0.366} = 491.8\ °\text{F abs} = 491.8\ °\text{R}$$

$\therefore \quad T_s = 491.8 + 180 = 671.8\ °F = 671.8\ °R$

We know that $\quad T°R = (t°F - 32) + 491.8 = t°\ F + 459.8$

$\therefore \quad T°\ R = 0°\ F + 459.8 = 459.8\ °\ R \approx 460\ °R$ **Ans.**

$T°R = 200 + 459.8 = 659.8\ °\ R \approx 660\ °R$ **Ans.**

Note the numerical value 459.8 K is in practice taken as 460 K.

Example 3.3 Two thermometers, one Centigrade and other Fahrenheit immersed in a fluid read the same numerical value. Find their temperature of the fluid expressed in °K and °R. Also, find the identical numerical value shown by the thermometer.

Solution $\quad t°\ C = T°\ K - 273.15$

$t°\ F = T°\ R - 459.67$

Hence, for $\quad t°\ C = t°\ F$

$T°\ K - 273.2 = T°\ R - 459.67$

$T°\ K - T°\ R = 459 - 273.15 = 186.52$

We know that numerical figure on the Rankine is 1.8 times that on Kelvin scale.

Hence, $\quad \frac{T°\,R}{T°\,K} = 1.8$

$\therefore \quad 1.8\ (T°\ K) - T°\ K = 186.52$

or $\quad T°\ K = \frac{186.52}{0.8} = 233.15\ °K$

or $\quad T°\ R = 1.87°\ K = 1.8 \times 233.16 = 419.68\ T°R$ **Ans.**

$T°\ F = 419.68 - 459.68 = -\ 50°\ F$ **Ans.**

Example 3.4 An exponential scale is chosen between 0°C and 100°C such that $T = ae^x - b$. On this new scale, what temperature corresponds to – 40°C.

Solution Let the property x that depends on temperature be x_0, x_{100} and x_i at 0°C, 100°C and at any temperature t.

As per given equation

At $\quad T = 0°C, \quad 0 = ae^{x_0} - b$

or $\quad b = ae^{x_0}$ ...(i)

At $\quad T = 100°C,\ 100 = ae^{x_{100}} - b = ae^{x_{100}} - ae^{x_0}$

or $\quad a = \frac{100}{e^{x_{100}} - e^{x_0}}$ ...(ii)

Hence, at temperature $T = t$, we have

$$t = a\,e^{x_t} - b = a\,e^{x_t} - a\,e^{x_0} = a\,(e^{x_t} - e^{x_0}) \qquad \text{...(iii)}$$

From Eqs. (ii) and (iii), we have

$$t = \frac{100\,(e^{x_t} - e^{x_0})}{(e^{x_{100}} - e^{x_0})} \qquad \text{...(iv)}$$

Hence, at $t = -40°C$, we have from Eq. (iv)

$$t_{(-40°C)} = \frac{100\,(e^{x_{-40}} - e^{x_0})}{(e^{x_{100}} - e^{x_0})} \qquad \textbf{Ans.}$$

Example 3.5 A certain thermometer using pressure as a thermometric property yields the values of pressure, p as 1.9 and 6.8 at the ice and steam point respectively. The temperature of the ice and steam point are assigned the number 32 and 212 respectively. Calculate the temperature corresponding to $p = 3$, if the temperature T is expressed as $T = a \ln p + b$ where a and b are constants.

Solution Let the pressure at ice and steam point be p_i and p_s respectively. Thus, $p_i = 1.9$ and $p_s = 6.8$.

$$\text{At ice point : } T = 32 = a \ln p_i + b = a \ln (1.9) + b \qquad \text{...(i)}$$

$$\text{At steam point : } T = 212 = a \ln p_s + b = a \ln (6.8) + b \qquad \text{...(ii)}$$

Subtracting Eq. (i) from Eq. (ii), we have

$$212 - 32 = a \ln 68 - a \ln 1.9 = \ln \left(\frac{6.8}{1.9}\right)$$

or
$$180 = a \ln \left(\frac{6.8}{1.9}\right)$$

or
$$a = 141.168$$

Substituting the value of a in Eq. (i), we have

$$32 = 141.68 \ln 1.9 + b$$

or
$$b = -58.609$$

Substituting the value of a and b in the temperature expression,

$$T = 141.168 \ln p - 58.609$$

The temperature at pressure $p = 3$, we have

$$T = 141.168 \ln 3 - 58.609 = 96.479 \text{ unit} \qquad \textbf{Ans.}$$

3.8 ELECTRICAL RESISTANCE THERMOMETERS

In the resistance thermometer, the change in resistance of a metal wire due to its change in temperature is the thermometric property. The relation between the resistance of wire and temperature is given by

$$R_t = R_0 (1 + \alpha t + \beta t^2) \qquad \text{... (3.19)}$$

where α and β are the constant of the wire used.

Figure 3.6 shows a modern type of platinum resistance thermometer. It consists of a pure, well annealed platinum wire (melting point 176°C and coefficient of thermal expansion 9 × 10¯6/°C,) doubled on itself to avoid induction effects, wound on a thin flat strip of insulating mica *M* and placed at the bottom of hard glass or glazed porcelain tube *T*, provided with an ebonite cap *E*. The free ends of the platinum wire are attached to long leads which pass through holes in several mica sheets *S* closely fixed along the upper part of the tube and connected to binding screw *PP* fixed in the ebonite cap. To eliminate the effect of variation in the resistance of the leads a pair of compensating leads of the same material (platinum) are used. In order to prevent moisture depositing on the mica and rendering the insulation less effective, the tube *T* is evacuated and sealed. To measure the changes in the resistance. Callender and Griffiths Bridge (Fig. 3.7) is used, the platinum resistance leads are connected to arm *S* while the compensating leads *CC* is connected to variable arm *R*. The two arms *P* and *Q* are of equal resistance.

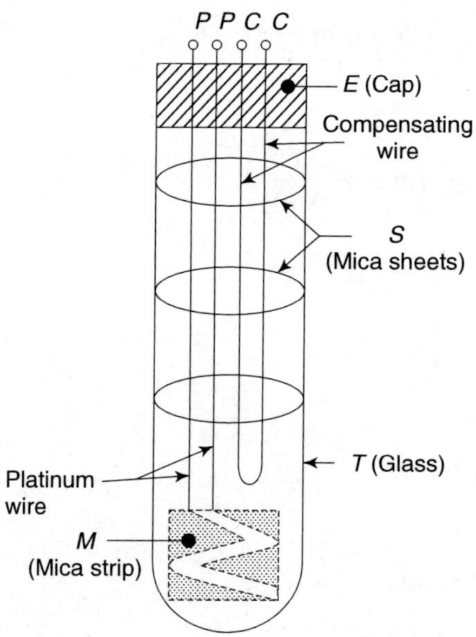

Fig. 3.6 Platinum resistance

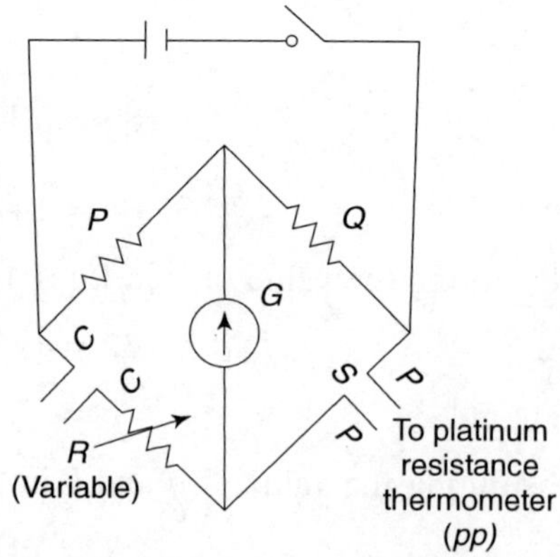

Fig. 3.7 Callender and Griffiths bridge

In order to measure the temperature, the platinum resistance thermometer is placed in the bath whose temperature is to be measured. A suitable resistance is introduced in the arm *R*. By adjusting the position of jockey on the wire *MN*, the balance point is obtained, indicated by no deflection in the galvanometer *G*. The unknown temperature *t* is given by

$$t = \frac{R_t - R_0}{R_{100} - R_0} = 100 \quad \text{...(3.20)}$$

where R_t = resistance of wire at temperature t

R_0 = resistance of wire at ice point, 0°C

R_{100} = resistance of wire at steam point, 100°C

The advantages of platinum resistance thermometers are (i) it is convenient to use, (ii) the temperature range is varying from 200°C to 1200°C, (iii) it reads correct value up to 0.01°C, (iv) its accuracy is as good as that of gas thermometer.

The main drawbacks are : (i) it does not give quick reading, (ii) it takes time to balance the bridge.

Example 3.6 A platinum resistance thermometer has a resistance of 2.8 ohm at 0°C and 4 ohm at 100°C. Calculate the temperature when the resistance indicated will be 6 ohm.

Solution Let resistance at 0°C, 100°C and t°C be represented as R_0, R_{100} and R_t respectively. The governing equation of platinum resistance thermometer is

$$R_t = R_0 (1 + \alpha \Delta t)$$

at

$$t = 100°\text{C}$$

$$R_{100} = R_0 (1 + \alpha \Delta t)$$

or

$$4 = 2.8 [1 + \alpha (100 - 0)] = 2.8 [1 + 100 \alpha]$$

or

$$\alpha = \left(\frac{4}{2.8} - 1\right)\Big/100 = 4.285 \times 10^{-3}/°\text{C}$$

Now,

$$R_t = R_0 (1 + \alpha \Delta t)$$

or

$$6 = 2.8 [1 + 4.285 \times 10^{-3} (t - 0)]$$

or

$$t = \left(\frac{6}{2.8} - 1\right)\Big/4.285 \times 10^{-3} = 266.71°\text{C}$$ **Ans.**

3.9 THERMOELECTRIC THERMOMETERS

The principle of thermoelectric thermometer is based on Seeback effect. According to this principle, when one of junctions of two different metals is kept at higher temperature (called hot junction) and other at lower temperature (called cold junction), an e.m.f. is set up; this causes a current to flow in the circuit. *The e.m.f.* (ε) *produced is the thermometric effect.*

Figure 3.8 shows a thermocouple. The e.m.f. can be measured by a micro voltmeter or digital thermometer or potentiometer to a high degree of accuracy. The choice of metal depends on the temperature range to be investigated, however copper-constantan, chromel-alumel and platinum-platinum-rhodium are typical combinations in use.

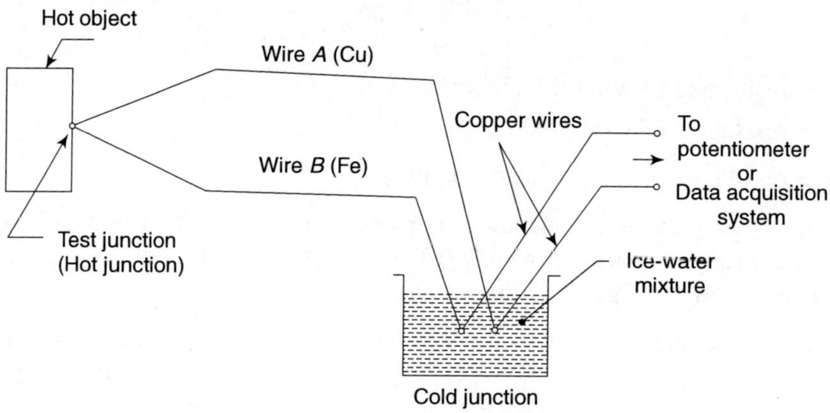

Fig. 3.8 Thermocouple

A thermocouple is calibrated with the help of constant volume gas thermometer by measuring the e.m.f. at various known temperatures, the reference junction being kept at 0°C. The results may be represented by the equation

$$\varepsilon = a + bt + ct^2 + dt^3$$

where ε = thermal e.m.f.

$a, b, c, d,$ = constants

The advantages of thermocouple are:

(i) It is especially adapted for giving continuous records of rapidly changing temperature as in the cooling of alloys because of its small thermal capacity and, consequently, negligible tag.

(ii) The range is very wide varying from – 200°C to 1600°C.

(iii) It can be easily made direct reading and may be recorded.

The disadvantage include: It is less accurate than platinum reistance thermometer.

Example 3.7 The e.m.f. produced in a thermocouple with the test (hot) junction at t°C on gas thermometer scale and reference junction at ice point is given by the expression.

$$\varepsilon = 0.22\, t - 6 \times 10^{-4}\, t^2 \text{ mV}$$

The millivoltmeter is calibrated at ice and steam point. Find the temperature reading by this thermometer in a place where the gas thermometer reads 80°C.

Solution At ice point, when t = 0°C

$$\varepsilon = 0 \text{ mV}$$

At steam point, when t = 100°C, e.m.f. produced is given by

$$\varepsilon = 0.22 \times 100 - 6 \times 10^{-4} (100)^2 = 16 \text{ mV}$$

At 80°C, the e.m.f. produced

$$\varepsilon = 0.22 \times 80 - 6 \times 10^{-4} (80)^2 = 13.76 \text{ mV}$$

When gas thermometer reads 80°C, the thermocouple will read

$$= \frac{100}{16} \times 13.75 = 86°\text{C}$$ **Ans.**

3.10 RADIATION PYROMETER

The radiation pyrometer utilises the heating effect of energy radiated from a hot object as an index of its temperature. It consists of the following various parts shown in Fig. 3.9.

(i) A cylindrical tube which contains a concave mirror

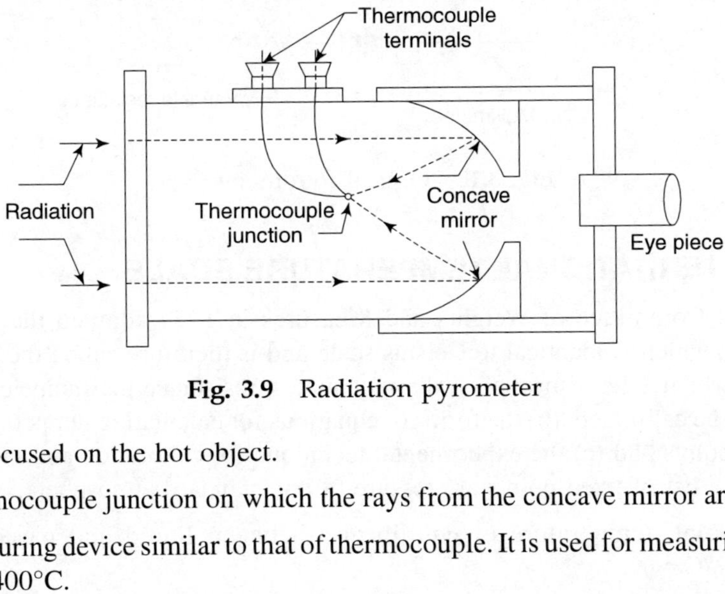

Fig. 3.9 Radiation pyrometer

(ii) Lens focused on the hot object.

(iii) A thermocouple junction on which the rays from the concave mirror are concentrated.

(iv) A measuring device similar to that of thermocouple. It is used for measuring temperature above 400°C.

3.11 SEGAR CONE

This is based on the principle of fusion and is suitable for measuring temperature between 600°C to 2000°C. Generally, the furnace temperature is measured from this instrument.

3.12 OPTICAL PYROMETER

This is based on monochromatic radiation of the hot object where temperature is to be measured. It is suited only for high temperature. Figure 3.10 shows the schematic diagram of an optical pyrometer. The hot body whose temperature has to be measured is viewed through a telescope. A small electric filament lamp is ranged inside this telescope such that the image of the filament is superimposed on it of the variation of hot body. In order to vary the temperature of the filament, a variable resistance is arranged in the filament circuit, by adjusting this resistance, the image of the filament merges or disappears into the image of the hot body. When this occurs, the temperature are similar. By calibrating controlled resistance, the temperature of the body can readily be found, This device is suitable for measuring temperature above 650°C.

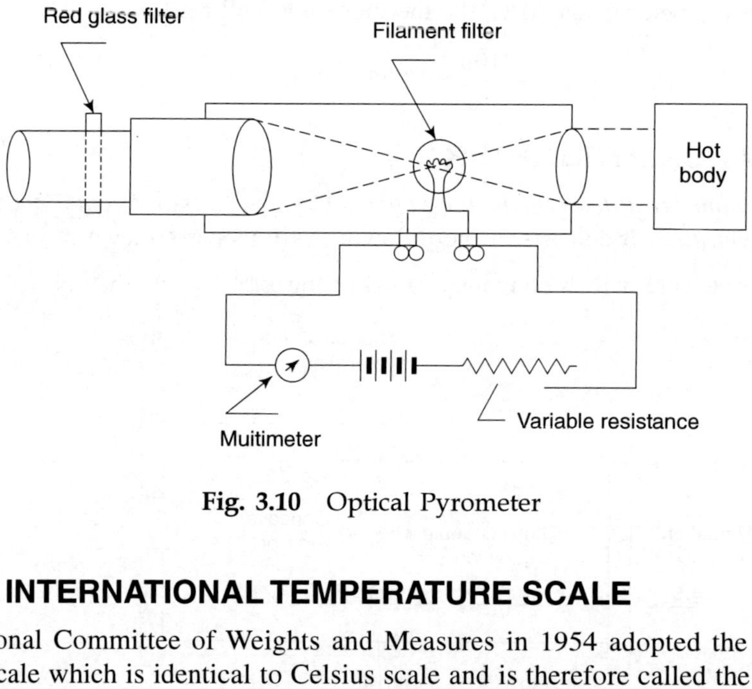

Fig. 3.10 Optical Pyrometer

3.13 THE INTERNATIONAL TEMPERATURE SCALE

The International Committee of Weights and Measures in 1954 adopted the international temperature scale which is identical to Celsius scale and is therefore called the International Celsius point at which the instruments (thermocouples, resistance thermometer and optical pyrometer) are to be calibrated, (b) the forms of equations for calculating temperature at points other than fixed points and (c) the experimental technique for calibration and measurements. The following is a list of fixed points at pressure of one standard atmosphere.

(a) Oxygen point, temperature of equilibrium between liquid and gaseous, oxygen = – 182.92°C.

(b) Ice point, temperature of equilibrium between ice and air-saturated water = 0.00°C.

(c) Steam point temperature of equilibrium between liquid water and ice vapour = 100.00°C.

(d) Sulphur point, temperature of equilibrium between liquid sulphur and its vapour = 444.60°C.

(e) Silver point = temperature of equilibrium between solid silver and liquid silver = 960.8°C.

(f) Gold point, temperature of equilibrium between solid gold and liquid = 1063°C.

For temperature interpolation, the International scale is divided into four regions. Each region uses an interpolation equation chosen to determine the temperature.

1. From – 182.97°C to 0°C, the temperature is measured by a platinum resistance thermometer by relation.

$$R_t = R_0\ [1 + \alpha t + \beta t^2 + \gamma\,(t - 100)\ t^2] \qquad ...(3.21)$$

where the four constants, R_0, α. β. γ are determined by calibration by a platinum resistance thermometer but a different equation is used.

2. From 600°C to 1083°C, the platinum-rhodium thermocouple is used.

The equation used is

$$E = a + b\ .\ t + c\ .\ t^2\ ... \qquad ...(3.22)$$

where the constants *a*, *b* and *c* are determined by calibration at freezing point of antimony, the silver point and the gold point.

3. Above 1063°C, the optical pyrometer calibration at the gold point is used. The formula used is

$$\frac{J_t}{J_{au}} = \frac{\exp\{C_2 / (t_{au} - T_0)\,\lambda\} - 1}{\exp\{C_2 / (t_{au} + T_0)\,\lambda\} - 1} \qquad ...(3.23)$$

where suffix au refers to the gold point temperature and *J* is the intensity of radiation. The radiation must be visible body spectrum and must be emitted by a black body. The constant C_2 = 1.438 cm *K*; λ is wavelength and T_0 is ice point temperature.

3.14 PRACTICAL ASPECT OF TEMPERATURE MEASUREMENTS

The various modern temperature measuring devices and their applications are listed in Tables 3.3 and 3.4 respectively. The selection of a particular temperature measuring device depends on the temperature range and the accuracy desired.

Table 3.3 Modern temperature meauring devices

Devices	Practical temperature range in °C	Effects utilised
Mercury in glass thermometers	- 39 to 525	Expansion of fluid
Mercury thermometers fused silica	- 36 to 640	Expansion of fluid
Gas expansion thermometer (N_2)	- 200 to 1500	Expansion of gas
Metal expansion thermometer	0 to 500	Expansion of solid

Contd.

Table Contd. 3.3

Electrical resistance	- 200 to 1200	Change in resistance
Thermoelectric (le-Chatelier)	0–1500	Thermoelectric effect
Thermoelectric (base metal)	0–1000	Thermoelectric effect
Radiation pyrometer	400 upwards	Radiation (Total)
Optical pyrometer	65 upwards	Radiation (Mono)
Segar cones	600 to 2000	Fusion

Table 3.4 Applications

Items	Instrument
Boiler combustion space	Optical pyrometer
Economiser, feed water heaters and chimney gases	Base metal pyrometer
Incandescent filaments	Optical pyrometer
Incandescent gas mantels	Radiation pyrometer
Condenser	Mercury thermometer
Cooler	Mercury thermometer

3.15 SUMMARY OF FORMULAE

1. $t = a\,L + b$ and $t = a\,L$
2. $t(L) = 273.16\ L/L_t$, $t\ (\varepsilon) = 273.16\ \varepsilon/\varepsilon_t$, $t\ (R) = 273.16\ R/R_t$

   $t\ (V) = 273.16\ V/V_t$ and $t\ (p) = 273.16\ p/p_t$
3. $C/5 = (F - 32)/9 = \text{Re}/4$
4. $\dfrac{T_s}{T_i} = \lim\limits_{p_t \to 0} \dfrac{p_s}{p_i} = 1.366$ and $T_s - T_i = 100$
5. $R_t = R_0\ (1 + \alpha t + \beta t^2)$
6. $t = \dfrac{R_t - R_0}{R_{100} - R_0}$
7. $\varepsilon = a + bt + ct^2 + dt^2$

PROBLEMS

Things to Think About

3.1. Does 1°R represent a larger or smaller temperature unit or interval than 1 K? How does one convert a temperature on the Kelvin scale to the Rankine scale?

3.2. What difficulties might be encountered if water were used as the thermometric substance in the liquid-in-glass thermometer?

3.3. How will you measure the temperature of sun and other heavenly stars?

3.4. How is the temperature of human body maintained?

Short and Long Theoretical Questions.

3.5. Give the concept of temperature and differentiate between heat, temperature and internal energy.

3.6. What do you mean by equality of temperature?

3.7. Define and explain "Zeroth Law of Thermodynamics".

3.8. What are the requirements of a temperature scale?

3.9. Explain the method of establishing the temperature scale in Centigrade and Fahrenheit.

3.10. What are the requirements of a standard scale of temperature?

3.11. What are the different scales of temperature? Establish the mathematical relations among each other.

3.12. Give the method establishing the absolute temperature scale?

3.13. What are the advantages of gases over liquids as a thermometric substances?

3.14. Explain the working principle of electrical resistance thermometer.

3.15. Explain the working principle of thermo-electric thermometer.

3.16. Explain the working principle of radiation and optical pyrometers.

3.17. What do you mean by International Temperature scale?

3.18. What instrument would you use to measure the temperature of (a) a boiler furnace and (b) a metal furnace. Describe with the help suitable sketches, the construction of one of these instrument and indicate the method of using it.

3.19. Describe briefly the appropriate methods of measuring the temperatures of (a) the flue gas from a boiler, (b) the material in a melting furnace, (c) a visible heat of an exhaust valve, (d) the exhaust from an internal combustion engine and (e) a gas system with a fluctuating temperature, the fluctuations having a cycle time of 20 sec.

3.20. What types of temperature measuring instruments would you effect in the following applications.

(i) A heat treatment bath or furnace (400–1000°C).

(ii) A boiler furnace (1660°C).

(iii) A boiler flue (370°C)

(iv) A fuel calorimeter water system (20°C).

(v) Superheat steam temperature distant reading (320°C). Indicate a few words underlying physical principle on which each of the instrument is based.

Numerical Questions

3.21. Invent a thermometer preferably of the type you have never heard of. Define the temperature scale of this thermometer if freezing and boiling points of water are taken as 100° and 300° respectively. Correlate this scale with the Celcius scale. If °M reading of this scale is a certain number of degree on a corresponding absolute temperature scale, what is absolute temperature at °M? [*Ans.* t°M = $2t$°C + C100, 446°M]

3.22. Convert 63°F in °C, K, °Re and °R.

3.23. If temperature scale is graduated according to the equation $t = 100 + 3t_c$ where t is the temperature reading on the scale and t_c is Celsius temperature. Find (a) freezing and boiling point of the thermometric substance and (b) the absolute temperature corresponding to 20° temperature reading on the scale.

3.24. A certain temperature scale A is arbitrarily defined by identifying the boiling point of a particular substance (whose boiling point is 160°C) as 85°A. The freezing point of the same substance is -70° A. If the absolute zero temperature on the same scale is – 109°A, calculate the freezing point of the substance on Rankine scale. Pressure may be taken as 1 atm.

[*Ans.* 3798°R]

3.25. A platinum resistance thermometer have a resistance of 2.56 at 0°C and 3.56 ohm at 100°C. Calculate the temperature when the resistance indicate is 5.56 ohm. [*Ans.* 300°C]

3.26. In a mercury thermometer, the length of the mercury column is 5 cm when the thermometer is at the ice point and is 25 cm when the thermometer is at the steam point. Assume this length as a thermometric property and let t be the temperature on linear Ceisius scale as determined by this thermometer.

(a) A Celsius temperature t^* is defined by a quadratic equation $t = ax^2 + b$ where $t^* = 0°$ at the ice point at $t^* = 100°$ at the steam point. Find the temperature t^*, when $t = 50°$ and sketch graph between t^* and t.

(b) A Celsius scale in logarithmic form is defined as $t^* = a \ln X + b$ where $t^* = 0°$ at the ice point and $t^* = 100°$ at the steam point. Find the temperature t^* when $t = 50°$ and sketch graph of t^* versus t.

3.27. A certain thermometer is caiibiated using the ice and steam point as fixed point, at temperature 0° and 100° respectively. The function of thermometric substance is taken as $t = a \ln X + b$ instead of the usual linear function $t = aX + b$. Show that the new scale is given by

$$t = 100\left\{\left(\ln\frac{X}{X_i}\right)\Big/\left(\ln\frac{X_s}{X_i}\right)\right\}$$

3.28. The volume of a liquid at different temperature is given by

$$V = V_0\,(1 - \alpha t + \beta t^2)$$

where $\alpha = 0.001$, $\beta = 0.000003$ and t is the temperature in degree centigrade measured on the scale of a constant volume air thermometer. A thermometer is constructed with this liquid, the fixed points are determined in the same manner, and the thermometer uniformly graduated as a centigrade instrument. What temperature will this thermometer read when $t = 50°C$?

3.29. An exponential scale is chosen between 0°C and 100°C such as

$$t = ae^x + b$$

On this scale, what temperature corresponds to – 50°C.

3.30. If one junction of a thermocouple is kept at the ice point and the other junction at a temperature t as measured by Celsius mercury in glass thermometer, the e.m.f. (ε) of the thermocouple is given by a quadratic function of t

$$\varepsilon = \alpha\tau + \beta\tau^2$$

where ε is millivolts, $\alpha = 0.2$ and $\beta = 5 \times 10^{-4}$

(a) Calculate the e.m.f. when t = 100°C, 200°C, 400°C and 500°C. Also sketch a graph of a graph of ε versus t.

(b) Suppose the e.m.f. (ε) is taken as a thermometric property and that a temperature scale t is defined by the linear equation

$$t^* = a\varepsilon + b$$

where $t^* = 0$ at the ice point and $t^* = 100°$ at the steam point. Calculate the numerical value of a and b and sketch a graph of ε versus t.

(c) Calculate the value of t^* when t = – 100°C, 400°C. Also sketch a graph of t^* versus t over this range.

(d) Is the t^* scale Centigrade scale? Is it a linear scale?

3.31. A platinum resistance thermometer has a resistance of 12.78 ohm in ice, 17.77 ohm in steam and 25.67 ohm when placed in a furnace. Calculate the fundamental interval of the thermometer and the platinum resistance temperature of the furnace. If the difference between the gas scale temperature and the temperature measured by the above thermometer at the boiling point of sulphur (445°C) was 23°C, calculate the gas scale temperature of the furnace.

[*Ans.* 4.99 ohm, 258.3°C, 265°C]

3.32. The resistance R of platinum varies with temperature t°C as measured on the gas scale according to the equation

$$R_t = R_0\,(1 + at - bt^2)$$

where $a = 3.8 \times 10^{-4}$ $b = 5.6 \times 10^{-7}$, calculate the platinum temperature corresponding to a gas scale temperature of 400°C. Why is there a difference between these two values?

3.33. The relation between resistance R and temperature T for a thermistor closely follows the relation

$$R_t = R_0 \exp\left[\beta\left(\frac{1}{T} - \frac{1}{T_0}\right)\right]$$

where R_0 is the resistance in ohms (Ω), measured at temperature T_0(K) and β is a material constant with unit of K. Take R_0 = 2.2 Ω and T_0 = 310 K. The calibration test results in R_t = 0.31 Ω at T = 422 K. Calcutate the value of β for the thermistor and make a plot of resistance versus temperature.

3.34. The relation between electrical resistance R_t and temperature T for a resistance temperature indicator is expressed as

$$R_t = R_0\,(1 + \alpha\,(T - T_0)$$

where R_0 is the resistance, in ohms (Ω), measured at reference temperature t_0 (in °C) and α is a material constant with units of $(°C)^{-1}$. During test, R_t = 51.39 Ω at 0°C and 51.72 Ω at 90°C. What resistance reading would correspond to a temperature of 50°C on this thermometer?

Software Based Questions

3.35 Develop a software for devising a new temperature scale and compare this with °C and Kelvin scale.

3.36. Develop a software to estimate the temperature of gas at inlet of gas turbine by measuring the exhaust temperature of gas turbine.

3.37. Develop a software to plot a graph between resistance and temperature for a thermistor.

Design and Open Ended Questions

3.38. Temperature measurement of flowing gases and liquids is to be made in a nuclear power plant facility. In this venture, valves are to be tested. Using the standard for temperature measurement prepare a report discussing what standards must be met in the design of testing facility and what requirements these standards place on the design?

3.39. In a combined cycle power plant, the temperature of gases passing through heat recovery steam generator is to be measured at various locations. Using the standards, what instruments with their specification will be used.

Multiple-Choice Answer Type Questions

3.40. In a constant volume gas thermometer, the thermometric property used is

(a) mass (b) length (c) volume (d) pressure

3.41. Steam point and ice point temperature in Fahrenheit scale can be given by

(a) 100 and 0 (b) 373,15 and 273.15 (c) 212 and 32 (d) 491.67 and 671.67

3.42. Triple point of water temperature in Rankine scale is

(a) 32.02 (b) 273.16 (c) 0.01 (d) 491.69

3.43. As per the zeroth law of thermodynamics if the two bodies *A* and *B* are in thermal equilibrium with a third body *C* separately then the two bodies *A* and *B* shall be also in thermal equilibrium with each other. Using this law as basis of temperature measurement, which body will refer to thermometer.

(a) Body A (b) Body B (c) Body C (d) None of these

3.44. In a new temperature scale say M, the boiling and freezing points of water at one atmosphere are 100° M and 300°M respectively. Correlate thus scale with centigrade scale. The reading of 0°M respectively. Correlate this scale with centigrade scale. The reading of 0°M on the centigrade scale is

(a) 0°C (b) 50°C (c) 100°C (d) 150°C

3.45. A mercury thermometer cannot be used to measure a temperature of

(a) 100°F (b) 380°C (c) 212°C (d) 212°F

3.46. The thermometric property of electrical resistance thermometer is

(a) Current (b) Potential difference (c) Magnetic (d) Resistance

Answers to Multiple-Choice Questions

3.40. (d) 3.41. (c) 3.42. (d) 3.43. (c) 3.44. (d) 3.45. (b)

3.46. (d)

CHAPTER

4

Behaviour of Ideal (Perfect) and Real Gases

OVERVIEW

Many thermodynamic devices such as petrol, diesel engines, gas turbines, jet engines and rockets, etc. use gases as their working fluid. It is the changes in the properties of the working fluid that operate these devices such as air or gas compressor, gas turbines, internal combustion engines, etc. In order to design these thermodynamic devices, it is essential to study the governing laws of gases which are discussed in this vary chapter.

4.1 CONCEPT OF AN IDEAL GAS (MACRO ANALYSIS)

For practical purposes, in thermodynamical problems, we are concerned with fluids as a system in its various phases such as liquid, vapour and gas. *As long as a fluid remains homogeneous in composition and chemical form, and retains its chemical identity, it is termed as a pure substance.*

During thermodynamic processes, a fluid (system) may retain or change its phase. In other words, a gas may remain a gas or a liquid may turn into gaseous phase under suitable conditions of pressure and temperature. In thermodynamics, we generally use the term gas for pure substance which fills up a given space completely without changing its phase during any thermodynamic process. *If the working fluid (system) condenses or evaporates during the thermodynamic process, it is called vapour.*

Many gases which do not change their phase during a thermodynamic process and obey a set of common rules governing change of their properties are called ideal (or perfect) gases.

To understand the behaviour of an ideal gas, consider the interrelationship between the various properties of the substance. There exists a relationship between pressure p, specific volume v and temperature T of a substance and this is given by

$$F\ (p, \mathrm{v}, T,) = 0$$

$$p = f(\text{v}, T) \quad \text{...(4.1)}$$

Equation (4.1) applies widely for gases because in gaseous state where there is a change in pressure, there is an appreciable change in specific volume. There exists a condition for all real gases at which they behave in a similar manner. This condition is

$$p\text{v} = (RT)_{p \to 0, \text{ real gas}}$$

or

$$p = \left(\frac{RT}{\text{v}}\right)_{p \to 0, \text{ real gas}} \quad \text{...(4.2)}$$

This identical behaviour is called the ideal behaviour and the state of zero pressure is called the ideal state.

The above behaviour for real gases at the ideal, i.e., state of $p \to 0$ gives the concept of ideal gas – a hypothetical gas that will behave in an ideal manner at all pressures. An ideal gas is not a real a gas but a hypothetical gas which displays at all pressures the same behaviour which the real gases displays only at the state $p \to 0$. For an ideal gas, i.e., perfect gas, the following conditions are used.

(i) An ideal gas obeys the law $p\text{v} = RT$.

(ii) The internal energy is a function of temperature alone, i.e., $U = f(T)$ only.

In strict sense, no gas which exists in nature is perfect. But according to definitions, many gases such as air, nitrogen, oxygen, helium, hydrogen, carbon dioxide, etc., are treated as perfect gases becaue the obey the gas laws at normal temperature and pressure.

4.2 BOYLE'S LAW

This law was enunciated by Robert Boyle in 1662 A.D. and stated as :

When temperature remains constant, the volume of a perfect gas is inversely proportional to the absolute pressure.

Let

p = absolute pressure of gas

V = volume of gas at pressure p

T = absolute temperature of gas

Thus, $V \alpha \left(\frac{1}{p}\right)_{T = \text{constant}}$

or

$$V = \left(\frac{c}{p}\right)_{T = \text{constant}}$$

where c is constant of proportionality.

or

$$pV = \text{constant}$$

This reveals that the product of absolute pressure and volume of a given quantity of gas is constant when the temperature remains constant.

If the gas is made to change its state from 1 to 2, without change in temperature, then

$$p_1 V_1 = p_2 V_2 \qquad ...(4.3)$$

or

$$\frac{V_1}{V_2} = \frac{p_2}{p_1} \qquad ...(4.4)$$

All real gases obey this law very neraly and a perfect gas, by definition, obeys it exactly.

This law can be verified by means of a simple apparatus shown in Fig. 4.1. It consists of a glass tube *A* sealed at one end while the other end is connected with a reservoir filled with mercury by a flexible rubber pipe. The reservoir *R* can be raised or lowered and the pressure inside *A* is due to the column of mercury of height *h* together with atmospheric pressure acting on the reservoir. The volume of air inside *A* will be corresponding to the length in *A* occupied by air. The experimental results when plotted on a graph give the rectangular hyperbola shown in Fig. 4.2. The temperature remains constant during operation. Similar readings may be taken at another temperatures of the system.

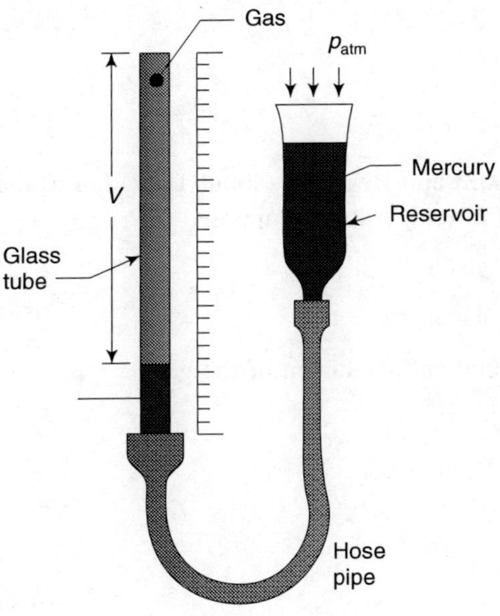

Fig. 4.1 Apparatus to prove Boyle's law

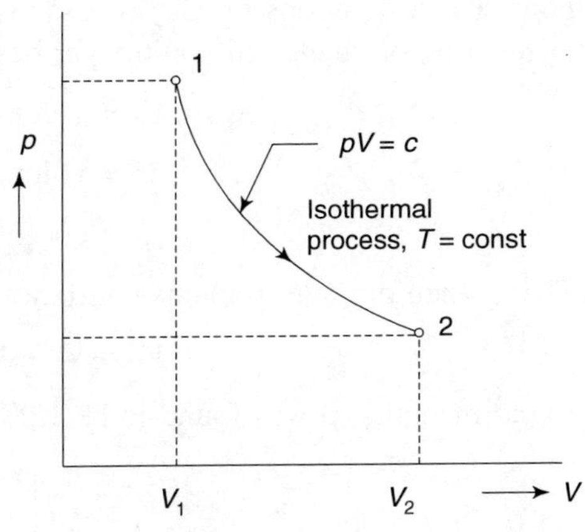

Fig. 4.2 Boyel's law on *p-V* diagram

On per unit mass basis, the Eq. (4.4) may be written as

$$\frac{v_1}{v_2} = \frac{p_2}{p_1} \qquad ...(4.5)$$

Example 4.1 Air is compressed in a vessel from pressure 1 bar and volume 8 m^3 to 2 m^3 at constant temperature. Find the final pressure.

Solution We know, $\quad p_1 V_1 = p_2 V_2$

or $$10^2 \times 8 = p_2 \times 10^2 \times 2$$

or $$p_2 = 4 \text{ bar}$$ **Ans.**

4.3 CHARLES' LAW

This law was stated by Charles' in 1887 A.D. It may be stated in two ways:

(a) *If pressure remains constant, the volume of a given mass of gas varies directly as the absolute temperature. Thus*

$$V \alpha (T)_{p = \text{constant}}$$

or $$\frac{V}{T} = \text{constant}$$

If the process occurs from state 1 to 2, then

$$\frac{V_1}{T_1} = \frac{V_2}{T_2}$$

or $$\frac{V_2}{V_1} = \frac{T_2}{T_1} \qquad ...(4.6)$$

Further investigations by Gay–Lussac (about 1802) and Regnault (about 1847) led to the determination of change in volume per degree rise in temperature. Suppose

V_0 = volume of the gas at 0°C

V_t = volume of the gas at t°C

β = coefficient of cubical expansion of the gas

Then, when pressure remains constant

$$V_1 = V_0 + V_0 \beta t = V_0 (1 + \beta t)$$

Experimentally, β was found to be 1/273 per °C

$\therefore$ $$V_t = V_0 \left(1 + \frac{t}{273}\right) \qquad ...(4.7)$$

From this question, it follows that *at a temperature – 273°C, the volume of the gas will be zero. This is considered to be the point of absolute zero temperature.*

$$V_1 = V_0 \left(\frac{273 + t}{273}\right) = \frac{V_0 T_t}{T_0}$$

or $$\frac{V_t}{V_0} = \frac{T_t}{T_0} \quad \text{or} \quad \frac{V_t}{T_t} = \frac{V_0}{T_0} \qquad ...(4.8)$$

An experiment to prove Charles' law can be performed with the apparatus shown in Fig. 4.3. It consists of a definite mass of gas, kept at atmospheric pressure by the mercury levels in the U-tube, is raised through a series of temperature and the corresponding volumes at each temperatures are noted by means of the graduated scale. Experimental results when plotted give the straight line shown in Fig. 4.4.

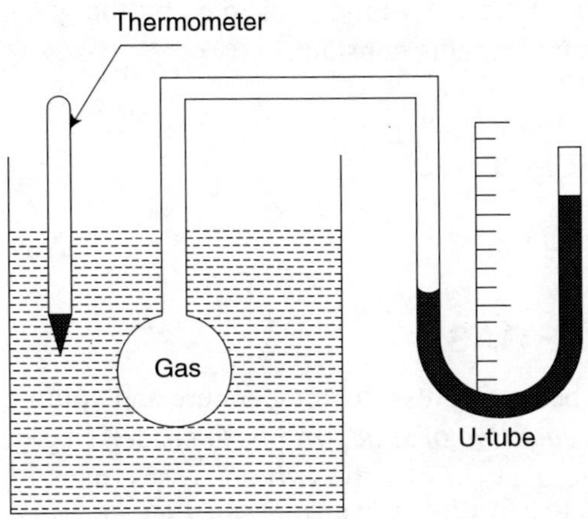

Fig. 4.3 Apparatus to prove Charles' law

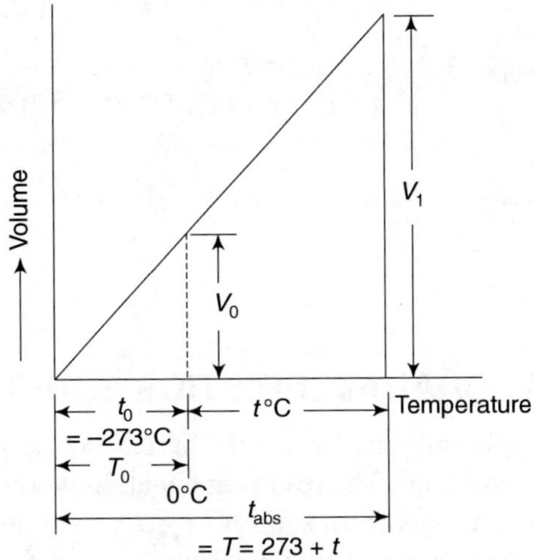

Fig. 4.4 Charles's law on *p*-*V* diagram

$$V_t = S.t + V_0 \quad \text{where } S \text{ is the slope of the line}$$

$$= V_0 (1 + \beta t) \quad \text{where } \beta = \frac{S}{V_0} = \frac{1}{273} \quad \text{if } t \text{ is in °C}$$

$\therefore$
$$V_t = V_0 \left(\frac{273 + t}{273}\right) = V_0 \frac{T_t}{T_0}$$

or
$$\frac{V_t}{273 + t} = \frac{V_0}{273} = \text{constant}$$

If the straight line is produced back until it cuts the horizontal axis it cuts at a temperature of – 273 °C. At – 273 °C, the volume of the gas is zero and *this is called* **absolute zero of temperature**. Thus, when the temperatures are measured from absolute zero the straight line law will be valid.

$$\frac{V_t}{T} = \frac{V_0}{T_0}$$

Thus, this experiment proves the Charles' law.

(b) *If the volume remains constant, the pressure of a given mass of gas varies directly with temperature.* Thus,

$$p \propto (T)_{v = \text{constant}}$$

or
$$p = c\,(T)_{v = \text{constant}}$$

If the process occurs between two states

$$\frac{p_2}{p_1} = \left(\frac{T_2}{T_1}\right)_{v = \text{constant}} \quad \text{...(4.9)}$$

Example 4.2 A mass of air occupies 3 m^3 at 20°C. Find the volume occupied by the given mass of air at a temperature of 200°C if the pressure remains constant.

Solution We know that
$$\frac{V_2}{V_1} = \frac{T_2}{T_1} = \frac{V_2}{3} = \frac{273+200}{273+20} = \frac{473}{293}$$

or
$$V_2 = \frac{473}{293} \times 3 = 4.843 \text{ m}^3$$ **Ans.**

4.4 CHARACTERISTICS EQUATION OF GAS

By combining Boyle's and Charles' law, a relation between pressure, temperature and volume may be found. The resulting equation would be an *equation of state for gas*. Consider 1 kg of mass changes from state p_1, v_1, T_1, to another state p_2, v_2, T_2 by the two processes, namely, constant pressure and constant temperature shown in Fig. 4.5. During constant pressure,

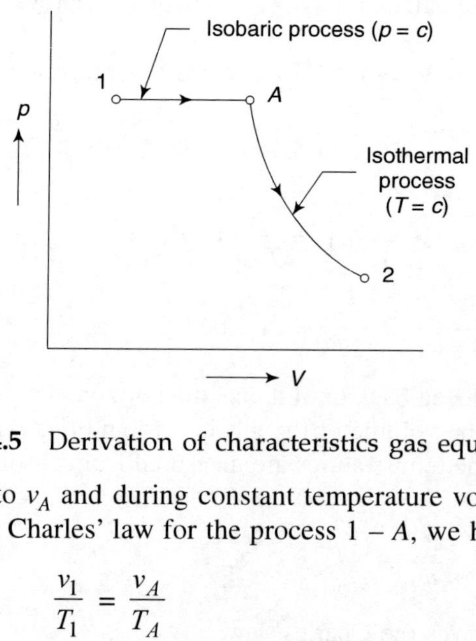

Fig. 4.5 Derivation of characteristics gas equation

volume changes from v_1 to v_A and during constant temperature volume changes from v_A to final volume v_2. Applying Charles' law for the process 1 – A, we have

$$\frac{v_1}{T_1} = \frac{v_A}{T_A}$$

or
$$v_A = \frac{v_1}{T_1} T_A = \frac{v_1 T_2}{T_1} \quad (\text{since } T_A = T_2) \quad \text{...(4.10a)}$$

For constant temperature process $A - 2$, Boyle's law shows

$$v_A\, p_A = v_2\, p_2$$

or
$$v_A = v_2 \frac{p_2}{p_A} = v_2 \frac{p_2}{p_1} \quad (\text{since } p_A = p_1) \qquad \text{...(4.10b)}$$

Equating the Eqs. (4.10 a) and (4.10 b), we have

or
$$v_1 \frac{T_2}{T_1} = v_2 \frac{p_2}{p_1} \quad \text{or} \quad \frac{p_1 v_1}{T_1} = \frac{p_2 v_2}{T_2}$$

or
$$\frac{pv}{T} = \text{constant} = R \qquad \text{...(4.11)}$$

or
$$\frac{p_1 v_1}{T_1} = \frac{p_2 v_2}{T_2} = R \qquad \text{...(4.12)}$$

where R is a constant and called the ***characteristics gas constant***. The unit of R is

$$R = \frac{pv}{T} = \frac{N}{m^2} \times \frac{m^3}{\text{kg} \cdot \text{K}} = \text{J/kg K}$$

where p = pressure, N/m^2

v = specific volume, m^3/kg

T = temperature, K

Equation (4.12) is the eqution of state or characteristic equation of 1 kg mass of the gas. Table 4.1 gives the gas constant R for various gases.

Table 4.1 Gas constant

Gas	Molecular weight M	Characteristics gas constant R (J/kg K)	Universal gas constant $R_0 = \bar{R} = MR$, as (M = m/n) J/kg mole K
Air	29	287	
O_2	32	259.859	
N_2	28.016	296.729	8314
H_2	2.016	4124.4	
C_2	44.01	1887.65	

Note: 1 kg mole = 10^5 mole

Note: Here Boyle' law, Charles' law and characteristics of gas law are known as ***perfect gas laws***.

For a gas having m kg mass, $pV = mRT$ ...(4.13)

4.5 AVOGADRO'S HYPOTHESIS AND UNIVERSAL GAS CONSTANT

(a) Mole. *A mole of a substance has a mass numerically equal to the molecular weight of the susbstance.* Thus, 1 gm mole (mole) oxygen has a mass of 32 gm, 1 kg mole of oxygen has a mass of 32 kg and 1 kg mole of carbon dioxide has mass of 44 kg and so on.

For a certain gas, if m is its mass in kg, and M its molecular weight in kg/kg mole, then the number of kg moles of the gas, n is given by

$$n = \frac{\text{m (kg)}}{M\,\text{(kg/kg mole)}} = \frac{m}{M} \text{ kg mole} \quad ...(4.14)$$

The molar volume, v_{mole} is given by

$$v_{mole} = \frac{V}{n} \text{ m}^3\text{/kg mole} \quad ...(4.15)$$

where V is the total volume of gas in m^3

(b) Avogadro's Number. There are 6.02×10^{26} molecules in a kg mole of a substance. This is known as Avogadro's number (A). If N be the total number of molecules, then

$$N = nA \quad ...(4.16)$$

(c) Avogadro's Hypothesis. *Avogrado's law (established in 1811) states that equal volume of all perfect gases at the same temperature and pressure contain an equal number of molecules.* Therefore, if two ideal gases having masses m_A and m_B and molecular weight M_A and M_B are contained in two vessel of equal volume at the same temperature and pressure, hence each gas will contain the same number of molecules say N. Thus,

$$m_A = nM_A = \frac{N}{A} M_A \quad \text{and } m_B = nM_B = \frac{N}{A} M_B$$

or

$$\frac{m_A}{m_B} = \frac{M_A}{M_B}$$

But $m = \rho V$ where volume V is the same for both gases and ρ is density.

$$\frac{\rho_A}{\rho_B} = \frac{v_B}{v_A} = \frac{M_A}{M_B} \quad \text{since } \rho = \frac{1}{v}$$

$$\therefore \quad M_A\, v_A = M_B\, v_B = v_{mole} = \text{constant} \quad ...(4.17)$$

Equation (4.17) tells us that *the product of molecular weight and specific volume is constant for all gases at the same temperature and pressure and is called the molar volume,* i.e., volume of 1 kg mole of gas.

Avogardo's law may also be stated as the volume of 1 kg mole of all gases at normal temperature and pressure is the same and is equal to 22.4 m^3. Thus, for hydrogen

$$v_{mole} = \frac{M}{\rho} = \frac{2.015}{0.09} = 22.4 \text{ m}^3\text{/kg mole}$$

or $$\text{Density} = \rho = \frac{\text{Molecular weight}}{22.4} \text{ kg/m}^3 \qquad \text{...(4.17a)}$$

(d) Universal Gas Constant

$$pV = mRT = n\,(MR)\,T$$

or $$p\left(\frac{V}{n}\right) = p\,v_{\text{mole}} = (MR)\,T$$

$$MR = \frac{p\,v_{\text{mole}}}{T} \qquad \text{...(4.18)}$$

In the case of several gases A, B, C, etc., at constant pressure p and temperature T, the Eq. (4.18) may be written as

$$M_A\,R_A = \frac{p\,(v_{\text{mole}})_A}{T}, \quad M_B\,R_B = \frac{p\,(v_{\text{mole}})_B}{T}, \text{ etc.,}$$

But $$(v_{\text{mole}})_A = (v_{\text{mole}})_B = (v_{\text{mole}})_C$$

∴ $$M_A\,R_A = M_B\,R_B = MR = R_0 = \bar{R} = \text{constant} \qquad \text{...(4.19)}$$

The product MR represented by R_0 (or $\bar{\ }$) is known as the ***Universal Gas Constant*** and it is the same for all gases (Table 4.1). Thus, from Eq. (4.18).

$$pv_{\text{mole}} = R_0\,T \text{ and } pV = nR_0T$$

The value of R_0 or R_{mole} or $\bar{R}$ will be given by ...(4.20)

$$R_0 = \bar{R} = \frac{1.10325 \times 10^5 \times 22.4}{273} \approx 8314 \text{ J/kg mole K.}$$

or $$R_0 = \bar{R} = 8.314 \text{ kJ/kg mole } K = 8.314 \text{ J/mole K.}$$

The unit of R_0 is J/mole. K or kJ/kg mole K.

Example 4.3 Find the volume of a perfect gas at pressure 0.4 MPa and temperature 20°C. Assume the atmospheric pressure as 1 bar.

Solution We know $$pv_{\text{mole}} = MRT = R_0\,T$$

But $$R_0 = 8314 \text{ J/kg mole K}, \; p = 0.4 \times 10^6 \; Pa = 4 \times 10^5 \text{ N/m}^2$$

$$T = 273 + 20 = 293 \text{ K}$$

∴ $$v_{\text{mol}} = \frac{R_0\,T}{p} = \frac{8314 \times 293}{4 \times 10^5} = 6.09 \text{ m}^3$$ **Ans.**

Example 4.4 A tank having molecular volume 4 m^3 contains 30 kg of an ideal gas of molecular weight 29. The temperature is 20°C. Find the pressure.

Solution

$$p = \frac{R_0\,T}{v_{\text{mol}}} = \frac{8314 \times (273 + 20)}{4} = 6.09 \times 10^5\ \frac{\text{N}}{\text{m}^2}$$

$$= 0.609 \text{ MPa}$$

Example 4.5 In a pressure vessel of 20 m^3, O_2, H_2 and CO_2 are stored at the pressure of 1 MPa and temperature 27°C. Find the mass of each gas stored in the tank.

Solution We have

$$R = \frac{R_0}{M} \text{ J/kg K}$$

∴

$$R \text{ for } O_2 = \frac{8314}{32} = 259.81 \text{ J/kg K}$$

$$R \text{ for } N_2 = \frac{8314}{28.016} = 296.758 \text{ J/kg K}$$

$$R \text{ for } CO_2 = \frac{8314}{44} = 188.954 \text{ J/kg K}$$

Now,

$$m = \frac{pV}{RT}$$

$$(m)_{O_2} = \frac{10^6 \times 20}{259.81 \times 300} = 256.59 \text{ kg}$$

$$(m)_{H_2} = \frac{10^6 \times 20}{296.758 \times 300} = 224.64 \text{ kg}$$

$$(m)_{CO_2} = \frac{10^6 \times 20}{188.954 \times 300} = 352.81 \text{ kg}$$

Ans.

Example 4.6 In a pressure vessel, 8 m^3 of N_2 gas is stored at pressure of 50 bar and a temperature of 27°C. Determine: (a) the mass of the gas, (b) the molecular volume, and (c) the density of the gas. Assume M for N_2 = 28.016.

Solution

$$R = \frac{R_0}{M} = \frac{8314}{28.016} = 296.758 \text{ J/kg K}$$

We have,

$$m = \frac{pV}{RT} = \frac{50 \times 10^5 \times 8}{296.758 \times 300} = 449.299 \text{ kg}$$ **Ans.**

(b) Molecular volume $= \dfrac{R_0\,T}{p} = \dfrac{8314 \times 300}{50 \times 10^5} = 0.4988 \text{ m}^3$ **Ans.**

(c) Density of the gas $= \dfrac{\text{Mass}}{\text{Volume}} = \dfrac{449.299}{8} = 56.16 \text{ kg/m}^3$ **Ans.**

Example 4.7 Nitrogen gas has to be stored at 140 bar and 30°C in a steel cylinder having 0.5 m^3 capacity. The cylinder has to be protected against excessive pressure by a fusible plug which will melt and allow the gas to escape if the temperature rises too high. How many kg of nitrogen will the cylinder hold at the desired condition? At what temperature must the fusible plug melt in order to limit the pressure in the receiver to 155 bar.

Solution We have $$pV = mRT$$

$$R = \frac{R_0}{M} = \frac{8314}{28} = 296.928 \text{ J/kg K}$$

$\therefore$ $$140 \times 10^5 \times 0.5 = m \times 296.928 \times (273 + 30)$$

or $$m = \frac{140 \times 10^5 \times 0.5}{296.928 \times 303} = 77.8 \text{ kg}$$ **Ans.**

Also, $$\frac{p_1 V_1}{T_1} = \frac{p_2 V_2}{T_2}$$

where $p_1 = 140 \times 10^5$ N/m^2,

$V_1 = 0.5$ m^3

$T_1 = 303$ K

$p_2 = 155 \times 10^5$ N/m^2

$V_2 = 0.5$ m^3

T_2 =?

$\therefore$ $$\frac{140 \times 10^5 \times 0.5}{303} = \frac{155 \times 10^5 \times 0.5}{T_2}$$

or $$T_2 = 335.464 \text{ K}$$

and $$t_2 = 333.464 - 273 = 62.46°\text{C}$$ **Ans.**

Example 4.8 Calculate the mass and number of moles of air displaced by a special balloon having a radius of 2 m. The atmospheric pressure and temperature are 1 bar and 27°C.

If instead of air, hydrogen gas fills the balloon under the same conditions of pressure and temperature, find the mass and the number of moles of hydrogen. Assume molecular weight of air = 28.97 and that of hydrogen = 2.

Solution The volume of hydrogen $= \frac{4}{3}(\pi r^3) = \frac{4}{3}\pi (2)^3 = 33.51$ m

The balloon will displace the air volume equivalent to its own volume. Thus, the volume of air displaced, $V = 33.51$ m^3.

$\therefore$ Mass of air, $m = \frac{pV}{RT} = \frac{1 \times 10^5 \times 33.51}{287 \times 300} = 38.91$ kg

But Mass = Number of moles × Molecular weight

$\therefore$ Moles of air displaced $= \frac{38.91}{28.97} = 1.34$ moles **Ans.**

But Avogadro's hypothesis states that equal volume of ideal gases contain the same number of molecules at the same conditions of pressure and temperature, so

Moles of hydrogn = Moles of air = 1.34 **Ans.**

Moles of hydrogen = 1.34 × 2 = 2.64 kg **Ans.**

4.6 ENTHALPY OF AN IDEAL GAS

The property, enthalpy, H is defined as the sum of internal energy and flow work. Thus,

$$H = U + pV \quad \text{...(4.21)}$$

The combination $(U + pV)$ often plays an important role in the analysis of open system.

On per unit mass basis,

$$h = u + pv;\ \text{kJ/kg} \quad \text{...(4.22)}$$

For an ideal gas, $pv = RT$

$$\therefore \quad h = u + RT \quad \text{...(4.23)}$$

Because the internal energy, u is a function of temperature alone [i.e., $u = f(T)$ of ideal gass], hence the *enthalpy of an ideal gas is a function or temperature alone. However, this is not true in the case of vapour.*

4.7 SPECIFIC HEATS OF AN IDEAL GAS

The specific heat of a gas is defined as the quantity of heat required to raise the temperature of unit mass of gas by one degree. This is denoted by c.

The average specific heat c is equal to the amount of heat transferred per unit mass divided by the accompanying increase of its temperature. That is,

$$c_{acv} = \frac{Q}{m\,\Delta T} = \frac{q}{\Delta T} \quad \text{...(4.24)}$$

where c_{cav} = specific heat (kJ/kg . K)

Q = heat interaction (kJ)

ΔT = temperature difference (K)

m = mass (kg)

q = heat interaction per unit mass (kJ/kg) $\left(= \frac{\delta Q}{m}\right)$

where q and T are very small, then the ratio of $q/\Delta T$ tends to limit which indicates the instantaneous specific heat at temperature T. Hence,

$$c = \lim_{\Delta T \to 0} \frac{q}{\Delta T} = \frac{\delta q}{d T} \quad \text{...(4.25)}$$

In order to find the mean specific heat over a wide temperature range, the specific heat is integrated and

$$c_{\text{acv}} = \frac{\int_{T_1}^{T_2} c\, dT}{\int_{T_1}^{T_2} dT} = \frac{\int_{T_1}^{T_2} c\, dT}{T_2 - T_1} \quad \text{...(4.26)}$$

The specific heat of gas changes with temperature in a complex manner and various equations have been developed for this. The specific heat of solids and liquids depends upon the temperature but unlike gases it is not sensitive to the type of heating processes (constant pressure and constant volume) involved.

It is an established fact that when heat is supplied to a given quantity of gas, energy will be stored in it. But to know the quantity of heat supplied, the conditions under which heat is supplied must be known. Normally, *heating is done at constant volume and constant pressure.* In order to understand these, let us consider the following cases.

Consider a quantity of gas contained in a fixed container shown in Fig. 4.6 (a). When heat is supplied to the gas, the volume of the gas will not change because the size of the container is fixed but there will be an increase in pressure and temperature of the gas.

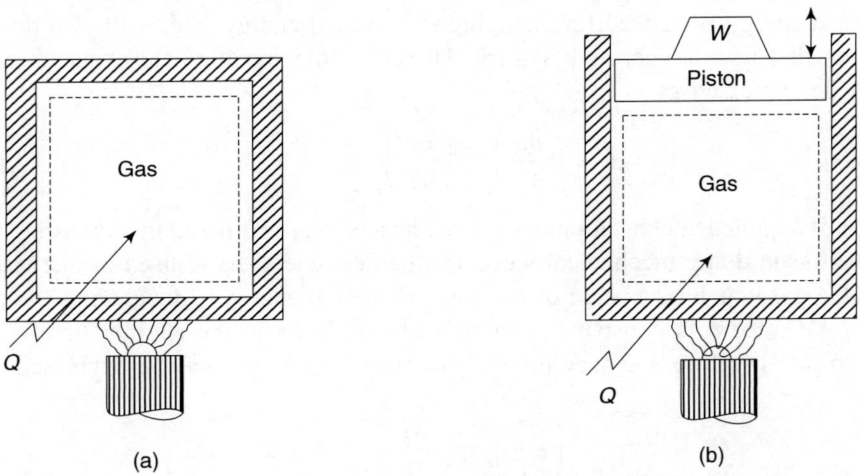

Fig. 4.6 (a) Heating at constant volume **Fig. 4.6** (b) Heating at constant pressure

This increase in pressure and temperature may be given by the expression $p_1/T_1 = p_2/T_2$ since volume is constant. No work is done in this case. Hence, the whole of the heat energy is utilised in increasing the kinetic energy of the gas molecules which will be indicated by the rise in temperature.

When the volume of the gas is constant the quantity of heat required to raise the temperature of unit mass of gas to one degree is known as the specific heat of gas at constant volume. It is generally denoted by c_v.

Now, consider a gas contained in a cylinder tightly fitted with a frictionless piston. The piston carries a load as shown in Fig. 4.6 (b). When heat is supplied to the gas, the kinetic energy of the gas molecules will increase but the tendency for the pressure to rise will be counterbalanced by the piston rising upward and thereby lifting the load and increasing the volume of the gas. Here, the energy supplied increases the temperature and volume of the gas and does work. The pressure remains constant throughout and is equal to the weight on the piston divided by the area of the piston.

When heat is supplied at constant pressure, the quantity of heat required to raise the temperature of unit mass of a gas to one degree is known as the specific heat at constant pressure, It is generally denoted by c_p.

Thus, the gases have two specific heats, (a) at constant pressure, c_p and (b) at constant volume, c_v. It is interesting to note that the specific heat at constant pressure, c_p is greater than the specific heat at constant volume, c_v, *because the heat supplied to a gas at constant pressure is utilised for two purposes, namely*:

(a) for increasing the internal energy of the gas, and

(b) for doing external work.

Thus, we can say that when heat is supplied at constant volume, no external work is done and all the heat energy is utilised for changing the internal energy. This will also be seen in the chapter on first law of Thermodynamics. Therefore, the specific heat of gas at constant volume may be expressed as

$$c_v = \left(\frac{\partial u}{\partial T}\right)_v = \left(\frac{\delta q}{dT}\right)_v \quad \text{...(4.27)}$$

When heat is supplied at constant pressure, the heat energy is utilized in increasing internal energy as well as in doing mechanical work, or, the heat energy is utilised in increasing the enthalpy of the gas which takes care of increase in internal energy as mechanical work (i.e., $\Delta h = \Delta u + p\Delta V$ since p is constant). This will also be seen in the chapter of first law of thermodynamics. Therefore, the specific heat at constant pressure may be expressed as

$$c_p = \left(\frac{\partial h}{\partial T}\right)_p = \left(\frac{\delta q}{dT}\right)_p \quad \text{...(4.28)}$$

As internal energy u and enthalpy h of an ideal gas are functions of temperature alone, the specific heats of an ideal gas are also functions of temperature alone. Thus,

$$c_p = \left(\frac{dh}{dT}\right)_p = \left(\frac{\delta q}{dT}\right)_p \qquad ...(4.29)$$

$$c_v = \left(\frac{du}{dT}\right)_p = \left(\frac{\delta q}{dT}\right)_v \qquad ...(4.30)$$

From the definition of enthalpy, we have, $h = u + pv$

and from characteristic gas equation, $pv = RT$

$$\therefore \qquad h = u + RT$$

Differentiating with respect to T, we get $\frac{dh}{dT} = \frac{du}{dT} + R$

Putting the value of dh/dT and du/dT in Eqs. (4.29) and (4.30), we get

$$c_p - c_v = R \text{ kJ/K} \qquad ...(4.31)$$

$$c_p - c_v = \frac{R}{J} \text{ kcal/K in M.K.S.} \qquad ...(4.32)$$

From the above equation, we find that the difference between the specific heat at constant pressure and constant volume for an ideal gas is equal to the characteristic gas constant, R. For real gases, the difference $(c_p - c_v)$ is never less than R but is greater than R.

The ratio of c_p/c_v is denoted by γ and this is called the ***adiabatic index***.

The value of γ is found experimentally.

For air $\gamma = 1.4$, $c_p = 1.005$ kJ/kg K and $c_v = 0.718$ kJ/kg K

In Table 4.2, the values of c_p and c_v for different gases are given.

Table 4.2 Zero pressure specific heats for various gases in kJ/kg K

Temp., °C	Air			Carbon dioxide, CO_2			Carbon monoxide, CO		
	c_p	c_v	γ	c_p	c_v	γ	c_p	c_v	γ
0	1.005	0.716	1.401	0.816	0.628	1.3	1.038	0.741	1.4
38	1.005	0.72	1.4	0.858	0.67	1.283	1.042	0.745	1.399
93	1.009	0.7243	1.397	0.908	0.72	1.262	1.042	0.745	1.399
149	1.017	0.728	1.394	0.958	0.77	1.246	1.05	0.753	1.394
260	1.038	0.7494	1.383	1.034	0.845	1.223	1.071	0.774	1.384
400	1.067	0.7829	1.368	1.113	0.925	1.204	1.105	0.808	1.367
1090	1.197	0.908	1.316	1.306	1.117	1.169	1.243	0.947	1.314
2200	1.276	0.988	1.291	1.394	1.205	1.156	1.314	1.017	1.292

Contd.

Contd. Table 4.2

Temp., °C	Hydrogen, H_2			Nitrogen, N_2			Oxygen, O_2		
	c_p	c_v	γ	c_p	c_v	γ	c_p	c_v	γ
0	14.197	10.073	1.409	1.038	0.741	1.4	0.9169	0.71	1.397
38	14.343	10.219	1.404	1.038	0.745	1.399	0.921	0.661	1.394
93	14.448	10.324	1.399	1.0425	0.745	1.398	0.933	0.674	1.387
149	14.490	10.366	1.398	1.046	0.749	1.396	0.946	0.686	1.378
260	14.528	10.4	1.397	1.063	0.766	1.388	0.983	0.724	1.36
400	14.582	10.458	1.394	1.092	0.795	1.374	1.025	0.761	1.34
1090	15.733	11.609	1.355	1.226	0.922	1.319	1.13	0.87	1.298
2200	17.743	13.619	1.303	1.323	1.009	1.294	1.2134	0.954	1.272

$$\gamma = \frac{c_p}{c_v} = \frac{c_v + R}{c_v} = 1 + \frac{R}{c_v} \qquad ...(4.33)$$

∴

$$c_v = \frac{R}{\gamma - 1} \text{ kJ/kg K} \qquad ...(4.34)$$

or

$$c_v = \frac{R}{J(\gamma - 1)} \text{ kcal/kg/K in M.K.S.} \qquad ...(4.35)$$

and

$$c_p = c_v \times \gamma = \frac{R\gamma}{(\gamma - 1)} \text{ kJ/kg K} \qquad ...(4.36)$$

or

$$c_p = \frac{\gamma R}{J(\gamma - 1)} \text{ ... kcal/kg K in M.K.S.} \qquad ...(4.37)$$

If M is the molecular weight of the gas, then the molar specific heat (kJ/kg mole K) of the gas will be given by

$$(c_p)_{mol} = M.c_p \text{ and } (c_v)_{mol} = M.c_v \qquad ...(4.38)$$

$$(c_p)_{mol} - (c_v)_{mol} = R_0 = \overline{R}\text{; kJ/kg mol K}$$

Molar specific heat is defined as the quantity of heat required to raise the temperature of one mole (22.4 m^3) of a substance by one degree centigrade.

From Eq. (4.30), we have, $du = c_v\, dT$

or

$$\Delta u = u_2 - u_1 = (q_{1-2})_{v=c} = \int_{T_1}^{T_2} c_v\, dT \qquad ...(4.39)$$

If c_v remains constant then $\Delta u = u_2 - u_1 = (q_{1-2})_{v=c} = c_v\,(T_2 - T_1)$ ...(4.40)

$$U_2 - U_1 = (Q_{1-2})_{v=c} = mc_v\,(T_2 - T_1) \qquad ...(4.41)$$

This is the desired expression for change in internal energy and heat transferred.

From Eq. (4.30), we have $dh = c_p\, dT$

or
$$\Delta h = h_2 - h_1 = (q_{1-2})_{p\,=\,c} = \int_{T_1}^{T_2} c_p\, dT \qquad ...(4.42)$$

If c_p remains constant, then $\delta h = h_2 - h_1 = (q_{1-2})_{p\,=\,c} = c_p\,(T_2 - T_1)$

or
$$H_2 - H_1 = (Q_{1-2})_{p\,=\,c} = mc_p\,(T_2 - T_1) \qquad ...(4.43)$$

In general, $\delta q = c\, dT$

or
$$Q_{1-2} = \int_1^2 m\, c\, dT \qquad ...(4.44)$$

It is interesting to note that numerical value of specific heat ratio of gas in all the units is the same at the same conditions.

We have seen that specific heats in general are not constant. There is a significantly large variation of specific heats with temperature, but no significant change with pressure except at very high value. As pressure decreases, the behaviour of real gas approaches to ideal gases, and so the specific heat of real gases, measured at very low pressure are called ***either ideal gas specific heats or zero pressure specific heats***. The symbols c_{po} and c_{poc} are often used to indicate that the values concerned are those for zero pressure or for very large specific volume. When dealing only with an ideal gas, we omit the subscript o and ∞ but we must remember that c_p and c_v values we use for any ideal gas are actually the c_{po} and $c_{p\infty}$ values of the corresponding real gas.

Figure 4.7 (a) and (b) show the zero pressure specific heat data of several gases as a function of temperature.

It is worthwhile noting that the specific heat values for diatomic and triatomic gases shown in Fig. 4.7 (a) and (b) are mathematically well behaved, monotonically increasing function of temperature and for the pressure indicated. This fact makes it possible to fit algebraic equations, which require only a minimum number of terms for the experimental curves. Typical equations for specific heat as a function of temperature are

$$c_p = a + bT + cT^2 + dT^3 + ...$$
$$c_p = a_1 + b_1\, T^{-1/2} + c_1\, T^{-1} + ...$$
$$c_p = a_2 + b_2\, T^{-1} + c_2\, T^{-2} + ...$$

where the constants a, b, c, d, etc., are known for any particular substances. The units must be specified. Table 4.3 shows the relationship of specific heat for various gases.

Although, we are primarily interested in low-pressure specific heat data applicable to ideal gases, it is appropriate at this point to examine the influence of pressure and temperature on specific heat data. Figure 4.7 (c) shows the dependency of specific heat values on pressure and temperature. The function increases monotonically at all temperatures. But, at higher

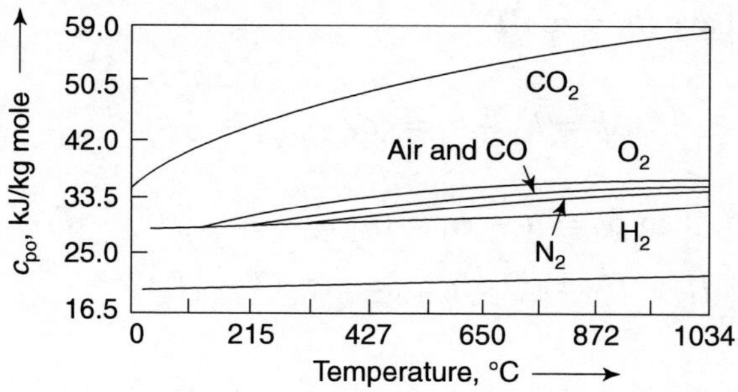

(a) Molar c_{po} for six gases as a function of temperature

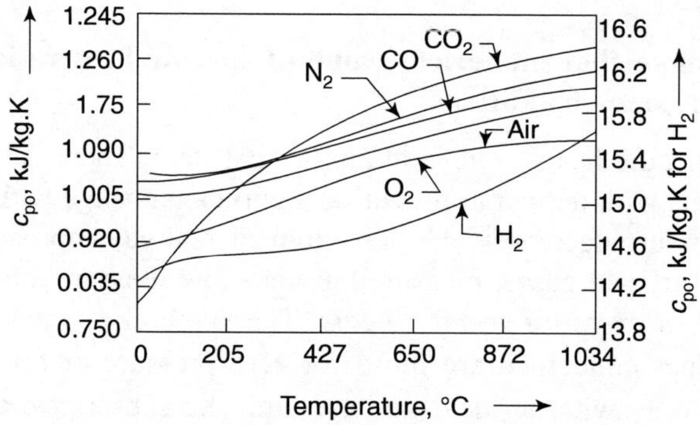

(b) c_{po} for seven gases as function of temperature

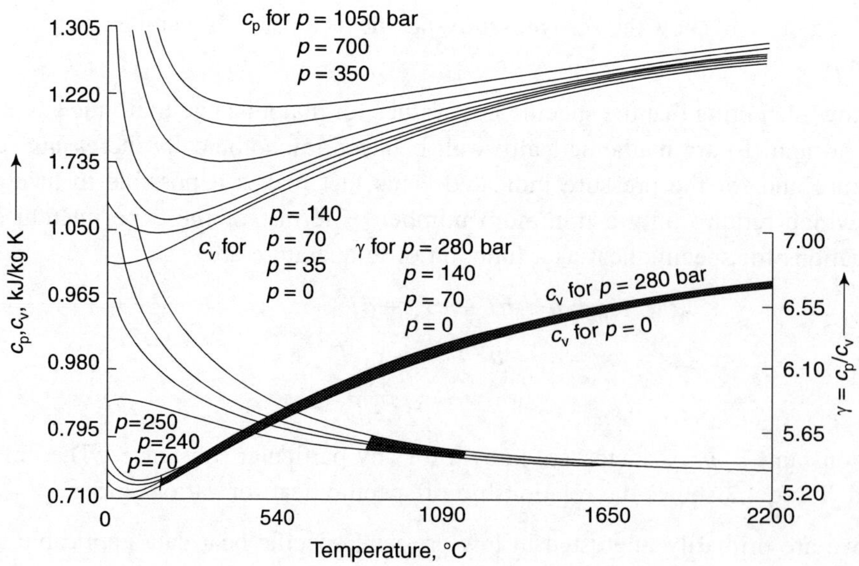

(c) Effect of pressure and temperature on c_p, c_v and γ

Fig. 4.7

pressure both the specific heats increase with decreasing temperature below certain values of temperature. It is also interesting to note that c_v is affected only slightly by pressure where c_p is more adversely affected. Nevertheless, the increase in c_p is roughly, only 20% at 94°C when raised from 0.07 to 350 bar; under similar conditions c_v increases by about 2% for dry air. At much higher temperature, the percentage increase due to pressure is much smaller. *That is why at pressure where a gas still behaves essentially like an ideal gas, it is usually safe to use the zero-pressure specific heat data for practical calculation.* The gas which deviates more from the ideal gas equation of state, it is to be expected that the specific heats, especially c, will become more pressure dependent.

Example 4.9 Calculate the specific heat of a substance during a process where 5 kg of substance receives 1500 kJ of heat and undergoes a temperature change from 50°C to 200°C.

Solution

$$Q_{1-2} = m \int_{t=50}^{t=200} c\, dt = 5 \times c\,(200 - 50)$$

or

$$c = \frac{1500}{150 \times 5} = 2 \text{ kJ/kg K}$$ **Ans.**

Table 4.3 Variable specific heats zero pressure based on spectroscopic data

Substances with temperature range	Molecular weight M	$(c_p)_{mol}$, kJ/kg mole K
Air (280 to 1500 K)	29	$26.62 + 41.53 \times 10^{-4}\, T - 35.67 \times 10^{-8}\, T^2$
O_2 (280 to 2800 K)	32	$48.21 - \frac{724.3}{\sqrt{T}} + \frac{6405.8}{T}$
N_2 (280 to 500 K)	28.016	$39.64 - \frac{14528\, T}{10^4} + 350.56 \times 10^4/T^2$
H_2 (300 to 2200 K)	2.016	$24.11 + \frac{24.19\, T}{10^4} + 83.73 \times \frac{10^4}{\sqrt{T}}$
H_2O (steam) (280 to 2500 K)	18.016	$83.14 - \frac{2499.5}{\sqrt{T}} + \frac{31401}{T}$
CO_2 (280 to 3500 K)	44	$67.8 - \frac{27339.8}{T} + 590.3 \times \frac{10^4}{T^2}$
C_2H_{10} (*n*-butane) (280 to 1800 K)	55.08	$18.25 + \frac{1687.2\, T}{10^4} - \frac{2872.1\, T^2}{10^8}$
C_3H_8 (propane) (280 to 1800 K)	44.06	$9.45 + \frac{1339.7\, T}{10^4} - 2273.4 \frac{T^2}{10^8}$

Contd.

Table Contd. 4.3

CH_4 (methane) (280 to 1800 K)	16.032	$14.15 + \frac{419.5\,T}{10^4} - 555.58\frac{T^2}{10^8}$
C_2H_6 (280 to 1800 K)	30.048	$9.19 + \frac{890.5\,T}{10^4} - 1423.5\frac{T^2}{10^4}$
C_2H_4 (280 to 1800 K)	28.032	$11.32 + \frac{678.2\,T}{10^4} - 1423.5\frac{T^2}{10^4}$
C_2H_2 (280 to 1800 K)	26.016	$49.99 + \frac{102\,T}{10^4} - \frac{3148.4 \times 10^4}{T^2}$

Example 4.10 The specific heat of a gas at constant pressure is 1.0299 kJ/kg and the value of the specific heat at constant volume is 0.753 kJ/kg. Find (a) the adiabatic index and (b) the gas constant.

Solution

$$\gamma = \frac{c_p}{c_v} = \frac{1.0299}{0.753} = 1.3677 \qquad \textbf{Ans.}$$

We know,

$$R = c_p - c_v = 1.0299 - 0.753 = 0.2769 \text{ kJ/kg K} \qquad \textbf{Ans.}$$

Example 4.11 The characteristic equation of a certain gas is $pV = m\,RT$ where T is in K. Calculate the specific heat of the gas at constant pressure if the specific heat of the gas at constant volume is 0.6489 kJ/kg K. Take R = 259.96 J/kg K.

If 0.4 kg of this gas occupies a volume of 0.3 m^3 at a pressure of 2 bar, calculate the internal energy of the gas reckoned from – 40°C.

Solution We have

$$c_p = c_v + R = 0.6489 + \frac{259.96}{1000} = 0.9088 \text{ kJ/kg K}$$

Again,

$$pV = mRT$$

or

$$2 \times 10^5 \times 0.3 = 0.4 \times 259.6 \times T$$

or

$$T = \frac{2 \times 10^5 \times 0.3}{0.4 \times 259.96} = 577 \text{ K} = 304°\text{C}$$

$$\Delta U = mc_v\,[t - (-40)] = 0.4 \times 0.6489\,[300 + 40]$$

$$= 89.288 \text{ kJ} \qquad \textbf{Ans.}$$

Example 4.12 A gas having mass 4.5 kg undergoes a process from 50°C to 100°C. If the specific heat of the gas is a function of temperature alone, calculate the heat transfer during the process for the relation

$$c_p = 1.256 + 80/(T + 50) \text{ kJ/kg}$$

Solution We know,

$$Q_{1-2} = \int_{T_1}^{T_2} m \cdot c_p \cdot dT = 4.5 \int_{323}^{373} \left[1.256 + \frac{80}{T+50}\right] dT$$

$$= 4.5 \left[\int_{323}^{373} 1.256\, dT + \int_{323}^{373} \frac{dT}{0.0125\,T + 0.625}\right]$$

$$= 4.5 \left[(1.256 \times 50) + \left(\frac{1}{0.0125} \ln (0.0125\,T + 0.625\right)_{323}^{373}\right]$$

$$= 4.5 \left[62.8 + \frac{1}{0.0125} \{\ln (4.662 + 0.625) - \ln (4.037 + 0.625\}\right]$$

$$= 328.03 \text{ kJ}$$

Example 4.13 Find the change in enthalpy when 1 kg of oxygen is heated from 300 K to 1000 K. The specific heat at constant pressure c_p is given by

$$c_p = 45 + \frac{700}{\sqrt{T}} + \frac{6000}{T} \text{ kJ/kg mole K}$$

Solution We know,

$$h_2 - h_1 = \int_{T_1}^{T_2} c_p\, dT = \int_{T_1}^{T_2} \left(45 + \frac{700}{\sqrt{T}} + \frac{6000}{T}\right) dT$$

$$= 45 (T_2 - T_1) + 700 \times 2 \left(T_2^{1/2} - T_1^{1/2}\right) + 6000 \ln \frac{T_2}{T_1}$$

$$= 45 (1000 - 300) + 1400 (1000^{1/2} - 300^{1/2}) + 6000 \ln \frac{1000}{300} = 58738 \text{ kJ/kg mole}$$

$$\therefore \quad h_2 - h_1 = \frac{58738}{32} = 1835.56 \text{ kJ/kg}$$ **Ans.**

4.8 KINETIC THEORY OF GASES (MICROSCOPIC ANALYSIS OF PERFECT GASES)

We have discussed the concept of an ideal gas on the basis of its behaviour from the macroscopic point of view. Now, we shall develop the concept of perfect gas from the microscopic point of view, considering individual molecules of the gas and their behaviour. The kinetic theory of gases deals with the behaviour of molecules constituting the gases. Bernoulli and Boyle are considered to be the originators of the kinetic theory. From the middle of the 19th century the kinetic theory of gases, was greatly developed by Clausius, Maxwell, Boltzmann, Jeans, van der Waals, Lorentz and Raleigh covering almost the entire field of the different phenomena related with gases.

According to the kinetic theory, the molecules of all gases are in continuous motion. As a result of this, they possess kinetic energy which is transferred from molecule to molecule during their collision. This transferred energy produces a change in velocities of individual molecules. Thus, we see that the complete phenomenon of molecular behaviour is very complex. So for analysis of ideal gas the following assumptions are made.

Assumptions (Mathematical Modeling)

(a) The molecules of gases are assumed to be rigid, perfectly elastic, solid spheres.

(b) The mean distance between molecules is about 2×10^{-10} m while the distance between molecules is about 3×10^{-9} m at S.T.P., i.e., the distance is about 10 times the diameter.

(c) The molecules are in a state of incessant random motion moving in all directions with all possible velocities and the gas is said to be in a state of molecular chaos.

(d) The collisions between molecules in one kg mole, known as Avogadro's number is 6.03×10^{26}. Under standard conditions, 1 kg mole occupies 22.4 m^3, it means there will be 3×10^{25} molecules in 1 m^3 volume of gas. Due to collisions, the number of molecules per unit volume of the gas remains constant.

(e) The time spent in collision is negligible as compared to the time during which the molecules are moving independently.

(f) Between collisions, the molecules move in a straight line with uniform velocity because of frictionless effect between molecules. *The distance between two collisions is called the free path of the molecule* ; the average distance travelled by a molecule between successive collisions is known as the *mean free path.*

(g) The volume of a molecule is so small that it is negligible as compared to the total volume of the gas.

(h) The motion imparted to the gas molecules by gravity is negligible in comparison to the effect of the continued collisions between them.

(i) The average kinetic energy of the gas molecules is proportional to the absolute temperature.

4.8.1 Governing Equation

Imagine a quantity of gas to be contained in a cubical vessel of side l with perfectly clastic walls and let n represent the very large number of molecules in the vessel as shown in Fig. 4.8 (a).

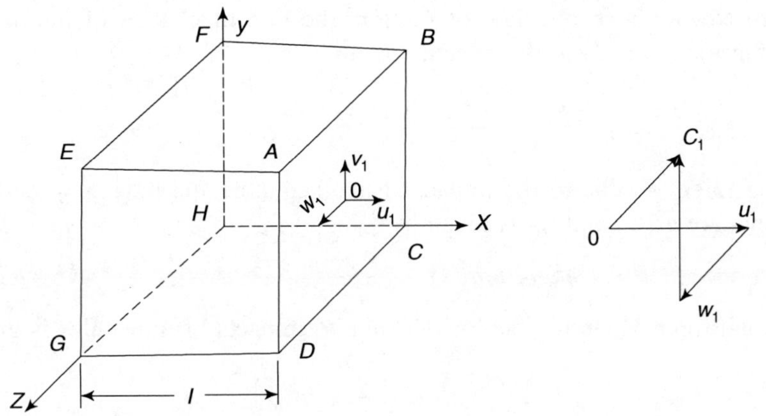

Fig 4.8 (a) Kinetic theory analysis

Now, these molecules may be assumed to have a vessel of side l with perfectly elastic walls and let n represent a very large number of molecules in a certain direction. This velocity c_1 can be resolved into three components u_1, v_1 and w_1 parallel to three co-ordinate axes X, Y and Z which are again assumed parallel to the sides of the cube as mentioned in the figure. Thus,

$$c_1^2 = u_1^2 + v_1^2 + w_1^2 \qquad ...(4.45)$$

Now, suppose one molecule of gas having mass m strikes the wall surface $ABCD$ of the cube with velocity u_1. Since the collision is perfectly elastic, the molecule will rebound from this face with the same velocity $- u_1$.

$\therefore$ The momentum of the molecule before it strikes the face $ABCD = mu_1$.

The momentum of the molecule after the impact $= mu_1$.

Hence, the change in momentum at each impact in direction normal to the surface $ABCD$

$$= mu_1 - (mu_1) = 2mu_1 \qquad ...(4.46)$$

After striking the face $ABCD$, the molecule rebounds and travels to the face $EFGH$, collides with it, rebounds and travels back again to the face $ABCD$ covering $2l$ distance.

This means heat the molecule covers $2l$ distance to hit the same face again. Hence, the time taken by the same molecule to strike the same face $ABCD$ again is

$$= \frac{2l}{u_1}$$

Therefore, the rate of change of momentum for one molecule of the gas

$$= \frac{2\,mu_1}{2l / u_1} = \frac{mu_1^2}{l} \qquad ...(4.47)$$

According to Newton's second law of motion, the rate of change of momentum is the force. If F_1 is the force due to ones molecule, then

$$F_1 = \frac{mu^2}{l}$$

Similarly, the force, F_2 due to the impact of another molecule having velocity c_2 whose three components are u_2, v_2 and w_2 is given by

$$F_2 = mu_2^2 / l$$

Hence, the total force F_x on the face *ABCD* due to impacts of molecules is given by

$$F = \frac{m}{l}\left(u_1^2 + u_2^2 + \ldots\ldots\, u_n^2\right) \qquad \ldots(4.48)$$

Since pressure is force per unit area, hence pressure exerted on the wall *ABCD* is given by

$$p_x = \frac{F_x}{l^3} = \frac{m}{l^3}\left(u_1^2 + u_2^2 + u_3^2 + \ldots.\, u_n^2\right) \qquad \ldots(4.49)$$

Similarly, if p_y and p_z represent the pressure on other faces which are perpendicular to the *Y* and *Z* axes respectively, we have

$$p_y = \frac{m}{l^3}\left(v_1^2 + v_2^2 + \ldots + v_n^2\right) \qquad \ldots(4.50)$$

$$p_z = \frac{m}{l^3}\left(w_1^2 + w_2^2 + \ldots + w_n^2\right) \qquad \ldots(4.51)$$

Since the pressure exerted by the gas is the same in all directions, i.e., $p_x = p_y = p_z$ the average pressure p of the gas is given by

$$p = \frac{p_x + p_y + p_z}{3}$$

or

$$p = \frac{m}{3\,l^3}\left[\left(u_1^2 + v_1^2 + w_1^2\right) + \left(u_2^2 + v_2^2 + w_2^2\right) + \ldots + \left(u_n^2 + v_n^2 + w_n^2\right)\right] \qquad \ldots(4.52)$$

But

$$c_1^2 = u_1^2 + v_1^2 + w_1^2,\; c_2^2 = c_2^2 + u_2^2 + w_2^2 \;\ldots \text{ and so on}$$

$$l^3 = V = \text{volume of the gas, } m^3$$

∴

$$p = \left(\frac{1}{3}\right)\frac{m}{V}\left(c_1^2 + c_2^2 + \ldots\, c_n^2\right)$$

or

$$p = \left(\frac{1}{3}\right)\frac{mn}{V}\,\overline{C}^2 \qquad \ldots(4.53)$$

or

$$pV = \left(\frac{1}{3}\right) mn\,\overline{C}^2 \ldots \qquad \ldots(4.54)$$

where $\bar{C}^2 = \dfrac{c_1^2 + c_1^2 + ... + c_n^2}{n}$ known as the mean square velocity

or
$$\bar{C} = \sqrt{\frac{c_1^2 + c_2^2 + ...c_n^2}{n}}$$

$\bar{C}$ is called the *root mean square (rms) velocity of the molecules* and is equal to the square root of the mean of the square of the velocities of individual molecules which is evidently not the same as mean of the velocities of the different molecules.

Equation (4.54) is the fundamental equation of the kinetic theory of gases and is often referred to as kinetic equation of gases. Equation (4.54) may also be written as

$$pV = \frac{2}{3} \cdot \frac{1}{2} mn\,\bar{C}^2 = \frac{1}{3} mn\,\bar{C}^2 \qquad ...(4.55)$$

where (1/2) $mn\bar{C}^2$ is the average translation, or linear kinetic energy of the system of particles.

or
$$pv = \frac{2}{3} \cdot \frac{1}{2} \bar{C}^2 \qquad ...(4.56)$$

where v = specific volume of the gas = $\dfrac{V}{mn}$

From Eq. (4.56), we have

$$\bar{C} = \sqrt{\frac{3p}{\rho}} = \frac{\sqrt{3pV}}{mn} \qquad ...(4.57)$$

where $\rho = \dfrac{1}{v}$ = density of the gas

m = mass of one molecule

and n = number of molecules

(i) Interpretation of Temperature. From kinetic equation, we have, $pV = \left(\dfrac{1}{3}\right) mn\,\bar{C}^2$

If v_{mole} is the volume occupied by 1 gram molecule (i.e., 1 mole) of gas and N is the number of molecules in 1 gram molecule of gas, then

$$pv_{mol} = \frac{1}{3} m\,N\,\bar{C}^2 = \frac{2}{3} \cdot \frac{1}{2} \cdot mN\,\bar{C}^2$$

But M = molecular weight = $m \,.\, N$

From Boyle's law, $pv_{mol} = R_0\,T$

Equating these two, $\dfrac{2}{3} \cdot \dfrac{1}{2} mN\,\bar{C}^2 = R_0\,T$

or $$\text{K.E.} = \frac{1}{2} mN\bar{C}^2 = \frac{3}{2} R_0\, T \qquad \text{...(4.58)}$$

or $$\frac{1}{2} m\,\bar{C}^2 = \left(\frac{3}{2}\right)\frac{R_0}{N} T = \left(\frac{3}{2}\right) KT$$

where $K = \dfrac{R_0}{N}$ = Boltzmann constant

Thus, the mean *kinetic energy per molecule in a given mass of gas is proportional to the absolute temperature of the gas.*

From Eq. (4.58), we have

$$\bar{C} = \sqrt{3\frac{R_0}{M} T} \quad \alpha\ \sqrt{T} \qquad \text{...(4.59)}$$

This suggests that the *rms velocity is directly proportional to square root of absolute temperature of gas.*

(ii) Derivation of Gas Laws. Gas laws may be derived from the kinetic theory of gases.

(a) Perfect Gas Equation and Boyle's Law. From kinetic equation

$$pV = \frac{2}{3}\cdot\frac{1}{2} mn\,\bar{C}^2$$

or $$pv_{\text{mol}} = \frac{2}{3}\cdot\frac{1}{2} mN\,\bar{C}^2$$

According to kinetic theory, there is direct proportionality between absolute temperature and kinetic energy of the molecules, that is, $\dfrac{1}{2} m\,\bar{C}^2 = \dfrac{3}{2} KT$ ...(4.60)

or $$pv_{\text{mol}} = KNT \qquad \text{...(4.61)}$$

From Eq. (4.61), we may write

$$pv_{\text{mole}} = R_0 T$$

where $R_0 = KN$ constant

If T remains constant, then pv = constant and this is the Boyle's law. *This is the perfect gas equation of state.*

(b) Charles' Law. From the kinetic equation, $pv = \dfrac{1}{2} mn\,\bar{C}^2$

or $$pv_{\text{mole}} = \frac{2}{3}\cdot\frac{1}{2} mN\,\bar{C}^2 = KNT \qquad \text{...(4.62)}$$

At constant pressure, $v_{\text{mole}}\ \alpha\ T$

This is Charles's law.

(c) Avogadro's Hypothesis. From the kinetic equation

$$pV = \frac{2}{3}\cdot\frac{1}{2}\, mn\,\overline{C^2}$$

If equal volume of two gases is considered at the same pressure, then kinetic equation applied to one gas is

$$pV = \frac{2}{3}\cdot\frac{1}{2}\, m_1\, n_1\, \overline{C_1^2}$$

Kinetic equation applied to second gas,

$$pV = \frac{2}{3}\cdot\frac{2}{3}\, m_2\, n_2\, \overline{C_2^2}$$

or $$\frac{2}{3}\cdot\frac{2}{3}\cdot m_1\, n_1\, \overline{C_1^2} = \frac{2}{3}\cdot\frac{1}{2}\, m_2\, n_2\, \overline{C_2^2}$$

or $$\frac{1}{2}\, m_1\, n_1\, \overline{C_1^2} = \frac{1}{2}\, m_2\, n_2\, \overline{C_2^2} \qquad ...(4.63)$$

When the temperature of both the gases is the same, their mean kinetic energy per molecule will also be the same, so that

$$\frac{1}{2}\, m_1\, \overline{C_1^2} = \frac{1}{2}\, m_2\, \overline{C_2^2} \qquad ...(4.64)$$

With the help of Eq. (4.64), we have

$$n_1 = n_2 \qquad ...(4.65)$$

Under the same conditions of temperature and pressure, equal volumes of two gases contain the same number of molecules.

This is the Avogadro's Hypothesis.

(d) Graham's Law of Diffusion. From kinetic equations

$$pV = \left(\frac{1}{3}\right) mn\,\overline{C^2}$$

or $$pv = \left(\frac{1}{3}\right)\overline{C^2}$$

or $$\overline{C^2} = \frac{3p}{\rho} \text{ where } \rho \text{ is the density}$$

or $$\overline{C} = \sqrt{(3p/\rho)}$$

The rate of the diffusion (r) of a gas depends on the velocity of molecules, i.e., the greater the velocity, the higher will be the rate of diffusion, i.e., $r \infty C$. Thus, we have

$$r \propto \sqrt{(3p/\rho)}$$

or $$r \propto \sqrt{(1/\rho)} \text{ where } \rho \text{ is constant.} \quad ...(4.66)$$

This is known as Graham's law of diffusion.

(iii) Specific Heat of Ideal Gases from Kinetic Theory. The kinetic theory of gases is also useful in providing some insight into the specific heats of monatomic and diatomic gases. The approach presented here is not particularly useful for gases more complex than these. The internal energy of an ideal gas is the sum of molecular kinetic energy and potential energies. The molecular potential energy of an ideal gas is constant. Therefore, the molecular potential energy of an ideal gas may be assumed to be zero. In monatomic gases, the only form of molecular kinetic energy is translational one. Hence, on the basis of the kinetic theory of gases, the internal energy of an ideal is given by Eq. (4.58) i.e.,

$$M{\cdot}u = \frac{1}{2} M\bar{C}^2 = \frac{3}{2} R_0 T$$

where $M{\cdot}u$ is the internal energy per mole gas

Then, the molar specific heat at constant volume is

$$(c_v)_{mol} = \left(\frac{\partial M\cdot u}{\partial T}\right)_v = \left(\frac{3}{2}\right) R_0$$

The molar specific heat at constant pressure is given by

$$(c_p)_{mol} = (c_v)_{mol} + R_0 = \left(\frac{3}{2}\right) R_0 + R_0 = \left(\frac{5}{2}\right) R_0$$

∴ The ratio of specific heats for a monatomic ideal gas is given by

$$\gamma = \frac{(c_p)_{mol}}{(c_v)_{mol}} = \frac{5/2}{3/2} = \frac{5}{3}$$

On a molar basis, since R_0 = 8.314 kJ/kg mole K, we find that $(c_p)_{mol}$ = 20.785 kJ/kg mole and $(c_p)_{mol}$ = 12.47 kJ/kg mole K, for a monatomic gas like argon or helium.

In order to extend this discussion to diatomic molecules, it is important, first, to consider what is meant by a degree of freedom with relation to an energy mode. Energy modes are those physically distinguishable means by which a particle can have energy, e.g., in translation, in rotation, in vibration, etc. *A degree of freedom is defined as any independent quantity which must be specified to determine the energy of a molecule.* For example, a monatomic molecule constrained to move in x direction has one degree of freedom and the velocity in x direction must be specified. If the molecule is free to move in any direction, the velocity in x, y and z directions must be specified, and the molecule has three degrees of freedom. If the moment of inertia of the molecule about the x axis is not zero and the molecule rotates about this axis, then there is a rotational kinetic energy and the angular velocity must be specified. Rotation about other axis adds other degree of freedom. An important principle in kinetic theory development is the ***Equipartition of Energy Principle*** which states that the total energy of a molecule is divided equally among their degree of freedom. Therefore, the total kinetic energy per mole of monatomic molecules which involves translational energy only and have three degrees of freedom is 3 $R_0\,T/2$, the energy per degree of freedom is (1/2) $R_0\,T$. Thus, the *equipartition of energy principle* clearly states that each additional degree of freedom involves additional energy in the amount (1/2) $R_0\,T$ per mole. The internal energy per mole, having particles with f degree of a freedom, is given by

$$Mu = \frac{f}{2} R_0\, T$$

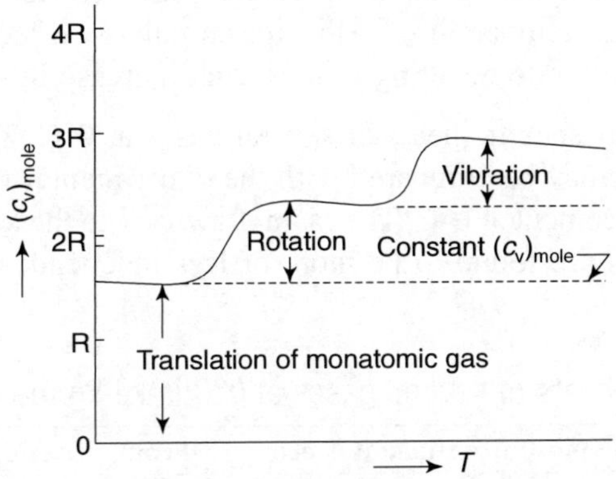

Fig. 4.8 (b) Variation of c_v based on kinetic theory

$$\therefore \qquad (c_v)_{mol} = \left(\frac{\partial\, M \cdot u}{\partial\, T}\right)_v = \frac{f}{2} R_0$$

and

$$(c_p)_{mol} = (c_v)_{mol} + R_0 = \frac{f}{2} R_0 + R_0 = \frac{f+2}{f} R_0$$

$$\gamma = \frac{(c_p)_{mol}}{(c_v)_{mol}} = \frac{f+2}{f}$$

Let us consider now a diatomic molecule consisting of two point masses. In addition to translator, kinetic energy, the diatomic molecule possesses rotational kinetic energy by virtue of rotation about the two mutually perpendicular axes which are perpendicular to the line joining the two masses. The energy of rotation about the axis connecting the atom is zero because the moment of inertia about this axis is zero for point particles. It is also possible that atoms may vibrate along the line connecting them. This vibration of atoms involves two additional degrees of freedom because both potential and kinetic energy are involved. In order to determine the vibrational energy, both position and velocity of the atoms must be specified. In other words, a diatomic molecule may have seven degrees of freedoms—three associated with translation, two with rotation and two with vibration. It is to be kept in mind that vibration occurs only at higher temperature. Thus, at normal room temperature, for example, a diatomic gas usually has only five degrees of freedom, further, at very low temperature, there is no rotation of the molecule, hence $f = 3$ of a diatomic gas. Thus, according to kinetic theory of gases, for diatomic gases we come to the conclusion that

At low temperature : $(c_v)_{mol} = (3/2)\ R_0$, $(c_p)_{mol} = (5/2)\ R_0$

At intermediate temperature : $(c_v)_{mol} = (5/2)\ R_0$, $(c_p)_{mol} = (7/2)\ R_0$

At high temperature : $(c_v)_{mol} = (7/2)\ R_0$, $(c_p)_{mol} = (9/2)\ R_0$

The variation $(c_v)_{mol}$ with temperature is shown in Fig. 4.8 (b). Reason for the gradual change in the value of $(c_v)_{mol}$ is that as the temperature increases, not the all molecules begin to rotate or vibrate at the same temperature, so that the cumulative effect when more molecules begin to rotate and then to vibrate resulting in a gradual increase in specific heat.

Table 4.4 gives the molar specific heats of several gases at 0°C. The specific heat values predicted by kinetic theory may be compared with the experimental values presented earlier. Both values are in good agreement at 0°C. The main drawback to the kinetic theory of specific heat is that these quantities are found to be more or less independent of temperature while this is not true.

Table 4.4 Molar specific heats of several gases at 0°C in kJ/kg mole K

Gas	Experimental values at 1 atm		From kinetic theory (No vibration)	
	$(c_v)_{mol}$	$(c_p)_{mol}$	$(c_v)_{mol}$	$(c_v)_{mol}$
A	12.456	20.732	$(3/2)\ R_0 = 12.456$	$(5/8)\ R_0 = 20.774$
He	12.623	20.94	$(3/2)\ R_0 = 12.456$	$(5/8)\ R_0 = 20.774$
Air	20.732	29	$(3/2)\ R_0 = 20.774$	$(5/8)\ R_0 = 29.05$
CO	20.774	29.05	$(3/2)\ R_0 = 20.774$	$(5/8)\ R_0 = 29.05$
N_2	20.774	29.05	$(3/2)\ R_0 = 20.774$	$(5/8)\ R_0 = 29.05$
O_2	20.9	29.218	$(3/2)\ R_0 = 20.774$	$(5/8)\ R_0 = 29.05$

Example 4.14 Find the root mean square speed of oxygen and hydrogen molecules at 70°C.

Solution Assume O_2 and H_2 as perfect gases. From Eq. 4.59, we have

$$(\bar{C})_{O_2} = \sqrt{3\left(\frac{R_0\,T}{M}\right)} = \sqrt{\left(3 \times \frac{8314}{32} \times 343\right)} = 517.056 \text{ m/s} \qquad \textbf{Ans.}$$

$$(\bar{C})_{H_2} = \sqrt{\left(3 \times \frac{8314}{2} \times 343\right)} = 2068.22 \text{ m/s} \qquad \textbf{Ans.}$$

Example 4.15 At what temperature will the root mean square velocity of hydrogen be double of its value at N.T.P. pressure remaining constant.

Solution We have $\bar{C} \propto \sqrt{(T)}$

Let the velocity at N.T.P., i.e., at T_0 °C be C. If T (K) is the temperature required, the velcoity $\overline{C}$ at that temperature = 2 C_0

$$\therefore \qquad \frac{\overline{C}}{\overline{C}_0} = \frac{2\,\overline{C}_0}{\overline{C}_0} = \sqrt{\left(\frac{T}{T_0}\right)}$$

or
$$4 = \frac{T}{T_0}$$

or
$$T = 4\ T_0 = 4 \times 273 = 1092 \text{ K}$$

4.9 MEAN FREE PATH

In the assumptions on perfect gases, the molecules were assumed as having point masses and the effect of collisions were neglected. But in actual practice, this is not a fact. The molecules have very small finite sizes and there are collisions of molecules. These two factors, (molecules having finite sizes and collisions between molecules) are interdependent since the molecules could never collide with each other if they were point masses without dimensions. As a molecule, with finite size moves forward, it is sure to collide with another molecule after a short interval and its velocity and direction of motion will be changed so that its path will consist of a series of short zig-zag lines (Fig. 4.8 c) limited to the confines of the vessel that contains the gas. The same process holds good for all the molecules of the gas.

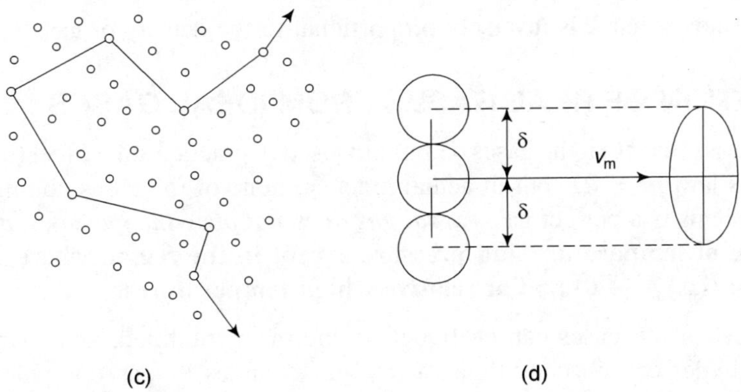

Fig. 4.8 (c) and (d) Mean Free, Path

Assuming, therefore, collisions between finite-sized molecules of a gas, the path travelled by a molecule between two consecutive collisions will be a straight line with uniform velocity since molecules exert no force over one another except during collisions, hence it is called the ***free path***. As these paths of a molecule between collisions will not be of the same length, the mean free path is a more useful quantity to be considered. *It is defined as the average distance travelled by a molecule between two successive collisions.* If the total path travelled in N collisions is S, then mean free path is given by $\lambda = S/N$.

Now, consider a gas which contains n molecules per c.c. Let σ be the diameter of each of the molecule. It is assumed here that the molecule under consideration is only in motion while all others are at rest. The moving molecule will collide with all these molecules whose centre lies within a distance σ from the centre (Fig. 4.8 d). Let u_m be the mean velocity of the molecule in one second, it will collide with all the molecules, the centre of molecules in this cylinder volume will be $\pi\sigma^2 u_m \cdot n$ and this will be also the number of collisions N. Thus,

$$\lambda = \frac{\overline{S}}{N} = \frac{u_m}{\pi \sigma^2 u_m \cdot n} = \frac{1}{\pi \sigma^2 n} \quad \text{...(4.67)}$$

This equation is known as **Clausius equation.**

On the basis of more realistic assumption that all the molecules have the same average speed, Boltzmann found that

$$\lambda = \frac{1}{\sqrt{2 \pi \sigma^2 n}}$$

In all the gases, the mean free path is inversely proportional to the square of the molecular diameter.

The value of λ for O_2 is 8×10^{-6} cm. If m is the mass of one molecule,

$$\lambda = \frac{m}{\pi \sigma^2 \rho} \quad \text{...(4.67a)}$$

The equation shows that λ is inversely proportional to the density of gas.

4.10 DEVIATION OF REAL GASES FROM IDEAL GASES

Though, we have assumed that the gases commonly used in practice are perfect (or ideal) gases that obey the gas law, $pv = RT$, but in actual practice, none of the gases commonly used in engineering problems is a perfect gas. ***A real gas does not obey the gas laws, $pv = RT$* over the entire range of temperature and pressure except in the region where the pressure approaches zero (i.e., $p \to 0$) and at relatively high temperatures.**

Normally, most of the gases can be liquefied and even solidified, i.e., its phase can be changed. Thus, under certain conditions, a real gas changes its gaseous phase to another state, and at the same time it does not accurately obey the ideal gas law. In other words, ***all real gases are vapour forms of a substance***. It will be seen that the assumption of ideal gas behaviour for the real gases is not sufficiently accurate at *low temperature or at very high pressure*. There is appreciable deviation in the real and ideal gas laws under such a condition. Hence, it is essential to study the actual behaviour of real gases. This is possible on the basis of kinetic theory of gases.

According to the kinetic theory of gases the ideal gas equation of state, $pv = RT$ may be derived provided the following conditions are satisfied.

(i) There is a large number of molecules in a small volume of gas.

(ii) Molecules collisions with one another and with the walls of the container are perfectly elastic.

(iii) The molecules are separated by distance, large compared to their own dimensions, so that there is no molecular attraction between the particles of the gas and that they are in a state of continuous motion.

(iv) There are no intermolecular forces with one another except when they collide. Under the above assumptions, the behaviour of the real gas approaches closely to the ideal gas. But if the temperature and pressure are such that the above assumptions are not satisfied, the behaviour of the real gas deviates from that of ideal gas, the main reason for this behaviour are given below.

1. Shape Factors and Intermolecular Forces In fact, molecules are not point masses and have many shape factors, i.e., there exists a shape factor for each molecule. At high pressure, the volume fraction occupied by the atoms is significant. The molecules attract or repel each other depending upon the distance of separation.

Figure 4.9 shows the variation of intermolecular forces F and energy of attraction E with the intermolecular distance, r. The following conclusions may be drawn from Fig. 4.9.

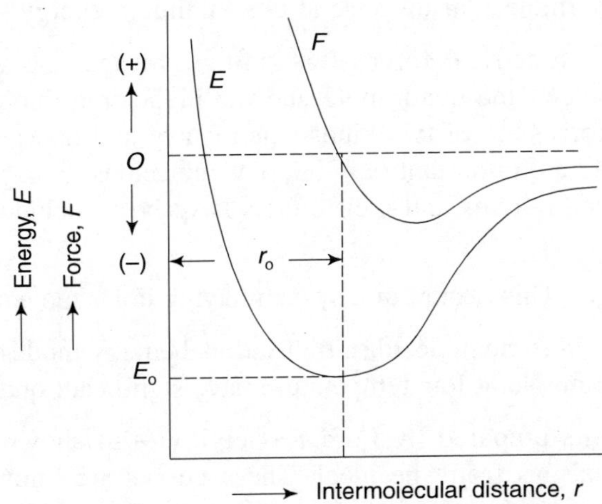

Fig. 4.9 Intermolecular forces and energy

(a) When the intermolecular distance is large, i.e., $r/r_0 >> 1$, the energy of attraction E is negligible and the intermolecular force F is nearly zero.

(b) When the intermolecular distance is small the molecules start attracting one another and the force and potential energy of the system decrease and become negative. These forces are known as ***long-range attractive forces***. As the intermolecular distance decreases the attractive force increases and attains a maximum value. At about $r/r_0 = 1$, the potential energy is the minimum and is denoted by $- E_0$.

(c) As the intermolcular distance still decreases, the attraction of energy decreases, becomes zero and then becomes repulsive. As the molecules come closer, the repulsion keeps on increasing resulting in the increase of potential energy. These forces are known as ***short range repulsive forces***.

For the analysis, molecules are classified as simple molecules (Ar, Kr, Xe) may be defined as having spherical shape and an inverse 12th power repulsive potential. Such molecules may well be expressed by a Lennard – Jones 6–12 potential.

$$E = E_0\left[\left(\frac{r_0}{r}\right)^{12} - 2\left(\frac{r_0}{r}\right)^{6}\right] \qquad ...(4.68)$$

where r_0 = distance of separation for minimum potential energy

$-E_0$ = the maximum energy of attraction or the minimum potential energy

For the minimum energy, $\frac{dE}{dr} = 0$ which gives $r = r_0$

Putting this value in Eq. (4.68), we get $E = -E_0$

This means that the intermolecular distance at this minimum energy $-E_0$ is r_0.

2. Electrical Forces. Electrical forces (excluding hydrogen bonding) depend upon the molecular dipole moment, the quadrupole and the higher multipole moments and the polarizability. Electrical forces are of two kinds—permanent and non-permanent. Electrical forces of a permanent kind are important only in polar molecules. Non-permanent electrical forces are found in all real systems and their effects have been included in intermolecular forces.

3. Hydrogen Bonding. This occurs in a system having polar molecules.

4. Quantum Effects. In some molecules, translational energy modes must be quantized. For molecules of H_2, He and Ne at low temperature have significant quantum effects.

The experimental results prepared by Lord Rayleigh (1905) shown in Fig. 4.10 gives the deviation of the actual gas from the ideal. These curves are known as ***isotherms***. *If Boyle's law is obeyed, pv should be constant at all pressures and the graph should be a straight horizontal line. But this is only an ideal behaviour which no real gas will show. At low pressure* the deviations are represented by the left hand end of the curves. For H_2, pv increases as p increases while for N_2 and CO_2, pv decreases as p increases. This suggests that at low pressure pv for N_2 and CO_2 has lower value while for H_2 it has higher value than expected of an ideal gas.

At high pressures the deviations are indicated by the right–hand ends of the curves. The diagram shows that pv for H_2 increases continuously with the rise of pressure. In the case of N_2 and CO_2, however, pv first decreases and then rises. At very high pressure, the curves are seen to pass the ideal gas curves. Thus, at high pressure, gases H_2, N_2 and CO_2 have higher values of pv than that expected of an ideal gas.

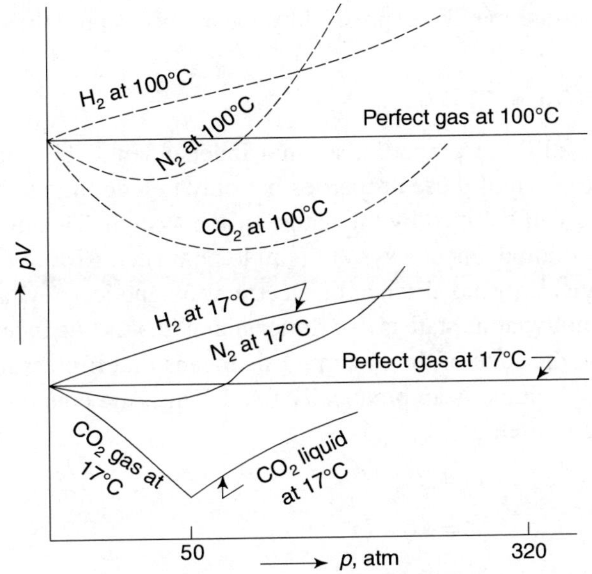

Fig. 4.10 Deviation of real gases from ideal gas

Effect of temperature on deviations becomes apparent from the relative study of the two sets of curves at 17°C and 100°C. At 17°C, the curve for N_2 has a more clearly marked deviation than at 100°C. This effect is still more clear in the case of CO_2 which shows a small depression at 100°C while at 17°C a sudden fall occurs at 55 atm, at which pressure the gas liquefies. Thus, at lower temperature, the deviation becomes more pronounced with the fall in temperature till the approach of the conditions of liquefaction the curve records a sharp break. ***This shows Boyle's law is valid only at very low pressure and moderately at high temperature.***

4.11 CRITICAL STATE

*The limiting state at which no discontinuities are noticeable between liquid and vapour phase of a pure substance is called the **critical state**.* In other words, it is the meeting point of saturated liquid line and saturated vapour line of a pure substance. Broadly speaking, critical state is the limit beyond which a liquid vapour transformation is not possible (See Sec. 9.4).

4.12 EQUATION OF STATE

Properties such as pressure, temperature, specific volume, internal energy, electrical resistance, etc., may be used to specify the state of thermodynamic system in equilibrium. The relationship between various properties provides us with a set of equations, each of which may be called *the equation of state of the system. However, the particular relationship between pressure, specific volume and temperature is conventionally known as the equation of state of the system*. The equation of state may be expressed by means of graph, tables of properties or a equation of the form

$$f(p, v, T) = 0 \quad ...(4.69)$$

It is to be noted that all these properties are not independent. This suggests that we can assign arbitrary values to each of these properties but only to a definite number, the values of the remaining properties will be fixed by the nature of the system. This may be illustrated by an example. Consider an homogeneous system (single phase) consisting of a pure substance (single chemical substance). In the absence of electrical magnetic, gravitational surface and motion effects, the thermodynamic state is two dimensional, that is two independent properties are sufficient to describe the state of the system. This means that if pressure and temperature are specified the specific volume is automatically fixed. Thus, the equation of state for these type of systems may be written as

$$p = f(v, T) \quad ...(4.70)$$

$$v = \phi(p, T) \quad ...(4.71)$$

$$T = \psi(p, v) \quad ...(4.72)$$

In the case of a perfect gas, the equation of state relating to pressure temperature and specific volume is expressed by the equation

$$pv = RT \quad ...(4.73)$$

The equation of state of a real gas may also be expressed by the following form

$$pv = A + \frac{B}{v} + \frac{C}{v^2} + ... \quad ...(4.74)$$

where A, B, C, etc., are called the virial coefficient and are function of temperature alone.

In terms of p,

$$pv = RT\,(A + B'p + C'p^2\ ...) \quad ...(4.75)$$

For a perfect gas, therefore, $A = RT$ in Eq. (4.74) and $A = 1$ in Eq. (4.75) and other virial coefficients are zero. We have already seen that the perfect gas law becomes inadequate at low molal volumes or high pressure interaction cannot be neglected at high particle density. *Realising the inherent inaccuracies in describing real gases by the perfect gas laws investigators have proposed numerous equations of state for real gases. The derivations are either empirical based on experimental data or have physical interpretation based on kinetic theory.*

Whatever be the equation of state, the following conditions have to be satisfied:

(i) There should not be any difficulty in solving p, v or T.

(ii) It should yield partial derivatives such as $((\partial u/\partial p)_T$, etc., without difficulty, so that, other properties may be calculated.

The various types of equation of states are mentioned below.

4.13 VAN DER WAALS EQUATION OF STATE (1873)

In 1873, van der Waals proposed an equation of state which was *an attempt to correct the ideal gas equation of state so that it would be applicable to real gases.* While deriving the ideal gas equation of state on the basis of kinetic theory in which it was assumed that (i) there is no intermolecular forces between particles and (ii) the volume of the molecules is negligible in comparison with the gas (i.e., molecules are of point masses). But as the pressure increases, there exists an intermolecular force between particles and the volume occupied by the particles may become a significant part of the total volume. Hence, real gases have intermolecular attraction forces and the volume occupied by the particles is significant.

4.13.1 Correction for the Intermolecular Attractions

In order to find a correction term to compensate for the attraction forces between molecules, consider the forces of attraction on molecules hitting the wall of the confining vessel. The molecules which are interior of the mass of the gas are unaffected by intermolecular attractions since it is being attracted equally in all directions and the forces cancel on another. But the molecules which are close enough to wall shown by B in Fig. 4.11 exert an attractive force which causes the molecules to strike the wall with less velocity, i.e., less momentum than that with which it would have struck in the absence of intermolecular attraction. When the momentum is less, the pressure exerted by the gas is also less. This reduction of pressure is known as *internal pressure*. It is evident that the decrease in pressure (i.e., internal pressure) is proportional to the number of the attracting molecules per unit volume and to the number of attracting molecules that strike the unit area of the wall per unit time. Both these factors are proportional to the number of molecules in unit volume of the gas or the density of gas. van der Waals suggested that the decrease in pressure is proportional to the square of molal density and, in turn, is equal to a/v^2 where v *is the molal volume*, and a is proportionality constant. That is, decrease in pressure $\propto$ (density of gas)^2 = a/v^2.

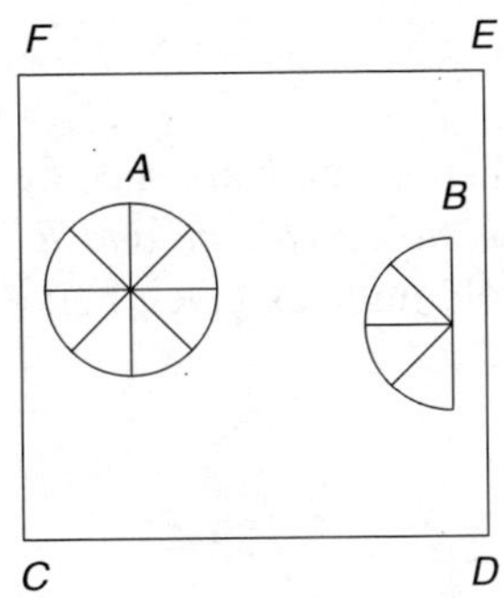

Fig. 4.11 Intermolcular attraction's correction

Thus, the pressure exerted by the gas p is equal to kinetic pressure say p, as calculated by the molecular change of momentum minus the correction term to compensate for the attraction forces (i.e., internal pressure). Thus,

$$p = \text{Kinetic pressure} - \text{Internal pressure}$$

or
$$p = p' - a/v^2$$

or
$$p' = p - a/v^2 \qquad \text{...(4.76)}$$

Here the constant a accounts for the reduction in total pressure due to long range attractive force.

4.13.2 Correction for Finite Size of Molecules

The second correction is due to the size of the molecule. To account for the volume occupied by molecules, van der Waals proposed that the specific volume in the ideal gas equation of state v must be replaced by the term $(v - b)$ where the constant b accounts for the reduction in volume due to the finite size of the molecule of a real gas. This reduction in volume causes more molecular collisions and hence an increase in pressure.

It is to be noted that the constant b also takes into account the increase in pressure due to short range repulsive force between the molecules.

Thus, van der Waals realising the behaviour of real gas and its deviation from the prefect gas proposed the following equation of state for a real gas

$$\left(p + \frac{a}{v^2}\right)(v - b) = RT \qquad ...(4.77)$$

or

$$p = \frac{RT}{v - b} - \frac{a}{v^2} \qquad ...(4.78)$$

where a and b are specific constants and depend upon the nature of the substance. *v represents the molar volume and R is the universal gas constant.* Normally, Eq. (4.78) is expressed in mole units as given below

p = pressure, N/m^2

v = molar volume, m^3/kg mole K

T = temperature, K

R = gas constant = 8314 J/kg mole K

a = constant, $N.m^4/(kg\ mole)^2$

b = constant, m^3/kg mole

Table 4.6. shows the values of constants a and b for some important substances.

On comparing the van der Waals equation of state with that of perfect gas, we find the two differences, namely:

(a) A new term (a/v^2) has been added to the term pressure, p.

(b) A new term b has been subtracted from the volume term, v.

It is to be noted that as the pressure approaches zero and the specific volume approaches infinity, the correction terms (a/v^2) and b are negligible and the equation reduces to $pv = RT$.

At low pressure where the mean free path is large in comparison to the molecule's dimensions, the quantity b in van der Waals equation may be neglected. Hence, Eq. (4.77) reduces to

or

$$\left(p + \frac{a}{v^2}\right)v = RT$$

or $$pv = RT - \frac{a}{v^2} \qquad ...(4.79)$$

Equation (4.79) states that the product pv is less than RT. This accounts for the shape of the curves of Fig. 4.12 at low pressure. Similarly, at high pressure, the term a/v^2 can be neglected compared to p and van der Waals equation reduces to

$$p(v - b) = RT$$

or $$pv = RT + bp \qquad ...(4.80)$$

Equation (4.80) is known as ***Clausius equation of state***. It obvious that the product pv is greater than RT and increases linearly with pressure.

The van der Waals equation can be expanded in the form

$$pv^3 - (pb + RT)\, v^2 + av - ab = 0 \qquad ...(4.81)$$

The above equation suggests that van der Waals equation of state is cubic in volume v and therefore, has three roots at each pressure. Figure 4.12 shows several isotherms (T = constant) for van der Waals gas on p–v diagram. Only one value of volume is obtained at temperatures equal to or above the critical isotherms, while three values of specific volumes can be obtained at temperatures below the critical temperature.

4.13.3 Determination of Constants *a* and *b*

The van der Waals equation, though, widely used for the real gases, has a fundamental weakness that the constants a and b actually vary with temperature. Hence, their values should be determined empirically for particular regions of pressure and temperature. One specific method of calculating the two constants a and b is based on the experimental result that the critical isotherm on a p – v diagram has a zero slope at the critical state for gases. In addition, this isotherm goes through a point of inflection at the critical state. This is shown in Fig. 4.12 and the curve marked T_c is the **critical isotherm**. On the basis that the critical isotherm goes through a point of inflection at the critical state, we may write the following equations which are valid at the critical state.

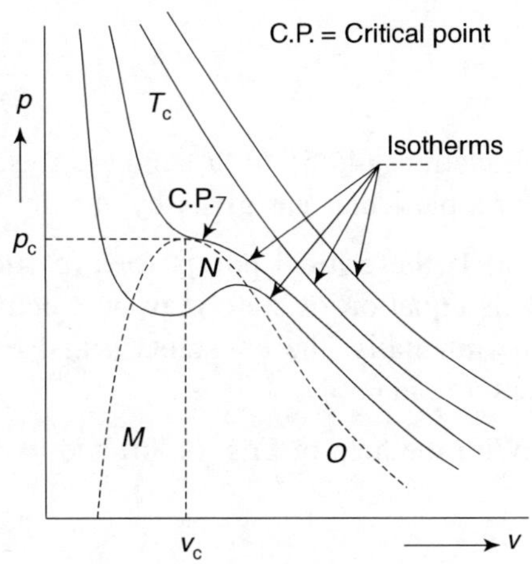

Fig. 4.12 Isotherms on p – v diagram

$$\left(\frac{\partial p}{\partial v}\right)_c = 0 \quad \text{and} \quad \left(\frac{\partial^2 p}{\partial v^2}\right)_c = 0$$

where subscript c denotes critical point. The van der Waals equation at the critical point is

$$p_c = \frac{RT_c}{v_c - b} - \frac{a}{v_c^2} \qquad ...(4.82)$$

This equation represents the curve *MNO* shown by dotted line. Differentiating with respect to v_c and noting that T is a constant gives

$$\left(\frac{\partial p}{\partial v}\right)_c = -\frac{RT}{(v_c - b)^2} + \frac{2a}{v_c^3} = 0$$

or
$$RT_c = \frac{2a}{v_c^3}(v_c - b)^2 \qquad ...(4.83)$$

Again,
$$\left(\frac{\partial^2 p}{\partial v^2}\right) = \frac{2\,RT}{(v_c - b)^3} - \frac{6a}{v_c^4} = 0$$

or
$$RT_c = \frac{3a}{v_c^4}(v_c - b)^3 \qquad ...(4.84)$$

or
$$v_c = 3\,b \qquad ...(4.85)$$

Putting the value of v_c in Eq. (4.84), we get

$$T_c = \frac{a}{27\,Rb} \qquad ...(4.86)$$

Putting the value v_c and T_c in Eq. (4.82), we get

$$p_c = \frac{a}{27\,b^2} \qquad ...(4.87)$$

Equations (4.85), (4.86) and (4.87) show the critical properties in terms of constants a and b for a particular gas given by van der Waals equation. It leads to the following results.

(a) If the critical points for a particular gas are known, the constants for the van der Waals equation of state may be calculated. As the critical volume is difficult to measure experimentally, the constants a and b are expressed in terms of critical temperature and pressure only.

With the help of Eqs. (4.86) and (4.87), we have

$$a = \frac{27}{64}\frac{R^2\,T_c^2}{p_c} \qquad ...(4.88)$$

and
$$b = \frac{RT_c}{8\,p_c} \qquad ...(4.89)$$

(b) The van der Waals equation may be compared with that of and ideal gas equation by writing $pv = RT$ at the critical state.

$$\frac{p_c\, v_c}{RT_c} = \frac{3}{8} \qquad \text{...(4.90)}$$

It is obvious from Eq. (4.90), that the ratio (pv/RT) at the critical point is a constant 3/8 and is therefore, same for all substances. For an ideal gas, its value is unity.

The experimental value of $p_c\, v_c/RT$ given in Table 4.6 which is significantly different from 3/8 suggests the limited accuracy of van der Waals equation. It may be noted that the deviation is not very large.

Table 4.5 Constants *a* and *b*

Substance	a, (N m^4/(kg mole)^2	b, (m^3/kg mole)	$p_c\, v_c/RT_c$
Air	13.537×10^4	0.0362	—
CO	3.629×10^4	0.0423	0.277
He	317.193×10^4	0.0228	0.327
H_2	2.511×10^4	0.0262	0.306
H_2O (vapour)	35.51×10^4	0.0300	0.233
O_2	13.93×10^4	0.0314	0.292
Hg (vapour)	203.26×10^4	0.0657	0.909

4.13.4 Limitation of van der Waals Equation

We have seen in spite of the general agreement deductions from the van der Waals equation show appreciable deviation from experimental results. The following are the factors which limit this equation from the most general use:

(a) *The values of a and b* obtained by different methods do not agree. Not only this, the values obtained by a single method vary considerably at different temperatures. This clearly suggests a fundamental drawback in the equation which is not easily remedied by suitable corrections.

(b) *Critical coefficient.* The expression $RT_c/p_c v_c$ is known as the ***critical coefficient***. For a perfect gas, it is equal to unity. This critical coefficient can be found out from the critical properties p_c, v_c and T_c of actual gases obeying the van der Waals equation.

$$\frac{RT_c}{p_c\, v_c} = R \times \frac{8a}{27\, R\, b} \times \frac{27\, b^2}{a} \times \frac{1}{3\, b} = \frac{8}{32} = 2.66$$

It is a well-known fact that the critical coefficient must have the same value for all the gases if van der Waals equation is correct for actual gases. But this vaiue is found to vary from gas to gas, the values ranging from 3.28 for hydrogen to 4.79 for acetic acid. This means, that it depends on the molecule's structure of the gas. In addition, none of the gases examined has a value less than 3, which is obviously greater than the theoretical value of 2.66.

(c) *Critical value.* According to van der Waals equation $v_c = 3b$ for all gases. The experimental values of gases show the volume of v_c varies from 1.46 for argon to 2.86 for hydrogen. Thus, *the van der Waals equation is not obeyed exactly by any gas near its critical point.*

Summarizing the above situation, we may say that the van der Waals equation succeeds in explaining the deviations from the perfect gas common to all actual gases but it fails when applied to the difference between the various gases. Therefore, other equations have been suggested by Dietetics, Berthlot, Redlich-Kowng and others to account for these differences but they are also not very correct, and so no more useful than the simpler van der Waals equation. However, the van der Waals equation is more used due to its simplicity.

Example 4.16 Find the pressure exerted by 1 kg of oxygen at 150°C if the specific volume is 0.3 m^3/kg using (a) van der Waals equation of state, and (b) perfect gas law.

Solution (a) According to van der Waals equation

$$p = \frac{RT}{v-b} - \frac{a}{v^2}$$

where $v = 0.3 \times 9.6$ m^3/kg mole. since $M_{02} = 32$, $T = 150 + 273 = 423$ K

$a = 13.93 \times 10^4$ N·m^4 (kg mole)^2

$b = 0.0314$ m^3 kg mole, $R = 8314$ J/kg mole K

Putting the value,
$$p = \frac{8314 \times 423}{9.6 - 0.0314} - \frac{13.93 \times 10^4}{(9.6)^2} = 3.6602 \times 10^5 \text{ N/m}^2$$

$$= 3.6602 \text{ bar}$$ **Ans.**

(b) According to perfect gas law

$$p = \frac{8314 \times 423}{9.6} = 3.6633 \times 10^5 \text{ N/m}^2 = 3.6633 \text{ bar}$$ **Ans.**

4.14 REDUCED COORDINATES

Related coordinates are the non-dimensional coordinates which are obtained by dividing the coordinates with the corresponding coordinates at the critical state. Thus, the equations

$$p_r = \frac{p}{p_c} \quad \text{...(4.91)}$$

$$v_r = \frac{v}{v_c} \quad \text{...(4.92)}$$

$$T_r = \frac{T}{T_c} \quad \text{...(4.93)}$$

are known as the ***reduced coordinates***. It means that the reduced pressure p_r, reduced volume v_r and reduced temperature T_r at a given state are obtained by dividing the pressure p, volume v and temperature T at the state by the value of the same property at the critical state. According to van der Waals

$$v_r = 3b$$

$$T_c = \frac{8a}{27\,Rb}$$

and
$$p_c = \frac{a}{2762}$$

Now, with the help of Eqs. (4.91), (4.92), (4.93), we have

$$v = v_r . v_c = 3b . v_r \qquad ...(4.94)$$

$$T = T_r\, T_c = \frac{8\,a}{27\,Rb} T_r \qquad ...(4.95)$$

$$p = p_r\, p_c = \frac{a}{27\,b^2}\, p_r \qquad ...(4.96)$$

Putting these values in the van der Waals equations, we get

$$\left(p_r + \frac{3}{v_r^2}\right)\left(v_r - \frac{1}{3}\right) = \frac{8}{3} T_r \qquad ...(4.97)$$

Equation (4.97) is the van der Waals equation of state expressed in reduced coordinates. It is to be noted that this equation is independent of R, a and b which are constants specific to one gas. Thus, the van der Waals equation expressed in reduced coordinates should be valid for all gases.

4.15 COMPRESSIBILITY FACTOR AND LAW OF CORRESPONDING STATES

4.15.1 Compressibility Factor

For real gases, especially at high pressure and low temperature, the ideal gas equation of state is not a good approximation. It is because of the fact that at high pressure and low temperature, the specific volume becomes very large. The departure of real gas behaviour from that of ideal gas is frequently characterised by a **compressibility factor** Z, which is defined as

$$Z = \frac{pv}{RT} \qquad ...(4.98)$$

Since RT/p is the ideal gas specific volume (v_{ideal}) of a gas, the compressibility factor may be considered a measure of the ratio of the actual specific volume to the ideal gas specific volume. That is,

$$Z = \frac{v_{actual}}{v_{ideal}}$$

For an ideal gas, the compressibility factor Z is unity. Also, as the pressure is reduced, the compressibility factor of any gas approached unity, since a gas acts more like a perfect gas as the pressure is lowered.

$$\lim_{p \to 0} Z = 1$$

The value of Z for any real gas may be less or more than unity, depending upon the pressure and temperature of the gas which must be finite. Hence, we can say that the compressibility factor measures the percentage deviation of any actual gas from ideal gas behaviour. The compressibility factor for several common gases is shown in Fig. 4.13 as a function of pressure for three temperatures. For pressure below 50 atm, the deviation of nitrogen and argon from the ideal gas equation is 3% or less at all three reported temperatures. However, both carbon dioxide at 300 K and water vapour at 600 K (both triatomic gases) rapidly deviate from ideal gas conditions as the pressure is raised. Nevertheless, if the temperature of CO_2 is raised to 600 K, the compressibility factor of the gas remains within about 1% of unity for pressure up to 100 atm. If CO_2 is maintained at relatively high temperatures, the assumption of ideal gas behaviour is quite good. Modern steam power plants operate at pressures above 100 atm. and at temperatures around 840 K. Apparently, at these conditions of steam, the ideal gas equation of state will not give answers within the permissible errors for engineering calculations.

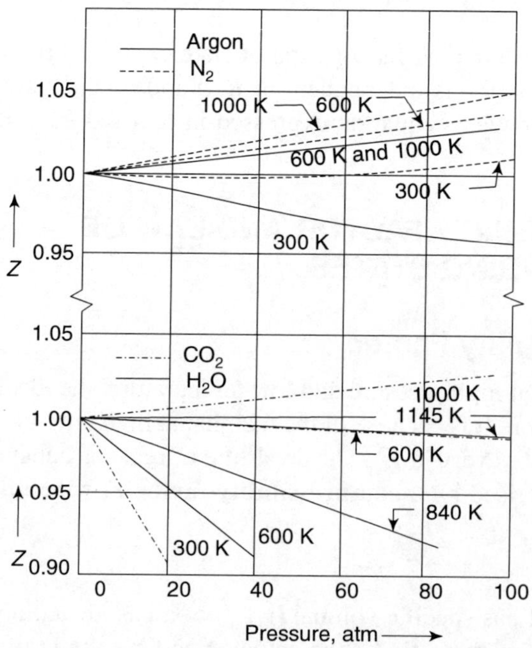

Fig. 4.13 Compressibility factor for gases

We have seen that a plot of Z versus p for different isotherms as shown in Fig. 4.13 is very useful for studying the behaviour of real gases. This chart is generally called a ***specific compressibility chart*** showing isotherms and isometrics, the word specific is used here because one isotherm takes care of one substance only at that temperature.

4.15.2 Law of Corresponding States

Though specific compressibility charts are different for different substances but they have the same trends in common. Thus, it is possible to determine values of Z from other properties of a gas based on only limited amount of data by taking the help of common trends of gases. *These common trends when stated in the form of law is called the **law of corresponding states** which reads as*:

If the properties p, v and T are properly selected, then a single equation of state or single compressibility chart may be possible which will show the behaviour of all substances.

(a) van der Waals Principle. It was van der Waals first who proposed the law of corresponding states. According to van der Waals, since all substances exist in a gas, liquid or solid phase and all display common trends such as critical point, triple point, etc., then p, v, T relationship of all substances could be possible by a single model. He took the critical point to be the base reference and the critical properties p_c, v_c and T_c to be the reducing properties such that

$$p_r = \frac{p}{p_c}; T_r = \frac{T}{T_c} \text{ and } v_r = \frac{v}{v_c}$$

Thus, the principle of corresponding state may be written as

$$f(p_r, v_r, T_r) = 0 \qquad \text{...(4.99)}$$

We have

$$Z = \frac{pv}{RT} = \frac{p_c p_r}{R}, \frac{v_c v_r}{T_c T_r}$$

or

$$Z = \frac{p_c v_c}{RT_c} \cdot \frac{p_r v_r}{T_r} = \frac{p_r v_r}{T_r} Z_c \qquad \text{...(4.100)}$$

where $Z_c = \frac{p_c v_c}{RT_c} = \frac{3}{8}$ according to van der Walls

$$\therefore \qquad Z = \left(\frac{3}{8}\right)\frac{p_r v_r}{T_r} \qquad \text{...(4.101)}$$

With the help of Eqs. (4.97), and (4.100), we have

$$\left(p_r + \frac{27 p_r^2}{64 T_r^2 Z^2}\right)\left(\frac{8 T_r Z}{3 p_r} - \frac{1}{3}\right) = \left(\frac{8}{3}\right) T_r$$

or $$\left(Z^3 - \frac{p_r}{8T_r} - 1\right)Z^2 + \left(\frac{27p_r}{64T_r^2}\right)Z - \frac{27p_r^2}{512T_r^2} = 0 \quad ...(4.102)$$

The equation is another form of the van der Waals equation. Now, we can write this equation as

$$Z = f(p_r, T_r) \quad ...(4.103)$$

Equation (4.103) is another form of the law of corresponding states.

The law of corresponding states** is the basis of generalised compressibility chart.* A plot of Z versus T_r for different values isotherms ($T_r = c$) for different gases shows that all gases have almost the same chart with an accuracy within 5%. This *chart is known as the **generalized compressibility chart shown in Fig. 4.14. *A gas that has a compressibility factor less than unity is more compressible than a perfect gas.* If is obvious from the figure that all the curves converge at $Z = 1$ as the pressure approaches zero, suggesting that all the gases behave as ideal gas ($Z = 1$) at low pressure. With the help of this chart, various properties of real substances may be deteimined. Credit goes to Weber (1933) and then Nelson and Obert who drew up this chart.

Equation (4.99) may also be put in different form.

Since, $$v_r = \frac{v}{v_c} = \frac{ZRT/p}{Z_cRT_r/p_c} = \frac{Z}{Z_c} \cdot \frac{T}{T_r} \cdot \frac{p_c}{p}$$

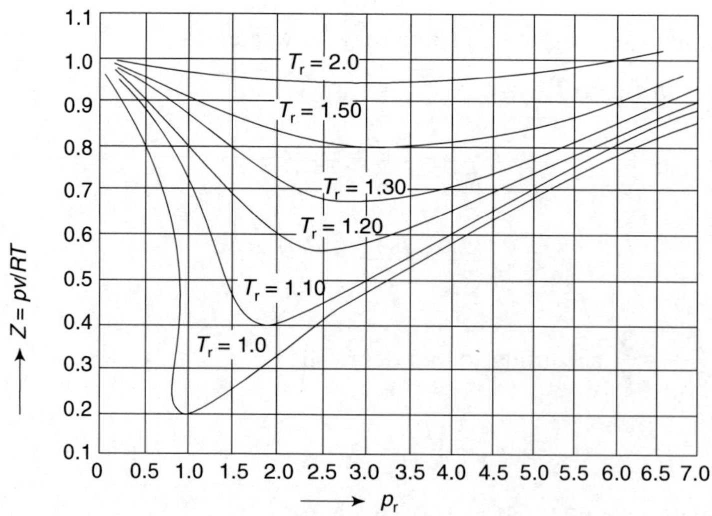

Fig. 4.14 Generalised compressibility chart

Hence, Eq. (4.99) is modified as

$$f\left(p_r, T_r, \frac{Z}{Z_c}\right) = 0 \qquad \text{...(4.104)}$$

From Eqs. (4.97) and (4.104), it follows that if reduced isotherms are constructed on a diagram having p_r and v_r as coordinates, correlation for all substances would be exact at the critical point, since $p_r = T_r = v_r = Z/Z_c = 1$ at the critical point. This shows that van der Waals principle of corresponding states is exactly valid for all substances at the critical point. Let us examine its validity at other condition. At low pressure, where p_r tends to zero and Z tends to unity, Z/Z_r tends to a limit $(1/Z_c)$. Note that $(1/Z_c)$ has different values for different gases since Z_c varies from 0.2 to 0.3. This means that the van der Waals principle gives more error at low pressure if the gases have different Z_c values.

(b) Su's Principle. In order to correct the fault of van der Waals principle of corresponding states, Kamerlingh Onnes suggested that a new set of reduced parameters should be derived by combining Z_c, with either p_r, v_r or T_r. On the basis of this suggestion, Su proposed a principle of corresponding states given as

$$f(p_r, T_r, v_r, Z_c) = f(p_r, T_r, v_r) = 0 \qquad \text{...(4.105)}$$

But $$(v_r Z_c) = \left(\frac{ZT_r}{p_r}\right)$$

$$f(p_r, T_r, Z) = 0 \qquad \text{...(4.106)}$$

Su's principle of corresponding states is very accurate at low pressure but gives error at critical point if the substances have different value of Z_c. The two principles are equivalent if Z_c is the same for all substances.

Thus, *we have seen that neither principle is accurate in all regions if the substances have different values of Z_e*. However, Su's principle is preferable since it is more exact in the low and medium pressure regions and approximately accurate in all regions except critical point.

(c) Other Principles. In has been seen that the failure of the above two principle of corresponding states is the variation of Z_c. This fault can be removed by restricting the correlation to substances having essentially the same Z_c values. Missener and Saferian proposed that

$$f(p_r, T_r, Z_c) = 0 \qquad \text{...(4.107)}$$

Pitzer proposed that

$$Z = Z(T_r, p_r, w)$$

or $$f(p_r, T_r, Z, w) = 0 \qquad \text{...(4.108)}$$

Pitzer defined w as

$$w = \frac{\log_{10}(p_c/p_s) - \log_{10}(p_c'/p_s')}{\log_{10}(p_c'/p_s')}$$

where p_s = vapour pressure at $T_r = 0.7$

p'_s = vapour pressure of simple fluid at $T_r = 0.7$

The choice for $T_r = 0.7$ was taken in the interest of simplicity because for gases such as methane, argon; krypton and the xenon

$$\frac{p'_r}{p'_c} = 0.1 \text{ at } Tr = 0.7$$

Thus,
$$w = \log_{10}\left(\frac{p_s}{p_c}\right) - 1 = -\left[\log_{10}(p_r)_{T_r=0.7}{}^{+1}\right]$$

This w is called the ***acentric factor***

Pitzer further suggested that

$$Z = z^{\circ}(p_r, T_r) + wz'(p_r, T_r) \quad ...(4.109)$$

The above equations show that Z is the sum, a universal function of p_r and T_r and the product of acentric factor and another universal function of p_r and T_r

We have,
$$a = \frac{d(\ln p_{rv})}{d(\ln T_r)} = \frac{dp_{rv}}{dT_r} \text{ at critical point}$$

But according to Riedel

$$\ln P_{rv} = \alpha \ln T_r + 0.0838(\alpha - 3.75)(36T_r^{-1} - 35 - T_r^6 + 42 \ln T_r)$$

$$\therefore \quad \alpha = 5.808 + 4.93\, w$$

If Z_c is expressed in terms of α then

$$\frac{1}{Z_c} = 3.72 + 0.26\,(\alpha - 7)$$

The above result is valid for many substances.

Example 4.17 Determine the specific volume of CO_2 at 200°C and 60 bar by the use of (a) perfect gas law, (b) van der Waals equation, and (c) compressibility chart.

Solution (a) According to the perfect gas law

$$v = \frac{RT}{p} = \frac{R_0\, T}{M\, p} \text{ m}^3/\text{kg} = \frac{8314}{44} \times \frac{(200 + 273)}{60 \times 10^5}$$

$$= 0.014895 \text{ m}^3/\text{kg}$$ **Ans.**

According to van der Waals equation

$$\left(p + \frac{a}{v^2}\right)(v - b) = RT$$

where $a = 363 \times 10^3 \text{ N.m}^4/(\text{kg mole})^2 = \dfrac{363 \times 10^3}{(44)^2} \text{ N.m}^4/\text{kg}^2$

$$b = 0.0423 \text{ m}^3/\text{kg mole} = \frac{0.0423}{(44)^2} \text{ m}^3/\text{kg}$$

$$R = \frac{R_0}{M} = \frac{8314}{44} \text{ J/kg K}$$

$\therefore$ $$\left(60 \times 10^5 + \frac{363 \times 10^3}{(44)^2 \times v^2}\right)\left(v - \frac{0.0423}{44}\right) = \frac{8314}{44} \times 473$$

or $$v = 0.014 \text{ m}^3/\text{kg}$$ **Ans.**

(c) According to the compressibility chart

$$T_r = \frac{T}{T_c} = \frac{473}{304.2} = 1.5548, \quad p_r = \frac{p}{p_c} = \frac{60}{72.9} = 0.823$$

$\therefore$ $Z = 0.99$ from the compressibility chart

$$v = \frac{ZRT}{p} = 0.99 \times \frac{8314}{44} \times \frac{473}{60 \times 10^5} = 0.01474 \text{ m}^3/\text{kg}$$ **Ans.**

4.16 OTHER EQUATIONS OF STATE FOR REAL GASES

After the appearance of van der Waals equation of state, about 125 equations of state have been proposed by various investigators. Some of them are given below:

4.16.1 Dieterici Equation (1899) (Two Constants)

In 1899, Dieterici proposed an equation of state

$$p = \frac{RT}{v - b} e^{-a/RTv} \quad \text{...(4.110)}$$

Later on, he modified the equation as

or $$p = \frac{RT}{v - b} e^{-k} \quad \text{...(4.111)}$$

where $k = \dfrac{a\, T_c^{1/2}}{v\, RT^{3/2}}$

Still later Joffe modified Eq. (4.110) to

$$p = \frac{RT}{v-b} - \frac{R(T \cdot T_c)^{1/2}}{v-b}\left(1 - e^{-a/RTv}\right) \qquad \text{...(4.112)}$$

In the above equation, the constants a and b may be calculated by using the conditions

$$\left(\frac{\partial p}{\partial v}\right)_T = 0 \text{ at C.P. and } \left(\frac{\partial^2 p}{\partial^2 v}\right)_T = 0 \text{ at C.P.}$$

This result in

$$a = \frac{4\,R^2\,T_c^2}{p_c\,e^2}$$

and $$b = \frac{RT}{p_c\,e^2}$$

The Dieterici equation is in good agreement with the compressibility chart along the critical isotherm but in other regions it gives error.

There are many other equations of state for real gases such as:

(a) Redlich–Kowng equation (1909) (Two constants)

(b) Whole equation (1914) (Three constants)

(c) Macleod equation (1944) (Four constants)

(d) The Beattie–Bridgeman equation of state (1928) (Five constants)

Example 4.18 Find the second and third virial coefficients of van der Waals equation of state when expressred in the form

$$\left(p + \frac{a\,n^2}{V^2}\right)(V - n\,b) = nRT$$

where n = number of moles

Solution The van der Waals equation of state is given by

$$\left(p + \frac{a\,n^2}{V^2}\right)(V - nb) = nRT$$

$$pV + \frac{a\,n^2}{V} + p\,n\,b - \frac{a\,b\,n^2}{V^2} = nRT$$

or $$pV = nRT\left(1 + \frac{pb}{RT} - \frac{an}{VRT} + \frac{abn^2}{V^2 RT}\right)$$

But $$p = \frac{nRT}{V - nb} - \frac{an^2}{V^2}$$

$$\therefore \quad pV = nRT\left[1+\frac{b}{RT}\left(\frac{nRT}{V-nb}-\frac{an^2}{V^2}\right)-\frac{an}{VRT}+\frac{abn^2}{V^2RT}\right]$$

$$= nRT\left[1+\frac{nb}{V-nb}-\frac{an}{VRT}\right] = nRT\left[1+\frac{nb}{V}\left(1-\frac{nb}{V}\right)^{-1}-\frac{an}{VRT}\right]$$

$$= nRT\left[1+\frac{nb}{V}\left(1+\frac{nb}{V}+\frac{n^2b^2}{V}+\ldots\right)-\frac{an}{VRT}\right]$$

$$= nRT\left[1+\frac{n}{V}\left(b-\frac{a}{RT}\right)+\frac{n^2b^2}{V^2}+\frac{n^3b^3}{V^3}+\ldots\right]$$

Comparing with virial form of equation, second coefficient $B = b - \frac{a}{RT}$.

Third coefficient $\qquad C = b^2$

Example 4.19 The volume occupied by one mole of 1-butene at 249°C is 0.83 m³/kg mole. Calculate the pressure using (i) Dieterici equation, (ii) Redlich–Kwong equation and (iii) ideal gas equation. Assume that for 1-butene, T_c = 419.6 K, p_c = 39.7 bar and R = 0.082 bar m³/kg mole K.

Solution (i) The values of constants used in Dieterici's equation are found to be a = 16 and b = 0.117.

Using Dieterici equation as Joffe

$$p = 40.45 \text{ bar}$$ **Ans.**

(ii) The values of constant used in Redlich–Kwong equation are found to be a = 260 and b = 0.075. Using Redlich Kwong equation, p = 43.13 bar **Ans.**

(iii) Using ideal gas equation

$$p = \frac{RT}{v} = \frac{0.082 \times 522}{0.83} = 51.57 \text{ bar}$$ **Ans.**

The results show the importance of equation of state for real gases.

4.17 *p-v-T* SURFACE

We have seen that various euations of state whether for ideal or real substance have relationship between three variables p, v and T. It is possible to plot these variables along three mutually perpendicular axes resulting in surface representation of equation of state graphically. Such a surface is called the p–v–T surface, whether these surfaces are drawn to the scale or without scale but in any case they will present fundamental properties of the individual substance and will serve as a tool in the study of properties and processes of thermodynamics.

4.18 *p-v-T* SURFACE OF AN IDEAL GAS

Figure 4.15 shows a p–v–T surface of an ideal gas. The constant temperature, constant volume and constant pressure lines are shown. It is to be noted that each point in this surface represents a possible equilibrium state and line on the surface represents a quasi-static process.

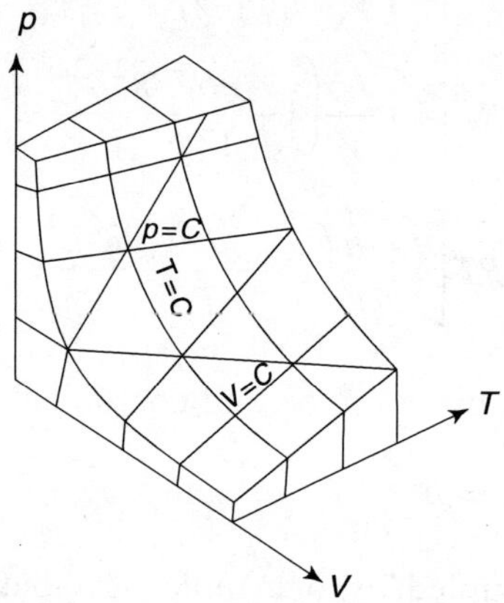

Fig. 4.15 p–v–T surface of an ideal gas

It is possible to draw p–v and p–T planes with the help of p, v, T planes. The p–v and p–T planes for a ideal gas are shown in Fig. 4.16. Figure 4.16 (a) is the graphical representation of Boyle's law due to isothermal lines and Fig. 4.16 (b) is the graphical representation of Charles' law due to isobaric lines.

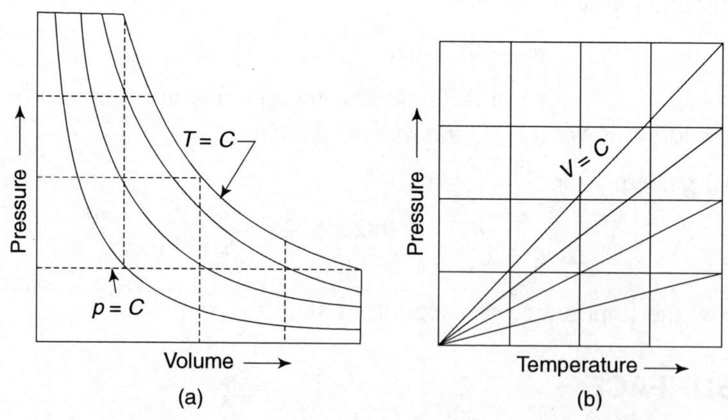

Fig. 4.16 p–v and p–T planes for an ideal gas

4.19 REPRESENTATION OF PARTIAL DERIVATES ON A *p-v-T* SURFACE

It is essential to understand the geometrical meaning of partial derivates of a substance such as $\left(\frac{\partial p}{\partial v}\right)_T$, $\left(\frac{\partial v}{\partial T}\right)_p$ and $\left(\frac{\partial p}{\partial T}\right)_v$. A p–v–T surface of an ideal gas is shown in Fig. 4.17. The

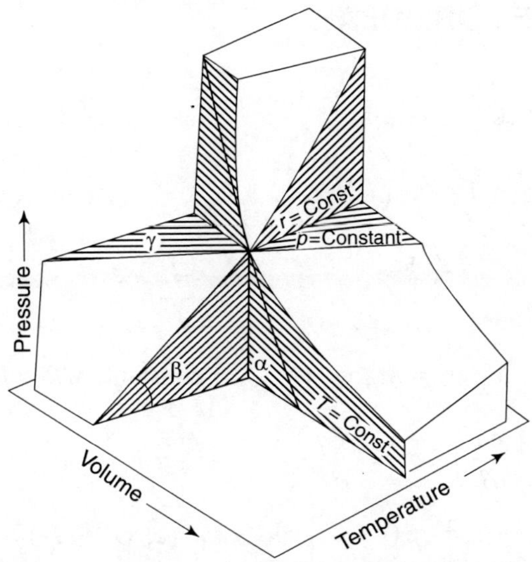

Fig. 4.17 Representation of partial derivatives on p–v–T surface

temperature, pressure and volume planes intersect at a point on the surface. Due to Boyle's law, the isothermal plane intersects the surface in a hyperbola while due to Charles' law, the other two planes, namely, constant pressure and constant volume intersect the surface in lines.

In the partial derivative $\left(\frac{\partial p}{\partial v}\right)_T$, the suffix T refers to the constant temperature and hence the value of derivative is represented by the slope of the curve obtained by the intersection of p–v–T surface by a constant temperature plane, i.e., by slope of the hyperbola curve. That is,

$$\left(\frac{\partial p}{\partial v}\right)_T = \tan \alpha \qquad ...(4.113)$$

$$\left(\frac{\partial p}{\partial T}\right)_v = \tan \beta \text{ [constant volume process]} \qquad ...(4.114)$$

$$\left(\frac{\partial v}{\partial T}\right)_p = \tan \gamma \text{ [constant pressure process)} \qquad ...(4.115)$$

It is obvious from Fig. 4.17 that the slope of the hyperbola (i.e., tan α) has a different value along the whole length of the curve. In other words, the value of partial derivative $\left(\frac{\partial p}{\partial v}\right)_T$ will have different values at different points. For example, the partial derivative of perfect gas equation of state ($pv = RT$) will be

$$\left(\frac{\partial p}{\partial v}\right)_T = \frac{-RT}{v^2} \qquad ...(4.116)$$

At high pressure, the volume of the gas is very small, thus the value of slope will be large. This statement will be confirmed if the value of α is found at the high pressure.

4.20 SUMMARY OF FORMULAE

1. $p = \left(\dfrac{RT}{v}\right)_{p \to 0,\ \text{real gas}}$

2. $pV = \text{constant},\ p_1 V_1 = p_2 V_2,\ \left(p \,\alpha\, \dfrac{1}{V}\right)_{T=c}$

3. $V \alpha (T)_{p = \text{constant}},\ V_2/V_1 = T_2/T_1$

4. $p_1 V_1/T_1 = p_2 V_2/T_2$

5. $v_{\text{mole}} = \dfrac{V}{n},\ N = nA,\ M_A v_A = M_B v_B = v_{\text{mole}} = \text{constant},\ MR = R_0$

6. $c_{\text{ave}} = \dfrac{Q}{m \,\Delta T}, c_p = \left(\dfrac{\delta q}{dT}\right)_p, c_v = \left(\dfrac{\delta q}{dT}\right)_v, c_p - c_v = R, \gamma = c_p / c_v$

7. $Q_{1-2} = \int mc\, dT, h_2 - h_1 = c_p (T_2 - T_1), u_2 - u_1 = c_v (T_2 - T_1)$

8. $c_p = a + bT + cT^2 + dT^3 + \ldots$

9. $pV = \left(\dfrac{1}{3}\right) mn\, \bar{C}^2 = \left(\dfrac{2}{3}\right)\left(\dfrac{1}{2}\right) mn\, \bar{C}^2$

10. $p = \dfrac{RT}{v-b} - \dfrac{a}{v^2}$ where $a = \dfrac{27}{64} \dfrac{R^2 T_c^2}{p_c}$ and $b = \dfrac{RT_c}{8\,p_c}$ (van der Waals equation)

11. $Z = \dfrac{pv}{RT},\ \ Z = \dfrac{v_{\text{act}}}{v_{\text{ideal}}}$ (compressibility factor)

12. $p = \dfrac{RT}{v-b} e^{-a/RTv}$ (Dieterici equation)

PROBLEMS

Things to Think About

4.1. Why does food cook more quickly in a pressure cooker than in water boiling in an open container?

4.2. Why do frozen water pipes tend to burst?

4.3. Discuss implications on aquatic life if water contracts on freezing.

4.4. With the help of phase diagram explain why a film of liquid water forms under the blade of an ice skate,

4.5. Estimate the mass of air contained in a bicycle tyre.

4.6. If a block of aluminium and a block of steel having equal volumes each receive the same energy input by heat transfer, which block would experience the greater temperature increase?

Short and Long Theoretical Questions

4.7. What do you mean by a perfect gas? Give the answer according to macroscopic point of view. Under what conditions does it behave as a real gas?

4.8. Define Boyle's, Charles' and gas laws and give their expressions.

4.9. Define specific heat? Why are there two types of specific heat?

4.10. Derive the expressions:

(a) $c_p - c_v = R$, (b) $(Q_{1-2})_p = mc_p\,\Delta T$, (c) $(Q_{1-2})_v = mc_v\,\Delta T$

4.11. What do you mean by enthalpy of gas?

4.12. Give the concept of an ideal gas from the kinetic theory of gases (i.e., microscopic point of view).

4.13. Derive the expression for pressure exerted by a gas according to kinetic theory of gases. From the kinetic theory of gases, deduce the gas laws.

4.14. Discuss the deviations of real gases from ideal gases.

4.15. What is the van der Waals equation of state? What is the significance of the various constants involved in this equation? What are its limitations?

4.16. Explain the significance of the van der Waals equation of state

$$p = \frac{RT}{v-b} - \frac{a}{v^2}$$

Find the values of the constants a and b in terms of critical parameters p_c, v_c and T_c.

4.17. What do you mean by a critical state of real gas?

4.18. What do you mean by compressibility factor and law of corresponding states? Give their physical significance.

4.19. Write the virial equation of state.

4.20. How many equations of stats of real gases do you know? Give their expressions.

4.21. What is geometrical meaning of partial derivatives?

4.22. Show the critical constants for a substance given by Dieterici equation of state

$$p = \frac{RT}{v-b}e^{-(a/RTv)} \text{ and given by } p_c = \frac{a}{4\,e^3\,b^2};\ v_c = 2b \text{ and } T_c = \frac{a}{4\,Rb}$$

4.23. Prove that the values of a and b for the Berthelot equation of state

$$p = \frac{RT}{v-b} - \frac{a}{Tv^2} \text{ where } a = \frac{27\,R^3\,T_c^3}{64\,p_c},\ b = \frac{1}{8}\frac{RT_c}{p_c}$$

4.24. Show that the approximation $v \approx b$ in van der Waals equation of state can be reasonably made for making it applicable to liquid, whereby it may be written as

$$v = b + \frac{RT\,b^2}{a + pb^2}$$

4.25. (a) Explain the van der Waals equation of state and describe it limitations.

(b) Distinguish between the characteristic gas constant and molar gas constant.

Numerical Questions

4.26. Find the values of characteristic gas constants R for the following gases.

(a) Hydrogen, (b) Oxygen (c) CO_2 (d) CO

(e) CH_4

[*Ans.* (a) 4.124, (b) 0.2598, (c) 1.887, (d) 0.2969, (e) 0.5196 kJ/kg K]

4.27. An air tank has 1.8 metre inner diameter and 3 m length. Calculate the mass of air in the receiver when the pressure is 7 bar and temperature is 293 K. [*Ans.* 141.21 kg]

4.28. Find the mass of oxygen in a cylinder having a capacity of 100 litres at a temperature of 20°C. The pressure in the cylinder is maintained at 80 bar, R = 0.259859 kg/kg K

4.29. 10 kg mole of a gas occupies 683.6 m^3 volume at a temperature of 13°C. If the density of the gas under these conditions is 0.644 kg/m^3 ; find the gas constant, molecular weight of the gas and pressure of the gas. [*Ans.* 0.576 kJ/kg K, 14.42, 0.03395 bar]

4.30. A tank having volume of 10 m^3 has to be used for storing oxygen, nitrogen, and carbon dioxide at 15 bar and 40°C. Calculate the mass of the each gas that can be stored in that tank.

[*Ans.* O_2 = 184.5 kg, N_2 = 161.39 kg, CO_2 = 253.62 kg]

4.31. Calculate the volume of 6 kg of air at a pressure of 4 bar and temperature of 50°C. If c of air is 1.005 kJ/kg K and c_v is 0.72 kJ/kg K. [*Ans.* 0.251 m^3]

4.32. An unknown gas has been developed that the physical and chemical properties which make it useful for various processes industry. The gas has been sent to a chemical laboratory, where its specific heat at constant pressure has been determined as a function of temperature. An equation representing a mean curve through the experimental results is of the from

$$c_p = \left[e^{T/1000} - \text{In}\frac{T}{120} - \frac{T^2}{10^7} \right] \times 4.18$$

where c_p is in kJ/kg K. and T is in K. The equation is valid between 250 K and 500 K. If during a certain constant pressure process the temperature is found to have increased from 25°C to 225°C, calculate the change in enthalpy per kg of the gas.

4.33. The empirical equation for c_p at zero pressure for oxygen may be expressed as

$$c_p = 48.2 - 720\,T^{-1/2} + \frac{6406}{T}$$

where T is in K and c_p has units of kJ/(kg mole) K. Calculate the change in the enthalpy when the temperature of oxygen is increased at constant pressure from 38°C to 1095°C. What per cent error will be introduced by using the arithmetic mean of specific heat between the given temperature limit?

4.34. Nitrogen gas is heated at constant pressure of 1.36 bar from 166°C to 200°C. The empirical specific heat at constant pressure of this gas at low pressure is given by

$$c_p = 39.64 - \frac{14.5 \times 10^3}{T} + \frac{4.85 \times 10^6}{T^2}$$

where c_p is in kJ/(kg mole) K and T is in K. Calculate the quantity of heat added in kJ/kg and compute the change in internal energy in kJ/kg. List any necessary assumptions.

4.35. Calculate from the van der Waals equations

(a) the pressure exerted by 1 kg of CO_2 at a temperature of I00°C if the specific volume is 9.013 m^3/kg,

(b) the pressure exerted by 1 kg of air at 35°C occupying 0.86 m^3,

(c) the temperature of steam at 8 bar having specific volume 0.27 m^3/kg, and

(d) the specific volume of air at 25°C and 50 bar.

4.36. The density of hydrogen at 0°C and 760 mm pressure is 0.00009 gm per c.c. Find the root mean square velocity of hydrogen molecule.

4.37. Calculate the root mean square velocity of a molecule of nitrogen at N.T.P.

4.38. Calculate the velocity of chlorine molecule at 12°C and 7 cm pressure.

4.39. Calculate the velocity of N_2 molecule 12°C and 78 cm pressure.

4.40. Given that the critical pressure, temperature and volume of a gas are 73 bar, 31°C and 0.2066 m^3 of its volume at 15°C and 1 atm respectively, calculate the constants of the van der Waals equation.

4.41. Express the value on the critical temperature in terms of a, b and R. Calculate its value of CO_2 where a = 0.00874 and b = 0.0023. [*Ans.* $T_c = -\frac{89}{27\,Rb}$; t_c = 32.5°C]

4.42. Find the specific volume of CO_2 at 64°C and 71 bar pressure using (c) the generalised compressibility chart.

4.43. Calculate the pressure exerted by ammonia using the generalised compressibility chart for the conditions of 170°C and specific volume 13 c.c./gm.

4.44. Assuming the principle of corresponding state to be valid in the form of $v_r = f(p_r, T_r)$ for gases H_2, N_2, CO_2, and water, calculate the actual pressure and temperature for each gas that will place it in the reduced state defined by $T_r = 4$ and $p_r = 30$.

4.45. The heat required to heat a certain perfect gas at constant pressure from 15°C to 95°C is 1134 kJ/kg. When the same gas is heated at constant volume between the same temperature, the heat required is 808 kJ/kg. Calculate c_p c_v, R and molecular weight of the gas.

[*Ans.* 14.2 kJ/kg K, 10.1 kJ/kg K, 1.405. 41 kJ/kg K, 2.038]

4.46. Calculate the heat rejected per kg of a perfect gas which has a molecular weight of 26 and a value of γ = 1,26 for the following cases:

(a) When the gas is contained in rigid vessel at 3 bar and 315°C and is cooled until the pressure falls to 1.5 bar.

(b) When the gas expands at constant pressure till the temperature falls from 280°C to 20°C.

[*Ans.* (a) 36l kJ/kg; (b) 403 kJ/kg]

4.47. A steel cylinder having a volume of 0.01653 m^3 contains 5.6 kg of ethylene gas ($C_2 H_4$) Molecular weight is 28. Calculate the temperature to which the cylinder may be heated without the pressure exceeding 200 bar; given the compressibility factor $Z = 0.505$.

4.48. A rigid vessel is connected to a spherical elastic balloon through a small bore pipe containing a valve. Both contain air at the ambient temperature of 30°C. The volume of the rigid vessel is 0.1 m^3 and the initial pressure is 300 kPa. This initial diameter of the balloon is 0.5 m and air rushes to fill the balloon till pressure equilibrium is reached. It may be assumed that the pressure inside the balloon is directly proportional to its diameter and also that the final temperature of the air is uniform throughout at 30°C. Calculate the final pressure in the system and the final volume of the balloon.

4.49. Calculate the compressibility factor for water at 100 bar and 400°C, using

(a) data from the compressibility chart, and

(b) data from the tables [*Ans.* (a) 0.86, (b) 0.85]

4.50. Calculate the volume in m^3, occupied by 40 kg of nitrogen (N_2) (at) 17 MPa and 180 K.

[*Ans.* 0.103]

Software Based Questions

4.51. Develop a software for predicting the thermodynamic behaviour of ideal and real gases using various expressions for real gases.

4.52. Develop a software to calculate enthalpy of air at the end of compression process in a diesel engine and axial flow compressor.

4.53. Assuming ideal gas behaviour for air, develop a software to plot to scale the isothermal 300, 500, 1000 and 2000 K on a p–v diagram.

Design and Open Ended Questions

4.54. There is a chance that frazil ice forming upstream of hydroelectric power plant may block the flow of water to the turbine. Prepare a report summarizing the mechanism of frazil ice formation and alternative means of eliminating frazil ice block of power plants. For one of the alternatives, prepare an estimate of the cost of maintaining a 60 MW power plant frazil ice-free.

4.55. In order to study allotropic transformation metallurgists use phase diagrams. Discuss the features of the phase behaviour of solids which are important in the fields of metallurgy and material processing.

4.56. During research, many new substances have been found as potential working fluids for power plants or refrigeration systems or heat pumps. Prepare thermodynamic property data needed to assess the feasibility of a candidate substance for possible use as a working fluid. Write a paper discussing your findings.

Multiple-Choice Answer Type Questions

4.57. At a given temperature and pressure all molecules occupy the same volume. This is called

(a) Charles' law (b) Boyle's law, (c) Avogadro's law (d) Regnault's law

4.58. The characteristic gas constant (R) and universal gas constant (R_0) is related through molecular weight (M) is as follows.

(a) $R = MR_0$ (b) $R = R_0/M$ (c) $R_0 = R + M$ (d) $R = R_0 - M$

4.59. c_p and c_v are related as

(a) $c_p + c_v = R$ (b) $c_p - c_v = R$ (c) $c_p - c_v = 0$ (d) c_p/c_v

4.60. The compressibility factor Z for real gases depends upon

(a) $Z > 1$ (b) $Z \geq 1$ (c) $Z \leq 1$ (d) $Z = 1$

4.61. The van der Waals equation of state for real gases may be given by

(a) $p=\dfrac{RT}{v+b}+\dfrac{a}{v^2}$ (b) $p=\dfrac{RT}{v-b}-\dfrac{a}{v^2}$ (c) $p=\dfrac{RT}{v^2}+\dfrac{a}{v^2}$ (d) $p=\dfrac{RT}{v+b}-\dfrac{a}{v^2}$

4.62. According to van der Waals equation the critical molar volume (v_c) is given by

(a) $v_c = b$ (b) $v_c = 2\,b$ (c) $v_c = 3\,b$ (d) $v_c = 4\,b$

4.63. The real gas behaviour is expressed as

(a) $p=\dfrac{RT}{v}$ (b) $p=\left(\dfrac{RT}{v}\right)$ (c) $p=\left(\dfrac{RT}{v}\right)_{p\to\infty}$ (d) $p=\left(\dfrac{RT}{v}\right)_{p\to 0}$

Answers to Multiple-Choice Questions

4.57. (c) 4.58 (b) 4.59 (b) 4.60 (b) 4.61 (b) 4.62 (c)

4.63. (d)

CHAPTER

5

First Law of Thermodynamics in General and for Non-flow Processes

OVERVIEW

In the past century between 1843 and 1848, Joule conducted experiments which were the first step in the analysis of thermodynamic system leading to the discovery of the First Law of Thermodynamics. This law is the law of conservation of energy (i.e., energy can neither be created nor destroyed but only converted from one form to another) when it is used in a system involving heat and work transfer. In this chapter, the first law of thermodynamics will be applied to *non-flow processes*.

5.1 FIRST LAW OF THERMODYNAMICS UNDERGOING CYCLIC PROCESS

The first law of thermodynamics may be stated in many ways:

(a) *If a system undergoes a cyclic change (the end state being precisely the same as the initial state), then the algebraic sum of the work delivered to the surrounding is proportional to the algebraic sum of heat taken from the surrounding*. Using the same unit for work and heat, the first law is expressed as

$$\oint \delta Q = \oint \delta W \qquad ...(5.1)$$

(b) *Heat and work are mutually convertible one into other*. But according to the law of conservation of energy, energy can neither be created nor destroyed hence it follows that the total energy associated with the energy conversion remains constant. Thus,

Creation of energy = 0

or Rate of creation of energy = 0

(c) *It is impossible to construct a perpetual motion machine of first kind PMM-I. PMM-I is capable of producing energy without corresponding expenditure of the energy* (Fig. 5.1). The converse of the above statement is also true. It is impossible to construct a machine which would continuously consume work without some other form of energy appearing simultaneously (Fig. 5.2).

Verification of First Law. In order to verify the first law of thermodynamics, let us consider Joule's experiment as shown in Fig. 5.3. In Fig. 5.3(a), a quantity of liquid is kept in an insulated vessel. Work is done on the fluid by stirring of the paddle wheel. This work input to the fluid causes a rise in the temperature of the fluid. The amount of work done on the fluid is calculated by the product of the weight and the vertical height through which the weight descends.

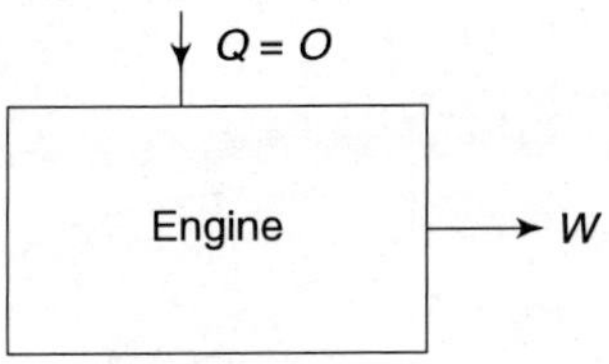

Fig. 5.1 PPM–1 (Power producing)

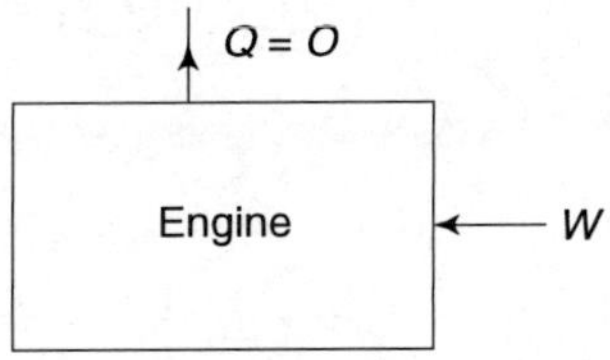

Fig. 5.2 PPM–1 (Power absorbing)

Next by removing the insulation in the system the system is immersed in a water bath as shown in Fig. 5.3(b) such that the heat is transferred from the fluid till the original state of the fluid is reached which will be indicated by the temperature and pressure. In this way, the system undergoes a complete cycle. The amount of heat rejected by the fluid is equal to the increase of energy of the water bath and may be determined by measuring the rise in temperature of the water bath. Joule conducted many such experiments involving different work interactions in a variety of systems. For example, Joule used work in the form of electrical energy after measuring the heat transfer by a heating coil, completed the cycle by restoring the system to its initial state. These experiments led to an important conclusion: that the net work input W was directly proportional to the net heat Q transferred from the system regardless of the kind of work done, the rate at which the work was done and the scheme used for transferring it into heat. *Thus, for closed system undergoing cyclic process,*

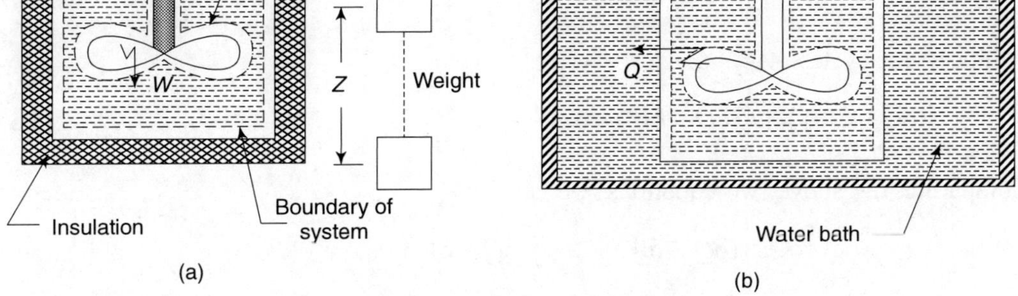

Fig. 5.3 Joule's experiment

$$\oint \delta Q \propto \oint \delta W$$

If heat and work are expressed in the same unit, then

or
$$\oint \delta Q = \oint \delta W \qquad ...(5.2)$$

In a cycle (Fig. 5.4) involving two processes, the first law may be expressed as

$$Q_{1-2} + Q_{2-1} = W_{1-2} + W_{2-1} \qquad ...(5.3)$$

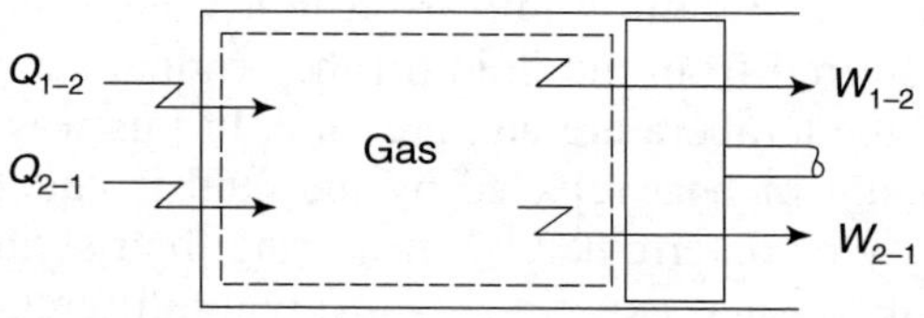

Fig. 5.4 Cyclic processes

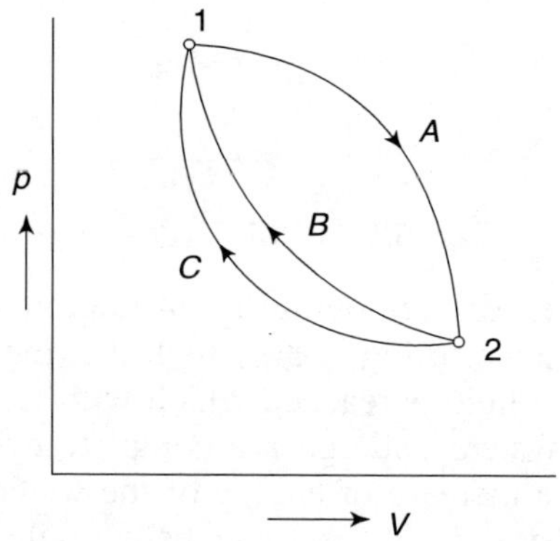

Fig. 5.5 Existence of internal energy

5.2 FIRST LAW OF THERMODYNAMICS UNDERGOING A PROCESS (Change of State Leads to the Concept Internal Energy)

One very important consequence of the first law of thermodynamics is that the *energy of a system is a property*. In order to prove this, consider a closed system which undetgoes a change of state from 1 to 2 as shown in Fig. 5.5. The system may proceed from state 1 to state 2 along the path A and then back to the original state 1 along two paths say B and C. If the system operates over the cycle 1–A–2–B–1 first, then the first law yields

$$\oint (\delta Q - \delta W) = 0$$

or
$$\int_1^2 (\delta W - \delta W)_A + \int_2^1 (\delta Q - \delta W)_B = 0 \qquad ...(5.4)$$

Similarly, for cycle 1–A–2–C–1

$$\int_1^2 (\delta Q - \delta W)_A + \int_2^1 (\delta Q - \delta W)_C = 0 \qquad ...(5.5)$$

Comparing, the above two equations, we get

$$\int_2^1 (\delta Q - \delta W)_B = \int_2^1 (\delta Q - \delta W)_C \qquad ...(5.6)$$

Equation (5.6) suggests that when a system operates between states 1 and 2, quantity $\int_1^2 (\delta Q - \delta W)$ is a constant irrespective of the path, i.e., whether the system proceeds along path *B* or path *C*. Thus, the value of $\int_1^2 (\delta Q - \delta W)$ is solely fixed by the end states and is independent of the path of process.

Therefore, $(\delta Q - \delta W)$ *is an exact differential* and its integral *is a property because it satisfies the condition of a property. This is a called total energy* (*E*) of the system. Thus, for non-flow and non-cyclic process, we have

$$\int_1^2 (\delta Q - \delta W) = \int_1^2 dE = E_2 - E_1 \qquad \text{...(5.7)}$$

or

$$\delta Q - \delta W = dE \qquad \text{...(5.8)}$$

where $\delta W = \delta W_{pdv} + \delta W_{\text{shaft}} + \delta W_{\text{elect}} + ...$

Equation (5.8) is a differential form of the first law of thermodynamics applied to a closed system. For an isolated $\delta Q = 0$ and $\delta W = 0$, hence $dE = 0$. Therefore, E = constant. Thus, it follows that in *an isolated system energy remains constant.*

The energy *E* includes all forms of energy in the system,

i.e., $$E = U + \text{KE} + \text{PE} + \text{chemical energy} + ...$$

where energy *U* is the *internal energy* due to molecular kinetics and is independent of macroscopic motion, gravity electricity and magnetism. Internal energy is sometimes also known as ***intrinsic*** or ***macroscopic energy***. With all forms of energy, except chemical, electrical and magnetic, etc., the Eq. (5.7) becomes:

$$\int_1^2 (\delta Q - \delta W) = \int_1^2 [dU + d(\text{KE}) + d(\text{PE})]$$

or

$$\delta Q - \delta W = dU + d\left(\frac{mC^2}{2}\right) + d(mgz)$$

or

$$Q_{1-2} - W_{1-2} = (U_2 - U_1) + \left(\frac{m(C_2^2 - C_1^2)}{2}\right) + mg(z_2 - z_1) \qquad \text{...(5.9)}$$

In the absence of kinetic energy, potential energy, chemical, electrical and magnetic effects, which is possible in a closed system the above differential form of equation becomes

$$\int_1^2 (\delta Q - \delta W) = \int_1^2 dU$$

For cyclic process, $dU = 0$, i.e., $\oint dU = 0$

or

$$\delta Q - \delta W = 0 \qquad \text{...(5.10a)}$$

or

$$Q_{1-2} - W_{1-2} = U_2 - U_1 \qquad \text{...(5.10b)}$$

For non-flow process, $\delta W = pdV$

Hence, $$\delta Q = dU + pdV \qquad ...(5.11)$$

On per unit mass basis, for non-flow and non-cyclic process, the first law of thermodynamics is

$$\delta q - \delta w = du \qquad ...(5.12)$$

On the basis of what has been said it can be stated that the first law enables us to define a property of the system called internal energy just as the zeroth law enables us to define a property of matter called temperature.

It is to be noted that Eq. (5.10) gives the change in internal energy and not the absolute value of it. To evaluate the absolute value of energy, *some reference point or datum is assumed at which the internal energy is zero,* Further, internal energy is invisible and only its effects can be sensed.

Heat and work interaction for states 1 to 2 may be written either Q_{1-2} or Q simply and W_{1-2} or W simply.

5.2.1 Time Rate Form of Energy Balance

The energy balance given by Eq. (5.8) an also be written on time rate basis as follows. By dividing by the time interval $\Delta\tau$, an expression is obtained by the average rate of change of energy in terms of *average rates* of energy transfer by heat and work during time interval, $\Delta\tau$. Thus,

$$\frac{\Delta E}{\Delta \tau} = \frac{Q}{\Delta \tau} - \frac{W}{\Delta \tau}$$

In the limit, when $\Delta\tau$ tends to zero

$$\lim_{\Delta\tau \to 0}\left(\frac{\Delta E}{\Delta \tau}\right) = \lim_{\Delta\tau \to 0}\left(\frac{Q}{\Delta \tau}\right) - \lim_{\Delta\tau \to 0}\left(\frac{W}{\Delta \tau}\right)$$

or $$\frac{dE}{d\tau} = \dot{Q} - \dot{W} \qquad ...(5.13)$$

where $\dot{Q}$ and $\dot{W}$ are expressed in kW. Eq. (5.13) is known as the ***instantaneous time rate form of the energy balance***.

Since the time rate of change of energy is given by

$$\frac{dE}{d\tau} = \frac{dKE}{d\tau} + \frac{dPE}{d\tau} + \frac{dU}{d\tau}$$

Thus, Eq. (5.13) can be expressed alternatively as

$$\frac{dKE}{d\tau} + \frac{dPE}{d\tau} + \frac{dU}{d\tau} = \dot{Q} - \dot{W} \qquad ...(5.14)$$

5.3 INTERNAL ENERGY OF A PERFECT GAS

Joule demonstrated experimentally that the internal energy of a perfect gas is a function of absolute temperature and is independent of pressure and volume changes. The apparatus he used is shown in Fig. 5.6. It consists of a chamber *A* containing gas in thermodynamic equilibrium and an evacuated chamber *B*. Both of them are connected by a stopcock. The two chambers constitute a closed system and are immersed in a water bath which is insulated from outside. Suppose that initially the system is in thermal equilibrium with its environment at uniform temperature, t_1. When the stopcock between chambers *A* and *B* is opened, the gas in *A* expands freely into *B*, occupying both chambers. During expansion process the temperature of the gas in *A* decreases below its initial value while the temperature in *B* increases. These temperature changes develop temperature gradients so that heat flows from the water bath to chamber *A* and from chamber *B* to the water bath. Due to this, there is an additional flow of gas from *A* to *B*. After sometimes, there is complete thermodynamic equilibrium between system and environment. As a result of this process, the heat transferred to the system from the environment is given by

$$Q_{\text{system}} = -Q_{\text{enviroment}} = mc\,(t_2 - t_1) = mc\,(T_2 - T_1)$$

where *m* and *c* are the mass and average specific heat of the environment and $(t_2 - t_1) = (T_2 - T_1)$ is the temperature change in the environment. As there is no movement of the boundary of the system, hence no external work is done during the free expansion of the gas. Applying the first law to this system, we have

$$\Delta U = Q_{\text{system}}$$

Substituting for Q_{system} from the above equation

$$\Delta U = mc_v\,(t_1 - t_2) = mc_c\,(t_2 - t_1) \qquad \ldots(5.15)$$

or

$$U_2 - U_1 = mc_v\,(t_2 - t_1) = mc_v\,(T_2 - T_1)$$

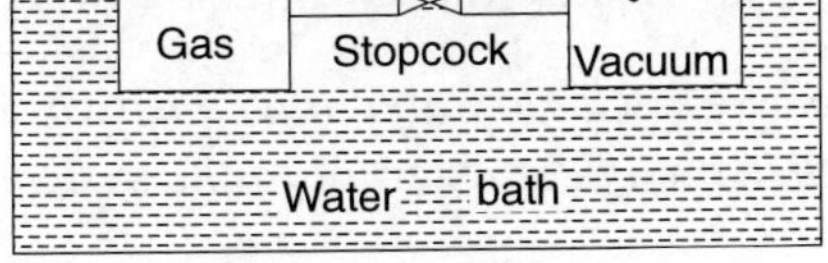

Fig. 5.6 Internal energy of perfect gas

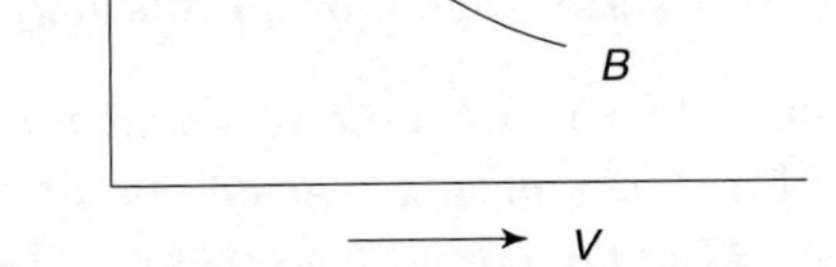

Fig. 5.7 Internal energy only a function of temperature

It is possible to measures the specific heat and temperature t_1 and t_2. Measurements reveal that the temperature of the water bath remains practically constant. This impiles that chamber *A* loses as much energy as chamber *B* gains, and there is no net heat interaction between the system and the environment. The absence to both heat interaction and work interaction shows that internal energy of this system remains constant, even when work involving a volume change was done between parts of the system. Thus, it is concluded that internal energy of a perfect gas is independent of volume and depends only on temperature. This statement is called Joule's law which is expressed as

$$\left(\frac{\partial u}{\partial v}\right) = 0 \qquad \text{...(5.16)}$$

This is also shown on the $p - V$ diagram of Fig. 5.7 that internal energy is independent of pressure and volume and is only a function of temperature. The curve *AB* shows the constants internal energy *U* or constant temperature *T*, while curve *CD* shows the constant internal energy $(U + dU)$ or constant $(T + dT)$ line.

Example 5.1 The following amount of heat transfer occurs during a cycle comprising four processes. Calculate the work done of the cycle and indicate about the type of work 120 kJ, – 20 kJ, + 16 kJ and + 24 kJ.

Solution

$$\oint \delta Q = 120 - 20 + 16 + 24 = 140 \text{ kJ}$$

$$\oint \delta W = \oint \delta Q = 140 \text{ kJ}$$

Here work is done by the system and hence work transfer is positive.

Example 5.2 For the following processes indicate signs in tabular form for Q, W and ΔE in terms of positive, negative or zero. The systems to be considered are given in italic.

(a) *Water and its vapour in a rigid boiler* The boiler is fired and steam is generated at high pressure and temperature.

(b) *Water and steam* circulating through various components of steam power plants.

(c) *Hydrogen and oxygen in the form of combustible mixture* within a rigid and insulated wall. A spark, which may be considered infinitesimal and negligible, causes the combustible mixture combine chemically to form water.

(d) *The gasoline-air mixture* The mixture enters an automobile engine and leaves as products of combustion.

(e) *An electric capacitor* It discharges through an electric motor to raise a weight.

Solution *It is worthwhile to recall the definition work which states that positive work is done by a system during a given operation, when the sole effect external to the system could be reduced to the rise of a weight.*

(a) In the case of steam generation in the boiler, there is no work done either by the system or on the system as the boiler is rigid, i.e., $pdV = 0$. If we allow the steam to expand either in a steam engine or turbine, then as per second law of thermodynamics, the sole effect external to the system cannot be a rise of weight only but also either the surroundings become warmer or the steam does not return to the original state. Therefore, the energy transfer to the system is heat. Thus,

$$W = 0, Q = +\text{ve and hence } \Delta E = Q - W = +\text{ve}$$

(b) The working fluid passing through various elements (boiler, turbine, condenser and pump) of steam power plant constitutes a closed system. Heat added in the boiler is greater than the heat rejected in the condenser. This means that there is a net positive heat transfer to the system. Similarly, work developed by turbine is much greater than the work input to the boiler feed pump, hence there is a positive work transfer by the system. Since the system is going through a cyclic change, hence the change in the energy of the system is zero. Thus,

$$Q = +\text{ve}, \quad W = +\text{ve} \quad \text{and} \quad \Delta E = 0$$

(c) Since the system is surrounded by an insulated surface hence the heat exchange either from the system to the surrounding or vice versa is zero. This combustible mixture is contained in a rigid vessel, so the boundary work (*pdV*) is zero. The electrical energy required to initiate the spark is negligible. Therefore, the work transfer to or form the system is zero. Using non-flow energy equation, $\Delta E = Q - W = 0$. However, there is only transformation of chemical energy into internal energy manifested by the increase in temperature. This system is a good example of an isolated system. Thus,

$$Q = 0, W = 0 \quad \text{and} \quad \Delta E = 0$$

(d) In automobile engine, the gasoline-air mixture burns and the chemical energy of the fuel is converted into heat energy which increases the temperature and thus its internal energy. Gas while expanding inside the engine does work and finally leaves as combustion products at lower energy. The balance of the energy appears partly a work transfer and partly as heat transfer, both energy flow from the system to the surroundings. Thus, by convention

$$W = +\text{ve} \quad Q = -\text{ve} \quad \text{and} \quad \Delta E = -\text{ve}$$

(e) The capacitor being the system discharges into an electric motor which may be assumed frictionless and that the copper loss in motor is negligible. The electrical energy associated with current flow can be converted solely into rise of a weight external to the boundary of the system. Hence, the work transfer is positive. Since the discharge timing is very short, the process can be assumed adiabatic as no heat transfer can occur across the capacitor boundary during this short time. Thus,

$$W = +\text{ve}, \quad Q = 0 \quad \text{and} \quad \Delta E = -\text{ve}$$

The result of various processes are tabulated as follows:

Process	*a*	*b*	*c*	*d*	*e*
Q	+ve	+ve	0	–ve	0
W	0	+ve	0	+ve	+ve
ΔE	+ve	0	0	–ve	–ve

Example 5.3 **Case (a)** A wooden block (non-conduccor of heat) slides down a rough inclined plane made of copper (conductor of heat) and thus the temperature of plane rises, State whether the heat, work and change in energy are positive, negative or zero as a result of sliding down for each of the following systems (i) the block, (ii) the plane, and (iii) the block and the plane.

Case (b) State the above by considering the block made of copper and plane made of wood. In this case, due to sliding the temperature of the block increases.

Tabulate the results in both the cases.

Solution **Case (a) :** Refer to Fig. 5 8 and recall the definition of work. Further, when a system does positive work its surroundings do an equal amount of negative work and vice versa.

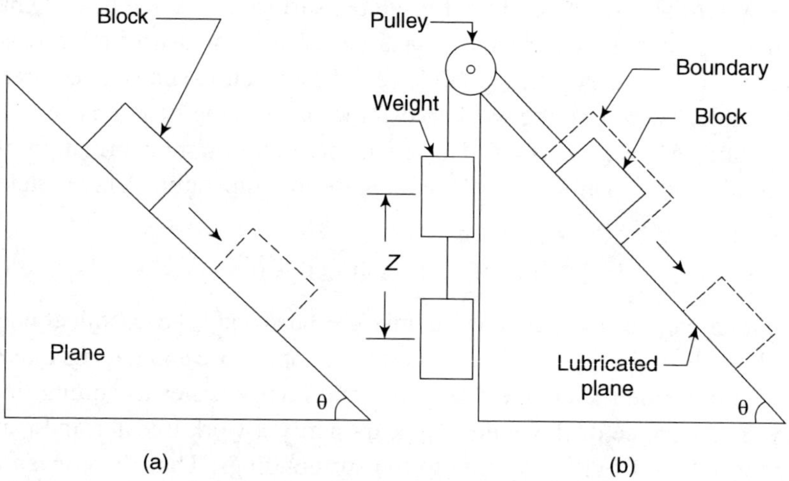

Fig. 5.8 Plane and block (a) actual (b) equivalent or reduced to

(i) *Block as a system and plane as surrounding* When wooden block slides down on copper plane, heat is generated on the plane surface due to sliding friction which results in the rise in temperature of the plane but there is no transfer of heat to the block as it is non-conductor of heat.

Hence, for block $Q = 0$. For the block, $W = +\text{ve}$ because although the actual external effect to itself is the rise in temperature of the plane, it could be reduced to the sole external effect of rise of weight by connecting the block by a suitable frictionless pulley mechanism and lubricating the plane. Using first law

$$\Delta E = Q - W = 0 - W = -\text{ve}$$

From above, it follows that for the block $\Delta E = -\text{ve}$. This may be explained by the fact that there is no change in the block except due to sliding, the potential energy of the block has decreased.

(ii) *Plane as system and block as surrounding* If plane is assumed as the system $Q = 0$ because, though the temperature of the plane has risen, energy due to temperature difference (i.e., heat energy) does not cross the boundary of system. Here, the block is the surrounding which is non-conducting. Since for block, $W = +\text{ve}$, hence for plane, $W = -\text{ve}$. Applying first law

$$\Delta E = Q - W = 0 - (-W) = +W = +\text{ve}$$

This may be attributed to fact that the internal energy of the plane has increased as a result of increase, in its temperature.

(iii) *Block and plane as system* If block and plane is considered a system then atmosphere becomes the surroundings. In this case, the system does not react with the surroundings in any way. Thus, $Q = 0$ and $W = 0$ for the system. Using first law, $\Delta E = Q - W = 0 - 0 = 0$. The above findings may be explained by the fact that the decrease in the potential energy of the part to the system (i.e., the block) is exactly equal to the increase in the potential energy of the other part of the system (i.e., the plane).

System	Q	W	ΔE
(i) Block	0	+ve	–ve
(ii) Plane	0	–ve	+ve
(iii) Block and plane	0	0	0

Case (b) The plane is non-conductor of heat while the block is conductor of heat.

(i) *Block as a system and plane as surroundings* For the block, $Q = 0$ because even though the block has become warmer as a result of sliding friction but there is no heat interaction with the surrounding (i.e., plane) as the plane is non-conducting. Interestingly for block, $W = 0$ because due to sliding down the block there is no effect "external" to itself but there is only internal effect, i.e., the system (block) has become warmer which cannot be compared with the case (a) where there is an external effect (i.e., warming the plane) which could be reduced to rise in weight external to boundary resulting in positive work done. Using first law, $\Delta E = Q - W = 0$. The finding can be explained by the fact that the decrease in potential energy of the block is exactly compensated by the increase in its internal energy due to rise in its temperature.

(ii) *Plane as a system* For plane as a system, $Q = 0$ as it is non-conducting. $W = 0$ as it is zero for block. Therefore, $\Delta E = Q - W = 0$. This may be explained by the fact that the plane suffers no change at all.

(iii) *Block and plane as a system* In this case, $Q = 0$, $W = 0$ and hence symbol $\Delta E = 0$ as the combined system does not interact with the surroundings.

System	Q	W	ΔE
(i) Block	0	0	0
(ii) Plane	0	0	0
(iii) Block and plane	0	0	0

Example 5.4 A system consists of 50 kg of copper block initially kept 5 m above the concrete floor and both are at the same temperature. The copper block is allowed to fall. Find Q, W, ΔU, ΔKE, and ΔE for the following conditions:

(a) at the instant the block is about is to hit the floor,

(b) just after the block has come to rest on the floor, and

(c) after enough heat has been transferred so that the block and floor are essentially at the same temperature as they were initially.

Solution (a) Since system and surroundings are at the same temperature, hence $Q = 0$. As the falling body does not produce any work, hence. $W = 0$. Thus, from first law

$$\Delta E = Q - W = 0 - 0 = 0$$

But $$\Delta E = \Delta U + \Delta \text{KE} + \Delta \text{PE} = 0$$

Since $\Delta U = 0$ as there is no change in the temperature of the system, hence

$$\Delta \text{KE} = -\Delta \text{PE} = -mg(z_{\text{final}} - z_{\text{initial}}) = -50 \times 9.81\ (0 - 5) = 2452.5 \text{ J}$$ **Ans.**

(b) Once the block comes to rest on the floor, $\Delta \text{KE} = 0$

$$\Delta \text{PE} = -2452.5 \text{ J and by first law, } \Delta E = 0$$

But $$\Delta E = \Delta U + \Delta KE + \Delta PE = 0$$

$$\Delta U = \Delta \text{PE} = -2452.5 \text{ J}$$ **Ans.**

(c) When the block and floor are at the same temperature as they were initially means

$$\Delta U = 0. \text{ Also } \Delta \text{KE} = 0 \text{ and } \Delta \text{PE} = -2452.5 \text{ J}$$

Heat released to surroundings is

$$\Delta Q = \Delta E = \Delta U + \Delta \text{PE} + \Delta \text{KE} = 0 - 2452.5 + 0 = -2452.5 \text{ J}$$ **Ans.**

Example 5.5 In a piston-cylinder arrangement in which the system is air, calculate the quantity of heat added to the cycle, when the cycle is complete by two processes : compression and expansion as follows:

(a) During compression process, piston does 2,00,000 J of work on air while 80 kJ of heat is rejected to the surrounding.

(b) During expansion process, air does 3,00,000 J of work on the piston.

Solution Let processes 1–2 and 2–1 show the compression and expansion processes respectively and thus the cycle is complete. During compression process 1–2, we have, from first law.

$$Q_{1-2} - W_{1-2} = \Delta U = U_2 - U_1$$

or $$-80 - \frac{(-200000)}{1000} = U_2 - U_1 = 120 \text{ kJ}$$

It is to be noted that Q is positive when heat is taken from surrounding to system and W is positive when the work is delivered from system to surroundings. Hence, for process 1–2, Q and W are negative.

$$U_2 - U_1 = 120 \text{ kJ}$$

Again, for expansion process 2–1, we have $Q_{2-1} - W_{2-1} = U_1 - U_2$

or $$Q_{2-1} = \frac{300000}{1000} - 120 = 180 \text{ kJ}$$ **Ans.**

Example 5.6 Calculate the distance moved by a diesel locomotive train consuming 1000 kg of oil if 30 per cent of the heat generated by the oil is converted into mechanical work. The tractive effort required is 40 N/tonne dead mass of the locomotive where the mass of the locomotive is 2000 tonnes. Assume 1 kg of oil liberates 43000 kJ of heat on burning.

Solution Since only 30 per cent of heat generated is converted into mechanical work or useful work, hence the quantity of 1 tonne (i.e., 1000 kg) of oil is given by

$$W = 0.3 \times 1000 \times 43000 = 129 \times 10^5 \text{ kJ}$$

Since the tractive effort is 40 N/tonne of dead mass, hence the force required to move the train

$$\text{Force} = 40 \times 2000 = 80000 \text{ N}$$

$$\text{Work done} = \text{Force} \times \text{Distance}$$

Let x be the distance moved by the train, hence $129 \times 10^5 \times 10^3 = 80000\, x$

$$x = 161.25 \times 10^3 \text{ m} = 161.25 \text{ km}$$ **Ans.**

Example 5.7 The following data refers to a closed system which undergoes a thermodynamic cycle consisting of four processes.

Process	Heat transfer (kJ/min)	Work transfer (kJ/min)
1-2	NIL	- 1000
2-3	40,000	NIL
3-4	- 4000	+ 26,000
4-1	- 12,000	- 1000

Show that the data is consistent with the first law of thermodynamics and calculate:

(a) Net rate of work output in kW.

(b) Efficiency of the cycle.

(c) Change in internal energy for each process.

Solution Algebraic sum of heat transfer is

$$\oint \delta Q = 0 + 40000 - 4000 - 12000 = 24000 \text{ kJ/min}$$

Algebraic sum of work transfer is

$$\oint \delta W = -1000 + 26000 - 1000 = 24000 \text{ kJ/min}$$

$$\oint \delta Q = \oint \delta W \text{ which is according to first law.}$$

(a) Net rate of work $= 24{,}000$ kJ/min $= \dfrac{24000}{60} = 400$ kW **Ans.**

(b) Thermal efficiency $= \dfrac{\text{Net work done}}{\text{Heat supplied}} = \dfrac{24000}{40000} \times 100 = 60$ per cent **Ans.**

(c) We have, $\delta Q - \delta W = dU$, hence change in internal energy may be tabulated as follows

Process	δQ	ΔW	$dU = (\Delta Q - \Delta W)$ kJ/min
1-2	0	- 1000	1000
2-3	40000	NIL	40000
3-4	- 4000	26000	- 30000
4-1	- 12000	- 1000	- 11000
Sum	24000	24000	0

Thus, we see that for the complete cycle, the *total change in internal energy is zero.*

Example 5.8 During a non-flow process, the temperature of the system changes from 20°C to 40°C. The work done by the system and heat transfer per degree rise in temperature, at each temperature reached is given by

$$\frac{\delta W}{dt} = (4 - 0.08\, t) \text{ kJ/°C and } \frac{\delta Q}{dt} = 1.003 \text{ kJ/°C}$$

Determine change in internal energy of the system during the process.

Soluitonne We have, $\int \delta Q - \int \delta W = \int dU = \Delta U$

or

$$\Delta U = \int_{t_1}^{t_2} 1.003\, dt - \int_{t_1}^{t_2} (4 - 0.08)\, dt$$

$$= 1.003\,(t_2 - t_1) - \left[4\,(t_2 - t_1) - 0.08\,\frac{(t_2^2 - t_1^2)}{2}\right]$$

$$= 1.003 \times 20 - \left[80 - \frac{0.08 \times 1200}{2}\right] = 20.06 - 32 = -11.94 \text{ kJ}$$ **Ans.**

Example 5.9 A paddle fixed to the shaft of an engine rotates in closed hollow vessel containing water. This closed vessel is mounted freely on the shaft and restraint to its turning moment is proved by weight attached to its side. Find the temperature rise for the following data.

Engine speed = 400 rpm

Brake load = 50 kg at a leverage of 1 m

Quantity of water = 220 kg

Duration of test = 30 minutes

Solution Work done = 2πNT J/min

where Torque = T = Brake load × leverage = 500 × 1 = 500 J

Work done in 30 minutes = 2π × (400 × 30) × 500 = 37699111.84 J

Heat gained by water in 30 minutes

$$Q = m_w \, c_{pw} \, \Delta t_w$$

or
$$\Delta t_w = \frac{37699111.84}{10^3 \times 220 \times 4.18} = 40.99°\text{C} \quad \textbf{Ans.}$$

5.4 APPLICATION OF FIRST LAW TO A CLOSED SYSTEM (NON-FLOW PROCESS) INVOLVING PERFECT GASES

We have seen that the governing equation of first law of thermodynamics for a closed system is given by

$$Q_{1-2} - W_{1-2} = (\Delta E)_{\text{system}}$$

where $(\Delta E)_{\text{system}} = (\Delta U + \Delta KE + \Delta PE)_{\text{sysem}}$

If ΔKE and ΔPE are negligible, then $Q_{1-2} - W_{1-2} = \Delta U$ ...(5.17)

The aim of this analysis is to apply the first law of thermodynamics to various useful processes of a closed system (i.e., non-flow system) which changes from one state to another and to develop some relations for thermodynamic analysis as given below.

5.4.1 Constant Volume or Isochoric Process

As the name implies, the volume of the system remains constant in constant volume process.

Figure 5.9 shows a completely closed vessel filled with a perfect gas being heated from outside. This heating results in an increase in pressure and temperature of the system at constant volume as represented by process 1–2 on the p–V diagram shown in Fig. 5.10. Let Q units of heat be supplied during the process in which the property of the system is changed from state 1 to state 2. Since no expansion or compression of the gas takes place, hence the work done by the system is zero and it is also obvious from the p–V diagram that the area under the diagram is zero. Thus, all the heat supplied in the system is stored, i.e., the internal energy will be increased.

$$\therefore \quad \delta W = 0 \quad \text{...(5.18)}$$

$$\therefore \quad Q_{1-2} = U_2 - U_1; \text{ kJ}$$

In differential form $\delta Q = dU$ ...(5.19)

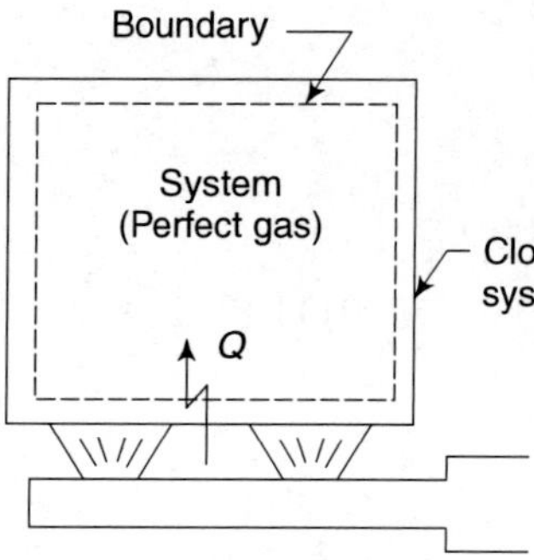

Fig. 5.9 Arrangement for constant volume heat addition system

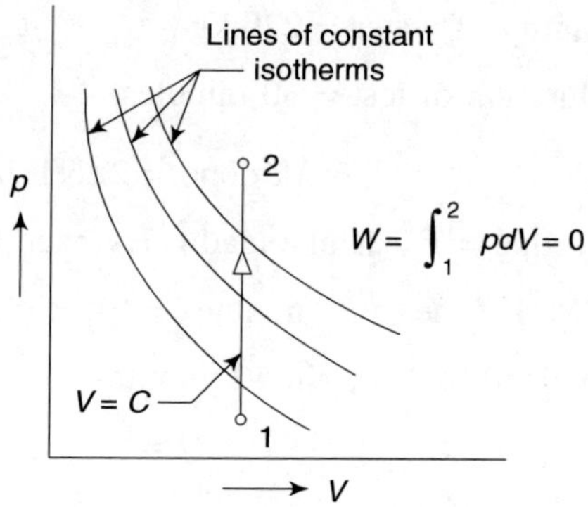

Fig. 5.10 Isochoric process on p–V diagam

The Eq. (5.19) suggests *that during constant volume process, the heat transfer is equal to the change in internal energy of the system.* Thus,

$$(Q_{1-2})_v = mc_v\,(T_2 - T_1);\ \text{kJ} \qquad ...(5.20)$$

5.4.2 Constant Pressure or Isobaric Process

In this process, the pressure of the system remains constant throughout. Figure 5.11 shows how heat energy can be supplied to a system at constant pressure. It is illustrated by the perfect gas contained in the cylinder while the piston with weight is placed on the gas. The piston can move along the cylinder wall. If heat is supplied to the system, its temperature will rise and it will expand which will force the piston to move upward. Thus, a displacement work done by the system against a constant force. Assume that this process is carried out in a thermodynamic equilibrium (i.e., quasi-static process). The above process is represented on p–V diagram shown in Fig. 5.12. Lines of constant temperature lines are shown for clarity and understanding regarding the increase of temperature of the system.

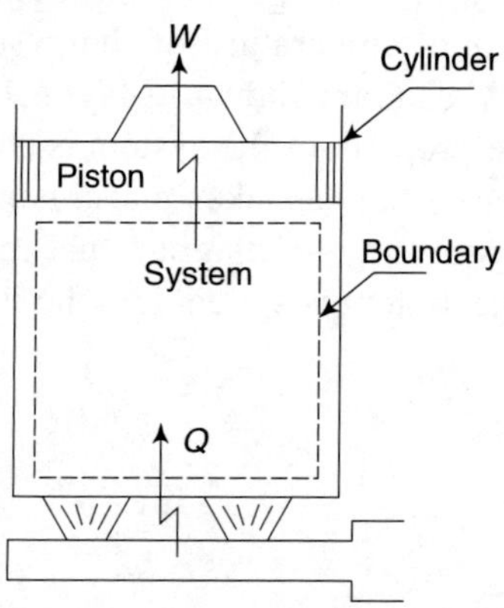

Fig. 5.11 Heat addition at constant pressure

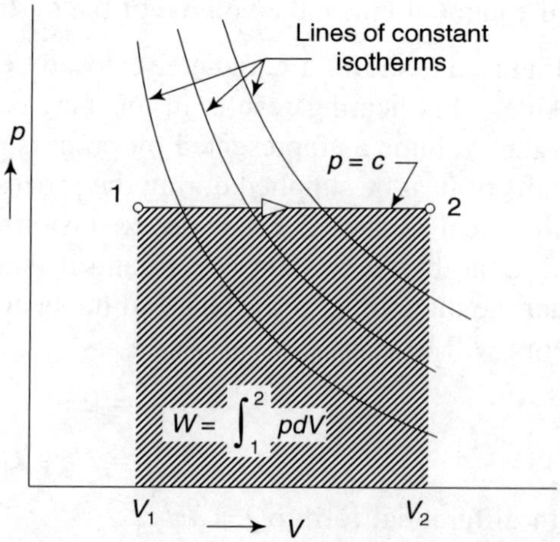

Fig. 5.12 Isobaric process on p–V diagram

Let Q be the amount of heat to be supplied to the system during the process of the system from state 1 to 2. Then the work done by the system is

$$\delta W = pdV$$

or $$W_{1-2} = \int_1^2 pdV = p(V_2 - V_1), \text{ kJ} \quad \text{...(5.21)}$$

The first law of thermodynamics can be written as

$$\delta Q = \delta W + dU$$

or $$Q_{1-2} = (U_2 - U_1) + p\,(V_2 - V_1) \quad \text{...(5.22)}$$

Since p is constant throughout the process, Eq. (5.22) may be written as

$$\delta Q = dU + d\,(pV) = d\,(U + pV) = dH$$

or $$Q_{1-2} = H_2 - H_1 = mc_p\,(T_2 - T_1); \text{ kJ} \quad \text{...(5.23a)}$$

or $$q_{1-2} = h_2 - h_1 = c_p\,(T_2 - T_1); \text{ kJ/kg} \quad \text{...(5.23b)}$$

where h is the enthalpy of the gas, kJ/kg

Thus, during constant pressure process, heat transfer is equal to the change in enthalpy.

5.4.3 Constant Temperature or Isothermal Process

In this process, the temperature of system remains constant. It is rather difficult to execute the isothermal process in practice. Figure 5.13 shows an arrangement for performing isothermal process under ideal circumstances. A constant temperature, T_1 reservoir surrounds a piston cylinder arrangement. Assume that a perfect gas is at any instant at temperature T_1 inside the cylinder. At this equilibrium state, the two are in thermal equilibrium because the temperature of the system and surrounding is the same. Hence, there is no transfer of heat energy across the boundary of the system in either direction.

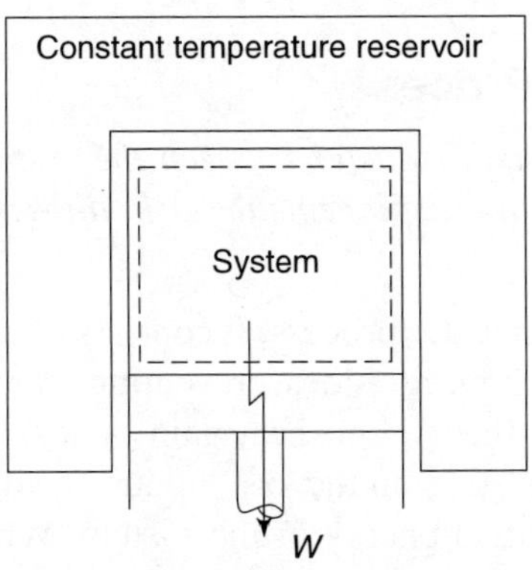

Fig. 5.13 Arrangement for isothermal process

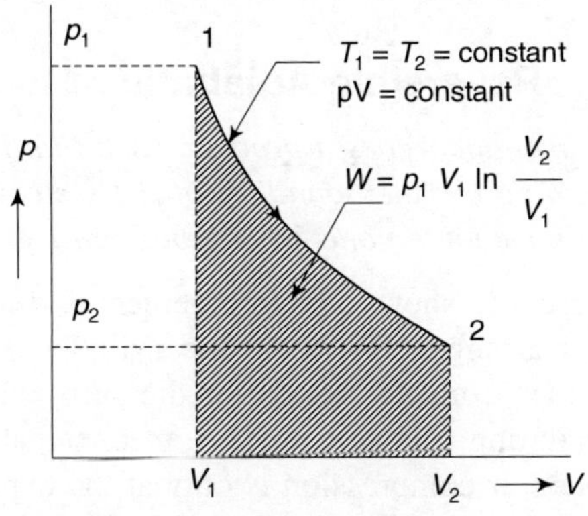

Fig. 5.14 Isothermal process on p–V diagram

Suppose, now the piston moves slightly downwards. This results in expansion of gas by volume dV and consequently the pressure and temperature of the system drop by the amount dp and dT, respectively. Since the temperature of the system has dropped by the dT hence the heat will flow from the surrounding till the system reaches the original temperature T_1. If the piston moves again slightly downwards, similar events would happen. Thus, the isothermal process will be possible only when the process is quasi-static and the system and the surrounding is maintained at the same temperature. The isothermal expansion process form state 1 to 2 is shown on p–V diagram in Fig. 5.14.

The isothermal process may also be realised from the concept of Boyle's law in a closed system. That is,

$$pV = \text{constant} = c$$

Since there is no change in the temperature during the process, there is no change in the internal energy of the system. Applying first law of thermodynamics

$$\delta Q - \delta W = dU$$

But $$dU = 0 \text{ or } U_1 = U_2 \text{ since temperature is constant} \quad ...(5.24)$$

∴ $$\delta Q = \delta W \quad ...(5.25)$$

But $$\delta W = pdV$$

$$Q_{1-2} = W_{1-2} = \int_1^2 pdV$$

or $$W_{1-2} = c\int_1^2 \frac{dV}{V} = c\left[\ln \frac{V_2}{V_1}\right] \quad ...(5.26)$$

But $$p_1 V_1 = p_2 V_2 = c$$

∴ $$W_{1-2} = p_1 V_1 \ln \frac{V_2}{V_1} = p_1 V_1 \ln \frac{p_1}{p_2} \quad ...(527)$$

5.4.4 Reversible Adiabatic or Isentropic Process

If a system undergoes a process, such that there is no heat transfer between the system and surrounding but the boundary of the system moves giving displacement work, the system is said to have undergone an adiabatic process.

Figure 5.15 shows and arrangement performing an adiabatic process. It consists of a piston cylinder arrangement where the cylinder is insulated from all sides preventing transfer of heat either from the system or the surrounding. Since there is no heat gain or loss by the system during adiabatic process, mechanical work done by or on the system during adiabatic expansion or compression is only at the expense of internal energy of the system. When the piston makes downward stroke, the adiabatic expansion process occurs and this is represented on p–V diagram as shown in Fig. 5.16.

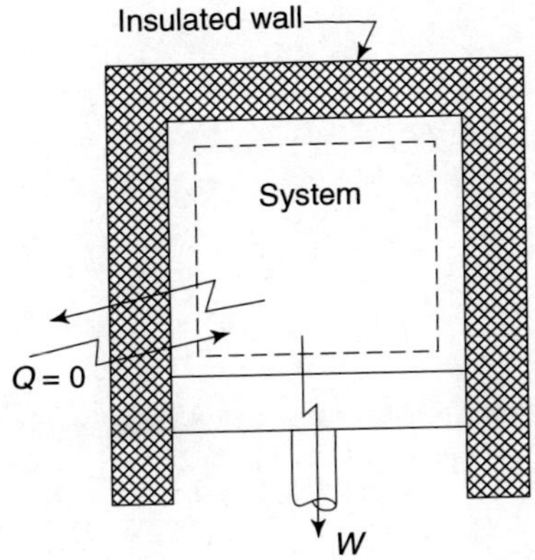

Fig. 5.15 Arrangement for adiabatic process

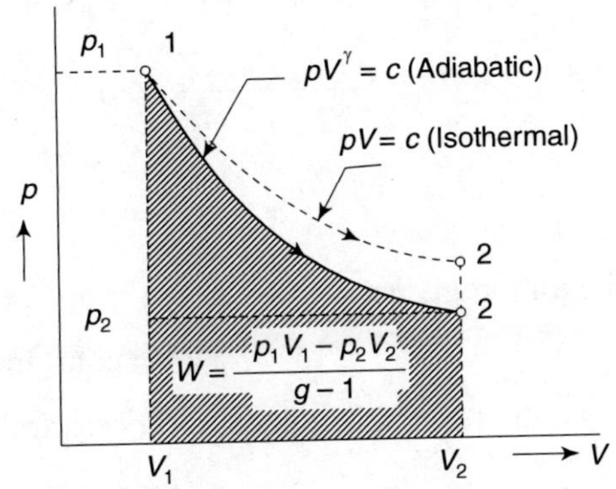

Fig. 5.16 Adiabatic process on p–V diagram

(A) Governing equation. Applying first law of thermodynamics

$$\delta Q - \delta W = dU$$

But $$dQ = 0$$

$$\delta W = - dU = - mc_v\, dT \quad \text{...(5.28)}$$

This equation suggests that the work done by the system is at the cost of internal energy. If the expansion or compression process is quasi-static (i.e., frictionless)

$$\delta W = pdV \quad \text{...(5.29)}$$

$\therefore$ $$pdV = - dU$$

or $$pdV + dU = 0 \quad \text{...(5.30)}$$

For perfect gas, $$pdV + mc_v\, dT = 0 \quad \text{...(5.31)}$$

But $$pV = mRT$$

So, after differenting $$pdV + Vdp = mRdT \quad \text{...(5.32)}$$

It is to be noted that here pdV, Vdp and RdT are all expressed in the same unit

or $$dT = \frac{pdV + Vdp}{mR}$$

Substituting the value of dT from the above equation into Eq. (5.31), we have

$\therefore$ $$c_v \left(\frac{pdV + Vdp}{R} \right) + pdV = 0$$

or $$c_v\, (pdV + Vdp) + pdV \times R = 0$$

But $$R = c_p - c_v$$

$\therefore$ $$c_v\, (pdV + Vdp) + (c_p - c_v)\, pdV = 0$$

Dividing both sides by $c_v\, p\,.\,V$, we get

$$\frac{dp}{P}+\frac{c_p}{c_v}\frac{dV}{V}=0$$

or

$$\frac{dp}{P}+\gamma\frac{dV}{V}=0$$

Integration gives: $\ln p+\gamma\ln V=\text{constant}=c$

$\therefore$ $pV^{\gamma}=\text{constant}$

Thus, the reversible adiabatic process follows the law

$$pV^{\gamma}=\text{constant} \qquad ...(5.33)$$

On per unit mass basis, $pv^{\gamma}=\text{constant}$

(B) *p-V-T* Relationship. Equation (5.33) may also be written in the following forms for state point 1 and 2.

(i) $$p_1V_1^{\gamma}=p_2V_2^{\gamma}$$

or $$\frac{V_2}{V_1}=\left(\frac{p_1}{p_2}\right)^{1/\gamma} \qquad ...(5.34)$$

For perfect gas, $$\frac{p_1V_1}{T_1}=\frac{p_2V_2}{T_2}$$

(ii) or $$\frac{T_2}{T_1}=\frac{p_2}{p_1}\cdot\frac{V_2}{V_1}=\left(\frac{V_1}{V_2}\right)^{\gamma-1} \qquad ...(5.35)$$

(iii) $$\frac{T_2}{T_1}=\left(\frac{p_2}{p_1}\right)^{(\gamma-1)/\gamma} \qquad ...(5.36)$$

(C) Work done. To find the work done during adiabatic process, Eq. (5.29) is helpful.

Work done during the process

$$W_{1-2}=\int_1^2 pdV$$

But $pV^{\gamma}=c$ or $p=c\cdot V^{-\gamma}$

$\therefore$ $$W_{1-2}=c\int_1^2 V^{-\gamma}\,dV=c\left[\frac{V_2^{-\gamma+1}-V_1^{-\gamma+1}}{1-\gamma}\right]$$

But $$c=p_1V_1^{\gamma}=p_2V_2^{\gamma}$$

or $$W_{1-2}=\frac{p_2V_2-p_1V_1}{1-\gamma}=\frac{p_1V_1-p_2V_2}{\gamma-1};\ \text{kJ} \qquad ...(5.37)$$

$$= mR\,\frac{(T_1 - T_2)}{(\gamma - 1)}\,;\ \text{kJ} \qquad ...(5.38)$$

On per unit mass basis, $w_{1-2} = \dfrac{p_1 v_1 - p_2 v_2}{\gamma - 1} = \dfrac{R\,(T_1 - T_2)}{(\gamma - 1)}$; kJ/kg

5.4.5 Polytropic Process

All processes described earlier, namely: (i) isochoric process, (ii) isobaric process, (iii) isothermal process, (iv) isentropic processes are quasistatic processes under ideal conditions, and may be described by a single equation, called the *polytropic law* which is

$$pV^n = \text{constant} \qquad ...(5.39)$$

It is essential that the value of index 'n' remains constant throughout in any particular process. It will be shown that the above processes are nothing but some particular forms of polytropic process, each having a definite value of index n. These are

(a) $n = 0$ for isobaric process

(b) $n = \infty$ for isochoric process

(c) $n = 1$ for isothermal

(d) $n = \gamma = \dfrac{c_p}{c_v}$ for isentropic process

(e) $n > 1 < \gamma$ most practical index, generally called *polytropic process*

The amount of the work done either for compression or expansion may also be compared for various processes from the p–V diagram because the *area under the diagram shows the work done.*

(A) Displacement or Mechanical Work. Mechanical work done during polytropic process can be calculated in a similar way as isentropic process.

$$W_{1-2} = c\int_1^2 pdV^{-n}\,dV = c\left(\frac{V_2^{1-n} - V_1^{1-n}}{1-n}\right)$$

But $\qquad c = p_1 V_1^n = p_2 V_2^n$

$$W_{1-2} = \frac{p_2 V_2^n \,.\, V_2^{1-n} - p_1 V_1^n V_1^{1-n}}{1-n} = \frac{p_1 V_1 - p_2 V_2}{n-1}\,;\ \text{kJ} \qquad ...(5.40)$$

For perfect gas, $p_1 V_1 = mRT$ and $p_2 V_2 = mRT_2$

$$W_{1-2} = mR\,\frac{(T_1 - T_2)}{(n-1)}\,;\ \text{kJ}$$

On per unit mass basis

$$w_{1-2} = \frac{p_1 v_1 - p_2 v_2}{(n-1)} \text{ kJ/kg} = \frac{R(T_1 - T_2)}{(n-1)} \quad ...(5.41)$$

(B) Change in Internal Energy

$$U_2 - U_1 = mc_v (T_2 - T_1) \text{ kJ} \quad ...(5.42)$$

or

$$u_2 - u_1 = c_v (T_2 - T_1), \text{ kJ/kg} \quad ...(5.43)$$

(C) Heat Transfer. Heat transfer during polytropic process may be calculated by applying the first law of thermodynamics.

$$Q_{1-2} = (U_2 - U_1) + \int_1^2 pdV$$

From gas laws, we have

$$Q_{1-2} = mc_v (T_2 - T_1) + \frac{mR(T_2 - T_1)}{1-n} = m\left(c_v + \frac{R}{1-n}\right)(T_2 - T_1) \quad ...(5.44)$$

$$= m\left(c_v + \frac{c_p - c_v}{1-n}\right)(T_2 - T_1) = m\frac{c_p - nc_v}{1-n}(T_2 - T_1) \text{ kJ} = mc_n (T_2 - T_1), \text{ kJ} \quad ...(5.45)$$

where $c_n = \dfrac{c_p - nc_v}{1-n}$

On per unit basis, $q_{1-2} = c_n (T_2 - T_1)$, kJ/kg ...(5.46)

Sometimes c_n is called the ***polytropic specific heat.***

Also $\dfrac{c_p}{c_v} = \gamma$, then

$$c_n = \frac{\gamma - n}{1-n} c_v$$

Putting the value of c_n in Eq. (5.45), we have

$$Q_{1-2} = m\left(\frac{\gamma - n}{1-n}\right) c_v (T_2 - T_1), \text{ kJ} \quad ...(5.47)$$

or

$$q_{1-2} = \frac{\gamma - n}{1-n} c_v (T_2 - T_1), \text{ kJ/kg} \quad ...(5.48)$$

But $c_v = R/(\gamma - 1)$

$\therefore$

$$Q_{1-2} = \left(\frac{\gamma - n}{\gamma - 1}\right) \times \text{Polytropic work done, kJ} \quad ...(5.49)$$

To analyse the difference in work done between polytropic and adiabatic expansion, we have

$$(W_{1-2})_p - (W_{1-2})_a = mR\,(T_2 - T_1)\left[\frac{1}{1-n} - \frac{1}{1-\gamma}\right]$$

$$= \frac{m\,(\gamma - n)\,R}{(\gamma-1)\,(1-n)}\,(T_2 - T_1) = c_n\,(T_2 - T_1) \qquad ...(5.50)$$

The difference in work between polytropic and adiabatic processes according to the first law of thermodynamics, must be equal to the transfer of heat to or from the system. Now, with the help of the polytropic specific heat c_n, the different processes may be distinguished as given in Table 5.1.

Table 5.1 Various forms of polytropic process, $pV^n = C$

(a) Isobaric process	$n = 0$	$c_n = c_p$
(b) Isochoric process	$n = +\infty$	$c_n = c_v$
(c) Isothermal process	$n = 1$	$c_n = \infty$
(d) Isentropic process	$n = \gamma$	$c_n = 0$
(e) Process in which	$1 > n > 0$	$c_n > c_p$
(f) Process in which	$\gamma > n > 1$	$c_n < 0$
(g) Process in which	$\infty > n > \gamma$	$c_v > c_n > 0$

Since $c_p > c_v$ for perfect, gas hence $\gamma > 1$. Therefore, the process in which $\gamma > n > 1$, the value of c_n will be negative. Since this process is very important from practical point of view, the *negative value of c_n* is explained further. In spite of the heat supplied to the system during such an expansion process, the temperature of the system falls. This happens because of the fact that the work is done by the gas due to the change of internal energy with a consequent decrease in temperature.

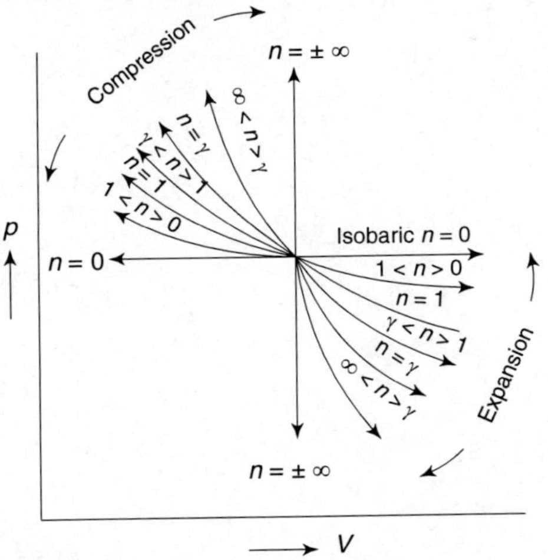

Fig. 5.17 Polytropic processes on p–V diagram

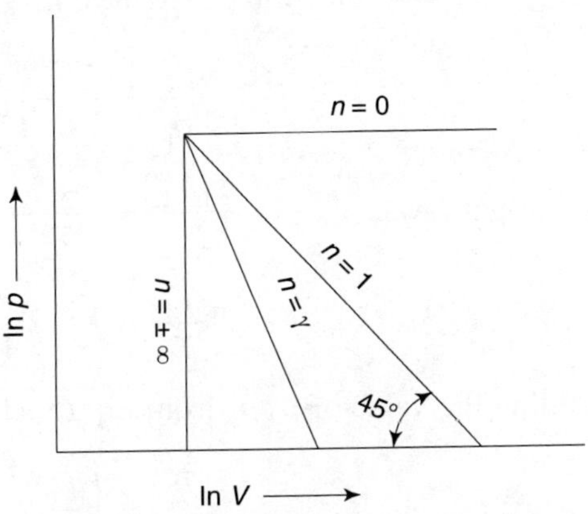

Fig. 5.18 Polytropic process on log graph

In order to compare these various forms of polytropic processes they are shown on p–V diagram of Fig 5.17. Assuming the starting point as (i) the fourth quadrant will contain expansion process and the second quadrant will contain compression. For a perfect gas, if various values of n are substituted in the polytropic equation, then we will get the corresponding equations. In Fig. 5.17, the value of index n increases from $n = 0$ at isobaric process to $n = \infty$ at isochoric process. *From practical point of view, the positions between isothermal and isentropic are very important since most of the processes belong to this category.*

A logarithmic graph of presure and volume can be drawn for the different polytropic processes and would make an interesting study.

Taking logarithmic of $pV^n = c$ gives

$$\ln p + n \ln V = \text{Constant}$$

The graph between $\ln p$ vs $\ln V$ is shown in Fig. 5.18. On this graph, each process takes the shape of a straight line. The isothermal line ($n = 1$) is inclined at 45° to the horizontal.

The various forms of polytropic process are summarised in Table 5.1

(D) Polytropic Index, *n*. The value of the polytropic index (n) during a change of state is found as follows.

We have
$$pV^n = \text{Constant}$$

$$p_1 V_1^n = p_2 V_2^n$$

Taking log both sides, we have

$$\ln p_1 + n \ln V_1 = \ln p_2 + n \ln V_2$$

or
$$n = \frac{\ln p_2 - \ln p_1}{\ln V_1 - \ln V_2} = \frac{\ln \dfrac{p_2}{p_1}}{\ln \dfrac{V_1}{V_2}} = \frac{\ln \dfrac{p_1}{p_2}}{\ln \dfrac{V_2}{V_1}} \qquad \text{...(5.51)}$$

(E) *p-V-T* Relationship. We have $p_1 V_1^n = p_2 V_2^n$

or
$$\frac{p_2}{p_1} = \left(\frac{V_1}{V_2}\right)^n \qquad \text{...(5.52)}$$

From gas laws,
$$\frac{p_1 V_1}{T_1} = \frac{p_2 V_2}{T_2} \qquad \text{...(5.53)}$$

or
$$\frac{T_2}{T_1} = \frac{p_2}{p_1} \times \frac{V_2}{V_1} \qquad \text{...(5.54)}$$

Putting the value of p_2/p_1 from Eq. (5.52) into Eq. (5.54)

$$\frac{T_2}{T_1} = \left(\frac{V_1}{V_2}\right)^{n-1} \qquad \text{...(5.55)}$$

We have,

$$\frac{V_1}{V_2} = \left(\frac{p_2}{p_1}\right)^{1/n}$$

$$\frac{T_2}{T_1} = \left(\frac{p_2}{p_1}\right)^{(n-1)/n} \qquad ...(5.56)$$

5.4.6 Constant Internal Energy or Free Expansion Process

As the name implies, the internal energy of the system during this process remains constant. The constant internal energy process is possible in a reversible process when at each stage heat transfer and work are equal but it is difficult to achieve. In order to analyse this process, consider a system shown in Fig. 5.19. It consists of an insulated tank in which there are two compartments separated by a partition wall. In one compartment, there is a perfect gas and other compartment is in state of perfect vacuum. Now, if the partition is removed or the partition wall is punctured, the system will expand and fill up the whole space as shown in Fig. 5.19 (b). The free expansion process is represented on *V–U* diagram as shown in Fig. 5.20. **The process is adiabatic but irreversible.**

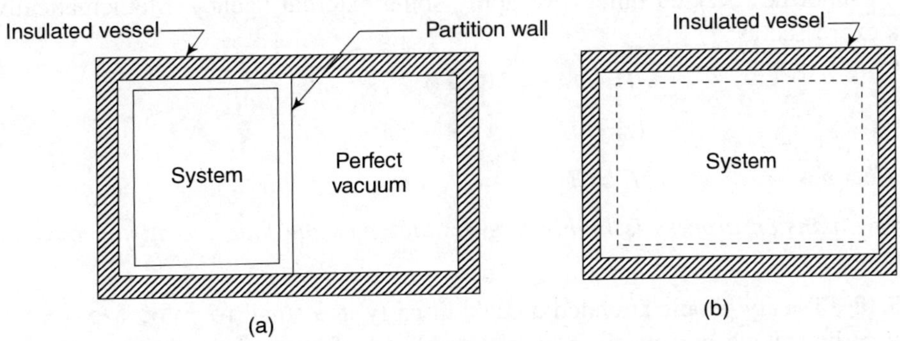

Fig. 5.19 Free expansion process on *V–U* diagram

Since the tank is insulated there is no heat transfer from or into the system. Thus, we may call this expansion adiabatic expansion but ths difference between this expansion and adiabatic expansion lies in the fact that *this expansion has no resistance to overcome whereas the adiabatic expansion is against the resistance offered by the piston.* Therefore, this expansion is called the ***free expansion process***. Since the expansion is free from any resistance, no work is done by the system. In a free expansion process, neither heat nor work is transferred. Consequently, according to first law of thermodynamic, *the internal energy of the system remains constant*, i.e., *temperature of the system remains constant.* Because the temperature of the system remains constant, this process is called **constant internal energy process**. It is also called the constant enthalpy process as temperature, i.e., enthalpy of the system remains constant. *This process cannot be called adiabatic or isothermal process due to fundamental difference between them.*

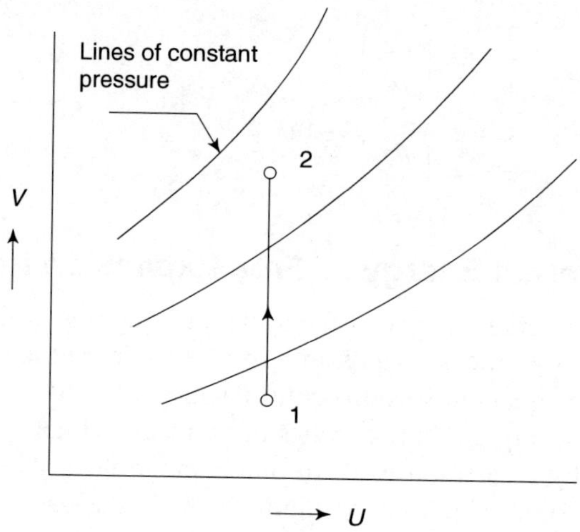

Fig. 5.20 *V–U* diagram of free expansion

It is also to be noted that adiabatic or isothermal process may be reversible but the free expansion cannot be reversed unless we apply some external agency. Mathematically, this process is expressed as

$$dU = \delta Q - \delta W = 0$$

or $$U_2 = U_1 \qquad ...(5.57)$$

For perfect gas, $$T_1 = T_2$$

The free expansion process is highly irreversible since the fluid is eddying continuously during the process.

Example 5.10 Twenty people attended a cocktail party in a small room of size 6 × 8 m and has a 3 m ceiling. Each person gives up about 400 kJ of heat per hour. Assuming that the room is completely sealed off and insulated, calculate the air temperature rise in 15 minutes. Assume c_v of air 0.718 kJ/kg and R = 0.287 kJ/kg K and each person occupies a volume of 0.07 m^3. Give your comment on the result.

Solution Refer to Fig 5.21.

We take the air in the room as a thermodycamic system and assume that the people are adding heat to the air in a constant volume process. The room volume is $V_{room} = 8 \times 6 \times 3 = 144$ m^3.

The volume of air is obtained by subtracting the volume occupied by the people

$$V_{air} = 144 - 0.07 \times 20 = 142.6 \text{ m}^3$$

Assuming that the constant volume heat addition starts at standard atmospheric conditions of 1.303 bar and 20°C, the mass of air is

$$m = \frac{pV}{RT} = \frac{1.013 \times 10^2 \times 142.6}{0.287 \times 293} = 171.78 \text{ kg}$$

From first law, $Q - W = \Delta U$

No work is done on the air, hence, $Q = \Delta U = 400 \times 20 = 8000$ kJ/h

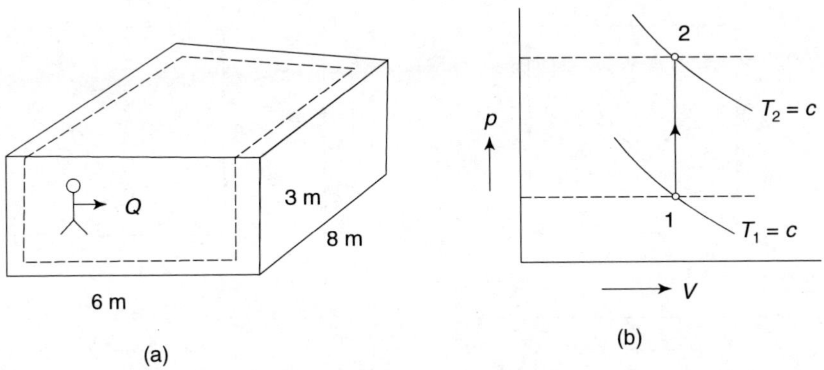

Fig. 5.21 Example 5.10

For 15 minutes period, $\Delta U = 8000 \times \frac{15}{60} = 2000$ kJ

But $\Delta U = mc_v \, \Delta T$

or $\Delta T = \frac{2000}{171.78 \times 0.718} = 16.31°\text{C}$ **Ans.**

Comment *This problem illustrates rather the obvious need for adequate ventialtion when a large number of people are confined in a small space.*

Example 5.11 A perfect gas changes from equilibrium state 1 to equilibrium state 2 by three possible quasi-static processes, namely, *A*, *B* and *C*. Path *A* consists of constant volume process followed by constant pressure; path *B* consists of a process of constant pressure followed by constant volume and process *C* is adiabatic process. The internal energy at state 1 is 750 J. The work transfer during paths *A*, *B* and *C* are 120, 200 and 150 J. Calculate the internal energy of the system at state 2 and the heat energy transferred during three processes.

Solution Refer to Fig. 5.22. Since internal energy is a property, it does not depend on path. It will be the same for all three cases. **Ans.**

For path *C*, $Q_{1-2} = 0$ **Ans.**

$$-\Delta U = W_{1-2}$$

Hencc, $U_1 - U_2 = 150$ J

or $U_2 = U_1 - 150 = 750 - 150 = 600$ J **Ans.**

Path A $\quad Q_{1-2} = (U_2 - U_1) + W_{1-2} = -150 + 120 = -30$ J **Ans.**

Path B, $\quad Q_{1-2} = (U_1 - U_1) + W_{1-2} = -150 + 200 = 50$ J **Ans.**

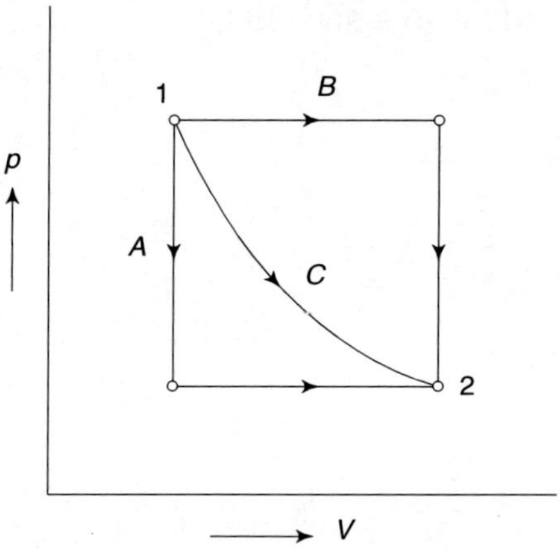

Fig. 5.22 Example 5.11

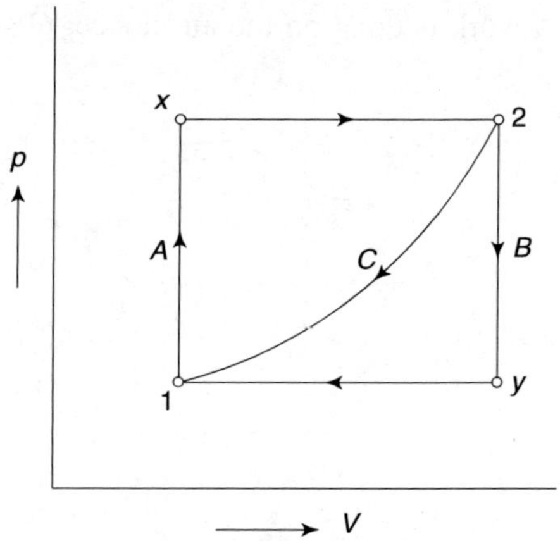

Fig. 5.23 Example 5.12

Example 5.12 A system undergoes a process from state 1 to state 2 by path A as shown in Fig. 5.23. During this process A, system does 200 kJ of work and receives 600 kJ of heat from the surrounding. Now, the system may be returned to the original state 1 along either path B and C. Along path B and C 80 and 120 kJ work is done on the system what will be heat exchange during 2–y if the system has an internal energy of zero at state point 1 and 360 kJ at state y.

Solution Refer to Fig. 5.23.

Process A, $\quad Q_{1-2} = (U_2 - U_1) + W_{1-2}$ or $\quad 600 = (U_2 - U_1) + 200$

or $\quad U_2 - U_1 = 400$ kJ

Process B, $\quad Q_{2-1} = (U_1 - U_2) + W_{2-1} = -400 + (-80) = -480$ kJ

The negative sign indicates the heat flows out from the system.

Process C, $\quad Q_{2-1} = (U_1 - U_2) + W_{2-1} = 400 + (-120) = -520$ kJ

Since $\quad U_1 = 0$, hence $U_2 = 400$ kJ, $U_y = 360$ kJ

Process 2–y. It is a constant volume process in which the work done is zero. Thus,

$$Q_{2-y} = (U_y - U_2) + W_{2-y} = (360 - 400) + 0 = -40 \text{ kJ} \quad \textbf{Ans.}$$

Negative sign shows that heat transfers from the system.

Example 5.13 During winter season, a room measuring $5 \times 6 \times 3$ m^3 is heated electrically from initial temperature of 0°C to 20°C. The air pressure inside the room is the same as that of surroundings and is equal to 74 cm of Hg. The pressure remains constant during the heating process. The heat capacity of furniture and walls is 32 kJ/K. The specific heat of air is 1005 kJ/kg K. Calculate the amount of electric energy needed for heating the room. How much air escapes through gaps, and windows during this period?

Solution Refer to Fig. 5.24. 76 cm Hg = 1.013 bar

$$\therefore \quad 74 \text{ cm of Hg} = \frac{1.013}{76} \times 74 = 0.9863 \text{ bar}$$

Neglecting the space occupied by furniture, the volume of air = 5 × 6 × 3 = 90 m^3.

$$\text{Mass of air before heating, } m_1 = \frac{p_1 V_1}{RT_1} = \frac{0.9863 \times 90 \times 10^2}{0.287 \times (273 + 0)} = 113.29 \text{ kg}$$

$$\text{Mass of air after heating, } m_2 = \frac{p_2 V_2}{RT_2} = \frac{0.9863 \times 10^2 \times 90}{0.287 \times (273 + 20)} = 105.56 \text{ kg}$$

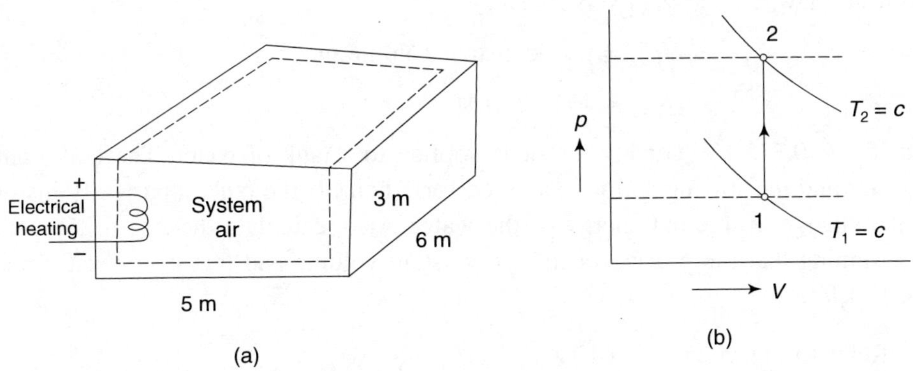

Fig. 5.24 Example 5.13

Thus, the amount of air escaping during heating = 113.29 – 105.56 = 7.73 kg **Ans.**

Heat required to raise the temperature of air from °C to 20°C

$$= Q_1 = mc_p (T_2 - T_1)$$

$$\text{During heating period, mean value of } m = \frac{113.29 + 105.56}{2} = 109.44 \text{ kg}$$

$$\therefore \quad Q_1 = 109.42 \times 1.005 \times 20 = 2199.74 \text{ kJ}$$

Heat required to heat walls and furniture = Q_2 = 32 × 20 = 640 kJ

Total heat supplied = 2199.74 + 640 = 2839.74 kJ

$$\text{The electric energy needed, } P_e = \frac{2839.74}{3600} = 0.788 \text{ kWh.}$$ **Ans.**

Example 5.14 In a vessel, 10 kg of O_2 is heated in a reversible, non-flow, constant volume process so that the pressure of O_2 is increased two times that of the initial value. The initial temperature is 20°C. Calculate (a) the final temperature, (b) the change in internal energy, (c) the change in enthalpy and (d) the heat transfer. Take R = 0.259 kJ/kg K and c_v = 0.652 kJ/kg K for oxygen.

Solution Refer to Fig. 5.9 and Fig. 5.10 for the constant volume process.

(a) $$T_2 = \left(\frac{p_2}{p_1}\right) T_1 = 2 \times (273 + 20) = 586 \text{ K}$$ **Ans.**

(b) Change in internal energy

$$\Delta U = mc_v (T_2 - T_1) = 10 \times 0.652 (586 - 293) = 1910.36 \text{ kJ}$$

(c) Change in enthalpy = $\Delta H = mc_p (T_2 - T_1) = m \times (R + c_v) (T_2 - T_1)$

$$= 10 \times (0.259 + 0.652) (586 - 293) = 2669.23 \text{ kJ}$$ **Ans.**

(d) From first law, $Q_{1-2} = \Delta U + W_{1-2}$

But $W_{1-2} = 0$ in constant volume process

$Q_{1-2} = 1910.36$ kJ **Ans.**

Example 5.15 A 0.735 kW stirring motor is applied to a tank of water. The tank contains 22 kg of water and the stirring action is applied for 1 hour. If the tank is perfectly insulated, calculate the change in internal energy of the water. Also, calculate the rise in temperature of water, assuming that the process occurs at constant volume and that c_v for water may be taken as 4.18 kJ/kg K.

Solution Refer to Fig. 5.25.

The increase in the internal energy of water is clearly the work added during the 1 hour time period

$$\Delta U = W_{1-2} = 0.735 \times 3600 = 2644 \text{ kJ}$$ **Ans.**

$$\Delta U = mc_v \, \Delta T$$

or $$\Delta T = \frac{\Delta U}{mc_v} = \frac{2646}{22 \times 4.18} = 28.77°\text{C}$$ **Ans.**

Example 5.16 A system receives 200 kJ of heat at constant volume process and rejects 220 kJ of heat at constant pressure and 40 kJ of work is done on the system. The system is brought to its original state by an adiabatic process. Calculate the adiabatic work. If the initial internal energy is 240 kJ, then calculate the value of internal energy at all points.

Solution Refer to Fig. 5.26.

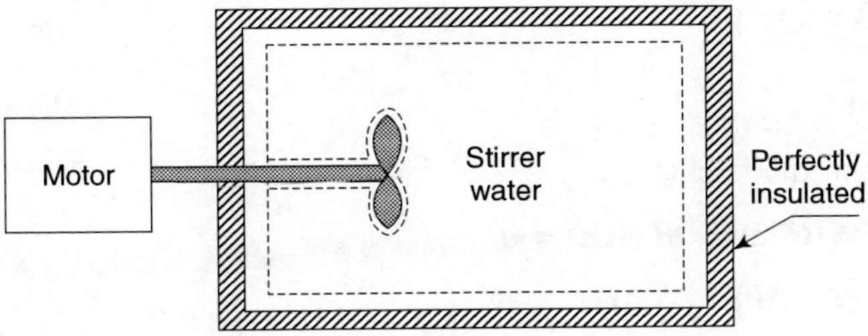

Fig. 5.25 Example 5.15

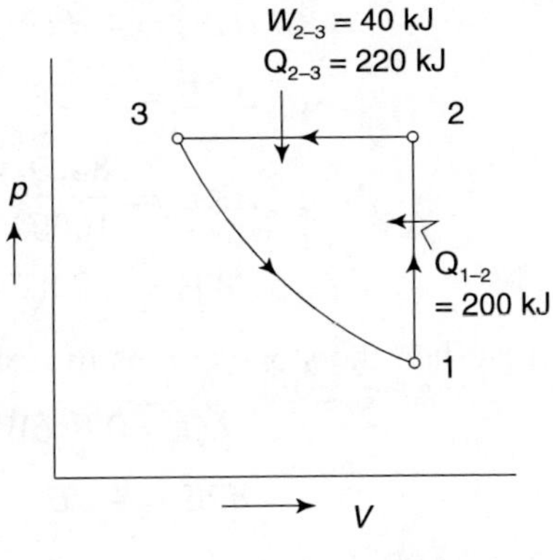

Fig. 5.26 Example 5.16

Process 1 – 2 : $Q_{1-2} = 200 = U_2 - U_1$

$\therefore$ $U_2 = 200 + 240 = 440$ kJ **Ans.**

Procee 2 – 3 : $Q_{2-3} = (U_3 - U_2) + W_{2-3}$ or $-220 = (U_3 - 440) - 40$

or $U_3 = 260$ kJ **Ans.**

Process 3 – 1 : $Q_{3-1} = 0$ because the process is adiabatic

$\therefore$ $\oint \delta Q = Q_{1-2} + Q_{2-3} + Q_{3-1} = 200 - 220 + 0 = -20$ kJ

According to First Law, $\oint \delta Q = \oint \delta W$ or $-20 = -40 + W_{3-1}$

or $W_{3-1} = 20$ kJ **Ans.**

Example 5.17 A system consists of stone having a mass of 10 kJ and a bucket containing 200 kg of water. Initially the stone and water are at the same temperature and then stone falls into water at a height of 90 m. Calculate (i) the change in internal energy, (ii) the change in kinetic energy, (iii) the change in potential energy, (iv) the heat transferred, and (v) the work transferred for the following cases:

(a) At the instant the stone is about to enter the water.

(b) Just after the stone comes to rest in the bucket.

(c) Heat is transferred to the surroundings at such a rate that stone and water remain at a temperature at which they were initially.

Solution (a) According to first law

$$\delta Q - \delta W = dU + d\,(\text{KE}) + d\,(\text{PE})$$

Let us assume that as the stone falls, there is no heat transfer to or from the stone, i.e., $\delta Q = 0$ and $\delta W = 0$.

Thus, $dU + d\,(\text{KE}) + d\,(\text{PE}) = 0$

But a cyclic non-flow energy equation gives $dU = 0$ **Ans.**

$$d\,(\text{KE}) = -\,d\,(\text{PE})$$

or $\Delta\text{KE} = -\Delta\text{PE} = -\text{mg}\,(z_2 - z_1) = -10 \times 9.81(0 - 90) = 8829 \text{ J}$ **Ans.**

or $\Delta\text{KE} = \dfrac{8829}{1000} = 8.829 \text{ kJ}$ **Ans.**

$\therefore$ $\Delta\text{PE} = -8.829 \text{ kJ}$ **Ans.**

(b) When the stone comes to rest in the bucket

$$\delta Q = 0, \quad \delta W = 0 \quad \text{and} \quad d\,(\text{KE}) = 0$$

$\therefore$ $d\text{PE} = -dU \quad \text{or} \quad \Delta\text{PE} = -\Delta U$

But $\Delta\text{PE} = mg\,(z_2 - z_1) = 10 \times 9.81\,(0 - 90) = -8829 \text{ J} = -8829 \text{ J} = -8.829 \text{ kJ}$ **Ans.**

$\Delta U = 8.829 \text{ kJ}$ **Ans.**

(c) When heat is transferred to the surroundings, after sometime, water and stone will acquire the same temperature as initial one.

Hence, $dU = 0, \quad d(\text{KE}) = 0 \quad \text{and} \quad \delta W = 0$ **Ans.**

From first law, $Q_{1-2} = \Delta\text{PE} = mg\,(z_2 - z_1) = -8.829 \text{ kJ}$ **Ans.**

Example 5.18 A cylinder is equipped with a spring loaded piston. The inside diameter of the cylinder is 8 cm and the stiffness of the spring is 150 N/cm of compression. Calculate the amount of heat added to the system so that the piston moves by 4 cm if the initial pressure, volume and temperature of air in the cylinder is 3 bar, 0.00006 m^3 and 27°C respectively.

Solution In this example, the external resistance to the piston is the spring force is shown in Fig. 5.27 and the gas between the piston and cylinder is at high pressure, hence work will be done by the gas in compressing the spring. The work done by the gas is

$$W = \frac{1}{2} Sx^2$$

where S = spring stifness, N/m

x = spring compression, m

The final volume of the gas in the cylinder

$$V_2 = V_1 + \frac{\pi}{4} D^2 \times x = 0.00006 + \frac{\pi}{4}\left(\frac{8}{100}\right)^2 \times \frac{4}{100} \text{ m}^3 = 0.00026 \text{ m}^3$$

$$W = \frac{1}{2} \times 150 \times 10^2 \times \left(\frac{4}{100}\right)^2 = 12 \text{ Nm} = 12 \text{ J}$$

Similarly, the final pressure of the gas is given by

$$p_2 = p_1 + \frac{S \cdot x}{(\pi/4)\,D^2} = 3 + \frac{150 \times 10^2 \times 4}{(\pi/4)\,(8/100)^2 \times 10^2 \times 10^2} = 4.1936 \text{ bar}$$

For closed system, i.e., non-flow process, $W_{1-2} = \dfrac{p_1 V_1 - p_2 V_2}{n-1}$

or $$12 = \frac{3\times10^5 \times 0.00006 - 4.1936\times10^5 \times 0.00026}{n-1}$$

or $$n = -\ 6.586$$

The negative value of n is very surprising but at the same time very interesting. This is because of the fact that as the heat is added, the compression of the spring takes place and both pressure and volume increase simultaneously.

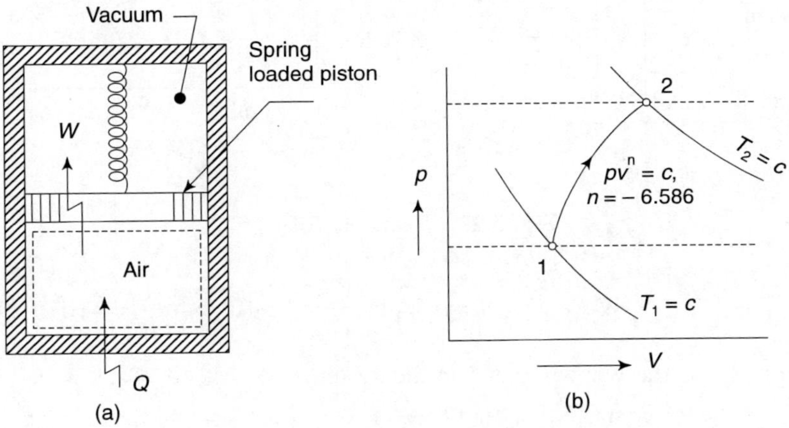

Fig. 5.27 Example 5.18

$$Q_{1-2} = mc_n\ (T_2 - T_1)$$

$$m = \frac{p_1 V_1}{RT_1} = \frac{2\times10^5 \times 0.00006}{287\times300} = 0.000139 \text{ kJ}$$

and $$c_n = c_v\left(\frac{\gamma - n}{1-n}\right) = \frac{0.718\ (1.4 + 6.586)}{1 + 6.586} = 0.7558$$

Also $$T_2 = T_1\left(\frac{p_2}{p_1}\right)^{(n-1)/n} = 300\left(\frac{4.1936}{3}\right)^{(6.586-1)/(-6.586)} = 441.23 \text{ K}$$

$$Q_{1-2} = 0.000139 \times 0.7558 \times (441.23 - 300) = 0.01483 \text{ kJ}$$ **Ans.**

Example 5.19 Calculate the final temperature, pressure, work done and heat transfer if the fluid is compressed reversibly from volume 6 m^3 to 1 m^3 in a cylinder. Comment on the result when the initial temeprature and pressure of fluid as 20°C and 1 bar. The index of compression may be assumed as 1.3 and 1.4 respectively. Take c_p = 1.005 and c_v = 0.718 kJ/ kg K and R = 0.287 kJ/kg K.

Solution Refer to Fig. 5.28.

(a) For $n = 1$: $p_1 V_1 = p_1 V_2$ or $1 \times 10^4 \times 6 = p_2 \times 10^4 \times 1$

or $p_2 = 6$ bar and $T_1 = T_2 = 20 + 273 = 293$ K **Ans.**

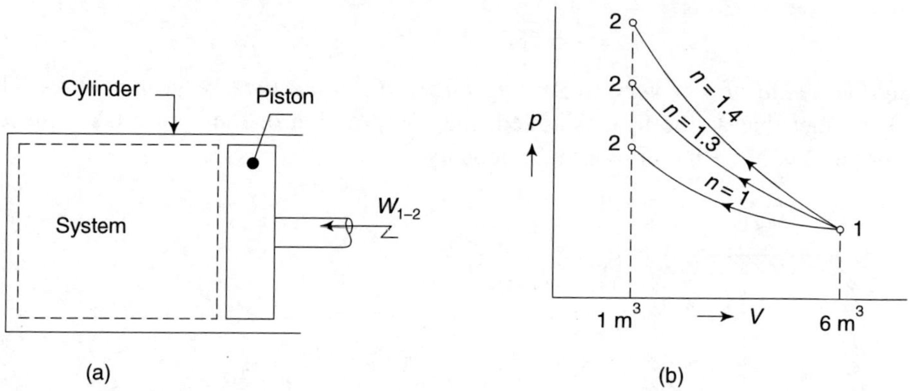

Fig. 5.28 Example 5.19

$$W_{1-2} = Q_{1-2} = p_1 V_1 \ln \frac{V_2}{V_1} = 1 \times 10^2 \; 6 \ln \frac{1}{6} = -1075.29 \text{ kJ} \quad \textbf{Ans.}$$

Here –ve sign shows the work is done on the system,

(b) For $n = 1.3$: $p_1 V_1^{1.3} = p_2 V_2^{1.3}$

or $1 \times 10^2 \times 6^{1.3} = p_2 \times 10^2 \times 11^{1.3}$

or $p_2 = 6^{1.3} = 10.27$ bar

Also, $$T_2 = T_1 \left(\frac{V_1}{V_2}\right)^{n-1} = 293 \times 6^{0.3} = 501.54 \text{ K} \quad \textbf{Ans.}$$

or $$W_{1-2} = \frac{p_1 V_1 - p_2 V_2}{n-1} = \frac{1 \times 10^2 \times 6 - 10.27 \times 10^2 \times 1}{0.3} = -1423.33 \text{ kJ} \quad \textbf{Ans.}$$

$$Q_{1-2} = mc_n (T_2 - T_1) = \frac{m(c_p - nc_v)}{1-n}(T_2 - T_1)$$

But $$m = \frac{p_1 V_1}{RT_1} = \frac{1 \times 10^2 \times 6}{0.287 \times 293} = 7.135 \text{ kg}$$

$\therefore$ $$Q_{1-2} = \frac{7.135\,(1.005 - 1.3 \times 0.718)}{1 - 1.3} \times (501.54 - 293) = -356.60 \text{ kJ} \quad \textbf{Ans.}$$

(c) For $n = 1.4$, the process is reversible adiabatic.

$$p_1 V_1^{1.4} = p_2 V_2^{1.4} \quad \text{or} \quad 1 \times 10^2 \times 6^{1.4} = p_2 \times 10^2 \times 1^{1.4}$$

or
$$p_2 = 6^{1.4} = 12.286 \text{ bar}$$

Also,
$$T_2 = T_1 \left(\frac{V_2}{V_1}\right)^{\gamma - 1} = 293 \times 6^{0.4} = 599.968 \text{ K}$$

$$W_{1-2} = \frac{p_1 V_1 - p_2 V_2}{\gamma - 1} = \frac{1 \times 10^2 \times 6 - 12.286 \times 10^2 \times 1}{0.4} = -1571.5 \text{ kJ} \quad \textbf{Ans.}$$

In adiabatic process, $Q_{1-2} = 0$ **Ans.**

Example 5.20 A system containing 0.2 m³ of air at a pressure of 4 bar and 160°C expands isentropically to pressure of 1.06 bar and after this the gas is heated at the constant pressure till the enthalpy increases by 65 kJ. Calculate the work done. Now, imagine that these processes are replaced by a single reversible polytropic process producing the same work between initial and final state; find the index of expansion in this case, c_p of air = 1.005 kJ/kg K.

Solution Refer to Fig. 5.29.

For isentropic process,
$$V_2 = V_1 \left(\frac{p_1}{p_2}\right)^{1/\gamma} = 0.2\left(\frac{4}{1.06}\right)^{1/1.4} = 0.5164 \text{ m}^3$$

Also,
$$T_2 = T_1 \left\{\frac{p_2}{p_1}\right\}^{(\gamma-1)/\gamma} = 433\left(\frac{1.06}{4}\right)^{0.4/1.4} = 296.27 \text{ K}$$

$$m = \frac{p_1 V_1}{R T_1} = \frac{4 \times 10^2 \times 0.2}{0.287 \times 433} = 0.6437 \text{ kg}$$

Let Q_{2-3} be the heat addition at constant pressure.

∴
$$Q_{2-3} = m c_p (T_3 - T_2) \quad \text{or} \quad 65 = 0.6437 \times 1.005\,(T_3 - 296.27)$$

or
$$T_3 = 396.74 \text{ K}$$

Also,
$$V_3 = V_2 \frac{T_3}{T_2} = 0.5164 \times \frac{396.74}{296.27} = 0.6915 \text{ m}^3$$

The total work done by these processes 1–2–3 is given by

$$W_{1-2} + W_{2-3} = \frac{p_1 V_1 - p_2 V_2}{\gamma - 1} + p_2 (V_3 - V_2)$$

$$= \frac{4 \times 10^2 \times 0.2 - 1.06 \times 10^2 \times 0.5164}{0.4} + 1.06 \times 10^2 (0.6915 - 0.5164)$$

$$= 81.71 \text{ kJ}$$ **Ans.**

According to the question, $W_{1-2} + W_{2-3} = W_{1-3} = \dfrac{p_1 V_1 - p_3 V_3}{n - 1}$

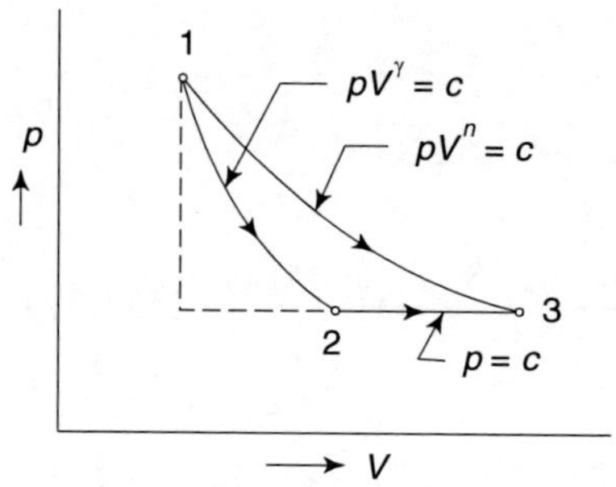

Fig. 5.29 Example 5.20

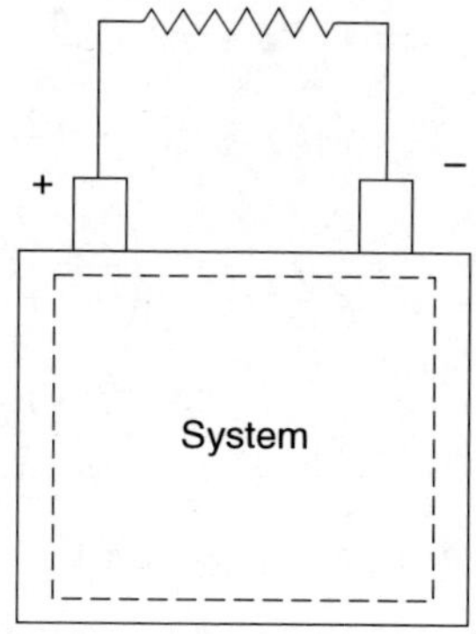

Fig. 5.30 Example 5.21

or $$81.71 = \frac{4 \times 10^2 \times 0.2 - 1.06 \times 10^2 \times 0.6915}{}$$

or $$n = 1.082$$ **Ans.**

Example 5.21 In order to test a thermally isolated battery, it is discharged at atmospheric pressure and constant volume. During a 1 hour test, it has been found that a current of 60 amp and 3 volts flows while the temperature increases from 20°C to 35°C. Calculate the change in internal energy of the cell during the period of operation.

Solution Refer to Fig. 5.30.

$$W_{1-2} = -\int_{q_1}^{q_2} E\, dq = \int_{\theta_1}^{\theta_2} EI\, d\theta = EI\, \Delta\theta$$

$$= 3 \times 60 \times 3600 = 648 \times 10^3 \text{ J}$$

Since heat transfer due to thermally insulated battery is zero, i.e., $Q_{1-2} = 0$

or $$\Delta U = -648 \times 103 \text{ J}$$ **Ans.**

Example 5.22 A cylinder closed at its two ends is divided into two compartments by means of free piston. Both piston and cylinder are perfectly insulated. One compartment contains air and the second one contains nitrogen. Initially, both air and nitrogen have pressure of 1 bar, volume 0.02 m^3 and temperature 10°C. Heat is supplied from the side of air compartment by an electric heater till the volume of nitrogen is reduced to 0.01 m^3. Find (a) the final temperature of air and (b) the heat supplied to air. Assume c_p for air = 1.005 kJ/kg K, R_{air} = 0.287 kJ/kg K and γ = 14 for nitrogen.

Solution. Refer to Fig. 5.31.

Due to free movement of the piston, the final pressure in two compartments must be the same, After heating, the final volume of air allowing nitrogen volume of 0.01 m^3

$$= (0.02 + 0.02) - 0.01 = 0.03 \text{ m}^3$$

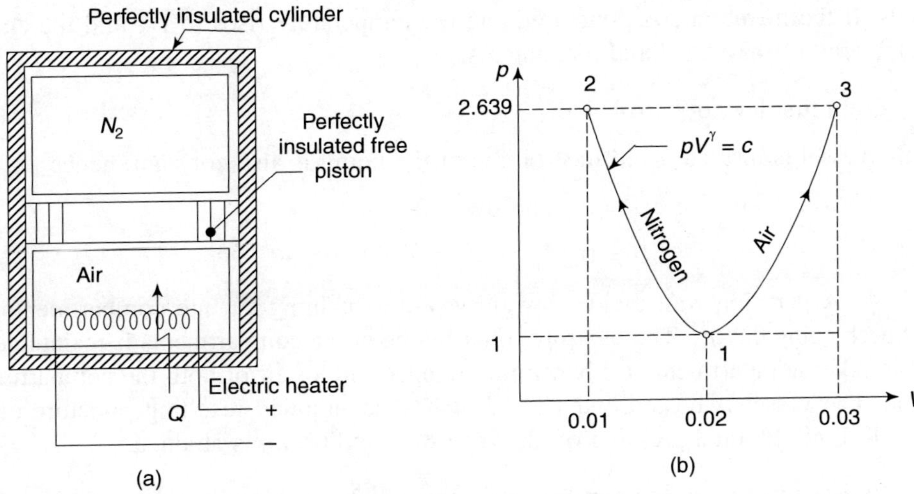

Fig. 5.31 Example 5.22

On heating the air will expand and do work on nitrogen which is being compressed adiabatically.

$$p_1 V_1^{\gamma} = p_2 V_2^{\gamma} \quad \text{or} \quad p_2 = 1\left(\frac{0.02}{0.01}\right)^{1.4} = 2.639 \text{ bar} = p_3$$

$$(W_{1-2})_{N_2} = \frac{p_1 V_1 - p_3 V_3}{\gamma - 1} = \frac{1\times10^2 \times 0.02 - 2.639\times10^2 \times 0.01}{0.4} = -\ 1.5975 \text{ kJ}$$

For air, the relation is $\dfrac{p_1 V_1}{T_1} = \dfrac{p_3 V_3}{T_3}$

(a) $$T_3 = \frac{p_2 V_3 T_1}{p_1 V_1} = \frac{2.639\times10^2 \times 0.03\times 283}{1\times10^2 \times 0.02} = 1120.68 \text{ K}$$

The change in internal energy of the air = $(\Delta U)_A = mc_v\,(T_2 - T_1)$

But $$m = \frac{p_1 V_1}{RT_1} = \frac{1\times10^2 \times 0.02}{0.287\times 283} = 0.02462 \text{ kg}$$

$\therefore$ $$(\Delta U)_A = 0.02462 \times 0.716\,(1120.68 - 283) = 14.766 \text{ kJ}$$

Since the initial conditions are the same and piston is of free type the work done by the air is equal to the work done on nitrogen. Hence, for a closed system, the heat added to air is

(b) $$Q_{1-2} = \Delta U + W_{1-2} = 14.766 - 1.5975 = 13.1085 \text{ kJ}$$ **Ans.**

Example 5.23 A membrane divides a closed isolated vessel into two compartments. One compartment contains water and the other nitric acid. The initial temperature T_1, is same in both liquids. If the membrane is punctured and the temperature rises to T_2, find the change in internal energy of the vessel and its contents.

Solution From first law, $\delta Q - \delta W = dU$

Since the vessel is insulated and the volume of the liquid is also constant, hence

$$\delta Q = 0 \quad \text{and } \delta W = 0$$

$$\delta U = 0$$ **Ans.**

Example 5.24 A partition wall divides a rigid vessel containing air into compartments, the volume of each being 0.2 m^3. The pressure of air in one of the compartments is maintained at 2 bar and in other compartment at 4 bar. Initial temperature of air of both the compartments is the same. The vessel is heated and 125 kJ of heat is supplied resulting puncture of the partition wall. Find the final pressure of air when the equilibrium is attained.

Solution Before puncture of the partition wall, the initial conditions are shown in Fig. 5.32.

$$m_1 = \frac{p_1 V_1}{RT_1} = \frac{2 \times 10^2 \times 0.2}{0.287 \times T_1} = \frac{139.37}{T_1} \text{ kg}$$

$$m_2 = \frac{p_2 V_2}{RT_2} = \frac{4 \times 10^2 \times 0.2}{0.287 \times T_1} = \frac{278.74}{T_1} \text{ kg}$$

$\therefore$ $$\text{Total mass of air} = m = \frac{139.37 + 278.74}{T_1} = \frac{418.11}{T_1} \text{ kg}$$

$$\text{Heat added} = Q = mc_v (T_{\text{final}} - T_1) = 125 \text{ kJ}$$

or $$T_{\text{final}} = \frac{125\, T_1}{418.11 \times 0.718} + T_1 = 1.4163\, T_1$$ **Ans.**

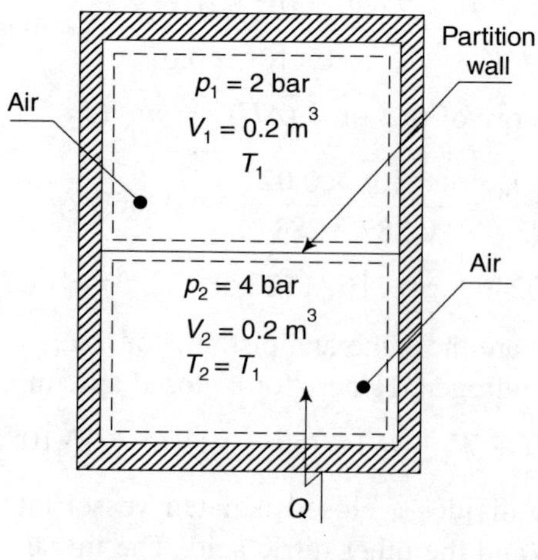

Fig. 5.32 Example 5.24

Average pressure when the partition wall is punctured = $\dfrac{2+4}{2}$ = 3 bar

It is to be noted that the vessel is rigid, pressure are directly proportional to temperature.

$$\frac{p}{3} = \frac{1.4163\, T_1}{T_1}$$

or $$p = 4.2489 \text{ bar}$$ **Ans.**

Example 5.25 Prove that the formula $T^b\, V^{a-b}\, e^{kT}$ = constant for the adiabatic expansion of the gas if $c_p = a + kT$ and $c_v = b + kT$, where a, b and k are constants and T in K.

Calculate the work done if 2 kg of gas originally occupying 0.2 m^3 at 40 bar expands adiabatically until the temperature is 260°C, given that $a = 0.948$, $b = 0.656$ and $k = 0.001$.

Solution Due to adiabatic process, $\delta Q = 0$. From the first law.

$$\delta Q = dU + \delta W$$

Based on 1 kg of gas, $0 = c_v\, dT + pdv = (b + kT)\, dT + \dfrac{RT}{v} dv$

$$R = c_p - c_v = a - b$$

$$0 = b\frac{dT}{T} + kdT + (a-b)\frac{dv}{v}$$

Integrating this, we have: Constant = $b \ln T + kT + (a - b) \ln v$

or $$T^b\, v^{a-b}\, e^{kT} = \text{Constant}$$ **Ans.**

For adiabatic process, $W_{1-2} = \Delta U = -\, mc_v\,(T_2 - T_1)$

Non $$p_1\, V_1 = mRT_1$$

Here $$R = a - b = (0.948 - 0.656) = 0.292 \text{ kJ/kg K}$$

$$40 \times 10^2 \times 0.2 = 2 \times 0.292 \times T_1$$

or $$T_1 = 1369.86 \text{ K} \quad \text{and} \quad T_2 = 533 \text{ K}$$

$\therefore$ $$W_{1-2} = -\int_{1369.86}^{533} 2 \times (0.656 + 0.0001\, T)\, dT$$

$$= 2\left[0.656\, T + \frac{0.0001\, T^2}{2}\right]_{533}^{1369.86} = 1257.2 \text{ kJ}$$ **Ans.**

Example 5.26 Monatomic ideal gas at 12 bar is contained in a tank on capacity 0.028 m^3. The tank is placed in a room where the pressure and temperature are 1 bar and 20°C respectively. Find the amount of heat transferred, the work done and the change in internal energy for each of the following processes:

(i) Leakage occurs which may be considered to be sufficiently slow so that the tank and the contents remain at a constant temperature.

(ii) The slow leakage process may be considered reversible adiabatic by assuming the tank thermally insulated.

(iii) The tank is thermally insulated and the leakage process is to be considered. However, the leakage is sudden and the tank pressure instantaneously drops to the room pressure.

(iv) The leakage process may be assumed as polytropic process having n =1.3

Assume $c_{p\,\text{mol}}$ = 20 kJ/kg mole K, $c_{v\,\text{mol}}$ = 12 kJ/kg mole K

Solution (i) Since the leakage process occurs due to the difference in pressure, hence the leakage will stop when the tank pressure is equal to room pressure, i.e., 1 bar. Thus, the mass

$$m = \frac{p_2 V_2}{RT_2} = \frac{1\times 10^2 \times 0.028}{R \times 293} \text{ kg} = \frac{0.009556}{R} \text{ kg}$$

Since during the process temperature remains constant, hence the work done

$$W_{1-2} = p_2 V_2 \ln \frac{p_1}{p_2} = mRT_2 \ln \frac{p_1}{p_2} = \frac{0.009556}{R} R \times 293 \ \ln 12 = 6.9575 \text{ kJ} \quad \textbf{Ans.}$$

Since the temperature remains constant, hence $dU = 0$

$$\delta Q = \delta W + dU$$

or

$$\delta Q = \delta W = 6.9575 \text{ kJ}$$

(ii)

$$\gamma = \frac{c_p}{c_v} = 20/12 \quad \textbf{Ans.}$$

$$T_2 = T_1 \left(\frac{p_2}{p_1}\right)^{(\gamma-1)/\gamma} = 293 \left(\frac{1}{12}\right)^{[20/12-1]/(20/12)} = 108.44 \text{ K}$$

$$\text{Work done, } W = \frac{mR(T_1 - T_2)}{\gamma - 1}$$

$$\text{But } m \text{ at the final condition} = m = \frac{1\times 10^2 \times 0.028}{R \times 108.44} \text{ kg}$$

$$\text{Work done} = \frac{1\times 10^2 \times 0.028}{R\times 108.44} \times \frac{R}{\frac{20}{12} - 1} (293 - 108.44) = 7.14 \text{ kJ} \quad \textbf{Ans.}$$

Since the process is adiabatic, there is no interaction of heat

$$Q = 0$$

From first law, the change in internal energy = $\Delta U = -\ W = -\ 7.14$ kJ **Ans.**

(iii) If p_a is constant external pressure acting during the process, then work done

$$W_{1-2} = p_a \times \text{change in volume} = p_a \left(\frac{RT_2}{p_2} - \frac{RT_1}{p_1} \right)$$

As the process is adiabatic, work is numerically equal to change of internal energy, thus

$$W_{1-2} = p_a \left(\frac{RT_2}{p_2} - \frac{RT_1}{p_1} \right) = -\Delta U = -c_v (T_2 - T_1)$$

or
$$\left(\frac{p_a T_2}{p_2 T_1} - \frac{p_a}{p_1} \right) = \frac{c_v}{R} - \frac{c_v T_2}{RT_1}$$

or
$$\frac{T_2}{T_1} = \frac{(p_a/p_1) + (c_v/R)}{(p_a/p_2) + (c_v/R)} = \frac{1/12 + 12/8}{1 + 12/8} = 0.6333$$

Thus,
$$T_2 = 0.6333 \times 293 = 185.56 \text{ K}$$

Work done $= \dfrac{mR(T_1 - T_2)}{\gamma - 1} = \dfrac{1 \times 10^2 \times 0.028}{R \times 185.56} \times R \dfrac{(293 - 185.56)}{(20/12) - 1} = 2.434$ kJ **Ans.**

Change in internal energy $= \Delta U = -W = -2.434$ kJ **Ans.**

(iv)
$$T_2 = T_1 \left(\frac{p_2}{p_1} \right)^{(n-1)/n} = \left(\frac{1}{12} \right)^{0.3/1.3} = 165.129 \text{ K}$$

$$\text{Work done} = \frac{m\,R(T_1 - T_2)}{n-1} = \frac{1 \times 10^2 \times 0.028}{R \times 165.129} R \frac{(293 - 165.129)}{0.3} = 7.227 \text{ kJ}$$

$\Delta U = mc_v (T_2 - T_1) = \dfrac{1 \times 10^2 \times 0.028}{8 \times 165.129} \times 12 \times (165.129 - 293) = -3.248$ kJ **Ans.**

From first law, heat supplied $= W + \Delta U = 7.227 - 3.248 = 3.979$ kJ **Ans.**

Example 5.27 A piston cylinder arrangement contains a gas of amount 2 kg and expands from an initial pressure of 6 bar and volume of 0.2 m^3 to a final pressure of 1 bar. The process of expansion follows the law $pV^{1.2}$ = constant. The internal energy is related by $u = 240\,pv + 1840$.

where u is in kJ/kg, p is in bar, v is in m^3/kg

(a) If the expansion is frictionless, find the changes in heat transfer, internal energy and work transfer.

(b) If during expansion 33 kJ of heat is transferred to the system, find the work done. Explain the difference in work for the above two cases.

Solution Refer to Fig. 5.33. We have $u = 240\,pv + 1840$. Putting the initial and final conditions

$$u_1 = 240\,p_1\,v_1 + 1840 \quad \text{and} \quad u_2 = 240\,p_2\,v_2 + 1840$$

$$\therefore \qquad u_2 - u_1 = 240\ (p_2\, v_2 - p_1\, v_1)$$

$$U_2 - U_1 = m \times 240\ (p_2\, v_2 - p_1\, v_1) = 240\ (p_2\, V_2 - p_1\, V_1)$$

But
$$V_2 = V_1 \left(\frac{p_1}{p_2}\right)^{1/n} = 0.2\ (6)^{(1/1.2)} = 0.8902\ \text{m}^3$$

$$\therefore \qquad U_2 - U_1 = 240\ (1 \times 0.8902 - 6 \times 0.2) = -\ 74.35\ \text{kJ} \qquad \textbf{Ans.}$$

$$\text{Work done} = \frac{p_1\, V_1 - p_2\, V_2}{n-1} = \frac{6 \times 10^2 \times 0.2 - 1 \times 10^2 \times 0.8902}{0.2} = 154.9\ \text{kJ} \qquad \textbf{Ans.}$$

$$Q_{1-2} = \Delta U - W_{1-2} = -\ 74.35 + 154.9 = 80.55\ \text{kJ} \qquad \textbf{Ans.}$$

(b) Since the initial and final conditions are the same hence the change in internal energy is the same, i.e.,

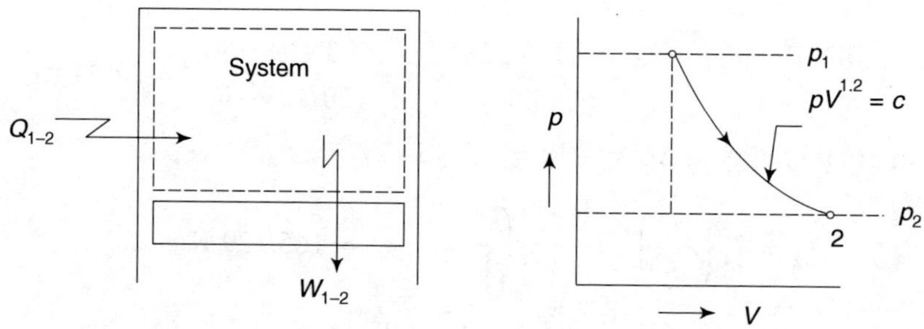

Fig. 5.33 Example 5.27

$$\Delta U = -\ 74.35\ \text{kJ} \qquad \textbf{Ans.}$$

$$\therefore \qquad W_{1-2} = Q_{1-2} - \Delta U = 33 - (-\ 74.35) = 107.35\ \text{kJ} \qquad \textbf{Ans.}$$

In the case (b) $\int pdV$ is the same as the case (a) since path is the same. But the work input in part (b) is not equal $\int pdV$. This simply means that process (b) is not a frictionless process.

Example 5.28 2 m^3 of hydrogen at a pressure of 1 bar and 20°C is compressed isentropically to 4 bar. The same gas is expanded and restored to original volume by constant volume heat rejection process. Determine (a) pressure, volume and temperature at each end of operation (b) the heat added during the isothermal process; (c) the heat rejected during constant volume process, and (d) change in internal energy during each process. Assume R = 4.206 kJ/kg K, c_p = 14.25 kJ/kg K.

Solution Refer to Fig. 5.34. Since the value of c_v and γ are not given, we have to calculate these.

We have
$$c_p - c_v = R = (14.25 - c_v) = 4.206 \text{ or } c_v = 10.044\ \text{kJ/kg K}$$

$\therefore$
$$\gamma = \frac{c_p}{c_v} = \frac{14.25}{10.044} = 1.4187$$

Isentropic compression process 1 – 2:

$$V_2 = V_1 \left(\frac{p_1}{p_2}\right)^{1/\gamma} = 2 \times \left(\frac{1}{4}\right)^{1/1.4187} = 0.7527 \text{ m}^3 \qquad \textbf{Ans.}$$

$$T_2 = T_1 \left(\frac{p_2}{p_1}\right)^{(\gamma-1)/\gamma} = 293\,(4)^{0.4187/1.4187} = 441.11 \text{ K} \qquad \textbf{Ans.}$$

Given $p_1 = 1$ bar and $p_2 = 4$ bar **Ans.**

$$\text{Work done} = \frac{p_1 V_1 - p_2 V_2}{\gamma - 1} = \frac{1\times10^2 \times 1 - 4 \times 10^2 \times 0.7527}{0.4187} = -\,480.24 \text{ kJ} \qquad \textbf{Ans.}$$

Since the process is isentropic, $\delta Q = 0$

Applying first law : $\delta Q - \delta W = dU$

or $-\,W_{1-2} = \Delta U = 480.24$ kJ **Ans.**

Isothermal expansion 2 – 3 : $p_2\, V_2 = p_3\, V_3 = p_3\, V_1$ since $V_3 = V_1$

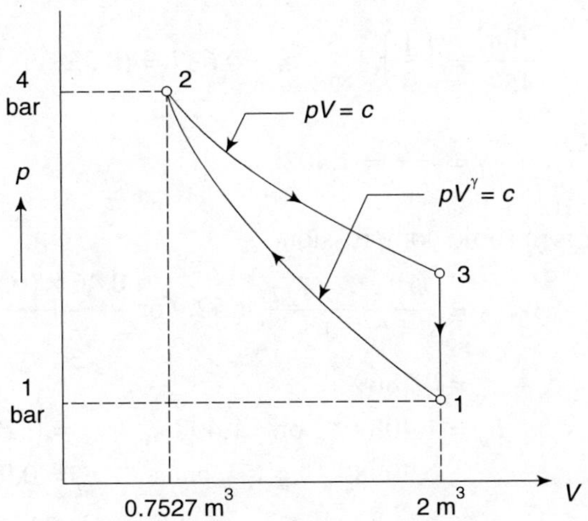

Fig. 5.34 Example 5.28

or
$$p_3 = p_2 \times \frac{V_2}{V_1} = 4 \times 10^2 \times \frac{0.7527}{2} = 150.54 \text{ kN/m}^2$$

or
$$p_3 = 150.64 \text{ kN/m}^2 = 1.5054 \qquad \textbf{Ans.}$$

Now
$$V_3 = V_1 = 2 \text{ m}^3$$

and $$T_2 = T_3 = 441.11 \text{ K}$$ **Ans.**

In isothermal process, the change in internal energy, $\Delta U = 0$ **Ans.**

Hence, from first law, $Q_{1-2} = W_{1-2}$

$$\therefore \quad Q_{1-2} = p_2 V_2 \ln \frac{V_3}{V_2} = 4 \times 10^2 \times 0.7527 \ln \frac{2}{0.7527} = 294.22 \text{ kJ}$$

Heat rejected at constant volume = $Q_{3-1} = mc_v (T_1 - T_3)$

But $$m = \frac{p_1 V_1}{RT_1} = \frac{1 \times 10^2 \times 2}{4.206 \times 193} = 0.1622 \text{ kg}$$

$$\therefore \quad Q_{3-1} = 0.1622 \times 4.206 (293 - 441.11) = -101.94 \text{ kJ}$$ **Ans.**

During constant volume process, $\delta W = 0$,

Hence, from first law, $Q_{1-2} = \Delta U = -101.04$ kJ **Ans.**

Example 5.29 Calculate c_p and c_v of air if 0.46 kg of air at 179°C expands isentropically to three times its original volume. The work done by the air is 52.9 kJ. Take final temperature 288 K.

Solution For isentropic process, $\left(\frac{T_2}{T_1}\right) = \left(\frac{V_1}{V_2}\right)^{\gamma-1}$

or $$\frac{288}{452} = \left(\frac{1}{3}\right)^{\gamma-1} \quad \text{or} \quad 0.637 = (0.333)^{\gamma-1}$$

$\therefore$ Taking log, we get, $$\gamma = \frac{c_p}{c_v} = 1.408$$

The work done during isentropic compression

$$W_{1-2} = \frac{mR(T_1 - T_2)}{\gamma - 1} \text{ or } 52.9 \text{ or } \frac{0.46 \times (c_p - c_v) \times (452 - 288)}{0.408}$$

or $$c_p - c_v = 0.2869$$

But $$c_p = 1.408\, c_v \quad \text{or} \quad 1.408\, c_v - c_v = 0.2896$$

or $$c_v = 0.7098 \text{ kJ/kg K} \quad \text{and} \quad c_p = 0.99 \text{ kJ/kg K}$$ **Ans.**

Example 5.30 A certain amount of gas at 304°C expands from 0.0361 m^3 according to law pV^n = constant up to volume 0.21 m^3. The initial pressure is 14 bar and the final pressure is 1 bar. The value of specific heat at constant volume is 0.748 kJ/kg K at 304°C and at this temperature, the value of $\gamma = 1.39$. The specific heats at constant pressure and constant volume are given by

$$c_p = a + 0.00012\, T \text{ and } c_v = 8 + 0.00012\, T$$

Find (a) final temperature, (b) work done (c) change in internal energy, (d) heat transfer and (e) polytropic specific heat

Solution We have, $p_1 V_1{}^n = p_2 V_2{}^n$

or $$14 \times 10^2 \times 0.0361^n = 1 \times 10^2 \times 0.21^n \text{ or } n = 1.498$$

From the gas laws (a) $T_2 = T_1 \dfrac{p_2 V_2}{p_1 V_1} = 577 \times \dfrac{1}{14} \times \dfrac{0.21}{0.0361} = 293.75 \text{ K}$

The value of T_2 can also be calculated by the equation

$$T_2 = T_1 \left(\frac{p_2}{p_1}\right)^{(n-1)/n} = 293.75 \text{ K}$$

(b) $$W_{1-2} = \frac{p_1 V_1 - p_2 V_2}{n-1} = \frac{14 \times 10^2 \times 0.0361 - 1 \times 10^2 \times 0.21}{0.498} = 59.31 \text{kJ}$$

$$= 69.31 \text{ kJ}$$

(c) The change in internal energy

$$\Delta U = m c_{v\text{mean}} (T_2 - T_1) = m \left\{ b + 0.00012 \left(\frac{T_1 + T_2}{2} \right) \right\} (T_2 - T_1)$$

$$R = c_v (\gamma - 1) = 0.748\ (1.39 - 1) = 0.291 \text{ kJ/kg K}$$

But $$m = \frac{p_1 V_1}{R T_1} = \frac{14 \times 10^2 \times 0.0361}{0.291 \times 577} = 0.3 \text{ kg}$$

At $$577 \text{ K}, c_v = 0.748 = b + 0.00012 \times 577$$

or $$b = 0.678$$

But $$c_p = \gamma c_v = 1.39 \times 0.748 = 1.0397 \text{ kJ/kg K}$$ **Ans.**

$\therefore$ $$\Delta U = 0.3 \left\{ 0.678 + \frac{0.00012\,(577 + 293.75)}{2} \right\} (293.75 - 577)$$

$$= -\ 62.05 \text{ kJ, i.e., loss}$$ **Ans.**

From first law, $Q_{1-2} - W_{1-2} = \Delta U$

$\therefore$ $$Q_{1-2} = -\ 62.05 + 59.31 = -\ 2.74 \text{ kJ, i.e., heat rejected}$$ **Ans.**

Example 5.31 A cylinder contains 0.115 m^3 of gas at 1 bar and 90°C. This is compressed to volume 0.0288 m^3, the final pressure being 5.67 bar. Calculate (a) the mass of the gas, (b) energy of the gas, and (d) and heat transfer during compression.

If, after the above compression, the gas is cooled at constant pressure to its original temperature of 90°C, find the further work of compression required. Assume $\gamma = 1.4$, $R =$ 0.3 kJ/kg K.

Solution (a) $m = \dfrac{pV}{RT} = \dfrac{1 \times 10^2 \times 0.115}{0.30 \times 363} = 0.1056 \text{ kg}$ **Ans.**

(b) $$p_1 V_1^n = p_2 V_2^n$$

or $$\left(\frac{0.115}{0.0288}\right)^n = \frac{5.67}{1} \quad \text{or} \quad (3.993)^n = 5.67$$

$\therefore$ $$n = 1.253$$

$$T_1 = T_1 (V_1/V_2)^{n-1} = 363\ (0.115/0.0288)^{0.253} = 515.27 \text{ K}$$

(c) $$\Delta U = mc_v (T_2 - T_1) = m \frac{R}{\gamma - 1}(T_2 - T_1)$$

$\therefore$ $$\Delta U = \frac{0.1056 \times 0.3\ (515.27 - 363)}{0.4} = 12.059 \text{ kJ}$$ **Ans.**

(d) $$W_{1-2} = \frac{p_1 V_1 - p_2 V_2}{n-1} = \frac{mR(T_1 - T_2)}{n-1} = \frac{0.1056 \times 0.3\ (363 - 515.27)}{0.253} = -19.066 \text{ kJ}$$

From first law,

$$Q_{1-2} = \Delta U + W_{1-2} = 12.059 - 19.066 = -7.007 \text{ kJ}$$ **Ans.**

$$Q_{1-2} = mc_n (T_2 - T_1) = m \frac{R(\gamma - n)}{(\gamma - 1)(1 - n)}(T_2 - T_1) = -7.007 \text{ kJ}$$

Negative sign shows that the heat is rejected from the system.

(e) Further, work done at constant pressure

$$= p\ (V_2 - V_1) = mR\ (T_2 - T_1) = 0.1056 \times 0.3\ (363 - 515.27) = -4.823 \text{ kJ}$$ **Ans.**

Example 5.32 Prove the following equation for a adiabatic process.

$$T^{a_2} e^{(bT + (c/2)T^2} v^{(a_1 - a_2)} = \text{constant}$$

where $c_p = a_1 + bT + cT^2$ and $c_v = a_2 + bT + cT^2$

Solution We have, $$c_p - c_v = R = a_1 - a_2$$

$$c_v\, dT + p dv = \delta q = 0 \text{ for adiabatic process}$$

or $$c_v \frac{dT}{T} + R\frac{dv}{v} = 0$$

Putting the values of c_v and R, we have, $a_2 \dfrac{dT}{T} + bdT + cT\ dT + (a_1 - a_2)\dfrac{dv}{v} = 0$

Integrating, we have, $a_2 \ln T + bT + \;-T\; + (a_1 - a_2) \ln v = \text{constant}$

$$T^{a_2} e^{(bT)(c/2T^2)} v^{(a_1 - a_2)} = \text{constant}$$

Example 5.33 A vertical piston-cylinder assembly fitted with an electrical resistor contains air. The atmosphere exerts a pressure of 1.0133 bar on the top of the piston which has a mass of

50 kg and a face area of 0.1 m^2. With the passing of electric current through the resistor, the volume of the air slowly increases by 0.03 m^3 while pressure remains constant. The mass of the air is 0.3 kg and its specific internal energy increases by 40 kJ/kg. The air and the piston are at rest initially and finally. The piston-cylinder is made of ceramic composite material and serves as good insulator. Assume that there is no friction between piston and cylinder wall. Draw the system and calculate the work done by the system and heat transfer from the resistor to the air for a system consisting of (a) the air alone, and (b) the air and the piston.

A Model Solution Refer to Fig. 5.35.

Data given $p_{atm} = 1.0133$ bar, $m_{piston} = 50$ kg, $A_{piston} = 0.1$ m^2, $m_{air} = 0.3$ kg, $V_2 - V_1 = 0.03$ m^3, $\Delta u_{air} = 40$ kJ/kg.

To find W and Q for both gases.

Assumptions (i) Two closed systems under consideration are shown in Fig. 5.35, (ii). The heat transfer from resistor to air is such that air expands slowly and its pressure remains constant, (iii) The changes in kinetic and potential energies are negligible, (iv) The internal energy of the piston is not effected by the heat transfer as piston material is a good conductor, (v) The friction between piston and cylinder wall is negligible.

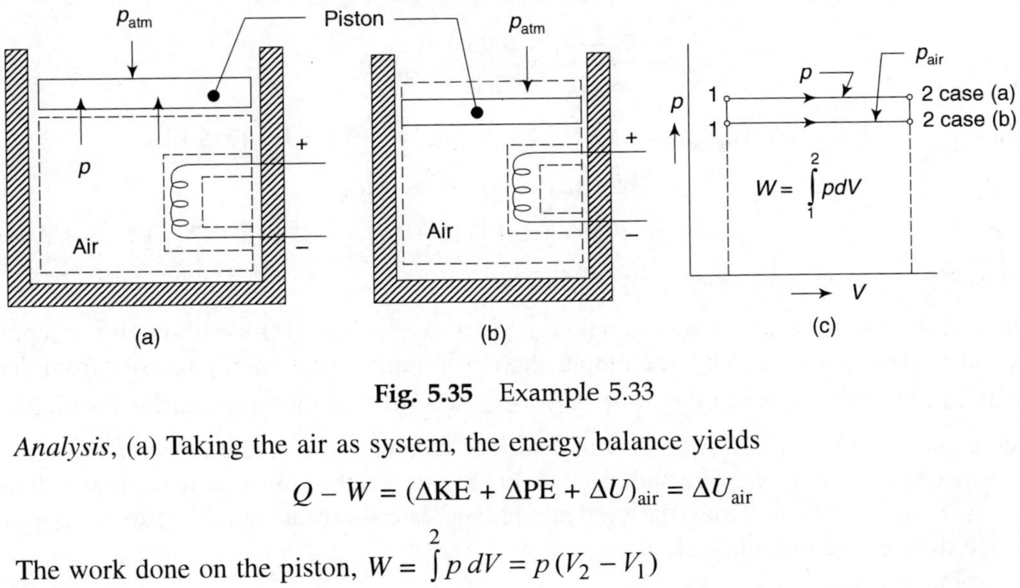

Fig. 5.35 Example 5.33

Analysis, (a) Taking the air as system, the energy balance yields

$$Q - W = (\Delta KE + \Delta PE + \Delta U)_{air} = \Delta U_{air}$$

The work done on the piston, $W = \int_1^2 p\,dV = p(V_2 - V_1)$

In this case, the work is done on the bottom of piston. In order to determine the pressure p acting on the bottom of piston, a force balance on slowly moving frictionless piston is used. The upward force exerted by the air on the bottom of piston is balanced by the weight of the piston and the downward force exerted by atmospheric pressure. Thus,

$$pA_{piston} = m_{piston} \cdot g + p_{atm} \cdot A_{piston}$$

or $$p = \frac{m_{\text{piston}} \cdot g + p_{\text{atm}} \cdot A_{\text{piston}}}{A_{\text{piston}}} = \frac{50 \times 9.81 + 1.0133 \times 10^5 \times 0.1}{0.1}$$

$$= 106235 \text{ N/m}^2 = 1.06235 \text{ bar}$$

Thus, $W = 1.06235 \times 10^2 \times 0.03 = 3.18705$ kJ **Ans.**

And $Q = W + \Delta U_{\text{air}} = W + m_{\text{air}}\, \Delta u_{\text{air}} = 3.18705 + 0.3 \times 40 = 15.18705$ kJ **Ans.**

(b) Taking air and piston as the system, the energy balance yields,

$$Q - W = (\Delta\text{KE} + \Delta\text{PE} + \Delta U)_{\text{air}} + (\Delta\text{KE} + \Delta\text{PE} + \Delta U)_{\text{piston}}$$

Here, ΔKE and ΔPE = 0 for air, ΔKE and $\Delta U = 0$ for piston, thus

$$Q - W = (\Delta\text{PE})_{\text{piston}} + (\Delta U)_{\text{air}}$$

For this system, work is done at the top of piston as it pushes aside the surrounding atmosphere. Thus,

$$W = \int_1^2 p\, dV = p_{\text{atm}}\,(V_2 - V_1) = 1.0133 \times 10^2 \times 0.03 = 3.0399 \text{ kJ}$$ **Ans.**

The potential energy change of the piston, $(\Delta\text{PE})_{\text{piston}} = mg\, \Delta z$

Here, $$\Delta z = \frac{V_2 - V_1}{A_{\text{piston}}} = \frac{0.03}{0.1} = 0.3 \text{ m}$$

Thus, $(\Delta\text{PE})_{\text{piston}} = 50 \times 9.81 \times 0.3 = 147.15 \text{ J} = 0.14715$ kJ

Thus, $$Q = W + (\Delta\text{PE})_{\text{piston}} + m_a\, \Delta u_{\text{air}}$$

$$= 3.0399 + 0.14715 + 0.3 \times 40 = 15.18705 \text{ kJ}$$ **Ans.**

Thus, Q is same in both cases, only W differs.

Example 5.34 Under steady state operation, a gearbox receives 100 kW through the input shaft and delivers power through the output shaft. The rate of heat energy transfer from the gearbox as a system is given by $Q = \bar{h}\, A_s\, (t_b - t_a)$ where $\bar{h}$ is the heat transfer coefficient and equal to 0.175 kW/m^2 K, A_s is the outer surface area of gearbox and equal to 1.0 m^2, t_b is the temperature of outer surface and equal to 303 K and t_a is the temperature of surrounding air and is equal to 293 K. Draw the system and calculate the heat transfer rate and power delivered through the output shaft.

A Model Solution Refer to Fig. 5.36

Data given $\dot{W}_1 = -100$ kW, $\bar{h} = 0.175$ kW/m^2 K, $A_s = 1.0$ m^2, $t_b = 303$ K, $t_a = 293$ K

To find $\dot{Q}$ and $\dot{W}_2$

Assumptions The gearbox is a closed system and at steady state.

Analysis The rate of heat energy transfer by convection from the gearbox (–ve sign convention)

$$Q = -\bar{h}\,A_s\,(t_b - t_a) = -0.175 \times 1.0\,(303 - 293) = -1.75 \text{ kW} \quad \textbf{Ans.}$$

The negative sign in Q shows that the heat is transferred from the system to surrounding.

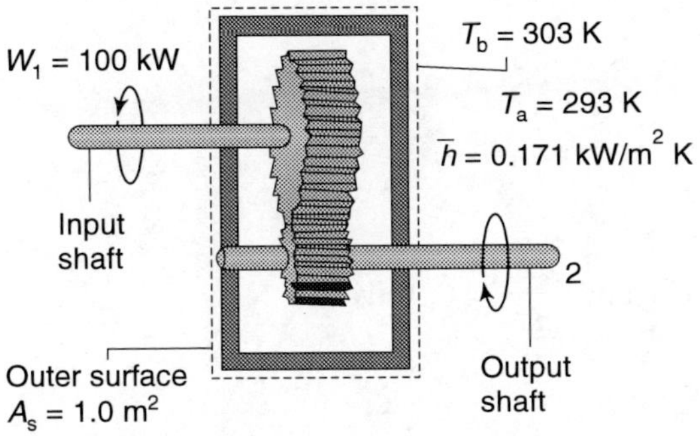

Fig. 5.36 Example 5.34

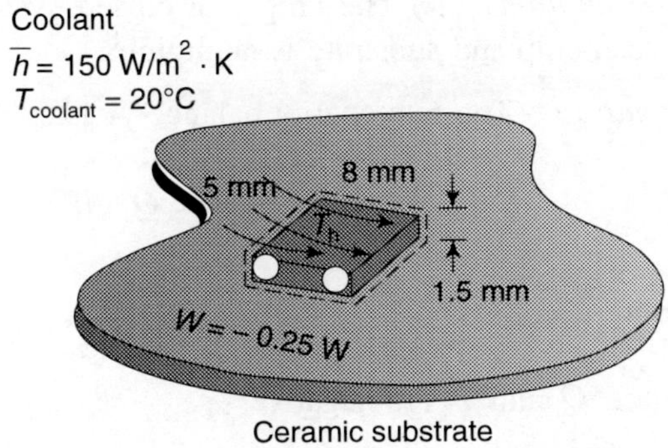

Fig. 5.37 Example 5.35

The energy rate balance equation yields

$$\frac{dE}{d\tau} = \dot{Q} - \dot{W}$$

Here, $$\frac{dE}{d\tau} = 0 \quad \text{so} \quad \dot{W} = \dot{Q}$$

Here $\dot{W}$ represents the net power which is the sum of input power $\dot{W}_1$ and output power $\dot{W}_2$.

Here, $\dot{W}_1 = -\,100$ kW as per sign conversion so

Thus, $\dot{W} = \dot{W}_1 + \dot{W}_2 = \dot{Q}$

$\therefore$ $\dot{W}_2 = \dot{Q} - \dot{W}_1 = -\,1.75 - (-\,100) = 98.25$ kW **Ans.**

The positive sign of $\dot{W}_2$ indicates that work is transferred from the system to the surroundings. Further, the input energy is equal to the output energy.

Example 5.35 An electronic device employs a silicon chip measuring 8 mm on a side and 1.5 mm in thickness is embedded in a ceramic substrate. Under steady state condition, the electrical power input to the silicon chip is 0.25 W. The top surface of the chip is exposed to coolant (air) whose temperature is 20°C. The convective heat transfer coefficient between the chip and the coolant is 150 W/m^2 K. Calculate the surface temperature neglecting the heat transfer due to conduction between the chip and substrate.

A Model Solution Refer to Fig. 5.37.

Data given Chip size 8 mm on one side, thickness 1.5 m, $\dot{W} = -2.5$ W, $t_{\text{coolant}} = 20°\text{C}$, $\bar{h} = 150 \text{ W/m}^2 \text{K}$.

To find t_{surface}

Assumptions (i) The chip is a closed system at steady state and (ii) the heat transfer between chip and substrate is negligible.

Analysis The rate energy balance yields

$$\frac{dE}{d\tau} = \dot{Q} - \dot{W}$$

But $$\dot{Q} = -\bar{h} A_s (t_s - t_c) \text{ and } \frac{dE}{d\tau} = 0, \text{ thus}$$

Here $\dot{Q}$ and $\dot{W}$ are negative

$$0 = -\bar{h} A_s (t_s - t_c) - (-0.25)$$

or $$-150 \times 8 \times 8 \times 10^{-6} (t_s - 20) + 0.25 = 0$$

or $$t_s = \frac{0.25}{150 \times 8 \times 8 \times 10^{-6}} + 20 = 46.04°\text{C}$$ **Ans.**

Eaxmple 5.36 The rate of heat transfer between an electric motor and its surrounding varies with time and is expressed as $\dot{Q} = -0.2 [1 - e^{-0.05\tau}]$ where τ is in seconds and $\dot{Q}$ is in kW. The motor runs at $\omega = 110$ rad/s and applies a constant torque of 25 Nm to an external load. The input power to the electric motor is 3 kW. Derive the expression for the change in energy.

A Model Solution Refer to Fig. 5.38.

Data given $\omega = 100$ rad/s, $T = 25$ Nm, $\dot{W}_{\text{elect}} = -3$ kW.

To find ΔE expression

Assumptions The system is a closed one.

Analysis The instantaneous time rate form of energy balance yields

$$\frac{dE}{d\tau} = \dot{Q} - \dot{W}$$

Here $\dot{W}$ represents the net power from the system, which is

$$\dot{W} = \dot{W}_{\text{shaft}} + \dot{W}_{\text{elect}} = T\omega + \dot{W}_{\text{elect}}$$

$$\dot{W} = 25 \times 110 \times 10^{-3} + (-3) = -0.25 \text{ kW}$$

The negative sign shows that the electrical power input is greater than that the shaft power transferred. Thus,

$$\frac{dE}{d\tau} = -0.2\,[1 - e^{0.05\,\tau}] - (-0.25)$$

$$\frac{dE}{d\tau} = 0.25 - 0.2\,[1 - e^{-0.05\,\tau}] = 0.05 + 0.2\,e^{-0.05\,\tau}$$

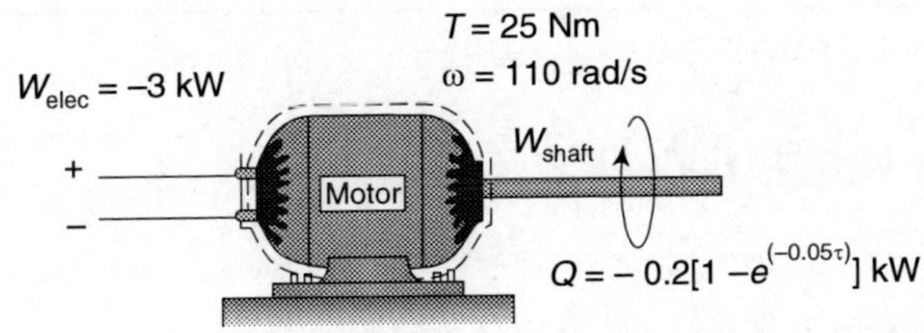

Fig. 5.38 Example 5.36

or $$\Delta E = \int_0^{\tau} [0.05 + 0.2\,e^{-0.05\,\tau}]$$ **Ans.**

5.5 SUMMARY OF FORMULAE

1. $\oint \delta Q = \oint \delta W$ (First law of thermodynamics for cyclic process)
2. $\delta Q - \delta W = dE = dU + d\,(\text{KE}) + d\,(\text{PE}) + \ldots$ and $\delta Q - \delta W = dU$ for a process
3. $\dot{Q} - \dot{W} = \frac{d\,(\text{KE})}{d\tau} + \frac{d\,(\text{PE})}{d\tau} + \frac{dU}{d\tau}$ (Rate equation) and $\frac{dE}{d\tau} = \dot{Q} - \dot{W}$
4. $\Delta U = mc_v\,(t_2 - t_1) = mc_v\,(T_2 - T_1)$

Adiabatic		Polytropic
5. $p_1 V_1^{\gamma} = p_2 V_2^{\gamma}$ (adiabatic)	:	$p_1 V_1^{n} = p_2 V_2^{n}$ (polytropic)
6. $\frac{V_2}{V_1} = \left(\frac{p_1}{p_2}\right)^{1/\gamma}$	:	$\frac{V_2}{V_1} = \left(\frac{p_1}{p_2}\right)^{1/n}$
7. $\frac{p_1 V_1}{T_1} = \frac{p_2 V_2}{T_2}$	:	$n = \frac{\ln (p_2/p_1)}{\ln (V_1/V_2)}$
8. $\frac{T_2}{T_1} = \left(\frac{V_1}{V_2}\right)^{\gamma - 1}$	:	$\frac{T_2}{T_1} = \left(\frac{V_1}{V_2}\right)^{n - 1}$
9. $\frac{T_2}{T_1} = \left(\frac{p_2}{p_1}\right)^{(\gamma - 1)/\gamma}$	:	$\frac{T_2}{T_1} = \left(\frac{p_2}{p_1}\right)^{(n - 1)/n}$
10. $(W_{1-2})_{\text{adta}} = \frac{p_1 V_1 - p_2 V_2}{\gamma - 1}$	:	$(W_{1-2})_{\text{poly}} = \frac{p_1 V_1 - p_1 V_2}{n - 1}$

$$= \frac{mR(T_1 - T_2)}{\gamma - 1} \quad : \quad = \frac{mR(T_1 - T_2)}{n - 1}$$

11. $Q_{\text{adiabatic}} = 0 \quad : \quad (Q_{1-2})_{\text{poly}} = m\left(\frac{\gamma - n}{1 - n}\right) c_v (T_2 - T_1)$

12. $p_1 V_1 = p_2 V_2$ (Isothermal)
13. $(W_{1-2})_{\text{iso}} = p_1 V_1 \ln \frac{V_2}{V_1} = p_1 V_1 \ln \frac{p_1}{p_2}$
14. $U_1 = U_2$ for free expansion process
15. $W = 0$ for free expansion process and $U = c$
16. $\delta W_{nf} = pdV$

PROBLEMS

Things to Think About

5.1. Two power cycles each receive the same energy input Q_{in} and discharges energy O_{out} to the same lake. If the cycles have different efficiencies, think which discharges the greater amount Q_{out}. Discuss the implications of discharge on the environment.

5.2. What is the source of energy released from the sun? Has the sun a certain life? If so, what is the estimate of life?

5.3. In what form and the amount the energy released from the sun reaches the earth?

5.4. The symbols ΔU, ΔKE ΔPE are used to denote the energy change during a process but the work and heat transfer for the process represented respectively, simply as W and Q. Why is it so?

5.5. How will you harness the solar, wind, geothermal and ocean energy for the use of mankind?

5.6. What energy transfers by work and heat can you identify for a flying aeroplane and rocket?

5.7. Using nanotechnology, what devices you think of which will give you work or heat transfer.

5.8. Identify the energy wastage in a moving automobile, train, ship and plane. How will you reduce them? Is there any possibility to use them?

Short and Long Theoretical Questions

5.9. Define first law of thermodynamics.

5.10. Define internal energy of the system. Show that the internal energy is a property of the system.

5.11. List the conditions for which the equation $\delta Q - \delta W = dU$ is true.

5.12. Derive an expression for the first law of thermodynamics applied to a closed system. Also derive rate equation.

5.13. Derive the p, V, T relationship and expression for work done for closed system for all possible thermodynamic processes with reference to work done.

5.14. If the internal energy of a perfect gas is zero at absolute zero temperature, show that when its volume is v and its pressure p, its internal energy is given by

$$u = \frac{pv}{\gamma - 1} \text{ where } \gamma = \frac{c_p}{c_v}$$

5.15. Represent the various thermodynamic processes on p-V diagram for compression and expansion. Comment on the variations.

Numerical Questions

5.16. In a certain cyclic process, the heat interactions are + 44 kJ, – 108 kJ, – 32 J and 136 kJ. Find the net work during the cyclic process. [*Ans*. + 40 kJ]

5.17. A car having mass 1500 kg is running at a speed of 40 km/h. In order to save a man, the brakes are applied and the vehicle is suddenly brought to rest. Calculate the rise in temperatures of brakes. The mass of the brake is 15 kg and the specific heat of the material of the brakes is 0.46 kJ/kg K. [*Ans*. 13.41°C]

5.18. The volume of receiver, containing a gas whose specific heats at constant pressure and constant volume are 0.238 and 0.169 kJ/kg K respectively, is 0.566 m^3. The temperature and pressure in the receiver are 20°C and 3.5 bar, respectively. Calculate the pressure when 0.91 kg of gas been pumped out.

5.19. A steam locomotive having weight of 2500 tonnes experiences the resistance to the motion of a locomotive train equal to 49 N per tonne weight of the train. If 1000 kg of coal is burnt, calculate the distance covered by the locomotive. The calorific value of the coal is 3372 kJ/kg. It may be assumed that only 10 per cent of the heat generated by burning coal in the fire box is converted into work. [*Ans*. 26.77 km]

5.20. In a vessel containing 0.002 m^3 of water kept initially at 20°C is heated to its boiling point of 100°C by an 0.5 kW electric heater. Find the time required to achieve the desired object assuming that the system does not exchange heat with the surrounding. [*Ans*. 22.29 minutes]

5.21. 1 kg of air at 3.5 bar and occupying 0.35 m^3 is heated at constant volume until its temperature has risen to 316°C. Find (a) initial temperature of air, (b) the final pressure of air, (c) heat added, (d) gain in internal energy per kg. c_v = 0.715 kJ/kg K.

[*Ans*. 153.82°C, (b) 4.82 bar, (c) 116.44 kJ/kg (d) 116.44 kJ/kg (e)116.44 kJ/kg]

5.22. One kg of air having an initial volume of 0.3 m^3 is heated at constant pressure of 3.2 bar until the volume is doubled. Find (a) heat added, (b) work done, (c) the initial and final temperature of air. [*Ans*. 336.16 kJ, (b) 96 kJ, (c) 61.49°C and 395.98°C]

5.23. The cold air of the oven flows through a pipe of 50 cm diameter and enters the air heater at a temperature of 30°C. In the air, heat temperature of air rises to 800°C at constant pressure. Calculate heat supplied.

5.24. A system receives 800 kJ of heat while work of 980 kJ is transferred. Is this possible?

[*Ans*. Impossible, because this violates first law]

5.25. A household refrigerator contains fresh food and it is closed. 1 kWh of electric energy is consumed in cooling the food and internal energy of the system decreases by 500 kJ as the temperature drops. Find the magnitude and sign of heat transferred for the process. Assume the entire refrigerator and its contents as a system. [*Ans.* – 4100 kJ]

5.26. The average heat transfer from a person to the surrounding when he is at rest is about 314 kJ/h. Assume that the ventilation system of an auditorium seating 2000 persons fails. Find (a) the change in the internal energy of the air in the auditorium during the first 15 minutes after the ventilation fails and (b) the change in internal energy of the system considering the auditorium and all people inside it as system and no heat transfers to the surrounding.

[*Ans.* (a) 1,57,000 kJ (b) 0]

5.27. 5 m^3 of gas at 8 bar at 180°C is heated keeping the pressure same throughout till the volume is doubled. Find the heat added, the external work done and the change in internal energy during the process. [*Ans.* (a) 14006.68 kJ, 4000 kJ, 10006.05 kJ]

5.28. Calculate the difference in nonflow work, pdV and flow work, Vdp if 1 kg of air having specific volume of 0.4 m^3 at 1.35 bar is compressed according to the law $pV^{1.3} = c$ to a pressure of 10 bar.

5.29. During a nonflow process, the work per degree temperature increase in $\delta W/dt = 300$ J/K and the internal energy is given by $U = (100 + 05\ t)$ J. Calculate the heat transferred from the system if temperature changes from 50°C to 100°C.

5.30. A gas undergoes the following reversible nonflow processes during its change from initial state, pressure $p_1 = 15$ bar and $v_1 = 1$ m^3/kg to final state having pressure 6 bar and specific volume 2 m^3/kg.

(a) polytropic process, pv^n = constant.

(b) constant volume followed by a constant pressure.

(c) constant pressure followed by constant volume. Sketch the processes on p–V co-ordinates and determine the work done by the gas during each process.

[*Ans.* (a) 9.319×10^2 kJ/kg, (b)6×10^5 kJ/kg, (c) 15×10^2 kJ/kg]

5.31. A cycle pump forces air into a tyre against a pressure of 5 bar . The movement of the piston is arranged in such a manner that it may move (i) slowly or (ii) quickly Calculate the swept length of the stroke in these two cases before the air begins to enter the tyre. Assume that the pump stroke is 25 cm and atmospheric pressure is 1 bar. [*Ans.* 20 cm, 17.191 cm

5.32. A piston cylinder arrangement contains 3 kg of working fluid. When the piston moves out the temperature of the system falls from 95°C to 40°C and the net heat rejected from the cylinder wall is equal to 20 kJ. The specific internal energy is given by the expression $u = u_0 + t/5.23$ where u_0 is a constant and t is the temperature in °C. Calculate the work done.

5.33. An electric generator coupled to a windmill produces an average electric power output of 25 kW. The power is used to charge a storage battery. Heat transfer from the battery to the surroundings occurs at a constant rate of 2 kW. Calculate for 12 h of operation

(a) the total amount of energy stored in the battery, in kJ.

(b) the value of stored energy in \$, if electricity is valued at \$ 0.08 per kWh.

5.34. A certain gas occupies 0.1 m^3 at 1 bar pressure. It is compressed adiabatically to pressure of 7 bar. Find the new pressure, new volume and change in internal energy. The density of the gas at 0°C and 1 bar is 1.5 kg/m^3. Take c_v = 0.728 kJ/kg K.

5.35. One m^3 of air at pressure of 0.15 MPa and temperature of 30°C is compressed to a volume of 0.15 m^3 according to the law $pV^{1.3}$ = Const. Find the work done during the process.

5.36. A system executing a certain process, the work done by it per degree temperature increase at each temperature attained is given by the relation $\delta W/dt = (1 - 0.05\ T)$ kJ/K and the heat transferred is given by $(\delta Q/dT)$ = 1.05 kJ/K. Find the increase or decrease of internal energy of the system.

5.37. A cylinder-piston arrangement contains 3 kg of steam at 15 bar, dry and saturated. The piston is restrained by a spring which is fully extended when the volume inside the cylinder is zero. The spring force is proportional to the spring displacement. The weight of the piston may be neglected. Heat is transferred to steam until its volume is 15% of the initial volume. Calculate (a) final pressure, (b) final state of steam, and (c) heat transfer.

5.38. A cylinder fitted with a piston contains a gas. A number of small weights are placed on the piston. The initial pressure and volume of the gases are 1 bar and 1 m^3. A bunsen burner supplies heat to the cylinder. Find the work done by the system for the following cases:

(a) Pressure remains constant while volume increases five time the initial volume.

(b) When the weights are removed from the system such that the following processes occur.

(i) pV = constant, (ii) $pV^{1.4}$ = constant and (iii) $p = (40/V^2) + 2$. The volume is 5 times the initial volume in all the above three cases, (c) the piston is held constant by means of a pin resulting drop pressure to 0.5 bar.

[*Ans.* (a) 39.25 × 10^4J, (b) (i) 6.73 × 10^4 J, (ii) 11.68 × 10^4 J,(iii) 3.92 × 10^4J, (c) zero]

5.39. A gas having volume 0.15 m^3 of air and mass 1 kg at a pressure of 6 bar gauge expands to five times its original volume.

Find (a) the final temperature, (b) work done, and (c) heat interchange for the following expansion process (i) adiabatic (ii) hyperbolic, (iii) constant pressure (iv) polytropic with n = 1.3. Assume c_p = 1.004 kJ/kg K and c_v = 0.72 kJ/kg K. In each case, sketch the process on p–v diagram.

[*Ans.* (a) (i) 492.18 K (ii) 361.95 K, (iii)1811 K, (iv) 223.2 K, (b) (i) 124400 J (ii) 169040 J, (iii) 42000 J, (iv) 134200 J, (c) (i) 0 (ii) 166 kJ, (iii) 1663 kJ]

5.40. A volume of water equal to 2.002 m^3 initially at 20°C is heated to boiling point. A 0.5 kW electric heater is supplied for this. Assuming that the system does not exchange heat with the surrounding, calculate the time required to achieve the desired object.

5.41. A paddle wheel stirs a fluid contained in a tank. During the process of operation, 7209 kJ of work are input to the paddle wheel and 1569 kJ of heat are transferred from the tank. Determine the change in internal energy of the system considering the system to be composed of tank and the fluid.

5.42. Determine the ratio of isothermal work to adiabatic work when a gas expands reversibly from the same initial state isothermally and adiabatically. The pressure ratio is 6 in both and cases. The adiabatic index of the gas is 1.6.

5.43. Two unequal vessels '*A*' and '*B*' are connected by a pipe of negligible volume. Vessel '*A*' contains 0.15 m^3 of air at 28 bar and 95°C. Vessel '*B*' of unknown volume contains air at 1 bar and 5°C. The valve is opened and when the properties are determined in equilibrium state, it is found that the final pressure and temperature are 15 bar and 45°C respectively. Compute volume of vessel '*B*' and determine the change in internal energy of air (both vessel) if any.

5.44. A fluid contained in a constant volume calorimeter experiences a temperature rise of 30°C. When a certain process occurs, the heat supplied to the fluid during the process is 146 kJ. If the system contains 2 kg of fluid whose specific heat at constant volume is 5 kJ/kg K, determine the change in internal energy of the fluid and work transfer, if any.

5.45. 1 kg of a perfect gas is compressed from 1.1 bar, 27°C according to law $pV^{1.3}$ = constant, until the pressure is 6.6 bar. Find the heat flow to or from the cylinder walls for the following cases:

(a) when the gas is ethane (M = 30) having c = 1.75 kJ/kg K.

(b) when the gas is argon (M = 40) having c_p = 0.515 kJ/kg K.

[*Ans.* (a) 84.5 kJ/kg; (b) 59 kJ/kg]

5.46. 0.05 kg of carbon dioxide (M = 44) with a volume of volume of 0.03 m^3 at 1.025 bar, is compressed reversibly until the pressure is 6.15 bar. Determine the final temperature, the work done and the heat flow, for the following cases:

(a) process $pv^{1.4}$ = constant

(b) isothermal process

(c) adiabatic process

Take γ = 1.3.

[*Ans.* (a) 270°C, 5.138 kJ, 1.713 kJ; (b) 52.6°C, 5.51 kJ; – 5.51 kJ; (c) 219°C, 5.25 kJ, 0 kJ]

5.47. A piston-cylinder arrangement as shown in Fig. 5.39, the frictionless piston is free to travel between two sets of stop. When the piston is resting on the lower stops, the enclosed volume is 0.5 m^3, while the enclosed volume is 0.75 m^3 when the piston encounters the upper stop. Initially, the cylinder is filled with air at 0.1 MP and 20°C. The air is then heated until the piston touches the upper stop. If the mass of the piston is such that 0.3 MPa pressure is required to move it against atmospheric pressure, calculate (a) the final pressure in the cylinder, (b) the heat transfer, and (c) work transfer during the process.

5.48. A piston-cylinder arrangement shown in Fig. 5.40 contains 0.1 kg of air at 40°C. At this state the spring touches the piston but exerts no force on it. Heat is then transferred to the air causing the piston to rise during which process the spring resisting force is proportional to the distance moved, with a spring constant of 55 kN/m. The piston area is 0.06 m^2. Calculate the following:

(a) the temperature in the cylinder when the pressure reaches 0.3 MPa; and

(b) the heat transfer.

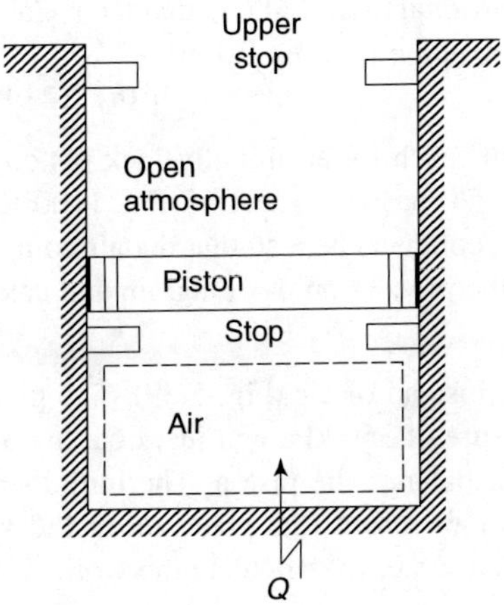

Fig. 5.39 Example 5.47

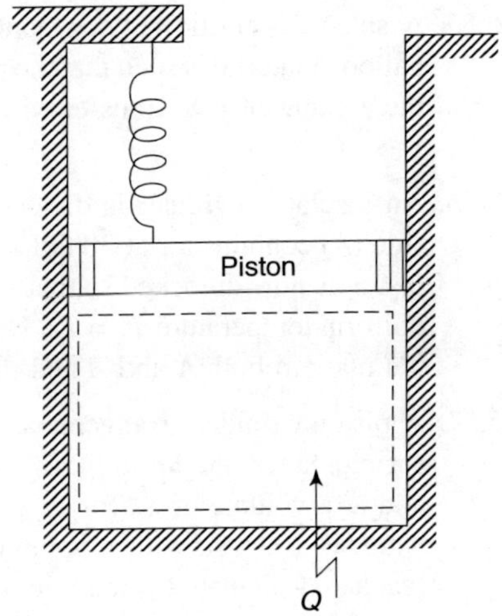

Fig. 5.40 Example 5.48

5.49. In a piston-cylinder arrangement shown in Fig. 5.40, the initial volume inside the cylinder is 0.12 m^3. At this state the pressure inside is 100 kPa, which just balances the atmospheric pressure outside plus the piston weight, the spring is touching but exerts no force on the piston at this state. The air is now heated until the final volume of air becomes two and half times. The final pressure of the gas is 350 kPa, and during the process the spring force is proportional to the displacement of the piston from the initial position. By showing the process on *p-V* diagram and considering the gas inside as the system, calculate the work done by the system. Also calculate the percentage of this work done against the spring.

5.50. A thermoelectric refrigerator maintains a constant temperature in a cold space. This device operates on a current of 30 A and a voltage of 25 V, and rejects heat to the surrounding at the rate of 5 kW. If the temperature in the cold space remains constant, calculate the rate of heat leak into the cold space.

5.51. A small elevator is being designed for a construction site to carry four 70 kg workers to the top of 110 m tall building in less than 2 min. The elevator cage will have a counter-weight to balance the mass. Calculate the smallest size (power) electric motor that can drive this unit.

5.52. The rate of heat transfer to the surroundings from a person at rest is about 400 kJ/h. If the ventilation system fails in an auditorium containing 100 people, calculate the rate (degrees/minute) of the air temperature change. Assume the energy goes into the air of volume 1500 m^3 initially at 300 K and 101 kPa. [*Ans.* 0.53°C/min]

5.53. A computer in a closed room of 350 m^3 dissipates energy at a rate of 10 kW. The room has 50 kg wood, 25 kg steel, and air with all material at 300 K and 101 kPa. If all mass heats up uniformly, calculate the time taken to increase the temperature of room by 10°C.

5.54. The heater in a spacecraft suddenly fails due to some problem. Heat is lost by radiation at the rate of 100 kJ/h and the electric instruments generate 75 kJ/h. At the initial condition of the spacecraft, the air is at 100 kPa, 20°C with a volume of 10 m^3. Calculate the time required to reach an air temperature of – 20°C in the spacecraft. [*Ans.* 15 h]

5.55. A spherical elastic balloon contains (N_2) at 20°, 500 kPa. The initial volume is 0.5 m^3. The balloon material is such that the pressure inside is proportional to the balloon diameter. Calculate the amount of heat transferred to the balloon until its volume reaches 1.0 m^3.

[*Ans.* 1239.5 kJ]

5.56. An insulated cylinder is divided into two parts of 2 m^3 each by an initially lock piston. One side (*A*) contains air at 300 kPa, 350 K, and other side (*B*) has an air at 1.5 MPa, 1100 K. The piston is now unlocked so that it is free to move, and it conducts heat so that the air comes to a uniform temperature $T_A = T_B$. Draw the system, show the process on p–v diagram and calculate the mass in both *A* and *B* and the final *T* and *P*.

5.57. A piston cylinder arrangement contains 1 kg of water instead of air (Fig. 5.39). The piston is spring loaded and initially rests on some stops. A pressure of 350 kPa will just float the piston, and at a volume of 1.5 m^3, a pressure of 500 kPa will balance the piston. The initial state of water is 100 kPa with a volume of 0.5 m^3. Heat is supplied until its pressure of 450 kPa is reached. Calculate (a) the initial temperature and final volume, (b) calculate the work and heat transfer and plot p-v diagram.

Software Based Questions

5.58. Develop a software to calculate the work done and heat exchange for various non-flow processes.

5.59. A gas is contained in a closed tank. An electric resistor in the tank transfers energy to the gas at constant rate of 2000 W. Heat transfers between the gas and the surroundings occurs at a rate of $Q = 60\ t$ where $\dot{Q}$ is in watt and t is time, in min Develop a software for the following:

(a) Plot the time rate of change of energy of the gas for $0 < t < 20$ min, in watts

(b) Determine the net change in energy of the gas after 20 min, in kJ

(c) If the cost of electricity is \$ 0.1 per kWh, what is the cost of the electrical input to the resistor for 20 minutes for operation?

5.60. Using polytropic process during compression and expansion processes in a water cooled diesel engine, develop a software which will give pressure and temperature variation in the cylinder for a linear variation in volume. Also calculate the work done in each process.

Design and Open Ended Questions

5.61. Use of hydrogen energy as a fuel is talk of the 21st century. Investigate the issues surrounding the so-called hydrogen economy covering various methods of production and write a report considering possible uses of hydrogen and obstacles in using it as a primary fuel source.

5.62. Electric car engine, is one of the options for keeping the environment pollution free. Now it is a reality. Write a report of future scope of such engine with advantages, disadvantages and limitations.

5.63. Energy conservation is gaining popularity due to depletion of fossil fuel and increasing cost. Write a report involving all possible energy conservation measures in various fields of production and utilisation, you may consult various research papers.

Multiple-Choice Answer Type Questions

5.64. The equation $\delta Q - \Delta W = dU$ is valid for (a) Flow process, (b) Non-flow cyclic process (c) Non-flow, non-cyclic process, (d) Unsteady flow process.

5.65. Work done in free expansion process is

(a) Zero (b) Negative (c) Positive (d) Maximum

5.66. In a free expansion process involving ideal gas

(a) $p = c$ (b) $v = c$ (c) $u = c$ (d) $w = c$

5.67. The internal energy for a perfect gas is expressed as

(a) $\left(\frac{\partial u}{\partial v}\right) = c$ (b) $\frac{\partial u}{\partial v} = 0$ (c) $\frac{\partial u}{\partial v} < 0$ (d) $\frac{\partial u}{\partial v} > 0$

5.68. The following amount of heat transfer occurs during a cycle comprising of four processes.

120 kJ, – 20 kJ, 16 kJ and 24 kJ

The work is during the cycle is given by

(a) 100 kJ (b) 120 kJ (d) 130 kJ (d) 140 kJ

5.69. The specific heat of polytropic processes (c_n) is given by

(a) $c_v\left(\frac{\gamma - n}{1 - n}\right)$ (b) $c_v\left(\frac{n - \gamma}{1 - n}\right)$ (c) $c_v\left(\frac{1 - n}{\gamma - n}\right)$ (d) $c_v\left(\frac{n - 1}{\gamma - n}\right)$

5.70. The exponent of polytropic process is given by

(a) $\frac{p_2/p_1}{V_1/V_2}$ (b) $\frac{\ln (V_1/V_2)}{\ln (p_2/p_1)}$ (c) $\frac{\ln (p_2/p_1)}{\ln (V_1/V_2)}$ (d) $\frac{p_1 - p_2}{V_1 - V_2}$

Answers to Multiple-Choice Questions

5.64. (c) 5 65. (a) 5.66. (c) 5.67. (b) 5.68. (d) 5.69. (a)

5.70. (c)

CHAPTER

6

First Law of Thermodynamics for Flow Process

OVERVIEW

In the preceding chapter, the first law analysis (energy balance) for non-flow process (i.e., control mass going through a process) has been developed. Many applications in thermodynamics in which flow enters and leaves the devices such as pumps, compressors, turbines, heat exchangers are handled more conveniently by control volume approach.

The objective of this chapter is to develop and illustrate the use of control volume forms of the conservation of mass and conservation of energy principles to solve the problems related to flow processes.

6.1 FLOW PROCESSES AND CONTROL VOLUME

We have discussed the application of the first law of thermodynamics to a closed system in the preceding section. But in most engineering plants such as internal combustion engines, steam engines, boilers, turbines, pumps, compressors, jet engines, etc., the working fluid continuously flows in and out of the plant components. **Such a process is called the flow process and the system is called open system**. For example, consider a portion of a gas turbine plant as shown in Fig. 6.1. The air at high temperature and high pressure enters the gas turbine at section 1 and after doing work by the rotor due to expansion comes out at low pressure at section 2. The heat and work interactions are shown in the diagram. In flow process, the analysis like in closed system, for finding the heat and work interactions.is not feasible; hence a concept of ***control volume and control surface is used which is very convenient and most feasible***.

The control volume is a volume of fixed shape and position. Like a closed system, the control volume is also defined by a boundary which is called control surface. But there is a difference between system and control volume. *The boundary of the closed system may change shape, position and orientation, but the control boundary is fixed and the matter flows across the boundary of the control volume. In the control volume concept, a careful selection of control surface is necessary.*

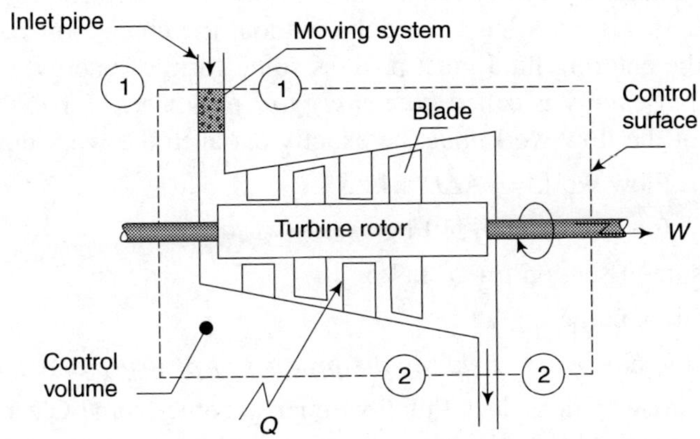

Fig. 6.1 Control volume concept

The main advantage of control volume is that there is no need to bother about the numerous changes of state and work and heat interactions experienced by the gas during its passage through the different parts of the open system. With the control volume concept, the average change of state of gas from entrance to exit can be related to the shaft work delivered by the turbine (Fig. 6.1) and the heat flow from the turbine across the control volume boundary.

6.2 FLOW ENERGY OR FLOW WORK

Flow energy or flow work is associated with the flow process, hence in a closed system this term in absent. *The flow work is the work required to cause the flow of fluid in any passage. In other words, it represents the work necessary to advance a fluid against the existing pressure.* Consider a fluid flowing through a pipe of uniform cross-sectional area A as shown in Fig. 6.2.

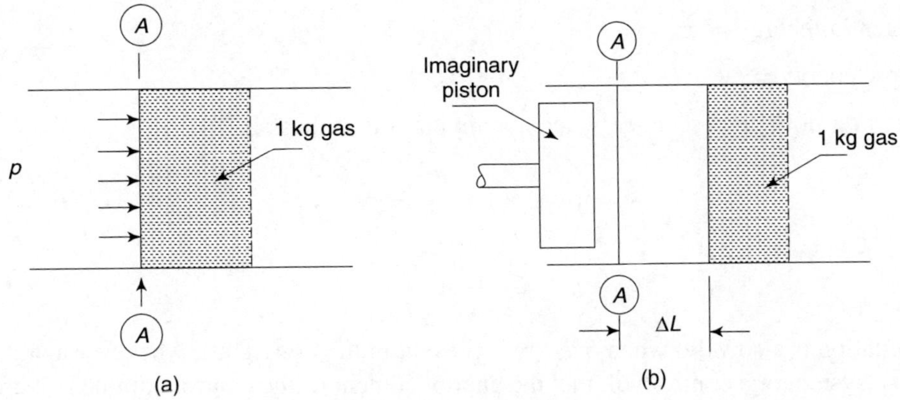

Fig. 6.2 Flow work

Let 1 kg of fluid move a distance ΔL. It is assumed that the pressure p and velocity C at the section a–a are uniform. In order to calculate the work done required in covering the distance ΔL, let us imagine an imaginary piston behind the section a–a which forces the fluid and thus moved a distance ΔL. If this work has been done without the change of the internal, kinetic or potential energy, the entering fluid must possess some form of energy to account for this work. This new form of energy is called ***flow energy or flow work***. Since the above process is ideal, the amount of the flow work must be exactly equal to the work done by the piston.

$$\text{Flow work} = pA\Delta L;\ \text{kJ/kg}$$

or

$$\text{Flow work} = pv;\ \text{kJ/kg} \qquad ...(6.1)$$

where p = pressure at section $a - a$, kN/m^2

v = specific volume, m^3/kg

Thus, we see that the *flow work is simply the product of pressure and specific volume.*

Now, the question arises where does this flow energy come from? Clearly, it is from *the pump located in the surrounding. It is also evident that when a mass of fluid enters, work must be done on a system and when the mass leaves the boundary of a system, flow work must be done on the environment.*

6.3 ENERGY ACCOMPANYING MASS IN A FLOW PROCESS

We have seen that in flow process, there in an extra term of energy called *flow energy* which comes into the picture besides heat and work. Thus, in a flow process the energy contained within the control volume is influenced by (i) heat transfer, (ii) work transfer, and (iii) energy flow accompanying mass flow at entrance and exist of the control volume. It is, therefore, essential to determine the *energy flow per unit mass* flowing at various position coordinates of the control volume.

The various forms of energies in a flow process per unit mass are:

(i) Internal energy = u

(ii) Kinetic energy = $C^2/2$

(iii) Potential energy = gz

(iv) Flow energy = pv

Therefore, in a flow process, energy accompanying a unit mass is

$$e_f = (u + pv) + \frac{C^2}{2} + gz$$

or

$$e_f = h + \frac{C^2}{2} + gz \qquad ...(6.2)$$

This equation is also valid when *friction* is present in the flow. Thus, whenever a unit mass flows into a system or a control volume, the energy content of the control volume is increased by the internal, kinetic and potential energy of the mass and also by flow energy to the extent, *pv. This flow energy (pv) is not the energy contained within the mass by virtue to its internal composition, velocity or elevation but it is a form of energy transmitted with the flow which must be supplied to the flow stream for the correct energy balance in the open system. If the working fluid leaves the system, the flow energy again appears, but this leaving flow energy which forces the working fluid out of this system, is supplied by transformation of internal, potential or kinetic energies of the mass stored within the system.*

6.4 STEADY AND UNSTEADY FLOW PROCESS

Steady flow process is defined as a process in which the conditions within the control volume do not vary with time. The following conditions exist in a steady flow process.

(i) The *mass flow rates* at entrance and exit of the control volume are the same and do not vary with time, i.e., the mass flow rate remains constant within the system.

(ii) The state of energy of the fluid at the entrance or exit of the control volume does not vary with time.

(iii) The state of the fluid at any point within the control volume is the same at all times.

(iv) The rate at which heat and work transfer across the control surface, does not vary with time.

(v) There is no change in the chemical composition of the system so no change in chemical energy is involved.

Any open system such as a steam or gas turbine would constitute a steady flow process operating at full load, constant rate of steam or gas flow and constant rate of heat transfer.

An **unsteady process** is one which does not satisfy the conditions of steady flow, i.e., ***conditions vary with time.***

6.5 FIRST LAW OF THERMODYNAMICS APPLIED TO OPEN SYSTEM (FLOW PROCESS)

6.5.1 General Steady Flow Energy Equation (SFEE)

Consider a control volume of a thermodynamic open system as shown in Fig. 6.3 (a) through which a working fluid is passing at a steady rate. Let the fluid enters the system at location 1–1 and leaves the system at location 2–2. Let the subscripts 1 and 2 refer to the inlet and outlet conditions in the control volume. *The energy balance required in the open system for the steady flow process can be written as*:

Flow energy entering the control volume at section 1 + Heat entering the control volume –Work transferred from the control volume – Flow energy leaving the control volume at section 2

= Increase of energy stored within the control volume. ...(6.3a)

In steady flow process, there is no increase of energy stored within the control volume, hence

Energy entering the control volume = energy leaving the control volume. ...(6.3b)

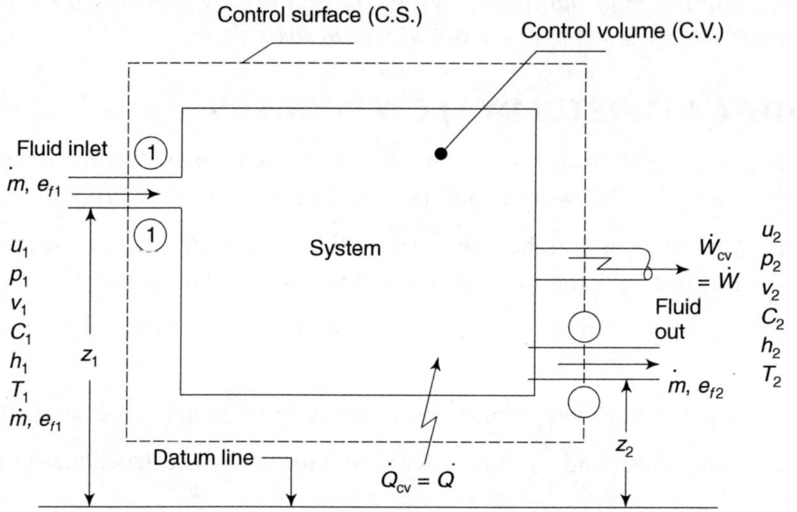

Fig. 6.3 Analysis of open system

Let $\dot{m}$ = mass flowing through the control volume, kg/s

e_{f1} and e_{f2} = energy accompanying unit mass at entrance and exit of the control volume, respectively, kJ/g

$\dot{Q}$ = heat entering the control volume, kJ/s or kW

$q = \dot{Q}/\dot{m}$ = heat entering the control volume per unit mass basis, kJ/kg for simplicity

$\dot{W}$ = work transferred from the control volume, kJ/s or kW

$w = \dot{W}/\dot{m}$ = work transferred from the control volume per unit mass, kJ/kg for simplicity

C_1 and C_2 = velocity of fluid at inlet and outlet respectively, m/s

z_1 and z_2 = elevation of entrance and exit section, respectively, m

6.5.1.1 One Stream

(A) Based on Time Rate

Now, according to Eq. (6.3b)

or $$\dot{m}\, e_{f1} + \dot{Q}_{cv} = \dot{m} \cdot e_{f2} + \dot{W}_{cv}$$

where $$e_f = u + pv + \frac{C^2}{2} + gz = h + \frac{C^2}{2} + gz$$

or $$\dot{Q}_{cv} - \dot{W}_{cv} = \dot{m}\,(e_{f2} - e_{f1}) \qquad ...(6.4)$$

Let $$\dot{Q}_{cv} = \dot{Q} \text{ and } \dot{W}_{cv} = \dot{W} \text{ for simplicity}$$

or $$\dot{Q} - \dot{W} = \dot{m}\left[\left(h_2 + \frac{C_2^2}{2} + gz_2\right) - \left(h_1 + \frac{C_1^2}{2} + gz_1\right)\right] \qquad ...(6.5)$$

*Equation (6.5) is called the steady flow energy equation (**SFEE**). This is also the first law of thermodynamics applied to open system for steady flow process.*

Equation (6.5) can be written in the form of energy entering equal to energy leaving

$$\dot{m}\left(h_1 + \frac{C_1^2}{2} + gz_1\right) + \dot{Q} = \dot{m}\left(h_2 + \frac{C_2^2}{2} + gz_2\right) + \dot{W} \qquad ...(6.6)$$

Equations (6.5) or (6.6) is often known as *Rate Equation*.

Equation (6.4) can also be written in **differential form**

$$\delta\dot{Q} - \delta\dot{W} = d\dot{E}_f \qquad ...(6.7)$$

(B) Based on Unit Mass. For unit mass, Eq. (6.6) is written as

$$h_1 + \frac{C_1^2}{2} + gz_1 + q = h_2 + \frac{C_2^2}{2} + gz_2 + w \qquad ...(6.8)$$

Important Note: Heat exchange (q) and work exchange (w) are not the property of system. For simplicity, expressed in kJ/kg are abbreviated as q and w, otherwise, it should be written as $\frac{\delta Q}{d\dot{m}}$ and $\frac{\delta W}{d\dot{m}}$ respectively.

or $$q - w = (e_{f2} - e_{f1})$$

While for unit mass in differential form, it becomes

$$\delta q - \delta w = de_f = d\left(h + \frac{C^2}{2} + gz\right) \qquad ...(6.9)$$

Equations (6.7) and (6.9) are called differential forms of the steady flow energy equations or first law of thermodynamics in an open system for steady flow process.

6.5.1.2 More than One Stream

When more than one fluid enters and leaves the system in a steady flow, the continuity equation becomes

$$\dot{m}_a + \dot{m}_b + ... = \dot{m}_c + \dot{m}_d + ... \qquad ...(6.10)$$

where $\dot{m}_a, \dot{m}_b, \dot{m}_c, \dot{m}_d,$ are masses of different constituents entering and leaving the system (Fig. 6.4) in a given time interval, and the steady flow process becomes

$$\dot{Q} + \dot{m}_a \left[h + \frac{C^2}{2} + gz \right]_a + \dot{m}_b \left[h + \frac{C^2}{2} + gz \right]_b$$

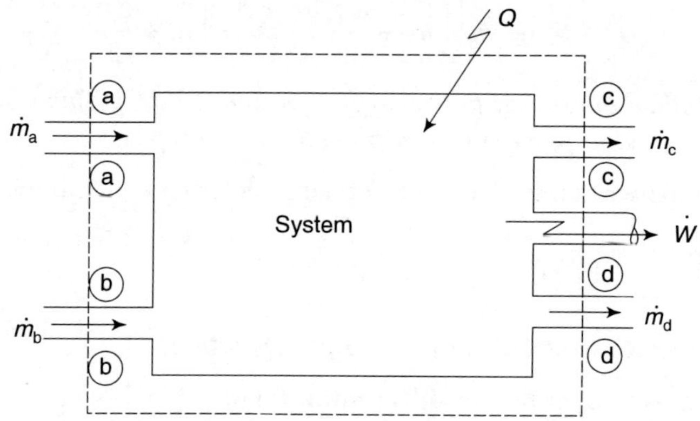

Fig. 6.4 Steady flow process involving more than one fluid stream

$$= \dot{m}_c \left[h + \frac{C^2}{2} + gz \right]_c + \dot{m}_d \left[h + \frac{C^2}{2} + gz \right]_d + \dot{W} \qquad \text{...(6.11)}$$

where subscripts *a*, *b*, etc., refer to entering flow and *c*, *d* etc., refer to flow leaving the system.

Note *If all the terms of energy equation have to be expressed per unit mass* (i.e., kJ/kg), then Eq. 6.8 may be written as

$$h + \frac{}{2 \times 1000} + \frac{gz}{1000} + q = h_2 + \frac{C_2^2}{2 \times 1000} + \frac{gz_2}{1000} + w \qquad \text{...(6.12)}$$

The equation based on units mass is more convenient in analysing the open system handling one stream.

If all the terms of energy equation have to be *expressed per unit time* (i.e., kW), *the rate Eq. (6.6) may be written as*

$$\dot{m}\left(h_1 + \frac{C_1^2}{2 \times 1000} + \frac{gz_1}{1000} \right) + \dot{Q} = \dot{m}\left(h_2 + \frac{C_2^2}{2\ 1000} + \frac{gz_2}{1000} \right) + \dot{W} \qquad \text{...(6.13)}$$

The rate equation is more convenient in analysing the open system handling more than one stream.

Adiabatic Mixing of Perfect Gases *The mixing of two fluid streams is quite a common in engineering*. It can be assumed to occur adiabatically. Figure 6.5 shows the mixing of two streams. Let the stream have mass flow rate $\dot{m}_1$ and $\dot{m}_2$ and temperature T_1 and T_2. Let the resulting mixed stream have temperature T_3. Since $\dot{Q} = 0$, $\dot{W} = 0$, $\Delta KE = 0$ and $\Delta PE = 0$, hence the energy equation gives

$$H_1 + H_2 = H_3$$

or $$\dot{m}_1 h_1 + \dot{m}_2 h_2 = \dot{m}_3 h_3$$

For a perfect gas, $h = c_p T$, hence for mixing of same gases at different temperatures,

$$\dot{m}_1 T_1 + \dot{m}_2 T_2 = \dot{m}_3 T_3 = (\dot{m}_1 + \dot{m}_2) T_3 \qquad ...(6.14)$$

It is to be noted that the mixing process is ***highly irreversible*** *due to the large amount of eddying and churning of the fluid that takes place.*

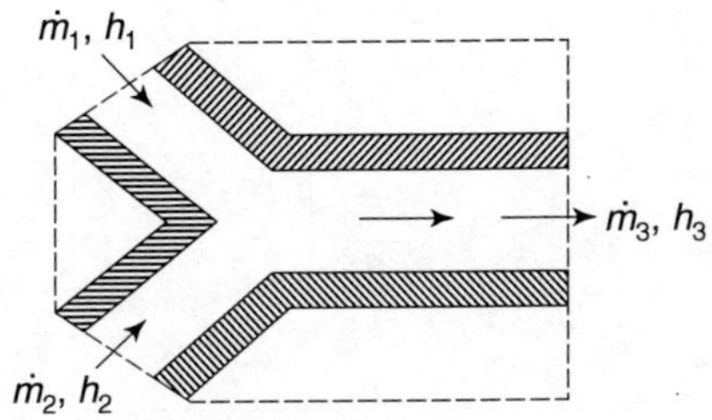

Fig. 6.5 Adiabatic mixing of two streams

6.5.2 General Unsteady Flow Process

In the unsteady flow processes, the mass flow rates at entrance and exit may not be equal and may vary with time. Similarly, the energy entering, leaving and stored in the system may also

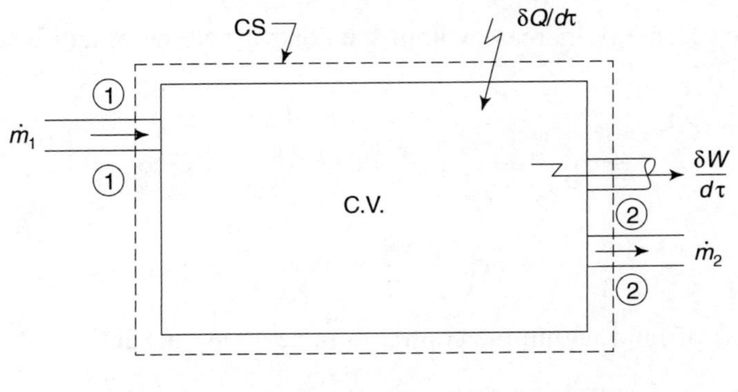

Fig. 6.6 Variable mass flow in unsteady flow process

vary with time. Consider an open system through which a fluid is flowing under unsteady state condition (Fig. 6.6). The rate at which the mass of fluid within the control volume is accumulated is equal to the net rate of mass flow across the control surface. Thus,

$$\frac{dm_{cv}}{d\tau} = \dot{m}_1 - \dot{m}_2 = \frac{dm_1}{d\tau} - \frac{dm_2}{d\tau} \qquad ...(6.15)$$

where m_{cv} is the mass of fluid within the control volume at any instant.

Over any finite period of time.

$$\Delta m_{cv} = \Delta m_1 - \Delta m_2 \qquad ...(6.16)$$

The rate of accumulation of energy within the control volume is equal to the rate of energy flow across the control surface. Thus, (Fig. 6.7)

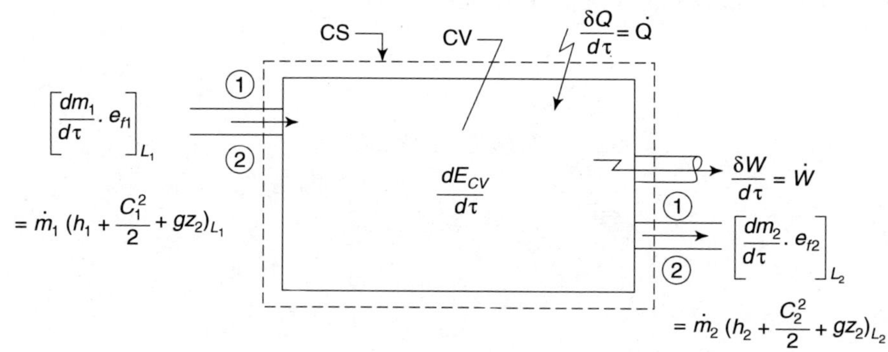

Fig. 6.7 Unsteady flow process (Rate equation)

Rate of energy increases within the control volume

= Rate of flow energy entering the control volume at section 1 + Rate of heat flow entering the control volume–Rate of work transferred from the control volume – Rate of flow energy leaving the control volume. ...(6.17)

If E_{cv} is the rate of energy increase within the control volume at any instant, then from Eq. (6.17), we have

$$\frac{dE_{cv}}{d\tau} = \dot{m}_1\left(h_1 + \frac{C_1^2}{2} + gz_1\right)_{L_1} + \left(\frac{\delta Q}{d\tau}\right)_{cv} - \dot{m}_2\left(h_2 + \frac{C_2^2}{2} + gz_2\right)_{L_2} - \left(\frac{\delta W}{d\tau}\right)_{cv} \qquad ...(6.18)$$

Since $$E_{cv} = m\left(u + \frac{C^2}{2} + gz\right)_{cv}$$

where m is the mass of fluid within the control volume at any instant.

Thus, $$\frac{dE_{cv}}{d\tau} = \frac{d}{d\tau}\left(mu + \frac{mC^2}{2} + mgz\right)_{cv}$$

$$= \left[\frac{dm_1}{d\tau}\left(h_1 + \frac{C_1^2}{2} + gz_1\right)\right]_{L_1} + \left(\frac{\delta Q}{d\tau}\right)_{cv} - \left[\frac{dm_2}{d\tau}\left(h_2 + \frac{C_2^2}{2} + gz_2\right)\right]_{L_2} - \left(\frac{\delta W}{d\tau}\right)_{cv} \quad ...(6.19)$$

where suffixes L_1 and L_2 show the inlet and outlet sections. **Equation (6.19) is the general form of unsteady flow energy equation.**

or $$\frac{dE_{cv}}{d\tau} = \dot{Q}_{cv} - \dot{W}_{cv} + \left[\dot{m}_1\left(h_1 + \frac{C_1^2}{2} + gz_1\right)\right]_{L_1} - \left[\dot{m}_2\left(h_2 + \frac{C_2^2}{2} + gz_2\right)\right]_{L_2} \quad ...(6.20)$$

Equation (6.20) is another general form of equation for unsteady flow process. This is also called ***Unsteady Flow Rate Equation*** for control volume.

Neglecting changes in KE and PE, Eq. (6.20) becomes after removing suffix *cv*:

$$\frac{dE}{d\tau} = \dot{Q} - \dot{W} + \dot{m}_f\,(h_i - h_e) \quad ...(6.20a)$$

In many situations in the control volume, we may have several entering or leaving mass flow rates so a summation over those terms is often needed. **Thus, the final form of first law of thermodynamics applied to flow processes becomes**

$$\frac{dE_{cv}}{d\tau} = \dot{Q}_{cv} - \dot{W}_{cv} + \left[\Sigma\dot{m}_1\left(h_1 + \frac{C_1^2}{2} + gz_1\right)\right]_{L_1} - \left[\Sigma\dot{m}_2\left(h_2 + \frac{C_2^2}{2} + gz_2\right)\right]_{L_2} \quad ...(6.21)$$

For any finite interval $\Delta\tau$ (i.e., τ_1 to τ_2), the Eq. (6.19) may be written in integral form as (Fig. 6.8).

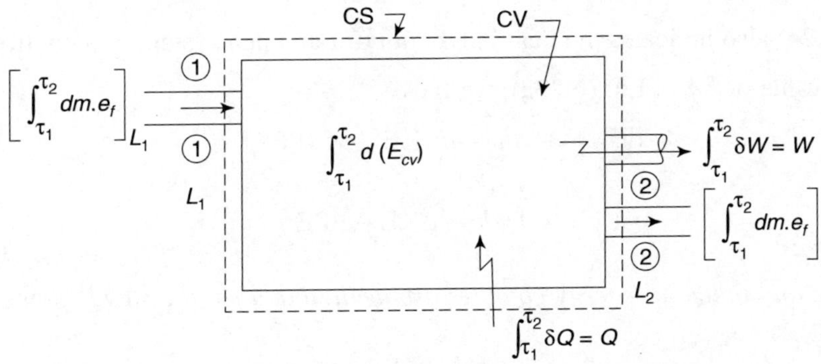

Fig. 6.8 Unsteady flow process (Integral form)

$$\int_{\tau_1}^{\tau_2} dE_{cv} = Q - W + \left[\int_{\tau_1}^{\tau_2} dm\, e_f\right]_{L_1} - \left[\int_{\tau_1}^{\tau_2} dm\, e_f\right]_{L_2} \quad ...(6.22)$$

where $e_f = \left(h + \frac{C^2}{2} + gz\right)$ and $E_{cv} = \left(mu + \frac{mC^2}{2} + mgz\right)_{cv}$

Equation (6.22) is the general equation for unsteady flow for control volume in integral form.

For **steady flow process**, $dE_{cv}/d\tau = 0$ and $\dot{m}_1 = \dot{m}_2$ then Eq. (6.20) becomes the same Steady Flow Energy Equation (SFEE).

$$\dot{Q}_{cv} - \dot{W}_{cv} = \dot{m}\left[\left(h_2 + \frac{C_2^2}{2} + gz_2\right) - \left(h_1 + \frac{C_1^2}{2} + gz_1\right)\right] \qquad ...(6.23)$$

6.6 MECHANICAL OR DISPLACEMENT WORK IN A STEADY FLOW SYSTEM

We have seen that in a non-flow process (i.e., in a closed system), the work done = $\delta w_{nf} = pdv$. For a steady flow reversible process, a similar expression for work done is also feasible. The steady flow energy equation per unit mass is

$$\delta q - \delta w_{sf} = dh + d\,(\text{KE}) + d\,(\text{PE})$$

Here, suffix *sf* is used to indicate the steady flow process.

The work interaction between system and enviornment in steady flow is given by

$$-\,\delta w_{sf} = dh + d\,(\text{KE}) + d\,(\text{PE}) - \delta q \qquad ...(6.23a)$$

In a closed system, the reversible heat interaction δq is given as

$$\delta q = du + pdv = du + pdv + vdp - vdp = d\,(u + pv) - vdp$$

or

$$\delta q = dh - vdp \qquad ...(6.24)$$

Equation (6.24) also indicates the heat interaction for an open system or a control volume.

Putting the value of δq in Eq. (6.23a), we have

$$\delta w_{sf} = -\,vdp - d\,(\text{KE}) - d\,(\text{PE})$$

or

$$w_{sf} = -\int_1^2 vdp - \Delta\text{KE} - \Delta\text{PE}$$

If the kinetic and potential energy changes are negligible, then the steady flow work

$$w_{sf} = -\int_1^2 vdp \qquad ...(6.25)$$

Equation (6.25) is a very important deduction. It not only gives the value of work of a steady flow process but also points out the basic difference between the work term for flow process and non-flow process. In non-flow process if ΔKE and ΔPE are zero, then

$$w_{nf} = \int_1^2 pdv \qquad ...(6.26)$$

Equations (6.25) and (6.26) may be compared with the *p–v* diagram of flow work and non-flow work as shown in Fig. 6.9.

Consider a reversible expansion curve [Fig. 6.9 (a)]. Area *m*–1–2–*n* shows the work done in a non-flow process. Thus,

$$\text{Area } (m - 1 - 2 - n) = \int_1^2 p dv \quad \text{...(6.27)}$$

Similarly, the area (*b* – 2 – 1 – *a*) in Fig. 6.9 (b) shows the work done in a steady flow process. Thus,

$$\text{Area } (b - 2 - 1 - a) = -\int_1^2 v dp \quad \text{...(6.28)}$$

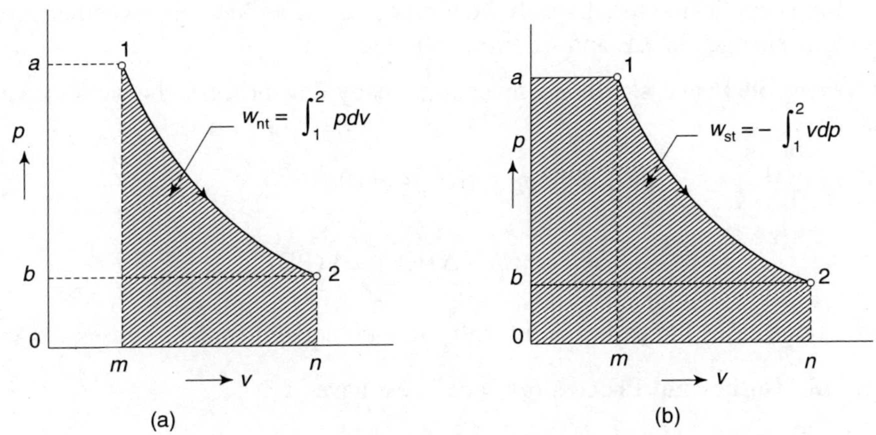

Fig. 6.9 (a) Non-flow work (b) Flow work

But area (*b* – 2 – 1 – *a*) = area (*a* – 1 – *m* – 0) + area (*m* – 1 – 2 – *n*) – area (*b* – 2 – *n* – 0)

$$\therefore \qquad w_{sf} = p_1 v_1 + \int_1^2 pdv - p_2 v_2 = -\int_1^2 vdp \quad \text{...(6.29)}$$

where $p_1 v_1$ = flow work brought into the control of volume by virtue of incoming fluid.

$\int pdv$ = work within the control volume (i.e., non-flow work).

$p_2 v_2$ = flow work leaving the control volume by virtue of outgoing fluid.

Thus, the sum of these terms represents the *work done in a steady flow process.*

6.7 WORK DONE IN STEADY FLOW PROCESSES FOR AN IDEAL GAS

If we assume that a working fluid undergoes a steady flow process from state 1 to state 2, the equation for steady flow based on per kg of flow may be written as

$$\delta q - \delta w_{sf} = dh + d \text{ (KE)} + d \text{ (PE)} \quad \text{...(6.30)}$$

(i) Adiabatic Process. In the case of adiabatic, reversible or irreversible process, $\delta q = 0$, hence

$$\delta w_{sf} = -\, dh - d\,(\text{KE}) - d\,(\text{PE}) = -\, vdp - d\,(\text{KE}) - d\,(\text{PE})$$

For an ideal gas,

$$\therefore \quad w_{sf} = -\int_1^2 c_p\, dT + \frac{C_1^2 - C_2^2}{2} + g\,(z_1 - z_1)$$

If changes in KE and PE are negligible, then

$$w_{sf} = -\,\Delta h = h_1 - h_2 = -\int_1^2 c_p\, dT \; ; \text{kJ/kg} \qquad ...(6.31)$$

This equation is applicable to all steady flow processes of an ideal gas whether reversible or irresversible if changes in KE and PE are negligible.

(ii) Any Reversible Process. In any reversible steady flow process, the work is expressed by the equation

$$\delta w_{sf} = -\, vdp - (\text{KE}) - d\,(\text{PE})$$

or

$$w_{sf} = -\int_1^2 vdp - \Delta\,(\text{KE}) - \Delta\,(\text{PE})$$

Now, for various cases the expression for w_{sf} may be derived for ideal gases

(a) Reversible Isothermal Process (*pv* = *c*). We have

$$-\int_1^2 vdp = -\int \frac{RT_1}{p}\, dp = RT_1 \ln \frac{p_1}{p_2} = RT_1 \ln \frac{v_2}{v_1}$$

$$\therefore \quad w_{sf} = RT_1 \ln \frac{p_1}{p_2} + \frac{C_1^2 - C_2^2}{2} + g\,(z_1 - z_2) \qquad ...(6.32)$$

If changes in kinetic and potential energies are negligible, then

$$w_{sf} = RT_1 \ln \frac{p_1}{p_2} = RT_1 \ln \frac{v_2}{v_1} = p_1\, v_1 \ln \frac{v_2}{v_1}, \ \text{kJ/kg} \qquad ...(6.33)$$

(b) Reversible Polytropic Process (*pv*^{n} = *c*)

$$-\int_1^2 vdp = -\int_1^2 \left(\frac{c}{p}\right)^{1/n} dp = \frac{n}{n-1}(p_1\, v_1 - p_2\, v_2)$$

$$\therefore \quad w_{sf} = \frac{n}{n-1}[p_1\, v_1 - p_2\, v_2] = \frac{n}{n-1} R\,(T_1 - T_2) \text{ neglecting KE and PE} \qquad ...(6.34)$$

Now, from first law, the heat transferred is

or
$$q = (h_2 - h_1) + \frac{n}{n-1} R\,(T_1 - T_2) \qquad \text{...(6.35)}$$

Equations (6.34) and (6.35) can be used to find out work and heat flow respectively in reversible polytropic process.

(c) Reversible Adiabatic Process ($pv^\gamma = c$). The work interaction in this process is calculated in the same manner as $pv^\gamma = c$ be simply putting $n = \gamma$. Thus,

$$w_{sf} = \frac{\gamma}{\gamma-1}(p_1 v_1 - p_2 v_2) = \frac{\gamma}{\gamma-1} R\,(T_1 - T_2) \qquad \text{...(6.36)}$$

(d) Reversible Constant Volume Process ($v = c$).

$$-\int_1^2 v dp = -v\,(p_2 - p_1) = v\,(p_1 - p_2)$$

$\therefore$
$$w_{sf} = v\,(p_1 - p_2) + \frac{C_1^2 - C_2^2}{2} + g\,(z_1 - z_2) \qquad \text{...(6.37)}$$

$w_{sf} = v\,(p_1 - p_2)$ kJ/kg. Neglecting changes in KE and PE. ...(6.38)

(e) Reversible Constant Pressure ($p = c$).

$$-\int v dp = 0$$

$\therefore$
$$w_{sf} = 0 + \frac{C_1^2 - C_2^2}{2} + g\,(z_1 - z_2) \qquad \text{...(6.39)}$$

If the change in the kinetic and potential energies are negligible, then

$$w_{sf} = 0 \qquad \text{...(6.40)}$$

In order to differentiate the work done in a non-flow and steady flow process of an ideal gas, they may be tabulated as follows in Table 6.1.

Table 6.1 Comparison of work done

Process	$w_{sf} = -\int vdp$	$w_{nf} = \int pdv$
Constant volume	$v\,(p_1 - p_2)$	0
Constant press	0	$p\,(v_2 - v_1)$
Isothermal	$p_1 v_1 \ln \frac{v_2}{v_1}$	$p_1 v_1 \ln \frac{v_2}{v_1}$
Polytropic	$\frac{n}{n-1}(p_1 v_1 - p_2 v_2)$	$\frac{p_1 v_1 - p_2 v_2}{n-1}$
Adiabatic	$\frac{\gamma}{\gamma-1}(p_1 v_1 - p_2 v_2)$	$\frac{(p_1 v_1 - p_2 v_2)}{\gamma-1}$

6.8 CONTINUITY EQUATION

The continuity equation is the equation of conservation of mass. Consider a control volume fixed in a one-dimensional fluid flow shown in Fig. 6.10.

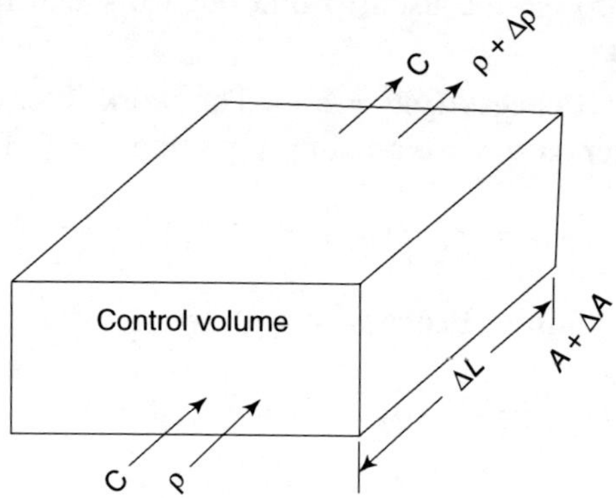

Fig. 6.10 Control volume

Rate of mass of flow into the control volume	−	Rate of mass flow out of the control volume	=	Rate of increase of mass within the control volume.

(i) Rate of mass flow into the control volume = ρAC

(ii) Rate of mass flow into of the control volume = $(\rho + \Delta\rho)\ (A + \Delta A)\ (C + \Delta C)$

(iii) Rate of increase of mass within the control volume

$$= -\frac{\partial}{\partial\tau}\left[\rho\left(A + \frac{\Delta A}{\tau}\right)\Delta L\right]$$

$$\therefore \qquad \rho\, AC - [(\rho + \Delta\rho)\ (A + \Delta A)\ (C + \Delta C)] = \frac{\partial}{\partial\tau}\left[\rho\left(A + \frac{\Delta A}{\tau}\right)\Delta L\right]$$

or $$[-(\rho A\Delta C) - (\rho C\Delta A) - (AC\Delta\rho) - (\rho\Delta A\Delta C) - (A\Delta\rho\Delta C) - (C\Delta\rho\Delta A) - (\Delta A\Delta\rho\Delta C)]$$

$$= \frac{\partial}{\partial\tau}\left[\rho\left(A + \frac{\Delta A}{\tau}\right)\Delta L\right]$$

Dividing by ΔL and taking the limit when $\Delta L \to 0$, we get

$$\left[-\rho A\frac{\partial C}{\partial L} + \rho\, C\frac{\partial A}{\partial L} + AC\frac{\partial\rho}{\partial\tau}\right] = \frac{\partial}{\partial\tau}(\rho\, A)$$

or $$-\frac{\partial}{\partial L}(\rho AC) = A\left(\frac{\partial\rho}{\partial\tau}\right) \qquad (6.41)$$

Equation (6.41) is a general one-dimensional continuity equation. Now, if the flow is steady, then $\frac{\partial\rho}{\partial\tau} = 0$ and hence

$$\frac{d}{dL}(\rho AC) = 0 \tag{6.42}$$

$$\rho A C = \text{constant} = m \tag{6.43}$$

Differentiating and dividing by $\rho A C$, we get

$$\frac{d\rho}{\rho} + \frac{dA}{A} + \frac{dC}{C} = 0 \tag{6.44}$$

Equations (6.42), (6.43) and (6.44) are the *various forms of continuity equations* for steady flow.

For several possible flows

$$\frac{dm_{cv}}{d\tau} = \Sigma \dot{m}_t - \Sigma \dot{m}_e \tag{6.45}$$

6.9 EULER'S AND BERNOULLI'S EQUATIONS

Euler's equation is the basic equation in fluid mechanics. It can be derived from Newton's second law of motion. *According to Newton's second law of motion, the sum of the forces acting in a certain direction on a particle at rest or in motion, is equal to the time rate of change of momentum in the same direction.*

Consider a frictionless steady flow of a fluid in a stream tube of infinitesimal length as shown in Fig. 6.11.

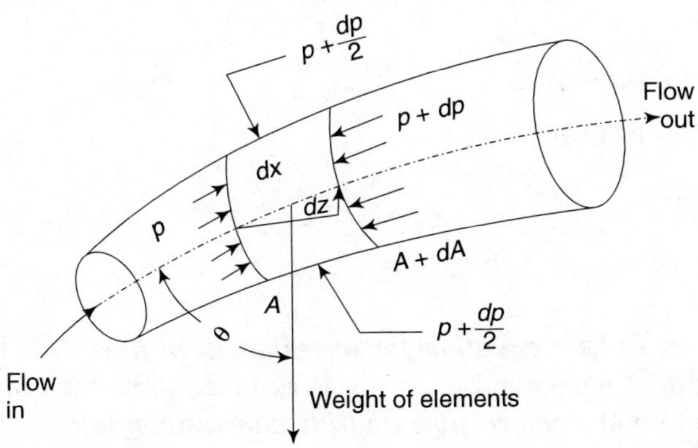

Fig. 6.11 Flow in a stream tube

The centre stream line is inclined at angle of θ with the vertical. Let dx be the distance along the centre between two adjacent sections and the cross-sectional area and the pressures at two sections are A, $A + dA$ and p, $p + dp$ respectively. The following forces are acting:

(i) Pressure force acting at section of flow $= pA$.

(ii) The component of the force due to pressure that is exerted on the sides of the fluid element in the direction of flow

$$= \left(p + \frac{dp}{2}\right) dA$$

(iii) The pressure force acting at section $(A + dA)$ in the opposite direction of flow $(p + dp)\,(A + dA)$.

(iv) The weight of this fluid element in the stream tube

$$= \text{specific weight} \times \left(\frac{A + dA}{2}\right) dx$$

where $\left(A + \frac{dA}{2}\right)$ is the average cross-sectional area between the two sections. The component of the weight in the x direction is

$$-\rho\left(A + \frac{dA}{2}\right) dx \cos\theta = -\rho g\left(A + \frac{dA}{2}\right) dx \frac{dz}{dx} = -\rho g\left(A + \frac{dA}{2}\right) dz$$

where $\cos\theta = \dfrac{dz}{dx}$

Thus, applying the momentum equation in the x direction gives

$$pA + \left(p + \frac{dp}{2}\right) dA - (p + dp)\,(A + dA) - \rho g\left(A + \frac{dA}{2}\right) dz = \rho ACdC$$

where C is velocity

Neglecting the term dA/dp, we get

$$Adp + \rho gAdz + \rho AC\,dC = 0$$

Dividing by $A\rho$ results

$$\frac{dp}{\rho} + C\,dC + g\,dz = 0 \qquad \text{...(6.46)}$$

Equation (6.46) *is an* ***Euler's equation for one-dimension flow***. If it is to be noted that the fluid was assumed frictionless and gravity was assumed to be the only field force. This equation is applicable to both ***compressible and incompressible fluid***.

Euler's equation can be integrated directly if the flow is incompressible because density remains constant, hence

$$\frac{p_2 - p_1}{\rho} + \frac{C_1^2 - C_1^2}{2} + g\,(z_1 - z_1) = 0$$

or

$$\frac{p_1}{\rho} + \frac{C_1^2}{2} + gz_1 = \frac{p_2}{\rho} + \frac{C_2^2}{2} + gz_1 \qquad \text{...(6.47)}$$

or $$\frac{p}{\rho}+\frac{C^2}{2}+gz = \text{constant} \qquad ...(6.48)$$

***Equations (6.47) and (6.48) are the various forms of Bernoulli equation* applicable to steady, one dimensional, frictionless flow of an incompressible fluid neglecting magnetic or electric effects.**

6.9.1 Compression of SFEE, Euler's and Bernoulli's Equation

It will be noteworthy to compare this result with that of the steady flow energy equation. The steady flow energy equation can be written as

$$\delta q - \delta w_{sf} = dh + \frac{CdC}{2} + gdz$$

Since the process is revesible, hence

$$\delta w_{sf} = -\,vdp - \frac{CdC}{2} - gdz$$

If no work is transferred, $(dw)_{sf} = 0$, then

$$vdp + CdC + gz = 0$$

or $$\frac{dp}{\rho} + CdC + gz = 0 \qquad ...(6.49)$$

Thus, Eq. (6.49) is the same as Euler's equation, and Euler's equation is the steady flow energy equation if the following conditions are satisfied.

(i) The process is reversible.

(ii) There is no work interaction.

Further, Bernoulli's equation is restricted to frictionless incompressible fluid but SFEE is not restricted and is valid for viscous compressible fluid as well.

6.10 THROTTLING PROCESS (JOULE–THOMSON COEFFICIENT)

Throttling process is an irreversible process in which a fluid flows across a restriction in such a manner that there is a drop of pressure without any change in enthalpy of the fluid, kinetic energy, potential energy and there is no work or no heat transfer. Such a process occurs in the flow through *porous plug, a partially closed valve or a small orifice*. Figure 6.12 shows the Joule–Thomson experiment for the throttling process. Joule–Thomson performed the basic throttling process in 1852 and so it has come to be known as Joule-Thomson porous plug experiment.

In this experiment, a steady stream of gas at pressure p, and temperature T, flows through a porous plug contained in a horizontal tube which is thermally insulated. The fluid comes out of the porous plug at a lower pressure, p_2. The system does not exchange work with the surrounding.

Since the process is steady flow process, applying the first law of steady flow energy equation, we get

$$u_1 + p_1 v_1 + \frac{C_1^2}{2} + gz_1 + q = u_2 + p_2 v_2 + \frac{C_2^2}{2} + gz_2 + w$$

or $$h_1 + \frac{C_1^2}{2} + gz_1 + q = h_2 + \frac{C_2^2}{2} + gz_2 + w$$

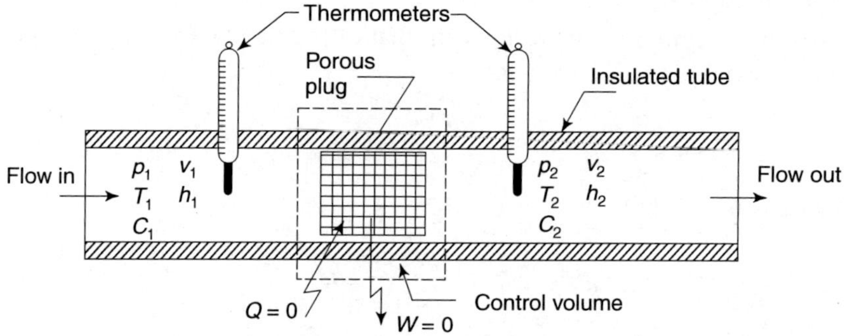

Fig. 6.12 Porous plug experiment

If ΔKE = 0, ΔPE = 0, $q = 0$ and $w = 0$ hence

$$h_1 = h_2 \qquad ...(6.50)$$

Thus, enthalpy before throttling is equal to the enthalpy after throttling.

The question arises whether the temperature and internal energy change in a throttling process or not. The answer of this question depends on whether the fluid **behaves as an ideal gas or not**. *As we know that enthalpy of an ideal gas is a function of temperature alone*, hence

$$T_1 = T_2 \qquad ...(6.51)$$

and therefore $$u_1 = u_2 \qquad ...(6.52)$$

Thus, we see that for an ideal gas, throttling process takes place at constant enthalpy, constant temperature and constant internal energy.

Real Gas Case But the **enthalpy of the real gas** is not the function of temperature alone, hence $T_1 \neq T_2$. Thus, change in internal energy is not zero in real gas and is given by

$$u_2 - u_1 = p_1 v_1 - p_2 v_2$$

Assume a series of experiments performed on a real gas keeping the initial pressure p_1 and temperature T_1 constant, but with a different downstream pressure of the plug say p_2, p_3, p_4 etc., Let T_2, T_3, T_4, etc., be the corresponding temperatures after throttling. The results from these experiments can be plotted as a constant enthalpy curve on a $T - p$ plane. If several different rates of flow are established at each condition of T_1 and p_1, a series of constant enthalpy curves is obtained as shown in Fig. 6.13.

The maximum point on each curve is called the ***inversion point*** and the locus of the inversion point is called the ***inversion curve***. The slope of the constant enthalpy curve on $p - T$ diagram at a particular state may be ***positive, zero or negative in value***. This slope is called the **Joule–Thomson coefficient** (μ_j) and is expressed as

$$\mu_J = \left(\frac{\partial T}{\partial p}\right)_h \qquad ...(6.53)$$

On the left side of an inversion point, μ_J is positive, at the right side of the inversion point, μ_J is negative, at the inversion point μ_J is zero. Since, there is always a pressure drop in throttling process Δp is always negative. Thus, when μ_J is positive, the temperature change is negative, therefore throttling process produces ***cooling effect***. Likewise, when μ_J is negative, the temperature change is positive, therefore, throttling produces ***heating effect***. *It is to be noted that at any particular pressure, μ_J is positive only within a certain temperature range*. These temperatures are called the ***upper and lower inversion points*** indicated in Fig. 6.13. Between these two temperatures, throttling causes a drop in the temperature of the gas and outside of this temperature range, throttling results in a temperature rise. *Cooling can take place only if the initial temperature before throttling is below the maximum inversion temperature*. This temperature is **5 T_c** where ***T_c* is the critical temperature**. *Put differently, above a certain pressure, throttling can cause only a* ***heating effect***. *This is* ***point 'a'*** *in the figure. Similarly, above a certain temperature throttling can cause a cooling effect. This is the* ***point 'b'*** *in the figure*.

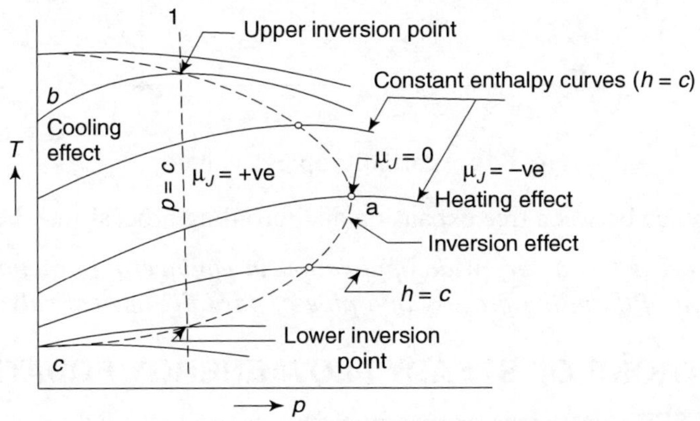

Fig. 6.13 Constant enthalpy and inversion curves of real gas

Evidently in a perfect gas, where the temperature remains constant in throttling process

$$\mu_J = \left(\frac{\partial T}{\partial p}\right)_h = 0 \text{ for a perfect gas} \qquad ...(6.54)$$

This experiment can be used to find the specific heat of gas. If h is expressed in terms of p and T, then

$$dh = \left(\frac{\partial h}{\partial p}\right)_T dh + \left(\frac{\partial h}{\partial T}\right)_P dT$$

For adiabatic throttling process, $dh = 0$, hence

$$0 = \left(\frac{\partial h}{\partial p}\right)_T \left(\frac{\partial p}{\partial T}\right)_h + \left(\frac{\partial h}{\partial T}\right)_p$$

or
$$\left(\frac{\partial h}{\partial T}\right)_p = -\left(\frac{\partial h}{\partial p}\right)_T \left(\frac{\partial p}{\partial T}\right)_h = -\frac{1}{\mu_J}\left(\frac{\partial h}{\partial p}\right)_T \quad ...(6.55)$$

The property $(\partial h/\partial p)_T$ is called the ***constant temperature coefficient***. Equation (6.55) may be used to determine c_p and consequently other properties.

We have already discussed the *free expansion process. This process is also a **Joule expansion not Joule–Thomson expansion***. The *Joule–coefficient* is defined as

$$\eta = -\left(\frac{\partial T}{\partial v}\right)_u \quad ...(6.56)$$

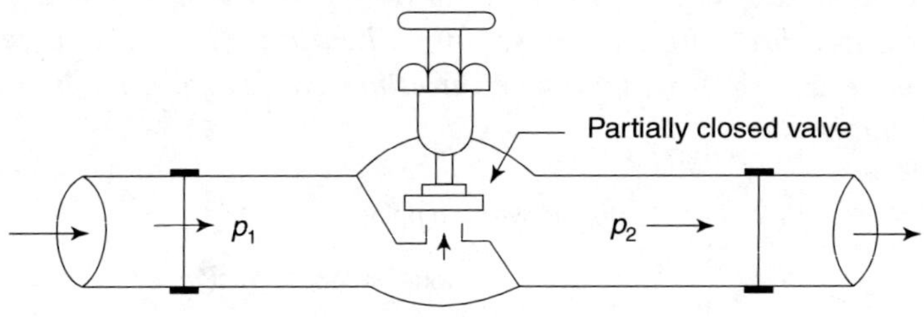

Fig. 6.14 Partially opend valve

Hence, the difference between free expansion and throttling process may be distinguished.

The throttling process has a very wide application in engineering practice. Figure. 6.14 shows an application of throttling process in a pipe of partially opened valve.

6.11 APPLICATIONS OF STEADY FLOW ENERGY EQUATION (SFEE)

The applications of SFEE in various engineering devices are given below.

6.11.1 Nozzle

Nozzle is a passage of varying cross section by means of which pressure energy of flowing fluid is changed into kinetic energy. The chief use of a nozzle is to produce a jet of high velocity of stream to drive turbine or to produce thrust in jet engines. Figure 6.15 shows a nozzle. From the steady flow energy equation considering per unit mass

$$h_1 + \frac{C_1^2}{2} + gz_1 + q = h_2 + \frac{C_2^2}{2} + gz_2 + w$$

The work done w is zero, there is no heat transfer, i.e., $q = 0$, and $z_1 = z_2$, hence

$$C_2 = \sqrt{[2(h_1 - h_2) + C_1^2]} \quad ...(6.57)$$

For perfect gas,

$$C_2 = \sqrt{\left[2c_p(T_1 - T_2) + C_1^2\right]} = \sqrt{\left[2c_p\, T_1 \left\{1 - (p_2/p_1)^{(\gamma-1)/\gamma}\right\} + C_1^2\right]} \quad ...(6.58)$$

If $C_1 = 0$, then Eq. (6.58) reduces to

$$C_2 = \sqrt{2(h_1 - h_2)} = \sqrt{2c_p\, T_1 \left\{1 - (p_2/p_1)^{(\gamma-1)/(\gamma)}\right\}} \quad ...(6.59)$$

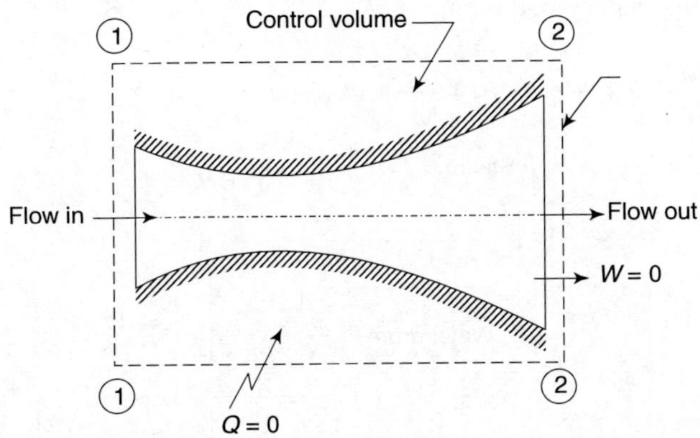

Fig. 6.15 Flow through a nozzle (convergent–divergent)

6.11.2 Diffuser

The function of a diffuser is to convert kinetic energy into pressure energy of fluid. Its chief use is at the entrance of jet engines, combustion chamber of jet engines, at the exhaust hood of water turbines. Similarly, in the case of diffuser (Fig. 6.16) where the *kinetic energy changes to pressure energy*, the energy equation can be applied and similar expression may be obtained

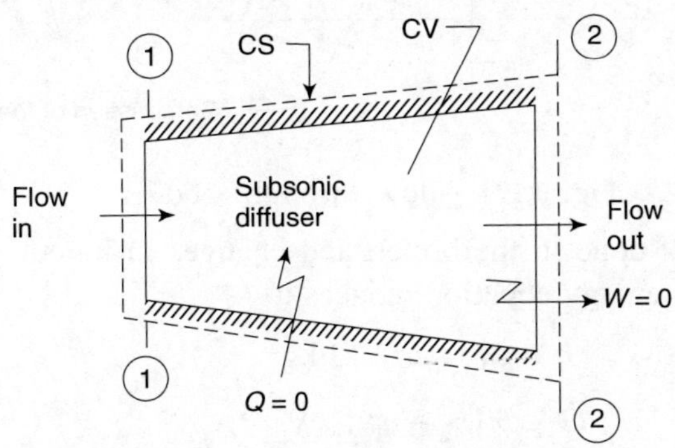

Fig. 6.16 Flow through a diffuser

$$h_2 - h_1 = \left(C_1^2 - C_2^2\right)/2$$

or $$c_p T_1 \left\{\left(\frac{p_2}{p_1}\right)^{(\gamma-1)/\gamma} - 1\right\} = \frac{C_1^2 - C_2^2}{2} \quad ...(6.59a)$$

6.11.3 Boiler

In the boiler shown in Fig. 6.17, chemical energy of the fuel is released by the combustion process in the form of heat. This energy is used to generate steam from water which is pumped by a centrifugal pump at pressure higher than steam generated pressure at constant pressure which may be used to drive steam turbine and for process work. In the boiler, only phase of water changes. Applying energy equation,

$$h_1 + \frac{C_1^2}{2} + gz_1 + q = h_2 + \frac{C_2^2}{2} + gz_2 + w$$

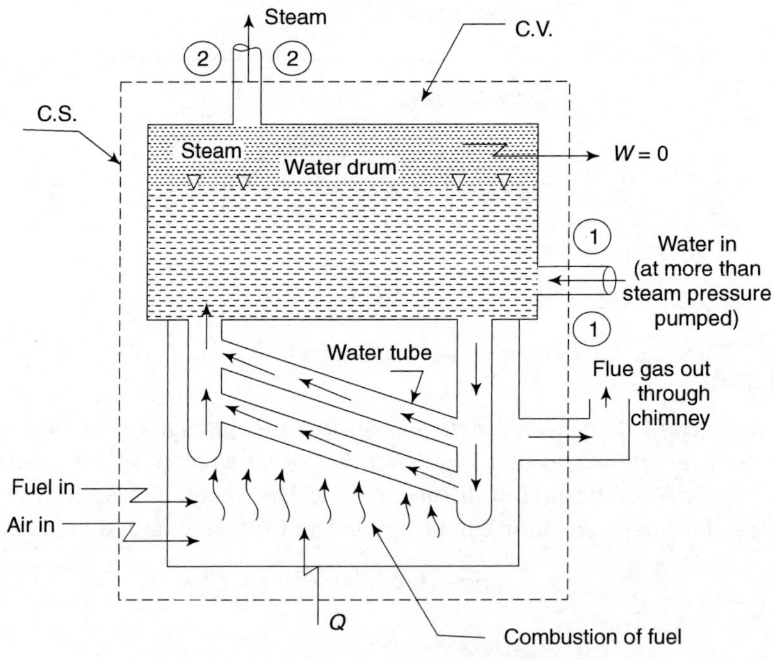

Fig. 6.17 Flow through a boiler

Since there is no work done in the boiler, and changes in kinetic energy and potential energy are negligible, the energy equation reduces to

$$q = (h_2 - h_1);\ \text{kJ/kg} \quad ...(6.60a)$$

$$\dot{Q} = \dot{m}\left(h_2 - h_1\right);\ \text{kW} \quad ...(6.60b)$$

where $\dot{m}$ = mass flow rate of steam, kg/s

Rate equation may also be used to find out $\dot{Q}$.

6.11.4 Turbine

A turbine converts the heat energy of fluid into useful work. The energy level of the working fluid decreases continuously along the flow stream while passing through blades and expansion process occurs. This is shown in Fig. 6.18. In this system:

(a) change in kinetic energy is negligible,

(b) change in potential energy is negligible, and

(c) heat transfer is negligible, since the turbine is thermally insulated.

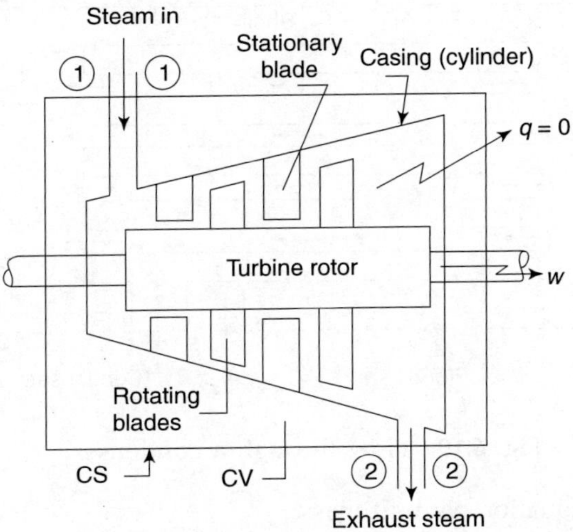

Fig. 6.18 Flow through a turbine

Applying energy equation per unit mass

$$h_1 + \frac{C_1^2}{2} + gz_1 + q = h_2 + \frac{C_2^2}{2} + gz_2 + w$$

or
$$w = (h_1 - h_2) + \frac{C_1^2 - C_2^2}{2} + (z_1 - z_2)\, g + q$$

If
$$\frac{C_1^2 - C_2^2}{2} = 0,\ g\ (z_1 - z_1) = 0 \quad \text{and} \quad q = 0, \text{ hence}$$

$$w = (h_1 - h_2), \text{ kJ/kg} \qquad ...(6.61a)$$

$$\dot{W} = \dot{m}_s\,(h_1 - h_2);\ \text{kW} \qquad ...(6.61b)$$

The rate equation may also be applied to find out $\dot{W}$.

6.11.5 Condenser

The function of a condenser is to condense the steam coming out from steam turbine by rejecting heat from the steam to cooling water circulating in the condenser. It is one of the main elements of steam power plant. *About 55% energy is lost in condenser.* In this case, there are two steady flow mass streams : one for steam and the other for cooling water. In the condenser shown in Fig. 6.19.

(a) The work done is zero.

(b) There is no change in kinetic energy.

(c) There is neligible change in potential energy.

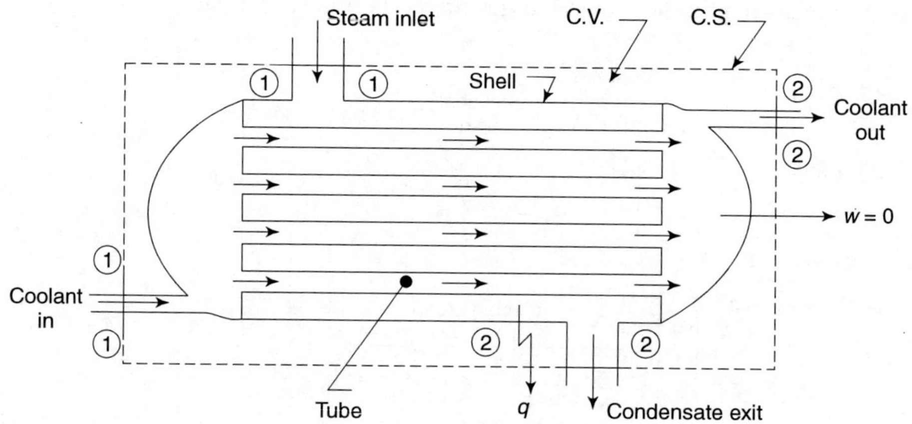

Fig. 6.19 Flow through a condenser

Applying the energy equation per unit mass

$$q = (h_1 - h_2) + \frac{C_1^2 - C_2^2}{2} + (z_1 - z_2)\,g - w$$

But $\dfrac{C_1^2 - C_2^2}{2} = 0, \quad g\,(z_1 - z_2) = 0, \quad w = 0$, hence

$$q = (h_1 - h_2) \text{ kJ/kg} \qquad ...(6.62a)$$

$$\dot{Q} = \dot{m}\,(h_1 - h_2) \text{ kW} \qquad ...(6.62b)$$

6.11.6 Evaporator

Evaporator is one of the main elements of a refrigerating system. In the evaporator, heat is taken from the surrounding (enclosed space) for evaporating of the working fluid (refrigerant) and as a result the space gets cooled. Evaporation takes place at constant temperature and pressure. Figure 6.20 shows a evaporator. In the evaporator.

(a) The work done is zero.

(b) There is no change in kinetic and potential energy.

Applying energy equation per unit mass

$$q = (h_2 - h_1);\ \text{kJ/kg} \qquad \text{...(6.63a)}$$

$$\dot{Q} = \dot{m}\,(h_2 - h_1);\ \text{kW} \qquad \text{...(6.63b)}$$

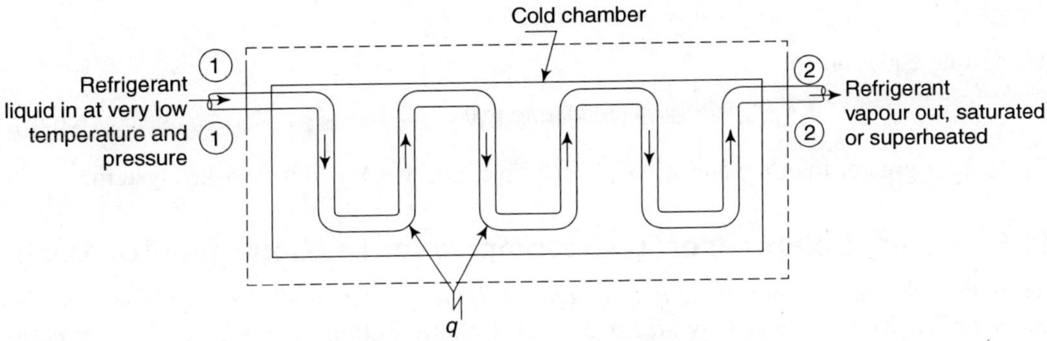

Fig. 6.20 Flow through an evaporator

6.11.7 Reciprocating Compressors

The function of a compressor is to compress the air or gas from low pressure to high pressure with the help of work input. A reciprocating compressor is shown in Fig. 6.21 which includes a piston-cylinder arrangement with a receiver. The receiver is a vessel which is sufficiently large such that there is no fluctuation of flow in the control volume. The compressor may be air or water cooled. Applying steady flow energy equation per unit mass basis.

$$h_1 + \frac{C_1^2}{2} = gz_1 + q = h_2 + \frac{C_2^2}{2} + gz_2 + w$$

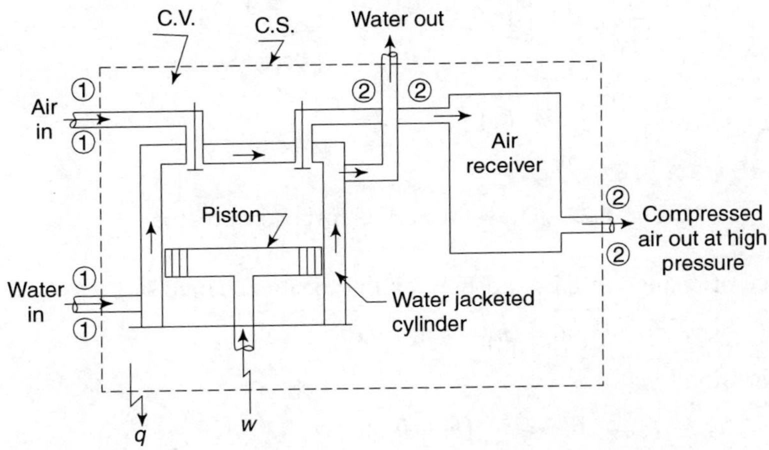

Fig. 6.21 Flow through a reciprocating compressor

The changes in kinetic and potential energies are negligible. Here q is negative since heat is rejected from the system, hence

$$w = (h_1 - h_2) - q \text{ kJ/kg} \qquad \text{...(6.64a)}$$

In the absence of heat exchange, i.e., $q = 0$

$$w = (h_1 - h_2); \text{ kJ/kg}$$

Using rate equation

or
$$\dot{W} = \dot{m}\,(h_1 - h_2); \text{ kW} \qquad \text{...(6.64b)}$$

Since h_2 is greater than h_1, hence, $\dot{W}$ is negative, i.e., work is done on the system.

6.11.8 Rotary Compressor (Turbocompressor, i.e., Centrifugal or Axial)

The function of a turbocompressor is to compress the air or gas with the help of a rotating element called rotor followed by stator due to dynamic action of working fluid through centrifugal and diffusion action or only diffusion action. This increases the energy level of fluid in the direction of flow stream. Here the work done on the system is to rotate the rotor/impeller causing pressure rise in the compressor (centrifugal or axial) as shown in Fig. 6.22. From steady flow energy equation

$$h_1 + \frac{C_1^2}{2} + gz_1 + q = h_2 + \frac{C_2^2}{2} + gz_2 + w$$

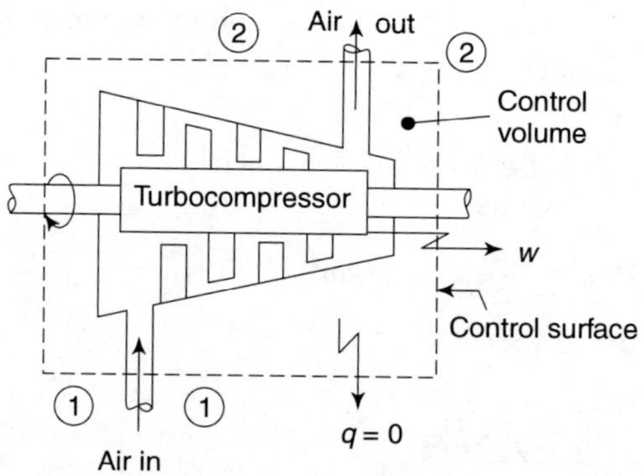

Fig. 6.22 Flow through a turbocompressor

In the absence of change in P.E. and K.E. and no heat exchange.

$$w = (h_1 - h_2); \text{ kJ/kg} \qquad \text{...(6.65a)}$$

Using rate equation,

or
$$\dot{W} = \dot{m}_a\,(h_1 - h_2); \text{ kW} \qquad \text{...(6.65b)}$$

where $\dot{m}_a$ = mass flow rate of air, kg/s

Since h_2 is greater than h_1, hence work done is negative, i.e., work is done on the gas.

Example 6.1 A steam turbine operates under steady-flow conditions. It receives 7200 kg/h of steam from the boiler. The steam enters the turbine at enthalpy of 2800 kJ/kg, a velocity of 4000 m/min and an elevation of 4 m. The steam leaves the turbine at enthalpy of 2000 kJ/kg, a velocity of 8000 m/min and an elevation of 1 m. Due to radiation, heat losses from the turbine to the surrounding amount of 1580 kJ/h. Calculate the output of the turbine.

Solution Refer to Fig. 6.18,

The steady flow energy equation when expressed each term in kJ/kg may be written as

$$h_1 + \frac{C_1^2}{2\times1000} \times \frac{gz_1}{1000} + q = h_2 + \frac{C_2^2}{2\times1000} + \frac{gz_2}{1000} + w$$

Here, $\dot{m} = 7200 \text{ kg/h} = 2 \text{ kg/s}$

$$q = \frac{1580}{2\times3600} = -\ 0.2194 \text{ kJ/kg}$$

$$\therefore \quad 2800 + \frac{1}{2\times1000}\left(\frac{4000}{60}\right)^2 + \frac{9.81\times4}{1000} - 0.2194$$

$$= 2000 + \frac{1}{2\times1000}\left(\frac{8000}{60}\right)^2 + \frac{9.81\times1}{1000} + w$$

or $\quad w = 793.363, \text{ kJ/kg}$

or $\quad \dot{W} = 793.363 \times 2 = 1586.72; \text{ kW}$ **Ans.**

This example may also be solved by **rate Eq. (6.5 or 6.6)**

Example 6.2 Steam enters a nozzle at a pressure of 7 bar and 20°C (i.e., initial enthalpy 2850 kJ/kg) and leaves at a pressure of 1.5 bar. The initial velocity of steam at the entrance is 40 m/s and the exit velocity from the nozzle is 700 m/s. The mass flow rate through the nozzle is 1400 kg/h. The heat loss from the nozzle is 11705 kJ/h. Determine the final enthalpy of steam and the nozzle area if the specific volume is 1.24 m^3/kg.

Solution Refer to Fig. 6.16.

The steady flow energy equation when expressed each term kJ/kg can be written as

$$h_1 + \frac{C_1^2}{2\times1000} + \frac{gz_1}{1000} + q = h_2 + \frac{C_2^2}{2\times1000} + \frac{gz_2}{1000} + w$$

Since $\quad z_1 - z_2 = 0 \quad \text{and} \quad w = 0$

Since the heat loss is from the system to surrounding, hence

$$q = -\frac{\dot{Q}}{\dot{m}} = -\frac{11705 \text{ kJ/h}}{\dot{m}} = -\frac{11705}{1400} = -\ 8.36 \text{ kJ/kg}$$

$$\therefore \qquad h_2 = 2850 + \frac{40^2}{2 \times 1000} - \frac{700^2}{2 \times 1000} - 8.36 = 2597.44 \text{ kJ/kg} \qquad \textbf{Ans.}$$

From the continuity equation,

$$A_2 = \frac{\dot{m}v_2}{C_2} = \frac{1400}{3600} \times \frac{1.24}{700} = 0.000688 \text{ m}^2 = 6.88 \text{ cm}^2 \qquad \textbf{Ans.}$$

The problem may also be solved by **rate Eq. (6.6)**.

Example 6.3 In a boiler, water enters with an enthalpy of 168 kJ/kg and steam leaves with enthalpy of 2925 kJ/kg. Find the heat transferred per kg of steam. The changes in kinetic and potential energies may be neglected.

Solution Refer to Fig. 6.17.

The energy equation expressed in kJ/kg is given by

$$h_1 + \frac{C_1^2}{2 \times 1000} + \frac{gz_1}{1000} + q = h_2 + \frac{C_2^2}{2 \times 1000} + \times \frac{gz_2}{1000} + w$$

Here $\qquad \Delta\text{KE} = 0, \quad \Delta\text{PE} = 0, \quad w = 0.$ Hence,

$$h_1 + q = h_2$$

$$q = h_2 - h_1 = 2925 - 168 = 2757 \text{ kJ/kg} \qquad \textbf{Ans.}$$

Hence, 2757 kJ/kg of heat has to be transferred to the water in the boiler.

Example 6.4 A reciprocating water jacketed ammonia compressor installed in cold storage delivers 3 kg/min of ammonia with an enthalpy of 1670 kJ/kg. At the suction line the enthalpy of ammonia is 1465 kJ/kg. If the power input of the compressor is 16 kW, find the rate of heat transfer to the water jacket. Changes in kinetic and potential energies may be neglected.

Solution Refer to Fig. 6.21.

In the absence of potential and kinetic energy, the energy equation for steady flow process may be written as

$$h_1 + q = h_2 + w$$

But $\qquad \dot{W} = -16 \text{ kW} = -16 \text{ kJ/s}$

or $\qquad w = \dfrac{-16 \times 60}{3} = -320 \text{ kJ/kg}$

Hence $\qquad q = (1670 - 1465) - 320 = -115 \text{ k/kg}$

or $\qquad \dot{Q} = \dot{m}q = -\dfrac{3}{60} \times 115 = -5.75 \text{ kW} \qquad$ **Ans.**

Negative sign shows the rejection of heat from the system.

The example may also be solved by rate Eq. (6.6).

Example 6.5 A reciprocating air compressor installed in a fertiliser factory takes in air at 1 bar and 20°C and delivers at 6 bar. Calculate the work done, heat transfer and change in internal energy per kg of air compressed if the compression process follows:

(a) isothermal, (b) reversible adiabatic and (c) polytropic $pv^{1.25}$ = constant. The change in potential and kinetic energies may be neglected.

Solution (a) Refer to Fig. 6.21. During isothermal process, the steady flow work done is given by

$$(q_{1-2})_{sf} = (w_{1-2})_{sf} = p_1 v_1 \ln \frac{v_2}{v_1} = RT_1 \ln \frac{p_1}{p_2}$$

$$= 0.287 \times 293 \text{ In } \frac{1}{6} = -\ 150.67 \text{ kJ/kg}$$

The –ve sign shows that heat transfer takes place from the system.

Note, $q - w = \Delta h = 0$ since T = constant

or $$q = w = -\ 150.76 \text{ kJ/kg}$$ **Ans.**

Since T is constant, hence $\Delta U = 0$ **Ans.**

(b) During adiabatic process the steady flow work is given by

$$(w)_{sf} = \frac{\gamma}{\gamma - 1}(p_1 v_1 - p_2 v_2) = \frac{\gamma}{\gamma - 1} R(T_1 - T_2)$$

But $$T_2 = T_1 \left(\frac{p_2}{p_1}\right)^{\gamma - 1/\gamma} = 293 \times (6)^{0.286} = 489.1 \text{ K}$$

$\therefore$ $$(w)_{sf} = \frac{1.4}{1.4 - 1} \times 0.287\ (293 - 489.1) = -\ 196.98 \text{ kJ/kg}$$ **Ans.**

The –ve sign shows the work is done on the air.

Since process is adiabatic, hence $q = 0$ **Ans.**

The change in internal energy is

$$\Delta u = c_v\ (T_2 - T_1) = \frac{R}{(\gamma - 1)}(T_2 - T_1) = \frac{0.287}{0.4}\ (489.1 - 293) = 140.7 \text{ kJ/kg}$$ **Ans.**

(c) During polytropic process, the steady flow work is

$$(w)_{sf} = \frac{n}{n - 1}(p_1 v_1 - p_2 v_2) = \frac{n}{n - 1} R(T_1 - T_2)$$

But $$T_2 = T_1 \left(\frac{p_2}{p_1}\right)^{(n-1)/n} = 293\,(6)^{0.25/1.25} = 419.27 \text{ K}$$

$\therefore$ $$(w)_{sf} = \frac{1.25}{0.25} \times 0.287\,(293 - 419.27) = -\,181.19 \text{ kJ/kg} \quad \textbf{Ans.}$$

Now, $$(q)_{sf} = (w)_{sf} + \Delta h = -\,181.19 + c_p\,(T_2 - T_1)$$

$$= -\,181.19 + \frac{0.287}{\left(1 - \dfrac{1}{1.4}\right)}\,(419.27 - 293) = -\,54.35 \text{ kJ/kg} \quad \textbf{Ans.}$$

$$(\Delta u)_{sf} = c_v\,(T_2 - T_1) = \frac{R}{(\gamma - 1)}(T_2 - T_1) = \frac{0.287\,(419.27 - 293)}{0.4} = 90.59 \text{ kJ/kg} \quad \textbf{Ans.}$$

Example 6.6 A boiler used in a paper factory is fed with 160000 kg/h of water having an enthalpy of 836 kJ/kg. The steam leaves the boiler for utilisation at an enthalpy of 3200 kJ/kg. The amount of coal burnt having calorific value as 30930 kJ/kg is 15000 kg/h. For the combustion process, 13 kg of air is supplied per kg of coal in the boiler at an enthalpy of 67 kJ/kg, For each kg of coal fed, 13.6 kg of gases leave the boiler plant at an enthalpy of 293 kJ/kg. Calculate the energy that is not accounted for per kg of coal if the changes in kinetic and potential energies are negligible.

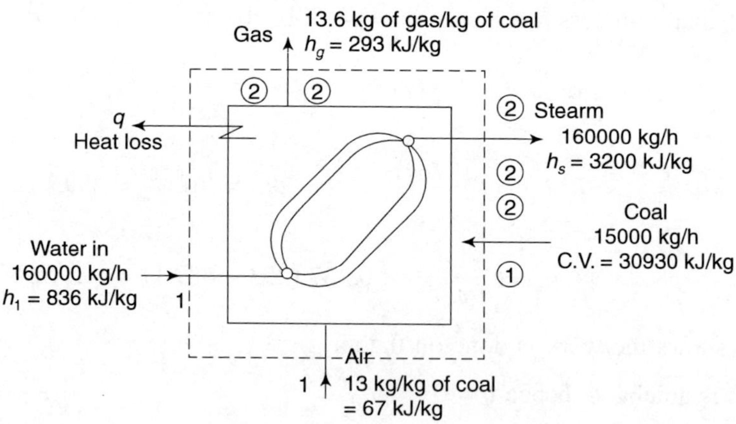

Fig. 6.23 Example 6.6

Solution Refer to Fig. 6.23.

This problem may be solved by applying the principle of energy equation. It is obvious from the data that the mass entering the control volume is greater than the mass leaving the control volume. This is because of the fact that heat unaccounted for due to radiation, etc., has to be considered.

Applying the energy equation neglecting KE and PE gives

$(h_1)_{water} + (h_1)_{coal} + (h_1)_{air} + q = (h_2)_{steam} + (h_2)_{gas} + w$

In the case of boiler, $w = 0$

$(h_1)_{water} + (h_1)_{coal} + (h_1)_{air} + q = (h_2)_{steam} + (h_2)_{gas}$

Now, writing each term as the ratio per kg of coal, we have

$$q = \frac{160000 \times 3200}{15000} + \frac{13.6 \times 15000 \times 293}{15000} - \frac{160000 \times 836}{15000} - \frac{15000 \times 30930}{15000} - \frac{13 \times 15000 \times 61}{15000}$$

$$= -\ 2522.19 \text{ kJ/kg of coal.}$$ **Ans.**

This means that 2522.19 kJ/kg of coal is rejected out from the boiler as unaccounted.

Example 6.7 An axial flow compressor of a gas turbine plant receives air from atmosphere at a pressure 1 bar, temperature 300 K and velocity 60 m/s. At the discharge of compressor, the pressure is 5 bar and the velocity is 100 m/s. The mass flow rate through the compressor is 20 kg/s. Assuming isentropic compression, calculate the power required to drive the compressor. Also calculate the inlet and outlet pipe diameters.

Soiution Refer to Fig 6.22. From first law of thermodynamics for steady flow process neglecting heat loss and changes in P.E. and each term expressed in kJ/kg, we have

$$h_1 + \frac{C_1^2}{2 \times 1000} = h_2 + \frac{C_2^2}{2 \times 1000} + w$$

or $$-\,w = (h_2 - h_1) + \frac{C_2^2 - C_1^2}{2 \times 1000} = c_p\,(T_2 - T_1) + \frac{C_2^2 - C_1^2}{2 \times 1000}$$

The final temperature is given by

$$T_2 = T_1 \left(\frac{p_2}{p_1}\right)^{(\gamma - 1)/\gamma} = 300\,(5)^{0.4/1.4} = 475.14 \text{ K}$$

$$-\,w = 1.005\,(475.14 - 300) + \frac{100^2 - 60^2}{2 \times 1000} = 179.21 \text{ kJ/kg}$$

$\therefore$ $$-\,\dot{W} = 20 \times 179.21 = 3584.2 \text{ kW}$$ **Ans.**

From continuity Eq., $A = \dfrac{\dot{m}}{\rho C}$ **Ans.**

$\therefore$ $$A_1 = \frac{20 \times 0.287 \times 300}{1 \times 10^2 \times 60} = 0.287 \text{ m}^2$$

$\therefore$ $$D_1 = \sqrt{\frac{4\,A_1}{\pi}} = 0.604 \text{ m}$$ **Ans.**

$$A_2 = \frac{20 \times 0.287 \times 475.14}{5 \times 10^2 \times 100} = 5.45 \times 10^{-2} \text{ m}^2$$

$$\therefore \quad D_2 = \sqrt{\frac{4 A_2}{\pi}} = 0.263 \text{ m}$$

Note The above example may also be solved by expressing SFEE as rate equation, i.e.,

$$\dot{m}\left(h_1 + \frac{C_1^2}{2} + gz_1\right) + \dot{Q} = \dot{m}\left(h_2 + \frac{C_2^2}{2} + gz_2\right) + \dot{W}$$

or

$$-\dot{W} = \dot{m}\left[(h_2 - h_1) + \frac{C_2^2 - C_1^2}{2 \times 10^3} + g\frac{(z_2 - z_1)}{10^3}\right]$$

$$= 20\left[1.005\,(475.14 - 300) + \frac{100^2 - 60^2}{2 \times 10^3} + 0\right] = 3584.2 \text{ kW} \qquad \textbf{Ans.}$$

Example 6.8 A metal extrudes slowly through a die under a pressure of 2200 bar. The density of the metal is 9400 kg/m^3 and its specific heat is 0.21 kJ/K. The metal may be assumed as incompressible material. Calculate the rise in temperature of the metal if there is no cooling of the metal during extruding process.

Solution Apply the general energy equation in a extruding process

$$u_1 + p_1 v_1 + \frac{C_1^2}{2} + gz_1 + q = u_2 + p_2 v_2 + \frac{C_2^2}{2} + gz_2 + w$$

The changes in potential energy, $g\,(z_2 - z_1)$ is assumed to be zero. The system is considered well insulated, hence $q = 0$.

Since the metal extruding process is very slow, the ΔKE is assumed zero.

With the above assumptions, the energy equation reduces to

$$u_1 + p_1 v_1 = u_2 + p_2 v_2$$

or

$$u_2 - u_1 = p_1 v_1 - p_2 v_2$$

Since the metal is incompressible, hence, $v_1 = v_2 = \frac{1}{9400}$ m^3/kg, $p_2 = 1$ bar

$$\therefore \quad u_2 - u_1 = (p_1 - p_2)\, v_1 = \frac{2199 \times 10^2}{9400} = 23.41 \text{ kJ/kg}$$

But

$$u_2 - u_1 = c_v\,(T_2 - T_1)$$

$$\therefore \quad T_2 - T_1 = \frac{23.4}{0.21} = 111.442°\text{C} \qquad \textbf{Ans.}$$

Example 6.9 Saturated water at enthalpy 600 kJ/kg is injected into saturated steam at an enthalpy 2400 kJ/kg. If the mixing process is accomplished at constant pressure and the mixing ratio of water to steam 1 : 10 by mass, find the enthalpy of the mixture.

Solution Neglecting the changes in kinetic and potential energies and assuming that there is no work except flow work, the first law may be written as

$$m_s h_s + m_w h_w = m_{\text{mix}} h_{\text{mix.}} = (m_s + m_w) h_{\text{mix.}}$$

$$h_{\text{mix.}} = \frac{h_s + (m_w/m_s) h_w}{1 + (m_w/m_s)} = \frac{2400 + 0.1 \times 600}{1 + 0.1} = 2236.36 \text{ kJ} \quad \textbf{Ans.}$$

Example 6.10 A small nuclear reactor is being cooled by passing liquid sodium through it. The liquid sodium leaves the reactor at 1.6 bar and 350°C. The liquid sodium is cooled to 300°C by passing through at heat exchanger before returning to the reactor.

In the heat exchanger, heat is transferred from the liquid sodium to water which enters the equipment at 70 bar and 40°C (h = 167 kJ/kg) and leaves at the same pressure as a saturated vapour (h = 2772 kJ/kg). The mass flow rate of the sodium, is 10000 kg/h and its specific heat is constant at 1.2 kJ/kg K. Calculate the mass flow rate of water evaporated in the exchanger in kg/h and the heat transfer rate between the fluid in kJ/h.

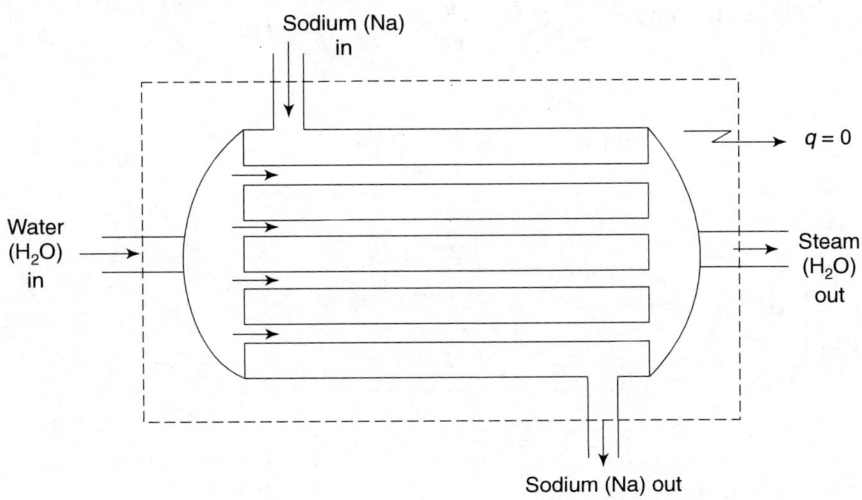

Fig. 6.24 Example 6.10

Solution Refer to Fig. 6.24. Writing the energy equation applied to the heat exchanger (Fig. 6.24) in which changes in KE and potential energies are zero and there is interaction of heat and work.

$$\dot{m}_{\text{H 20}} (h_{\text{out}} - h_{\text{in}})_{\text{H 20}} = \dot{m}_{\text{No}} (h_{\text{in}} - h_{\text{out}})_{\text{no}} = \dot{m}_{No}\, c_p\, (T_{\text{in}} - T_{\text{out}})_{\text{No}}$$

$$\dot{m}_{H_2O}\,(2772 - 167) = 10{,}000 \times 1.2\,(350 - 300)$$

or $$\dot{m}_{H_2O} = \dot{m}_{water} = 230.32 \text{ kg/h}$$ **Ans.**

Example 6.11 A turbojet engine in flight takes in air at a velocity of 280 m/s. The ambient air temperature is – 10°C. Gas temperature at the outlet of nozzle is 610°C. The air-fuel ratio is 55. The calorific value of the fluid 44 MJ/kg of fuel. As a result of incomplete combustion, 3% of chemical energy of the fuel is not released in the reaction. Heat loss from the turbojet engine is 20 kJ/kg of air. Calculate the velocity of the exhaust gas jet. Take c_p or air = 1.005 kJ/kg and c_p of gas = 1.06 kJ/kg.

Solution Refer to Fig. 6.25. The steady flow process energy equation for a turbojet engine is given by rate equation by neglecting changes in PE.

$$\dot{m}_a\left(h_{a1} + \frac{C_{a1}^2}{2}\right) + \dot{E}_{fc} + \dot{Q} = \dot{m}_g\left(h_{g2} + \frac{C_{g2}^2}{2}\right) + \dot{W}$$

where $\dot{E}_{fc}$ = actual chemical energy of fuel. Here $\dot{Q}$ is negative as there is heat loss.

Let $\dot{m}_g = \dot{m}_a + \dot{m}_f$ and $h = c_p\,(T - 0)$, and dividing throughout by $\dot{m}_a$, then

$$\left(c_{pa}\,T_{a1} + \frac{C_{a1}^2}{2}\right) + (1 - 0.03) \times \frac{\dot{m}_f}{\dot{m}_a}\text{C.V.} - \frac{\dot{Q}}{\dot{m}_a} = \frac{\dot{m}_g}{\dot{m}_a}\left(c_{pg}\,T_{g2} + \frac{C_{g2}^2}{2}\right)$$

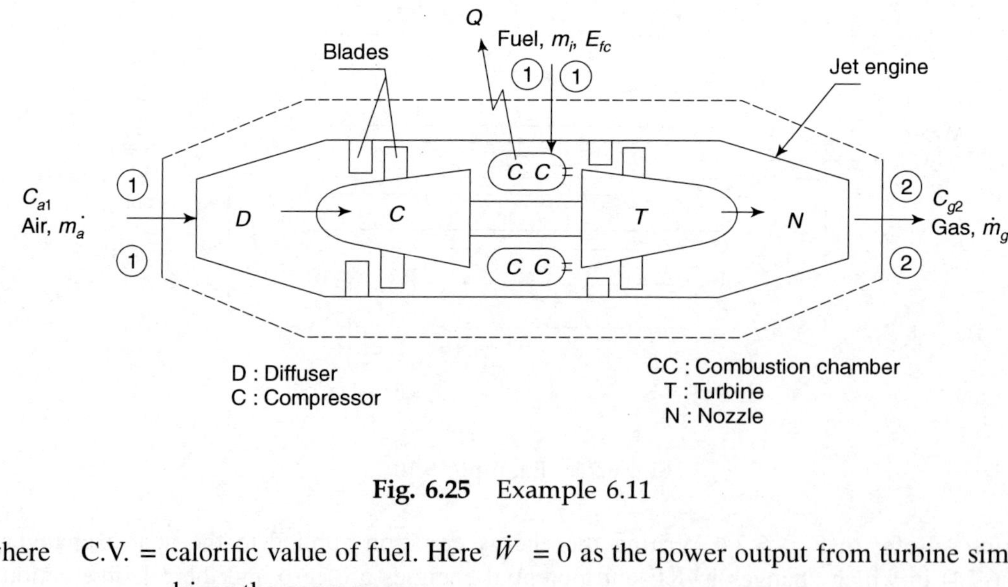

Fig. 6.25 Example 6.11

where C.V. = calorific value of fuel. Here $\dot{W}$ = 0 as the power output from turbine simply drives the compressor.

or $$1\left(1.005 \times 263 + \frac{280^2}{2 \times 1000}\right) + 0.97 \times \frac{1}{55} \times 44.44 \times 10^3 - 20$$

$$= \left(1 + \frac{1}{55}\right)\left(1.06 \times 883 + \frac{C_{g2}^2}{2 \times 1000}\right)$$

$$c_{g2} = 457.41 \text{ m/s}$$ **Ans.**

Example 6.12 An air stream of 2 kg/min with a velocity of 20 m/s and temperature 30°C is mixed with another air stream of 4 kg/min with velocity 30 m/s and temperature 40°C. 30 kJ/kg of mixture of heat is lost to the surroundings during the mixing. Draw the system with control volume and find temperature of the stream mixture leaving the system.

Take $h = c_p (T - 0)$ where $c_p = 1.04$ kJ/kg K

Solution Refer to Fig. 6.26.

From continuity equation, $\dot{m}_1 + \dot{m}_2 = \dot{m}_3$ or $\frac{2}{60} + \frac{4}{60} = \dot{m}_3$

or $$\dot{m}_3 = \frac{1}{10} \text{ kg/s}$$

Applying steady flow rate energy equation to the system neglecting changes in P.E.

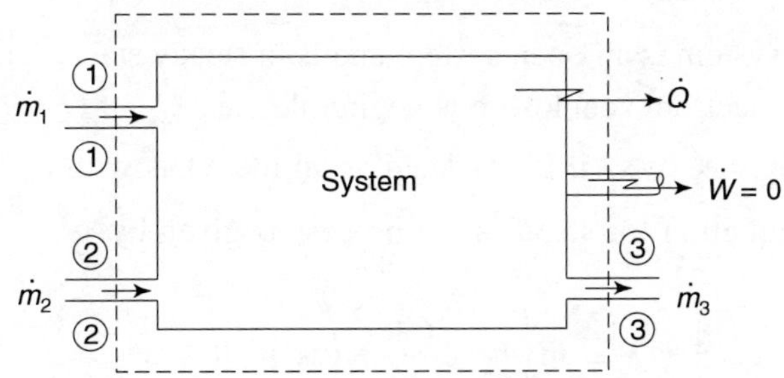

Fig. 6.26 Example 6.12

$$\dot{m}_1\left(h_1 + \frac{C_1^2}{2}\right) + \dot{m}_2\left(h_2 + \frac{C_2^2}{2}\right) + \dot{Q} = \dot{m}_3\left(h_3 + \frac{C_3^2}{2}\right) + \dot{W}$$

Here, $\dot{W} = 0$ and $\dot{Q} = -30 \times \dot{m}_3 = -30 \times \frac{1}{10}$ kW

or $$\dot{m}_1\left(c_p\, T_1 + \frac{C_1^2}{2}\right) + \dot{m}_2\left(c_p\, T_2 + \frac{C_2^2}{2}\right) + \dot{Q} = \dot{m}_3\left(c_p\, T_3 + \frac{C_3^2}{2}\right)$$

Expressing each term in kW, we have

$$\frac{2}{60}\left(1.04 \times 303 + \frac{20^2}{2 \times 1000}\right) + \frac{4}{60}\left(1.04 \times 313 + \frac{30^2}{2 \times 1000}\right) - 30 \times \frac{1}{10}$$

$$= \frac{1}{10}\left(1.04\,T_3 + \frac{15^2}{2\times 1000}\right)$$

or $$T_3 = 281.04 \text{ K}$$

or $$t_3 = 8.04°\text{C}$$

Example 6.13 For reliable and safe working of the electronic components (ECs) of a computer, they are cooled by air flowing through a fan mounted at the inlet of electronic enclosure. During steady state condition, air enters at 20°C and 1 atm. For noise control, the velocity of entering air is limited only up to 1.4 m/s. For temperature control, the temperature of the air at the exit is limited up to 33°C. The electric power input to electronic components and fan are 85 W and 20 W respectively. Calculate the smallest fan inlet diameter for which the limits on the entering air velocity and exit air temperature are met.

A Model Solution Refer to Fig. 6.27.

Data given : p_1 = 1 atm, T_1 = 20°C, $C_1 \leq$ 1.4 m/s, $T_2 \leq$ 33 °C,

$$\dot{W}_{e,\,\text{comp}} = -\,85\ W,\ \dot{W}_{\text{fan}} = -\,20\ W$$

To find : Fan size, D_1

Assumptions (i) The system is an open system and is in steady state, (ii) Heat transfer from the outer surface of the electronic enclosure is negligible, i.e., $\dot{Q}_{cv} = \dot{Q} = 0$ (iii) The changes, in PE and KE are negligible, (iv) Air is modeled as an ideal gas with c_p = 1.005 kJ/kg K.

Analysis : The rate equation for steady-flow process is given by

$$\dot{m}\left(h_1 + \frac{C_1^2}{2} + gz_1\right) + \dot{Q} = \dot{m}\left(h_2 + \frac{C_2^2}{2} + gz_2\right) + \dot{W}$$

Since $\dot{Q} = 0$, ΔKE = 0, ΔPE = 0, so

$$\dot{m}h_1 = \dot{m}h_2 + \dot{W}$$

Here $$\dot{W} = (-\,85) + (-\,20) = -\,105\ W \text{ as per convention, so}$$

$$\dot{m}h_1 = \dot{m}h_2 - 105$$

or $$\dot{m} = \frac{105}{c_p\,(T_2 - T_1)} \qquad ...(1)$$

From continuity equation at inlet

$$\dot{m}v_1 = A_1\,C_1 \quad \text{or} \quad \dot{m} = \frac{A_1\,C_1}{v_1} = \frac{A_1\,C_1 \cdot p_1}{RT_1}$$

Substituting the expression of $\dot{m}$ in Eq. (1), we have

Thus, $$\frac{A_1 C_1 p_1}{RT_1} = \frac{1.01325 \times 10^5}{c_p (T_2 - T_1)}$$

or $$A_1 = \frac{1.01325 \times 10^5}{c_p (T_2 - T_1)} \times \frac{RT_1}{p_1 C_1}$$

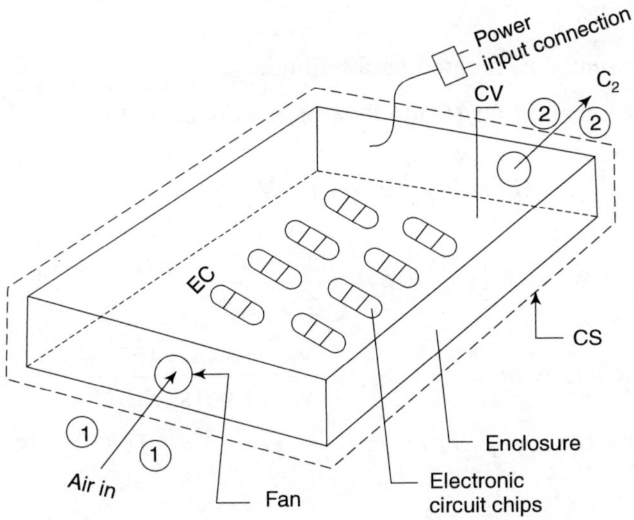

Fig. 6.27 Example 6.13

$$= \frac{1.01325 \times 10^5 \times 297 \times (273 + 20)}{1.005 \times 10^3 (33 - 20) \times 1.01325 \times 10^5 \times 1.4} = 4.757 \times 10^{-3} \text{ m}^2$$

$$A_1 = 4.757 \times 10^{-3} = \left(\frac{\pi}{4}\right) D_1^2$$

or $$D_1 = \left(\frac{4 \times 4.757 \times 10^{-3}}{\pi}\right)^{1/2} = 0.0778 \text{ m} = 7.78 \text{ cm}$$ **Ans.**

Thus, the minimum fan inlet diamter is 7.78 cm.

Example 6.14 In order to clean the sidings of a house, a power washer machine is used. Water enters at 20°C, 1 atm with a volumetric flow rate of 0.15 L/s through a 2.8 cm diameter hose. A jet of water comes out at 24°C, 1 atm, with a velocity of 60 m/s at an elevation of 5 m. At the steady state condition, heat loss from the power washer to the surrounding is 10% of power input. Calculate the power input to the motor of power washer. Take specific volume of water equal to 1.0018 × 10 J, m^3/kg.

A Model Solution Refer to Fig. 6.28.

Data given p_1 = 1 atm, T_1, = 20°C, = 0.15 L/s, D_1, = 2.8 cm, T_2 = 24°C, p_2 = 1 atm, C_2 = 60 m/s, z_2 = 5 m, z_1 = 0, $\dot{Q} = -0.1\ \dot{W}_{input}$, $v_1 = 1.0018 \times 10^{-3}\ m^3/kg$

To find Power input, $\dot{W}$ in kW

Assumptions (i) A control volume enclosing the power input and the delivering hose is at steady state (Fig. 6.28).

(ii) The water is assumed as incompressible fluid.

Analysis : The rate equation for steady flow process is given by

$$\dot{m}\left(h_1 + \frac{C_1^2}{2} + gz_1\right) + \dot{Q} = \dot{m}\left(h_2 + \frac{C_2^2}{2} + gz_2\right) + \dot{W}$$

or
$$\dot{W} = \dot{m}\left[(h_1 - h_2) + \frac{C_1^2 - C_2^2}{2 \times 10^3} + \frac{g(z_1 - z_2)}{10^3}\right];\ kW$$

$$\text{The mass flow rate} = \dot{m} = A_1\, C_1\, \rho_1 = \frac{\dot{V}_1}{v_1} = \frac{0.15 \times 10^{-3}}{1.0018 \times 10^{-3}} = 0.1497\ kg/s$$

$$h_1 - h_2 = (u_1 - u_2) + (p_1\, v_1 - p_2\, v_2) = c_v\,(T_1 - T_2) + 0 = 4.18\,(20 - 24) = -\ 16.72\ kJ/kg$$

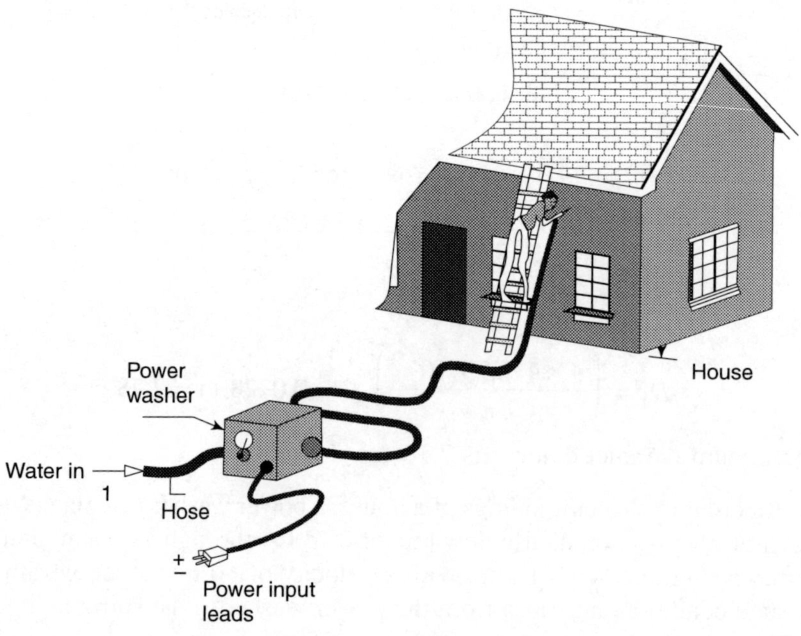

Fig. 6.28 Example 6.14

Here, $$C_1 = \frac{V_1}{A_1} = \frac{0.15 \times 10^{-3}}{(\pi/4)(2.8/100)^2} = 0.243 \text{ m/s}$$

$$\therefore \quad \frac{C_1^2 - C_2^2}{2 \times 10^3} = \frac{0.243^2 - 60^2}{2 \times 10^3} = -1.7999 \text{ kJ/kg}$$

and $$\frac{g(z_1 - z_2)}{10^3} = \frac{9.81\,(0 - 5)}{10^3} = -0.04905 \text{ kJ/kg}$$

Substituting values, we have

$\dot{W} = 0.1497\,[(-16.72) + (-1.7999) + (-0.04905)] = -2.779$ kW **Ans.**

Here, the minus sign indicates the work input to the power washer.

Example 6.15 There exists a combination of a heat exchanger, a turbine and a nozzle in series. Air at a temperature of 20°C passes through a heat exchanger at a velocity of 40 m/s where its temperature is raised to 1000°C. The heated air then enters a turbine with the same velocity of 40 m/s and expands until temperature falls to 700°C. On leaving the turbine, the exhaust air from the turbine enters a nozzle with velocity 70 m/s where it expands until the temperature falls to 500°C. The mass flow rate through each unit is 10 kg/s. Draw the system and calculate (a) the rate of heat transfer to the air in the heat exchanger, (b) the power output from the turbine assuming no heat loss, and (c) the velocity of air at the exit of the nozzle assuming no heat loss.

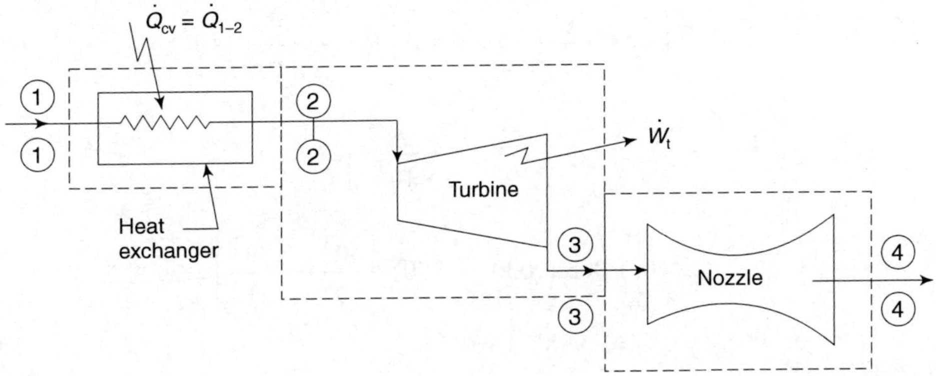

Fig. 6.29 Example 6.15

Solution. Refer to Fig. 6.29.

Data given $t_1 = 20$°C, $C_1 = 40$ m/s, $t_2 = 1000$°C, $C_2 = 40$ m/s, $t_3 = 700$°C, $C_3 = 70$ m/s, $t_4 = 500$°C, $\dot{m} = 10$ kg/s.

To find $\dot{Q}_{\text{each}}$, $\dot{W}_t$ and $C_{nj} = C_4$

Assumptions (i) The flow is in steady state and the control volume encloses all elements separately.

(ii) The heat loss from turbine and nozzle is negligible, i.e., $\dot{Q}_{cv} = \dot{Q} = 0$ and there is no work done from heat exchanger and nozzle.

(iii) The air is modeled as an ideal gas.

(iv) Changes in PE is negligible.

Analysis SFEE in rate form applied over heat exchanger yields.

$$\dot{m}\left(h_1 + \frac{C_1^2}{2} + gz_1\right) + \dot{Q}_{t-2} = \dot{m}\left(h_2 + \frac{C_2^2}{2} + gz_2\right) + \dot{W}_{1-2}$$

Using assumptions, the above equation yields,

$$\dot{m}h_1 + \dot{Q}_{1-2} = \dot{m}h_2 \text{ as } C_1 = C_2$$

(a) $\dot{Q}_{1-2} = \dot{m}\,(h_2 - h_1) = \dot{m}\,c_p\,(T_2 - T_1) = 10 \times 1.005\,(1000 - 20) = 9844$ kW **Ans.**

(b) Applying SFEE in rate form across turbine, we have

$$\dot{m}\left(h_2 + \frac{C_2^2}{2} + gz_2\right) + \dot{Q}_{2-3} = \dot{m}\left(h_3 + \frac{C_2^2}{2} + gz_3\right) + \dot{W}_{2-3}$$

Using assumptions, the above equation takes the form

$$\dot{m}\left(h_2 + \frac{C_2^2}{2\times10^3}\right) + 0 = \dot{m}\left(h_3 + \frac{C_3^2}{2\times10^3}\right) + \dot{W}_{2-3}$$

or

$$\dot{W}_{2-3} = \dot{m}\left[c_p\,(T_2 - T_3) + \frac{C_2^2 - C_3^2}{2\times10^3}\right]$$

$$= 10\left[1.005\,(1000 - 700) + \frac{40^2 - 70^2}{2 \times 10^3}\right] = 2998.5 \text{ kW}$$

or

$$\dot{W}_{2-3} = \dot{W}_t = 2998.5 \text{ kW}$$ **Ans.**

(c) SFEE in rate form applied over nozzle yields

$$\dot{m}\left(h_3 + \frac{C_3^2}{2} + gz_3\right) + \dot{Q}_{3-4} = \dot{m}\left(h_4 + \frac{C_4^2}{2} + gz_4\right) + \dot{W}_{3-4}$$

Using assumptions, we have

$$\dot{m}\left(h_3 + \frac{C_3^2}{2}\right) + 0 = \dot{m}\left(h_4 + \frac{C_4^2}{2}\right) + 0$$

or $$\frac{C_4^2}{2} = (h_3 - h_4) + \frac{C_3^2}{2}$$

or $$C_4 = \left[2c_p(T_3 - T_4) + C_3^2\right]^{1/2}$$

$$= [\,2 \times 1.005 \times 10^3\,(\,700 - 500) + 70^2]^{1/2} = 637.88 \text{ m/s}$$ **Ans.**

6.12 UNSTEADY FLOW PROCESSES

It is very difficult to solve the general energy equation for unsteady flow process. But in some special cases, its solution is easily available. For such cases, the conditions necessary are given below.

(i) There is only one flow boundary

(ii) e_f = constant.

(iii) $e = e_{cv} = u$ (i.e., the energy is only in the from of internal energy)

Case 1 : Flow boundary L_2 Absent This is a case of **filling process**. The unsteady flow energy Eq. (6.22) reduces to

$$Q_{cv} - W_{cv} = -\left[\int_{\tau_1}^{\tau_2} dm \cdot e_f\right]_{L_1} + (m_2 e_2)_{\tau_2} - (m_1 e_1)_{\tau_1} \qquad ...(6.66)$$

For simplicity, $Q_{ev} = Q$ and $W_{cv} = W$

Since e_f = constant and $e_2 = u_2$, $e_1 = u_1$ hence

$$\therefore \quad \left[\int_{\tau_1}^{\tau_2} dm\right]_{L_2} = (m_2 - m_1) \qquad ...(6.67)$$

Equation (6.67) suggests that the mass flowing to the control volume from time τ_1 to τ_2 is equal to the increase in the mass within the control volume from time τ_1 to τ_2. Thus,

$$\therefore \quad Q - W = m_2\, u_2 - m_1\, u_1 - (m_2 - m_1)\, e_f \qquad ...(6.68)$$

Case 2 : Flow Boundary L_1 Absent This is the case of **emptying process**. The unsteady flow energy Eq. (6.22) reduces to

$$Q_{cv} - W_{cv} = -\left[\int_{\tau_1}^{\tau_2} dm \cdot e_f\right]_{L_2} + (m_2\, e_2)_{\tau_1} - (m_1\, e_1)_{\tau_2} \qquad ...(6.69)$$

e_f = Constant, $e_1 = u_1$, $e_2 = u_2$ hence

$$\int_{\tau_1}^{\tau_2} [dm]_{L_1} = (m_1 - m_2) \qquad ...(6.70)$$

$$\therefore \qquad Q - W = m_2\, u_2 - m_1\, u_1 + (m_1 - m_2)\, e_f$$

$$Q - W = (m_2\, u_2 - m_1\, u_1) - (m_2 - m_1)\, e_f \qquad ...(6.71)$$

Comparing Eqs. (6.68) and (6.71), we find that the same equation is obtained for both the above cases, no matter whether the flow boundary L_2 *or* L_1 *is absent. These two cases are found in practice, for example, filling and emptying processes.*

6.13 FILLING PROCESS

Let us consider a *filling process in which a bottle is to be filled from a large reservoir* as shown in Fig. 6.30. The work interaction is possible by change in volume of the bottle or by internal electrical or mechanical devices. Also, heat interaction may be possible between system and the surrounding. But, if the reservoir is too large compared to the bottle, the properties of the entering fluid stream are essentially constant. Then e_f is also constant. In this filling process, the flow boundary L_2 is absent, hence the Eq. (6.68) is applicable, i.e.,

$$Q - W = m_2\, u_2 - m_1\, u_1 - (m_2 - m_1)\, e_f \qquad ...(6.72)$$

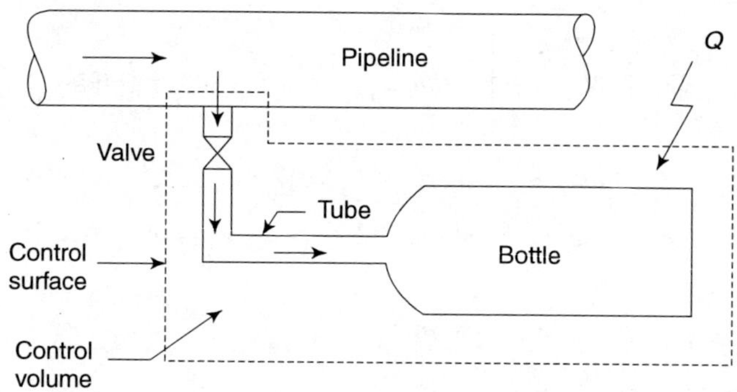

Fig. 6.30 Filling process

6.14 EMPTYING PROCESS

In the emptying process, the fluid flows from the system to the surrounding. This process is reverse of filling process. If the surrounding is too large compared to the system, e_f is essentially constant. Also, the flow boundary L_1 is absent here. Thus, Eq. (6.71) may be used to describe the flow.

$$Q - W = m_2\, u_2 - m_1\, u_1 - (m_2 - m_1)\, e_f \qquad ...(6.73)$$

But in most of the practical cases, the velocity of flowing stream C_f is not constant with time and thus e_f is not a constant. *Therefore, Eq. (6.71 or 6.73) is not directly applicable for such practical processes.*

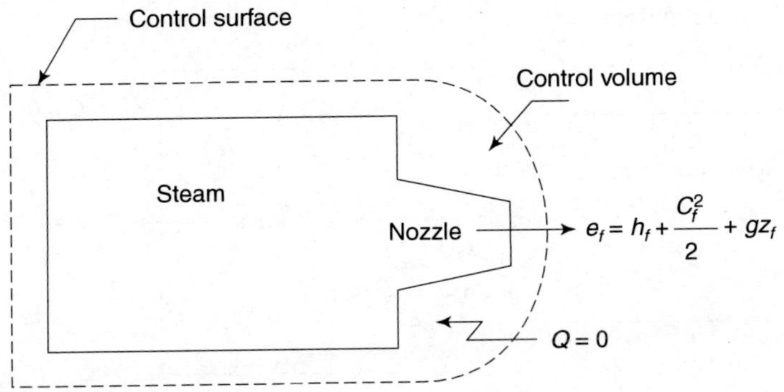

Fig. 6.31 Emptying process

As an example, consider a nozzle exhausting the gas in the atmosphere (Fig. 6.31). In this case, the fluid within the system undergoes an isentropic expansion and flows to the surrounding at constant pressure. It is to be noted here that all elements of fluid leaving the system have the same values of pressure and temperature but each element has a different velocity. The reasons of this are given below.

Let us consider the first element of the fluid leaving the system. This element undergoes expansion in reversible manner in the nozzle from pressure p_1 to p_f resulting in a high velocity and drop of temperature from T_1 to T_f.

But the second element may undergo reversible adiabatic expansion in two phases–partly within the system itself from p_1 to a slightly lower pressure and then in the nozzle. Thus, the second element leaves with small velocity at the exit of the nozzle compared to the first element due to the lower pressure range in the nozzle. Since the overall expansion is the same, i.e., p_1 to p_f, hence the exit temperature will also be T_f. Thus, the last fluid element undergoes the entire expansion within the system, and therefore, has negligible enthalpy at exit from the nozzle.

Let $\quad h_f$ = enthalpy of flow at the nozzle exit.

$\quad e_f$ = energy of flow at the nozzle exit.

From the above statement, it is clear that although h_f = constant, e_f is variable, this suggests that Eq. (6.68) is not applicable in this case.

However, if the control volume consists of a impulse turbine adjacent to the nozzle as shown in Fig. 6.32, then Eq. (6.68) is applicable because the kinetic energy from the nozzle can be converted completely into work without changing the enthalpy of the fluid, hence

$$Q - W_{rev} = m_2 u_2 - m_1 u_1 - (m_2 - m_1) e_f$$

As $\quad Q = 0,$

Thus, $$-W_{rev} = (m_2 u_2 - m_1 u_1) - (m_2 - m_1) e_f \qquad \text{...(6.74)}$$

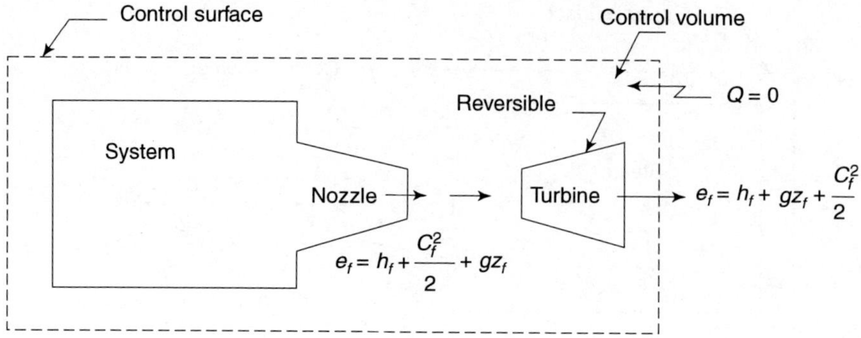

Fig. 6.32 Turbine fed by fluid being emptied

Example 6.16 An insulated rigid tank with zero heat capacity contains 10 kg of air at 2 bar and 40°C. It is quickly filled with air from a large reservoir at 15 bar and 60°C. Find the final state of air in the tank and the amount added, c_p =1.03 and c_v = 0.71 kJ/kg K.

Solution This is a filling process example. Let suffixes 1 and 2 denote the conditions of air in tank before and after filling process. The energy equation for filling process is given by

$$Q - W = m_2 u_2 - m_1 u_1 - (m_2 - m_1) e_f$$

Here, condition 1 and 2 refer to before and after the filling process which is quick.

Here, $$m_1 = \frac{p_1 V_1}{RT_1} \text{ and } m_2 = \frac{p_2 V_2}{RT_2}$$

$$\therefore \quad \frac{m_2}{m_1} = \frac{p_2 V_2}{RT_2} \cdot \frac{RT_1}{p_1 V_1} = \frac{p_2}{p_1} \frac{T_1}{T_2} \quad \text{since } V_1 = V_2$$

$$= 7.5 \frac{T_1}{T_2}$$

Since $Q = 0$ and $W = 0$, hence the above energy equation becomes

$$m_2 u_2 - m_1 u_1 - (m_2 - m_1) e_f = 0$$

But $u_2 = c_v T_2$, $u_1 = c_v T_1$ and $e_f = c_p T_{\text{reservoir}}$

$$\left(\frac{7.5 T_1}{T_2} c_v T_2\right) - c_v T_1 - \left(\frac{7.5 T_1}{T_2} - 1\right) c_p T_{\text{reservoir}} = 0$$

$$\therefore \quad 6.5 c_v T_1 - \left(\frac{7.5 T_1}{T_2} - 1\right) c_p T_{\text{reervoir}} = 0$$

$$6.5 \times 0.71 (273 + 40) - \left(\frac{7.5 (273 + 40)}{T_2} - 1\right) \times 1.03 \times (273 + 60) = 0$$

or $$T_2 = 444.56 \text{ K}$$ **Ans.**

Since $$\frac{m_2}{m_1} = 7.5 \frac{T_1}{T_2}$$

or $$m_2 = \frac{7.5 \times 313 \times 10}{444.56} = 52.80 \text{ kg}$$

∴ $$\text{Mass added} = 52.80 - 10 = 42.80 \text{ kg}$$ **Ans.**

Example 6.17 In an internal combustion engine at the end of the exhaust stroke, its combustion chamber is filled with 0.000 11 kg of exhaust gas at 750 K and atmosphere pressure. The piston then descends on the intake stroke sucking 0.0016 kg of air at 293 K in the cylinder. If the suction process is assumed to take place at constant atmospheric pressure and if the heat transfer is negligible, find the temperature and pressure of the mixture at the end of the intake stroke.

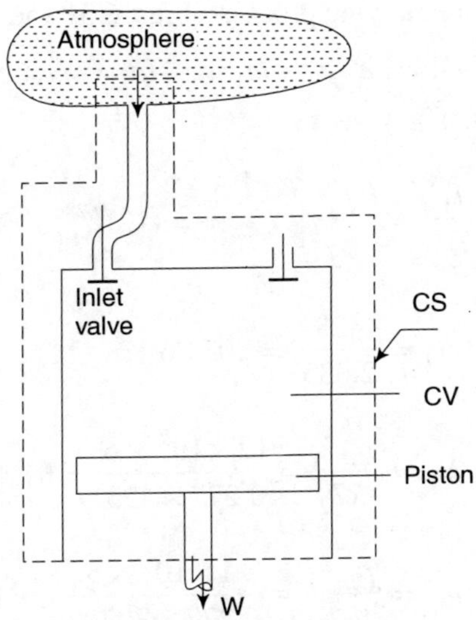

Fig. 6.33 Example 6.17

Solution Refer to Fig. 6.33. This problem is a filling process one. Let suffixes 1 and 2 denote the condition of the working fluid before and after the intake stroke. The energy equation may be written as

$$m_2 u_2 - m_1 u_1 - (m_2 - m_1) e_f = Q - W$$

Since there is no heat interaction, hence $Q = 0$

Thus, suction occurs at constant pressure, hence

$$W = \int_1^2 pdV = p (V_2 - V_1) = p (m_2 v_2 - m_1 v_1)$$

$$\therefore \quad m_2 u_2 - m_1 u_1 - (m_2 - m_1) e_f = -p (m_2 v_2 - m_1 v_1)$$

or
$$pm_1 v_1 + m_1 u_1 + (m_2 - m_1) e_f = p . m_2 v_2 + m_2 u_2$$

or
$$m_1 h_1 + (m_2 - m_1) e_f = m_2 h_2$$

or
$$m_1 c_p T_1 + (m_2 - m_1) c_p T_{\text{reservoir}} = m_2 c_p T_2$$

or
$$0.00011 \times 1.005 \times 750 + (0.0016 - 0.00011) \times 1.005 \times 293 = 0.0016 \times 1.005 \times T_2$$

or
$$T_2 = 324.41 \text{ K} \quad \text{and } p_2 = 1 \text{ atm} = \text{suction pressure}$$ **Ans.**

Example 6.18 An insulated and rigid tank contains 6 m^3 of air at 12 bar and 150°C. This air is allowed to expand to 1 bar. Find the maximum work that can be obtained from the escaping air in a diabatic process. Take $\gamma = 1.4$ and $R = 0.277$ kJ/kg K, $c_p = 1.005$ and $c_v = 0.728$ kJ/kg K.

Solution This is an example of emptying process. (Figs. 6.31 and 6.32). The energy equation is

$$Q - W = m_2 u_2 - m_1 u_1 - (m_2 - m_1) e_f$$

Since the tank is insulated, $Q = 0$

$$\frac{T_1}{T_f} = \left(\frac{p_1}{p_f}\right)^{(\gamma - 1)/\gamma} = (12)^{0.286} = 2.035$$

$$T_f = \frac{423}{2.035} = 207.86 \text{ K}$$

$$m_1 = \frac{p_1 V_1}{RT_2} = \frac{12 \times 10^2 \times 6}{0.277 \times 423} = 61.448 \text{ kg}$$

$$m_2 = \frac{p_2 V_2}{RT_1} = \frac{1 \times 10^2 \times 6}{0.277 \times 207.86} = 10.42 \text{ kg}$$

Also
$$e_f = c_p T_f$$

where, T_f is the final or reservoir temperature, putting all these values

$$-W = m_2 c_v T_f - m_1 c_v T_1 - (m_2 - m_1) c_p T_f$$

$$-W = 10.42 \times 0.728 \times 207.86 - 61.448 \times 0.728 \times 423 - (10.42 - 61.448) \times 1.005 \times 207.86$$

or
$$W = 6686.06 \text{ kJ}$$ **Ans.**

Example 6.19 A pressure vessel of volume 0.45 m^3 contains an electric heating coil but it is thermally insulated. The vessel is equipped with a release valve which allows the gas to escape when the pressure in the vessel reaches 2 bar. The pressure vessel is filled with nitrogen at a pressure of 1.35 bar and a temperature of 15°C and the gas is then heated until the temperature in the vessel is 160°C. Find the heat transfer (a) up to the time when the release valve opens and (b) after this time.

Assume $c_p = 1.045$ and $c_v = 0748$ kJ/kg K.

Solution Refer to Fig. 6.34. We have, $R = c_p - c_v = 1.045 - 0.748 = 0.297$ kJ/kg K

$$m = \frac{p_1 V_1}{RT_1} = \frac{1.35 \times 10^2 \times 0.45}{0.297 \times 288} = 0.71 \text{ kg}$$

(a) Till the release valve opens, the heating is at constant volume with constant mass, Thus,

$$T_2 = T_1 \frac{p_2}{p_1} = 288 \times \frac{2}{1.35} = 426.66 \text{ K}$$

$$Q_{1-2} = mc_v (T_2 - T_1) = 0.71 \times 0.748 \times (426.66 - 288) = 73.639 \text{ kJ} \quad \textbf{Ans.}$$

(b) After the release of valve, due to heating of the gas, temperature will exceed T_2 very slightly, causing a corresponding rise in pressure, and hence an escape of the gas. For a steady state of heating, there will be corresponding steady flow of gas from the vessel.

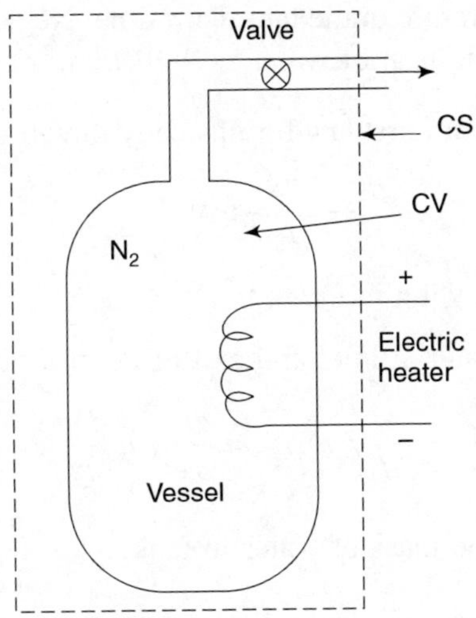

Fig. 6.34 Example 6.19

Avoiding sign complications, the principle of conservation of energy gives.

$$U_1 + Q_{1-2} + (U)_{\text{incoming}} + (W)_{\text{incoming mass.}} = U_2$$

where suffices 1 and 2 refer before and after release. The sum of the internal energy and the work of introduction of the incoming mass is the enthalpy of incoming mass. Hence, for a mass (dm) entering the vessel.

$$m\, c_v\, T + \delta Q + dm\, c_p\, (T + dT) = (m + dm)\, c_v\, (T + dT)$$

Multiplying out and neglecting second order terms,

$$m\, c_v\, T + \delta Q + c_p\, T\, dm = m\, c_v\, T + m\, c_v\, dT + c_v\, T\, dm$$

or

$$\delta Q + c_p\, T\, dm = m\, c_v\, dT + c_v\, T\, dm$$

But

$$m = \frac{pV}{RT} \quad \therefore \quad dm = -\frac{pV}{RT^2}\, dT$$

$\therefore$

$$\delta Q = -c_p\, T\left(\frac{-pV}{RT^2}\, dT\right) + c_v\left(\frac{pV}{RT}\, dT - T\,\frac{pV}{RT^2}\, dT\right) = c_p\,\frac{pV}{R}\cdot\frac{dT}{T}$$

$$Q_{1-2} = c_p\,\frac{pV}{R}\ln\frac{T_2}{T_1} = \frac{1.045 \times 2 \times 10^2 \times 0.45}{0.297}\ln\frac{433}{426.66} = 4.67 \text{ kJ} \quad \textbf{Ans.}$$

Example 6.20 A tank containing 50 kg of water, initially at 40°C, has one inlet and one exit with equal mass flow rate. Water enters the tank at 40°C at the rate of 200 kg/h. Heat energy is removed from the water at the rate of 8 kW by means of a cooling coil immersed in water. The water is thoroughly mixed by a mechanical stirrer so that the water temperature in the tank is uniform. The power input to the water from the stirrer is 0.3 kW. Derive an expression for the variation of water temperature in the tank with time. Neglect the effects of kinetic and potential energy. Assume specific heat of water as 4.2 kJ/kg.

Solution Refer to Fig. 6.35. From first law for unsteady flow process, we have rate equation

$$\dot{Q} - \dot{W} + \dot{m}_f\,(h_1 - h_e) = \frac{dE}{d\tau} = \frac{dU}{d\tau}\text{; kW}$$

where $\dot{m}_f$ = mass of fluid flowing in kg/s

h_1 and h_e = enthalpies of water at inlet and exit of control volume in kJ/kg

Here,

$$\frac{dU}{d\tau} = \frac{d}{d\tau}(mu) = m\frac{du}{d\tau} + u\frac{dm}{d\tau}$$

Since $\frac{dU}{d\tau} = 0$, where m is the mass of water in tank.

Let T be the temperature of water in tank at time τ. Assuming T to be uniform,

We have,

$$mc_p\,\frac{dT}{d\tau} = \dot{Q} - \dot{W} + m_f\, c_p\,(T_i - T)$$

or

$$(50 \times 4.2)\frac{dT}{d\tau} = -8 - (-0.3) + \frac{200}{3600} \times 4.2\,(T_i - T)$$

Here, T_i = 40°C, so the above equation becomes

$$(50 \times 4.2)\frac{dT}{d\tau} = -7.7 + \frac{200}{3600} \times 4.2\,(40 - T)$$

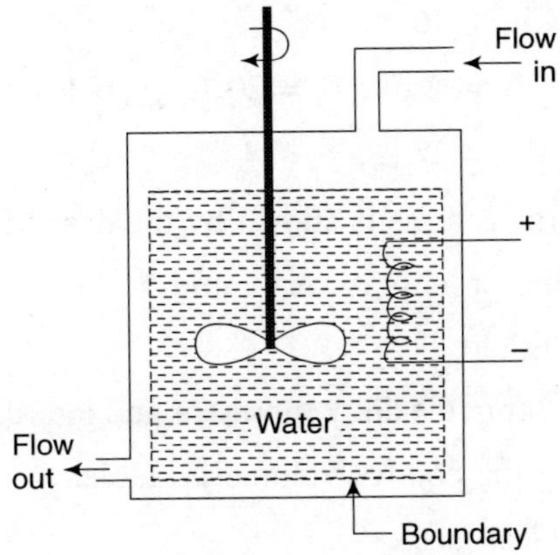

Fig. 6.35 Example 6.20

or
$$\frac{dT}{d\tau} = (0.0084 - 0.00111\,T)$$

Integrating the above equation and using boundary conditions, $T = 40°C$ to T and $\tau = 0$ to τ, we have

$$\int_{T=40°C}^{T=T} \frac{dT}{(0.0084 - 0.00111\,T)} = \int_{\tau=0}^{\tau=\tau} d\tau$$

or
$$T = 7.56 + 32.462\, e^{-0.0111\tau}$$ **Ans.**

Example 6.21 An air compressor is used to fill rapidly a 2 m^3 tank at 20°C, 1 atm. The filling process is governed by the law $pv^{1.4}$ = constant. The kinetic and potential energy effects are negligible. If the ratio of the final to the initial mass of air in the tank is 3, calculate the work input to the compressor.

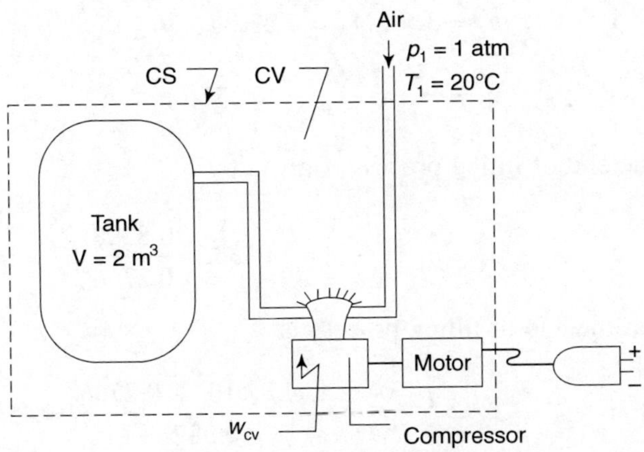

Fig. 6.36 Example 6.21

A Model Solution Refer to Fig. 6.36.

Data given $V = 2\ \text{m}^3$, $T_1 = 20°\text{C}$, $p_1 = 1$ atm, $pv^{1.4}$ = constant.

To find $W_{cv} = W_{\text{input}}$

Assumptions (i) The control volume is shown by the dashed line enclosing the diagram.

(ii) $Q_{cv} = Q = 0$ as the filling process is very rapid.

(iii) ΔKE and ΔPE are negligible.

(iv) The mass of air stored within the air compressor and interconnecting pipe is negligible.

(v) The state of air entering the control volume is stable.

(vi) The air behaves as an ideal gas.

(vii) The reference for evaluating u and h is taken as 0°C.

Analysis The unsteady flow filling process is

$$Q - W = m_2\, u_2 - m_1\, u_1 - (m_2 - m_1)\, e_f$$

Here conditions 1 and 2 refer to before and after the filling process in the tank.

Here
$$e_f = \left(h + \frac{C^2}{2} + gz\right)_{\text{reserve}} = h_{\text{atm}} = c_p\,(20 - 0) = 1.005 \times 20 = 20.1 \text{ kJ/kg}$$

$$v_1 = \frac{RT_1}{p_1} = \frac{0.287 \times (273 + 20)}{1.01332 \times 10^2} = 0.8298 \text{ m}^3\text{/kg}$$

The initial mass in the tank= $m_1 = \dfrac{V}{v_1} = \dfrac{2}{0.8298} = 2.41$ kg

At the final condition, $\dfrac{m_2}{m_1} = 2$

or
$$m_2 = 3 \times 2.41 = 7.23 \text{ kg}$$

∴
$$v_2 = \frac{V}{m_2} = \frac{2}{7.23} = 0.2766 \text{ m}^3\text{/kg}$$

The pressure at the end of filling process is given by

$$p_2 = p_1 \left(\frac{v_1}{v_2}\right)^{\gamma} = 1.01335 \left(\frac{0.8298}{0.2766}\right)^{1.4} = 4.717 \text{ bar}$$

The temperature at the end of filling process is

$$T_2 = \frac{p_2\, v_2}{R} = \frac{4.717 \times 10^2 \times 0.2766}{0.287} = 454.6 \text{ K} = 181.6°\text{C}$$

$$u_2 = c_v\,(T_2 - 0) = 0.718\,(181.6 - 0) = 130.388 \text{ kJ/kg}$$

Substituting the values in energy equation

$$-W = 7.23 \times 130.388 - 2.41 \times 14.36 - (7.23 - 2.41) \times 20.1 = 811.21 \text{ kJ} \quad \textbf{Ans.}$$

Negative sign shows the work input to compressor during filling process.

Example 6.22 A mass of air is stirred by a paddle wheel in an insulated constant volume tank. There is a negligible contribution of velocity due to stirring to the internal energy of air. Air flows out through a small valve in the tank at a rate such that the temperature of the air in the tank remains constant. At a certain instant, the conditions are as follows : Tank volume = 0.2 m^3, pressure = 1 MPa, temperature 200°C, the power input to paddle wheel is 0.15 kW. Calculate the rate of flow of air out of tank at this instant by assuming internal energy of air, $u = u_0 + 0.72t$ where u is in kJ/kg, u_0 is in kJ/kg and t is in °C.

Solution Refer to Fig. 6.37.

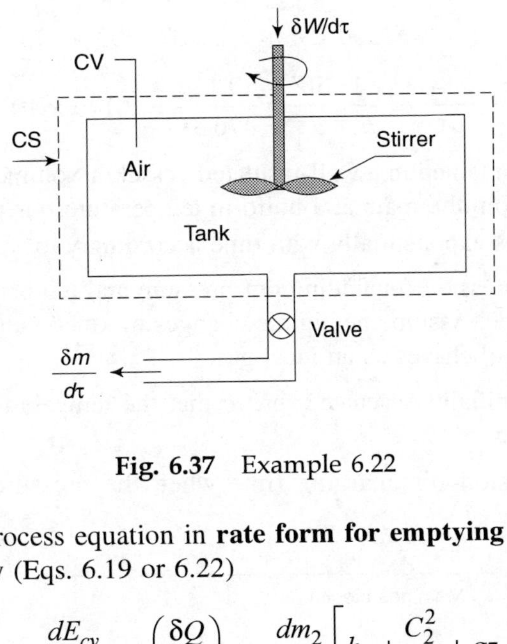

Fig. 6.37 Example 6.22

The unsteady flow process equation in **rate form for emptying process** (flow boundary L_1 is absent) is given by (Eqs. 6.19 or 6.22)

$$\frac{dE_{cv}}{d\tau} = \left(\frac{\delta Q}{d\tau}\right)_{cv} - \frac{dm_2}{d\tau}\left[h_2 + \frac{C_2^2}{2} + gz_2\right]_{L_2} - \left(\frac{\delta W}{d\tau}\right)_{cv}$$

Here, $$\left(\frac{\delta Q}{d\tau}\right)_{cv} = 0, \quad C_2 = 0, \quad z_2 = 0$$

Hence, using –ve sign convention for work input, we have

$$\frac{dE_{cv}}{d\tau} = \left(\frac{\delta W}{d\tau}\right)_{cv} - \frac{dm_2}{d\tau}[h_2]$$

Here the condition 2 refers to any instant in the tank, so, since there is no change in the internal energy in the tank, so $dE_{cv}/d\tau = 0$.

Thus, $$\frac{\delta W}{d\tau} = \frac{dm_2}{d\tau} h_2$$

Since $$h = u + pv$$

Let $$u = 0, \text{ at } t = 0 \text{ K (reference temperature, i.e., } -273°\text{C)}$$

Since $$u = u_0 + 0.72\, t$$

$\therefore$ $$0 = u_0 + 0.72\,(-273)$$

or $$u_0 = 0.72 \times 273$$

At t°C : $$u = 0.72 \times 273 + 0.72\, t = 0.72\,(t + 273) \text{ kJ/kg}$$

Thus, $$h_2 = 0.72\,(t + 273) + 0.287\,(t + 273) = 1.007\,(t + 273)$$

At $$t = 200°\text{C} : h_2 = 1.007\,(200 + 273) = 476.31 \text{ kJ/kg}$$

Thus, $$\frac{dm_2}{d\tau} = \frac{1}{h_2}\frac{\delta W}{d\tau} = \frac{0.15}{476.31} = 3.149 \times 10^{-4} = 1.1337 \text{ kg/h}$$ **Ans.**

Example 6.23 A gas is contained in a well-insulated vessel of volume V at a pressure p_0 and temperature t_0. The gas from the main at a uniform temperature t_1 is pumped into the vessel and the flow rate decreases exponentially with time accordingly to $\dot{m} = \dot{m}_0\, e^{-a\tau}$ where a is constant. Develop the expression for finding out pressure and temperature of the gas in the vessel as a function of time. Assume negligible changes in kinetic and potential energies of the gas in the tank and gas behaves as an ideal gas.

(a) If the vessel was initially evacuated, prove that the temperature inside the vessel is independent of time.

(b) Derive the expression for charging time when the pressure inside the vessel is independent of time.

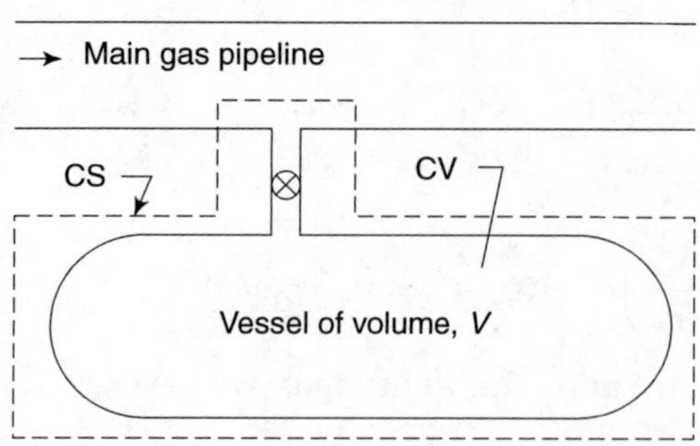

Fig. 6.38 Example 6.23

Solution Refer to Fig. 6.38.

The unsteady flow energy equation in rate form for filling process (L_2 is absent) is given by from Eq. 6.19 or Eq. 6.22,

$$\frac{dE_{cv}}{d\tau} = \left(\frac{\delta Q}{d\tau}\right)_{cv} + \frac{dm_1}{d\tau}\left[h_1 + \frac{C_1^2}{2} + gz_1\right]_{L_1}$$

Neglecting potential and kinetic energies, and let $(\delta Q/d\tau)_{cv} = 0$, we have

$$\frac{dE_{cv}}{d\tau} = \frac{dm_1}{d\tau} h_1 = h_1 \dot{m}_0 e^{-a\tau}$$

where h_1 is the enthalpy of the gas in the main.

Integration of the above equation results in

$$E_{cv} = E = E_0 + \frac{h_1 \dot{m}_0}{a}(1 - e^{-a\tau})$$

where E_0 is the initial energy of the vessel at the beginning of charging process (i.e., $E = E_0$ at $\tau = 0$)

Neglecting changes in KE and PE, we have

$$Mu = M_0 u_0 + \frac{\dot{m}_0}{a}(1 - e^{-a\tau})(u_1 + p_1 v_1) \qquad \text{...(i)}$$

where M_0 is the initial mass of the air in the vessel at the beginning of charging process ($M = M_0$ at $\tau = 0$). Here M and M_0 are used to differentiate between $\dot{m}_0$ and $\dot{m}$. Note that M and M_0 are mass of gas whereas $\dot{m}_0$ and $\dot{m}$ are mass flow rate.

We have, $$\dot{m} = \frac{dm}{d\tau} = \dot{m}_0 e^{-a\tau} = \text{flow rate in vessel from main}$$

Integration of above equation yields

$$M = M_0 + \frac{\dot{m}_0}{a}(1 - e^{-a\tau}) \qquad \text{...(ii)}$$

where M is the mass of gas in vessel at any instant

From Eq. (i) and (ii), we have

$$\left[M_0 + \frac{\dot{m}_0}{a}(1 - e^{-a\tau})\right]u - M_0 u_0 = \frac{\dot{m}_0}{a}(1 - e^{-a\tau})(u_1 + RT_1)$$

or $$M_0 c_v (T - T_0) = \frac{\dot{m}_0}{a}[(1 - e^{-at}) c_v (T_1 - T) + RT_1]$$

or $$T = \frac{M_0 c_v T_0 + \dfrac{\dot{m}_0}{a}(1 - e^{-a\tau}) c_v T_1}{\left[M_0 + \dfrac{\dot{m}_0}{a}(1 - e^{-a\tau})\right] c_v} \qquad \text{...(iii)}$$ **Ans.**

and $$p = \frac{MRT}{V} = \frac{R}{V \cdot c_v}\left[M_0\, c_v\, T_0 + \frac{\dot{m}_0}{a}(1 - e^{-a\tau})\, c_p\, T_1\right]$$

or $$p = p_0 + \frac{\dot{m}_0\, R}{aV}(1 - e^{-a\tau})\, \gamma T_1 \quad \text{...(iv)}$$ **Ans.**

Equations (iii) and (iv) show that the temperature and pressure of the gas in the vessel are the functions of time. Now, consider the cases of questions:

(a) If the vessel was initially evacuated, i.e., $M_0 = 0$, then $T = \gamma\, T_1$, i.e., the temperature of the air inside the vessel becomes independent of time and is equal to $\gamma\, T_1$ throughout the charging process.

(b) The charging process will cease when the pressure inside the vessel reaches that of the main. The expression for charging time can be found by letting $p = p_1$ in the equation (iv) of the pressure relation. Thus,

$$p_1 - p_0 = \frac{\dot{m}_0\, R\, \gamma\, T_1}{aV} - \frac{\dot{m}_0\, R}{aV} e^{-a\tau}\, \gamma\, T_1$$

After rearranging, we have

$$e^{-a\tau} = \frac{\dot{m}_0\, R\, \gamma\, T_1 / aV}{\dfrac{\dot{m}_0\, R\, \gamma\, T_1}{aV} - (p_1 - p_0)}$$

or $$\tau = -\frac{1}{a}\ln\left[1 - (p_1 - p_0)\frac{aV}{m_0\, R\, \gamma\, T_1}\right]$$ **Ans.**

Example 6.24 A very large vessel containing steam is used to supply steam for emergency power generation. The steam is contained in the vessel at 20 bar and 350°C. The vessel is connected with turbine through a valve and the turbine is connected to small initially evacuated tank with a volume of 0.8 m^3. When the situation arises for the requirement of emergency power, the valve is opened and the tank fills with steam until the pressure is 20 bar. At this situation, the temperature of the tank is 400°C. This filling process takes place adiabatically and kinetic and potential energy effects are negligible. Calculate the amount of work done by the turbine. Take internal energy of steam at 20 bar and 400°C as 3137 kJ/kg and enthalpy at 20 bar and 350°C as 2945.2 kJ/kg and specific volume at 20 bar and 400°C is 0.15120 m^3/kg.

A Model Solution Refer to Fig. 6.39.

Data given Steam at 20 bar and 350°C, $v = 0{,}8\ m^3$ of initially evacuated tank, $(m_1 = 0)$. After filling, p = 20 bar and 400°C.

To find Work done by turbine in kJ/kg.

Assumption (i) It is a filling process from a large reservoir to tank through turbine and the control volume is shown by dashed line in Fig. 6.39.

(ii) For the control volume, $Q_{cv} = 0$ and ΔKE and ΔPE are negligible.

(iii) The state of steam in the large vessel remains constant while the final state is the smaller tank is an equilibrium state.

(iv) The amount of steam stored within the turbine and the interconnecting pipe, the end of filling process is negligible.

Analysis For the filling process, the unsteady flow energy equation is

$$Q - W = (m_2 u_2 - m_1 u_1) - (m_2 - m_1) e_f$$

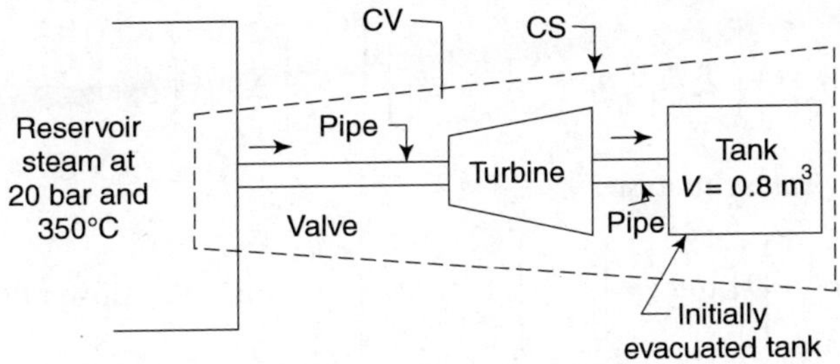

Fig. 6.39 Example 6.24

Here condition 1 and 2 refer to before and after the filling process in the tank. Since before the filling process, the tank was evacuated, so $m_1 = 0$. Further, e_f refers to the flow energy of the reservoir.

Thus,
$$e_f = \left(h + \frac{C^2}{2} + gz\right)_{\text{reser}} = h_{\text{reser}} \text{ as } C = 0, \quad z = 0 \text{ and } Q_{cv} = 0$$

Substituting these values, we have

$$-W_{cv} = -W = m_2 u_2 - m_2 h_{\text{res}}$$

or
$$W = m_2 h_{\text{res}} - m_2 u_2$$

But
$$m_2 = \frac{V}{v_2} = \frac{0.8}{0.15120} = 5.291 \text{ kg} \quad \text{and} \quad u_2 = h_2 - p_2 v_2$$

$\therefore$
$$W = [5.291 \times 3137 - 5.291 \times 2945.2] = 1014.86 \text{ kJ}$$ **Ans.**

6.15 SUMMARY OF FORMULAE

1. $e_f = h + \dfrac{C^2}{2} + gz$

2. $\dot{m}\left(h_1 + \dfrac{C_1^2}{2} + gz_1\right) + \dot{Q} = \dot{m}\left(h_2 + \dfrac{C_2^2}{2} + gz_2\right) + \dot{W}$ (Rate equation) of SFEE

3. $\delta\dot{Q} - \delta W = d\dot{E}_f$ (Differential rate form of SFEE)

4. $h_1 + \frac{C_1^2}{2} + gz_1 + q = h_2 + \frac{C_2^2}{2} + gz_2 + w$ (SFEE based on unit mass)

5. $\dot{Q} + \dot{m}_a \left[h + \frac{C^2}{2} + gz \right]_a + \dot{m}_b \left[h + \frac{C^2}{2} + gz \right]_b$

$= \dot{m}_c \left[h + \frac{C^2}{2} + gz \right]_c + \dot{m}_d \left[h + \frac{C^2}{2} + gz \right]_d + \dot{W}$ for (Multi-stream)

6. $\frac{dE_{cv}}{d\tau} = \dot{Q}_{cv} - \dot{W}_{cv} + \left[\Sigma\, \dot{m}_1 \left\{ h_1 + \frac{C_1^2}{2} + gz_1 \right\} \right]_{L_1} - \left[\Sigma\, \dot{m}_2 \left(h_2 + \frac{C_2^2}{2} + gz_2 \right) \right]_{L_2}$

(Unsteady flow rate equation)

7. $\int_{\tau_1}^{\tau_2} dE_{cv} = Q - W + \left[\int_{\tau_1}^{\tau_2} dm\, e_f \right]_{L_1} - \left[\int_{\tau_1}^{\tau_2} dm\, e_f \right]_{L_2}$ (Unsteady flow rate equation)

where $e_f = \left(h + \frac{C^2}{2} + gz \right)$ and $E_{cv} = \left(mh + \frac{mC^2}{2} + mgz \right)$

8. $w_{sf} = -\int_1^2 vdp - \Delta\text{KE} - \Delta\text{PE} = -\int_1^2 vdp$ if $\Delta\text{KE} = \Delta\text{PE} = 0$

9. $w_{nf} = \int_1^2 pdv$

10. $w_{sf} = \int_1^2 vdp = v\,(p_1 - p_2)$ for $v = c$

11. $w_{sf} = 0$ for $p = c$

12. $w_{sf} = p_1\, v_1 \ln \frac{v_2}{v_1}$ for $T = c$

13. $w_{sf} = \frac{\gamma}{\gamma - 1}\,(p_1\, v_1 - p_2\, v_2)$ for $pv^\gamma = c$

14. $w_{sf} = \frac{n}{n-1}\,(p_1\, v_1 - p_2\, v_2)$ for $pv^n = c$

15. $\rho AC = \dot{m}$ (continuity equation)

16. $\frac{d\rho}{\rho} + \frac{dA}{A} + \frac{dC}{C} = 0$ (continuity equation in differential form)

17. $\frac{dp}{\rho} + C\, dC + gdz = 0$ (Euler's equation)

18. $\frac{p}{\rho}+\frac{C^2}{2}+gz$ = constant (Bernoulli equation)

19. $h_1 = h_2$ = constant (Throttling process)

20. $\mu_J = \left(\frac{\partial T}{\partial p}\right)_h$ (Joule–Thomson coefficient)

21. $\eta = -\left(\frac{\partial T}{\partial v}\right)_u$ (Joule coefficient)

22. $Q - W = (m_2 u_2 - m_1 u_1) - (m_2 - m_1) e_f$ (Filling and emptying process)

PROBLEMS

Things to Think About

6.1. What is the reason that a computer cooled by a constant speed fan operates satisfactorily at sea level but overheats at high altitude?

6.2. List an example where the inlet and exit mass flow rates for a control volume are equal, yet the control volume is not in steady state.

6.3. Using the mass and energy balance rate, describe the process of blowing a balloon through modeling assumptions.

6.4. When air enters a diffuser and decelerates, what is the effect on the pressure of air?

6 5. Discuss the desirability for a coolant circulating inside the engine of an automobile to have a large or small specific heat.

6.6. List the commonly encountered devices that undergo periods of transient operation.

6 7. How will achieve cooling and heating effect from the same device?

6.8. In the space, whether the energy requirement for flying a vehicle will be less or more than in atmospheric air? Why?

Short and Long Theoretical Questions

6.9. Differentiate between flow and non-flow work.

6.10. What are the requirements of a steady flow process?

6.11. What do you mean by continuity equation?

6.12. What is difference between Euler's and Bernoulli's equation?

6.13. What is difference between SFEE and Euler's equation?

6.14. What do you mean by throttling process?

6.15. What do you mean by inversion curve?

6.16. Derive the expression for work done for closed system for all possible processes.

6.17. Derive the general energy equation for steady flow process. Explain the physical significance of the several terms of the equations.

6.18. Derive the continuity equation.

6.19. Derive the Euler's equation.

6.20. Derive an expression for emptying and filling process.

6.21. Write down the general energy equation for unsteady flow process. How is it modified for closed system?

Numerical Questions

6.22. A jet plane is flying at a speed of 6500 km/h at an altitude where the pressure is 0.5 bar and temperature is –12°C. Air passes through the diffuser of the jet plane and leaves with velocity of 100 m/s. If the flow is isentropic, calculate the temperature and pressure at the outlet of the diffuser. [*Ans.* –1.9S5°C: 0 57 bar]

6.23. An evacuated tank of 0.4 m^3 volume is connected to a pipeline by means of small valve. Air at temperature 70°C and pressure 0.3 MPa flows from the pipe into the tank until the pressure in the tank equals line pressure. Calculate the mass of that air in tank. The pipe and the tank may be assumed thermally insulated. Take c_p = 1.005 and c_v = 0.718 kJ/kg K. [*Ans.* 2.281 kg]

6.24. Steam enters a steam turbine at a velocity of 40 m/s, and a specific enthalpy of 2400 kJ/kg and leaves it at velocity of 90 m/s, and a specific enthalpy of 1950 kJ/kg. The heat losses from the turbine to the surrounding are limited to be 280 kJ/min and the steam consumption is 5000 kg/h. Determine the power developed by turbine.

6.25. An air compressor is provided with a water jacket for cooling. A test indicates that compressor requires 180 kJ of work per kg of air flow through the compressor while the enthalpy of the air increases from 73 kJ per kg as it passes through the compressor. The enthalpy of the circulating water increases by 60 kJ per kg of air. Calculate the heat transfer from the compressor to the surrounding.

6.26. 325 kJ of heat is supplied to a system at constant volume. The system rejects 365 kJ of heat at constant pressure and 100 kJ of work is done on it. The system is brought to its initial state by an adiabatic work. Calculate all values of internal energy at all the end states if initial value is 400 J.

6.27. A rigid and insulated tank with zero heat capacity contains 8 kg of air at 3 bar and 30°C. The tank is quickly filled with air from a very large reservoir at 18 bar and 65°C. Find the final temperature and mass added in the tank assuming c_p = 1.005 and c_v = 0.718 kJ/kg K.

[*Ans.* 434 K and 25.6 kg]

6.28. An adiabatic diffuser is employed to reduce the velocity of stream of air from 238 m/s to 50.4 m/s. The inlet pressure is 1.02 bar and the inlet temperature is 394°C. Determine the final pressure and the required outlet area if the mass flow rate is 6.83 kg/s. [*Ans.* 1.25 bar, 62 cm^2]

6.29. Steam enters heat exchanger at 13.6 bar and 261°C, where it condenses on the outside of heat-exchanger tubes. The condensed steam leaves the exchanger as a saturated liquid at 13.6 bar and at a flow rate of 455 kg/h. The steam is condensed by passing cold water through the inside of the tubes. The cooling water enters at a temperature of 10°C and experiences a temperature rise of 7.24°C before leaving the heat exchanger. What flow rate of cooling water is required in kg/h? Draw the system.

Hint See the value of enthalpies from Steam Table. [*Ans.* 925000 kJ/h]

6.30. A lake has its surface 305 m above sea level. Water flows through 10 cm diameter pipe to a turbine situated above sea level. If no heat energy transfers across to the water flowing in the pipe, calculate (a) the velocity of water on entering the turbine, (b) the flow rate of water entering the turbine in kg/s, (c) the temperature rise that would occur if the water hits a non-conducting slab and not the turbine.

6.31. A gas expands through a nozzle following a reversible polytropic law $pV^{1.3} = c$. There is no heat transfer and there is no change in potential energy but the pressure drops from 1.5 MPa to 0.15 MPa and the specific volume increases from 0.1 m^3 to 0.6 m^3. If the entrance velocity to the nozzle is 70 m/s, calculate the exit velocity.

6.32. A reciprocating air compressor with water jacket compresses 2830 litres/min of air with an initial specific volume of 730 litres/kg. The enthalpy of the air is increased by 105 kJ/kg. Transfer of heat to cooling water and surrounding is at the rate of 189 kJ/min. Assuming steady flow and neglecting changes in K.E. and P.E., find power required to drive the compressor. Draw the system.

6.33. In a gas turbine, the mass flow rate of gas through the turbine is 17 kg/s. The power developed by the turbine is 1400 kW. The enthalpies of the gases at inlet and outlet are 1200 kJ/kg and 360 kJ/kg, respectively. The velocity of gas at inlet and outlet are 60 m/s and 150 m/s, respectively. Calculate the rate at which heat is rejected from the turbine. Calculate also the area of inlet pipe if specific volume of gas at inlet is 0.5 m^3/ kg. [*Ans.* Q_R = 119.3 kW, A_1 = 0.142 m^3]

6.34. The mass flow rate of air through a reciprocating compressor is 0.4 kg/s. The air enters at 6 m/s with pressure of 1 bar and a specific volume of 0.85 m^3/kg and leaves at 4.5 m/s with a pressure of 6.9 bar and a specific volume of 0.16 m^3/kg. The internal energy of the air leaving the compressor is 88 kJ/kg greater than that of air entering. Cooling water of cylinder jacket absorbs heat at the rate of 59 kW. Determine the power required to drive the compressor and the inlet and outlet pipe cross-sectional areas.

6.35. A gas turbine receives gases from the combustion chamber at 7 bar and 650°C with a velocity of 90 m/s. The gases leave the turbine at 1 bar with a velocity of 45 m/s. Determine the work done if the expansion is isentropic. Assume n = 1.333, c_p = 1.11 kJ/kg K. [*Ans.* 394.2 kJ/kg]

6.36. Air at 1 bar, 20°C is taken into a gas turbine power plant at a velocity of 120 m/s through an opening of 0.15 m^2 cross-sectional area. The air is compressed in a compressor, heated in a combustion chamber, expanded through the turbine and exhausted at 1.2 bar, 300°C through an opening of 0.1 m^2. The power output is 1 MW. Calculate the net amount of heat added to the air in kW/kg.

6.37. In a water cooling tower air enters at a height of 1.5 m above the ground level and leaves at a height of 8 m. The air inlet and outlet velocities are 25 m/s and 35 m/s respectively. Air temperatures are 30°C and 70°C at the entry and exit respectively. The hot water at a temperature 80°C enters the cooling tower at a height of 8.5 m with velocity 3 m/s. The cooled water leaves at a temperature of 45°C with velocity 1 m/s. The cooling tower is well insulated and a fan of 2.5 kW drives the air through the cooler. Calculate the amount of air flow rate required for 1 kg/s of water flow through cooler.

6.38. In an oil cooler, oil flows steadily through a bundle of metal tubes submerged in a steady stream of cooling water. The oil enters the cooler at 90°C and leaves at 40°C while water enters at 25°C and leaves at 70°C. Calculate the cooling water flow rate for cooling 1 kg/s of oil. The enthalpy of the oil is given by

$$h = 17\,t + 11 \times 10^{-4}\,t^2 \text{ kJ/kg.}$$

6.39. During a steady-flow process 5000 kg/h of fluid passes through an apparatus in which the exit pipe is 2 m below the level of the inlet connection. Pertinent data regarding the working fluid are : pressure decreases from 700 to 140 kPa; velocity increases from 60 to 360 m/s; internal energy decreases by 50 kJ/kg; specific volume increases from 0.06 to 0.36 m^3/kg. If the heat radiated from the apparatus is 10,000 kJ/h, calculate the work exchanged with the surroundings. Indicate the direction of the work. [*Ans.* – 32.5 kW, work is done on the system]

6.40. A condenser brings 1 kg/s of steam flow at 10 kPa from 35°C to saturated liquid at 10 kPa. The cooling is done by lake water at 20°C that returns to the lake at 30°C. For an insulated condenser, calculate the flow rate of cooling water.

6.41. A diffuser receives an ideal gas at 100 kPa, 300 K with a velocity of 250 m/s and the exit velocity is 2.5 m/s. Calculate the exit temperature if the gas is (a) argon and (b) helium.

[*Ans.* (a) 360 K, 306 K]

6.42. The front of a jet engine acts as a diffuser receiving air at 900 km/h, – 5°C, 50 kPa during flight bringing it to 100 m/s relation to the engine before entering the compressor. If the flow area is reduced to 70% of the inlet flow area, calculate the temperature and pressure in the compressor inlet.

6.43. Helium gas is throttled from 1.5 MPa, 20°C to a pressure of 100 kPa. Due to larger diameter of exit pipe than that of inlet, the inlet and exit velocities are equal. Calculate the exit temperature of helium and the ratio of pipe diameters.

6.44. A small, high speed turbine working on compressed air produces a power output of 100 W. The inlet condition of air is 400 kPa, 50°C and exit condition is 150 kPa, – 30°C. Assuming the changes in kinetic and potential energies negligible and the process is adiabatic, calculate the required mass flow rate of air through the turbine. [*Ans.* 0.00125 kg/s]

6.45. Air enters an insulated diffuser operating at steady state with a pressure of 1 bar a temperature of 60°C and a velocity of 220 m/s. At the exit, pressure is 1.4 bar and the temperature is 71°C. Calculate the ratio of exit flow area to the inlet flow area and the exit velocity in m/s.

6.46. The inlet ducting ahead of to a jet engine forms a diffuser that decelerates the entering air to zero velocity relative to the engine before the air enters the compressor. Jet airplane flies at a speed of 1000 km/h where the local atmospheric pressure is 0.6 bar and the air temperature is 8°C. Neglecting heat transfer and changes in potential energy, calculate the temperature, in °C, of the air entering the compressor. [*Ans.* 319.65 K]

6.47. An air turbine receives air at 10 bar and 900 K and expands up to 1 bar and 500 K. The inlet velocity is small compared to the exit velocity of 100 m/s. The turbine operates at steady state and develops a power output of 3.2 MW. Neglecting heat exchange between turbine and surroundings and changes in PE, calculate the mass flow rate of air and exit area of turbine.

[*Ans.* 7.53 kg/s and 0.108 m^2]

6.48. The intake to a hydraulic turbine installed in a flood control dam is located at an elevation of 10 m above the turbine exit. Water enters the turbine at 20°C with negligible velocity and comes out at 10 m/s. If the change in temperature and the heat exchange is negligible, calculate the mass flow of water for the 500 kW turbine output. [*Ans.* 10, 440 kg/s]

6.49. A water cooled reciprocating air compressor compresses air at steady state from 1 bar, 300 K to 7 bar with a mass flow rate of 5 kg/s. The compression process is polytropic and is governed by $pv^{1.27}$ = constant. Heat transfer occurs at a rate of 50 kJ/kg of air flowing to cooling water circulating in a water jacket enclosing the compressor. If the changes in kinetic and potential energy of the air from inlet to exit are negligible, calculate the compressor power input if mechanical efficiency is 90%.

6.50. A solar collector panel with a surface area of 3 m^2 receives energy from the sun at a rate of 1880 kJ/h m^2 of collector surface. 36% of the incoming energy is lost to the surrounding and the remainder is used to heat liquid water from 43°C to 50°C. The water passes to the solar collector with negligible pressure drop. Neglecting changes in kinetic and potential energy, calculate at steady state the mass flow rate of water.

6.51. Electronic components are mounted on the inner surface of a horizontal cylindrical duct having inner diameter 0.2 m. In order to prevent overheating of the electronics, the cylinder is cooled by a stream of air flowing through it and by convection from its outer surface. Air enters the duct at 25°C, 1 bar and a velocity of 0.3 m/s and comes out with negligible changes kinetic energy and pressure at a temperature that cannot exceed 40°C. The electronic components require 0.2 kW of electric power. Calculate the minimum rate of heat transfer by convection from the cylinder's outer surface in order to meet the limit on the temperature of exiting air.

6.52. A rigid tank having a volume of 0.5 m^3 is initially evacuated. A hole develops in the wall and the air from the surrounding at 1 bar, 21°C flows in until its pressure in the tank reaches 1 bar. Neglecting the heat transfer in the contents of the tank and the surroundings, calculate the final temperature in the tank. [*Ans.* 138°C]

6.53. A tank having 1 m^3 of volume initially contains air at 300 kPa, 300 K. Air slowly escapes from the tank until the pressure drops to 100 kPa. The remaining air in the tank undergoes a polytropic process ($pv^{1.2}$ = constant). For a control volume enclosing the tank, calculate the heat transfer.

6.54. A rigid tank of 1 m^3 contains nitrogen initially at 10 bar 300 K. Heat transfer to the contents of the tank occurs until the temperature increases to 400 K. During the process, a pressure-relief valve allows nitrogen to escape, maintaining constant pressure in the tank. Neglecting changes in kinetic and potential energy and taking specific heat value at 350 K for nitrogen, calculate (a) the amount of escaped nitrogen, (b) the amount of energy transfer by heat.

[*Ans.* (a) 2.9 kg, (b) 1043 kJ]

Software Based Questions

6.55. The mass flow rate of water at the inlet of a control volume varies with time and is given by $\dot{m} = 200\,(1 - e^{-2\tau})$ where $\dot{m}$ is in kg/h and τ is in *h*. At the exit mass flow rate is constant at 200 kg/h. The initial mass in the control volume is 100 kg.

With the above data, develop a software (a) to plot the inlet and exit mass flow rates and the amount of mass contained in the control volume as a function of time, for τ ranging from 0 to 5h. (b) calculate the time, in *h*, when the tank is nearly empty.

6.56. Develop a software to plot the variation of exit temperature of gas with exit velocity in a gas nozzle. For specific data, consider methane (CH_4) gas enters a horizontal, well insulated nozzle operating at steady state at 100°C and a velocity of 10 m/s. For plot, consider the variation of exit velocity ranging from 500 to 600 m/s.

6.57. Develop a software to find the temperature rise in cooling water of water-jacket reciprocating compressor. For specific data, consider such a compressor in which air enters with a volumetric flow rate of 37 m^3/min at 136 kPa, 305 K and exits at a pressure of 680 kPa and a temperature of 400 K. The power input to the compressor is 155 kW. Using the software (a) calculate the temperature increase of the cooling water, if cooling water mass flow rate is 85 kg/min, (b) plot the temperature increase of the cooling water versus the cooling water mass flow rate which varies from 80 to 100 kg/min.

6.58. Develop a software to calculate the exit velocity of gas from a jet engine. Take manufacturer's data for validation.

6.59. Develop a software to calculate brake-specific fuel consumption, brake thermal efficiency and exhaust gas temperature for passenger cars. Take manufacturer's data of leading passenger cars of all segments for validation of results.

6.60. Develop a software for thermodynamic analysis of filling and emptying processes. Take standard data such as bottling plant for LPG.

6.61. Develop a software for calculating the mass flow rate of air required in a gas turbine plant which may develop power ranging from 100 MW to 300 MW.

Design and Open Ended Questions

6.62. Practical measures should be taken by worldwide manufacturers to use energy resources more efficiently. Prepare a comprehensive report of these possible practical measures listing several specific opportunities and discussing their potential impact on profitability and productivity.

6.63. A factory has been taking the service of a 85 kW reciprocating compressor delivering 30 m^3/min at 3 bar (gauge) for about 8000 operating hours annually and now its useful life is over. Investigate options replacing the old air compressor and prepare a report containing options and recommendations.

6.64. Prepare a report providing guidelines for selecting fan for cooling electronic circuit. The option includes type of fans and its location at inlet or at exit.

6.65. For measuring mass flow rates of gases and liquid flowing in pipe and ducts include rotometers, turbine flow meters, orifice-type flow meters, thermal flow meters and Coriolis type flow meters. Prepare a report involving basic principles of operation of each with suitability and limitations.

6.66. At many places windmills or wind turbines are used to develop power from wind. There are various types of windmills such as Mandaras, Darriens and propeller. Prepare a report with the sketches of each including working principles and limitations.

6.67. Prepare a report for economic feasibility of installing a heat exchanger in the stack of industrial paint-drying oven discharging 40 m^3/min of gases combustion product at 250°C to heat air that would provide for some of the space heating needs of the plant.

6.68. Based on the manufacturer's data of commercial planes, prepare a report for comparative study of design features, brake-specific fuel consumption, propulsive efficiency, thrust, cruising speed, comfort, etc., for the major commercial planes such as Boieng, Air Bus, etc. which may be helpful to various airlines during selection.

6.69. Based on the manufacturer's data of passenger cars, prepare a report for comparative study of design features, brake-specific fuel consumption, power output, acceleration, brake thermal efficiency, emission, comfort, etc. for B-segment (air conditioned) passenger cars such as Maruti Zen, Santro, Palio, Indica, SUV, etc., for the use of customers while deciding to purchase.

Multiple-Choice Answer Type Questions

6.70. Law of energy conservation is stated by

(a) Zeroth law of thermodynamics (b) First law of thermodynamics

(c) Second law of thermodynamics (d) Third law of thermodynamics

6.71. The concept of internal energy is given by

(a) Zeroth law of thermodynamics (b) First law of thermodynamics

(c) Second law of thermodynamics (d) Third law of thermodynamics

6.72. Steady flow energy equation when applied to a boiler shall yield

(a) $q = h_2 - h_1$ (b) $q = W$ (c) $q = 0$ (d) $q = u$

6.73. The steady flow work done (w_{sf}) is given by

(a) pv (b) $\int_1^2 pdv$ (c) $-\int_1^1 vdp$ (d) $\frac{p_1 v_1 - p_2 v_2}{n-1}$

6.74. Joule–Thomson coefficient (μ_J) is expressed by

(a) $-\left(\frac{\partial T}{\partial p}\right)_v$ (b) $\left(\frac{\partial T}{\partial p}\right)_h$ (c) $-\left(\frac{\partial T}{\partial p}\right)_u$ (d) $\left(\frac{\partial T}{\partial p}\right)_v$

6.75. Joule-coefficient (η) is expressed by

(a) $-\left(\frac{\partial T}{\partial u}\right)_v$ (b) $\left(\frac{\partial T}{\partial v}\right)_u$ (c) $-\left(\frac{\partial T}{\partial v}\right)_u$ (d) $\left(\frac{\partial T}{\partial T}\right)_v$

6.76. Throttling process occurs when the flow passes through

(a) Nozzle (b) Diffuser (c) Turbine (d) Valve

6.77. The energy equation for filling and emptying processes are expressed by

(a) $Q - W = (m_2 u_2 + m_1 u_1) - (m_2 - m_1) e_f$

(b) $Q - W = (m_2 u_2 + m_1 u_1) + (m_2 - m_1) e_f$

(c) $Q - W = (m_2 u_2 - m_1 u_1) + (m_2 - m_1) e_f$

(d) $Q - W = (m_2 u_2 - m_1 u_1) - (m_2 - m_1) e_f$

Answers to Multiple-Choice Questions

6.70. (b) 6.71. (b) 6.72. (a) 6.73. (c) 6.74. (b) 6.75. (a)

6.76. (d) 6.77. (d)

CHAPTER

7

Second Law of Thermodynamics

OVERVIEW

This chapter deals with the essence and statement of the second law of thermodynamics with its all corollaries for designing a heat engine or a heat pump. As the first law gives the concept of **internal energy**, similarly the second law gives the concept of **entropy** which is discussed in Chapter 8.

7.1 LIMITATIONS OF FIRST LAW AND ESSENCE OF SECOND LAW

The first law of thermodynamics states that heat and work are mutually convertible. In other words, it is the conservation of energy, i.e., every process is possible, provided energy is neither created nor destroyed. *The first law places no restrictions on the direction of flow of heat and work.* Thus, according to first law, in every cyclic process, work is completely converted into heat or heat is completely converted into work; potential energy can be transformed into kinetic energy or kinetic energy can be transformed into potential energy, heat flows from hot to cold or from cold to hot, gas expands from high pressure to low or from low pressure to high pressure. But from experience, we know that this is not true. In natural way, heat is not completely converted into work, potential energy is converted into kinetic energy but the reverse is not automatically true, heat flows from hot to cold but the reverse is not automatically true. *These spontaneous processes can be reversed but they will not be reversed automatically. They require some external source for reverse process.*

For more examples consider the following cases:

(i) Consider two heat reservoirs, one at a higher temperature (T_1) and the other at lower temperature T_2. There is a flow of heat from higher temperature body to lower temperature body. When they are brought in contact with each other, the amount of energy lost by the higher temperature reservoir will be equal to the amount of energy gained by the lower temperature reservoir. The flow of heat from lower temperature reservoir to higher temperature reservoir does not violate the first law of thermodynamics if heat coming out from one is equal to heat entering into the other. But in actual practice this is not possible. Thus, the *transfer of energy is governed by some directional law.*

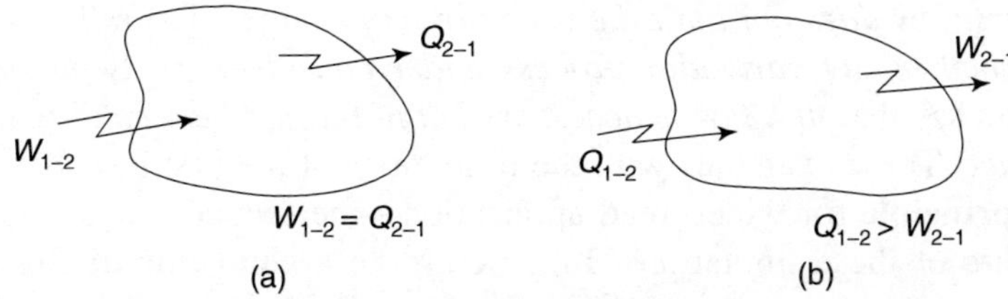

Fig. 7.1 Qualitative distinction between work and heat

(ii) Consider a rotating flywheel brought to rest by friction in its bearing. As a result of this, the temperature of the wheel and bearings rises and the increase in their internal energy is equal to the original energy of the flywheel. The first law is valid here. Now, by cooling down of the flywheel and bearings to get the kinetic energy of the flywheel also satisfies the first law, but this does not happen in actual practice. Thus, there exists a *directional law which governs the transformation of energy.*

(iii) Consider an ideal gas undergoing free expansion process through a valve into an evacuated vessel. We notice that the temperature remains constant but there is an increase in volume and fall in pressure. Now suppose the reverse process takes place, i.e., gas at low pressure should rush back through the valve and compresses itself into its original states spontaneously. Though, this reverse process is true according to the first law, but in actual practice it is not possible.

(iv) Consider a chemical reaction initiated by a spark plug in an insulated rigid vessel containing hydrogen and oxygen in proper proportions. After the reaction, the system consists of water vapour at high temperature and pressure. The final internal energy is the same as the initial one. Now, suppose the reverse process takes place, i.e., the water vapour could spontaneously dissociate into hydrogen and oxygen at a lower temperature and pressure. Though, from the first law, this reverse process is true but in actual practice it is not true.

(v) From Joule's experiment (Sec. 5.1), we conclude that work can be completely converted into heat but from experience we know that the complete conversion of heat into work in a cycle is not possible. Thus, for a cyclic process

$$W = \underset{\rightarrow}{Q} \text{ and } \underset{\rightarrow}{Q} > W$$

The arrow indicates the direction of energy transformation. This is also illustrated in Fig. 7.1. In Fig. 7.1(a), the system is taken from state 1 to 2 by work transfer W_{1-2} and then Q_{2-1} to complete cycle. It is always found that $W_{1-2} = Q_{2-1}$. But if the system is taken from state 1 to 2 by heat transfer Q_{1-2} but this cannot be back by work transfer (Fig. 7.1 b) from state 2 to 1, i.e., $Q_{1-2} > W_{2-1}$. Some heat has to be rejected. **Work is said to be a high grade energy and heat a low grade energy. The complete conversion of low grade energy into high grade energy in a cycle is impossible.**

From the above *examples, it is obvious that the first law does not help the engineer to predict whether the system will or will not undergo a change, i.e., whether a certain process is possible or not*. We have seen that a spontaneous process can proceed in a particular direction only. *Hence, the first law does not limit the extent of any energy conversion nor it gives any indication as to whether any particular process under consideration is possible or not.* In other words, we can say that *first law is necessary but not a sufficient condition for a certain process to take place*. These examples with the limitations of the first law show the **need for another general principle** for widespread application—the second law of thermodynamics which will take care of these limitations. In general, the **second law of thermodynamics** in the light of the above discussion may be defined as "*Heat cannot flow itself from colder body to a hotter body*" *and it can flow only when some external work is supplied to it, for example, in refrigeration system.*

7.2 ASPECTS OF SECOND LAW

From the preceding discussions, it may be summarized that the second law and deductions from it are useful because they provide means to study the following aspects:

(i) Predicting the direction of processes.

(ii) Establishing conditions for equilibrium of a process.

(iii) Determining the best theoretical performance of cycles, engines and other devices.

(iv) Evaluating quantitatively the factors that preclude the attainment of the best theoretical performance level.

(v) Defining a temperature scale independent of the properties of any thermometric substance.

(vi) Developing means for evaluating properties such as u and h in terms of properties that are more readily obtained experimentally.

(vii) Quantifying irreversibilities in the system and processes.

(viii) Giving the concept of entropy and entropy production (generation).

7.3 THERMAL RESERVOIR

A thermal reservoir is defined as a sufficiently large system in stable equilibrium to which and from which finite amount of heat can be transferred without any change in its temperature. Generally, the high temperature reservoir is called the ***heat source*** *and the low temperature one is called the* ***heat sink****. Boiler furnace, combustion chamber, nuclear reactor, sun, etc., may be treated as heat source while atmospheric air, river and ocean water may be treated as heat sink.*

7.4 HEAT ENGINE

A heat engine is defined as a thermodynamic system operating in a cycle in which net heat is transferred and from which net work is delivered. To constitute a heat engine cycle, the system has to undergo a series of processes. Figure 7.2 shows the principle of a heat engine. Thus, the boiler, the turbine, the condenser and the pump separately in a power plant cannot be regarded as a heat engine because each does not complete a cycle. A combination of these is a heat engine since it completes a cycle as shown in Fig. 7.3. Steam generated in a boiler enters the turbine and does work, the exhaust steam is condensed in a condenser and is returned to the boiler by a feed pump. In this way, the working fluid is returned to its original state and thus the cycle is completed.

$$\oint \delta Q = \oint \delta W$$

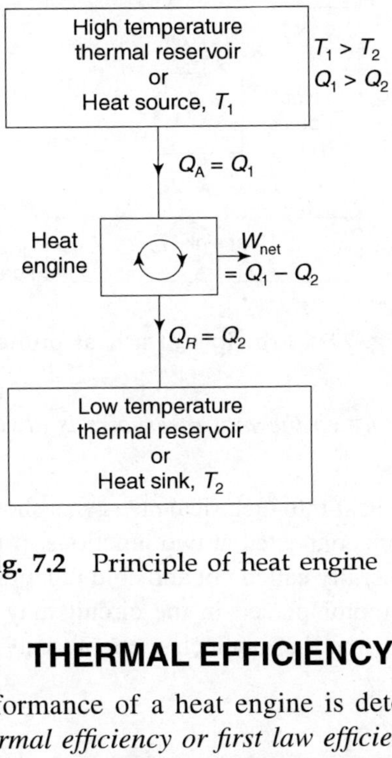

Fig. 7.2 Principle of heat engine

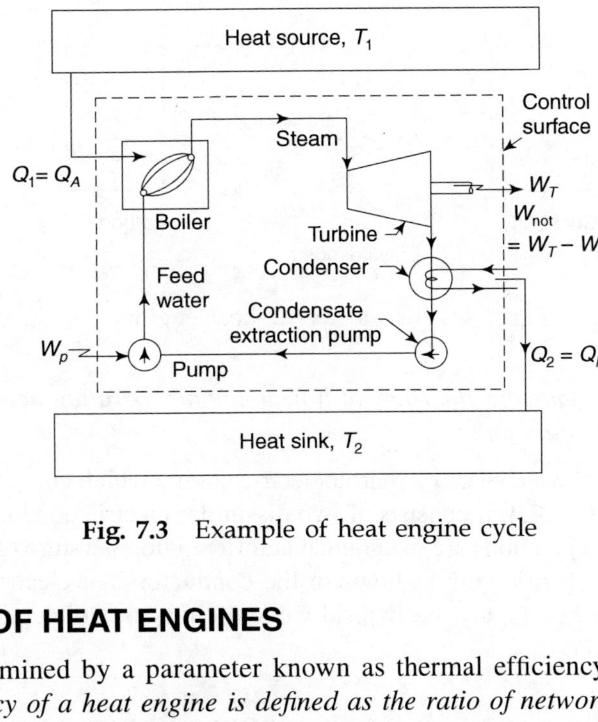

Fig. 7.3 Example of heat engine cycle

7.5 THERMAL EFFICIENCY OF HEAT ENGINES

Performance of a heat engine is determined by a parameter known as thermal efficiency. *Thermal efficiency or first law efficiency of a heat engine is defined as the ratio of network output to heat input.* Only a fraction of the heat input is converted into work and the rest is rejected. Fig. 7.2).

Let, $Q = Q_A$ = Heat added or transferred to a system, kJ

Q_2 = Heat rejected from a system, kJ

W_{net} = Net work done by a system, kJ

$$\text{Then, thermal efficiency, } \eta_{th} = \frac{\text{Output}}{\text{Input}} = \frac{W_{net}}{Q_1} = \frac{W_{net}}{Q_A} \quad ...(7.1)$$

From the first law of thermodynamics applied to a heat engine cycle,

$$\oint \delta Q = \oint \delta W$$

or $$Q_1 - Q_2 = W_{net}$$

$\therefore$ $$\eta_{th} = \frac{Q_1 - Q_2}{Q_1} = 1 - \frac{Q_2}{Q_1} \quad ...(7.2)$$

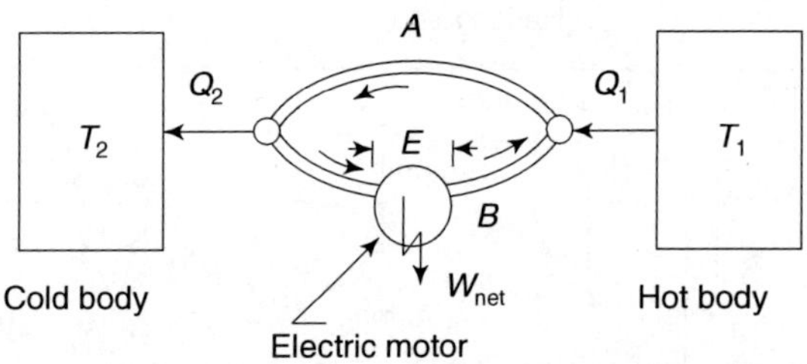

Fig. 7.4 Thermoelectric heat engine

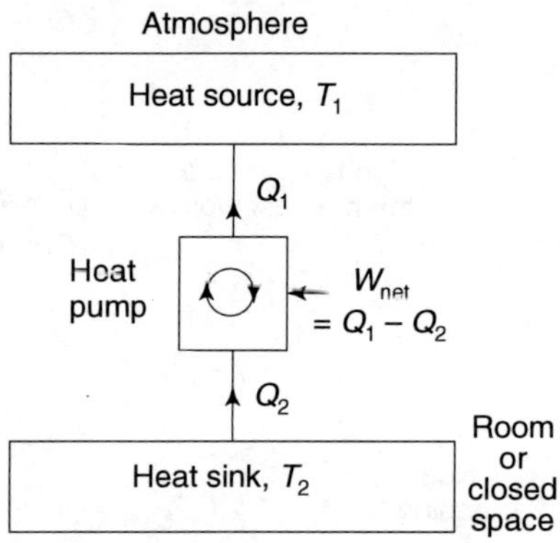

Fig. 7.5 Principle of a heat pump

Thus, the efficiency of a heat engine operating between two thermal reservoirs is always less than unity.

Now, consider a thermoelectric engine which converts heat into electrical energy as shown in Fig. 7.4. It consists of two dissimilar electric conductors connected at two junctions. If the two junctions are maintained at different temperature (generally called hot and cold junctions) an electric current flows in the conductors An electric motor placed in the circuit may be used to supply mechanical work. Hence, the efficiency of the thermoelectric engine may be given as

$$\eta = \frac{W_{net}}{Q_1} = \frac{EI}{Q_1} = \frac{EI}{Q_A} \quad ...(7.3)$$

where E and I are the electric potential and current, respectively. Here, efficiency is less than 100%. But according to the first law, it should be 100%. The question arises why this difference? The answer is given by second law.

7.6 HEAT PUMP AND COEFFICIENT OF PERFORMANCE

A heat pump is defined as a thermodynamic system operating in a cycle which removes heat from a low-temperature body and delivers into a high temperature body taking external energy in the form of work input. The working fluid undergoes a series of processes and thus constitutes a heat pump cycle shown in Fig. 7.5.

The heat pump may be used in a refrigeration system, such as refrigerator, air-conditioner or a cold storage plant, etc. The primary function of these systems is to transfer heat from the low temperature space or system. For example, heat is pumped out so as to cool the houses in summer. A heating system utilises a heat pump to transfer heat to a high temperature system. For example, heat is pumped in so as to heat the house in winter season. Various parts of heat pump are shown in Fig. 7.6.

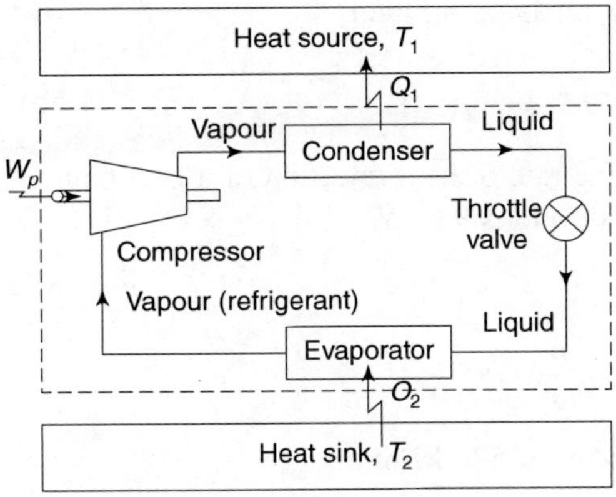

Fig. 7.6 Heat pump

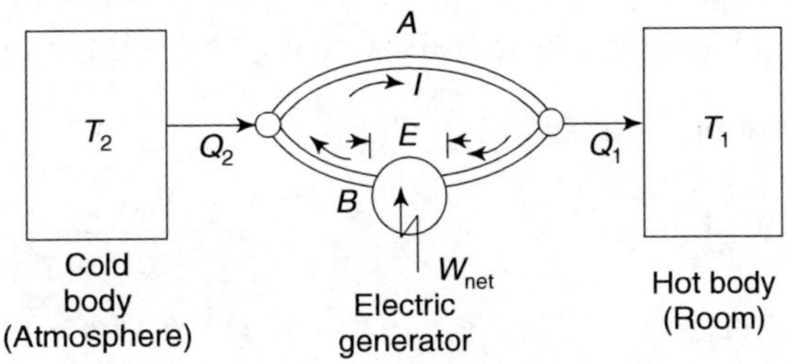

Fig. 7.7 Thermoelectric heat pump

In the case of heat pump, the term efficiency is replaced by ***coefficient of performance*** (***COP***) as index of performance of heat pump to distinguish it from heat engine. The definition of (COP) depends on the use of heat pump whether the system is operating as heat pump or refrigerator.

Let Q_1 = heat rejected by a system, kJ

Q_2 = heat supplied to a system, kJ

W_{net} = network required by the system, kJ

Applying the first law to the cycle.

$$\oint \delta Q = \oint \delta W$$

or

$$Q_1 - Q_2 = W_{net}$$

COPs of refrigerator and heat pump are given by

$$(\text{COP})_{\text{refrigerator}} = \frac{\text{Refrigerating effect}}{\text{Work done input}} = \frac{Q_2}{W_{net}} = \frac{Q_2}{Q_1 - Q_2} \quad ...(7.4)$$

For heat pump

$$(\text{COP})_{\text{heat pump}} = \frac{\text{Heating effect}}{\text{Work input}} = \frac{Q_1}{W_{net}} = \frac{Q_1}{Q_1 - Q_2} = 1 + (\text{COP})_{\text{Ref}} \quad ...(7.5)$$

A thermoelectric heat pump transfers heat from a low to a high temperature thermal reservoir by supplying electric energy as shown in Fig. 7.7. Q_2 amount of heat is transferred from the low temperature reservoir and Q_1 amount of heat is transferred to the high-temperature. Therefore, COP of this pump is given by

$$(\text{COP})_{\text{Heat pump}} = \frac{Q_1}{W} = \frac{Q_1}{EI} \quad \text{...(7.5a)}$$

If the thermoelectric heat pump is operating as a refrigerator, then

$$(\text{COP})_{\text{Ref}} = \frac{Q_2}{W} = \frac{Q_2}{EI} \quad \text{...(7.5b)}$$

Example 7.1 Calculate the thermal efficiency and the rate of heat rejection of a heat engine which receives heat at the rate of 2000 kJ/min and develops 9.6 kW.

Solution Refer to Fig. 7.2

We know, $$\eta_{th} = \frac{\text{Work done}}{\text{Heat supplied}} = \frac{W}{Q_A}$$

But $$W = 9.6 \text{ kJ/s} = 9.6 \times 60 = 576 \text{ kJ/min.}$$

$$\therefore \quad \eta_{th} = \frac{576}{2000} = 28.8\%$$ **Ans.**

The heat rejected $= Q_R = Q_A - W = 2000 - 576 = 1424$ kJ/min **Ans.**

Example 7.2 An engine is to be designed to produce 0.735 kW by taking continuously 6000 kJ/h from heat source which is at 505 K. The heat sink is atmosphere, i.e., sink temperature is atmospheric. The atmospheric temperature varies from 38°C in summer to 0°C in winter. Comment on the performance of the engine.

Solution $$W = 0.735 \text{ kW} = 0.735 \times 3600 = 2646 \text{ kJ/h}$$

$$\therefore \quad \text{Efficiency of the proposed engine} = \frac{W}{Q_A} = \frac{2646}{6000} = 44.1\%$$

In order to comment on the performance of the proposed engine, let us calculate the efficiency of the Carnot engine (see Sec. 7.13) in summer and winter. Carnot engine's efficiency when source temperature is 505 K and sink temperature 38°C in summer is given by

$$\eta_{c_1} = \frac{T_1 - T_2}{T_1} = \frac{505 - 311}{505} = 38.41\%$$

Carnot engine efficiency working between 505 K and sink temperature 273 K in the winter is given by

$$\eta_{c_2} = \frac{T_1 - T_2}{T_1} = \frac{505 - 273}{505} = 45.94\%$$

According to second law, no engine can be more efficient than the Carnot engine hence the proposed engine will work satisfactorily in the summer but for the winter it will not work and have to be redesigned.

Example 7.3 For heating a house in winter, it requires 204820 kJ/h. The heat is supplied by a heat pump which absorbs heat from cold air outside in winter to supply heat to the house 28424 kJ/h is needed by the heat pump for its operation. Calculate the COP of heat pump and the heat abstracted from outside. What will be COP of the heat pump when it serves to cool the house in the summer requiring same 204820 kJ/h of heat rejection?

Solution Refer to Fig. 7.5.

$$\text{Work input} = W = Q_1 - Q_2$$

or

$$Q_2 = Q_1 - W = 204820 - 28424 = 176396 \text{ kJ/h}$$

$$(\text{C.O.P})_{\text{Heat pump}} = \frac{Q_1}{Q_1 - Q_2} = \frac{Q_1}{W} = \frac{204820}{28424} = 7.205 \quad \textbf{Ans.}$$

$$(\text{C,O,P,})_{\text{Ref.}} = \frac{Q_2}{W} = \frac{176396}{28424} = 6.205 \quad \textbf{Ans.}$$

These results suggest that the same device has two values of COP depending upon the uses.

7.7 AVAILABLE AND UNAVAILABLE ENERGY

In the operation of any reversible heat engine, a portion of heat supplied from a high constant temperature source must be rejected to a sink at a lower temperature. We have already seen that the amount of heat rejected for a given heat input is reduced by lowering the sink temperature. For all practical purposes, there is a **minimum value of sink temperature which can be considered in most cases**. *This normally corresponds to the temperature of atmospheric air or a large body of water,* etc. Although, this ambient temperature is a variable, the variation is very small. Sometimes very low temperature of the heat sink is used but this requires an additional expenditure of work to attain it. In space applications, the solar system may have very low heat sink, even then, the sink condition may be treated as invariable in all the cases.

Generally speaking, *any system employed by the engineers must be operated within the earth's atmosphere and the temperature and pressure of the atmosphere impose a limit over the extent to which a given form of energy can be converted into work.* Some forms of energies (e.g., potential) can be efficiently converted into work and some less efficiently. Thus, the energy of a system can be divided into two forms as the following.

(i) Available Energy. *This is defined as that part of energy which can be converted into work by ideal processes which reduce the system to a dead state, a state in equilibrium with the earth and atmosphere.*

(ii) Unavailable Energy. *It is defined as that part of energy which cannot be converted into work even by ideal processes which reduce the system to a dead state, a state in equilibrium with the earth and atmosphere.*

According to the definitions cited above, the potential and kinetic energies are completely available form of energies because the elevation and velocity are measured from the surface of earth and work is done on the earth. In the circumstances of changed datum of zero potential i.e., if the datum is the centre of the earth, then a large part of potential energy would be unavailable, *thus the datum of zero energy decides the availability and unavailability of energies. J.W. Gibbs is credited as the originator of availability concept.*

7.8 THE DEAD STATE

Generally, the atmosphere or low temperature heat sink is regarded as the most stable state having the following characteristics:

(i) It is homogeneous in temperature, pressure and composition.

(ii) There is no relative motion in its various parts.

(iii) All parts are at the same level.

Under such a condition, the atmosphere alone cannot produce any work.

Now, suppose that any system employed by an engineer within this atmosphere has the following characteristics.

(i) It is at rest relative to the atmosphere.

(ii) It is in equilibrium with the atmosphere regarding pressure, temperature, electricity and magnetism.

Under the above conditions, no work will be obtained from interaction between the system and the atmosphere. Further, even if the system is in a state of stable equilibrium, we cannot obtain work from either the system or the atmosphere of by interaction between the two. In other words, it is the dead state of the system.

Therefore, *the dead state of a system is defined as the state of stable equilibrium of the system within the atmosphere, i.e., its capacity to do work ceases and its available energy is zero. Earth and its atmosphere are regarded to be in stable state.* The following conditions specify the dead state of a system:

(a) The system should be in a stable and inert phase, thus any phase change or chemical reaction is not possible.

(b) The system should have pressure and temperature equal to that of atmosphere and its elevation and velocity should be zero. Mathematically,

$$(p)_{\text{final}} = p_0, \quad (z)_{\text{final}} = 0, \quad (T)_{\text{final}} = T_0, \quad (C)_{\text{final}} = 0$$

When p_0 and T_0 are the atmospheric pressure and temperature of the pressure and temperature of a system which is now in dead state or pressure and temperature of the heat sink.

7.9 STATEMENT OF THE SECOND LAW

There are several statements of the second law of thermodynamics.

7.9.1 Kelvin–Planck Statement

(i) *It is impossible to construct an engine which will work in a complete cycle and produce no other effect except the raising of a weight and cooling of heat reservoir.*

or

(ii) *No process is possible whose sole effect is the removal of heat from a single thermal reservoir at an uniform temperature and the performance of an equal amount of work.*

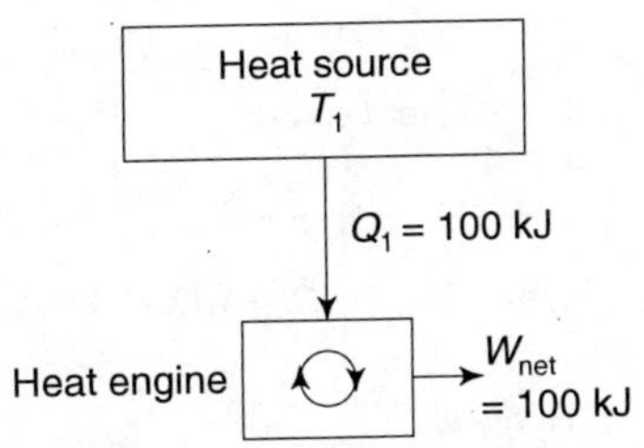

Fig. 7.8 Satisfying only first law and violating second law (PMM–II)

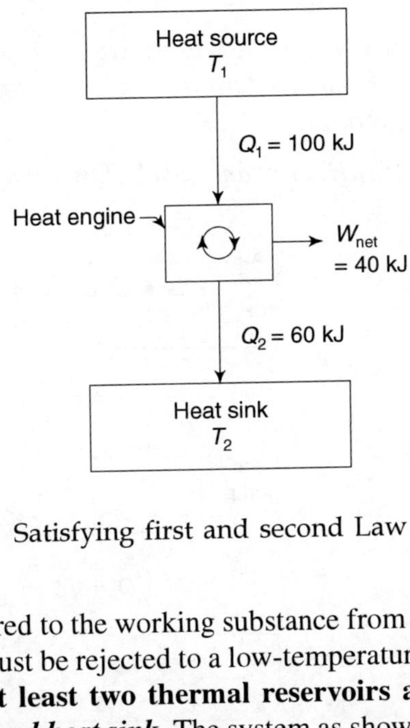

Fig. 7.9 Satisfying first and second Law

This statement means that only part of the heat transferred to the working substance from a high temperature reservoir is converted to work, the rest must be rejected to a low-temperature reservoir. Hence, in **order to operate a heat engine, at least two thermal reservoirs at different temperatures are essential**, i.e., ***a heat source and heat sink***. The system as shown in Fig. 7.8 is taking heat from a heat source and continuously converting an equal amount of work. Though, this system is satisfying the first law but violating the second law because there is an absence of lower temperature heat reservoir, i.e., heat sink and heat received is equal to the work done. *This type of an engine is not possible to construct.*

The statement of the second law further means that ***no heat engine can have a 100 per cent efficiency***. Hence, the engine shown in Fig. 7.9 is possible to construct because it is operating between two heat reservoirs – one heat source and the other heat sink giving efficiency less than 100 per cent. This system satisfies both first and second laws.

The word cycle in this statement has an important implication. In order to understand this, consider the isothermal expansion of a perfect gas in a piston-cylinder arrangement as a result of heat interaction. According to first law, the heat is completely converted into work since the internal energy (i.e., temperature) is constant. Due to the motion of the piston, there will be an increase in volume and decrease in the pressure of the gas. The process continues till the end of the stroke at which point no further work can be done. Hence, it is obvious from this example that ***continuous conversion of heat into work requires a cyclic process***.

7.9.2 Clausius Statement

(i) *It is impossible to construct a heat pump which operating in a cycle will remove heat continuously from a lower temperature reservoir and transfer to a higher temperature reservoir without any amount of external work being done on it.*

or

(ii) *No process is possible whose sole effect is the removal of heat from a reservoir at lower temperature and the absorption of equal amount of heat by a reservoir at a higher temperature.*

(iii) *Heat cannot itself flow from a colder body to a hotter body.*

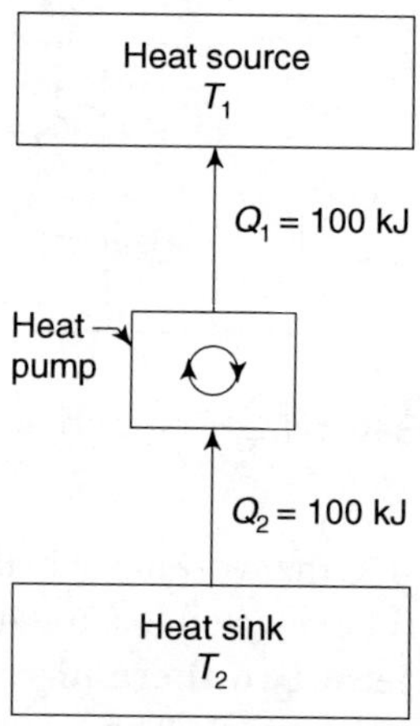

Fig. 7.10 Violating second law

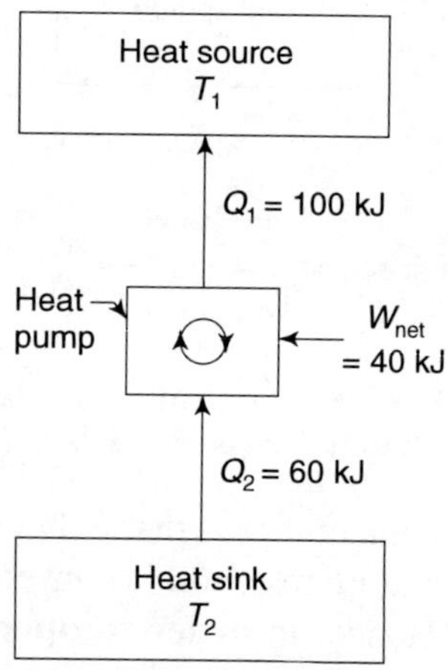

Fig. 7.11 Satisfying second law

or

This statement means that energy in the form of work must be supplied to the heat pump in order to transfer heat from a cold body to hot body. Thus, the coefficient of performance can never be infinity.

In order understand the Clausius statement, Fig. 7.10 may be helpful. In Fig. 7.10, a heat pump transfers heat from a lower temperature body (i.e., heat sink T_2) to a higher temperature body (i.e., heat source T_1) without any external work. This *violates the second law, hence such a heat pump is impossible to construct though, it satisfies the first law.*

Figure 7.11 shows a heat pump in which external work is supplied to transfer heat from a low temperature body to a high temperature body giving COP less than infinity, thus satisfying the first and second laws.

7.9.3 Perpetual Motion Machine of Second Kind

It is impossible to construct a perpetual motion machine of the second kind (PMM-II).

A perpetual motion machine of the second kind (PMM-II) is one which operates in a cycle and delivers an amount of work equal to heat extracted from a single reservoir at a uniform temperature as shown Fig. 7.10.

7.9.4 Principle of Degradation of Energy

(i) *The available energy of an isolated system decreases in all real processes and is conserved in ideal processes.*

(ii) *As a result of natural process, energy steadily becomes less available for the performance of work.*

7.9.5 Principle of Increase of Entropy

(i) *The entropy of an isolated system increases in all real processes and is conserved in ideal processes.*

or

(ii) *As a result of natural process, energy steadily becomes less available for the performance of work.*

Among all types of statement, the statement given in sections (7. 9.1) and (7.9.2) are widely used. In fact, these statements are the basic statements and others are derived from the corollaries of second law.

7.10 EQUIVALENCE OF KELVIN AND CLAUSIUS STATEMENT

Although the Kelvin and Clausius statements appear to be different, but they are really equivalent in the sense that a violation of either statement means the violation of the other.

Consider a heat engine shown in Fig. 7.12 as violating the Kelvin–Planck statement by absorbing heat Q_1 from a single reservoir and producing an equal amount of work $W = Q_1$. The work delivered (W) from the engine is used to drive a heat pump which takes in Q_2 amount of heat from a low temperature reservoir and transfers ($Q_1 + Q_2$) amount of heat to a high temperature reservoir. Out of the heat ($Q_1 + Q_2$) transferred to the high temperature reservoir, a part Q_1 can be delivered to furnish the heat input to the engine. Thus, the combined system of the heat engine and the heat pump then acts like a heat pump transferring an amount of heat Q_2 from a low temperature to a high temperature reservoir without any external work. This violates Clausius statement.

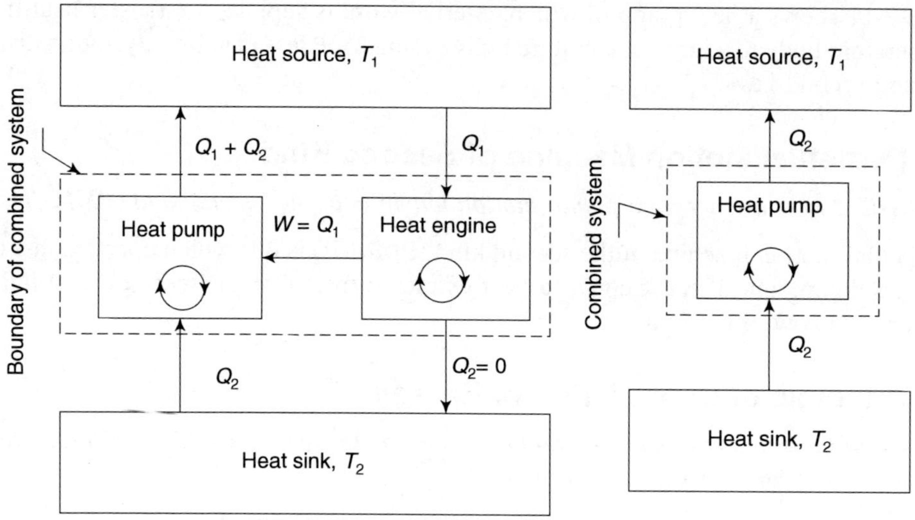

Fig. 7.12 Violation of Kelvin statement

Now consider heat pump as shown in Fig. 7.13 that is violating the Clausius statement by receiving no work and transferring an amount of heat Q_2 from a low temperature to a high temperature reservoir. Let a amount of heat Q_1 which is greater than Q_2, be transferred from the high temperature reservoir to a heat engine which is developing a network $W = Q_1 - Q_2$, and rejects Q_2 amount of heat to the low temperature reservoir. We see that there is no heat interaction with the low temperature reservoir hence it can be neglected. The combined system of heat engine and heat pump acts like a heat engine exchanging heat with a single reservoir and producing an equal amount of work. This violates the Kelvin statement.

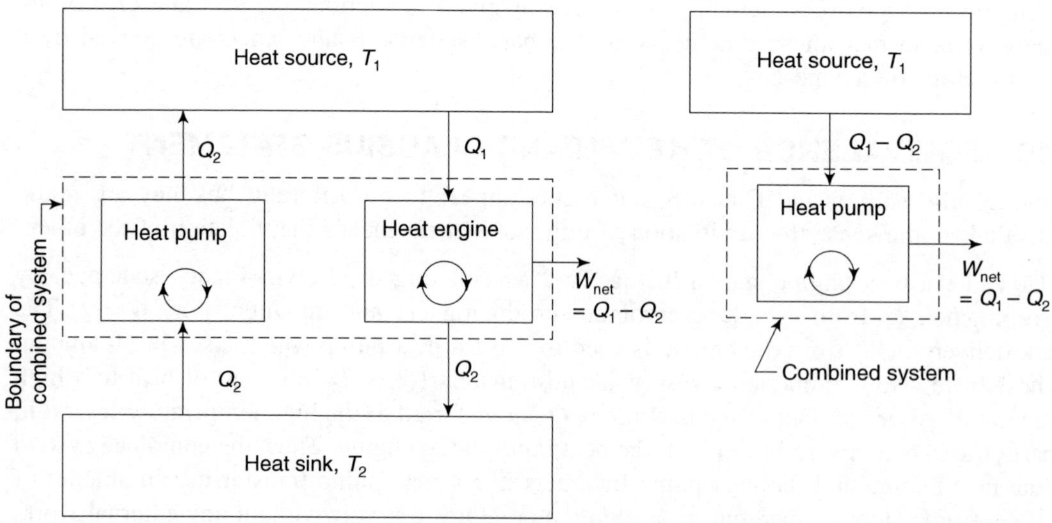

Fig. 7.13 Violation of Clausius statement

Since the violation of Kelvin statement implies also a violation of Clausius statement and vice versa, so it is concluded that the **two statements are equivalent**.

7.11 REASONS FOR IRREVERSIBILITY

As mentioned earlier that the thermodynamic processes may be divided in two classes.

(i) Reversible or ideal processes

(ii) Irreversible or natural processes

In chapter one, both processes have been defined with some examples. However, in this chapter, a detailed study of reasons for irreversibility with examples will be discussed.

The irreversibility for a process may be caused due to either one or both of the following:

(i) Lack of thermodynamic equilibrium during the process, and (ii) involvement of dissipative effects during the process.

7.11.1 Irreversibility due to Lack of Thermodynamic Equilibrium

If there is a lack of thermodynamic equilibrium (mechanical, electrical, thermal or chemical) between the system and its surrounding or between two systems or two parts of the same system, a process is caused due to spontaneous change which is irreversible. The following examples will reveal this fact.

(a) *Heat transfer through a finite temperature difference* As per definition, a reversible heat transfer process is one in which heat is transferred through an infinitesimal temperature

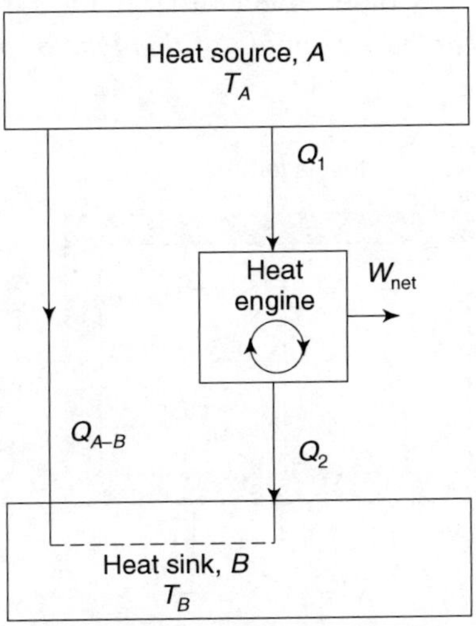

Fig. 7.14 Heat transfer through a finite temperature difference

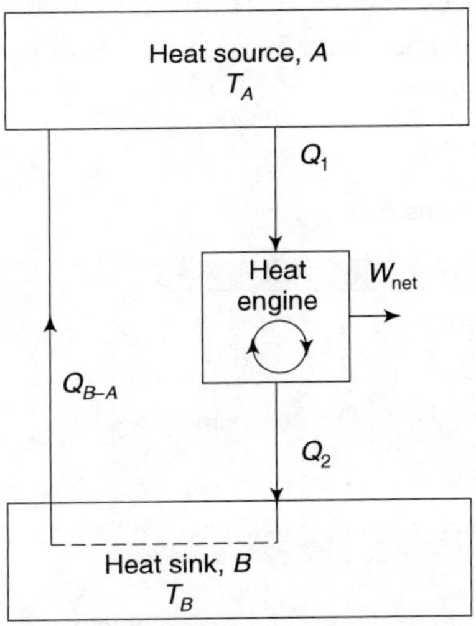

Fig. 7.15 Irreversible heat transfer through a finite temperature difference

difference in the direction of high temperature to low temperature. In order to transfer a finite amount of heat through an infinitesimal temperature difference, an infinite area or infinite amount of time is needed. But all actual processes are through finite temperature difference employing finite area and are therefore irreversible due to lack of thermal equilibrium. ***The greater temperature difference is the greater irreversibility***. The process cannot be reversed automatically.

It is possible to show by the second law of thermodynamics that heat transfer through a finite temperature difference is irreversible. Consider a heat source (*A*) at temperature T_A and a heat sink (*B*) at temperature T_B ($T_A > T_B$). Let Q_{A-B} be the amount of heat flowing from *A* to *B* as a result of temperature difference (Fig. 7.14). Further, let us assume that an engine operating between *A* and *B* taking heat Q_1 from *A* and discharging Q_2 amount of heat to *B* and delivering work, W_{net}. Now reverse the heat transfer process and so Q_{B-A} amount of heat will be flowing from *B* to *A* and suppose that the working of the engine is such that (Fig. 7.15).

$$Q_2 = Q_{B-A}$$

Under this condition, the heat sink *B* is eliminated and the net result is that the engine exchanging heat with a single reservoir and producing work W_{net}. This violates the Kelvin–Planck statement. *This means that this heat transfer process* (Q_{A-B}) *cannot be reversed and hence this is an irreversible process.*

(b) *Free expansion* The lack of pressure equilibrium within the interim of the system or between the system and surroundings causes both the system and its surrounding or the system alone to undergo an spontaneous change of state which will cease only when pressure (mechanical) equilibrium will be established.

Such process cannot be reversed without producing any other effect either in the system or surroundings and hence it is an irreversible process. *The best example of this type of process is free expansion.*

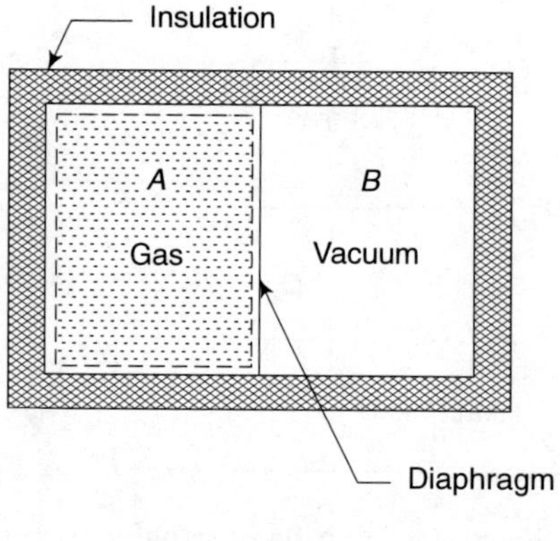

Fig. 7.16 Free expansion process

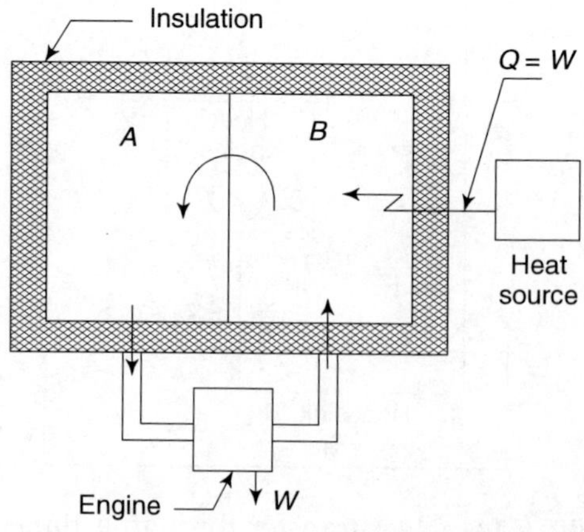

Fig. 7.17 Irreversible free expansion process

Consider an insulated container (Fig. 7.16) having two compartments A and B separated by a thin diaphragm. Compartment A contains a mass of gas while compartment B is completely evacuated. If the diaphragm is punctured, the gas in A will expand into B till pressure equilibrium is reached. As discussed earlier, such process is called a free or unrestricted expansion process. If the process cannot be reversed, hence it is a irreversible process. This can be demonstrated by second law as below.

First of all let us assume that free expansion is a reversible process and that the gas in B returns to A with an increase in pressure, and B attains the same evacuated condition as before (Fig. 7.16). There is no after events as per reversible condition. Now install an engine (simply a machine, not a cyclic heat engine) between A and B and permit the gas to expand through the engine from A to B. As a result of this expansion, the engine develops a work output W at the expense of the internal energy of the gas (system). In order to restore the internal energy of gas in B to its initial value, a heat source transfers heat $Q = W$. Now with the assumption of reversed free expansion process, the system can be restored to the initial state of high pressure in A and vacuum in B. The net result is a cycle in which the net work output W is accomplished by exchanging heat with a single reservoir as the reservoir B gets eliminated. This violates the Kelvin–Planck statement. The assumption of reversed free expansion process gets violated and hence it is an irreversible process.

7.11.2 Irreversibility due to Dissipative Effect

Dissipative effects associated with a process include the effects due to friction, viscosity, inelasticity, electrical resistance and magnetic hysteresis. Friction is more predominant among all. The irreversibility of a process may be due to the dissipative effects in which work is done by or on the system without producing an equivalent decrease or increase in the kinetic energy or potential energy or internal energy of the system. Since the transformation of work into the other form of energy is in the presence of dissipative effects, so the work is said to be dissipated. The following are specific examples in this regard.

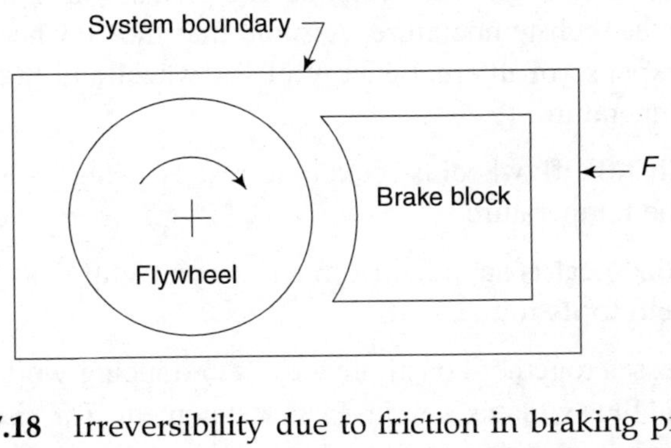

Fig. 7.18 Irreversibility due to friction in braking process

(a) *Braking of rotating flywheel* Friction is always present in moving devices. Consider a system consisting of a flywheel and a brake block (Fig. 7.18). Suppose that flywheel is rotating with a certain speed and it is brought to rest by applying the force on friction brake. The design of the braking system is such that the distance moved by the brake block is very small, so work transfer is negligible. In most of the cases, the braking process occurs very rapidly so there is a little heat transfer which may be neglected. Let suffixes 1 and 2 refer to conditions before and after braking respectively. By the first law, we have

$$Q_{1-2} - W_{1-2} = E_2 - E_1$$

or

$$O - O = E_2 - E_1$$

or

$$E_2 = E_1 \qquad \text{...(7.6)}$$

Equation (7.6) suggests that energy of the system remains constant. The energy may exist in the form of internal, kinetic and potential energy, so

$$U_2 + \frac{m\,C_2^2}{2} + mgz_2 = U_1 + \frac{m\,C_1^2}{2} + mgz_1$$

After braking, the wheel comes to rest, hence $C_2 = 0$ and there is no change in potential energy, so

$$U_2 = U_1 + \frac{m\,C_1^2}{2} \qquad \text{...(7.7)}$$

From Eq. (7.7), we conclude that the molecular internal energy of the system (i.e., the brake and the wheel) after applying the brake increases by the absorption of the kinetic energy of the reverse process, the conversion of this increase in molecular internal energy into KE within the system to cause the wheel to rotate is not possible, hence the ***process is irreversible*** which can be also demonstrated by second law.

Let us assume that the process is reversible and assume the following cycle consisting of three processes *A*, *B* and *C*.

Process A As a result of the absorption of the KE of the wheel while at rest, initially the wheel and brake are at high temperature. Assume that the flywheel starts rotating at a particular speed at the expense of internal energy of the wheel and brake which will result in the decrease of its temperature.

Process B Assume that the flywheel is brought to rest by using its KE in raising weights without any change in the temperature.

Process C Assume that heat is supplied from an external source to the flywheel and the brake to restore the system to its initial state.

The above three processes together constitute a cycle producing work by exchanging heat with a single reservoir and thus violates Kelvin–Planck statement. *This means that assumption of reversible braking possible was wrong and hence braking process (i.e., transformation of KE into molecular internal energy) is irreversible.*

(b) *Paddle-wheel work transfer* Figure 7.19 shows the paddle-wheel work (also known as stirring work) transfer to a system in insulated container. This work transferred is dissipated adiabatically into an increase in the molecular internal energy of the system. The irreversibility of the process may be proved as follows by second law.

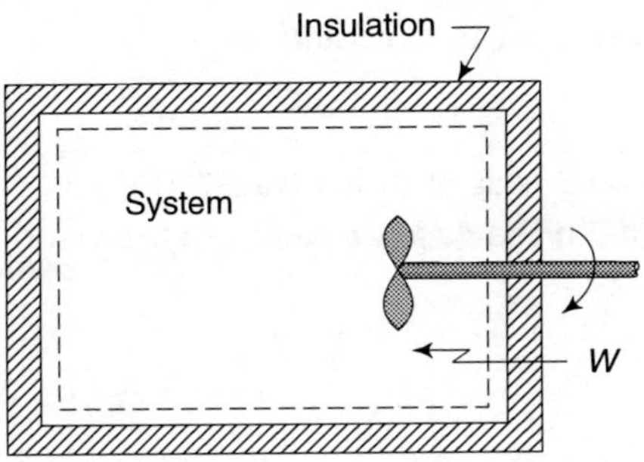

Fig. 7.19 Paddle wheel work transfer

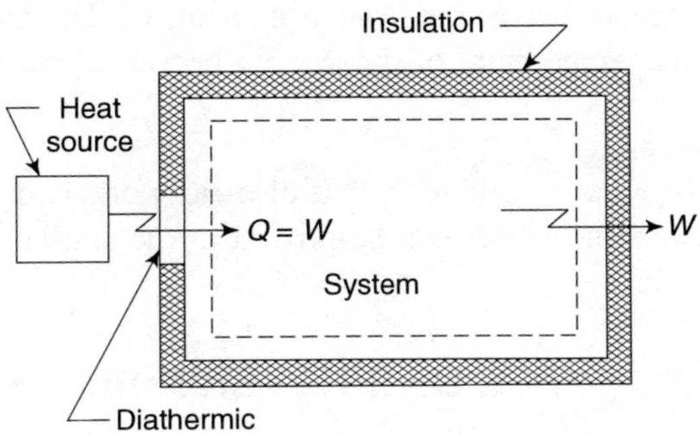

Fig. 7.20 Irreversible paddle wheel work transfer

First of all assume that the process is reversible and so the same amount of work is delivered by the system at the expanse of its molecular internal energy and temperature of the system falls down (Fig. 7.20). In order to restore the system to initial state, the heat is transferred to the system from an external source. These two processes together constitute a cycle producing work by exchanging heat with a single reservoir and this violates Kelvin–Planck statement and it will become PPM-I1. *So the paddle-wheel work transfer (i.e., dissipation of stirring work into molecular internal energy) is an irreversible process.*

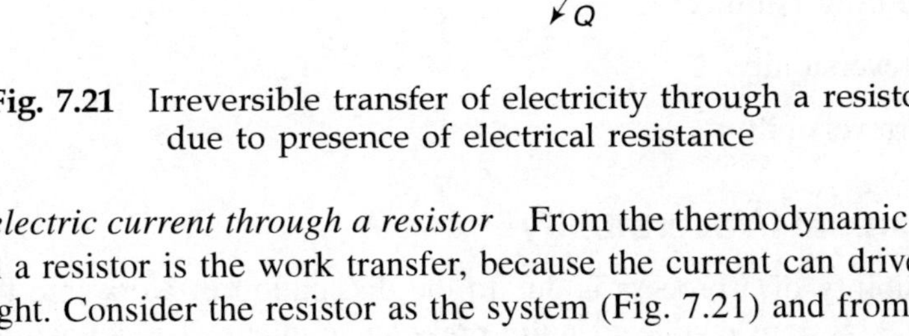

Fig. 7.21 Irreversible transfer of electricity through a resistor due to presence of electrical resistance

(c) *Flow of electric current through a resistor* From the thermodynamic point of electric current through a resistor is the work transfer, because the current can drive a motor which can raise a weight. Consider the resistor as the system (Fig. 7.21) and from first law.

$$Q_{1-2} = W_{1-2} = U_2 - U_1$$

Since Q_{1-2} and W_{1-2} are both negative here, hence

$$W_{1-2} = (U_2 - U_1) + Q_{1-2} \quad ...(7.8)$$

From Eq. (7.8), it follows that a part of the work transfer is stored in the form of an increase in internal energy of the system resulting in the increase in its temperature and the remainder leaves the system as heat. Under steady state conditions, the internal energy and hence temperature of the system becomes constant with respect to time and thus

$$W_{1-2} = Q_{1-2}$$

This equation suggests that electric work is dissipated into heat and this transformation is irreversible which can be proved in the similar fashion as in the case of paddle-wheel work transfer.

7.12 TYPES OF IRREVERSIBILITIES

We have discussed in detail the causes of irreversibilities. Irreversibilities can be of two types namely:

1. Internal irreversibilities
2. External irreversibilities

7.12.1 Internal irreversibilities

It is caused by the internal dissipative effects like friction, turbulence, electrical resistance, magnetic hysterisis, etc., within the system.

7.12.2 External Irreversibilities

It refers to the irreversibility occurring at the system boundary just away from the system boundary like heat interaction with the surroundings due to finite temperature gradient. In the system boundary extended to cover the location of external irreversibilities, then it will be only internal irreversibilities.

However, there are some other types of irreversibilities. They include:

(i) Mechanical irreversibility

(ii) Thermal irreversibility

(iii) Chemical irreversibility

7.12.3 Mechanical Irreversibility

When the irreversibility of a process is due to the dissipation of work into the increase in internal energy of a system or due to finite pressure gradient, it is known as mechanical irreversibility.

7.12.4 Thermal Irreversibility

When the irreversibility is caused by finite temperature gradient, it is known as thermal irreversibility. For example, in a heat engine cycle in which there is a temperature difference (i) between the source and the working fluid during the heat supply and (ii) between the working fluid and sink during heat rejection, they exhibit *external thermal irreversibility.*

7.12.5 Chemical Irreversibility

If the irreversibility is caused due to a finite concentration gradient or a chemical reaction, it is known as chemical irreversibility.

Engineers should be able to recognize the irreversibilities, evaluate their influence and develop practical means for reducing them. Steps taken in reducing irreversibilities are constrained by a number of factors.

7.13 INTERNALLY REVERSIBLE PROCESSES

In an irreversible process, irreversibilities are present within the system, its surrounding or both. *A reversible process is one in which there are no internal or external irreversibilities.* **An internally reversible process** is one in which there are no internal irreversibilities. But, the irreversibilities may be located within surroundings.

At every intermediate state of internally reversible process of a closed system, all intensive properties are uniform throughout each phase present. This means that the temperature, pressure, specific volume, and other intensive properties do not vary with positions. For reversibility, no spontaneous processes can be present. From these considerations, it can be concluded that the internally reversible process consists of a series of equilibrium states. It is quasi-equilibrium process. However, the internally reversible process term is more common in use.

The internally reversible concept can be employed to redefine the definition of thermal reservoir. In subsequent discussions, we assume that no internal irreversibility are present within the thermal reservoir. *Accordingly, every process of a thermal reservoir is an internally reversible process.*

7.14 CARNOT CYCLE

A French military engineer Nicolas Sadi Carnot was among the first to study the principles of the second law of thermodynamic. In 1824, he introduced a classical ideal cycle named after him in which the working fluid, after passing through a sequences of events, is brought back to its initial state. On the basis of the Carnot cycle, Clausius and Kelvin were able to define the second law of thermodynamics.

The Carnot cycle shown in Fig. 7.22 consists of an alternate series of two reversible isothermal processes and two reversible adiabatic processes. Since all processes are individually reversible, hence the Carnot cycle as a whole is a *reversible cycle.*

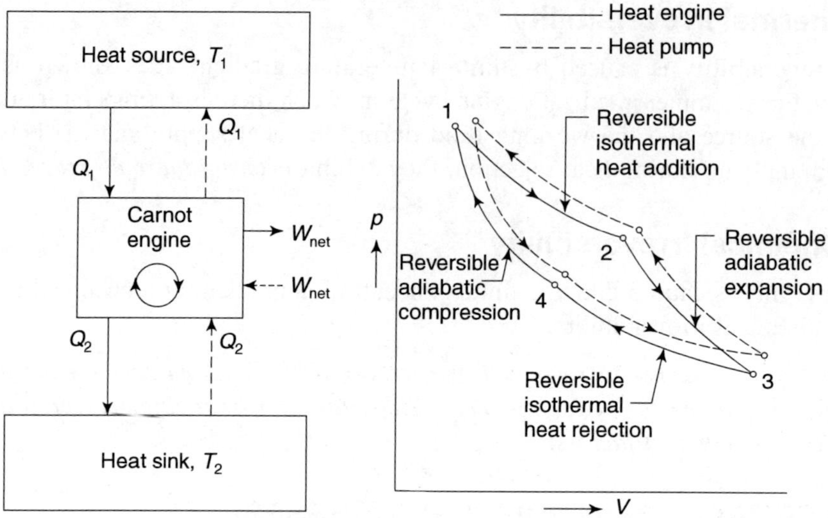

Fig. 7.22 The Carnot cycle

7.14.1 Heat Engine

In order to understand the working of an engine on Carnot cycle, let us assume that the working fluid is enclosed in a cylinder having a frictionless piston (Fig. 7.23). The walls of the cylinder and the piston are taken as perfect insulators of heat. The cylinder head is arranged such that it acts like a perfect conductor as well as a perfect insulator. There are a heat source at temperature T_1, a heat sink at temperature T_2 and a work reservoir. Starting from an initial state 1, the working of the Carnot cycle for a heat engine is explained in the following manner:

(a) *Process 1–2* During the process 1–2, heat is transferred reversibly and isothermally to the working substance from a heat source, T_1. This is accomplished by placing the cylinder

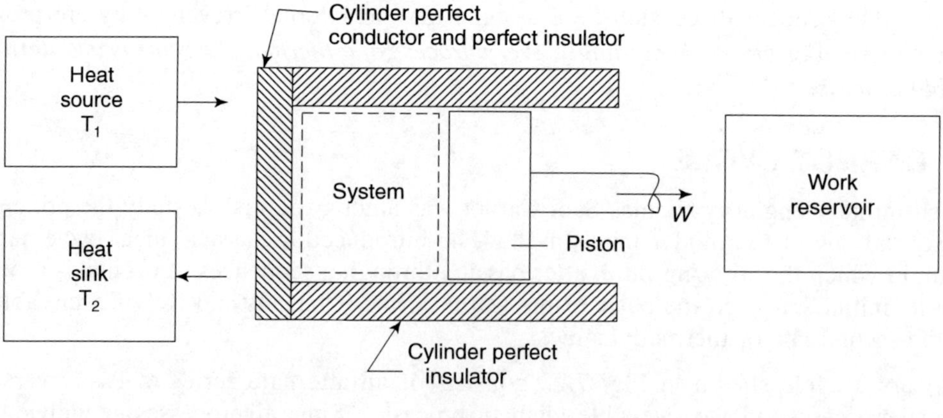

Fig. 7.23 Working of a Carnot cycle

in contact with the heat source T_1 and the fluid is allowed to expand slowly. Thus, some work is delivered and the quantity of heat Q_1 is withdrawn from the heat source. The work done is given by the area underneath path 1–2 of the *p*–*V* diagram.

(b) *Process 2–3* During the process 2–3, reversible adiabatic expansion takes place. Since the system (cylinder-piston arrangement) is thermally insulted, no heat transfer takes place The fluid is allowed to expand still further adiabatically until its temperature drops from T_1 to T_2. Work is delivered to the work reservoir and is represented by the area underneath path 2–3 of the *p*–*V* diagram.

(c) *Process 3–4* During the process 3–4, reversible isothermal heat rejection at constant temperature T_2 takes place. This is accomplished by placing the cylinder in contact with heat sink T_2. The fluid is slowly compressed. Work is supplied by the work reservoir while Q_1 amount of heat flows into the heat sink. The work done in the process is represented by the area underneath path 3–4 of the *p*–*V* diagram.

(e) *Process 4–1* During the process 4–1 which completes the cycle, reversible adiabatic compression takes place so that the working substance is returned to the original state 1. This is accomplished by supplying work from the work reservoir and is represented by the area underneath path 4–1 of the *p*–*V* diagram. Due to compression, the temperature of the working fluid is raised from low temperature T_2 to high temperature T_1, and since the cylinder is thermally insulated, no transfer of heat takes place.

7.14.2 Heat Pump

If a *Carnot cycle for a heat engine is carried out in a reversed sequence, the result will be Carnot cycle for a heat pump.* Such a cycle is shown by dotted line in Fig. 7.22.

The various processes are:

(a) *Process 4–3* During process 4–3, heat is transferred from a low temperature reservoir to the working fluid in a reversible isothermal manner.

(b) *Process 3–2* During process 3–2, reversible adiabatic compression takes place and the temperature of the working fluid is raised from a low temperature T_2 to a higher temperature T_1.

(c) *Process 2–1* During process 2–1, heat is transferred reversibly and isothermally from the working fluid to the high temperature reservoir.

(d) *Process 1–4* During process 1–4, the working fluid is returned to initial state 4 in a reversible adiabatic manner. The temperature of the working fluid decreases from T_1 to T_2.

As in the heat engine cycle, the net work done is represented by the area 1–2–3–4 and is equal to $\oint \delta W$.

Since the Carnot cycle is a reversible cycle and the reversible process does not occur in nature, hence the *Carnot cycle is an ideal cycle*.

So far dealing with the Carnot cycle is concerned, no specific characteristics were imposed regarding the working substance. We will discuss now, that efficiency or COP of the Carnot cycle is independent of the amount or nature of the working fluid.

7.14.3 Efficiency of Carnot Cycle as an Engine

The efficiency of an engine is defined as

$$\eta_{\text{engine}} = \frac{\text{Work done per cycle}}{\text{Heat supplied per cycle}} = \frac{W}{Q} = \frac{\oint \delta W}{Q_1}$$

To find an expression of efficiency, we have to find out work done per cycle and the heat supplied per cycle.

(i) Work done by the system during reversible isothermal heat addition from 1 to 2 is

$$W_{1-2} = p_1 V_1 \ln \frac{V_2}{V_1} = m R T_1 \ln \frac{V_2}{V_1}$$

(ii) Work done by the system during reversible adiabatic expansion from 2 to 3 is

$$W_{2-3} = \frac{p_2 V_2 - p_3 V_3}{\gamma - 1} = mc_v (T_1 - T_2)$$

(iii) Work done by the system during reversible adiabatic compression proces from 3–4 is

$$W_{3-4} = -p_3 V_3 \ln \frac{V_3}{V_4} = -m R T_2 \ln \frac{V_3}{V_4}$$

(iv) Work done by the system during reversible adiabatic compression process from to 4–1 is

$$W_{4-1} = \frac{p_4 V_4 - p_1 V_1}{\gamma - 1} = -mc_v (T_1 - T_2)$$

Hence, network per cycle 1 – 2 – 3 – 4 = $W_{1-2} + W_{2-3} + W_{3-4} + W_{4-1}$

$$= m R T_1 \ln \frac{V_2}{V_1} + mc_v (T_1 - T_2) - mRT_2 \ln \frac{V_3}{V_4} - mc_v (T_1 - T_2)$$

$$= m R T_1 \ln \frac{V_1}{V_2} - m R T_2 \ln \frac{V_3}{V_4}$$

or net work done per cycle $= mR \left[T_1 \ln \frac{V_2}{V_1} - T_2 \ln \frac{V_3}{V_4} \right] = Q_1 - Q_2$

Since there is no change in internal energy in an isothermal process, the heat suppled (Q_1) is equal to the work done by the system during that process. That is,

$$Q_1 = W_{1-2} = mRT_1 \ln \frac{V_2}{V_1}$$

$$\therefore \quad \eta_{engine} = \frac{mR\left[T_1 \ln \frac{V_2}{V_1} - T_2 \ln \frac{V_3}{V_4}\right]}{m\,RT_1\,T_1 \ln \frac{V_2}{V_1}} = \frac{T_1 \ln \frac{V_2}{V_1} - T_2 \ln \frac{V_3}{V_4}}{T_1 \ln \frac{V_2}{V_1}} \qquad ...(7.9)$$

But during adiabatic proces 2–3.

$$\frac{V_2}{V_3} = \left(\frac{T_2}{T_1}\right)^{1/(\gamma - 1)} \qquad ...(7.10)$$

and during process 4–1,

$$\frac{V_1}{V_4} = \left(\frac{T_2}{T_1}\right)^{1/(\gamma - 1)} \qquad ...(7.11)$$

Comparison of the Eqs. (7.10) and (7.11), reveals that

$$\frac{V_2}{V_3} = \frac{V_1}{V_4} \quad \text{or} \quad \frac{V_2}{V_1} = \frac{V_3}{V_4} \qquad ...(7.12)$$

Putting Eq. (7.12) in Eq. (7.9), we get

$$\eta_{engine} = \frac{T_1 - T_2}{T_1} = 1 - \frac{T_2}{T_1} = 1 - \frac{T_{min}}{T_{max}} \qquad ...(7.13)$$

Hence, from Eq. (7.13) we can draw the following *conclusions*.

(a) Equation (7.13) suggests that the efficiency of such an engine is independent of the working fluid and limited by temperature T_1 and T_2 only.

(b) If $T_2 = 0$, efficiency of the engine will be maximum, i.e., 100%, but this means the absence of the heat sink which is impossible according to the second law.

(c) When there is a large temperature difference ($T_1 - T_2$), a higher value of efficiency can be attained. **In other words, efficiency increases with higher value of T_1 and lower value of T_2. T_1 is limited by metallurgical condition and T_2 is limited by atmospheric or sink condition.**

(d) If $T_1 = T_2$, no work will be done, and efficiency will be zero.

7.14.4 Coefficient of Performance as a Refrigerator

$$(\text{COP})_{Ref} = \frac{Q_2}{W_{net}} = \frac{Q_2}{Q_1 - Q_2} = \frac{m\,RT_2 \ln \frac{V_3}{V_4}}{mR\left[T_1 \ln \frac{V_2}{V_1} - T_2 \ln \frac{V_3}{V_4}\right]}$$

or

$$(\text{COP})_{Ref} = \frac{T_2}{T_1 - T_2} = \frac{T_{min}}{T_{max} - T_{min}} \qquad ...(7.14)$$

For **higher value of (COP)$_{\text{Ref}}$, (T_1 – T_2) should be as minimum as possible. In other words, (COP)$_{\text{ref}}$ increases with lower value of T_1 and higher value of T_2.**

7.14.5 Coefficient of Performance as a Heat Pump

$$(\text{COP})_{\text{Heat pump}} = \frac{Q_1}{W_{\text{net}}} = \frac{Q_1}{Q_1 - Q_2} = \frac{T_1}{T_1 - T_2} \qquad ...(7.15)$$

From the above discussions, it can be stated that the magnitude of the quantity of heat absorbed or rejected is proportional to the absolute temperatures of the heat reservoirs in all the cycles.

Being an ideal cycle, the Carnot cycle gives the maximum possible thermal efficiency attainable between any two given temperature limits. But the cycle is, however, impractical for the following reasons.

7.14.6 Reasons for the Impracticability of Carnot Cycle

(i) The cycle demands that all four processes should be reversible. This necessitates that the movement of the piston should be frictionless, which is an impossible condition.

(ii) Isothermal heat addition is not possible in practice, because when heat is supplied temperature cannot remain constant, it must increase as it does in practice.

(iii) In order to achieve isothermal operation, the condition necessary for the cycle is that the piston should move very slowly, while achieving the adiabatic operation, the piston movement should be very fast. The variation of speed during the different processes in one cycle of the practical engine is not possible.

(iv) Isothermals and adiabatics of the gas have a very insignificant difference in slope, in other words, the *p–V* plot is greatly extended both in the horizontal and the vertical directions. This is only possible under conditions when the cylinder involves greater pressure and volume.

Example 7.4 A Carnot heat engine works between two thermal reservoirs *A* and *B*. *A* is at constant temperature 600°C and *B* is at 250°C. Half of the power developed by the Carnot engines is used to derive a generator to produce electricity and the other half is used to drive a heat pump which receives heat from thermal reservoir *B* and rejects heat to a thermal reservoir *C* which is at temperature 400°C. Calculate the heat rejected to thermal reservoir *C* by the heat pump as percentage of heat from thermal reservoir *A* to the Caraot engine. Also calculate the heat rejected per hour to thermal reservoir *C* if 480 kW are generated by generator assuming 100% of generator efficiency.

Solution Refer to Fig. 7.24.

Efficiency of the Carnot heat engine is

$$\eta = \frac{T_A - T_B}{T_A} = \frac{873 - 523}{873} = \frac{330}{873} = 40.09\%$$

Let Q_A be the heat supplied by the reservoir to the Carnot heat engine. Then, the work done by the engine.

$$\eta \times Q_A = \frac{350}{873} Q_A$$

$\therefore$ Work done supplied to heat pump $= \frac{1}{2} \cdot \frac{350}{873} Q_A = \frac{350}{1756} Q_A$

$$\therefore \quad (\text{C.O.P.})_{\text{Heat pump}} = \frac{T_C}{T_C - T_B} = \frac{673}{673 - 523} = 4.486 \qquad \textbf{Ans.}$$

By definition, $(\text{C.O.P.})_{\text{Heat pump}} = \frac{\text{Heat Rejected to } C}{\text{Work input}}$

or
$$4.486 = \frac{Q_c \times 1756}{350\, Q_A}$$

or
$$\frac{Q_C}{Q_A} = 4.486 \times \frac{350}{1756} = 89.41\% \qquad \textbf{Ans.}$$

Since power output of generator = 480 kW

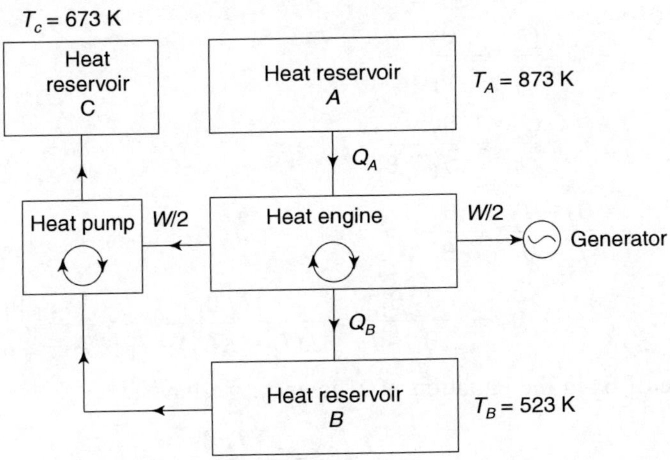

Fig. 7.24 Example 7.4

$\therefore$ Total power developed by the Carnot engine = 480 × 2 = 960 kW

$$\therefore \quad 960 = \frac{350}{873} Q_A$$

or
$$Q_A = \frac{960 \times 873}{350} = 2394.5 \text{ kW}$$

But $$\frac{Q_c}{Q_A} = 0.8941$$

$\therefore$ $$Q_c = 0.8941 \times 2394.5 = 2140.92 \text{ kW}$$ **Ans.**

Example 7.5 A working fluid goes through a Carnot cycle of operations, the upper absolute temperature of the fluid being θ_1, and the lower absolute temperature being θ_2. The amounts of heat taken in and rejected by the working fluid are Q_1 and Q_2 respectively. On account of losses of heat, the source temperature T_1 is higher than θ_1 and the heat sink temperature T_2 is lower than θ_2. If

$$T_1 = \theta_1 + KQ_1 \quad \text{and} \quad T_2 = \theta_2 + KQ_2$$

where K is the same constant for both the equations, show that the efficiency of the plant is given by

$$\eta = 1 - \frac{T_1}{(T_2 - 2KQ_1)}$$

Solution Efficiency of the Carnot cycle is given by

$$\eta = \frac{Q_1 - Q_2}{Q_1} = \frac{\theta_1 - \theta_2}{\theta_1}$$

or $$\frac{Q_2}{Q_1} = \frac{\theta_2}{\theta_1}$$

But $$\frac{Q_2}{Q_1} = \frac{\theta_2 - T_2}{T_1 - \theta_1}$$

$\therefore$ $$\frac{\theta_2 - T_2}{T_1 - \theta_1} = \frac{\theta_2}{\theta_1}$$

or $$\theta_2 = \frac{T_2\,\theta_1}{2\,\theta_1 - T_1} = \frac{T_2\,\theta_1}{2\,(T_1 - KQ_1) - T_1} = \frac{T_2\,\theta_1}{T_1 - 2\,KQ_1}$$

Putting the value of θ_2 in the equation of efficiency, we have

$$\eta = \frac{\theta_1 - \theta_2}{\theta_1} = \frac{\theta_1 - \dfrac{T_2\,\theta_1}{T_1 - 2\,KQ_1}}{\theta_1} = 1 - \frac{T_1}{T_1 - 2\,KQ_1}$$ **Proved**

Example 7.6 What is the more effective way to increase the efficiency of a Carnot engine? Comment on the results.

Solution The efficiency of a Carnot engine is given by

$$\eta = 1 - \frac{T_2}{T_1}$$

There are two possible ways to increase the efficieny of a Carnot engine. One way is to decrease T_2 keeping T_1 constant while the other is to increase T_1 keeping T_2 constant. Let T_2 decrease by ΔT with T_1 remaining the same, hence

$$\eta_1 = 1 - \frac{T_2 - \Delta T}{T_1}$$

If T_1 is increased by the same ΔT, T_2 remaining the same

$$\eta_2 = 1 - \frac{T_2}{T_1 + \Delta T}$$

Difference between two efficiencies yield

$$\eta_1 - \eta_2 = \frac{T_2}{T_1 + \Delta T} - \frac{T_2 - \Delta T}{T_1} = \frac{(T_1 - T_2)\,\Delta T + \Delta T^2}{T_1\,(T_1 + \Delta T)}$$

Since $\quad T_1 > T_2, \quad \text{so} \quad (\eta_1 - \eta_2) > 0$

This shows that the more effiective way to increase the Carnot cycle efficiency is to decrease T_2.

There is a limit for the lower value of T_2 *as it depends on the environmental condition. The increae in the value of* T_1 *is limited by the metallurgical condition of the material for the engine.*

Example 7.7 A satellite incorporates a reversible heat engine which operates between a hot reservoir at T_1 and a radiating panel at T_2. The radiation from the panel, $Q_2 = KAT_2^{\,4}$ where K is Stefan–Boltzmann constant and A is the area of the panel. For a given work output and a value of T_1, prove that the area of the panel will be the minimum when $T_2/T_1 = 3/4$. Further prove that $A_{min} = 256\ W/27\ KT_1^{\,4}$ where W is the output of the engine.

Solution Refer to Fig. 7.2.

The efficiency of a reversible engine is given by

$$\eta = \frac{W}{Q_1} = \frac{T_1 - T_2}{T_1} = 1 - \frac{Q_2}{Q_1}$$

or

$$\frac{W}{T_1 - T_2} = \frac{Q_1}{T_1} = \frac{Q_2}{T_2} = \frac{K\,AT_2^4}{T_2}$$

or

$$A = \frac{W}{K\,T_2^3\,(T_1 - T_2)} = \frac{W}{K\,(T_1\,T_2^3 - T_2^4)}$$

For a given value of W and T_1, A will be the minimum when

$$\frac{dA}{dT_2} = \frac{-W}{K}(3\,T_1\,T_2^2 - 4\,T_2^3)\,(T_1\,T_2^3 - T_2^4)^{-2} = 0$$

Since $\frac{W}{K}(T_1 T_2^3 - T_2^4)^{-2} \neq 0$

Hence $3 T_1 T_2^2 - 4 T_2^3 = 0$

or $\frac{T_2}{T_1} = \frac{3}{4}$ **Proved**

Thus, $$A = A_{min} = \frac{W}{K (3/4)^3 T_1^3 [T_1 - (3/4) T_1]}$$

or $$A_{min} = \frac{256 W}{27 K T_1^4}$$ **Proved**

Example 7.8 What is the minimum amount of work input necessary to a refrigerator which will convert 1 kg of water at 20°C into ice at –5°C?

Solution Refer to Fig. 7.6. Removal of heat from 1 kg of water at 20°C so that it becomes ice at –5°C is given by

$$Q_2 = m_w c_{pw} (0 - 20) + (-335) + m_i c_{pi} (-5 - 0)$$

$$= -1 \times 4.18 \times 20 - 335 - 1 \times 2.09 \times 5 = -429.05 \text{ kJ}$$

–ve sign shows the removal of heat from water.

COP of a Carnot heat pump working as a refrigerator is

$$\text{COP} = (\text{COP})_{max} = \frac{T_2}{T_1 - T_2} = \frac{268}{293 - 268} = 10.72$$

The COP of an actual refrigerator working between the two temperature limits will always be less than the reversible one.

Hence, $$W_{min} = \frac{Q_2}{(\text{C.O.P.})_{max}} = \frac{429.05}{10.72} = 40.02 \text{ kJ}$$ **Ans.**

Example 7.9 A heat engines receives heat from a solar collector plate. Solar radiation of amount 1880 kJ/m$^2 \cdot h$ falls on the solar collector plate and raises its temperature equal to 95°C. The engine rejects heat to atmosphere where temperature is assumed to be 20°C. Calculate the minimum solar collector area that would be required for a plant producing 1 kW of useful shaft power.

Solution Refer to Fig. 7.25. The maximum efficiency of a heat engine (Carnot) operation between the collector plate temperature and the atmospheric temperature is given by

$$\eta_{max} = 1 - \frac{T_2}{T_1} = 1 - \frac{293}{368} = 0.2039$$

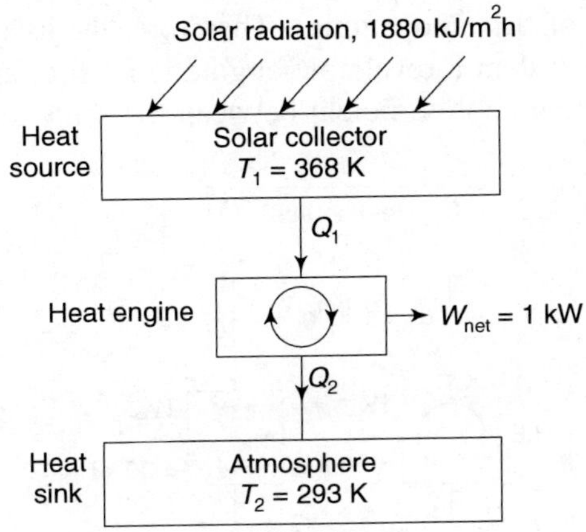

Fig. 7.25 Example 7.9

The efficiency of any actual heat engine operating between these temperature limits would be less than this efficiency. Thus, the minimum heat supplied to the engine is

$$Q_{min} = Q_1 = \frac{W_{net}}{\eta_{max}} = \frac{1}{0.2039} = 4.904 \text{ kW} = 4.904 \times 3600 \text{ kJ/h}$$

The minimum area required for the solar collector plate is

$$A_{min} = \frac{4.904 \times 3600}{1880} = 9.3906 \text{ m}^2$$

7.15 COROLLARY 1 : MAXIMUM REVERSIBLE HEAT ENGINE EFFICIENCY (CARNOT'S THEOREM)

No heat engine working in a cycle between two constant temperatures reservoirs can be more efficient than a reversible engine working between the same two reservoirs.

Let a reversible engine R and an irreversible engine I work between the same temperature limits of heat source T_1 and heat sink T_2 as shown in Fig. 7.26.

Let the heat engine I be more efficient than the reversible engine R and let the heat supplied to the two engines be the same say 100 kJ. Since irreversible engine I is assumed more efficient than the reversible engine R, hence the work done by this engine (say W_I = 30 kJ) will be more than the work done by reversible engine (say W_R = 20 kJ). Now, reverse the reversible engine R so that it work as a heat pump by taking work input from the engine I as shown in the Fig. 7.27. By doing this, the irreversible engine, besides being able to drive the reversible heat pump, gives an additional network $W_{net} = W_I - W_R$, i.e., (30 – 20 = 10 kJ) as shown in Fig. 7.28. If heat equivalent of $W_I - W_R$ = 10 kJ is fed to the heat source T_1, we find that the combined system of reversible heat pump and irreversible heat engine has the net effect of transferring 10 kJ (i.e., $Q_{2R} - Q_{2I}$) of heat from a low temperature heat sink at T_2 to a high temperature heat source at T_1 without any external work. This is obviously a violation of the second law of thermodynamics. Therefore, the assumption that irreversible engine has a higher efficiency than a reversible engine is invalid, and the reversible engine operating on the Carnot cycle is more efficient between the same temperature limits.

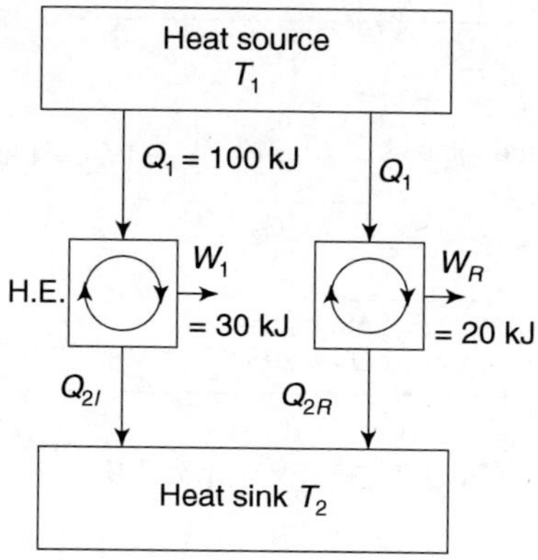

Fig. 7.26 Carnot's theorem-I

Another argument leading to the violation of Kelvin–Planck statement of the second law is

$$\eta_R = \frac{W_R}{Q_{1R}} = \frac{W_R}{W_R + Q_{2R}} = \frac{1}{1 + (Q_{2R})/W_R}$$

and

$$\eta_1 = \frac{W_1}{Q_{1I}} = \frac{W_1}{W_1 + Q_{2I}} = \frac{1}{1 + (Q_{2I})/W_1}$$

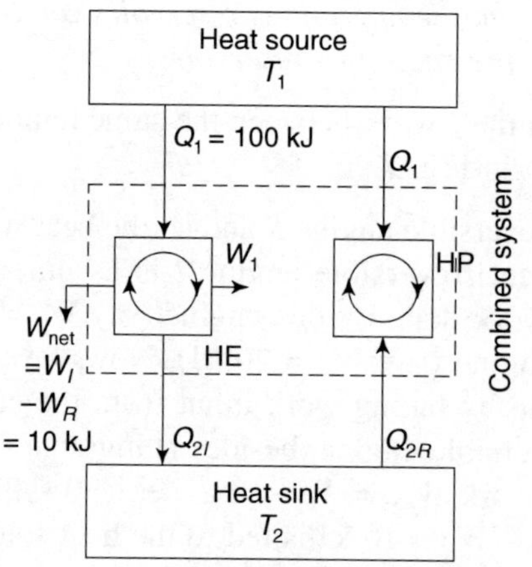

Fig. 7.27 Carnot's theorem-I

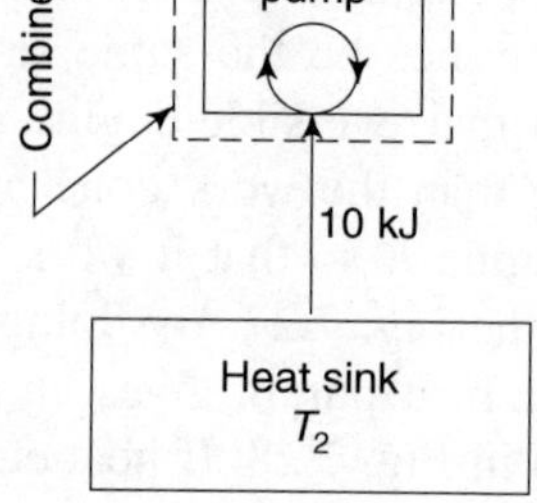

Fig. 7.28 Combined system

Since $\eta_I > \eta_R$ is already assumed, hence

$$\frac{Q_{2R}}{W_R} > \frac{Q_{2I}}{W_I}$$

If $$Q_{2R} = Q_{2I}, \text{ then } W_I > W_R$$

Now assume that the reversible engine R is revered to operate as a heat pump. W_1 can be used to supply W_R to the heat pump and the network output is $W_I - W_R$. Since $Q_{2R} = Q_{2I}$, the low temperature heat reservoir can be eliminated and the net heat leaving the high temperature reservoir is

$$Q_{1I} - Q_{1R} = (W_I + Q_{2I}) - (W_R + Q_{2R}) = W_I - W_R$$

This suggests that heat is transferred with a single reservoir and an equal amount of work is performed which is violating the Kelvin–Planck statement.

Hence, the reversible engine is more efficient than any engine working between the same temperature limits.

7.16 COROLLARY 2 : ALL REVERSIBLE ENGINES HAVE THE SAME EFFICIENCY (CARNOT'S THEOREM)

All reversible engines working between the two constant temperature reservoirs have the same efficiency.

Suppose two reversible engines R_1 and R_2 are working between the same two thermal reservoirs as shown in Fig. 7.29. Let R_1 be more efficient than R_2. Let the same amount of heat Q_1 be supplied from the heat source at T_1. Similar to Corollary 1, it can be shown that if R_2 is reversed to run as a heat pump using the output of R_1 as shown in Fig. 7.30, we come

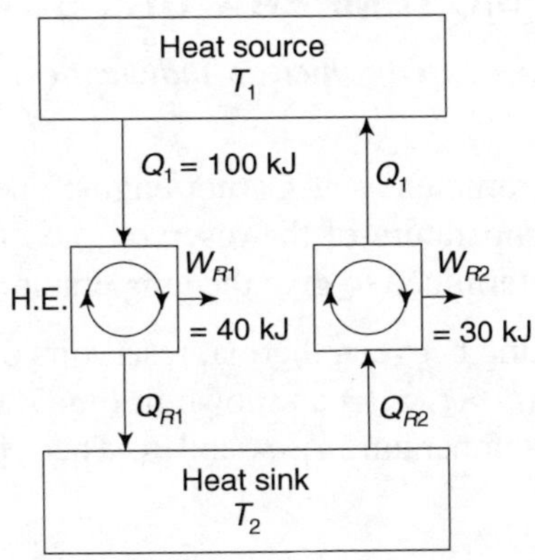

Fig. 7.29 Carnot's theorem–II

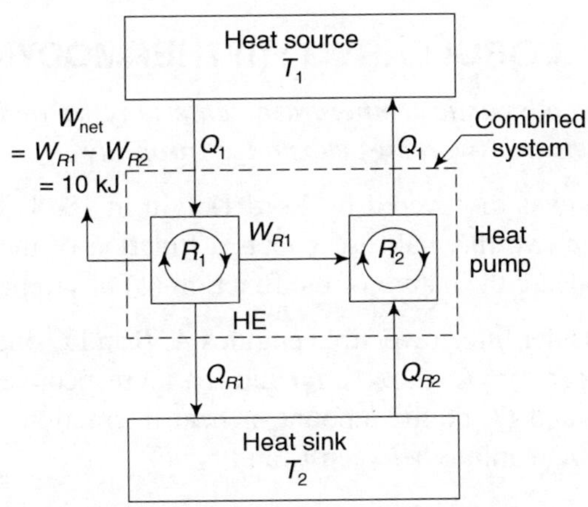

Fig. 7.30 Carnot's theorem–II

to the conclusion that the combined system of heat pump and engine R_1 is removing heat $(Q_{R2} - Q_{R1})$ continuously from a low temperature reservoir without any external work. This violates the Clausius statement of second law:

This can be proved in other way also. Since it has been assumed that the efficiency of R_1 is more than R_2, then by assumption.

$$W_{R1} > W_{R2}$$

$\therefore$ $$(Q_1 - Q_{R1} > (Q_1 - Q_{R2})$$

or $$Q_{R2} > Q_{R1}$$

Now, R_2 is reversed to run as heat pump. The heat pump abstracts heat Q_{R2} from a reservoir at T_2 and delivers heat Q_1 to reservoir T_1. Since the heat required by R_1 is also Q_1, we can eliminate the high temperature reservoir, the net heat thus leaving the low temperature reservoir is

$$Q_{R2} - Q_{R1} = W_{R1} = W_{\text{net}}$$

This means that heat is exchanged with a single reservoir and an equal amount of work is performed, i.e., the new system is a perpetual motion machine of the second kind which is contradictory to Kelvin statement, hence all reversible engines have the same efficiency working between the same temperature limits.

It has been proved already that the efficiency of the Carnot engine is independent of the working fluid and function of temperature alone.

$$\eta_R = 1 - \frac{Q_2}{Q_1} = 1 - \frac{1}{Q_1/Q_2} = 1 - \frac{1}{\phi\,(t_1, t_2)}$$

or $$\eta_R = \phi\ (t_1, t_2)$$

7.17 COROLLARY 3 : (I) THERMODYNAMIC TEMPERATURE SCALE

A thermodynamic temperature scale may be defined as that which is independent of the thermometric (working) substance (property).

This was discovered by Lord Kelvin in 1848. The efficiency of Carnot engine operating between two thermal reservoir is a function of the temperature of the reservoir, i.e., $\eta_R = \phi\ (t_1, t_2)$. Here the nature of the function (ϕ) has to be determined to give the temperature scale.

Consider three reversible engines A, B and C operating between thermal reservoirs of pairs of temperature (t_1, t_2), (t_2, t_3) and (t_1, t_3) respectively $(t_1 > t_2 > t_3)$ as shown in Fig. 7.31. Let Q_1, Q_2 and Q_3 be the amount of heat interactions at temperature t_1, t_2 and t_3. Then, for the engine A working between t_1 and t_2.

$$\eta_A = 1 - \frac{Q_2}{Q_1} = 1 - \frac{1}{Q_1/Q_2} = 1 - \frac{1}{\phi\,(t_1, t_2)}$$

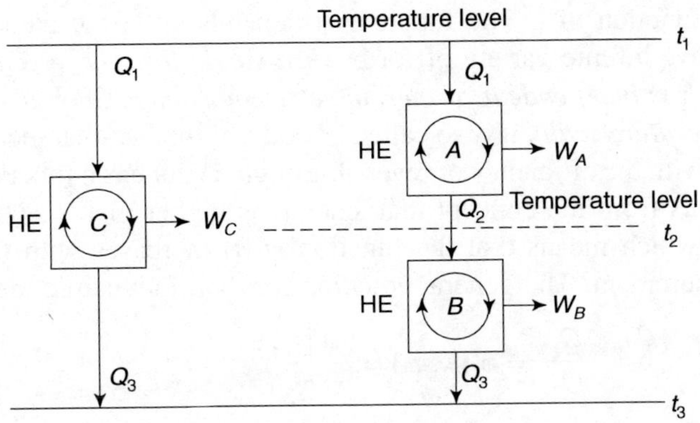

Fig. 7.31 Concept of thermodynamic temperature scale

$$\therefore \quad \frac{Q_1}{Q_2} = \phi\,(t_1, t_2) \qquad ...(7.16)$$

where ϕ is any function

For heat engines B working between temperature limits t_2 and t_3

$$\eta_B = 1 - \frac{Q_3}{Q_2} = 1 - \frac{1}{Q_2/Q_3} = 1 - \frac{1}{\phi\,(t_2, t_3)} \qquad ...(7.17)$$

or $$\frac{Q_2}{Q_3} = \phi\,(t_2, t_3)$$

Again, for heat engine C working between temperature limits t_1 and t_3

$$\eta_c = 1 - \frac{Q_3}{Q_1} = 1 - \frac{1}{Q_1/Q_3} = 1 - \frac{1}{\phi\,(t_1, t_3)}$$

or $$\frac{Q_1}{Q_3} = \phi\,(t_1, t_3) \qquad ...(7.18)$$

But $$\frac{Q_1}{Q_2} = \frac{Q_1/Q_3}{Q_2/Q_3}$$

or $$\phi\,(t_1, t_2) = \frac{\phi\,(t_1, t_3)}{\phi\,(t_2, t_3)} \qquad ...(7.19)$$

In Eq. (7.19), the left hand side has no term containing t_3, therefore, the right-hand side of the equation should be also free from term t_3, and hence, it may be written as

$$\phi\,(t_1, t_2) = \frac{\psi\,(t_1)}{\psi\,(t_2)} \qquad ...(7.20)$$

where ψ is another function of t. The function ψ depends upon the nature of temperature and may therefore have infinite variety of forms. *If a single form is selected, a temperature scale is defined which is independent of thermometric substance. Such a scale is called the thermodynamic temperature scale.* It is so called because of the fact that the temperature scale has been established with the efficiency of reversible engine which is a function of temperature only. It is also obvious from the concept that since t_1 is higher than t_2, Q_1 is large than Q_2 and therefore ψ (t_1) which means that the function ψ (t) increases with t and can be used for temperature measurement. The general equation for thermodynamic temperature scale is

$$\frac{Q_1}{Q_2} = \phi\,(t_1, t_2) = \frac{\psi\,(t_1)}{\psi\,(t_2)} \qquad ...(7.21)$$

7.18 COROLLARY 3 : (II) ABSOLUTE TEMPERATURE SCALE OR KELVIN TEMPERATURE SCALE

Kelvin had chosen the following form of the function ψ

$$\psi\,(t) = T \qquad ...(7.22)$$

Thus, we may write Eq. (7.21) as

$$\frac{Q_1}{Q_2} = \phi\,(t_1, t_2) = \frac{T_1}{T_2} \qquad ...(7.23)$$

Equation (7.23) defines *the absolute or Kelvin temperature scale.*

It **is independent of thermometric substance**.

A thermometric scale for the temperature function (T) can be established by considering a string of Carnot engines as shown in Fig. 7.32. The engines are so arranged that the heat rejected by first engine is absorbed by the engine following it and the temperature limits are so chosen that each engine develops the same amount of work. Hence,

$$W = Q_1 - Q_2 = Q_2 - Q_3 = Q_3 - Q_4 \qquad ...(7.24)$$

or

$$W = Q_1\left(1 - \frac{Q_2}{Q_1}\right) = Q_2\left(1 - \frac{Q_3}{Q_2}\right) = Q_3\left(1 - \frac{Q_4}{Q_3}\right)$$

or

$$W = Q_1\left(1 - \frac{T_2}{T_1}\right) = Q_2\left(1 - \frac{T_3}{T_2}\right) = Q_3\left(1 - \frac{T_4}{T_3}\right)$$

or

$$W = \frac{Q_1}{T_1}(T_1 - T_2) = \frac{Q_2}{T_2}(T_2 - T_3) = \frac{Q_3}{T_3}(T_3 - T_4)$$

But from Eq. (7.23)

$$\frac{Q_1}{T_1} = \frac{Q_2}{T_2} = \frac{Q_3}{T_3}$$

Hence,

$$T_1 - T_2 = T_2 - T_3 = T_3 - T_4 \qquad ...(7.25)$$

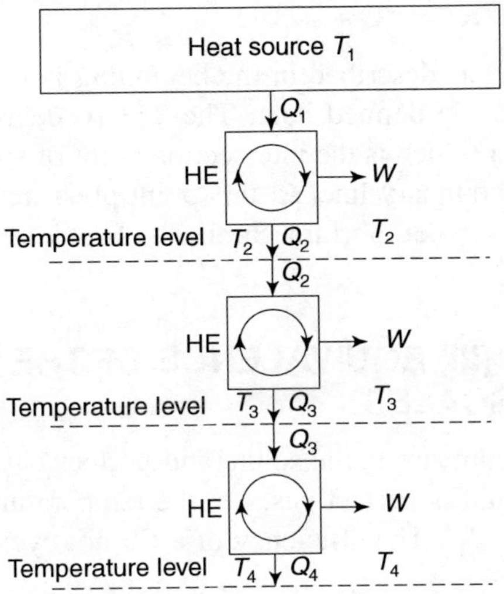

Fig. 7.32 Establishment of absolute scale

This means that equal temperature intervals on the absolute scale are the temperature intervals between which Carnot engines operate in such a way that each engine absorbs the heat rejected by the preceding one and all engines develop the same amount of work.

Now, we can assign the numerical values to standard fixed points of temperature on the absolute scale. Usually, **steam point and ice point are chosen as fixed reference temperature** and the interval between these two points is taken as 100 degrees corresponding to centigrade system. Thus

$$T_s - T_i = 100 \qquad \text{...(7.26)}$$

where T_s = temperature of steam point

T_i = temperature of ice point

Now, if the Carnot engine works between a heat source at temperature of steam point and a heat sink at the temperature of ice point, the **efficiency of the engine will be 26.8 per cent.**

$$\therefore \qquad \eta_c = \eta_{th} = 1 - \frac{Q_2}{Q_1} = 1 - \frac{T_i}{T_s} = 0.268 \qquad \text{...(7.26a)}$$

Solving the Eqs. (7.26) and (7.26a) gives

or $$\frac{T_i}{T_s} = 0.372$$

$$T_i = 273.15 \quad \text{and} \quad T_s = 373.15$$

For practical use, these values are rounded to $T_i = 273$ and $T_s = 373$. **These are called the absolute temperature values of the absolute scale (i.e., Kelvin scale)**. Hence, the relation

$$T_i\ (K) = t°C + 273$$

The temperature scale which was described in the beginning is in close agreement with the thermodynamic temperature scale defined here. The 273.16 degree on the Kelvin scale is called the **triple point** of water which is the intersection point of solid-liquid phase boundary line and liquid-vapour phase boundary line. At this point pressure and temperature are such that solid, liquid and gas exists in equilibrium. Similar to Kelvin scale, other scales may also be devised.

7.19 COROLLARY 3 : (III) EQUIVALENCE OF THE ABSOLUTE AND PERFECT GAS SCALES

It will be interesting here to compare the absolute and perfect gas scale. Consider a Carnot cycle in which the working fluid is perfect gas. Let the temperature limits within which the Carnot cycle works be T_1' and T_2'. The efficiency of a Carnot cycle is given as

$$\eta = 1 - \frac{Q_2}{Q_1} = 1 - \frac{T_2'}{T_1'}$$

or $$\frac{Q_2}{Q_1} = \frac{T_2'}{T_1'} \qquad ...(7.27)$$

But the absolute temperature scale is defined by

$$\frac{Q_1}{Q_2} = \frac{T_1}{T_2}$$

∴ $$\frac{T_1'}{T_2'} = \frac{T_1}{T_2} \qquad ...(7.28)$$

Since the numerical value assigned to the *fixed point* (*triple point*) *may be chosen to be* the same on both scales, then $T' = T$. In other words, the absolute temperature is mumericaliy equal to the absolute temperature as measured by the perfect gas thermometer. *Hence, absolute scale and perfect gas temperature scale are identical.*

7.20 COROLLARY 3 (IV): UNATTAINABILITY OF NEGATIVE ABSOLUTE TEMPERATURE

It is impossible to attain a negative temperature on absolute scale.

In order to prove this, let us examine the expresion.

$$\eta_{\text{constant}} = 1 - \frac{Q_2}{Q_1} = 1 - \frac{T_2}{T_1} \qquad ...(7.29)$$

If sink temperature T_2 becomes zero, the efficiency of reversible engine becomes 100 per cent. Since $T_2 = 0$, it means that no heat is rejected in the heat sink or there is an absence of heat sink. This means the engine will work with a single reservoir at a fixed temperature continuously producing work. But this violates the second law of thermodynamics. Thus, it may be concluded that absolute zero is definitely the lower limit on the therrnodynamic scale and a lower value than this cannot be expected. Therefore, it is impossible to attain negative temperature on the absolute scale.

7.21 COROLLARY 3 (V) : UNATTAINABILITY OF ABSOLUTE ZERO TEMPERATURE

It is impossible (in the absence of perfect heat insulator) for a finite system to attain the absolute zero of temperature.

In order to prove this, let us consider a finite system of temperature, T. Suppose, we want to reduce the absolute temperature T of finite system to zero. A heat pump as shown in Fig. 7.33 is employed to withdraw heat from this system and discharge the same to the surrounding which is at some higher level of temperature T_1. In the absence of perfect insulation of the system, it will receive Q quantity of heat at a finite rate from the surrounding. The work required by the heat pump is given by

$$W = Q_1 - Q = \frac{Q}{T}(T_1 - T) \qquad ...(7.30)$$

as $$\frac{Q_1}{Q} = \frac{T_1}{T}$$

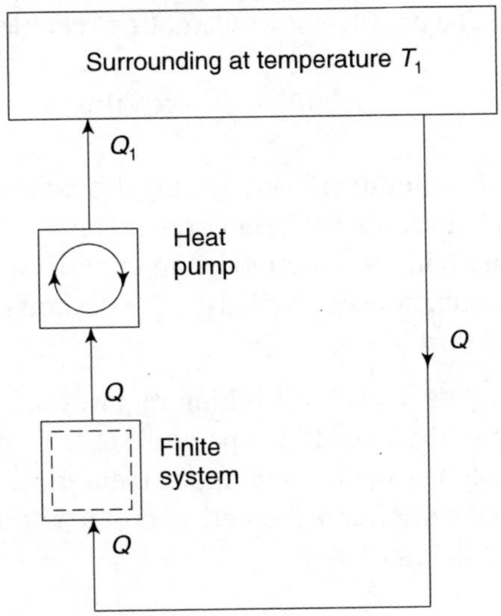

Fig. 7.33 Unattainability of absolute zero

Since Q is finite in the absence of perfect insulation and if T becomes zero, the work required by the heat pump would by infinite. This is impossible, hence absolute zero of temperature in the absence of perfect heat insulation is unattainable.

Example 7.10 In a freezer compartment of a refrigerator, a temperature of – 5°C is maintained by steadily circulating a refrigerant at low temperature through passages in the walls of freezer compartment. The air surrounding the refrigerator is at 27°C. The rate of heat transfer from the freezer compartment to the refrigerant is 8100 kJ/h and the power input required to operate the refrigerator is 3300 kJ/h. Draw the system and calculate the coefficient of performance (COP) of the refrigerator and compare with the coefficient of performance of a reversible refrigeration cycle operating between the reservoirs at the same temperature limits. Comment on the results.

A Model Solution Refer to Fig. 7.34.

Data given $T_2 = 273 - 5 = 268$ K, $T_1 = 273 + 27 = 300$ K, $Q_2 = Q_{abs} = 8100$ kJ/h, $\dot{W}_{cycle} = 3300$ k/h

To find COP of refrigerator and to compare with reversible refrigerator.

Assumptions (i) The system shown in Fig. 7.34 is in steady state.

(ii) The freezer compartment and the surrounding air play the role of low temperature and high temperature thermal reservoirs, respectively.

Analysis The coefficient of performance of refrigerator is given

$$\text{COP} = \frac{\text{Refrigerating effect}}{\text{Work input}} = \frac{8100}{3300} = 2.45 \qquad \textbf{Ans.}$$

The coefficient of Carnot (reversible) refrigerator is

$$(\text{COP})_{\text{Carnot}} = (\text{COP})_{\text{Rev.}} = \frac{T_2}{T_1 - T_2} = \frac{T_L}{T_H - T_L} = \frac{268}{300 - 268} = 8.375 \qquad \textbf{Ans.}$$

Comments : There is a big difference between the coefficient of performance of refrigerator and reversible (Carnot) refrigerator which suggests that there may be some potential for improving the thermodynamic performance of refrigerator. The objective should be carried out judiciously. Obviously, the desired improvement may require increase in size, complexity and cost.

Example 7.11 A dwelling requires 6.5×10^5 kJ per day to maintain the temperature at 21°C when the outside temperature at 0°C. An electric heat pump is used to supply this energy. Calculate (a) the minimum theoretical work input for one day of operation in kJ/day, and (b) the minimum theoretical cost to operate the heat pump in ₹ per day of the cost of electricity is ₹ 2.5 per kW/h.

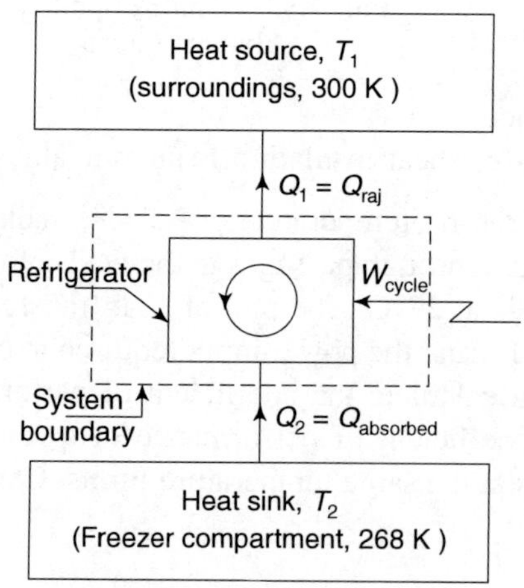

Fig. 7.34 Example 7.10

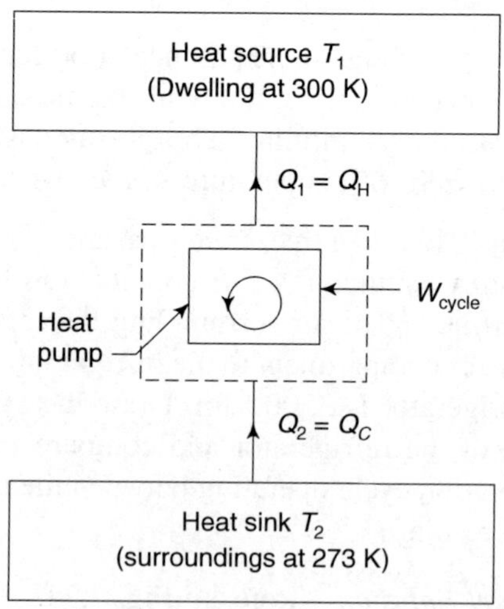

Fig. 7.35 Example 7.11

A Model Solution Refer to Fig. 7.35.

Data given $Q_1 = Q_{Ref.} = 6.5 \times 10^5$ kJ/day, $T_2 = T_C = 273$ K, $T_1 = 294$ K, cost of electricity = ₹ 2.5 kWh.

To find $\dot{W}_{cycle} = \dot{W}_{net}$ and cost of electricity in ₹ per day.

Assumptions (i) The system boundary is shown in Fig. 7.35. (ii) The dwelling and surrounding air play the role of heat source and heat sink respectively.

Analysis For heat pump, the work input, $W_{cycle} = W_{net}$ is given

$$W_{cycle} = \frac{Q_1}{\text{COP}} = \frac{Q_H}{\text{COP}}$$

The COP of heat pump is less than or equal to the $(\text{COP})_{max}$ of a reversible heat pump when each operates between the same two thermal reservoirs, i.e., COP $\leq (\text{COP})_{max}$.

$$W_{cycle} \geq \frac{Q_1}{(\text{COP})_{max}} \geq \left(1 - \frac{T_2}{T_1}\right) Q_1$$

or

$$W_{cycle} \geq \left(1 - \frac{273}{294}\right) \times 6.5 \times 10^5 = 0.46 \times 10^5 = 4.64 \times 10^4 \text{ kJ/day}$$

Thus, the minimum theoretical work input is 4.64×10^4 kJ/day. The minimum theoretical cost will be

$$= (W_{cycle})_{min} = 4.64 \times 10^4 \left(\frac{\text{kJ}}{\text{day}}\right) \times 2.5 \left(\frac{₹\ s}{\text{kJ} \times 3600\ s}\right)$$

$$= \frac{4.64 \times 10^4 \times 2.5}{3600} = ₹\ 32.22/\text{day}$$ **Ans.**

Example 7.12 Prove the theory proposed by Kelvin in pointing out the thermodynamic wastefulness of burning fuel for the direct heating of a house. It is much more economical to use the high temperature heat produced by combustion of fuel in a heat engine and then to use the work so developed to pump heat from outdoors (surroundings) up to the temperature desired in the house including the heat rejected from the engine. Take temperature of high temperature reservoirs, house and surroundings as 500 K, 300 K and 273 K respectively.

Solution Refer to Fig. 7.36.

Let a boiler furnish heat Q_1 at high temperature T_1 by burning the fuel. The heat is absorbed by a heat engine which develops work output W and rejects the waste heat Q_2 into the house at T_2. The work W so produced is, in turn, used to operate a mechanical refrigerator or heat pump which extracts heat Q_3 from outdoors (surroundings) at temperature T_3 and rejects Q_2' amount where $Q_2' = Q_3 + W$ into the house. As a result of this cycle of operations, a total quantity of heat equal to $(Q_2 + Q_2')$ is provided in the house, against Q_1 which would be provided directly by the ordinary combustion of fuel.

Data given $T_1 = 500$ K, $T_2 = 300$ K and $T_3 = 273$ K.

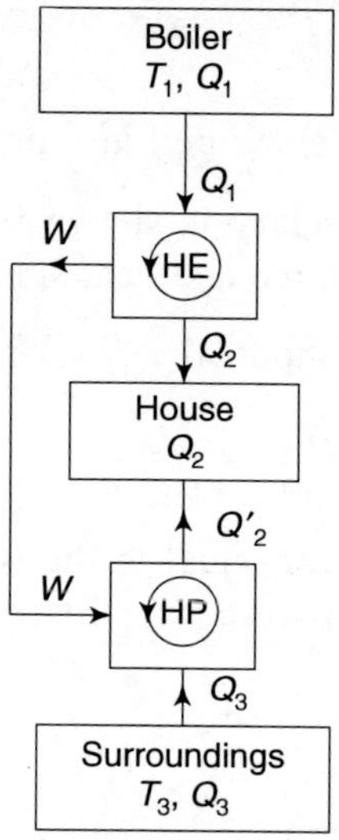

Fig. 7.36 Example 7.12

For the reversible heat engine,

$$\frac{Q_2}{Q_1} = \frac{T_2}{T_1} \quad \text{or} \quad Q_2 = Q_1\left(\frac{T_2}{T_1}\right) \qquad ...(1)$$

The efficiency of heat engine is given by

$$\eta = \frac{W}{Q_1} = \frac{T_1 - T_2}{T_1}$$

or

$$W = \left(\frac{T_1 - T_2}{T_1}\right)Q_1$$

For reversible heat pump, COP is given by

$$\text{COP} = \frac{Q_2'}{W} = \frac{T_2}{T_2 - T_3}$$

or

$$Q_2' = \left(\frac{T_2}{T_2 - T_3}\right)W = \left(\frac{T_2}{T_2 - T_3}\right)\left(\frac{T_1 - T_2}{T_1}\right)Q_1 \qquad ...(2)$$

The ratio of heat supplied in the house with the use of heat engine and heat pump to directly in the house is generally called multiplication factor (MF) is given by

$$\text{MF} = \frac{Q_2 + Q_2'}{Q_1} = \frac{Q_1\left(\frac{T_2}{T_1}\right) + \left(\frac{T_2}{T_2 - T_3}\right)\left(\frac{T_1 - T_2}{T_1}\right)Q_1}{Q_1}$$

$$= \frac{T_2}{T_1} + \left(\frac{T_2}{T_2 - T_3}\right)\left(\frac{T_1 - T_2}{T_1}\right) = \frac{T_2^2 - T_2 T_3 + T_2 T_1 - T_2^2}{T_1 (T_2 - T_3)}$$

$$= \frac{T_2 (T_1 - T_3)}{T_1 (T_2 - T_3)} = \frac{300\,(500 - 273)}{500\,(300 - 273)} = 5.044 \qquad \textbf{Ans.}$$

Comment : The above result shows that every kg coal burnt and utilised in the proposed system would deliver heat equivalent to over 5 kg. Even considering the efficiency of real engine to 50%, the saving is quite significant and thus Kelvin's preposition is proved.

7.22 SUMMARY OF FORMULAE

1. $\eta_{th} = \dfrac{Q_1 - Q_2}{Q_1} = \dfrac{W_{net}}{Q_1} = 1 - \dfrac{Q_2}{Q_1}$

2. $(h_{th})_{elec} = \dfrac{EI}{Q_A}$ (Efficiency of thermoelectric engine)

3. $(COP)_{Ref} = \dfrac{Q_2}{W_{net}} = \dfrac{Q_2}{Q_1 - Q_2}$

4. $(COP)_{Heat\ pump} = \dfrac{Q_1}{W_{net}} = \dfrac{Q_1}{Q_1 - Q_2}$

5. $(COP)_{Heat\ pump} = \dfrac{Q_1}{EI}$ (For thermoelectric heat pump)

6. $(COP)_{Ref} = \dfrac{Q_2}{EI}$ (For thermoelectric refrigerator)

7. $\eta_{Carnot\ engine} = \dfrac{T_1 - T_2}{T_1} = 1 - \dfrac{T_2}{T_1}$

8. $(COP)_{Carnot\ Ref} = \dfrac{T_2}{T_1 - T_2}$

9. $(COP)_{Carnot\ Heat\ pump} = \dfrac{T_1}{T_1 - T_2}$

10. $\eta_R = \phi\,(t_1, t_2)$

11. $\dfrac{Q_1}{Q_2} = \phi\,(t_1, t_2) = \dfrac{T_1}{T_2}$

PROBLEMS

Things to Think About

7.1. Describe the processes that would satisfy the conservation of energy principle, but does not actually occur in nature.

7.2. List some of the principal irreversibilities present during operation of (a) an aeroplane engine, (b) an automobile engine, (c) a household refrigerator, (d) an oiled fired boiler, and (e) an electric water heater.

7.3. List the principal irreversibilities in the operation of a gearbox and classify them in terms of internal and external.

7.4. Whether the power output from steam turbine will be more in reversible expansion or in actual expansion.

7.5. Whether the power input to a compressor will be less in reversible compression or in actual compression?

7.6. If a window air conditioner were placed on a table in a room and operated, whether the room temperature would increase, decrease or remain the same? Justify your answer.

7.7. Specify whether the coefficient of performance of a refrigeration cycle will be less than one, or greater than one.

7.8. Steam power plant typically rejects energy by heat transfer to a body of water or the atmosphere. Will you advise to reject this energy instead to a large blocks of ice maintained by a refrigeration system?

7.9. In order to increase the thermal efficiency of a reversible power cycle, would it be better to increase high temperature reservoir (T_H) or decrease low temperature reservoir (T_C) by an equal amount?

7.10. Can you think of a heat engine more efficient than Carnot engine. Is second law a directional law or not?

Short and Long Theoretical Questions

7.11. What are the limitations of the first law of thermodynamics and essence of second law of thermodynamics?

7.12. What is the qualitative difference between heat and work? Why are heat and work not completely interchangeable forms of energy?

7.13. Define a thermal reservoir, a heat engine, and a heat pump.

7.14. With the help of schematic diagram of a heat engine, derive the expression for thermal efficiency.

7.15. With the help of schematic diagram of a heat pump, derive the expression for coefficient of performance as a heat pump.

7.16. Prove that $(COP)_{HP} = 1 + (COP)_{Ref}$.

7.17. State and explain the various statements of second law of thermodynamics with examples.

7.18. Why is direct heating thermodynamically wasteful?

7.19. Discuss the scope of second law of thermodynamics.

7.20. Justify the following statements:

It is impossible to construct an engine to work between two reservoirs, each having a fixed and uniform temperature, which will exceed in efficiency a reversible engine working between the same reservoirs.

7.21. Show the temperature scale is independent of the working fluid.

7.22. Show the equivalence of Clausius and Kelvin statement of second law of thermodynamics.

7 23. Show the negative absolute temperature is impossible.

7.24. All spontaneous processes are reversible. Explain.

7 25. What are the causes of irreversibility?

7.26. Discuss the various causes of irreversibility in detail.

7.27. Discuss irreversibiliiy due to the following:

(a) Heat transfer through a finite temperature difference

(b) Free expansion process

(c) Braking of a rotating flywheel

(d) Paddle wheel work transfer

(e) Flow of electric current through a resistor

7.28. What is a internally reversible process?

7.29. Discuss the types of irreversibilities.

7.30. Define external, internal, thermal, mechanical and chemical irreversibilities.

7.31. What is a Carnot cycle? Derive the expression of thermal efficiency and COP of a Carnot cycle working as engine and heat pump, respectively.

7.32. Why is Carnot cycle a not practical cycle?

7.33. Prove the corollary of second law. "No heat engine working in a cycle between two constant temperature reservoirs can be more efficient than a reversible engine working between the same two reservoirs".

7.34. Prove that all reversible engines have the same efficiency (Carnot's theorem)

7.35. Derive absolute temperature scale from Carnot.

7.36. Show the equivalence of absolute and perfect gas scales.

7.37. Show the unattainability of negative absolute temperature.

7.38. A reversible heat engine works between three heat reservoirs A, B and C. The engine receives equal amount of heat from reservoirs A and B at temperature T_A and T_B respectively and rejects heat to reservoir C at temperature T_C. Show that the efficiency of the cycle is given by

$$\frac{T_A}{T_B} = (2x - 1) + 2\,(1 - x)\,\frac{T_A}{T_C}$$

where x is the ratio of efficiency of engine to that of reversible engine.

7.39. (a) The specific heats of a perfect gas are $c_p = \beta + \lambda T$ and $c_v = \alpha + \lambda T$ where α, β and λ are constant. Prove that along an adiabatic.

$$p^{\alpha}\, V^{\beta}\, c^{\lambda T} = \text{constant}$$

(b) A charge of this gas is compressed adiabatically in an engine from the initial parameters p_1, T_1 to final parameters p_2, T_2. If the quantity $(T_1 - T_2)$ is negligible as compared to T_1 or T_2, prove that

$$T_2 = \frac{T_1\,(\beta + \gamma\, T_1)}{\beta\left(\dfrac{p_2}{p_1}\right)^{(\alpha - \beta)/\beta} + \gamma\, T_1}$$

7.40 Define the term absolute zero of temperature and show how a scale of temperature can be devised from the operation of reversible heat engines.

7.41. An ideal engine operates on the Carnot cycle, using a perfect gas as the working fluid. The ratio of the greatest to the least volume is fixed and is n to 1, the lower temperature of the cycle is also fixed, but the volume compression ratio r of the reversible adiabatic compression process is variable. The ratio of specific heat is λ. Show that if the work done in the cycle is a minimum, then

$$(\gamma - 1)\ \ln\ \frac{n}{r} + \frac{1}{r^{\gamma - 1}} - 1\ = 0$$

7.42. A heat engine receives heat Q_{1a} at temperature T_{1a} and Q_{1b} at temperature T_{b1}. The engine rejects heat Q_{2p} at temperature T_{2p} and Q_{2q} at temperature T_{2q}. Prove that

$$\frac{Q_{1a}}{T_{1a}} + \frac{Q_{b1}}{T_{b1}} - \frac{Q_{2p}}{T_{2p}} - \frac{Q_{2q}}{T_{2q}}\ \le 0$$

Numerical Questions

7.43. A Carnot engine with an efficiency of 40 per cent receives 400 kJ/h from a high temperature source and rejects heat to a sink at 26.7°C. What is the power output of the engine and the temperature of the source in °C. [*Ans.* 0.444 kW. 226.5°C]

7.44. A patent application claims that an engine which receives heat at 160°C and rejects to sink at 4.54°C is capable of delivering 0.1 kWh of energy for energy 1050 kJ received by the engine. Is this a valid claim? [*Ans.* Yes]

7.45. A refrigerator operating on a reversed Carnot cycle is required to transfer 348.33 kW from a reservoir at – 30°C to the atmosphere at 25°C. Calculate power input. [*Ans.* 78.84. kW]

7.46. The efficiency of a Carnot engine rejecting heat to a cooling pond at 27°C is 30 per cent. The cooling pond receives 13.91 kW. Calculate the power of the engine and the source temperature. [*Ans.* 5 96 kW 155.7°C]

7.47. A heat engine drives a heat pump. The heat delivered by the heat engine as well as by the pump is used to heat the water circulating through the heat radiators of a building. The efficiency of the heat engine is 27% and the coefficient of performance of the heat pump is 4. Calculate ratio of the heat transferred to the circulating water to the heat taken by the heat engine.

[*Ans.* 1.81]

7.48. A Carnot refrigerator is maintaining foodstuffs in a refrigerator area at 4.44°C by rejecting heat to atmosphere at 27°C. It is desired to maintain some frozen foods at – 17.75°C with the same sink temperature of 27°C. What per cent increase in work input will be required for the frozen food unit over the refrigerated unit for the same quantity of heat removed?

[*Ans.* 115.7%]

7.49. An inventor claims to have a device which receives 1000 watts of energy as heat but puts out only 750 watts of electrical power. The rest of the energy is put out as mechanical work and dissipated outside his device by friction. Discuss the validity of these claims.

7.50. A power plant using air as the working fluid is limited to a maximum temperature of 850°C. If it could operate on a Carnot cycle with an overall volume ratio of 30 and a volume ratio for isothermal expansion of 4, find the maximum efficiency possible and the temperature of heat rejection. [*Ans.* 53.35%, 222°C]

7.51. In a certain process, a vapour while condensing at 430°C transfers heat to water at 250°C. The resulting steam is used in a power cycle which rejects heat at 40°C. What fraction of the available energy in the heat transferred from the process vapour at 430°C is lost due to irreversible heat transfer to water at 250°C?

7.52. An office building in Srinagar is to be heated by means of a heat pump. The mean air temperature within the building is maintained at 20°C when the outside mean air temperature is – 4°C. The heat picked up by the heat pump from outside air is 10,460 kJ/h and the power required to drive the pump is 10.29 kW. Evaluate (a) the heat transfer rate from heat pump to the building and the coefficient of performance of the pump, (ii) the maximum possible coefficient of performance for a heat pump under these conditions and the maximum power input to satisfy the given heating requirement. [*Ans.* 15.91 kW; 1.224; 12.21; 1.3 kW]

7.53. An engineer claims to have developed a refrigerator which takes in 10460 kJ of thermal energy as heat at a temperature of 23°C and rejects 12426 kJ to surroundings at 25°C, receives 1905 kJ of energy as work during the refrigerating process. Examine the claims of the engineer on the basis of the first and second law of thermodynamics and state whether such a machine can possibly be built and whether you would be willing to purchase such a machine.

7.54. A solar-powered heat pump employed in a space vehicle receives heat from a solar collector at T_h, rejects heat to the atmosphere at T_a and pumps heat from a cold space at T_c. The heat transfer rates are Q_h, Q_a, and Q_c, respectively. Derive an expression for the minimum ratio Q_h/Q_c in terms of the three temperatures. If $T_h = 500$ K, $T_a = 300$ K, $T_c = 200$ K, $Q_c = 10$ kW, calculate the minimum value of Q_h.

7.55. A domestic refrigerator is maintained at a temperature of 2°C. Every time the door is opened, warm material is placed inside, introducing an average of 420 kJ, but making only a small change in a temperature of the refrigerator. The door is opened 20 times a day, and the refrigerator operates at 15% of the ideal COP. The electric work input the compressor costs ₹ 6.4 per kWh.

Calculate the monthly bill of the refrigerator assuming the atmospheric temperature as 30°C.

[*Ans.* ₹ 304.00]

7.56. Two reversible heat engines *A* and *B* are arranged in series. *A* rejects heat directly to *B*. Engine *A* receives 30 kJ at a temperature of 500°C from a heat source, while engine *B* is in communication with a cold sink at a temperature of 5°C. If the work output of *A* is twice that of *B*, calculate (a) the intermediate temperature between *A* and *B*, (b) the thermal efficiency of each engine.

7.57. An ice-making plant is to be designed to produce ice at atmospheric pressure and at 0°C from water. The mean temperature of cooling water circulating through the condenser of the refrigerating machine is 20°C. Calculate the minimum electrical work input in kWh to produce 1 tonne of ice. If the COP of the actual machine is 40% of the ideal, recalculate the electrical work input.

7.58. A heat pump is used to maintain a house at a temperature of 20°C during winter by pumping heat from atmosphere which is at temperature of –10°C. The heat losses through the walls of the house are estimated to be 0.65 kW per unit of temperature difference between the inside of the house and the atmosphere. Calculate the minimum power required to drive the heat pump.

If the same heat pump is to be used to cool the house in summer provided the room temperature, heat loss and work input remaining the same, calculate the maximum possible atmospheric temperature.

[*Ans.* 2 kW, 50°C]

7.59. The heat capacity at constant volume of a system is given by $c_v = A + 2\,BT$ where $A = 0.014$ J/K and $B = 4.3 \times 10^{-4}$ J/K. A low temperature reservoir at 0°C is available. Calculate the maximum amount of work that can be obtained if the system is cooled down from 100°C to the temperature of the environment.

7.60. A heat pump is used for cooling in summer and heating in winter. The heat loss 0.44 kW per degree difference between outside and inside temperatures. The average outside temperature is 32°C in summer and – 4°C in winter. Determine (i) the power requirement for both heating and cooling, (ii) which condition determines the unit size that must be purchased?

7.61. A car engine burns 5 kg fuel at 1600 K and rejects energy to the radiator and the exhaust at an average temperature of 750 K. If the calorific value of fuel is 40×10^3 kJ/kg, calculate the maximum amount of work the engine can provide.

7.62. In a steam power plant 1 MW is added at 700°C in a boiler, 0.58 MW is taken out at 40°C in the condenser and the pump work is 0.02 MW. Calculate the plant thermal efficiency. Assuming the same pump work and heat transfer to the boiler is given, calculate the power produced by turbine if the plant were running on the Carnot cycle.

[*Ans.* 0.7 MW]

7.63. An air-conditioner provides 1 kg/s of air at 15°C cooled from outside atmospheric air at 35°C. Calculate the amount of power needed to operate the air-conditioner.

[*Ans.* 1.39 kW]

7.64. It is proposed to heat a house in the winter with a heat pump. The house is to be maintained at 20°C at all times. If the ambient temperature outside drops to –10°C and the rate at which the heat lost from the house is estimated to be 25 kW, calculate the minimum electrical power required to drive the heat pump.

[*Ans.* 4.9 kW]

7.65. Electrical solar cells are capable of producing power with 20% efficiency. Assume a heat engine with a low temperature heat rejection at 30°C driving an electric generator with 80% efficiency.

Calculate the effective high temperature in the heat engine to give the same overall efficiency as the solar cells.

7.66. A household freezer operates in a room at 20°C. Heat must be transferred from the cold space at a rate of 2 kW to maintain its temperature at – 30°C. Calculate smallest power of the motor required to operate this freezer. [*Ans.* 0.41 kW]

7.67. An inventor has developed a refrigeration unit that maintains the cold space at – 10°C while operating in a room of 25°C. The coefficient of performance is 8.5, verify the claim.

7.68. A certain solar-energy collector produces a maximum temperature of 100°C. The energy is used in a cyclic heat engine that operates in a 10°C environment. Calculate the maximum thermal efficiency. If the collector is redesigned to focus the incoming light to produce a maximum temperature of 300°C, calculate the maximum thermal efficiency.

7.69. A furnace can deliver heat Q_{H1} at T_{H1} and is proposed to use this to drive a heat engine with a rejection at T_{atm} instead of direct room heating. The heat engine drives a heat pump that delivers Q_{H2} at T_{room} using the atmosphere as the cold reservoir. Calculate the ratio Q_{H2}/Q_{H1} as a function of temperatures. Is this a better set-up than direct room heating from the furnace?

7.70. A heat engine has a solar collector receiving 0.2 kW/m^2 inside which a transfer media is heated to 450 K. The collected energy powers a heat engine which rejects heat at 40°C. If the heat engine deliver 3 kW, what will be the size of solar collector.

7.71. A household refrigerator is maintained at a temperature of 2°C. Every time the door is opened, warm material is placed inside, introducing an average of 420 kJ, but making only a small change in the temperature of refrigerator. The door is opened 20 times a day and the refrigerator operates at 15% of the ideal COP. The electric tariff is ₹ 1.60 per kWh. Calculate the monthly bill for this refrigerator if atmospheric air temperature is at 30°C. [*Ans.* ₹ 76.00]

7.72. An ice-making plant produces ice at atmospheric pressure and at 0°C from water. The mean temperature of cooling water circulating through the condenser of the refrigerating machine is 18°C. Calculate the minimum electric work in kWh, required to produce 1 tonne of ice. The enthalpy of fusion of ice at atmospheric pressure is 333.5 kJ/kg.

7.73. An engine in outer space operates on the Carnot cycle. The only way in which heat can be transferred from the engine is by radiation. The rate at which heat is radiated in proportional to the fourth power of the absolute temperature and to the area of the radiating surface. Prove that for a given power output and a given T_1, the area of the radiator will be minimum when

$$\frac{T_2}{T_1} = \frac{3}{4}$$

7.74. Two reversible heat engines are arranged in series, *A* is rejecting heat directly to *B*. Engine *A* receives 200 kJ at a temperature of 421°C from a hot source, while engine *B* is in communication with cold sink at a temperature of 4.4°C. If the work output of it is twice that of *B*, calculate:

(a) the intermediate temperature between *A* and *B*,

(b) the efficiency of each engine,

(c) the heat rejected to the cold sink. [*Ans.* 143.45°C 40%, 33.52%, 80 kJ]

7.75. A heat engine is used to drive a heat pump. The heat transfer from the heat engine and heat pump are used to heat the water circulating through the radiators of a building. The efficiency of the heat engine and COP of the heat pump are 27% and 4 respectively. Calculate the ratio of the heat transfer to the circulating water to the heat transfer to the heat engine. [*Ans.* 1.811]

7.76. A power cycle operates between a reservoir at a temperature T_H and lower temperature reservoir at 280 K. At steady state, the cycle develops 40 kW of power while rejecting 1000 kJ/min of energy by heat transfer to the cold reservoir. Calculate the minimum theoretical value for T_H.

[*Ans.* 952 K]

7.77. During steady state condition, a power cycle having a thermal efficiency of 38% generates 100 MW of electricity while discharging energy by heat transfer to cooling water at an average temperature of 20°C. The average temperature of steam passing through the boiler is 500°C. Calculate (a) the rate at which energy is discharged to the cooling water in kJ/h, (b) the minimum theoretical rate at which energy couid be discharged to the cooling water, in kJ/h. Comment on the results.

7.78. Ocean thermal energy conversion (OTEC) power plants generate power by utilising the naturally occurring decrease with depth of the temperature of ocean water. At a certain location, the ocean surface temperature is 27°C while at a depth of 700 m, the temperature is 7°C. Calculate the maximum thermal efficiency for any power cycle operating between these temperatures.

[*Ans.* 6.666%]

7.79. Geothermal power plants harness underground sources of hot water or steam for the production of electricity. One such plant receives a supply of hot water at 170°C and rejects heat energy to the atmosphere at 15°C. Calculate the maximum possible thermal efficiency for any power cycle operating between these temperatures.

7.80. A solar collector receives solar radiation at the rate of 0.315 kW/m^2 of area and provides energy to a storage unit whose temperature remains constant at 230°C. The power cycle receives energy by heat transfer from the storage unit, generates electricity at the rate of 0.6 MW, and discharges energy by heat transfer to the surrounding at 20°C. Under steady state condition, calculate the minimum collector area required in m^2.

7.81. At steady state, a refrigerator having COP equal to 3, removes energy by heat transfer from a freezer compartment at 0°C at the rate of 6100 kJ/h and discharge energy by heat transfer to the surrounding, which are at 20°C, calculate (a) the power input to the refrigerator and compare with power input required by reversible refrigeration cycle operating between the same reservoirs, (b) if the cost of electricity is ₹ 3.00 kWh, calculate the actual and minimum theoretical operating costs, each in ₹ per day.

Software Based Questions

7.82. A heat pump maintains a dwelling at temperature T when the outside temperature is 5°C. The heat transfer rate through the walls and roof is 2000 kJ/h per degree of temperature difference between the inside and outside. If electricity costs ₹ 2.0 per kWh, develop a software to work out the following.

(a) Calculate the minimum theoretical operating cost for each day of operation when $T = 22$°C.

(b) Plot the minimum theoretical operating cost for each day of operation as a function of T ranging from 18 to 23°C.

7.83. Develop a software to ascertain the claim of the inventor of heat engine and heat pump. Validate the results with realistic data.

7.84. Develop a software to calculate the minimum theoretical callector area required to generate electricity by a power cycle utilising the solar radiation. Assume that the solar collector receives solar radiation at the rate of 0.315 kW/m^2 and provides energy to a storage unit whose temperature remains constant at 220°C. The power cycle receives energy by heat transfer from the storage unit. The power cycle discharges energy by heat transfer to the surrounding at 20°C. Plot the graph between power output and collector area required as a function of T_H.

7.85. A reversible power cycle receives Q_H from a heat reservoir at temperature T_H and rejects energy by heat transfer to the surroundings at the temperature T_0. The work delivered by the power cycle is used to drive a refrigeration cycle that removes Q_C from a cold reservoir at temperature T_C and discharge energy by heat transfer to the same surroundings at T_0. Develop a software to plot Q_C/Q_H vessus T_H/T_0 for various values of T_C/T_0 = 0.85, 0.9 and 0.95 and versus T_C/T_0 for various values of T_H/T_0 = 2, 3, and 4.

Design and Open Ended Questions

7.86. A 2000 cc engine passenger car having a seating capacity for five persons is to be provided air-conditioning facility. The car may run up to 100 km/h. Estimate the air-conditioning capacity requirement and extra cost on this, in the form of capital and operating cost.

7.87. Prepare a comprehensive paper outlining the contribution made by Carnot, Clausius, Kelvin–Planck, Gibbs to the development of second law of thermodynamics.

7.88. A 1000 MW steam power plant located on the river side rejects heat in the cooling water of the condenser drawn from the river and hot water again meets in the river through open channel. Prepare a report for alternative means of cooling the condenser with environmental issues related to each year alternatives.

7.89. Prepare a comparative report on daily cost for heating a building and maintaining at 20°C by means of an electric heat pump, direct electrical resistance heating and a conventionsl gas-fired furnace. Assume the heat transfer rate through the walls and roof a building is 3066 kJ/h per degree temperature difference between the inside and outside. Take outdoor temperature ranging from 15°C to – 30°C.

7.90. An article published by Fnimanuel (1988) states that hurricanes are an example of a natural Carnot engine. Prepare a report about the scientific and engineering merit to harness the energy.

7.91. A US patent No. 4,980,572 describes a method for generating electricity using gravitational theory. In this method, massive spinning wheels located underground that serve as the primemover of an alternator for generating electricity. Each wheel is to be kept in motion by torque pulses transmitted to it via a suitable mechanism from vehicles passing overhead. Prepare a comprehensive report about the practical difficulties in implementing such a method to generate electricity.

Multiple-Choice Answer Type Questions

7.92. The efficiency of a Carnot engine is given as 0.75. If the cycle direction is reversed, what will be the value of COP of the Carnot refrigerator?

(a) 0.271 (b) 0.33 (c) 1.27 (d) 2.3

7.93. A Carnot cycle is having an efficiency of 0.75. If the temperature reservoir is 727°C, what is the temperature of low temperature reservoir?

(a) 230.5°C (b) 181.75°C (c) 150.8°C (d) 250.5°C

7.94. Two reversible engines operate between temperature limits of 900 K and T_2 400 K respectively. For both to be equally efficient, the value of T_2 will be

(a) 600 K (b) 650 K (c) 700 K (d) 750 K

7.95. The coefficient of performance (COP) of a heat pump and refrigerator may be

(a) slightly more than 0 (b) equal to 1

(c) greater than 1 (d) All of these

7.96. The difference between COP of a heat pump and COP of refrigerator is

(a) 0 (b) 1 (c) > 1 (d) < 1

7.97. For a reversible heat engine operating between source and sink having temperatures T_1 and T_2 and heat interactions being Q_1 and Q_2 respectively, the thermodynamic temperature scale states

(a) $\frac{Q_1}{Q_2} = \frac{T_1}{T_2}$ (b) $\frac{Q_1}{Q_2} = \frac{T_2}{T_1}$ (c) $\frac{Q_1}{T_1} + \frac{Q_1}{T_2} = 0$ (d) None of these

7.98. COP of reversible heat pump is 1.2. If it is reversed to run as reversible heat engine then its efficiency shall be

(a) 0.233 (b) 0.833 (c) 1.23 (d) 1.43

Answers to Multiple-Choice Questions

7.92. (b) 7.93. (b) 7.94. (a) 7.95. (d) 7.96. (c) 7.97. (a)

7.98. (b)

CHAPTER

8

Entropy and Third Law of Thermodynamics

OVERVIEW

While dealing with the first law, we initially stated the law in terms of cycle but then defined a property called *internal energy*, which enabled us to use the first law quantitatively for processes. Similarly, we have stated the second law for a cycle, and we now find that the second law leads to a property called *entropy*, which enables us to treat the second law quantitatively for processes. *It is worth to emphasise that energy and entropy are both abstract concepts that help to describe certain observations*. Therefore, thermodynamics can be described as the science of three E_s, namely, **energy, entropy and equilibrium** as mentioned in Chapter 1.

This chapter deals with the various remaining corollaries dealing with concept of entropy, entropy production and application of entropy. In the last, the third law of thermodynamics has also been discussed which also relate to entropy under certain conditions.

8.1 COROLLARY 4 : CLAUSIUS INEQUALITY

When a system undergoes a complete cyclic process, the integral of $\delta Q/T$ around the cycle is less than zero or equal to zero. The equality and unequality signs hold for reversible or irreversible cycles respectively. This theorem is known as Clausius inequality. Mathematically,

$$\oint \frac{\delta Q}{T} \leq 0 \qquad \text{...(8.1)}$$

where δQ represents, the heat transfer at a part of the system boundary during a portion of the cycle and T *is the absolute temperature at that part of boundary*. It is to be noted that integrant is evaluated at the boundary of the system executing the cycle.

In order to prove this, consider a reversible cycle *ABCD* on *p–V* diagram as shown in Fig. 8.1. The cycle is subdivided by a family of reversible adiabatics and reversible isothermals. Every two adjacent adiabatics are connected by two reversible isothermals so that the heat transferred during all isothermal processes are equal to the heat transferred during the original cycle. Since no two adiabatics cross each other, the cycle can be subdivided into a large number of Carnot cycles which provide the same amount of work.

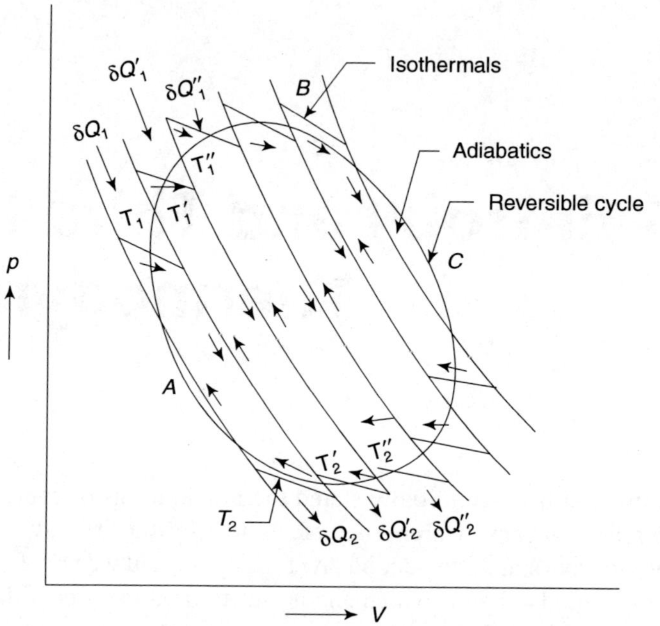

Fig. 8.1 Clausius inequality

Since Carnot cycle is reversible cycle so, for the first small Carnot cycle, the efficiency is given as

$$\eta_1 = 1 - \frac{\delta Q_2}{\delta Q_1} = 1 - \frac{T_2}{T_1}$$

or
$$\frac{T_2}{T_1} = 1 - \frac{\delta Q_2}{\delta Q_1}$$

or
$$\frac{\delta Q_1}{T_1} = \frac{\delta Q_2}{T_2}$$

But considering the direction of heat interaction, +ve for heat added to the cycle, and, –ve for heat rejected from cycle, the above equation becomes

$$\frac{\delta Q_1}{T_1} = -\frac{\delta Q_2}{T_2} \quad \text{or} \quad \frac{\delta Q_1}{T_1} + \frac{\delta Q_2}{T_2} = 0$$

Similarly, for the second Carnot cycle

$$\frac{\delta Q_1'}{T_1'} + \frac{\delta Q_2'}{T_2'} = 0$$

For other Carnot cycles, similar equations are drawn.

Thus, for the reversible cycle *ABCD*

$$\left(\frac{\delta Q_1}{T_1}+\frac{\delta Q_2}{T_2}\right)+\left(\frac{\delta Q_1'}{T_1'}+\frac{\delta Q_2'}{T_2'}\right)+........ = 0$$

or
$$\oint\left(\frac{\delta Q}{T}\right)_R = 0 \text{ for reversible process.} \qquad ...(8.2)$$

In the case of an irreversible engine *I*, it has been shown that $\eta_1 < \eta_R$, hence

$$1-\frac{\delta Q_{2I}}{\delta Q_{1I}} < 1-\frac{\delta Q_{2R}}{\delta Q_{1R}}$$

or
$$1-\frac{\delta Q_{2I}}{\delta Q_{1I}} < 1-\frac{T_2}{T_1}$$

or
$$\frac{\delta Q_{1I}}{T_1}-\frac{\delta Q_{2I}}{T_2} < 0$$

Consider the direction of heat interaction with engines

$$\frac{\delta Q_{1I}}{T_1}+\frac{\delta Q_{2I}}{T_2} < 0$$

Similarly, for other irreversible engines

$$\frac{\delta Q_{1I}'}{T_1'}+\frac{\delta Q_{2I}'}{T_2'} < 0$$

Similarly, up all the small irreversible engines

$$\oint\left(\frac{\delta Q}{T}\right)_I < 0 \text{ for irreversible process} \qquad ...(8.3)$$

Thus, for a cycle of process-reversible and irreversible, we have

$$\oint\frac{\delta Q}{T} \le 0 \qquad ...(8.4)$$

Note that the temmperature (T) in the Calusius unequality pertains to reservoirs and not systems.

The Calusius inequality not only gives the mathematical expression to the second law of thermodynamics, but it also throws light on the irreversibility of the system. The Clausius Inequality for irreversible process may also be written as

$$\oint\frac{\delta Q}{T}+I = 0 \qquad ...(8.5)$$

where I indicates the amount by which the given cycle is irreversible. If $I = 0$, the cycle is reversible.

Equation (8.1) can be expressed equivalently as

$$\oint \left(\frac{\delta Q}{T}\right)_b = -(S_{gen})_{cycle}$$

where $(S_{gen})_{cycle}$ is identified as the entropy generation (i.e., production) in subsequent discussion and can be viewed as representing the strength of inequality. *The suffix b refers to the boundary of the system. In summary.*

$(S_{gen})_{cycle} = 0$ no irreversibility present within the system.

$(S_{gen})_{cycle} > 0$ irreversibility present within the system

$(S_{gen})_{cycle} < 0$ impossible

8.2 COROLLARY 5 : ENTROPY

The cyclic integral of quantity δQ/T for a reversible cycle being equal to zero suggests that the quantity δQ/T is a point function and is therefore property of a system. This property is called entropy denoted by S. Mathematically,

$$\int_1^2 \left(\frac{\delta Q}{T}\right)_{rev} = \int_1^2 dS$$

or

$$\left(\frac{\delta Q}{T}\right)_{rev} = dS$$

Here reversible cycle means internally reversible.

Considering mathematical and physical aspects, the entropy is defined as follows:

The amount of heat contained in a substance and the degree to which it is ordered are described by the property of state known as entropy. The higher the disorder, the greater is the increase in entropy.

In order to prove entropy is a property, let us consider arbitrary internally reversible cycles as shown in Fig. 8.2.

We know that for an internally reversible cycle

$$\oint \left(\frac{\delta Q}{T}\right)_{rev} = 0$$

Thus, for cycles 1–*A*–2–*B*–1 and 1–*A*–2–*C*–1, we have

$$\oint \left(\frac{\delta Q}{T}\right)_{rev} = \underbrace{\int_1^2 \left(\frac{\delta Q}{T}\right)_{rev}}_{\text{Path } A} + \underbrace{\int_2^1 \left(\frac{\delta Q}{T}\right)_{rev}}_{\text{Path } B} = 0$$

and
$$\oint\left(\frac{\delta Q}{T}\right)_{rev} = \underset{\text{Path } A}{\int_1^2\left(\frac{\delta Q}{T}\right)_{rev}} + \underset{\text{Path } C}{\int_2^1\left(\frac{\delta Q}{T}\right)_{rev}} = 0$$

Comparison of above two equation results in

$$\underset{\text{Path } B}{\int_2^1\left(\frac{\delta Q}{T}\right)_{rev}} = \underset{\text{Path } C}{\int_2^1\left(\frac{\delta Q}{T}\right)_{rev}} \quad \text{...(8.6)}$$

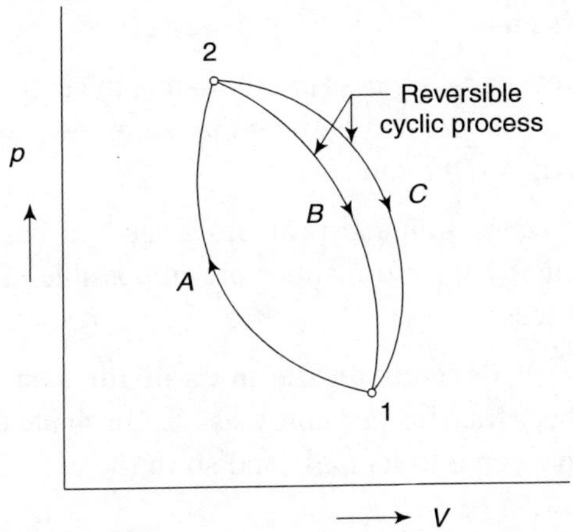

Fig. 8.2 Entropy is a point function

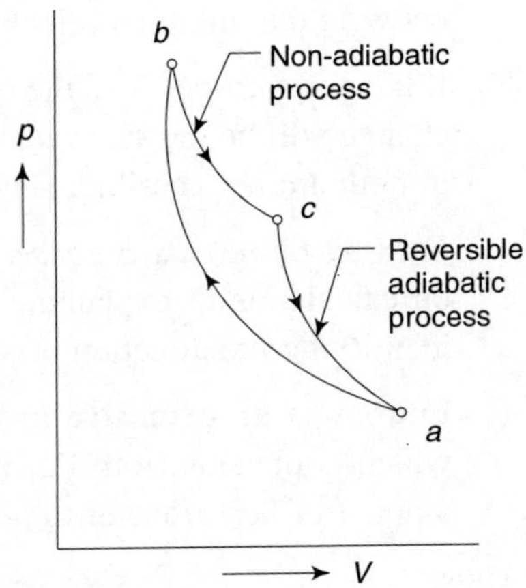

Fig. 8.3 Concept of entropy

Hence, integral of $\delta Q/T$ is independent of path. It is a point function (i.e., exact differential), hence it is a property. This property as stated above is the *entropy*. Thus,

$$dS = \left(\frac{\delta Q}{T}\right)_{rev} \quad \text{...(8.7)}$$

In the case of thermodynamic reversible process from state 1 to state 2, the change in entropy

$$S_2 - S_1 = \int_1^2\left(\frac{\delta Q}{T}\right)_{rev} \quad \text{...(8.8)}$$

For reversible adiabatic, $\delta Q = 0$, hence

$$\Delta S = S_2 - S_1 = 0, \quad \text{i.e., } S_1 = S_2 = \text{constant} \quad \text{...(8.9)}$$

We see that for reversible adiabatic process, the change in entropy is zero. *Thus, entropy is a property of a system that remains constant during reversible adiabatic process.*

From the above discussion, it follows that:

(i) The method of *defining entropy* is *analogous* to the method of *defining internal energy*. The amount of work interaction or heat interaction in a process having two fixed states depends on how the process is carried out. On the other hand, the change in entropy and the change in internal energy are invariant.

(ii) From Eq. (8.7), the change in entropy can be calculated, only if the path connecting the end states is reversible. The change in entropy for an irreversible process may also be calculated by devising a reversible process (or a series of reversible processes) between the initial and final state of the system.

(iii) If the *end states* of both the reversible or irreversible process *are* the same, the entropy change will be the same for the two processes but it will be seen that common states in both are not possible, hence the entropy is not the same.

(iv) *Entropy represents a consequence of the second law* and provides a method based on calculation of explaining why certain energy transformations are impossible, and identifying the direction of spontaneous process.

(v) Entropy is **an extensive property because it depends on the mass of the system** which is obvious from Eq. (8.10). If the heat transfer per unit mass is constant, the amount of heat transfer to a system is proportional to its mass and so in the entropy.

Thus, $$\Delta S = m\,\Delta s \qquad ...(8.10)$$

where m is the mass of the system. The unit of entropy is kJ/K while the unit of specific entropy is kJ/Kg K.

The *concept of entropy* can also be developed in another way. According to the second law, two or more reversible adiabatic processes cannot have a common state. In order to prove this, consider two reversible adiabatic processes *ab* and *ca* having a common state *a* as shown in Fig. 8.3. In order to complete the cycle, let a non-adiabatic reversible process joins states *b* and *c*. For this cycle

$$(\delta Q)_{bc} - (\delta W)_{\text{cycle}} = (dE)_{\text{cycle}} = 0$$

or $$(\delta W)_{\text{cycle}} = (\delta Q)_{bc}$$

Here the net effect of the cycle is to convert heat completely into work. This violates the second law.

Thus, we conclude that ***two reversible adiabatics cannot have a common state***. This suggests that there must be a property of matter that remains constant during reversible adiabatic process. This new property is called **entropy**. For example, two constant temperature processes cannot have a common state because of property of matter temperature remaining constant. *Thus, entropy is a property of matter that remains constant during reversible adiabatic process.*

8.3 COROLLARY 6 : GENERAL STATEMENT OF SECOND LAW

The entropy of an isolated system, increases in all real processes and is conserved in reversible process. **Planck calls this statement as the most general statement of second law.** This is called entropy principle.

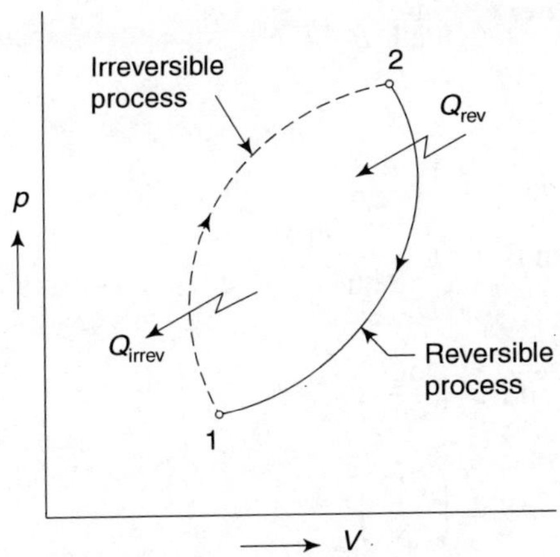

Fig. 8.4 Planck's statement

In order to prove this, let us consider an irreversible process 1–2 along path *A* shown in Fig. 8.4. Let the cycle be completed by a reversible process 2–1 along pathe *B*, so that processes 1–*A*–2 and 2–*B*–1 together form an irreversible *cycle*. From the Clausius Inequality.

$$\oint \frac{\delta Q}{T} < 0 \text{ for irreversible process}$$

or
$$\underset{\text{Path } A}{\int_1^2 \left(\frac{\delta Q}{T}\right)_{irrev}} + \underset{\text{Path } B}{\int_2^1 \left(\frac{\delta Q}{T}\right)_{rev}} < 0$$

But
$$\underset{\text{Path } B}{\int_2^1 \left(\frac{\delta Q}{T}\right)_{rev}} = \underset{\text{Path } B}{\int_2^1 dS}$$

But since path B is reversible, the limits of the integral can be reversed and therefore

$$\underset{\text{Path } B}{\int_{2}^{1}\left(\frac{\delta Q}{T}\right)_{\text{rev}}} = -\underset{\text{Path } B}{\int_{1}^{2} dS}$$

$$\therefore \qquad \underset{\text{Path } A}{\int_{1}^{2}\left(\frac{\delta Q}{T}\right)_{\text{irrev}}} = \underset{\text{Path } B}{\int_{1}^{2} dS}$$

or

$$\underset{\text{Path } B}{\int_{1}^{2} dS} > \underset{\text{Path } A}{\int_{1}^{2}\left(\frac{\delta Q}{T}\right)_{\text{irrev}}}$$

or

$$dS > \left(\frac{\delta Q}{T}\right)_{\text{irrev}} \qquad ...(8.11)$$

But

$$dS = \left(\frac{\delta Q}{T}\right)_{\text{rev}}$$

Hence, in general, we may write

$$dS \geq \frac{\delta Q}{T} \qquad ...(8.12)$$

or

$$\Delta S = S_2 - S_1 \geq \int_{1}^{2} \frac{\delta Q}{T} \qquad ...(8.13)$$

where inequality and equality signs show the irreversibility and reversibility respectively. The Eq. (8.13) suggests that entropy increases in irreversible process and is conserved in reversible process. ***Note that δQ is the heat interaction with the system and T is the temperature of environment.***

For adiabatic process whether reversible or irreversible, $\delta Q = 0$, hence

$$\Delta S \geq 0 \qquad ...(8.14)$$

This means that entropy is zero in reversible adiabatic and greather than zero in irreversible adiabatic.

Although, the entropy changes between *two fixed states* is the same for any path, reversible or irreversible, the heat transfer to the system is less (more, if heat is transferred from the system) for the irreversible path than for the reversible one. In other words, *these two paths, irreversible and reversible, will not cause the same change of state in the envrionment starting from the same initial state* and hence the change in entropy is different. This is obvious from the fact that entropy increases in irreversible processes and is conserved in reversible processes.

8.3.1 Entropy Principle

In the case of isolated system, there is no heat and work interaction with the environment, and so the total energy of all possible states remains constant. Thus, $\delta Q = 0$ for isolated system.

$$dS_{\text{isolated}} > 0 \qquad \text{...(8.15)}$$

Since process 1–2 is an arbitrary process, the entropy of the isolated system increases with every irreversible process.

Equation (8.15) is called the ***principle of increase of entropy*** which may be considered of another form to describe the second law.

Following the principle of increase of entropy, the entropy of the universe increases owing to the natural process (irreversible process) such that

$$\Delta S_{\text{univ}} = \Delta S_{\text{sys}} + \Delta S_{\text{envir}} = (S_{\text{gen}})_{\text{isolated}} > 0 \qquad \text{...(8.16)}$$

For reversible process,

$$\Delta S_{\text{system}} = \left(\frac{Q}{T}\right)_{\text{rev}} \quad \text{and} \quad \Delta S_{\text{envi}} = -\left(\frac{Q}{T}\right)_{\text{rev}}$$

$$\therefore \quad \Delta S_{\text{univ}} = 0 \text{ for reversible process.}$$

In general, we may write

$$\Delta S_{\text{univ}} \geq 0 \qquad \text{...(8.17)}$$

The Eq. (8.17) is an another form of Second Law, i.e., *entropy of the universe increases in all irreversible processes and is conserved in reversible process.*

Rudolf Clausius summarized the first and second law in the following words:

(a) *Die Energie der Welt ist konstant.*

(b) *Die Entropie der Welt Strebt einen Maximum zu. The translation of above two* statements in English is:

(i) *The energy of the world (universe) is constant.*

(ii) *The entropy of the world tends towards a maximum.*

We have seen that entropy of an isolated system always increases and it **will become the maximum at the state of *equilibrium***. In other words, if the entropy of an isolated system varies with some parameters say x, then there is certain value of x_e which maximizes the entropy, i.e., $(ds/dx) = 0$ and it represents the equilibrium state (Fig. 8.5). The system is then said to exist at the peak of entropy hill and $ds = 0$. In **conclusion, we may say that *when the system is at equilibrium, any conceivable change in entropy would be zero.***

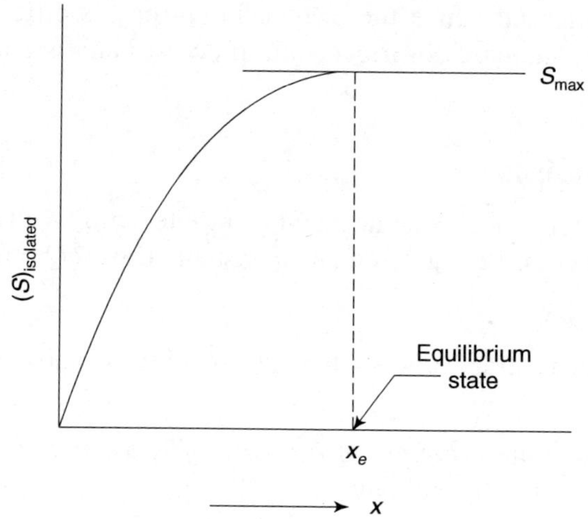

Fig. 8.5 Equilibrium state of an isolated system

8.4 ENTROPY GENERATION (PRODUCTION) IN A CLOSED SYSTEM (CONTROL MASS)

The conclusion from the previous considerations is that the entropy change for an irreversible process is larger than the change in a reversible process for the same δQ and T. This may be expressed in the form of an *equality or entropy balance*

$$dS = \frac{\delta Q}{T} + \delta S_{gen} \quad ...(8.18)$$

where $\delta S_{gen} \geq 0$ ...(8.19)

The amount of entropy, δS_{gen} is the entropy generation in the process due to irreversibilities occurring inside the closed system, a control mass (Fig. 8.6 a). This internal generation can be caused by the processes mentioned in Chapter 7, Section 7.11 such as friction, unrestrained expansion, heat transfer due to finite temperature difference, etc.

Equation (8.19) is then valid with the equal sign for a reversible process and greater than sign for an irreversible process. Since the entropy generation is always positive, and the smallest in a reversible process, nearly zero, we may deduce some limits for the work and heat transfer.

Let us consider a reversible process for which the entropy generation is zero, that the heat and work transfer may be expressed as

$$\delta Q = TdS \quad \text{and} \quad \delta W = pdV \quad ...(8.20)$$

For an irreversible process with non-zero entropy generation, the heat transfer from Eq. (8.18) becomes

$$\delta Q_{irrev} = TdS - T \cdot \delta S_{gen} \qquad ...(8.21)$$

Equation (8.21) shows that the heat transfer in irreversible case is less than that of reversible.

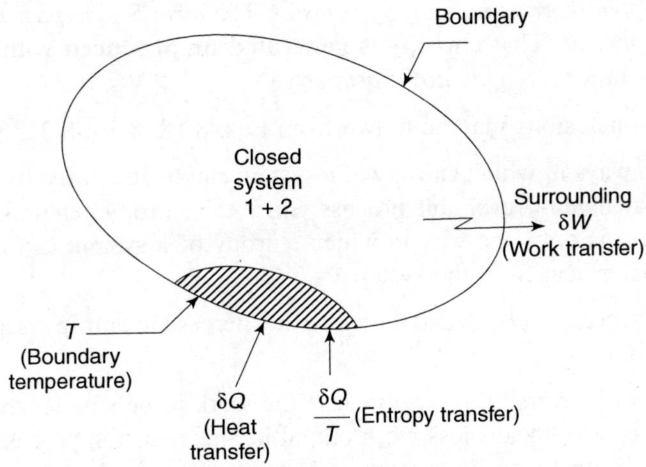

Fig. 8.6 (a) A Closed system interacting with surrounding

From the first law, the work transfer is given by

$$\delta Q_{irrev} = dU + \delta W_{irrev} \qquad ...(8.22)$$

The work in irreversible process is no larger equal to pdV but it is smaller, but the property relation is valid. Thus,

$$TdS = dU + pdV$$

Thus,
$$\delta W_{irrev} = pdV - T \cdot \delta S_{gen} \qquad ...(8.23)$$

Equation (8.23) shows that work is reduced by an anount proportional to the entropy generation. That is why $T \cdot \delta S_{gen}$ is often called "*lost work*", although it is not a real work or energy quantity lost, *but a lost opportunity to extract work.*

If Eq. (8.18) is integrated between the initial and final limits.

$$\underbrace{\int_1^2 dS}_{\text{Entropy change}} = S_2 - S_1 = \underbrace{\int_1^2 \left(\frac{\delta Q}{T}\right)_b}_{\text{Entropy transfer}} + \underbrace{S_{gen}}_{\text{Entropy generation}} \qquad ...(8.24)$$

Equation (8.24) represents the entropy balance for closed (control mass) system. Here **subscript *b* represents the boundary of the surface at which heat transfer occurs**. The term on the right side of Eq. (8.24) is associated with the heat transfer to or from the system during the process. This term can be interpreted as the **entropy transfer accompanying heat**

transfer. The direction of entropy transfer is the same as the direction of the heat transfer and the same sign convention applies as for heat transfer. *A positive value implies that entropy is transferred into the system, and a negative value means that entropy is transferred out. When there is no heat transfer, there is no entropy transfer.* The *term* S_{gen} *is positive when internal irreversibilities are present.* That **entropy is generated or produced** within the system by the action of irreversibilities. $\boldsymbol{S_{gen}}$ ***is not a property***.

Some important conclusions may be drawn from Eqs. 8.18, 8.19, 8.21, 8.23 and 8.24.

(i) There are two ways in which entropy of a system can be increased by transferring heat to it and by having an irreversible process. Since the entropy generation cannot be less than zero, there is only one way in which entropy of a system can be decreased and that is to transfer heat from the system.

(ii) For adiabatic process, $dQ = 0$, and therefore the increase in entropy is always associated with irreversibilities.

(iii) The presence of irreversibilities will cause the work to be smaller than the reversible process. This clearly means less work output in an expansion process and more work input into the control mass in a compression process.

When applying the entropy balance to a closed system, it is essential to remember the requirements imposed by the second law on entropy production (or generation). The second law requires that entropy production be positive or zero in value.

$$S_{gen} : \begin{cases} > 0 \text{ Irreversibilities present within the system} \\ = 0 \text{ No irreversibilities present within the system} \end{cases}$$

It is worth to note that the value of entropy production cannot be negative. By contrast, the change of entropy of the system may be positive, negative or zero.

$$(S_2 - S_1) : \begin{cases} > 0 \\ = 0 \\ < 0 \end{cases} \qquad \text{...(8.25)}$$

Like other properties, entropy change can be determined without the knowledge of the details of the process.

In order to illustrate the *entropy transfer and entropy generation* (*or production*) *concepts*, as well as the accounting nature of the entropy balance, consider the example shown in Fig. 8.6 (b). Here, the example in consideration shows a system consisting of a gas or liquid in a rigid container stirred by a paddle-wheel while receiving the heat transfer Q from a reservoir. The boundary surface temperature where heat transfer occurs in at T_b. By definition, the reservoir is free of irreversibilities, however, the system is not without irreversibilities as the fluid friction is obviously present. Further, there may be other irreversibilities within the system.

If we apply the entropy balance to the system and reservoir, Eq. (8.24) takes the form of

$$S_2 - S_2 = \frac{Q}{T_b} + S_{gen} \text{ as } T_b \text{ is constant.}$$

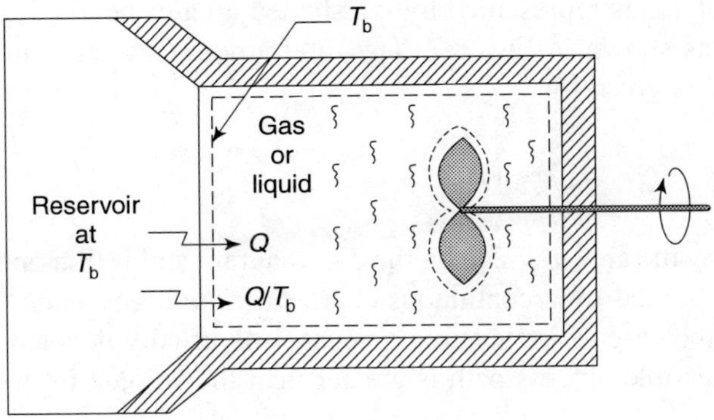

Fig. 8.6 (b) Example of entropy transfer and entropy generation or production

Here, Q/T_b accounts for entropy transfer into the system accompanying heat transfer Q. The entropy balance for the reservoir takes the form

$$\Delta S_{res} = \frac{Q_{res}}{T_b} + 0 = \frac{Q_{res}}{T_b}$$

where, $S_{gen} = 0$ **because the reservoir is without irreversibilities. Since $Q_{res} = -Q$, the last equation is expressed as**

$$\Delta S_{res} = -\frac{Q}{T_b}$$

Here, the minus sign shows the entropy is carried out of the reservoir accompanying heat transfer. It is worth to note that entropy of the reservoir decreases by an amount equal to the entropy transferred from it to the system.

In order to evaluate the entropy transfer term of the entropy balance requires information regarding both the heat transfer and temperature of the boundary where the heat transfer occurs. Sometimes, boundary surface is either unknown or not defined, such as when system passes through states sufficiently far from equilibrium. In such a situation, it is conveniently, therefore, to enlarge the system to include enough of the immediate surroundings that the temperature of the boundary of the enlarged system corresponds to the temperature away from the immediate vicinity of the system, T_f. Thus, entropy transfer = Q/T_f.

8.5 TEMPERATURE ENTROPY DIAGRAM

Selection of T and S as two parameters to describe a thermodynamic system is very beneficial. On the T–S diagram, isotherms and reversible adiabatics are represented by horizontal and vertical lines because during isothermal process, temperature remains constant but entropy changes, while for reversible adiabatic entropy remains constant but temperature changes. In the case of reversible process, the heat interaction is given by

$$(\delta Q)_{rev} = TdS$$

The amount of heat $(\delta Q)_{rev}$ is represented by the shaded area of height T and width dS under the process path 1–2 as shown in Fig. 8.7. The total amount of heat interaction during the reversible process 1–2 is given by

$$(Q_{1-2})_{rev} = \int_1^2 TdS \quad \text{...(8.26)}$$

This is represented by the area *a*–1–2–*b* on the *T–S* diagram and is true only when the process is reversible. It is represented by a continuous curve *thus the area under the T–S diagram is the heat exchange.* In the case of irreversible process, it is usually denoted by dotted line and the area under an irreversible process path is greater than the amount of heat interaction.

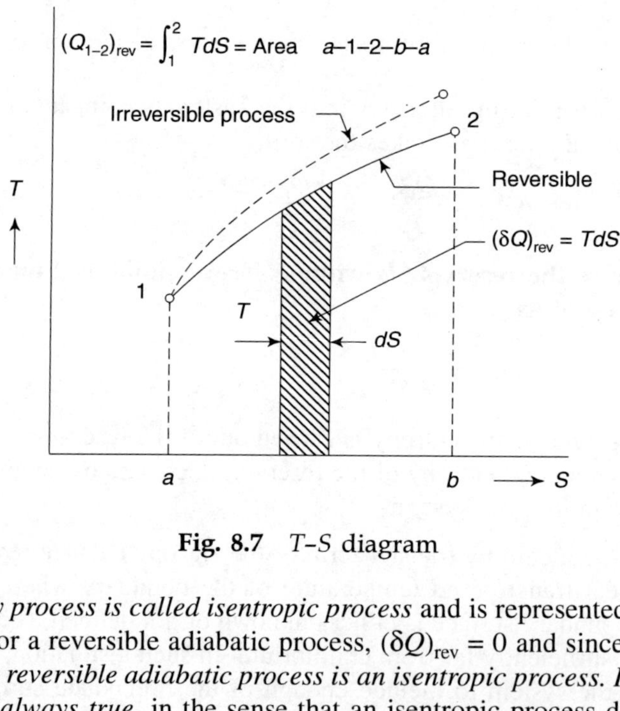

Fig. 8.7 *T-S* diagram

Constant entropy process is called isentropic process and is represented by vertical line on the *T–S* diagram. For a reversible adiabatic process, $(\delta Q)_{rev} = 0$ and since $dS = (\delta Q/T)_{rev}$. So $dS = 0$. *Therefore, a reversible adiabatic process is an isentropic process. It is to be noted that the opposite is not always true*, in the sense that an isentropic process does not necessarily mean a reversible adiabatic process. For example, a fluid flow accompanied by friction results in an increase in entropy. If at the same time, heat is transferred from the system at such a rate that the *entropy of the system is maintained constant*, then the process is iseptropic. Such a process is neither reversible nor adiabatic and is accompanied by an increase in entropy of the surroundings. This is clear from the fact that since $dS = 0$ in constant entropy process but $\delta Q/T < 0$, which means $\delta Q < 0$. *Thus an irreversible-adiabatic process requires heat flow from the system.*

8.6 ENTROPY AS A RATE EQUATION FOR CLOSED SYSTEM

In some case, entropy balance equation is needed in a rate form, so a given process can be treated in time. The rate form is also the basis for the development of the entropy balance equation in the general control volume analysis for an unsteady state.

By taking an incremental change in S from Eq. (8.18), we have

$$\frac{dS}{\delta\tau} = \frac{1}{T}\frac{\delta Q}{\delta\tau} + \frac{\delta S_{gen}}{\delta\tau} \quad \text{...(8.27)}$$

It is possible that for a given control volume, we may have more than one source of heat transfer. Thus, the final rate form of entropy balance equation in the limit is expressed as

$$\frac{dS_{nf}}{d\tau} = \frac{dS_{cm}}{\delta\tau} = \sum_j \frac{1}{T_j}\dot{Q}_j + \dot{S}_{gen} \quad \text{...(8.28)}$$

Equation (8.28) expresses the rate of entropy change as due to the flux of entropy into the control mass from heat transfer, there is an increase in entropy due to irreversible process inside the control mass. The term, $\dot{Q}_j / T_j$ represents the time rate of entropy transfer through the portion of boundary whose instantaneous temperature is T_j. It is sometimes convenient to use the entropy balance expressed in differential form

$$dS = \left(\frac{\delta Q}{T}\right)_b + \delta S_{gen} \quad \text{...(8.29)}$$

Note that the differentials of the nonproperties Q and S_{gen} are shown, respectively, as δQ and δS_{gen}. When there are no internal irreversibilities δS_{gen} vanishes and Eq. (8.29) reduces to Eq. (8.7).

8.7 APPLICATIONS OF ENTROPY PRINCIPLE

In fact, principle of increase of entropy is one of the most important laws of physical science. It is the quantitative statement of second law of thermodynamics and according to many, it is the best definition of second law. All the real processes are irreversible processes which are accompanied by entropy increase of the universe quantifying the extent of ineversibility in the process. The higher the entropy increase of the universe, the higher will be the irreversibility of the process and it is the maximum at equilibrium state. Some of the applications of entropy principle arc discussed below.

8.7.1 Transfer of Heat through a Finite Temperature Difference

Consider the case of transfer of heat through a finite temperature difference between two reservoirs (Fig. 8.8). Let Q be the rate of heat transfer from reservoir A at T_1 to reservoir B at T_2 where $T_1 > T_2$.

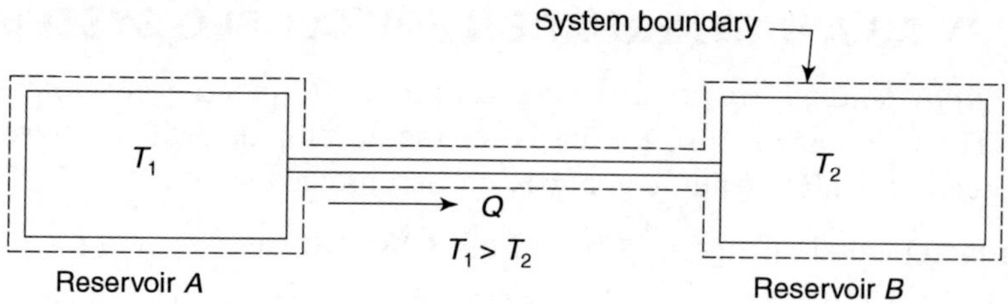

Fig. 8.8 Heat transfer through a finite temperature difference

For reservoir A, the change in entropy is $\Delta S_a = -\,Q/T_1$. It is negative because heat (Q) flows out of reservoir. For reservoir B, $\Delta S_B = Q/T_2$ and it is positive because heat flows into the reservoir. The rod connecting the reservoirs suffers no entropy change, because, once in the steady state, its coordinators do not change.

Therefore, for isolated system comprising of the two reservoirs and rod and since entropy is the additive property, so

Thus,
$$\Delta S = \Delta S_A + \Delta S_B$$

or
$$\Delta S_{\text{univ}} = \Delta S_A + \Delta S_B = -\frac{Q}{T_1} + \frac{Q}{T_2} = Q\frac{(T_1 - T_2)}{T_1\, T_2} \quad \text{...(8.30)}$$

It is worth to analyse Eq. 8.30. Since $T_1 > T_2$, ΔS_{univ} is positive and the process is irreversible and possible. If $T_1 = T_2$, $\Delta S_{\text{univ}} = 0$, which signifies that the process is isentropic. If $T_1 < T_2$, ΔS_{univ} is negative and the process is impossible as heat cannot flow itself from lower to higher temperature.

8.7.2 Mixing of Two Fluids at Different Temperatures

In order to check validity of the second law, consider the case of mixing of two fluids having different mass, different specific heat and at different temperatures. Let m kg of fluid of specific heat c_1 at temperature T_1, is isobarically and adiabatically mixed with m_2 kg of fluid of specific heat c_2 at temperature T_2 ($T_1 > T_2$) by removing the partition wall (Fig. 8.9). Let at equilibrium, T_f be the final temperature and $T_2 < T_f < T_1$. Since energy interaction is exclusively confined to the two fluids, so the system is an isolated system. Taking energy balance,

$$m_1\, c_1\, (T_1 - T_f) = m_2\, c_2\, (T_f - T_2)$$

or
$$T_f = \frac{m_1\, c_1\, T_1 + m_2\, c_2\, T_2}{m_1\, c_1 + m_2\, c_2} \quad \text{...(8.31)}$$

Entropy change of the fluid in subsystem 1

$$\Delta S_1 = \int_{T_1}^{T_f} \frac{\Delta Q_{\text{rev}}}{T} = \int_{T_1}^{T_f} \frac{m_1\, c_1\, dT}{T} = m_1\, c_1 \ln \frac{T_f}{T_1} \quad \text{...(8.32)}$$

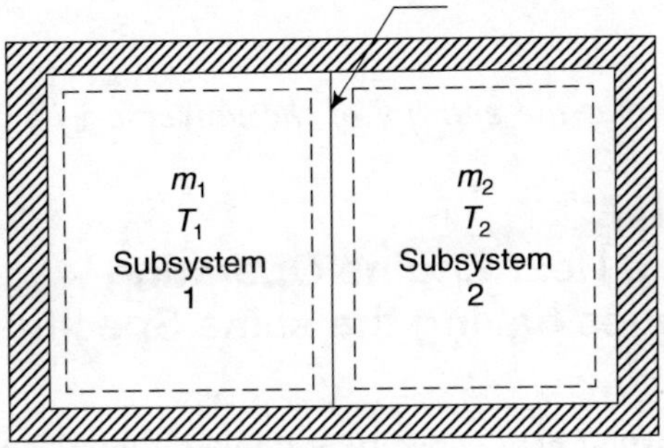

Fig. 8.9 Mixing of two fluids

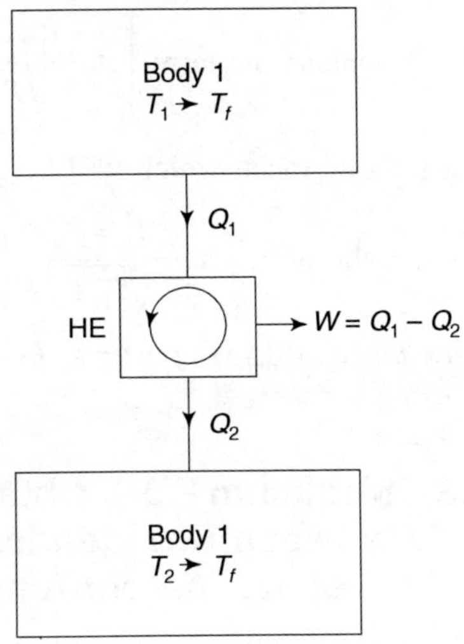

Fig. 8.10 Maximum work obtainable from heat engines

This will be onviously negative as $T_1 > T_f$

Entropy change of the fluid in subsystem 2

$$\Delta S_2 = \int_{T_2}^{T_f} \frac{\delta Q_{rev}}{T} = \int_{T_2}^{T_f} \frac{m_2\, c_2\, dT}{T} = m_2\, c_2 \ln \frac{T_f}{T_2} \qquad ...(8.33)$$

This entropy change is positive, since $T_2 < T_f$

Thus,
$$\Delta S_{univ} = \Delta S_1 + \Delta S_2 = m_1\, c_1 \ln \frac{T_f}{T_1} + m_2\, c_2 \ln \frac{T_f}{T_2} \qquad ...(8.34)$$

Now substituting the value of T_f from Eq. (8.31), we have

$$\Delta S_{univ} = m_1\, c_1 \ln \frac{(m_1\, c_1\, T_1 + m_2\, c_2\, T_2)}{(m_1\, c_1 + m_2\, c_2)\, T_1} + m_2\, c_2 \ln \frac{(m_1\, c_1\, T_1 + m_2\, c_2\, T_2)}{(m_1\, c_1 + m_2\, c_2)\, T_2}$$

If $m_1 = m_2 = m$ and $c_1 = c_2 = c$, then the above equation becomes

$$\Delta S_{univ} = m.c \left[\ln \frac{T_1 + T_2}{2\, T_1} + m.c \ln \left(\frac{T_1 + T_2}{2\, T_2} \right) = m \cdot c \ln \left(\frac{T_1 + T_2}{2\sqrt{T_1 + T_2}} \right) \right]^2$$

or
$$\Delta S_{univ} = 2\, m.c \ln \frac{T_1 + T_2}{2\sqrt{T_1\, T_2}} \qquad ...(8.35)$$

If we analyse the term $\left[\ln \frac{T_1 + T_2}{2\sqrt{T_1 T_2}}\right]$ or the above Eq. (8.35), the term ln $(T_1 + T_2)/2$ is the arithmetic mean which will be greater than the geometrical mean, $\sqrt{T_1 T_2}/2$.

Hence, the term $\left[\ln \frac{T_1 + T_2}{2\sqrt{T_1 + T_2}}\right]$ is positive

This mean ΔS_{univ} is positive, hence it is proved that the entropy of the universe will be increasing.

8.7.3 Maximum Work Obtainable from a Heat Engine Operating between Two Identical Finite Bodies having the same Specific Heat at Different Temperatures

Consider two identical finite bodies of the same constant heat capacity at temperature T_1 and T_2 respectively ($T_1 > T_2$). If a heat engine is operated between the two bodies acting as a thermal energy reservoirs (Fig. 8.10), the engine receives Q_1 amount of heat and rejects Q_2 amount of heat by developing W amount of work.

As work is delivered by the heat engine, the temperature of the body 1 having mass m will be decreasing and that of body 2 having mass m will be increasing. When both bodies attain the final temperature T_f, the engine will stop operating. The maximum work is possible only in the case of reversible process. Let c_p be the specific heat of two bodies at constant pressure.

$$\text{Work done} = W = Q_1 - Q_2 = mc_p\,(T_1 - T_f) - mc_p\,(T_f - T_2)$$

or

$$W = mc_p\,(T_1 + T_2 - 2T_f) \qquad \text{...(8.36)}$$

For the maximum work output or maximum efficiency, the heat engine cycle must be reversible (i.e., Carnot cycle). According to Clausius inequality for reversible cycle

$$\oint \frac{\delta Q}{T} = 0$$

The above equation signifies that the change in entropy in the complete cycle will be zero, Thus

$$\Delta S_{\text{univ}} = \Delta S_1 + \Delta S_2 = 0$$

or

$$\Delta S_{\text{univ}} = mc_p \ln \frac{T_f}{T_1} + m\,c_p \ln \frac{T_f}{T_2} = 0$$

or

$$mc_p \left[\ln \frac{T_f}{T_1} + \ln \frac{T_f}{T_2}\right] = 0$$

or

$$T_f = \sqrt{T_1 T_2} \qquad \text{...(8.37)}$$

Thus, for maximum work done, $T_f = \sqrt{T_1 T_2}$

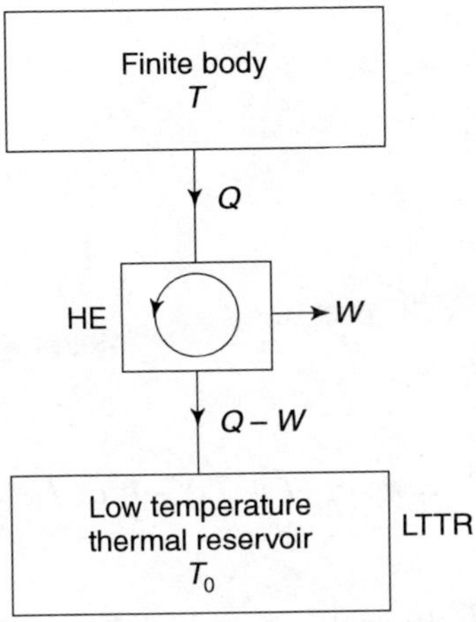

Fig. 8.11 Maximum work obtainable when one body is LTTR at T_0

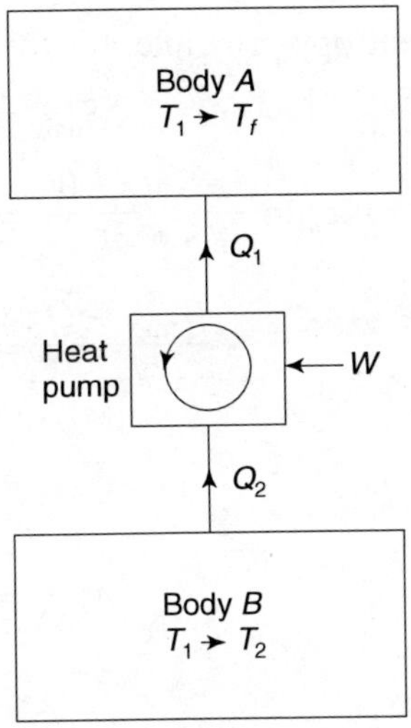

Fig. 8.12 Minimum work input to heat pump

Substituting the value of T_f in Eq. (8.36), we have

$$W_{\max} = m\, c_p \left[T_1 + T_2 - \sqrt{T_1 T_2}\right]$$

or

$$W_{\max} = m\, c_p \left[\sqrt{T_1} - \sqrt{T_2}\right]^2 \quad ...(8.38)$$

8.7.4 The Maximum Work Obtainable from a Heat Engine Operating between a Finite Body and a Thermal Reservoir at Dead State

Consider a heat engine operating between a finite body at temperature T and a low temperature thermal reservoir (LTTR) at dead state T_0 where $T > T_0$ (Fig. 8.11). Let m be the mass of the finite body and c_p be its specific heat. During operation, as heat is withdrawn from finite body, its temperature decreases but the temperature of themal reservoir will not change and it will always T. The engine will stop working when the temperature of finite body reaches T_0. During the working period, the amount of work delivered is W and the heat rejected to low temperature thermal reservoir (LTTR) is $Q - W$.

$$\Delta S_{\text{Body}} = \int_T^{T_0} m\, c_p \frac{dT}{T} = m\, c_p \ln \frac{T_0}{T}$$

$$\Delta S_{\text{LTTR}} = \int \frac{\delta Q}{T_0} = \frac{Q - W}{T_0}$$

$$\therefore \qquad \Delta S_{\text{univ}} = \Delta S_{\text{Body}} + \Delta S_{\text{LTTR}} = mc_p \ln \frac{T_0}{T} + \frac{Q - W}{T_0}$$

By the entropy principle, for the maximum work done

$$\Delta S_{\text{univ}} = 0$$

Thus, $$mc_p \ln \frac{}{T} \frac{Q \quad W}{T} = 0 \quad \text{and } W = W_{\max}$$

or $$\frac{W_{\max} - Q}{T_0} = mc_p \ln \frac{T_0}{T}$$

or $$W_{\max} = mc_p\, T_0 \ln \frac{T_0}{T} + Q$$

or $$W_{\max} = Q + mc_p\, T_0 \ln \frac{T_0}{T} = m\, c_p\, (T - T_0) - mc_p\, T_0 \ln \frac{T}{T_0}$$

or $$W_{\max} = mc_p \left[(T - T_0) - T_0 \ln \frac{T}{T_0} \right] \qquad \text{...(8.39)}$$

8.7.5 The Minimum Work Input to a Heat Pump Operating between Two Identical Finite Bodies having the same Mass and Specific Heat

Consider the case of a heat pump operating between two identical finite bodies *A* and *B* (Fig. 8.12). Let *m* and c_p be the mass and specific heat of the two bodies.

In the beginning, both the bodies are at the same temperature T_1 but operation of heat pump cools down one body (i.e., *B*) to temperature T_2. Let T_f be the final temperature of the body *A*.

From first law of thermodynamics,

$$\oint \delta W = \oint \delta Q$$

or $$-W = Q_2 - Q_1$$

or $$W = Q_1 - Q_2$$

Heat removed from body *A* to cool it from T_1 to T_2

$$Q_2 = mc_p\, (T_1 - T_2)$$

Heat discharged to body $A = Q_1 = Q_2 + W = mc_p\, (T_f - T_1)$

Thus, $$W = mc_p\, [(T_f - T_1) - (T_1 - T_2) = mc_p\, [T_f - 2T_1 + T_2) \qquad \text{...(8.40)}$$

We know that for the minimum work input, the heat pump should operate on Carnot cycle, i.e., reversible cycle. Thus,

$$\oint \frac{\delta Q}{T} = 0 \quad \text{or} \quad \Delta S_{\text{univ}} = 0$$

$$\Delta S_A = \int_{T_1}^{T_2} m\,c_p \frac{dT}{T} = m\,c_p \ln \frac{T_2}{T_1} \quad \text{(negative)}$$

$$\Delta S_B = \int_{T}^{T_f} m\,c_p \frac{dT}{T} = m\,c_p \ln \frac{T_f}{T_1} \quad \text{(positive)}$$

Thus, $m c_p \ln \dfrac{T_2}{T_1} + m\,c_p \ln \dfrac{T_f}{T_1} = 0$

or $$\ln \frac{T_1}{T_2} = \ln \frac{T_f}{T_1}$$

or $$T_f = \frac{T_1^2}{T_2} \qquad \text{...(8.41)}$$

Substituting the values of T_f in Eq. (8.40), the minimum work input to the heat pump is

$$W_{\min} = m\,c_p \left[\frac{T_1^2}{T_2} - 2\,T_1 + T_2\right] \qquad \text{...(8.42)}$$

Note ***The Articles 8.7.1 to 8.7.5 may also be put in question form to prove the results obtained.***

8.7.6 Processes Involved in External Mechanical Irreversibility

8.7.6.1 Isothermal Dissipation of Work

Consider the case of isothermal dissipation of work through a system into the internal energy of a reservoir. The electric current I flows through a resistor which is constant with a reservoir. At steady state, the internal energy of the resistor is constant and so its temperature. According to first law

$$W = Q$$

It is to be noted that the flow of current represents work transfer. According to the system arrangement, the work is dissipated isothermally into heat transfer to the surroundings. Since the surroundings absorb Q units of heat at temperature, so

$$\Delta S_{\text{surv}} = \frac{Q}{T} = \frac{W}{T}$$

At steady state, $\Delta S_{\text{system}} = 0$

$$\therefore \quad \Delta S_{\text{univ}} = \Delta S_{\text{system}} + \Delta S_{\text{surv}} = \frac{W}{T} \qquad \text{...(8.43)}$$

8.7.6.2 Adiabatic Dissipation of Work

Consider the case of adiabatic dissipation of work in which the stirring work is supplied to a viscous thermally insulated liquid which is dissipated adiabatically into internal energy increase of the liquid (Fig. 8.13). Due to stirring work, the temperature of the liquid increases from T_1 to T_f. Since the process is adiabatic, so

$$\Delta S_{\text{surv}} = 0$$

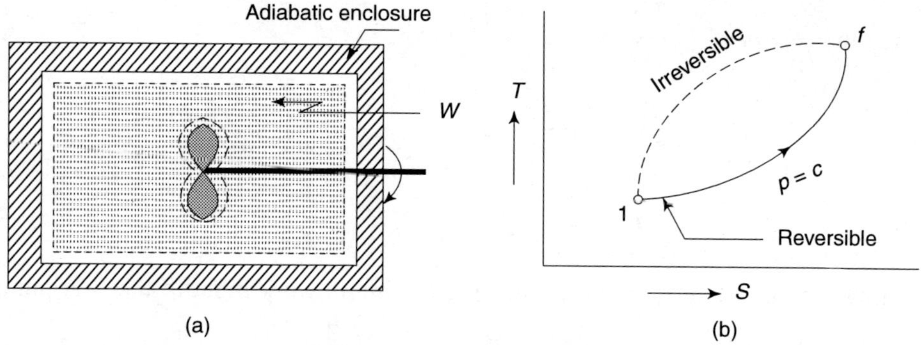

Fig. 8.13 Adiabatic dissipation of work

In order to calculate the entropy change of the system, the original irreversible path shown by dotted line must be replaced by a reversible path between the same end states, 1 and f. Further, the irreversible performance of work is replaced by a reversible isobaric flow of heat from a series of reservoirs ranging from T_1 to T_f to cause the same change in the state of the system. Thus, the entropy change of the system is expressed as

$$\Delta S_{\text{sys}} = \int_1^f \left(\frac{\delta Q}{T}\right)_R = \int_1^f \left(m\, c_p \frac{dT}{T}\right)_R = m\, c_p \ln \frac{T_f}{T_1}$$

Thus, $$\Delta S_{\text{univ}} = \Delta S_{\text{sys}} + \Delta S_{\text{surr}} = m c_p \ln \frac{T_f}{T_1} \quad \text{(positive)} \qquad ...(8.44)$$

8.7.6.3 Entropy Transfer with Heat Flow

As per sign convention, when heat is added to the system, it is positive, so the entropy of the system, $dS = \delta Q_{\text{rev}}/T$ increases. When heat is removed from the system, δQ is given negative sign conversion, and so the entropy of the system decreases.

When heat is transferred to the system of fixed mass (i.e., closed system), its internal energy increases and as a result of which the molecules of gas will move with higher kinetic energy and collide more frequently, so the *disorder of the system increases*. In other words, we can say that *heat is thus regarded as disorganised or disordered energy transfer which increases molecular chaos*. Consider the heat flows from the system to surroundings at T_0 as shown in Fig. 8.14.

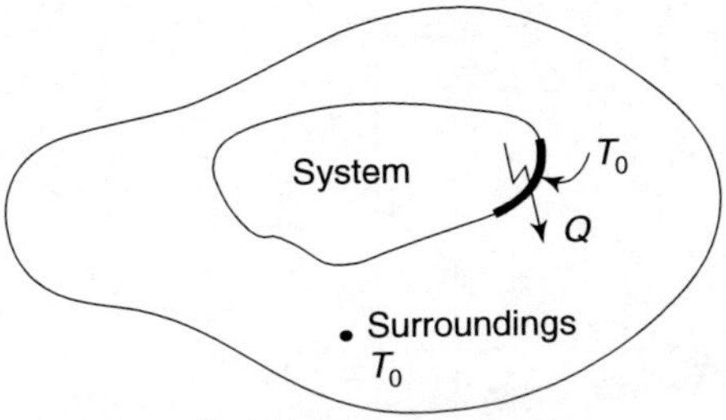

Fig. 8.14 Entropy transfer with heat flow

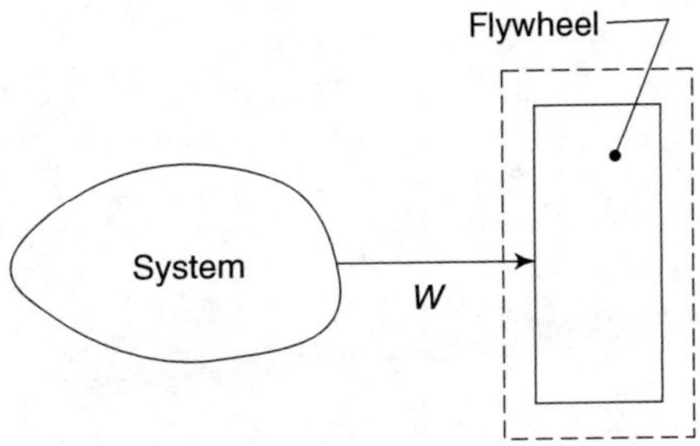

Fig. 8.15 No entropy transfer with work transfer

The entropy increase of the surrounding, $\Delta S_{surr} = \dfrac{Q}{T_0}$

The entropy of the system is reduced by $\Delta S_{syst} = -\dfrac{Q}{T_0}$

It is to be noted that the temperature of the boundary where heat transfer occurs is the constant temperature, T_0. In other words, it may be said that the system has lost entropy to the surroundings, or surroundings have gained entropy from the system. This clearly states that there is a entropy transfer from the system to the surroundings along with heat flow. Alternatively, we may say that *since the heat inflow increases the molecular disorder, there is a flow of disorder along with heat.*

In contrast to this, there is no entropy transfer associated with work transfer which may be visualised from the arrangement shown in Fig. 8.15 in which system delivers work to a flywheel where energy is stored in a fully recoverable form.

This is because of the fact that the flywheel molecules are simply put into rotation around the axis in a perfectly organised manner, and there is no dissipation and hence no entropy increase of the flywheel. The same argument holds good for the compression of a ring or raising of a weight to a certain height. *Thus, there is no entropy transfer with work transfer.*

8.8 ENTROPY GENERATION (PRODUCTION) IN AN OPEN SYSTEM (CONTROL VOLUME)

The second law of thermodynamics can be applied to an open system (control volume) by a procedure similar to the application of first law of control volume. Figure 8.16 shows an open system. The control volume and control surface are also shown. In an open system, *there is a transfer of three quantities, namely, mass, energy and entropy*. The control surface can have one or more openings for mass transfer. If the control surface is rigid, energy in the form of shaft work transfers across it.

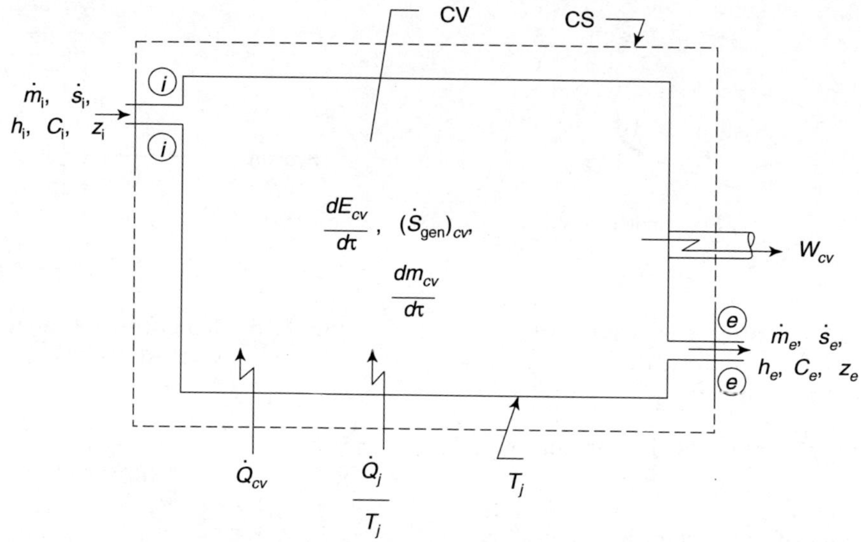

Fig. 8.16 The entropy balance for a control volume on a rate form

The mass balance (continuity equation) gives

$$\sum_i \dot{m}_i - \sum_e \dot{m}_e = \frac{dm_{cv}}{d\tau} \qquad \text{...(8.45)}$$

(Net mass transfer rate) (Rate of mass assumulation in CV)

The energy balance gives

$$\frac{dE_{cv}}{d\tau} = \dot{Q}_{cv} - \dot{W}_{cv} + \sum_i \dot{m}_i \left(h_i + \frac{C_i^2}{2} + gz_i \right) - \sum_e \dot{m}_e \left(h_e + \frac{C_e^2}{2} + gz_e \right) \qquad \text{...(8.46)}$$

(Rate of energy accumulation in CV) (Net rate of energy transfer)

The entropy rate balance gives

$$\frac{dS_{cv}}{d\tau} = \sum_j \frac{\dot{Q}_j}{T_j} + \sum_i \dot{m}_i \, s_i - \sum_e \dot{m}_e \, s_e + (\dot{S}_{gen})_{cv} \qquad \text{...(8.47)}$$

(Rate of entropy change) (Rate of entropy transfer) (Rate of entropy generation)

where $dS_{cv}/d\tau$ represents the time rate of change of entropy within the control volume. The terms $\dot{m}_i \cdot s_i$ and $\dot{m}_e \cdot s_e$ account respectively, for rates of entropy transfer into and out of the control volume accompanying mass flow. The term $\dot{Q}_j$ represents the time rate of heat transfer at the location of boundary where the instantaneous temperature is T_j. The ratio $\dot{Q}_j/T_j$ represents the accompanying rate of entropy transfer. The term $(\dot{S}_{\text{gen}})_{cv}$ denotes the time rate of entropy generation (i.e., production) due to irreversibilities within the control volume.

8.8.1 Integral Form

The rate balance can be expressed in terms of local properties to obtain forms that are more generally applicable. Thus, the terms $S_{cv}(\tau)$ representing the total entropy associated with control volume at time τ can be written as a volume integral

$$S_{cv}(\tau) = \int_V \rho s dv$$

where ρ and s denote, respectively, the total density and specific entropy. The rate of entropy transfer accompanying heat transfer can be expressed more generally as on integral over the surface of the control volume. Thus,

$$\text{Time rate of entropy transfer accompanying heat transfer} = \int_A \left(\frac{\dot{q}}{T}\right)_b dA$$

where $\dot{q}$ is the heat flux; the time rate of heat transfer per unit surface area through the location of the boundary where the instantaneous temperature is T_j. The subscript "b" is added as a reminder that the integral is evaluated on the boundary of the control volume. Thus, finally

$$\frac{d}{d\tau}\int_V \rho s dv = \int\left(-\right) dA + \sum_i\left(\int_A s\rho\, C_n\, dA\right)_i - \sum_e\left(\int_A s\rho\, C_n\, dA\right)_e + (\dot{S}_{\text{gen}})_{cv} \qquad \text{...(8.48)}$$

where C_n denotes the normal component in the direction of flow of the velocity relative to the flow area, (dA).

8.8.2 Analysis of Control Volume at Steady State

It is worth to note that a great number of engineering analyses involve control volume at steady state.

At steady state, the conservation of mass principles takes the form by letting $dm_{cv}/d\tau = 0$,

$$\sum_i \dot{m}_i = \sum_e \dot{m}_e \qquad \text{...(8.49)}$$

The energy rate balance at steady state is given by putting $dE_{cv}/d\tau = 0$

$$0 = \dot{Q}_{cv} - \dot{W}_{cv} + \sum_i \dot{m}_i\left(h_i + \frac{C_2^2}{2} + gz_i\right) - \sum_e \dot{m}_e\left(h_e + \frac{C_e^2}{2} + gz_e\right) \qquad \text{...(8.50)}$$

The steady state of the entropy rate balance is obtained by letting $dS_{cv}/d\tau = 0$

$$0 = \sum \frac{\dot{Q}_j}{T_j} + \sum_i \dot{m}_i s_i - \sum_e \dot{m}_e s_e + (\dot{S}_{gen})_{cv} \qquad ...(8.51)$$

For the analysis of control volume problems, these equations must be solved simultaneously.

8.8.2.1 One-Inlet and One-Exit Control Volume

Many applications in engineering problems involve one-inlet, one-exit control volumes at steady state. For such case, the steady state entropy rate balance is given by

$$0 = \sum_j \frac{\dot{Q}_j}{T_j} = \dot{m}(s_1 - s_2) + \frac{(\dot{S}_{gen})_{cv}}{\dot{m}}$$

or

$$s_2 - s_1 = \frac{1}{\dot{m}}\left(\sum_j \frac{\dot{Q}_j}{T_j}\right) + \frac{(\dot{S}_{gen})_{cv}}{\dot{m}} \qquad ...(8.52)$$

Equation (8.52) *is the expression for entropy rate balance for unit mass flowing through the control volume.* From this equation, it is concluded that the entropy of a unit mass passing from inlet to exit can increase, decrease or remain the same.

In the special case, where there is no entropy transfer accompanying heat transfer, Eq. (8.52) reduces to

$$s_2 - s_1 = \frac{(\dot{S}_{gen})_{cv}}{\dot{m}} \qquad ...(8.53)$$

8.9 FIRST AND SECOND LAWS COMBINED

From the second law of thermodynamics, we have

$$\delta Q_{rev} = TdS$$

From the first law, for closed system (control mass)

$$\delta Q = dU + pdV$$

From the above two equations

$$Tds = dU + pdV \qquad ...(8.54)$$

We know that, $H = U + pV$

or

$$dH = dU + pdV + Vdp = Tds + Vdp$$

or

$$Tds = dH - Vdp \qquad ...(8.55)$$

Equations (8.54) and (8.55) are the thermodynamic equations obtained by the combination of the first and second laws.

It is worth to analyse the various equations related to the first and second laws and their combinations.

(i) $\delta Q = dE + \delta W$: This equation holds good for any process, reversible or irreversible and for any system. Here energy (E) and work (W) take all types of possible forms.

(ii) $\delta Q = dU + \delta W$: This equation holds good for any process undergone by a closed system. Here work (W) takes all possible forms.

(iii) $\delta Q = dU + pdV$: This equation holds good for a closed system when only pdV work is present and valid for reversible or quasi-static process.

(iv) $\delta Q = TdS$: This equation is valid for only reversible processes.

(v) $TdS = dU - Vdp$: This equation is valid for closed system and true for any process reversible or irreversible. Further, this equation is independent of path.

(vi) $TdS = dH - Vdp$: This equation is also true for closed system and valid for any process reversible or irreversible. Further, this equation is also independent of path.

It is to be noted that the term irreversible process is doubtful, so no irreversible path or process can be plotted on thermodynamic coordinates. It is more logical to state that the change of state is an irreversible proces rather than to say the process is irreversible. *However, to differentiate the state caused by irreversible process, we show the irreversible process on the thermodynamic scale by a path.*

8.10 ABSOLUTE VALUE OF ENTROPY

All the governing equations discussed above give the change in entropy. In order to know the absolute value of entropy, we have to choose the reference state. This reference state may be 0°C (for air and steam) where the initial value of entropy will be zero so that the absolute value of entropy at the end of state may be found. In refrigeration, the reference state may be –100°C at which initial entropy value will be zero. In gas turbine analysis, the ambient condition may be taken as the reference state. For example, if $\Delta s = 0.05$ kJ/kg K, means if $s_1 = 0$ at (say 0°C), so $s_2 = s = 0.05$ kJ/kg K.

8.11 ENTROPY CHANGES FOR AN IDEAL GAS DURING VARIOUS PROCESSES

The first law is valid for reversible or irreversible process and in the absence of potential and kinetic energies,

$$\delta Q - \delta W = dU$$

If the process is reversible, then

$$(\delta Q)_{rev} - \delta W = dU$$

Dividing by T, we have

$$\left(\frac{\delta Q}{T}\right)_{rev} - \frac{\delta W}{T} = \frac{dU}{T}$$

or $$dS = \frac{pdV}{T} + \frac{dU}{T} \qquad ...(8.56)$$

or $$TdS = pdV + dU \qquad ...(8.57)$$

or $Tds = pdv + du$ per unit mass basis

Equation (8.57) is valid for both reversible and irreversible processes because they express the relationships between properties and are therefore, independent of the path

Equation (8.57) may also be written as

$$Tds = pdV + d\,(H - pV) = pdV + dH - Vdp - pdV$$

or $$TdS = dH - VdP \qquad ...(8.58)$$

or $$dS = \frac{dH}{T} - \frac{Vdp}{T} \qquad ...(8.59)$$

Equations (8.57) and (8.58) represent the first and second laws and are applicable to both reversible or irreversible process of a closed or open system.

Integrating Eqs. (8.56) and (8.57), we have

$$S_2 - S_1 = \int_1^2 \frac{pdV}{T} + \int_1^2 \frac{dU}{T} \qquad ...(8.60)$$

and $$S_2 - S_1 = \int_1^2 \frac{dH}{T} - \int_1^2 \frac{Vdp}{T} \qquad ...(8.61)$$

For a perfect gas,

$$dU = mc_v\, dT,\ dH = mc_p\, dT \quad \text{and} \quad \frac{p}{T} = \frac{mR}{V}, \frac{V}{T} = \frac{mR}{p}$$

Hence, Eq. (8.60) becomes

$$dS = mc_v \frac{dT}{T} + mR\frac{dV}{V}$$

or $$S_2 - S_1 = m\int_{T_1}^{T_2} c_v \frac{dT}{T} + mR\int_{v_1}^{v_2} \frac{dV}{V} \qquad ...(8.62)$$

If c_v remains constant

$$S_2 - S_1 = mc_v \ln\frac{T_2}{T_1} + mR\ln\frac{V_2}{V_1} \qquad ...(8.63)$$

Now Eq. (8.61) may be written as

$$S_2 - S_1 = m \int_{T_1}^{T_2} c_p \frac{dT}{T} - mR \int_{p_1}^{p_2} \frac{dp}{p} \quad ...(8.64)$$

If c_p remains constant, then

$$S_2 - S_1 = m\, c_p \ln \frac{T_2}{T_1} - mR \ln \frac{p_2}{p_1} \quad ...(8.65)$$

From the concept of entropy,

$$\delta Q = TdS \text{ for a reversible process} \quad ...(8.66)$$

(i) Constant Volume Process During constant volume process, volume V remains constant.

i.e., $\int dV = 0$, hence from the Eq. (8.63)

$$S_2 - S_1 = m \int_{T_1}^{T_2} c_v \frac{dT}{T} = m \cdot c_v \ln \frac{T_2}{T_1}; \text{ kJ/K} \quad ...(8.67)$$

On the basis of per unit mass

$$s_2 - s_1 = c_v \ln \frac{T_2}{T_1}; \text{ kJ/kg K} \quad ...(8.68)$$

The process when plotted on T–S diagram will have the nature as shown in Fig. 8.17.

The area under the curve 1–2 (i.e., area a–1–2–b) shows the amount of heat supplied during constant volume process carried from state point 1 to state point 2 and is mathematically

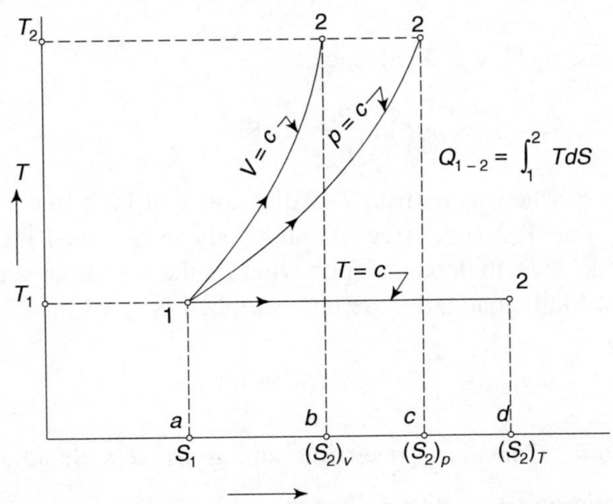

Fig. 8.17 Constant volume, constant pressure and constant temperature process

$$Q_{1-2} = \int_1^2 TdS \quad ...(8.69)$$

(ii) Constant Pressure Process During the constant pressure process, p = constant and thus $\int dp = 0$, hence from the Eq. (8.6), we have

$$S_2 - S_1 = m\, c_p \int_1^2 \frac{dT}{T} = m \cdot c_p \ln \frac{T_2}{T_1} \quad ...(8.70)$$

The process when plotted on T–S diagram will have the nature as shown in Fig. 8.17. Since c_p is greater than c_v, hence a line of constant pressure slopes less steeply than a line of constant volume. The area under the curve 1–2 (i.e., area a–1–2–c) shows the amount of heat supplied (since $\delta Q = TdS$) during constant pressure process from state point 1 to state point 2. Mathematically,

$$Q_{1-2} = \int_1^2 TdS$$

(iii) Isothermal Process $(pV = c = mRT)$

From Eq. (8.62), we have

$$S_2 - S_1 = m \int c \frac{dT}{T} \quad \int mR \frac{dV}{V}$$

But $dT = 0$, hence

$$S_2 - S_1 = mR \int_1^2 \frac{dV}{V} = mR \ln \frac{V_2}{V_1} \quad ...(8.71)$$

For isothermal process, $p_1 V_1 = p_2 V_2$. Hence,

$$S_2 - S_1 = mR \ln \frac{p_1}{p_2}; \text{ kJ/K} \quad ...(8.72)$$

The isothermal process when plotted on T–S diagram will look like as in Fig. 8.17. The area under the straight line 1–2 (i.e., area 1–2–d–a) shows the heat interaction. The heat supplied or rejected by the system depends upon whether the system undergoes an expansion or compression process. Mathematically, the heat supplied is

$$Q_{1-2} = \int_1^2 TdS = T(S_2 - S_1)$$

(iv) Adiabatic Process Consider a reversible and an irreversible adiabatic process.

For reversible adiabatic process, $\delta Q = 0$, hence

$$\Delta S = S_2 - S_1 = 0 \quad \text{or} \quad S_1 = S_2 \qquad \text{...(8.73)}$$

For irreversible process, we have seen that

$$\Delta S = S_2 - S_1 > 0 \qquad \text{...(8.74)}$$

Plotting the adiabatic process on T–S chart, it will look as shown in Fig. 8.18.

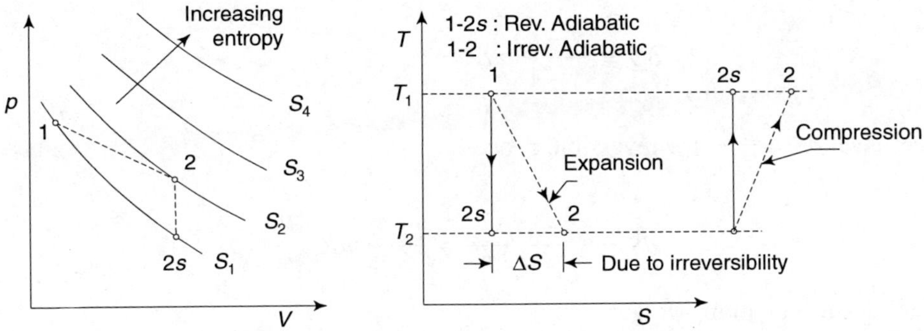

Fig. 8.18 Adiabatic process

First consider a reversible adiabatic expansion, it will be represented as a straight line l–2*s* on T–S diagram and the change in entropy is zero which is obvious. If an irreversible adiabatic expansion process takes place between the same temperature limits T_1 and T_2, then due to internal friction, internal energy of the system at the end of expansion will be more than at point 2 and the properties of the system at the end of state will be represented by state point 2 to the right of 2*s*. Let due to the friction $\delta Q'$ be the amount of heat gained by the system, then the increase of entropy will be given by $dS = \delta Q'/T$.

This confirms the earlier statement that $dS > 0$ for irreversible process. Hence, there is an increase in entropy due to irreversible process.

Similarly, the reversible and irreversible adiabatic compression processes will be represented by l–2*s* and 1–2 respectively.

The reversibie and irreversible adiabatic process may also be explained on the p–V diagram (Fig. 8.18) with the help of constant entropy lines. Let the ratio of expansion for reversible and irreversible be the same. The reversible adiabatic process will be represented by 1–2*s*, i.e., by constant entropy curve, S_1. But in the case of irreversible process, at the end of expansion, the temperature of the system and hence the internal energy of system will be more than at point 2*s* due to friction. Thus, the end point will shift from 2*s* to 2 vertically from point 2 lying on another entropy curve 2. We have seen that $dS > 0$ in the irreversible process, hence it is clear that $S_1 > S_{1s}$ and $S_2 > S_{2s}$, etc.

During the non-flow adiabatic process work is done at the cost of internal energy. The change in internal energy in the case of irreversible process, due to friction, is more than that of isentropic, i.e., $(U_1 - U_2)_{irrev} > (U_1 - U_2)_{rev}$, in other words, the amount of work done during irreversible expansion process is less than that done in isentropic process. *Therefore, we can conclude that higher entropy at the end of expansion means less availability of heat energy for conversion into work*, and *that entropy is an index for availability or non-availability of heat energy for conversion into work.*

(v) Polytropic Process : ($pV^n = c$) In the previous article, for polylropic process, the heat interaction was given by

$$\delta Q = \frac{\gamma - n}{\gamma - 1} pdV$$

But we have, $dS = \frac{\delta Q}{T}$ for reversible process

$$\therefore \qquad dS = \frac{\gamma - n}{\gamma - 1} \frac{pdV}{T} = \frac{\gamma - n}{\gamma - 1} mR \frac{dV}{V}$$

Integrating this equation, we get

$$S_2 - S_1 = \frac{\gamma - n}{\gamma - 1} mR \ln \frac{V_2}{V} ; \text{ kJ/K} \qquad ...(8.75)$$

On per unit mass basis

$$\text{or} \quad s_2 - s_1 = c_n \ln \frac{T_2}{T_1} = c_v \left(\frac{\gamma - n}{1 - n}\right) \ln \frac{T_2}{T_1} = \frac{\gamma - n}{(\gamma - 1)(n - 1)} R \ln \frac{T_2}{T_1} \qquad ...(8.76)$$

$$s_2 - s_1 = \frac{\gamma - n}{\gamma - 1} R \ln \frac{V_2}{V_1} = \frac{\gamma - n}{\gamma - 1} R \ln \frac{v_2}{v_1} \text{ kJ/kg K} \qquad ...(8.77)$$

These equations are valid for both reversible and irreversible process because the entropy change is expressed in the forms of systems properties. The change in entropy for polytropic process may be found from Eqs. 8.63 and 8.65 also.

8.12 CARNOT CYCLE ON *T–S* DIAGRAM

Carnot cycle on *T–S* diagram looks like a rectangle as shown in Fig. 8.19. The various heat exchanges and entropy changes are as follows:

$$Q_{1-2} = T_1 (S_2 - S_1) \quad ; \quad \Delta S = S_2 - S_1$$

$$Q_{1-2} = 0 \quad ; \quad \Delta S = 0$$

$$Q_{1-2} = T_2 (S_4 - S_3) \quad ; \quad \Delta S = S_4 - S_3$$

$$Q_{1-2} = 0 \quad ; \quad \Delta S = 0$$

Here $S_2 - S_1 = S_4 - S_3$, i.e., *the increase in entropy during the process* 1–2 is equal to the decrease in entropy during process 3–4.

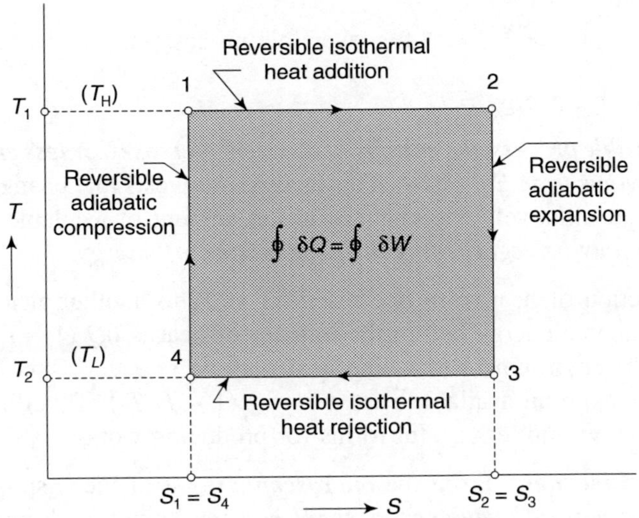

Fig. 8.19 Carnot on *T–S* diagram

From first law,

$$\oint \delta Q = \oint \delta W$$

or

$$\oint TdS = \oint pdV \text{ for reversible process}$$

The above equation means that heat or work interaction during the cycle may be represented by the area 1–2–3–4 on *T–S* diagram.

8.13 PHYSICAL INTERPRETATION OF ENTROPY

Discussing entropy only in mathematical parameter is not sufficient. It is essential to understand the significance of entropy and its continuous creation in irreversible processes, hence the physical interpretations of entropy are important.

(i) Entropy and Degradation of Energy According to second law of thermodynamics, when heat δQ is transferred to a system operating in a cycle, only a portion of the heat is available for work. This portion is known as available energy. We know that the efficiency of a reversible engines operating between two thermal reservoirs, heat source (T) and heat sink T_0 is given by $(1 - T_0/T)$ and the work done will be the product of efficiency and heat input. Thus, the available energy will be

$$\text{Available energy} \leq \delta Q\left(1 - \frac{T_0}{T}\right)$$

The inequality and equality signs signify irreversible and reversible processes respectively. The unavailable energy in the reversible engine is given by

$$\text{Unavailable energy} = \delta Q\left(\frac{T_0}{T}\right)$$

But $$\frac{\delta Q}{T} = dS \text{ for reversible process}$$

$\therefore$ $$\text{Unavailable energy} = T_0\, dS \qquad \text{...(8.78)}$$

During the irreversible process, the entropy of a system increases, or less energy is available for work which is obvious from Eq. (8.78). This analysis suggests that change in entropy may be used to describe quantitatively any changes in the amount of available energy that may occur. *Thus,* ***entropy may be regarded as the degradation of energy.***

During the interaction of heat from one reservoir at T_1 to another at lower temperature T_2, the maximum available energy before the transfer of heat is $\delta Q\ (1 - T_0/T_1)$ where T_0 is the temperature of the environment to which heat may be rejected. Once the heat transfer process is over, the maximum available energy is $\delta Q\ (1 - T_0/T_2)$. This difference is called the degradation of energy into less useful forms for producing work.

(ii) Entropy and Disorder From the *microscopic* point of view, spontaneous process can be analysed. *As a natural phenomenon, there is a tendency for all physical systems to proceed to states in which the particles become less ordered.* It means that ***entropy may be regarded as measure of the disorder of the system.*** Now, imagine the transfer of heat to a homogeneous system. As temperature rises, the molecules of the system move in a larger variety of ways in each of the various vibrational, rotational, translational, electronic, etc. modes. Or, we may say that as the rejection of heat takes place from a system, disorder of molecules will take place although an equivalent amount of disorder or greater disorder results in the environment. Thus, molecular disorder of the universe, is therefore, continuously increasing due to irreversible processes of nature, and for that reason the ***Second Law is also called, law of increase of entropy. The entropy of the system will be the maximum when it will be in thermodynamic equilibrium with the environment as disorder will be the maximum.***

(iii) Entropy and Probability According to the concept of *probability*, a system is less likely to be in a orderly state than in a disorderly state. Note that spontaneous change in the nature proceeds from the less probable to the more probable states. *In the state of thermodynamic equilibrium,* ***the degree of molecular disorder is the maximum and the state of the system is the state of the highest probability***. Therefore, *entropy* being a *function of molecular disorder*, may be used to measure the *statistical probability* of a macrostate of a system. Thus,

$$dS_{\text{isolated}} \geq 0$$

suggests that an *isolated system tends to approach complete randomness.* In the state of equilibrium of an isolated system, there is no change in the properties of the system because the system is in a state corresponding to the macrostate of maximum probability.

In order to understand the meaning of thermodynamic probability, consider the free expansion of a gas of volume V_a into an evacuated chamber. After expansion, let the total volume be V_b. If W denotes probability, then the probability that a certain molecule of the gas will be in volume, V_b is unity, i.e., $W_b = 1$. Also, the probability that a certain molecule of the gas will be in volume V_a is proportional to the ratio of the two volumes, such that $W_a = V_a/V_b$. Taking into consideration the several independent events, the probability that they will all occur simultaneously is equal to the product of their individual probabilities. Thus, for N molecules, the probability that they all are located in V_a at any particular time is $(V_a/V_b)^n$.

or
$$\frac{W_a}{W_b} = \left(\frac{V_a}{V_b}\right)^N \qquad ...(8.79)$$

According to **Boltzmann**, *there exists a relationship between probability and entropy.* From the *concept of probability, the state of equilibrium is the most probable state* and, from the *concept of classical thermodynamics*, the *state of equilibrium represents the state of maximum entropy. In the irreversible process, both the thermodynamic probability and the entropy of a system increase.* Entropy is however, an extensive (i.e., *additive*) property, whereas *thermodynamic probability is a multiplicative property*, Boltzmann suggested that

$$S = k \ln W \qquad ...(8.80)$$

where $k = R/N_0$ = Boltzmann's constant. This constant represents the gas constant based on one molecule. If the system is made of several parts, each having fixed energy, volume and composition, then the entropy of such a system is equal to $S = S_1 + S_2 + ...$ *The thermodynamic probability of a state of a system (being multiplicative property) is equal to the product of the individual probabilities of the composition of the system.* Mathematically, $W = W_1 . W_2$.

Therefore, entropy of a system according to Boltzmann relation becomes

$$S = S_1 + S_2 + ... = k \ln W_1 + k \ln W_2 + ... = k \ln W$$

The entropy change during free expansion process is given by

$$dS = mc_v \frac{dT}{T} + mR\frac{dV}{V}$$

Since temperature remains constant during free expansion process, hence

$$S_b - S_a = mR \ln\frac{V_b}{V_a} = N\frac{R}{N_0}\ln\frac{V_b}{V_a} = k \ln\frac{W_b}{W_a} \qquad ...(8.81)$$

Equation (8.81) *suggest that the change of entropy is proportional to the logarithm of the ratio of the statistical probabilities of the end states of the system. The probability of the final state of the process is greater than initial state.*

(iv) Thermal Death There is also a *philosophical aspect of entropy.* ***The first and second laws of thermodynamics suggest that the energy of the universe is constant but the entropy of the universe is increasing continuously.*** Now, one may ask "can the entropy of the universe increase indefinitely or is there a maximum possible entropy for the universe? If such a limit exists, when will it be reached? We know that all real processes are irreversible processes and in all real processes, the entropy increases. It means that the energy available to mankind is continuously decreasing and in near future, no energy will be available for use. When such a condition is reached, the entropy of the universe will be at its maximum and complete disorder and randomness will prevail. Also, at that stage of the universe, all matter will be at the same temperature level and the universe will according to Clausius reach the state of **thermal "death"**. In these circumstances, no energy will be available for use. But the total energy of the universe will not be different from its present value.

Similar philosophical argument may be given about the *theory of probability*. From the standpoint of probability, all possible microstates of a macrostate are equally probable. Putting it in other way, a system can exist in a particular microstate at any time, so that, no particular time for its existence is more probable than any other time. That is, all possible events and states of the universe occur randomly. But according to second law, only those processes which result in an increase in the entropy of the universe can occur. ***Hence, the second law may be interpreted to mean as one of very high statistical probability but not one of absolute certainty***.

Example 8.1 Verify the Clausius inequality in a steam power plant cycle shown in Fig. 8.20. Dry and saturated steam leaves the boiler at enthalpy of 2632 kJ/kg at a constant temperature of 163°C. The steam expands in the turbine and it leaves the condenser at an enthalpy of 2352 kJ/kg. In the condenser, where the temperature is 50°C, heat is rejected to circulating cooling water and the condensate leaves the condenser at 432 kJ/kg. The feed water is pumped back to boiler having enthalpy 652 kJ/kg.

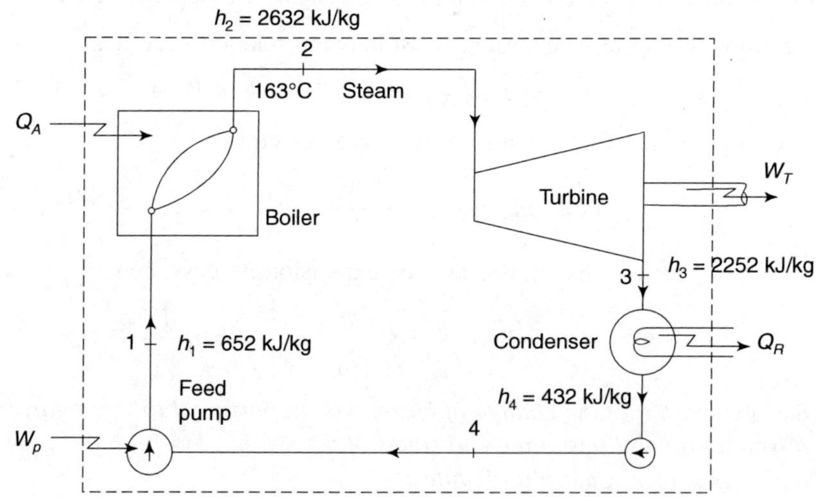

Fig. 8.20 Example 8.1

Solution Refer to Fig. 8.20.

In the steam power plant, the boiler serves as a heat source while the condenser serves as a heat sink. Hence,

$$\oint \frac{\delta Q}{T} = \int \left(\frac{\delta Q}{T}\right)_{\text{Boiler}} + \int \left(\frac{\delta Q}{T}\right)_{\text{Condenser}}$$

Here the temperature is constant both in the boiler as well as condenser, hence

$$\oint \frac{\delta Q}{T} = \frac{1}{T_{\text{boil}}} \int_1^2 \delta Q + \frac{1}{T_{\text{cond}}} \int_3^4 \delta Q = \frac{Q_{1-2}}{T_{\text{boil}}} + \frac{Q_{3-4}}{T_{\text{cond}}}$$

$$= \frac{2632 - 652}{163 + 273} + \frac{432 - 2252}{50 + 273} = -\,1.09 < 0$$ **Ans.**

The Clausius inequality is valid in this steam power plant.

Example 8.2 A heat engine receives 1000 kW of heat at constant temperature of 285°C. The heat is rejected at 5°C. The possible heats rejected are:

(a) 840 kW, (b) 498 kW, (c) 300 kW

Comment on the results.

Solution Apply the Clausius inequality in each case with proper sign convention

(a) $$\oint \frac{\delta Q}{T} = \int \left(\frac{\delta Q}{T}\right)_{\text{Add}} - \int \left(\frac{\delta Q}{T}\right)_{\text{Rej}} = \frac{1000}{25 + 273} - \frac{840}{5 + 273} = -\,1.312 < 0$$

Hence, the cycle is irreversible **Ans.**

(b) $$\oint \frac{\delta Q}{T} = \frac{1000}{285 + 273} - \frac{498}{5 + 273} = 0.0$$

Hence, the cycle is reversible **Ans.**

(c) $$\oint \frac{\delta Q}{T} = \frac{1000}{285 + 273} - \frac{300}{5 + 273} = 0.713$$

The Clausius inequality does not hold good, hence, the cycle is not possible.

Example 8.3 During the braking test of an electric motor of 11 kW for one hour the heat generated by the friction forces was found to be transferred to surroundings at 25°C. Calculate resulting entropy increase.

Solution Heat generated by friction force

$$Q = 11 \times 3600 = 39600 \text{ kJ/h.}$$

The change in entropy $= S_2 - S_1 = \dfrac{Q}{T} = \dfrac{39600}{25 + 273} = 132.88$ kJ/K

Example 8.4 During summer to cool water for drinking purpose 1 kg of ice at – 2°C in mixed with 5 kg of water at 27°C in an insulated container which is subjected to atmospheric pressure of 1 bar. Calculate the temperature of the mixture and the change in entropy for the spontaneous process. Assume specific heat of ice as 2.093 kJ/kgK and latent heat of ice as 335 kJ/kg.

Solution Refer to Fig. 8.21. After mixing the ice with water the mixture temperature will obviously be greater than – 2°C and less than 27°C. Let t_f be the final mixture temperature.

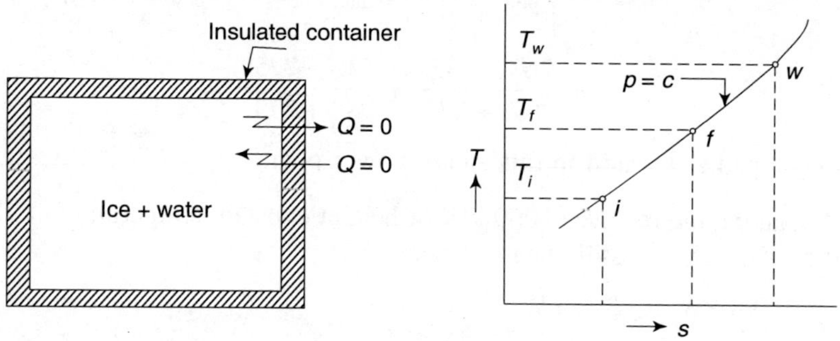

Fig. 8.21 Example 8.4

Heat gained by ice = Heat lost by water

Heat gained by ice = Heat gained in solid phase + Latent heat of ice + Heat gained by liquid phase

$$m_i\, c_{pi}\, (0 + 2) + m_i\, h_{fgi} + m_{iw}\, c_{piw}\, (t_f - 0) = m_w\, ,\, c_{pw}\, (27 - t_f)$$

or $$1 \times 2.093 \times 2 + 1 \times 335 + 1 \times 4.18\, t_f = 5 \times 4.18\, (27 - t_f)$$

or $$t_f = 8.976°\text{C}$$

Change in entropy of ice will be given by $(\Delta S)_{\text{ice}}$ = Entropy change due to heating of ice from – 2°C to 0°C at constant pressure + entropy change due to melting of ice at constant temperature during phase change + entropy change due to heating of ice-water from 0°C to 8.976°C at constant pressure.

$$(\Delta S)_{\text{ice}} = m_i\ .c_{pi} \int_{271}^{273} \frac{dT}{T} + \frac{m_i\, h_{fgi}}{273} + m_{iw}\, c_{piw} \int_{273}^{281.976} \frac{dT}{T}$$

$$= 1 \times 2.093 \ln \frac{273}{271} + \frac{1 \times 335}{273} + 1 \times 4.18 \ln \frac{281.976}{173} = 1.377 \text{ kJ/K}$$

Change in entropy of water will be given by

$(\Delta S)_{\text{water}}$ = Change in entropy due to cooling water from 27°C to 8.976°C at constant pressure

$$= m_w\, c_{pw} \int_{300}^{281/976} \frac{dT}{T} = 5 \times 4.18 \ln \frac{281.976}{300} = -\ 1.294 \text{ kJ/K}$$

Hence, change in entropy of the system is given by

$$(\Delta S)_{system} = (\Delta S)_{ice} + (\Delta S)_{water} = 1.377 - 1.294 = 0.083 \text{ kJ/K} \quad \textbf{Ans.}$$

The reader is advised to plot the whole process on $T - Q$ diagram.

$(\Delta S)_{system} > 0$ means that the process is irreversible.

Example 8.5 A gearbox of centrifugal compressor receives 78 kW from an electric motor and delivers 74 kW to compressor operating in a steady state of 55°C. Temperature of the surrounding is at 60°C. Calculate the entropy growth in the gearbox and the rate of entropy increase of the surroundings.

Solution There is a loss of 4 kW of work in the gearbox. This loss will generate heat in the gearbox. Hence, the rate at which heat is being generated due to the loss of work is given by

$$Q_{gen} = 4 \times 60 = 240 \text{ kJ/min}$$

∴ Increase in entropy due to loss of work is the entropy growth in the gearbox and this is given as

$$(\Delta S)_{gear\ box} = \frac{240}{55 + 273} = 0.7317 \text{ kJ/K. min.} = 0.0121 \text{ kW/K} \quad \textbf{Ans.}$$

The surrounding receives 240 kJ/min of heat at 60°C and there will be corresponding increase in entropy given as

$$(\Delta S)_{surr} = \frac{240}{60 + 273} = 0.72 \text{ kJ/K. min.} = 0.012 \text{ kW/K} \quad \textbf{Ans.}$$

Example 8.6 A heat exchanger equipped in a factory handles 40 kg/min of water which is heated from 20°C to 70°C by a hot gas which enters the heat exchanger at 150°C with the rate of 80 kg/min. Determine the change in entropy. Assume specific heat for water and gases as 4.18 and 1.045 kJ/kg K.

Solution Heat gained by water = Heat lost by gas

or $$m_w . c_{pw}\,(t_2 - t_1) = m_g\, c_{pg}\,(t_3 - t_4)_g$$

or $$40 \times 4.18 \times (70 - 20) = 80 \times 1.045\,(150 - t_4)$$

or $$t_4 = 50°\text{C}$$

Entropy change of gas during cooling at constant pressure is

$$(\Delta S)_{gas} = m_g\, c_{pg} \ln \frac{T_4}{T_3} = \frac{80}{60} \times 1.045 \ln \frac{50 + 273}{150 + 273} = -\ 0.375 \text{ kW/K}$$

Entropy change of water during heating at constant pressure.

$$(\Delta S)_{water} = m_w \cdot c_{pw} \ln \frac{T_2}{T_1} = \frac{40}{60} \times 4.18 \ln \frac{70+273}{20+273} = 0.439 \text{ kW/K}$$

Net change in entropy is

$$(\Delta S)_{net} = (\Delta S)_{water} + (\Delta S)_{gas} = 0.439 - 0.375 = 0.064 \text{ kW/K} \quad \textbf{Ans.}$$

Example 8.7 (a) For a system of perfect gas, prove that two reversible adiabatic lines plotted on any thermodynamic coordinates cannot intersect.

(b) Prove that reversible adiabatic lines are steeper than isothermal lines on a p–V plane.

(c) Prove that the constant volume lines on T–S plane are steeper than constant pressure line working between the same temperature limits.

Solution (a) We know that in reversible adiabatic (isentropic) process, $dS = \delta Q/T = 0$, i.e., no change is in entropy. Therefore, the two constant entropy curves S_1 and S_2 are drawn on a p–V diagram (Fig. 8.22). First assume that both of them intersect. But at the point of intersections there will be two values of entropy. Since entropy is a property and it is a point function, therefore, it cannot have two values at point of intersection. Thus, two reversible adiabatic (isentropic) curves do not intersect (Fig. 8.22).

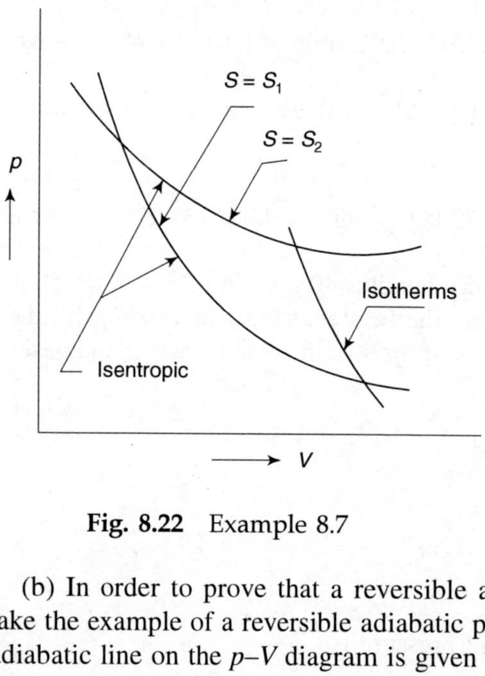

Fig. 8.22 Example 8.7

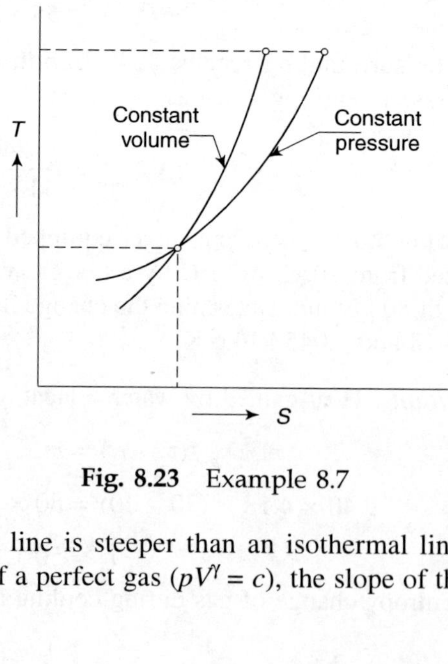

Fig. 8.23 Example 8.7

(b) In order to prove that a reversible adiabatic line is steeper than an isothermal line, take the example of a reversible adiabatic process of a perfect gas ($pV^\gamma = c$), the slope of the adiabatic line on the p–V diagram is given by

$$\frac{dp}{dV} = -\gamma \frac{p}{V}$$

For isothermal process ($pv = c$), the slope is

$$\frac{dp}{dV} = -\frac{p}{V}$$

It is clear from the above two equations that the adiabatic line is steeper than the isothermal line on p–V diagram because $\gamma > 1$. This is shown in Fig. 8.22. Further, the isothermal and adiabatic lines cannot touch without intesection.

(c) We know that for a constant pressure process

$$ds = \frac{\delta q}{T} = \frac{c_p\, dT}{T}$$

or

$$\frac{dT}{ds} = \frac{T}{c_p} \;\alpha\; T \text{ since } c_p \text{ is constant}$$

Hence, dT/ds is proportional to absolute temperature. Now, for constant pressure process $dT/ds = T/c_p$ and for constant volume process $dT/ds = T/c_v$. From these two equations, it is obvious that constant volume line will be steeper than constant pressure as c_v is smaller than c_p. This is shown in Fig. 8.23.

Example 8.8 A household refrigerator absorbs heat at 1°C and rejects heat at 48°C. A 2 kW electric motor drives the compressor of refrigerator. 40000 kJ/h of heat is absorbed by the refrigerator at low temperature. Determine the amount of heat rejected per hour and irreversibility.

Solution

$$2 \text{ kW} = 2 \times 3600 = 7200 \text{ kJ/h} = W_{net}$$

From the first law of thermodynamics, heat rejected to higher temperature

$$Q_1 = W_{net} + Q_2 = 7200 + 40000 = 47200 \text{ kJ/h}$$

Inirreversible cycle, taking care of direction of heat

$$\oint \frac{\delta Q}{T} + I = -\frac{Q_1}{T_1} + \frac{Q_2}{T_2} + I = 0 \text{ where } I \text{ is irreversibility}$$

or

$$I = \frac{47200}{48 + 273} - \frac{40000}{1 + 273} = 1.02 \text{ kJ/h}$$ **Ans.**

Example 8.9 A valve connects two vessels one exactly twice the volume of the other. The whole arrangements is immersed in a constant temperature bath of the water.

The small vessel contains air at 15 and the other is evacuated. When the valve is opened, the air expands to fill both the vessels. Calculate the final pressure of air and change in entropy when the conditions have settled.

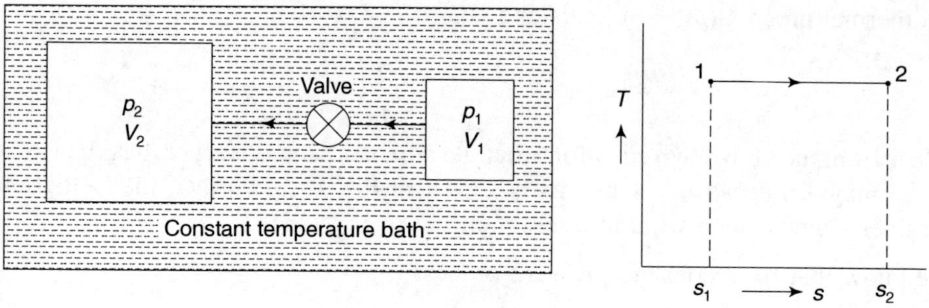

Fig. 8.24 Example 8.9

Solution Refer to Fig. 8.24. This is a throttling process, hence temperature and enthalpy remain constant.

$$\therefore \qquad p_2 = \frac{p_1 V_1}{V_2} = \frac{15 \times 1}{2} = 7.5 \text{ bar} \qquad \textbf{Ans.}$$

For constant temperature process,

$$s_2 - s_1 = R \ln \frac{v_2}{v_1} = R \ln \frac{V_2}{V_1} = 0.287 \ln 2 = 0.1989 \text{ kJ/kg} \qquad \textbf{Ans.}$$

Example 8.10 One kg of air is compressed according to the law $pV^{1.25}$ = const. from a pressure of 1.03 bar and a temperature of 15°C to a pressure of 17 bar. Calculate (a) the temperature at the end of compression, (b) the heat rejected or received by the air during the process and (c) the change in entropy. Sketch the operation on T–S diagram. Compare the results if the compression process is isothermal.

Take $\qquad c_p = 1.005$ and $c_v = 0.715$ kJ/kg K

Solution Refer to Fig. 8.25

(a) We have
$$T_2 = T_1 \left(\frac{p_2}{T_1}\right)^{(n-1)/n} = 288 \left(\frac{17}{1.03}\right)^{0.25/1.25} = 504.56 \text{ K}$$

Work done during polytropic process

$$W = \frac{p_1 V_1 - p_2 V_2}{(n-1)} = \frac{mR(T_1 - T_2)}{(n-1)} = \frac{m(c_p - c_v)}{n-1}(T_1 - T_2)$$

$$= \frac{(1.005 - 0.715)}{0.25}(288 - 504.56) = -251.2 \text{ kJ/kg}$$

The negative sign shows that work is done on the gas. The heat interaction during the process is given by

$$Q = \frac{\gamma - n}{\gamma - 1} \times \text{Polytropic work done and } \gamma = \frac{c_p}{c_v} = \frac{1.005}{0.715} = 1.4055$$

$$\therefore \qquad Q = \frac{1.4055 - 1.25}{1.4055 - 1} \times (-251.2) = -96.32 \text{ kJ/kg}$$

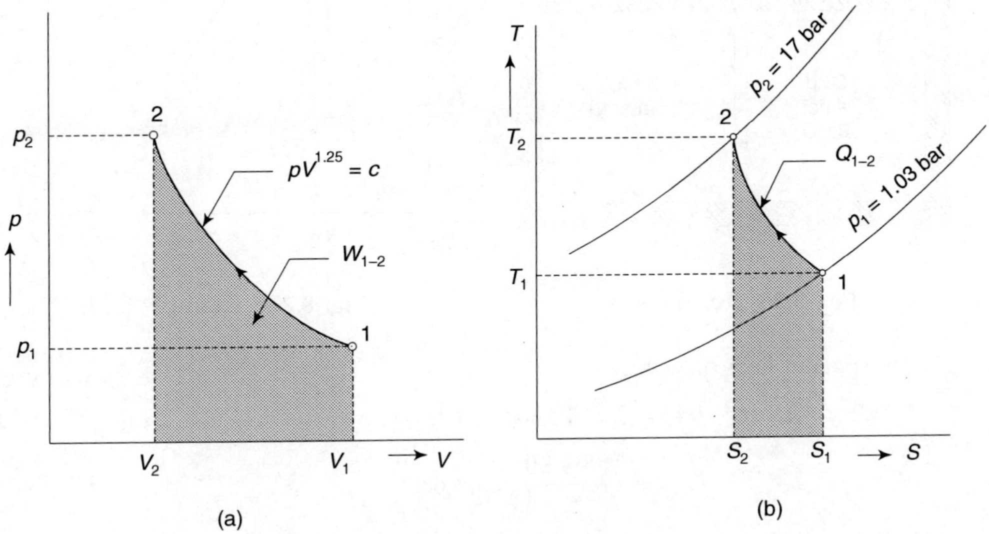

Fig. 8.25 Example 8.10

The negative sign indicates that heat will be rejected from the system. This is shown by hatched area on T–S diagram.

The entropy change during polytropic process based on per kg of fluid is given by

$$s_2 - s_1 = \frac{\gamma - n}{\gamma - 1} R \ln \frac{V_2}{V_1}. \quad \text{But } \frac{V_2}{V_1} = \left(\frac{T_1}{T_2}\right)^{1/(n-1)}$$

$$\therefore \qquad s_2 - s_1 = \frac{\gamma - n}{(\gamma - 1)(n - 1)} \times R \ln \frac{T_2}{T_2} = \frac{1.4055 - 1.25}{(1.4055 - 1)(1.25 - 1)} 0.29 \ln \frac{288}{504.56}$$

$$= -0.2494 \text{ kJ/kg K.}$$ **Ans.**

The negative sign indicates the decrease of entropy.

Example 8.11 2 kg of water at 94°C is mixed with 3 kg of water at 10°C in an isolated system. Calculate the change of entropy due to mixing process.

Solution Referring to the isolated system shown in Fig. 8.26, there is no heat or work interaction with the surroundings, hence from first law, the change of internal energy of the system is zero. Let t_f be the final temperature, then

$$m_1 \, c_p \, (t_1 - t_f) = m_2 \, c_p \, (t_f - t_2)$$

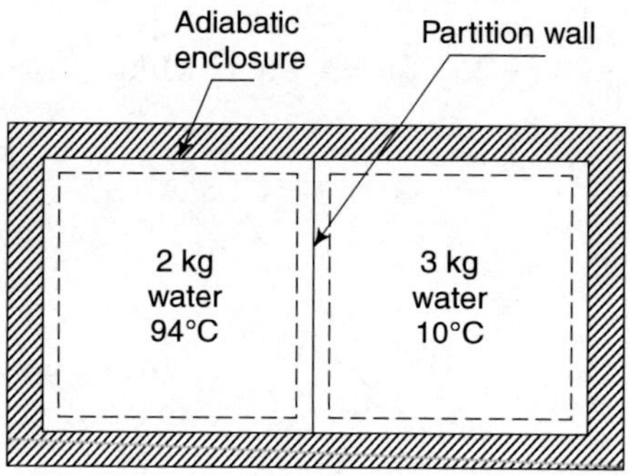

Fig. 8.26 Example 8.11

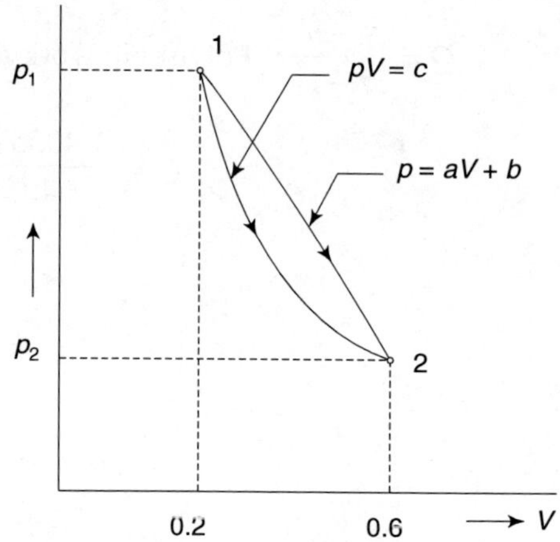

Fig. 8.27 Example 8.12

where c_p is the specific heat of water

$$\therefore \quad 2 \times 4.18\ (t_f - 94) = 3 \times 4.18\ (t_f - 10)$$

or
$$t_f = \frac{188 + 30}{5} = 43.6°\text{C}$$

The total change of the entropy of the system is the sum of the entropy changes of its components, i.e.,

$$\Delta S = \Delta S_1 + \Delta S_2 = m_1\ c_p\ \ln \frac{T_f}{T_1} + m_2\ c_p\ \ln \frac{T_f}{T_2}$$

$$= 2 \times 4.18 \times \ln \frac{316.6}{367} + 3 \times 4.18 \times \ln \frac{316.6}{283} = 0.1719 \text{ kJ/K} \quad \textbf{Ans.}$$

Example 8.12 The expansion of a perfect gas is so controlled that the pressure changes according to the law, $p = aV + b$, where a and b are constant and V is the volume. The mass of gas is 1.4 kg and the initial and final pressure are 6 bar and 2 bar. The corresponding volumes are 0.2 and 0.6 m^2. Assume $\gamma = 1.39$. $R = 0.28$ kJ/kg K. Find (a) the change in entropy per kg during the expansion, (b) the maximum value of internal energy per kg reckoned from 0°C, and (c) the net heat removed or added during the cycle.

Solution Refer to Fig. 8.27. According to question, the process follows a straight path such that

$$p = aV + b$$

Putting the values, $\quad 6 = 0.2\ a + b$ and $2 = 0.6\ a + b$

These equations give $\quad a = -10$ and $b = 8$

(a) We know that change in entropy per kg of fluid is

$$s_2 - s_1 = c_v \ln \frac{T_2}{T_1} + R \ln \frac{V_2}{V_1} \text{ kJ/kg K}$$

But the data of questions tell us $p_1 v_1 = p_2 V_2$, this means that $T_1 = T_2$ (i.e., process is isothermal).

$\therefore$ $$s_2 - s_1 = 0.28 \ln \frac{0.6}{0.2} = 0.3076 \text{ kJ/kg K}$$ **Ans.**

(b) Although $p_1 V_1 = p_2 V_2$, i.e., $T_1 = T_2$, it is not true that the temperature of the gas remains constant throughout, as such a path would be represented by a hyperbola pV = constant and not a straight line, $p = AV + b$. Thus, the temperature will vary as a function of pressure or volume where

$$p = aV + b = a\frac{mRT}{p} + b \text{ or } p^2 - bp = amRT$$

Differentiating we get $2pdp - bdp = amRdT$

T is the maximum when

$$\frac{dT}{dp} = \frac{2b - b}{amR} = 0$$

or $$p = \frac{b}{2}$$

or $$p = \frac{8}{2} = 4 \text{ bar and } V = \frac{p - b}{a} = 0.4 \text{ m}^3$$

Thus, $$T_{max} = \frac{p \times 10^3\ V}{mR} = \frac{4 \times 10^2 \times 0.4}{1.4 \times 0.28} = 408.16 \text{ K}$$

$$u_{max} = c_v (T_{max} - T_0) = \frac{R}{(\gamma - 1)}(T_{max} - T_0) \text{ kJ/kg}$$

$$= \frac{0.28\,(408.16 - 273)}{0.39} = 97.03 \text{ kJ/kg}$$ **Ans.**

(c) $$Q_{1-2} = \Delta U + W_{1-2}. \quad \text{Since } T_1 = T_2 \text{ hence } \Delta U = 0$$

$$Q_{1-2} = 0 + W_{1-2} = 0 + 4 \times 10^2 \times (0.6 - 0.2) = 160 \text{ kJ}$$ **Ans.**

It is interesting to note that the heat flowing into the gas during the first portion of the expansion (until T_{max} = 408.16 K) will be greater than 160 kJ due to the increase in internal energy but for remainder of the operation, heat is rejected again giving the net heat added to 160 kJ.

Example 8.13 Show that when a perfect gas changes from a state p_1, V_1, T to another state p_2, V_2, T_2, the increase in entropy per unit mass is given by

$$s_2 - s_1 = c_v \ln \frac{p_2}{p_1} + c_p \ln \frac{V_2}{V_1}$$

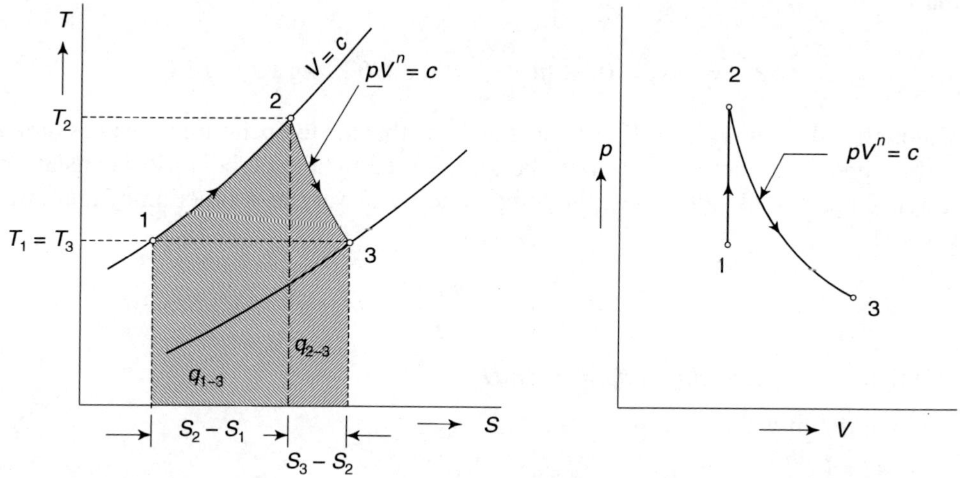

Fig. 8.28 Example 8.13

Find the value of index n so that the gain of entropy during the heating of the gas at constant volume between temperatures T_1 and T_2 will be the same as that during an expansion according to the law pV^n = constant between the same temperatures. Show also that the heat supplied per unit mass of gas will be the same in each case.

Solution Refer to Fig. 8.28.

$$\delta Q = dU + pdV \text{ or } \frac{\delta Q}{T} = \frac{dU}{T} + \frac{pdV}{T}$$

or

$$dS = mc_v \frac{dT}{T} + mR \frac{dV}{V}$$

Based on per unit mass, $s_2 - s_1 = c_v \ln \dfrac{T_2}{T_1} + (c_p - c_v) \ln \dfrac{V_2}{V_1}$

But

$$\frac{T_2}{T_1} = \frac{p_2}{p_1} \frac{V_2}{V_1}$$

$\therefore$

$$s_2 - s_1 = c_v \ln \frac{p_2}{p_1} + c_p \ln \frac{V_2}{V_1}$$

Proved

It is to be noted that $T_1 = T_3$ but $T_2 > T_1$. During constant volume process 1–2, the change in entropy per unit mass is given by

$$s_2 - s_1 = c_v \ln \frac{T_2}{T_1} \text{ kJ/kg K} \quad \text{...(a)}$$

We have $\quad pV^n = c$

$$\therefore \quad \frac{T_2}{T_3} = \left(\frac{p_2}{p_3}\right)^{(n-1)/n} \quad \text{and} \quad \frac{T_2}{T_3} = \left(\frac{V_3}{V_2}\right)^{n-1}$$

But we have during process 2–3,

$$s_3 - s_2 = c_v \left(\frac{\gamma - n}{n-1}\right) \ln \frac{T_2}{T_3} \quad \text{...(b)}$$

Equating the Eqs. (a) and (b) and using $T_3 = T_1$, we have

$$(n - 1)\, c_v = c_p - nc_v$$

or
$$n = \frac{c_p + c_v}{2\, c_v} = \frac{\gamma + 1}{2}$$

At constant volume $\quad q_{1-2} = c_v\,(T_2 - T_1)$ kJ/kg and $q_{2-3} = (u_3 - u_2) + w_{2-3}$

But
$$w_{2-3} = \frac{p_2\, v_2 - p_3\, v_3}{(n-1)} = \frac{R\,(T_2 - T_3)}{n-1}$$

$$q_{2-3} = c_v(T_3 - T_2) + (c_p - c_v)\left(\frac{T_2 - T_3}{(n-1)}\right) = (T_2 - T_3)\left[\frac{c_p - c_v}{n-1} - c_v\right]$$

But $\quad n = \dfrac{\gamma + 1}{2}$ and $T_3 = T_1$, hence

$$q_{2-3} = (T_2 - T_1)\left[\frac{2\,(c_p - c_v)}{\gamma - 1} - c_v\right] = (T_2 - T_1)\left(\frac{2\, c_p - c_v - \gamma\, c_v}{\gamma - 1}\right)$$

But $\quad c_p = \gamma c_v$

$$\therefore \quad q_{2-3} = c_v\,(T_2 - T_1)\,\frac{(\gamma - 1)}{(\gamma - 1)} = c_v\,(T_2 - T_1)$$

$$\therefore \quad q_{1-2} = q_{2-3}$$

Proved

Example 8.14 4 kg of air is compressed in a reversible steady flow polytropic process ($pV^{1.25} = c$) from 1 bar and 30°C to 10 bar. Calculate the work input, heat transferred and the change in the entropy.

Solution $\quad T_1 = T_1 \left(\dfrac{p_2}{p_1}\right)^{(n-1)/n} = (273 + 300)\,(10)^{0.25/1.25} = 480.22$ K

$$W_{sf} = -\int_1^2 Vdp = -\frac{n}{n-1} mR(T_2 - T_1) = -\frac{1.25}{0.25} \times 4 \times 0.287\,(480.22 - 303) \quad \textbf{Ans.}$$

$$= -\ 1017.24 \text{ kJ}$$

The negative sign indicates that work is done on the gas.

In the absence of potential and kinectic energy, the steady flow energy equation

$$H_1 + Q_{1-2} = H_2 + W_{1-2}$$

$$Q_{1-2} = (H_2 - H_1) + W_{1-2} = mc_p\,(T_2 - T_1) + W_{1-2}$$

$$= 4 \times 1.005\,(480.22 - 303) - 1017.24 = -\ 304.81 \text{ kJ (Rejected)} \quad \textbf{Ans.}$$

Change in entropy during polytropic process is given by

$$S_2 - S_1 = \frac{\gamma - n}{\gamma - 1} mR \ln \frac{V_2}{V_1} = \frac{\gamma - n}{(\gamma - 1)(n - 1)} mR \ln \frac{T_1}{T_2}$$

$$= \frac{1.4 - 1.25}{0.4 \times 0.25} \times 4 \times 0.287 \ln \frac{303}{480.22} = -\ 0.79245 \text{ kJ/K} \quad \textbf{Ans.}$$

This may also be solved in another way

$$S_2 - S_1 = \int_1^2 \frac{dH}{T} - \int_1^2 \frac{Vdp}{T} = mc_v \int_1^2 \frac{dT}{T} - mR \int_1^2 \frac{dp}{p}$$

$$= mc_v \ln \frac{T_2}{T_1} - mR \ln \frac{p_2}{p_1} = -\ 0.79245 \text{ kJ/K} \quad \textbf{Ans.}$$

Example 8.15 A gas undergoes a non-flow process according to the law $p = (0.15/V + 2)$ bar, where V is volume in m^3. Calculate the change in enthalpy during the process if 20 kJ of heat is rejected from the system. The initial volume is 0.8 m^3 and at the end of the process, the volume is 0.3 m^3.

Solution Enthalpy is expressed by

$$H = U + pV \quad \text{or} \quad dH = dU + Vdp$$

But from first law,

$$dH = \delta Q + Vdp \qquad ...(1)$$

According to data,

$$p = \frac{0.15}{V} + 2 \quad \text{or } dp = \frac{-\ 0.15}{V^2} dV \text{ bar}$$

Substituting the value of dp in the Eq. (1), we have

$$dH = \delta Q - 0.15 \times 10^2 \frac{dV}{V}$$

or

$$H_2 - H_1 = Q_{1-2} - 0.5 \times 10^2 \ln \frac{V_2}{V_1} = 20 - 0.15 \times 10^2 \ln \frac{0.3}{0.8}$$

$$= -20 + 14.7 = -5.3 \text{ kJ} \qquad \textbf{Ans.}$$

Negative sign indicates decrease of enthalpy.

Example 8.16 0.2 m^3 of air at 1 bar and 60°C is compressed to 0.05 m^3 according to the law $pV^{1.3} = c$. Now, the heat is added at constant volume until its pressure is 10 bar. Calculate the change in entropy in each process undergone.

Solution Refer to Fig. 8.29.

$$\text{Mass of air} = \frac{p_1 V_1}{RT_1} = \frac{1\times 10^2 \times 0.2}{0.287\times 333} = 0.2092 \text{ kg}$$

During compression process, the final pressure is

$$p_2 = p_1 \left(\frac{V_1}{V_2}\right)^n = 1\times\left(\frac{0.2}{0.05}\right)^{1.3} = 6.062 \text{ bar}$$

$$T_2 = T_1 \left(\frac{p_2}{p_1}\right)^{(n-1)/n} = 333\left(\frac{6.062}{1}\right)^{0.3/1.3} = 504.71 \text{ K}$$

Change of entropy during polytropic process

$$S_2 - S_1 = m\left(\frac{\gamma - n}{\gamma - 1}\right) R \ln\frac{V_2}{V_1} = 0.2092\left(\frac{1.4-1.3}{1.4-1}\right)\times 0.287 \ln\frac{0.05}{0.2} = -0.0208 \text{ kJ} \qquad \textbf{Ans.}$$

Entropy change during constant volume heat addition process is given by

$$S_3 - S_2 = mc_v \ln\frac{p_3}{p_2} = m\frac{R}{(\gamma-1)}\ln\frac{p_3}{p_2} = 0.2092\times\frac{0.287}{0.4}\ln\frac{10}{6.062} = 0.0751 \text{ kJ/K} \qquad \textbf{Ans.}$$

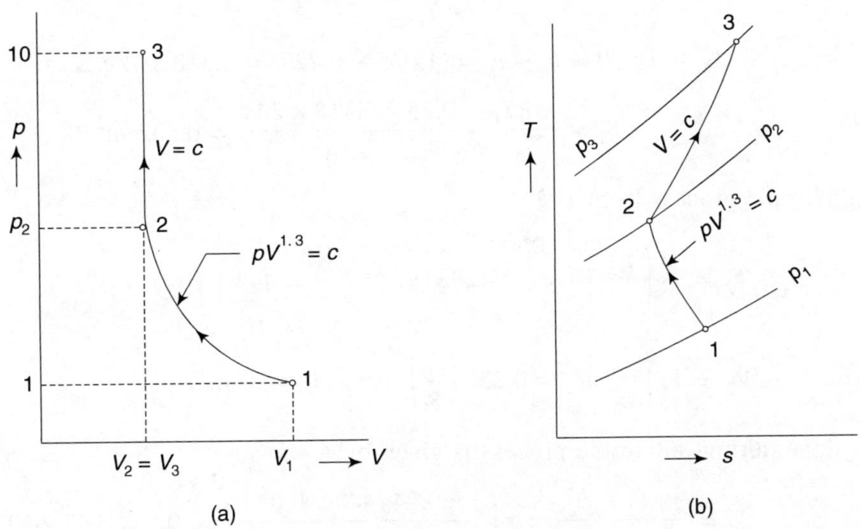

Fig. 8.29 Example 8.16

Example 8.17 (a) Show that the approximate change of entropy during a polytropic process equals the quantity of heat transferred divided by the mean absolute temperature.

(b) 0.25 kg of air at pressure 1 bar and temperature 20°C is compressed to a pressure of 8 bar according to the law $pV^{1.25}$ = constant. Calculate the following:

(i) Initial and final conditions of the gas, (ii) Work interaction, (iii) Heat interaction, (iv) Change of entropy, (v) Approximate change of entropy, (vi) Percentage error if the entropy is calculated by the approximate method.

Assume $c_p = 1.045$ and $c_v = 0.727$ kJ/kg K

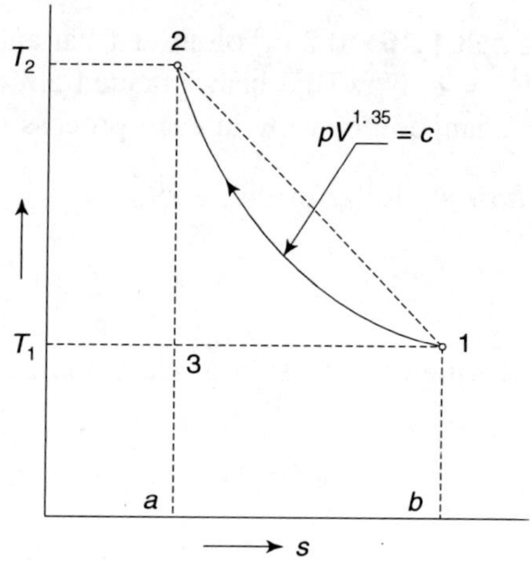

Fig. 8.30 Example 8.17

Solution Refer to Fig. 8.30. Process 1–2 shows the polytropic process from 1 to 2. The area under curve 1–2 shows the heat supplied during the process. The curve 1–2 has got small curvature so for approximation we consider this as a straight line.

(a) Heat absorbed = Triangular area 123 + Rectangular area *a* 31 *b*

or
$$Q_{app} = \frac{1}{2}(S_2 - S_1)(T_2 - T_1) + (S_2 - S_1)\,T_1$$

$$= (S_2 - S_1)\left[\frac{T_1 + T_2}{2}\right] = (S_2 - S_1) \times T_{mean}$$

or
$$= S_2 - S_1 = \frac{Q_{app}}{T_{mean}}$$ **Proved**

(b)
$$R = c_p - c_v = (1.045 - 0.727) = 0.318 \text{ kJ/kg K}$$

(i)
$$V_1 = \frac{mRT_1}{p_1} = \frac{0.25 \times 0.318 \times 293}{1 \times 10^2} = 0.232 \text{ m}^3$$ **Ans.**

During polytropic process from 1–2

$$T_2 = T_1\left(\frac{p_2}{p_1}\right)^{(n-1)/n} = 293\,(8)^{0.25/1.25} = 444.1 \text{ K}$$ **Ans.**

$$V_2 = V_1\left(\frac{p_1}{p_2}\right)^{1/n} = 0.232\left(\frac{1}{8}\right)^{1/1.25} = 0.0439 \text{ m}^3$$ **Ans.**

(ii) Work done during polytropic process is given by

$$= \frac{p_1 V_1 - p_2 V_2}{n-1} = \left(\frac{1 \times 0.242 - 8 \times 0.045}{0.25}\right) \times 10^2 = -47.2 \text{ kJ}$$ **Ans.**

The –ve sign shows the work done on the air.

(iii) $$\text{Heat transferred} = \frac{\gamma - n}{\gamma - 1} \times \text{work done}$$

But $$\gamma = \frac{c_p}{c_v} = \frac{1.045}{0.727} = 1.437$$

$$\therefore \quad Q = \frac{1.437 - 1.25}{1.437 - 1} \times (-47.2) = -20.19 \text{ kJ}$$

The –ve sign shows that heat transfers from air

(iv) $$S_2 - S_1 = m_{cv} \frac{\gamma - n}{n} \ln \frac{p_1}{p_2} = 0.25 \times 0.727 \times \frac{1.437 - 1.25}{1.25} \ln \frac{1}{8} = -0.0565 \text{ kJ}$$ **Ans.**

(v) $$T_m = \frac{293 + 444.1}{2} = 268.55 \text{ K}$$

$$\text{Approximate change in entropy} = \frac{Q_{\text{app}}}{T_m} = -\frac{20.19}{371} = -3.544 \text{ per cent}$$ **Ans.**

(vi) $$\text{Percentage error} = \frac{0.0564 - 0.0544}{0.0564} = 3.546 \text{ per cent}$$

Example 8.18 (a) Deduce an expression for the change in entropy of 1 kg of gas in terms of the temperature T_1 and T_2 before and after compression, if the law of compression is pV^n = constant and the specific heats are of the form

$$c_p = a + KT \quad \text{and} \quad c_v = b + KT$$

(b) The compression ratio of an engine is 9 to 1, and the initial pressure and temperature are 1 bar and 120°C and the volume is 860 cm^3. At the end of compression, the temperature is 560°C. The values of the constants for the specific heats are a = 0.948, b = 0.672, K = 0.0001 and T is the absolute temperature. Find:

(i) the mass of charge undergoing compression, the final pressure and the index of compression, (ii) the change of entropy per kg of charge, and (iii) the change in internal energy per kg of charge.

Solution (a) For per unit mass, first law for a non-flow process is

$$\delta q = du + dw = c_v \, dT + pdv$$

or $$\left(\frac{\delta q}{T}\right)_{\text{rev}} = c_v \frac{dT}{T} + \frac{pdv}{T}$$

or $$ds = c_v \frac{dT}{T} + \frac{pdv}{T} \quad \text{...(1)}$$

We have $$R = c_p - c_v = (a + KT) - (b + KT) = a - b$$

Also $$\frac{p}{T} = \frac{R}{v} = \frac{a-b}{v}$$

Putting the value $\frac{p}{T}$ in Eq. (1), we have

$$ds = (b + KT)\frac{dT}{T} + (a-b)\frac{dv}{v} \qquad ...(2)$$

Integrating this equation, we get

$$s_2 - s_1 = b \ln \frac{T_2}{T_1} + K(T_2 - T_1) + (a-b) \ln \frac{v_2}{v_1} \qquad ...(3)$$

But from gas equation, $$\frac{v_2}{v_1} = \frac{T_2}{T_1}\frac{p_1}{p_2}$$

or $$\ln \frac{v_2}{v_1} = \ln \frac{T_2}{T_1} + \ln \frac{p_1}{p_2} = \ln \frac{T_2}{T_1} + \ln \left(\frac{v_2}{v_1}\right)^n = \ln \frac{T_2}{T_1} + n \ln \frac{v_2}{v_1}$$

or $$\ln \frac{v_2}{v_1} = \frac{\ln \frac{T_2}{T_1}}{1-n} \qquad ...(4)$$

With the help of Eqs. (2) and (3), we have

$$s_2 - s_1 = K(T_2 - T_1) + b \ln \frac{T_2}{T_1} - \frac{a-b}{n-1} \ln \frac{T_2}{T_1}$$

or $$s_2 - s_1 = K(T_2 - T_1) + \left(b - \frac{a-b}{n-1}\right) \ln \frac{T_2}{T_1}$$ **Ans.**

(b) $$c_p = a + KT = 0.948 + 0.0001\,T \text{ and } c_v = b + KT = 0.672 + 0.0001\,T$$

$$R = c_p - c_v = 0.948 - 0.672 = 0.276 \text{ kJ/kg K}$$

Now, from the gas equation

$$m = \frac{p_1 V_1}{RT_1} = \frac{1 \times 10^2 \times 860 \times 10^{-6}}{0.276 \times 293} = 0.001063 \text{ kg}$$ **Ans.**

Also $$p_2 = \frac{p_1 V_1 T_2}{T_1 V_2} = \frac{1 \times 10^2 \times 833 \times 9}{393} = 19.07 \times 10^2 \text{ kN/m}^2 = 19.07 \text{ bar}$$ **Ans.**

$$n = -\frac{\ln \frac{p_2}{p_1}}{\ln \frac{V_1}{V_2}} = \frac{\ln 19.07}{\ln 9} = 1.341$$ **Ans.**

(c) $$s_2 - s_1 = K\,(T_2 - T_1) + \left(b - \frac{a-b}{n-1}\right)\ln\frac{T_2}{T_1}$$

$$= 0.0001\,(833 - 393) + \left(0.672 - \frac{0.948 - 0.672}{1.341 - 1}\right)\ln\frac{833}{393} = -\,0.0591 \text{ kJ/kg K} \quad \textbf{Ans.}$$

The –ve sign indicates the decrease in entropy.

Example 8.19 1 kg of hydrogen having molecular weight 2.016 undergoes free expansion process from a volume of 1 m^3 to 4 m^3. Find the change in entropy of the gas and the surroundings. What will be the change in entropy if the process of expansion between two states is brought by reversible isothermal expansion process? Compare the net change in the entropy of the universe.

Solution $$R_{H_2} = \frac{8314}{2.016} = 4124 \text{ J/kg K} = 4.124 \text{ kJ/kg K}$$

In the free expansion process, temperature remains constant, hence the process may be considered as isothermal. Thus, the change in entropy of the system (gas) is given by

$$S_2 - S_1 = m\,R\ln\frac{V_2}{V_1} = 1 \times 4.124 \ln 4 = 5.717 \text{ kJ/kg}$$

In free expansion process, there is no interchange of heat with the surroundings, hence entropy of the surroundings remains unchanged.

$$\therefore \qquad (S_2 - S_1)_{surr} = 0$$

Entropy of the universe is

$$(\Delta S)_{univ} = (S_2 - S_1)_{system} + (\Delta S)_{surr} = 5.717 + 0 = 5.717 \text{ kJ/kg} \quad \textbf{Ans.}$$

If the process is isothermal, then there is an exchange of heat between the gas and the surroundings. The gas gains heat while and equal amount of heat is lost by the surroundings.

$$(\Delta S)_{univ} = (S_2 - S_1)_{sys} + (S_2 - S_1)_{surr} = 5.717 - 5.717 = 0 \quad \textbf{Ans.}$$

Example 8.20 Calculate the change in entropy when nitrogen gas is heated in a steady flow process from 5 bar and 27°C to 6 bar and 917°C.

$$\text{Assume } c_p = \left[39.3 - 7.64 \times \frac{10^3}{T} + \frac{1.5 \times 10^6}{T^2}\right] \text{kJ/kg mole K}$$

and $$R = 0.3 \text{ kJ/kg K},$$

Molecular weight of nitrogen = 28.016

Solution We have, $$s_2 - s_1 = \int_{T_1}^{T_2} c_p \frac{dT}{T} - R\ln\frac{p_2}{p_1}$$

Substituting the values by taking care of unit, we have

$$s_2 - s_1 = \frac{1}{28.016}\left[\int_{300}^{1190}\left(39.3 - \frac{7.64\times10^3}{T} + \frac{1.5\times10^6}{T^2}\right)\frac{dT}{T}\right] - 0.3\ \ln\left(\frac{6}{5}\right)$$

$$= \frac{1}{28.016}\left[39.3\ \ln T + \frac{7.64\times10^3}{T} - \frac{1.5\times10^6}{2T^2}\right]_{300}^{1190} - 0.3\ \ln\left(\frac{6}{5}\right)$$

$$= \frac{1}{28.016}\left[39.3\ \ln\frac{1190}{300} + 7640\left(\frac{1}{1190} - \frac{1}{300}\right) - 75\times10^4\left(\frac{1}{1190^2} - \frac{1}{300^2}\right)\right] - 0.3\ \ln\frac{6}{3}$$

$= 1.212$ kJ/kg K **Ans.**

Example 8.21 A thermodynamic system has a heat capacity at constant volume expressed as $c_v = AT^2$ where $A = 0.043$ J/K^3. The system is originally at 300 K. A low temperature thermal reservoir at 200K is available so that a heat engine can be operated. Calculate the maximum amount of work that can be obtained as the system is cooled down to the temperature of the reservoir.

Solution Refer to Fig. 8.31.

A heat engine may receive heat from the system and may reject it to the low temperature thermal reservoir. Heat removed from the system is given by

$$Q_1 = \int c\ \ dT = \int_{300}^{200} 0.043\,T\ \ dT = 0.043\left[-\right]_{300}^{200} = -\ 272.33 \times 10^3\ \text{J}$$

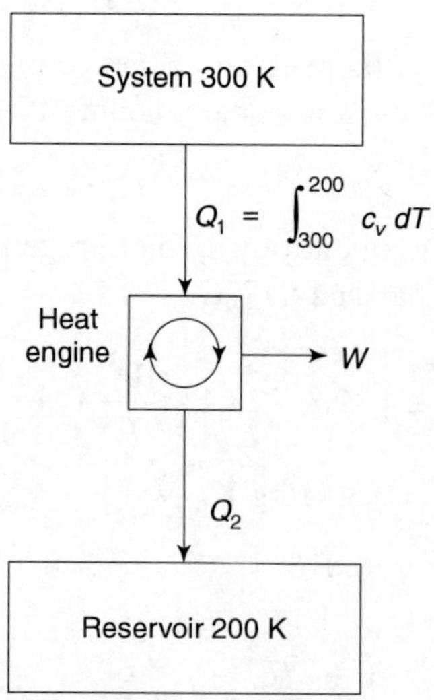

Fig. 8.31 Example 8.21

The change in the entropy of the system,

$$(\Delta S)_{\text{system}} = \int_{300}^{200} c_v \frac{dT}{T} = \int_{300}^{200} \frac{0.043\, T^2\, dT}{T} = \frac{0.043}{2}[200^2 - 300^2] = -\,1075 \text{ J/K}$$

$$(\Delta S)_{\text{reser}} = \frac{Q_1 - W}{T_{\text{reser}}} = \frac{272.33 \times 10^3 - W}{200}$$

We have, $(\Delta S)_{HE} = 0$, hence

$$(\Delta S)_{\text{univ}} = (\Delta S)_{\text{system}} + (\Delta S)_{\text{reser}} = -\,1075 + \frac{272.33 \times 10^3 - W}{200}$$

Since $(\Delta S)_{\text{univ}} \geq 0$

or
$$-\,1075 + \frac{272.33 \times 10^3 - W}{200} > 0$$

or
$$-\,1075 + 1316.65 - \frac{W}{200} \geq 0$$

or
$$\frac{W}{200} \leq 286.65$$

For the maximum work done,

$$W_{\max} = 286.65 \times 200 = 57330 \text{ J}$$ **Ans.**

Example 8.22 Three identical finite bodies *A*, *B*, and *C* of constant heat capacity are at temperature 400, 200 and 400 K, respectively. A heat engine is operated between *A* and *B* and a heat pump working as refrigerator is operated between *B* and *C*. The heat pump is operated by the output of heat engine. If no work or heat supplied from outside, find the highest temperature to which any one of the bodies can be raised by the operation of heat engine or refrigerator.

Solution A heat engine and a refrigerator are operated between three temperature limits 400, 200 and 400 K as shown in Fig. 8.32.

Let T_f be final temperature of bodies *A* and *B* and T'_f be the final temperature of body *C*.

Since $(\Delta S)_{\text{univ}} \geq 0$, hence

$$(\Delta S)_A + (\Delta S)_B + (\Delta S)_C + (\Delta S)_{HE} + (\Delta S)_{\text{Ref}} \geq 0$$

As $(\Delta S)_{HE} = 0$ and $(\Delta S)_{\text{Ref}} = 0$, hence

$$(\Delta S)_A + (\Delta S)_B + (\Delta S)_C \geq 0$$

Let C be the heat capacity of finite system, hence

$$\left(C \ln \frac{T_f}{400} + C \ln \frac{T_f}{200} + C \ln \frac{T'_f}{400} \right) \geq 0$$

or
$$C \ln \frac{T'_f \,.\, T'_f}{32 \times 10^6} \geq 0$$

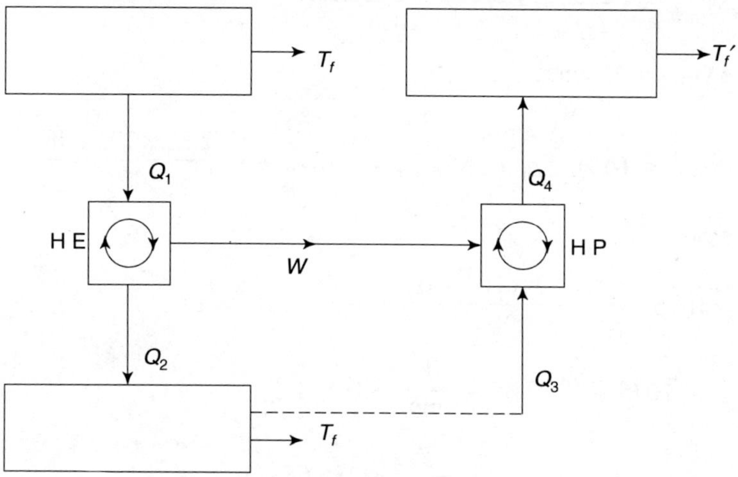

Fig. 8.32 Example 8.22

For the minimum value of T_f,

$$C \ln \frac{T_f^2 \,.\, T'_f}{32 \times 10^6} = 0 = \ln 1$$

or
$$T_f^2 \,.\, T'_f = 32 \times 10^6 \qquad \text{...(a)}$$

The above equation involves two unkowns. When T_f is minimum, T'_f will be maximum.

Further,
$$Q_1 = Q_2 + Q_4$$

or
$$C\,(400 - T_f) = C\,(T_f - 200) + C\,(T'_f - 400)$$

or
$$T'_f = 1000 - 2T_f \qquad \text{...(b)}$$

From Eqs. (a) and (b), we get

$$T_f^2\,(1000 - 2T_f) = 32 \times 10^6$$

or
$$T_f \approx 250 \text{ K by trial and error method}$$

∴
$$T'_f = 100 - 2 \times 250 = 500 \text{ K}$$

Example 8.23 A closed system has been designed to execute a reversible cycle 1–2–3–4–5–6–1 consisting of six processes. During processes 1–2 and 3–4 the system receives 1000 kJ and 800 kJ of heat respectively at constant temperature of 500 K and 400 K respectively. Processes 2–3 and 4–5 are isentropic expansions in which the system temperature gets reduced from 500 K to 400 K and from 400 K to 300 K respectively. During process 5–6, the system rejects heat at constant temperature of 300 K. Process 6–1 is an isentropic compression process. Calculate the work done by the system during the cycle and the thermal efficiency of the cycle.

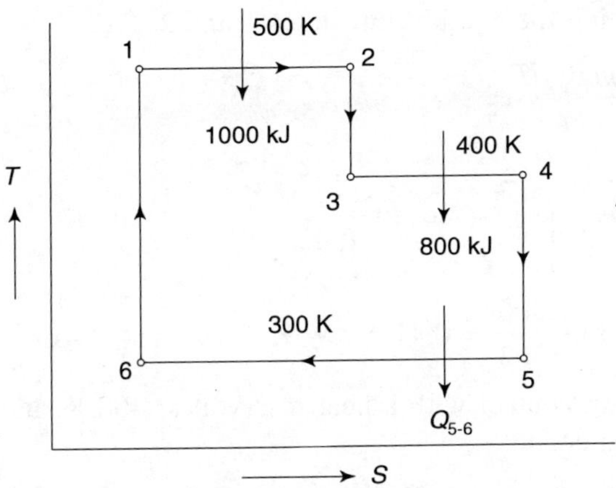

Fig. 8.33 Example 8.23

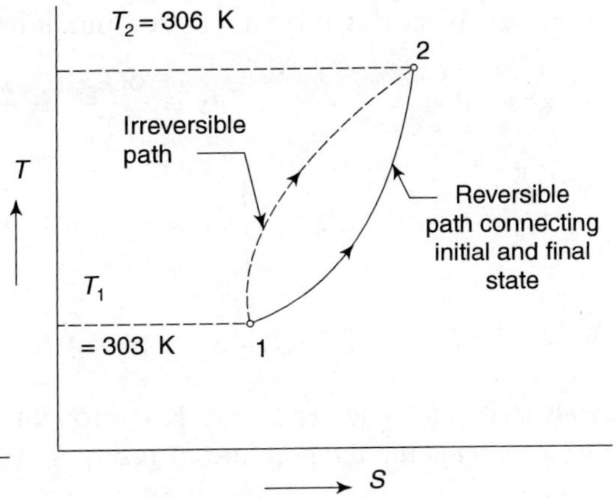

Fig. 8.34 Example 8.24

Solution Refer to Fig. 8.33

$$Q_{5-6} = T_{5-6} \cdot (\Delta S)_{5-6}$$

From process 5–6, we have

$$(\Delta S)_{5-6} = -\left[(\Delta S)_{1-2} + (\Delta S)_{3-4}\right]$$

or

$$(\Delta S)_{5-6} = -\left[\frac{Q_{1-2}}{T_{1-2}} + \frac{Q_{3-4}}{T_{3-4}}\right] = -\left[\frac{1000}{500} + \frac{800}{400}\right] = -4 \text{ kJ/K}$$

$$Q_{5-6} = 300 \times (-4) = -1200 \text{ kJ}$$

For cyclic process,

$$\oint \delta Q = \oint \delta W$$

$$W_{\text{cycle}} = \sum_{\text{cycle}} Q = Q_{1-2} + Q_{3-4} - Q_{5-6}$$

or

$$W_{\text{cycle}} = 1000 + 800 - 1200 = 600 \text{ kJ}$$ **Ans.**

Thermal efficiency is given by

$$\eta_{\text{th}} = \frac{W_{\text{cycle}}}{Q_{\text{added}}} = \frac{W_{\text{cycle}}}{Q_{1-2} + Q_{3-4}} = \frac{600}{1800} = 33.3\%$$ **Ans.**

Example 8.24 In a hydraulic turbine, water flows through blade passages in which friction occurs which causes the water temperature to rise from 30°C to 33°C. If there is no heat transfer, calculate the change of entropy of water passing through turbine. Assume water as incompressible fluid and the process occurs at constant volume. Assume 10 kg/s of flow flowing through turbine.

Solution Refer to Fig. 8.34. The presence of friction makes the process irreversible. The flow process is indicated by the dotted line joining the equilibrium states 1 and 2.

We know that
$$dS = \frac{\delta Q_{rev}}{T} = \frac{m\, c_v\, dT}{T}$$

or
$$\int_1^2 dS = S_2 - S_1 = mc_v \int_1^2 \frac{dT}{T} = mc_v \ln \frac{T_2}{T_1}$$

or
$$S_2 - S_1 = 10 \times 4.18 \times \ln \frac{306}{303} = 0.4118 \text{ kW/K} \qquad \textbf{Ans.}$$

Example 8.25 5 kg water at 288 K is brought in contact with a heat reservoir at 400 K. If the water attains the temperature of reservoir, calculate:

(a) The entropy change of water, of the heat reservoir, and of the universe.

(b) If the water is heated in two stages first from 288 K to 340 K by first bringing it in contact with a reservoir at 350 K and then with a reservoir at 400 K, calculate the change of entropy of the universe.

(c) Suggest the method of heating from 255 K to 400 K so that there is no change in the entropy of the universe.

Solution Refer to Fig. 8.35. The heating process is assumed to take place at constant volume.

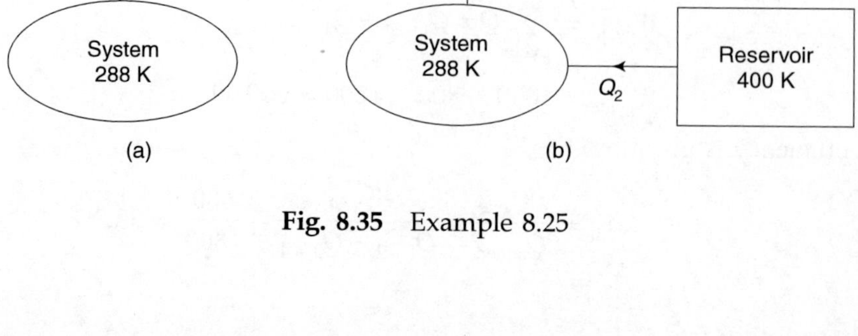

Fig. 8.35 Example 8.25

or $$(\Delta S)_{system} = m \cdot c_v \ \ln \frac{T_2}{T_1} = 5 \times 4.18 \times \ln \frac{400}{288} = 6.865 \text{ kJ/K}$$ **Ans.**

It is assumed here that reservoir temperature remains constant irrespective of amount of heat withdrawn from it.

The amount of heat absorbed by the system from the reservoir

$$= Q = mc \ \Delta T = 5 \times 4.18 \times (400 - 288) = 2344.8 \text{ kJ}$$

Entropy change of the reservoir, $(\Delta S)_{res} = \frac{-Q}{T} = -\frac{2344.8}{400} = -5.861$ kJ/K

or $$(\Delta S)_{univ} = 6.865 - 5.861 = 1.004 \text{ kJ/K}$$ **Ans.**

Entropy change of universe, $(\Delta S)_{univ} = (\Delta S)_{system} + (\Delta S)_{reser}$

or $$(\Delta S)_{univ} = 6.865 - 5.861 = 1.004 \text{ kJ/K}$$ **Ans.**

(b) In this case, water is heated in two stages, first from 288 K to 340 K and then from 340 K to 400 K, thus

$$(\Delta S)_{water} = (\Delta S)_{system} = \int_{288 \text{ K}}^{340 \text{ K}} \frac{m\,c \cdot dT}{T} + \int_{340 \text{ K}}^{400 \text{ K}} m \cdot c \cdot \frac{dT}{T}$$

$$= 5 \times 4.187 \left[\ln \frac{340}{288} + \ln \frac{400}{340} \right] = 6.877 \text{ kJ/K}$$

$$(\Delta S)_{res\ I} = -\frac{5 \times 4.18\,(340 - 288)}{340} = -\frac{-1086.8}{340} = -3.1964 \text{ kJ/kg}$$

$$(\Delta S)_{res\ II} = \frac{-5 \times 4.18\,(400 - 350)}{400} = -\frac{1045}{400} = -2.613 \text{ kJ/kg}$$

$$(\Delta S)_{univ} = (\Delta S)_{sys} + (\Delta S)_{res\ I} + (\Delta S)_{res\ II} = 6.877 - 3.1964 - 2.6125$$

$$= 1.0681 \text{ kJ/K}$$ **Ans.**

From the above result, it is concluded that the entropy of the universe decreases with two-stage heating as compared to single-stage heating.

(c) *The change of entropy of the universe decreases with the increase of number of small stage heating. If the number of heating stage is infinite, obviously, there will no change of entropy and the heating process may be treated as isentropic.*

Example 8.26 During a steady state operation, a gearbox receives 70 kW through the input shaft and delivers power through the output shaft. Taking gear as system, the heat energy transfer by convection is $Q = \hbar A_s\,(T_b - T_a)$ where $\hbar$ = heat transfer coefficient = 0.171 kW/m^2. K; A_s = outer surface area = 1.0 m^2, T_b = temperature of the outer surface 310 K and T_a = temperature of the surrounding air = 300K. Calculate the energy transferred from output shaft and the rate of entropy production for

(a) the gearbox as the system

(b) the enlarged system consisting of the gearbox and enough of its surroundings that heat transfer occurs at the temperature the surrounding away from immediate vicinity of gearbox where T_a = 300 K.

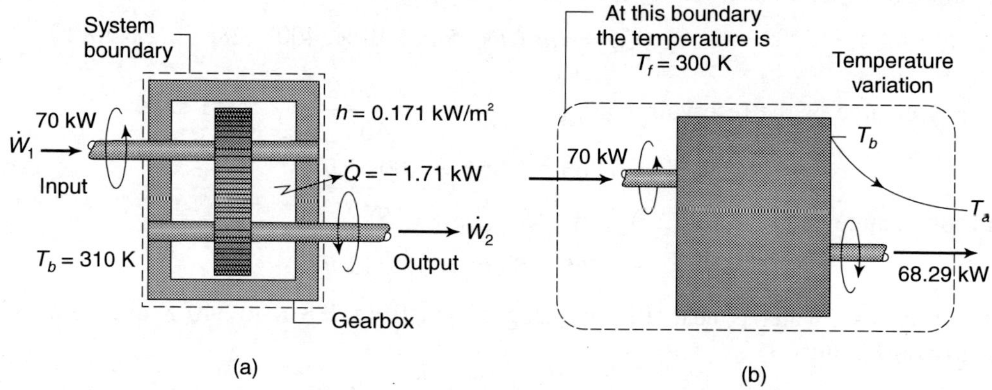

Fig. 8.36 Example 8.26

A Model Solution Refer to Fig. 8.36.

Data given : $\dot{W}_{\text{input}} = \dot{W}_1 = 70$ kW, $\hbar = 0.171$ kW/m² K, $A_s = 1$ m², $T_b = 310$ K

$T_a = 300$ K

To find : Rate of entropy generation ($\dot{S}_{\text{gen}}$) and $\dot{W}_2 = \dot{W}_{\text{output}}$

Assumptions : (i) In part (a) the gearbox is taken as the closed system operating at steady state, (ii) In part (b), the gear box and a portion of its surrounding are taken as closed system, (iii) The temperatures of surroundings are each uniform.

Analysis : The heat transfer from the gearbox taking sign convention,

$$\dot{Q} = \hbar A_s (T_b - T_a) = -0.171 \times 1 (310 - 300) = -1.71 \text{ kW}$$

From the energy rate balance,

$$\frac{dE}{d\tau} = \dot{Q} - \dot{W} = 0 \quad \text{or} \quad \dot{Q} = \dot{W} = \dot{W}_1 + \dot{W}_2$$

$$\dot{W}_2 = \dot{Q} - \dot{W}_1 = -1.71 - (-70) = 68.29 \text{ kW}$$ **Ans.**

The entropy rate balance yields

$$\frac{dS}{d\tau} = \frac{\dot{Q}}{T_b} + \dot{S}_{\text{gen}} = 0$$

(a) $$\dot{S}_{gen} = -\frac{\dot{Q}}{T_b} = -\frac{(-1.71)}{310} = 5.516 \times 10^{-3} \text{ kW/K}$$

(b) Since the heat transfer takes place at temperature T_a for the enlarged system, the entropy rate balance yields.

$$\frac{dS}{d\tau} = \frac{\dot{Q}}{T_a} + \dot{S}_{gen} = 0$$

or $$\dot{S}_{gen} = \frac{\dot{Q}}{T_a} = -\frac{(-1.71)}{300} = 5.7 \times 10^{-3} \text{ kW/K}$$ **Ans.**

Example 8.27 A 0.5 kg metal bar initially at 2000°C is removed from an oven and quenched by immersing it in a closed tank containing 10 kg of water initially at 600°C. Each substance can be modelled as incompressible. An approximate constant specific heat for the water is 4.18 kJ/kg K and for the metal is 0.5 kJ/kg K. Heat transfer from the tank is neglected calculate (a) the final equilibrium temperature of the metal bar and water, and (b) the amount of entropy produced.

A Model Solution Refer to Fig. 8.37.

To find T_a and S_{gen}

Data given $m_m = 0.5$ kg, $T_{mi} = 2273$ K, $m_w = 10$ kg

$T_{wi} = 873$ K, $c_m = 0.5$ kJ/kg K, $c_w = 4.18$ kJ/kg K.

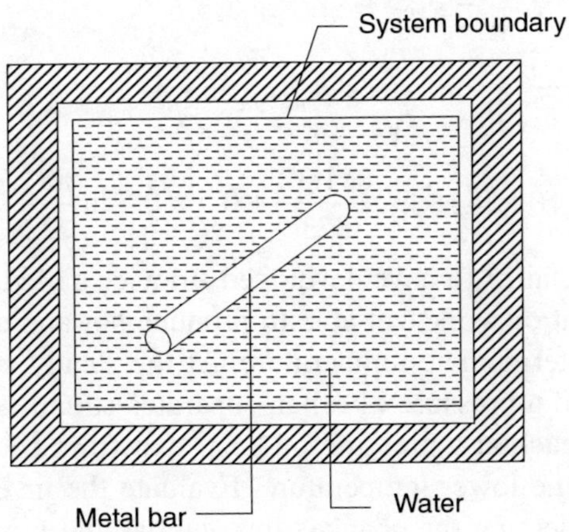

Fig. 8.37 Example 8.27

Assumptions (i) The metal bar and water within the tank form a closed system as shown in Fig. 8.37.

(ii) There is no energy transfer by heat or work. The system is isolated.

(iii) There is no change in KE and PE

(iv) The water and metal bar are each modeled as incompressible.

Analysis The energy balance yields

$$Q - W = \Delta U + \Delta PE + \Delta KE$$

Since $Q = W = 0$, $\Delta PE = \Delta KE = 0$. So, $\Delta U = 0$

or $(\Delta U)_{metal} + (\Delta U)_{water} = 0$

$$m_m\, c_m\, (T_f - T_{mi}) + m_w\, c_w\, (T_f - T_{wi}) = 0$$

(a)

$$T_f = \frac{m_w\,(c_w/c_m)\,T_{wi} + m_m T_{mi}}{m_w\,(c_w/c_m) + m_m}$$

$$= \frac{10\,(4.18/0.5)\,873 + 0.5 \times 2273}{10\,(4.18/0.5) + 0.5} = 881.32 \text{ K} \qquad \textbf{Ans.}$$

(b) The entropy balance for the system yields

$$\Delta S = \int_1^2 \left(\frac{\delta Q}{T}\right)_b + S_{gen}$$

But $\left(\dfrac{\delta Q}{T}\right)_b = 0$ as system is isolated, so

So, $(\Delta S)_{water} + (\Delta S)_{metal} = S_{gen}$

But

$$S_{gen} = m_w\, c_w \ln \frac{T_f}{T_{wi}} + m_m\, c_m \ln \frac{T_f}{T_{mi}}$$

$$= 10 \times 4.18 \ln \frac{881.32}{873} + 0.5 \times 0.5 \ln \frac{881.32}{2273} = 0.1596 \text{ kJ/K} \qquad \textbf{Ans.}$$

Example 8.28 An inventor claims to have developed a device called vortex tube requiring no energy transfer either by heat or work, yet able to produce hot and cold stream of air from a single stream of air at an intermediate temperature. At the steady state condition, air enters at a temperature of 20°C and a pressure of 6 bar, separates stream at temperature of – 20°C and 80°C respectively, and each at a pressure of 1 bar exit from each end. 60% of the mass entering the device exits at the lower temperature. Evaluate the inventor's claim, employing the ideal gas model and neglecting the changes in potential and kinetic energies.

A Model Solution Refer to Fig. 8.38.

Data given $T_1 = 20°\text{C}$, $p_1 = 6$ bar, $T_3 = -20°\text{C}$, $p_3 = 6$ bar, $T_2 = 80°\text{C}$,

$p_2 = 1$ bar, $\dot{m}_3 = 0.6\ \dot{m}_1$

To find Validate the claim.

Assumptions (i) The system is an open system and the control volume is shown in Fig. 8.38.

(ii) $$\dot{W}_{cv} = 0 \quad \text{and} \quad \dot{Q}_{cv} = 0$$

(iii) $$\Delta PE = 0, \quad \Delta KE = 0$$

(iv) The air is modeled as ideal gas with constant c_p = 1.005 kJ/kg K

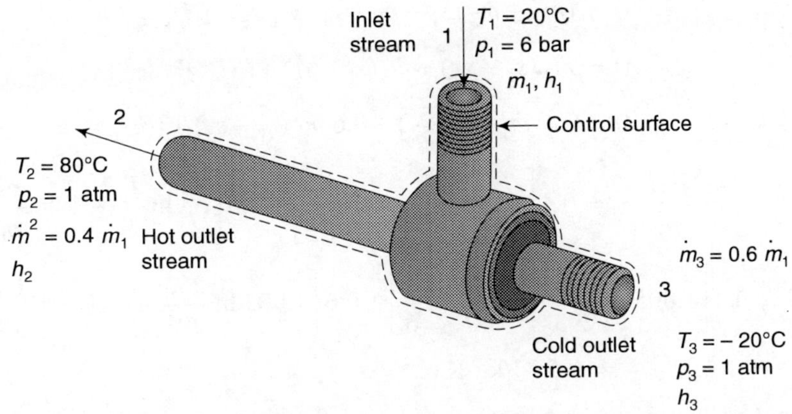

Fig. 8.38 Example 8.28

Analysis In order to evaluate the claim, the conservation mass and energy principles must satisfy the second law of thermodynamics, and in particular, the rate of entropy generation cannot be negative.

$$\text{Mass balance : } \dot{m}_1 = \dot{m}_2 + \dot{m}_3 \qquad ...(1)$$

$$\text{Energy balance : } \dot{m}_1 h_1 - \dot{m}_2 h_2 - \dot{m}_3 h_3 = 0 \qquad ...(2)$$

As per data, $\dot{m}_3 = 0.6\ \dot{m}_1$, so by combining mass and energy balance.

We have

$$(\dot{m}_2 + \dot{m}_3) h_1 - \dot{m}_2 h_2 - \dot{m}_3 h_3 = 0$$

or $$\dot{m}_2(h_1 - h_2) + \dot{m}_3(h_1 - h_3) = 0$$

or $$0.4\ m_1\ [c_{[}\ (T - T_2) + 0.6\ m_1\ c_p\ (T_1 - T_3)] = 0$$

or $$0.4\ (20 - 80) + 0.6\ (20 - (-20)) = 0$$

or $$0 = 0$$

Thus, the energy and mass balance are satisfied, i.e., first law is satisfied.

The entropy rate balance at steady state is expressed as

$$0 = \sum_j \frac{\dot{Q}_j}{T_j} + \dot{m}_1 s_1 - \dot{m}_2 s_2 - \dot{m}_3 s_3 + (\dot{S}_{gen})_{cv} \quad ...(3)$$

Since there is no heat tansfer, so $\dot{Q}_j = 0$. Thus,

$$0 = \dot{m}_1 s_1 - \dot{m}_2 s_2 - \dot{m}_3 s_3 + (\dot{S}_{gen})_{cv}$$

or

$$0 = (m_2 + \dot{m}_3) s_1 - \dot{m}_2 s_2 - \dot{m}_3 s_3 + (\dot{S}_{gen})_{cv}$$

or

$$0 = \dot{m}_2 (s_1 - s_2) + \dot{m}_3 (s_1 - s_3) + (\dot{S}_{gen})_{cv}$$

or

$$0 = 0.4\ \dot{m}_1 (s_1 - s_2) + 0.6\ \dot{m}_1 (s_1 - s_3) + (\dot{S}_{geb})_{cv}$$

or

$$\frac{(\dot{S}_{gen})_{cv}}{\dot{m}_1} = 0.4\left[c_p \ln\frac{T_2}{T_1} - R\ln\frac{p_2}{p_1}\right] + 0.6\left[c_p \ln\frac{T_3}{T_1} - R\ln\frac{p_3}{p_1}\right]$$

or

$$\frac{(\dot{S}_{gen})_{cv}}{\dot{m}_1} = 0.4\left[1.005\ln\frac{353}{293} - 0.287\ln\frac{1}{6}\right] + 0.6\left[1.005\ln\frac{253}{293} - 0.287\ln\frac{1}{6}\right] \quad \textbf{Ans.}$$

$$= 0.5 \text{ kJ/kg K}$$

Thus, the second law of thermodynamics is also satisfied based on this evaluation, the inventor's claim does not violate principles of thermodynamics.

Example 8.29 10 kg of ice at – 7°C is exposed to the atmosphere which is at 27°C. The ice melts and comes into thermal equilibrium with the atmosphere.

(a) Calculate the increase of entropy of the universe.

(b) Also calculate the minimum amount of work necessary to convert the water back to ice at –7°C. Assume c_p of ice as 2.093 kJ/kg K and latent heat of fusion of ice as 333.3 kJ/kg.

Solution Refer to Fig. 8.39.

(a) Heat absorbed by ice from atmosphere is given by

Q = Heat absorbed in solid phase + Latent heat of ice + Heat absorbed in liquid phase to attain equilibrium with atmospheric temperature

or

$$Q = m.c_{ice}\ [0 - T_{ice}) + m_{ice}\ h_{fg} + m_{iw}\ c_{pw}\ (T_a - 0)$$

$$= 10 \times 2.093\ [0 - (-7)] + 10 \times 333.3 + 10 \times 4.18\ (27 - 0)$$

$$= 4608.11 \text{ kJ}$$

Entropy change of the atmosphere

$$(\Delta S)_{atm} = \frac{-Q}{T} = -\frac{4608.11}{300} = -15.36 \text{ kJ/K}$$

Entropy change of ice (system) as it gets heated from – 7°C to 0°C

$$(\Delta S_{\text{I}})_{\text{sys}} = \int_{266}^{273} m\, c_p \frac{dT}{T} = 10 \times 2.093 \ln \frac{273}{266} = 0.5436 \text{ kJ/K}$$

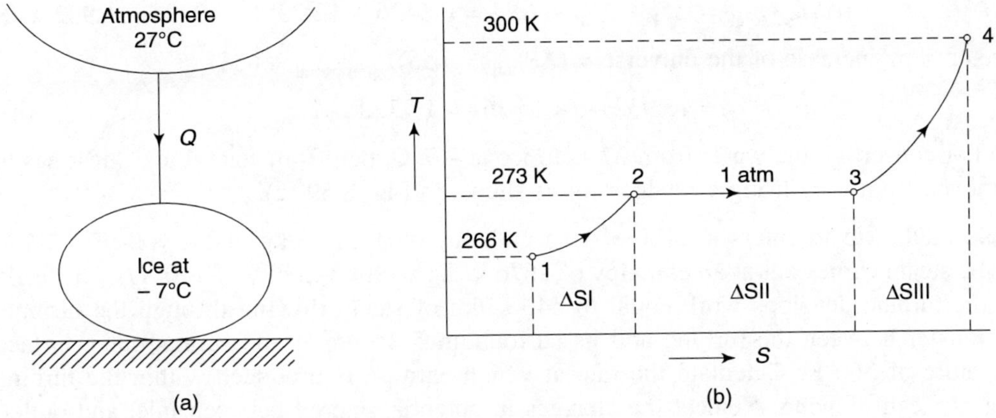

Fig. 8.39 (a, b) Example 8.29

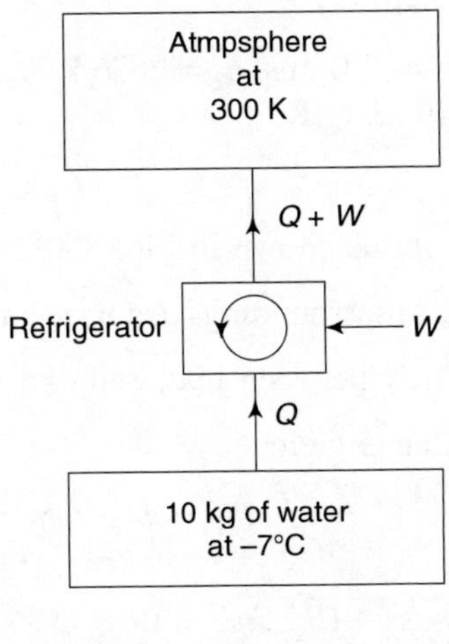

Fig. 8.39 (c) Example 8.29

Entropy change of system as it melts at 0°C to become water, i.e., entropy change during melting

$$(\Delta S_{\text{II}})_{\text{sys}} = \frac{Q_{\text{fusion}}}{T_a} = \frac{m \cdot h_{fg}}{T_a} = \frac{10 \times 333.3}{273} = 12.208 \text{ kJ/K}$$

Entropy change of water as it gets heated from 0°C to 27°C.

$$(\Delta S_{III})_{sys} = \int_{273}^{300} m_w \, c_{pw} \ln \frac{dT}{T} = 10 \times 4.18 \ln \frac{300}{273} = 3.942 \text{ kJ/K}$$

Thus, total change in entropy as ice melts into water and attains equilibrium with atmosphere.

$$(\Delta S)_{sys,\ total} = \Delta S_I + \Delta S_{II} + \Delta S_{III} = 0.5436 + 12.208 + 3.942 = 16.932 \text{ kJ/K}$$

The entropy increase of the universe = $(\Delta S)_{univ} = (\Delta S)_{system,\ total} + (\Delta S)_{atm}$

$$= 16.932 + (-15.36) = 1.572 \text{ kJ/K}$$ **Ans.**

(b) For converting the water from 27°C to ice at – 7°C, i.e., from state 4 to 1, heat has to be removed from it by using a refrigerator as shown in Fig. 8.39 (c).

Example 8.30 Steam enters a turbine at an enthalpy of 3231 kJ/kg and a velocity of 170 m/s. The steam comes out at an enthalpy of 2676 kJ/kg with a velocity of 110 m/s. At steady state, the turbine develops work equal to 540 kJ/kg of steam flowing through the turbine. Heat transfer between the turbine and its surroundings occurs at an average outer surface temperature of 360 K. Calculate the rate at which entropy is produced within the turbine per kg of steam flowing. Neglect the changes in potential energy between inlet and outlet. The entropy of steam at inlet and exit of turbine are 6.9212 kJ/kg K and 7.3549 kJ/kg K respectively.

A Model Solution Refer to Fig. 8.40 (a).

Data given h_1 = 3231 kJ/kg, C_1 = 170 mis, h_2 = 2676 kJ/kg, C_2 = 110 m/s, w = 540 kJ/kg, T_{bs} = 360 K. s_1 = 6.9212, s_2 = 7.3549 kJ/kg K.

To find S_{gen}

Assumptions (i) The control volume shown in Fig. 8.40 is in steady state.

(ii) Heat transfer from the turbine to surroundings occurs at an specified average temperature.

(iii) The change in potential energy between inlet and exit is neglected.

Analysis Mass and entropy balance yield

$$\dot{m}_1 = \dot{m}_2 = \dot{m}$$

$$\sum_j \frac{\dot{Q}_j}{T_j} + \dot{m}_1 \cdot s_1 - \dot{m}_2 \, s_2 + (\dot{S}_{gen})_{cv} = 0$$

Since the heat transfer occurs only at T_{bs} = 369 K the term $\sum_j \dot{Q}_j / T_j$ reduces to $\dot{Q}_{cv}/T_{bs}$.

Thus,

$$\frac{\dot{Q}}{T_{bs}} + \dot{m}\,(s_1 - s_2) + (\dot{S}_{gen})_{cv} = 0$$

or $$\frac{(\dot{S}_{gen})cv}{\dot{m}} = -\frac{\dot{Q}_{cv}/\dot{m}}{T_{bs}} + (s_2 - s_1) \qquad ...(i)$$

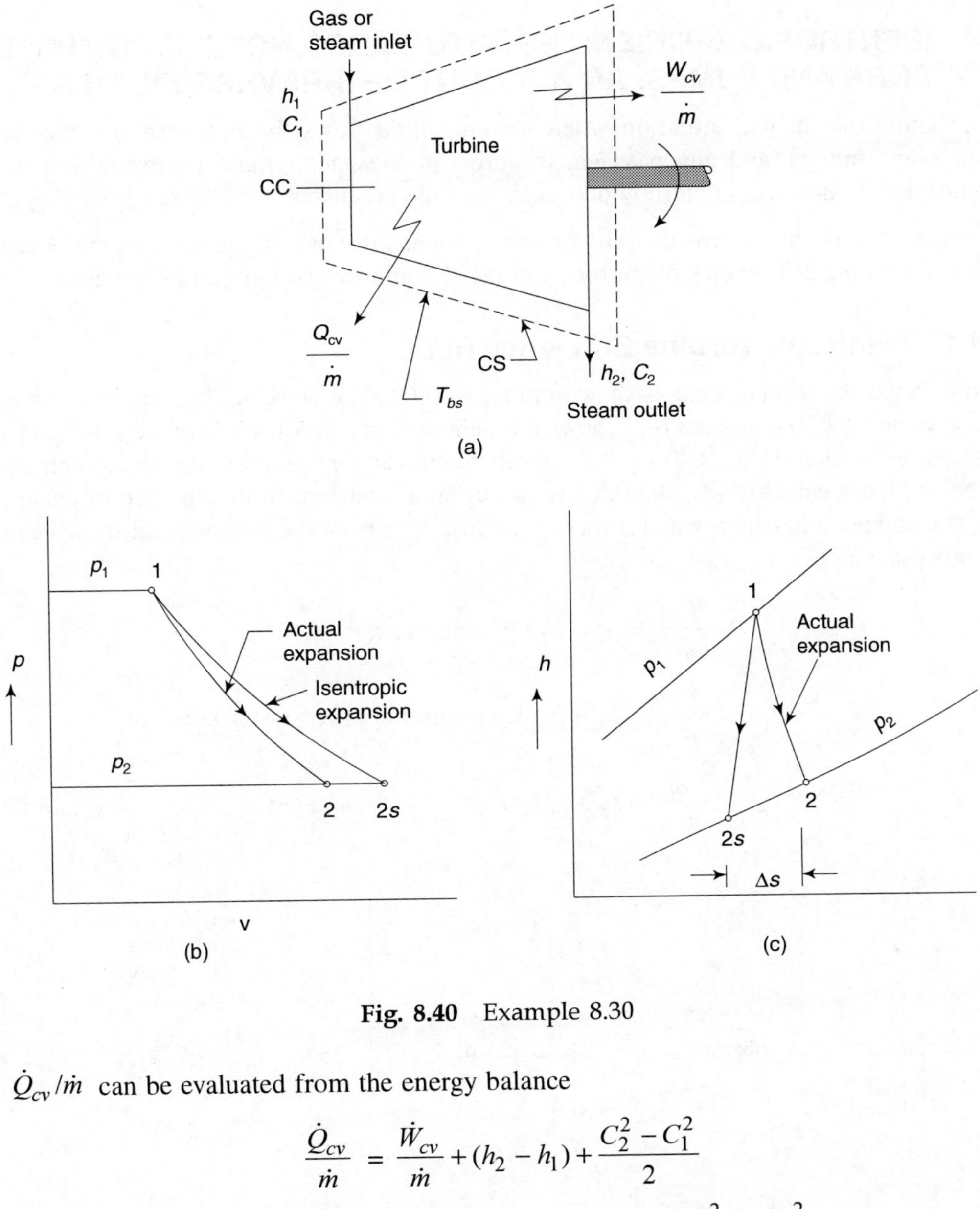

Fig. 8.40 Example 8.30

$\dot{Q}_{cv}/\dot{m}$ can be evaluated from the energy balance

$$\frac{\dot{Q}_{cv}}{\dot{m}} = \frac{\dot{W}_{cv}}{\dot{m}} + (h_2 - h_1) + \frac{C_2^2 - C_1^2}{2}$$

$$\frac{\dot{Q}_{cv}}{\dot{m}} = 540 + (2676 - 3231) + \frac{110^2 - 170^2}{2 \times 10^3} = -23.4 \text{ kJ/kg}$$

Substituting the value of $\dot{Q}_{cv}/\dot{m}$ in Eq. (i), we have

$$\frac{(\dot{S}_{gen})_{cv}}{\dot{m}} = -\frac{(-23.4)}{360} + (7.3549 - 6.9212) = 0.4987 \text{ kJ/kg K} \quad \textbf{Ans.}$$

8.14 ISENTROPIC EFFICIENCIES OF TURBINES, NOZZLE, COMPRESSORS AND PUMPS: AN ACCOUNT OF IRREVERSIBILITIES

As we know that in real situation when working fluid flows through turbines, nozzles, compressors, pumps, and any passage, the process does not remain isentropic but it is accompanied by increase in entropy due to irreversible processes.

In order to account the irreversibility, engineers generally use efficiencies. In the present section, isentropic efficiencies for turbines, nozzles, compressors and pumps are used.

8.14.1 Isentropic Turbine Efficiency (η_t)

Consider a gas or steam turbine as shown in Fig. 8.40 (a). The working fluid enters at state 1 and comes out at 2. The process l–2*s* shows the isentropic expansion while process 1–2 shows the actual expansion. (Fig. 8.40 c). As a result of irreversible process, the actual enthalpy will be as higher enthalpy (h_2) than due to isentropic expansion. In the absence of changes in kinetic and potential energy and if there is no heat loss from the turbine, then from steady flow process

$$\frac{\dot{W}_{cv}}{\dot{m}} = h_1 - h_2 \text{ for actual flow}$$

and

$$\frac{\dot{W}_{cv}}{\dot{m}} = h_1 - h_{2s} \text{ for isentropic process } (s = \text{constant})$$

Here

$$\frac{\dot{S}_{gen}}{\dot{m}} = s_2 - s_1 \geq 0 \qquad \text{...(8.82)}$$

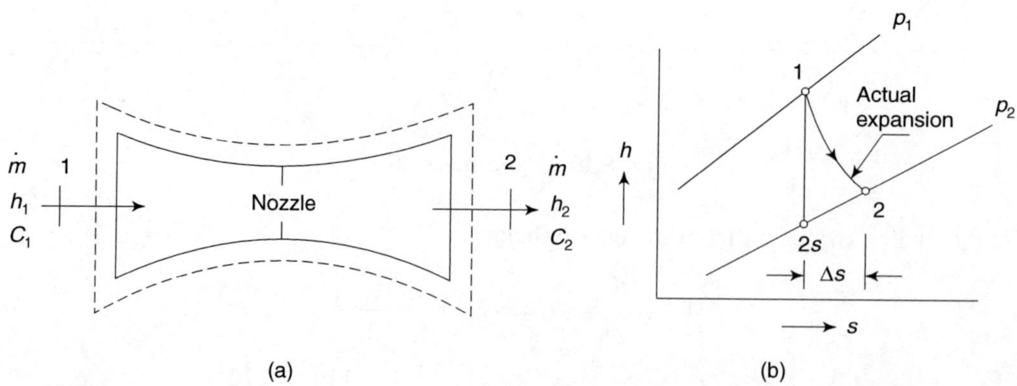

Fig. 8.41 Comparison of actual and isentropic expansion through turbine

The isentropic turbine efficiency (η_t) is defined as the ratio of actual turbine work to that due to isentropic expansion. Thus,

$$\eta_t = \frac{(\dot{W}_{cv}/\dot{m})}{(\dot{W}_{cv}/\dot{m})_s} = \frac{h_1 - h_2}{h_1 - h_{2s}} \qquad \text{...(8.83)}$$

In other words, isentropic turbine efficiency is defined as the ratio of actual enthalpy drop to the isentropic enthalpy drop.

If η_t is known, the actual work done may be calculated. Alternatively, if actual work done is known, η_t may be evaluated.

8.14.2 Isentropic Nozzle Efficiency (η_n)

Consider a nozzle through which a working fluid gas or steam flows. The isentropic expansion is represented by the process 1–2*s* while actual expansion process is 1–2 (Fig. 8.41). If there is no heat exchange from the nozzle and no changes in potential energy, and $W = 0$, the steady flow equation results in

$$h_1 + \frac{C_1^2}{2} = h_{2s} + \frac{C_{2s}^2}{2} \quad \text{for isentropic process}$$

or

$$\frac{C_{2s}^2 - C_1^2}{2} = h_1 - h_2$$

and

$$h_1 + \frac{C_1^2}{2} = h_2 + \frac{C_2^2}{2} \quad \text{for actual process}$$

or

$$\frac{C_2^2 - C_1^2}{2} = h_1 - h_{2s}$$

The isentropic nozzle efficiency (η_n) is defined as the ratio of actual gain in kinetic energy to isentropic gain in kinetic energy. Thus,

$$\eta_n = \frac{C_2^2 - C_1^2}{C_{2s}^2 - C_1^2} = \frac{h_1 - h_2}{h_1 - h_{2s}} \qquad \text{...(8.84)}$$

If η_n is known, actual velocity at the exit may be known.

8.14.3 Isentropic Compressor and Pump Efficiencies

The flow through a compressor is shown in Fig. 8.42. The isentropic compression is represented by 1–2*s* while actual compression is given by 1–2. If there is no change in KE and PE, and no heat exchange, SFEE equation yields.

$$\left(\frac{-\dot{W}_{cv}}{\dot{m}}\right)_s = h_{2s} - h_1 \quad \text{for isentropic compression}$$

and $$\left(\frac{-\dot{W}_{cv}}{\dot{m}}\right)_s = h_2 - h_1 \text{ for actual compression}$$

The isentropic compression efficiency (η_c) *is defined as the ratio of isentropic work input to actual work input.* Thus,

$$\eta_c = \frac{(-\dot{W}_{cv}/\dot{m})_s}{(-\dot{W}_{cv}/\dot{m})} = \frac{h_{2s} - h_1}{h_2 - h_1} \quad ...(8.85)$$

In other words, *isentropic compression efficiency is defined as the ratio of isentropic enthalpy rise to the actual enthalpy rise.*

If η_c is known, actual work input may be calculated.

Similarly, isentropic pump efficiency may be defined.

Example 8.31 A gas turbine operates at steady state with inlet condition at 15 bar and 1500 K and gas leaves the turbine at 1 bar. If the isentropic efficiency of the turbine is 85% calculate the work developed per unit mass. Neglect changes in KE and PE. Assume c_p of gas = 1.005 kJ/kg K. If mass flow rate of gas is 100 kg/s, calculate power developed, in MW.

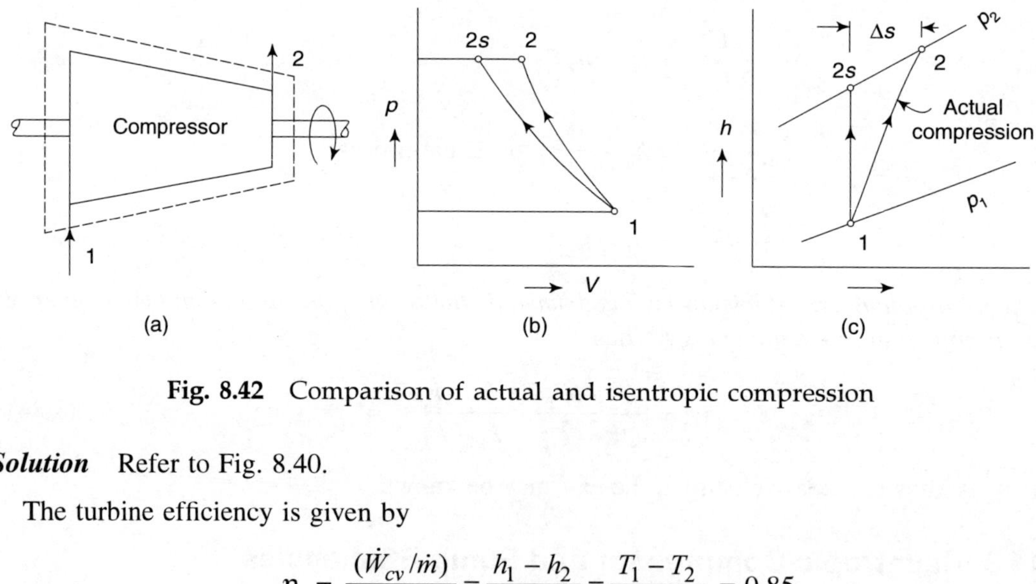

Fig. 8.42 Comparison of actual and isentropic compression

Solution Refer to Fig. 8.40.

The turbine efficiency is given by

$$\eta_t = \frac{(\dot{W}_{cv}/\dot{m})}{(\dot{W}_{cv}/\dot{m})_s} = \frac{h_1 - h_2}{h_1 - h_{2s}} = \frac{T_1 - T_2}{T_1 - T_{2s}} = 0.85$$

We have $$T_{2s} = T_1\left(\frac{p_2}{p_1}\right)^{(\gamma-1)/\gamma} = 1500\left(\frac{1}{15}\right)^{0.4/1.4} = 691.93 \text{ K}$$

$$T_2 = T_1 - \eta_t (T_1 - T_{2s}) = 1500 - 0.85 (1500 - 691.93) = 813.14 \text{ K}$$

$$\dot{W}_{cv}/\dot{m} = h_1 - h_2 = c_p\,(T_1 - T_2) = 1.005\,(1500 - 813.14) = 690.29 \text{ kJ/kg} \quad \textbf{Ans.}$$

Power developed = $\dot{W}_{cv}$ = 690.29 × 100 × 10^{-3} = 690.29 MW **Ans.**

Example 8.32 An air compressor compresses air from 1 bar 288 K to 15 bar with an isentropic efficiency of 84 per cant. Calculate the work input per unit mass. Neglect changes in KE and PE. Assume c_{pa} = 1.005 kJ/kg K. If mass flow rate of air is 100 kg/s calculate the power input, in MW.

Solution Refer to Fig. 8.42.

Isentropic efficiency of compressor is given by

$$\eta_c = \frac{-(\dot{W}_{cv}/\dot{m})_s}{-(\dot{W}_{cv}/m)} = \frac{h_{2s} - h_1}{h_2 - h_1} = \frac{T_{2s} - T_1}{T_2 - T_1} = 0.84$$

$$T_{2s} = T_1\left(\frac{p_2}{p_1}\right)^{(\gamma-1)/\gamma} = 288\,(15)^{0.4/1.4} = 624.33 \text{ K}$$

$$T_2 = T_1 + \frac{T_{2s} - T_1}{\eta_c} = 288 + \frac{624.33 - 288}{0.84} = 688.39 \text{ K}$$

$$-\dot{W}_{cv}/\dot{m} = h_2 - h_1 = c_p\,(T_2 - T_1) = 1.005\,(688.39 - 288)$$

$$= 402.39 \text{ kJ/kg} \quad \textbf{Ans.}$$

$$-\dot{W}_{cv} = 402.39 \times 100 = 40239 \text{ kW} = 402.39 \text{ MW} \quad \textbf{Ans.}$$

Example 8.33 Gas enters a nozzle at 15 bar and 1500 K with a velocity of 30 m/s. The pressure at the exit of nozzle is 5 bar. If the nozzle efficiency is 90%, calculate the actual exit velocity. Neglect changes in PE and heat exchange between nozzle and surroundings. Take c_p = 1.005 kJ/kg K.

Solution Refer to Fig. 8.41.

Nozzle efficiency is expressed as

$$\eta_n = \frac{C_2^2 - C_1^2}{C_{2s}^2 - C_1^2} = \frac{h_1 - h_2}{h_1 - h_{2s}} = \frac{T_1 - T_2}{T_1 - T_{2s}} = 0.9$$

$$T_{2s} = T_1\left(\frac{p_2}{p_1}\right)^{(\gamma-1)/\gamma} = 1500\left(\frac{5}{15}\right)^{0.4/1.4} = 1095.89 \text{ K}$$

$\therefore$

$$T_2 = T_1 - (T_1 - T_{2s}) \times \eta_n = 1500 - (1500 - 1095.89) \times 0.9$$

$$= 1136.3 \text{ K}$$

From SFEE,

$$h_1 = \frac{C_1^2}{2} = h_2 + \frac{C_2^2}{2}$$

Thus, $$C_2 = \sqrt{2\,(h_1 - h_2) + C_1^2} = \sqrt{2\,c_p\,(T_1 - T_2) + C_1^2}$$

$$= \sqrt{2 \times 1.005 \times 10^3\,(1500 - 1136.3) + 30^2} = 855.53 \text{ m/s}$$ **Ans.**

8.15 THIRD LAW OF THERMODYNAMICS

At extremely low temperatures, all substances become solid, and as absolute zero temperature is approached, the degree of atomic or molecular activity decreases. This reveals the possibility that motion ceases at absolute zero. Although, some energy would be required for the atoms to be held in a crystalline lattice, each atom would then be in the same state. Consequently, probability would be equal to 1 and the *entropy as defined by statistical thermodynamics, S = K ln W, would be equal to zero. If variations in geometrical configurations are possible, W may be greater than 1 and S greater than 0.* This phenomenon leads to another law known as third law of thermodynamics.

This third law of thermodynamics states that:

The entropy of any pure substance in thermoaynamic equilibrium approaches zero as the absolute temperature approaches zero. When expressed mathematically,

$$\lim_{T \to 0} S = 0 \qquad \text{...(8.86)}$$

The third law was developed in the early twentieth century primarily through the efforts of Nernst, and consequently it is often called the Nernst theorem. The third law enables us to calculate the absolute entropy of a substance from the definition of change in entropy.

$$S - S_0 = \int_0^T \frac{\delta Q_{rev}}{T}$$

and since a state exists at the absolute zero where $S_0 = 0$, hence

$$S = \int_0^T \frac{\Delta Q_{rev}}{T} \qquad \text{...(8.87)}$$

The **term "zero entropy" must be understood carefully**. A state of zero entropy is one for which one single system quantum state is always observed. *It is to be noted that a temperature of absolute zero is unattainable measurements and cannot be made to prove that the entropy value is zero at zero degree absolute.* It can merely be assumed on the basis of extrapolation, that this is true.

It is again interesting to note that the third law does not limit the substance to any given crystalline form. The substance may, at the lowest temperature measurement, be in any of the several possible crystalline form and each one to the different crystalline forms can conform to the requirements of the third law.

In general, any substance obtainable in pure crystalline from at equilibrium is likely to conform to the requirement of third law. Hence, the absolute value of entropy can be calculated. *Non-conformity with the third law occurs because the substance is not in a quantum state at 0 K, due to factors such as (i) the presence of two, or more isotopes, (ii) randomness in the structure of the crystal, (iii) non-equilibrium distribution of molecules among the quantum states of energy, (iv) presence of different molecules in a solution or in a mixture, and (v) existence of subcooled state rather than crystalline from.*

The importance of third law is now clear. It is useful for calculating thermodynamic properties and in analysing chemical and phase equilibria. *It is also helpful in explaining the behaviour of solids at very low temperature.*

8.16 SUMMARY OF FORMULAE

1. $\oint\left(\frac{\delta Q}{T}\right) \leq 0$: Clausius inequality

2. $\left(\frac{\delta Q}{T}\right)_{\text{rev}} = dS$: Entropy definition

3. $(\Delta S)_{\text{isen}} = 0$, i.e., S = constant

4. $(\Delta S)_{\text{univ}} = (\Delta S)_{\text{sys}} + (\Delta S)_{\text{surr}} = (S_{\text{gen}})_{\text{isolated}} \geq 0$

5. $S_2 - S_1 = \int_1^2 \left(\frac{\delta Q}{T}\right) + S_{\text{gen}}$

6. $(Q_{1-2})_{\text{rev}} = \int_1^2 TdS$

7. $\frac{dS_{nf}}{d\tau} = \frac{1}{T}\frac{\delta Q}{\delta\tau} + \frac{\delta S_{\text{gen}}}{\delta\tau}$: Rate equation for closed system

8. $\frac{dS_{nf}}{d\tau} = \sum_j \frac{1}{T_j} Q_j + \dot{S}_{\text{gen}}$: Rate equation for more than one source

9. $\sum_i \dot{m}_i - \sum_e \dot{m}_e = \frac{dm_{cv}}{d\tau}$: Continuity equation

10. $\frac{dS_{cv}}{d\tau} = \sum_j \frac{\dot{Q}_j}{T_J} + \sum_i \dot{m}_i\, s_i - \sum \dot{m}_e\, s_e + (\dot{S}_{\text{gen}})_{cv}$: Entropy balance for c.v.

11. $0 = \sum_j \frac{\dot{Q}_j}{T_j} + \dot{m}\,(s_1 - s_2) + \frac{(S_{\text{gen}})}{\dot{m}}$: One inlet and outlet, steady state

12. $S_2 - S_1 = m\, c_v \ln \frac{T_2}{T_1} + mR \ln \frac{V_2}{V_1}$ Valid for any process

13. $S_2 - S_1 = mc_p \ln \dfrac{T_2}{T_1} - mR \ln \dfrac{p_2}{p_1}$ Valid for any process

14. $S_2 - S_1 = \dfrac{\gamma - n}{\gamma - 1} mR \ln \dfrac{V_2}{V_1}$ for polytropic process

15. $\delta Q = dE + \delta W$ and $\delta Q = dU + \delta W$

16. $TdS = dU + pdV$

17. $Tds = dH - Vdp$

18. $S = k \ln W$ where W is probability

PROBLEMS

Things to Think About

8.1. Among mass, energy and entropy, which are conserved?

8.2. Suppose you have to explain entropy production to a child, how will you explain it?

8.3. Think a process of a closed system for which the entropy of both the system and its surroundings increase.

8.4. Is it possible for the entropy of both a closed system and its surroundings to decrease during a process?

8.5. Discuss the transfer of entropy into or out of, a closed system.

8.6. How will you calculate the entropy production in the sun?

8.7. How will you calculate the entropy production in a nuclear reactor?

8.8. How will you calculate the entropy production during a storm?

8.9. All states of an adiabatic and internally reversible process of a closed system have the same entropy, but is a process between two states having same entropy necessarily adiabatic and internally reversible?

8.10. It is a well-known fact that reduction of irreversibilities within a system can be improved by its thermodynamic performance. But steps taken in this direction are usually constrained by other considerations. Discuss some of these.

Short and Long Theoretical Questions

8.11. Define Clausius inequality and prove it.

8.12. Define entropy and show that it is a property of the system.

8.13. Give a physical explanation of entropy.

8.14. Why is the Carnot cycle on T–s plot a rectangle?

8.15. What do you understand by entropy principle?

8.16. Show that the entropy of an isolated system increases in all real processes and is conserved in the reversible process.

8.17. Why is the entropy increase of an isolated system a measure of the extent of irreversibility of the process undergone by the system?

8.18. State the summary given by Rudolf Clausius about first and second laws of thermodynamics.

8.19. Show that the transfer of heat through a finite temperature difference is irreversible.

8.20. Show that the adiabatic mixing of two fluids is irreversible.

8.21. What causes an increase in entropy?

8.22. Why is the second law called a directional law of nature?

8.23. Derive the expression for entropy generation (production) in a closed system.

8.24. Prove that the maximum work obtainable from a heat engine operating between two finite bodies at different temperatures is given by

$$W_{max} = m\, c_p \left[\sqrt{T_1} - \sqrt{T_2}\right]^2$$

8.25. Prove that the maximum work obtainable from a heat engine operating between a finite body and a thermal reservoir at dead state is given by

$$W_{max} = m\, c_p \left[(T - T_0) - T_o \ln \frac{T}{T_0}\right]$$

8.26. Prove that the minimum work input to a heat pump operating between two identical finite bodies having same mass and specific heat is given by

$$W_{min} = m\, c_p \left[\frac{T_1^2}{T_2} - 2\,T_1 + T_2\right]$$

8.27. Derive the expression for entropy generation in an open system (control volume).

8.28. What do you mean by absolute value of entropy?

8.29. Define and explain third law of thermodynamics.

8.30. Answer the following True and False. If the answer is False, explain why?

(a) A process that violates the second law of thermodynamics violates the first law of thermodynamics.

(b) The change in entropy of a closed system must be greater than zero or equal to zero.

(c) A closed system can experience an increase in entropy only when irreversibilities are present within the system during the process.

(d) Entropy is produced in every internally reversible process of a closed system.

(e) The entropy remains constant in an adiabatic and internally reversible process.

(f) When a net amount of work is done on a closed system undergoing an internally reversible process, a net heat transfer of energy from the system also takes place.

(g) The energy of an isolated system must remain constant, but the entropy can only decrease.

8.31. By mixing two equal masses of the same liquid initially at the temperature T_1 and T_2 an isolated system of total mass m is formed. Eventually, the system attains an equilibrium state. Each mass is incompressible with constant specific heat, c.

(a) Show that the entropy produced is given by

$$S_{gen} = mc\ln\left[\frac{T_1 + T_2}{2(T_1T_2)^{1/2}}\right]$$

(b) Prove that S_{gen} must be positive.

8.32. A system undergoing a thermodynamic cycle receives Q_H at temperature T'_H and discharges Q_c at temperature T'_c. There are no other heat transfers. Show that the net work developed per cycle is given by

$$W_{cycle} = Q_H\left(1 - \frac{T'_c}{T'_H}\right) - T'_c \,.\, S_{gen}$$

where S_{gen} is the amount of entropy produced per cycle due to irreversibilities within the system.

8.33. A closed system undergoes a process in which work is done on the system and the heat transfer Q occurs only at temperature T_b. For each care, determine whether the entropy change of the system is positive, negative, zero or indeterminate.

(a) internally reversible process, $Q > 0$

(b) internally reversible process, $Q = 0$

(c) internally reversible process $Q < 0$

(d) internally irreversibilities present, $Q > 0$

(e) internal irreversibilities present, $Q = 0$

(f) internal irreversibilities present, $Q < 0$

8.34. A system maintained at constant volume is initially at temperature T_1, and a heat reservoir at lower temperature T_0 is available. Show that the maximum work recoverable as the system is cooled to T_0 is

$$W_{max} = c_v\left[(T_1 - T_0) - T_0 \ln\frac{T_1}{T_0}\right]$$

8.35. A closed system consists of an ideal gas with constant specific heat ratio γ. The gas undergoes a process in which pressure increases from p_1 to p_2. Show that the ratio of entropy change for an isothermal process to the entropy change for a constant volume process is $(1 - \gamma)$. Show the process on p–v and T–s co-ordinates.

8.36. A quantity of perfect gas at an absolute temperature T is heated at constant pressure until the absolute temperature is $2T$. The gas is then expanded, according to the law pV = constant until the temperature has fallen to the initial temperature, T. Show that the total quantity of heat transferred to the gas from the surrounding during the whole process is given by

$$Q = \frac{n}{n-1} mRT$$

Also show that the change of entropy per kg of gas during the polytropic expansion only is

$$\Delta s = \frac{(c_p - nc_v)}{n-1} \ln \frac{T_2}{T_1}$$

8.37. During a reversible process, the temperature and entropy are related by

$$T = a \, . \, S + b \, . \, S^2 + c \, . \, S^3$$

where a, b and c are constants. Find the heat added between T_1 and T_2.

8.38. Show that the change of entropy during a polytropic process for a perfect gas on per unit mass is given by

$$s_2 - s_1 = c_v \left(\frac{\gamma - n}{n-1}\right) \ln \frac{T_1}{T_2}$$

8.39. Derive the expression for change in entropy during isothermal process, constant volume process and constant pressure processes.

8.40. The specific heat of a gas at constant pressure is expressed as $c_p = a + bt$, where a and b are constant and T is the absolute temperature. Show that the change of entropy per kg of gas is given by

$$\Delta S = a \ln \frac{T_2}{T_1} + b\,(T_2 - T_1) + R \ln \frac{p_1}{p_2}$$

When the absolute pressure and temperature alter from p_1, T_1 to p_2, T_2.

8.41. (a) Using the second law of thermodynamics as a postulate show that the entropy of the universe always increases in any real process.

(b) State whether the following processes are reversible or irreversible. In each reversible process, show how all the effects of the processes can be completely undone. In each irreversible process, devise a perpetual motion machine of the second kind which would use consequences if the process were reversible.

(c) Some water confined under a piston is stirred rapidly with a stirrer till its temperature rises.

(d) A rigid chamber contains a partition on the one side of which is air at a high pressure and the other side completely vacuum. The partition is suddenly removed and the air expands.

Numerical Questions

8.42. A certain process occurs in the following manner: (a) Addition of heat at constant volume so that the pressure rises from 6 bar to 15 bar. (b) Polytropic expansion $pV^{1.3} = c$ till the system reaches original temperature of 110°C. (c) Isothermal compression till the system reaches original state.

Find (i) temperature at the end of constant volume process, (ii) pressure at the end of polytropic expansion, and (iii) change in entropy during each of the three stages per kg of air.

[*Ans.* (i) 957.51 K, (ii) 0.2834 bar, (iii) $(\Delta s)_{v=0} = 0.652$ kJ/kg·K $(\Delta s)_{poly} = 0.17$ kJ/kg·K $(\Delta s)_{iso} = -0.872$ kJ/kg·K]

8.43. 1 m^3 of nitrogen contained in a cylinder behind a piston is initially at 1 bar and 15°C. The gas is compressed isothermally and reversibly until the pressure is 4 bar. Find (a) the change of entropy, (b) heat exchange, and (c) the work done.

[*Ans.* (a) – 0.47 kJ/K, (b) 135.6 kJ, (c) 135.6 kJ]

8.44. One kg of air is compressed according to the law $pV^{1.2}$ = constant from 1.05 bar and 15°C to 16.8 bar. Find the change in entropy. For c_p = 1.005 and c_v = 0.6673 kJ/kg K.

[*Ans.* 0.728 kJ/K]

8.45. One kg of air expands according to law $pV^{1.2}$ = constant from initial condition of 7 bar and 300°C to a final pressure of 1.4 bar. Find the change in entropy during the process.

8.46. Find the value of c_p and γ for oxygen which has a value of 0.264 for gas constant and c_v = 0.65 kJ/kg K. If 3 kg of oxygen undergoes a non-flow process during which the pressure remains constant at 7 bar from an initial volume 1200 litres to a state where the temperature is 520 K. Find (a) change in internal energy, (b) change of enthalpy, (c) heat transferred, and (d) the work done.

8.47. 2 kg of ice at 0°C is mixed with 5 kg of water at temperature 20°C. Determine the change in entropy of the adiabatic system. [*Ans.* 0.0533 kJ/K. Only a portion of ice melts]

8.48. From the height of 1000 metres, a stone weighing 50 kg falls in vacuum and strikes a 100 kg steel plate. Find the value of change of entropy assuming stone and plate to be an adiabatic system. Take specific heats of both stone and plate as 0.836 kg/kg K. Stone and plate both experience a temperature rise of 15°C. [*Ans.* 1.255 kJ/K]

8.49. A wagon having 1 tonne mass moving at 100 km/h is brought to rest with the application of brakes. Calculate the increase of entropy. [*Ans.* 0.1406 kJ/K]

8.50. A piece of iron rod in red hot condition is plunged suddenly into a mass of water weighing 10 kg. Calculate the change in entropy if the initial temperature of water 20°C is raised to 55°C. [*Ans.* 4.619 kJ/K]

8.51. During a polytropic compression ($pV^n = c$), a gas of 0.2 m^3 volume at 1 bar and 100°C contained in a cylinder is compressed to a final volume of 0.05 m^3 and pressure of 6.5 bar. Find, (a) mass of gas contained is the cylinder, (b) value of index n, (c) increase in internal energy of the gas, (d) work done during compression, (e) heat interaction, and (f) change in entropy. Show the process on p–v and T–s diagrams. Take γ = 1.4 and R = 0.287 kJ/K.

[*Ans.* (a) 0.179 kg (b) 1.35, (c) 30.54 kJ (d) – 35 kJ, (e) 4.435 kJ]

8.52. A perfectly insulated cylindrical vessel with closed insulated ends is fitted with a piston which is mechanically free and is a perfect insulator. One side of the piston encloses 0.03 m^3 of air at 3 bar and 47°C and the other side encloses 0.06 m^3 of hydrogen at 3 bar at – 15°C. An electric heater is fitted in the air side of the cylinder but not in the hydrogen side. The piston moves without friction till the air occupies 0.06 m^3. γ for air as well as H_2 may be taken as 1.4 and specific heat at constant volume for H_2 is 10.07 kJ/kg K. Find (a) final temperature and pressure of both gases, (b) work done on hydrogen, (c) quantity of heat received by air, and (d) change of entropy of hydrogen. [*Ans.* (a) 242.5 K, 10137 bar, 340 K, 7.9 bar

(b) – 14.126 kJ (c) 333.4 kJ (d) 0.0014 kJ/K]

8.53. Thermal power systems for use in space normally reject energy as heat by radiating to space. Since fluid filled radiators are generally quite heavy, it has been suggested that this energy be converted to electricity and the current run through lighter resistors to dissipate the energy as heat. What do you think of this idea? Energy transfer as heat from a space vehicle must take place by radiation. The weight of space radiator is proportional to its area, which is determined by the rate at which the energy must be radiated as heat, the rate of energy radiation is proportional to the product of area and the fourth power of the radiator temperature. Consider a reversible 2*T* engine giving a fixed amount of power and operating with fixed source temperature. Show that the least radiator weight is obtained when the radiator temperature is 0.75 times the source temperature.

8.54. Calculate the changes of entropy per kg of air in the following cases:

(i) Air expands isothermally from 6 bar to 3 bar.

(ii) Air is compressed to half the volume at constant pressure.

(iii) Heat is supplied to air at constant volume till the pressure becomes threefold. What would be the change in entropy if the air undergoes the above three processes in sequence. Take $c_p = 1.005$, $c_v = 0.720$ kJ/kg K. Plot the process on *p–v* and *T–s* charts.

[*Ans.* (a) 0.198 kJ/kg K, (b) 0.693 kJ/kg K, (c) 0.788 kJ/kg K]

8.55. 2 kg of air at 15 bar and 260°C expands reversibly and polytropically to 1.5 bar and 64°C. Determine the index of expansion, work transfer, change in internal energy, heat transferred and change in entropy. Represent the process or, *p–v* and *T–s* diagram.

8.56. In an air turbine, the expansion of air takes place from 6.8 bar 430°C to 1.013 bar and 150°C. The heat loss from the turbine is negligible. Show that the process is irreversible and calculate the change of entropy per kg of air. [*Ans.* 0.0355 kJ/kg K]

8.57. Air at 15°C and 1.05 bar occupies a volume of 0.02 m^3. The air is heated at constant volume until the pressure is 4.2 bar and then cooled at constant pressure back to the original temperature. Calculate the net heat flow to or from the air or the net entropy change. Sketch the process on *T–S* diagram

$$Q = -6.3 \text{ kJ}, \Delta S = -0.0101 \text{ kJ/K}$$

8.58. 0.03 m^3 of nitrogen having molecular weight 28 contained in a cylinder behind a piston is initially at 1.05 bar and 15°C. The gas is compressed isothermally and reversible until the pressure is 4.7 bar. Find the change in entropy, the heat flow and the work done. Sketch the process on *p–v* air *T–S* diagram. [*Ans.* $\Delta S = -0.01516$ kJ/K, $Q = 4.37$ kJ, $W = 4.37$ kJ

8.59. 1 kW electric heater is placed in a 8.4 m^3 insulated room filled with air at 1 bar and 20°C. The heater is allowed to operate for 15 minutes. Calculate the change in entropy. Sketch the process on *T–S* diagram. [*Ans.* 2.556 kJ/K]

8.60. 1 m^3 of air is heated reversibly at constant pressure from 5 bar and 15°C to 300°C, and is then cooled reversible at constant volume back to pressure of 1.03 bar. Determine the net heat flow and the overall change entropy. Sketch the process on *p–V* and *T–S* diagram.

8.61. 1 m^3 air expands reversibly in a cylinder behind a piston in such a way that the temperature remains constant at 260°C while the volume is doubled. The piston is then moved in, and heat is rejected by air reversibly at constant pressure until the volume is the same as it was initially. Determine the net heat flow and the overall change of entropy. Show the process on *T–S* diagram. [*Ans.* – 161.9 kJ/kg, – 0.497 kJ/kg K]

8.62. In a gas turbine plant, the hot gases enter the turbine at 550°C and 5 bar and leave at 1 bar. The process is approximately adiabatic but the entropy changes by 0.174 kJ/kg K. Find the exit temperature of the gases. Assume $\gamma = 1.333$, $c_p = 1.11$ kJ/kg K. Show the process on *T–S* diagram. [*Ans.* 370°C]

8.63. A lead ball of 20 kg having specific heat 0.13 kJ/kg K and at a temperature of 40°C falls from a height of 5000 m and hits a cake of ice kept on the ground at – 1°C in an adiabatic enclosure. Assume that all the energies possessed by the ball is fully utilised to just melt the ice and to bring the ice-water and ball at 0°C. Calculate the mass of ice and the change is entropy in the system comprising of ice and ball. [*Ans.* 3.21 kg, 0.449 kJ/K]

8.64. An isothermal compression process of 50 kg/s of nitrogen from 305 K and 1 bar occurs in a large, water-cooled compressor. Determine (a) the change is entropy and (b) the power required.

8.65. Water at 25°C is pumped from a lake to an elevated storage tank having volume 60 m^3. The average elevation of the water in the tank is 50 m above the surface of the lake. Initially, the tank contains air at 110 kPa, 25°C, and the tank. The pump is operated until the tank is three-quarters full of water. The temperature of the water and air in the tank remains at 25°C. Calculate the work input to the pump and the change in entropy during the process.

8.66. Calculate the change in entropy of the universe in the case of following processes.

(a) A copper block of 500 g mass and with c_p of 150 J/K at 100°C is placed in a lake at 10°C.

(b) The same block at 10°C is dropped from a height of 100 m into the lake.

(c) Two such blocks at 15°C and 5°C are joined together.

8.67. 2 kg of water undergoes a process from an initial state where the pressure is 2.5 Mpa, and 100°C. Calculate the entropy change of the water, in kJ/K, assuming the process is (a) irreversible, (b) internally reversible. [*Ans.* – 11.438 kJ/K]

8.68. Air enters a compressor operating at steady state at 1 bar. 22 °C with a volumetric flow rate of 1 m^3/min and is compressed to 4 bar, 177°C. The power input is 3.5 MW. Work out the following ignoring kinetic and potential energy effects.

(a) For a control volume enclosing the compressor only, calculate the heat transfer rate, in kW and the change in specific entropy from inlet to exit, in kJ/kg K. What additional information would be required to evaluate the rate of entropy generation?

(b) determine the rate of entropy generation, if kW/K, for an enlarged control volume enclosing the compressor and a portion of immediate surrounding so that the heat transfer occurs at the ambient temperature of 22°C.

8.69. An axial flow compressor operating at steady state receives air at 27°C, 1 bar and compressor it to a pressure of 2.1 bar. The work input required is 95 kJ/kg. Heat transfer from compressor occurs at the rate of 14 kJ/kg at the location on the compressor's surface where the temperature is 40°C. Neglect the changes is KE and PE. Calculate (a) the temperature of the air at the exit, in °C, and (b) the rate at which entropy is produced within the compressor, in kJ/kg K.

8.70. Air enters a turbine operating at steady state at 6 bar and 1100 K and expands up to 1 bar with isentropic efficiency of 85%. Calculate the power output and increase in entropy.

Software Based Questions

8.71. Develop a software to calculate the work input and entropy production in a water cooled reciprocating compressor using air, N_2, natural gas as the working fluid. Tabulate the results for a real input data.

8.72. Develop a software to calculate the work input and entropy production in a centrifugal/axial flow air compressor. Tabulate the results for a real input data.

8.73. Develop a software to calculate the work output from a gas turbine and entropy production. Tabulate the results for a real input data.

8.74. Develop a software to calculate the exit velocity from a gas nozzle and entropy production. Tabulate the results for a real data.

8.75. Develop a software to calculate delivery pressure of a diffuser and entropy production. Tabulate the results for an real data.

Design and Open Ended Questions

8.76. Many leading economists have called for the use of principles of thermodynamics in economics. According to this view, entropy and second law of thermodynamics are relevant in assessing not only the exploitation of natural resources for industrial and agricultural production but also the impact on the natural environment of waste from such production. Prepare a report arguing in favour or against this proposition.

8.77. Write a paper to study the growth of living organisms using the entropy and entropy production.

8.78. In the era of nanotechnology and micromachines, there is increasing interest in producing micro turbines, pumps and heat exchangers that weigh less than 1 gram and have volumes of 1 cm^3 or less. The designers have to give a special emphasis on irreversibilities on the performance of such tiny devices. Prepare a report discussing the influence of irreversibilities related to heat transfer and friction on the design and operation of these devices.

8.79. Prepare a report covering the energy loss, entropy production, environmental impact, property loss, business loss, human life loss due to terrorist attack on twin world trade towers of USA, called as 9/11.

8.80. Prepare a report covering the fuel requirement, energy requirement, entropy production caused by launching a space rocket.

8.81. Prepare a report on energy loss and entropy production on a volcanic eruption. Is there any possibility to the use of energy released in volcanic eruption?

8.82. Prepare a report on the theory of thermal death caused by entropy production.

Multiple-Choice Answers Type Questions

8.83. The first law of thermodynamics gives $dU = \delta Q - \delta W$ and second law tells that $dS > \delta Q/T$. In which of the way these two laws can be combined and written.

(a) $dU \geq TdS - pdV$ (b) $dU \leq TdS - pdV$

(c) $dU \leq TdS + pdV$ (d) $dU = TdS - pdV$

8.84. For a reversible engine cycle, the Clausius inequality says

(a) $\oint \frac{\delta Q}{T} > 0$ (b) $\oint \frac{\delta Q}{T} < 0$ (c) $\oint \frac{\delta Q}{T} = 0$ (d) None of these

8.85. For an irreversibie process in an isolated system the change in entropy is expressed as

(a) $dS_{\text{isolated}} < 0$ (b) $dS_{\text{isolated}} > 0$ (c) $dS_{\text{isolated}} = 0$ (d) None of these

8.86. The entropy generated or total change in entropy for a process to be impossible will be

(a) $S_{gen} > 0$ (b) $S_{gen} = 0$ (c) $S_{gen} < 0$ (d) $S_{gen} \geq 0$

8.87. The change in entropy in polytrcpic process is given by

(a) $\left(\frac{n-1}{n-\gamma}\right) c_v \ln\left(\frac{T_2}{T_1}\right)$ (b) $c_v \ln \frac{T_2}{T_1}$ (c) $c_p \ln \frac{T_2}{T_1}$ (d) $\left(\frac{n-\gamma}{n-1}\right) c_v \ln \frac{T_2}{T_1}$

8.88. Write the correct relation

(a) $TdS = dU + pdV$ (b) $dH = TdS + Vdp$

(c) $TdS = c_p\, dT - Vdp$ (d) All of these

8.89. On T–s diagram, the slope of constant pressure line, $(\tan \theta)_p$ and the slope of constant volume line $(\tan \theta)_v$ is related as

(a) $(\tan \theta)_p = (\tan \theta)_v$ (b) $(\tan \theta)_v > (\tan \theta)_p$

(c) $(\tan \theta)_p > (\tan \theta)_v$ (d) $(\tan \theta)_p \times (\tan \theta)_v = 1$

Answers to Multiple-Choice Questions

8.83. (a) 8.84. (c) 8.85. (b) 8.86. (c) 8.87. (d) 8.88. (d)
8.89. (b)

CHAPTER

9

Properties of Pure Substance (Steam) and Applications

OVERVIEW

Thermal power plants operate with the help of some working fluids (water/steam). These working fluids are generally *pure or real substances*. It is the changes in properties of the working fluid, i.e., pure substance which operates the power plants. Further, the working fluid used in the refrigeration and air-conditioning system is also a pure substance called *refrigerant such as Fr–12, Fr–22, NH_3, etc. It is the changes in the properties of refrigerant that cause cooling effect*. Therefore, it is essential to study the properties of pure substances.

This chapter deals with the methodology to calculate properties of pure substances and the applications of the first and second laws of thermodynamics to devices dealing with steam.

9.1 PURE SUBSTANCE (STEAM)

The pure substance is sometimes also called **one component system**. *The important characteristics of a pure substance or one component system is that it has a homogeneous and invariable chemical composition irrespective of the phase or phases in which it exists*. A mixture of ice, water and steam is an example of a pure substance or one component system since the chemical composition of the three phases is the same. Similarly, a system consisting of O_2 as a vapour, as a liquid, as a solid, or a combination of these is also pure substance. A mixture of different gases may be treated as pure substance provided that its composition is uniform. If one of the gases condenses, the mixture can no longer be considered to be a pure substance since the condensed phase has a different composition than the original mixture. For example, air is a pure substance only as long as it is all gas or all liquid. A mixture of dry gaseous air and liquid air is not a pure substance because the chemical composition of the liquid phase (liquid phase will be richer in N_2) is different from that of the vapour.

Note The following steam tables and Mollier charts have been used in this chapter.

(i) Mathew and Ragers, Black Well (UK)

(ii) Nam Chand and Brothers, (India)

Steam (water vapour) is regarded as the most important pure substance in engineering practice because it is used as a working fluid in steam engines and turbines. It is also used for process heating work. In this section, we will discuss the properties of steam. The properties of steam were first investigated experimentally by Regent and subsequently by Prof. Callender.

9.2 PHASE

A pure substance may exist in different aspects known as phases. *The phase of a substance is the homogeneous, chemical and physical state of aggregation of molecules.* This does not mean that substance may have one phase only of one type. For example, ice is found in a number of solid phases, namely, ice I, ice II, and ice III and so on, one of such phases may be amorphous and others may be different crystalline forms.

In general, a pure substance exists in three phases namely:

(i) Solid, (ii) liquid and (iii) vapour or gaseous phase. At low temperature, the molecules exist as solids, when the temperature increases they may melt into the liquid phase, at still higher temperature they evaporate and become vapour. For example, *water has three phases namely, ice, water and steam.* Under certain conditions, all these three phases may co-exist. The co-existence of two phases is very common in engineering practice. When ice melts, there is a transformation of phase from solid to liquid which is called the ***melting or fusion of ice***. If state transformation of phase takes place from the liquid to solid, it is called the *freezing or solidification.* If the transformation takes place from the liquid phase to vapour phase, it is called ***vaporization***. If the transformation takes place in opposite direction, i.e., from vapour to liquid, it is called ***condensation***.

It is to be noted the *boundary between different phases is unusually sharp and clearly defined except* in some cases. For example, between vapour and liquid phase above the critical point, one phase blends into the other so that a definite boundary cannot be detected.

9.3 PHASE TRANSFORMATION (ICE INTO STEAM) AT CONSTANT PRESSURE

To see the change of phase of water under different conditions, consider 1 kg of ice at –10°C contained in a cylinder under a pressure p say 1 atm exerted by a weight W placed on the frictionless piston as shown in Fig. 9.1 (a). Now, let heat be supplied to the cylinder continuously and slowly. This state of ice corresponds to state A shown in Figs. 9.2 and 9.3.

Due to heat supplied, the ice gets warmer and the temperature of ice rises and approaches 0°C as recorded by thermometer. The warming process is represented by AB in Fig. 9.2. At point B, the temperature is 0°C.

On further heating, the temperature remains constant at 0°C and ice melts, i.e., a change of phase takes place from solid to liquid state. The two phases (i.e., mixture of ice and water) exist in equilibrium. This process is shown by BC. At point C all the ice has melted and there is only one phase, i.e., liquid water as shown in Fig. 9.1 (b). It is to be noted that the liquid water is under the same atmospheric pressure. The quantity of heat required to transform ice into water at constant temperature (i.e., 0°C) is called ***latent heat of fusion or enthalpy of fusion***. *The latent heat at 1 atm is numerically equal to 335 kJ/kg*. However, there is a small decrease in volume during this melting process but this is practically negligible, as shown in Fig. 9.3.

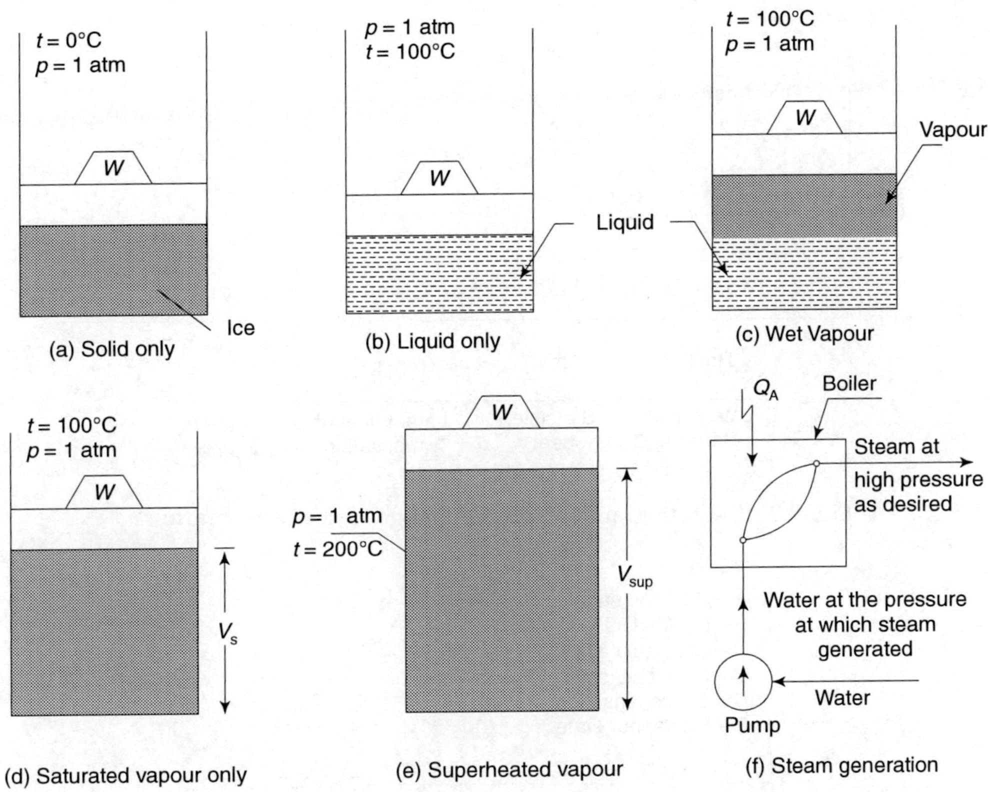

Fig. 9.1 Formation of steam

With further addition of heat, the temperature of water rises till temperature of vaporization (i.e., boiling) is reached. This process is represented by *CD* in Fig. 9.2. The point *D* corresponds to the vaporization or saturation temperature. ***This boiling temperature is a function of pressure***. Since the pressure within the cylinder is 1 atm, the temperature of water rises from 0°C to 100°C. The state of water at t_D = 100°C is called the ***saturated liquid*** state because any further addition of heat causes vaporization to start. The heat required for change from liquid state to vaporization point is called the ***sensible heat***. During the process, from 0 to 4°C, the volume slightly decreases but after that it increases as shown in Fig. 9.3.

When more heat is added, the liquid water starts evaporating and once again another change of phases occurs at constant temperature. This process is represented by *DE* in Fig. 9.2. In

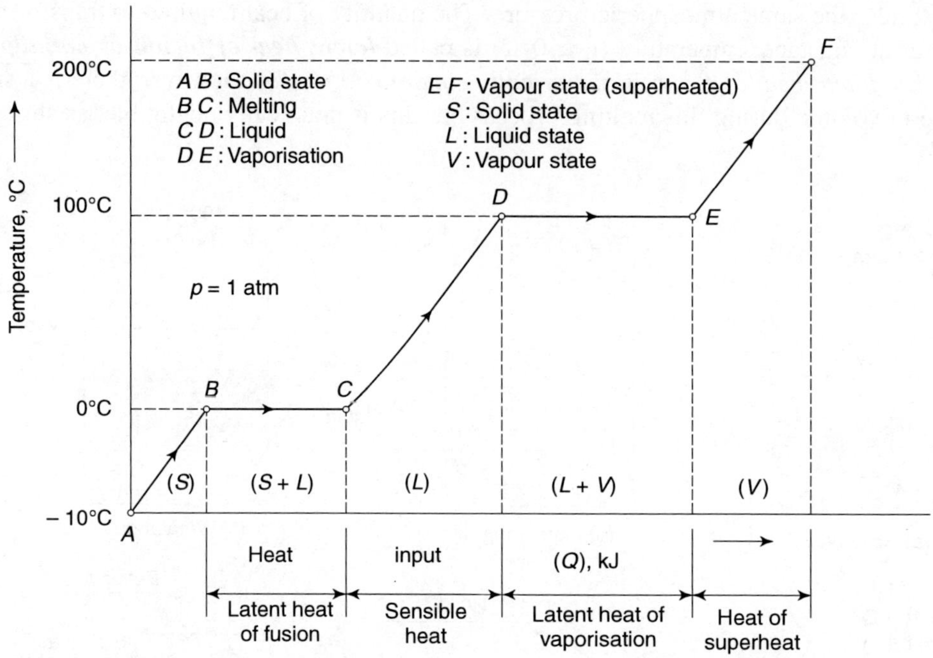

Fig. 9.2 *T–Q* diagram for phase change of ice into steam

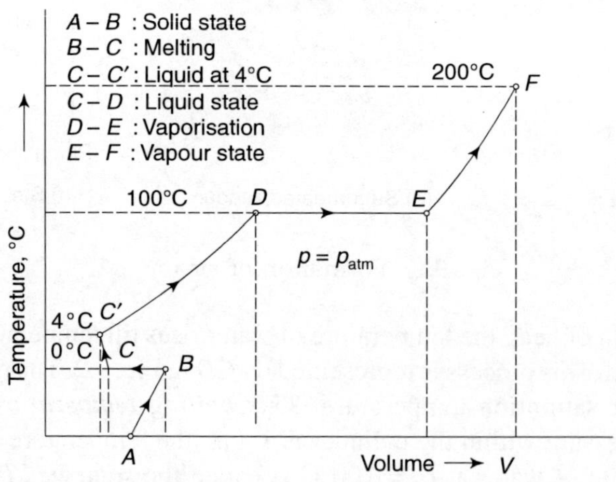

Fig. 9.3 *T–V* diagram for phase change of ice into steam

this state there exists a two-phase mixture of water and its vapour as shown in Fig. 9.1 (c). There is a considerable increase in volume as shown in Fig. 9.3. The piston, therefore, rises up. The rate of evaporation depends upon the rate of heat supply. At point *D*, all the water has vaporized and the state of steam at this point is called ***dry and saturated steam*** and its corresponding temperature is called ***saturation temperature***. The heat required to vaporise state at constant temperature is called the ***latent heat of vaporization***. The numerical value of latent heat of vaporization at 1 atm is equal to 2256.9 kJ/kg.

With further addition of heat to the saturated steam, the temperature of the steam rises as represented by process *EF* in Fig. 9.2. There will also be an increase in volume as shown in Fig. 9.3. The piston, therefore, will rise again as shown in 9.1 (e). This process is called the superheating of steam and the resulting steam is called the ***superheated steam***.

It is to be noted that the formation of steam in the present study is under 1 atm. pressure. *At different pressures (below critical), the nature of the various processes will be the same but the saturation temperature of steam will be different.* The formation of 1 kg of water at 0°C into steam at 1 atm has little practical significance. ***In practice, the water is pumped into a boiler at desired high pressure (say 10 to 200 bar or more) and then it is transformed into steam (saturated or superheated) with the addition of heat at constant pressure (Fig. 9.1.f.).***

9.4 *T-v* DIAGRAM FOR WATER (CRITICAL STATE)

There exists an important state known as ***critical state*** of the substance which may be illustrated by *T–v* diagram (Fig. 9.4. of water). Consider 1 kg of water contained in a vessel and heated at constant pressure. Let us plot the constant pressure lines that represent the state through which water passes as it is heated from initial state. The line *ABCD* shown in Fig. 9.4 represents such a constant pressure line of 1 atm. The various portions of this line represent different states of water as follows:

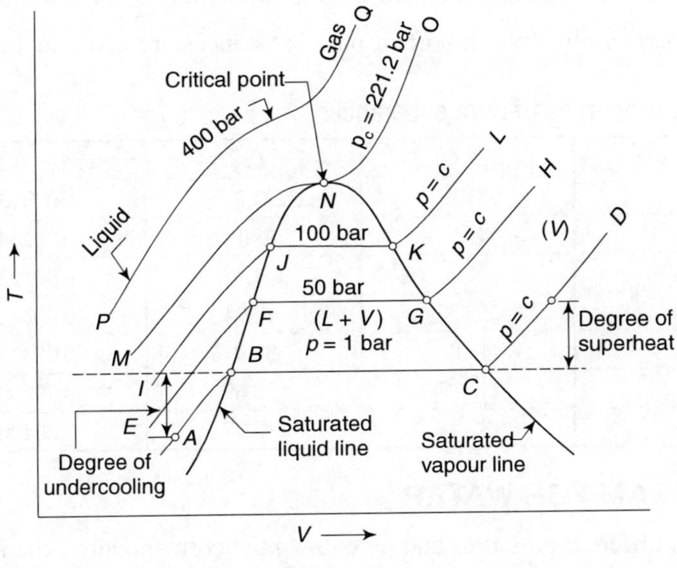

Fig. 9.4 *T–v* diagram of water

AB : *Heating of liquid to saturation temperature*. During process *AB*, heat transfer produces an increase in temperature and small increase in volume. If the temperature of the liquid is less than saturated temperature at a given pressure, the liquid is said to be *sub-cooled* or *compressed*.

BC : *Change of phase at constant temperature from liquid to vapour state*. When point *C* is reached, all the liquid has been converted into vapour. The point *B* and *C* are called the ***saturated liquid state*** and ***saturated vapour state***.

CD : **Superheating of vapour**. At point *C*, the change of phase from liquid to vapour is complete and further transfer of heat to the system, i.e., from *C* to *D* increases both temperature and volume. When the temperature of the vapour is higher than the saturation temperature at a given pressure, it is called ***superheated vapour***.

If the heating process alone is repeated at different pressures such as 50 bar, 100 bar, curves such as *EFGH* and *ITKL* respectively are obtained. The locus of points like *E, F, J*, etc., is known as **saturated liquid line** and that of *C, G, K*, etc., is known as ***saturated vapour line***.

It is to be noted that there are two discontinuities in the slope of each of constant-pressure lines (e.g., points *B* and *C*) with a constant temperature portion between them. If the pressure is increased beyond a certain value called the ***critical pressure***, the discontinuities an well as constant temperature portion disappear. This is shown in constant ***pressure line of 221.2 bar*** represented by *MNO* and other constant pressure line above it. ***The limiting state at which no discontinuities are noticeable is called the critical state***. The critical state is represented by point *N* in Fig. 9.4. **The point *N* is actually a point of inflection and change of phase occurs at this point abruptly**. In other words, *there exists a limit beyond which a liquid vapour transformation is not possible called the critical state*. At **critical state, v_{fg} = 0.**

The critical properties for some important pure substances are given in Table 9.1.

Table 9.1 Critical properties of pure substances

Substance	T_c (°C)	p_c (bar)	v_c (m^3/kg)
Water	374.15	221.2	0.00310
Air	- 141.5	38.1	0.00284
CO_2	31.2	71.48	0.00122
H_2	- 239.76	12.94	0.003220
N_2	- 146.8	33.9	0.00320
O_2	- 118.12	50.69	0.00230
NH_3	132.4	112.77	0.00420

9.5 *P–v* DIAGRAM FOR WATER

If we heat the ice at different pressures and note down the corresponding changes in volumes for different phases, a phase equilibrium diagram shown in Fig. 9.5 (a) will be obtained. ***A saturation state is a state from which a change of phase may occur without a change of***

pressure or temperature. Saturated solid line is obtained by joining the saturated solid sates at various pressure. Similarly, all the saturated liquid states with respect to solidification and all the saturated liquid states with respect to vaporization and all the saturated vapour states are joined together.

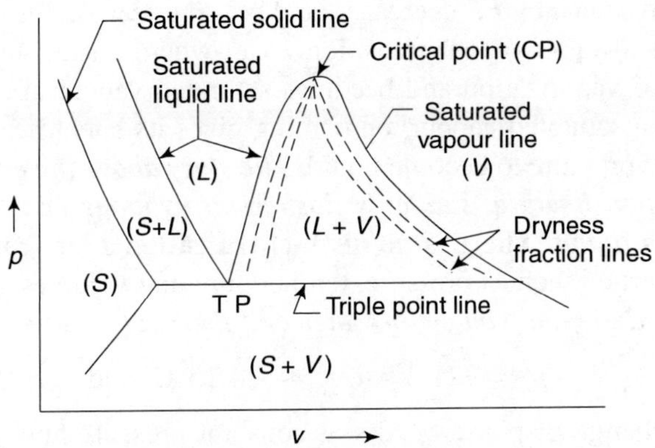

Fig. 9.5 (a) Phase equilibrium diagram of water on p–v co-ordinates

Liquid is generally the working fluid in power cycles and interest is often centred to the liquid-vapour region only. Figure 9.5 (b) shows p–v diagram for water and other pure substance indicating only liquid and vapour phase. AB is the liquid saturated line while BC is the saturated vapour line. State B is the critical point which is the meeting point of saturated liquid line and dry saturated vapour line. Consider a typial isotherm $DEFG$ a pure substance on p–v diagram.

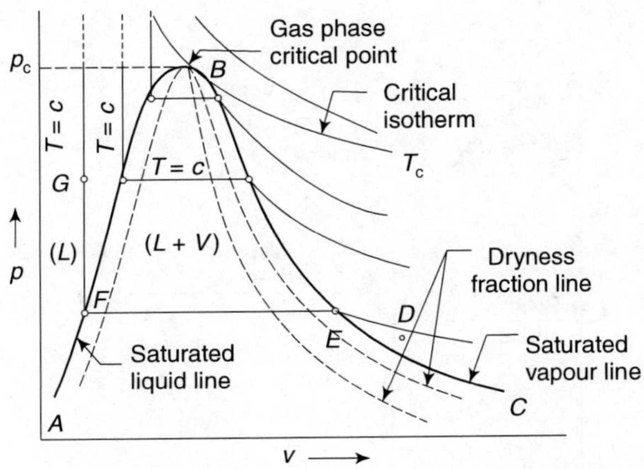

Fig. 9.5 (b) p–v diagram for water

If the vapour at state *D* is compressed slowly and isothermally, the pressure will rise until there is saturated vapour at *E*. If the compression is continued, condensation takes place, the pressure remaining constant, so long as the temperature remains constant. At any point between *E* and *F*, the liquid and vapour are in equilibrium. Since a very large increase in pressure is needed to compress the liquid, line *FG* is almost vertical. With the increase in temperature the liquid-vapour transition, as represented by *EF* decreases, and becomes zero at the critical point. Below, the critical point only is a liquid-vapour transition zone, where a saturated liquid, on heating absorbs the latent heat of vaporization and becomes saturated vapour at constant temperature and pressure. Likewise, a saturated vapour, on cooling gives its latent heat of condensation at constant temperature and pressure to become liquid. ***The story above the critical point is totally different as a liquid, upon heating, suddenly flashes into vapour or a vapour, on cooling suddenly condenses into liquid*. There is no distinct transition zone from liquid to vapour or vice versa**. The properties such as pressure, temperature and volumes at this point is called ***critical pressure***, *temperature and volume of water and their values are*

$$p_c = 221.2 \text{ bar}, \ t_c = 374.15°\text{C, and } v_c = 0.00317 \text{ m}^3/\text{kg}$$

As mentioned earlier, change of phase occurs at constant pressure and temperature. A pure liquid at a given pressure will transform into vapour only at a particular temperature known as ***saturation temperature*, (t_{sat}), which is a function of pressure**. Likewise, at a given temperature, the liquid will boil or condense only at a particular pressure called ***saturation pressure* (p_{sat}), which is a function of temperature**.

9.6 *p–T* DIAGRAM FOR WATER (TRIPLE POINT)

The triple point plays an important role in the study of the general change of phase of a substance. It may be well understood by the following example of ice-water-steam on *p–T* plane an shown in Fig. 9.6. Along the sublimation line, the solid and vapour phases are in

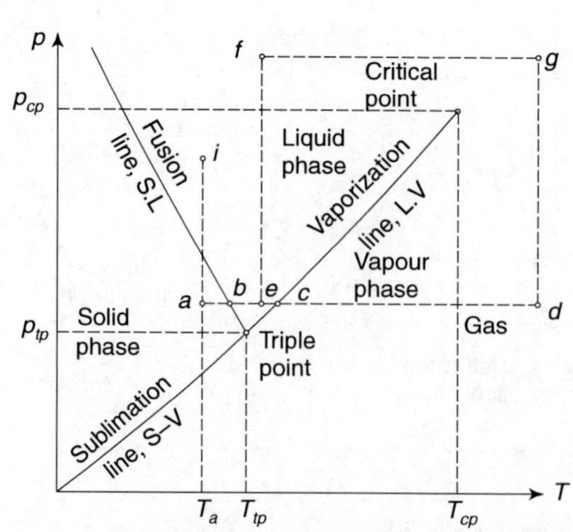

Fig. 9.6 Triple point

equilibrium along the fusion line, the solid and liquid phases are in equilibrium and along vaporization line the liquid and vapour phases are in equilibrium. ***The only point at which all three phases may exist in equilibrium is the triple point***. The vaporization line ends at the critical point because there is no distinct change from the liquid phase to the vapour phase above the critical point. The fusion curve, the vaporisation curve and the sublimating meet at triple point. Thus, the triple point is defined as follows—*The triple point is such a state in which the solid, liquid and vapour phases of a substance co-exist in equilibrium.*

It is to be noted that the slopes of the sublimation and vaporization curves for all substances are positive. The slope of fusion curve for most substances is positive, but for water, it is negative. The boiling (saturation) temperature is very sensitive to pressure change as indicated by the vaporization curve which gives the saturation temperature at different pressures but the temperature at which a solid melts is not much sensitive to pressure change as indicated by the small slope of the fusion curve.

Consider a solid in state *a*, which is at a pressure higher than triple point pressure. When the temperature is increased while pressure remaining constant (line *ab*) the substance passes from the solid to the liquid phase. Along the constant-pressure line *bcd*, the substance passes from the liquid phase to the vapour phase and then gaseous phase as temperature increases. At the triple point, three phases co-exist in equilibrium. At a pressure above the critical point such as *fg* there is no sharp distinction between liquid and vapour phase.

Triple point data for some common substances is shown in Table 9.2.

Table 9.2 Triple point data

Substance	t_{tp} (°C)	p_{tp} (bar)
H_2O	0.01	0.006113
CO_2	– 56.2	5.2156
O_2	– 218.4	0.0015
N_2	– 209.7	0.1282
H_2	– 39	0.13×10^{-8}

9.7 *p-v-T* SURFACE FOR PURE SUBSTANCE

The *p-v-T* surface of three real substances exhibiting distinct characteristics such as water which expands on freezing, carbon dioxide which contracts on freezing and helium which consists of two forms of liquid phase in the neighbourhood of 4 K as shown in Figs. 9.7 to 9.9.

In these presentations, the pressure, specific volume and temperature are plotted on mutually perpendicular co-ordinates and each possible equilibrium state is thus represented by a point on the surface. *Any point on the p-v-T surface is determined by any two independent properties such as p and T*. The regions of the surface that represent a single phase, namely, the solid, liquid and vapour phases, are labelled, these surfaces being curved. The two-phase regions namely, the solid-liquid, solid-vapour and liquid-vapour regions are ruled surfaces and are made up of straight lines parallel to the specific volume axis. It is to be noted that in the two-phase regions, lines of constant pressure are also lines of constant temperature though the specific volume may change. Further, the triple point actually appears as the triple line on the *p-v-T* surface as the pressure and temperature of the triple point are fixed, but the specific volume may vary depending on the proportion of each phase. The various lines of constant temperature (isotherms) are depicted and *critical isotherm has a point of inflection at the critical point.*

9.7.1 H_2O and CO_2

There is one distinction between the behaviour of CO_2 and H_2O which is noticeable from the slopes of *S–L* line of *p-T* plane of CO_2 and H_2O. In the case of CO_2, *S–L* line slopes towards the right, whereas in H_2O, it slopes left. This may be explained as H_2O contract on freezing to solid state while water expands on freezing to solid state. In other words, in the case of water, the freezing temperature decreases with an increase in pressure while for CO_2, the freezing temperature increases as the pressure increases. *Consider an isotherm abcdef in Figs. 9.7 and 9.8.* As the pressure of vapour is increased along the isotherm *abcdef*, water first becomes solid and then liquid while CO_2 first becomes liquid and then solid.

9.7.2. He

In the neighbourhood of 4 K, He exhibits an unique behaviour (Fig. 9.9.). When isothermal compression of helium in vapour phase at temperature between 5.19 K (critical temperature) and 2.16 K is taking place, it condenses to a liquid phase called Helium I, the isothermal compression below 2.16 K results in the liquid phase of Helium II.

An *interesting point about this p-v-T surface is the existence of two triple lines.* At one of these two lines, the two forms of liquids are in equilibrium with the vapour while at the other they are in equilibrium with solid. Solid helium does not exist at pressure lower than 25 bar nor does it exist in equilibrium point with the vapour phase.

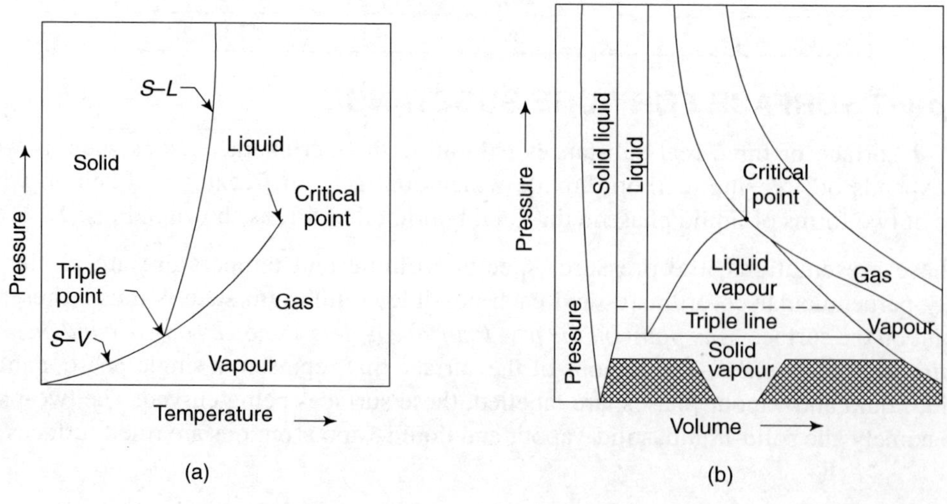

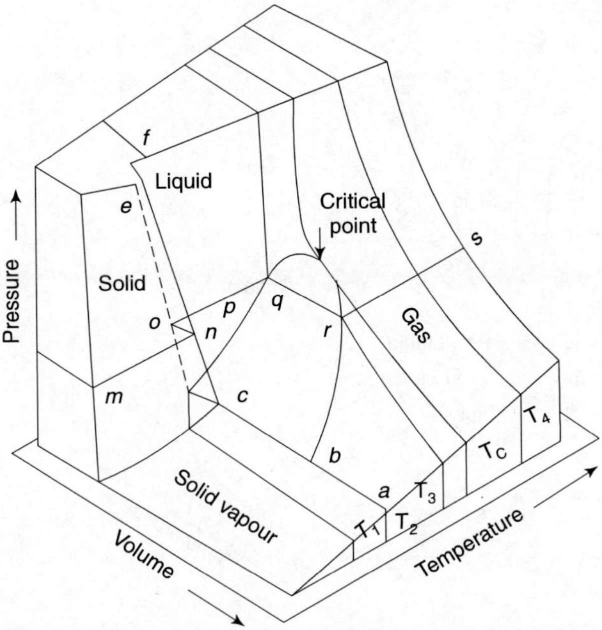

(c)

Fig. 9.7 p-v-T surface of H_2O

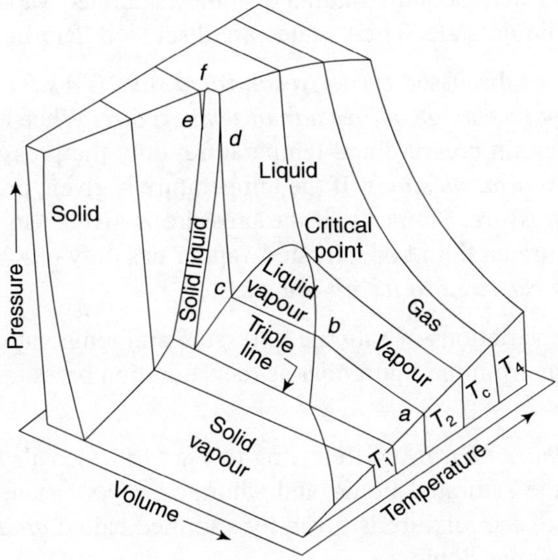

Fig. 9.8 p-v-T surface of CO_2

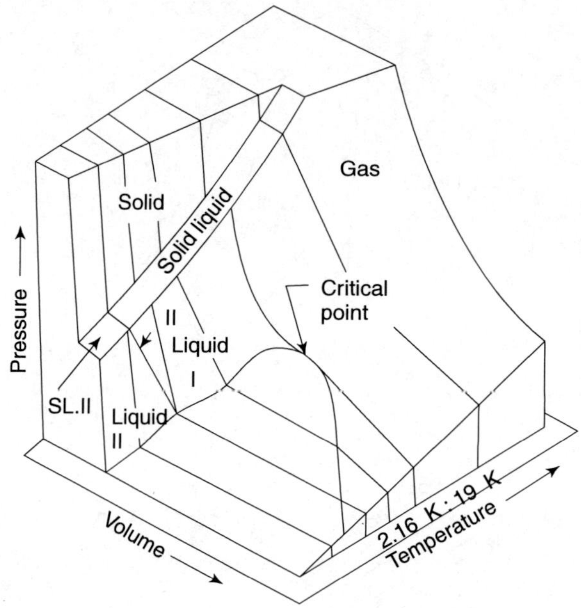

Fig. 9.9 *p-v-T* surface of He

9.8 VARIOUS STATES OF WATER-STEAM

Water-steam combination may acquire saturated liquid, saturated vapour, wet, superheated vapour and compressed liquid state. These states are discussed here in brief.

(a) *Saturated States* As discussed earlier, *a saturated state is a state from which a change of phase may occur without a change of pressure or temperature*. When a liquid and its vapour are in equilibrium at a certain pressure and temperature, only the pressure or temperature is sufficient to identify the *saturation states*. If the temperature is given, pressure is fixed, which is known as *saturation pressure*. Similarly, if the pressure is given, saturation temperature is fixed. In other words, saturated liquid or saturated vapour has only *one independent variable, i.e.,* ***only one property is required to fix up the state***.

Figure 9.10 shows the variations of saturated pressure and temperature of steam up to the critical point. It is obvious from the figure that as the saturation pressure increases saturation temperature increases.

(b) *Wet Steam* (*Two-Phase Mixture*) Referring to Fig. 9.11 (see also Fig. 9.15) when the substance lies between the saturated liquid and saturated vapour state, it is termed as ***wet steam***. The composition of the mixture is given by a termed called ***quality of steam*** and the state will be within the vapour dome.

(c) *Superheated Vapour* If the temperature of the vapour is greater than the saturation temperature corresponding to the given pressure, the vapour is said to be in ***superheated***

state. (State 1 in Fig. 9.11). The difference between the temperature of superheat vapour (t_1) and the saturation temperature (t_{sat}) at that pressure is called the ***degree of superheat***. Thus, degree of superheat = $t_1 - t_{sat}$.

(d) *Compressed or Subcooled Liquid* If the temperature of a liquid is less than the saturation temperature at the given pressure, the liquid is said to be *compressed liquid* (state 2 in Fig. 9.11). Since the *properties of liquids vary little with pressure*, hence properties of compressed liquids are taken from the *saturation tables at the temperature of compressed liquids*.

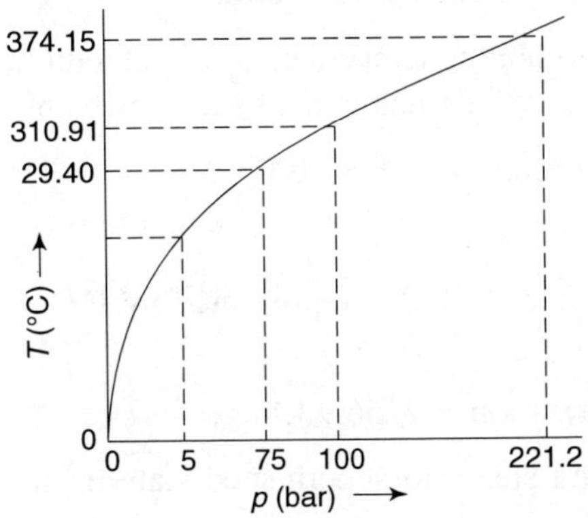

Fig. 9.10 Saturation pressure versus temperature

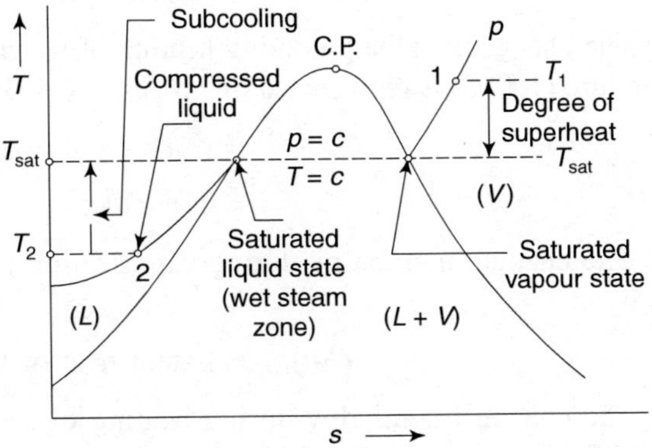

Fig. 9.11 Saturated, wet superheat and subcooling

If the liquid is cooled below its saturation temperature at a certain pressure, it is called *subcooled*. The difference between the saturation temperature and actual liquid temperature (i.e., $t_s - t_2$) is termed as the **degree of subcooling** (Fig. 9.11).

Note *It is to be observed that steam as a vapour will not obey the laws of perfect gases unless it is in a highly dried condition (i.e., highly superheated steam).* Although, steam is not considered as a perfect gas on account of it being a mixture of dry steam and moisture, it possesses properties like those of gases namely, *pressure, volume, temperature, internal energy, enthalpy and entropy*. But the pressure, volume, temperature of steam as a vapour are not connected by a simple relationship expressed by the characteristics equation for a perfect gas.

9.9 ENTHALPY CHANGES DURING FORMATION OF STEAM

The representation of formation of steam is shown in Fig. 9.2. *Since the heating process is at constant pressure, hence the heat input may be regarded as* ***the change in enthalpy***. The enthalpy changes during processes at 1 atm are as follows:

(a) The change in enthalpy from –10°C to 0°C as represented by process AB

$$= (\Delta h)_{AB} = c \times \Delta t = c \times \Delta T$$

where c = specific heat of ice = 2.09 kJ/kg K

$$(\Delta t) = \Delta T = \text{temperature difference}$$

or
$$(\Delta h)_{Ag} = 2.09\ (0 + 10) = 20.9 \text{ kJ/kg}$$

(b) The change in enthalpy during melting of ice at 0°C represented by process BC is given by

$$= (\Delta h)_{BC} = \text{Latent heat of fusion} = \mathbf{335\ kJ/kg}$$

(c) The change in enthalpy during heating of water up to boiling temperature, i.e., saturation temperature (100°C) is represented by process CD is known as the **sensible heat** and is given by

$$= (\Delta h)_{CD} = c \times \Delta t = c \times \Delta T = 4.18 \times (100 - 0)$$

$$= 418 \text{ kJ/kg}$$

(d) The change in enthalpy during vaporization process at 100°C is represented by DE is given by

$$(\Delta h)_{DE} = \text{Latent heat of vaporization} = 2256.9 \text{ kJ/kg},$$

(e) The change in enthalpy during heating of saturated steam to superheated state up to say 200°C is represented by EF is given by

$$(\Delta h)_{EF} = c_p \times \Delta t = c \times \Delta T = c_p\ (t_{sup} - t_g)$$

where c_p = sp. heat of superheated steam = 1.967 kJ/kg K

Δt temperature difference = ΔT = 200 – 100 = 100°C

$\therefore$
$$(\Delta h)_{EF} = 1.967 \times 100 = 196.7 \text{ kJ/kg}$$

It may be observed here that *enthalpy quantities are always measured* **from some datum.** This suggests that enthalpy may **be negative in magnitude. The negative sign does not mean that the substance does not contain any energy.** In order to determine the properties of steam, ***liquid water at 0°C is arbitrarily assigned the value of zero enthalpy***. This means that the enthalpy at point equal to – 335 kJ/kg is referred to as zero enthalpy such as at point C, in Fig. 9.2.

9.10 WET STEAM (TWO-PHASE MIXTURES)

Wet steam is defined as a mixture of dry steam and suspended water particles. Thus, it is a two-phase mixture. The steam ordinarily generated in any boiling vessel is wet steam. By applying heat to wet steam, it becomes dry and saturated as all the suspended water particles evaporate.

The relative amount of each phase present in wet steam (i.e., **in the two-phase mixture**) determines the ***quality or dryness fraction of steam***. *The quality or dryness fraction of steam* *x is defined as the ratio of the mass of dry steam to the total mass of mixture* (i.e., *wet steam*). Thus, for any wet vapour

$$x = \frac{\text{Mass of dry steam}}{\text{Mass of mixture}} = \frac{\text{Mass of dry vapour}}{\text{Mass of mixture}}$$

If m_g = mass of dry saturated steam per kg of mixture

m_f = mass of suspended water or fluid particles per kg of mixture

$$\therefore \quad x = \frac{m_g}{m_g + m_f} \quad \text{...(9.1)}$$

The above equation shows that dryness fraction of saturated liquid is zero because vapour phase is absent whereas the dryness fraction of dry and saturated steam is unity because water particles are absent.

If the dryness fraction of steam is 0.9, it means that in one kg of wet steam, 0.9 kg is the dry steam and 0.1 kg is the suspended water particles.

It is to be noted that *quality of steam* (*x*) *is regarded as the intensive property of steam.*

9.11 PROPERTIES OF PURE SUBSTANCE (STEAM)

The thermodynamic properties of steam includes temperature, pressure, specific volume, enthalpy, internal energy, entropy, etc. We have already discussed the variations of temperature and pressure of steam. Now, we will study the rest of the properties.

(a) Specific Volume *The volume of one kg of steam is called the specific volume of steam.* The specific volume of dry and saturated steam is denoted by v_g. If the steam is wet having dryness fraction *x*, then (1 – *x*) kg of water particles are suspended, in *x* kg of dry steam. Hence, specific volume of wet steam is given by

$$v = xv_g + (1 - x)\, v_f \quad \text{...(9.2)}$$

where v = specific volume of wet steam, m^3/kg

v_f = specific volume of saturated water or fluid, m^3/kg.

But v_f is very small except at very high pressure, hence $(1 - x)\, v_f$ is generally neglected.

$$\therefore \quad v = xv_g \;\; m^3/kg \quad \text{...(9.3)}$$

It is to be noted that volume of water at high pressure cannot be neglected.

For superheated steam, the specific volume may be found with reasonable accuracy by applying the general gas law pv/T = constant, since it is assumed that superheated steam behaves as a perfect gas from dry saturated condition, i.e., from the beginning of superheating.

Thus,

$$\frac{p_g \cdot v_g}{T_g} = \frac{p_s \cdot v_s}{T_s} = \frac{p_{sup} \cdot v_{sup}}{T_{sup}}$$

But superheating is done at constant pressure, hence $p_g = p_f = p_{sup}$. Therefore,

$$\frac{v_g}{T_g} = \frac{v_{sup}}{T_{sup}}$$

or
$$v_{sup} = v_g \frac{T_{sup}}{T_g} \quad \text{...(9.4)}$$

If we know the value of specific volume for saturated steam v_g, temperature of superheated steam T_{sup} and temperature of dry and saturated steam T_g, we can find the value v_{sup}.

According to Prof. Linds, the specific volume for superheated steam valid up to pressure of 20 atm is given by

$$v_{sup} = \frac{0.417\, T_{sup}}{p_{sup}} - 0.016 \text{ m}^3\text{/kg}$$

where p_{sup} is the pressure of superheating, kN/m^2

(b) Enthalpy of Steam The transformation of water from 0°C into steam needs heat quantities given below:

(i) Stensible heat of water (h_f) *It is the quantity of the heat required to increase the temperature of 1 kg of water from 0°C to boiling point or saturation temperature at a given pressure. Sensible heat increase as the pressure increases.*

(ii) Latent heat (h_{fg}) *It is the quantity of heat required to convert 1 kg of water at saturation temperature for a given pressure into dry and saturated steam at that temperature and pressure. The value of latent heat of vaporization decreases as the pressure increases and it becomes zero when the critical pressure is reached.*

Thus, the enthalpy of wet steam = $h = h_f + x\, h_{fg}$ kJ/kg ...(9.5)

where h_f = enthalpy of water, kJ/kg

The value of sensible heat (h_f) can be calculated by the equation given below.

$$h_f = m \cdot c_p\,(t_g - 0) = 1 \times 4.18 \times (t_g) = 4.18\, t_g \text{ kJ/kg}$$

Thus, the sensible heat of water is equal to the 4.18 t_g, i.e., 4.18 × boiling temperature at a given pressure. Its validity can be confirmed from the fact that for a wet steam having dryness fraction x, only x part mass of the water absorbs latent heat, and $(1 - x)$ part remains in water form with an enthalpy h_f. Since the saturation temperature increases with pressure, hence the sensible heat will also increase with pressure. The enthalpy of the wet steam will also increase with pressure.

For dry and saturated steam ($x = 1$), the enthalpy (h_g) is given by

$$h_g = h_f + h_{fg} \text{ kJ/kg} \quad \text{...(9.6)}$$

Thus, the enthalpy of dry and saturated steam is the heat which must be supplied to 1 kg of water at 0°C to convert it at constant pressure into dry and saturated steam at the pressure and temperature considered. The enthalpy of dry and saturated steam increases with the pressure.

For superheated steam, the enthalpy = $h_{sup} = h_g + c_p\ (t_{sup} - t_g)$, kJ/kg ...(9.7)

where h_{sup} = enthalpy of superheated steam, kJ/kg

c_p = mean specific heat of superheated steam at constant pressure, kJ/kg K

h_g = enthalpy of saturated steam, kJ/kg

t_{sup} = temperature of superheat steam, °C

t_g = saturation temperature of steam, °C

Hence, $(h_{sup} - t_g)$ is called ***degree of superheat***. The value of mean specific heat of superheat steam (c_p) depends on the degree of superheat and the pressure of steam generation.

(c) Internal Energy of Steam The internal energy of steam is

$$u = h - pv, \text{ kJ/kg}$$

where h = enthalpy of steam, kJ/kg

p = pressure of steam, kN/m^2

v = sp. volume of steam, m^3/kg

Hence, the internal energy of saturated liquid, $u_f = h_f - p_f\, v_f$ ...(9.8)

Internal energy of wet steam, $u = (h_f + xh_{fg}) - p_g \cdot x \cdot v_g$ ...(9.9)

Internal energy of saturated steam, $u_g = h_g - p_g\, v_g$ ...(9.10)

The internal energy of superheated steam, $u_{sup} = u_{sup} - p_{sup} \cdot v_{sup}$ ...(9.11)

The change in internal energy during conversion of water into steam is

$$u_2 - u_1 = (h_2 - h_1) - (v_2 - v_1)\ p;\ \text{kJ/kg} \quad ...(9.12)$$

This suggests that the change in internal energy is the difference of

(i) change in enthalpy $(h_2 - h_1)$, and

(ii) heat required to expand the volume from v_1 to v_2 at constant pressure, p.

In some cases, the change in volume $(v_2 - v_1)$ is negligible, hence the change in internal energy is equal to change in enthalpy.

But in general, the change in internal energy differs from change in enthalpy by an amount $p\ (v_2 - v_1)$. The value of $p\ (v_2 - v_1)$ is known as ***external work of expansion***.

We know that during evaporation of steam at 1 atm and 100°C, the change in enthalpy is equal to latent heat of vapourization h_{fg}. Thus,

$$h_g - h_f = h_2 - h_1 = h_{fg}$$

$\therefore \quad h_2 - h_1 = h_{fg} = (u_2 - u_1) + p\,(v_2 - v_1) = h'_{fg} + w_{evp}$ ...(9.13)

where h'_{fg} = internal latent, heat kJ/kg

$w_{evp} = p\,(v_2 - v_1)$ = external work of evaporation

(d) Entropy of Steam The change in entropy of steam is

$$ds = \left(\frac{\delta q}{T}\right)_{rev}; \text{ kJ/kg K}$$

When heating is at constant pressure, hence

$\delta q = c_p\, dT$ where c_p = specific heat at constant pressure

$$ds = \frac{c_p\, dT}{T}$$

$$\Delta s = s_2 - s_1 = \int_{T_1}^{T_2} \frac{c_p\, dT}{T} = c_p \ln \frac{T_2}{T_1}$$

If 0°C is chosen as the datum for entropy, the entropy of steam is

$$s = c_p \ln \frac{T}{273}; \text{ kJ/kg K} \quad \text{...(9.14)}$$

Entropy of water at saturation temperature, $s_f = c_{pf} \ln \dfrac{T_g}{273}$ ...(9.15)

where c_{pf} = 4.18 kJ/kg K

During evaporation of 1 kg of water at saturation temperature, the heat added is equal to the latent heat of vaporization = h_{fg}. Hence, the entropy of vaporization is

$$s_{fg} = \frac{h_{fg}}{T_g}$$

$\therefore$ Entropy of dry and saturated steam, $s_g = s_f + s_{fg}$ ...(9.16)

Entropy of wet steam, $s = c_{pf} \ln \dfrac{T_g}{273} + \dfrac{x h_{fg}}{T_g} = s_f + x\, s_{fg}$ ...(9.17)

Entropy of superheat steam, $s_{sup} = s_f + s_{fg} + (c_p) \ln \dfrac{T_{sup}}{T_g}$ ...(9.18)

or

$$s_{sup} = c_{pf} \ln \frac{T_g}{273} + \frac{h_{fg}}{T_g} + (c_p)_{sup} \ln \frac{T_{sup}}{T_g} \quad \text{...(9.19)}$$

where $(c_p)_{sup}$ = 2 to 2.7 or even more kJ/kg K depending upon the degree of superheat and superheating pressure.

(e) Steam Tables Properties of steam, such as pressure, temperature, specific volume, enthalpy, entropy, etc., have been experimentally determined and tabulated and are available in a form called ***steam table***. For engineering calculations, these values are picked up from steam tables and not calculated each time from the equations which are complex in nature, and also because of the complex behaviour of specific heat of water and steam and the complex relationship between pressure, saturation temperature and specific volume.

In steam tables, properties of dry and saturated steam from 1 kg of steam are given as shown in Table 9.3. For wet steam, the properties may easily be calculated from Steam Tables of dry and saturated steam.

The properties of superheated steam are tabulated (Table 9.4) separately. They depend not only on pressure but also on superheating temperature (T_{sup}) or degree of superheat ($T_{sup} - T_g$).

(f) Interpolation For properties which are not tabulated exactly in the tables, it is necessary to interpolate between the values tabulated. For example, to find *t*, *u*, *h*, *s* of dry saturated steam at 15.5 bar, it is necessary to interpolate by the governing linear equation for the desired property say *y* is

$$y = y_1 + \frac{p - p_1}{p_2 - p_1}(y_2 - y_1)$$

$$h_g = \text{Saturated enthalpy at 15 bar} + \frac{(15.5 - 15)}{(16 - 15)} \times \text{(Saturation enthalpy at 16 bars, saturation enthalpy at 15 bar)}$$

$$h_g = 2792 + \frac{(15.5 - 5)}{(16 - 15)}(2794 - 2792) = 2793 \text{ kJ/kg}$$

The interpolation is shown in Fig. 9.12. Sometimes double interpolation is also done.

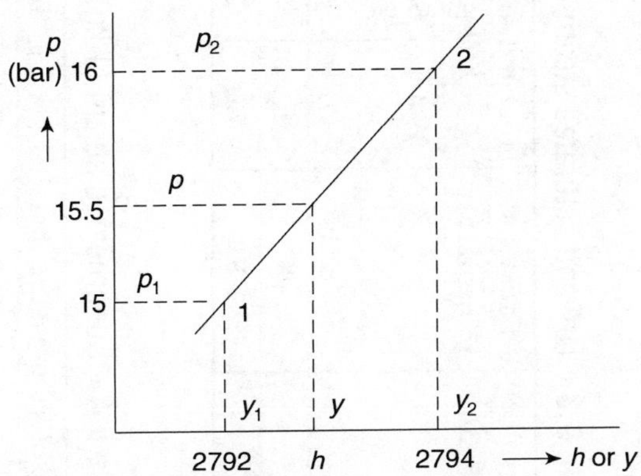

Fig. 9.12 Interpolation

Suffixes used :

f : stands for fluid

w : stands for water

fg : stands for fluid and gaseous state, L or h_{fg} : stands for latent heat

g : stands for gaseous (vapour) state (dry and saturated)

s : stands for dry and saturated steam (gaseous phase)

Sup : Superheated steam

Table 9.3 Dry and saturated steam (1 kg)

Absolute pressure	Sauration temeprature	Internal energy kJ/kg		Enthalpy, kJ/kg			Entropy, kJ/kg K			Specific volume m³/kg	
		Water	Steam	Water	Evaporation	Dry steam	Water	Evaporation	Dry steam	Water	Dry steam
		(u_f)	(u_g)	(h_f)	(h_{fg})	(h_g)	(s_j)	(s_{fg})	(s_g)	(v_f)	(v_g)
bar	t°C	u_w	u_s	h_w	L	h_s	s_w	s_L	s_g	v_w	v_s
10	179.9	762	2584	763	2015	2778	2.138	4.448	6.586	0.001128	0.1944

Table 9.4 Superheated steam (1 kg)

Absolute pressure, bar		Temperature of superheated steam, t_{sup}, °C				
		200°C	300°	400°	500°	600°C
10	h	2829	3052	3264	3478	3698
(179.9°C)	v	0.2061	0.258	0.3065	0.354	0.4010
	s	6.695	7.124	7.464	7.76	8.028

(g) Properties of Other Pure Substances Articles (9.8) and (9.11) valid for steam are also valid for other pure substances such as **refrigerant like NH_3, R–12 or R–22, etc.** *With the help of these equations, the properties may be tabulated. Separate refrigerant tables and charts are prepared for the use in refrigeration and air conditioning problems.* The reference state for refrigerants are taken generally as –100°C where *h*, *u*, *s* have zero values.

Example 9.1 Find the dryness fraction of steam if 2 kg of water is in suspension with 90 kg of dry and saturated steam.

Solution Mass of dry and saturated steam, $m_g = 90$ kg

Mass of water in suspension, $m_f = 2$ kg

$\therefore$ Dryness fraction, $x = \dfrac{m_g}{m_g + m_f} = \dfrac{90}{90+2} = \dfrac{90}{92} = 0.9782$ **Ans.**

Example 9.2 Calculate the state of steam, i.e., whether it is wet, dry or superheated for the following cases:

(i) Steam has a pressure of 15 bar and specific volume of 0.12 m^3/kg.

(ii) Steam has a pressure of 10 bar and temperature 200°C.

(iii) Steam has a pressure of 30 bar and if 2700 kJ/kg of heat is required to generate steam.

Solution (i) At 15 bar, the specific volume of dry and saturated steam is 0.13177 m^3/kg. But the specific volume of steam given is 0.2 m^3/kg, this indicates that the steam is wet. We known that

$$v = 0.12 = (1 - x)\, v_f + x \,.\, v_g$$

At 15 bar from steam table, $v_g = 0.001154$ m^3/kg

$$\therefore \quad 0.12 = (1 - x)\, 0.001154 + x \times 0.13167$$

or $$x = \frac{0.1188}{0.1305} = 0.9103 \quad \textbf{Ans.}$$

Since the specific volume of water is negligible, we may neglect this. By neglecting v_f, we have

$$v = xv_g \quad \text{or} \quad 0.12 = x \times 0.1316$$

or $$x = \frac{0.12}{0.13167} = 0.91136 \quad \textbf{Ans.}$$

(ii) The saturation temperature corresponding to 10 bar is 179.0°C from Steam Tables. But the steam in question has a temperature of 200°C, which means that the steam is superheated. Thus, the degree of superheat, $t_{sup} - t_g = 200 - 179.9 = 20.1$°C. **Ans.**

(iii) At 30 bar, the dry and saturated steam has enthalpy equal to 2803 kJ/kg as obtained from steam tables. But the enthalpy of the given steam is 2700 kJ/kg which is less than the saturated steam at that pressure, this indicates that the steam is wet

Thus, $h = 2700 = h_f + xh_{fg}$

At 30 bar, $h_f = 1008$ kJ/kg and $h_{fg} = 1795$ kJ/kg

$\therefore$ $2700 = 1008 + x \times 1795$

$x = 0.9426$ **Ans.**

Example 9.3 Two kg of steam at a pressure of 20 bar exists in the following conditions : (i) wet steam with a dryness fraction of 0.9, (ii) dry and saturated steam, (iii) superheated steam with temperature 250°C. Calculate (a) enthalpy, (b) volume, (c) entropy and (d) internal energy. Assume $c_p = 2.302$ kJ/kg K for superheated steam.

Solution From Steam Tables, at 20 bar

$h_f = 909$ kJ/kg $\qquad s_f = 2.447$ kJ/kg K

$h_{fg} = 1890$ kJ/kg $\qquad s_g = 6.340$ kJ/kg K

$h_g = 2799$ kJ/kg $\qquad s_{fg} = s_g - s_f = 6.3 - 2.447 = 3.893$ kJ/kg

$v_g = 0.09957$ m^3/kg $\qquad t_g = 212.4$°C

(a) (i) Enthalpy of wet steam $= h = h_f + x\,h_{fg} = 909 + 0.9 \times 1890 = 2610$ kJ/kg

$\therefore$ Total enthalpy of wet steam $= H = 2 \times 52201$ kJ **Ans.**

(ii) Enthalpy of dry steam $= h_g = 2799$ kJ/kg **Ans.**

$\therefore$ Total enthalpy of dry steam $= H_g = 2 \times 2799 = 5598$ kJ

(iii) Enthalpy of superheated steam is

$$h_{sup} = h_g + c_p\,(t_{sup} - h_g) = 2799 + 2.023\,(250 - 212.4) = 2875.06 \text{ kJ/kg}$$

Total enthalpy of superheated steam $= H_{sup} = 2 \times 2875.06 = 5750.12$ kJ **Ans.**

(b) (i) The specific volume, $v = (1 - x)\,v_f + xv_g$

Neglecting specific volume of water, $v = xv_g = 0.9 \times 0.09957 = 0.089613$ m^3/kg

$\therefore$ Total volume of wet steam, $V = 2 \times 0.08961 = 0.179226$ m^3 **Ans.**

(ii) Total volume of dry steam, $V_g = 2 \times v_g = 2 \times 0.09957 = 0.19914$ m^3 **Ans.**

(iii) The specific volume of superheated steam,

$$v_{sup} = v_g \frac{T_{sup}}{T_g} = 0.09957 \times \frac{523}{485.4} = 0.10728 \text{ m}^3\text{/kg}$$

Total volume of superheated steam, $V_{sup} = 2 \times 0.10728 = 0.21456$ m^3 **Ans.**

(c) (i) The entropy of wet steam is

$$s = s_f + x\,s_{fg} = 2.447 + 0.9 \times 2.893 = 5.9507 \text{ kJ/kg K}$$

$\therefore$ Total entropy of wet steam = $S = 2 \times 5.9507 = 11.9014$ kJ/kg K **Ans.**

(ii) The total entropy of dry steam = $m \times s_g = 2 \times 6.34 = 12.68$ kJ/K **Ans.**

(iii) The entropy of superheated steam

$$= s_{sup} = s_g + c_p \ln \frac{T_{sup}}{T_g}$$

$$= 6.34 + 2.202 \ln \frac{523}{485.4} = 6.5042 \text{ kg/K}$$

The total entropy of superheated steam = $2 \times 6.5042 = 13.0008$ kJ/K **Ans.**

The enthalpy and entropy of superheated steam may also be read directly from steam tables.

(d) (i) The internal energy of wet steam is

$$u = h - pv = 2610 - 20 \times 10^2 \times 0.089613 = 2430.77 \text{ kJ/kg}$$

Total internal energy of wet steam = $U = 2 \times 2430.77 = 4861.54$ kJ **Ans.**

(ii) $$u_g = h_g - pv_g = 2799 - 20 \times 10^2 \times 0.09957 = 2599.86 \text{ kJ/kg}$$

Total internal energy of dry steam, $U_g = 2 \times 2599.86 = 5199.72$ kJ **Ans.**

(iii) The internal energy of superheated steam is given by

$$u_{sup} = h_{sup} - pv_{sup} = 2875.403 - 20 \times 0.10728 \times 10^2 = 2660.84 \text{ kJ/kg}$$

The total internal energy of superheated steam, $U_{sup} = 2 \times 2660.84 = 5321.68$ kJ **Ans.**

Example 9.4 Two boilers, *A* and *B*, discharge equal amount of steam into a common main steam pipe. The steam from boiler *A* is at 10 bar and 350°C and from *B* at 10 bar and 0.94 dryness fraction. Calculate (a) the equilibrium condition after mixing in the main, (b) the loss of entropy by high temperature steam, (c) gain of entropy of the low pressure steam, and (d) net increase or decrease of entropy.

Solution From Steam Tables at 10 bar and 350°C (For boiler *A*)

$$h_{sup} = 3188 \text{ kJ/kg}, \; s_{sup} = 7.301 \text{ kJ/kg K}$$

At 10 bar for dry and saturated steam (from Steam Tables)

$$h_f = 763 \text{ kJ/kg}, \; h_{fg} = 2015 \text{ kJ/kg}, \; h_g = 2778 \text{ kJ/kg}$$

$$s_f = 2.138 \text{ kJ/kg K}, \; s_g = 6.586 \text{ kJ/kg K}$$

$$s_{fg} = s_g - s_f = 6.586 - 2.138 = 4.448 \text{ kJ/kg K}$$

To simplify let us assume that 1 kg of steam is delivered by both boilers.

Hence, the enthalpy of mixture, $h_{mix} = \dfrac{h_A + h_B}{2} = \dfrac{(h_f + x\,h_{fg}) + h_{sup}}{2}$

$$= \frac{(763 + 0.94 \times 2015) + 3158}{2} = 2907.55 \text{ kJ/kg}$$

The enthalpy of dry and saturated steam at 10 bar = 2778 kJ/kg. This means that mixture steam in the main is superheated.

Corresponding to mixture enthalpy, h_{mix} = 2907.55 kJ/kg and pressure at 10 bar from Steam Tables for superheated conditions using interpolation

$$(t_{sup})_{mix} = 234.15°\text{C}, (v_{sup})_{mix} = 0.22433 \text{ m}^3/\text{kg}, (s_{sup})_{mix} = 6.8527 \text{ kJ/kg}$$

(b) Loss of entropy by high temperature steam

$$= (s_{sup})_A - (s_{sup})_{mix} = 7.301 - 6.8527 = 0.4483 \text{ kJ/kg}$$ **Ans.**

(c) Gain of entropy of the low pressure steam

$$= (s_{sup})_{mix} - (s_f \_\ x\,s_{fg}) = 6.8527 - (2.138 + 0.94 \times 4.498)$$

$$= 0.48676 \text{ kJ/kg K}$$ **Ans.**

(d) The net gain in entropy = 0.48676 – 0.4483 = 0.03846 kJ/kg K **Ans.**

Example 9.5 A metallic drum of volume 2 m^3 is to be completely filled with dry and saturated steam at 8 bar. For this, the drum is evacuated and the necessary amount of water is placed. Now this water is completely converted into steam by heating. Calculate (a) the necessary amount of water to be placed in the drum and the resulting temperature of steam, and (b) the amount of water if the steam generated is 0.9 dryness fraction.

Solution (a) From Steam Tables, at 8 bar, the saturation temperature

$$t_g = 170.4°\text{C}$$ **Ans.**

The specific volume of dry and saturated steam at 8 bar = v_g = 0.2403 m^3/kg

$$\text{Mass of water} = \text{Mass of steam} = \frac{\text{Volume of drum}}{\text{Sp. volume of steam}} = \frac{2}{0.2403}$$

$$= 8.3229 \text{ kg}$$ **Ans.**

(b) The specific volume of wet steam is, $v = (1 - x)\,v_f + x\,v_g$

Neglecting the sp. volume of water = $v = x_g = 0.9 \times 0.2403 = 0.21627$ m^3/kg

$$\therefore \quad \text{Mass of water} = \frac{\text{Volume of drum}}{\text{Sp. volume of steam}} = \frac{2}{0.2627} = 7.6132 \text{ kg}$$ **Ans.**

Example 9.6 A vessel of volume 0.05 m^3 contains a mixture of saturated water and saturated steam at a temperature 250°C. The mass of liquid present is 10 kg. Calculate the pressure, the mass, the specific volume, the enthalpy, and the entropy of steam.

Solution From Steam Tables at 250°C. p_{sat} = 39.73 bar

$$v_f = 0.0012512 \text{ m}^3/\text{kg},\ v_g = 0.05013 \text{ m}^3/\text{kg}$$
$$h_f = 1085.36 \text{ kJ/kg},\ h_{fg} = 1716.2 \text{ kJ/kg}$$
$$s_f = 2.7927 \text{ kJ/kg K},\ s_{fg} = 3.2802 \text{ kJ/kg K}$$

Volume of liquid, $V_f = m_f\, v_f = 10 \times 0.0012512 = 0.012512 \text{ m}^3$

Volume of vapour, $V_g = 0.05 - 0.012512 = 0.037488 \text{ m}^3$

Mass of vapour, $m_g = \dfrac{V_g}{v_g} = \dfrac{0.037488}{0.05013} = 0.7478$ kg

Total mass of mixture $= m_f + m_g = 10 + 0.7478 = 10.7478$ kg **Ans.**

Quality of mixture, $x = \dfrac{m_g}{m_f + m_g} = \dfrac{0.7478}{10.7478} = 0.0695$ **Ans.**

$\therefore$
$$v = v_f + x\,(v_g - v_f) = (1 - x)\, v_f + x\, v_g$$
$$= 0.012512 + 0.0695\ (0.05013 - 0.012512)$$
$$= 0.01512 \text{ m}^3/\text{kg}$$ **Ans.**
$$h = h_f + x\, h_{fg} = 1085.36 + 0.0695 \times 1716.2$$
$$= 1204.63 \text{ kJ/kg K}$$ **Ans.**
$$s = s_f + x\, s_{fg} = 2.7927 + 0.0695 \times 3.2802$$
$$= 3.02067 \text{ kJ/kg K}$$ **Ans.**

Example 9.7 The pressure of steam in a condenser is 0.16 bar and the dryness fraction is 0.85. Calculate the amount of heat that must be abstracted from steam in order to condense (a) 1 kg of steam, and (b) 1 m^3 of steam.

Solution In order to condense steam, its latent heat must be abstracted.

From Steam Tables, at 0.16 bar, latent heat, h_{fg} = 2369 kJ/kg

(a) Heat to be removed $= x \cdot h_{fg} = 0.85 \times 2369 = 2013.65$ kJ/kg **Ans.**

(b) From Steam Tables, specific volume of dry and saturated steam, $v_g = 9.432 \text{ m}^3/\text{kg}$

$\therefore$ Specific volume of steam of 0.85 dryness fraction neglecting volume of water,

$$xv_g = 0.85 \times 9.423 - 8.00955 \text{ m}^3/\text{kg}$$

$\therefore$ Heat to be removed per m^3 of steam $= \dfrac{2013.65}{8.00955} = 251.40$ kJ **Ans.**

Example 9.8 In a piston cylinder arrangement as shown in Fig. 9.13, when throttle valve is closed 3 kg of dry steam at pressure 0.8 bar is retained in the cylinder. The throttle valve is opened till the pressure in the cylinder rises to 6 bar. During this opening period of throttle valve, 5 kg of dry steam enters the cylinder and 230 kJ of heat is transferred to the cylinder and the work done on the spring is 1860 kJ. Calculate the final quality of steam.

Solution Refer to Fig. 9.13.

Let Q = heat transferred to the cylinder = 230 kJ

m_1 = mass of steam retained in the cylinder before operation = 3 kg

m = mass of steam transferred to the cylinder = 5 kg

m_2 = mass of steam in the cylinder at the end of operation = $m_1 + m$ = 8 kg

h_1 = enthalpy of steam retained in the cylinder before operation

h = enthalpy of steam transferred to the cylinder

h_2 = enthalpy of steam at the end of operation

U_1 = internal energy of steam at the end of operation

U = internal energy of steam transferred to the cylinder

U_2 = internal energy of steam after the operation

W = work done by the steam on piston = 1860 kJ

From first law (unsteady flow), $Q + me_f = (m_2\, u_2 - m_1\, u_1) + W$ ...(i)

Neglecting PE and KE of entering steam, $e_f = h$, so the above equation becomes

$$Q + m\,(u + pv) = (m_2\, u_2 - m_1\, u_1) + W$$

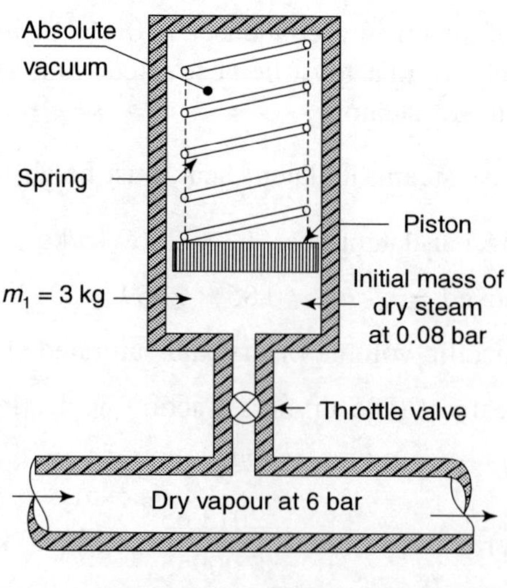

Fig. 9.13 Example 9.8

where pv is the flow energy associated with entering steam. From Steam Tables at 6 bar

$$h = h_g = 2757 \text{ kJ/kg}$$

At $\quad 0.8 \text{ bar}, h_{fg} = 2665 \text{ kJ/kg}, \quad v_g = 2.087 \text{ m}^3\text{/kg}$

$\therefore \quad u_1 = h_g - p_1 v_g = 2665 - 0.8 \times 10^2 \times 2.087 = 2498.04 \text{ kJ/kg}$

Substituting the value in Eq. (1)

Now, $\quad 230 + 5 \times 2757 = (8\, u_2 - 3 \times 2498.04) + 1360$

or $\quad u_2 = 2456.14 \text{ kJ/kg}$

But $\quad u_2 = 2456.14 = h_2 - p_2\, v_2 = (h_{f2} + x_2\, h_{fg2}) - p_2\, x_2\, v_{g2}$

or $\quad 2456.14 = 670 + x_2 \times 2087 - 6 \times 10^2 \times 0.3156\, x_2$

or $\quad x_2 = 0.941$ **Ans.**

Example 9.9 In a piston-cylinder arrangement, a constant pressure of 7 bar maintained in the cylinder by the weight in the piston. The steam flowing from the main pipe at 7 bar, 340°C enters into the cylinder. Determine the final temperature.

Solution From first law

$$Q = (U_2 - U_1) + W = (U_2 - U_1) + p_2\,(v_2 - v_1) \times m$$

Since there is no steam in the cylinder in the beginning, hence $U_1 = 0, \quad v_1 = 0$

$$Q = mh_1$$

$\therefore \quad mh_1 = mu_2 + p_2\, v_2 \times m = mh_2 \quad \text{or} \quad h_1 = h_2$

$t_2 = 340°\text{C}$ **Ans.**

9.12 STEAM PROPERTY DIAGRAMS

The properties of steam are given in tabular form for engineering calculations. To make the procedure still simpler and to acquire information needed readily *property data of steam are presented in graphical forms called diagrams or charts.* Various diagrams may be prepared such as *T v*, *p–h*, *p–v*, *T–s* and *h–s*. Among these diagrams, the two most important diagrams are (i) temperature-entropy (*T–s*) diagram and (ii) enthalpy-entropy (*h–s*) diagram, also known as Mollier diagram after the name of its inventor Dr. R. Mollier. Taking the various properties of steam from Steam Tables, these diagrams are drawn. *T–v* and *p–v* diagrams have been discussed earlier.

9.12.1 Temperature-Entropy (*T–s*) Diagram

Figure 9.14 shows a phase equilibrium diagram on *T*–s coordinates. This gives the idea of behavioural changes taking place between the three phases (*S* + *L* + *V*) of a pure substance on *T–s* coordinates. The isobars 1–2–3–4–5–6 at two different values are depicted in this figure.

This could be easily obtained by calculating the entropy changes taking place from assumed initial temperature to final temperature through different regimes of heating.

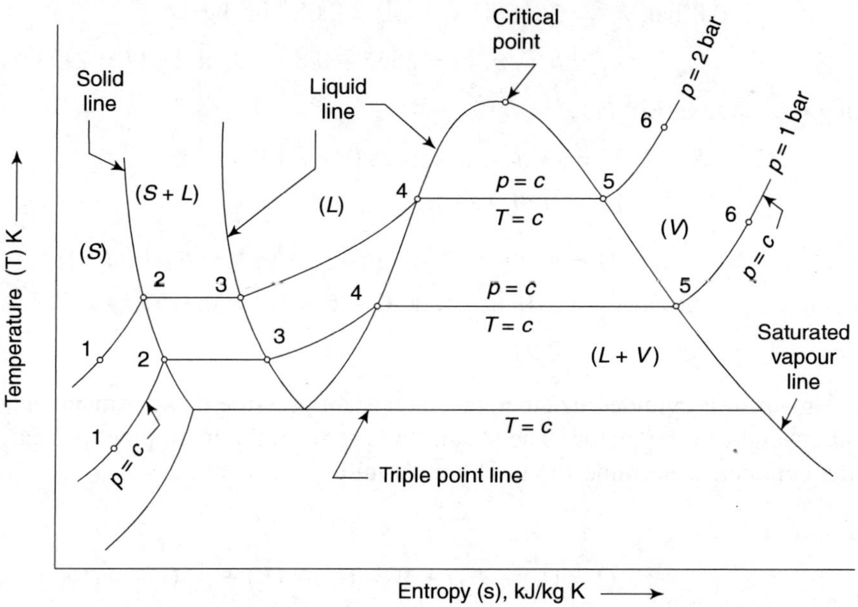

Fig. 9.14 Phase equilibrium diagram on *T*–*s* co-ordinates

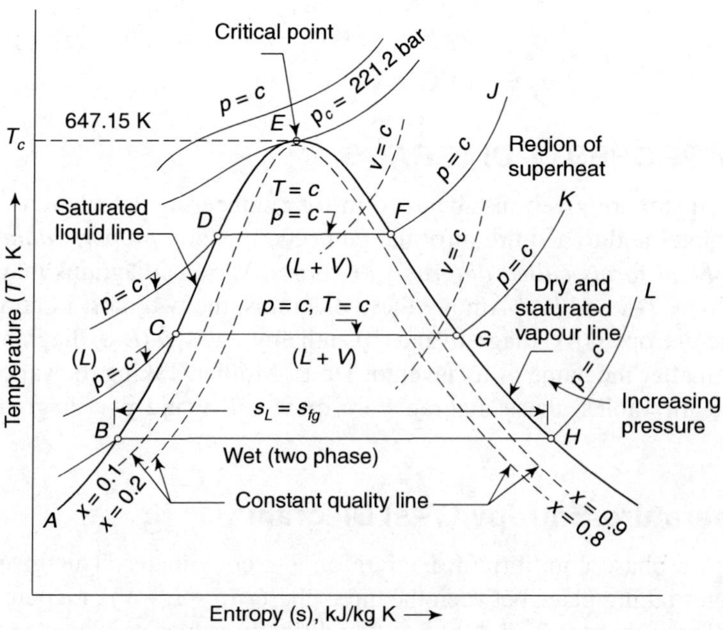

Fig. 9.15 *T*–*s* diagram for steam

Most often liquid-vapour transformation only are of interest and Fig. 9.15 shows *T–s* diagram for water indicating only liquid and vapour. In the diagram, *ABCDE* is the saturated water line where *EFGH* is dry and saturated vapour line. The two lines meet at a point *E* called the ***critical point of water*** where temperature and pressure are 647.31 K and 221.2 bar, respectively. In addition, constant pressure lines such as *EI*, *DFJ*, *CGK* and *BHL* are drawn. The nature of the constant pressure line in liquid, we and superheat region is not the same. Slopes are governed by equations.

$$\left(\frac{\partial T}{\partial s}\right)_v = \frac{T}{c_v} \quad \text{and} \quad \left(\frac{\partial T}{\partial s}\right)_p = \frac{T}{c_p} \qquad \text{...(9.20)}$$

The lines of constant quality (i.e., dryness fraction) lines are also shown. These lines may be drawn by dividing into equal number of parts by the various intercepts of constant pressure lines between saturated liquid and saturated vapour line such as *DF*, *CG*, etc., and drawing the locii of corresponding dividing points. The constant specific volume line is shown by dotted line. For better usefulness of diagram, the constant enthalpy lines may also be drawn.

9.12.2 Enthalpy-Entropy (*h–s*) or Mollier Diagram

In engineering problems, it is mostly required to find out the amount of heat supplied or change in enthalpy during a process. From *T–s* diagram, the amount of heat supplied or change in enthalpy is not readily available. Therefore, *h–s* diagram becomes a necessity. Such a *h–s* diagram was first prepared by Dr. Mollier and hence it is known as Mollier diagram.

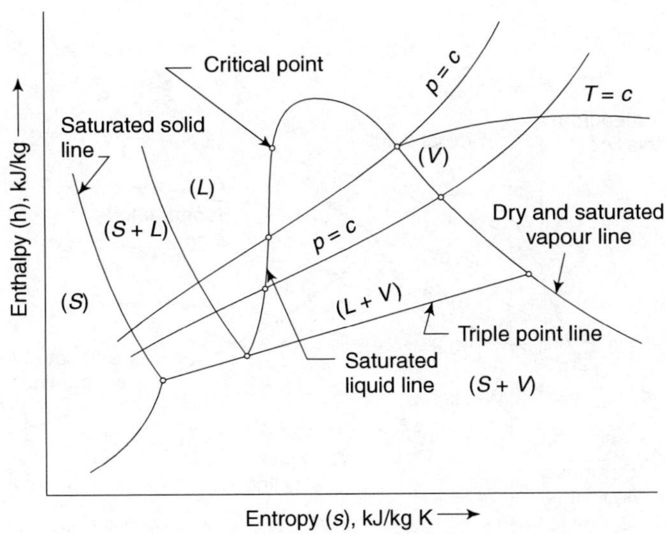

Fig. 9.16 Phase equilibrium diagram of a pure substance on *h-s* co-ordinates

From the first and second laws, the following Tds relation is obtained

$$Tds = dh - vdp$$

or
$$\left(\frac{\partial h}{\partial s}\right)_p = T \qquad ...(9.21)$$

This equation forms the basis of h–s diagram of a pure substance.

Figure 9.16 shows the phase equilibrium diagram of a pure substance on h–s coordinates indicating the saturated solid line, saturated liquid line, saturated vapour line, various phases and transition zones.

Figure 9.17 shows the h–s or Mollier diagram indicating only the liquid and vapour phases. Line AB represents the saturated liquid line. BCD represents the dry and saturated vapour line. Point B indicates the critical point. Since $(\partial h/\partial s)_p = T$, so inclination of constant pressure line will be equal to absolute temperature. In the wet region $(L + V)$, temperature is constant. Thus, the *constant pressure lines in this very zone will be straight line and will merge with constant temperature line*. As the pressure increases, the saturation temperature increases and so the slope of the isobar also increases. Hence, the constant pressure lines diverge from one another and the critical isobar is a tangent at the critical point. In the superheated region, these pressure lines bend upwards because temperature increases beyond the point of saturation. It is obvious from the diagram that the constant pressure line diverges as the value of entropy increases. At the vapour saturation point the lines of constant temperature will naturally diverge from lines of constant pressure and at low pressure and temperature, these lines becomes almost horizontal.

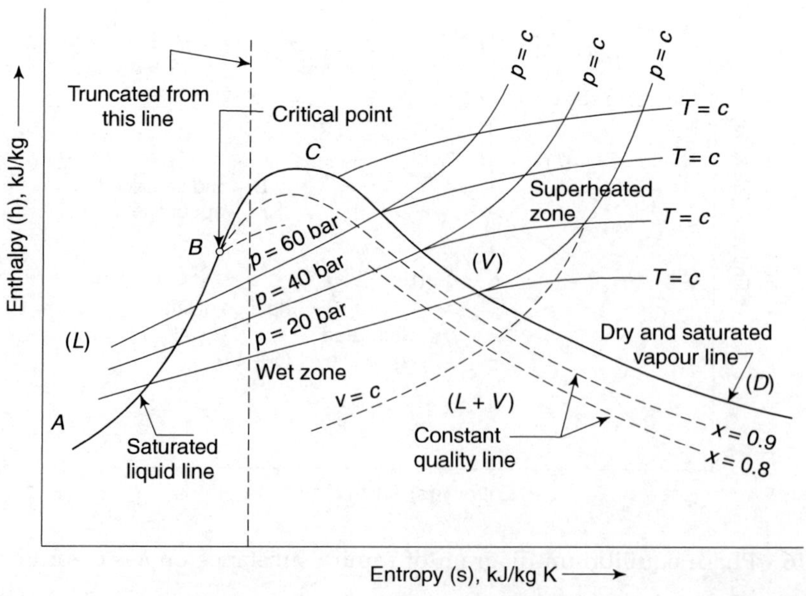

Fig. 9.17 h–s diagram for steam

Lines of constant quality (x) are drawn and in some chart lines of constant specific volume are drawn. *The commercial Mollier diagram is truncated from a point beyond the critical and so in such case, enthalpy of liquid cannot be read.*

9.12.3 Pressure Enthalpy (*p–h*) Diagram

Figure 9.18 is a pressure-enthalpy diagram. *AB* represents the saturated liquid line and *BC* the dry and saturated vapour line. The liquid line and the dry and saturated vapour line meet at the critical point *B*. The lines of constant temperature and constant entropy are drawn. The constant quality lines are also drawn in the diagram. **The *p–h* diagram of refrigerants are suitable for refrigeration cycles and throttling process.** *The nature of p–h diagram for refrigerant is more or less the same.*

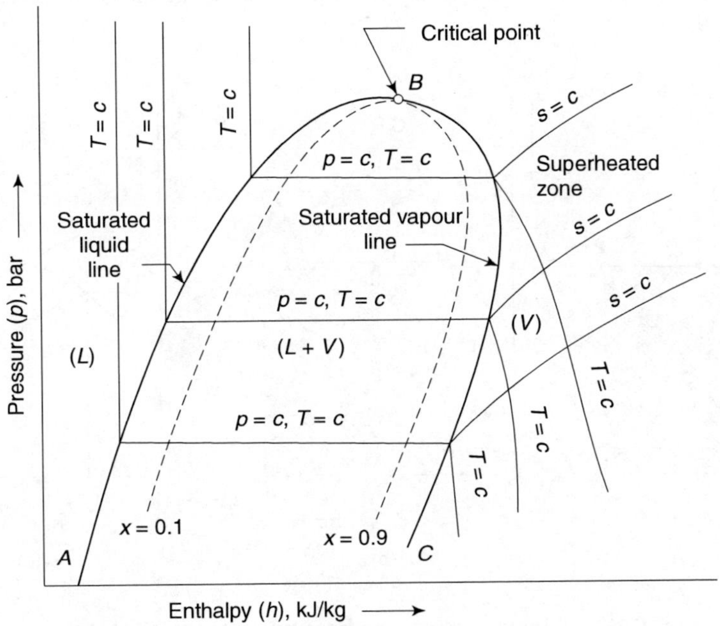

Fig. 9.18 *p-h* diagram for steam

9.13 APPLICATIONS OF FIRST AND SECOND LAW INVOLVING STEAM AS WORKING FLUID

Since steam is used as a working fluid in prime movers like steam engines and turbines, it undergoes various processes such as constant pressure, constant volume, adiabatic, isothermal, and polytropic processes. In some cases, steam may be compressed and in some cases it may be expanded. Thus, the analysis of these various processes for steam will be discussed **because steam does not behave as ideal gas and thus does not obey $pv = RT$.** However, the following basic energy equations derived from first and second laws hold good for steam.

$$\delta q = du + pdv, \delta q = dh - vdp \text{ and } \delta q = Tds$$

The expressions for work done during nonflow process $\int pdv$, flow process $-\int vdp$, SFEE, entropy balance, etc. also hold good. *The applications of various thermodynamic processes are given in* **subsequent subheadings.**

9.14 CONSTANT PRESSURE PROCESS

(a) *p–v*, *T–s*, and *h–s* Representation Assume that steam undergoes a constant pressure process. Then let the initial condition of steam be in the wet region to be represented as point 1 and the final condition in the superheat region by point 2 shown in Fig. 9.19 on *p–v*, *T–s* and *h–s* diagrams. The initial condition is set corresponding to initial pressure, and dryness fraction while the final condition is set corresponding to the pressure and final temperature. Thus, the process 1–2 shows the constant pressure heating process on *p–v*, *T–s* and *h–s* diagrams in Fig. 9.19.

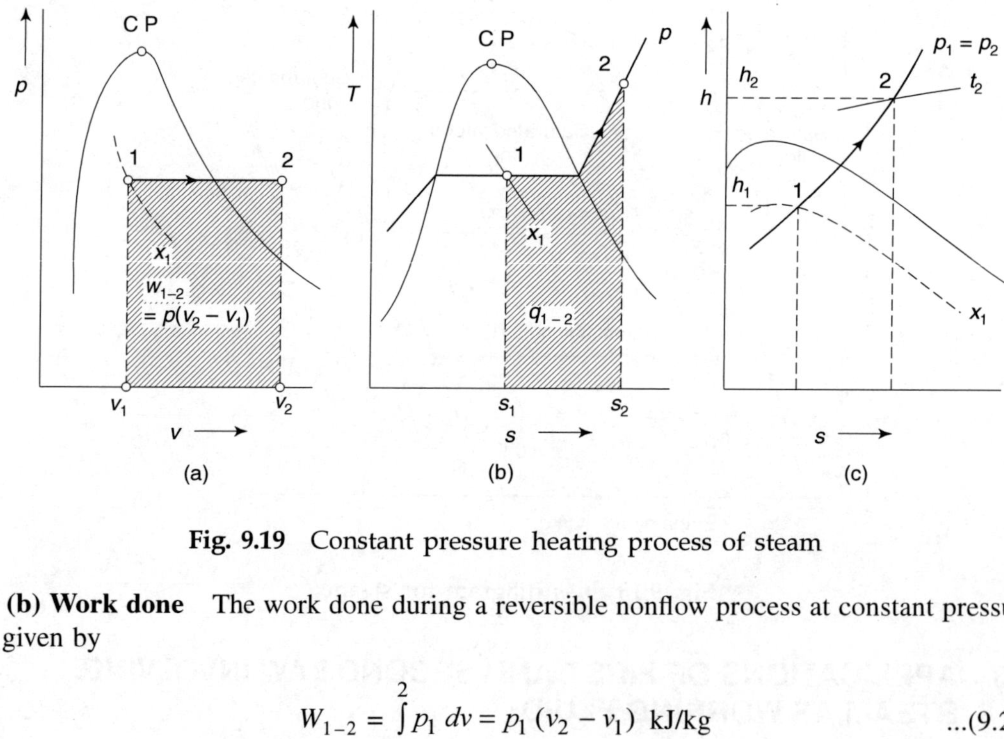

Fig. 9.19 Constant pressure heating process of steam

(b) Work done The work done during a reversible nonflow process at constant pressure is given by

$$W_{1-2} = \int_1^2 p_1 \, dv = p_1 (v_2 - v_1) \text{ kJ/kg} \qquad ...(9.22)$$

where $v_1 = xv_{g1} + (1 - x)v_{f1}$ at pressure p_1 taken from the Steam Tables.

v_2 = volume of superheated steam corresponding to pressure $p_1 = p_2$ and temperature T_2 taken from Steam Tables.

(c) Heat Transferred The heat transferred during nonflow process is given by

$$\delta q = du + pdv$$

or
$$q_{1-2} = (u_2 - u_1) + \int pdv = (u_2 - u_1) + p_2 v_2 - p_1 v_1 = h_2 - h_1, \text{ kJ/kg} \quad ...(9.23)$$

Thus, the heat transferred at constant pressure is equal to change in enthalpy. The value of h_1 is taken from Steam Tables corresponding to pressure p_1 and initial dryness fraction (i.e., $h_1 = h_{f1} + x_1 h_{fg1}$). The value of h_2 is taken from Steam Tables corresponding to pressure $p_1 = p_2$ and final temperature T_2. The same equation holds good for steady flow process.

(d) Change in Internal Energy The change in internal energy,

$$u_2 - u_1 = (h_2 - h_1) - (p_2 v_2 - p_1 v_1) \quad ...(9.24)$$

At constant pressure, for cooling process, the nomenclature used for states in Fig. 9.19 will be interchanged.

(e) Change in Entropy $\Delta s = s_2 - s_1$ directly available from chart or table.

Example 9.10 5000 kg of steam enters a surface condenser per hour at pressure 0.07 bar. 2,50,000 litres of cooling water are passed through the tubes per hour and its temperature rises from 21°C to 32°C. The condensed steam leaves the condenser with a temperature of 30°C. Calculate the condition of steam entering the condenser.

Solution Refer to Fig. 9.20. Volume flow rate of cooling water = 2.5×10^5 litres/h

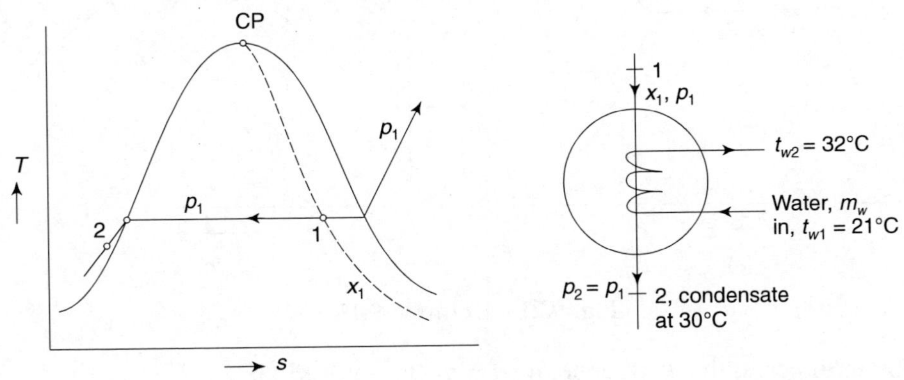

Fig. 9.20 Example 9.10

$$\text{We know that 1 litre} = 10^{-3} \text{ m}^3$$

$$\text{Volume flow rate of cooling water} = 2.5 \times 10^5 \times 10^{-3} = 250 \text{ m}^3\text{/h}$$

$$\text{We know that density of water at 20°C} = 1000 \text{ kg/m}^3$$

$\therefore$ Mass flow rate of cooling water = (250×1000) kg/h

$$\therefore \quad \text{Mass of cooling water per kg of steam} = \frac{250 \times 1000}{5000} = 50 \text{ kg/kg of steam}$$

At 0.07 bar from Steam Tables, $h_{f1} = 163$ kJ/kg, $h_{fg1} = 2409$ kJ/kg

$\therefore$ Enthalpy of steam, $h_1 = h_{f1} + x_1 h_{fg1} = 163 + x_1 \times 2409$

Enthalpy of condensate at 30°C, $h_{f2} = 125.7$ kJ/kg

Heat lost by steam = Heat gained by cooling water

or $$h_1 - h_{f2} = m_w (t_{w2} - t_{w1}) c_p$$

or $$(163 + x_1 \times 2409) - 125.7 = 50 (32 - 21) \times 4.18$$

or $$x_1 = \frac{2261.7}{2409} = 0.9388$$ **Ans.**

Example 9.11 3 kg of steam at 10 bar and 250°C undergoes a constant cooling pressure process. The resulting steam is wet having dryness fraction 0.6. Calculate (a) work done, (b) change in enthalpy, and (c) heat transferred assuming the nonflow process.

Solution Refer to Fig. 9.21.

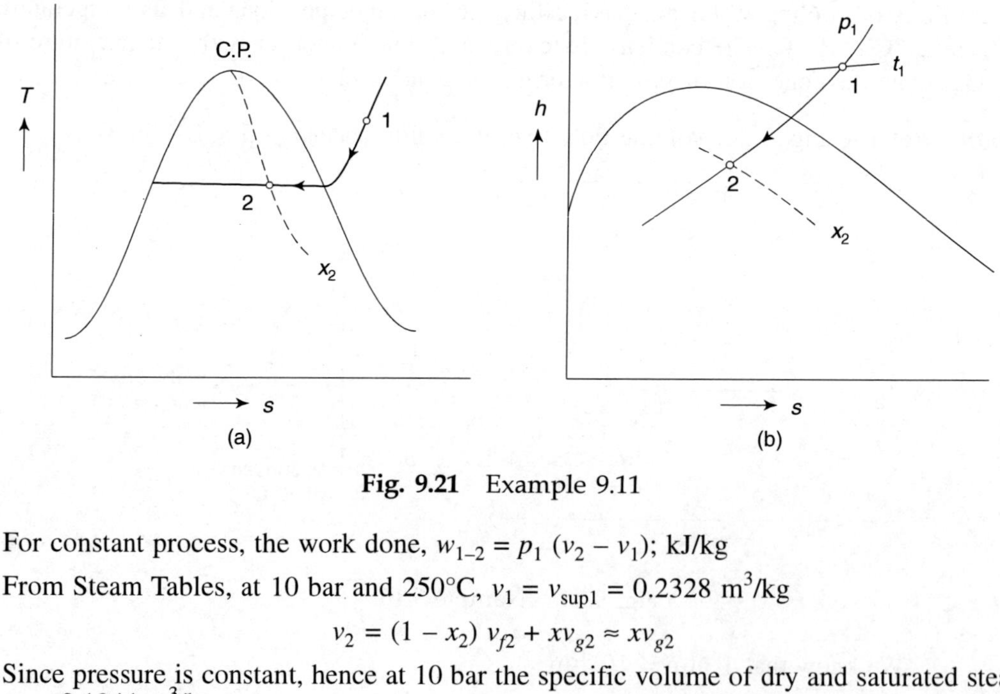

Fig. 9.21 Example 9.11

For constant process, the work done, $w_{1-2} = p_1 (v_2 - v_1)$; kJ/kg

From Steam Tables, at 10 bar and 250°C, $v_1 = v_{sup1} = 0.2328$ m^3/kg

$$v_2 = (1 - x_2) v_{f2} + xv_{g2} \approx xv_{g2}$$

Since pressure is constant, hence at 10 bar the specific volume of dry and saturated steam $= v_{g2} = 0.1944$ m^3/kg

$$w_{1-2} = 10 \times 10^2 (0.1944 \times 0.6 - 0.2328) = - 116.16 \text{ kJ/kg}$$

(a) Total work done, $W_{1-2} = 3 \times (- 116.16) = - 348.48$ kJ/kg **Ans.**

(b) Heat transfer, $q_{1-2} = h_2 - h_1$

From Steam Tables, at 10 bar and 250°C, $h_1 = h_{sup1} = 2944$ kJ/kg

$$h_2 = h_{f2} + x_2\, h_{fg2} = 763 + 0.6 \times 2015 = 1972 \text{ kJ/kg}$$

$\therefore$ $$Q_{1-2} = m\,(h_2 - h_1) = 3 \times (1972 - 2944) = -2916 \text{ kJ}$$ **Ans.**

Example 9.12 During a steay flow process, 4 kg of steam at 15 bar and 260°C loses heat of amount 3770 kJ at constant pressure. Find the final condition of steam.

Solution Refer to Fig. 9.21. Heat transfer during steady flow at constant pressure

$$Q_{1-2} = m\,(h_2 - h_1), \text{ kJ} = -3770 = 4\,(h_2 - h_1)$$

From Steam Tables, at 15 bar and 260°C, $h_1 = h_{sup1} = 2929.56$ kJ/kg

$\therefore$ $$h_2 = \frac{-3770}{4} + 2929.56 = 1987.06 \text{ kJ/kg}$$

But $$h_2 = h_{f2} + x_2\, h_{fg2}$$

At $$15 \text{ bar}, h_{f2} = 845 \text{ kJ/kg}, \quad h_{fg2} = 1947 \text{ kJ/kg}$$

$\therefore$ $$1987.06 = 845 + x_2 \times 1947$$

or $$x_2 = 0.5865$$ **Ans.**

Example 9.13 A cylinder-piston assembly contains 3 kg of water inside it at a temperature 20°C. The cylinder is well insulated and its cross-sectional area is 0.4 m^2. A pressure equivalent of 8 bar is applied on the water by the piston. Heat is supplied to cylinder by an electric heater at the rate of 1000 W. Calculate the time elapsed if the piston rises through a distance of 1.5 metre. Also calculate the quality of steam.

Solution Refer to Fig. 9.22 with proper location of states as per data. From Steam Tables, specific volume of water at 20°C is $v_f = 0.0010018$ m^3/kg.

$\therefore$ Total initial volume of water $= m \times v_f = 3 \times 0.0010018 = 0.0030054$ m^3

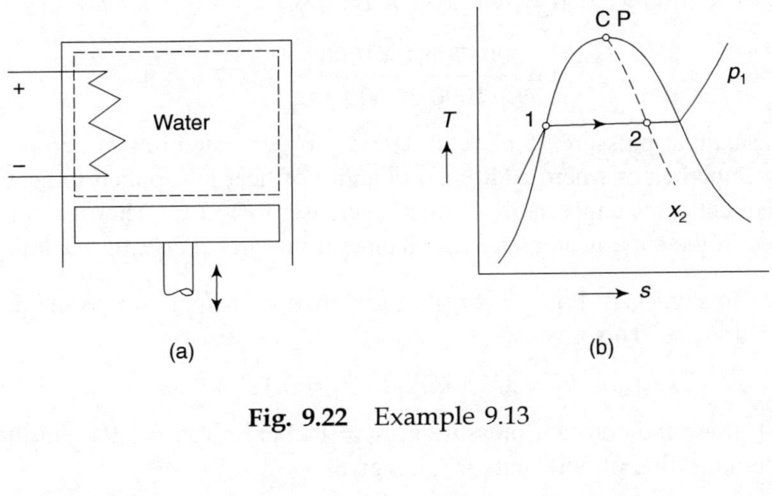

Fig. 9.22 Example 9.13

This piston will only move upwards when steam is generated due to heating. Change in volume due to rise of piston.

$$= \text{Distance moved} \times \text{Cross-sectional area}$$

$$= 1.5 \times 0.4 = 0.6 \text{ m}^3$$

$\therefore$ Final volume of steam formed = 0.0030054 + 0.6 = 0.6030054 m^3

$$\text{Specific volume of steam, } v_2 = \frac{0.6030054}{3} = 0.2010018 \text{ m}^3/\text{kg}$$

At 8 bar, specific volume of dry saturated steam, $v_{g2} = 0.2403$ m^3/kg

Hence, $v_2 = x_2\, v_{g2}$ or $0.2010018 = x_2 \times 0.2403$

or $x_2 = 0.836$ **Ans.**

During heating, pressure remains constant and equal to 8 bar.

Enthalpy of water at 20°C, $h_{f1} = 83.9$ kJ/kg

Enthalpy of steam at 8 bar, $h_2 = h_{f2} + x_2\, h_{f2}$

At 8 bar, $h_{f2} = 721$ kJ/kg, $h_{fg2} = 2048$ kJ/kg

$\therefore$ $h_2 = 721 + 0.836 \times 2048 = 2433.128$ kJ/kg

During constant pressure, the heat needed is equal to change in enthalpy. Hence,

$$q_{1-2} = h_2 - h_{f1} = 2433.128 - 83.9 = 2349.228 \text{ kJ/kg}$$

Total heat added by the heater = 3 × 2349.288 = 7047.684 kJ

Let τ be the time elapsed in hour during heating, hence,

$$\tau \times 1000 \times 3600 = 7047.684 \times 1000$$

or
$$\tau = \frac{7047.684 \times 1000}{1000 \times 3600} = 1.957 \text{ h}$$ **Ans.**

Example 9.14 Steam at pressure 15 bar and 0.95 dry is generated in a boiler and is made to pass through its superheater where additional quantity of heat is supplied to heat at constant pressure. Consequently, the temperature of steam increases to 350°C. Determine the following: (a) heat supplied in the superheater, and (b) change in internal energy. Use Mollier diagram.

Solution Refer to Fig. 9.23. Point 1 is set on the chart corresponding to the pressure p_1 = 15 bar and 0.95 dryness. This gives

$$h_1 = 2595 \text{ kJ/kg}, \quad v_1 = 0.127 \text{ m}^3/\text{kg}$$

From point 1 along the constant pressure at p_1 = 15 bar and x_1 = 0.95 and the point 2 is set where temperature line of 350° cuts it. This gives

$$h_2 = 3150 \text{ kJ/kg}$$
$$v_2 = 0.19 \text{ m}^3\text{/kg}$$

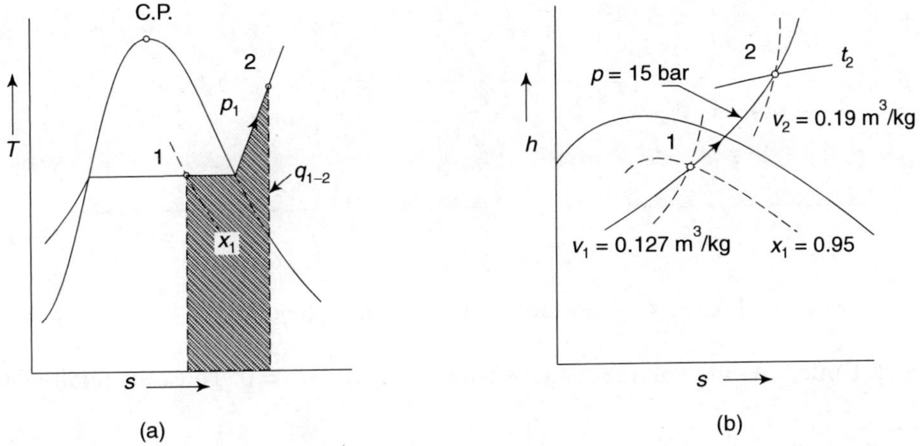

Fig. 9.23 Example 9.14

(a) Quantity of heat supplied in the superheater,

$$h_2 - h_1 = 3150 - 2595 = 555 \text{ kJ/kg} \quad \textbf{Ans.}$$

(b) Change in internal energy, $u_2 - u_1 = (h_2 - p_2 v_2) - (h_1 - p_1 v_1)$

$$= (3150 - 15 \times 10^2 \times 0.19) - (2595 - 15 \times 10^2 \times 0.127)$$
$$= 460.5 \text{ kJ/kg} \quad \textbf{Ans.}$$

9.15 CONSTANT VOLUME PROCESS

(a) *p–v*, *T–s* and *h–s* Representation Assume that steam undergoes a constant volume cooling process. Let the initial condition of superheated steam be defined by pressure p_1 and superheating temperature T_1 and the final condition of steam as pressure p_2 and dryness fraction x_2. This process is shown as 1–2 on *p–v* and *T–s* diagrams in Fig. 9.24.

Since, $\quad v_1 = v_{sup}$

$$v_2 = v_{f2}(1 - x_2) + x_2 v_{g2} \text{ at pressure } p_2$$

But $\quad v_1 = v_2 = v_{sup1} = v_{f2} + x_2 v_{g2} - x_2 v_{f2}$

$$\therefore \quad x_2 = \frac{v_{sup\,1} - v_{f2}}{v_{g2} - v_{f2}} = \frac{v_1 - v_{f2}}{v_{g2} - v_{f2}} \approx \frac{v_1}{v_{g2}} \quad \text{...(9.25)}$$

This gives the quality of steam at the final condition 2, hence the point 2 on *T–s* diagram can early be located.

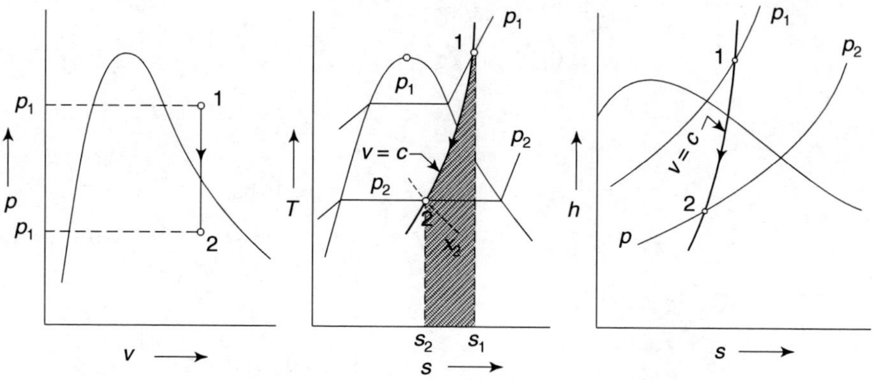

Fig. 9.24 Constant volume process (cooling)

(b) Work Done As the volume remains constant, hence $dv = 0$. Thus, the nonflow work, $\int pdv = 0$.

(c) Heat Transferred From first law,

$$dq = du + pdv \quad \text{or} \quad q_{1-2} = (u_2 - u_1) + \int pdv$$

Since $\qquad dv = 0$, hence, $q_{1-2} = u_2 - u_1$

or
$$q_{1-2} = (h_2 - p_2 v_2) + (h_1 - p_1 v_1), \text{ kJ/kg} \qquad ...(9.26)$$

The value of h_1, and h_2 can be found corresponding to initial and final condition from Steam Tables.

We see that during constant volume process, the heat transferred is equal to change in internal energy.

(d) Change in Entropy $\Delta s = s_2 - s_1$ from Table or Chart.

For heating process at constant volume the nomenclature used for states in Fig. 9.24 will be interchanged.

Example 9.15 Water and steam at atmospheric pressure are contained in a boiler drum. The air is expelled and stop valve is closed. Calculate (a) the heat required to convert 1 kg of water from this condition to dry saturated steam at 6 bar.

(b) When the stop valve is opened and boiler supplies steam at 6 bar, calculate the heat required per kg of steam formed from feed water at 48°C.

Solution Refer to Fig. 9.25. Let suffixes 1 and 2 denote the initial and final conditions in each process.

From Steam Tables at 1 bar, $h_{f1} = 417$ kJ/kg, $\quad v_{f1} = 0{,}001044\ \text{m}^3/\text{kg}$

At 6 bar, $h_{f2} = 670$ kJ/kg, $h_{fg2} = 2087$ kJ/kg, $h_{g2} = 2757$ kJ/kg

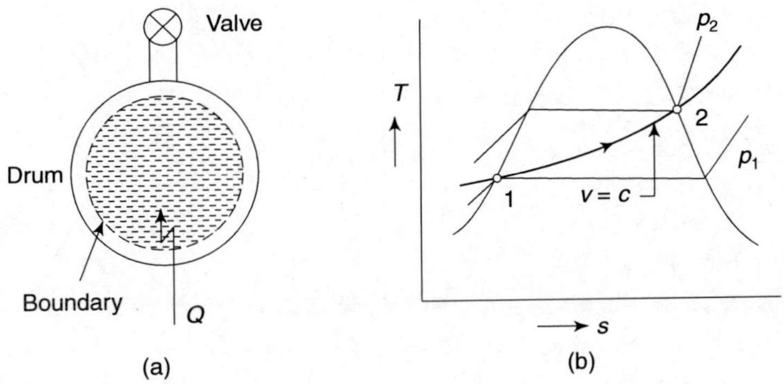

Fig. 9.25 Example 9.15

Since the stop valve is closed, the generation of steam takes place at constant volume (non-flow) process. The heat transferred during constant volume (non-flow) process is given by

$$q_{1-2} = (u_2 - u_1) + w_{1-2}$$

But $\quad w_{1-2} = 0, \quad$ hence $q_{1-2} = u_2 - u_1$ kJ/kg

For 1 kg of water, $\quad u_1 = h_{f1} - p_1 v_{f1} = 417 - 1 \times 10^2 \times 0.001044 = 416.895$ kJ/kg

For steam at 6 bar, $\quad u_2 = h_2 - p_2 v_2 = 2757 - 6 \times 10^2 \times 0.3156 = 2567.64$ kJ/kg

$\therefore \quad q_{1-2} = 2567.64 - 416.895 = 2150.745$ kJ/kg **Ans.**

(b) When the stop valve is opened, there is a flow process at constant pressure and the heat transfer is given by

$$q_{1-2} = h_2 - h_1 = h_{g2} - h_{f1}$$

At 48°C, the enthalpy of water from the Steam Tables, $h_{f1} = 200.9$ kJ/kg

$\therefore \quad q_{1-2} = 2757 - 200.9 = 2556.1$ kJ/kg **Ans.**

Example 9.16 A household pressure cooker contains 2 kg of dry and saturated steam at 4 bar while in operation. Calculate the quantity of heat that must be rejected so that the steam becomes wet with 0.7 dryness fraction. Also, calculate the pressure and temperature of the steam that exists in the cooker after the heat rejection.

Solution Refer to Fig. 9.26 with proper location of states.

The volume of steam in the pressure cooker before and after the heat rejection is the same. Hence, this process is a constant volume process. The heat transfer during constant volume process is given by

$$q_{1-2} = u_2 - u_1 = (h_2 - p_2 v_2) - (h_1 - p_1 v_1)$$

From Steam Tables at 4 bar, $h_{f1} = 605$ kJ/kg, $\quad h_{fg1} = 2134$ kJ/kg, $\quad h_{g1} = 2739$ kJ/kg

$$v_{f1} = 0.001085 \text{ m}^3/\text{kg}, \quad v_{g1} = 0.4623 \text{ m}^3/\text{kg}$$

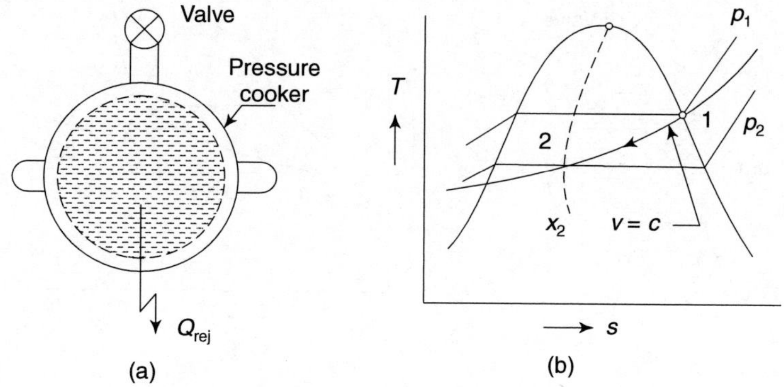

Fig. 9.26 Example 9.16

Volume of pressure cooker = Volume of steam

$$V_1 = mv_1 = mv_{g1} = 2 \times 0.4623 = 0.9246 \text{ m}^3$$

Since volume remains constant, hence $v_1 = v_{g1} = v_2 = v_{f2}(1 - x_2) + x_2 v_{g2}$

Neglecting volume of water, $v_1 = v_{g1} = x_2 v_{g2} = 0.4623 = 0.7 \times v_{g2}$

or
$$v_{g2} = \frac{0.4623}{0.7} = 0.6604 \text{ m}^3/\text{kg}$$

Corresponding to $v_{g2} = 0.6604$ m^3/kg, from the Steam Tables, we have by interpolation

$$p_2 = 2.73 \text{ bar}, t_{g2} = 130.2°\text{C}, h_{f2} = 548 \text{ kJ/kg}, h_{fg2} = 2173 \text{ kJ/kg}$$

$$h_2 = h_{f2} + x_2 h_{fg2} = 548 + 0.7 \times 2173 = 2069.1 \text{ kJ/kg}$$

$$v_2 = v_1 = v_{g1} = 0.4623 \text{ m}^3/\text{kg and } h_1 = h_{g1} = 2739 \text{ kJ/kg}$$

$\therefore$
$$q_{1-2} = (2069.1 - 2.73 \times 10^2 \times 1 \times 0.6604) - (2739 - 4 \times 10^2 \times 0.4623)$$
$$= -665.26 \text{ kJ/kg}$$

$\therefore$ Total heat rejected, $Q_{1-2} = m \times q_{1-2} = -2 \times 665.26 = -1330.52$ kJ **Ans.**

Example 9.17 A sealed and evacuated drum of volume 0.2 m^3 contains saturated water at 64.56°C. It is heated till the pressure and temperature of steam becomes 15 bar and 260°C respectively. Calculate (a) the mass of water contained in the drum, and (b) the heat required.

Solution Refer to Fig. 9.27. The steam generated will fill the drum completely.

From Steam Tables at 15 bar and 260°C by interpolation, $v_3 = v_{sup3} = 0.1555$ m^3/kg

(a) Mass of water = Mass of steam = $\dfrac{\text{Total volume, m}^3}{\text{Sp. volume m}^3/\text{kg}} = \dfrac{0.2}{0.1555} = 1.286$ kg **Ans.**

(b) From Steam Tables, corresponding to saturated water at 64.56°C

$$p_1 = 0.25 \text{ bar}, h_{f1} = h_{f2} = 271.5 \text{ kJ/kg}$$

$$h_{fg1} = 2345.7 \text{ kJ/kg}, h_{g1} = 2617.7 \text{ kJ/kg}, v_{g1} = v_{g2} = 6.322 \text{ m}^3\text{/kg}$$

Let us assume that first constant pressure heating takes place till whole volume of 0.2 m^3 is filled by wet steam at 0.25 bar. Thus,

$$V = 0.2 = m\,(x_2\, v_{g2}) = 1.286 \times x_2 \times 6.322$$

or

$$x_2 = \frac{0.2}{1.286 \times 6.322} = 0.02459$$

Heat added during constant pressure process

$$= h_{f2} + x_2\, h_{fg2} - h_{f1} = x_2\, h_{fg2} = 0.0259 \times 2345 = 60.735 \text{ kJ/kg}$$

Once the drum is filled with wet steam at 0.25 bar further addition of heat takes place at constant volume till the required condition is reached.

From Steam Tables of 15 bar and 260°C. $h_{supl} = h_3 = 2947.8$ kJ/kg

Heat added at constant volume is given by

$$q_{2-3} = u_3 - u_2 = (h_3 - p_3\, v_3) - (h_2 - p_2\, v_2) = (2947.8 - 15 \times 10^2 \times 0.1555)$$

$$- (271.5 + 0.02459 \times 2345.7 - 0.25 \times 10^2 \times 6.322 \times 0.02459)$$

$$= 2389.25 \text{ kJ/kg}$$

Heat added during generation of steam = 60.735 + 2389.25 = 2449.8 kJ/kg

Hence, total heat added = 1.285 × 2449.98 = 3148.22 kJ **Ans.**

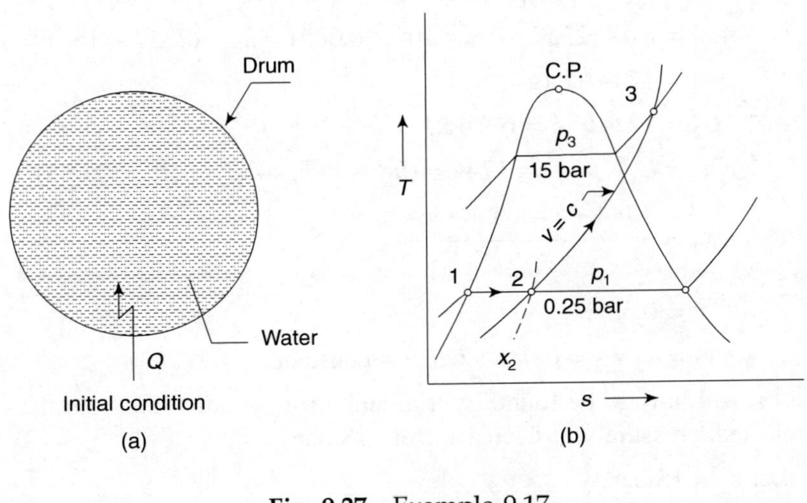

Fig. 9.27 Example 9.17

Example 9.18 A vessel of volume 0.34 m^3 contains 4 kg of steam at a pressure of 18 bar. 1840 kJ/kg of heat is rejected from the vessel during a constant volume process. Calculate (a) final internal energy, (b) final dryness fraction, (c) pressure, (d) change in entropy, and (e) work done.

Solution Refer to Fig. 9.24 with proper **location of states as per data as state 1 will be in wet zone**. Specific volume of steam contained in the vessel in the beginning,

$$v_1 = \frac{\text{Volume of vessel, m}^3}{\text{Mass of steam, kg}} = \frac{0.34}{4} = 0.085 \text{ m}^3\text{/kg}$$

From Steam Tables at 18 bar,

$$v_{g1} = 0.1104 \text{ m}^3\text{/kg}, t_{s1} = 207.1°\text{C}, h_{f1} = 885 \text{ kJ/kg}, h_{fg1} = 1912 \text{ kJ/kg}$$

$$h_{g1} = 2797 \text{ kJ/kg}, s_{f1} = 2398 \text{ kJ/kg K}, s_{g1} = 6.379 \text{ kJ/kg K}, s_{fg1} = 3.981 \text{ kJ/kg K}$$

The specific volume of the steam contained in the vessel $v_1 = 0.085$ m^3/kg is less than v_{g1} = 0.1104 m^3/kg, hence the steam is wet. Neglecting volume of water

$$v_1 = x_1\, x_{g1} \quad \text{or} \quad 0.085 = x_1 \times 0.1104$$

or
$$x_1 = \frac{0.085}{0.1104} = 0.7699$$

(a) During constant volume process, heat transferred is given by

$$q_{1-2} = u_2 - u_1 = (h_2 - p_2\, v_2) - (h_1 - p_1\, v_1)$$

or
$$-\frac{1840}{4} = u_2 - (h_1 - p_1\, v_1), \text{ here –ve sign is used to indicate heat rejection}$$

But
$$h_1 = h_{f1} + x_1\, h_{fg1} = 885 + 0.7699 \times 1912 = 2357 \text{ kJ/kg}$$

$\therefore$
$$-460 = u_2 - (2357 - 18 \times 10^2 \times 0.085) = u_2 - (2357 - 153) = u_2 - 2204$$

or
$$u_2 = 1744 \text{ kJ/kg}$$

$\therefore$
$$U_2 = 1744 \times 4 = 6976 \text{ kJ}$$ **Ans.**

(b)
$$u_2 = h_2 - p_2\, v_2 = 1744 = (h_{f2} + x_2\, h_{fg2} - p_2 \times 10^2 \times 0.085)$$

or
$$x_2 = \frac{1744 + p_2 \times 8.5 - h_{f2}}{h_{fg2}} \qquad \text{...(i)}$$

$$= 0.5276$$

Also
$$v_2 = x_2\, v_{g2} = 0.085 \text{ since } v = \text{constant} \qquad \text{...(ii)}$$

The value of x_2 will have to be found by trial and error by satisfying both the equations. Since heat is rejected, pressure will decrease from 18 bar.

First trial Let p_2 = 15 bar

$\therefore$
$$\frac{1744 + 15 \times 8.5 - 845}{1947} = x_2$$

or $\qquad x_2 = 0.5272$ and $x_2 v_{g2} = 0.5272 \times 0.1317 = 0.0694 < 0.085$

Second trial Let $p_2 = 12$ bar

Hence, $\dfrac{1744 + 12 \times 8.5 - 798}{1986} = x_2 = 0.5276$

and $x_2 v_{g2} = 0.5276 \times 0.1632 = 0.086 \approx 0.085$

(c) Hence, $p_2 = 12$ bar **Ans.**

(d) From Steam Tables at 12 bar

$$s_{f2} = 2.216 \text{ kJ/kg K}, s_{fg2} = 4.307 \text{ kJ/kg K}, s_{g2} = 6.523 \text{ kJ/kg K}$$

The chance in entropy, $s_2 - s_1 = (s_{f2} + x_2 s_{fg2}) - (s_{fg1} + x_1 s_{fg1})$

$$= (2.216 + 0.5276 \times 4.307 - (2.398 + 0.7699 \times 3.981)$$

$$= -0.9746 \text{ kJ/kg K}$$ **Ans.**

Hence, the entropy decreases

(e) $\qquad$ Work done $= \int pdv = 0$ since there is no change in volume.

Example 9.19 A closed metallic boiler drum of capacity 0.24 m^3 contains steam at a pressure of 11 bar and a temperature of 200°C. Calculate the quantity of steam in the vessel. At what pressure in the vessel will the steam be dry and saturated if the vessel is cooled? The cooling is further continued till the temperature of steam in the vessel becomes 116.9°C. Calculate the pressure and condition of steam at the end of the process.

Solution Refer to Fig. 9.24. From the Steam Tables, the saturation temperature corresponding to 11 bar is 184.1°C. Thus, the steam is superheated. The volume of the superheated steam is given by

$$v_{sup\,1} = \frac{v_{g1} T_{sup\,1}}{T_{g1}}$$

At $\qquad$ 11 bar, $v_{g1} = 0.1774$ m^3/kg

$$\therefore \qquad v_{sup\,1} = \frac{0.1774 \times (273 + 200)}{(273 + 184.2)} = 0.1835 \text{ m}^3\text{/kg}$$

This value may be directly read from Steam Tables.

Since the vessel is closed, the volume of steam will remain constant.

$$\text{Mass of steam, } m = \frac{\text{Volume of drum, m}^3}{\text{Sp. volume of steam, m}^3\text{/kg}} = \frac{0.24}{0.1835} = 1.3079 \text{ kg}$$ **Ans.**

From Steam Tables, corresponding to $v_{sup\,1} = v_{g1} = 0.1835\ m^3/kg$ the saturation pressure, $p_{s1} = 10.545$ bar. **Ans.**

After cooling, the temperature becomes 116.9°C, the saturation pressure is 1.8 bar and specific volume of dry and saturated steam is 0.9774 m^3/kg

∴ $$v_{sup\,1} = v_2 = x_2\, v_2 \quad \text{or} \quad 0.1835 = x_2 \times 0.9774$$

or $$x_2 = \frac{0.1835}{0.9774} = 0.1877$$ **Ans.**

Example 9.20 A vessel of 1.5 m^3 capacity is full of steam at a pressure of 14 bar having dryness fraction 0.85. Water at 17.2°C is injected under a constant head of 150 m. How much water will it be necessary to inject in order to reduce the pressure in the vessel to 2 bar. Also calculate the final percentage of water and dry steam respectively in the vessel. Assume that there are no radiation and conduction losses.

Solution Refer to Fig. 9.24 with proper location of states as per data. From Steam Tables at 14 bar, $h_{f1} = 830$ kJ/kg, $h_{fg} = 1960$ kJ/kg, $v_{g1} = 0.1408\ m^3/kg$.

At 2 bar, $h_{f2} = 505$ kJ/kg, $h_{fg} = 2202$ kJ/kg, $v_{g2} = 0.8850\ m^3/kg$, $v_{f2} = 0.0011495\ m^3/kg$

The original mass of steam, $V_1 = 1.5 = mx_1 \times v_{g1} = m \times 0.85 \times 0.1408$

or $$m = \frac{1.5}{0.85 \times 0.1408} = 12.533 \text{ kg}$$

Let m_f be the mass of water injected, neglecting the volume occupied by the water in the wet steam at 14 bar and assuming that there are A kg of water and B kg of dry steam in the vessel finally. The final dryness fraction is

$$x = \frac{B}{A + B}$$

We have, $$V = 1.5 = A\, v_{f2} + B\, v_{g2} \text{ or } 1.5 = A \times 0.0011495 + B \times 0.8856$$

or $$1.6937 = A \times 0.001297 + B$$

Also, $$A + B = 12.533 + m_f$$

Putting the value of B from the above equation, we have

$$A + 1.6937 - A \times 0.001297 = 12.533 + m_f$$

or $$A = 10.835 + 1.00129 \times m_f \quad ...(1)$$

and $$B = 1.698 - 0.00129\, m_f \quad ...(2)$$

Energy of water injected, $q = h_f + pv$

150 m of water = 14.8 bar since 10.2 m of water = 1.013 bar

$\therefore$ Energy of water $= 67.94 + 14.8 \times 10^2 \times 0.001 = 69.42$ kJ/kg

$\therefore$ Total energy of injected water $= Q = 69.42 \times m_f$ kJ

Now $u_1 = h_1 - p_1 v_1 = h_{f1} + x_1 h_{fg1} - p_1 v_1$

$= 830 + 0.85 \times 1960 - 14 \times 10^2 \times 0.85 \times 0.1408 = 2328.44$ kJ/kg

or $U_1 = 2328.44 \times 12.533 = 29182.33$ kJ

$U_2 = H_2 - p_2 V_2 = 505 \times A + (2202 - 2 \times 10^2 \times 0.8856)\ B$

or $U_2 = 505\ A + 2024.88\ B$

$\therefore$ At constant volume process, heat transfer $= Q = U_2 - U_1$

$$69.42 \times m_f = 505\ A + 2024.88\ B - 2918.33 \qquad ...(3)$$

There are three equations and three unknowns, hence from Eqs. (1), (2) and (3), we get

$m_f = 47.47$ kg, $A =$ 58.14 kg of water, i.e., 97.4%, and $B =$ 1.6775 kg of dry steam, i.e., 2.6% **Ans.**

9.16 ADIABATIC PROCESS (REVERSIBLE AND IRREVERSIBLE)

(a) *T–s* and *h–s* Representation Assume that steam undergoes an adiabatic expansion process. The initial condition of steam is superheated and the final condition is wet. If the process is reversible adiabatic (i.e., isentropic), then the process is represented as 1–2*s* on *h–s* and *T–s* diagram shown in Fig. 9.28.

If the process is irreversible adiabatic, the process represented by 1–2 showing the increase in entropy. Since no heat transfer takes place during irreversible process, the area under the curve 1–2 on *T–s* diagram has no significance to the heat transfer because such significance is attached only with reversible process.

9.16.1 Non-Flow Process

(b) Work Done From first law, for non-flow process (say expansion in steam engine with no mass transfer), the work done is given by

$$\delta w = \delta q = du$$

But since the process is adiabatic, hence $\delta q = 0$

$$\therefore \quad (w_{1-2})_{s=c} = (u_1 - u_{2s}) \text{ for reversible process} \qquad ...(9.27)$$

$$\text{and} \quad w_{1-2} = (u_1 - u_2) \text{ for irreversible process} \qquad ...(9.28)$$

Hence, ideal work lost due to irreversibility is given by

$$(w_{\text{lost}})_{\text{non-flow}} = (w_{1-2})_s - w_{1-2} = u_2 - u_{2s} \qquad ...(9.29)$$

where u_1, u_{2s} and u_2 are found according to initial and final conditions.

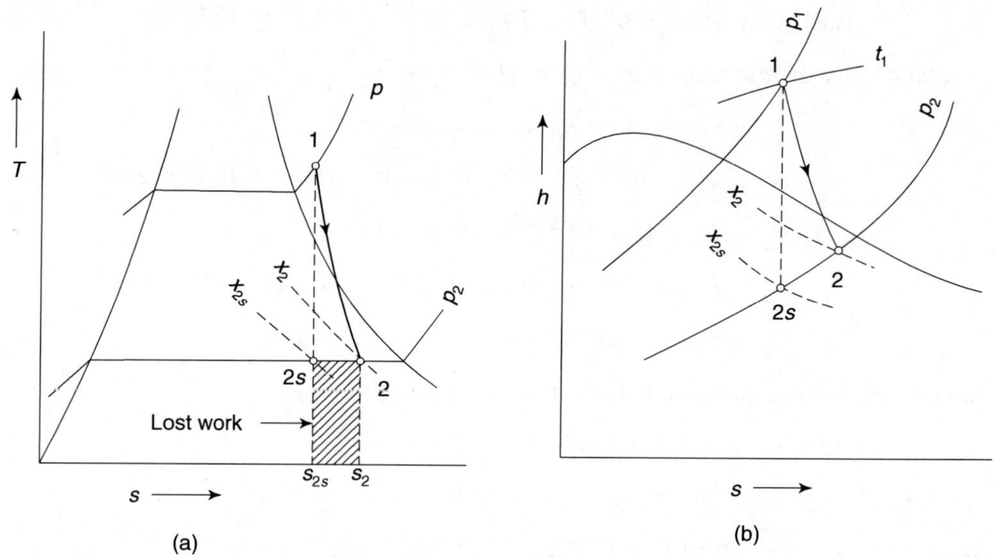

Fig. 9.28 Adiabatic process (Expansion)

9.16.2 Flow Process

For steady flow process (say expansion in steam turbine) neglecting ΔKE and ΔPE, work done is given by

$$\delta w = \delta q - dh = - dh \text{ since } \delta q = 0$$

∴ $$(w_{1-2})_s = (h_1 - h_{2s}) \text{ for reversible process} \quad ...(9.30)$$

and $$w_{1-2} = (h_1 - h_2) \text{ for irreversible process} \quad ...(9.31)$$

Hence, ideal work lost due to irreversibility is given by

$$(w_{\text{lost}})_{\text{steady flow}} = (w_{1-2})_s - w_{1-2} = (h_2 - h_{2s}) \quad ...(9.32)$$

Actual enthalpy h_2 may be found from the concept of isentropic turbine expansion efficiency.

Thus, $$\eta_{\text{exp}} = \frac{h_1 - h_2}{h_1 - h_{2s}} \quad ...(9.33)$$

The entropy generation may be calculated following the expressions given in Chapter 8.

On the T–s diagram the work lost due to irreversibility is shown by shaded area.

Note *There are many ways to represent reversible and irreversible processes. Adiabatic reversible process may be represented as 1–2s (meaning s = c) while irreversible as 1–2. In* ***another nomenclature****, isentropic and irreversible processes are represented as 1–2 and 1–2′ respectively. Here, the former is used.*

Example 9.21 In a steam turbine, 3 kg/s of steam expands isentropically from initial pressure 20 bar and 350°C to a final temperature of 85°C. Calculate (a) the quality of the steam at the end to the procees, and (b) steady flow work done.

Solution Refer to Fig. 9.28 for the representation of the process. Process 1–2*s* shows the isentropic expansion of steam. We know that for isentropic process.

$$s_1 = s_{2s} = s_{f2} + s_{2s}\, s_{fg2}$$

From the Steam Tables, at 20 bar and 350°C,

$$s_1 = s_{\text{sup}\,1} = 6.957 \text{ kJ/kg K}, \quad h_{\text{sup}\,1} = 3138 \text{ kJ/kg} = h_1$$

At 85°C, from the Steam Tables, $s_{f2} = 1.134$ kJ.kg K, $s_{g2} = 7.544$ kJ/kg K,

$$s_{fg} = 641 \text{ kJ/kg K}, \quad h_{f2} = 355.9 \text{ kJ/kg}, \quad h_{fg2} = 2295.6 \text{ kJ/kg}$$

$$\therefore \quad 6.957 = 1.134 + x_{2s} \times 6.41$$

(a) or

$$x_{2s} = \frac{5.823}{6.41} = 0.9084$$

(b) The work done during steady flow process is given by

$$w_{1-2} = h_1 - h_{2s}$$

But

$$h_{2s} = h_{f2} + x_{2s}\, h_{fg2} = 355.9 + 0.9084 \times 2295.6 = 2441.22 \text{ kJ/kg}$$

$$\therefore \quad (w_{1-2})_{s=c} = 3138 - 2441.22 = 696.78 \text{ kJ/kg}$$

$\therefore$ Total work done, $(W_{1-2})_{s=c} = 3 \times 696.78 = 2090.34$ kW **Ans.**

Example 9.22 In a perfectly insulated cylinder, 3 kg/s of steam expands isentropically from an initial pressure of 20 bar and 300°C to a final temperature of 85°C. Calculate (a) the quantity of steam at the end of the process, and (b) non-flow work done.

Solution Refer to Fig. 9.28. For isentropic process, $s_1 = s_{2s} = s_{f2} + x_{2s}\, s_{fg2}$

From Steam Tables, at 20 bar and 300°C, $h_1 = h_{\text{sup}\,1} = 3025$ kJ/kg

$$v_1 = v_{\text{sup}\,1} = 0.1225 \text{ m}^3\text{/kg}, \quad s_1 = s_{\text{sup}\,1} = 6.768 \text{ kJ/kg K}$$

At 85°C from Steam Tables,

$$s_{f2} = 1.134 \text{ kJ/kg K}, \quad s_{g2} = 7.544 \text{ kJ/kg K}, \quad s_{f2} = 6.41 \text{ kJ/kg K}$$

$$h_{f2} = 335.9 \text{ kg}, \quad h_{fg2} = 2295.6 \text{ kJ/kg}, \quad v_{g2} = 2.282 \text{ m}^3\text{/kg},$$

$$p_2 = p_{s2} = 0.578 \text{ bar}$$

$$\therefore \quad 6.768 = 1.134 + x_{2s} \times 6.41$$

(a) or

$$x_{2s} = \frac{5.634}{6.41} = 0.8789$$

Ans.

(b) For non-flow isentropic process, the work done is given by

$$(\delta w)_{nf} = \delta q - du \text{ . As } q = 0, \text{ so}$$

or $$(w_{1-2})_{nf} = -(u_2 - u_1) = u_1 - u_2 = (h_1 - p_1 v_1) - (h_2 - p_2 v_2)$$

But $$h_{2s} = h_{f2} + x_{2s} h_{fg2} = 355.9 + 0.8789 \times 2295.6 = 2373.5 \text{ kJ/kg}$$

$$\therefore \quad (w_{1-2})_{nf} = (3025 - 20 \times 10^2 \times 0.1225) - (2373.5 - 0.578 \times 10^2 \times 0.8789 \times 2.228)$$

$$= 519.97 \text{ kJ/kg}$$

Total work done = $(W)_{nf} = 3 \times 519.97 = 1559.91$ **Ans.**

Example 9.23 What is the final condition steam in each of the following operations if the initial condition in each case is 0.95 dry and 10 bar?

(a) The steam loses 145 kJ/kg at constant pressure, (b) its temperature is reduced at constant volume to 165°C, and (c) it does 230 kJ/kg of work isentropically in a steady flow process. Use of Mollier chart is not permitted in this example.

Solution Refer to Fig. 9.29. Let suffixes 1 and 2 refer to the initial and final conditions of steam respectively.

(a) During constant pressure heat transferred, $q_{1-2} = h_2 - h_1 = -145$ kJ/kg

– ve sign in used, since steam loses heat

$$\therefore \quad h_{f2} + x_2 h_{fg2} = h_{f1} + x_1 h_{fg1} - 145$$

But at constant pressure of 10 bar from the Steam Tables.

$$h_{f1} = h_{f2} = 763 \text{ kJ/kg} \quad \text{and} \quad h_{fg1} = h_{fg2} = 2015 \text{ kJ/kg}$$

$$\therefore \quad x_2 . h_{fg2} = x_1 h_{fg1} - 145$$

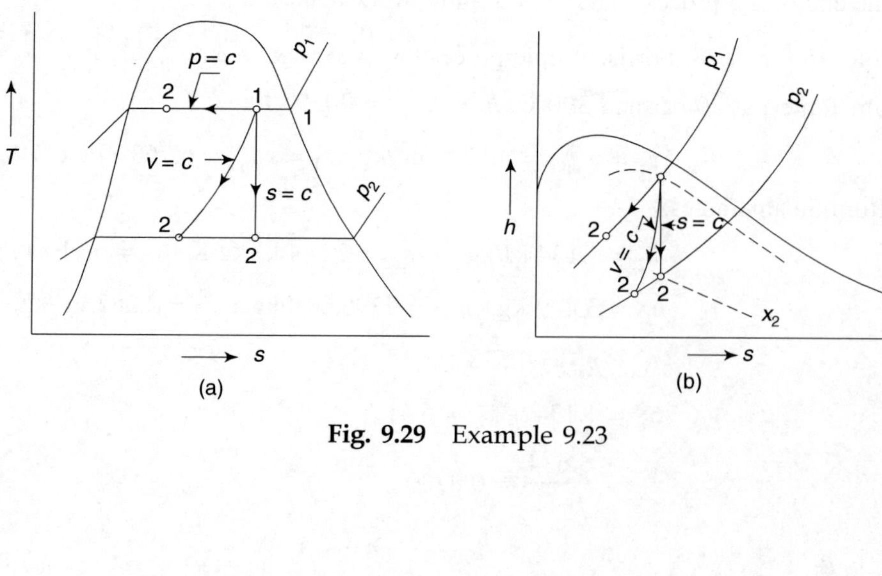

Fig. 9.29 Example 9.23

or $$x_2 = \frac{x_1\, h_{fg1} - 145}{h_{fg2}} = \frac{0.95 \times 2015 - 145}{2015} = 0.878$$ **Ans.**

(b) Since volume remains constant, so $v_1 = v_2$

But $$v_1 = (1 - x_1)\, v_{f1} + x_1\, v_{g1} \approx x_1\, v_{g1}$$

and $$v_2 = (1 - x_2)\, v_{f2} + x_2\, v_{g2} \approx x_2\, v_{g2}$$

From Steam Tables at 10 bar, $v_{g1} = 0.1944$ m^3/kg and 165°C, $v_{g2} = 0.2728$ m^3/kg

∴ $$0.95 \times 0.1944 = x_2 \times 0.2728$$

or $$x_2 = \frac{0.95 \times 0.1944}{0.2728} = 0.6769$$ **Ans.**

(c) During isentropic expansion, entropy remains constant, hence

$$s_1 = s_2 = s_{f2} + x_2\, s_{fg}$$

But $$s_1 = s_{f1} + x_1\, s_{fg1}$$

From the Steam Tables, at 10 bar, $s_{f1} = 2.138$ kJ/kg K, $s_{fg1} = 4.448$ kJ/kg K

∴ $$s_1 = 2.138 + 0.95 \times 4.448 = 6.3636 \text{ kJ/kg K}$$

Now, $$s_1 = 6.3636 = s_{f2} + x_2\, s_{fg2}$$

Also, $$h_1 = h_{f1} + x_1\, h_{fg1} = 763 + 0.95 \times 2015 = 2677.25 \text{ kJ/kg}$$

∴ $$h_2 = h_1 - w_{1-2} = 2677.25 - 230 = 2447.25 \text{ kJ/kg}$$

The work done by the steam is possible only when it expands, i.e., pressure goes down from 10 bar.

$$h_2 = 2447.25 = h_{f2} + x_2\, h_{fg2} \qquad ...(1)$$

Also $$s_2 = s_{f2} + x_2\, s_{fg2} = 6.3636 \text{ kJ/kg K} \qquad ...(2)$$

The value of x_2 will be found by trial and error assuming the value p_2 say 9. 8, 7 etc., bar and solving above equations

First trial Let $p_{g2} = 7$ bar

$$2447.25 = 697 + x_2 \times 2067$$

or $$x_2 = 0.9467$$

and $1.992 + 0.8467 \times 4.717 = 5.9858 < 6.3636$

Second trial Let $p_2 = 2.5$ bar

$$2447.25 = 535 + x_2 \times 2182$$

or $$x_2 = 0.8763$$

and $1.607 + 0.8763 \times 5.446 = 6.379 \approx 6.3636$

Hence, the final quality of steam, $x_2 = 0.8763$

Note *This example will be solved easily from Mollier chart or if the final pressure is given, then there is no need of trial and error.*

Example 9.24 4 kg of steam expands adiabatically from 16 bar and 250°C to 0.6 bar in a steam turbine such that the steam is just dry and saturated at the end of expansion. Calculate (a) the work done by the steam, and (b) work lost due to irreversibility.

Solution Refer to Fig. 9.30. Let processes 1–2*s* and 1–2 represent the isentropic and adiabatic irreversible processes respectively.

Neglecting changes in PE and KE, the actual work done by turbine is

$$w_{1-2} = (h_1 - h_2);\ \text{kJ/kg}$$

The isentropic work done, $(w_{1-2})_{s=c} = (h_1 - h_{2s})$; kJ/kg

From Steam Tables, at 16 bar and 250°C, $h_1 = h_{\text{sup}\,1} = 2920.8$ kJ/kg

At 0.6 bar for dry and saturated steam, $h_2 = 2653$ kJ/kg

Hence, the actual work done by the turbine, $w_{1-2} = 2920.8 - 2653 - 267.8$ kJ/kg

(a) $$W_{1-2} = 4 \times 267.8 = 1071.2\ \text{kJ}$$

(b) If the flow is isentropic, then $h_{2s} = h_{f2} + x_{2s}\, h_{fg2}$

At 0.6 bar, $h_{f2} = 360$ kJ/kg, $h_{fg2} = 2293$ kJ/kg. For isentropic expansion

$$s_1 = s_{2s} = s_{f2} + x_{2s}\, s_{fg2}$$

At 16 bar and 250°C, $s_1 = s_{\text{sup}\,1} = 6.6782$ kJ/kg K

At 0.6 bar, $s_{f2} = 1.145$ kJ/kg K, $s_{fg2} = 6.86$ kJ/kg K

$\therefore$ $$6.6782 = 1.145 + x_{2s} \times 6.386$$

or $$x_{2s} = 0.8664$$

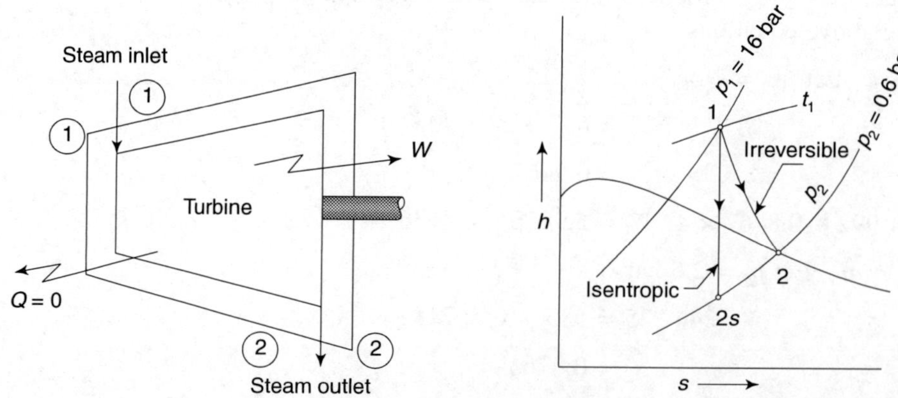

Fig. 9.30 Example 9.24

$$h_{2s} = h_{f2} + x_{2s}\, h_{fg2} = 360 + 0.8664 \times 2293 = 2346.65 \text{ kJ/kg}$$

Work done due to isentropic expansion

$$(w_{1-2})_s = h_1 - h_{2s} = 2920.8 - 2346.65 = 574.14 \text{ kJ/kg}$$

Work done lost due to irreversibility

$$(w_{1-2})_s - w_{1-2} = h_2\, h_{2s} = 574.15 - 267.8 = 306.35 \text{ kJ/kg}$$

Total work done lost, $W_{lost} = 4 \times 306.35 = 1225.36$ kJ **Ans.**

$$\text{Percentage of ideal work lost} = \frac{w_{1-2s} - w_{1-2}}{w_{1-2}} \times 100 = \frac{306.35}{574.15} \times 100 = 53.35\% \quad \textbf{Ans.}$$

9.17 ISOTHERMAL PROCESS

(a) *p–v*, *T–s* and *h–s* Representation Assume that steam undergoes an isothermal expansion process. Let the initial condition of steam be wet and the final condition be superheated. The initial condition 1 is represented on the chart as shown in Fig. 9.31 corresponding to the initial pressure, temperature and dryness fraction. The final condition 2 is represented, corresponding to final pressure p_2.

(b) Heat Transferred The heat transferred is given by

$$q_{1-2} = T \int_1^2 ds = T\,(s_2 - s_1) \text{ kJ/kg} \quad \text{...(9.34)}$$

Thus, the area under *T–s* is diagram shown by cross-hatched area gives the heat transferred.

(c) Work Done For non-flow process, the work done is given by

$$(\delta w)_{nf} = \delta q - du \quad \text{or } (w_{1-2})_{bf} = q_{1-2} - (u_2 - u_1) \quad \text{...(9.35)}$$

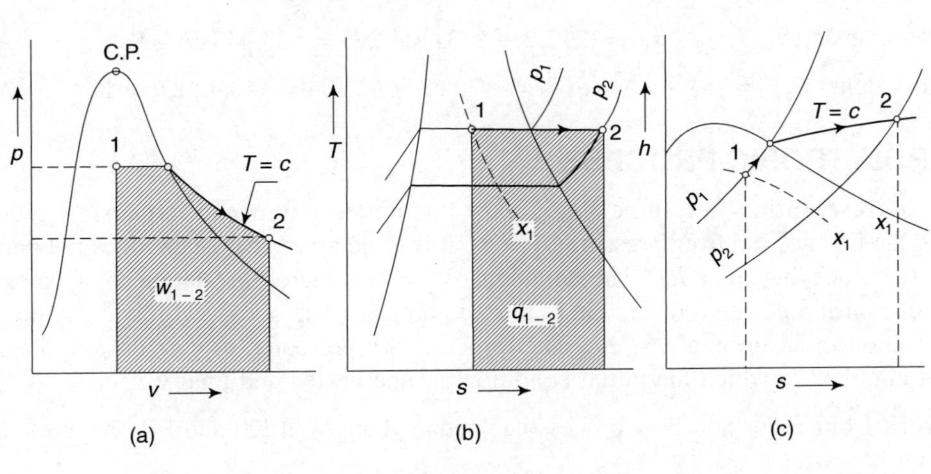

Fig. 9.31 Isothermal process

For steady flow process, the work done is given by

$$(\delta w)_{sf} = \delta q - dh \quad \text{or } (w_{1-2})_{sf} = q_{1-2} - (h_2 - h_1) \qquad ...(9.36)$$

The value u_1, u_2 and h_2 are found corresponding to initial and final conditions from Steam Tables or chart.

Example 9.25 In a steam engine cylinder, dry and saturated steam expands from 22 bar to 2 bar isothermally. Calculate (a) change in enthalpy, (b) change in internal energy, (c) change in entropy, (d) heat transferred, and (e) work done. Assume the non-flow process in the cylinder.

Solution Refer to Fig. 9.31 with proper location of states with initial condition at dry saturated line. From Steam Tables corresponding to 22 bar

$$t_{g1} = 217.2°\text{C},\ h_{f1} = 931 \text{ kJ/kg},\ h_{fg1} = 1870 \text{ kJ/kg},$$

$$h_{g1} = 2801 \text{ kJ/kg},\ v_{s1} = 0.09067 \text{ m}^3\text{/kg},\ s_{fg1} = 2.492 \text{ kJ/kg K}$$

$$s_{fg1} = 3.813 \text{ kJ/kg K},\ s_{g1} = 6.305 \text{ kJ/kg K}$$

Since the process is isothermal, therefore, $t_{g1} = t_1 = t_2 = 217.2°\text{C}$

From Steam Tables, at 2 bar and 217.2°C

$$h_2 = h_{\text{sup}2} = 2905 \text{ kJ/kg},\ v_{\text{sup}2} = v_2 = 1.1215 \text{ m}^3\text{/kg},$$

$$s_2 = s_{\text{sup}1} = 7.576 \text{ kJ/kg K}$$

(a) Change in enthalpy, $h_2 - h_1 = 2905.4 - 2801 = 104.4$ kJ/kg **Ans.**

(b) The change in internal energy, $u_2 - u_1 = (h_2 - p_2 v_2) - (h_1 - p_1 v_1)$

$= (2905.4 - 2 \times 10^2 \times 1.1215) - (2801 - 22 \times 10^2 \times 0.09069) = 79.62$ kJ/kg **Ans.**

(c) Change in entropy, $s_2 - s_1 = 7.576 - 6.305 = 1.271$ kJ/kg K **Ans.**

(d) Heat transferred, $T(s_2 - s_1) = (217.2 + 273) \times 1.271 = 623.04$ kJ/kg K **Ans.**

(e) Work done, $w_{1-2} = (u_1 - u_2) + q_{1-2} = -79.62 + 623.04 = 543.42$ kJ/kg K **Ans.**

9.18 POLYTROPIC PROCESS

(i) *p–v* Representation Assume that steam undergoes polytropic expansion process following the law $pv^n = c$ from pressure p_1 to p_2. It is to be noted that steam does not behave as perfect gas obeying $pv = RT$. The equation $pV^n = c$ is merely a statement of pressure-volume relationship during a reversible polytropic process. The non-flow polytropic process has application in steam engines, etc. This process is represented as 1–2 on *p–v* diagram shown in Fig. 9.32 in which the initial condition is superheated and final wet.

(b) Work Done For non-flow process neglecting changes in KE and PE, the work done is given by

$$(w_{2-2})_{nf} = \int_1^2 p\,dv = \frac{p_1 v_1 - p_2 v_2}{n - 1} \text{ kJ/kg} \qquad ...(9.37)$$

For steady flow process neglecting change in KE and PE

$$(w_{1-2})_{sf} = -\int_1^2 v dp = \frac{n}{n-1}(p_1 v_1 - p_2 v_2) \quad ...(9.38)$$

(c) Heat Transferred From first law, for non-flow process

$$(\delta q)_{nf} = du + \delta w = du + pdv$$

or

$$(q_{1-2})_{nf} = (u_2 - u_1) + \frac{p_1 v_1 - p_2 v_2}{(n-1)}; \text{ kJ/kg} \quad ...(9.39)$$

For steady flow, neglecting ΔKE and ΔPE

$$(\delta q)_{sf} = dh + \delta w = dh - \int_1^2 v dp$$

or

$$(q_{1-2})_{sf} = (h_2 - h_1) + \frac{n}{n-1}(p_1 v_1 - p_2 v_2); \text{ kJ/kg} \quad ...(9.40)$$

$$(q_{1-2})_{nf} = (u_2 - u_1) + \frac{p_1 v_1 - p_2 v_2}{n-1}; \text{ kJ/kg} \quad ...(9.41)$$

Thus, heat transferred during flow and non-flow is the same.

Example 9.26 Steam at a pressure of 5 bar and 0.8 dry expands in a cylinder according to the law $pv^{1.35} = c$ to 2 bar. Find the interchange of heat between the steam and the cylinder per kg of steam and work done.

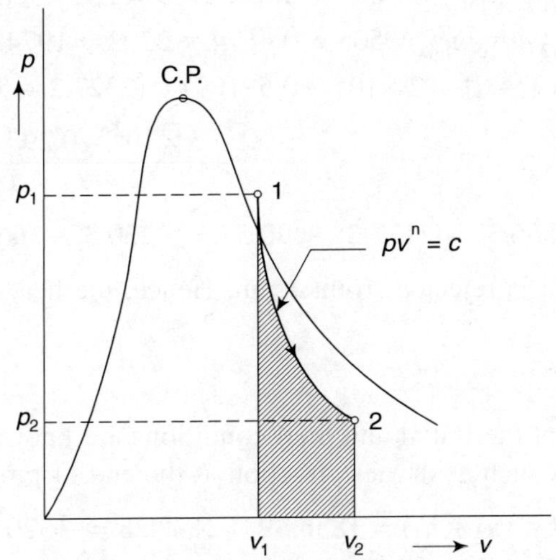

Fig. 9.32 Polytropic process

Solution Refer to Fig. 9.32 with proper location of steam. The expansion of steam in the cylinder is a non-flow process. From the Steam Tables at 5 bar

$$t_{g1} = 151.8°\text{C},\ h_{f1} = 640 \text{ kJ/kg},\ h_{fg1} = 2109 \text{ kJ/kg},\ h_{g1} = 2749 \text{ kJ/kg}$$

$$v_{g1} = 0.3748 \text{ m}^3\text{/kg},\ s_{f1} = 1086 \text{ kJ/kg K},\ s_{g1} = 6.822 \text{ kJ/kg K}$$

and $$s_{fg1} = 4.902 \text{ kJ/kg K}$$

At 2 bar, $$t_{g1} = 120°\text{C},\ h_{f2} = 505 \text{ kJ/kg},\ h_{fg2} = 2202 \text{ kJ/kg},\ v_{g2} = 0.8856 \text{ m}^3\text{/kg}$$

The equation for the law of expansion is $pv^{1.35} = c$, hence, $p_1\ v_1^{1.35} = p_2\ v_2^{1.35}$

or $$v_2 = v_1\left(\frac{p_1}{p_2}\right)^{1/1.35}$$

But $$v_1 = x_1\ v_{g1} \text{ neglecting volume of water} = 0.8 \times 0.3748 = 0.29984 \text{ m}^3\text{/kg}$$

$\therefore$ $$v_2 = 0.29984 \times \left(\frac{5}{2}\right)^{1/1.35} = 0.59109 \text{ m}^3\text{/kg}$$

But $$v_2 = 0.59109 = x_2\ v_{g2} = x_2 \times 0.8856$$

or $$x_2 = \frac{0.59109}{0.8856} = 0.6674$$

Hence, the final dryness fraction of steam, $x_2 = 0.6674$

During polytropic process, the heat transferred in non-flow system is given by

$$q_{1-2} = (u_2 - u_1) + \frac{p_1v_1 - p_2v_2}{(n-1)} = (h_2 - p_2v_2) - (h_1 - p_1v_1) + \frac{p_1v_1 - p_2v_2}{(n-1)}$$

But $$h_1 = h_{f1} + x_1\ h_{fg2} = 640 + 0.8 \times 2109 = 2327.3 \text{ kJ/kg}$$

and $$h_2 = h_{f2} + x_2\ h_{fg2} = 505 + 0.6674 \times 2202 = 1974.61 \text{ kJ/kg}$$

$\therefore$ $$q_{1-2} = (1974.61 - 2 \times 10^2 \times 0.59109) - (2327.2 - 5 \times 10^2 \times 0.29984) + \frac{5 \times 10^2 \times 0.29984 - 2 \times 10^2 \times 0.59109}{(1.35 - 1)}$$

$$= 1856.39 - 2177.28 + 90.57 = -230.32 \text{ kJ/kg}$$

–ve sign shows that heat is rejected from steam. Hence, the heat transfer from steam to cylinder will be

$$= -230.32 \text{ kJ/kg}$$ **Ans.**

Since all the properties at the initial and final conditions are known, we can calculate the other important parameters such as dryness fraction at the end of process = $x_2 = 0.6674$

Change in internal energy, $(u_2 - u_1) = 1856.39 - 2177.28 = -320.89$ kJ/kg

$$\text{Work done} = \frac{p_1\ v_1 - p_2\ v_2}{(n-1)} = \frac{5 \times 10^2 \times 0.29984 - 2 \times 10^2 \times 0.59109}{1.35 - 1} = 90.57 \text{ kJ/kg}$$

Example 9.27 Dry and saturated steam at 12 bar expands in cylinder to 2 bar having 0.8 dryness fraction according to the law $pv^n = c$. Calculate (a) the index of expansion and (b) work done.

Solution Refer to Fig. 9.32 with proper location of states. From Steam Tables at 12 bar, $v_1 = v'_{g1} = 0.8856\ m^3/kg$

At 2 bar, $v_{g2} = 0.8856\ m^3/kg$, hence $v_2 = x_2 \times v_{g2} = 0.8856 \times 0.8 = 0.7084\ m^3/kg$

We have, $p_1 v^n{}_1 = p_2 v^n{}_2$. Taking log of both sides, we have

$$n = \frac{\ln p_1 - \ln p_2}{\ln v_2 - \ln v_1} = \frac{\ln p_1/p_2}{\ln v_2/v_1} = \frac{\ln (12/2)}{\ln (0.7084/0.1632)} = 1.22 \qquad \textbf{Ans.}$$

$$\text{Work done} = \frac{p_1 v_1 - p_2 v_2}{(n-1)} = \frac{10^2 (12 \times 0.1632 - 2 \times 0.7084)}{(1.22 - 1)} = 246.18 \text{ kJ/kg} \qquad \textbf{Ans.}$$

Example 9.28 Steam at a pressure of 6 bar and with 41.2°C of superheat is expanded adiabatically and reversibly in a cylinder to 1 bar. Using Steam Tables, calculate the work done per kg of steam during the expansion and deduce the mean adiabatic index of expansion over this range.

Solution Refer to Fig. 9.28. The expansion of steam in the cylinder is a non-flow process.

$$\delta q - \delta w = du$$

But $\delta q = 0$ since the process is adiabatic

$$\therefore \qquad w_{1-2} = -(u_2 - u_1) = u_1 - u_2 = (h_1 - p_1 v_1) - (h_2 - p_2 v_2)$$

where suffixes 1 and 2 refer to initial and final conditions

From the Steam Tables at 6 bar and 41.2°C,

$$h_1 = h_{\text{sup}1} = 2851 \text{ kJ/kg}, \; v_1 = v_{\text{sup}1} = 0.3522 \text{ m}^3\text{/kg},$$

$$s_1 = s_{\text{sup}1} = 6.966 \text{ kJ/kg K}$$

From Steam Tables at 1 bar,

$$h_{f2} = 417 \text{ kJ/kg}, \; h_{fg2} = 2258 \text{ kJ/kg}, \; h_{f2} = 2675 \text{ kJ/kg},$$

$$v_{f2} = 1.649 \text{ m}^3\text{/kg}$$

$$s_{f2} = 1.303 \text{ kJ/kg K}, \; s_{fg2} = 6.05 \text{ kJ/kg K}$$

Since the process is isentropic, therefore

$$s_1 = s_2 = s_{f2} + x_2 \times s_{fg2} \quad \text{or} \quad 6.986 = 1.303 + x_2 \times 6.056$$

or $$x_2 = 0.9354$$

$$h_2 = h_{f2} + x_2\, h_{fg2} = 417 + 0.9354 \times 2258 = 2529.13 \text{ kJ/kg}$$

and $$v_2 = x_2\, v_{g2} = 0.9354 \times 1.694 = 1.5845 \text{ m}^3\text{/kg} \qquad \textbf{Ans.}$$

$$w_{1-2} = (2851 - 6 \times 10^2 \times 0.3522) - (2529.13 - 1 \times 10^2 \times 1.5845) = 269 \text{ kJ/kg}$$

Also $$w_{1-2} = \frac{p_1 v_1 - p_2 v_2}{(n-1)} = 269 = \frac{10^2}{(n-1)} (6 \times 0.3522 - 1 \times 1.5845)$$

or $$n = 1.1965$$ **Ans.**

Example 9.29 Steam at 3 bar and 0.9 dry expands in a cylinder till the volume is four times that at the commencement. The law of expansion is $pv^{1.1} = c$. Determine the quantity of heat which must pass into or out of 1 kg of steam during the expansion. Also, calculate the change in internal energy, amount of work done and condition of steam at the end of process.

Solution Refer to Fig. 9.33. The expansion in the cylinder is a non-flow process. From Steam Tables, at 3 bar, $h_{f1} = 561$ kJ/kg, $h_{fg1} = 2164$ kJ/kg, $v_{g1} = 0.6057$ m^3/kg.

Initial volume of steam at the commencement of expansion

$$v_1 = x_1 v_{f1} = 0.9 \times 0.6057 - 0.545 \text{ m}^3\text{/kg}$$

Final volume of steam, $v_2 = 4 \times v_1 = 4 \times 0.545 = 2.18$ m^3/kg

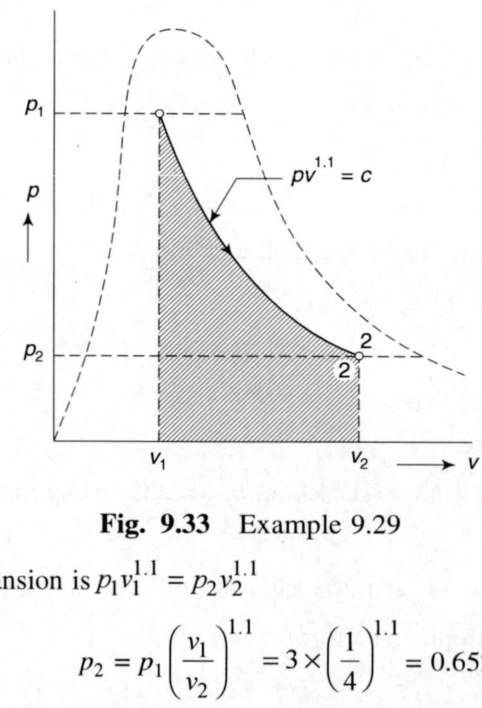

Fig. 9.33 Example 9.29

Since the law of expansion is $p_1 v_1^{1.1} = p_2 v_2^{1.1}$

or $$p_2 = p_1 \left(\frac{v_1}{v_2}\right)^{1.1} = 3 \times \left(\frac{1}{4}\right)^{1.1} = 0.6529 \text{ bar}$$

From Steam Tables at 0.6529 bar, $h_{f2} = 369$ kJ/kg, $h_{fg} = 2288$ kJ/kg, $v_{g2} = 2.535$ m^3/kg

Now, $$v_2 = 2.180 = x_2 v_{g2} = x_2 \times 2.35$$

or $$x_2 = \frac{2.180}{2.535} = 0.8599$$ **Ans.**

The change in internal energy, $\Delta u = u_2 - u_1 = (h_2 - p_2v_2) - (h_1 - p_1v_1)$

$$= (h_2 + x_2\, h_{fg2} - p_2v_2) - (h_{f1} + x_1 h_{fg1} - p_1v_1)$$

$$= (369 + 0.8599 \times 2288 - 0.6529 \times 10^2 \times 2.18)$$

$$- (561 + 0.9 \times 2164 - 3 \times 10^2 \times 0.18)$$

$$= 301.61 \text{kJ/kg}$$ **Ans.**

$$w_{1-2} = \frac{p_1v_1 - p_2v_2}{(n-1)}$$

$$= \frac{3\times10^2 \times 0.545 - 0.6529\times10^2 \times 2.18}{0.1}$$

$$= 211.67 \text{ kJ/kg}$$ **Ans.**

$$q_{1-2} = (u_2 - u_1) + w_{1-2} = -\ 301.61 + 211.678$$

$$= -\ 89.932 \text{ kJ/kg}$$ **Ans.**

Hence, 89.932 kJ/kg of heat is flowing out of cylinder.

9.19 THROTTLING PROCESS

The throttling process described earlier involves the passing of a higher pressure fluid through a narrow constriction resulting in (i) reduction in pressure and temperature, (ii) increase in volume, (iii) increases in entropy, and (iv) constant enthalpy, i.e., the important characteristics of the throttling process is that *enthalpy remains constant*. The process is adiabatic and no heat flow from or to the system but it is *not reversible*. Steam stop valve and throttle valve are installed at the entry of steam turbine in power plants. When flow of steam takes place through these valves, throttling process occurs.

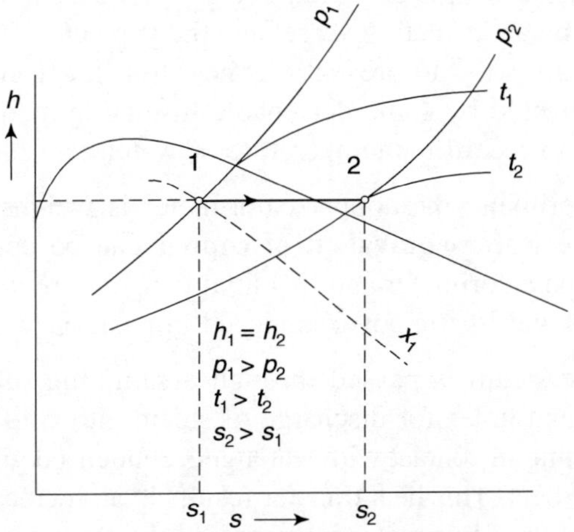

Fig. 9.34 Throttling of steam

Let us assume that the initial condition of the steam is wet represented as point 1 corresponding to pressure p_1 and dryness fraction x_1 unknown. (Fig. 9.34). This steam is allowed to throttle to pressure p_2 resulting in the superheating of steam (t_2). The final state 2 is set by p_2 and t_2. Since enthalpy remains constant, the initial state 1 is represented on *h–s* chart by drawing a horizontal line from point 2 to pressure p_1. The dryness fraction line passing through 1 gives the quality of steam (x_1). It is obvious from the *h–s* chart that entropy increases during throttling process. Therefore, for throttling process,

$$h_1 = h_2 = h_{sup\,2}$$

or

$$h_{f1} + x_1\, h_{fg1} = h_{g2} + c_{ps}\,(T_{sup\,2} - T_{g2}) \qquad \text{...(9.42)}$$

The value of h_{f1} , h_{fg} and $h_{sup\,2}$ may be found out from the Steam Tables, the value of dryness fraction x_1, may be calculated from the above equation. *The throttling process is, therefore, used in determining the dryness fraction of steam in power plant practice.*

9.20 METHODS OF DETERMINING THE DRYNESS FRACTION

There are four methods of determining the dryness fraction of steam experimentally.

They are as follows:

(i) Bucket or barrel calorimeter

(ii) Separating calorimeter

(iii) Throttling calorimeter

(iv) Combined separating and throttling calorimeter.

9.20.1 Bucket or Barrel Calorimeter

The bucket calorimeter used for determining the dryness fraction of steam as shown in Fig. 9.35. It consists of a copper calorimeter which contains a measured quality of water. This calorimeter is surrounding by an insulating vessel and the gap between the two vessel is filled with some insulating medium so as to prevent any heat transfer from or to the system. The top of the vessel is also covered by some detachable insulating medium. A thermometer is inserted in the calorimeter to record the temperature of water.

Before starting the experiment, the copper calorimeter is weighed and by knowing the specific heat of copper, the **water equivalent of copper** can be determined. The required amount of water is put in the calorimeter and its initial temperature is noted. The calorimeter is then placed in the vessel and its top cover is placed in position.

Now, the sample of the steam is passed through a sampling tube via the control and pipe terminating into fine exit holes for discharge of steam into cold water contained in the calorimeter. The steam coming in contact with water gets condensed giving out its entire latent heat and part of its sensible heat. This heat transfer results in an increase in the temperature of water. The mass of water in the calorimeter increases by the amount of steam condensed. As a precautionary measure, sufficient quantity of steem should be passed into the calorimeter so that there is a marked rise in the temperature so as to minimising the error in the experiment.

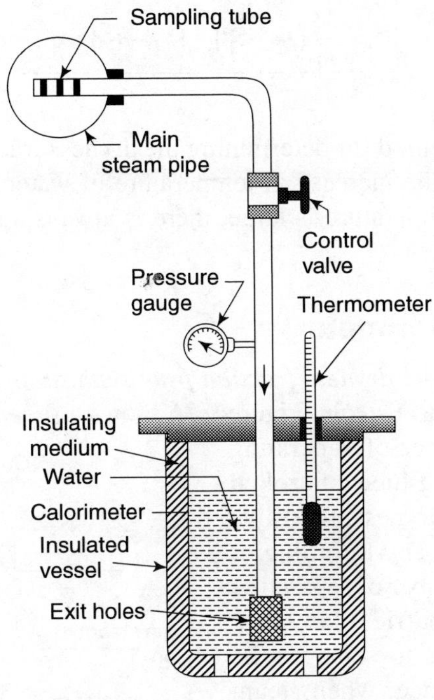

Fig. 9.35 Bucket calorimeter

Now, the supply of steam is closed. The temperature of the water is recorded. The calorimeter is again weighed. The increase in the water will give the weight of steam condensed and the increase in temperature will give amount of heat transferred.

Let p = pressure of steam in the main pipe, bar

t = saturation temperature corresponding to pressure, p°C

h_{fg} = latent heat of vaporization corresponding to p, kJ/kg

x = dryness fraction of steam in the main pipe

m_1 = mass of cold water before steam supply, kg

m_2 = mass of water after steam supply

M_1 = water equivalent of the vessel. kJ/K

∴ $M = M_1 + m_1 \cdot c_{pf}$ = water equivalent of calorimeter, kJ/K

$m = m_2 - m_1$ = mass of steam condensed, kg

t_1 and t_2 = initial and final temperature of water

If there is no heat lost, then according to conservation of energy.

$$m[c_{pf}(t_g - t_2) + xh_{fg}] = M(t_2 - t_1)$$

$$\therefore \qquad x = \frac{\frac{M}{m}(t_2 - t_1) - (t_g - t_2) \times c_{pf}}{h_{fg}} \qquad ...(9.43)$$

It is to be noted that this method for determining the dryness fraction is approximate. There are many sources of errors. The increase in temperature of water should neither be too low nor too high. If the rise in temperature is large, there is always a chance for heat losses due to radiation.

9.20.2 Separating Calorimeter

The principle of determining the dryness fraction from separating calorimeter is to separate the moisture content from the wet steam. Figure 9.36 shows a separating calorimeter. The wet steam enters the inner chamber of separating calorimeter through a sampling tube controlled by a valve. While testing for dryness, the valve should be fully open avoiding any throttling or wiredrawing. In the inner chamber of separating calorimeter, separators or baffle plates are arranged in such a manner as to separate the moisture from the steam. Hence, when steam enters the separating calorimeter, it undergoes a sudden reversal of direction of motion when it strikes the baffle plates. This causes the water particles which have greater inertia to separate from the wet steam. The water is collected in the bottom portion of the inner chamber of the separating calorimeter which is connected to a gauge glass having a graduated scale showing the quantity of water separated. The dry steam moves upwards (shown by firm arrows) and enters the annular space between the inner and outer surface of the separating calorimeter and goes downward.

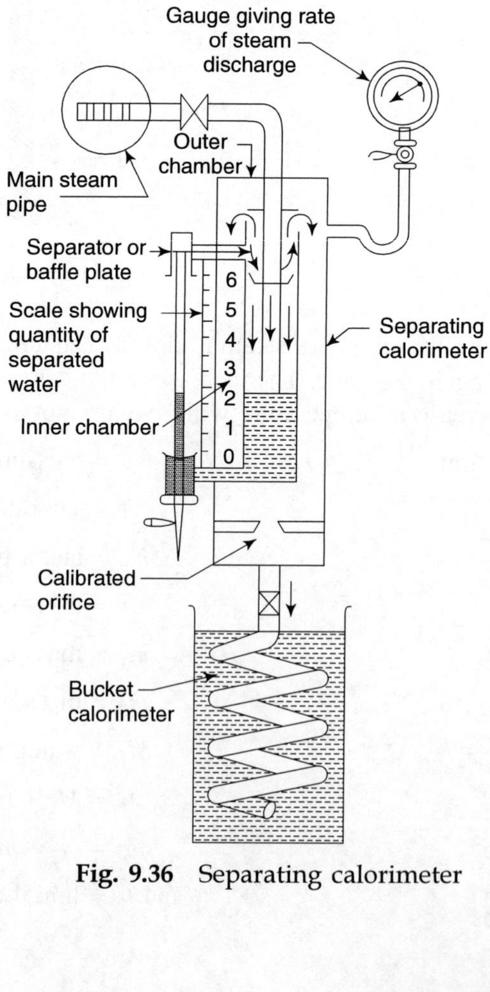

Fig. 9.36 Separating calorimeter

The discharge of dry and saturated steam may be measured by fitting a gauge in the outer chamber of the separating calorimeter. This gauge also gives the pressure of dry and saturated steam. Another way to measure the discharge of the dry and saturated steam is to pass it in a bucket calorimeter through a calibrated orifice. The condensed steam is then weighed.

Let m = mass of water separated in certain time interval

M = mass of dry steam read from the gauge or condensed and weighed in that time interval, kg

$$\therefore \quad x = \text{dryness fraction of wet steam} = \frac{M}{m+M} \qquad \text{...(9.44)}$$

This method is also approximate because of incomplete separation of moisture and inaccuracy of gauge.

9.20.3 Throttling Calorimeter

The principle of the throttling calorimeter is to throttle the wet steam so that it becomes superheated. Since the enthalpy remains constant, the dryness fraction can be found. Figure 9.37 shows such a throttling calorimeter. The wet steam whose dryness fraction has to be determined enters a sampling tube from the main pipe of steam. The sampled steam will represent the main wet steam. In order to ensure this, the sampling tube which is kept in the main steam pipe with the holes facing upstream, should be as far as possible downstream from the elbows and valves. This sampled steam while passing through the orifice or throttle valve is throttled. This throttled steam is led to the inner chamber of the throttling calorimeter. For the accurate result from the throttling calorimeter, *the necessary condition is that the steam should be in a superheated state after throttling*. In general, 5°C of superheat is most desirable. The manometer attached to the outer casing gives the pressure of steam after throttling. Normally, the pressure after throttling is slightly more by a few mm of H_g above atmospheric pressure. The thermometer attached gives the temperature of the throttled steam. It is put in oil pocket to give correct value of temperature of the throttled steam. If the temperature recorded by the thermometer is more than the saturation temperature corresponding to recorded pressure, it means it is in superheated state after throttling.

The steam of the inner chamber flows upwards and turns downwards in the annular space between the inner and outer chamber. In order to prevent any heat transfer from and to the system, the outer chamber is insulated. As a precautionary measure before taking the temperature of throttled steam, the steam should flow enough to ensure constant temperature.

Let p_1 = initial pressure of wet steam, bar

x_1 = dryness fraction of wet steam

h_{f1} = enthalpy of water at p_1, kJ/kg

h_{fg1} = latent heat of vaporization at p_1, kJ/kg

p_2 = pressure of throttled steam, bar

c_p = sp. heat of superheated temparature of throttled steam, kJ/kg K

t_2 = saturation temperature of throttled steam corresponding to pressure p_2,°C

$t_{sup\,2}$ = temperature of a superheated throttled steam as recorded by thermometer,°C.

We know that during throttling process, the enthalpy remains constant, hence

Enthalpy before throttling at pressure p_1 = Enthalpy after throttling at pressure, p_2

$$\therefore \qquad h_1 = h_{sup\,2}$$

or
$$h_{f1} + x_1 . h_{fg1} = h_{sup\,2} \qquad ...(9.45)$$

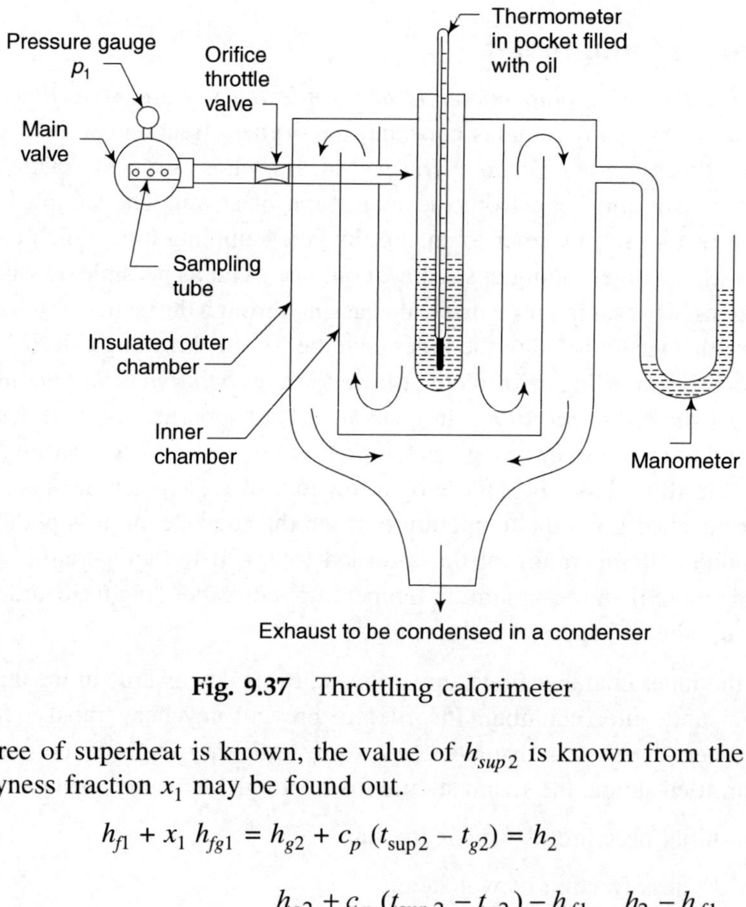

Fig. 9.37 Throttling calorimeter

If the degree of superheat is known, the value of $h_{sup\,2}$ is known from the Steam Tables, hence the dryness fraction x_1 may be found out.

Also,
$$h_{f1} + x_1\, h_{fg1} = h_{g2} + c_p\,(t_{sup\,2} - t_{g2}) = h_2$$

or
$$x_1 = \frac{h_{g2} + c_p\,(t_{sup\,2} - t_{g2}) - h_{f1}}{h_{fg1}} = \frac{h_2 - h_{f1}}{h_{fg1}} \qquad ...(9.46)$$

where $\qquad c_p$ = 2.607 kJ/kg K is normally used.

With Eqs. (9.45) or (9.46), we can easily calculate the value of the dryness fraction.

This process has got some limitations. Since the necessary condition for this process is that steam must be superheated after throttling, this method cannot be used for finding out the dryness fraction of very wet steam. As the final pressure is near about atmospheric, the minimum dryness fraction of steam that can be found out reliably from this process depends upon the initial pressure of wet steam. This minimum quality that can be measured by the throttling calorimeter varies from 93.4% for 40 bar to 98.5% for 3 bar of initial pressure. Thus, steam having dryness fraction less than 0.934 cannot be measured by a throttling calorimeter accurately for steam pressure below 20 bar.

9.20.4 Combined Separating and Throttling Calorimeter

The separating calorimeter suffers from the drawback that water particles from wet steam are not fully separated thus resulting in higher value of dryness fraction, whereas the throttling calorimeter suffers from the drawback that it is not suitable for very wet steam because throttled steam will not be superheated. If a combination of separating and throttling calorimeter is used, it results in the removal of the drawbacks of both the calorimeters.

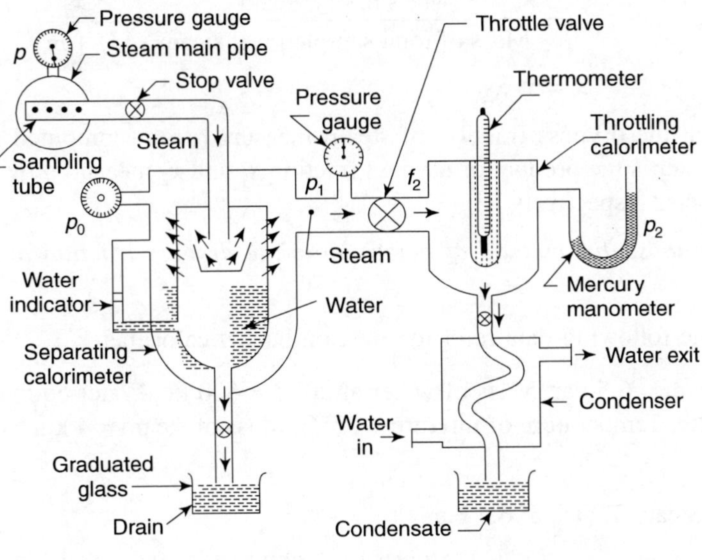

Fig. 9.38 Combined separating and throttling calorimeter

Figure 9.38 shows such a combined separating and throttling calorimeter. A sample of steam passing through the sampling tube controlled by a valve first enters the separating calorimeter where most of the moisture is separated and measured by an indicator attached to it. Now, comparatively dry steam from separating calorimeter is allowed to pass through the throttling calorimeter where it is throttled and the exhaust steam is condensed and measured.

Let $\qquad x$ = dryness fraction of steam

M = mass of steam coming from separating calorimeter and passing through throttling calorimeter, kg

m = mass of water separated in the separating calorimeter, kg

$x_1 = x_{sup}$ = dryness fraction of steam measured by separating calorimeter alone

$$= \frac{M}{M+m}$$

p_1 and p_2 = pressures before and after throttling

$x_2 = x_{sup}$ = dryness fraction of steam measured by throttling calorimeter

$$= \frac{h_{g2} + c_p\,(t_{sup\,2} - t_{g2}) - h_{f1}}{h_{fg1}}$$

Thus, mass of dry steam entering the throttling calorimeter = $x_2 \cdot M$

$$\therefore \quad x = \frac{\text{Mass of dry steam}}{\text{Mass of total sampled wet steam}} = \frac{Mx_2}{M+m}$$

or $$x = x_1\, x_2 \qquad \text{...(9.47)}$$

This shows that the dryness fraction of steam measured from combined separating and throttling calorimeter is the product of dryness fraction x_1 and x_2 measured by separating and throttling calorimeter respectively.

Among all methods, this method of determining the dryness fraction of steam is most satisfactory.

Example 9.30 The following data refer to a test on bucket calorimeter.

Pressure of steam = 6.5 bar. Mass of water at 20°C = 500 kg. Water equivalent of bucket calorimeter = 30 kg, Temperature of mixture = 38°C. Mass of steam = 4 kg. Find the dryness fraction of steam.

Solution From Steam Tables at 6.5 bar

$$t_g = 161.9°\text{C},\ h_{fg} = 2077 \text{ kJ/kg},\ M = 500 + 30 = 530 \text{ kg}$$

The dryness fraction for bucket calorimeter is given by

$$x = \frac{\frac{M}{m}(t_2 - t_1) - (t_g - t_2) \times c_{pf}}{h_{fg}} = \frac{\frac{530}{4}(28 - 20) - (161.19 - 38) \times 4.18}{2077}$$

or $x = 0.9$ **Ans.**

Example 9.31 Steam at 9 bar and 0.93 dry is contained inside a vessel of 0.9 m^3 volume. The delivery valve is opened and the steam is blown off until the pressure reduces to 4 bar.

The delivery valve is then closed and the vessel cooled until the pressure is 3 bar. Assume that enthalpy per kg of steam remains constant during blowing off period. Calculate, (a) the mass of steam blown off, and (b) quality of steam in the vessel after cooling.

Solution Let suffixes 1, 2 and 3 represent initial, blowing off and cooling conditions respectively. From Steam Tables at 9 bar

$$h_{f1} = 743 \text{ kJ/kg}, h_{fg1} = 2031 \text{ kJ/kg}, v_{g1} = 0.2149 \text{ m}^3\text{/kg}$$

At 4 bar, from Steam Tables, $h_{f2} = 605$ kJ/kg, $h_{fg2} = 2134$ kJ/kg, $v_{g2} = 00.4623$ m^3/kg

At 3 bar, $v_{f3} = 0.6057$ m^3/kg, $h_{f3} = 561$ kJ/kg, $h_{fg3} = 2164$ kJ/kg

Since during blowing off period enthalpy remains constant, therefore, $h_1 = h_2$

But $$h_1 = h_{f1} + x_1 h_{fg} = 743 + 0.93 \times 2031 = 2631 \text{ kJ/kg}$$

and $$h_2 = h_{f2} + x_2 h_{fg2} = 605 + x_2 \times 2134 = 2631.83$$

or $$x_2 = 0.9497$$

$$V_1 = 0.9 = x_1 \times v_{g1} \times m_1 = 0.93 \times 0.2149 \times m_1$$

or $$m_1 = \frac{0.9}{0.93 \times 0.2149} = 4.503 \text{ kg}$$

$\therefore$ Mass of steam contained in the vessel at the initial condition, $m_1 = 4.503$ kg.

Even after blowing-off steam, the volume of steam in the vessel remains constant and equal to 0.9 m^3. Thus,

$$V_1 = 0.9 = V_2 = x_2 \times v_{g2} \times m_2$$

or $$m_2 = \frac{V_2}{x_2 \times v_{g2}} = \frac{0.9}{0.9497 \times 0.4623} = 2.0498 \text{ kg}$$

(a) Hence, the mass of steam blown off, $m_1 - m_2 = 4.503 - 2.0498 = 2.453$ kg **Ans.**

(b) At 3 bar also the total volume of steam is the same 0.9 m^3. Since the valve is closed so mass $m_2 = m_3 = 2.0496$ kg

$\therefore$ $$V_1 = 0.9 = V_3 = x_3 \times v_{g3} \times m_3 = x_3 \times 0.6057 \times 2.0496$$

or $$x_3 = \frac{0.9}{0.6057 \times 2.0496} = 0.7248$$ **Ans.**

(c) Since the vessel is closed while cooling, hence this cooling is a non-flow process. Heat transferred at constant volume and non-flow process,

$$q_{1-2} = u_2 - u_2 = (h_3 - p_3 v_3) - (h_2 - p_2 v_2)$$

But $$h_2 = h_1 = 2631.83 \text{ kJ/kg}$$

$$h_3 = h_{f3} + x_3 h_{fg} = 561 + 0.7248 \times 2164 = 2129.46 \text{ kJ/kg}$$

$$q_{2-3} = (2129.46 - 3 \times 10^2 \times 0.7248 \times 0.6057) - (2631.83 - 4 \times 10^2)$$
$$= -\ 339.93 \text{ kJ/kg}$$

Heat transferred during cooling. The –ve sign is to show the heat rejection.

Example 9.32 During a test on separating calorimeter, the following observations were taken.

Mass of water separated = 0.3 kg/min, mass of steam passing through calorimeter = 4 kg/min. Calculate the dryness fraction.

Solution From separating calorimeter, $x = \dfrac{M}{M+m} = \dfrac{3}{3+0.3} = 0.909 = 90.0\%$ **Ans.**

Example 9.33 The following data refer to a throttling calorimeter.

Pressure in the main steam pipe = 10 bar, pressure after throttling = 1.2 bar, temperature after throttling = 120°C. Assuming c_p = 2.303 kJ/kg K for steam after throttling, calculate the dryness fraction. Also, calculate the maximum degree of wetness which can be shown by this calorimeter for this steam. If the wetness is beyond the range of this calorimeter, suggest another method.

Solution Refer to Figs. 9.34 and 9.37.

During throttling process, enthalpy remains constant. Hence, $h_1 = h_2$

$$\therefore \quad h_{f1} + x_1\, h_{fg1} = h_{g2} + c_p\,(t_{\text{sup}\,2} - t_{g2}) \quad ...(1)$$

From the Steam Tables at 10 bar, h_{f1} = 763 kJ/kg, h_{fg1} = 2015 kJ/kg

At 1.2 bar, h_{g2} = 2683 kJ/kg, t_{g2} = 104.8°C

$$\therefore \quad 763 + x_1 \times 2015 = 2683 + 2.093\,(120 - 104.8)$$

or $x_1 = 0.9686$ **Ans.**

The maximum degree of wetness occurs when $t_2 = t_g$ in Eq. (1). Thus,

$$763 + x_1 \times 2015 = 2683$$

or $x_1 = 0.9582$ **Ans.**

If the dryness fraction is lower than 0.9528, a combination of separating and throttling calorimeter will be used.

Example 9.34 In a test on a combined separating and throttling calorimeter, the following data were obtained:

Pressure in the steam main = 14 bar, pressure of steam after throttling = 13.2 cm of Hg gauge, mass of steam collected in the separator = 0.8 kg/min, discharge from throttling calorimeter = 10 kg/min, temperature of steam after throttling = 115°C, barometer = 76 cm of Hg. Calculate (a) the dryness fraction of steam in the main, (b) the velocity of flow in the steam main if the diameter is 12 cm and the flow of steam is 130 kg/min, (c) the mass of water required to condense the steam after the throttling calorimeter if the temperature rise of the cooling water is 12°C.

Solution Refer to Fig. 9.38.

The dryness fraction of steam from separating and throttling calorimeter is calculated by the equation.

$$x = x_1 . x_2$$

where $x_1 = \frac{M}{M+m}$ = dryness fraction of steam measured from separating calorimeter.

$$= \frac{10}{10+0.8} = 0.9259$$

x_2 = dryness fraction of steam measured from throttling calorimeter

$$= \frac{h_{g2} + c_p\left(t_{\sup 2} - t_{g2}\right) - h_{f1}}{h_{fg1}}$$

Now, pressure after throttling

$$p_2 = 76 + 13.2 = 89.2 \text{ cm of Hg} = \frac{89.2}{76} \times 1.013$$

$$= 1.1889 \text{ bar}$$

From Steam Tables at 11889 bar, t_{g2} = 104.8°C, h_{g2} = 2683 kJ/kg

At 14 bar, h_{f1} = 830 kJ/kg, h_{fg1} = 1960 kJ/kg

$$\therefore \quad x_2 = \frac{2683 + 2.093(115 - 104.8) - 830}{1960}$$

$\therefore$ (a) $x = x_1 \cdot x_2 = 0.926 \times 0.956 = 0.885$ **Ans.**

(b) Specific volume of steam in the main pipe, $v = x. v_g$

$$= 0.885 \times 0.1408 = 0.1246 \text{ m}^3\text{/kg}$$

But from continuity equation, $\dot{m}\ v = A \cdot C$

or Velocity, $C = \frac{\dot{m}v}{A} = \frac{130 \times 0.1246}{(\pi/4)(0.12)^2} = 1432.2$ m/min **Ans.**

(c) If $\dot{m}$ kg of water is required in the condenser to remove the superheat and latent heat, then

$$10\ (2704.\ 3 - 104.8) = \dot{m} \times 12 \times 4.18$$

or $$\dot{m} = 518.24 \text{ kg/min}$$ **Ans.**

Example 9.35 Steam at 18 bar and 0.95 dry is throttled to 14 bar and passed to an engine which expands it isentropically to 0.3 bar and exhausts at this pressure. Determine (a) entropy per kg of steam entering the engine (b) steam consumption of the engine in kg/kWh, (c) the thermal energy per kg of steam leaving the boiler. Use Mollier chart.

Solution Refer to Fig. 9.39. The point 1 is set on *h–s* chart corresponding to pressure 18 bar and 0.95 dry. This gives h_1 = 2698 kJ/kg.

Since during throttling process, enthalpy remains constant, hence $h_1 = h_2$ = 2698 kJ/kg.

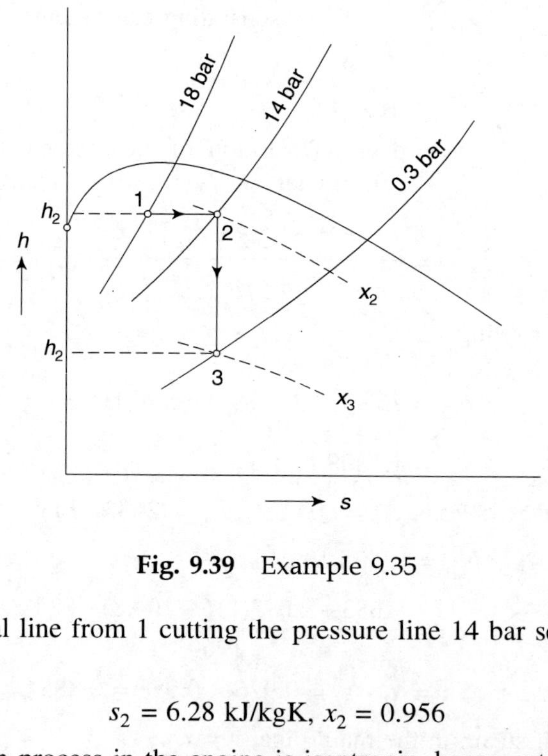

Fig. 9.39 Example 9.35

Thus, the horizontal line from 1 cutting the pressure line 14 bar sets the point 2. Hence, from chart,

$$s_2 = 6.28 \text{ kJ/kgK},\ x_2 = 0.956$$

Since the expansion process in the engine is isentropic, hence entropy remains constant. The vertical line from point 2 cutting the pressure line 0.3 bar sets the point 3. This gives

$$h_3 = 2096 \text{ kJ.kg},\ x_3 = 0.777,\ v_3 = 4.14 \text{ m}^3\text{/kg}$$

(a) Entropy per kg of steam entering the engine, s_2 = 6.28 kg/kg K **Ans.**

(b) Work done in flow process, $w = h_2 - h_3$ = 2698 – 2096 = 602 kJ/kg **Ans.**

(c) Specific steam consumption, $SSC = \dfrac{3600}{602}$ = 5.98 kg/kWh

Example 9.36 A drum contains saturated liquid and saturated vapour at 30°C. Calculate (a) the fraction of the initial mass pumped off so that the contents in the tank reach a temperature equal to 0°C. (b) The fraction of the initial mass that can be solidified. Assume no loss of heat to the tank and neglect the mass of vapour in the tank.

Solution Let m_1 = initial mass, m_2 = mass when 0°C is reached and m_3 = mass when solid and liquid is in equilibrium.

At 30°C, h_{f1} = 125.66 kJ/kg h_{g1} = 2554.5 kJ/kg At 0°C, $h_{f2} = 0$, h_{g2} = 2501.6 kJ/kg

(a) Energy transferred = Change in liquid energy

or $$(m_1 - m_2) \times \text{Average enthalpy} = m_1\, h_{f1} - m_2\, h_{f2}$$

or $$(m_1 - m_2) \times \frac{2554.5 + 2501.6}{2} = m_1 \times 125.66 - m_2 \times 0$$

or $$\frac{m_2}{m_1} = \frac{2428.84}{2528} = 0.96$$

or $$\frac{m_1 - m_2}{m_1} = \frac{1 - 0.96}{1} = 0.04 \text{ or } 4\%$$

∴ Fraction of initial mass removed = 4% **Ans.**

(b) Energy transferred = Change in liquid energy at 0°C

or $$(m_2 - m_3) \times \text{enthalpy at 0°C} = m_3\, h_{f3} = m_3 \times 335$$

or $$(m_2 - m_3) \times 2501.6 = m_3 \times 335$$

or $$\frac{m_3}{m_2} = \frac{2501.6}{2836.6} = 0.8819$$

or $$m_3 = 0.8819,\ m_2 = 0.8819 \times 0.96\ m_1 = 0.8466\ m_1$$

This means 84.66% of the initial mass is solidified. **Ans.**

Example 9.37 A steam boiler initially contains 10 m^3 of steam and 10 m^3 of water at 20 bar. Dry and saturated steam is taken out at constant pressure until 8 m^3 of water is left. Calculate the heat transferred to the boiler during this process.

Solution Refer to Fig. 9.40.

This is the case of an *emptying process* (unsteady flow). Let suffixes 1 and 2 denote the initial and final conditions. At 20 bar from Steam Tables.

$$v_{f1} = v_{f2} = 0.001176 \text{ m}^3/\text{kg},\quad v_{g1} = v_{g2} = 0.099623 \text{ m}^3/\text{kg}$$

$$h_{g1} = h_{g2} = 2797.3 \text{ kJ/kg};\quad h_{f1} = h_{f2} = 908.4 \text{ kJ/kg}$$

$$u_{f1} = u_{f2} = h_{f1} - p_1\, v_{f1} = 908.4 - 20 \times 10^2 \times 0.00176 = 904.88 \text{ kJ/kg}$$

$$u_{g1} = u_{g2} = h_{g1} - p_1 v_{g1} = 2797.3 - 20 \times 10^2 \times 0.099623 = 2598 \text{ kJ/kg}$$

The initial mass of saturated water, $m_{f1} = \dfrac{V_{f1}}{v_{f1}} = \dfrac{10}{0.001127} = 8873.11$ kg

The initial mass of saturated steam, $m_{g1} = \dfrac{V_{g1}}{v_{g1}} = \dfrac{10}{0.099623} = 100.37$ kg

The initial mass of saturated water and steam,

$$= m_1 = m_{f1} + m_{g1} = 8873.11 + 100.37 = 8973.48 \text{ kg}$$

The final mass of saturated water, $m_{f2} = \dfrac{V_{f2}}{v_{f2}} = \dfrac{8}{0.001127} = 7098.49$ kg

The final mass of saturated steam, $m_{g2} = \dfrac{12}{0.099623} = 120.45$ kg

The initial mass of saturated water and steam,

$$m_2 = m_{f2} + m_{g2} = 7098.49 + 120.45 = 7218.91 \text{ kg}$$

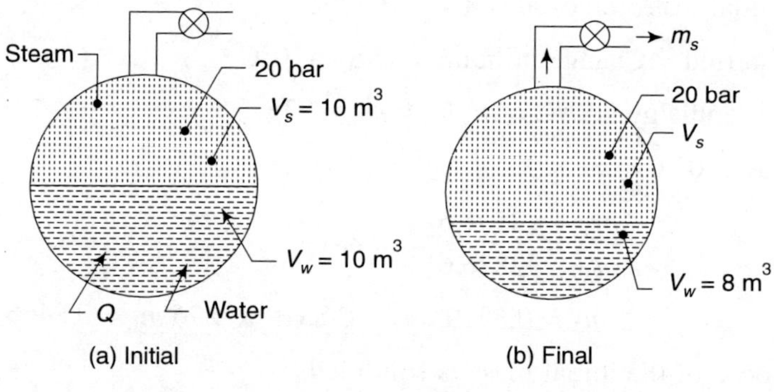

Fig. 9.40 (a) Initial (b) Final Example 9.37

Mass of dry steam taken out of boiler, $m_s = m_1 - m_2 = 8973.48 - 7218.91 = 1754.47$ kg

The energy gives, $U_1 + Q = u_2 + m_g h_{g2}$

or $\quad (m_{f1} u_{f1} + m_{g1} u_{g1}) + Q = (m_{f2} u_{f2} + m_{g2} u_{g2}) + m_{g2} h_{g2}$

or $\quad 8873.11 \times 904.88 + 100.37 \times 2598 + Q$

$$= 7098.49 \times 906 + 120.45 \times 2598 + 1754.57 \times 2797.3$$

or $\quad 8029.099 \times 10^3 + 260.761 \times 10^3 + Q$

$$= 6431.23 \times 10^3 + 312.929 \times 10^3 + 4908 \times 10^3$$ **Ans.**

$$Q = 3362.35 \times 10^3 \text{ kJ} = 3362.35 \text{ MJ}$$ **Ans.**

Example 9.38 A large insulated vessel is divided in two chambers with the help of a partition, one containing 10 kg of dry and saturated steam at 3 bar and other 20 kg of steam having dryness fraction 0.85 at 6 bar. If the partition is removed and the steam is allowed to mix thoroughly and to settle, calculate the mixture pressure, quality of the mixture and change in entropy during the mixing process.

Solution Refer to Fig. 9.41. At 3 bar, $v_1 = v_{g1} = 0.606274 \text{ m}^3/\text{kg}$

$$\therefore \quad V_1 = m_1 v_1 = 10 \times 0.606274 \text{ m}^3 = 6.06274 \text{ m}^3$$

At 6 bar, $v_2 = x_2 v_{g2} + (1 - x_2) v_{f2} = 0.85 \times 0.315901 + 0.15 \times 0.001101 = 0.26868 \text{ m}^3 \text{ kg}$

$$\therefore \quad V_2 = m_2 v_2 = 20 \times 0.26868 = 5.37362 \text{ m}^3$$

Total volume, $V_m = V_3 = V_1 + V_2 = 6.06274 + 5.37362 = 11.43636 \text{ m}^3$

Total mass of the mixture, $m_m = m_3 = m_1 + m_2 = 30$ kg

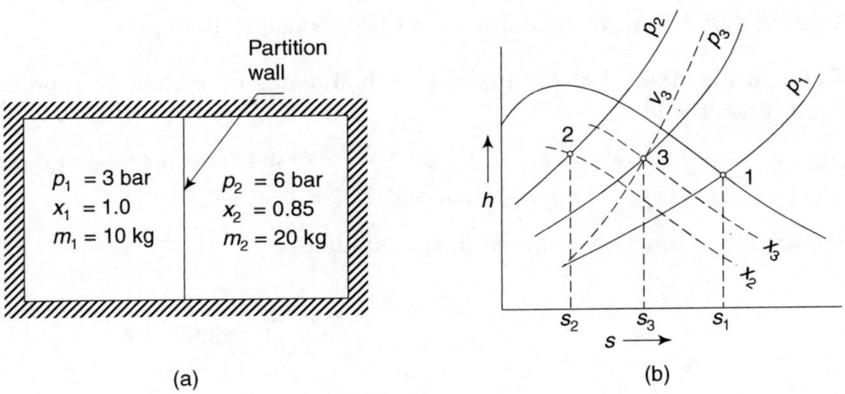

Fig. 9.41 Example 9.38

$$\therefore \quad \text{Specific volume of mixture, } v_m = v_3 = \frac{V_3}{m_3} = \frac{11.43636}{30} = 0.381212 \text{ m}^3/\text{kg}$$

Energy balance yields, $m_1 u_1 + m_2 u_2 = m_3 u_3$

$$u_1 = h_1 - p_1 v_1 = 2724.71 - 3 \times 10^2 \times 0.606274 = 2542.82 \text{ kJ/kg}$$

$$u_2 = h_2 - p_2 v_2 = (h_{f2} + x_2 h_{f2}) - p_2 x_2 v_{g2}$$

$$= (670.3 + 0.85 \times 2085.24) - 6 \times 10^2 \times 0.85 \times 0.315901$$

$$= 2281.6 \text{ kJ/kg}$$

Now $\quad 10 \times 2542.82 + 20 \times 2281.6 = 30 \times u_3$

or

$$u_3 = \frac{10 \times 2542.82 + 20 \times 2281.6}{30} = 2368.67 \text{ kJ/kg}$$

For simplicity, let $h_3 = 2539.9$ kJ/kg and $v_3 = 0.38116$ m^3/kg the state 3 is fixed on Mollier chart which gives

$$p_3 = 5 \text{ bar}, s_1 = 7.0 \text{ kJ/kg K}, x_3 = 0.91, s_2 = 6.00 \text{ kJ/kg K},$$
$$s_3 = 6.4 \text{ kg/kg K}$$

The second calculation may be made for more accurate results. Entropy changes during the mixing process is given by

$$\Delta S = m_3 s_3 - (m_1 s_1 + m_2 s_2) = 30 \times 6.4 - (10 \times 7 + 20 \times 6)$$
$$= 2 \text{ kJ/kg}$$

Example 9.39 A frictional, perfectly conducting piston divides a rigid vessel in two parts, *A* and *B*. Initially, part *A* contains 0.4 m^3 of air at a pressure of 10 bar and part *B* contains 0.4 m^3 of wet steam at a pressure of 10 bar. Heat is now added to both parts until all the water in part *B* is evaporated. At this condition, the pressure in part *B* is 15 bar. Determine the initial quality of steam in part *B* and the total amount of heat added to both parts.

Solution Refer to Fig. 9.42. Let the initial and final states of system be represented by suffixes 1 and 2 respectively.

At per data, $p_{A1} = p_{g1} = 10$ bar, $T_{A1} = T_{g1} = T_{B1} = 452.9$ K (from Steam Tables) as the piston is perfect conductor, $p_{A2} = p_{B2} = 15$ bar and $T_{2.g} = 471.3$ K

From gas laws, the volume of air at the final condition is

$$V_{A2} = V_{A1} \times \frac{T_{A2}}{T_{A1}} \times \frac{p_{A1}}{p_{A2}} = 0.4 \times \frac{471.3}{452.9} \times \frac{10}{15} = 0.2775 \text{ m}^3$$

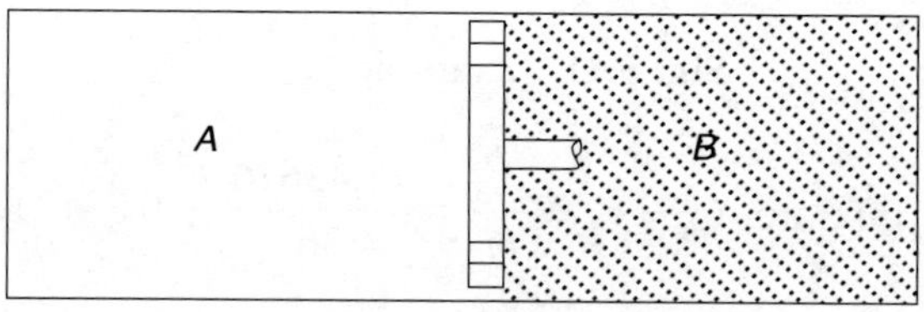

Fig. 9.42 Example 9.39

The volume of steam at the final condition, $V_{B2} = 0.4 + 0.4 - 0.2775 = 0.5225$ m^3

At 15 bar, specific volume of dry saturated steam (from Steam Tables)

$$V_{s2} = 0.1317 \text{ m}^3/\text{kg}$$

$$\text{The mass of steam, } m_g = \frac{V_{B2}}{v_{g2}} = \frac{0.5225}{0.1317} = 3.967 \text{ kg}$$

The specific volume of steam at initial state, $v_1 = \dfrac{V_{B1}}{m_g} = \dfrac{0.4}{3.967} = 0.10083 \text{ m}^3/\text{kg}$

Let x_1 be the dryness fraction of steam at the initial state, so $= v_1 = v_{f1} = x_1 (v_{g1} - v_{f1})$

or $$0.10083 = 0.001127 + x_1 (0.1943 - 0.001127)$$

or $$x_1 = 0.516$$ **Ans.**

From the energy equation for closed system, we have $Q - W = \Delta U$, here, $W = 0$

$\therefore$ $$Q = \Delta U = (\Delta U)_a + (\Delta U)_s = m_a c_{va} \Delta T_a + m_s \Delta h_s + (p_1 v_{g1} - p_2 v_{g2})$$

From Steam Tables, $$h_2 = h_{g2} = 2789.9 \text{ kJ/kg}$$

$$h_1 = h_{f1} + x_1 h_{fg} = 762.61 + 0.516 \times 20136 = 1801.6 \text{ kJ/kg}$$

$$m_a = \frac{p_a V_A}{RT_A} = \frac{10 \times 10^5 \times 0.4}{287 \times 452.88} = 3.0775 \text{ kg}$$

$\therefore$ $$Q = 3.0775 \times 0.718 (471.29 - 452.88) + 3.967 (2789.9) - 1801.6$$

$$+ \frac{10 \times 10^5 \times 0.4}{1000} - \frac{15 \times 10^5 \times 0.5225}{1000} = 3576.6 \text{ kJ}$$ **Ans.**

Example 9.40 Steam enters a nozzle operating at steady state with pressure at 40 bar, temperature 400°C with a velocity of 10 m/s. Neglect heat transfer from the nozzle to the surrounding and changes in potential energy. At the exit, the pressure is 15 bar and the velocity is 665 m/s. The mass flow rate through the nozzle is 5 kg/s. Calculate the exit area of the nozzle.

Solution Refer Fig. 9.43. From continuity equation, $A_2 = \dfrac{\dot{m}\, v_2}{C_2}$

v_2 is unknown here which has to be find out.

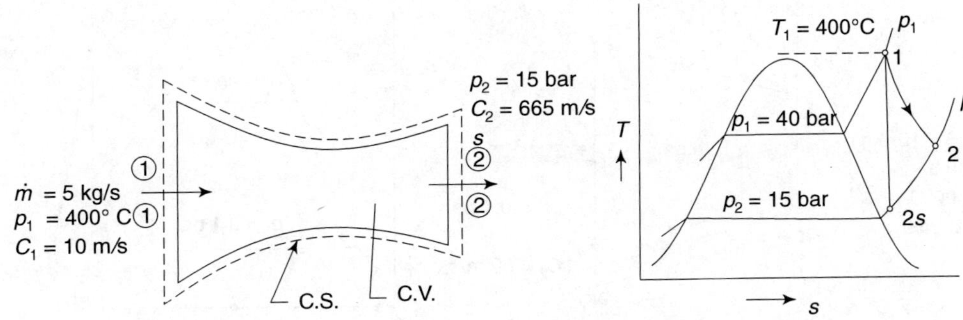

(a) (b)

Fig. 9.43 Example 9.40

From SFEE, we have, $\dot{Q}_{cv} - \dot{W}_{cv} = \dot{m}\left(h_2 + \frac{C_2^2}{2} + gz_2\right) - \dot{m}\left(h_1 + \frac{C_1^2}{2} + gz_1\right)$

Here $\dot{Q}_{cv} = 0,\ \dot{W}_{cv} = 0,\ \Delta PE = 0$, hence $h_2 = h_1 + \frac{C_1^2 - C_2^2}{2}$

From Steam Table, $p_1 = 40$ bar and $T_2 = 400°C$, $h_1 = 3213.6$ kJ/kg

$$h_2 = 3213.6 + \frac{10^2 - 665^2}{2 \times 10^3} = 2992.52 \text{ kJ/kg}$$

From Steam Table at $p_2 = 15$ bar and $h_2 = 2992.52$ kJ/kg, $v_2 = 0.1627$ m³/kg

Thus, $A_2 = \frac{5 \times 0.1627}{665} = 1.2233 \times 10^{-3} = 12.233 \text{ cm}^3$ **Ans.**

Example 9.41 Steam enters a turbine operating at steady state with a mass flow rate of 46×10^3 kg/s having pressure 60 bar, temperature 400°C and velocity 10 m/s. Turbine develops 10 MW of power. At the exit, the pressure is 0.1 bar, the quality of steam is 0.9 and the velocity is 60 m/s. Determine the rate of heat transfer between the turbine and the surrounding. Also calculate the rate of entropy production if heat transfer occurs at 350 K. Neglect changes in potential energy.

Solution Refer to Fig. 9.44. From the steady flow energy equation (SFEE), we have

$$\dot{Q}_{cv} - \dot{W}_{cv} = \dot{m}\left(h_2 + \frac{C_2^2}{2} + gz_2\right) - \dot{m}\left(h_1 + \frac{C_1^2}{2} + gz_1\right)$$

Let $\Delta PE = 0$, so

$$\dot{Q}_{cv} = \dot{W}_{cv} + \dot{m}\left[(h_2 - h_1) + \frac{C_2^2 - C_1^2}{2}\right]$$

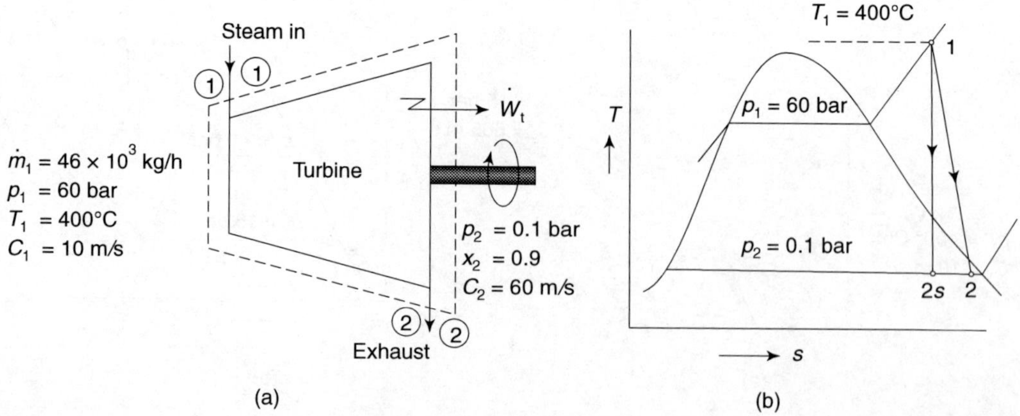

Fig. 9.44 Example 9.41

From Steam Tables, at $p_1 = 60$ bar, $T_1 = 400°C$, $h_1 = 3177.3$ kJ/kg

At $p_2 = 0.1$ bar, $h_{f2} = 191.83$ kJ/kg, $h_{fg2} = 2392.8$ kJ/kg so

$$h_2 = h_{f2} + x_2\, h_{fg2} = 191.83 + 0.9 \times 2392.8 = 2345.35 \text{ kJ/kg}$$

$$\dot{Q}_{cv} = 10 \times 10^3 + \frac{46 \times 10^3}{3600}\left[(2345.35 - 3177.3 + \frac{60^2 - 10^2}{2 \times 10^3}\right]$$

$$= 10 \times 10^3 + \frac{46 \times 10^3}{3600}[(-)\,831.95 + 1.75]$$

$$= -\,601.65 \text{ kW}$$ **Ans.**

From Steam Tables, at $p_1 = 60$ bar and $T_1 = 400°C$, $s_1 = 6.5408$ kJ/kg K

At $p_2 = 0.1$ bar

$$s_2 = s_{f2} + x_2\,(s_{g2} - s_{f2}) = 0.6493 + 0.9\,(8.1502 - 0.6493)$$

$$= 7.4 \text{ kJ/kg K}$$ **Ans.**

From the entropy balance, $\sum_j \frac{\dot{Q}_j}{T_j} + \dot{m}_1\, s_1 - \dot{m}_2\, s_2 + (\dot{S}_{\text{gen}})_{cv} = 0$

or $$\frac{\dot{Q}_{cv}}{T_b} + \dot{m}\,(s_1 - s_2) + (\dot{S}_{\text{gen}})_{cv} = 0$$

or $$(\dot{S}_{\text{gen}})_{cv} = -\frac{\dot{Q}_{cv}}{T_b} + \dot{m}\,(s_2 - s_1) = -\frac{(-\,601.65)}{350} + \frac{46 \times 10^3}{3600}(7.4 - 6.5408)$$

$$= 12.697 \text{ kW/K}$$ **Ans.**

Example 9.42 A steam condenser operating at steady state receives steam from the exhaust of steam turbine at 0.1 bar with a quality of 0.95 and the condensate exits at 0.1 bar and 45°C. Cooling water enters the condenser in a separate stream through water tubes as a liquid at 25°C and exits as a liquid at 35°C without any changes in pressure. Heat transfer from the outer surface of the condenser is ignored and changes in the kinetic and potential energies of the flowing stream are negligible. Calculate the following:

(a) The ratio of the mass flow rate of the cooling water to the mass flow rate of the condensing steam.

(b) The rate of energy transfer from the condensing steam to the cooling water, in kJ/kg of steam passing through the condenser.

A Model Solution Refer to Fig. 9.45.

Data given Steam $p_1 = 0.1$ bar $x_1 = 0.93$, $p_2 = 0.1$ bar, $t_2 = 45°C$

Cooling water $t_3 = 25°C$, $t_4 = 35°C$

To find $\dot{m}_3/\dot{m}_1$ and $(\dot{Q}_{cv}/\dot{m}_1)$ steam side

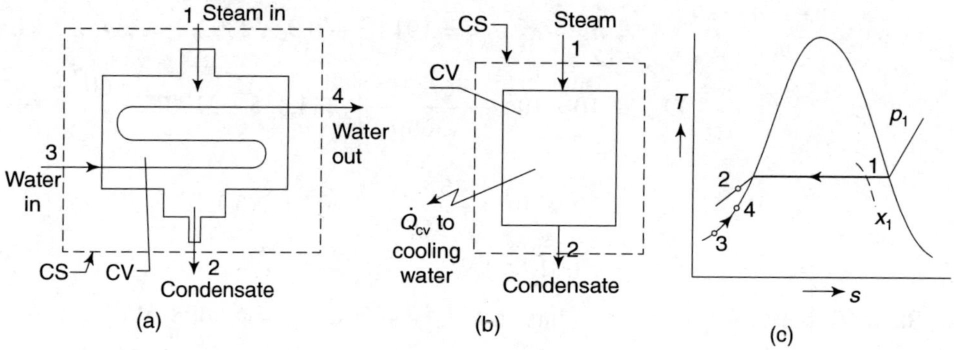

Fig. 9.45 Example 9.42

Assumptions (i) In each case, the control volume shown in Fig. 9.45 is in steady state, (ii) There is no heat transfer from condenser to surroundings, and $\dot{W}_{cv} = 0$, (iii) $\Delta KE = 0$, $\Delta PE = 0$, (iv) cooling water is incompressible, (v) water and steam do not mix.

Analysis Since water and steam do not mix, so the mass balance gives

$$\dot{m}_1 = \dot{m}_2 \quad \text{and} \quad \dot{m}_3 = \dot{m}_4$$

(a) Applying the steady state form of energy rate balance applied to the overall condenser, we have

$$\dot{m}_1 (h_1 - h_2) + \dot{m}_3 (h_3 - h_4) = 0$$

or

$$\frac{\dot{m}_3}{\dot{m}_1} = \frac{h_1 - h_2}{h_4 - h_3}$$

From Steam Tables, at 0.1 bar, $h_{f1} = 191.83$ kJ/kg $h_{g1} = 2584.7$ kJ/kg

$$h_1 = h_{f1} + x_1 (h_{g1} - h_{f1}) = 191.83 + 0.93 (2584.7 - 191.83)$$
$$= 2417.19 \text{ kJ/kg}$$

At 45°C,

$$h_2 = h_f (t_2) = 188.45 \text{ kJkg}$$

$$h_4 - h_3 = c_{pw} (T_4 - T_3)_w = 4.18 \times 10 = 41.8 \text{ kJ/kg}$$

Thus,

$$\frac{\dot{m}_3}{\dot{m}_1} = \frac{2417.9 - 188.45}{41.8} = 54.46$$ **Ans.**

(b) For a control volume enclosing the steam side of the condenser only, the steady state form of the enrgy balance yields

$$\dot{Q}_{cv} - \dot{W}_{cv} = \dot{m}_2 \left(h_2 + \frac{C_2^2}{2} + gz_2 \right) - \dot{m}_1 \left(h_1 + \frac{C_1^2}{2} + gz_1 \right)$$

Since $\dot{W}_{cv} = 0, \quad \Delta KE = 0, \quad \Delta PE = 0$, so

$$\frac{\dot{Q}_{cv}}{\dot{m}_1} = (h_2 - h_1) = 188.45 - 2417.1 = -2228.74 \text{ kJ/kg} \qquad \textbf{Ans.}$$

Here, the negative sign indicates that heat is transferred from the condensing steam to cooling water.

9.21 SUPERCRITICAL PRESSURE OF STEAM

So far we have discussed only the properties of steam generated below the critical pressure of water, i.e., 221.2 bar. For practical reasons, engineers were constrained to work with pressure below the critical pressure for a long time, but this difficulty has now been overcome and steam can now be generated much above the critical pressure. This is possible only when generation of steam takes place without the boiling or bubbling of water. The method of generation of supercritical pressure steam is completely different from that of below critical pressure. The properties of supercritical pressure of steam are separate and have separate tables.

9.22 SUMMARY OF FORMULAE

1. $x = \dfrac{m_g}{m_g + m_f}$
2. $v = (1 - x)\, v_f + x v_g \sim x v_g$
3. $v_{sup} = v_g \cdot \dfrac{T_{sup}}{T_g}$
4. $h = h_f + x\, h_{fg}$
5. $h_g = h_f + h_{fg}$
6. $h_{sup} = h_g + c_p\,(T_{sup} - t_g)$
7. $u = h - pv$
8. $u_f = h_f - p_f\, v_f$
9. $u = \left(h_f + x\, h_f\right) - p_g \cdot x \cdot v_g$
10. $u_g = h_g - p_g\, v_g$
11. $s_f = c_{pf} \ln \dfrac{T_g}{273}$
12. $s_{fg} = \dfrac{h_{fg}}{T_g}$
13. $s_g = s_g + s_{fg}$
14. $s = s_f + x\, s_{fg}$
15. $s_{sup} = s_g + c_p \ln \dfrac{T_{sup}}{T_g} = c_{pf} \ln \dfrac{T_g}{273} + \dfrac{h_{fg}}{T_g} + c_{p\,sup} \ln \dfrac{T_{sup}}{T_g}$
16. $y = y_1 + \dfrac{p - p_1}{p_2 - p_1}(y_2 - y_1)$ (Linear interpolation)
17. $h_1 = h_2$ (Throttling process)
18. $x = \dfrac{\dfrac{M}{m}(t_2 - t_1) - (t_g - t_2)\, c_{pf}}{h_{fg}}$ (Bucket calorimeter)

19. $x = \dfrac{M}{M+m}$ (Separating calorimeter)

20. $x_1 = \dfrac{h_{g2} + c_p\,(t_{\sup 2} - t_{g2}) - h_{f1}}{h_{fg1}}$ (Throttling calorimeter)

21. $x = x_1 . x_2$ (Separating and throttling calorimeter)

PROBLEMS

Things to Think About

9.1. At critical temperature of water, is water a two-phase mixture?

9.2. Is it true that the saturation temperature increases with pressure?

9.3. Is it true that h_{fg} decreases with increase of pressure? Justify your answer.

9.4. List the thermodynamic cycle in which a pure substance takes the shape of liquid phase and liquid-vapour phase.

9.5. What happens to the level of ocean and sea due to evaporation of water and fall of rains?

9.6. Whether the steam formation in a boiler is due to constant pressure heating or constant volume heating?

9.7. Can water at – 40°C exist as a vapour or as a liquid?

9.8. Is it true that steam bath is beneficial for health? Give the reasoning of your answer.

Short and Long Theoretical Questions

9.9. Define a pure substance. Is water a pure substance?

9.10. What do you mean by a phase of substance?

9.11. Discuss the generation of steam at constant pressure. Show the various processes on temperature-enthalpy diagram.

9.12. Draw the following charts for steam

(a) *p–v* (b) *T–v* (c) *h–s* (d) *T–s* and

(e) *p–h* : Discuss the various lines.

9.13. Represent the following process on *T–s*, *h–s* and *p–v* diagrams for expansion of steam (a) constant pressure, (b) constant volume, (c) isothermal, (d) adiabatic, (e) polytropic and throttling.

9.14. Discuss ths various methods of determining the dryness fraction of steam with relative merits and demerits.

9.15. Describe with a net sketch the working of a combined separating and throttling calorimeter.

9.16. What do you mean by super-critical pressure of steam?

9.17. Explain the term (a) latent heat, (b) sensible heat, (c) total heat, (d) dryness fraction, (e) saturation temperature and (f) specific volume as applied to steam.

Numerical Questions

9.18. Calculate the dryness fraction of steam if 0.8 kg of water is in suspension with 45 kg of dry steam. [*Ans.* $x = 0.982$]

9.19. Calculate the enthalpy, entropy and internal energy for each of the following:

(a) 5 kg of steam at 15 bar and 0.95 dry

(b) 5 kg of steam at 1 bar and 0.95 dry

(c) 5 kg of steam at 15 bar and 400°C.

[*Ans.* (a) 13456 kJ/kg 21.171 kJ/kg, 12822.96 kJ
(b) 12412.5 kJ, 35.28 kJ/K, 12007.85 kJ
(c) 17360.5 kJ, 38.104 kJ/kg, 15589.75 kJ]

9.20. 15000 kg of water and 15 kg of dry saturated steam at 10 bar are contained in the boiler drum. Determine the total heat content and the internal volume of the boiler. When the pressure in the boiler is 2 bar, calculate the mass of steam and water. Also, calculate the temperature of the water in the boiler. [*Ans.* 3045647.5 kJ, 19.8195 m^3]

9.21. Dry and saturated steam at a pressure of 11 bar is contained in a closed metallic drum. Due to cooling, the temperature in the drum reduces to 105°C. Calculate: (a) the pressure in the vessel, and (b) heat lost in cooling. [*Ans.* (a) 1.208 bar, (b) 2059.25 kJ/kg]

9.22. A boiler is supplied with feed water at a temperature of 44°C and converted into steam at a gauge pressure of 5 bar and a temperature of 302°C. If 900 kg of steam is generated by the boiler per hour, calculate the quantity of heat supplied to the water to produce steam per minute, assuming barometric pressure of 750 mm of Hg and specific heat of superheated steam as 2.28 kJ/kg K. [*Ans.* 40063.95 kJ]

9.23. Find the state of steam whether wet, dry or superheated for the following cases:

(a) $t = 90°C$, $s = 7.276$ kJ/kg

(b) $p = 6$ bar, $h = 2727$ kJ/kg

(c) $p = 15$ bar, $u = 2300$ kJ/kg

(d) $v = 0.23$ m^3/kg, $p = 8$ bar

(e) $t = 600°C$, $p = 75$ bar

[*Ans.* (a) wet, 0.964; (b) wet, 0.986; (c) superheated by 177°C, (d) wet, 0.95, (e) superheated by 309.51°C]

9.24. A vessel containing 5 kg of steam at 8 bar and 250°C is cooled by pouring water over the outer surface, till the inside pressure falls to 5 bar. Calculate (a) the final state of the steam, (b) heat loss, and (c) loss of internal energy. [*Ans.* (b) wet, 0.7795; (b) 33385.5 kJ; (c) 2900.38 kJ]

9.25. Calculate the external work of evaporation, internal latent heat and internal energy of one kg of steam at a pressure of 13 bar when the steam is (a) 0.8 dry, and (b) dry and saturated.

[*Ans.* (a) 157.2 kJ, 1421.46 kJ, 2229.6 kJ (b) 196.55 kJ, 1777 kJ, 2586 kJ]

9.26. A tank containing 180 kJ of water at 16°C is to be heated to 66°C by blowing dry and saturated steam at 2 bar into the tank. Determine the mass of steam that should enter the tank to heat the water to the required temperature. [*Ans.* 13.93 kg]

9.27. Feed water at 65°C enters a boiler and steam is generated at a pressure of 17 bar and 260°C. If 75% of the heat of the coal of calorific value of 29274 kJ/kg is utilised in producing steam, determine the mass of coal consumed in producing 5000 kg of steam per hour. Assume c_p = 2.09 kJ/kg K for superheated steam. [*Ans.* 601 kg]

9.28. Determine the amount of heat to be supplied to 2 kg of water at 40°C in order to convert it into steam at 14 bar and 0.9 dry. [*Ans.* 4843.59 kJ]

9.29. A rigid metallic vessel of capacity 0.055 m^3 contains steam at a pressure of 15 bar and 250°C. Calculate (a) the mass of steam contained in the vessel. If the vessel is cooled till the pressure in the vessel becomes 10.5 bar determine, (b) final dryness fraction of steam, and (c) amount of heat rejected between the initial and final states.[*Ans.* (a) 0.355 kg, (b) 0.824, (c) 153.5 kJ]

9.30. Determine the mass of steam of dryness fraction 0.95 required to move a piston of diameter of 40 cm through a distance of 1 metre at a constant pressure of 10 bar. [*Ans.* 0.668 kg]

9.31. A cylinder fitted with a movable piston contains 1 kg of steam at an initial pressure of 6 bar and volume of 0.26 m^3. Determine the dryness fraction of steam. The steam expands according to the law $pV^n = c$ to a final volume of 0.52 m^3 and the pressure is 2.5 bar. Calculate the value of index n and the work done during expansion. [*Ans.* 0.807, 1.263, 5333.7 kJ]

9.32. Determine the internal energy of 1 kg of steam at 1 bar and 0.9 dry. If this steam is compressed to 10 bar according to law $pv^{1.1}$ = const., calculate the final dryness of steam and the change in internal energy, stating whether the latter is an increase or decrease.

[*Ans.* 2294.66 kJ, 0.964, 2520 kJ (increase)]

9.33. A small steam turbine exhausts into a condenser at the rate of 5 kg/min. The pressure inside the condenser is 0.15 bar. The steam from the condenser leaves water at 40°C. Cooling water in circulated through the condenser tubes at the rate of 115 kg/min with a rise in temperature from inlet to outlet of 20°C. Determine the dryness fraction of steam entering the condenser. Draw the system, show the process an p–v, T–s and h–s chart. [*Ans.* (a) 0.787]

9.34. Superheated steam at 5.5 bar and 250°C is contained in a cylinder fitted with piston. The initial volume of steam is 0.028 m^3. The expansion of steam takes place according to the law $pv^{4/3}$ = Constant. Determine (a) final volume and temperature (b) work done, (c) change in internal energy, and (d) heat transferred. Take final pressure equal to 1.5 bar.

[*Ans.* (a) 0.07826 m^3, 173.8°C; (c) – 13.7 kJ (decrease)
(d) – 0.627 kJ (heat transfer from steam]

9.35. Steam at 7 bar having dryness fraction 0.9 expands reversibly at constant pressure until the temperature is 200°C. Find the work done and heat supplied per kg of steam during the process.

[*Ans.* 38.2 kJ/kg; 288.7 kJ/kg]

9.36. Dry and saturated steam expands reversibly in a cylinder behind a piston until the pressure is 0.1 bar. Calculate the change in internal energy per kg of steam if the heat is supplied continuously during the process in order to keep the temperature constant. [*Ans.* 37.2 kJ/kg]

9.37. 1 kg of steam at 20 bar having dryness fraction 0.9 is heated reversibly at constant pressure to a temperature of 300°C. Determine the heat supplied, the change in entropy and show the process on *T–s* diagram. [*Ans.* 415 kJ/kg; 0.8173 kf/kg K]

938. A steam engine receives steam at 4 bar having dryness fraction 0.8. The steam expands according to $pv^{1.05}$ = constant to a condenser pressure of 1 bar. Calculate the change of entropy per kg of steam during the expansion and sketch the process on *T–s* diagram. [*Ans.* 0.381 kJ/kg K]

9.39. Two boilers generating steam deliver equal amount of steam into a common main. One boiler is fitted with a superheater and other without a superheater. Both the boilers operate at a pressure of 15 bar which is also the pressure in the main. The temperature of steam in the superheated boiler is 300°C. If it is desired to maintain a temperature of 225°C in the main, what will be quality of steam being raised in the boiler without superheated. Also, calculate the net increase or decrease of entropy. [*Ans.* 0.944, 0.10078 kJ/kg K]

9.40. A rigid metallic vessel is divided into two parts by a partition. Part *A* contains 1 kg of steam at 3.4 bar dry and saturated. Part *B* contains 2 kg of steam at 6.8 bar having 0.8 dryness fraction. The partition is removed and the pressure in the vessel after sometime is found to be 4.8 bar. Calculate (a) volume of vessel, (b) state of steam at 4.8 bar, and (c) heat interchange between steam and the surrounding walls of the vessel. [*Ans.* 1.0055 m^3, 0.7649 and – 11709.6 kJ]

9.41. The following data refer to a test on combined separating and throttling calorimeter.

Mass of steam entering the separating calorimeter = 5200 gm. Water drained from separator = 270 gm. Pressure of steam before throttling = 8.5 bar. Pressure of steam after throttling = 1.2 bar. Temperature of steam after throttling = 105°C. Assume c_p for superheated steam as 2.091 kJ/kg K. Calculate the dryness fraction of steam on entry. [*Ans.* 0.911]

9.42. Steam at a pressure of 8 bar and 0.96 dry passes through a throttle valve and emerges just dry and saturated. Calculate the lower pressure. [*Ans.* 1.35 bar]

9.43. Steam is available at a pressure of 21 bar and 78°C superheat but its pressure must be reduced to 1.05 bar before it can be used in a heating apparatus. This reduction can be brought about either (a) by allowing the steam to do 3735 kJ of work per kg in an engine exhausting at 1.05 bar or (b) by throttling the steam to 1.05 bar and the injecting water at a temperature of 66.6°C into it. The heating apparatus requires 2785 kW. Assume that the condensate leaves the heater at 79°C and find, (a) The quality of high pressure steam required per hour in both the cases, (b) the power of the engine, (c) the mass of injection water required per hour in order to yield same quality of steam for heating as obtained by process (a).

[*Ans.* (a) 1 kg (b) 3.71 kW (c) 1.030 kg/h]

9.44. A steam engine receives steam from a boiler through a long pipeline. The steam leaving the boiler is at 14 bar and 204°C. Some of the steam at the engine end of the pipeline is bled off and passes through a throttling calorimeter. The temperature of steam entering the calorimeter is 193°C while pressure and temperature of steam after passing through the calorimeter are 1.05 bar and 115°C. Calculate (a) the heat loss per kg of steam during passage through the pipes, (b) quality of the steam at the engine.

9.45. In a bucket colorimeter, 5 kg of steam is sampled at 7 bar. The water equivalent at 20°C is 10 kg. The bucket contains 140 kg of water at 20°C. The temperature of water when steam condensed into it was measured as 40°C. Calculate the dryness fraction of steam. [*Ans.* 0.963]

9.46. Steam at a pressure of 8.4 bar but of unknown quality flows into a separator at the rate of 4.2 kg/s. Water is removed from the separator at the rate of 0.1 kg/s while the remainder of the steam passes through a throttle valve causing a pressure drop to 1.4 bar. Due to throttling, temperature of steam increases to 114.5°C. Assume c_p = 2.07 kg/kg K for superheated steam. Calculate the dryness fraction of steam (a) entering the valve and (b) entering the separator, and (c) the change in specific volume after throttling.

[*Ans.* (a) 0.965, (b) 0.942, (c) 1.023 m^3]

9.47. Steam at 10.5 bar and 0.98 dry is throttled to 1.4 bar. If the velocity of steam in the pipe after throttling is 61 m/s and flow rate is 5.45 kg/min, calculate: (a) temperature of steam in pipe after throttling and (b) diameter of pipe of low pressure side. [*Ans.* (a) 114.8°C (b) 4.92 cm^2]

9.48. Steam at 10 bar and 200°C is supplied to feed water heater. The steam raises the temperature of water from 90°C to 5°C below the saturation temperature of heating steam. During heat transfer process, the steam is condensed and the condensate leaves at saturation temperature. Determine (a) the amount of steam to be delivered to feed water heater for heating 40 kg/s of feed water through the requisite temperature range, (b) diameter of the steam pipe supplying to heater if the velocity or flow is 30 m/s. [*Ans.* (a) 6.86 kg/s., (b) 24.8 cm]

9.49. A closed vessel having a volume of 2.84 m^3 contains 1.365 kg of mixture of water and steam at a pressure of 7 bar. Find how much heat must be added to the vessel to make the steam 0.9 dry and find the resulting steam pressure. [*Ans.* 3826.5 kJ, 7.47 bar]

9.50. 5 kg of steam per sec enters a steam turbine at 26 bar and 400°C. It expands adiabatically to 1 bar such that condition after expansion is dry and saturated. State if the process is reversible or not and calculate ideal work lost if the process is not reversible. [*Ans.* Irreversible. 18.7%]

9.51. In a cylinder, 5 kg of steam at a pressure of 14 bar and 280°C expands to 1.4 bar according to law, $pv^{1.25}$ = constant. Calculate (a) the final temperature, (b) the work done, (c) heat transferred, (d) the change in entropy. [*Ans.* (a) 108.75°C, (b) 1807.4 kJ(c) 466.29 (d) 1.28kJ/K]

9.52. A certain quantity of saturated water at 59.6°C is contained in a evacuated drum of 0.9 m^3 capacity. It is heated till the pressure and temperature is 20 bar and 240°C. Calculate (a) mass of water (b) heat required. [*Ans.* (a) 8.1 kg, (b) 19265.6 kJ]

9.53. Steam at 0.05 bar and 100°C is to be condensed completely by a reversible constant pressure process. Find the heat to be removed per kg of steam and the change of entropy. Sketch the process on *T–s* diagram. [*Ans.* 25.50 kJ/kg: 8292 kJ/kg K]

9.54. A turbine in supplied with steam at 40 bar, 400°C, the steam expands through the turbine in steady flow condition to an exit pressure of 0.2 bar and a dryness fraction of 0.93. The inlet velocity is negligible but the steam leaves at high velocity through a duct of 0.14 m^2 cross-sectional area. Determine the power output of the turbine if the mass flow rate is 3 kg/s and the mechanical efficiency is 90%. Calculate the change of entropy. Assume negligible heat loss.

[*Ans.* 2098 kW, 0.643 kJ/kg K]

9.55. A rigid steel tank contains a mixture of vapour and liquid water at a temperature of 50°C. The tank has a volume of 0.8 m^3, the liquid phase occupying 40% of the volume. Determine the amount of heat added to the system to raise the pressure to 20 bar.

9.56. In order to operate a steam turbine at part load conditions, a throttle valve is installed on the steam line before steam turbine. The line conditions are 1.5 MPa, 300°C, and the turbine exhaust pressure is fixed at 10 kPa. Assuming isentropic expansion through the turbine. Calculate the following:

(a) The full load power output of the turbine per kg of steam.

(b) The pressure to which the steam must be throttled to produce 70% of the full load output. Show both processes on a *T–s* diagram.

9.57. A direct contact feed water heater is used to preheat the water before entering a boiler by direct mixing of steam withdrawn from turbine with feed water coming from condenser. Steam at 800 kPa having dryness fraction 0.95 enters the feed water heater and feed water at 800 kPa and at 35°C also enters the heater. The final feed water leaves at 800 kPa and temperature 150°C with mass flow rate of 2 kg/s. Calculate the rate of increase in entropy per second if the process is steady flow and adiabatic.

9.58. Steam enters a turbines at 700 kPa, 250°C and exhausts at 15 kPa. The efficiency of the turbine is 80%. Calculate (a) the power output per kg of steam, (b) the quality of steam leaving the turbine.

Software Based Questions

9.59. Develop a software in database form for the properties of steam for the following case by taking data from table:

(a) Temperature based (up to 100°C) for saturated steam

(b) Pressure based for dry and saturated steam

(c) Superheated steam properties, pressure and temperature based

(d) Supercritical steam properties, pressure and temperature based.

9.60. Develop a software to calculate work done, heat transfer, change in internal energy and change in entropy for the various non-flow processes. Print the results for one real specific case in each.

9.61. Develop a software to calculate the work done from steam turbine, kinetic energy gain from steam nozzle, cooling water requirement for steam condenser, fuel requirement for a boiler and dryness fraction of steam from a separating and throttling calorimeter. Print the results for one real specific case in each.

9.62. Develop a software in database form for various refrigerants presently used in practice by taking data from refrigrants tables.

Design and Open Ended Questions

9.63. Collect the various governing equations on the basis of which steam tables and charts have been prepared.

9.64. There are three methods, namely, separating, throttling and separating-throttling methods which are common to calculate dryness fraction of steam. Device some new methods to calculate more accurate dryness fraction of steam.

9.65. From the research papers, collect the empirical equations in polynomial forms which may be used to calculate properties of steam. Compare these values with the table.

9.66. Develop an experiment to visualise the sequence of events as a two-phase liquid-vapour mixture which is heated at constant volume near its critical point. Prepare a report regarding your observations separating the two phases where the average specific volume is less and greater than critical. What happens to the meniscus in the vicinity of the critical point?

9.67. Prepare a report about the depletion of earth in stratospheric zone due to CFC (chloroflurocarbon) refrigerant such as *Fr*-12 and *Fr*-22. Now what type of refrigerants have been suggested by world bodies.

9.68. With the help of research papers write a comprehensive report about two-phase flow.

9.69. Writc a report starting from the realisation of steam power from inception till today development.

Multiple-Choice Answer Type Questions

9.70. If the dryness fraction of a given sample of stean is 0.9, the mass of water particles contain in 5 kg of steam will be

(a) 0.2 kg (b) 0.3 kg (c) 0.4 kg (d) 0.5 kg

9.71. The values of critical pressure and temperature of steam are

(a) 374.15 bar, 221.297°C (b) 221.297 bar, 374.15°C

(c) 274.15 bar, 32I.297°C (d) 321.297 bar, 274.15°C

9.72. With increase of pressure, the latent heat of steam

(a) decreases (b) increase

(c) remains ihe same (d) behaves unpredictably

9.73. Consider the following statements.

(A) Boiling process is a constant pressure process

(B) Boiling process is a constant temperature process involving heat transfer with infinitesimal temperature difference

(C) Boiling process is a reversible process

(D) Boiling process is an isentropic process

(a) A, B and D (b) A, C and D (c) B, C and D (d) A, B, C

9.74. On *T*–*s* diagram the change of state during transformation of water from 20°C to 20°C steam at atmospheric pressure due to addition of heat can be represented by

(a) slant line (b) vertical line (c) horizontal line (d) none of these

9.75. Slope of constant pressure line on enthalpy-entropy diagram of steam with increasing pressure

(a) increases (b) decreases (c) remains unaltered (d) none of these

9.76. During dryness fraction measurement of steam using throttling calorimeter, the wet state of steam is throttled so as to lie in

(a) wet state (b) dry and saturated state

(c) superheated states (d) super saturated state

9.77. By increasing pressure, the specific volume of steam

(a) remains the same (b) increase

(c) decreases (d) increases and then decreases

Answers to Multiple-Choice Questions

9.70. (d) 9.71. (b) 9.72. (a) 9.73. (d) 9.74. (c) 9.75. (a)

9.76. (c) 9.77. (c)

CHAPTER

10

General Thermodynamic Relations

OVERVIEW

In this chapter, we will establish some general thermodynamic relations which may be applied to the development of the equations of state for different substances which do not undergo a chemical reaction or a change in phase. These relations will be helpful to compute the changes in those properties which cannot be easily measured experimentally, for example, energy and entropy. Among all relations, the most popular are the Maxwell relations named after the discoverer James Clerk Maxwell.

10.1 SOME MATHEMATICAL PRELIMINARIES

It will be worthwhile to review a few principles of partial differential calculus to make the most efficient use of thermodynamic relations. Let z be a function of two independent variables x and y, then

$$z = f(x, y) \qquad ...(10.1)$$

and, according to calculus,

$$dz = \left(\frac{\partial z}{\partial x}\right)_y dx + \left(\frac{\partial z}{\partial y}\right)_x dy \qquad ...(10.2)$$

where dz is called an ***exact differential***. Here, the subscripts are employed to show which variable is held constant.

Let

$$\left(\frac{\partial z}{\partial x}\right)_y = M \quad \text{and} \quad \left(\frac{\partial z}{\partial y}\right)_x = N \qquad ...(10.3)$$

The second mixed partial derivatives are equal, i.e.,

$$\left(\frac{\partial^2 z}{\partial x\, \partial y}\right) = \left(\frac{\partial^2 z}{\partial y\, \partial x}\right) \qquad ...(10.4)$$

or $$\left(\frac{\partial M}{\partial y}\right) = \left(\frac{\partial N}{\partial x}\right) \quad ...(10.5)$$

Hence, Eq. (10.2) becomes $dz = M\,dx + N\,dy$ ...(10.6)

It is obvious from Eq. (10.1) that, in principle, either x or y could be expressed implicitly in terms of the other variables. In other words, there exists relation of the form

$$x = \Psi\,(y, z) \text{ and } y = \phi\,(x, z)$$

Now considering, $$x = \Psi\,(y, z),\ dx = \left(\frac{\partial x}{\partial y}\right)_z dy + \left(\frac{\partial x}{\partial z}\right)_y dz \quad ...(10.7)$$

Substituting the value of dx in the Eq. (10.2), we have

$$dz = \left(\frac{\partial z}{\partial x}\right)_y \left\{\left(\frac{\partial x}{\partial y}\right)_z dy + \left(\frac{\partial x}{\partial z}\right)_y dz\right\} + \left(\frac{\partial z}{\partial y}\right)_x dy$$

or $$dz = \left(\frac{\partial z}{\partial x}\right)_y \left(\frac{\partial x}{\partial y}\right)_z dy + \left(\frac{\partial z}{\partial x}\right)_y \left(\frac{\partial x}{\partial z}\right)_y dz + \left(\frac{\partial z}{\partial y}\right)_x dy$$

Since there are only two independent varibales, we can choose y and z at our will. For example, let $dy = 0$ and $dz \neq 0$, then

$$\left(\frac{\partial z}{\partial x}\right)_y \left(\frac{\partial x}{\partial z}\right)_y dz = 1 \quad ...(10.8)$$

Likewise, let $dz = 0$ and $dy \neq 0$, hence, $\left(\frac{\partial z}{\partial x}\right)_y \left(\frac{\partial x}{\partial y}\right)_z = -\left(\frac{\partial z}{\partial y}\right)_x dy$

or $$\left(\frac{\partial x}{\partial y}\right)_z \left(\frac{\partial y}{\partial z}\right)_x \left(\frac{\partial z}{\partial x}\right)_y = -1 \quad ...(10.9)$$

Equations (10.8) and (10.9) are useful equations. It is also possible to write it as

$$\left(\frac{\partial z}{\partial x}\right)_y = \frac{1}{(\partial x/\partial z)_y}.$$

10.2 DEFINITIONS

Let us define some new important thermodynamic terms which will be useful for generalising the thermodynamic relations.

10.2.1 Helmholtz Free Energy, *F*

It is defined as the difference between the internal energy and the product of the temperature and entropy. Hence,

$$F = U - T \cdot S;\ \text{kJ}$$

and $$f = u - Ts;\ \text{kJ/kg} \quad ...(10.10)$$

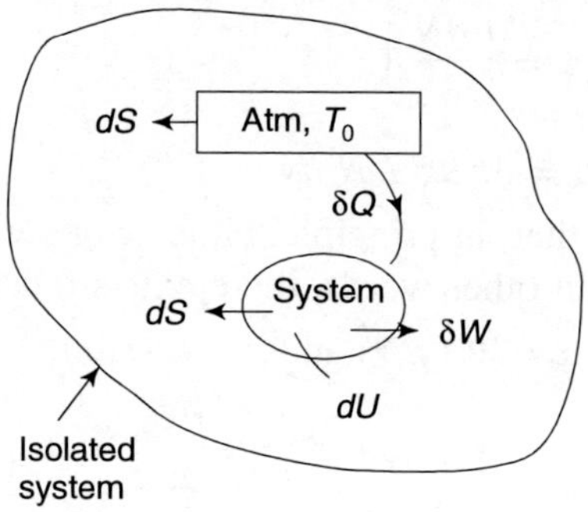

Fig. 10.1 Work exchange equal to decrease in Helmholtz function

This equation applies to a closed system. For example, consider the heat interaction between a closed system and its environment at a pressure p_0 and temperature T_0 (Fig. 10.1). The work done on this, in the absence of kinetic, potential, electrical, magnetic and surface energies is given by this first law:

$$-\delta W = dU - \delta Q \qquad ...(10.11)$$

Now suppose that the system exchanges heat with the environment only, the change of entropy of the system from the second law will be

$$dS \geq \frac{\delta Q}{T_0}$$

Putting the value of δQ in the first law, we have work done on the system

$$-\delta W \geq dU - T_0\, dS \qquad ...(10.12)$$

Integrating between two equilibrium states, we have

$$-W_{1-2} \geq (U_2 - U_1) - T_0\,(S_2 - S_1) \qquad ...(10.13)$$

If the initial and final temperatures of the system are equal to the temperature of the environment, then the above can be written as

$$-W_T \geq [(U_2 - T_2\,S_2) - (U_1 - T_1\,S_1)]_T \geq (F_2 - F_1)_T \qquad ...(10.14)$$

Equation (10.14.) states that in a closed system, when initial and final temperatures are equal to that of the environment and the system exhanges heat with the enviorment only the work done is either equal to or greater than the change in the ***Helmholtz function****. In the case of work done by the system.*

$$W_T \leq (F_1 - F_2)_T \qquad ...(10.15)$$

This shows that the work done by the system will be equal to or less than decrease in Helmholtz function.

10.2.2 Gibbs Free Energy

It is defined as the difference between the enthalpy and the product of temperature and entropy. Hence,

$$G = H - TS; \text{ kJ}$$

and
$$g = h - Ts; \text{ kJ/kg} \quad ...(10.16)$$

This equation applies to a steady flow process. In this system too, heat is exchanged with the environment only, and the system exists in thermal equilibrium with the environment both initially and ultimately. Thus, the above equation may be written as

$$G = U + pV - TS \quad ...(10.17)$$

In the steady flow process, the net flow work ($p_1 V_1 - p_2 V_2$) at entry to and exit from the system affects the amount of work done, hence the expression for work done on the system is

$$-\delta W = dH - \delta Q$$

or
$$-W \geq (U_2 - U_1) - T_0 (S_2 - S_1) - (p_1 V_1 - p_2 V_2)$$

or
$$-W \geq (U_2 + p_2 V_2 - T_0 S_2) - (U_1 + p_1 V_1 - T_0 S_1) \quad ...(10.18)$$

Since the initial and final temperature of the system were specified to be equal to the environmental temperature, therefore

$$-W_T > (U_2 + p_2 V_2 - T_2 S_2)_T - (U_1 + p_1 V_1 - T_1 S_1)_T$$

or
$$-W_T \geq (G_2 - G_1)_T \quad ...(10.19)$$

Equation (10.19) suggests that in a steady flow process when the initial and final temperatures are equal to that environment and when the system exchanges heat with the environment only the work done on the system is equal to or greater than the change in the Gibbs function. The equality and inequality sign indicate reversible and irreversible process. In the case of work done by the system,

$$W_T \leq (G_1 - G_2)_T$$

If the system is also in pressure equilibrium with the environment, then

$$W_{T.p} \leq (G_1 - G_2)_{T.p} \quad ...(10.20)$$

Therefore, both the Gibbs and the Helmholtz free function establish a criteria for thermodynamic equilibrium. At equilibrium, these functions are at their minimum values.

10.2.3 Coefficient of Volume Expansion, β

It is defined as the ratio of the change in volume with temperature to the volume at constant pressure when the change in temperature and the corresponding change in volume become infinitesimal. Hence,

$$\beta = \frac{1}{V}\left(\frac{\partial V}{\partial T}\right)_p = \frac{1}{v}\left(\frac{\partial v}{\partial T}\right)_p = [T^{-1}]$$

In the case of perfect gas, $\beta = \frac{1}{v}\left(\frac{\partial v}{\partial T}\right)_p = \frac{p}{RT}\left(\frac{R}{p}\right) = \frac{1}{T}$ ...(10.21)

Thus, the coefficient of volumetric expansion of a perfect gas varies inversely with absolute temperature and is independent of both pressure and volume. The coefficient of real gas, however, depends on both pressure and temperature.

10.2.4 Isothermal Compressibility, *K*

It is defined as the negative of the ratio of the change in volume with pressure, to the volume at constant temperature, when the change in pressure and the corresponding change in volume become infinitesimal.

$$K = -\frac{1}{V}\left(\frac{\partial V}{\partial p}\right)_T = \frac{-1}{v}\left(\frac{\partial v}{\partial p}\right)_T = [p^{-1}] \quad \text{...(10.22)}$$

The negative sign is included because an increase in pressure generally results in a decrease in volume. When the system consists of a perfect gas, the coefficient of compressibility becomes

$$K = \frac{-1}{v}\left(\frac{\partial v}{\partial p}\right)_T = \frac{1}{v}\left(\frac{-RT}{p^2}\right) = \frac{1}{p} \quad \text{...(10.22a)}$$

The compressibility of a substance is measured by its change of volume when the external pressure changes. Let the pressure on a certain substance of volume V be increases by dp resulting in diminishing volume dV. The bulk modulus of elasticity E is given by

$$E = -\frac{dp}{dV/V} = -V\frac{dp}{dV} \quad \text{...(10.22b)}$$

where dV/V *is the volumetric strain.* E is not a property since the derivative dp/dV depends on the process. Hence, the above equation becomes

$$E = \frac{1}{K - \beta\, dT/dp}$$

In solids and liquids, change in pressure produces negligible change in temperature, hence dT/dp is zero. Therefore, the reciprocal of the coefficient of compressibility of solids and liquids is the bulk modulus of elasticity. *For the gases, the terms dT/dp cannot be neglected and so the bulk modulus of elasticity is not a property.*

Internally, a substance is not compressible or incompressible. It is the process which determines whether the substance is incompressible or compressible. If there is an appreciable change in volume (i.e., density), the substance is said to be compressible.

Now, we can relate the volume change with β and K, in a single component, single phase system changes in volume may occur due to changes in pressure or temperature hence,

$$dV = \left(\frac{\partial V}{\partial T}\right)_p dT + \left(\frac{\partial V}{\partial p}\right)_T dp$$

Dividing by V, this becomes

$$\frac{dV}{V} = \frac{1}{V}\left(\frac{\partial V}{\partial T}\right)_p dT + \frac{1}{V}\left(\frac{\partial V}{\partial p}\right)_T dp = \beta\, dT - K\, dp$$

Thus, we see that the volume change can be expressed in terms of β and K. Hence, *both β and K are intensive properties independent of the volume of the system.*

10.3 DIFFERENTIAL RELATIONS FOR *U, H, G* AND *F*

With the help of first and second laws of thermodynamics we can find the differential relations between U, H, G and F for closed or open system.

$$dU = TdS - pdV \quad \text{...(10.23)}$$

$$dH = TdS + Vdp \quad \text{...(10.24)}$$

For unit mass,

$$du = Tds - pdv \quad \text{...(20.23a)}$$

$$dh = Tds + vdp \quad \text{...(10.24a)}$$

Here, g and f are expressed as

$$dg = dh - Tds - sdT \quad \text{...(10.25)}$$

$$df = du - Tds - sdT \quad \text{...(10.26)}$$

With the help of Eq. (10.23) and (10.24), we get

$$dg = vdp - sdT \quad \text{...(10.27)}$$

$$df = -\,pdv - sdT \quad \text{...(10.28)}$$

10.4 SOME IMPORTANT PROPERTIES EXPRESSED IN DIFFERENTIAL FORM

(i) From Eq. (10.23 a), $u = f(s, v)$

Hence,
$$du = \left(\frac{\partial u}{\partial s}\right)_v ds + \left(\frac{\partial u}{\partial v}\right)_s dv = Tds - pdv$$

$$\therefore \quad \left(\frac{\partial u}{\partial s}\right)_v = T \text{ and } \left(\frac{\partial u}{\partial v}\right)_s = -\,p \quad \text{...(10.29)}$$

(ii) From Eq. (10.24 a), $h = f(s, p)$, hence,

$$dh = \left(\frac{\partial h}{\partial s}\right)_p ds + \left(\frac{\partial h}{\partial p}\right)_s dp = Tds + vdp$$

$$\therefore \quad \left(\frac{\partial h}{\partial s}\right)_p = T \text{ and } \left(\frac{\partial h}{\partial p}\right)_s = v \quad \text{...(10.30)}$$

(iii) From Eq. (10.25), $g = f(s, p)$, hence

$$\left(\frac{\partial g}{\partial p}\right)_T = v \text{ and } \left(\frac{\partial g}{\partial T}\right)_p = -s \qquad ...(10.31)$$

(iv) From Eq. (10.26), $f = f(p, T)$, hence

$$df = \left(\frac{\partial f}{\partial v}\right)_T dv + \left(\frac{\partial f}{\partial T}\right)_v dT = -pdv - sdT$$

$$\left(\frac{\partial f}{\partial v}\right)_T = -p \text{ and } \left(\frac{\partial f}{\partial T}\right)_v = -s \qquad ...(10.32)$$

Comparing the above equations, we get

(a) $$\left(\frac{\partial u}{\partial s}\right)_v = \left(\frac{\partial h}{\partial s}\right)_p = T \qquad ...(10.33)$$

(b) $$\left(\frac{\partial u}{\partial v}\right)_s = \left(\frac{\partial f}{\partial v}\right)_T = -p \qquad ...(10.34)$$

(c) $$\left(\frac{\partial g}{\partial T}\right)_p = \left(\frac{\partial f}{\partial T}\right)_v = -s \qquad ...(10.35)$$

(d) $$\left(\frac{\partial g}{\partial T}\right)_T = \left(\frac{\partial h}{\partial p}\right)_s = v \qquad ...(10.36)$$

10.5 MAXWELL RELATIONS

(i) **From u:** We have from Eqs. (10.33) and (10.34),

$$\left(\frac{\partial u}{\partial s}\right)_v = T \text{ and } \left(\frac{\partial u}{\partial v}\right)_s = -p$$

Differentiating, we have, $\dfrac{\partial^2 u}{\partial s\, \partial v} = \left(\dfrac{\partial T}{\partial v}\right)_s = \dfrac{\partial^2 u}{\partial v\, \partial s} = -\left(\dfrac{\partial p}{\partial s}\right)_v$

or $$\left(\frac{\partial T}{\partial v}\right)_s = -\left(\frac{\partial p}{\partial s}\right)_v \qquad ...(10.37)$$

Equation (10.37) is called the *Maxwell relation from u*. Likewise, Maxwell relations of h, g and f can also be obtained.

(ii) **From h:** We have, $\left(\dfrac{\partial h}{\partial s}\right)_p = T$ and $\left(\dfrac{\partial h}{\partial p}\right)_s = v$

$$\therefore \quad \frac{\partial^2 h}{\partial s\, \partial p} = \left(\frac{\partial T}{\partial p}\right)_s = \frac{\partial^2 h}{\partial p\, \partial s} = \left(\frac{\partial v}{\partial s}\right)_p$$

or $$\left(\frac{\partial T}{\partial p}\right)_s = \left(\frac{\partial v}{\partial s}\right)_p \qquad ...(10.38)$$

(ii) **From g:** We have, $\left(\frac{\partial g}{\partial p}\right)_T = v$ amd $\left(\frac{\partial g}{\partial T}\right)_p = -s$

$$\therefore \quad \frac{\partial^2 g}{\partial p\,\partial T} = \left(\frac{dv}{\partial T}\right)_p = \frac{\partial^2 g}{\partial T\,\partial p} = -\left(\frac{\partial s}{\partial p}\right)_T$$

or $$\left(\frac{\partial v}{\partial T}\right)_p = -\left(\frac{\partial s}{\partial p}\right)_T \qquad ...(10.39)$$

(iii) **From f:** We have, $\left(\frac{\partial f}{\partial v}\right)_T = -p$ and $\left(\frac{\partial f}{\partial T}\right)_p = -s$

or $$\frac{\partial^2 f}{\partial v\,\partial T} = -\left(\frac{\partial p}{\partial T}\right)_v = \frac{\partial^2 f}{\partial T\,\partial v} = -\left(\frac{\partial s}{\partial v}\right)_T$$

or $$\left(\frac{\partial p}{\partial T}\right)_v = \left(\frac{\partial s}{\partial v}\right)_T \qquad ...(10.40)$$

Equations (10.37), (10.38), (10.39) and (10.40) are called *Maxwell realtions*.

10.6 GENERALISED RELATIONS FOR c_p, c_v, K AND β

(i) We have already established the relation for c_p and c_v.

$$c_p = \left(\frac{\partial h}{\partial T}\right)_p \quad \text{and} \quad c_v = \left(\frac{\partial u}{\partial T}\right)_v$$

Using the realtions between partial derivatives of h, T and s, we can express c_p, c_v in many forms. We have,

$$\left(\frac{\partial h}{\partial T}\right)_p \left(\frac{\partial T}{\partial s}\right)_p \left(\frac{\partial s}{\partial h}\right)_p = 1 \quad \text{so} \quad \left(\frac{\partial h}{\partial T}\right)_p = \left(\frac{\partial h}{\partial s}\right)_p \left(\frac{\partial s}{\partial T}\right)_p$$

But we have, $$\left(\frac{\partial h}{\partial s}\right)_p = T$$

$$\therefore \quad c_p = \left(\frac{\partial h}{\partial T}\right)_p = T\left(\frac{\partial s}{\partial T}\right)_p \qquad ...(10.41)$$

Differentiating, $$\left(\frac{\partial c_p}{\partial p}\right)_T = T\left(\frac{\partial^2 s}{\partial T\,\partial p}\right) = T\frac{\partial}{\partial T}\left(\frac{\partial s}{\partial p}\right)_T$$

But $$-\left(\frac{\partial s}{\partial p}\right)_T = \left(\frac{\partial v}{\partial T}\right)_p \qquad ...(10.42)$$

$$\therefore \quad \left(\frac{\partial c_p}{\partial p}\right)_T = T\frac{\partial}{\partial T}\left(-\frac{\partial v}{\partial T}\right)_p$$

or
$$\left(\frac{\partial c_p}{\partial p}\right)_T = -T\left(\frac{\partial^2 v}{\partial T^2}\right)_p \qquad ...(10.43)$$

(ii) Similarly, $\left(\frac{\partial u}{\partial T}\right)_v\left(\frac{\partial T}{\partial s}\right)_v\left(\frac{\partial s}{\partial u}\right)_v = 1$ or $\left(\frac{\partial u}{\partial T}\right)_v = \left(\frac{\partial u}{\partial s}\right)_v\left(\frac{\partial s}{\partial T}\right)_v$

But
$$\left(\frac{\partial u}{\partial s}\right)_v = T$$

$$\therefore \quad c_v = \left(\frac{\partial u}{\partial T}\right)_v = T\left(\frac{\partial s}{\partial T}\right)_v \qquad ...(10.44)$$

Differentiation of c_v results, $\left(\frac{\partial c_v}{\partial v}\right)_T = T\frac{\partial^2 s}{\partial u\,\partial T} = T\frac{\partial}{\partial T}\left(\frac{\partial s}{\partial v}\right)_T$

But
$$\left(\frac{\partial s}{\partial v}\right)_T = \left(\frac{\partial p}{\partial T}\right)_v \qquad \therefore \quad \left(\frac{\partial c_v}{\partial v}\right)_T = T\frac{\partial}{\partial T}\left(\frac{\partial p}{\partial T}\right)_v$$

or
$$\left(\frac{\partial c_v}{\partial v}\right)_T = T\left(\frac{\partial^2 p}{\partial T^2}\right)_v \qquad ...(10.45)$$

(iii) With the help of Eqs. (10.41) and (10.44), we may write

$$c_p - c_v = T\left(\frac{\partial s}{\partial T}\right)_p - T\left(\frac{\partial s}{\partial T}\right)_v$$

Assume $s = f(T, p)$, hence, $ds = \left(\frac{\partial s}{\partial T}\right)_p \cdot dT + \left(\frac{\partial s}{\partial p}\right)_T dp$

$$\therefore \quad \left(\frac{\partial s}{\partial T}\right)_v = \left(\frac{\partial s}{\partial T}\right)_p + \left(\frac{\partial s}{\partial p}\right)_T\left(\frac{\partial p}{\partial T}\right)_v$$

or
$$\left(\frac{\partial s}{\partial T}\right)_p - \left(\frac{\partial s}{\partial T}\right)_v = -\left(\frac{\partial s}{\partial p}\right)_T\left(\frac{\partial p}{\partial T}\right)_v$$

$$\therefore \quad c_p - c_v = -T\left\{\left(\frac{\partial s}{\partial p}\right)_T\left(\frac{\partial p}{\partial T}\right)_v\right\}. \quad \text{But} \quad \left(\frac{\partial s}{\partial p}\right)_T = -\left(\frac{\partial v}{\partial T}\right)_p$$

$$\therefore \quad c_p - c_v = T\left(\frac{\partial v}{\partial T}\right)_p\left(\frac{\partial p}{\partial T}\right)_v \qquad ...(10.46)$$

Also, from the partial derivation for the variables p, v, T, we have

$$\left(\frac{\partial p}{\partial T}\right)_v\left(\frac{\partial T}{\partial v}\right)_p\left(\frac{\partial v}{\partial p}\right)_T = -1 \quad \text{or} \quad \left(\frac{\partial p}{\partial T}\right)_v = -\left(\frac{\partial v}{\partial T}\right)_p\left(\frac{\partial p}{\partial v}\right)_T$$

Putting this in Eq. (10.46), we have

$$c_p - c_v = -T\left(\frac{\partial v}{\partial T}\right)_v^2\left(\frac{\partial p}{\partial v}\right)_T \qquad ...(10.47)$$

This is a very important relation in thermodynamics. The following facts are derived from this equation.

(a) Since $\left(\frac{\partial v}{\partial t}\right)^2$ is always positive, and $\left(\frac{\partial p}{\partial v}\right)$ for any substance is negative, $(c_v - c_v)$ is always positive. Therefore, c_p is always, greater than c_v.

(b) As $T \to 0\ K$, $c_p \to c_v$ or at absolute zero, $c_p = c_v$

(c) When $\left(\frac{\partial v}{\partial T}\right)_p = 0$, (e.g., for water at 4°C), when density is the maximum, or specific volume the minimum, $c_p = c_v$.

(d) For an ideal gas, $pv = RT$, thus

$$\left(\frac{\partial v}{\partial T}\right)_p = \frac{R}{p} = \frac{v}{T} \quad \text{and} \quad \left(\frac{\partial p}{\partial v}\right) = -\frac{RT}{v^2} \quad \text{so} \quad c_p - c_v = R$$

(iv) We have $\beta = \frac{1}{v}\left(\frac{\partial v}{\partial T}\right)_p$, and $K = -\frac{1}{v}\left(\frac{\partial v}{\partial p}\right)_v T$

$$\therefore \qquad \left(\frac{\partial v}{\partial T}\right)_p = \beta \cdot v \quad \text{and} \quad \left(\frac{\partial p}{\partial v}\right)_T = -\left(\frac{1}{K \cdot v}\right)$$

Putting this value in Eq. (10.47), we have $c_p - c_v = (-T)\,(\beta v)^2\left(\frac{-1}{K \cdot v}\right)$

or

$$c_p - c_v = \frac{T\beta^2 v}{K} \qquad ...(10.48)$$

(v) The ratio of c_p and c_v is given as

$$\gamma = \frac{c_p}{c_v} = \frac{T\left(\frac{\partial s}{\partial T}\right)_p}{T\left(\frac{\partial s}{\partial T}\right)_v} = \frac{\left(\frac{\partial s}{\partial T}\right)_p}{\left(\frac{\partial s}{\partial T}\right)_v}$$

We have, $\left(\frac{\partial s}{\partial T}\right)_p\left(\frac{\partial T}{\partial p}\right)_s\left(\frac{\partial p}{\partial s}\right)_T = -1$ and $\left(\frac{\partial s}{\partial T}\right)_v\left(\frac{\partial T}{\partial v}\right)_s\left(\frac{\partial v}{\partial s}\right)_T = -1$

$$\gamma = \frac{c_p}{c_v} = \frac{\left(\frac{\partial s}{\partial T}\right)_p}{\left(\frac{\partial s}{\partial T}\right)_v} = \frac{\left(\frac{\partial p}{\partial T}\right)_s\left(\frac{\partial s}{\partial p}\right)_T}{\left(\frac{\partial v}{\partial T}\right)_s\left(\frac{\partial s}{\partial v}\right)_T} = \left(\frac{\partial p}{\partial v}\right)_s\left(\frac{\partial v}{\partial p}\right)_T$$

or
$$\gamma = \frac{(\partial p/\partial v)_s}{(\partial p/\partial v)_T} \qquad ...(10.49)$$

This equation is an important deduction. As c_p cannot be less than c_v in any case, the slope of an isentropic process at any state on the p–v diagram cannot be less than that of the isothermal process.

10.7 RELATIONS FOR INTERNAL ENERGY AND ENTHALPY

(i) Various forms of internal energy and enthalpy may be derived for general use.

If
$$u = f(T, v)$$

$\therefore$
$$du = \left(\frac{\partial u}{\partial T}\right)_v dT + \left(\frac{\partial u}{\partial v}\right)_T dv = c_v\, dT + \left(\frac{\partial u}{\partial v}\right)_T dv \qquad ...(10.50)$$

In order to find $\left(\frac{\partial u}{\partial T}\right)_T$, let us assume $u = \Psi\,(s, v)$

$\therefore$
$$du = \left(\frac{\partial u}{\partial s}\right)_v ds + \left(\frac{\partial u}{\partial v}\right)_s dv$$

or
$$\left(\frac{\partial u}{\partial v}\right)_T = \left(\frac{\partial u}{\partial s}\right)_v \left(\frac{\partial s}{\partial v}\right)_T + \left(\frac{\partial u}{\partial v}\right)_s$$

But, we have from Eqs. (10.33), (10.40) and (10.35)

$$\left(\frac{\partial u}{\partial s}\right)_v = T, \quad \left(\frac{\partial s}{\partial v}\right)_T = \left(\frac{\partial p}{\partial T}\right)_v, \quad \left(\frac{\partial u}{\partial v}\right)_s = -\,p$$

$\therefore$
$$\left(\frac{\partial u}{\partial v}\right)_T = T\left(\frac{\partial p}{\partial T}\right)_v - p \qquad ...(10.51)$$

Sometimes, Eq. (10.51) known as energy equation. Putting the value of $\left(\frac{\partial u}{\partial v}\right)_T$ in Eq. (10.50), we get

$$du = c_v\, dT + \left[T\left(\frac{\partial p}{\partial T}\right)_v - p\right] \qquad ...(10.52)$$

(ii) Similarly, for finding dh, let $h = f\,(T, p)$

or
$$dh = \left(\frac{\partial h}{\partial T}\right)_p dT + \left(\frac{\partial h}{\partial p}\right)_T dp = c_p\, dT + \left(\frac{\partial h}{\partial p}\right)_T dp \qquad ...(10.53)$$

In order to find $\left(\frac{\partial h}{\partial p}\right)_T$, let $h = \Psi\,(s, p)$

$$\therefore \qquad dh = \left(\frac{\partial h}{\partial s}\right)_p ds + \left(\frac{\partial h}{\partial p}\right)_s dp$$

or
$$\left(\frac{\partial h}{\partial p}\right)_T = \left(\frac{\partial h}{\partial s}\right)_p \left(\frac{\partial s}{\partial p}\right)_T + \left(\frac{\partial h}{\partial p}\right)_s$$

We have from Eqs. (10.30 and (10.39)

$$\left(\frac{\partial h}{\partial s}\right)_p = T, \left(\frac{\partial s}{\partial p}\right)_T = -\left(\frac{\partial v}{\partial T}\right)_p, \left(\frac{\partial h}{\partial p}\right)_s = v$$

$$\therefore \qquad \left(\frac{\partial h}{\partial p}\right)_T = v - T\left(\frac{\partial v}{\partial T}\right)_p \qquad \text{...(10.54)}$$

Putting the value of $\left(\frac{\partial h}{\partial p}\right)_T$ in the Eq. (10.54), we have

$$dh = c_p\, dT + \left\{v - T\left(\frac{\partial v}{\partial T}\right)_p\right\} dp \qquad \text{...(10.55)}$$

(iii) Other forms of enery Eq. (10.51) and (10.54) may be derived with the help of K and β.

We have,
$$\left(\frac{\partial u}{\partial v}\right)_T = T\left(\frac{\partial p}{\partial T}\right)_v - p$$

But
$$\left(\frac{\partial p}{\partial T}\right)_v \left(\frac{\partial T}{\partial v}\right)_p \left(\frac{\partial v}{\partial p}\right)_T = -1 \quad \text{or} \quad \left(\frac{\partial p}{\partial T}\right)_v = -\left(\frac{\partial v}{\partial T}\right)_p \left(\frac{\partial p}{\partial v}\right)_T = \frac{\beta v}{Kv} = \frac{\beta}{K}$$

Putting this in Eq. (10.51), we have $\left(\frac{\partial u}{\partial v}\right)_T = \frac{T\beta}{K} - p$ ...(10.56)

Similarly, we have, $\left(\frac{\partial h}{\partial p}\right)_T = v - T\left(\frac{\partial v}{\partial T}\right)_p$

But $\left(\frac{\partial v}{\partial T}\right)_p = \beta v$, putting this value in the above Eq. (10.54), we have

$$\left(\frac{\partial h}{\partial p}\right)_T = v\,(1 - \beta\, T) \qquad \text{...(10.57)}$$

(iv) We know $\qquad u = h - pv$

Differentiating, we have, $\left(\frac{\partial u}{\partial p}\right)_T = \left(\frac{\partial h}{\partial p}\right)_T - p\left(\frac{\partial v}{\partial p}\right)_T - v = v - v\,\beta\, T + pKv - v$

or
$$\left(\frac{\partial u}{\partial p}\right)_T = pKv - v\beta T \qquad \text{...(10.58)}$$

10.8 VARIOUS *Tds* EQUATIONS

Various forms of *Tds* equations are derived for general use.

(i) Let $s = f(T, v)$

Differentiating, we have, $ds = \left(\frac{\partial s}{\partial T}\right)_v dT + \left(\frac{\partial s}{\partial v}\right)_T dv$

From Eqs. (10.44) and (10.40), we have $\left(\frac{\partial s}{\partial T}\right)_v = \frac{c_v}{T}$ and $\left(\frac{\partial s}{\partial v}\right)_T = \left(\frac{\partial p}{\partial T}\right)_v$

Putting these values, we have, $ds = \frac{c_v}{T} dt + \left(\frac{\partial p}{\partial T}\right)_v dv$

or
$$Tds = c_v\, dT + \left(\frac{\partial p}{\partial T}\right)_v dv \qquad ...(10.59)$$

Equation (10.59) is known as *first Tds equation.*

(ii) If $s = f(T, p)$, then after differentiating, $ds = \left(\frac{\partial s}{\partial T}\right)_p dT + \left(\frac{\partial s}{\partial p}\right)_T dp$

From Eqs. (10.41), (10.42), we have, $\left(\frac{\partial s}{\partial T}\right)_p = \frac{c_p}{T}$ and $\left(\frac{\partial s}{\partial p}\right)_T = -\left(\frac{\partial v}{\partial T}\right)_p$

Putting these values, we have, $ds = \frac{c_p}{T} dT - \left(\frac{\partial v}{\partial T}\right) dp$

or
$$Tds = c_p\, dT - T\left(\frac{\partial v}{\partial T}\right)_p dp \qquad ...(10.60)$$

Equation (10.60) is known as *second Tds equation.*

(iii) Again let us assume, $s = f(p, v)$

Differentiating, we have

$$ds = \left(\frac{\partial s}{\partial p}\right)_v dp + \left(\frac{\partial s}{\partial v}\right)_p dv$$

Also, $\left(\frac{\partial s}{\partial p}\right)_v \left(\frac{\partial p}{\partial T}\right)_v \left(\frac{\partial T}{\partial s}\right)_v = 1$

or
$$\left(\frac{\partial s}{\partial p}\right)_v = \left(\frac{\partial T}{\partial p}\right)_v \left(\frac{\partial s}{\partial T}\right)_v = \left(\frac{\partial T}{\partial p}\right)_v \frac{c_v}{T}$$

Also, $\left(\frac{\partial s}{\partial v}\right)_p \left(\frac{\partial v}{\partial T}\right)_p \left(\frac{\partial T}{\partial s}\right)_p = 1$ or $\left(\frac{\partial s}{\partial v}\right)_p = \left(\frac{\partial T}{\partial v}\right)_p \left(\frac{\partial s}{\partial T}\right)_p = \left(\frac{\partial T}{\partial v}\right)_p \frac{c_p}{T}$

$$\therefore \qquad ds = \frac{c_v}{T}\left(\frac{\partial T}{\partial p}\right)_v dp + \frac{c_v}{T}\left(\frac{\partial T}{\partial v}\right)_p dv$$

or
$$Tds = c_v\left(\frac{\partial T}{\partial p}\right)_v dp + c_p\left(\frac{\partial T}{\partial v}\right)_p dv \qquad \text{...(10.61)}$$

Equation (10.61) is known as *third Tds equation.*

(iv) With the help of β and K, the three Tds equations can also be expressed in alternative forms as follows.

(a) From Eq. (10.21), we have, $\beta v = \left(\frac{\partial v}{\partial T}\right)_p$

Also from Eqs. (10.22) and (10.21),

$$\frac{\beta}{K} = -\left(\frac{\partial v}{\partial T}\right)_p \Big/ \left(\frac{\partial v}{\partial p}\right)_T = -\left(\frac{\partial v}{\partial T}\right)_p \left(\frac{\partial p}{\partial v}\right)_T = \left(\frac{\partial p}{\partial T}\right)_v$$

Puttng the value of β and β/K, we have, $Tds = c_v\, dT + \dfrac{T\beta}{K} dv$ ...(10.62)

$$Tds = c_p\, dT - T\,\beta v dp \qquad \text{...(10.63)}$$

$$Tds = \left(\frac{c_v K}{\beta}\right) dp + \left(\frac{c_p}{\beta\, v}\right) dv \qquad \text{...(10.64)}$$

Note *Joule–Thomson coefficient has already been discussed in chapter 6.*

Example 10.1 The isothermal coefficient of compressibility of water at 15°C and atmospheric pressure is 50 × 10^{-6} bar. What absolute pressure must be exerted on certain mass of water to decrease its volume 5% at the same temperature.

Solution
$$K = 50 \times 10^{-6} = -\frac{1}{v}\left(\frac{\partial v}{\partial p}\right)_T$$

Integrating,
$$\int_{p=1\text{ bar}}^{p} dp = -2 \times 10^4 \int_{v}^{0.950\, v} \frac{dv}{v} = -2 \times 10^4 \ln 0.95 = 1024$$

$$p = 1024 + 1 = 1025 \text{ atm}$$ **Ans.**

Example 10.2 Show that c_v of a gas is a function of temperature alone according to van der Waals equation of state.

Solution The van der Waals equation of state, $p = \dfrac{RT}{v-b} - \dfrac{a}{v^2}$

Differentiating this, we have, $\left(\frac{\partial p}{\partial T}\right)_v = \frac{R}{v-b}$ or $\left(\frac{\partial^2 p}{\partial T^2}\right)_v = 0$ ...(i)

We have for c_v, $\left(\frac{\partial c_v}{\partial v}\right)_T = T\left(\frac{\partial^2 p}{\partial T^2}\right)_v$ ...(ii)

From Eq. (i) and Eq. (ii), we get $\left(\frac{\partial c_v}{\partial v}\right)_T = 0$ ...(10.65)

This clearly shows that c_v of a van der Waals gas is independent of volume and therefore of pressure. Thus, it is a function of temperature alone.

Example 10.3 Prove that internal energy of an ideal gas is a function of temperature alone.

Solution The ideal gas equation is $p = \frac{RT}{v}$

But for internal energy Eq. (10.51) gives

$$\left(\frac{\partial u}{\partial v}\right)_T = T\left(\frac{\partial p}{\partial T}\right)_v - p = T\frac{R}{v} - p = p - p = 0 \qquad \textbf{Proved.}$$

This is also obvious from the definition of the Boyle's law that if the temperature remains constant, there is no change in internal energy with volume or pressure. Therefore, u is a function of temperature alone.

Example 10.4 Find the value of $(\Delta h)_T$ for a fluid that obeys the equation of state

$$p = \frac{RT}{v} - \frac{a}{v^2}$$

Solution From Eq. (10.51), we have

$$\left(\frac{\partial u}{\partial v}\right)_T = T\left(\frac{\partial p}{\partial T}\right)_v - p \quad \text{or} \quad (du)_T = \left[T\left(\frac{\partial p}{\partial T}\right)_v - pdv\right]$$

$$\left(\frac{RT}{v} - p\right)dv = \left(\frac{a^2}{v^2}\right)dv \quad \text{so} \quad (\Delta u)_T = \int_{v_1}^{v_2} \frac{a}{v^2}\,dv = \frac{a}{v_1} - \frac{a}{v}$$

But $$(\Delta h)_T = (\Delta u)_T + (p_2 v_2 - p_1 v_1)_T$$

$$= \left(\frac{a}{v_1} - \frac{a}{v_2}\right) + \left[\left(\frac{RT}{v_2} - \frac{a}{v_2^2}\right)v_2 - \left(\frac{RT}{v_1} - \frac{a}{v_1^2}\right)v_1\right] = \left(\frac{a}{v_1} - \frac{a}{v_2}\right) + \left(\frac{a}{v_1} - \frac{a}{v_2}\right)$$

or $$(\Delta h)_T = 2a\left(\frac{1}{v_1} - \frac{1}{v_2}\right) \qquad \textbf{Ans.}$$

Another Method From the Eq. (10.54), we have

$$\left(\frac{\partial h}{\partial p}\right)_p = v - T\left(\frac{\partial v}{\partial T}\right)_p \quad \text{or} \quad (dh)_T = \left[v - T\left(\frac{\partial v}{\partial T}\right)_p\right] dp$$

The given equation of state is not explicit in v, hence, it is difficult to find $\left(\frac{\partial v}{\partial T}\right)_p$. For finding this, let $p = f(v, T)$.

$$\therefore \quad dp = \left(\frac{\partial p}{\partial v}\right)_T dv + \left(\frac{\partial p}{\partial T}\right)_v dT$$

or $$(dh)_T = \left[v - T\left(\frac{\partial v}{\partial T}\right)_p\right]\left[\left(\frac{\partial p}{\partial v}\right)_T dv + \left(\frac{\partial p}{\partial T}\right)_v dT\right] = \left[v - T\left(\frac{\partial v}{\partial T}\right)_p\right]\left[\left(\frac{\partial p}{\partial v}\right)_T dv\right]$$

since $dT = 0$

or $$(dh)_T = v\left(\frac{\partial p}{\partial v}\right)_T dv - T\left(\frac{\partial v}{\partial T}\right)_T \left(\frac{\partial p}{\partial v}\right)_T dv$$

But $$\left(\frac{\partial v}{\partial T}\right)_p \left(\frac{\partial T}{\partial p}\right)_v \left(\frac{\partial p}{\partial v}\right)_T = -1 \quad \text{or} \quad \left(\frac{\partial v}{\partial T}\right)_p \left(\frac{\partial p}{\partial v}\right)_T = -\left(\frac{\partial p}{\partial T}\right)_v$$

$$\therefore \quad (dh)_T = v\left(\frac{\partial p}{\partial v}\right)_T dv + T\left(\frac{\partial p}{\partial T}\right)_v dv = \left[v\left(\frac{\partial p}{\partial v}\right)_T + T\left(\frac{\partial p}{\partial T}\right)_v\right] dv$$

Differentiating the given equation of state in the example 10.4, we have

$$\left(\frac{\partial p}{\partial v}\right)_T = -\frac{RT}{v^2} + \frac{2a}{v^3} \quad \text{or} \quad \left(\frac{\partial p}{\partial v}\right)_T = \frac{R}{v}$$

$$\therefore \quad (dh)_T = \left(-\frac{RT}{v} + \frac{2a}{v^2} + \frac{RT}{v}\right) dv = \frac{2a}{v^2} dv$$

or $$(\Delta h)_T = \int_{v_1}^{v_2} \frac{2a}{v^2} = 2a\left(\frac{1}{v_1} - \frac{1}{v_2}\right)$$ **Ans.**

Example 10.5 For van der Waals equation of state, prove that

$$c_p - c_v = \frac{R}{1 - 2a(v-b)^2 / RT\, v^3}$$

Solution From generalised realtion for c_p and c_v, from Eq. (10.46), we have

$$c_p - c_v = T\left(\frac{\partial p}{\partial T}\right)_v \left(\frac{\partial v}{\partial T}\right)_p$$

From energy Eq. (10.51), we have

$$\left(\frac{\partial u}{\partial v}\right)_T = T\left(\frac{\partial p}{\partial T}\right)_v - p \quad \therefore \quad c_p - c_v = \left[\left(\frac{\partial u}{\partial v}\right)_T + p\right]\left(\frac{\partial v}{\partial T}\right)_p$$

From van der Waals equation of state,

$$p = \frac{RT}{v-b} - \frac{a}{v^2} \quad \therefore \quad \left(\frac{\partial u}{\partial v}\right)_T = \frac{a}{v^2}$$

Hence,

$$c_p - c_v = \left(\frac{a}{v^2} + p\right)\left(\frac{\partial v}{\partial T}\right)_p = \left(\frac{RT}{v-b}\right)\left(\frac{\partial v}{\partial T}\right)_p$$

Again from van der Waals equation of state, $\left(p + \frac{a}{v^2}\right)(v-b) = RT$

Differentiating w.r.t. T, we have $(v-b)(-2\,av^{-3})\left(\frac{\partial v}{\partial T}\right)_p + \left(p + \frac{a}{v^2}\right)\left(\frac{\partial v}{\partial T}\right)_p = R$

$$\therefore \quad \left(\frac{\partial v}{\partial T}\right)_p = \frac{R/(v-b)}{\dfrac{RT}{(v-b)} - \dfrac{2a}{v^3}}$$

Substituting the above value, we have

$$c_p - c_v = \frac{R}{1 - 2a\,(v-b)^2/RT\,v^3}$$ **Proved.**

Example 10.6 By using Maxwell's relation of thermodynamics, prove that Joule–Thomson coefficient, μ_j of a gas is given by the following expression

$$\mu_j = \left(\frac{\partial T}{\partial p}\right)_h = \frac{T^2}{c_p}\left[\frac{\partial}{\partial T}\left(\frac{v}{T}\right)\right]_p.$$

Solution From Maxwell's relations, we have

$$\left(\frac{\partial v}{\partial s}\right)_p = \left(\frac{\partial T}{\partial p}\right)_s \qquad ...(i)$$

$$-\left(\frac{\partial s}{\partial p}\right)_T = \left(\frac{\partial v}{\partial T}\right)_p \qquad ...(ii)$$

$$\left(\frac{\partial p}{\partial T}\right)_v = \left(\frac{\partial s}{\partial v}\right)_T \qquad ...(iii)$$

$$\left(\frac{\partial T}{\partial v}\right)_s = -\left(\frac{\partial p}{\partial s}\right)_v \qquad ...(iv)$$

Joule–Thomson coefficient $= \mu_j = \left(\frac{\partial T}{\partial p}\right)_h$

We know $$dh = Tds + vdp \quad \text{...(a)}$$

Let $$s = s\,(p, T) \quad \text{so} \quad Tds = T\left(\frac{\partial s}{\partial p}\right)_T dp + T\left(\frac{\partial s}{\partial T}\right)_p dT$$

Again, we know $$T\left(\frac{\partial s}{\partial T}\right)_p = c_p$$

Thus, $$Tds = c_p\, dT + T\left[-\left(\frac{\partial v}{\partial T}\right)_p\right] dp \quad \text{...(b)}$$

Substituting the value of Tds from Eq. (b) into Eq. (a), we have

$$dh = c_p\, dT - T\left(\frac{\partial v}{\partial T}\right)_p dp + vdp = c_p\, dT - T\left[\left(\frac{\partial v}{\partial T}\right)_p - \frac{v}{T}\right] dp$$

At constant enthalpy, $dh = 0$ for satisfying Joule–Thomson coefficient

$$0 = c_p\,(dT)_h - T\left[\left(\frac{\partial v}{\partial T}\right)_p - \frac{v}{T}\right]_h dp$$

or $$\left(\frac{\partial T}{\partial p}\right)_h = \frac{T}{c_p}\left[\left(\frac{\partial v}{\partial T}\right)_p - \frac{v}{T}\right] \quad \text{or} \quad \left(\frac{\partial T}{\partial p}\right)_h = \frac{T^2}{c_p}\left[\frac{\partial}{\partial T}\left(\frac{v}{T}\right)_p\right]$$

Thus, $$\mu_j = \left(\frac{\partial T}{\partial p}\right)_h = \frac{T^2}{c_p}\left[\frac{\partial}{\partial T}\left(\frac{v}{T}\right)\right]_p$$ **Proved.**

Example 10.7 Prove that c_p of an ideal gas is a function of temperature only.

Solution For an ideal gas, $$v = \frac{RT}{p}$$

$$\left(\frac{\partial v}{\partial T}\right)_p = \frac{R}{p} \quad \text{and} \quad \left(\frac{\partial^2 v}{\partial T^2}\right)_p = 0 \quad \text{...(i)}$$

Since $$\left(\frac{\partial c_p}{\partial p}\right)_T = -T^2\left(\frac{\partial^2 v}{\partial T^2}\right)_p \quad \text{...(ii)}$$

From Eqs. (i) and (ii), we have $\left(\frac{\partial c_p}{\partial p}\right)_T = 0$, i.e., c_p is a function of temperature alone. **Ans.**

Example 10.8 In the case of gas obeying the equations of state $\frac{pv}{RT} = 1 + Bp$ where B is a function of T only.

Show that $c_p = - RTp \dfrac{d^2}{dT^2} (BT) + (c_p)_0$

where $(c_p)_0$ is the value at very low pressure.

Solution We have, $\dfrac{pv}{RT} = 1 + Bp$ or $Bp = \dfrac{pv}{RT} - 1$

or
$$BT = \frac{T}{p}\left(\frac{pv}{RT} - 1\right) = \left(\frac{v}{R} - \frac{T}{p}\right)$$

so
$$\left[\frac{\partial}{\partial T}(BT)\right]_p = \frac{1}{R}\left(\frac{\partial v}{\partial T}\right)_p - \frac{1}{p}$$

or
$$\left[\frac{\partial^2 (BT)}{\partial T^2}\right]_p = \frac{1}{R}\left[\frac{\partial^2 v}{\partial T^2}\right] = -\frac{1}{RT}\left(\frac{\partial c_p}{\partial p}\right)_T$$

Integration of above equation gives $c_p = RTp \dfrac{d^2}{dT^2} (BT +(c_p)_0$ **Proved.**

Example 10.9 The Joule–Kelvin coefficient μ_j is a measure of temperature change during throttling process. A similar measure of the temperature change produced by an isentropic change of pressure is provided by the coefficient μ_s, where

$$\mu_s = \left(\frac{\partial T}{\partial p}\right)_s$$

Prove that
$$\mu_J - \mu_s = \frac{v}{c_p}$$

Solution For the second *Tds* equation,

$$Tds = c_p\, dT - T\left(\frac{\partial v}{\partial T}\right)_p dp \quad \text{or} \quad \left(\frac{\partial T}{\partial p}\right)_s = \mu_s = \frac{T}{c_p}\left(\frac{\partial v}{\partial T}\right)_p$$

Now,
$$\mu_j = \frac{1}{c_p}\left(T\left(\frac{\partial v}{\partial T}\right)_p - v\right]$$

Therefore,
$$\mu_s - \mu_J = \frac{v}{c_p}$$
Proved.

Example 10.10 Show that for van der Waals gas

(a) $\left(\dfrac{\partial c_p}{\partial v}\right)_T = 0$

(b) $(s_2 - s_1)_T = R \ln \dfrac{v_2 - b}{v_1 - b}$

(c) $T(v-b)^{\gamma/c_p}$ = Constant for an isentropic process

(d) $(h_2 - h_1)_T = (p_2 v_2 - p_1 v_1) + a\left(\frac{1}{v_1} - \frac{1}{v_2}\right)$

Solution (a) From the energy Eq. (10.51), we have

$$\left(\frac{\partial u}{\partial s}\right)_T = T\left(\frac{\partial p}{\partial T}\right)_v - p$$

or
$$\left(\frac{\partial^2 u}{\partial u \cdot \partial T}\right) = T\left(\frac{\partial^2 p}{\partial T^2}\right)_v + \left(\frac{\partial p}{\partial T^2}\right)_v - \left(\frac{\partial p}{\partial T}\right)_v$$

or
$$\left(\frac{\partial^2 u}{\partial u \cdot \partial T}\right) = T\left(\frac{\partial^2 p}{\partial T^2}\right)_v$$

But $c_v = \left(\frac{\partial u}{\partial T}\right)_v$

$\therefore$
$$\left(\frac{\partial c_p}{\partial v}\right)_T = \frac{\partial^2 u}{\partial T \cdot \partial v} = T\left(\frac{\partial^2 p}{\partial T^2}\right)_v \qquad \text{...(i)}$$

For van der Waals gas, $p = \frac{RT}{v-b} - \frac{a}{v^2}$

or
$$\left(\frac{\partial^2 p}{\partial T^2}\right)_v = 0 \qquad \text{...(ii)}$$

From Eqs. (i) and (ii), we get

$\left(\frac{\partial c_v}{\partial v}\right)_T = 0$, i.e., c_v is independent of volume and a function of temperature alone. **Proved.**

(b) From the first *Tds* Eq. (10.59), we have; $Tds = c_v\, dT + T\left(\frac{\partial p}{\partial T}\right)_v dv$

and from energy Eq. (10.51), we have, $\left(\frac{\partial u}{\partial v}\right) = T\left(\frac{\partial p}{\partial T}\right)_u - p$

Thus,
$$ds = c_v \frac{dT}{T} + \frac{1}{T}\left[p + \left(\frac{\partial u}{\partial v}\right)_T\right] dv$$

For van der Walls gas, we have $\left(p + \frac{a}{v^2}\right)(v-b) = RT$,

so
$$\left(\frac{\partial u}{\partial v}\right)_T = \frac{a}{v^2}$$

Thus, $$ds = c_v \frac{dT}{T} + \frac{1}{T}\left(p + \frac{a}{v^2}\right) dv \quad \text{or} \quad ds = c_v \frac{dT}{T} + \frac{R}{v-b} dv$$

Integrating the above equation results in

or $$(s_2 - s_1)_T = R \ln \frac{v_2 - b}{v_1 - b}$$ **Proved.**

(c) We have $$ds = c_v \frac{dT}{T} + \frac{R}{v-b} dv = 0$$

For isentropic process, $c_v \frac{dT}{T} + \frac{R}{v-b} dv = 0$

or $$\frac{dT}{T} + \frac{R}{c_v} \frac{dv}{v-b} = 0$$

After integration, we get $T\,(v - b)^{R/c_v}$ = constant **Proved.**

(d) We have

$$\left(\frac{\partial v}{\partial u}\right) = T\left(\frac{\partial p}{\partial T}\right)_v - p = \frac{a}{v^2}$$

or $$[du]_T = \frac{a}{v^2}[dv]$$

After integration, we get $(u_2 - u_1)_T = a\left[\frac{1}{v_1} - \frac{1}{v_2}\right]$

or $$(h_2 - h_1)_T = (p_2 v_2 - p_1 v_1) + a\left(\frac{1}{v_1} - \frac{1}{v_2}\right)$$ **Proved.**

Example 10.11 The equation of state of a certain substance over a certain range of temperature and pressure is expressed by the relation.

$$v = \frac{RT}{p} - \frac{b}{T^3}$$

where b is a constant. Derive an expression for : (a) the change in enthalpy, (b) the change in entropy of the substance for an isothermal process.

Solution (a) We have $$dh = Tds + vdp$$

or $$Tds = c_p\, dT - T\left(\frac{\partial v}{\partial T}\right)_p dp \text{ (i.e., second } Tds \text{ equation)}$$

or $$dh = c_p\, dT + \left[v - T\left(\frac{\partial v}{\partial T}\right)_p\right] dp$$

or $$(h_2 - h_1)_T = \int_1^2 \left[v - T\left(\frac{\partial v}{\partial T}\right)_p \right] dp \quad ...(i)$$

As $v = \dfrac{RT}{p} - \dfrac{b}{T^3}$

$$\therefore \quad \left(\frac{\partial v}{\partial T}\right)_p = \frac{R}{p} + \frac{3b}{T^4} \quad \text{or} \quad T\left(\frac{\partial v}{\partial T}\right)_p = \frac{RT}{p} + \frac{3b}{T^3}$$

or $$v - T\left(\frac{\partial v}{\partial T}\right)_p = \frac{RT}{p} - \frac{b}{T^3} - \frac{RT}{p} - \frac{3b}{T^3} = -\frac{4b}{T^3} \quad ...(ii)$$

From Eqs. (i) and (ii), we have

$$(h_2 - h_1)_T = \int_{p_1}^{p_2} -\frac{4b}{T^3}\,dp = \frac{4b}{T^3}(p_1 - p_2)_T$$ **Ans.**

(b) From the second *Tds* equation, we have

$$Tds = c_p\,dT - T\left(\frac{\partial v}{\partial T}\right)_p dp$$

or $$(ds)_T = -\left(\frac{\partial v}{\partial T}\right)_p (dp)_T = -\left(\frac{R}{p} + \frac{3b}{T^4}\right)(dp)_T$$

or $$(s_2 - s_1)_T = R\ln\frac{p_1}{p_2} + \frac{3b}{T^4}(p_1 - p_2)_T$$ **Ans.**

Example 10.12 The virial equation of state of a gas is given by

$$pv = RT\,(1 + B'\,p + C'p^2 + ...)$$

Prove that (a) $\lim\limits_{p \to 0}\left[T\left(\frac{\partial v}{\partial T}\right)_p - v \right] = RT^2\frac{dB'}{dT}$

and (b) the inversion temperature of a van der Waals gas is twice the Boyle's temperature.

Solution (a) As per given equation of state, $pv = RT\,(1 + B'\,p + C'p^2 + ...)$

or $$v = \frac{RT}{p} + RTB' + RTp\,C' + ...$$

or $$\left(\frac{\partial v}{\partial T}\right)_p = \frac{R}{p} + RT\frac{dB'}{dT} + RB' + RT\,p\frac{dC'}{dT} + R\,pC' + ...$$

or $$T\left(\frac{\partial v}{\partial T}\right)_p - v = RT^2\frac{dB'}{dT} + RT^2\,p\frac{dC'}{dT} + ...$$

Joule–Thomson coefficient is expressed as

$$\mu_J = \frac{1}{c_p}\left[T\left(\frac{\partial v}{\partial T}\right)_p - v\right] = \frac{RT^2}{c_p}\left[\frac{dB'}{dT} + p\frac{dC'}{dT} + ...\right]$$

or
$$\lim_{p \to 0}\left[T\left(\frac{\partial v}{\partial T}\right)_p - v\right] = RT^2 \frac{dB'}{dT}$$ **Proved.**

(b) As per van der Waals gas, for the Boyle's temperature T_B, we have

$$B = b - \frac{a}{RT} = 0 \quad \text{or} \quad T_B = \frac{a}{bR}$$

and
$$B' = \frac{B}{RT} = \frac{b}{RT} - \frac{a}{R^2\,T^3}$$

$\therefore$
$$\frac{dB'}{dT} = \frac{-b}{RT^2} + \frac{2\,a}{R^2\,T^3}$$

$$\lim_{p \to 0} \mu_J = \frac{RT^2}{c_p}\left[\frac{-b}{RT^2} + \frac{2a}{R^2\,T^3}\right] = 0$$

or
$$\frac{b}{RT^2} = \frac{2\,a}{R^2\,T^3}$$

or
$$T_1 = \frac{2\,a}{bR} = 2\,T_B = 2 \times \text{Boyle's temperature}$$ **Proved.**

10.9 CLAUSIUS–CLAPEYRON EQUATION

The Clapeyron equation gives us valuable data pertaining to the change of phase of a pure substance. *This equation presents a relationship between the saturation pressure, temperature and enthalpy of evaporation and the specific volume of phases of a substance*. For example, consider the change of state of a pure substance from the saturated liquid to saturated vapour state. This process occurs at constant temperature. During the vaporization process, both the temperature and the pressure are independent of the volume, during the change of phase extensive property are linearly interrelated. From Maxwell relation,

$$\left(\frac{\partial p}{\partial T}\right)_v = \left(\frac{\partial s}{\partial v}\right)_T \qquad ...(10.66)$$

Let suffixes d, f and g refer to solid, fluid and gaseous phases, respectively.

For a phase change from the liquid to the gaseous state, the right hand side of the Eq. (10.66) may be written as

$$\left(\frac{\partial s}{\partial v}\right)_T = \frac{s_g - s_f}{v_g - v_f}$$

Hence, Eq. (10.66) becomes

$$\left(\frac{\partial p}{\partial T}\right)_v = \frac{s_g - s_f}{v_g - v_f} \qquad ...(10.67)$$

For saturated states (saturated solid, liquid or gas), the pressure is a function of temperature alone and it is independent of volume. Hence,

$$\left(\frac{\partial p}{\partial T}\right)_v = \frac{dp}{dT} \qquad ...(10.68)$$

Now, with the help of Eq. (10.67) and (10.68), we have $\dfrac{dp}{dT} = \dfrac{s_g - s_f}{v_g - v_f}$

The change of entropy $(s_g - s_f)$ is equal to the enthalpy of vaporization divided by the absolute temperature, or

$$(s_g - s_f) = \frac{h_g - h_f}{T}$$

Hence, the Eq. (10.67) becomes

$$\frac{dp}{dT} = \frac{h_g - h_f}{T\,(v_g - v_f)} = \frac{h_{fg}}{Tv_{fg}} \qquad ...(10.69)$$

In general, $\dfrac{dp}{dT} = \dfrac{\Delta h}{T\Delta v}$

Equation (10.69) is called the Clapeyron equation. This equation is conveniently used to determine the enthalpy of vaporization from the data of the rate of change of pressure with temperature, i.e., the slope of the *p–T* curve of the substance and the specific volumes of the saturated liquid and saturated vapour (Fig. 10.2).

For any other phase, similar equations are derived. For example, for change from saturated solid to saturated liquid state, we may write

$$\frac{dp}{dT} = \frac{h_f - h_d}{T\,(v_f - v_d)}$$

and for change from saturated solid to saturated vapour state (sublimation), we may write

$$\frac{dp}{dT} = \frac{h_g - h_d}{T\,(v_g - v_d)}$$

At low pressure, we may assume that $v_f << v_g$, i.e., v_f may be neglected. The ideal gas equation of state may be assumed to be valid for low pressure, hence, $v_g = RT/p$.

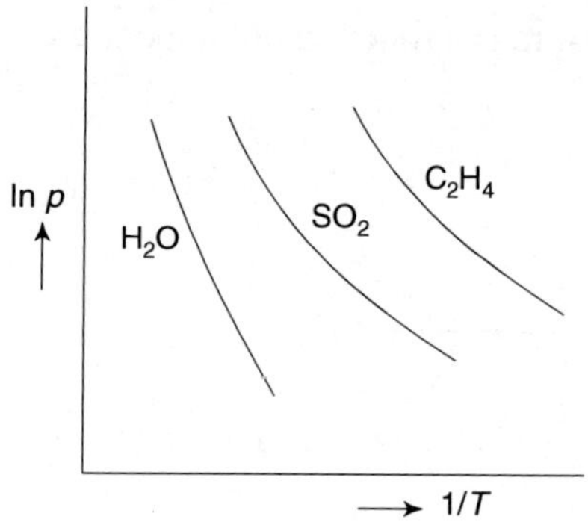

Fig. 10.2 ln p vs. $1/T$

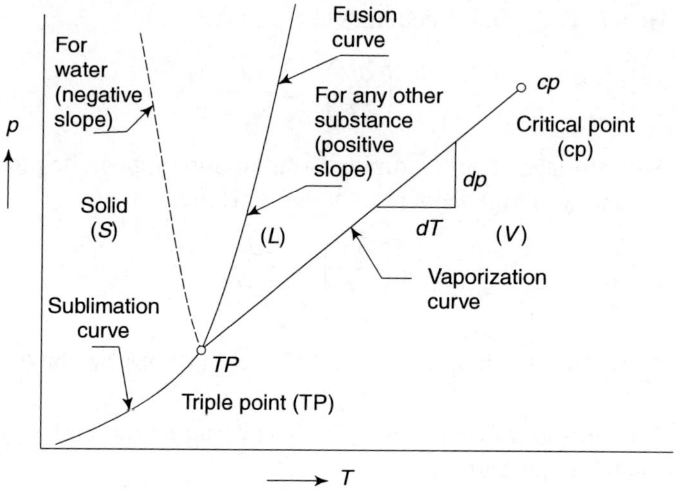

Fig. 10.3 Phase diagram for water and any other substance on p–T co-ordinates

Hence, the equation becomes $\dfrac{dp}{dT} = \dfrac{h_g - h_f}{T\,(v_g)} = \dfrac{(h_g - h_f)\,p}{RT^2}$

or
$$\frac{dp}{p} = \frac{(h_g - h_f)}{RT^2}\,dT = \frac{(h_{fg})}{RT^2}\,dT \qquad \text{...(10.70)}$$

Equation (10.70) is called the Clausius–Clapeyron equation and may also be written as

$$\frac{d\ \ln p}{d\,(1/T)} = -\frac{h_{fg}}{R} \qquad \text{...(10.71)}$$

Figure 10.2 shows a plot of a p versus $1/T$, and its slope can be readily used to determine the enthalpy of vaporization. The phase diagram for water and any other substance on p–T co-ordinates is shown n Fig. 10.3.

We can integrate the equation if the enthalpy of vaporization h_{fg} is assumed constant over a certan temperatures (or pressure range). We can integrate.

$$\int \frac{dp}{p} = \left[\frac{h_g - h_f}{R}\right] \times \int \frac{dT}{T^2}$$

or
$$\ln \frac{p_2}{p_1} = \left[\frac{h_g - h_f}{R}\left(\frac{1}{T_1} - \frac{1}{T_2}\right)\right] \qquad \text{...(10.72)}$$

With the help of this equation, we can find the vapour pressure p_2 corresponding to temperature T_2 if the initial conditions p_1 and T_1 are known.

Clapeyron's equation can also be used to estimate approximately the vapour pressure of a liquid at any arbitrary temperature in conjunction with a relation for the latent heat of a substance, known as **Trouton's rule which states that**

$$\frac{\bar{h}_{fg}}{T_B} = 88 \text{ kJ/kg mole K}$$

where $\bar{h}_{fg}$ is the latent heat of vaporization in kJ/kg mole K and T_B is the boiling point at 10.13 bar. We have

$$h_{fg} = \frac{\bar{R}T^2}{p}\frac{dp}{T}$$

Thus, $$\frac{dp}{dT} = \frac{88\,T_B}{\bar{R}\,T^2}\,p \quad \text{or} \quad \int_{101.335}^{p}\frac{dp}{p} = \frac{88\,T_B}{\bar{R}}\int_{T_B}^{T}\frac{dT}{T^2}$$

or $$\ln\frac{p}{101.325} = \frac{-88\,T_B}{R}\left(\frac{1}{T} - \frac{1}{T_B}\right)$$

or $$p = 101.325 \text{ exp.}\left[\frac{88}{R}\left(1 - \frac{T_B}{T}\right)\right] \qquad \text{...(10.73)}$$

Example 10.13 Estimate the enthalpy of vaporization of water at 100°C.

Solution From Steam Tables, at 95°C, 100°C and 105°C, the saturation pressures are 0.8453 bar, 1.0133 bar and 1.2080 bar, respectively.

Hence, $$\left(\frac{dp}{dT}\right)_{100°C} = \frac{1.2080 - 0.8453}{10} = 0.03627 \text{ bar/K}$$

At 100°C; $v_f = 0.00104$ m^3/kg; $v_g = 1.673$ m^3/kg

The Clapeyron equation is, $$\frac{dp}{dT} = \frac{h_g - h_f}{T\,(v_g - v_f)}$$

or $$0.03627 \times 10^2 = \frac{h_g - h_f}{(273 + 100)\,(1.673 - 0.00104)}$$

or $$h_g - h_f = h_{fg} = 3.627 \times 373 \times 1.67196 = 2261.94 \text{ kJ/kg}$$ **Ans.**

Example 10.14 If the boiling point (*TB*) of benzene at 1 atm pressure is 355 K, calculate the approximate value of vapour pressure of benzene at 305 K.

Solution Usng Clapeyron' and Trouton's equation, we have

$$p = 101.325 \text{ exp}\left[\frac{88}{\bar{R}}\left(1 - \frac{T_B}{T}\right)\right]$$

or $$= 101.325 \text{ exp}\left[\frac{88}{8.3143}\left(1 - \frac{355}{303}\right)\right] = 16.47 \text{ kPa}$$ **Ans.**

10.10 GAS TABLES

It is possible to express the thermodynamic properties of gases that obey the law of ideal gas in terms of temperature alone. These properties when tabulated are called the ***gas tables***. Hence, gas tables may be conveniently used to calculate property relations between different states of a system. The tabulated values take into consideration the variation of specific heat with temperature. The most widely used gas tables are of Keenan and Kaye. The gas tables are constructed in the following manner.

(a) Enthalpy and Internal Energy (*h* and *u*) We know from the previous article

If $h = h_0$ at $T = T_0$ then $C = h_0$ $\therefore h = \int c_p \, dT + h_0$

Here, the entliaipy is a function of temperature. The internal energy can be written as

$$u = h - pv = h - RT \qquad ...(10.74)$$

Therefore, with the help of Eqs. (10.73) and (10.74), h and u can be tabulated as a function of temperature only.

(b) Pressure Ratio (p_r) We can express the pressure ratio along an isentropic path as a function of temperature only. For an ideal gas, we have

$$ds = c_p \frac{dT}{T} - R\frac{dp}{p}$$

For an isentropic process, $ds = 0$, hence, $c_p \dfrac{dT}{T} = R\dfrac{dp}{p}$ ...(10.75)

Integrating this equation between temperature T_0 to T, we have

$$\ln \frac{p}{p_0} = \frac{1}{R}\int_{T_0}^{T} c_p \frac{dT}{T} \qquad ...(10.76)$$

The ratio (p/p_0) is called the relative pressure, p_r. In this equation, p_0 and p are pressures corresponding to T_0 and T respectively. If we select a certain value of T_0, the ratio (p/p_0) for an isentropic path is a function of T alone because $c_p = f(T)$ only. Hence,

$$p_r = \frac{p}{p_0} = \exp\left[\frac{1}{R}\int c_p \frac{dT}{T}\right]$$

Integrating Eq. (10.75), we get

$$\ln \frac{p_2}{p_1} = \frac{1}{R}\int_{T_1}^{T_2} c_p \frac{dT}{T} \qquad ...(10.77)$$

or

$$\ln \frac{p_2}{p_0} - \ln \frac{p_1}{p_0} = \frac{1}{R}\int_{T_1}^{T_2} c_p \frac{dT}{T} \quad \text{or} \quad \ln \frac{p_{r2}}{p_{r1}} = \frac{1}{R}\int_{T_1}^{T_2} c_p \frac{dT}{T} \qquad ...(10.78)$$

Comparing Eqs. (10.77) and (10.78), we get for isentropic process

$$\frac{p_2}{p_1} = \frac{p_{r2}}{p_{r1}} \qquad \text{...(10.79)}$$

(c) Valume Ratio (v_r) Following the same procedure as that of pressure ratio, the volume ratio along an isentropic path is

$$ds = 0 = c_v \frac{dT}{T} + R\frac{dv}{v} \qquad \text{...(10.80)}$$

or

$$\frac{dv}{v} = \frac{1}{R}\left(c_p \frac{dT}{T}\right) \quad \text{or} \quad \ln\frac{v}{v_0} = \frac{1}{R}\int_{T_0}^{T} c_v \frac{dT}{T} = \ln v_r$$

Here $v_r = v/v_0$ = relative volume. From abve equation v_r is a function of temperature alone. Similar to pressure ratio, we may get for volume ratio as

$$\frac{v_2}{v_1} = \frac{v_{r2}}{v_{r1}} \text{ for isentropic path} \qquad \text{...(10.81)}$$

(d) Change of Entropy We have

$$Tds = dh - vdp \quad \text{or} \quad ds = c_p \frac{dT}{T} - R\frac{dp}{p}$$

Integrating this, we get

$$s = \int c_p \frac{dT}{T} - R\ln p + c \qquad \text{...(10.82)}$$

Equation (10.82) reveals that entropy is a function of both T and p, hence in general, entropy cannot have a single value for any given value of T. Let us define a new function, Ψ.

$$\Psi = \int_{T_0}^{T} c_p \frac{dT}{T} \qquad \text{...(10.83)}$$

Here, temperature T_0 is an arbitrary reference temperature. For a constant pressure process, we have

$$s_2 - s_1 = \int_{T_1}^{T_2} c_p \frac{dT}{T} = \int_{T_0}^{T_2} c_p \frac{dT}{T} - \int_{T_0}^{T_1} c_p \frac{dT}{T} = \Psi_2 - \Psi_1$$

In the general case, we may write

$$s_2 - s_1 = \int_{T_1}^{T_2} c_p \frac{dT}{T} - R\ln\left(\frac{p_2}{p_1}\right) = (\Psi_2 - \Psi_1) - R\ln\frac{p_2}{p_1} \qquad \text{...(10.84)}$$

With the help of Eq. (10.84), we can claculate $s_2 - s_1$.

Example 10.15 In an air compressor, air is compressed from pressure 1 bar and temperature 310 K to a pressure of 4 bar. Calculate the changes in enthalpy and internal energy and the final temperature of air.

Solution From Gas Tables for air at temperature $T_1 = 310$ K

$$u_1 = 221.27 \text{ kJ/kg}, \quad h_1 = 310.24 \text{ kJ/kg}, \quad p_{r1} = 1.5546$$

We know $$\frac{p_2}{p_1} = \frac{p_{r2}}{p_{r1}} = 4 \quad \text{or} \quad p_{r2} = 4p_{r1} = 4 \times 1.5546 = 6.2184$$

Now, from Gas Tables corresponding to $p_{r2} = 6.2184$,

$$T_2 = 459.434 \text{ K}, h_2 - 461\ 43 \text{ kJ/kg}, u_2 = 329.57 \text{ kJ/kg}$$

$\therefore$ $$h_2 - h_1 = 461.43 - 310.24 = 151.19 \text{ kJ/kg}$$

and $$u_2 - u_1 = 329.57 - 221.27 = 108.3 \text{ kJ/kg}$$

In the Gas Tables, variation of specific heat with temperature is involved. If the specific heat is assumed to be constant and taking $\gamma = 1.4$ for air, then

$$T_2 = T_1\left(\frac{p_1}{p_2}\right)^{(\gamma-1)/\gamma} = 310\ (4)^{0.236} = 429.97 \text{ K}$$ **Ans.**

The difference between two methods may be marked.

Example 10.16 One kg of air expands reversible and adiabatically from an initial pressure of 1 bar. If the process is steady flow and the kinetic and potential energies are negligible, determine the final temperature volume and the work developed.

Solution Corresponding to 500 K from Gas Tables,

$$h_1 = 503.02 \text{ kJ/kg}, p_{r1} = 8.411, u_1 = 359.53 \text{ kJ/kg}, v_{r1} = 39.64$$

We know, $$p_{r2} = p_{r1} \times \frac{p_2}{p_1} = 8.411 \times \frac{1}{7} = 1.202$$

Now, corresponding to $p_{r2} = 1.202$ from Gas Tables,

$$T_2 = 287.95 \text{ K}, h_2 = 288.11 \text{ kJ/kg}, v_{r2} = 160.0$$

Now $$v_2 = v_1\left(\frac{v_{r2}}{v_{r1}}\right) = \frac{RT_1}{p_1}\left(\frac{v_{r2}}{v_{r1}}\right) = \frac{0.287 \times 500}{7 \times 10^2}\left(\frac{160.01}{39.64}\right) = 0.827 \text{ m}^3/\text{kg}$$ **Ans.**

The work interaction = $w = h_1 - h_2 = 503.02 - 288.11 = 214.91$ kJ/kg **Ans.**

Example 10.17 Calculate the change in entropy of air undergoing compression from 1 bar and 310°C to 700°C and 6 bar.

Solution The change in entropy, $s_2 - s_1 = \Psi_2 - \Psi_1 - R \ln \dfrac{p_2}{p_1}$

Corresponding $T_1 = 300 \text{ K}, \Psi_1 = 2.5482$

Corresponding $T_2 = 700 \text{ K}, \Psi_2 = 3.3861$

$$s_2 - s_1 = 3.3861 - 2.5483 - 0.287 \ln 6 = 0.32356 \text{ kJ/kg} \qquad \textbf{Ans.}$$

10.11 CHEMICAL POTENTIAL

In order to define the intensive state of a simple system, two independent intensive properties are essential. In the case of systems where changes in composition occur, or where phases are open for mass transfer, information is needed about concentration of components of the system. The composition of a homogeneous phase of a multicomponent system may be defined either by indicating the number of moles of each component or by indicating the relative molar concentration of each constituent and the total number of moles. Thus, we see that in either case of a system has k components, the composition is not adequately defined unless k, s independent variable are known. This clearly shows that if we want to define the thermodynamic state of a k component system adequately, it is necessary to specify a total $(k + 2)$ parameter. As an example, the internal energy of a homogeneous phase of a system having k components is defined by the equation

$$U = U(S, V, n_1, n_2, \ldots n_k) \qquad \ldots(10.85)$$

Differentiating this equation, we get

$$dU = \left(\frac{\partial U}{\partial S}\right)_{V, n_i} dS + \left(\frac{\partial V}{\partial V}\right)_{S, n_i} dV + \sum_{i=1}^{i=k}\left(\frac{\partial U}{\partial n_i}\right)_{S, V, n} dn_i \qquad \ldots(10.86)$$

where $n_1, n_2, \ldots, n_k$ are the number of moles of the components of the system. If the composition of the system does not change, the above equation becomes

$$dU = \left(\frac{\partial U}{\partial S}\right)_{v, n_i} ds + \left(\frac{\partial V}{\partial V}\right)_{S, n_i} dV$$

From first and second laws, $dU = TdS - pdV$

Equating the coefficient of independent variable S and V in the above equations, we get

$$\left(\frac{\partial U}{\partial S}\right)_{V, n_i} = T \qquad \ldots(10.87)$$

and

$$\left(\frac{\partial U}{\partial V}\right)_{s, n_i} = -p \qquad \ldots(10.88)$$

Now, we define a new term, the **molal chemical potential μ of component i as**

$$\mu_i = \left(\frac{\partial U}{\partial n_i}\right)_{S, V, n} \qquad \ldots(10.89)$$

Like pressure and temperature, the *chemical potential*, μ_i *is an intensive property.*

If changes in composition are considered in the statement of first and second laws, then the Eq. (10.86) may be written as

$$dU = TdS - pdV + \sum_{i=1}^{i=k} \mu_i \, dn_i \qquad ...(10.90)$$

Similarly, the Gibbs function may be written as

$$G = G\,(T, p, n_1, n_2, ...\, n_k) \qquad ...(10.91)$$

$$\therefore \quad dG = \left(\frac{\partial G}{\partial p}\right)_{T,\,n_i} dp + \left(\frac{\partial G}{\partial T}\right)_{p,\,n_i} dT + \sum_{i=1}^{i=k}\left(\frac{\partial G}{\partial n_i}\right)_{p,\,T,\,n} dn_i \qquad ...(10.92a)$$

The Gibbs free energy function, $dG = Vdp - SdT + \sum_{i=1}^{i=k} \mu_i \, dn_i$ ...(10.92b)

Equating the coefficients of dp; dT and dn_i result

$$\left(\frac{\partial G}{\partial p}\right)_{T,\,n_i} = V \;\; (10.93) \quad \left(\frac{\partial G}{\partial T}\right)_{p,\,n_i} = -\,S \;\; (10.94), \quad \left(\frac{\partial G}{\partial n_i}\right)_{p,\,T,\,n} = \mu_i \qquad ...(10.95)$$

Thus, we see that the chemical potential μ_i can be expressed in terms of either the internal energy of the Gibbs function as indicated by Eqs. (10.89) and (10.95)

Since $\qquad G = U + pV - TS,$

Hence, $d\,(U + pV - TS) = Vdp - SdT + \sum_{i=1}^{i=k} \mu_i \, dn_i$

$$dU + pdV + Vdp - TdS - SdT = Vdp - SdT + \sum^{i\;\;k} (\bar{\mu}_i \, dn_i)$$

or $\qquad dU = TdS - pdV + \sum_{i=1}^{i=k} \bar{\mu}_i \, dn_i$

which is the same equation as (10.67). This shows that functions

$$\left(\frac{\partial U}{\partial n_i}\right)_{s,\,V,\,n} \text{ and } \left(\frac{\partial G}{\partial n_i}\right)_{p,\,T,\,n} \text{ are equivalent.}$$

It is to be noted that the chemical potential, $\bar{\mu}$ *is used as an index of chemical equilibrium in the same manner as temperature and pressure are used as indices of thermal and mechanical equilibria. It is an intensive property and may be represented in terms of independent properties, such as*

$$\mu_i = \mu_i\,(u, V, n_1, n_2, ...\, n_k) \qquad ...(10.96)$$

Similar to U and G, the differentials of enthalpy and Helmholtz functions may be written as

$$dH = TdS + Vdp + \sum_i \bar{\mu}_i \, dn_i \qquad \text{...(10.97)}$$

$$dF = -SdT + pdV + \sum_i \bar{\mu}_i \, dn_i \qquad \text{...(10.98)}$$

Thus, the chemical potential μ_i

$$\mu_i = \left(\frac{\partial H}{\partial n_i}\right)_{S,p,n} = \left(\frac{\partial F}{\partial n_i}\right)_{T,V,n} = \left(\frac{\partial U}{\partial n_i}\right)_{V,S,n} = \left(\frac{\partial G}{\partial n_i}\right)_{T,p,n} \qquad \text{...(10.99)}$$

Equation (10.99) shows that the chemical potential of a species is the partial differential of the internal energy, enthalpy, Helmholtz function, or Gibbs function with respect to the number of moles of that species, subject to the conditions indicated by subscripts in the equation. It is expressed in energy per kg mole or energy per kg.

10.12 FUGACITY

The chemical potential plays an important role in describing multicomponent systems. In some situations, however, it is more convenient to work in terms of a related property called the ***fugacity***. The fugacity is introduced here for single and multiple component systems as given following:

10.12.1 Single-component System

Consider a system consisting of a single component. From one of the definition of chemical potential, we have

$$\mu_i = \left(\frac{\partial G}{\partial n_i}\right)_{T,p,n} . \quad \text{Thus,} \quad G = \sum_{i=1}^{j} n_i \, \mu_i$$

For single component, we have, $G = n\,\mu$ or $\mu = \dfrac{G}{n} = \bar{g}$

This shows that for a pure component, the chemical potential equals the Gibbs function per mole.

Property relations from exact differentials yield, $v = \left(\dfrac{\partial g}{\partial p}\right)_T$

$$\text{Thus,} \quad \left(\frac{\partial \mu}{\partial p}\right)_t = \bar{v} \qquad \text{...(10.100)}$$

For special case of ideal gas, $\bar{v} = \dfrac{\bar{R}T}{p}$ where $\bar{R} = R_0$.

Thus, the above equation takes the from of $\left(\dfrac{\partial \mu^*}{\partial p}\right)_T = \dfrac{\bar{R}T}{p}$

where asterisk denotes ideal gas. Integrating above equation at constant temperature, we get

$$\mu^* = \bar{R}\ T \ln p + C\ (T) \qquad \text{...(10.101)}$$

where $C\ (T)$ is a function of temperature. As the pressure p can take on values from zero to plus infinity, the ln p term of this expression, and thus the chemical potential, has an inconvenient range of values from minus infinity to plus infinity. The above Eq. (10.101) also suggests that the chemical potential can be determined only to within an arbitrary constant.

As a result of the above consideration, it is useful for many types of thermodynamic analyses to use *fugacity* (f) in place of thc chemical potenitial, which is defined as

$$\mu = \bar{R}\ T \ln f + C\ (T) \qquad \text{...(10.102)}$$

The comparison of Eq. (10.101) with Eq. (10.102) shows that *fugacity plays the same role in the general case as pressure plays in the ideal gas case*. Fugacity has the same units as pressure.

From Eq. (10.100) and Eq. (10.102), we get

$$\bar{R}\,T\left(\frac{\partial \ln f}{\partial p}\right)_T = \bar{v} \qquad \text{...(10.103)}$$

Integration of above Eq. (10.103) at constant temperature can determine the fugacity only to within a constant term. But the real gas behaviour is approached as pressure tends to zero, the constant term can be fixed by requiring that fugacity of a pure component equals the pressure in the limit of zero pressure. Thus,

$$\lim_{p \to 0}\left(\frac{f}{p}\right) = 1 \qquad \text{...(10.104)}$$

Evaluating Fugacity In order to evaluate fugacity consider the compressibility factor $Z = p\bar{v}/\bar{R}\,T$ and using it in Eq. (10.103) yields

$$\bar{R}\,T\left(\frac{\partial \ln f}{\partial p}\right) = \frac{\bar{R}\,T\,Z}{p} \quad \text{or} \quad \left(\frac{\partial \ln f}{\partial p}\right)_T = \frac{Z}{p}$$

Substracting $1/p$ from both sides and integrating from pressure p' to pressure p at a constant temperature gives

$$[\ln f - \ln p]_{p'}^{p} = \int_{p'}^{p} (Z-1)\, d \ln p \quad \text{or} \quad \left[\ln \frac{f}{p}\right]_{p'}^{p} = \int_{p'}^{p} (Z-1)\, d \ln p$$

Taking the limit as p' tends to zero and applying Eq. (10.104) results in

$$\ln\left(\frac{f}{p}\right) = \int_0^p (Z-1)\, d\ln p$$

Expressing the above equation in terms of reduced pressure, we get

$$\ln\left(\frac{f}{p}\right) = \int_0^{p_r} (Z-1)\, d\ln p_r \qquad \text{...(10.105)}$$

As the compressibility factor Z depends on the reduced temperature T_r and reduced pressure p_r, it follows that Eq. (10.105) depends on these property only. It is easier to calculate fugacity by using computer.

10.12.2 Multicomponent Systems

The fugacity of a component i of mixture can be defined by a procedure that parallels the definition for a pure component as given below:

$$\mu_i = \bar{R}T \ln \bar{f}_i + C_i(T) \qquad \text{...(10.106)}$$

where $\bar{f}_i$ is the fugacity of component, i

The *fugacity of a mixture of component,* $\bar{f}_i$ are evaluated as

$$\bar{R}T\left(\frac{\partial \ln \bar{f}_i}{\partial p}\right)_{T,n} = \bar{V}_1 \qquad \text{...(10.107)}$$

and

$$\lim_{p\to 0}\left(\frac{\bar{f}_i}{y_i\, p}\right) = 1 \qquad \text{...(10.108)}$$

The symbol $\bar{f}_i$ denotes of the component i in the mixture and it should be carefully distinguished from f_i which denotes the fugacity of pure, i.

The task of evaluating the fugacities of the components in a mixture is considerably simplified when the mixture can be modeled as an ideal solution. An ideal solution is a mixture for which

$$\bar{f}_i = y_i f_i \qquad \text{...(10.109)}$$

10.13 SUMMARY OF FORMULAE

1. $\left(\frac{\partial z}{\partial x}\right)_y \left(\frac{\partial x}{\partial z}\right)_y dz = 1$

2. $\left(\frac{\partial x}{\partial y}\right)_z \left(\frac{\partial y}{\partial z}\right)_x \left(\frac{\partial z}{\partial x}\right)_y = -1$

3. $F = U - TS$ and $f = u - Ts$ (Helmholtz free energy)
4. $-W_T \geq [(U_2 - T_2 S_2) - (U_1 - T_1 S_1]_T$
5. $G = H - TS$ and $g = h - Ts$ (Gibbs free energy)
6. $-W_T \geq (G_2 - G_1)_T$
7. $\beta = \frac{1}{v}\left(\frac{\partial v}{\partial T}\right)_p = \frac{1}{T}$ for perfect gas
8. $K = -\frac{1}{v}\left(\frac{\partial v}{\partial p}\right)_T = \left(\frac{-RT}{p^2}\right) = \frac{1}{p}$ (Isothermal compressibility)
9. $dg = vdp - sdT$ and $du = Tds - pdv$
10. $df = -vdp - sdT$ and $dh = Tds + vdp$

(9), (10) are (Differential relations)

11. $\left(\frac{\partial u}{\partial s}\right)_v = \left(\frac{\partial h}{\partial s}\right)_p = T$
12. $\left(\frac{\partial u}{\partial v}\right)_s = \left(\frac{\partial f}{\partial v}\right)_T = -p$
13. $\left(\frac{\partial g}{\partial T}\right)_p = \left(\frac{\partial f}{\partial T}\right)_v = -s$
14. $\left(\frac{\partial g}{\partial T}\right)_T = \left(\frac{\partial h}{\partial p}\right)_S = v$
15. $\left(\frac{\partial T}{\partial v}\right)_s = -\left(\frac{\partial p}{\partial s}\right)_v$
16. $\left(\frac{\partial T}{\partial p}\right)_s = \left(\frac{\partial v}{\partial s}\right)_p$
17. $\left(\frac{\partial v}{\partial T}\right)_p = -\left(\frac{\partial s}{\partial p}\right)_T$
18. $\left(\frac{\partial p}{\partial T}\right)_v = \left(\frac{\partial s}{\partial v}\right)_T$

(15), (16), (17), (18) Maxwell relations

19. $c_p = \left(\frac{\partial h}{\partial T}\right)_p = T\left(\frac{\partial s}{\partial T}\right)_p$
20. $c_v = \left(\frac{\partial v}{\partial T}\right)_v = T\left(\frac{\partial s}{\partial T}\right)_v$
21. $c_p - c_v = -T\left(\frac{\partial v}{\partial T}\right)_p^2\left(\frac{\partial p}{\partial v}\right)_T$

22. $\beta = \frac{1}{v}\left(\frac{\partial v}{\partial T}\right)_p$ (Coefficient of volume expansion)

23. $\gamma = \frac{(\partial p/\partial v)_s}{(\partial p/\partial v)_T}$

24. $\left(\frac{\partial u}{\partial v}\right)_T = T\left(\frac{\partial p}{\partial T}\right)_v - p$ (Energy equation)

25. $dh = c_p\, dT + \left\{v - T\left(\frac{\partial v}{\partial T}\right)_p\right\} dp$

26. $Tds = c_v\, dT + \left(\frac{\partial p}{\partial T}\right)_v dv$ (First Tds equation)

27. $Tds = c_p\, dT - T\left(\frac{\partial v}{\partial T}\right)_p dp$ (Second Tds equation)

28. $Tds = c_v \left(\frac{\partial T}{\partial p}\right)_v dp + c_p \left(\frac{\partial T}{\partial v}\right)_p dv$ (Third Tds equation)

29. $\mu = \bar{R}\, T \ln f + C\,(T)$ (where f is fugacity and μ is chemical potential)

30. $\lim\limits_{p \to 0}\left(\frac{f}{p}\right) = 1$

31. $\ln\left(\frac{f}{p}\right) = \int_0^{p_r} (Z - 1)\, d \ln p_r$

32. $\bar{f}_i = y_i f_i$

33. $\frac{dp}{p} = \frac{h_g - h_f}{RT^2} dT = \frac{h_{fg}}{RT^2} dT$ (Clausius–Clapeyron equation)

34. $p = 101.325 \exp\left[\frac{88}{R}\left(1 - \frac{T_B}{T}\right)\right]$ (Trouton's rule)

35. $\mu_i = \left(\frac{\partial G}{\partial n_i}\right)_{S,p,n} = \left(\frac{\partial F}{\partial n_i}\right)_{T,u,n} = \left(\frac{\partial U}{\partial n_i}\right)_{V,S,n} = \left(\frac{\partial G}{\partial n_i}\right)_{T,p,n}$ (Chemical potential)

36. $\mu_J = \left(\frac{\partial T}{\partial p}\right)_h$ and $\eta = -\left(\frac{\partial T}{\partial v}\right)_u$ (Discussed in chapter 6)

PROBLEMS

Things to Think About

10.1 If the function $p = p\ (T, v)$ is an equation of state, state whether $(\partial p/\partial T)_v$ is a property or not.

10.2. How is the Clausius–Clapeyron equation used?

10.3. If the equation of state of a gas is $p\bar{v} = \bar{R}T$, are the specific heats $\bar{c}_p$, and $\bar{c}_v$ necessarily functions of temperature alone?

10.4. For an ideal gas, list the value of Joule–Thomson coefficient.

10.5. Mention the states at which entropy departure is negligible. Is the fugacity coefficient, f/p closely equal to unity?

10.6. If two different liquids of known volumes are mixed, is the final volume necessarily equal to the sum of the original volumes?

10.7. For determining the specific volume of superheated water vapour at known pressure and temperature, under what cases you will use (i) Steam Tables, (ii) the generalised compressibility chart, (iii) an equation of state, and (iv) the ideal gas model?

10.8. In what ways, the generalised thermodynamic relations help in the study of thermodynamics?

Short and Long Theoretical Questions

10.9. Defining Helmholtz and Gibbs free energy. What is its physical significance?

10.10. Define coefficient of volume expansion and isothermal compressibility.

10.11. State the Kelvin–Clapeyron equation and its practical utility. Derive it.

10.12. Define chemical potential. Derive the general expression for it involving, U, H, F and G.

10.13. What do you mean by fugacity? Derive the expression for single and multi-component system. Is it extensive or intensive property?

10.14. Derive Maxwell equations.

10.15. Derive various Tds relations.

10.16. Derive relations for c_p, c_v and γ

10.17. Derive the relation $c_p = -\ T(\partial s/\partial T)_p$

10.18. Derive the expression $c_v = T\ (\partial s/\partial T)_v$

10.19. Derive the expression $c_p - c_v = -\ T\ (\partial v/\partial T)_p\ (\partial p/\partial T)_v$

10.20. Derive the expression

(a) $c_p - c_v = T_v\, \beta^2/K$ and (b) $\dfrac{c_p}{c_v} = \left(\dfrac{\partial p}{\partial v}\right)_s \left(\dfrac{\partial v}{\partial p}\right)_T$

10.21. Derive the following *Tds* equations

(a) $Tds = c_v\, dT + \beta T . dv/K$ (b) $Tds = c_p\, dT - v\beta T dp$

(c) $Tds = \frac{c_v K}{\beta} dp + \frac{c_p}{\beta \cdot v} dv$

10.22. Prove that for an ideal gas (a) c_v is a function of temperature only, (b) h is a function of temperature only, (c) u is a function of temperature only.

10.23. Show that $\left(\frac{\partial u}{\partial p}\right)_T = -T\left(\frac{\partial v}{\partial T}\right)_p - p\left(\frac{\partial v}{\partial p}\right)_T$

10.24. For gas obeying the van der Waals equation of state

$p = \frac{RT}{v-b} - \frac{a}{v^2}$ find the internal energy and entropy of the gas.

10.25. Derive the following equation for van der Waals equation of state

$$c_p - c_v = \frac{R}{\left[1 - \frac{2a(v-b)^2}{RT v^3}\right]}$$

10.26. Show that constant pressure lines on the Mollier diagram have equal slopes at the same temperature.

10.27. Derive the following expressions

(a) $\left(\frac{\partial c_p}{\partial p}\right)_T = -T\left(\frac{\partial^2 v}{\partial T^2}\right)_p$ and (b) $\frac{dV}{V} = \beta\, dT - K\, dp$

10.28. Show that the first law of thermodynamics of closed system may be written as

$$\delta Q = mc_v\, dT + \left\{p + \left(\frac{\partial U}{\partial V}\right)_T\right\} dV$$

10.29. Show that Clapeyron equation may be given as

$$\ln p = \frac{h_{fg}}{RT} + c \text{ where } c \text{ is a constant}$$

10.30. Derive the following equation

(a) $U = F - T\left(\frac{\partial F}{\partial T}\right)_v = -T^2\left(\frac{\partial F/T}{\partial T}\right)_v$ (b) $c_v = -T\left(\frac{\partial^2 F}{\partial T^2}\right)_v$

(c) $H = G - T\left(\frac{\partial G}{\partial T}\right)_p = -T^2\left(\frac{\partial G/T}{\partial T}\right)_p$ (d) $c_p = -T^2\left(\frac{\partial^2 G}{\partial T^2}\right)_p$

10.31. A gas obeys the equation of state as $\frac{pv}{RT} = 1 + \frac{B'}{v}$

where β' is a function of T only. Show that $c_v = -\frac{RT}{v}\frac{d^2}{dT^2}(B'T) + (c_v)_0$

where $(c_v)_0$ is the value of a very large volume.

10.32. Prove that the slope of a curve on a Mollier diagram reprsenting (a) a reversible isothermal process is equal to $T - 1/\beta$, and

(b) reversible isochoric process is equal to $T + \dfrac{\gamma - 1}{\beta}$

10.33. The following expressions for the equation of state and the specific heat, c_p are obeyed by certain gas

$$v = \frac{RT}{p} + \alpha T^2 \quad \text{and} \quad c_p = A + BT + Cp$$

where α, A, B and C are constants. Derive the expressions for (a) Joule–Thomson coefficient, μ_j, and (b) the specific heat c_v

$$\left[\textit{Ans.} \text{ (a) } \mu_J = \frac{\alpha T^2}{A + BT + Cp}, \text{ (b) } c_v = A + BT + \frac{CRT}{v - \alpha T^2} - R\left(\frac{v + \alpha T^2}{v - \alpha T^2}\right) \right]$$

10.34. Derive an expression for the Joule–Thomson coefficient as a function of T and v for a gas that obeys the van der Waals equations of state and where specific heat c_v is given by $c_v = A + BT + CT^2$, where A, B, C are constants. Estimate the temperature at the inversion states in terms of R, v and van der Walls constants a and b.

Numerical Questions

10.35. The vapour pressure of mercury at 399 K and 401 K is found to be 0.998 mm and 1.084 mm of mercury, respectively. Find the latent heat of vaporization of liquid mercury at 400 K.

[*Ans.* 61,634.98 kJ/kg mole]

10.36. Near the triple point, the vapour pressure of liquid ammonia (in atmospheres) is represented by

$$\ln p = 15.16 - \frac{3063}{T}$$

Calculate (a) temperature and pressure at the triple point, (b) latent heat of sublimation and vaporization, (c) latent heat of fusion at triple point.

[*Ans.* (a) 195 K, 0.585 atm, (b) 1498 kJ/kg, 1836 kJ/kg (c) 339 kJ/kg]

10.37. On the top of a hill, a certain liquid boils at a temperature 95°C, whereas it boils at a temperature of 105°C at the bottom. The latent heat is 4,187 kJ/kg mole. Calculate the approximate height of the hill. [*Ans.* 395 m^3]

10.38. The specific volumes of water and ice at 273.15 K are 0.001 and 0.001091 m^3/kg. Calculate the melting point increase due to increase of pressure by 1 atm. [*Ans.* – 0.00752 K]

10.39. Find the latent heat of vaporization of steam formed by boiling water under a pressure of 101.325 kPa. At a pressure near this, a rise of temperature of 1 K causes an increase of vapour pressure of 6.62 kPa. [*Ans.* 2257 kJ/kg]

10.40. Methane (CH_4) at 27°C, 10 MPa enters a turbine operating at steady state, expands adiabatically through 5 : 1 pressure ratio and exits at – 40°C. If $\bar{c}_{pv}$ = 35 kJ/kg mole K, calculate the work developed per kg of methane flowing through the turbine. Neglect changes in KE and PE. Compare with the value obtained using ideal gas model.

10.41. Nitrogen (N_2) enters a compressor operating at steady state at 1.1 MPa, 300 K and comes out at 8 MPa, 500 K. If the work input is 245 kJ/kg flowing, calculate the heat transfer. Neglect changes in KE and PE.

10.42. Oxygen (O_2) enters a compressor operating at steady with a mass flow rate of a 300 kg/min at 100 bar, 287 K. It is compressed adiabatically to 150 bar, 400 K. Calculate (a) power required and (b) the rate of entropy production. Neglect changes in PE and KE.

10.43. One kg mole of argon at 300 K is initially confined to one side of a rigid, insulated container divided into equal volumes of 0.3 m^3 by partition. The other side is initially evacuated. The partition is removed and the argon expands to fill the entire container. Using van der Walls equation of state, calculate the final temperature of the argon. Repeat the calculation using ideal gas equation.

Software Based Questions

10.44. Over a certain temperature interval, the saturation pressure-temperature curves of a substance is represented by an equation of the form as given below:

$\ln p_{sat} = A - B/T$ where A and B are empirically determined constants.

(a) Develop an expression for $h_g - h_f$ and $s_g - s_f$ in terms of p, v, T data and the constant B.

(b) Develop a software to calculate $(h_g - h_f)$ and $(s_g - s_f)$ for water vapour at different temperature and pressure and tabulate the results.

10.45. Develop a software to calculate the changes in specific enthalpy of air from atmospheric conditions to higher pressure and temperature by using van der Waals equation of state and using graphic package, plot the variation.

10.46. Develop a software to calculate the changes in specific enthalpy for ethylene from atmospheric conditions to higher pressure and temperature using Redlich–Kwong equation of state. Plot the variation.

10.47. Develop a software to calculate the fugacity of single and multicomponent systems.

Design and Open Ended Questions

10.48. Compressed natural gas (CNG) is being used as a fuel to replace gasoline for automobile engines. Prepare a comprehensive report considering light weight storage vessel (aluminium cylinders wrapped in fibers composite cane) to store up to a pressure of 200 bar to cover distance up to 100 km. The report should adhere to U.S. and Indian Department of Transportation guidelines specifying both the size weight and number of cylinders that would meet the above design standards.

10.49. Develop the preliminary design of a thermal storage system that would recover automobile engine waste heat for later use in improving the engine cold-start performance. Consider the weather conditions of UK and USA and storage period accordingly, specify the storage medium and determine when the medium should be charged with engine exhaust gases, the engine coolant or some combination. Discuss system configuration and its location in the automobile.

10.50. Prepare a critical report by evaluating the claim of an inventor who has prepared a new type of marine engine fueled by fresh water stored on board and seawater drawn in from the surrounding ocean, each entering the engine at the ambient pressure and temperature and exiting a single mixed stream at the ambient pressure and temperature.

10.51. Prepare a report on a *US patent concerning a portable refrigeration machine* requiring no external power supply and using carbon dioxide at its triple point. The report should involve its configuration and cost to cool a standard living room and feasibility of such a machine in practice.

Multiple-Choice Answer Type Questions

10.52. A characteristic function is one from which all properties of a substance can be obtained by differentiation alone, so that no arbitrary functions requiring supplementary data for their evaluation are required. The characteristics function is/are of the following

(a) $f(h, p, s) = 0$ (b) $f(u, v, s) = 0$ (c) $f(g, p, s) = 0$ (d) All of these

10.53. Which of the following is/are incorrect Maxwell equations?

(a) $\left(\frac{\partial T}{\partial v}\right)_s = -\left(\frac{\partial p}{\partial s}\right)_v$ (b) $\left(\frac{\partial T}{\partial p}\right)_s = \left(\frac{\partial v}{\partial s}\right)_p$ (c) $\left(\frac{\partial v}{\partial T}\right)_p = -\left(\frac{\partial s}{\partial p}\right)_v$ (d) $\left(\frac{\partial p}{\partial T}\right)_s = \left(\frac{\partial s}{\partial v}\right)_p$

10.54. The equation relating to the following three measurable properties is called (i) The latent heat, (ii) the slope of saturation pressure temperature line and (iii) the change in volume during phase transformation.

(a) Joule's equation (b) Maxwell relation (c) Clapeyron equation (d) Energy equation

10.55. Joule-Thomson coefficient (μ_j) is given by

(a) $\left(\frac{\partial T}{\partial p}\right)_u$ (b) $\left(\frac{\partial T}{\partial p}\right)_h$ (c) $\left(\frac{\partial T}{\partial v}\right)_u$ (d) $\left(\frac{\partial T}{\partial v}\right)_h$

10.56. Which of the following is/are incorrect *Tds* equations?

(a) $Tds = c_v\, dT + \left(\frac{\partial p}{\partial v}\right)_s dv$ (b) $Tds = c_p\, dT - T\left(\frac{\partial v}{\partial T}\right)_p dp$

(c) $Tds = c_v \left(\frac{\partial T}{\partial p}\right)_v dp + c_v \left(\frac{\partial T}{\partial v}\right)_p dv$ (d) $Tds = c_v\, dT + \left(\frac{\partial p}{\partial T}\right)_v dv$

10.57. Clausius–Clapeyron equation given by $dp/p = h_{fg} \cdot dT/RT^2$ is valid for

(a) high pressure phase transformation of solid-liquid type

(b) high pressure phase transformation of solid-vapour type

(c) low pressure phase transformation of liquid-vapour type

(d) low pressure phase transformation of solid-vapour types.

10.58. Expression given by $\frac{1}{v}\left(\frac{\partial v}{\partial T}\right)_p$ refers to

(a) Coefficient of volume expansion (b) Isothermal compressibility

(c) Isothermal bulk modulus (d) Isothermal in compressibility

10.59. (a) $\left(\frac{\partial u}{\partial s}\right)_v = \left(\frac{\partial h}{\partial s}\right)_p = T$ (b) $\left(\frac{\partial u}{\partial v}\right)_s = \left(\frac{\partial f}{\partial v}\right)_T = -p$

(c) $\left(\frac{\partial g}{\partial T}\right)_p = \left(\frac{\partial f}{\partial T}\right)_v = -s$ (d) $\left(\frac{\partial g}{\partial T}\right)_p = \left(\frac{\partial h}{\partial p}\right)_T = v$

10.60. Which of the following is or not correct thermodynamic relation?

(a) $c_p = T\left(\frac{\partial s}{\partial T}\right)_p$ (b) $c_v = T\left(\frac{\partial s}{\partial T}\right)_v$ (c) $c_p - c_v = \frac{T\,\beta\,v}{p\,K}$

(d) $c_p - c_v = -T\left(\frac{\partial v}{\partial T}\right)_p^2 \left(\frac{\partial p}{\partial v}\right)_T$

10.61. Clapeyron equation is given by

(a) $\frac{dp}{dT} = \frac{h_{fg}}{Tv_{fg}}$ (b) $\frac{dp}{dT} = \frac{h_{fg}}{sT}$ (c) $\frac{dp}{dT} = \frac{h_{fg}}{sv_g}$ (d) $\frac{dp}{dT} = \frac{h_g}{Tv}$

Answers to Multiple-Choice Questions

10.52. (d) 10.53. (d) 10.54. (c) 10.55. (b) 10.56. (a) 10.57. (c)

10.58. (a) 10.59 (d) 10.60. (c) 10.61. (a)

CHAPTER

11

Availability (Exergy) and Irreversibility

OVERVIEW

Energy is the driving force of civilization and thus its full utilisation is utmost important. The sources of energy can be divided into two groups, viz., **high grade energy and low grade energy**. High grade energy is the energy which may be totally transformed into (shaft) work neglecting friction, whereas low grade energy is the energy of which only a certain portion can be converted into work. In other words, the conversion of high grade energy into mechanical work is exempted from the limitation of second law while conversion of low grade energy is a subject to this law. The following are the examples of two kinds of energy.

High grade enery	Low grade energy
(a) Mechanical energy	(a) Heat or thermal energy
(b) Electrical energy	(b) Heat derived from the combustion of fossil fuels
(c) Water power	(c) Heat released from nuclear fission or fusion
(d) Wind power	
(e) K.E. of a jet	
(f) Tidal power	

Energy conservation idea alone is inadequate depicting some important aspects of energy resource utilisation. It is to be noted that the bulk of high grade energy (about 75%) is obtained from the sources of low grade energy such as heat derived from combustion of fossil fuels through the medium of cyclic heat engine. We have seen that complete conversion of low grade energy (heat) into high grade energy (shaft work) is impossible by second law of thermodynamics. *The part of low grade energy which is available for conversion is referred to as **available energy**, while the part, according to second law must be rejected to the sink is known as **unavailable energy***. The concept of availability comes from second law. We have also seen that maximum work can be done by Carnot engine only. ***Hence, the available energy is defined as that part of energy which can be converted into work by ideal (reversible)***

process which reduces the system to a dead state, a state in equilibrium with the earth and atmosphere. In other words, available energy is that portion of heat added to a system which instead could have been converted into work if added to reversible heat engine (Carnot engine) which rejects it to an environment of a dead state. Alternatively, *exergy is the maximum theoretical work obtainable as the system interacts to equilibrium with environment.*

In some books and journals instead of available work, **exergy is used**. *Exergy (E_x) is synonymous with available work.* Josiah Willard Gibbs is accredited with being the originator of the availability concept. He indicated that environment plays an important role in evaluating the available energy. Similarly, anergy or exergy loss is a alternative name of unavailable work Thus,

$$\text{Energy} = \text{Exergy} + \text{Anergy}$$

The unavailable energy is defined as that portion of energy which cannot be converted into work even by ideal (reversible) process which reduces the system to a dead state. In other words, unavailable energy is simply that portion of a quantity of heat which is not converted into work if added to reversible heat which rejects it to an environment which is in dead state. Thus, the *datum of zero energy decides the availability and unavailability of energy.* For availability two conditions must be fulfilled:

(i) The cycle must be reversible.

(ii) The cycle should reject heat at a temperature of the environment, i.e., dead state.

Availability represents a composite property that depends on the states of both system and the environment. At this state, no further interaction between and environment occurs. The concept of availability or exergy is used in determining the efficiency can be most effectively increased by reducing these losses. The exergy losses are generated throughout the system by the irreversible production of entropy caused by the non-ideal performance inherent in all systems, and components. Thus, exergy represents the quality of **energy**.

All the information obtained from exergy and exergy losses can be used to design thermal systems, guide efforts to reduce sources of inefficiency in existing systems and evaluate system economics. *The dead state chosen as p_0 = 1 atm and T_0 = 288 K, C_0 = 0, z_0 = 0.*

In this chapter, the primary interest is to derive expression for evaluating the availability or exergy and irreversibility of a given system or a given amount of energy and in processes.

11.1 DEAD STATE

The concept of *dead state* (as described earlier) is important for our understanding of the **property exergy**.

If the state of system departs from that of surroundings (environment), an opportunity exists for producing work which is obvious from Fig. 11.1. However, as the system changes state towards that of the surroundings, the opportunity diminishes, ceasing to exist when the two are in equilibrium with each other. This state of the system is known as ***dead state. At the dead state, the system is in equilibrium with surroundings in terms of pressure, temperature, motion, elevation and chemical equilibrium***. The dead state is designated by suffix 0 and p_0 =1 atm, T_0 = 288 K, C_0, = 0. z_0 = 0. At the dead state, both the system and environment possess energy,

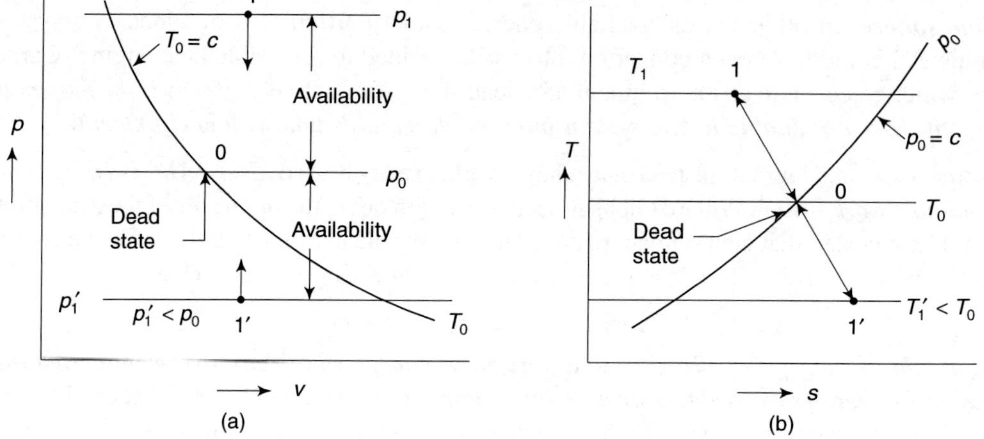

Fig. 11.1 Concept of dead state

but the *value of exergy is zero* because there is no possibility of spontaneous change within the system or environment, nor can there be an interaction between them. It is obvious from *Fig. 11.1 that when pressure and temperature of system are different* (*more or less*) *from the dead* state, there is an opportunity of available work or exergy.

11.2 ASPECTS OF EXERGY CONCEPTS

Exergy is the maximum theoretical work obtainable when the system interacts to equilibrium with environment, i.e., dead state. This is only possible when system goes through reversible processes. There are several aspects of exergy concepts as given below.

(i) ***Exergy is a measure of the departure of a system from that of environment***. This means it is an attribute of the system. However, once the environment is specified, a value can be assigned to exergy in terms of property values for the system only, thus ***exergy can be regarded as a property of the system***.

(ii) It is worth to note that the ***value of exergy cannot be negative***. If a system were at any state other than the dead state, the system would be able to change its condition spontaneously towards the dead state. As soon as the dead state is reached this tendency would cease. This implies that any change in state of the system to the dead state can be accomplished with at least zero work being developed and thus the maximum work (exergy) cannot be negative.

(iii) ***Exergy is not conserved but is destroyed by irreversibilities***. A limiting of exergy destruction may arise when exergy is completely destroyed, as would occur if a system were permitted to undergo a spontaneous change to the dead state with no provision to obtain work. This means the potential to develop work that existed originally would be a complete wastage in such a spontaneous process.

(iv) ***Exergy has been viewed as the maximum theoretical work obtainable from a system when it passes from a given state to the dead state while interacting with environment only.*** Alternatively, exergy may be viewed as minimum theoretical work input required to bring the system from the dead state to the given state.

(v) *At the dead state, both the system and environment possess energy, but the value of exergy is zero* because there is no possibility of a spontaneous change within the system or environment nor can there be an interaction between them. **All the spontaneous processes terminate at the dead state.**

(vi) Exergy is an *extensive property*.

(vii) The *dead state environment is regarded as free of irreversibilities*. All significant irreversibilities are located within the system and its immediate surroundings.

(viii) It is important to distinguish between the environment used for calculating exergy and a system's surroundings. Strictly speaking, the term surrounding refers to everything not included in the system. However, when concerning the exergy concept, we distinguish between the immediate surrounding where the intensive properties may vary during interactions with the system and the larger portion of surroundings at a distance, where the intensive properties are unaffected by process involving the system and its immediate surroundings. The term environment signifies the larger portion of surroundings, i.e., the dead state. The value of both **p_0 and T_0 of the environment are taken as p_0 = 1 bar, T_0 = 288 K, C_0 = 0 and z_0 = 0.**

11.3 AVAILABILITY (OR EXERGY) OF HEAT

Let us consider a certain amount of heat δQ is withdrawn from a heat reservoir or from a system of finite size. The question arises what is the availability or exergy of heat? In order to answer this question we have to find out the work that can be obtained when this heat δQ is transferred to reversible cycle which will reject that heat at temperature T_0 of the environment.

The heat δQ may be drawn by different ways. Two distinct cases may be possible as follows.

(a) Case 1: Heat δQ is withdrawn at Constant Temperature, T This method of drawing the heat is only possible when the heat is withdrawn from a heat reservoir. Under this condition, the temperature of heat reservoir remains constant. Figure 11.2 (a) shows the reversible cycle (i.e., Carnot cycle) in which the withdrawn heat is transferred. The cycle works between temperature T and T_0. According to first law

$$(\delta W_{\text{rew}})_{\text{cycle}} = (\delta Q)_{\text{add}} - (\delta Q)_{\text{rej}}$$

or
$$(W_{\text{cycle}}) = \int_1^2 \delta Q - \int_4^3 T_0\, dS = Q - T_0(S_3 - S_4) = T\Delta S - T_0 \Delta S$$

$$\text{Availability} = A = W_{\text{rev}} = E_x = Q - T_0\,\Delta S \tag{11.1}$$

Here, ΔS represents the change of entropy of the fluid during heat addition process.

Unavailability of heat = exergy loss = $UA = E_{x,\,\text{loss}} = Q - A = T_0\,\Delta S$ ...(11.2)

(b) Case 2: Heat δQ is Withdrawn not at Constant Temperature The method of drawing the heat is only possible when heat is drawn from a source of finite size whose temperature will vary after drawing the heat. The process 1–2 is shown in Fig. 11.2 (b) in which as the heat is withdrawn, the temperature of the source decreases while the temperature of the working fluid of the cycle increases. Since the process is reversible, at any instant during the process of heat transfer from the source to the cycle fluid, the temperature of two must be the same As we know that maximum work from δQ is possible when the cycle is reversible which rejects the heat at sink temperature T_0. It is possible by devising the reversible adiabatic processes 2–3 and 4–1 to complete the cycle,

$$(\delta W_{\text{rev}})_{\text{cycle}} = (\delta Q)_{\text{add}} - (\delta Q)_{\text{rej}}$$

$$(W_{\text{rev}})_{\text{cycle}} = Q - \int_4^3 T_0\,dS = Q - T_0(S_3 - S_4) = Q - T_0(S_b - S_a)$$

The reversible work of this cycle will be the availability of heat, hence

$$A = E_x = Q - T_0\Delta S \qquad \text{...(11.3)}$$

Here, ΔS represents the increase in entropy of the cycle fluid during the heat addition process.

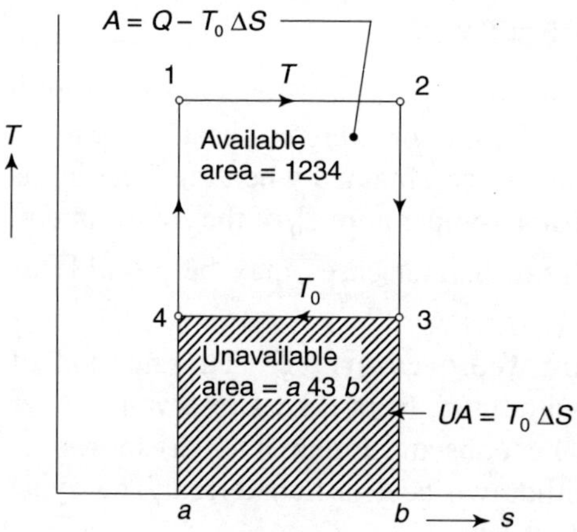

(a) Availability of heat

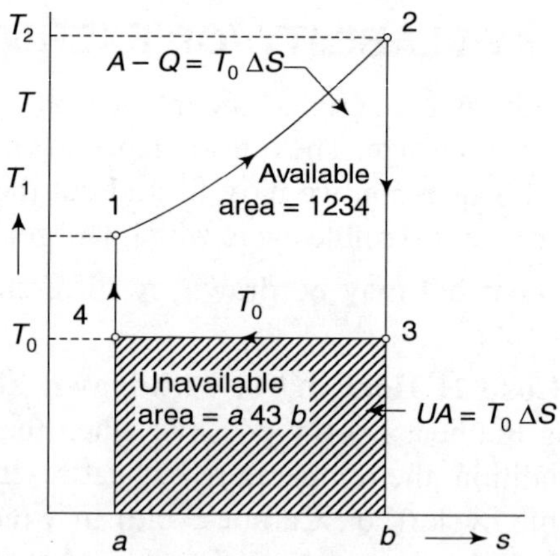

(b) Availability of heat. $T \neq$ constant

Fig. 11.2

We see that available part in both the cases is the same, hence the unavailable part will also be same.

Unvailable energy, $UA = E_{x,\,\text{loss}} = Q - A = Q - Q + T_0\,\Delta S = T_0\,\Delta S$ ...(11.4)

Example 11.1 From a heat reservoir 16,000 kJ of heat is withdrawn at 227°C while the sink temperature is at 1°C. Calculate the availability and unavailability of heat.

Solution The availability, $A = Q - T_0\,\Delta S$

Consider the heat 16,000 kJ to be transferred reversibility to a cycle at constant temperature of 227°C, the increase of entropy is

$$\Delta S = \frac{Q}{T} = \frac{16000}{227+273} = \frac{16000}{500} = 32 \text{ kJ/K}$$

$\therefore$ $$A = 16000 - 288 \times 32 = 6784 \text{ kJ}$$ **Ans.**

Unavailable part of heat, $T_0\,\Delta S = 288 \times 32 = 9216$ kJ **Ans.**

11.4 QUALITY OF ENERGY

As mentioned earlier, exergy signifies the quality of energy. In order to demonstrate this, consider the case of heat loss from a hot gas flowing through a pipeline (Fig. 11.3). Due to heat loss to the surroundings, the temperature of the gas decreases continuously from inlet at state *a* to the exit at state *b*. Although, the process is irreversible, but for the analysis consider a reversible isobaric path between the inlet and exit states of the gas as shown in Fig. 11.4. Now, consider an infinitesimal process on this irreversible isobaric process and for this infinitesimal change in entropy is given by

$$\Delta S = \frac{mc_p dT}{T} \quad \text{or} \quad \frac{dT}{dS} = \frac{T}{mc_p} \qquad \text{...(11.5)}$$

where m is the mass of gas flowing and c_p is the specific heat. The slop dT/dS depends on the gas temperature T. The decrement in T decreases the slope while increment in T will increase the slop.

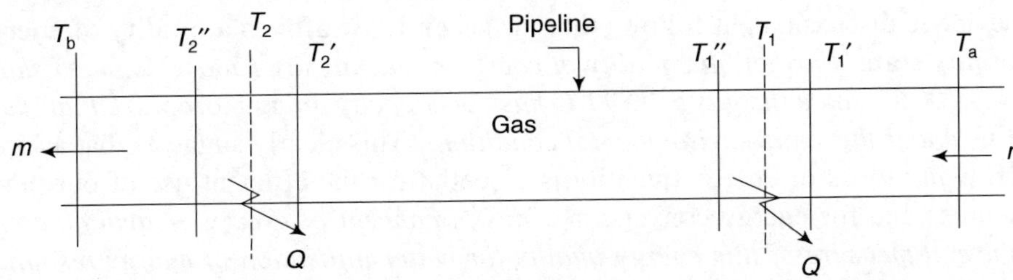

Fig. 11.3 Heat loss from a hot gas flowing through a pipeline

Let Q units of heat get lost to the surroundings during infinitesimal process as the temperature of the gas decreases from T_1' to T_1'', T_1, being the average of the two. Thus,

$$\text{Heat loss} = Q = mc_p\,(\;\;' - T_1''\,)\;T_1\,\Delta S_1 \qquad \text{...(11.6)}$$

Available energy lost with this heat at temperature T_1 is expressed as

$$W_1 = A_1 = Q - T_0\,\Delta S_1 \qquad \text{...(11.7)}$$

When the gas temperature has reached T_2 ($T_2 < T_1$), assume that the same heat loss Q occurs as the gas temperature decreases from T'_2 to T''_2, T_2 being the average temperature

$$\text{Heat loss} = mc_P\,(T'_2 - T''_2) = T_2\,\Delta S_2 \qquad \text{...(11.8)}$$

Thus, available energy lost with this heat loss at temperature T_2 is given by

$$W_2 = A_2 = Q - T_0\,\Delta S_2 \qquad \text{... (11.9)}$$

From Eqs. (11.6), and (11.8), we have $\Delta S_1 < \Delta S_2$ as $T_1 > T_2$

Thus, from Eqs. (11.7) and (11.9), we have. $W_1 > W_2$ ... (11.10)

The above result shows that the available energy lost during heat loss at higher temperature T_1 is higher as compared to the same heat loss at a lower temperature T_2. This means a heat loss of 1 kg, say at 1500°C is more harmful than the same heat loss of 1 kg at say 100°C. *From this it is concluded that adequate insulation must be provided for high temperature fluids ($T >> T_0$) to prevent precious heat loss.* This may not be important for low temperature fluids where T is slightly greater than T_0 since the available energy lost from such fluids would be low. Further, insulation must be provided adequately for very low temperature fluids ($T << T_0$) to prevent heat gain from surroundings and preserve available energy.

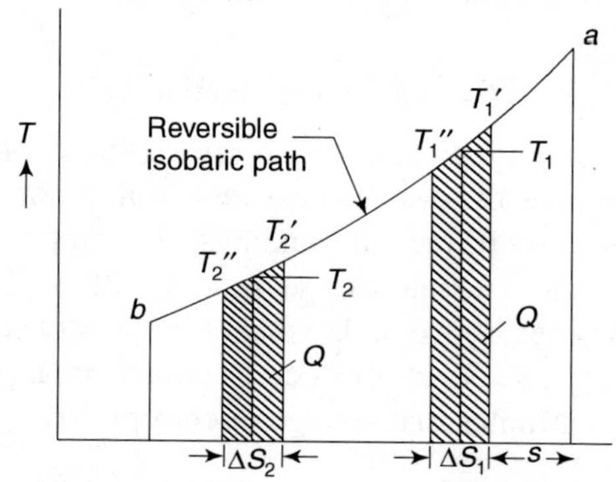

Fig. 11.4 Concept of energy quality

It is to be noted that the available energy or exergy of a fluid at a higher temperature T_1 is more than at a lower temperature T_2 and decreases with temperature.

From the above discussions, it follows that the **second law affixes a quality of energy of a system at any state**. *Further, the quality of energy of gas at, say 1500°C is much superior to that at, say 100°C, since the gas at 1500°C has the capacity doing more work than that the gas at 100°C, under the same environmental conditions.* This clearly suggests that awareness of this *energy quality* as of energy quantity is essential for the efficient use of our precious energy resources and for energy conversion. *Thus, a concept of exergy or available energy provides a useful measure of this energy quality for better utilisation of energy resources.*

11.5 AVAILABILITY OF A CLOSED SYSTEM

The availability of a closed system is the maximum useful work that can be obtained from the system in a non-flow process. Figure 11.5 shows a closed system at the state p_1, V_1 and T_1. Assume the atmosphere to be at the state p_0, V_0 and T_0, i.e., at the dead state.

Now, consider the system changes its state from (p_1, V_1, T_1) to dead state (p_0, V_0, T_0) during a reversible process. Let p, V, T represent the state of the system during an infinitesimal part of this reversible process and let δQ and δW_s represent the heat and work interaction between the system and the surroundings. It is to be noted that δQ and δW_S may be positive or negative depending upon the conditions.

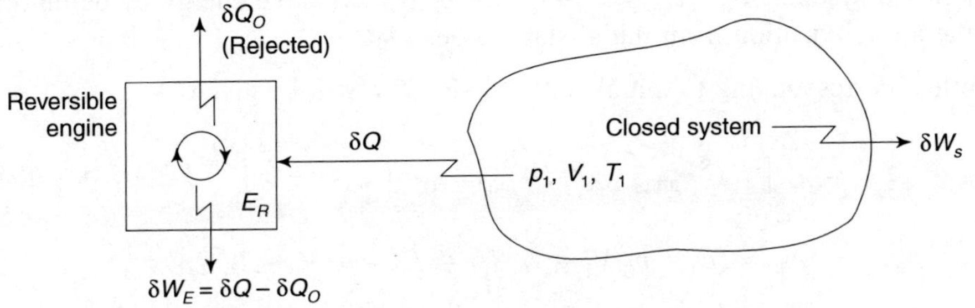

Fig. 11.5 Available energy of a closed system

Since there will be difference between temperature T and T_0, a reversible engine E_R working between T and T_0 may be assumed in order that the heat δQ is transferred either to or from the system. The work $(\delta W)_E$ of the engine E_R will always be greater than zero and it will be given as

$$(\delta W)_E = \left(\frac{T_0 - T}{T}\right)\delta Q \quad \text{Here, } \delta Q \text{ is negative by convention.} \qquad ...(11.11)$$

This work $(\delta W)_E$ is the maximum useful work available as a result of the heat exchange between the system and the surrounding during the infinitesimal process.

The work exchange (δW_s) between the system and the surrounding during this infinitesimal process is the gross work. A part of this work is spent in overcoming the resistance of the atmosphere to change dV in the volume of the system. Therefore, the maximum net available work due to the work exchange δW_s is given by

$$(\delta W)_{net} = (\delta W)_s - p_0\, dV \qquad ...(11.12)$$

The total maximum useful work,

$$(\delta W)_{max} = (\delta W)_E + (\delta W)_{net} = \left(\frac{T_0 - T}{T}\right)\delta Q + (\delta W)_s - p_0\, dV$$

$$= \{(\delta W)_s - dQ\} - p_0\, dV + T_0\left(\frac{\delta Q}{T}\right) \qquad ...(11.13)$$

According to first law applied to the closed system,

$$(\delta W)_s - \delta Q = -\,dE = -\,d\,(U + \text{KE} + \text{PE}) \qquad ...(11.14)$$

According to second law, $\dfrac{\delta Q}{T} = dS$ ...(11.15)

where dE and dS are change in total energy and change in entropy during the infinitesimal process. Putting the Eqs. (11.14) and (11.15) in (11.13), we get

$$(\delta W)_{max} = -dE - p_0\, dV + T_0\, dS = -d\,(E + p_0\, V - T_0\, S) \qquad ...(11.16)$$

In order to get$(W_{max})_{net}$, the Eq. (11.16) may be integrated from one equilibrium state to the order of a closed system. $(W_{max})_{net}$, i.e., W_{rev} will be the available energy by definition if we integrate the above equation from initial state to dead state.

Availability, corresponding to initial state (p_1, V_1, T_1), will be given by

$$A_1 = \int_{p_1,V_1,T_1}^{p_0,V_0,T_0} (\delta W_{max})_{net} = (W_{max})_{net} = W_{rev} = -\int_{E_1,V_1,S_1}^{E_0,V,S_0} d(E + p_0 V - T_0 S)$$

or $$A_1 = (E_1 + p_0\, V_1 - T_0\, S_1) - (E_0 + p_0\, V_0 - T_0\, S_0)$$

But $E_0 = U_0$ is dead state condition, hence

In order to generalise this equation, consider the initial state of the system as p, V. T, then

$$A_{nf} = W_{rev} = E_x = (E + p_0\, V - T_0\, S) - (U_0 + p_0\, V_0 - T_0\, S_0) \qquad ...(11.17)$$

Equation (11.17) gives the availability of the closed system having total energy E, entropy S and volume V.

If the kinetic and potential energies are neglected then Eq. (11.17) becomes

$$A_{nf} = W_{rev} = (U + p_0\, V - T_0\, S) - (U_0 + p_0\, V_0 - T_0\, S_0) \qquad ...(11.18)$$

or $$A_{nf} = E_{x,\,nf} = (U - U_0) + p_0\,(V - V_0) - T_0\,(S - S_0) \qquad ...(11.19)$$

Based on per unit mass,

$$a_{nf} = (e_x)_{nf} = (u - u_0) + p_0\,(v - v_0) - T_0\,(s - s_0)\ ;\ \text{kJ/kg} \qquad ...(11.20)$$

Now, it is possible to calculate change in availability of the system between two equilibrium states which is given as

$$\Delta A = A_{final} - A_{initial} = -(\Delta U + p_0\,\Delta V - T_0\,\Delta S) \qquad ...(11.21)$$

It is to be noted that the *availability of closed system is a property of the system.*

Though, it is a combination of properties but it is the only property when the state of the atmosphere (i.e., surrounding) is constant. Further, *it is an extensive property so it may be designated as* ***specific or molar availability***.

11.6 AVAILABILITY FUNCTION OF THE CLOSED SYSTEM

The availability function (Φ) for the closed is defined as

$$\Phi = U + p_0\, V - T_0\, S \qquad ...(11.22)$$

If the KE and PE energies are negligible, then the availability may be written in terms of availability function from Eq. (11.22) as

$$A_{nf} = W_{rev} = \Phi - \Phi_0 \qquad ...(11.23)$$

Now, from Eq. (11.23), $(\delta W_{rev}) = \delta W)_{max} = -\, dA = -\, d\Phi$

or $$A_{nf} = (W_{max})_{net} = -\,\Delta A = -\,A\Phi = A_1 - A_2 = \Phi_1 - \Phi_2. \quad ...(11.24)$$

It is obvious from Eq. (11.24), that if the change in kinetic and potential energies are negligible, the maximum useful work available from a closed system and its atmosphere during a change of states is equal to the negative change in the availability function of the closed system. Further, it also states that the change in available energy of the closed system between the two end states is equal to the change in availability function.

It will be worthwhile to note the difference between Gibbs function (G) and availability function, Φ.

$$G = U + pV - TS \text{ and } \Phi = U + p_0 V - T_0 S$$

We see that G involves only properties of the system while Φ involves properties of the system and atmosphere. If the system is in dead state, the availability function is equal to Gibbs function

$$\Phi_0 = G_0 \quad ...(11.25)$$

11.7 CLOSED SYSTEM EXERGY BALANCE

The exergy balance for a closed system may be developed by combining the closed system energy and entropy balance.

$$\text{Energy balance : } E_2 - E_1 = \int_1^2 \delta Q - W$$

$$\text{Entropy balance : } S_2 - S_1 = \int_1^2 \left(\frac{\delta Q}{T}\right)_b + S_{gen}$$

where W and Q represent, respectively, work and heat transfer between the system and surroundings. S_{gen} is the entropy generated or produced.

For the exergy balance multiply the entropy balance by T_0 and subtract the resulting expression from the energy balance which gives

$$(E_2 - E_1) - T_0\,(S_2 - S_1) = \int_1^2 \left(\frac{\delta Q}{T}\right)_b - W - T_0\, S_{gen}$$

or $$(E_{x2} - E_{x1}) - p_0(V_2 - V_1) = \int_1^2 \left(1 - \frac{T_0}{T_b}\right)\delta Q - W - T_0\, S_{gen}$$

Rearrangement of closed system exergy balance yields

$$\underbrace{(E_{x2}-E_{x1})}_{\text{exergy change}} = \underbrace{\int_1^2\left(1-\frac{T_0}{T_b}\right)\delta Q-\left[W-p_0(V_2-V_1)\right]}_{\text{exergy transfer}} \underbrace{-T_0 S_{gen}}_{\text{exergy destruction}} \qquad ...(11.26)$$

For specified end states and given values of p_0 and T_0, the exergy change $E_{x2} - E_{x1}$ on the left side of Eq. (11.26) can be evaluated from Eq. (11.19). The first underlined term of R.H.S. $\int_1^2\left(1-\frac{T_0}{T_b}\right)\delta Q$ represents the *exergy transfer accompanying heat.* The second underlined term $[W-p_0(V_2-V_1)]$ represents the *exergy transfer accompanying work.* The third underlined term $(T_0\,S_{gen})$ represents the *exergy destruction or exergy loss* $(E_{x,\,loss})$.

In summary, Eq. (11.26) states that the change in exergy of a closed system can be accounted for in terms of exergy transfers and the destruction or loss of exergy due to irreversibilities within the system.

As per second law of thermodynamics, the exergy destruction is positive when irreversibilities are present within the system during the processes and vanishes in the limiting case where there are no irreversibilities. Thus,

$$E_{x,loss}\begin{cases}>0 & \text{irreversibilities present in the system}\\ =0 & \text{no irreversibilities present within the system}\end{cases} \qquad ...(11.27)$$

As mentioned earlier, the value of *exergy destruction or loss cannot be negative.* Further, *exergy destruction or loss is not a property but the change in exergy of a system can be positive, negative or zero.*

$$(E_{x2}-E_{x1}) : \begin{bmatrix}>0\\ =0\\ <0\end{bmatrix} \qquad ...(11.28)$$

It is worth to note that most thermal systems are supplied with exergy inputs derived directly or indirectly from the consumption of fossil fuels. Accordingly, avoidable destructions and loss of exergy represent the waste of these resources. Thus, by devising ways to reduce such inefficiencies, better use can be made of fuels. It is the exergy balance that can be used to determine the locations, types, and magnitudes of energy resource waste and thus can play an important part in developing strategies for more effective fuel use.

11.7.1 Closed System Exergy Rate Balance

The closed system exergy rate balance is sometimes convenient to use. It is expressed as

$$\frac{dE_x}{d\tau} = \sum_j\left(1-\frac{T_0}{T_j}\right)\dot{Q}_j - \dot{W} - p_0\frac{dV}{d\tau} - \dot{E}_{x,\,loss} \qquad ...(11.29)$$

where $dE_x/d\tau$ is the time rate of change of exergy. The term $(1-T_0/T_j)\ \dot{Q}_j$ represents the time rate of exergy transfer accompanying heat transfer at the rate Q_j , occurring at the location on the boundary where the instantaneous temperature is T_j. W is the time rate of energy transfer by work. The accompanying rate of exergy transfer is expressed as $\left(\dot{Q} - p_0\ dV/d\tau\right)$ where $dV/d\tau$ is the time rate of change of system volume. The term $\dot{E}_{x,\,\text{loss}}$ signifies the time rate of exergy destruction due to irreversibilities within the system and $\dot{E}_{x,\,\text{loss}} = T_0\ \dot{S}_{\text{gen}}$.

For isolated system, the exergy balance yields $\left(\Delta E_x\right)_{\text{isolated}} = -\left(E_{x,\,\text{loss}}\right)_{\text{isolated}}$ ...(11.30)

11.7.2 Conceptualizing Exergy Transfer

In order to illustrate the use of closed system, exergy balance consider a simple experiment. Let us consider first a large metal part initially at the dead state which is put in a heating furnace vertically. The exergy of the metal part would increase because its elevation would be increased. As the metal part is heated, the exergy of the part would increase further as its temperature increases because of heat transfer from the hot furnace gases to metal part. If the metal part is subsequently quenched, it would experience a decrease of exergy as its temperature decreases due to heat transfer to the quenching medium. In each of these processes, the metal part would not actually interact with the environment used to assign exergy values. However, like the exergy values at the states attained by metal part, the exergy transfers taking place between the part and its surroundings would be evaluated *relative to the environment* used to define exergy.

11.7.2.1 Exergy Transfer Accompanying Heat

Let us consider a system undergoing a process in which heat transfer (Q) takes place across a portion of the **system boundary** where the temperature (T_b) is constant and $T_b > T_0$ (Fig. 11.6). The exergy transfer accompanying heat is

$$\left(1 - \frac{T_0}{T_b}\right) Q \qquad \text{...(11.31)}$$

The right side of this equation is recognized as the work. That could be developed by a reversible power cycle receiving (Q) amount of heat at temperature (T_b) and discharging energy by heat transfer to the environment at, T_0. We may interpret the magnitude of an exergy transfer accompanying heat transfer as the work that could be developed by supplying the heat transfer to a reversible power cycle operating between T_b and T_0. It is worth to note that this interpretation also applies for heat transfer below T_0, but then we think of the magnitude of an exergy transfer accompanying heat as the work that could be developed by a reversible power cycle receiving a heat transfer from the environment at T_0 and discharging Q at a temperature, $T_b < T_0$.

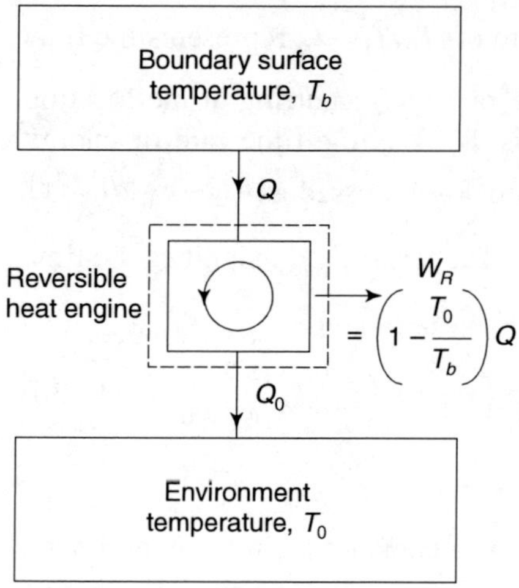

Fig. 11.6 Magnitude of energy transfer accompanying heat transfer

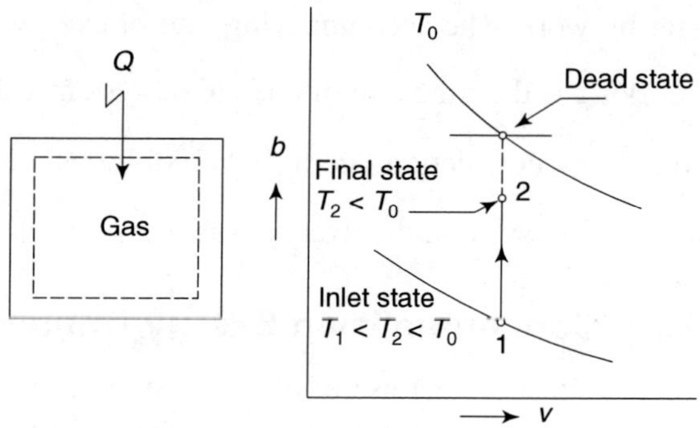

Fig. 11.7 Direction of energy transfer accompanying heat transfer when $T < T_0$

In the above discussion, we have only considered the magnitude of an exergy transfer accompanying heat. It is also necessary to account for direction. For the form of Eq. (11.31), it is clear that when T_b is greater than T_0, the heat transfer and accompanying exergy transfer would be in the same direction (i.e., from T_b to T_0). Both quantities could be positive or negative, but when T_b is less than T_0, (say refrigeration system), the sign of the exergy transfer would be opposite to the sign of heat transfer so the heat transfer and accompanying exergy transfer would be oppositely directed. For example, let us consider the system consisting of a gas heated at constant volume and its representation on p-v diagram (Fig. 11.7). Here, the initial and final temperatures of the gas are each less than T_0. Since the state of the system is brought closer to the dead state in this process, the exergy of the system must decrease as it is heated. Conversely, if the gas were cooled from state 2 to state 1, the exergy of the system would increase because the state of the system would be moved farther from the dead state.

11.7.2.2 Exergy Transfer Accompanying Work

In order to conceptualise the exergy transfer accompanying work, consider a closed system that does work (W) while undergoing a process in which system volume increases from V_1 to V_2 (Fig. 11.8). All the work (W) of the system in the process would not be available for delivery from a combined system plus environment because a portion would be spent in pushing aside the environment whose pressure is p_0. As the system does work on the surroundings equal to p_0 $(V_2 - V_1)$, the maximum amount of work that could be obtained from the combined system is

$$W_c = W_{max} = W - p_0 (V_2 - V_1)$$

which has the same exergy transfer accompanying work in Eq. (11.26). Further, the heat transfer, work and the accompanying exergy transfer can be in the same direction or oppositely directed. In the event of constant volume process, the exergy transfer will be equal to work transfer.

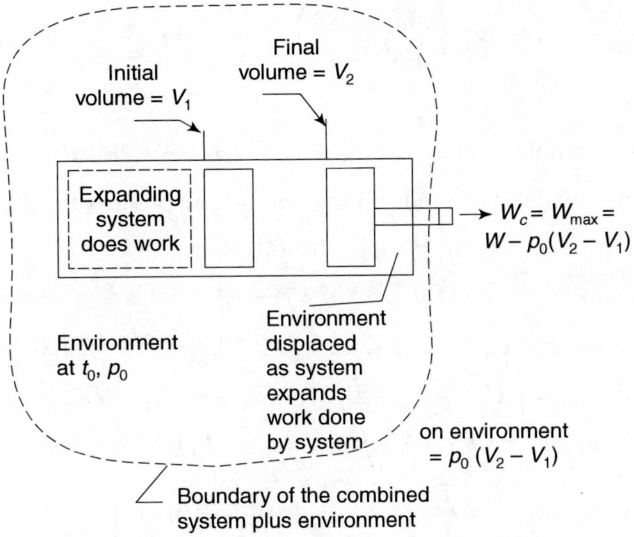

Fig. 11.8 Concept of exergy transfer from work

Example 11.2 Find the maximum work per kg of air that can be obtained from a piston-cylinder arrangement if the air expands from an initial pressure 9 bar, T_1 = 400 K to final state of p_2 = 1.5 bar, T_2 = 300 K. Neglect the changes in potential and kinetic energies. Assume T_0 = 288 K and p_0 = 1 bar. Also, calculate the availability at the initial and final states and as a whole.

Solution For non-flow process, $w_{rev} = w_{max} = (u_1 - u_2) + p_0 (v_1 - v_2) - T_0 (s_1 - s_2)$

$$= c_v.(T_1 - T_2) + p_0\left(\frac{RT_1}{p_1} - \frac{RT_2}{p_2}\right) - T_0\left(R\ln\frac{p_2}{p_1} - c_p \ln\frac{T_2}{T_1}\right)$$

$$= 0.71(400 - 300) + 1\left(\frac{0.287\times400}{9} - \frac{0.287\times300}{1.5}\right) - 287\left(0.287\ln\frac{1.5}{9} - 1.005\ln\frac{300}{400}\right)$$

$= 71 - 44.644 + 64.83 = 91.186$ kJ/kg **Ans.**

The availability at the initial state, $a_1 = (u_1 - u_0) + p_0 (v_1 - v_0) - T_0 (s_1 - s_0)$

$$= c_v(T_1 - T_0) + p_0\, R\left(\frac{T_1}{p_1} - \frac{T_0}{p_0}\right) - T_0\left(c_p \ln\frac{T_1}{T_0} - R\frac{p_1}{p_0}\right)$$

$$= 0.71(400 - 288) + 1 \times 0.287\left(\frac{400}{9} - \frac{288}{1}\right) - 288\left(1.005\ln\frac{400}{288} - 0.287\ln\frac{9}{1}\right)$$

$= 96.15$ kJ/ kg **Ans.**

The availability at the final state, $a_1 = (u_2 - u_0) + p_0 (v_2 - v_0) - T_0 (s_2 - s_0)$

$$= c_v(T_2 - T_0) + p_0\, R\left(\frac{T_2}{p_2} - \frac{T_0}{p_0}\right) - T_0\left(c_p \ln\frac{T_2}{T_0} - R\ln\frac{p_2}{p_0}\right)$$

$$= 0.71\ (300 - 288) + 1 \times 0.287\left(\frac{300}{1.5} - \frac{288}{1}\right) - 288\left(1.005 \ln\frac{300}{288} - 0.287 \ln\frac{1.5}{1}\right)$$

$= 4.552$ kJ/kg **Ans.**

The availability as a whole, $a = w_{rev} = w_{max} = a_1 - a_2 = 91.28$ kJ/ kg **Ans.**

Example 11.3 Find the availability of air during a non-flow process for the following cases :

(a) $p = 8$ bar, $T = T_0$, (b) $p = p_0$, $T = 150$ K

(c) $p = 8$ bar, $T = 400$ K, (d) $p = 0.5\ p_0$, $T = T_0$

Take $p_0 = 1.013$ bar, $T_0 = 288$ K, $R = 0.287$ kJ/kg K. Neglect PE of KE

Solution Availability, $A_{nf} = (U + p_0 V - T_0 S) - (U_0 + p_0 V_0 - T_0 S_0)$; kJ

On per unit mass basis, $a_{nf} = (u - u_0) + p_0 (v - v_0) - T_0 (s - s_0)$; kJ/kg

$$\text{(a) } a_{nf} = c_v (T - T_0) + p_0 R\left(\frac{T}{p} - \frac{T_0}{p_0}\right) - T_0\left(c_p \ln\frac{T}{T_0} - R \ln\frac{p}{p_0}\right)$$

Since $p = 8$ bar and $T = T_0$, hence

$$a_{nf} = c_v (T_0 - T_0) + \frac{1.013}{8} \times RT_0 - RT_0 + T_0\left(R \ln\frac{8}{1.103}\right) = 1.1077\ RT_0$$

$$= 1.1077 \times 0.287 \times 288 = 91.55 \text{ kJ/kg}$$ **Ans.**

$$\text{(b) } a_{nf} = c_v (T - T_0) + \frac{p_0\ RT}{p} - \frac{p_0\ RT}{p_0} - (s - s_0)$$

$$= (T - T_0)(c_v + R) - c_p\ T_0 \ln\frac{T}{T_0} = c_p (T - T_0) - c_p\ T_0 \ln\frac{T}{T_0}$$

Since $p = p_0$ and $T = 150$ K, hence

$$= 1.005\ (150 - 288) - 1.005 \times 288 \ln\frac{150}{288} = 50.11 \text{ kJ/ kg}$$ **Ans.**

$$\text{(c) } a_{nf} = c_v (T - T_0) + p_0 \frac{RT}{p} - RT_0 - T_0\left\{c_p \ln\frac{T}{T_0} - R \ln\frac{p}{p_0}\right\}$$

$$= 0.711\ (400 - 288) + \frac{1.013}{8}\ 0.287 \times 400 - 0.287 \times 288$$

$$- 288\left\{1.005 \ln\frac{400}{288} - 0.287 \ln\frac{8}{1.013}\right\} = 87.23 \text{ kJ/kg}$$ **Ans.**

$$\text{(d) } a_{nf} = c_v (T - T_0) + \frac{p_0 RT}{P} - RT_0 - T_0 (s - s_0)$$

$$= c_v(T_0 - T_0) + \frac{p_0 RT}{p} - RT_0 - T_0\left(-R\ln\frac{p}{p_0}\right) = RT_0 + RT_0 \ln 0.5 \text{ kJ/kg}$$

$$= 0.287 \times 288\,[1 + \ln 0.5] = 25.366 \text{ kJ/kg} \qquad \textbf{Ans.}$$

11.8 AVAILABILITY OF STEADY FLOW SYSTEM

The availability of steady flow system is the maximum useful work that can be obtained from the system in a steady flow process when (a) the flow is reversible and (b) at the exit, the flow stream is in its dead state. Let us consider a steady flow process as shown in Fig. 11.9. Let the entrance state of the flow stream be represented by p_1, V_1 and T_1. The surroundings which is in a dead state are represented as p_0, V_0 and T_0. Now, consider that during a reversible process the flow stream from initial state (p_1, V_1, T_1) goes to the exit state (p_0, V_0, T_0). During this reversible process the maximum useful work that can be obtained from the system consists of three parts as follows:

(a) work delivered due to reversible heat transfer,

(b) work delivered due to reversible work transfer, and

(c) work delivered due to steady reversible flow.

Let us consider an infinitesimal part of this flow process during which the flow stream is at the state (p, V, T) and it exchanges heat δQ is given by imagining a reversible engine as in the previous section.

$$(\delta W)_E = \left(\frac{T_0 - T}{T}\right)\delta Q$$

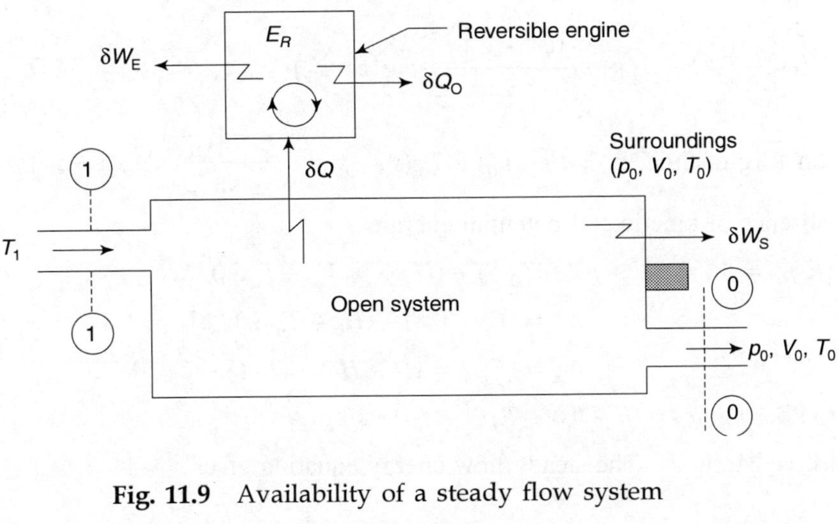

Fig. 11.9 Availability of a steady flow system

The maximum net useful work due to reversible work transfer δW_s is given by

$$(\delta W)_{net} = \delta W_s - pdV$$

The steady flow work available due to the flow process, $(\delta W)_{sf} = - V\,dp$

The total maximum useful work,

$$(\delta W)_{max} = (\delta W)_E + (\delta W)_{net} + (\delta W)_{sf} = \left\{\frac{T_0 - T}{T}\right\}\delta Q + (\delta W_s - pdV) - Vdp$$

$$= T_0\left(\frac{\delta Q}{T}\right) - (\delta Q - \delta W_s) - d(pV) \quad \text{...(11.32)}$$

But $\qquad \delta Q - \delta W_s = \delta E$ where $E = U + KE + PE$ and $\dfrac{\delta Q}{T} = dS$

$\therefore \qquad (\delta W)_{max} = - dE - d(pV) + T_0\, dS = - d\,[E + pV - T_0\, S]$

In order to get the value of $(W_{max})_{net,}$ the above equation will be integrated between entrance state $p_1\ V_1,\ T_1$ to dead state $p_0,\ V_0,\ T_0$. By definition, $(W_{max})_{net}$ will be the availability, hence

$$(A_{sf})_1 = \int_{E_1, p_1, V_1, S_1}^{E_0, p_0, V_0, S_0} -d(E + pV - T_0 S)$$

$$= (E_1 + p_1\, V_1 - T_0\, S_1) - (E_0 + p_0\, V_0 - T_0\, S_0)$$

In order to generalise the equation, let the state of the flow stream be defined by (p, V, T, S), then

$$A_{sf} = E_{x,sf} = \left(U + pV + \frac{mC^2}{2} + mgz - T_0 S\right) - \left(U_0 + p_0\, V_0 + \frac{mC_0^2}{2} + mgz_0 - T_0 S_0\right) \quad \text{...(11.33)}$$

$$= (H - H_0) - T_0(S - S_0) + m\frac{(C^2 - C_0^2)}{2} + mg(z - z_0) \quad \text{...(11.34)}$$

Based on 1 kg of fluid, $a_{sf} = (h - h_0) - T_0\,(s - s_0) + \left(\dfrac{C^2 - C_0^2}{2}\right) + g(z - z_0)$ ...(11.35)

In the absence of kinetic and potential energies,

$$(E_x)_{sf} = A_{sf} = (U + pV - T_0\, S) - (U_0 + p_0\, V_0 - T_0\, S_0) \quad \text{...(11.36)}$$

or $\qquad A_{sf} = (H - T_0 S) - (H_0 - T_0\, S_0)$; kJ ...(11.37)

or $\qquad A_{sf} = (E_x)_{sf} = (H - H_0) - T_0\,(S - S_0)$; kJ ...(11.38)

Based on per kg, $a_{sf} = (h - h_0) - T_0\,(s - s_0)$; kJ/kg ...(11.39)

Alternative Method The steady flow energy equation gives

$$W_{rev} = (W_{max})_{1-2} = A_{sf} = (H_1 - H_2) + Q_{1-2} + m\left(\frac{C_1^2 - C_2^2}{2}\right) + mg\,(z_1 - z_2)$$

Now considering system consisting of the surroundings, the work producing device, the second law gives $\Delta S + \Delta S_0 = 0 =$ for reversible process.

But the entropy change of the surroundings, $\Delta S_0 = \dfrac{-Q_{1-2}}{T_0} = -\Delta S$

$$\therefore \qquad T_0\, \Delta S = T_0 - (S_2 - S_1) = Q_{1-2}$$

Thus, $$A_{sf} = (W_{max})_{net} = (H_1 - H_2) - T_0\,(S_1 - S_2) + m\left(\frac{C_1^2 - C_2^2}{2}\right) + mg\,(z_1 - z_2)$$

Now, if the initial and final state be represented by (p, V, T) and (p_0, V_0, T_0), then

$$A_{sf} = (H - T_0\,S) + (H_0 - T_0\,S_0) + m\left(\frac{C^2 - C_0^2}{2}\right) + mg\,(z - z_0)$$

Neglecting KE and PE, $A_{sf} = (H - T_0\,S) - (H_0 - T_0\,S_0)$; kJ

Based on per kg, $a_{sf} = (h - T_0\,s) - (h_0 - T_0\,s_0)$; kJ/kg

It is to be noted that if the entering and exit pressure are both atmospheric, then Eq. (11.39) becomes identical to Eq. (11.20) for the closed system.

11.9 AVAILABILITY FUNCTION OF OPEN SYSTEM

For an open system, the availability function ψ is defined as

$$\psi = (U + pV - T_0\,S) = H - T_0\,S \qquad ...(1.l40)$$

$$\therefore \qquad A_{sf} = \psi + mgz + \frac{mC^2}{2} - \psi_0 \qquad ...(11.41)$$

In the absence of KE and PE, the availability is expressed as

$$(\delta W)_{max} = -\,d\,(H - T_0\,S) = -\,d\psi \qquad ...(11.42)$$

or $$(W_{max})_{net} = -\,\Delta\psi = \psi - \psi_0 \qquad ...(11.43)$$

$$\therefore \qquad A_{sf} = \psi - \psi_0 \qquad ...(11.44)$$

or $$(a)_{sf} = \psi - \psi_0 \qquad ...(11.45)$$

or $$(dA)_{sf} = -\,d\psi \qquad ...(11.46)$$

or $$(\Delta A)_{sf} = -\,\Delta\psi \qquad ...(11.47)$$

Equation (11.43) states that if changes in kinetic and potential energies are negligible, the maximum useful work available from a steady flow stream, when it changes from one state to another is equal to the negative of the change of availability function of the flow stream.

Equation (11.47) suggests that the changes in the availability of the flow stream is equal to the change in the availability function of the flow stream.

11.10 EXERGY RATE BALANCE FOR FLOW SYSTEM (CONTROL VOLUME)

The exergy rate balance for flow system (control volume) may be derived using an approach like steady flow energy equation. The result of the control volume exergy balance is given by

$$\underbrace{\frac{d(E_x)_{cv}}{d\tau}}_{\text{Rate of exergy change}} = \underbrace{\sum_j \left(1-\frac{T_0}{T_j}\right)\dot{Q}_j - \left(\dot{W}_{cv} - p_0\frac{dV_{cv}}{d\tau}\right) + \sum_i \dot{m}_i\, e_{xi} - \sum_e \dot{m}_e\, e_{xe}}_{\text{Rate of exergy transfer}} \underbrace{-\dot{E}_{x,\,loss}}_{\text{Rate of exergy destruction}} \quad ...(11.48)$$

where

$$e_{xi} = (h_i - h_0) - T_0\,(s_i - s_0) - \frac{C_i^2 - C_0^2}{2} + g\,(z_i - z_0) \quad ...(11.49)$$

$$e_{xe} = (h_e - h_0) - T_0\,(s_e - s_0) - \frac{C_e^2 - C_0^2}{2} + g\,(z_i - z_0) \quad ...(11.50)$$

$$\dot{E}_{x,\,loss} = T_0\,\dot{S}_{gen} \quad ...(11.51)$$

Here, the term $d(E)_{cv}/dx$ represents the time rate of change of the exergy of the control volume. "The term $\dot{Q}_j$ represents the time rate of heat transfer at the location on the boundary where the instantaneous temperature is T_j. The term $(1 - T_0/T_j)\,\dot{Q}_j$ represents the accompanying exergy transfer rate. The term $\dot{W}_{cv}$ represents the time rate of energy transfer by *work other than flow work*. The accompanying exergy transfer by work is given by $\left(\dot{W}_{cv} - p_0\dfrac{dV_{cv}}{d\tau}\right)$ where $dV_{cv}/d\tau$ represents the time rate of change of volume. The term $\dot{m}_i\, e_{xi}$ and $m_e\, e_{se}$ account for the time rate of exergy transfer accompanying mass flow and flow work at inlet i and exit e, respectively. The term $\dot{E}_{x,\,loss}$ accounts for the time rate of exergy destruction (or loss) due to irreversibilities within the control volume.

11.10.1 Steady State Condition

As a large number of engineering problems involve control volume at steady state, so steady rate of exergy rate balance has practical importance. At steady state, $d(E_x)_{cv}/d\tau = 0$, $dV_{cv}/d\tau = 0$, so Eq. (11.32) reduces to

$$0 = \sum_i \left(1-\frac{T_0}{T_j}\right)\dot{Q}_j - \dot{W}_{cv} + \sum_i \dot{m}_i\, e_{xi} - \sum_e \dot{m}_{ce} - \dot{E}_{x,\,loss} \quad ...(11.52)$$

The above equation signifies that the rate at which exergy is transferred into the control volume must exceed the rate at which exergy is transferred out, the difference being the rate at which exergy is destroyed within the control volume due to irreversibilities.

If there is a single inlet and single outlet represented by 1 and 2, respectively, Eq. (11.52) reduces to

$$0 = \sum_j \left(1 - \frac{T_0}{T_j}\right)\dot{Q}_j - \dot{W}_{cv} + \dot{m}\left(e_{x1} - e_{x2}\right) - \dot{E}_{x,\,loss} \qquad ...(11.53)$$

where $\dot{m}$ = mass flow rate

and $$e_{x1} - e_{x2} = \left(h_1 - h_2\right) - T_0\left(s_1 - s_2\right) + \frac{C_1^2 - C_2^2}{2} + g\left(z_1 - z_2\right) \qquad ...(11.54)$$

Example 11.4 Two kg of water at 50°C are mixed with 3 kg of water at 100°C in a steady flow process. Calculate (a) the temperature of the resulting mixture, (b) is the mixing isentropic? If not, what is the entropy change? (c) What is the unavailable energy with respect to the receiver at 50°C.

Solution From the first law, $2h_1 + 3h_2 = 5h_3$

Assuming the specific heat of water is constant, then

$$2c_p\,(50 - 0) + 3c_p\,(100 - 0) = 5c_p\,(t_3 - 0)$$

or $$100 + 300 = 5t_3$$

(a) $$t_3 = \text{temperature of mixture} = 80°\text{ C}$$

(b) The increase in entropy due to mixing process is

$$\Delta S = m_1\, c_p \ln \frac{T_3}{T_1} + m_2 c_p \ln\frac{T_3}{T_2} = 2 \times 4.18 \ln\frac{353}{323} + 3 \times 4.18 \ln\frac{353}{373} = 0.0514 \text{ kJ/K}$$ **Ans.**

(c) Unavailable energy, $T_0\, \Delta S = T_{\text{Rec}}\, \Delta S = (50 + 273) \times 0.0514 = 16.6$ kJ **Ans.**

Example 11.5 Water is evaporated at a constant temperature of 200°C as a result of heat interaction with combustion gases that are cooled from 1200°C to 260°C. The environment temperature is 40°C. Calculate the decrease in available energy per kg of gas due to heat interaction process. What is the availability of the energy of the gas before and after the heat interaction? Take c_p= 1.005 kJ/kg. K for gases, latent heat of vaporization of water is 1050 kJ/kg K.

Solution Refer Fig. 11.10. The heat transfer per kg of gas,

$$q = c_p\, \Delta t = 1.005(1200 - 260) = 944.7 \text{ kJ/kg}$$

Decrease in available energy, $-\,T_0\,(\Delta s_{\text{water}} + \Delta s_{\text{gas}}) = -\,T_0\left(\dfrac{q}{T_w} - c_p \dfrac{T_{gh}}{T_{gc}}\right)$

$$= -\,(40 + 273)\left(\frac{944.7}{200 + 273} - 1.005 \ln\frac{1200 + 273}{260 + 273}\right) = -305.37 \text{ kJ/kg of gas}$$ **Ans.**

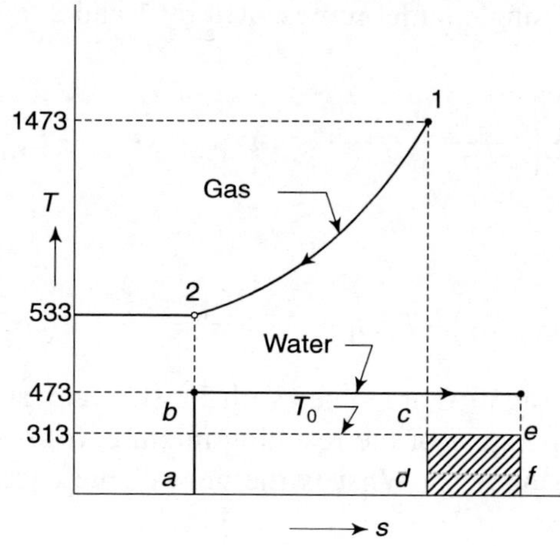

Fig. 11.10 Example 11.5

The unavailable energy before the heat transfer process is represented by area $a - b - c - d$ of Fig. 11.10 and is equal to $T_0 \, \Delta s_{gas} = 313 \times 1.021 = 319.766$ kJ/kg.

After the heat transfer process, the unavailable energy (area $a - b - e - f$) is

$$T_0 \, \Delta s_w = 313 \times 1.997 = 625.061 \text{ kJ/kg}$$

Therefore, the increase in the available energy due to the heat interaction (area $d - c - e - f$) is (625.061 – 319.766) = 305.295 kJ/kg. It is to be noted that the increase in the unavailable energy is the same as the decrease in available energy obtained before. The availability of energy per kg of gas before the heat transfer is

$$a_{sf1} = (h_1 - h_0) - T_0 (s_1 - s_0) = c_p (T_1 - T_0) - T_0 \left(c_p \cdot \ln \frac{T_1}{T_0} \right)$$

$$= 1.005 \, (1473 - 313) - 313 \left(1.005 \times \ln \frac{1473}{313} \right)$$

$$= 678.58 \text{ kJ/kg}$$ **Ans.**

The availability after the heat transfer $= c_p \, (T_2 - T_0) - T_0 \left(c_p \ln \frac{T_2}{T_0} \right)$

$$= 1.005(533 - 313) - 313 \times 1.005 \times \ln \frac{533}{313} = 53.65 \text{ kJ/kg}$$ **Ans.**

Example 11.6 What is the availability (i.e., maximum useful work) which can be produced per kg of saturated steam which enters steady flow system at 70 bar and leaves in equilibrium with the atmosphere at 1 bar and 15°C?

Solution Neglecting KE and PE energies, the availability, i.e., the maximum useful work per kg is given by

$$a_{sf} = (h - T_0\, s) - (h_0 - T_0\, s_0)$$

From Steam Tables, h = 2773.5 kJ/kg (corresponding to 70 bar), s = 5.8152 kJ/kg K

In determining the state of the water when it comes to equilibrium with the atmosphere (notice that the saturation temperature corresponding to 1 bar is 99.632°C), consequently, the water will have to be in compressed liquid region since its final temperature is lower than this temperature. In order to understand clearly, visualise an adiabattc expansion to 1 bar and 99.632°C, followed by condensation and subcooling to 15°C. The value of $(h - h_f)$ for compressing liquid from 0.017 bar (saturation pressure corresponding to 15°C) to 1 bar is found to be negligible. The change in entropy is also negligible. Therefore, $s_0 = s_f$ at 15°C = 0.2243 kJ/kg K, $h_0 = h_f$ at 15°C = 62.94 kJ/kg.

(a) s_f = (2773.5 – 288 × 5.8152) – (62.94 – 288 × 0.2243) = 1100.38 kJ/kg **Ans.**

Example 11.7 Find the availability of steam in steady flow for the following case:

p = 70 bar, t = 350°C, C = 200 m/s; z = 200 m

Take p_0 = 1 bar T_0 = 288 K

Solution (a) $$a_{sf} = \left(h + \frac{C^2}{2} + gz - T_0 s\right) - \left(h_0 - T_0 s_0\right)$$ **Ans.**

From Steam Tables, corresponding to 70 bar and 350°C

$$h = 3018.7 \text{ kJ/kg}, \; s = 6.2333 \text{ kJ/kg K}$$

At 288 K, $h_0 = h_{f0}$ = 62.94 kJ/kg K, $s_0 = s_{f0}$ = 0.2243 kJ/kg K

(a) $$a_{sf} = (3018.7 - 62.94) + \frac{200^2}{2 \times 1000} + \frac{200 \times 9.81}{1000} - 288(6.2333 - 0.2243)$$

$$= 1247.13 \text{ kJ/kg}$$ **Ans.**

11.11 IRREVERSIBILITY FOR CLOSED AND OPEN SYSTEM

In engineering, it is important to have some standard to which the actual performance can be compared. The concept of the maximum work has been introduced for this purpose. The maximum work is possible when the processes are totally reversible. But the actual work done by a system is always less than idealised reversible work. In other words, the **presence of irreversibilities leads to a decrease in the amount of work which could be produced and delivered to devices external to system and its local environment**. *The difference between maximum (idealised reversible) work obtainable for a given change state for the system and the actual work delivered is called the* ***irreversibility****, 1* ***of any process***. Thus,

$$I = E_{x,\, \text{loss}} = W_{\text{max}} - W_{\text{act}} = W_{\text{rev}} - W_{\text{act}} \quad ...(11.55)$$

In other words, irreversibility is defined as the amount by which the maximum possible increase in availability exceeds the actual increase.

11.11.1 Non-flow Process

For a non-flow process, we can put the value of W_{max} and W_{act} in Eq. (11.55), we get

$$\delta I = -(dU + p_0\, dV - T_0\, dS) - (\delta W_{total,\, act} - p_0\, dV)$$
$$= -dU + T_0\, dS - dW_{total,\, act}$$

From the conservation of energy, the quantity $\delta W_{total,\, act}$ may be replaced by $\delta Q - dU$.

Hence, $$\delta I = -dU + T_0\, dS + (dU - \delta Q) = T_0\, dS - \delta Q$$

Since we are discussing the irreversible case with heat transfer solely between system and environment, $\delta Q = -\delta Q_0$. But $\delta Q_0 = T_0\, dS = -\delta Q$

This is because of the fact that heat reservoir by concept is considered to be internally reversible.

$\therefore$ $$\delta I = T_0\, dS = T_0\,(dS_{sys} + dS_{envir})$$

or $$I = T_0\,(\Delta S_{sys} + \Delta S_{envir}) \quad ...(11.56)$$

or $$I = E_{x,\, loss} = T_0\ \Delta S_{univ} \quad ...(11.57)$$

Equation (11.56.) shows that the irreversibility of a process is the product of the environment temperature and the increase of entropy of both the system and the environment. Since neither the absolute temperature (T_0) nor the total change of entropy is negative, then

$$I \geq 0 \quad ...(11.58)$$

where the equality and inequality signs apply to **reversible and irreversible processes.**

Irreversibility. (I) may be interpreted as the magnitude of the decrease in available energy or the available energy loss or exergy loss ($E_{x\, loss}$). Thus, loss of available energy

$$E_{x,\, loss} = I = T_0\,(\Delta S_{sys} + \Delta S_{envir}) \quad ...(11.59)$$

If the system undergoes an **irreversible adiabatic process, $\delta Q_0 = 0$**, i.e., $\Delta S_{envir} = 0$, hence

$$I = T_0\ \Delta S_{sys} \quad ...(11.60)$$

11.11.2 Steady Flow Process

The value of W_{max} and W_{act} for steady flow process neglecting KE and PE are given by

$$\delta W_{max} = -(dH - T_0\, dS) \text{ and } \delta W_{act} = \delta Q_{act} = \delta Q - dH$$

Substituting the values of W_{max} and W_{act} in Eq. (11.55), we get

$$\delta I = -(dH - T_0\, dS) - \delta Q + dH = T_0\, dS - \delta Q$$

But $$\delta Q = -\delta Q_0 = -T_0\, dS_0$$

so $$\delta I = T_0\,(dS + dS_0) = T_0\,(dS_{sys} + dS_{envir})$$

or $$I = T_0\,(\,\Delta S_{sys} + \Delta S_{envir}) \qquad ...(11.61)$$

or $$I = E_{x,\,loss} = T_0\,\Delta S_{univ} \qquad ...(11.62)$$

The Gouya–Stodola theorem states that the rate of loss of available energy or exergy in a process is proportional to the rate of entropy generation. Thus, I is written as

$$I = T_0\,\Delta S_{univ} = T_0\,S_{gen} \qquad ...(11.63)$$

Thus, we see that same expression holds goods for irreversibility in both cases of non-flow and steady flow process.

A thermodynamically efficient process would involve the minimum exergy loss with the minimum rate of entropy generation.

11.11.3 Alternative Method for Finding Irreversibilities or Exergy Loss in Steady Flow Process

Irreversibility of exergy loss may be found by the difference of exergy in and exergy out.

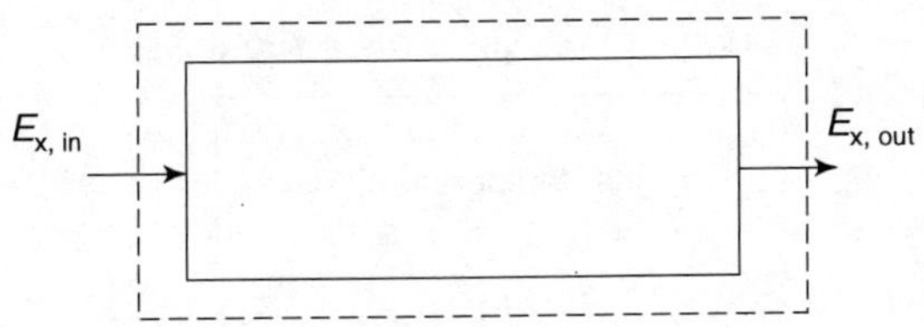

Fig. 11.11 Single stream system

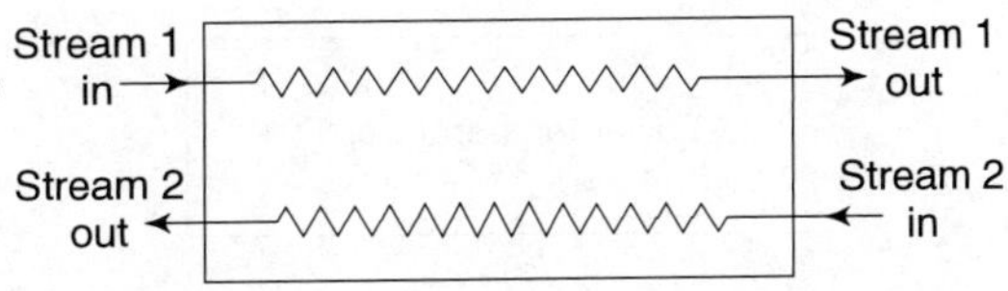

Fig. 11.12 Multi-stream system

Single stream. Consider the case of a single stream (Fig. 11.11)

The irreversibility I or $E_{x,\,loss} = I = E_{x,\,loss} = E_{x,\,in} - E_{x\,out}$ ...(11.64)

The final value will be the same as $T_0\,(\Delta S)_{univ}$

Multi-stream. Consider the case of multistream system (Fig. 11.12)

The irreversibility I or $E_{x,\,loss} = \Sigma\,E_{x\,in} - \Sigma\,E_{x\,out}$ ...(11.65)

It is to be noted that exergy destruction or loss may also be found from the exergy rate balance discussed earlier (Article 11.10).

11.11.4 Application of Gouya–Stodola Equation

11.11.4.1 Heat Transfer Through a Finite Temperature Difference

Consider the case of heat transfer $\dot{Q}$ from the hot reservoir at temperature T_0 to the cold reservoir at temperature T_2 (Fig. 11.13 a).

$$\dot{S}_{gen} = \frac{\dot{Q}}{T_2} - \frac{\dot{Q}}{T_1} = \dot{Q}\frac{T_1 - T_2}{T_1 T_2}$$

and $$\dot{E}_{x,\,loss} = \dot{I} = \dot{W}_{loss} = \dot{Q}\left(1 - \frac{T_2}{T_1}\right) = \dot{Q}\frac{(T_1 - T_2)}{T_1} = T_2\dot{S}_{gen} \quad ...(11.66)$$

If the heat transfer $\dot{Q}$ from T_1 to T_2 takes place through a reversible engine E, the entire work output is dissipated in the brake, from which an equal amount of heat is rejected to the reservoir at T_2 (Fig. 11.13b). Thus, the heat transfer through a finite temperature difference is equivalent to the destruction of its availability or exergy.

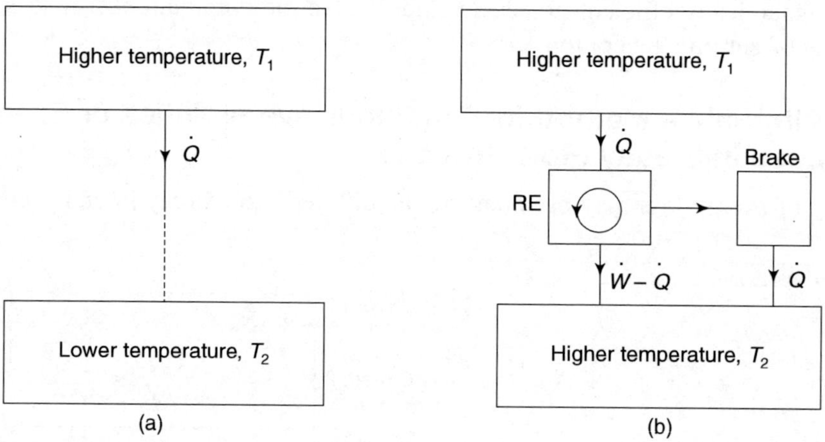

Fig. 11.13 Destruction of exergy by heat transfer through a finite temperature difference

11.11.4.2 Flow with Friction

Consider the case of adiabatic flow of an ideal gas under steady state condition through a segment of pipe. Due to friction, pressure drops from p_1 to p_2 where $\Delta p = p_1 - p_2$ (Fig. 11.14a).

From the first law, $h_1 = h_2$ and from second law, $Tds = dh - vdp$

or $$\int_1^2 ds = \int_1^2 \frac{dh}{T} - \int_1^2 \frac{v}{T}dp = 0 - \int_1^2 \frac{v}{T}dp = -\int_1^2 \frac{R}{p}dp$$

$$\dot{S}_{gen} = \int_1^2 \dot{m}dS = -\int_{p_1}^{p_2} \dot{m}R\frac{dp}{p} = -\dot{m}R\ln\frac{p_2}{p_1}$$

$$= -\dot{m}R\ln\left(1 - \frac{\Delta p}{p_1}\right) = -\,\dot{m}\,R\left(-\frac{\Delta p}{p_1}\right)$$

or $$\dot{S}_{gen} = \dot{m} R \frac{\Delta p}{p_1}$$

As $$\ln\left(1 - \frac{\Delta p}{p_1}\right) = -\frac{\Delta p}{p_1} \text{ since } \frac{\Delta p}{p_1} < 1 \text{ and negelecting higher terms}$$

$$\dot{I} = \dot{E}_{x,\,loss} = \dot{W}_{loss} = \dot{E}_{x,1} - E_{x,2} = \dot{m}\left[(h_1 - T_0 s_1) - (h_2 - T_0 s_2)\right]$$

$$= \dot{m}\, T_0 \,(s_2 - s_1) = \dot{m}\, T_0\, \dot{s}_{gen} = \dot{m}\, R\, T_0 \frac{\Delta p}{p_1} \qquad ...(11.67)$$

Equation (11.67) shows the exergy loss is proportional to the pressure drop and mass flow rate. It is shown by Grassmann diagram (Fig. 11.14 b). The width of the diagram is proportional to the exergy of the stream.

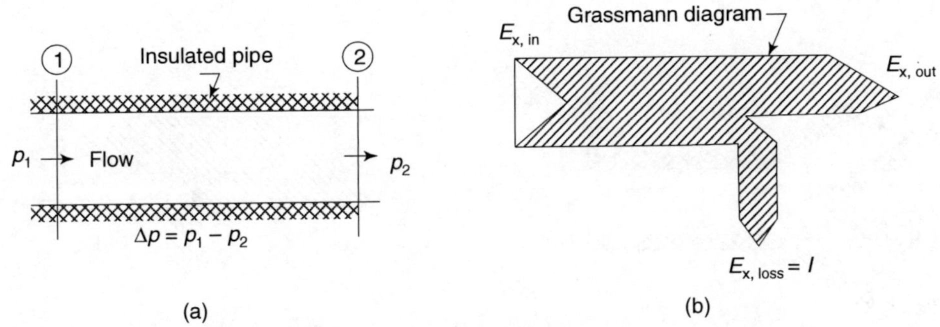

Fig. 11.14 Exergy loss in an insulated pipe due to friction

11.11.4.3 Mixing of Two Fluids

Consider the case of mixing of two streams of fluids 1 and 2 in an insulated pipe (Fig. 11.15a). Assume no pressure drop due to mixing.

Mass balance : $\dot{m}_1 + \dot{m}_2 = \dot{m}_3 = \dot{m}$ (say)

Let $$x = \frac{\dot{m}_1}{\dot{m}_1 + \dot{m}_2}$$

From the first law $\dot{m}_1 h_1 + \dot{m}_2 h_2 = (\dot{m}_1 + \dot{m}_2) h_3$

or $$x h_1 + (1 - x)\, h_2 = h_3$$

Since $h = f(T)$ only, so $x\, T_1 + (1 - x)\, T_2 = T_3$

or $$\frac{T_3}{T_1} = x + (1 - x)\, t \text{ where } t = \frac{T_2}{T_1}$$

From the second law $\dot{S}_{gen} = \dot{m}_3 s_3 - \dot{m}_1 s_1 - \dot{m}_2 s_2 = \dot{m} s_3 - x\dot{m} s_1 - (1-x)\dot{m} s_2$

or $$\frac{\dot{S}_s}{\dot{m}} = (s_3 - s_2) + x(s_2 - s_1) = c_p \ln\frac{T_3}{T_2} + xc_p \ln\frac{T_3}{T_2} \text{ as } p = \text{constant}$$

or $$\frac{\dot{S}_{gen}}{\dot{m}\,c_p} = \ln\left(\frac{T_3}{T_2}\right)\frac{T_1^{1-x}}{T_2^{1-x}} = \ln\left[\frac{T_3/T_1}{(T_2/T_1)^{1-x}}\right] = N_s \quad \text{...(11.68)}$$

where N_s = dimensionless number called, ***entropy generation number*** = $\dot{S}_{gen}/\dot{m}\,c_p$

or $$N_s = \ln\frac{x + t(1-x)}{t^{1\;x}}$$

If $$x = 1,\ t = 1, \text{ then } N_s = 0$$

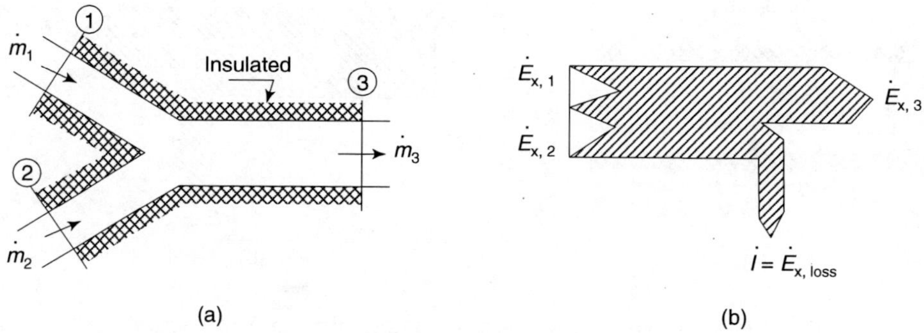

Fig. 11.15 Irreversibility due to mixing of two streams adiabatically

The irreversibility, $I = E_{x,\,loss} = T_0\,\dot{S}_{gen} = T_0\,\dot{m}c_p \ln\frac{x + t(1-x)}{t^{1-x}}$ ...(11.69)

The Grassmann diagram is shown in Fig. 11.15(b).

Example 11.8 Calculate the loss of available energy when air at a temperature of 600 K and 8 bar flows adiabatically across a restriction in a pipe and the pressure is reduced to 1.5 bar.

Solution The loss of available energy, $E_{x,\,loss} = T_0\,(\Delta S)_{system}$

Since $q = 0$, $w = 0$, hence $\Delta h = 0$ and $T_1 = T_2$, thus the process is isothermal.

Hence, $$\Delta s = -R \ln\frac{p_2}{p_1} = -0.287 \ln\frac{1.5}{8} = 0.4804 \text{ kJ/kg}$$

or $$E_{x,\,loss} = 0 + 288 \times 0.4804 = 138.355 \text{ kJ/kg}$$

Example 11.9 Calculate I for saturated steam entering in a steam turbine which is throttled adiabatically from 7 bar to 1.5 bar.

Solution $I = T_0 (S_{sys} + \Delta S_{surr})$. Since the process is adiabatic, $(\Delta S)_{surr} = 0$, hence,

$$I = T_0 (S_{sys}) = T_0 (\Delta S)_{steam}$$

From Steam Tables at 7 bar for dry steam,

$$h_1 = 2762 \text{ kJ/kg}, \; s_1 = 6.7052 \text{ kJ/kg K}$$

In throttling process, $h_2 = h_1 \quad \therefore \quad t_2 = 143.3°\text{C}$

From h–s chart, $s_2 = 7.451$ kJ/kg K

$\therefore$ $I = 288\ (7.451 - 6.7052) = 214.79$ kJ/kg **Ans.**

11.12 EXERGETIC OR SECOND LAW EFFICIENCY OR EFFECTIVENESS

The first law efficiency (η_I) is a common measure of use of energy. The first law or thermal efficiency is defined as the ratio of output energy of a device to input energy of the device. As mentioned earlier, the *first law is concerned only with the quantities of energy, and disregards the forms in which the energy exists*. Further, it does not also discriminate between the energies available at different temperatures.

It is the second law of efficiency (η_{II}) which provides a means of assigning a quality index to energy through the concept of available energy or exergy. Improved energy resource utilisation can be realised by reducing exergy destruction within the system and/or losses. Exergetic efficiencies also can be used to evaluate the effectiveness of engineering measures taken to improve the performance of a thermal system.

11.12.1 Effectiveness

The effectiveness of a cycle is defined as the ratio of increase in availability of the surroundings (due to work delivered by the cycle) to the decrease in availability of the surroundings (due to heat supplied to the cycle). Thus,

$$\eta_{II} = \varepsilon = \frac{|\text{ Increase in availability of surrounding }|}{|\text{Decrease in availability of surrounding }|} \qquad ...(11.70)$$

The algebraic sum of numerator and denominator of Eq. (11.70) represents the loss of available energy, $E_{x,\text{ loss}}$ because of the irreversibility of cycle.

(a) Effectiveness of Power Cycle The increase in the available energy of the surroundings in the case of power cycle is equal to the work delivered by the cycle and the decrease in the available energy of the surroundings is equal to the availability of heat supplied to the cycle. Thus,

$$\eta_{II} = \varepsilon_{\text{power cycle}} = \frac{|\text{ Work delivered by cycle }|}{|\text{ Availability of heat supplied }|} = \frac{W_{\text{output}}}{(E_x)_{\text{fuel}}} \qquad ...(11.71)$$

or
$$\eta_{II} = \eta_I / \eta_{\text{carnot}} \quad \text{...(11.72)}$$

(b) Effectiveness of Steady Flow Process Similarly, the effectiveness of steady flow process is given by

$$\eta_{II} = (\varepsilon)_{sf} = \frac{|\text{Increase in availability of surrounding}|}{|\text{Decrease in availability of the flow stream}|} \quad \text{...(11.73)}$$

(c) Effectiveness of a Turbine Figure 11.17 shows the process path in a turbine on T–s diagram. Due to friction the entropy increases during the expansion process while due to heat losses, entropy tends to decrease a little. The effectiveness of a turbine is given by

$$\eta_{II} = (\varepsilon)_{\text{turbine}} = \left|\frac{w}{(\Delta a)_{sf}}\right| = \left|\frac{w}{a_{sf}}\right| \quad \text{...(11.74)}$$

If changes in KE and PE energies are negligible, then the work delivered to the surroundings per unit mass is given by $w = (h_1 - h_2) - q$

$$a_{sf} = w_{\max} = (h_1 - h_2) - T_0 (s_1 - s_2)$$

$$\therefore \quad \eta_{II} = (\varepsilon)_{\text{turbine}} = \frac{h_1 - h_2 - q}{(h_1 - h_2) - T_0 (s_1 - s_2)} \quad \text{...(11.75)}$$

And, the loss of available energy, $I = a_{sf} - w = (h_1 - h_2) - T_0 (s_1 - s_2) - (h_1 - h_2) + q$

or
$$I = q - T_0 (s_1 - s_2) \quad \text{...(11.76)}$$

If the expansion process in the turbine is adiabatic, then $q = 0$, hence

$$\eta_{II} = (\varepsilon)_{\text{turbine}} = \left|\frac{h_1 - h_2}{(h_1 - h_2) - T_0 (s_1 - s_2)}\right| \quad \text{...(11.77)}$$

and
$$I = T_0 (s_2 - s_1) \quad \text{...(11.78)}$$

(d) Effectiveness of Pump or Compressor Since the work is supplied by the surroundings in the case of pump of compressor, there is an increase in available energy of the stream. The effectiveness of a pump or compressor is given by

$$(\varepsilon)_{\text{pump}} = \left|\frac{\text{Increase in available energy of the flow stream}}{\text{Decrease in availability of surroundings}}\right|$$

or
$$\eta_{II} = (\varepsilon)_{\text{pump}} = \left|\frac{a_{sf}}{w}\right| \quad \text{...(11.79)}$$

From first law, $w = q - \Delta h$. Taking sign convention of w in consideration

$$(\Delta a)_{sf} = \Delta h - T_0 \Delta s \text{ if K.E. and P.E.} = 0$$

$$\therefore \quad \eta_{II} = (\varepsilon)_{\text{pump}} = \left|\frac{\Delta h - T_0 \Delta s}{q - \Delta h}\right| \quad \text{...(11.80)}$$

and $$I = w - a_{sf} = q + T_0 \Delta s$$

If the compression process is adiabatic, $$\eta_{II} = (\varepsilon)_{pump} = \left|\frac{\Delta h - T_0 \Delta s}{-\Delta h}\right| \quad ...(11.81)$$

and $$I = T_0 \Delta s \text{ per unit mass} \quad ...(11.82)$$

(e) Effectiveness of a Heater There are two types of heaters open and closed. Figure 11.16 shows an open and closed heater, respectively. In the open heater, as the name implies, x kg of fluid enters at state 1 in the heater and is mixed with y kg of fluid at state 2. Thus, there is a direct contact and $(x + y)$ kg of fluid at state 3 leaves the heater. Now assume that due to irreversible mixing, the fluid entering at state 1 experiences an increase in available energy while the fluid entering at state 2 experiences decrease in available energy. In such a case, the increase in availability of x kg of fluid between states 1 and 3 is given by

$$= a_{sf} = x (\psi_3 - \psi_1) \text{ if KE = PE = 0}$$

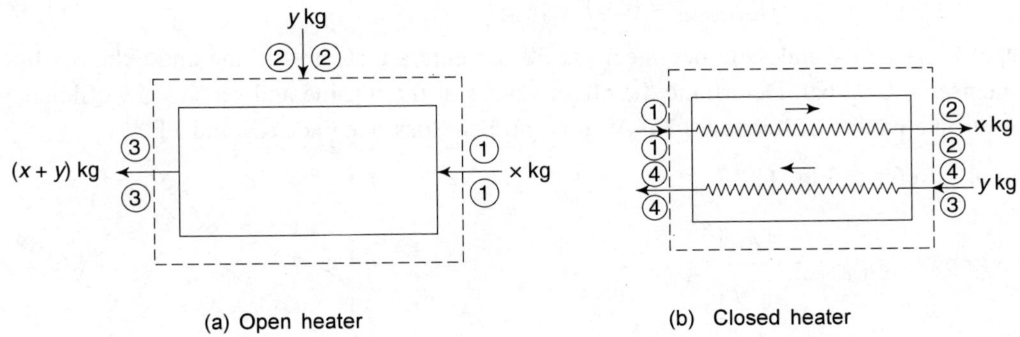

Fig. 11.16

Decrease in availability of y kg fluid between states 2 and 3 = $y (\psi_3 - \psi_2)$

$$\eta_{II} = (\varepsilon)_{heater} = \left|\frac{\text{Increase in availability}}{\text{Decrease in availability}}\right| = \left|\frac{x(\psi_3 - \psi_1)}{y(\psi_3 - \psi_2)}\right| \text{ for open heater} \quad ...(11.83)$$

Also,
$$\begin{aligned} I &= x (\psi_3 - \psi_1) + y (\psi_3 - \psi_2) \\ &= x [(h_3 - h_1) - T_0 (s_3 - s_1)] + y [(h_3 - h_2) - T_0 (s_3 - s_2)] \\ &= x (h_3 - h_1) + y (h_3 - h_2) - T_0 [x (s_3 - s_1) + y (s_3 - s_2)] \quad ...(11.84) \end{aligned}$$

In case of no heat loss from the heater, heat entering is equal to heat leaving. Thus,

$$x (h_3 - h_1) = y (h_2 - h_3) = y (h_3 - h_2)$$

or $$x (h_3 - h_1) + y (h_3 - h_2) = 0$$

$\therefore$ $$I = T_0 [x (s_3 - s_1) + y (s_3 - s_2)] = T_0 \Sigma \Delta s \quad ...(11.85)$$

From the closed heater, following the above procedure, the effectiveness is given by

$$\eta_{II} = (\varepsilon)_{\text{heater}} = \left|\frac{x(\psi_2 - \psi_1)}{y(\psi_4 - \psi_3)}\right| \quad ...(11.86)$$

(f) Effectiveness of a Condenser Though, there is a similarity between a heater and a condenser and difference lies in the temperature of the cooling water which is small due to the high flow. Therefore, the increase in available energy of the cooling water in negligible, hence

$$\eta_{II} = (\varepsilon)_{\text{heater}} = \left|\frac{\text{Increase in availability of cooling water}}{\text{Decrease in availability of cycle fluid}}\right| \quad ...(11.87)$$

$$\approx \left|\frac{0}{\text{Decrease availability of cycle fluid}}\right| \approx 0$$

$$(I)_{\text{condenser}} = (\Delta\psi)_{\text{cycle fluid}} \quad ...(11.88)$$

Example 11.10 Dry and saturated steam at 30 bar enters a steam turbine and exhausts into a condenser at 0.03 bar. Determine the effectiveness of the turbine and second law efficiency if the turbine having efficiency 0.75. Assume no heat loss. Neglect KE and PE

Solution Refer to Fig. 11.17.

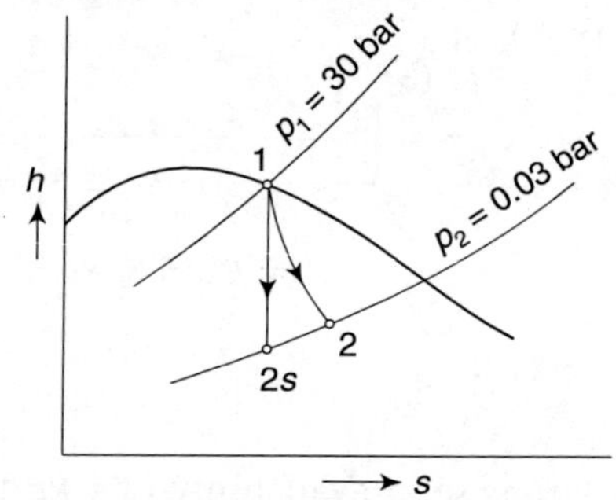

Fig. 11.17 Example 11.10

We know, $(\varepsilon)_{\text{turbine}} = \left|\dfrac{\Delta w}{a_{sf}}\right|$

We know, $(\eta)_{\text{turbine}} = \dfrac{\text{Actual enthalpy drop}}{\text{Isentropic enthalpy drop}}$

$$= \frac{h_1 - h_2}{h_1 - h_{2s}} = 0.75 \quad \textbf{Ans.}$$

From Steam Tables, at 30 bar, $h_1 = 2802.3$ kJ/kg, $s_1 = 6.1837$ kJ/kg K

At 0.03 bar, $h_f = 101$ kJ/kg, $h = 2545.6$ kJ/kg, $s_{fg} = 0.3544$ kJ/kg. K, $s = 8.5785$ kJ/kg K

$$T_2 = 297.1 \text{ K}$$

As $\quad s_1 = s_{2s} - 6.1837 = 0.3544 + x_{2s}\,(8.5785 - 0.3544)$

$\therefore \quad x_{2s} = 0.7088$

$\therefore \quad h_{2s} = 101 + 0.7088\,(2545.6 - 101) = 1833.73$ kJ/kg

Now $\quad 0.75 = \dfrac{2802.3 - h_2}{2802.3 - 1833.73}$

or $\quad h_2 = 2075.87$ kJ/kg

$$w = h_1 - h_2 = 2802.3 - 2075.87 = 726.43 \text{ kJ/kg}$$

$$a_{sf} = (h_1 - h_2) - T_0 (s_1 - s_2)$$

But $$s_2 = 0.3544 + \frac{2075.87 - 101}{297.1} = 7.0015 \text{ kJ/kg}$$

$\therefore$ $$a_{sf} = 726.43 - 288 (6.1837 - 7.0015) = 961.95 \text{ kJ/kg}$$

$$\eta_{II} = (\varepsilon)_{\text{turbine}} = \left|\frac{w}{a_{sf}}\right| = \left|\frac{726.43}{961.95}\right| = 0.755$$ **Ans.**

Example 11.11 Dry and saturated steam at 5 bar enters in direct contact heater (i.e., open heater) and mixes with water at 5 bar having enthalpy of 120 kJ/kg and entropy of 0.4 kJ/kg. It may be assumed that water leaves the heater as saturated liquid at 5 bar. Based on 1 kg of fluid, find effectiveness and loss of available energy.

Solution Refer to Fig. 11.16. Let suffixes 1, 2, and 3 denote the water at entrance, steam at entrance and water at exit of the heater respectively.

From Steam Tables, corresponding to 5 bar,

$$h_2 = 2747.5 \text{ kJ/kg} \qquad s_2 = 6.8192 \text{ kJ/kg K}$$

$$s_{3f} = 1.8604 \text{ kJ/kg K} \qquad h_{3f} = 640.12 \text{ kJ/kg}$$

If x kg and y kg represent the masses of water and steam at entrance, then

$$x + y = 1 \text{ or } x = 1 - y$$

Using the heat balance equation, $xh_{1f} + yh_2 = (x + y)\, h_{3f}$

or $$(1 - y)\, h_{1f} + yh_2 = h_{3f}$$

or $$y(h_2 - h_{1f}) = h_{3f} - h_{1f}$$

or $$y = \frac{h_{3f} - h_{1f}}{h_2 - h_{1f}} = \frac{640.12 - 120}{2747.5 - 120} = 0.1979 \text{ kg of steam}$$

$\therefore$ $$x = 1 - 0.1979 = 0.8021 \text{ kg of water}$$

$$(\varepsilon)_{\text{heater}} = \left|\frac{x(\psi_3 - \psi_1)}{y(\psi_3 - \psi_2)}\right| = \left|\frac{x\left[(h_{3f} - h_{1f}) - T_0 (s_{3f} - s_{1f})\right]}{y\left[(h_{3f} - h_{2f}) - T_0 (s_{2f} - s_2)\right]}\right|$$

$$= \left|\frac{0.8021[(640.12 - 120) - 288(1.8604 - 0.4)]}{0.1979[(640.12 - 2747.5) - 288(1.86.4 - 6.81920)]}\right| = 0.5939$$ **Ans.**

$$I = T_0\, [x(s_{3f} - s_{1f}) + y(s_{3f} - s_{2f})]$$

$$= 288\ [0.8021\ (1.8604 - 0.40) + 0.1979\ (\ 1.804 - 6.81920]$$

or $$I = 54.73 \text{ kJ/kg}$$ **Ans.**

11.13 SECOND LAW ANALYSIS OF THE POWER PLANT AND IC ENGINES

11.13.1 Power Plant

It is possible to do a second law analysis of a power plant by computing the changes in available energy for each process. The following quantities must be known for the complete analysis.

(i) Availability of the fuel air mixture

(ii) Availability of the furnace gas

(iii) Availability of heat received by the cycle

(iv) Availability of heat in the chimney gas

(v) Net work output of the plant

(vi) Loss of available energy for each element of the cycle such as

(a) turbine, (b) pumps, (c) heaters, (d) condenser, etc.

(vii) Second law efficiency, $\eta_{II} = \dfrac{(W_{\text{net}})_{\text{exergy}}}{(E_{x,f})_{\text{exergy}}}$ where $E_{x,f}$ = exergy of fuel

The following are the losses of available energy in the power plant.

(a) Losses of available energy due to combustion and due to the heat loss from the furnace = (i) – (ii)

(b) Loss of available energy due to temperature difference between the furnace and the cycle and due to chimney loss = (ii) – (iii)

(c) Loss of available energy due to temperature difference between furnace and cycle = (ii) – (iii) – (vi)

(d) Total loss of available energy, for the furnace and the cycle = (iii) – (vi)

(e) Loss of available energy within energy within the cycle = (iii) – (vi).

11.13.2 IC Engine

Figure 11.18 shows the availability of air-fuel mixture which enters a reciprocating spark ignition internal combustion engine and the irreversibility occurred during the process. This result has been obtained by calculating availability and irreversibility of each process. The crank angle is shown on abscissa, the left side of which represents the bottom dead centre when

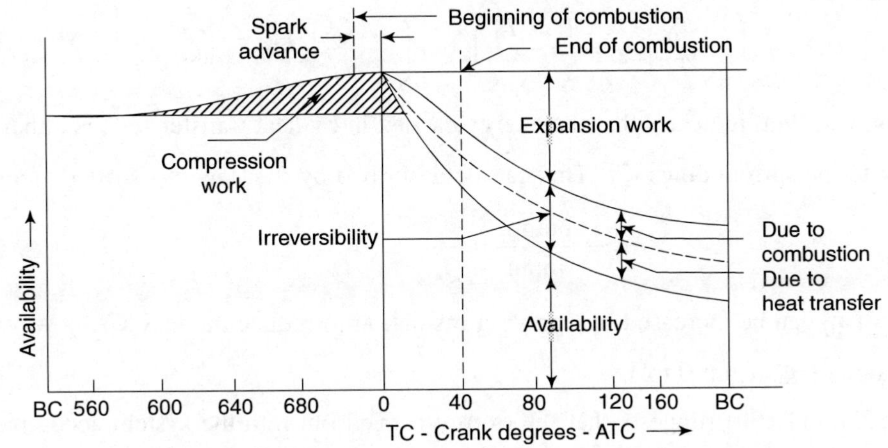

Fig. 11.18 Availability of S.I. internal combustion engine

the cylinder is assumed to be filled with an air-fuel mixture. The availability is represented by ordinate. As the compression process takes place, the availability of the mixture increases owing to the working in compressing the mixture. The expansion process takes place when the piston passes the top dead centre position. During the combustion and expansion process irreversibilities occur. The loss the irreversibility associated with a given change of state, the greater the work that will be done.

11.14 MATCHING END USE TO SOURCE

Thermal energy related tasks such as space heating in industrial furnace and process steam generation commonly involve the combustion of coal, oil or natural gas. In the situation, when the products of combustion are at a temperature significantly greater than required by a given task, the end use in not well matched to the source and the result is inefficient use of the fuel burnt. In order to illustrate this, consider a closed system receiving a heat transfer at the rate $\dot{Q}_s$ at a *source* temperature T_s, and delivering $\dot{Q}_u$ amount of heat at a *use* temperature T_u. Energy is lost to the surroundings by the heat transfer at the rate of $\dot{Q}_l$ across a portion of surface at T_l (Fig. 11.19). Assume that the system operates at steady state and there is no work transfer. Energy and exergy balance yield

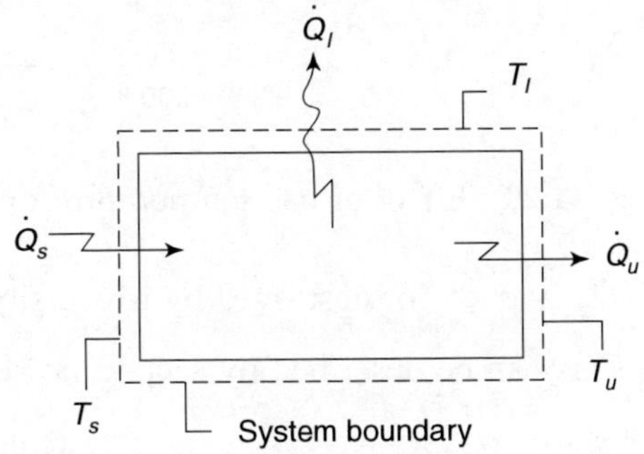

Fig. 11.19 System discussing efficient use of fuel

$$\dot{Q}_s = \dot{Q}_u + \dot{Q}_l \quad \text{...(11.89a)}$$

and $$\left(1-\frac{T_0}{T_s}\right)\dot{Q}_s = \left(1-\frac{T_0}{T_u}\right)\dot{Q}_u + \left(1-\frac{T_0}{T_l}\right)\dot{Q}_l + E_{x,\text{loss}} \quad \text{...(11.89b)}$$

Equation (11.89 a) indicates that the energy carried in by heat transfer, $\dot{Q}_s$, is either used, $\dot{Q}_u$ or lost to the surroundings $\dot{Q}_l$. This can be described by first law of thermodynamics as

$$\eta_I = \frac{\text{output}}{\text{input}} = \frac{\dot{Q}_u}{\dot{Q}_s} \quad \text{...(11.90)}$$

The value of η_I can be increased by applying insulation to reduce the loss $\dot{Q}_l$, however, the limiting value of $\dot{Q}_l$ from 0 to 1.

Further, Eq. (11.89b) suggests that the exergy carried out into the system accompanying the heat $\dot{Q}_s$ is either transferred from the system accompanying the heat transfers $\dot{Q}_u$ and $\dot{Q}_l$ or destroyed by irreversibilities within the system.

This can be described by seconds law efficiency (η_{II}).

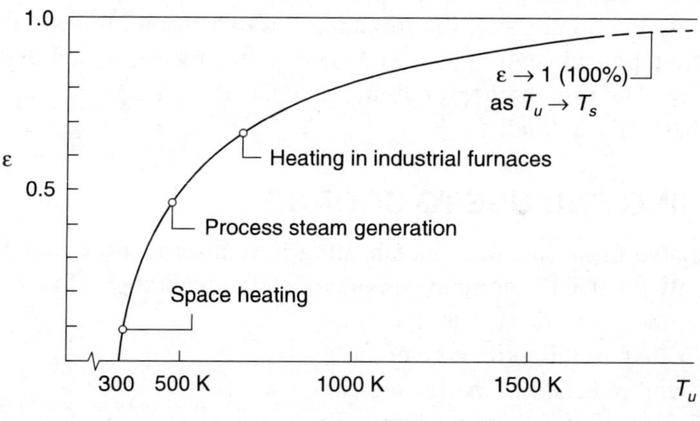

Fig. 11.20 Effect of use temperature on the exergetic efficiency (T_s = 2200 K, $\eta_{I'}$ = 100%)

$\dot{Q}_u$ and $\dot{Q}_l$ or destroyed by irreversibilities within the system.

This can be described by second law efficiency (η_{II}).

$$\eta_{II} = \frac{\text{Output}}{\text{Input}} = \frac{(1-T_0/T_u)\,\dot{Q}_u}{(1-T_0/T_s)\,\dot{Q}_s} \quad \text{...(11.91)}$$

From Eqs. (11.86) and Eq. (11.987), we can get

$$\eta_{II} = \eta_I\left(\frac{1-T_0/T_u}{(1-T_0/T_s)}\right) \quad \text{...(11.92)}$$

Equation (11.92) suggests that a value for η_I as close to unity as practical is important for proper utilisation of the exergy transferred from the hot combustion gas to the system. For proper utilisation of exergy, therefore, it is desirable to have a value for η_I as close to unity as practical and also a good match between the source and use temperature.

Figure 11.20 shows the central role of temperature, in exergetic efficiency considerations. The graph shows that η_{II} tends to unity (100%) as the use temperature approaches T_s. In most cases, the use temperature (T_u) is substantially below T_s. η_{II} for three applications namely space heating at $T_u = 320$ K, process steam generation at $T_u = 480$ K and heating in industrial furnaces at $T_u = 700$ K are shown in this figure. *These efficiency values suggest that fuel is used for more effectively in the higher use-temperature applications than in the lower use-temperature space heating.*

11.15 USING EXERGETIC EFFICIENCY

As mentioned earlier, exergetic efficiencies are useful for distinguishing means for utilising energy resources that are thermodynamically effective from those that are less so. It is worth to note that the limit of 100% exergetic efficiency should not be regarded as a practical objective. This theoretical limit could be attained only if there were no exergy destruction or losses. In practice, decisions are made on the basis of *total costs*. An increase in efficiency to reduce fuel consumption or otherwise utilise resources better normally requires additional expenditure for facilities and operations.

Various methods are now used to improve energy resource utilisation and all such methods must achieve their objective cost effectively. **One method is *co-generation***, which sequentially produces power and heat (or process steam) from the same utility for desired use. Two other methods employed to improve energy resource utilisation are ***power recovery*** and ***waste heat recovery***. Power recovery can be accomplished by inserting a turbine into a pressurized gas or liquid stream to capture some of the exergy which was going as a waste in spontaneous process. **Waste heat recovery is the utilisation of exergy from the exhaust of engine or turbines (say gas turbines) in the from of combined cycle**.

11.16 THERMOECONOMICS

Thermal systems appear in almost every industry and numerous examples are found in our day-to-day lives. Their design involves the applications of principle from thermodynamics, fluid mechanics, heat transfer, material science, manufacturing and mechanical design. *Their design also requires the explicit consideration of engineering economics, as cost is always a consideration. The term thermoeconomics may be a part of general application but in strict sense, it is often applied more narrowly to methodologies combining exergy and economics for optimizing the design and operation of thermal systems.*

In order to illustrate the use of exergy in the design, consider a thermal system consisting of electrical power generating unit and a heat recovery steam generator (Fig. 11.21). The combustion products entering the heat recovery steam generator can be regarded as having economic value.

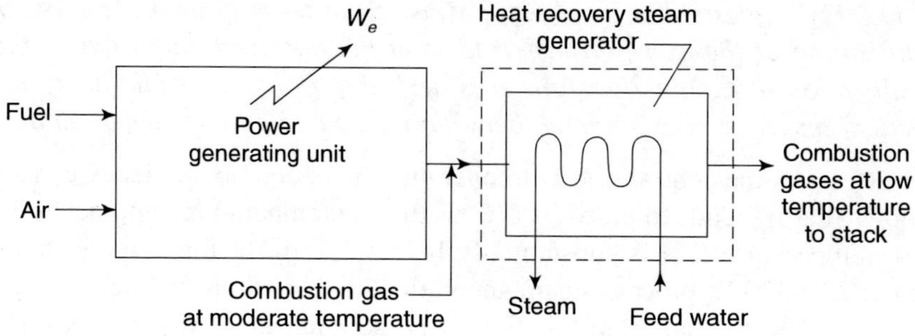

Fig. 11.21 Use of exergy in design

From the second law of thermodynamics, we know that average temperature difference, ΔT_{ave}, between two streams passing through a heat exchanger vanishes as the temperature difference approaches zero. For the heat recovery steam generator, this source of exergy destruction exacts an economic penalty in terms of fuel cost. Figure 11.22 shows the annual fuel cost attributed to the irreversibility of heat exchanger as a function of ΔT_{ave}. The fuel cost increases with ΔT_{ave} as irreversibility is directly related to the temperature difference.

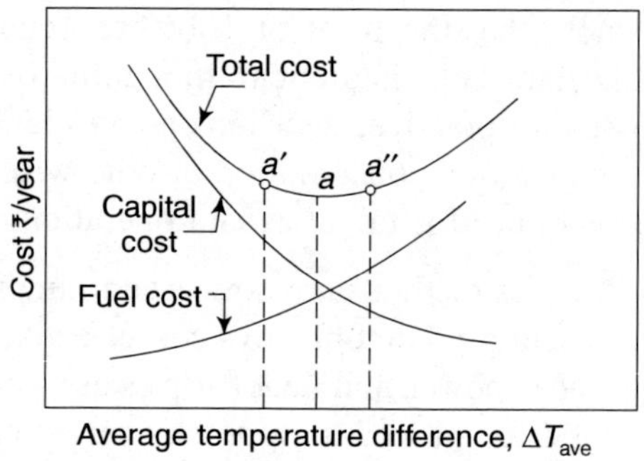

Fig. 11.22 Cost curve for a single heat exchanger

The heat transfer study reveals that there is an inverse relationship between ΔT_{ave} and the surface area required for a specified heat transfer rate. As a result, the capital cost decreases with increasing value of ΔT_{ave}. The total cost is the sum of capital cost and fuel cost. There is a small range of ΔT_{ave} which could be considered optimal in which the total cost could be considered as minimum. *If reducing the fuel cost were demand more important than minimizing the capital cost, we may choose a design corresponding to point a′. If capital cost is the main concern, then point a″ will be the desired operating point.*

Example 11.12 In order to generate steam, coal is burnt with excess air so that the maximum combustion temperature is 1900 K. The c_p of gas may be taken as 1.05 kJ/kg K. The heat evolved due to combustion process is used to generate dry and saturated steam at 15 bar that enters a steam turbine. The condenser pressure is atmospheric. The stack temperature is 600 K. Find per kg of coal.

(a) Loss of available energy due to combustion process, (b) loss of available energy for the stage, (c) loss of available energy due to temperature difference between furnace and cycle, (d) useful work of the cycle, (e) loss of available energy in condenser, (f) effectiveness of the process as a whole. Take, for coal, $\Delta G = -31395$ kJ/kg of coal, $T_0 = 288$ K.

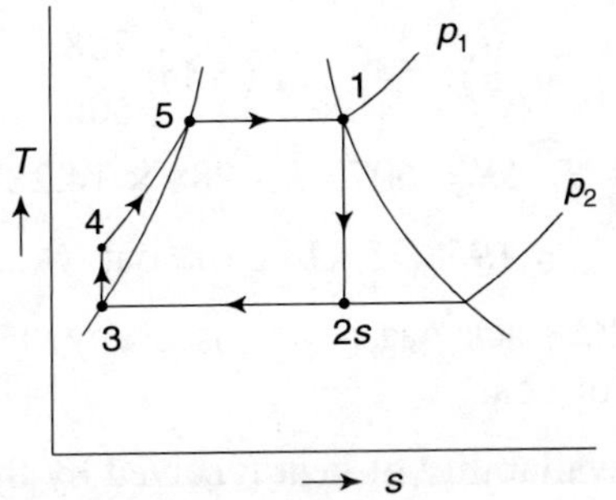

Fig. 11.23 Second law analysis of power plant

Solution Refer to Fig. 11.23. Since the availability of the coal-air mixture is – (ΔG), the availability of mixture before combustion = – ΔG = 31395 kJ/kg of coal.

In order to find the availability of combustion gas, we have to find the mass of the gas

$$\text{Heating value} = -(\Delta H) = mc_p\,(T_{gas} - T_0)$$

or $$31395 = m \times 1.05\,(1900 - 288)$$

or $$m = 18.54 \text{ kg/kg of coal.}$$

Assume that the gases are reversibly cooled from T_{gas} to T_0 by supplying heat to cycle. Thus,

$$\text{Heat supplied, } Q = m \cdot c_p\,(T_{gas} - T_0)$$

$$= 18.54 \times 1.05\,(1900 - 288) = 31380.8 \text{ kJ/kg coal.}$$

Change in entropy of the gas

$$(\Delta S)_{cycle} = m\,c_p \ln \frac{T_0}{T_{gas}} = 18.54 \times 1.05 \ln \frac{288}{1900} = -\,36.72 \text{ kJ/kg of coal K}$$

$\therefore$ $$(\Delta S)_{cycle} = -(\Delta S)_{gas} = 36.72 \text{ kJ/kg of coal K}$$

Now, availability of combustion process

= Availability of mixture before combustion – Availability of combustion gas

= 31395 – 20885.4 = 10509.6 kJ/kg of coal **Ans.**

(b) Similarly, the availability of stack gas may be found by cooling the stack gas reversible from 600 K to 288 K and finding Q and ΔS of the cycle.

Heat supplied, $Q = m \cdot c_p\,(T_{stack} - T_0) = 18.54 \times 1.05\,(600 - 288) = 6073.7$ kJ/kg coal

$$\Delta S = -(\Delta S)_{stack} = -\,m\,c_p \ln \frac{T_0}{T_{stack}}$$

$$= -\ 18.545 \times 1.05 \ln \frac{288}{600} = 14.288 \text{ kJ/kg of coal K}$$

$\therefore$ Availability of stack gas, $Q - T_0\ \Delta S = 6073.7 - 288 \times 14.288$

$$= 1958.75 \text{ kJ/kg of coal}$$

As the available energy in the stack gas is not used anywhere, hence all of this is lost. Thus, $(\Delta I)_{stack} = 1958.75$ kJ/kg of coal **Ans.**

(c) In order to claculate the availability of heat received by the Rankine cycle, we have to calculate Q and ΔS.

$$Q = m\ c_p\ (T_{gas} - T_{stack}) = 18.54 \times 1.05\ (1900 - 600) = 25307.1 \text{ kJ/kg of coal}$$

The process in turbine is shown on T–s diagram of Fig. 11.23. Let m_1 be the steam formed per kg of coal. Thus,

$Q = m_1\ (h_1 - h_{f4}) = m_1\ (h_1 - h_{f2})$ since $h_{f4} = h_{f3} = h_{f2}$ neglecting pump work.

From Steam Tables, at 15 bar = $h_1 = 2789.9$ kJ/kg of steam ; $s_1 = 6.4406$ kJ/kg K

$h_{f3} = h_{f2} = 417.51$ kJ/kg of steam (at 1 bar) ; $s_{f3} = s_{f2} = 1.3027$ kJ/kg K

$\therefore \quad 25307.1 = m_1\ (2789.9 - 417.51)$ or $m_1 = 10.667$ kg/kg of coal

$\therefore \quad \Delta S = m_1\ (s_1 - s_4) = m_1\ (s_1 - s_3) = 10.607 \times (6.4436 - 1.3007) = 54.55$ kJ/K

Availability of heat received by cycle

$$= Q - T_0\ \Delta S = 25307.1 - 288 \times 54.55 = 9596.7 \text{ kJ/kg of coal}$$

$\therefore$ Loss of available energy because of temperature difference between furnace and cycle.

= Availability of combustion gases-availability of heat received by the cycle-availability of stack gas

$$= 20805.4 - 9596.7 - 1958.75 = 9249.95 \text{ kJ/kg coal}$$ **Ans.**

(d) Neglecting the pump work, the output of the cycle is equal to the turbine work.

$$(W)_{cycle} = m_1\ (h_1 - h_2)$$

Due to isentropic expansion, $s_1 = s_{2s} = 6.4406 = 1.3027 + x_{2s}\ (7.3027 - 1.3027)$

or $\quad x_{2s} = 0.856$

$$h_{2s} = h_{2f} + x_{2s}\ h_{fg2} = 417.51 + 0.856 \times 2257.9 = 2350.27 \text{ kJ/kg}$$

$\therefore \quad (W)_{cycle} = 10.66\ (2789.9 - 2350.27) = 4869.53$ kJ/kg of coal. **Ans.**

(e) $(E_{x,\ loss})_{condenser} = -\ (\Delta\Psi)_{cyclic\ fluid} = -\ m_1\ (\Psi_3 - \Psi_2) = m_1\ [(h_{f3} - h_2) - T_0\ (s_{f3} - s_2]$ **Ans.**

$= -\ 10.667\ [(417.51 - 2350.27) - 288\ (1.3027 - 6.4406)] = 4832.63$ kJ/kg of coal **Ans.**

(f) $\varepsilon = \left|\dfrac{\text{Increase in availability}}{\text{Decrease in availability}}\right| = \left|\dfrac{(W)_{cycle}}{\Delta G}\right| = \left|\dfrac{4689.53}{31395}\right| = 0.14937$ **Ans.**

Now available energy balance can be found out.

Available energy(Exergy) balance

Item	Available energy kJ/kg of coal	Balance
1. Available energy of fuel-air mixture	–	31395(100%)
2. Loss of available energy		
(a) Due to combustion	10509.06	
(b) In stack gas	1958.75	
(c) Due to temperature dfference between furnace and cycle	9249.95	
(d) In condenser	4832.63	
3. Cycle work	4689	
4. Unaccounted	155.07	
	31395	31395 (100%)

Example 11.13 A machine element employs a parallel shaft gearbox which receives 800 kW through the high speed shaft but owing to friction, delivers 784 kW through the low speed shaft. Heat transfer takes place from the outer surface of the gearbox to the environment at a rate given by $\bar{h}\,A\,(T_b - T_0)$ where $\bar{h}$ = 0.2 kW/m^2 K, A = 2 m^2 and T_0 is the environment temperature which is equal to 288 K. The outer surface temperature of the gearbox, T_b, is uniform. Calculate the outer surface temperature and irreversibility rate for the gearbox in kW assuming steady state operation.

Solution The heat transfer from the outer surface of the gearbox to the environment is given by energy balance

$$\dot{Q} = \Sigma \dot{W} = -800 + 784 = -16 \text{ kW}$$

The negative sign shows the heat exchange from the system to the surroundings.

$$\dot{Q} = \bar{h}\,A\,(T_b - T_0)$$

or
$$16 = 0.2 \times 2\,(T_b - T_0) = 0.2 \times 2\,(T_b - 288)$$

or
$$T_b = 328 \text{ K}$$

The irreversibility rate, $\dot{I} = T_0\,\Delta S = T_0 \times \dfrac{\dot{Q}_{\text{rej}}}{T_b} = \dfrac{288}{328} \times 16 = 14.048 \text{ kW}$

Example 11.14 A cylinder of internal combustion engine contains 2500 cm^3 of gaseous combustion products at a pressure of 6 bar and a temperature of 850°C just before the exhaust valve opens. Assuming the combustion products as ideal gas, calculate the specific exergy of the gas. Take p_0 = 1.0 bar, T_0 = 288 K. Comment on the result.

Solution This is the case of a closed system. The specific exergy for a closed system is given by

$$(e_x)_{nf} = (u - u_0) + p_0 (v - v_0) - T_0 (s - s_0)$$

$$= c_v (T - T_0) + R \left(\frac{p_0 T}{p} - T_0 \right) - T_0 \left[c_p \ln \frac{T}{T_0} - R \ln \frac{p}{p_0} \right]$$

$$= 0.718 (1123 - 288) + 0.287 \left(\frac{1 \times 1123}{6} - 288 \right)$$

$$- 288 \left[1.005 \ln \frac{1123}{288} - 0.287 \ln \frac{6}{1} \right]$$

$$= 599.63 + (-28.939) - 245.76 = 324.83 \text{ kJ/kg}$$ **Ans.**

Comment The gases are at very high temperature just before the opening of exhaust valve and as a result the specific exergy is appreciable. This gas may be used to expand in gas turbines to get some further work done. Alternatively, this gas may be used in turbocharger which is an integral part of some internal combustion engines.

Example 11.15 A piston-cylinder assembly contains water initially as saturated liquid at 100°C. The water undergoes through two processes to the corresponding vapour state, during which the piston moves freely in the cylinder. The two processes are:

(a) The change in state is brought about by heating as it undergoes an internally reversible process at constant temperature and pressure.

(b) The change in state is brought about adiabatically by stirring action of a paddle wheel.

Calculate for each processes the change in exergy, the exergy transfer accompanying work, the exergy transfer accompanying heat and the exergy destruction or loss based on unit mass. Take p_0 = 1 bar, T_0 = 288 K.

A Model Solution Refer to Fig. 11.24, for case (a), and Fig. 11.25 for case (b).

Data given Initial state of saturated water at t_{s1} = 100°C, final state, saturated vapour.

To find Change in exergy, exergy transfer accompanying work and heat and exergy destruction in both processes.

Assumptions (a) (i) The water in the piston-cylinder assembly is a closed systems for both cases. (ii) The process is internally reversible. (iii) Temperature and pressure constant during process. (iv) Change in PE and KE are negligible for both cases. For case (b). (v) There is no heat transfer with the surroundings. (vi) The system is in equilibrium initially and finally.

(a) **Analysis** The change in specific exergy is given by

$$\Delta e_x = (u_2 - u_1) + p_0 (v_2 - v_1) - T_0 (s_2 - s_1)$$

$$= (u_g - u_f) + p_0 (v_g - v_f) - T_0 (s_g - s_f)$$

From Steam Tables at 100°C, $(u_2 - u_1) = 2087.56$ kJ/kg, $(v_g - v_f) = 1.672$ m^3/kg

and $$(s_g - s_f) = 6.048 \text{ kJ/kg K}$$

Thus, $$\Delta e_x = 2087.56 + 1 \times 10^2 \times 1.672 - 288 \times 6.048 = 512.936 \text{ kJ/kg}$$ **Ans.**

Exergy transfer accompanying work $= \dfrac{\dot{W}}{\dot{m}} - p_0\,(v_2 - v_1)$

$$= \frac{\dot{W}}{\dot{m}} - p_0\,(v_g - v_f) = (p - p_0)\,v_{fg} = 0$$ **Ans.**

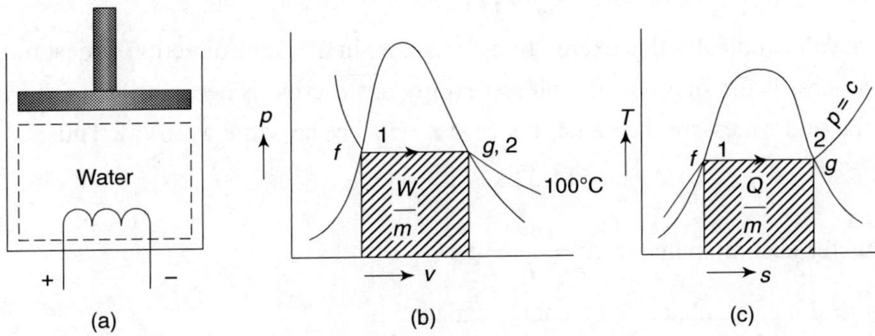

Fig. 11.24 Example 11.15, part (a)

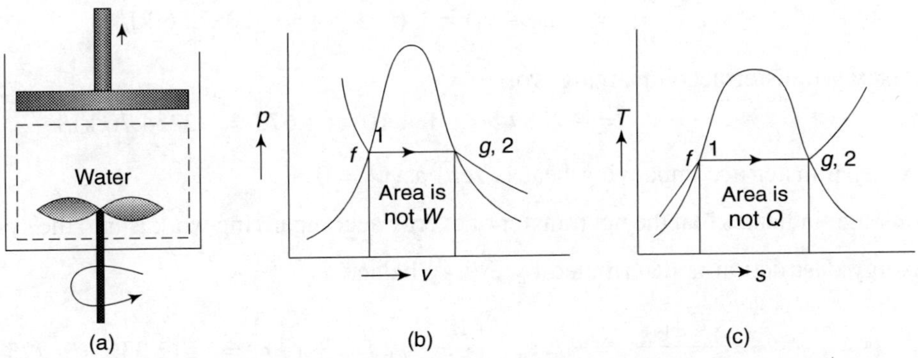

Fig. 11.25 Example 11.15 (b)

Although, the work has non-zero value, but exergy transfer is zero as $p = p_0$

The exergy transfer accompanying heat is given by $\left(1 - \dfrac{T_0}{T}\right)\dfrac{\dot{Q}}{\dot{m}}$

Since the process is internally reversible and at constant temperature, so

$$\dot{Q} = \int_1^2 TdS = \dot{m}\int_1^2 Tds \quad \text{or} \quad \frac{\dot{Q}}{\dot{m}} = (s_2 - s_1) = T\,(s_g - s_f)$$

From Steam Tables, $s_g - s_f = 7.3549 - 1.3069 = 6.048$ kJ/kg K

$$\therefore \qquad \frac{\dot{Q}}{\dot{m}} = 373 \times 6.048 = 2255.8 \text{ kJ/kg}$$

Thus, energy accompanying heat = $\left(1 - \dfrac{288}{373}\right) \times 2255.8 = 514.32$ kJ/kg **Ans.**

The positive value indicates that exergy transfer occurs in the same direction of heat transfer. Since the process is internally reversible, so exergy destruction is necessarily zero in value.

(b) Since the end states are the same, hence Δe_x will be the same as in (a). Thus,

$$\Delta e_x = 514.32 \text{ kJ/kg}$$ **Ans.**

Exergy transfer accompaying work = $\dfrac{\dot{W}}{\dot{m}} - p_0\,(v_g - v_f)$

But the work done calculated from energy balance is

$$\dot{Q} - \dot{W} = \Delta U + \Delta \text{KE} + \Delta \text{PE}$$

As $\dot{Q} = 0$, $\Delta\text{PE} = 0$, $\Delta\text{KE} = 0$, so

$$\frac{\dot{W}}{\dot{m}} = -(u_2 - u_1) = -(u_g - u_f) = -2087.56 \text{ kJ/kg}$$

Thus, exergy transfer accompanying work

$$= -2087.56 - 1 \times 10^2 \times 1.672 = -2254.76 \text{ kJ/kg}$$ **Ans.**

The exergy transfer accompanying heat is zero as $\dot{Q} = 0$.

Here minus sign indicates that the net transfer of exergy accompanying work is into the system.

The exergy destruction is determined by exergy balace

$$\frac{\dot{E}_{x,\,\text{loss}}}{\dot{m}} = -\Delta e_x - \left[\frac{\dot{W}}{\dot{m}} - p_0\,(v_g - v_f)\right] = -512.936 - (-2254.76)$$

$$= 1741.82 \text{ kJ/kg}$$ **Ans.**

Exergy destruction can also be calculated by

$$\frac{\dot{E}_{x,\,\text{loss}}}{\dot{m}} = T_0\, S_{\text{gen}}/\dot{m} = T_0\,(\Delta s_{\text{sys}} + \Delta s_{\text{sum}})$$

The above result is explained as follows.

2254.57 kJ/kg of energy is transferred into the accompanying work of this 1741.82 kJ/kg is destroyed by irreversibilities, having a net increase of ony 514.32 kJ/kg.

Example 11.16 A gearbox working under steady state condition receives 70 kW through the input shaft and delivers power through the output shaft. Taking gearbox as the system, the rate of energy transfer by convection is given by $Q = \bar{h}\ A_s\ (T_b - T_a)$ where $\bar{h}$ = heat transfer coefficient = 0.171 kW/m^2 K, A_s = 1.0 m is the outer surface area of the gearbox, T_b = temperature of outer boundary surface of gearbox = 300 K and T_a = temperature of surroundings away from the vicinity of gearbox = 288 K. Develop a full exergy accounting of gearbox. Take T_0 = 288 K, p_0 = 1 atm.

Solution Refer to *Fig. 5.36 for gearbox but data are different.*

The rate of energy transfer by heat transfer to surrounding taking sign convention is

$$\dot{Q} = -\ \bar{h}\, A_s\, (T_b - T_a) = -\ 0.171 \times 1.0\ (300 - 288)$$

$$= -\ 2.052 \text{ kW}$$

The energy balance for steady state yields, $\dot{Q} = \dot{W} = \dot{W}_1 + \dot{W}_2$.

or
$$\dot{W}_2 = \dot{Q} - \dot{W}_1 = -\ 2.052 - (-\ 70) = 67.948 \text{ kW}$$

Here minus sign of $\dot{W}_1$ indicates energy input while positive sign of $\dot{W}_2$ indicates energy transferred from output shaft.

The entropy rate balance yields, $\dfrac{dS}{d\tau} = \dfrac{\dot{Q}}{T_b} + \dot{S}_{gen}$

Since $dS/d\tau = 0$ for steady state, so, $\dot{S}_{gen} = \dfrac{\dot{Q}}{T_b} = -\dfrac{(-\ 2.052)}{300} = 6.84 \times 10^{-3}$ kW/K

$$\begin{bmatrix}\text{Time rate of exergy transfer}\\ \text{accompanying heat}\end{bmatrix} = \left(1 - \frac{T_0}{T_b}\right)\dot{Q} = \left(1 - \frac{288}{300}\right)(-\ 2.052)$$

$$= -\ 0.082 \text{ kW}$$

Here minus sign indicates the exergy transfer from the system.

$$\begin{bmatrix}\text{Time rate of exergy trasfer}\\ \text{accompanying work}\end{bmatrix} = \dot{W} - p_0\frac{dV}{d\tau}$$

Since the gearbox volume is constant, so $dV/d\tau = 0$, thus

Time rate of exergy transfer accompanying work, $\dot{W}$ = 70 kW

Accordingly, exergy is transferred to the gearbox via the high speed shaft at a rate equal to the power input, 70 kW, and exergy is transferred out via the low speed at a rate equal to the power output 67.948 kW.

The rate of exergy destruction, $E_{x,\ \text{loss}} = T_0\ \dot{S}_{gen} = 288 \times 6.84 \times 10^{-3} = 1.97$ kW

Thus, the exergy balance sheet is given below:

Rate of exergy in:		
• Through high speed shaft		70 kW (100%)
Disposition of exergy:		
• Rate of exergy out	low speed shaft	67.948 kW
	heat transfer	0.082 kW
• Rate of esergy destruction		1.97 kW
Total		70 kW (100%) **Ans.**

Example 11.17 Superheated steam enters a valve at 10 bar and 250°C and comes out at a pressure of 5 bar. The expansion of steam follows a throttling process. Calculate the specific flow exergy at inlet and exit and the exergy destruction. Take p_0 = 1 bar and T_0 = 288 K.

Solution Refer to Fig. 11.26. Here $\dot{Q}_{cv} = 0, \dot{W}_{cv} = 0$ and neglecting changes in KE and potential energy, the specific exergy at linlet is given by

$$e_{xf} = (h_1 - h_0) - (T_0 (s_1 - s_0)$$

From Steam Tables, at 10 bar and 250°C, h_1 = 2942.6 kJ/kg, s_1 = 6.9247 kJ/kg K

At 288 K, $h_0 = h_{f0}$ = 62.99 kJ/kg, $s_0 = s_{f0}$ = 0.2245 kJ/kg K

Thus, e_{xf1} = (2942.6 – 62.99) – 288 (6.9247 – 0.2245) = 949.95 kJ/kg K

Since the process is throttling, so $h_1 = h_2$ **Ans.**

Thus, as h_2 = 2942.6 kJ/kg and p_2 = 5 bar, s_2 = 7.23 kJ/kg K

The exergy at exit = $e_{xf2} = (h_2 - h_0) - T_0 (s_2 - s_0)$

= (2942.6 – 62.99) – 288 (7.23 – 0.2245) = 862.02 kJ/kg **Ans.**

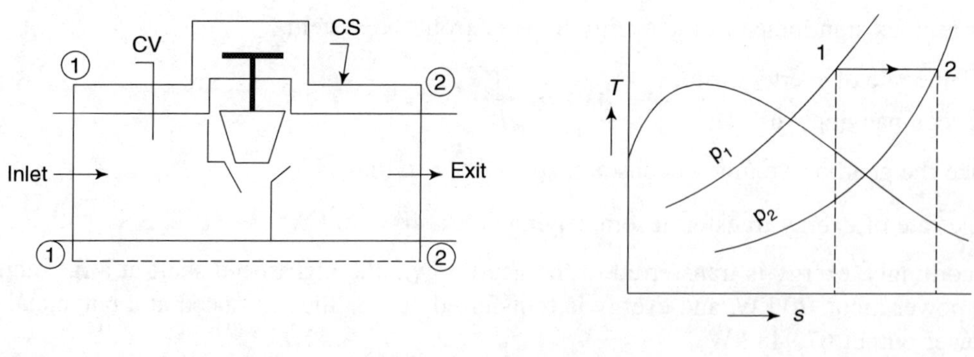

Fig. 11.26 Example 11.17

The steady state form of exergy rate balance is given by

$$0 = \sum_j \left(1 - \frac{T_0}{T_j}\right)\dot{Q}_j - \dot{W}_{cv} + \dot{m}\,(e_{xf1} - e_{xf2}) - \dot{E}_{x,\,loss}$$

Here, $$\dot{Q}_j = 0, \quad \dot{W}_{cv} = 0 \quad \text{so} \quad \frac{\dot{E}_{x,\,loss}}{\dot{m}} = T_0\,(\dot{S}_{gen})_{cv} = T_0\,(s_2 - s_1)$$

or $$\frac{E_{x,\,loss}}{\dot{m}} = 288\ (7.23 - 6.9247) = 89.9264 \text{ kJ/kg}$$ **Ans.**

Example 11.18 A counter flow heat exchanger acting as recuperator of gas turbine receives compressed air operating at steady state at 620 K, 10 bar, and exits at 870 K at 9.7 bar. Hot combustion gas enters as a separate stream at 1050 K, 1.1 bar and comes out at 1 bar. The mass flow rate of each stream is 100 kg/s. The heat transfer from the outer surface of the heat exchanger to the surrounding may be neglected. The changes in kinetic and potential energy are negligible. Assume ideal gas model for combustion gas, calculate the following:

(a) The exit temperature of combustion gas, (b) the net change in the flow exergy rate from inlet to outlet for each stream, and (c) the exergy destruction rate. Take p_0 = 1 bar and T_0 = 288 K. Comment on results.

Solution Refer to Fig. 11.27. For the steady state condition, the energy balance yields.

$$0 = \dot{Q}_{cv} - \dot{W}_{cv} + \dot{m}_0\left[(h_1 - h_2) + \frac{C_1^2 - C_2^2}{2} + g\,(z_1 - z_2)\right]$$

$$+ \dot{m}_g\left[(h_3 - h_4) + \frac{C_3^2 - C_4^2}{2} + g\,(z_3 - z_4)\right]$$

Here, $$\dot{Q}_{cv} = 0, \quad \dot{W}_{cv} = 0, \quad \Delta KE = 0, \quad \Delta PE = 0$$

and $$\dot{m}_a = \dot{m}_g, \quad \text{so} \quad h_2 - h_1 = h_3 - h_4$$

or $$(T_2 - T_1) = (T_3 - T_4) \quad \text{as} \quad c_{pa} = c_{pg}$$

(a) $$T_4 = T_3 - (T_2 - T_1) = 1050 - (870 - 620) = 800 \text{ K}$$ **Ans.**

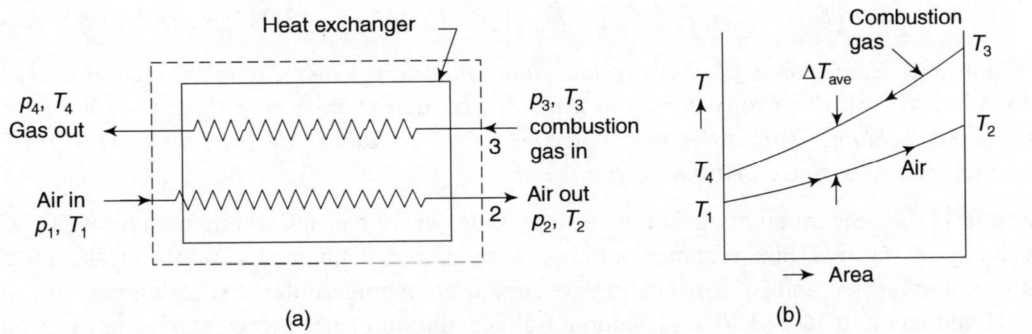

Fig. 11.27 Example 11.18

(b) The net change of exergy of flow stream of compressted air from inlet to outlet is given by

$$E_{xf2} - E_{xf1} = \dot{m}\,[(h_2 - h_1) - T_0\,(s_2 - s_1)]$$

$$= \dot{m}\left[c_p\,(T_2 - T_1) - T_0\left\{c_p \ln\frac{T_2}{T_1} - R\ln\frac{p_L}{p_1}\right\}\right]$$

$$= 100\left[1.005\,(870 - 620) - 288\left\{1.005\ln\frac{870}{620} - 0.287\ln\frac{9.7}{10}\right\}\right]$$

$$= 15067 \text{ kW} = 15.067 \text{ MW}$$ **Ans.**

Thus, the flow exergy of compressed air increases.

The net change of flow stream of hot combustion gas from inlet to outlet is given by

$$E_{xf4} - E_{xf3} = \dot{m}\left[(h_4 - h_3) - T_0\,(s_4 - s_3)\right]$$

$$= \dot{m}\left[c_p\,(T_4 - T_3) - T_0\left\{c_p \ln\frac{T_4}{T_3} - R\ln\frac{p_4}{p_3}\right\}\right]$$

$$= 100\left[1.005\,(800 - 1050) - 288\left\{1.005\ln\frac{800}{1050} - 0.287\ln\frac{1.0}{1.1}\right\}\right]$$

$$= -\ 180.42 \text{ kW} = -\ 18.042 \text{ MW}$$

Thus, the flow exergy of hot gas decreases.

(c) The rate of exergy destruction within control volume can be found from the exergy rate balance under steady state condition.

$$\frac{d\,(Ex)_{cv}}{d\tau} = 0 = \sum_j\left(1 - \frac{T_0}{T_j}\right)\dot{Q}_j - \dot{W}_{cv} + \dot{m}\,(e_{xf1} - e_{xf2}) + \dot{m}\,(e_{xf3} - e_{xf4}) - \dot{E}_{x,\,loss}$$

Since, $\dot{Q}_j = 0,\quad \dot{W}_{cv} = 0,$

so $$\dot{E}_{x,\,loss} = \dot{m}\,(e_{xf1} - e_{xf2}) + \dot{m}\,(e_{xf3} - e_{xf4}) = \Sigma\dot{E}_{xf,\,in} - \Sigma\dot{E}_{xf,\,out}$$

$$= -\left(\dot{E}_{xf2} - \dot{E}_{xf1}\right) - \left(\dot{E}_{xf4} - \dot{E}_{xf3}\right) = -\ 15.067 - (-\ 18.042) = 2.975 \text{ MW}$$ **Ans.**

Comment. *Comparison of above results shows that exergy increase of the compressed air stream is less than the exergy decrease of the combustion stream, even though each has the same energy change. This difference is accounted for by exergy destruction. It is worth to note that energy is conserved but exergy is not.*

Example 11.19 Steam enters a turbine at a pressure of 40 bar and a temperature of 500°C, a velocity of 160 m/s. Steam comes out at 40°C as saturated steam at 100 m/s. Heat transfer between the turbine and its surroundings occurs at an average outer surface temperature of 350 K and amounts to 24 kJ/kg. Develop a full accounting of net exergy carried in by steam per unit mass of steam flowing. Assume steady state condition and neglect changes in KE and PE. Take T_0 = 288 K and p_0 = 1 bar.

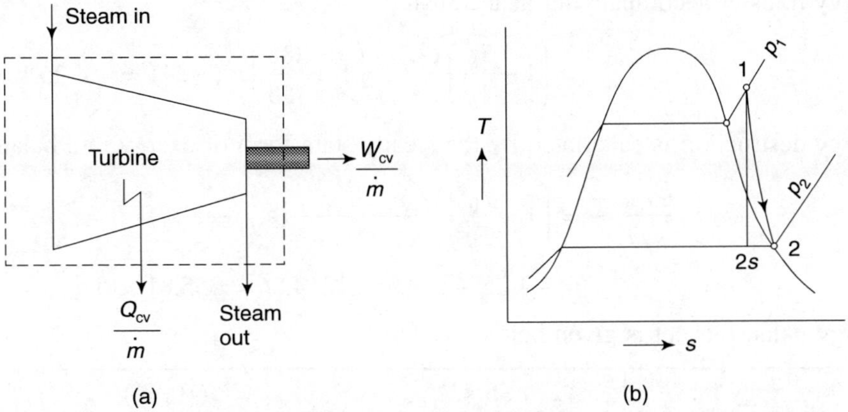

Fig. 11.28 Example 11.19

Solution Refer to Fig. 11.28. From Steam Tables, at 40 bar

and 500° C, h_1 = 3445.3 kJ/kg, s_1 = 7.0901 kJ/kg K

At 40°C, h_2 = 25743 kJ/kg, s_2 = 8.2570 kJ/kg K

At 15°C, $h_0 = h_{f0}$ = 62.99 kJ/kg, $s_0 = s_{f0}$ = 0.2245 kJ/kg K

The net exergy carried in per unit of steam flowing is given by

$$(e_{xf1} - e_{xf2}) = (h_1 - h_2) - T_0\,(s_1 - s_2) + \frac{C_1^2 - C_2^2}{2} + g\,(z_1 - z_1)$$

Here, $z_1 = z_2$, so

$$(e_{xf1} - e_{xf2}) = (3445.3 - 2574.3) - 288\,(7.0901 - 8.2570) + \frac{160^2 - 100^2}{2 \times 10^3}$$

$$= 871 - (-\,336.067) + 7.8 = 1214.86 \text{ kJ/kg}$$

From SFEE, $$h_1 + \frac{C_1^2}{2} - \frac{\dot{Q}_{cv}}{\dot{m}} = h_2 + \frac{C_2^2}{2} + \frac{\dot{W}_{cv}}{\dot{m}} h$$

or $$\frac{\dot{W}_{cv}}{\dot{m}} = (h_1 - h_2) + \frac{C_1^2 - C_2^2}{2} - \frac{\dot{Q}}{\dot{m}}$$

$$= (3445.3 - 2574.3) + \frac{160^2 - 100^2}{2} \times 10^3 - 24$$

$$= 871 + 7.8 - 24 = 854.8 \text{ kJ/kg}$$

At steady state, the exergy transfer accompanying work = $\dfrac{\dot{W}_{cv}}{\dot{m}}$ = 859.3 kJ/kg

The exergy transfer accompanying heat transfer

$$= \left(1 - \frac{T_0}{T_b}\right)\frac{\dot{Q}_{cv}}{\dot{m}} = \left(1 - \frac{288}{350}\right) \times (-24) = -4.25 \text{ kJ/kg}$$

The exergy destruction is calculated by the steady state form of exergy rate balance

$$\frac{E_{x,\,loss}}{\dot{m}} = \left(1 - \frac{T_0}{T_b}\right)\frac{\dot{Q}_{cv}}{\dot{m}} - \frac{\dot{W}_{cv}}{\dot{m}} + (e_{xf1} - e_{xf2})$$

$$= -4.25 - 854.8 + 1214.86 = 355.81 \text{ kJ/kg}$$

The exergy balance sheet is given below:

Net rate of exergy in:	1214.86 kJ/kg	(100%)
Disposition of exergy		
• Rate of exergy out		
work	854.8	(70.36)
heat transfer	4.25	(0.3486%)
• Rate of exergy destruction	355.81	(29.18%)
Total	1214.86	(100%)

Example 11.20 Air expands through a turbine from 6 bar, and 550°C to 1 bar 350°C. During expansion process, 15 kJ/kg of heat lost to surroundings. Neglect changes in KE and PE and take T_0 = 288 K and p_0 = 1 bar. Calculate per kg of air (a) decrease in availability or exergy, (b) the maximum work done, and (c) the irreversibility.

Solution The decrease in availability of exergy neglecting changes in KE and PE is given by

(a)
$$a_{sf1} - a_{sf2} = e_{xf1} - e_{xf2} = (h_1 - h_2) - T_0\,(s_1 - s_2)$$

$$= (h_1 - h_2) + T_0\,(s_2 - s_1)$$

$$= c_p\,(T_1 - T_2) + T_0\left[c_p \ln\frac{T_2}{T_1} - R \ln\frac{p_2}{p_1}\right]$$

$$= 1.005\,(873 - 623) + 288\left[1.005 \ln\frac{623}{873} - 0.287 \ln\frac{1}{6}\right]$$

$$= 301.7 \text{ kJ/kg}$$ **Ans.**

The maximum work, $\dfrac{\dot{W}_{max}}{\dot{m}}$ = decreases in availability = $e_{xf1} - e_{xf2}$ = 301.7 kJ/kg **Ans.**

From steady flow process neglecting ΔPE, $h_1 - \frac{\dot{Q}}{\dot{m}} = h_2 + \frac{\dot{W}}{\dot{m}}$

or
$$\frac{\dot{W}}{\dot{m}} = (h_1 - h_2) - \frac{\dot{Q}}{\dot{m}} = c_p (T_1 - T_2) - \frac{\dot{Q}}{\dot{m}}$$
$$= 1.005\,(873 - 623) - 15 = 236.25 \text{ kJ/kg}$$

(b) Irreversibility, $I = \frac{\dot{W}_{max} - \dot{W}}{\dot{m}} = 301.7 - 236 = 65.7$ kJ/kg **Ans.**

Alternatively, $I = E_{x,\,loss}/\dot{m} = T_0\,(\Delta s_{surr} + \Delta s_{surr}) = T_0\left[\left(c_p \ln\frac{T_2}{T_1} - R\ln\frac{p_2}{p_1}\right) + \frac{15}{288}\right]$

$$= 288\left[\left(1.005\ln\frac{623}{873} - 0.287\ln\frac{1}{6}\right) + \frac{15}{288}\right]$$
$$= 65.7 \text{ kJ/kg}$$ **Ans.**

Alternatively, I or $E_{x,\,loss}/\dot{m}$ may be found from exergy rate balance.

$$\frac{E_{x,\,loss}}{\dot{m}} = \left(1 - \frac{T_0}{T_b}\right)\frac{\dot{Q}_{cv}}{\dot{m}} - \frac{\dot{W}_{cv}}{\dot{m}} + (e_{xf1} - e_{xf2})$$

Here, T_b is not mentioned, so $T_b = T_0$. Thus,

$$\frac{E_{x,\,loss}}{\dot{m}} = \frac{\dot{Q}_{cv}}{\dot{m}} - \frac{\dot{W}_{cv}}{\dot{m}} + (e_{xf1} - e_{xf2}) = \frac{\dot{W}_{max}}{\dot{m}} - \frac{\dot{W}}{\dot{m}}$$
$$= 65.7 \text{ kJ/kg}$$ **Ans.**

Example 11.21 An ideal gas flows through an insulated pipe under steady state condition at the rate of 5 kg/s. There is 10% pressure drop from inlet to exit of long pipe. Calculate the exergy loss because of pressure drop due to friction. Take T_0 = 288 K and p_0 = 1 bar.

Solution The rate of generation in terms of pressure drop is given by

$$\dot{S}_{gen} = \dot{m}\,R\frac{\Delta p}{p_1} = 5 \times 0.287 \times \frac{0.1\,p_1}{p_1} = 0.1435 \text{ kW/K}$$

Rate of exergy loss, $\dot{I} = \dot{E}_{x,\,loss} = T_0\,\dot{S}_{gen}$ = 288 × 0.1435 = 41.328 kW **Ans.**

Note The example can also be solved from **general exergy rate balance equation**.

Example 11.22 A gas is flowing under steady state condition through a pipe at the rate of 10 kg/s. Owing to inadequate insulation of pipe, the gas temperature decreases from 900°C to 880°C between the two sections of the pipe. Neglecting pressure losses in pipe, calculate:

(a) The rate of exergy loss from the pipe. Take T_0 = 288 K, p_0 = 1 bar.

(b) For the same temperature drop of 20°C, when the gas is cooled from 120°C to 100°C, due to heat loss, calculate the rate of exergy loss from the pipe. Comment on these two results.

Solution (a) The rate of exergy loss is given by,

$$\dot{I} = \dot{E}_{x,\text{loss}} = T_0\,(\Delta\dot{S}_{\text{sys}} + \Delta\dot{S}_{\text{surr}}) = T_0\,\dot{S}_{\text{gen}}$$

Since the heat $\dot{Q}$ is loss from system to surrounding,

So $$\Delta\dot{S}_{\text{surr}} = -\frac{\dot{Q}}{T} = \frac{\dot{m}\,c_p\,(T_2 - T_1)}{T_0} \quad \text{and} \quad \Delta\dot{S}_{\text{system}} = \dot{m}\,c_p \ln\frac{T_2}{T_1}$$

Thus, $$\dot{I} = \dot{E}_{x,\text{loss}} = T_0\,\dot{m}\,c_p\left(\ln\frac{T_2}{T_1} - \frac{(T_2 - T_1)}{T_0}\right)$$

$$= 288 \times 10 \times 1.005\left[\ln\frac{1153}{1173} - \left(\frac{-20}{288}\right)\right] = 151.22 \text{ kW} \quad \textbf{Ans.}$$

(b) $$\dot{I} = \dot{E}_{x,\text{loss}} = 288 \times 10 \times 1.005\left[\ln\frac{373}{393} - \left(\frac{-20}{288}\right)\right]$$

$$= 48.822 \text{ kW} \quad \textbf{Ans.}$$

Note The above example may also be solved be *exergy rate balance for flow process.* In this equation, take $T_b = T_0$ and $W = 0$.

Comment *From the above results, it is concluded that the rate of exergy destruction due to heat loss at higher temperature is much higher than that at low temperature even for the same temperature drop. It suggests that the insulation must be provided to stop heat loss at higher temperature.*

Example 11.23 In a refrigerated fishing vessel, a mass of 1500 kg/h of fish initially at 1 bar and 300 K is to be cooled to – 30°C. The freezing point of fish is – 2.2°C and the specific heat of fish below and above freezing point are 1.7 and 3.2 kJ/kg respectively. The latent heat of fusion can be taken as 235 kJ/kg. Find the exergy produced in the chilling process. Take T_0 = 300 K and p_0 = 1 bar.

Solution Refer to Fig. 11.29.

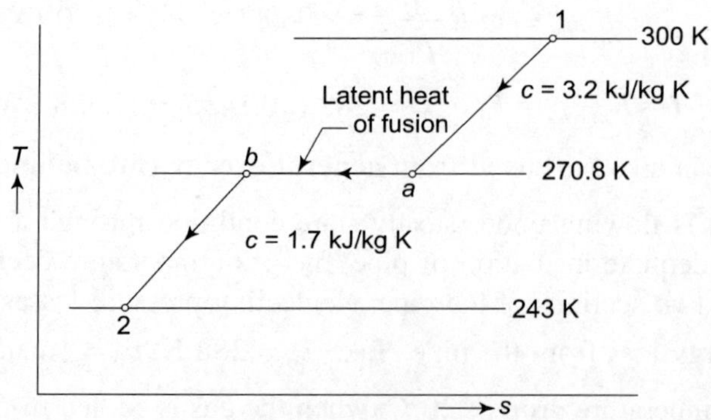

Fig. 11.29 Example 11.23

Exergy produced = $E_x = (H_2 - H_1) - T_0\,(S_2 - S_1)$

But $(H_1 - H_2) = 1500\,[1.7\,(270.8 - 243) + 235 + 3.2\,(300 - 270.8)]$

$= 563550$ kJ/h $= 563.55$ MJ/h

Hence, $H_2 - H_1 = -\,563.55$ MJ/h

Again, $$T_0\,(S_2 - S_1) = 1500 \times 300\left[1.7\ln\frac{243}{270.8} - \frac{235}{270.8} + 3.2\ln\frac{270.8}{300}\right]$$

$= -\,620{,}145$ kJ/h $= -\,620.145$ MJ/h

Hence, exergy produced $= -\,563.55 - (-\,620.145)$

$= 56.595$ MJ/h

$$= \frac{56.595 \times 10^3}{3600} = 15.72 \text{ kWh}$$ **Ans.**

11.17 SUMMARY OF FORMULAE

1. $A = Q - T_0\,\Delta S$ for heat
2. $UA = T_0\,\Delta S$ for heat
3. $A_{nf} = E_{x,\,nf} = (U - U_0) + p_0\,(V - V_0) - T_0\,(S - S_0)$
4. $a_{nf} = e_{x,\,nf} = (u - u_0) + p_0\,(v - v_0) - T_0\,(s - s_0)$
5. $E_{x2} - E_{x1} = \int_1^2 \left(1 - \frac{T_0}{T_b}\right)\delta Q - [W - p_0\,(V_2 - V_1)] - T_0\,S_{gen}$: For closed system
6. $(E_{x2} - E_{x1}) : \begin{cases} > 0 \\ = 0 \\ < 0 \end{cases}$
7. $E_{x,\,\text{loss}} \begin{cases} > 0 \text{ irreversibilities present in the system} \\ = 0 \text{ no irreversibilities present in the system} \end{cases}$
8. $\frac{dE_x}{d\tau} = \sum_j \left(1 - \frac{T_0}{T_j}\right)\dot{Q}_j - \dot{W} - p_0\,\frac{dV}{d\tau} - \dot{E}_{x,\,\text{loss}}$ (Closed system exergy rate balance)
9. $A_{st} = E_{x,\,sf} = (H - H_0) - T_0\,(S - S_0) + \frac{m\,(C^2 - C_0^2)}{2} + mg\,(z - z_0)$
10. $a_{sf} = e_{s,\,sf} = (h - h_0) - T_0\,(s - s_0) + \frac{C^2 - C_0^2}{2} + g\,(z - z_0)$

11. $$\frac{d(E_x)_{cv}}{d\tau} = \sum_j \left(1 - \frac{T_0}{T_j}\right)\dot{Q}_j - \left(\dot{W}_{cv} - p_0 \frac{dV_{cv}}{d\tau}\right) + \sum_i \dot{m}_1\, e_{xi} - \sum_e \dot{m}_e\, e_{xe} - \dot{E}_{x,\,loss}$$

(Exergy rate balance for unsteady state flow process)

12. $E_{x,\,loss} = T_0\, \dot{S}_{gen}$

13. $0 = \sum_j \left(1 - \frac{T_0}{T_j}\right)\dot{Q}_j - \dot{W}_{cv} + \sum_i \dot{m}_i\, e_{xi} - \sum_e \dot{m}_e\, e_{xe} - \dot{E}_{x,\,loss}$ (for steady state flow process)

14. $I = E_{x,\,loss} = W_{max} - W_{act} = W_{rev} - W_{act} = T_0\,(\Delta S_{sys} + \Delta S_{surr})$

15. $I = E_{x,\,loss} = E_{x,\,in} - E_{x,\,out}$ for single stream

16. $I = E_{x,\,loss} = \Sigma\, E_{x,\,in} - \Sigma\, E_{x,\,out}$

17. $\eta_{II} = \epsilon = \dfrac{W_{output}}{E_{x,\,f}} = \dfrac{\eta_f}{\eta_{Carnot}}$ for power cycle

18. $\eta_{II} = \epsilon_{sf} = \left|\dfrac{\text{Increase in availability of surroundings}}{\text{Decrease in availability of flow stream}}\right|$

PROBLEMS

Things to Think About

11.1. The world is passing through "energy crisis". Do you mean exergy crisis? Comment on this statement.

11.2. Whether a helium-filled balloon at temperature T_0 and pressure p_0 has exergy or not?

11.3. Do you think that exergy may have a negative value?

11.4. When a moving automobile is brought to rest by applying brake, what happens to the exergy associated with its motion?

11.5. Can an energy transfer by heat and the associated exergy transfer be in opposite direction?

11.6. Whether an evacuated space of volume has an exergy or not?

11.7. What is the significance of exergy destruction?

11.8. Is it possible for the flow exergy to be negative?

11.9. Exergy is calculated with reference to a datum? What is that datum? Potential energy has a 100% exergy and if the datum is changed from earth's surface to earth's centre, then what happens to exergy?

11.10. Whether the cost of electricity generated should be based on energy or exergy efficiency?

Short and Long Theoretical Questions

11.11. What do you mean by available and unavailable energy? What do you mean by dead state?

11.12. Derive an expression for availability or exergy of a non-flow process.

11.13. What do you mean by availability functions of a non-flow process?

11.14. Derive an expression for availability of a steady flow system.

11.15. What do you mean by availability function of an open system?

11.16. (a) What do you mean by irreversibiiity or exergy loss? Derive the expression.

(b) Define effectiveness of turbines, pumps, heaters.

11.17. What do you mean by high grade energy and low grade energy?

11.18. What do you mean by quality of energy?

11.19. Derive the expression of exergy rate balance for non-f!ow process.

11.20. Derive the expression of exergy rate balance for steady flow process.

11.21. What do you mean by matching end use to source?

11.22. Discuss exergetic efficiency and its use.

11.23. Discuss thermoeconomics.

11.24. Discuss the aspects of exergy concepts.

Numerical Questions

11.25. Determine the exergy, in MJ, of the contents of a 30 m^3 storage tank, if the tank is filled with (a) air as an ideal gas, at 500°C and 14 bar and (b) water vapour at 500°C dry and saturated at 14 bar.

11.26. In a rigid container, 80 kg of water at 60°C and 50 kg of water at 100°C are mixed together. Calculate (a) unavailable energy of each fluid before mixing, (b) unavailable portion of the mixture, and (c) change in available and unavailable energies due to mixing.

[*Ans.* (a) 13982 kJ, 15567 kJ, (b) 29796 kJ, (c) – 247 kJ, + 247 kJ]

11.27. Determine the availability of air at a pressure of 10 bar and a temperature of 100°C.

[*Ans.* 103.9 kJ/kg]

11.28. Compute availability of steam in a steady flow process at
(a) p = 20 bar. T = 303 K, C = 300 m/s, z = 400 m.
(b) at p = 20 bar, T = 333 K, C = 500 m/s, z = 0.

11.29. At constant atmospheric pressure 50 kg water is heated in a rigid container from 313 K to 370 K. Compute the available and unavailable energy if T_0 = 288 K.

[*Ans.* A = 1841.6 kJ, UA = 10071.4 kJ]

11.30. Calculate the availability of a gas (c_p = 1.09 kJ/kg K) at 1000°C with respect to an environment at 20°C. What will be the maximum efficiency for an engine using the energy of this gas and rejecting part of it to the environment? [*Ans.* 599.1 kJ/kg, 56.1%]

11.31. Find the change in the availability of 1 kg of oxygen at atmospheric pressure and 5°C when it expands isothermally to double its volume. [*Ans.* 20.36 kJ/kg]

11.32. 620 kJ of heat is exchanged between two constant-temperature reservoirs at temperature 800 K and 400 K. Evaluate (a) the total change in entropy due to heat transfer process, (b) the available energy before and after the heat transfer, (c) the increase in the unavailable energy. Show parts (a) and (b) on *T–s* diagram. [*Ans.* (a) 0.775 kJ/K; (b) 396 kJ 173 kJ, (c) 223 kJ]

11.33. One kg of water is heated at constant pressure 15°C to 150°C. The c_p of water is 4.18 kJ/kg K. Assuming steady flow, calculate (a) the heat interaction, (b) the unavailable energy based upon an environment temperature of – 10°C, and (c) show the pressure on a *T–s* diagram and indicate various energies designated.
[*Ans.* (a) 2717.3 kJ/kg; (b) 1940 kJ/kg]

11.34. One kg of oxygen and 1 kg of hydrogen are allowed to mix adiabatically without change in total volume. Find the decrease in availability and irreversibility. [*Ans.* –72.5 kJ, 72.27 kJ]

11.35. Evaluate the irreversibility when air is throttled adiabatically from 20 bar to 1 bar. Neglect changes in kinetic energies and assume steady state. [*Ans.* 247.7 kJ]

11.36. Dry and saturated steam at 30 bar enters a steam turbine and exhausts at 0.032 bar. The adiabatic work is 418.6 kJ. Calculate (e) efficiency of turbine, (b) effectiveness of the process, (c) loss of available energy of the process, and (d) loss of available energy in the condenser.

[*Ans.* (a) 0.782, (b) 0.79, (c) –203 kJ, (d) 0.66 kJ]

11.37. In axial flow compressor of a gas turbine, the air enters at 0.7 bar and – 7°C and leaves the compressor at 2.5 bar. The compression efficiency is 0.9. Evaluate per kg of air the compression work, the reversible work for the actual change of air and the irreversibility.

[*Ans.* – 144 kJ; – 129 kJ; 15 kJ]

11.38. In a gas turbine, air enters at 7 bar, 70°C with a velocity of 100 rn/s. It leaves the turbine at 1.4 bar, 5°C and 70 m/s velocity. The process is adiabatic. Evaluate per kg of air : (a) reversible work for the change in state of air, (b) actual work, (c) decrease in available energy of state of air, and (d) irreversibility of process. [*Ans.* (a) 142 kJ, (b) 68 kJ, (c) 142 kJ, (d) 74.5 kJ]

11.39. 1 kg of water at 100 bar undergoes a change in state from saturated liquid to saturated vapour. Calculate ΔI for the following cases.

(a) The change of state results from heat transfer from a reservoir at 600°C.

(b) The change of state is caused by an electrical resistance heater. [*Ans.* –1009 kJ,– 1741 kJ]

11.40. Air is heated at constant pressure from 4°C to 250°C. If the temperature of the environment is 4°C, what percentage of heat transfer to the air represents an increase in the available energy of the air?

11.41. Air at 20°C and 1 bar is compressed isentropically to a temperature of 150°C and then heated at constant pressure to 600°C. The specific heat at constant pressure c_p = 1.005 kJ/kg. The hot gas is then expanded isentropically to the initial pressure of 1 bar in a turbine to complete the cycle permitted to cool in the atmosphere to complete the cycle, (a) How much energy is supplied during the heating process from 150°C to 600°C? (b) How much of this energy is available? (c) How much of this energy is unavailable? (d) What is the thermal efficiency of this ideal cycle? (e) If 5 kg of gas per second passes through this machine, calculate the power developed.

11.42. A gas having c_p as 1.09 kJ/kg K at 7 bar and 500°C expands freely to atmospheric pressure. Calculate the decrease in availability (R = 0.284 kJ/kg K).

11.43. 1 kg of nitrogen is heated at constant volume from 50°C to 150°C. Calculate the available and unavailable portions of the heat added. [*Ans.* 45.2 kJ. 55.5 kJ]

11.44. Carnot cycle receives heat from the furnace gas which is cooled from 1773 K to 673 K. The cycle operates between 130 bar and 2.5 cm of Hg. Compute per kg of furnace gas: (a) stack loss, (b) condenser loss, (c) loss due to temperature difference between furnace and cycle, (d) effectiveness of the cycle and furnace combined, and (e) effectiveness of the cycle alone.

[*Ans.* (a) 146.5 kJ, (b) – 19.9 kJ, (c) 238.6 kJ, (d) 0.59 (c) 0.1]

11.45. Gas from the combustion chamber of a turbine plant enters the turbine at 700°C and 5 bar and expands to 1 bar. Assuming the gas can be treated as air, and the environment is at 20°C, find (a) the availability of gas entering the turbine, (b) the availability of gas after isentropic expansion, (c) the availability of the gas expanding with a expansion efficiency of 80%, (d) the irreversibility of expansion (a) to (b), and (e) the irreversibility of expansion (a) to (c).

[*Ans.* (a) 465.3 kJ/kg, (b) 104.9 kJ/kg, (c) 144.5 kJ/kg, (d) 0, (e) 32.6 kJ/kg]

11.46. 50 kg of water at 60°C is mixed with 80 kg of water at 100°C while the temperature of surroundings is 15°C. Calculate the decrease in available energy due to mixing. [*Ans.* 236 kJ]

11.47. Air enters an adiabatic compressor at 1 bar and 15°C. The compressed air leaves the compressor at 5.5 bar and is cooled to 40°C in an aftercooler. The mass flow rate flowing through the compressor is 0.01 kg/s. The adiabatic efficiency of the compressor is 75%. Determine (a) the power required to drive the compressor, and (b) the rate of exergy destruction for the overall process (compressor and cooler). [*Ans.* (a) 2.42 kW, (b) 1 kW]

11.48. In a steam generator, the hot gas from the furnace transfers heat to water which vaporizes at constant temperature and process of 242.6°C and 35 bar. The gases are cooled from 1100°C to 430°C and have average specific heat, c_p = 1.046 kJ/kg K over this temperature range. The latent heat of vaporization of steam at 35 bar is 1753.7 kJ/kg. The steam generation rate is 12.6 kg/s, and there is a negligible heat loss from the steam generator, determine (a) the rate of heat transfer, (b) the rate of loss of exergy of the gas, (c) the rate of gain of exergy of the steam, and (d) the rate of entropy generation. Take T_0 = 288K and p_0 = 1 bar.

[*Ans.* (a) 22.096 MW, (b) 15.606 MW, (c) 0.5 MW, (d) 20.76 kW/K]

11.49. The exhaust gases from the gas turbine are used to heat water is an adiabatic counter flow heat exchanger. The gases are cooled from 260°C to 120°C, while water enters the exchanger at 65°C. The mass flow rates of the gas and water are 0.38 kg/s and 0.5 kg/s, respectively. The constant pressure specific heat for gas and water as 1.09 and 4.18 kJ/kgK, respectively. Determine the rate of exergy loss due to heat transfer. Take T_0 = 35°C and p_0 = 1 bar. [*Ans.* 12.5 kW]

11.50. In a counter-flow heat exchanger, oil of specific heat 2.1 kJ/kgK is cooled from 440 K to 320 K while water of specific heat 4.2 kJ/kg K is heated from 290 K. to temperature T. The mass flow rates of water and oil are 3200 kg/h and 800 kg/h. Neglect pressure drop and changes in KE and PE. Calculate (a) the temperature at the exit water T, (b) the rate of exergy destruction, (c) the second law efficiency. Take, T_0 = 17°C and p_0 = 1 atm.

[*Ans.* (a) 305 K, (b) 414 MJ/h, (c) 10.9%]

11.51. Energy is received by a solar collector at the rate of 300 kW from a source temperature 2400 K. 60 kW of this energy is lost to the surroundings at a steady state and more temperature remains constant at 600 K. Calculate the first and second law efficiencies. Assume T_0 = 300K.

[*Ans.* 0.8 and 0.457]

11.52. 6 kg of air initially at 900 K, 3 bar fills a rigid tank. The air is cooled to 600 K. Work out the following. (a) Indicate the initial state, final state and dead state on a *T–v* diagram, (b) Determine the heat transfer, and (c) change in exergy. Take T_0 = 300 K, p_0 = 1 bar.

11.53. An electric motor operating at steady state draws current of 10 amp with a voltage of 220 V. The output shaft running at 1000 rpm produces a torque of 16 Nm applied to an external load. The rate of heat transfer from the motor to its surroundings is related to the surface temperature T_b and the ambient temperature T_0 by $\overline{h}\, A_s\,(T_0 - T_b)$ where $\overline{h}$ = 500 W/m^2 K, A_s = 0.195 m^2] T_0 = 293 K. Calculate the rate of exergy destruction and the rate of exergy transfer by heat, each in kW as a percentage of electric power supplied to the motor. Take T_0 = 293 K.

[*Ans.* 0.481 and 0.044]

11.54. Steam enters a steam turbine at 8.5 MPa, 500°C with a mass flow rate of 1200 kg/s. The steam exits from the turbine as saturated vapour at 10 kPa. Heat transfer from the turbine to its surroundings takes place at a rate of 30 MW at an average surface temperature of 180°C. Neglect changes in KE and PE. Work out the following:

(a) For the control volume enclosing the turbine, determine the power developed and the rate of exergy destruction, each in MW.

(b) If the turbine is located in a facility where the ambient temperature is 27°C, calculate the rate of exergy destruction for an enlarged control volume that includes the turbine and its immediate surroundings so that the heat transfer takes place from the control volume at the ambient temperature. Discuss why the exergy destruction values of parts (a) and (b) differ. Take p_0 = 100 kPa, T_0 = 300 K.

11.55. A reciprocating compressor filled with water jacket and operating at steady state takes in air with a volumetric flow rate of 0.2 m^3/s at 20°C, 100 kPa and discharges air at 165°C, 500 kPa. The cooling water enters the water jacket at 15°C, 100 kPa with a mass flow rate of 0.2 kg/s and comes out at 25°C and essentially the same pressure. The heat transfer from the water jacket to surroundings is negligible and changes in KE and PE are also negligible. For the water-jacketed compressor, prepare a full exergy accounting of the power input. Take T_0 = 288 K, p_0 = 1 atm.

11.56. A counter-flow heat exchanger operating at steady state has water entering as saturated steam at 1 bar with mass flow rate of 3 kg/s and exiting as saturated liquid at 1 bar. Air enters in a separate stream at 300K, 1 bar and comes out at 353K. Assume no pressure loss. Neglect heat transfer from heat exchanger to surroundings and changes in KE and PE are negligible. Calculate (a) the change in the flow exergy rate of each stream in, kW and (b) the rate of exergy destruction in the heat exchanges in kW. Take T_0 = 288 K, p_0 = 1 bar.

11.57. Air at 7 bar and 1000°C enters a turbine and expands to 1.5 bar and 665°C with a mass flow rate of 10 kg/s. The turbine operates at steady state with negligible heat transfer with its surrounding. For ideal gas, γ = 1.4 and neglecting KE and PE, calculate (a) the isentropic turbine efficiency, (b) the exergetic turbine efficiency.

Software Based Questions

11.58. Develop a software for the exergy analysis of all type of heat exchangers.

11.59. Develop a software for the exergy analysis of internal combustion engines.

11.60. Develop a software for the exergy analysis of gas turbine plant and steam turbine power plant.

11.61. Develop a software for the exergy analysis for the following cases : (a) reciprocating compressor, (b) centrifugal compressor, (c) axial flow compressor.

11.62. Develop a software for exergy analysis of refrigeration and air conditioning systems.

11.63. A geothermal source offers a stream of liquid water at temperature T ($\geq T_0$) and pressure p. Using the incompressible liquid model, develop a software to plot $(e_x)_f/c\ T_0$ where $(e_x)_f$ is the specific flow exergy and c is the specific heat, versus T/T_0 for p/p_0 = 1.0, 1.5 and 2.0. Take T_0 = 15°C, p_0 = 1 atm.

11.64. Steam at 60 bar and 450°C enters a valve operating at steady state and undergoes a throttling process. Develop a software to determine the following:

(a) Calculate the exit temperature, in °C and the exergy destruction rate, in kJ/kg of steam flowing, for an exit pressure of 35 bar.

(b) Plot the exit temperature, in °C, and the exergy destruction rate, in kJ/kg of steam flowing, each versus exit pressure ranging from 35 bar to 70 bar. Take T_0 = 15°C and p_0 = 1 bar.

11.65. Steam at 140 bar and 540°C with a mass flow rate of 500 kg/s enters a insulated turbine operating at steady state and exhaust at 0.05 bar. Develop a software to determine the following:

(a) Plot the temperature of exhaust steam in °C, the power developed by the turbine, in MW, and the rate of exergy destruction within the turbine, in MW, each versus the isentropic turbine efficiency varying from 0 to 100%. Neglect changes in KE and PE. Take T_0 = 20°C and p_0 = 1 bar.

11.66. Steam at 120 bar and 500°C enters a well-insulated turbine operating at steady state and exit at saturated steam at pressure p. Develop a software to find the following:

(a) For p = 4 bar, determine the exergy destruction rate, in kJ/kg of steam expanding through the turbine, and the turbine exergetic and isentropic efficiencies.

(b) Plot the exergy destruction rate in kJ/kg of steam flowing, and the exergetic efficiency and isentropic efficiency, each versus p ranging from 1 bar to 4 bar. Take T_0 = 288 K, p_0 = 1 atm.

11.67. Saturated steam at 30 bar enters an insulated turbine operating at steady state. A two-phase steam exits at 0.05 bar. Develop a software to plot each of the following versus the steam quality at the turbine exit ranging from 75 to 100%

(a) the power developed and the rate of exergy destruction each in kJ/kg,

(b) the isentropic turbine efficiency, and

(c) the exergetic turbine efficiency, take T_0 = 15°C, p_0 = 1 arm

Design and Open Ended Questions

11.68. A heat pump is used to heat a 300 m^2 dwelling in a locality. Using manufacturer's data for heat pump operation, calculate the exergetic efficiency.

11.69. In order to actuate pneumatic tools, a factory uses a 140 kW screw compressor that takes in 25 m^3/min of ambient air and delivers compressed air at 10 bar. One consultant has suggested that using compressed air is more expensive than the direct use of electricity for operating such tools. Prepare a report about your opinion in this regard on the basis of using exergy costing with electric rates in your locals.

11.70. Presently, pinch technology is a popular methodology for optimizing the design of heat exchanger networks in a complex thermal systems. Pinch analysis uses graphical approach of second law reasoning. Prepare a report discussing the use of pinch analysis within thermoeconomics.

11.71. Prepare a report on vortex tube based on manufacturer's data which have no moving mechanical part and converts an inlet stream of compressed air at an intermediate temperature in to two exiting steam one cold and other hot.

11.72. Take a existing 500 MW steam power plant and combined cycle power plant. Collecting the relevant input data, prepare a report discussing the exergy analysis of these plants and suggests the methods to reduce irreversibilities.

Multiple-Choice Answer Type Questions

11.73. Availability (exergy) of a system at any given state is

(a) a property of the system, (b) the maximum work obtainable as the system goes to dead state, (c) the total energy of the system, (d) the maximum useful work obtainable as the system goes to dead state.

11.74. Availability of a system at any given state 1 and dead state designated by subscript p_0 is given by

(a) $(E_1 - E_2) + p_0 (V_1 - V_2) - T_0 (S_1 - S_2)$

(b) $(E_1 - E_0) + p_0 (V_1 - V_0) - T_0 (S_1 - S_0)$

(c) $(h_1 - h_0) + g (z_1 - z_0) - T_0 (s_1 - s_0) + \frac{C_1^2}{2}$

(d) $(h_1 - h_2) + g (z_1 - z_2) - T_0 (s_1 - s_0) + \frac{C_1^2 - C_2^2}{2}$

11.75. The availability of a steady flow process for any state 1 and dead state represented by suffix 0 is given by

(a) $(E_1 - E_0) + p_0 (V_1 - V_0) - T_0 (S_1 - S_2)$

(b) $(h_1 - h_2) + g (z_1 - z_2) - T_0 (s_1 - s_2) + \frac{C_1^2 - C_2^2}{2}$

(c) $(h_1 - h_0) + g (z_1 - z_0) - T_0 (s_1 - s_0) + \frac{C_1^2 - C_0^2}{2}$

(d) $(E_1 - E_2) + p_0 (V_1 - V_2) - T_0 (S_1 - S_2)$

11.76. Irreversibility is given by

(a) $W_{act} - W_{rev}$ (b) $W_{rev} - W_{act}$ (c) W_{rev} (d) W_{act}

11.77. Exergetic (second law) efficiency of a thermal power plat with relation to first law efficiency

(a) more (b) less (c) equal (d) none of the above

11.78. The closed system exergy rate balance $dE/d\tau$ is given by

(a) $\sum_i \left(1-\frac{T_0}{T_j}\right)\dot{Q}_j - \left(\dot{W} + p_0 \frac{dV}{d\tau}\right) - \dot{E}_{x,\text{loss}}$ (b) $\sum_i \left(1-\frac{T_0}{T_j}\right)\dot{Q}_j + \left(\dot{W} - p_0 \frac{dV}{d\tau}\right) - \dot{E}_{x,\text{loss}}$

(c) $\sum_i \left(1+\frac{T_0}{T_j}\right)\dot{Q}_j + \left(\dot{W} + p_0 \frac{dV}{d\tau}\right) - \dot{E}_{x,\text{loss}}$ (d) $\sum_i \left(1-\frac{T_0}{T_j}\right)\dot{Q}_j + \left(\dot{W} + p_0 \frac{dV}{d\tau}\right) - \dot{E}_{x,\text{loss}}$

Answers to Multiple-Choice Questions

11.73. (a) 11.74. (b) 11.75. (c) 11.76. (b) 11.77. (b) 11.78. (a)

CHAPTER

12

Non-reactive Gas Mixtures

OVERVIEW

A mixture is defined as a collection of molecules, ions, electrons and so on. Each group of particles, distinguishable from the others by virtue of its chemical structure, is called a constituent (or species) of the mixture. In engineering problems, several thermodynamic systems are used which are a mixture of pure substances. *If a mixture remains homogeneous in composition throughout a process, it is considered a pure substance.* The following are a few examples of various mixtures which are encountered in engineering problems:

(i) *Dry air* It is a mixture of several gases like oxygen, nitrogen, carbon dioxide, etc. The approximate composition of dry air on the basis of mole fraction is N_2, 78.08%; O_2, 20.95%; Ar, 0.93%, CO_2, 0.03%, neon, helium, methane and others, 0.01%.

(ii) *Ordinary air* It is a mixture of dry air and water vapours.

(iii) *Products of combustion in a boiler or in combustion chamber of IC engines, etc.* It consists of a mixture of various gases like CO_2, SO_2, N_2, water vapour, etc.

Thus, the mixture of pure substances may be broadly classified in two groups:

(a) *mixture involving only the gases considered as ideal gases*, and

(b) *mixture involving ideal gases and vapours.*

The thermodynamic behaviour of a mixture of gases and vapours depends upon the individual properties of its constituent gases. Since substances in the gaseous phases are miscible with each other, there is no limit to the number of different gaseous mixtures possible; therefore a wide variation is possible in the properties of gaseous mixture. Thus, it is not practical to tabulate the properties of all the mixture except for those cases where there is constant chemical composition. Hence, it is more convenient to deduce the thermodynamic properties of mixture of gases and vapours. Furthermore, *individual constituents of gaseous mixture often react chemically with each other and these fractions introduce another factor that exerts a strong influence on the properties of gases. In this chapter, we will deal only with the mixture of inert (perfect) gases in which change in phase of constituents may take place.* The derived expressions apply, in general to gas mixtures over a wide range of temperature and pressure. Although, this study centres around perfect gases, mixture containing non-perfect gases often show only a negligible deviation from perfect gas behaviour.

12.1 SOME BASIC DEFINITIONS FOR GAS MIXTURES

In thermodynamics, there are in general two ways for measuring the mass within the system. One of these is to use units of mass such as **kg, pounds,** etc. The other basis for mass is the **mole unit**, i.e., **kg mole or gram moles** or **pound moles.**

When an analysis of a gas mixture is made on the basis of mass or weight, it is called ***gravimetric analysis***. For non-reacting gas mixture within a closed system, it is apparent that total mass of the mixture, m is the sum of the masses of each of the k components. Thus

$$m = m_a + m_b + m_c + \ldots = m_k = \sum_{i=a}^{k} m_i \qquad \ldots(12.1)$$

The ratio $(\dot{m}_i/m) = m_i / \Sigma m_i = y_i$ is defined as the mass fraction of the component of the mixture. Thus, the **mass fraction** of *a, b, c*, etc., component of gas is given by

$$y_a = \frac{m_a}{m},\quad y_b = \frac{m_b}{m},\quad y_c = \frac{m_c}{m},\quad y_k = \frac{m_k}{m}$$

$$\therefore \qquad y_a + y_b + y_c + \ldots + y_k = 1 \qquad \ldots(12.2)$$

This shows that the sum of mass fractions for all the components present in a mixture is unity.

If an analysis of gas mixture is based on the number of moles of each component present, the analysis is termed a ***molar analysis***. The total number of moles, n for a mixture is given by

$$n = n_a + n_b + n_c + \ldots n_k = \sum_{i=a}^{k} n_i$$

The **mole fraction** of any component, x_t is defined by the ratio $n_i / n = n_i / \Sigma n_i$. Thus,

$$x_a = \frac{n_c}{n},\quad x_b = \frac{n_b}{n},\quad x_c = \frac{n_c}{n},\quad x_k = \frac{n_k}{n}$$

$$\therefore \qquad x_a + x_b + x_c + \ldots + x_k = 1 \qquad \ldots(12.3)$$

From the above result, we conclude that the sum of the mole fractions for all the components present in the mixture is unity.

12.2 P-V-T RELATIONSHIP FOR MIXTURES OF IDEAL GASES

12.2.1 Dalton's Law of Additive Pressure

The PVT relationship for a mixture of gases is usually based on two rules. The first of these is known as ***Dalton's law of additive pressure***. *According to Dalton's law, the total pressure* *of a mixture of perfect gases is the sum of the partial pressure of the constituents, each measured alone at the temperature and volume of the mixture.* The partial pressure of a gas in mixture is the pressure it would exert if it alone occupied the whole volume of the mixture at the same temperature. Thus, Dalton's law can be written as

$$p = p_a + p_b + p_c + \ldots p_k = \sum_{i=a}^{k} p_i \Big|_{V,T} \qquad \ldots(12.4)$$

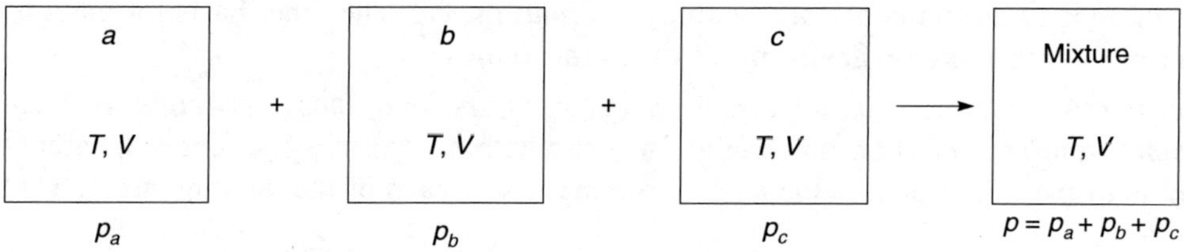

Fig. 12.1 Representation of Dalton's law

where p_i is the component pressure of the i component. A physical representation of the additive pressure rule is shown in Fig. 12.1 for the cases of three gases a, b and c.

$$p_aV = n_aR_0T,\ p_bV = n_b R_0 T,\ p_cV = n_c R_0 T, \text{ etc.}$$

Adding the above equations, we get

$$(p_a + p_b + p_c + \ldots)\ V = (n_a + n_0 + n_c + \ldots)\ R_0\ T$$

or

$$pV = nR_0\ T \qquad \ldots(12.5)$$

Equation (12.5) has the same form as the equation of state of a single-component perfect gas. *Thus, the mixture of perfect gases acts like a perfect gas.* When each individual equation of state is divided by the mixture's equation of state, the following equations for partial pressure are obtained

$$p_a = \frac{n_a}{n}p = x_a p, \qquad p_b = \frac{n_b}{n}p = x_b p, \qquad p_c = \frac{n_c}{n}p = x_c p, \text{etc.}$$

Therefore, for i^{th} component, the partial pressure p_i is

$$p_i = x_i p = \frac{n_i}{n}p \qquad \ldots(12.6)$$

where x_t is the mole fraction (or molar concentration) of constituent i

From Eq. (12.6), it is also obvious that the ratio of the partial pressure of any constituent to the total pressure is equal to the mole fraction of that consistent. As each constituent occupies the same volume and has the same temperature as the mixture, so we have

$$\frac{p_a}{n_a} = \frac{p_b}{n_b} = \frac{p_c}{n_c} = \frac{p_i}{n_i} = \frac{p}{n} \qquad \ldots(12.7)$$

12.2.2 Amagat-Leduc Law

This law states that the total volume of a mixture of gases is the sum of the partial volumes of constituents gases each measured alone at the pressure and temperature of mixture. This may be expressed by the relation

$$V = V_a + V_b + V_c + ... + V_k = \sum_{i=a}^{k} V_i \Big|_{p,T} \quad ...(12.8)$$

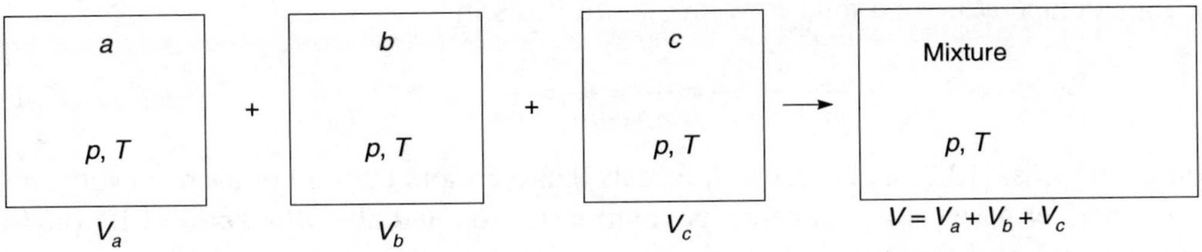

Fig. 12.2 Representation of Amagat's law of additive volumes

where V_i is the volume of ith component measured at the temperature and pressure of mixture. A physical interpretation of the Amagat–Leduc Law, can be obtained in the following way.

Let us consider a mixture of three ideal gases in a system of volume (V) with a total pressure (p) and a temperature, T. Theoretically, the gases could be separated so that one species occupies a certain part of the volume, the second species an another part and the third species the remaining volume. The temperature and pressure of each separate gas would still be identical. The actual volume filled by a given component gas would be its component volume and of course, the sum of the component volumes is the total volume of the closed system. A schematic representation of Amagat–Leduc Law of **additive volumes** is shown in Fig. 12.2 for the case of three gases.

When the mixture is analysed from the standpoint of Amagat's law, the perfect gas leads to

$$pV_a = n_a R_0 T, \quad pV_b = n_b R_0 T, \quad pV_c = n_c R_0 T, \quad \text{etc.}$$

Adding the above equations, we get

$$p(V_a + V_b + V_c + ...) = (n_a + n_b + n_c + ...) R_0 T$$

or

$$pV = nR_0 T \quad ...(12.9)$$

This equation also suggests that the mixture of gases behaves as an ideal gas. By dividing the perfect gas equation for each constituent by the equation for the entire mixture, we get

$$\frac{V_a}{V} = \frac{n_a}{n}$$

or

$$V_a = x_a V, \quad V_b = x_b V$$

and

$$V_c = x_c V$$

In general, for the ith component, the partial volume, V_i is expressed as

$$V_i = x_i V$$

or
$$x_i = \frac{V_i}{V} \quad ...(12.10)$$

This equation suggests that the ratio of partial volume of any constituent to the total is equal to the mole fraction of that constituent. As the partial volumes of the constituents are at the same temperature and total pressure, so it results in

$$\frac{V_a}{n_a} = \frac{V_b}{n_b} = \frac{V_c}{n_c} = \frac{V_i}{n_i} = \frac{V}{n} \quad ...(12.11)$$

Comparing Eqs. (12.7) and (12.11), it reveals that the mole fraction of each constituent in a mixture perfect gases is the same as its volume fraction and also, **the ratio of its partial pressure to the total pressure**. Thus,

$$x_t = \frac{n_i}{n} = \frac{V_i}{V} = \frac{p_i}{p} \quad ...(12.12)$$

The importance of this equation is understood when it is pointed out that many analysis of gas mixture are made experimentally on a volumetric basis. However, the evaluation of properties of mixtures is more often made on the basis of a gravimetric or mass basis. Thus, it is essential to be able to convert them from volumetric analysis to mass analysis. The relationship between mole fractions and volume fractions in an ideal gas mixtures provides the link of making these observations.

12.3 PROPERTIES OF MIXTURE OF IDEAL GASES

Thermodynamic properties of the individual gases and of the mixture may be obtained by the application of **Gibbs–Dalton‘s law** which is simply a generalisation of **Dalton’s law of additive pressure**. *According to the Gibbs–Dalton’s law, an extensive property of a mixture of perfect gases is the sum of the contributions of the individual constituents properties of mixture may be expressed on either a mass or a mole basis and measured alone at the temperature and volume of the mixture.*

Before dealing with extensive property, it is worthwhile to discuss the relation for characteristic gas constant, R and molecular weight, M of the mixture.

12.3.1 Gas Constant

For the mixture of gases, the perfect gas law gives

$$pV = mRT \quad ...(12.13)$$

and for each constituent at the same volume and temperature.

$$p_a V = m_a\, R_a\, T,\; p_b V = m_b\, R_0\, T,\; P_c V = m_c\, R_c\, T,$$

But $$p = p_a + p_b + p_c + ...$$

$$pV = (m_a R_a + m_b R_b + m_c R_c + ...) T$$

With the help of Eq. (12.13), the above equation gives

$$R = \frac{m_a}{m} R_a + \frac{m_b}{m} R_b + \frac{m_c}{m} R_b + ... = \sum \frac{m_i R_i}{m}$$

or $$R = y_a R_a + y_a R_b + ... = \sum y_i R_i \quad ...(12.14)$$

where y is the mass fraction, which is the ratio of mass of a constituent to the total mass of the mixture.

12.3.2 Molecular Weight

In order to determine the moelcular weight of the mixture, the mass of the mixture is divided by the total number of moles present in the mixture. Thus,

$$M = \frac{\sum_i m_i}{\sum_i n_i} = \frac{\sum_i m_i}{\sum_i (m_i / M_i)} = \frac{\sum_i (n M_i)}{\sum_i m_i} \quad ...(12.15)$$

The above equation suggests the mass of a constituent is equal to the number of moles of that constituent multiplied by its molecular weight. Thus, the mass of the individual components of the mixture are

$$m_a = n_a M_a, M_b = n_b M_b, m_c = n_c M_c$$

$$\therefore \quad m_a + m_b + m_c + ... = m = n_a M_a + n_a M_a + n_c M_c + ...$$

But the total mass of the mixture, m is given by

$$m = m_a + m_b + m_c + ... = nM$$

By combining the above two equations, we get

$$M = \frac{n_a}{n} M_a + \frac{n_b}{n} M_b + \frac{n_c}{n} M_c + ...$$

or $$M = x_a M_a + x_a M_b + ... = \sum x_i M_i \quad ...(12.16)$$

The average molecular weight of the gas mixture is the sum over all components of the mole fraction molecular weight. By putting the relation $R_i = R_0/M_i$ in Eq. (12.14), we get

$$\frac{R_0}{M} = y_a \frac{R_0}{m_a} + y_a \frac{R_0}{M_b} + y_c \frac{R_0}{M_c} + ...$$

This results, $$M = \frac{1}{(y_a / M_a) + (y_b / M_b) + (y_c / M_c) + ...} = \frac{1}{\sum_i (y_i / M_i)} \quad ...(12.17)$$

But $$R_0 = MR,$$

Hence, $$M = \frac{R_0}{R} \qquad ...(12.18)$$

Thus, the *expression of molecular weight of the mixture of gases is the same as ideal gas.*

12.3.3 Internal Energy

On the basis of the Gibbs–Dalton's law, the total internal energy of a mixture is given by

$$U = U_a + U_b + U_c + ... U_k = \sum_{i=a}^{k} U_i \Big|_{T,V} \qquad ...(12.19)$$

The total internal energy of each component may be expressed as mass basis or mole basis

$$mu = m_a u_a + m_b u_b + m_c u_c + ... m_k u_k \qquad ...(12.20)$$

and $$nu = n_a u_a + n_b u_b + n_c u_c + ... n_k u_k \qquad ...(12.21)$$

In Eq. (12.20), u is the specific internal energy whereas in Eq. (12.21), it is the molar specific internal energy.

If Eq. (12.21) is divided by n, the molar specific internal energy can also be found by

$$u = x_a u_a + x_b u_b + x_c u_c + ... x_k u_k = \sum_{i=a}^{k} x_i u_i; \text{ kJ/mole} \qquad ...(12.22)$$

where u_i is the molar specific internal energy of the ith component when found at the temperature and volume of mixture.

The internal energy per unit mass of mixture may also be claculated from Eq. (12.20).

$$u = y_a u_a + y_a u_a + y_c u_c + ... = \sum_i y_i u_i; \text{ kJ/kg} \qquad ...(12.23)$$

12.3.4 Enthalpy

The enthalpy of a mixture of ideal gaes is simply the sum of the enthalpies of the individual chemical species. Therefore,

$$H = H_a + H_b + H_c + ... + H_k = \sum_i^{k} H_i \Big|_{T,V} \qquad ...(12.24)$$

or $$mh = m_a h_a + m_b h_b + m_c h_c + ... \qquad ...(12.25)$$

or $$h = y_a h_a + y_b h_b + + ... = \sum_i y_i h_i ; \text{ kJ/kg} \qquad ...(12.26)$$

Also, $$nh = n_a h_a + n_b h_b + n_c h_c + ... \sum_i n_i h_i \qquad ...(12.27)$$

or $$h = x_a h_a + x_b h_b + x_c h_c + ... = \sum_i x_i h_i \qquad ...(12.28)$$

12.3.5 Specific Heats

An equation for constant volume specific heat and constant pressure specific heat may be found on either mass or a molar basis.

From Eq. (12.20), we have $mu = m_a\, u_a + m_b\, u_b + m_c\, u_c + ...$

Since $$u = c_v\, dT$$

Hence, $$mc_v\, dT = \{m_a\, c_{va} + m_b\, c_{vb} + m_a\, c_{vc} + ...\}\, dT$$

$$c_v = y_a\, c_{va} + y_b\, c_{vb} + y_c\, c_{vc} + ... = \sum_i y_i\, c_{vi};\ \text{kJ/kg} \quad ...(12.29)$$

A similar development for the molar specific heat at a mixture leads to

$$mc_v = n_a\, c_{va} + n_b\, c_{vb} + c_{vc} ...$$

or $$c_v = x_a\, c_{va} + x_b\, c_{vb} + x_c\, c_{vc} ...$$

$$= \sum_i x_i\, c_{vi};\ \text{kJ/kg mole} \quad ...(12.30)$$

The specific heat at constant pressure c_p may also be written for the mixture on similar basis. We have from Eq. (12.25).

$$mh = m_a\, h_a + m_a\, h_b + m_c\, h_c + ...$$

But $$h = c_p\, dT$$

Hence, $$mc_p\, dT = (m_a\, c_{pa} + m_b\, c_{pb} + m_c\, c_{pc} + ...)\, dT$$

or $$c_p = y_a\, c_{pa} + y_b\, c_{pb} + - ... = \sum_i y_i\, c_{\pi}\ ;\ \text{kJ/kg} \quad ...(12.31)$$

Also $$n\, c_p = n_a\, c_{pa} + n_b\, c_{pb} + n_c\, c_{pc} + ...$$

or $$c_p = x_a\, c_{pa} + x_b\, c_{pb} + x_c\, c_{pc} + ... = \sum_i x_i\, c_{pi}\ ;\ \text{kJ/kg mole} \quad ...(12.32)$$

12.3.6 Entropy

The entropy of a mixture of ideal gases can be determined on the basis of the Gibbs–Dalton's law. The entropy of a mixture of ideal gase is the sum of the entropies of the constituent gases when each gas is at the temperature and volume of the mixture

$$S = S_a + S_b + ...\, S_k = \sum_i S_i \Big|_{T,V} \quad ...(12.33)$$

or $$ms = m_a\, s_a + m_b\, s_b + m_c\, s_c + ...$$

or $$s = y_a\, s_a + y_b\, s_b + y_c\, s_c + ... = \sum_i y_i\, s_i\ \text{kJ/kg} \quad ...(12.34)$$

Also $$ns = n_a\, s_a + n_b\, s_b + n_c\, s_c + ...$$

or $$s = x_a\, s_a + x_b\, s_b + x_c\, s_c + ...\ \sum_i x_i\, x_i;\ \text{kJ/kg mole K} \quad ...(12.35)$$

The entropy change of a perfect gas mixture, expressed is terms of pressure and temperature is

$$s_2 - s_1 = \sum_i y_i (s_2 - s_1) = \sum_i y_i \, c_{pi} \int_{T_1}^{T_2} \frac{dT}{T} - \sum_i y_i \, R_i \ln \frac{p_{t2}}{p_{t1}}$$

$$s_2 - s_1 = c_p \ln \frac{T_2}{T_1} - R \ln \frac{p_2}{p_1} \qquad ...(12.36)$$

Also, $$s_2 - s_2 = c_v \ln \frac{T_2}{T_1} + R \ln \frac{v_2}{v_1} \qquad ...(12.37)$$

The change of entropy of perfect gas mixture may also be written on the mole basis

$$s_2 - s_1 = \sum_i x_i (s_2 - s_1) = \sum_i x_i \, c_{pi} \int_{T_1}^{T_2} \frac{dT}{T} - \sum_i x_i \, R_i \ln \frac{p_{t2}}{p_{t1}} \qquad ...(12.38)$$

$$= c_p \ln \frac{T_2}{T_1} - R \ln \frac{p_2}{p_1} \qquad ...(12.39)$$

12.4 ENTROPY CHANGE DUE TO ADIABATIC MIXING

Whenever there is a mixing of perfect gases, entropy changes as a result of the increase in the disorder of the system. Suppose the initial temperature of all the constituents are the same and the mixing process is adiabatic, the temperature does not change but the entropy increases. Let us assume that the initial pressure and temperature of the individual gases are p and T, equal to the total pressure and temperature of the mixture. In the mixing process, each gas may be considered to undergo a free expansion process from its initial pressure in the mixing, with no work interaction. Thus, mixing is an irreversible process. But from Second law.

$$dS = \sum_i (y_i \, ds_i) \geq 0 \qquad ...(12.40)$$

As the gases are perfect and the temperature remains constant, both internal energy and enthalpy are unchanged by the mixing process. Thus,

$$\Delta S = -\left(m_a \, R_a \ln \frac{p_a}{p} + m_b \, R_b \ln \frac{p_b}{p} + ... \right)$$

$$= -\sum_i m_i \, R_i \ln \frac{p_{i,2}}{p_{i,1}} \qquad ...(12.41)$$

On per unit mass basis, $\Delta s = -\sum_i y_i \, R_i \ln \frac{p_{i,2}}{p_{i,1}}$ ...(12.42)

In Eq. (12.42), the subscripts 1 and 2 refer to initial and final states of each component in the mixture. *Since each partial pressure is less than the total process, the change on entropy calculated from Eq. (12.42) is always a positive number.*

Example 12.1 An ideal-gas mixture consists of 10 per cent hydrogen, 48 per cent oxygen and 42 per cent carbon monoxide by weight. Calculate the volumetric analysis in per cent and the apparent molar mass.

Solution

1	2	3	4	5
Component	kg per 100 kg of mixture	Molecular weight (M_i) or (Molar mass), kg/mole	Moles per 100 kg of mixture	Moles per 100 moles of mixture
	(m_i)	(M_i)	$n_i = \frac{m_i}{M_i}$	$\frac{n_i}{\Sigma n_i} \times 100$
H_2	10	2	5	62.5
O_2	48	32	1.5	18.75
CO	42	28	1.5	18.75
		Total	8.00	100.00

In column 5, there are moles per 100 moles of mixture, or the volume per cent of each gas measured at the same pressure and temperature. It is hydrogen, which was present in the smallest amount in the gravimetric analysis, that has the largest percentage on a volumetric basis. There are 8 moles of mixture per 100 kg of mixture, consequently, the molar mass is 100/8 = 12.5 for the mixture.

Example 12.2 A gaseous fuel is composed of 20 per cent CH_4, 40 per cent C_2H_6 and 40 per cent C_3H_8, where all percentage are by volume. Calculate the gravimetric analysis of the fuel, the apparent molecular weight of the mixture and apparent gas constant for the mixture.

Solution It will be useful to select 100 moles as a basis for subsequent calculation.

1	2	3	4	5
Component	Moles per 100 moles of mixture (n_i)	Molecular weight (Molar mass kg/mole) (M_i)	kg per 100 mole of mixture $m_i = M_i \times n_i$	kg per 100 kg of mixture $\frac{m_i}{\Sigma m_i} \times 100$
CH_4	20	16	320	9.76
C_2H_6	40	30	1200	36.59
C_3H_6	40	44	1760	53.95
			$\Sigma m_i = 3280$	100.00

$$R = \frac{R_0}{M} = \frac{8314}{32.8} = 253.47 \text{ J/kg K}$$

Example 12.3 A mixture of gases having 2 kg of He and 5 kg of N_2 at 30°C and 1 bar is compressed in a reversible adiabatic process to 6 bar. Find (a) the final partial pressure of the constituents, (b) the final temperature and (c) change in internal energy of the mixture during the process. Assume $(c_v)_{N_2} = 0.744$, $(c_v)_{He} = 3.157$, $(c_p)_{N_2} = 1.049$, $(c_p)_{N_2} = 5.269$ kJ/kg K.

Solution (a) We have, $$n = \frac{m}{M}$$

$$(x)_{N_2} = \frac{(n)_{N_2}}{n} = \frac{5/28}{(2/4)+(5/28)} = 0.2625$$

$$(x)_{He} = \frac{(n)_{He}}{n} = \frac{2/4}{(2/4)+(5/28)} = 0.7374$$

We know that partial pressure ratio is equal to their mole fraction, hence

$$(p)_{N_2} = p(x)_{N_2} = 6 \times 0.2625 = 1.575 \text{ bar}$$ **Ans.**

$$(p)_{He} = p(x)_{He} = 6 \times 0.7374 = 4.424 \text{ bar}$$ **Ans.**

(b) We have $$\frac{T_2}{T_1} = \left(\frac{p_2}{p_1}\right)^{(\gamma-1)/\gamma}$$

$$(y)_{N_2} = \frac{(m)_{N_2}}{m} = \frac{5}{7} \quad \text{and } (y)_{He} = \frac{(m)_{He}}{m} = \frac{2}{7}$$

$$\therefore \quad c_p = (y)_{N_2} \cdot (c_p)_{N_2} + (y)_{He} \cdot (c_p)_{He} = \frac{5}{7}\times 1.049 + \frac{2}{7}\times 5.269 = 2.254 \text{ kJ/kg K}$$

$$c_v = (y)_{N_2} \cdot (c_v)_{N_2} + (y)_{He}\,(c_v)_{He} = \frac{5}{7}\times 0.744 + \frac{2}{7}\times 3.157 = 1.433 \text{ kJ/kg K}$$

$$\therefore \quad \gamma = \frac{c_p}{c_v} = \frac{2.254}{1.433} = 1.572$$

$$\therefore \quad T_2 = (273 + 30)\,(6)^{0.572/1.572} = 581.54 \text{ K}$$ **Ans.**

(c) The change in internal energy is expressed as

$$U_2 - U_1 = mc_v\,(T_2 - T_1) = 7 \times 1.433 \times (581.54 - 303) = 2794 \text{ kJ}$$ **Ans.**

Example 12.4 A perfect gas mixture consists of 3 kg of N_2 and 5 kg of CO_2 at a pressure of 3 bar and a temperature of 20°C. Calculate (a) the mole fraction of each constituent, (b) the equivalent molecular weight of the mixture, (c) the equivlaent gas constant of the mixture, (d) the partial pressure and partial volume, and (e) the volume and the density of mixture.

Solution We have $$x_i = \frac{n_i}{n}$$

$$\therefore \quad x_{N_2} = \frac{3/28}{(3/28)+(5/44)} = 0.485$$ **Ans.**

Similarly, $$x_{CO_2} = \frac{5/44}{(3/28)+(5/44)} = 0.5147$$ **Ans.**

(b) We know that $$M = x_a M_a + x_b M_b$$

or $$M = 0.485 \times 28 + 0.5147 \times 44 = 36.25 \text{ kg/kg mole}$$ **Ans.**

(c) We have $$R = y_a R_a + y_b R_b \text{ where } y_a = \frac{m_a}{m} \text{ and } y_b = \frac{m_b}{m}$$

$\therefore$ $$R = \frac{3}{8} \times \frac{8314}{28} + \frac{5}{8} \times \frac{8314}{44} = 229.39 \text{ J/kg K}$$ **Ans.**

(d) We have $$p_a = x_a p \text{ and } p_b = x_b p$$

$\therefore$ $$p_{N_2} = 0.485 \times 3 = 1.455 \text{ bar} \quad \text{and } p_{CO_2} = 0.515 \times 3 = 1.545 \text{ bar}$$ **Ans.**

$$V_{N_2} = \frac{m_{N_2} R_{N_2}}{p} T = \frac{3 \times (8314/28) \times 293}{3 \times 10^5} = 0.87 \text{ m}^3$$ **Ans.**

$$V_{CO_2} = \frac{m_{CO_2} R_{CO_2} T}{p} = \frac{5 \times (8314/44) \times 293}{3 \times 10^5} = 0.922 \text{ m}^3$$ **Ans.**

(e) We have $$V_i = x_i V \quad \text{or } V_{N_2} = x_{N_2}$$

or $$V = \frac{0.87}{0.485} = 1.793 \text{ m}^3$$

The density of mixture, $\rho = \rho_{N_2} + \rho_{CO_2} = \dfrac{3}{1.793} + \dfrac{5}{1.793} = 4.46 \text{ kg/m}^3$ **Ans.**

Example 12.5 A perfect gas mixture consists of 3 kg of N_2 and 5 kg CO_2 at pressure of 3 bar and a temperature of 20°C. Calculate the constant volume and constant change in internal energy, enthalpy and entropy of the mixture. Assume the specific heats remaining constant and the perfect gas law applies.

Also calculate the foregoing changes at constant pressure heating. Take $(c_v)_{N_2} = 0.74$, $(c_v)_{CO_2} = 0.66$, $(c_p)_{N_2} = 1.037$, $(c_p)_{CO_2} = 0.848$ kJ/kg K.

Solution For the constant volume process,

$$c_v = \sum_i y_i c_{vi} = y_{N_2} (c_v)_{N_2} + y_{CO_2} (c_v)_{CO_2}$$

or $$c_v = 38 \times 0.74 + \frac{5}{8} \times 0.66 = 0.69 \text{ kJ/kg K}$$

$$c_p = \sum_i y_i c_{pi} = y_{N_2} (c_p)_{N_2} + y_{CO_2} (c_p)_{CO_2}$$

or $$c_p = \frac{3}{8} \times 1.037 + \frac{5}{8} \times 0.848 = 0.9188 \text{ kJ/kg K}$$ **Ans.**

The change in internal energy

$$\Delta u = \int_{T_1}^{T_2} c_v \, dT = 0.69\ (40 - 20) = 13.8 \text{ kJ/kg of mixture}$$

or $$\Delta U = 8 \times 13.8 = 110.4 \text{ kJ}$$ **Ans.**

The change in entropy

$$\Delta h = \int_{T_1}^{T_2} c_p \, dT = 0.9188\,(40 - 20) = 18.376 \text{ kJ/kg of mixture}$$

or $$\Delta H = 8 \times 18.376 = 147 \text{ kJ}$$ **Ans.**

The change in entropy of the mixture, in terms of changes of temperature and volume is given by

$$\Delta s = c_v \ln \frac{T_2}{T_1} + R \ln \frac{v_2}{v_1}$$

But the volume is constant, hence, $\Delta s = 0.69 \ln \dfrac{313}{293} = 0.04556$ kJ/kg K

or $$\Delta s = 8 \times 0.04556 = 0.36448 \text{ kJ/K}$$ **Ans.**

For the constant pressure process, the internal energy and enthalpy values are the same since these are the function of temperature. The change of entropy is

$$\Delta s = c_p \ln \frac{T_2}{T_1} - R \ln \frac{p_2}{p_1}$$

For constant pressure process, $\Delta s = c_p \ln \dfrac{T_2}{T_1} = 0.9188 \ln \dfrac{313}{293} = 0.0606$ kJ/kg K

$$\Delta S = 0.0606 \times 8 = 0.4848 \text{ kg/K}$$ **Ans.**

Example 12.6 The following volumetric composition relate to a mixture of gases:

$$N_2 = 81\%,\ CO_2 = 11\%\ ,\ O_2 = 6\%,\ CO = 2\%$$

Determine (a) the gravimetric composition, (b) Molecular weight, and (c) Universal gas constant R for the mixture.

Solution (a) We know that volume fraction = Mole fraction x

And $M = \sum x_i M_i$ where M_i is the molecular weight of the constituent.

$$\text{Percentage of mass} = m\% = \frac{m_i}{m} \times 100$$

Using these relations, we can tabulate the results as given below.

Constituents	% by vol. or % by mole	kg mol per mole of mixture (x)	Molecular weight (M)	$M = x_i M_i$
N_2	81	0.81	28	22.62
CO_2	11	0.11	44	4.84
O_2	6	0.06	32	1.92
CO	2	0.02	28	0.56
				$\Sigma x_i M_i = 30$

(b) The molecular weight of the mixture, $M = \Sigma x_i m_i = 30$ **Ans.**

(c) $R = \dfrac{8.314}{30} = 0.277$ kJ/kg **Ans.**

12.5 MIXTURE OF PERFECT GASES AT DIFFERENT INITIAL PRESSURE AND TEMPERATURE

Consider three gases a, b and c at different initial pressure and temperature enclosed in a three containers as shown in Fig. 12.3. For simplicity, three gases have been considered, but the analysis will be applicable to any number of gases. When the connecting valves are opened so that the gases are allowed to diffuse into one another, the final temperature and pressure of the mixture will be at some intermediate value. Let us assume that no heat or work interaction takes place at the boundary of the system, hence the internal energy of the system (according to first law) will not be affected by the mixing and the internal energy of the mixture per unit mass will be given by

$$u = \frac{m_a}{m} u_a + \frac{m_b}{m} u_b + \frac{m_c}{m} u_c$$

But $$u = c_v T$$

Hence, $$c_v T = \frac{m_a}{m} c_{va} T_a + \frac{m_a}{m} c_{vb} T_b + \frac{m_b}{m} c_{vc} T_c = \sum \frac{m_i}{m} c_{vi} T_i$$

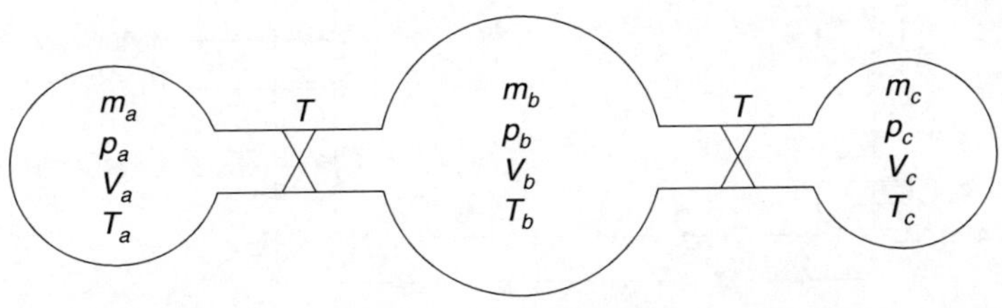

Fig. 12.3 Mixing of gases with different initial properties

If the gases are all monatomic, then c_v = (3/2) R and if they are all diatomic, then c_v = (5/2) R; and in both the cases, the above equation becomes

$$mRT = m_a R_a T_a + m_b R_b T_b + m_c R_c T_c \quad ...(12.43)$$

Thus, the temperature of mixture becomes

$$T = \frac{m_a R_a T_a + m_b R_b T_b + m_c R_c T_c}{mR} \quad ...(12.44)$$

But

$$mR = m_a R_a + m_b R_b + m_c R_c, \text{ hence}$$

$$T = \frac{m_a R_a T_a + m_b R_b T_b + m_c R_c T_c}{m_a R_a + m_b R_b + m_c R_c} \quad ...(12.45)$$

Using the perfect gas equation, the above equation may be expressed in terms of pressure, temperatures and volumes of the individual gases before mixing

$$T = \frac{p_a v_a + p_b V_b + p_c V_c}{(p_a v_a / T_a) + (p_b V_b / T_b) + (p_c V_c / T_c)} \quad ...(12.46)$$

Equation (12.43) may also be written (as in terms of mole numbers) as

$$nR_0 T = n_a R_0 T_a + n_b R_0 T_b + n_c R_0 T_c$$

Thus, the temperature of the mixture will be

$$T = \frac{n_a}{n} T_a + \frac{n_b}{n} T_b + \frac{n_c}{n} T_c = \Sigma x_i T \quad ...(12.47)$$

With the help of Eq. (12.44), the expression for pressure of mixture is given by

$$pV = p_a V_a + p_b V_b + p_c V_c$$

or

$$p = \frac{p_a V_a + p_b V_b + p_c V_c}{V} \quad ...(12.48)$$

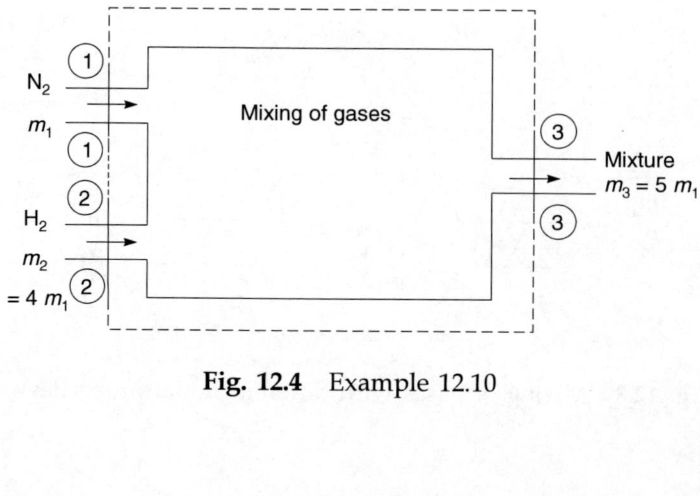

Fig. 12.4 Example 12.10

It is to be noted that the total volume of the mixture is the sum of the initial volume V_a, V_b and V_c. These latter volumes are not the partial volumes of components of the mixture; but the volume of the vessels which the component gases initially occupied.

Example 12.7 During a steady flow adiabatic process, N_2 and H_2 are mixed in the ratio of 4 kg of H_2 per kg of N_2. The hydrogen gas enters at 2 bar and 40°C and the nitrogen gas at 1.6 bar 250°C. After mixing, the pressure of the mixture is 1.8 bar. Calculate (a) the final temperature of the mixture, (b) the net entropy change per kg of mixture. Assume c_{pN_2} = 1.037 and c_{pH_2} = 14.344 kJ/kg K.

Solution We have, $$R = \frac{R_0}{M} = \frac{8314}{M}$$

$$\therefore \quad R_{N_2} = \frac{8314}{28} = 296.92 \text{ J/kg K}$$

and $$R_{H_2} = \frac{8314}{2} = 4157 \text{ J/kg K}$$

(a) From energy equation, $m_1 h_1 + m_2 h_2 = m_3 h_3$

and $$m_1 + m_2 = m_3 \quad \text{and } m_2 = 4\, m_1, \quad \text{hence } m_3 = 5\, m_1$$

Putting the value of m_2 and m_3 in the energy equation, we have

$$m_1 h_1 + 4\, m_1 h_2 = 5\, m_1 h_3 \text{ or } 0.2\, h_1 + 0.8\, h_2 = h_3$$

or $$0.2\, (c_p)_{N_2} T_{N_2} + 0.8\, (c_p)_{H_2} T_{H_2} = c_p T_3$$

But $$c_p = \frac{m_{N_2}}{m}(c_p)_{N_2} + \frac{m_{H_2}}{m}(c_p)_{H_2} = 0.2 \times 1.037 + 0.8 \times 14.344 = 11.682 \text{ kJ/kg K}$$

or $$0.2\, c_{pN_2} T_{N_2} + 0.8\, (c_p)_{H_2} T_{H_2} = \left[\frac{m_{N_2}}{m}(c_p)_{N_2} + \frac{m_{H_2}}{m}(c_p)_{H_2}\right]$$

or $0.2 \times 1.037 \times 523 + 0.8 \times 14.344 \times 313 = 11.682\, T_3$

or $$T_3 = 316.74 \text{ K}$$ **Ans.**

(b) The change in entropy of the mixture is

$$(\Delta s)_{N_2} + (\Delta s)_{H_2} = \left[\int_{T_1}^{T_2} c_p \frac{dT}{T} - R \ln \frac{p_2}{p_1}\right]_{N_2} + \left[\int_{T_2}^{T_1} c_p \frac{dT}{T} - R \ln \frac{p_2}{p_1}\right]_{H_2}$$

$$x_{N_2} = \frac{(n)_{N_2}}{n} = \frac{0.2/28}{(0.2/28) + (0.8/2)} = 0.01754$$

$$\therefore \quad (p_2)_{N_2} = p x_{N_2} = 1.8 \times 0.01754 = 0.03157$$

Now, $$x_{H_2} = 1 - 0.01754 = 0.9824$$

and $$(p_2)_{H_2} = 0.9824 \times 1.8 = 1.768 \text{ bar}$$

$\therefore$ $$\Delta s = 1.037 \ln \frac{316.74}{523} - \frac{8314}{28 \times 1000} \ln \frac{0.03157}{1.6}$$

$$+ 14.343 \ln \frac{316.74}{313} - \frac{8314}{2 \times 1000} \ln \frac{1.768}{2} = 0.1234 \text{ kJ/kg K} \quad \textbf{Ans.}$$

12.6 HETEROGENEOUS SYSTEM OF SEVERAL PHASES

Assume a heterogeneous system of volume (V) in which several homogeneous phases ($f = a, b, \ldots r$) exist in equilibrium. Assume, also that each phase consists of i (= 1, 2, ... k) components and that number of components in each phase is different from the others.

Within each phase, a change in internal energy which is accompanied by a change entropy, volume and composition is expressed as

$$dU_f = T_s\, dS_f - p_f\, dV_f = \sum_{i=1}^{i=k} (m_i\, dn_i)_f \qquad \text{...(12.49)}$$

Therefore, a change in internal energy of the entire system may be expressed as

$$\sum_{f=a}^{f=r} dU_f = \sum_{f=a}^{f=r} T_f\, dS_f - \sum_{f=a}^{f=r} p_f dV_f = \sum_{i=k}^{i=1} (\mu_i\, dn_i)_f \qquad \text{...(12.50)}$$

Also, a change in the internal energy of the entire system involves changes in internal energy of the constituent phases,

$$dU = dU_a + dU_b + \ldots + dU_r = \sum_{f=a}^{f=r} dU_f$$

Similarly, change in volume, entropy and chemical composition of the entire system may be expressed in terms of each of the constituent phases.

$$dV = dV_a + dV_b + \ldots + dV_r = \sum_{f=a}^{f=r} dV_f$$

$$dS = dS_a + dS_b + \ldots + dS_r = \sum_{f=a}^{f=r} dS_f$$

$$dn = dn_a + dn_b + \ldots + dn_r = \sum_{f=a}^{f=r} dn_f$$

In the cae of a closed system in equilibrium, the internal energy, volume, entropy and mass remain contant. Thus,

$$dU = dV = dS = dn = 0$$

In a closed system, when the system deviates slightly from the equilibrium state, involves change in the internal energy, entropy, volume and mass in each of the phases. Change in properties of one phase can be expressed in terms of the changes in properties of the remaining phases. Hence,

$$\left.\begin{aligned} dU_a &= -(dU_b + \ldots + dU_r) = -\sum_j dU_j \\ dV_a &= -\sum_j dV_j \text{ and } \quad dS_a = -\sum_j dS_j \end{aligned}\right\} \qquad \ldots(12.51)$$

where subscript j includes all phases except phase a. In the above expression, there are j independent variables and one dependent variable. Equation (12.51) may be written in the term of dependent and independent variable.

$$\left(T_a\, dS_a + \sum_j T_j\, S_j\right) - \left(p_a\, dV_a + \sum_j p_j\, dV_j\right) + \left[\sum_j (\mu_i\, dn_i) + \sum_j \sum_j (\mu_i\, dn_i)_j\right] = 0$$

Putting the value of Eq. (12.51) into the above equation, we have

$$\left(-T_a \sum_j dS_j + \sum_j T_j\, dS_j\right) - \left(-p_a \sum_j dV_j + \sum_j p_i\, dV_j\right) + \left[-\sum_i \sum_j \mu_i\, dn_i + \sum_j \sum_i \mu_i\, dn_j)_j\right] = 0$$

where subscript a refers to component i of phase a. Rearranging and combining the coefficients for the independent variables dS_J, dV_f and dn_i, yields

$$\sum_j (T_j - T_a)\, dS_j - \sum_j (p_j - p_a)\, dV_j + \sum_j \sum_i (\mu_{ij} - \mu_{ia})\, dn_{ij} = 0$$

Since dS_J, dV_j and dn_i are independent, their coefficients must vanish. Thus,

$$T_j = T_a,\ p_j = p_a \text{ and } \mu_{ij} = \mu_{ia} \qquad \ldots(12.52)$$

The above equations represent the condition that exists when the system is in thermal, mechanical and chemical equilibria. Obviously, a component that appears in several phases, must have the same temperature and pressure in all the phases. It is also obvious that the chemical potential of each of the component within one phase is equal to the chemical potential of the component in each of the remaining phases. For example, when the liquid phase of a non-component substance is in equilibrium with its vapour, the following equations hold good

$$T_j = T_g,\ p_j = p_g \text{ and } \mu_j = \mu_g$$

where subscripts f and g refer to liquid and vapour respectively.

12.7 SUMMARY OF FORMULAE

1. $y_a + y_b + y_c + \ldots y_k = 1$

2. $x_a + x_b + x_c + \ldots + x_k = 1$

3. $p = p_a + p_b + p_c + \ldots + p_k = \left.\sum_{i=a}^{k} p_i\right|_{V,T}$

4. $p_i = x_i\, p$

5. $\frac{p_a}{n_a} = \frac{p_b}{n_b} = \frac{p_c}{n_c} = ... \frac{p_i}{n_i} = \frac{p}{n}$

6. $V = V_a + V_b + V_c + ... + V_k = \sum_{i=a}^{k} V_i \Big|_{p,T}$

7. $v_i = x_i, v$

8. $\frac{v_a}{n_a} = \frac{v_b}{n_b} = \frac{v_c}{n_c} = ... \frac{v_i}{n_i} = \frac{v}{n}$

9. $x_i = \frac{n_i}{n} = \frac{v_i}{v} = \frac{p_i}{p}$

10. $R = y_a\, R_a + y_b\, R_b + y_c\, y_c = \Sigma\, y_i\, R_i$

11. $M = \frac{\Sigma(n\, m_i)}{\Sigma\, n_i}$

12. $M = \sum_i x_i\, M_i$

13. $M = \frac{1}{\Sigma\, y_i\, M_i}$

14. $U = \sum_{i=a}^{k} U_i \Big|_{V,T}$

15. $u = \sum_{i=a}^{k} x_i\, u_i$

16. $u = \sum_{i=a}^{k=0} y_i\, u_i$

17. $H = \sum_{i=a}^{k} H_i \Big|_{V,T}$

18. $h = \sum_{i=a}^{k} x_i\, h_i$

19. $h = \sum_{i=a}^{k} y_i\, h_i$

20. $S = \sum_{i\ \ a} \Big|_{v\ \ T}$

21. $s = \sum_{a}^{k} x_i\, s_i$

22. $s = \sum_{a}^{k} y_i\, s_i$

23. $s_2 - s_1 = x_i\, c_{pi} \int_{T_1}^{T_2} \frac{dT}{T} - \Sigma x_i\, T_i \ln \frac{p_{i,2}}{p_{i,1}}$

24. $s_2 - s_1 = y_i\, c_{pi} \int_{T_1}^{T_2} \frac{dT}{T} - \Sigma y_i\, R_i \ln \frac{p_{i,2}}{p_{i,1}}$

25. $T = \frac{m_a\, R_a\, T_a + m_b\, R_b\, T_b + m_c\, R_c\, T_c}{m_a\, R_a + m_b\, R_b + m_c\, R_c}$

26. $p = \dfrac{p_a V_a + p_b V_b + p_c V_c}{V}$

27. $dU = \sum_{j=a}^{j=r} dU_f$

28. $dV = \sum_{j=a}^{r} dV_f$

29. $dS = \sum_{j=a}^{r} dS_f$

30. $dn = \sum_{j=a}^{r} dn_f$

PROBLEMS

Things to Think About

12.1. "There exists a two-component mixture in which the mass and mole fractions are equal". Do you agree with this statement? Give reasoning about your answer.

12.2. In an equimolar mixture of O_2 and N_2, are the mass fractions equal?

12.3. If the molar analysis is known, how would you calculate the specific heat ratio γ, at 300 K of a mixture of N_2, O_2 and CO_2?

12.4. Whether entropy would be produced or not if two samples of the same gas at the same temperature and pressure were mixed adiabatically?

12.5. Which component of the fuel-air mixture in a cylinder of an automobile engine would have the greater mass fraction?

12.6. Why mixing of different energy stream in a container produces entropy production?

Short and Long Theoretical Questions

12.7. Explain Dalton's law of additive pressure.

12.8. Show that summation of mole and mass fraction is equal to unity.

12.9. Explain Amagat-Leduc law.

12.10. Derive the terms (a) partial pressure, (b) mole fraction, and (c) volume fraction of gas constituent in a mixture.

12.11. Derive expression for mixtures of ideal gases for the following : (a) internal energy, (b) enthalpy, (c) specific heats, (d) entropy, and (e) gas constant, (f) ΔS.

Numerical Questions

12.12. A gas mixture has the following volumetric analysis : N_2, 60%, CO_2, 33%, O_2, 7%.

(a) Determine the analysis on a mass basis.

(b) What is the mass of 28.3 m^3 of this gas when the pressure is 1.013 bar and the temperature is 27.7°C.?

[*Ans.* (a) 50, 43.3, 6.7, (b) 38.5 kg]

12.13. The volumetric analysis of a mixture of gases is as follows : N_2, 70%, CO_2, 20%, O_2, 10%. Determine:

(a) the gravimetric analysis,

(b) the component pressure of CO_2 if the pressure and temperature of the mixture are 1 bar and 60°C respectively, and

(c) kg of the mixture would be contained in 12.75 m^3 at the conditions given in part (b)?

[*Ans.* (a) 62, 279, 10.1; (b) 0.204 bar; (c) 14.55 Ig]

12.14. Hydrogen and nitrogen in separate streams enter an adiabatic mixing chamber in a 3 : 1 molar ratio. At the iniet the H_2 is at 1.7 bar and 60°C and the N_2 is at 1.7 bar and 280°C. The mixture leaves the chamber at 1.65 bar. Determine (a) the final temperature of the mixture, and (b) the entropy change of the hydrogen gas. [*Ans.* (a) 115°C, (b) 5.07 kJ/kg mole]

12.15. An insulated 0.085 m^3 tank is divided in two sections by a partition. One section is 0.0567 m^3 in volume and initially contains H_2 gas at 2.04 bar and 37.8°C. The remaining section initially holds N_2 gas at 3.4 partition bar and 10°C. The partition is then removed, and the gases are allowed to mix. Determine (a) temperature and pressure of the mixture, (b) change of entropy of the mixture, and (c) the entropy change for the process.

[*Ans.* (a) 25.4°C, (b) 2.36 bar, (c) 0.0685 kJ/K]

12.16. Two insulated containers are connected through an insulated pipe provided with a valve. One of the containers holds 2 moles of N_2 at 4 bar and 20°C. The other container hold 3 moles of N_2 at 0.5 bar and 55°C. The valve is opened and sufficient time is permitted to elapse for equilibrium to be attained. Calculate (a) temperature and pressure of the mixture, (b) change of entropy of the mixture, and (c) recalculate these values if the second container also holds O_2.

[*Ans.* (a) 41°C, 0.73 bar; (b) 18.97 kJ/kgK; (c) 41°C, 0.71 bar, 46.47 kJ/kg K]

12.17. A system consists of two tanks interconnected with a pipe and valve. One tank with a volume of 1.05 m^3 contains 2.27 kg of argon at 1.365 bar and 32.2°C. Then other tank has a volume of 0.227 m^3 and holds 1.82 kg of O_2 at 7 bar and 60°C. The valve is opened, and the gases are allowed to mix. The atmospheric temperature is 4.44°C, and during the mixing process 32.4 kJ of heat leaves the insulated system. Calculate (a) the final temperature, (b) the final mixture pressure, (c) the apparent molecular weight of the mixture (d) change in entropy of the argon in kJ/K, and (e) the entropy change of the atmospheric air in kJ/K.

[*Ans.* (a) 32.5°C, (b) 2.26 bar, (c) 36, (d) 0.0891 kJ/K, (e) 0.117 kJ/K]

12.18. A tank with a volume of 0.284 m^3 contains O_2 at 7 bar and 334°C. A second tank with a volume of 0.425 m^3 contains helium at 5.45 bar and 4.44°C. The tank walls are heavily insulated. The two tanks are interconnected and the gases are allowed to mix. Calculate (a) the final temperature in °C, (b) the final pressure in bar, and (c) the total entropy change in kJ/K.

[*Ans.* (a) 129°C, (b) 6.86 bar, (c) 0.8076 kJ/K]

12.19. A mixture of perfect gases at 20°C has the following composition by volume : N_2, 55 per cent, O_2, 20 per cent, CH_4, 25 per cent. If the partial pressure of CH_4 is 0.5 bar calculate (a) the partial pressure of N_2 and O_2 (b) the mass properties of the mixtures, and (c) gas constant of mixture and the volume per mole of the mixture.

[*Ans.* (a) 11.1 bar, 0.9 bar, (b) 25.8 kJ/kg mole, (c) 0.322 kJ/kg K]

12.20. The analysis by weight of a perfect gas mixture at 20°C and 1.2 bar is 10 per cent O_2, 70 per cent N_2, 15% CO_2 and 5 per cent CO. For a reference state of 0°C and atmospheric pressure, calculate (a) the partial pressure of the constituents at 20°C and 1.2 bar (b) gas constant of the mixture, (c) enthalpy and internal energy of the mixture, and (d) entropy of the mixture.

12.21. A rigid vessel has a volume of 0.01 m^3 and contains air at 1.013 bar and 20°C. Helium at 20°C is added to the vessel until the pressure is raised to 1.2 bar. Find (a) the partial pressure of the final components of the mixture composed of O_2, N_2 and H_2 (b) the masses of the components of the rnixiure, and (c) the gas constant of the mixtures.

12.22. 1 kg of O_2 at 0.7 bar and 40°C is mixed with 3 kg of He at 0.7 bar and 80°C. If the value of the mixture is equal to the sum of the volume of the O_2 and He before mixing find (a) the amount of heat necessary to raise the temperature of the mixture to 80°C at constant volume, (b) the entropy increase due to the mixing process only, and (c) the entropy increase due to the heating process. [*Ans.* (a) 25 kJ, (b) 1.084 kJ/K, (c) 0.071 kJ/K]

Software Based Questions

12.23. Develop a software to calculate the properties of non-reactive ideal gas mixture.

12.24. A gas mixture having a molar analysis of 60% O_2 and 49% N_2 enters an insulated compressor operating at steady state at 1 bar, 20°C with a mass flow rate of 0.5 kg/s and is compressed to 6 bar. Neglect changes in PE and KE. If the isentropic compressor efficiency is 80% develop a software to calculate the following :

(a) the temperature at the exit, in °C,

(b) the power required, in kW,

(c) the rate of entropy generation, in kW/K, and

(d) plot (a), (b) and (c) by varying delivery pressure from 3 bar to 6 bar.

12.25. A mixture having an analysis on a mass basis of 80% N_2, 20% CO_2 enters a nozzle operating at steady state with a velocity of 5 m/s and expands adiabatically through a pressure ratio of 8 : 1. If the isentropic efficiency of nozzle is 90%, develop a software to calculate the followings: (a) the exit velocity of mixture, in m/s, (b) plot the exit velocity variation by varying isentropic nozzle efficiency from 80 to 95%, and (c) plot the exit velocity variation by varying the expansion ratio from 4 : 1 to 8 : 1 having isentropic efficiency fixed at 90%.

Design and Open Ended Questions

12.26. A large number of gases at different pressure and temperature mixes in a mixing chamber. How will you calculate the mixture temperature and pressure and entropy generation. By taking a realistic data, give the results.

Multiple-Choice Answer Type Questions

12.27. Composition of dry air on mass basis is

(a) 0.232 kg O_2 and 0.76 kg N_2 (b) 0.21 kg O_2 and 0.79 kg N_2

(c) 0.31 kg O_2 and 0.69 kg N_2 (d) None of these

12.28. Dalton's law of partial pressure given by

(a) $p = p_a + p_b + p_c + ... + p_k \big|_T$ (b) $p = p_a + p_b + p_c + ... + p_k \big|_p$

(c) $p = p_a + p_b + p_c + ... + p_k \big|_V$ (d) $p = p_a + p_b + p_c + ... + p_k \big|_{V,T}$

12.29. Amagat–Leduc law is given by

(a) $V = V_a + V_b + V_c + ... + V_k \big|_p$ (b) $V = V_a + V_b + V_c + ... + V_k \big|_T$

(c) $V = V_a + V_b + V_c + ... + V_k \big|_v$ (d) $V = V_a + V_b + V_c + ... + V_k \big|_{p,T}$

Anwers to Multiple-Choice Questions

12.27 (a) 12.28 (d) 12.29 (d)

CHAPTER

13

Gas Power Cycles

OVERVIEW

Power or energy consumption is now one of the indexes of standard of living of mankind. The per capita energy consumption in terms of tonnes of oil equivalent (TOE) in USA was 6.95 in 2021 TOE against global average of 1.8 TOE. The energy consumption of India is 0.6 TOE while population of India is 17% of world population. The oil and gas reserves in India are 0.5 per cent of world reserves and coal reserves are 7% of world reserves. The per capita income of India is \$ 2191.00 in 2021, while of USA, it is \$ 69,288.00. Drinking water reserves are 2.4% of world reserves. All these data show the quality of life of the citizens of a country and one can guess the pace of development needed in India to achieve a reasonable quality of life of its citizens.

Today, there is an ever increasing demand of power, and in order to meet it, a large number of power producing devices have been developed. Broadly, they are classified as

(a) *Indirect energy converters.* and (b) *Direct energy converters.*

Power plants may also be classified as

(a) *Conventional power plants,* and (b) *Non-conventional power plants.*

In the indirect energy conversion system, the chemical energy of the fuel is first converted into heat, through combustion process, and then heat energy is converted into mechanical energy and then electrical energy as shown in Fig. 13.1. The indirect system includes the following :

(i) Internal combustion engines
(ii) Gas turbine plants
(iii) Steam power plants (engine, turbine)
(iv) Hydrogen fuel combustion turbine plants (to be ready by 2020)
(v) Nuclear power plants
(vi) Hydel power plants, (no fuel requirement), small size comes under renewals
(vii) Geothermal power plants, utilising earth's internal energy (no fuel requirement)

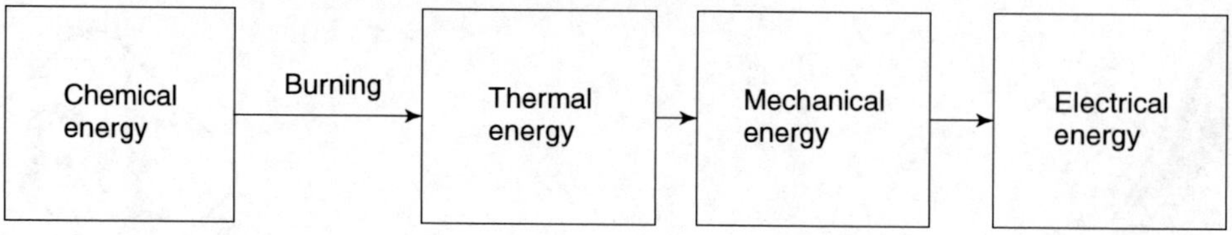

Fig. 13.1 Indirect energy converters

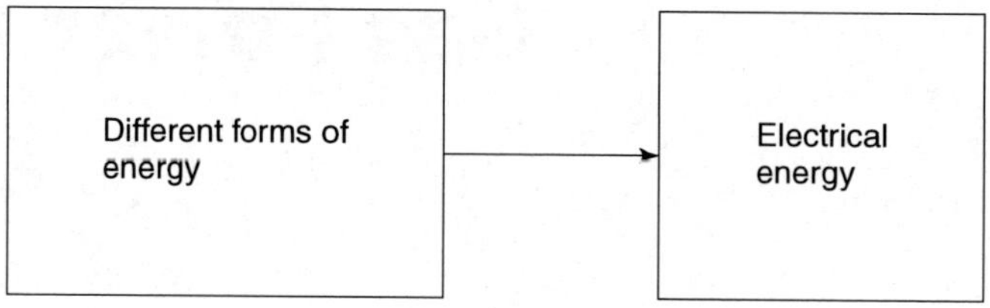

Fig. 13.2 Direct energy converters

(viii) Tidal and wave power plants (no fuel requirement), renewals

(ix) Ocean thermal energy power plants (no fuel requirement)

(x) Wind mills or turbines (no fuel requirement), renewals

(xi) Solar thermal power plants (no fuel requirement), renewals

The *first six fall* under the category of *conventional power plants* and the rest under non-conventional and renewals power plants.

In the direct energy conversion system, different types of energy are converted directly into electrical energy (Fig. 13.2). This includes:

(i) Fuel cells (ii) Thermoelectric generators

(iii) Thermoionic generators (iv) Magnetohydrodynamic power plant (MHD)

(v) Solar cells

Amongst the above devices, the ***working fluid*** **of some of them executes a cyclic process** and as we know, the essential elements of thermodynamic power cycles are:

(a) *Working fluid* This undergoes a change of state while receiving or rejecting heat and thus produces work.

(b) *Heat source and sink* These are essential for working fluid to receive heat from a heat source and reject in into a heat sink.

(c) *Arrangement* An arrangement is necessary for getting the work out of the working fluid. This may be done by *piston-cylinder or a turbine* or any other arrangement.

In this chapter, we will restrict ourselves to the analysis of the power plants which employ *gas as the working fluid* and executes a cyclic process, for example, internal combustion engines and gas turbines.

13.1 AIR STANDARD CYCLES (MODELLING OF GAS POWER CYCLES)

In gas power plants such as internal comustion engines and gas turbines, the working fluid remains gas throughout the cycle. The plants take in either a mixture of air and fuel or air and fuel separately, compress it to a high pressure, and cause it to burn in a combustion chamber liberating heat energy. The products of combustion at high pressure, temperature expand and do work and the exhaust leaves the engine or turbine.

In the most gas power cycles, the working fluid consists mainly of air. For example, in an internal combustion engine, the air-fuel ratio is generally 1015 : 1 and in the gas turbine 50 : 1 to 250 : 1. The properties of the working fluid closely follow the properties of air. Therefore, it is convenient to analyse the gas power cycles by devising an idealised cycle known as ***Air-Standard Cycle*** in which the working fluid is pure air.

The air-standard cycles are based on the following assumptions:

(i) The working fluid is air throughout the cycle and follows the perfect gas law $pV = mRT$.

(ii) The working fluid is homogeneous throughout at all times and no chemical reaction takes place.

(iii) The specific heats of air do not vary with temperature.

(iv) The mass of air in the cycle remains fixed, i.e., the engine operates as a closed cycle so that working fluid is restored to its initial state at the end of each cycle. Thus, there are no intake and exhaust processes.

(v) Combustion process which may appear in actual process is replaced by an equivalent heat addition process. This heat addition process is due to heat transfer alone.

(vi) Since there is no exhaust process, it is replaced by an equivalent heat rejection process. Further, this heat rejection process is due to heat transfer alone.

(vii) All processes are internally reversible, and compression and expansion processes are reversible.

The thermal efficiency, η_{th} of air standard cycle is given by

$$\text{Thermal efficieny} = \frac{\text{Work done}}{\text{Heat supplied}} = \frac{\text{Heat added} - \text{Heat rejected}}{\text{Heat supplied}}$$

or

$$\eta_{th} = \eta_{as} = \frac{Q_A - Q_R}{Q_A} = \frac{q_A - q_R}{q_A} \qquad \text{...(13.1)}$$

The thermal efficieny of the ideal air standard cycle is also known as the *air standard efficiency* (η_{as}). Sometimes, the terms *efficiency ratio* or *relative efficiency* is also used.

$$\text{Realtive efficiency, } \eta_{rel} = \frac{\text{Actual thermal efficiency}}{\text{Air standard efficiency}} = \frac{\eta_{tha}}{\eta_{as}} \qquad \text{...(13.2)}$$

Note: q_A or q_R have been used for convenience. Instead, it should be written as $\frac{Q_A}{m}$ or $\frac{Q_R}{m}$

The concept of air-standard cycle helps us in the following respects:

(i) To establish a standard of perfection by which the performance of actual engines can be estimated.

(ii) To study qualitatively the influence of a numnber of variables on the performance of actual power plants.

The air-standard cycles include the following :

(i) Carnot cycle, (ii) Otto cycle, (iii) Diesel cycle, (iv) Dual cycle, (v) Stirling cycle, (vi) Erricson cycle, (vii) Atkinson cycle, and (viii) Joule or Brayton cycle.

Usually, the devices which work on **gas power cycles are called heat engines.**

The passenger cars employ Otto (petrol) and Diesel cycle engines. Buses, trucks, rail engines employ diesel engine cycle. Now, in some buses and cars, LNGs are used. Aeroengines use petrol engines. Diesel engines are also employed in small size electricity generation plants varying from 1 kW to 10 MW. All gas turbine plants use Brayton cycle or their modifications. Jet engines employ gas turbine cycle with one stage of gas turbine followed by nozzle to produce jet which propels the vehicles in opposite direction based on Newton's third law of motion. They use aviation fuel (kerosene). **Due to pollution problems, electric vehicles are coming up.**

13.2 SOME DEFINITIONS FOR PISTON-CYLINDER ARRANGEMENT

A number of power producing devices (petrol and diesel engines) make use of the piston-cylinder arrangement in which the piston is observed to undergo cycles or revolutions. Figure 13.3 shows the various nomenclatures used in the piston-cylinder arrangement.

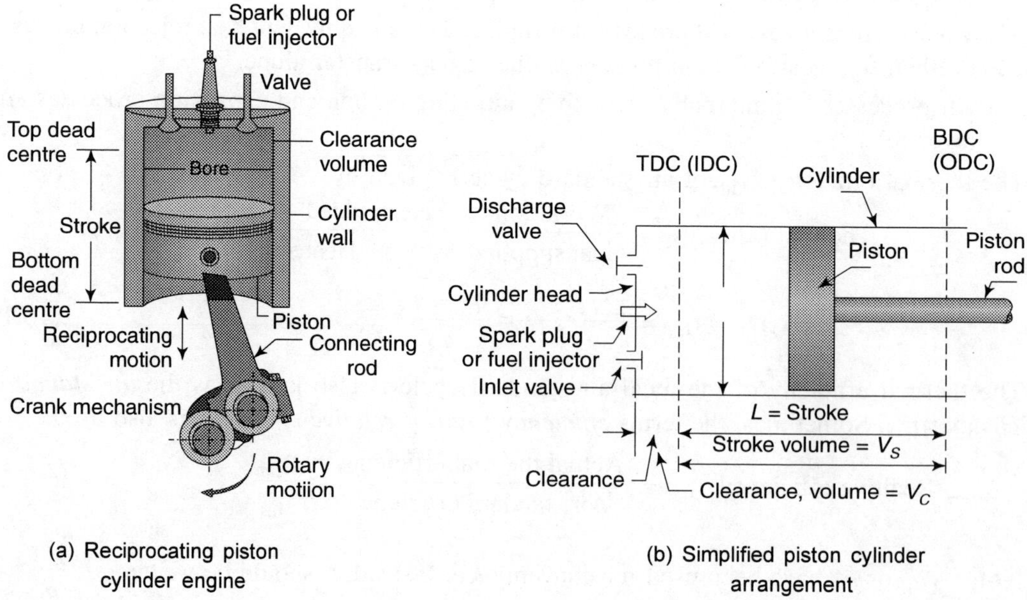

(a) Reciprocating piston cylinder engine

(b) Simplified piston cylinder arrangement

Fig. 13.3

(i) The cylinder or piston diameter d is known as *bore*. The extreme positions of the piston are called *top dead centre position* (TDC) *or inner dead centre position* (IDC) and bottom dead centre (BDC) *or outer deal centre position* (ODC). The distance that the piston moves in one direction is known as *stroke* (L), i.e., stroke is the distance between TDC and BDC position. Theoretically, the piston may touch the cylinder head but in actual practice, there is a clearance left between the cylinder, the minimum volume, i.e., volume of clearance, is called the *clearance volume* (V_c). Similarly, when the piston is at BDC, the fluid will occupy the maximum volume. The volume swept by the piston in moving from TDC to BDC or vice versa is called the *swept volume or piston displacement volume* (V_s). Hence,

(ii) Volume of cylinder = Clearance volume + Swept volume = $V_c + V_s$ ...(13.3)

(iii) *Compression ratio* (r) is defined as the ratio of cylinder volume to the clearance volume. Thus,

$$\text{Compression ratio} = r = \frac{\text{Full cylinder volume}}{\text{Clearance volume}} = \frac{V_1}{V_2} = \frac{V_c + V_s}{V_c} = 1 + \frac{V_s}{V_c} \qquad ...(13.4a)$$

$$\text{(iv) Clearance ratio} = C = \frac{??}{\text{Stroke volume}} = \frac{??}{V_s}$$

$$\therefore \qquad r = 1 + \frac{1}{c} = \frac{1+c}{c} \qquad ...(13.4b)$$

(v) *Mean effective pressure* (p_m) The mean effective pressure (m.e.p.) is the useful parameter in the study of the reciprocating devices that are used for power production. *It is defined as average pressure which, if acted on the piston during the entire power or outward stroke, would produce the same work output as the net work output for the actual cyclic process*. Thus,

$$\text{Work done per cycle} = W_{\text{cycle}} = p_m \times \text{Piston area} \times \text{Stroke} = p_m.A.L.;\ \text{kJ}$$

or

$$p_m = \frac{\text{Work done per cycle}}{\text{Stroke volume}} \qquad ...(13.5)$$

$$\text{Work done per minute} = p_m\ L\ A\ n$$

$$\text{Indicated power} = i.p. = \frac{p_m\ L\ A\ n}{60}\ \text{kW where } p_m \text{ is in kN/m}^2 \text{ or kPa} \qquad ...(13.6)$$

$$\text{Indicated power} = i.p. = \frac{100\ p_m\ L\ A\ n}{60}\ \text{kW} \qquad ...(13.7)$$

where p_m is in bar, n = number of cycle per minute, $n = \frac{N}{2}$ for four-stroke engines $n = N$ for two strokes engines, N = revolution per minute.

For reciprocating engines of comparable size, higher m.e.p. is an indication of better performance in terms of power produced at the same rated speed. The size of the internal combustion engines is denoted by volume capacity of cylinder expressed in cc, e.g., 800 cc, 1000 cc, 1500 cc engines. Reciprocating internal combustion engines may be ***four and two strokes engines***. *In four-stroke engines, there is one power stroke in two revolutions of crankshaft whereas in two-strokes engines, there is one power stroke in energy revolution of* *crankshaft. Therefore, theoretically two-stroke engines produces twice power than four-stroke engines but in actual practice, it is 1.6 times.*

13.3 CARNOT CYCLE

Carnot cycle is an ideal cycle (See Art. 7.10) which is not possible in practice and has the maximum efficiency compared to other cycles. Though, the analysis of Carnot cycle does not give the performance of actual cycle, it however permits the designer to arrive at the maximum possible efficiency that can be obtained in any heat engine under given conditions. The cycle is represented on *p-v* and *T-s* diagram shown in Fig. 13.4. The various processes are defined in section 7.12.

Consider 1 kg of working fluid.

From first law of thermodynamics, $q = w$ sincc $\Delta u = 0$ at constant temperature

Heated added = $q_A = q_{2-3} = RT_2 \ln v_3/v_2 = RT_H \ln v_3/v_2$

Heat rejected = $q_R = q_{2-3} = RT_2 \ln v_4/v_1 = RT_1 \ln v_4/v_1$

$$\text{Thermal efficiency, } \eta_{ih} = \frac{\text{Work done}}{\text{Heat added}} = \frac{q_A - q_R}{q_A} = \frac{T_H - T_L}{T_H} \qquad ...(13.8)$$

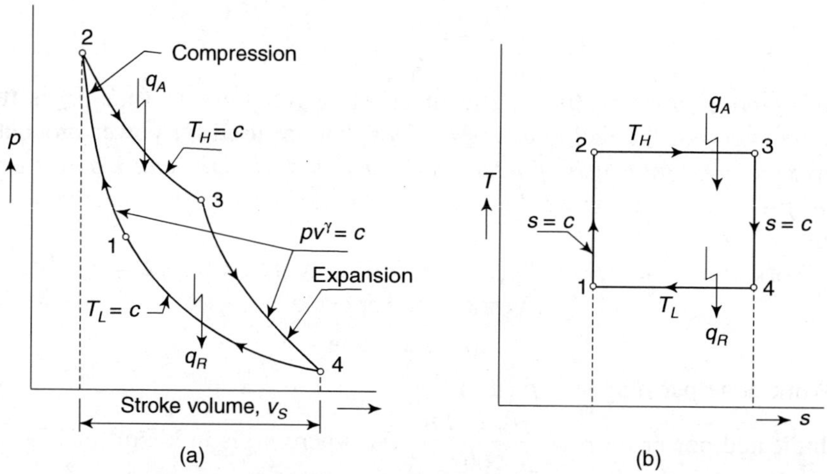

Fig. 13.4 Carnot cycle on *p-v* and *T-s* diagrams

Example 13.1 A Carnot cycle has lowest pressure and temperature equal to 1 bar and 20°C. Pressure after isothermal compression is 4 bar. Pressure after isentropic compression is 12 bar and after isothermal heat addition process is 6 bar. Calculate (a) the highest temperature in the cycle, (b) the change in entropy during isothermal expansion, (c) heat added to the cycle, (d) heat rejected by cycle, (e) power developed with 150 cycles per minute, (f) thermal efficiency of the cycle, and (g) mean effective pressure.

Important note: *For simplicity,* $\frac{Q}{m}$ *and* $\frac{W}{m}$ *are written as* q and w *respectively whereas* Q *and* W *are not the property of the system.*

Solution Refer to Fig. 13.4.

Given, $p_4 = 1$ bar, $T_L = T_4 = T_1 = 293$ K,

$p_2 = 12$ bar, $p_1 = 4$ bar and $p_3 = 6$ bar

Applying chracteristic gas equation at point 4, we have

$$v_4 = \frac{RT_L}{p_4} = \frac{RT_4}{p_4} = \frac{0.287 \times 293}{1 \times 10^2} = 0.8409 \text{ m}^3/\text{kg}$$

From process 4 – 1, $$v_1 = \frac{p_4\, v_4}{p_1} = \frac{1 \times 10^2 \times 0.8409}{4 \times 10^2} = 0.2102 \text{ m}^3/\text{kg}$$

From isentropic compression process 1 – 2, we have

$$v_2 = v_1 \left(\frac{p_1}{p_2}\right)^{1/\gamma} = 0.2102 \left(\frac{p_1}{p_2}\right)^{1/1.4} = 0.2102 \times \left(\frac{4}{12}\right)^{1/1.4} = 0.0959 \text{ m}^3/\text{kg}$$

$\therefore$ Stroke volume, $v_s = v_4 - v_2 = 0.8409 - 0.0959 = 0.745 \text{ m}^3/\text{kg}$

(a) Again, $$T_2 = T_H = T_1 \left(\frac{v_1}{v_2}\right)^{\gamma - 1} = T_L \left(\frac{v_1}{v_2}\right)^{\gamma - 1} = 293 \left(\frac{0.2102}{0.0959}\right)^{0.4} = 401.04 \text{ K}$$

(b) From process 2 – 3, $$v_3 = v_2 \frac{p_2}{p_3} = 0.0959 \times \frac{12}{6} = 0.1918 \text{ m}^3/\text{kg}$$

Change in entropy during isothermal expansion (heat addition) process is given

$$s_3 - s_2 = \frac{p_2\, v_2}{T_H} \ln \frac{v_3}{v_2} = \frac{12 \times 10^2 \times 0.0959}{401.04} \ln \frac{0.1918}{0.0959} = 0.1984 \text{ kJ/kg K}$$ **Ans.**

(c) Heat added, $q_A = p_2\, v_2 \ln \dfrac{v_3}{v_2} = T_H\,(s_3 - s_2) = 401.04 \times 0.1984 = 79.56$ kJ/kg **Ans.**

(d) Heat rejected, $q_R = p_4\, v_4 \ln \dfrac{v_1}{v_4} = T_L\,(s_1 - s_4) = -\,T_L\,(s_3 - s_2)$

$= -\,293 \times 0.1984 = -\,58.13$ kJ/kg **Ans.**

–ve sign is used to show the heat rejection.

(e) Work done per cycle, $q_A - q_R = 79.56 - 58.13 = 21.43$ kJ/kg

$$\text{Power developed} = \frac{21.4 \times 150}{60} = 53.57 \text{ kW}$$ **Ans.**

(f) Thermal efficiency $= 1 - \dfrac{T_L}{T_H} = 1 - \dfrac{293}{401.04} = 0.2695$ or 26.95% **Ans.**

(g) $\text{m.e.p.} = \dfrac{\text{Work done/kg}}{\text{Stroke volume}} = \dfrac{21.43}{0.8409 - 0.0959} = 28.76 \text{ kN/m}^2 = 0.2876 \text{ bar}$ **Ans.**

Example 13.2 It is claimed that an engine works on a Carnot cycle with air as the working fluid. The ratio of greatest to least volume is fixed and say x. The lower temperature of the cycle is fixed but the isentropic compression ratio (r) is variable. The ratio of specific heats is γ. Prove that for the maximum work done, the following equation holds good.

$$(\gamma - 1) \ln \left(\frac{x}{r}\right) + \frac{1}{r^{r-1}} - 1 = 0$$

Solution Refer to Fig. 13.4. Assume 1 kg of working fluid.

$$\text{Work done} = \text{Heat supplied} - \text{Heat rejected} = W = p_2 v_2 \ln \frac{v_3}{v_2} - p_4 v_4 \ln \frac{v_4}{v_1}$$

$$= p_2 v_2 \ln \frac{v_3}{v_2} + p_4 v_4 \ln \frac{v_1}{v_4} = RT_2 \ln \frac{v_3}{v_4} + RT_4 \ln \frac{v_1}{v_4} + RT_H \ln \frac{v_3}{v_4} + RT_L \ln \frac{v_1}{v_4}$$

But
$$T_2 = T_H = T_1 \left(\frac{v_1}{v_2}\right)^{\gamma - 1} = T_L \left(\frac{v_1}{v_2}\right)^{\gamma - 1} = T_L\, r^{\gamma - 1}$$

$$T_3 = T_H = T_4 \left(\frac{v_4}{v_3}\right)^{\gamma - 1} = T_L \left(\frac{v_4}{v_3}\right)^{\gamma - 1}$$

These two equations results in $\dfrac{v_1}{v_2} = \dfrac{v_4}{v_3}$ or $\dfrac{v_3}{v_2} = \dfrac{v_4}{v_1}$

We have,
$$\frac{v_4}{v_2} = x \quad \text{and} \quad \frac{v_1}{v_2} = r = \frac{v_4}{v_3} \quad \text{and} \quad \frac{v_3}{v_2} = \frac{v_3}{v_4} \times \frac{v_4}{v_2} = \frac{x}{r} \cdot \frac{v_1}{v_4}$$

$$\therefore \qquad w = RT_H \ln \frac{x}{r} + RT_L \ln \frac{r}{x} = RT_L (r^{\gamma - 1} - 1) \ln \frac{x}{r}$$

In this equation, the only variable is r, hence, for the maximum work done

$$\frac{dw}{dr} = 0 = RT_L \left[\left(\ln \frac{x}{r}\right)(\gamma - 1) r^{\gamma - 2} + (r^{\gamma - 1} - 1)\left(\frac{r}{x}\right)\frac{-x}{r^2}\right]$$

or
$$\left(\ln \frac{x}{r}\right)(\gamma - 1) r^{\gamma - 2} - r^{\gamma - 2} + \frac{1}{r} = 0 \quad \text{or} \quad \left(\ln \frac{x}{r}\right)(\gamma - 1) r^{\gamma - 1} - r^{\gamma - 1} - 1 = 0$$

or
$$(\gamma - 1) \ln \frac{x}{r} + \frac{1}{r^{\gamma - 1}} - 1 = 0$$ **Proved.**

13.4 OTTO CYCLE (CONSTANT VOLUME CYCLE)

The Otto cycle is named after its originator Dr. A.N. Otto (1876), a German scientist who devised this idealised cycle for the spark ignition internal combustion engine. Figure 13.5

shows the representation of this cycle on *p-v* and *T-s* diagrams. It is composed of four internally reversible processes, two adiabatic and two constant volume processes (hence the name constant volume cycle). The various processes are:

Process 1–2 Isentropic compression of air raising its pressure and temperature.

Process 2–3 Addition of heat at constant volume causing rise in pressure and temperature of air.

Process 3–4 Isentropic expansion of air from high pressure and temperature of air.

Process 4–1 Rejection of heat constant volume causing drop in pressure and temperature of air restoring initial state.

The proximity of air standard Otto cycle to the actual spark ignition cycle can be better understood by assuming a piston-cylinder as shown in Fig. 13.6. The actual engine consists of four strokes namely (i) *Suction*, (ii) *Compression*, (iii) *Expansion or Power and* (iv) *Exhaust stroke*. The actual engine operates as an open cycle. But as discussed earlier, this idealised cycle is free from suction and exhaust processes, being assumed a closed cycle. We will now see that the closed cycle assumption is not a wrong one.

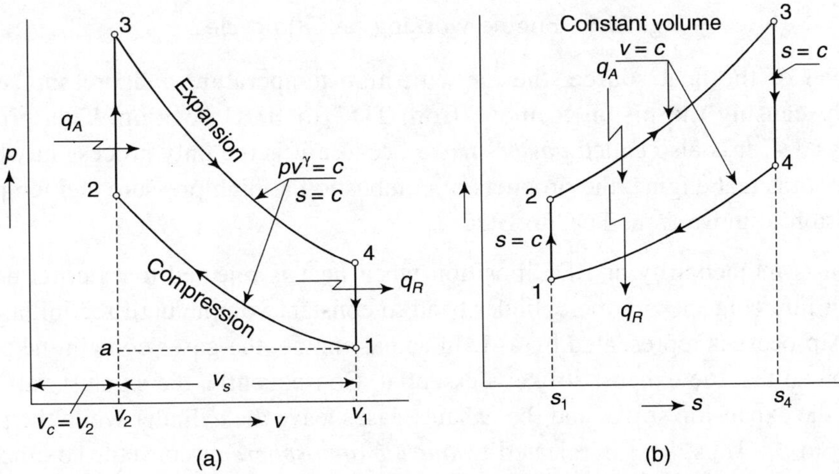

Fig. 13.5 Otto cycle on *p-v* and *T-s* diagrams

Consider that the piston is situated at BDC position and the cylinder is full of air. This is represented as point 1 on the diagram (Fig. 13.5.) As the piston moves from BDC position to TDC position, compression of the air occurs isentropically ending at 2 and thus raising the pressure and temperature air. In actual spark ignition engine, the mixture of air and fuel which has entered during suction stroke (i.e., movement of piston from TDC to BDC) through carburettor or MPFI is compressed (Fig. 13.6).

The piston is momentarily at rest at TDC position and heat is then added to air from an external source which is brought into contact with the cylinder head instantaneously so that both pressure and temperature rise to highest during a constant volume process 2–3. In an actual spark ignition engine, heat is added by igniting the fuel-air mixture in the combustion chamber of cylinder.

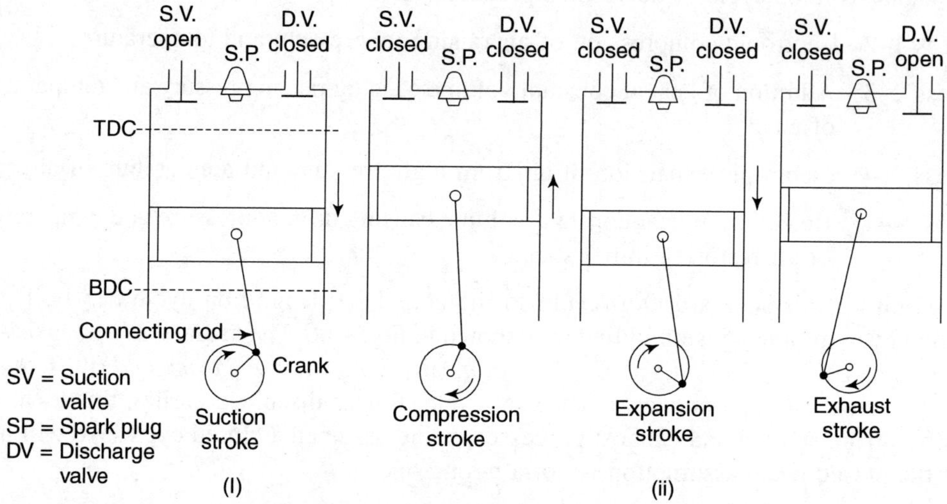

Fig. 13.6 Engine working on Otto cycle

On removal of the heat source, the air with high temperature and pressure expands isentropically causing the piston to move from TDC to BDC position. This process is represented as 3-4. It is also called ***power stroke*** because it is the only process in which the work is done. In actual engine, the products of combustion at high pressure and temperature forces the piston to move from TDC to BDC.

The piston is momentarily at BDC position when heat is rejected to external heat sink by bringing it into contact with the cylinder head to constant volume until the initial state is reached. This process is represented by 4–1. In actual engine, the gases contain the products of combustion so that the exhaust stroke is essential. Consequently, the exhaust valve opens at the end of the expansion stroke and the exhaust gases leave the cylinder while the pressure drop from p_4 to p_1. Thus, heat is rejected to *outside atmosphere* at constant volume during the process 4–1. The remaining exhaust gases are now expelled from the cylinder as the piston moves again from BDC to TDC. This process is called the **exhaust process**. In the case of actual engine, once the piston reaches the TDC position, the exhaust valve closes and the intake valve opens while the piston returns to BDC position. During suction stroke the cylinder is filled with fresh mixture or air and fuel shown by dotted line *a*–1.

In the air standard cycles, recharging of cylinder is not necessary since the same fluid continuously undergoes cyclic variation. It is to be noted that work required to push the charge out of cylinder is the same in magnitude but opposite in sign to that required to suck the new charge. Hence, these two functions, suction and exhaust, do not affect the new work done by the cycle and it is the only cycle 1–2–3–4–1 which is important in the theoretical cycle.

The *thermal efficiency* or air *standard efficiency* of the Otto cycle can be calculated as follows:

Consider 1 kg of air flowing through the cycle. Since the air in the cylinder acts as a closed system, hence, from first law for isentropic compression and expansion.

$$-w = \Delta u$$

For constant volume heat addition and rejection processes, since $w = 0$

$$\therefore \quad q = \Delta u = c_v \, \Delta T$$

$$\text{Heat added} = q_A = q_{2-3} = c_v (T_3 - T_2) \text{ kJ/kg}$$

$$\text{Heat rejected} = q_R = q_{1-4} = c_v (T_4 - T_1) \text{ kJ/kg}$$

$$\therefore \quad \text{Work done per cycle} = w = \text{Heat added} - \text{Heat rejected}$$

$$= c_v (T_3 - T_2) - c_v (T_4 - T_1) \text{ kJ/kg}$$

$$\text{Air standard efficiency} = \text{Thermal efficiency} = \frac{\text{Work done}}{\text{Heat added}}$$

or

$$\eta_{as} = \eta_{th} = \frac{c_v (T_3 - T_2) - c_v (T_4 - T_1)}{c_v (T_3 - T_2)} = 1 - \frac{T_4 - T_1}{T_3 - T_2} \quad \text{...(13.9)}$$

But

$$T_2 = T_1 \left(\frac{V_1}{V_2}\right)^{\gamma - 1} = T_1 \left(\frac{v_1}{v_2}\right)^{\gamma - 1} = T_1 \, r^{\gamma - 1} \quad \text{...(13.10a)}$$

$$T_3 = T_2 \frac{p_3}{p_2} = \alpha \, T_2 = \alpha \, T_1 \, r^{\gamma - 1} \quad \text{...(13.10b)}$$

where $\alpha = \dfrac{p_3}{p_2} = \dfrac{T_3}{T_2}$

$$T_4 = T_3 \left(\frac{1}{r}\right)^{\gamma - 1} = \alpha \, T_1 \quad \text{...(13.10c)}$$

Substituting the above values of T_2, T_3 and T_4 in Eq. (13.9), we get

$$\eta_{as} = \eta_{th} = 1 - \frac{\alpha \, T_1 - T_1}{\alpha \, T_1 \, r^{\gamma - 1} - T_1 \, r^{\gamma - 1}}$$

$$= 1 - \frac{1}{r^{\gamma - 1}} \quad \text{...(13.11)}$$

From Eq. (13.11), it is obvious that the air standard efficiency of the Otto cycle depends only on the compression ratio r and thus efficiency increases with compression ratio as shown in Fig. 13.7. In the beginning, the efficiency increases rapidly and after that it is slow.

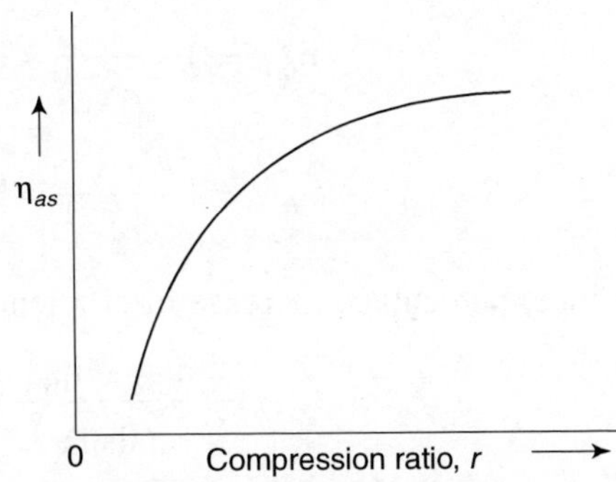

Fig. 13.7 Thermal efficiency of Otto cycle vs compression ratio

In actual engines working on Otto cycle, the **compression ratio varies from 5 to 8 and depends upon the quality of the fuel.**

Mean Effective Pressure (m.e.p.) *The mean effective pressure may be defined as the ratio of work done to the displacement volume of the piston.* Consider 1 kg of the working fluid.

Work done = c_v $[(T_3 - T_2) - (T_4 - T_1)]$ and displacement volume = $v_1 - v_2$

$$\text{m.e.p.} = \frac{\text{Work done}}{\text{Displacement volume}} = c_v \frac{[(T_3 - T_2) - (T_4 - T_1)]}{v_1 - v_2}$$

But
$$v_1 - v_2 = v_1\left(1 - \frac{1}{r}\right) = \frac{RT_1}{p_1}\left(\frac{r-1}{r}\right) = c_v\,\frac{(\gamma - 1)}{p_1}\left(\frac{r-1}{r}\right)T_1$$

Putting the values of T_2, T_3 and $(v_1 - v_2)$ we have

$$\therefore \quad \text{m.e.p.} = \frac{p_1 \cdot r\,[(\alpha\, r^{\gamma-1} - r^{\gamma-1}) - (\alpha - 1)]}{(\gamma - 1)(r - 1)} = p_1\, r\left[\frac{r^{\gamma-1}(\alpha - 1) - (\alpha - 1)}{(r-1)(\gamma-1)}\right]$$

or
$$\text{m.e.p.} = p_1\, r\left[\frac{(\alpha - 1)(r^{\gamma-1} - 1)}{(r-1)(\gamma-1)}\right] \text{ bar where } p_1 \text{ is in bar} \qquad \text{...(13.12)}$$

This is the required expression for m.e.p. for Otto cycle. The m.e.p. may also be expressed as follows :

$$\text{m.e.p.} = \left[\frac{(p_3\, v_3 - p_4\, v_4)}{(\gamma - 1)} - \frac{(p_2\, v_2 - p_1\, v_1)}{(\gamma - 1)}\right] \Big/ (v_1 - v_2)$$

Example 13.3 Compare the changes in efficiency of an Otto engine when the compression ratio changes from 4 to 5. Take $\gamma = 1.4$ for air.

Solution For Otto cycle, we know that

$$\therefore \quad \eta_{th1} = 1 - \frac{1}{r_1^{\gamma-1}} = 1 - \frac{1}{4^{0.4}} = 1 - 0.425 = 0.425 \text{ or } 42.5\%$$

$$\eta_{th2} = 1 - \frac{1}{r_2^{\gamma-1}} = 1 - \frac{1}{5^{0.4}} = 1 - 0.5253 = 0.4747 \text{ or } 47.47\%$$

Hence, percentage increase in efficiency

$$= \frac{\eta_{th2} - \eta_{th1}}{\eta_{th1}} \times 100 = \frac{0.4747 - 0.425}{0.425} \times 100 = 11.69\% \qquad \textbf{Ans.}$$

Example 13.4 The pressure limits in an Otto air cycle are 100 kN/m^2 and 2000 kN/m^2, respectively. The compression ratio is 4. Calculate (a) thermal efficiency, and (b) mean effective pressure. Take $\gamma = 1.4$ for air.

Solution Refer to Fig. 13.5 (a) Thermal efficiency for Otto cycle is

$$\eta_{th} = \eta_{as} = 1 - \frac{1}{r^{\gamma-1}} = 1 - \frac{1}{4^{0.4}} = 1 - 0.575 = 0.425 \text{ or } 42.5\%$$

(b) We have $p_1 = 100 \text{ kN/m}^2$ and $p_2 = 2 \times 10^3 \text{ kN/m}^2$

$$\frac{p_2}{p_1} = \left(\frac{v_1}{v_2}\right)^{\gamma} = (4)^{1.4} = 6.964$$

$\therefore$ $p_2 = p_1 \times 6.964 = 100 \times 6.964 = 6.964 \times 10^2 \text{ kN/m}^2$

and $p_3 = 2 \times 10^3 \text{ kN/m}^2$

$\therefore$ $$\frac{p_3}{p_2} = \alpha = \frac{2 \times 10^3}{6.964 \times 10^2} = 2.871$$

$$\text{m.e.p.} = p_1 r \left[\frac{(\alpha - 1)(r^{\gamma-1} - 1)}{(r-1)(\gamma - 1)}\right]$$

$$= 100 \times 4 \left[\frac{(2.871 - 1)(4^{0.4} - 1)}{(4-1) - (1.4 - 1)}\right] = 4.62 \times 10^2 \text{ kN/m}^2$$

$$= 4.62 \text{ bar}$$ **Ans.**

Mean effective pressure (m.e.p.) may also be calculated as follows:

$$\frac{p_4}{p_3} = \left(\frac{v_3}{v_4}\right)^{\gamma} = \left(\frac{1}{4}\right)^{1.4} = \frac{1}{6.964}$$

or $$p_4 = 20 \times 100 \times \frac{1}{6.964} = 2.871 \times 10^2 \text{ kN/m}^2$$

Let $v_3 = v_2 = v$ and $r = \dfrac{v_1}{v_2} = 4$, hence $v_1 = 4v$

$$\text{Work done} = w = \frac{(p_3 v_3 - p_4 v_4)}{\gamma - 1} - \frac{(p_2 v_2 - p_1 v_1)}{\gamma - 1}$$

$$= 10^2 \left[\frac{(20 \times v - 2.871 \times 4v)}{0.4} - \frac{(6.964 \times v - 1 \times 4v)}{0.4}\right]$$

$$= 13.88 \times v \times 10^2 \text{ kJ/kg}$$

$\therefore$ $$\text{m.e.p.} = \frac{w}{v_1 - v_2} = \frac{w}{3v} = \frac{13.88 \times v \times 10^2}{3v}$$

$$= 4.62 \times 10^2 \text{ kN/m}^2 = 4.62 \text{ bar}$$ **Ans.**

Example 13.5 An air standard Otto cycle has a compression ratio of 8. At the start of the compression process, the temperature is 26°C and the pressure is 1 bar. If the maximum temperature of the cycle is 1080°C, calculate (a) the heat supplied per kg of air, (b) the net work done per kg of air, and (c) the thermal efficiency of the cycle.

Solution Refer to Fig. 13.5 (a) 13.5 (b). From isentropic compression process 1 – 2

$$T_2 = T_1\left(\frac{v_1}{v_2}\right)^{\gamma-1} = T_1\left(\frac{V_1}{V_2}\right)^{\gamma-1} = 299 \times (8)^{0.4} = 686.89 \text{ K}$$

For isentropic expansion process 3 – 4,

$$T_4 = T_3\left(\frac{v_3}{v_4}\right)^{\gamma-1} = T_3\left(\frac{V_3}{V_4}\right)^{\gamma-1}$$

$$= (1080 + 273) \times \left(\frac{1}{8}\right)^{0.4} = 588.95 \text{ K}$$

(a) Heat added, $q_A = q_{2-3} = c_v\,(T_3 - T_2) = 0.718\,(1353 - 686.89) = 478.26$ kJ/kg **Ans.**

(b) Heat rejected, $q_R = q_{4-1} = c_v\,(T_4 - T_1) = 0.718\,(588.95 - 299) = 208.18$ kJ/kg **Ans.**

∴ Work done, $q_A - q_R = 478.26 - 208.18 = 270.08$ kJ/kg **Ans.**

(c) The thermal efficiency, $\eta_{th} = \dfrac{\text{Work done}}{\text{Heat supplied}} = \dfrac{270.08}{478.26} = 56.47\%$ **Ans.**

Example 13.6 In the constant volume cycle, the temperatures at the beginning and end of the compression are 43°C and 323°C respectively. Calculate the (a) air standard efficiency, and (b) the compression ratio. Assume $\gamma = 1.4$ for air.

A petrol engine, with above compression ratio develops 22 kW indicated power and consumes 7.5 litres of fuel per hour. The specific gravity of fuel is 0.78 and its calorific value is 44,000 kJ/kg. Calculate the indicated thermal efficiency and the relative efficiency.

Solution Refer to Fig. 3.5. $T = 273 + 43 = 316$ K and $T_3 = 273 + 323 = 596$ K

From isentropic compression process 1–2.

$$\frac{T_2}{T_1} = \left(\frac{v_1}{v_2}\right)^{\gamma-1} \quad \text{or} \quad \frac{v_1}{v_2} = r = \left(\frac{T_2}{T_1}\right)^{1/(\gamma-1)} = \left(\frac{596}{316}\right)^{1/0.4}$$ **Ans.**

(b) or $r = 4.885$

(a) Air standard efficiency, $\eta_{as} = 1 - \dfrac{1}{r^{\gamma-1}} = 1\ \dfrac{1}{4.885^{0.4}} = 56.98\%$ **Ans.**

Fuel consumption = 7.5 liters/h. We know that 1 litre of water = 1 kg of water

∴ Fuel consumption = 7.5 × 0.78 × 1 = 5.85 kg/h

$$\text{Indicated thermal efficiency} = \frac{\text{Indicated work}}{\text{Heat added}} = \frac{22 \times 3600}{44000 \times 5.85} = 30.76\%$$ **Ans.**

$$\text{Relative efficiency} = \frac{\text{Thermal efficiency}}{\text{Air standard efficiency}} = \frac{30.76}{46.985} \times 100 = 65.46\%$$ **Ans.**

Exampl 13.7 Show that for the maximum work to be done per kg of air in Otto cycle between given upper and lower limits of absolute tmperature T_3 and T_1 respectively, the ratio of comprssion should have the value $r = (T_3/T_1)^{1.25}$ when $\gamma = 1.4$.

Solution Refer to Fig. 13.5. Work done per cycle, w = Heat added – Heat rejected

$$= c_v\,(T_3 - T_2) - c_v\,(T_4 - T_1);\ \text{kJ/kg}$$

From isentropic compression process 1 – 2, $T_2 = T_1\, r^{\gamma - 1}$ where $r = \dfrac{v_1}{v_2} = \dfrac{v_4}{v_2}$

From isentropic expansion process 3 – 4, $T_4 = T_3 \left(\dfrac{1}{r}\right)^{\gamma - 1}$

$$\therefore \qquad w = c_v \left[T_3 - T_1\, r^{\gamma - 1} - T_3 \left(\frac{1}{r}\right)^{\gamma - 1} - T_1\right];\ \text{kJ/kg}$$

To obtain the maximum work, differentiate work done w.r.t. r and equate to zero. Thus,

$$\frac{dw}{dr} = 0 = c_v\,[-T_1\,(\gamma - 1)\, r^{\gamma - 2} - T_3\,(1 - \gamma)\, r]$$

or $$T_1\,(\gamma - 1)\, r^{\gamma - 2} = T_3\,(\gamma - 1)\, r^{-\gamma} \quad \text{or} \quad r^{2 - 2\gamma} = \frac{T_1}{T_3}$$

or $$r = \left(\frac{T_3}{T_1}\right)^{1/0.8} = \left(\frac{T_3}{T_1}\right)^{1.25}$$ **Proved.**

Example 13.8 A Fiat car working on Otto cycle has initial pressure 1 bar and pressure after compression 10 bar. Calculate (a) the compression ratio, (b) the percentage clearance, and (c) the thermal efficiency.

If the initial temperature is 26°C find (d) the temperature at the end of compression (e) the mean effective pressure if the maximum pressure in the cycle is 20 bar.

Solution Refer to Fig. 13.8. From isentropic compression process 1 – 2, we have

(a) $$\text{Compression ratio, } r = \frac{v_1}{v_2} = \frac{V_1}{V_2} = \left(\frac{p_2}{p_1}\right)^{1/\gamma} = \left(\frac{10}{1}\right)^{1/1.4} = 5.178$$ **Ans.**

(b) $$\text{Percentage clearance} = \frac{v_2}{v_1} \times 100 = \frac{100}{5.178} = 19.31\%$$ **Ans.**

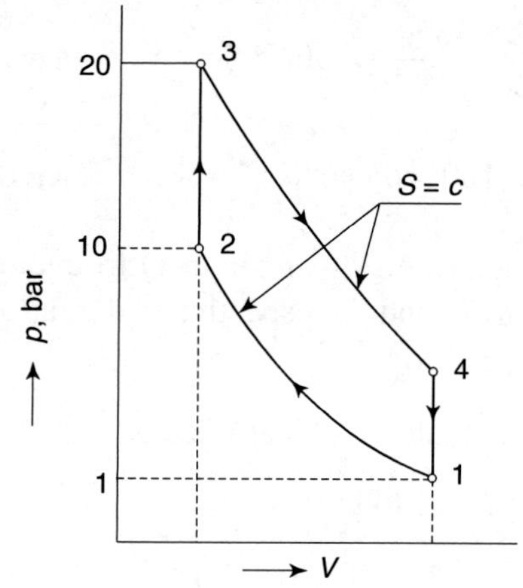

Fig. 13.8 Example 13.9

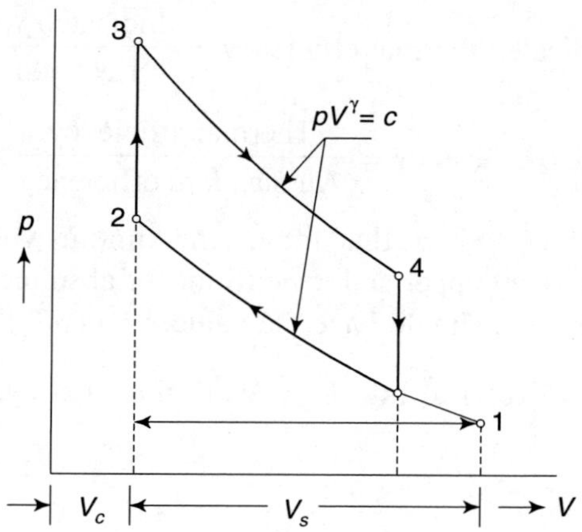

Fig. 13.9 Example 13.10

(c) $$\eta_{th} = 1 - \left(\frac{1}{r}\right)^{\gamma - 1} = 1 - \left(\frac{1}{5.178}\right)^{0.4} = 48.2\%$$ **Ans.**

(d) $$T_2 = T_1\,(r)^{\gamma - 1} = (273 + 26)\,(5.178)^{0.4} = 577.21\text{ K}$$ **Ans.**

(e) $$\text{m.e.p.} = p_1\,r\left[\frac{(\alpha - 1)\,(r^{\gamma - 1} - 1)}{(r - 1)\,(\gamma - 1)}\right],\ \text{here } \alpha = \frac{p_3}{p_2} = \frac{20}{10} = 2$$

$$\therefore \quad \text{m.e.p.} = 1 \times 5.178\left[\frac{(2 - 1)\,(5.178^{0.4} - 1)}{(5.178 - 1)\,(1.4 - 1)}\right] = 2.884\text{ bar}$$ **Ans.**

Example 13.9 A two-cylinder, two-stroke cycle engine working on the constant volume cycle has cylinder diameters of 35 cm and stroke 50 cm. The clearance volume is 1/15 of the swept volume the exhaust ports are closed by the piston when it has travelled 8 cm from the dead centre on the compression stroke. The indicated mean effective pressure is 4.5 bar and the specific fuel consumption is 0.27 kg/kWh, the calorific value of the fuel is 44,000 kJ/kg. Calculate (a) the indicated power at 180 rpm, and (b) the realtive efficiency based on the effective compression ratio of the engine.

Solution Refer to Fig. 13.9. Swept volume = Stroke volume = $V_s = V_1 - V_2$

$$\text{Clearance volume} = V_2 = V_3 = V_s/15 = V_c$$

$$\text{Stroke volume, } V_s = V_1 - V_2 = \frac{\pi}{4}d^2 \times L = \frac{\pi}{4} \times (0.35)^2 \times 0.5 = 0.0481\text{ m}^3$$

$$\therefore \quad \text{Clearance volume, } V_2 = \frac{V_s}{15} = \frac{0.0481}{15} = 0.0032\text{ m}^3$$

$$\therefore \quad \text{Cylinder volume, } V_1 = V_s + V_2 = 0.0481 + 0.0032 = 0.0513\text{ m}^3$$

$$\text{Cylinder area} = A = \frac{\pi}{4} d^2 = \frac{\pi}{4} (0.35)^2 = 0.0962 \text{ m}^3$$

$$\therefore \quad V_1 - V_2 = 0.0962 \times 0.08 = 0.007696 \text{ m}^3$$

$$\therefore \quad V_4 = 0.0513 - 0.00796 = 0.0433 \text{ m}^3$$

$$\therefore \text{ Effective compression ratio, } r_e = \frac{V_4}{V_2} = \frac{0.0433}{0.0032} = 13.53$$

(a)
$$i.p. = \frac{p_m \cdot LAn}{60} = \frac{4.5 \times 100 \times 0.5 \times 0.0962 \times 180 \times 2}{60}$$

$$= 129.87 \text{ kW}$$ **Ans.**

(b)
$$\text{Actual thermal efficiency} = \frac{\text{Work done}}{\text{Heat added}} = \frac{3600}{0.27 \times 44000} = 30.3\%$$

$$\text{Air standard efficiency, } 1 - \frac{1}{r^{\gamma - 1}} = 1 - \left(\frac{1}{13.53}\right)^{0.4} = 64.72\%$$

$$\therefore \quad \text{Relative efficiency} = \frac{\text{Actual thermal efficiency}}{\text{Air standard efficiency}} = \frac{30.3}{64.72} = 46.81\%$$ **Ans.**

Example 13.10 The bore and stroke of the cylinder of an engine working on Otto cycle are 18 cm and 32 cm, respectively. The clearance volume is 0.002 m³. Find compression ratio and the thermal efficiency of the engine. Take $\gamma = 1.4$ for air.

Solution Refer to Fig. 13.10. Clearance volume = $V_2 = V_3 = V_c = 0.002 \text{ m}^3$

$$\text{Stroke volume, } V_s = V_1 - V_2 = \frac{\pi}{4} d^2 \times L = \frac{\pi}{4} \times (0.18)^2 \times 0.32 = 0.00814 \text{ m}^3$$

$$\text{Cylinder volume, } V_1 = V_s + V_2 = 0.002 + 0.00814 = 0.01014 \text{ m}^3$$

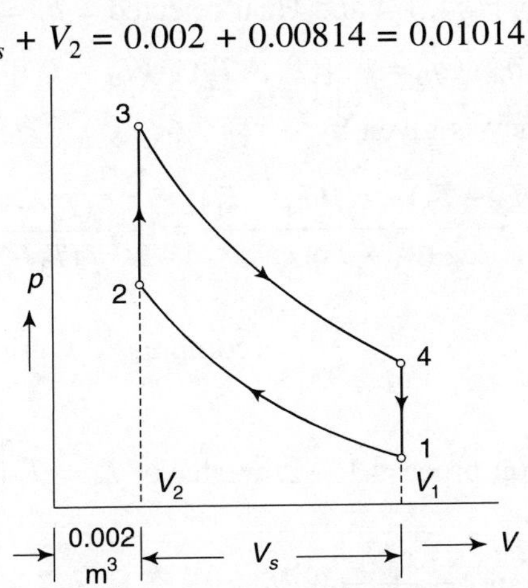

Fig. 13.10 Example 13.10

$\therefore$ Compression ratio, $= \text{—} = \dfrac{0.01014}{0.002} = 5.07$ **Ans.**

Thermal efficiency, $\eta_{th} = 1 - \dfrac{1}{r^{\gamma - 1}} = 1 - \dfrac{1}{5.07^{0.4}} = 47.77\%$ **Ans.**

13.5 DIESEL CYCLE (CONSTANT PRESSURE HEAT ADDITION CYCLE)

This cycle is named after its inventor Rudolph Diesel (1892) who devised this idealised cycle for the compression ignition internal combustion engines or the engines working on diesel oil. The cycle like the Otto cycle consists of four internally reversible processes, namely, two adiabatic, one constant pressure and other constant volume. Figure 13.11 shows the representation of Diesel cycle on *p-v* and *T-s* diagrams. The various processes are:

1–2 : *Isentropic compression of air* during which the piston moves from BDC to TDC position, thus raising pressure and temperature of air.

2–3 : *Addition of heat at constant pressure* from an external heat source which is brought into contact with cylinder head. During this process, pressure remains constant but volume and temperature increase. In actual diesel engine, the heat is added by injecting and burning of fuel (due to compression ignition) in the cylinder. At point 3, the heat addition is stopped and the volume ratio v_3/v_2 is called the ***cut-off ratio***.

3–4 : *Isentropic expansion of air at high pressure and temperature* to lower pressure and temperature till the piston reaches the BDC position. Due to expansion, work is done, and this stroke is called the *power stroke*. In actual engines, the products of combustion expands.

4–1 : *Rejection of heat at constant volume* to an external heat sink till the air restores its initial state and thus the cycle is completed.

Heat added, $q_A = q_{2-3} = c_p\,(T_3 - T_2)$ and Heat rejected $= q_R = q_{4-1} = c_p\,(T_4 - T_1)$

Work done per cycle, $w = q_A - q_R = c_p\,[(T_3 - T_2) - (T_4 - T_1)]$

Hence, the thermal efficiency is given by

$$\eta_{th} = \frac{\text{Work done}}{\text{Heat supplied}} = \frac{c_p\,(T_3 - T_2) - c_p\,(T_4 - T_1)}{c_p\,(T_3 - T_2)} = 1 - \frac{(T_4 - T_1)}{\gamma\,(T_3 - T_2)}$$

Let $$r = \frac{v_1}{v_2} = \frac{v_4}{v_2} = \text{compression ratio}$$

From isentropic compression process 1 – 2, we have, $T_2 = T_1 \left(\dfrac{v_1}{v_2}\right)^{\gamma - 1} = T_1\, r^{\gamma - 1}$

$$\text{Let cut-off ratio} = \frac{v_3}{v_2} = \beta$$

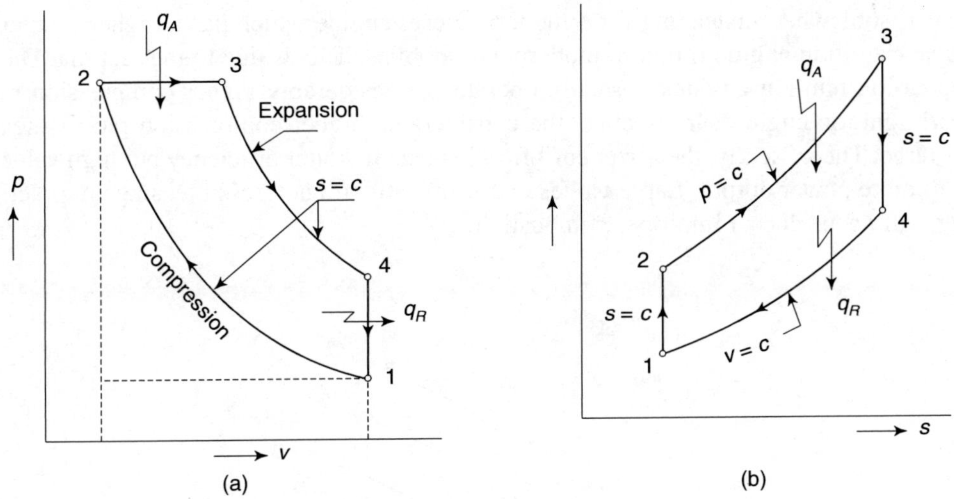

Fig. 13.11 Diesel cycle on *p-v* and *T-s* diagrams

From process 3 – 2, we have $\dfrac{T_3}{T_2} = \dfrac{v_3}{v_2} = \beta$ = cut-off ratio

or $$T_3 = T_2\,\beta = T_1 \,.\, \beta \,.\, r^{\gamma-1}$$

From isentropic expression process 3 – 4, we have $T_4 = T_3\left(\dfrac{v_3}{v_4}\right)^{\gamma-1} = T_3\left(\dfrac{v_3}{v_2}\cdot\dfrac{v_2}{v_4}\right)^{\gamma-1}$

or $$T_4 = T_3\left(\frac{\beta}{r}\right)^{\gamma-1} = T_1\,\beta\, r^{\gamma-1}\left(\frac{\beta}{r}\right)^{\gamma-1} = T_1\,\beta^{\gamma}$$

$\therefore$ $$\eta_{th} = \eta_{as} = 1 - \frac{1}{\gamma}\left[\frac{T_1\,\beta^{\gamma} - T_1}{\beta\, T_1\, r^{\gamma-1} - T_1\, r^{\gamma-1}}\right]$$

or $$\eta_{th} = \eta_{as} = 1 - \frac{1}{r^{\gamma-1}}\left[\frac{\beta^{\gamma}-1}{\gamma\,(\beta-1)}\right] \qquad ...(13.13)$$

Equation (13.13) gives the required expression for thermal efficiency in terms of compression ratio *r*, cut-off ratio β and adiabatic exponent γ. Figure 13.12 shows the variation of Diesel cycle efficiency with compression ratio (*r*) for various cut-off ratio β. It is obvious that the thermal efficiency of Diesel cycle increases with increase of compression ratio. It is also evident that as the cut-off ratio β increases, the value of terms in the bracket increases, hence the thermal efficiency decreases. When β = 1, the efficiency of the Diesel cycle is equal to the efficiency of Otto cycle. For a given compression ratio *r* under all circumstances when β is greater than unity, the efficiency of the Diesel cycle is less than that for Otto cycle. This is true because the term in the bracket in Eq. (13.13) is always positive. This comparative statement is somewhat misleading for a modern Diesel engine which has a higher efficiency than a spark ignition engine found in modern automobiles. This is due to the fact that Diesel or compression ignition engines normally operate at considerably higher compression ratio than spark ignition engines. In practice, the considerable higher compression ratio is nearly twice as large, Theoretically, the lower cut-off ratios lead to better efficiency but high value of β leads to more power output. Large values of cut-off ratio are not possible, since in practice, smoking will occur due to improper combustion.

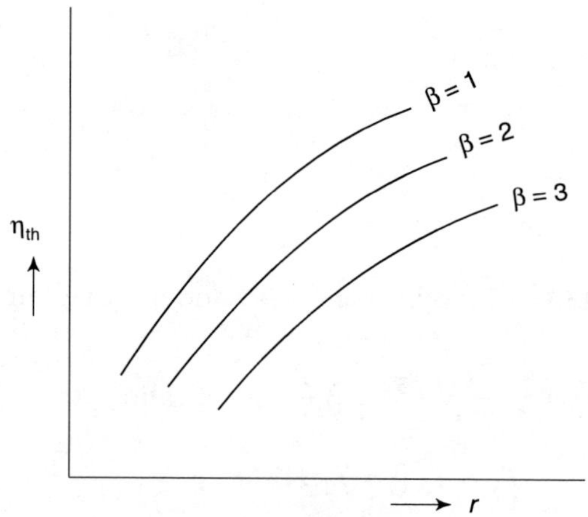

Fig. 13.12 Diesel cycle efficiency

$$\textbf{Mean effective pressure (m.e.p.)} = \frac{\text{Work done}}{\text{Displacement volume}} = \frac{c_p (T_3 - T_2) - c_v (T_4 - T_1)}{v_1 - v_2}$$

$$\text{But} \quad v_1 - v_2 = v_1\left(1 - \frac{v_2}{v_1}\right) = v_1\left(1 - \frac{1}{r}\right) = \frac{RT_1}{p_1}\left(\frac{r-1}{r}\right) = \frac{c_v(\gamma-1)T_1}{p_1}\left(\frac{r-1}{r}\right)$$

$$\therefore \quad \text{m.e.p.} = \frac{c_p (T_3 - T_2) - c_v (T_4 - T_1)}{c_v T_1 \dfrac{(\gamma-1)}{p_i}\left(\dfrac{r-1}{r}\right)} = p_1\left(\frac{r}{r-1}\right)\left(\frac{1}{\gamma-1}\right)\left[\gamma\left(\frac{T_3 - T_2}{T_1}\right) - \left(\frac{T_4 - T_1}{T_1}\right)\right]$$

or

$$\text{m.e.p.} = p_1 r\left[\frac{\gamma r^{\gamma-1}(\beta - 1) - (\beta^\gamma - 1)}{(r-1)(\gamma-1)}\right] \quad \text{...(13.14)}$$

Example 13.11 An air-standard Diesel cycle has a compression ratio of 16. The pressure at the beginning of the compression stroke is 1 and the temperature is 20°C. The maximum temperature is 1430°C. Determine the thermal efficiency and the mean effective pressure for this cycle.

Solution Refer to Fig. 13.11.

Given $r = \dfrac{v_1}{v_2} = 6$, $p_1 = 1$ bar, $T_2 = 273 + 20 = 293$ K, $T_3 = 273 + 1430 = 1703$ K

The thermal efficiency of diesel cycle, $\eta_{th} = 1 - \dfrac{1}{r^{\gamma-1}}\left[\dfrac{\beta^\gamma - 1}{\gamma(\beta - 1)}\right]$

Also $$T_2 = T_1\left(\frac{v_1}{v_2}\right)^{\gamma-1} = T_1\, r^{\gamma-1} = 293 \times 16^{0.4} = 887.79 \text{ K}$$

From process 2 – 3, $$\beta = \frac{v_3}{v_2} = \frac{T_3}{T_2} = \frac{1703}{887.79} = 1.918$$

$$\therefore \quad \eta_{th} = 1 - \frac{1}{16^{0.4}}\left[\frac{(1.918^{1.4} - 1)}{1.4\,(1.918 - 1)}\right] = 61.8\%$$ **Ans.**

Mean effective pressure (m.e.p.) $= p_1 \cdot r\left[\dfrac{\gamma \cdot r^{\gamma-1}(\beta - 1) - (\beta^\gamma - 1)}{(r-1)(\gamma-1)}\right]$

$$= 1 \times 16\left[\frac{1.4 \times 16^{0.4}\,(1.918 - 1) - (1.918^{1.4} - 1)}{(26-1)\,(1.4-1)}\right] = 3.851 \text{ bar}$$ **Ans.**

Example 13.12 A diesel engine operates on the air standard diesel cycle. The engine has six cylinders of 11 cm bore and 13 cm stroke. The engine runs at 2000 rpm. At the beginning of compression, the air is at 1 bar and 26°C. If the clearance volume is 12.5 per cent of the stroke volume, find (a) compression ratio, (b) pressure and temperature of the air after compression, and (c) thermal efficiency and power output if the air is heated to 1370°C.

Solution Refer to Fig. 13.11.

Stroke volume, $v_s = v_1 - v_2$ and Clearance volume $= v_c = v_2$

Hence, $$0.125 = \frac{v_2}{v_s} = \frac{v_2}{v_1 - v_2} = \frac{1}{(v_1/v_2) - 1} = \frac{1}{r-1}$$

(a) $\therefore$ Compression ratio, $r = 9$ **Ans.**

(b) $$v_1 = \frac{RT_1}{p_1} = \frac{0.287 \times 299}{1 \times 10^2} = 0.858 \text{ m}^3/\text{kg}$$

and $$v_2 = \frac{v_1}{r} = \frac{0.959}{9} = 0.0953 \text{ m}^2/\text{kg}$$

$$T_2 = T_1\left(\frac{v_1}{v_2}\right)^{\gamma-1} = 299\,(9)^{0.4} = 719.992 \text{ K}$$ **Ans.**

Also $$p_2 = p_1\left(\frac{v_1}{v_2}\right)^{\gamma} = (9)^{1.4} = 21.674 \text{ bar} \qquad \textbf{Ans.}$$

(c) $$\beta = \frac{v_3}{v_2} = \frac{T_3}{T_2} = \frac{1370 + 273}{719.992} = \frac{1643}{719.992} = 2.281$$

$$\eta_{th} = 1 - \frac{1}{r^{\gamma - 1}}\left[\frac{(\beta^{\gamma} - 1)}{\gamma(\beta - 1)}\right] = 1 - \frac{1}{9^{0.4}}\left[\frac{(2.281^{1.4} - 1)}{1.4\,(2.281 - 1)}\right]$$

$$= 0.497 \text{ or } 49.7\% \qquad \textbf{Ans.}$$

Stroke volume, $V_s = \frac{\pi}{4} \times (0.11)^2 \times 0.13 = 0.001235 \text{ m}^3$

$$V_1 = V_s + V_2 = V_s + 0.125\, V_s = 0.001235 \times 1.125$$

$$= 0.001389 \text{ m}^3$$

The mass of air for the six cylinder, $m = \frac{6\,V_1}{v_1} = \frac{6 \times 0.001389}{0.858} = 0.00971$ kg

Heat added $= Q_A = mc_p\,(T_3 - T_2) = 0.00971 \times 1.005\,(1643 - 719.992) = 9.007$ kJ

Now $$\frac{T_4}{T_3} = \left(\frac{v_3}{v_2}\right)^{\gamma - 1} = \left(\frac{v_3}{v_1}\right)^{\gamma - 1}$$

But $$v_3 = 2.281\, v_2 = \frac{2.281\, v_1}{r}$$

or $$\frac{v_3}{v_1} = \frac{2.281}{9} = 0.2534$$

$\therefore$ $$T_4 = 1643\,(0.2534)^{0.4} = 948.76 \text{ K}$$

Heat rejected, $Q_R = mc_v\,(T_4 - T_1) = 0.00971 \times 0.718\,(948.76 - 299) = 4.5200$ kJ

$$\text{Work done cycle} = Q_A - Q_R = 9.007 - 4.5299 = 4.4771 \text{ kJ}$$

$$\text{Power developed} = \frac{W \times N}{2 \times 60} = \frac{4.4771 \times 2000}{2 \times 60} = 74.618 \text{ kW} \qquad \textbf{Ans.}$$

Exampl 13.13 A diesel engine working on air standard diesel cycle takes in air at 1 bar and 15°C. The specific volume of air at the inlet is 0.8 m^3/kg. The compression ratio is 14 and 480 kJ/kg of heat is added at constant pressure. Determine (a) pressure, sp. volume and temperature at each point of the cycle, (b) thermal efficiency, (c) mean effective pressure, and (d) power developed by the engine if it consumes 0.2 kg/s of air.

Solution Refer to Fig. 13.11.

(a) **Point 1** $p_1 = 1 \text{ bar}, T_1 = 273 + 15 = 288 \text{ K}$

and $v_1 = 0.8 \text{ m}^3/\text{kg}$

Point 2 $v_2 = \frac{v_1}{r} = \frac{0.8}{14} = 0.05714 \text{ m}^3/\text{kg}$ **Ans.**

From isentropic compression 1 – 2, $p_2 = p_1 \left(\frac{v_1}{v_2}\right)^\gamma = 1 \times (14)^{1.4} = 40.23 \text{ bar}$ **Ans.**

$$T_2 = T_1 \left(\frac{v_1}{v_2}\right)^{\gamma - 1} = 288\,(14)^{0.4} = 827.64 \text{ K}$$ **Ans.**

Point 3 Heat added $= q_A = 840 = c_p\,(T_3 - T_2) = 1.005\,(T_3 - 827.64)$

or $T_3 = 1663.46 \text{ K}$ **Ans.**

From constant pressure heat addition process 2 – 3, $\frac{v_3}{v_2} = \frac{T_3}{T_2} = \frac{1663.46}{827.64} = 2.0098$

or $v_3 = 2.0098 \times 0.05714 = 0.1148 \text{ m}^3/\text{kg}$ **Ans.**

$p_3 = p_2 = 40 \text{ bar}$ **Ans.**

Point 4 From isentropic expansion process 3 – 4.

$$T_4 = \left(\frac{v_3}{v_4}\right)^{\gamma - 1} T_3 = \frac{T_3}{(v_4 / v_3)^{\gamma - 1}} = \frac{1663.46}{(0.8 / 0.1148)^{0.4}} = 765.16 \text{ K}$$ **Ans.**

From process 4 – 1, $p_4 = \frac{p_1 \times T_4}{T_1} = \frac{1 \times 765.16}{288} = 2.656 \text{ bar}$ **Ans.**

(b) Heat rejected $= q_R = c_v\,(T_4 - T_1) = 0.718\,(765.11 - 288) = 342.6 \text{ kJ/kg}$

∴ Work done $= w = q_A - q_R = 840 - 342.6 = 497.4 \text{ kJ/kg}$

∴ Thermal efficiency $= \frac{w}{q_A} = \frac{497.4}{840} = 59.2\%$ **Ans.**

(c) $$\text{m.e.p.} = \frac{\text{Work done}}{\text{Displacement volume}} = \frac{w}{v_1 - v_2} = \frac{497.4 \times 10^{-2}}{(0.8 - 0.05714)} = 6.6957 \text{ bar}$$ **Ans.**

Example 13.14 The German Mercedes "1900" car has a four stroke cylinder in line diesel engine with comprssion ratio 20 : 1 and expasion ratio 10 : 1. Calculate (a) the cur-off ratio and (b) air standard efficiency.

Solution Refer to Fig. 13.11. Cut-off ratio, $\beta = \frac{v_3}{v_2} = \frac{v_3}{v_4} \times \frac{v_1}{v_2}$

But compression ratio, $r = \dfrac{v_1}{v_2} = 20$ and expansion ratio $= \dfrac{v_4}{v_3} = 10$

(a) $\therefore$ $$\beta = \frac{20}{10} = 2$$ **Ans.**

(b) Air standard efficiency of diesel cycle is given by

$$\eta_{an} = 1 - \frac{1}{r^{\gamma-1}}\left[\frac{(\beta^{\gamma}-1)}{\gamma(\beta-1)}\right] = 1 - \frac{1}{20^{0.4}}\left[\frac{(2^{1.4}-1)}{1.4(2-1)}\right] = 0.6469 \text{ or } 64.69\%$$ **Ans.**

Example 13.15 A Diesel engine operating on a standard Diesel cycle has 20 cm bore and 30 cm stroke. The clearance volume is 420 cm^3. The fuel is injected at constant pressure for 5% of the stroke. Calculate the air-standard efficiency. If the cut-off is delayed from 5 to 8% what will be the percentage loss in efficiency. In both cases, the compression ratio is the same.

Solution Refer to Fig. 13.11.

$$\text{Clearance volume} = V_c = V_2 = 420 \text{ cm}^3$$

$$\text{Swept volume} = V_s = V_1 = V_2 = \frac{\pi}{4}d^2 \times L = \frac{\pi}{4} \times 20^2 \times 30 = 9424.778 \text{ cm}^2$$

$$\text{Cylinder volume} = V_1 = V_s + V_2 = 9423.778 + 420 = 9844.77 \text{ cm}^3$$

$$\text{Cut-off ratio} = \beta = \frac{V_3}{V_2} = \frac{891.238}{420} = 2.121$$

and $$\text{Compression ratio, } r = \frac{V_1}{V_2} = \frac{9844.778}{420} = 23.439$$

Efficiency of diesel cycle is given by

$$\eta_{th1} = 1 - \frac{1}{r^{\gamma-1}}\left[\frac{(\beta^{\gamma}-1)}{\gamma(\beta-1)}\right] = 1 - \frac{1}{23.449^{0.4}}\left[\frac{(2.121^{1.4}-1)}{1.4(2.121-1)}\right] = 66.34\%$$ **Ans.**

If the fuel is cut-off at 8% of the stroke

$$V_3 = 420 + 0.08 \times 9424.778 = 1173.982 \text{ m}^3$$

and $$\text{Cut-off ratio, } \beta = \frac{V_3}{V_2} = \frac{1173.982}{420} = 2.795$$

Hence, the new thermal efficiency is

$$\eta_{th2} = 1 - \frac{1}{r^{\gamma-1}}\left[\frac{(\beta^{\gamma}-1)}{\gamma(\beta-1)}\right] = 1 - \frac{1}{23.439^{0.4}}\left[\frac{(2.795^{1.4}-1)}{1.4 \times 1.795}\right] = 0.637590 \text{ or } 63.759\%$$ **Ans.**

$$\text{Percentage loss} = \frac{\eta_{th1} - \eta_{th2}}{\eta_{th1}} \times 100 = \frac{0.6634 - 0.63759}{0.6634} \times 100 = 3.89\%$$ **Ans.**

Example 13.16 A diesel engine has a compression ratio of 14. The fuel is cut-off at 0.08 stroke. The relative efficiency is 52%. Find the mass of the fuel of calorific value 41800 kJ/kg which would be required per kW per hour.

Solution Refer to Fig. 13.13.

$$\text{Compression ratio} = r = \frac{V_2}{V_1} = \frac{v_2}{v_1} = 14$$

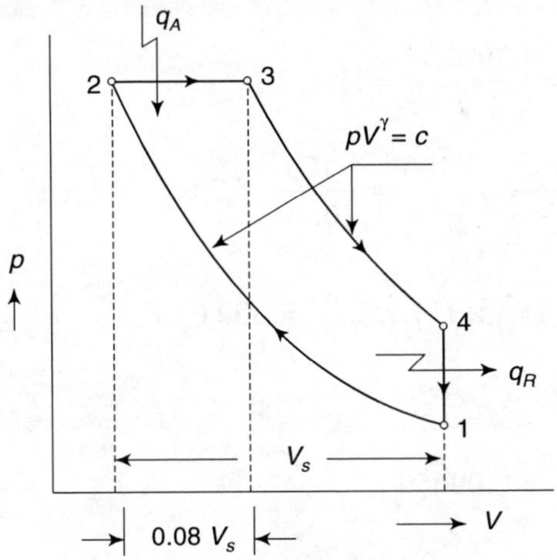

Fig. 13.13 Example 13.16

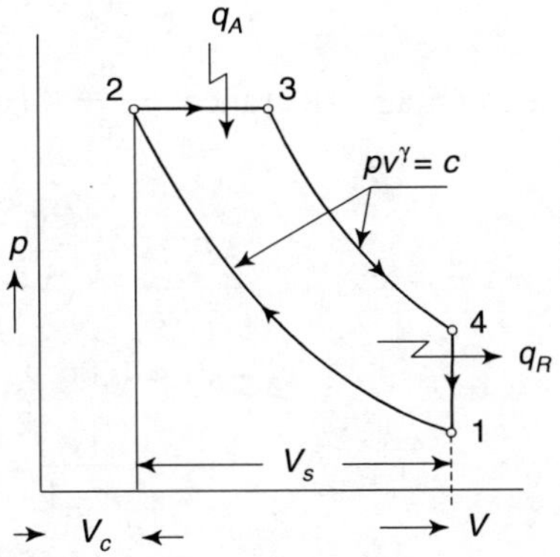

Fig. 13.14

$$V_s = \text{Stroke volume} = V_1 - V_2 \text{ and } V_3 - V_2 = 0.08\ V_s$$

$\therefore$
$$V_3 - V_2 = 0.08\ (V_1 - V_2) \text{ or } \frac{V_3}{V_2} - 1 = 0.08\left(\frac{V_1}{V_2} - 1\right)$$

or
$$\frac{V_3}{V_2} = 0.08\ (14 - 1) + 1 \text{ or } \beta = \frac{V_3}{V_2} = 2.04$$

$$\eta_{as} = 1 - \frac{1}{r^{\gamma-1}}\left[\frac{(\beta^\gamma - 1)}{\gamma\,(\beta - 1)}\right] = 1 - \frac{1}{14^{0.4}}\left[\frac{(2.04^{1.4} - 1)}{1.4\,(2.04 - 1)}\right]$$

$$= 0.5904 \text{ or } 59.04\%$$

or Actual thermal efficiency, $\eta_{tha} = \eta_{rel} \times \eta_{as} = 0.52 \times 0.5904 = 0.3071$

or
$$\eta_{tha} = \frac{\text{Actual work done, kJ}}{\text{Heat added, kJ}} = 0.3071 = \frac{\text{Work done}}{C.V. \times \dot{m}_f} = \frac{\text{Work done}}{41800 \times \dot{m}_f}$$

or
$$\frac{\text{Work done}}{\dot{m}_f} = 0.3071 \times 41800 \text{ kJ/kg of fuel} = \frac{0.3071 \times 41800}{3600} \text{ kWh/kg of fuel}$$

$\therefore$ Mass of fuel $= \dfrac{3600}{0.3071 \times 41800} = 0.2804$ kg/kWh. **Ans.**

Example 13.17 A diesel engine working on diesel cycle takes in air at pressure 1 bar and temperature 30°C. The pressure at the end of compresion is 30 bar and the cut-off is 6% of the stroke. Calculate (a) the compression ratio, (b) the percentage clearance, (c) the heat rejected, (d) the thermal efficiency, and (e) the mean effective pressure.

Solution Refer to Fig. 13.14. Assume 1 kg of working fluid.

(a) Compression ratio, $r = \dfrac{v_1}{v_2} = \left(\dfrac{p_2}{p_1}\right)^{1/\gamma} = \left(\dfrac{30}{1}\right)^{1/1.4} = 11.352$

(b) Percentage clearance $= \dfrac{v_2}{v_s} \times 100 = \dfrac{v_2}{v_1 - v_2} = \dfrac{100}{r-1} = \dfrac{100}{10.352} = 9.659\%$ **Ans.**

(c) $T_2 = T_1 \left(\dfrac{v_1}{v_2}\right)^{\gamma - 1} = 303 \times (11.352)^{0.4} = 800.67$ K

and $v_3 - v_2 = 0.06\, v_s$

But $v_2 = \dfrac{v_s}{r-1} = \dfrac{v_s}{10.352} = 0.09659\, v_s$

or $v_3 = 0.09659\, v_s + 0.06\, v_s = 0.1565\, v_s$

And $\dfrac{v_3}{v_2} = \beta = \dfrac{0.1565}{0.09650} = 1.6202$

From process 2 – 3, $T_3 = T_2 \dfrac{v_3}{v_2} = 800.67 \times 1.6202 = 1297.24$ K

Hear added $= q_A = c_p\,(T_3 - T_2) = 1.005\,(1297.24 - 800.67) = 499.05$ kJ/kg **Ans.**

From isentropic process 3 – 4,

$$T_4 = T_3 \left(\frac{v_3}{v_4}\right)^{\gamma - 1} = T_3 \left(\frac{v_3}{v_1}\right)^{0.4} = 1297.24 \left(\frac{0.1565\, v_s}{11.352 \times 0.09659\, v_s}\right)^{0.4} = 595.42 \text{ K}$$

Heat rejected $= q_A = c_v\,(T_4 - T_1) = 0.718\,(595.42 - 303) = 209.957$ kJ/kg **Ans.**

(d) $\eta_{th} = 1 - \dfrac{1}{r^{\gamma-1}} \left[\dfrac{(\beta^\gamma - 1)}{\gamma\,(\beta - 1)}\right] = 1 - \dfrac{1}{11.352^{0.4}} \left[\dfrac{(1.6202^{1.4} - 1)}{1.4\,(0.6202)}\right] = 0.5793$ or 57.93% **Ans.**

(e) Mean effective pressure is given by

$$p_1 r \left[\frac{\gamma r^{\gamma - 1}(\beta - 1) - (\beta^\gamma - 1)}{(r-1)(\gamma - 1)}\right]$$

$$= 1 \times 11.352\left[\frac{1.4 \times 11.352^{0.4}\,(1.6202 - 1) - (1.6202^{1.4} - 1)}{(11.352 - 1)\,(1.4 - 1)}\right] = 3.644 \text{ bar} \qquad \textbf{Ans.}$$

13.6 DUAL CYCLE (LIMITED PRESSURE CYCLE)

Since it is difficult to achieve either constant volume or constant pressure heat addition in actual spark ignition or compression ignition engines due to appreciable amount of time required for the completion of combustion process, therefore, a **dual cycle** has been proposed. This special cycle combines the features of the theoretical Otto and Diesel cycles and the end result is a theoretical dual cycle which probably comes a little closer to matching the actual performance of modern compression ignition engine. In other words, in dual cycle, part of heat addition takes place at constant volume and rest at constant pressure. The dual cycle is also called ***mixture or limited presure cycle***. Figure 13.15 shows the representation of fual cycle on p-v and T-s diagrams. The various processes are:

1–2 : Isentropic compression of air resulting in increase in pressure and temperature.

2–3 : Addition of heat at constant volume resulting in increase in pressure and temperature.

3–4 : Addition of heat at constant pressure resulting in increase in volume and temperature.

4–5 : Isentropic expansion of air from high temperature and pressure to lower pressure and temperature to do work.

5–1 : Rejection of heat at constant volume till the air is restored to its initial state.

Consider 1 kg of working fluid.

$$\text{Heat added} = q_A = c_v\,(T_3 - T_2) + c_p\,(T_4 - T_3)$$

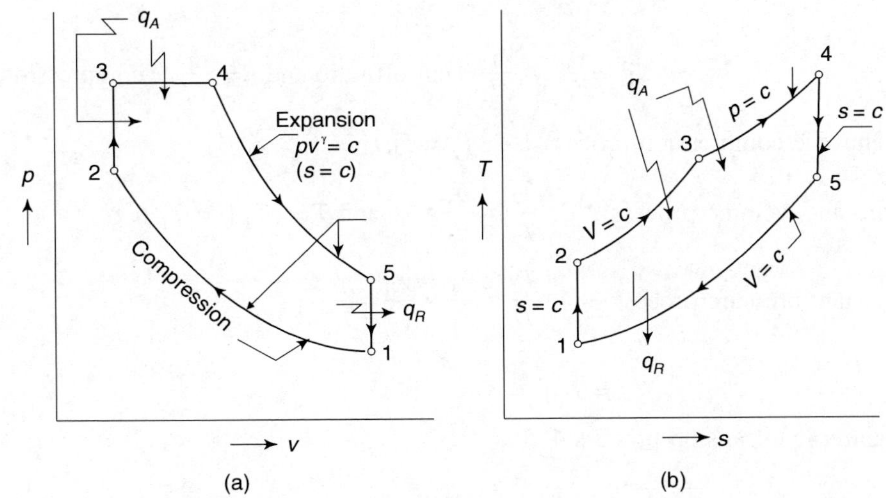

Fig. 13.15 Dual cycle on p-v and T-s diagrams

$$\text{Heat rejected} = q_R = c_v\,(T_5 - T_1)$$

$$\text{Work done per cycle} = q_A - q_R = c_v\,(T_3 - T_2) + c_p\,(T_4 - T_3) - c_v\,(T_5 - T_1)$$

$$\text{Thermal efficiency} = \frac{\text{Work done}}{\text{Heat supplied}} = \frac{c_v\,(T_3 - T_2) + c_p\,(T_4 - T_3) - c_v\,(T_5 - T_1)}{c_v\,(T_3 - T_2) + c_p\,(T_4 - T_3)}$$

$$= 1 - \frac{(T_5 - T_1)}{(T_3 - T_2) + \gamma\,(T_4 - T_3)} \qquad \text{...(13.15)}$$

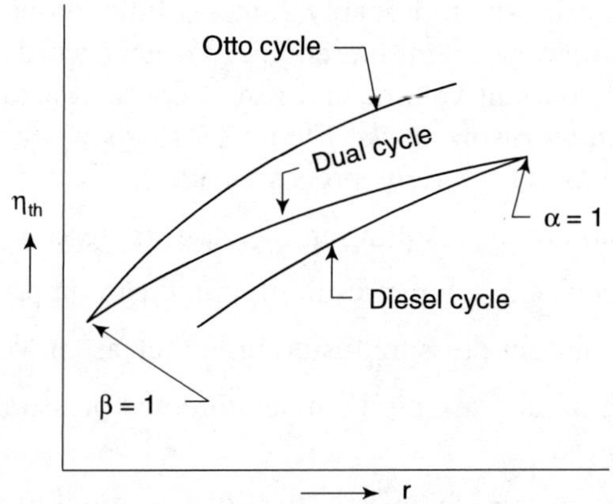

Fig. 13.16 η_{th} of dual cycle verus r

Let $\alpha = \dfrac{p_3}{p_2}$ = pressure ratio for constant volume heating

$\beta = \dfrac{v_4}{v_3} = \dfrac{v_4}{v_2}$ = cut-off ratio and $r = \dfrac{v_1}{v_2}$ = compression ratio

From isentropic compression process 1–2, $T_2 = T_1\, r^{\gamma - 1}$

From constants volume process 2–3, $\dfrac{T_3}{T_2} = \dfrac{p_3}{p_2} = \alpha$ and $T_3 = T_2\,\beta = T_1\,\alpha\, r^{\gamma - 1}$

From constant pressure process 3–4, $\dfrac{T_4}{T_3} = \dfrac{v_4}{v_3} = \dfrac{v_4}{v_2} = \beta$

or $$T_4 = T_3\,\beta = T_1 \cdot \alpha \cdot \beta \cdot r^{\gamma-1}$$

From isentropic expansion process 4–5

$$\frac{T_5}{T_4} = \left(\frac{v_4}{v_5}\right)^{\gamma-1} = \left(\frac{v_4}{v_2}\frac{v_2}{v_5}\right)^{\gamma-1} \left(\frac{v_4}{v_2}\cdot\frac{v_2}{v_1}\right)^{\gamma-1} = \left(\frac{\beta}{r}\right)^{\gamma-1}$$

or $$T_5 = T_4\left(\frac{\beta}{r}\right)^{\gamma-1} = T_1\cdot\alpha\cdot\beta\cdot r^{\gamma-1}\cdot\beta^{\gamma-1}\cdot r^{1-\gamma} = T_1\,\alpha\,\beta^{\gamma}$$

$\therefore$ $$\eta_{th} = 1-\frac{(\alpha\,\beta^{\gamma}-1)}{(\alpha\,r^{\gamma-1}-r^{\gamma-1})+\gamma\,(\alpha\,\beta\,r^{\gamma-1}-r^{\gamma-1})}$$

or $$\eta_{th} = 1-\frac{1}{r^{\gamma-1}}\left[\frac{(\alpha\,\beta^{\gamma}-1)}{(\alpha-1)+\gamma\,\alpha\,(\beta-1)}\right] \qquad ...(13.16)$$

Figure 13.16 shows the variation of dual cycle efficiency with compression ratio. At cut-off ratio, $\beta = 1$, the efficiency of dual cycle is equal to that of the Otto and at pressure ratio, $\alpha = 1$, it is equal to diesel cycle efficiency.

Mean effective pressure is calculated as follows:

$$\text{m.e.p.} = \frac{\text{Work done}}{\text{Displacement volume}} = \frac{w}{v_1 - v_2} = \frac{c_v\,(T_3 - T_2) + c_p\,(T_4 - T_3) - c_v\,(T_5 - T_1)}{v_1 - v_2}$$

But $$v_1 - v_2 = v_2\left(1-\frac{1}{r}\right) = v_1\left(\frac{r-1}{r}\right) = \frac{RT_1}{p_1}\left(\frac{r-1}{r}\right) = \frac{c_v\,(v-1)\,T_1}{p_1}\left(\frac{r-1}{r}\right)$$

$\therefore$ $$\text{m.e.p.} = \frac{p_1\cdot r}{(\gamma-1)\,(r-1)}\left[\frac{T_3-T_2}{T_1}+\frac{\gamma\,(T_4-T_3)}{T_1}-\frac{T_5-T_1}{T_1}\right]$$

$$= \frac{p_1\cdot r}{(\gamma-1)\,(r-1)}\,[r^{\gamma-1}\,(\alpha-1)+\alpha\,(\beta-1)\,\gamma\,r^{\gamma-1}-(\alpha\,\beta^{\gamma}-1)]$$

or $$\text{m.e.p.} = \frac{p_1\cdot r}{(\gamma-1)\,(r-1)}\,[r^{\gamma-1}\,\{\alpha-1)+\gamma\,\alpha\,(\beta-1)\}-(\alpha\,.\,\beta^{\gamma}-1)] \qquad ...(13.17)$$

This is the required expression for m.e.p. of dual cycle.

Example 13.18 An air standard dual cycle has pressure and temperature at the beginning of compression as 1 bar and 35°C, respectively. The compression ratio is 11, the pressure ratio during heat addition 1.6 and cut-off ratio 1.7. Calculate (a) the percentage clearance, (b) the pressure, temperature and volume at each point of the cycle, (c) heat added, (d) heat rejected, (e) work done, (f) thermal efficiency, and (g) mean effective pressure. Assume 0.5 kg of air flowing in the cycle.

Solution Refer to Fig. 13.15.

(a) Percentage clearance $= \dfrac{V_2}{V_s}\times 100 = \dfrac{V_2}{V_1 - V_2}\times 100 = \dfrac{100}{r-1} = \dfrac{100}{11-1} = 10\%$

(b) **Point 1** $\quad V_1 = \dfrac{mRT_1}{p_1} = \dfrac{0.5\times 0.287\times 308}{1\times 10^2} = 0.4419\ \text{m}^3$ **Ans.**

Point 2 $p_2 = p_1 \left(\frac{V_1}{V_2}\right)^{\gamma} = 1 \times (11)^{0.4} = 2.609$ bar **Ans.**

$$T_2 = T_1 \left(\frac{V_1}{V_2}\right)^{\gamma - 1} = 308\ (11)^{0.4} = 803.72 \text{ K}$$ **Ans.**

$$V_3 = \frac{V_1}{r} = \frac{0.4419}{11} = 0.0401 \text{ m}^3$$ **Ans.**

Point 3 $p_3 = p_2 \times \alpha = 28.7 \times 1.6 = 45.92$ bar **Ans.**

$$T_3 = T_2 \left(\frac{p_3}{p_2}\right) = 803.72 \times 1.6 = 1285.95 \text{ K}$$ **Ans.**

Point 4 $V_4 = V_2 \times \beta = 0.0401 \times 1.7 = 0.0681 \text{ m}^3$ **Ans.**

$$T_4 = T_3 \left(\frac{V_4}{V_3}\right) = 1285.95 \left(\frac{0.0681}{0.0401}\right) = 2183.87 \text{ K}$$ **Ans.**

Point 5 $T_5 = T_4 \left(\frac{V_4}{V_5}\right)^{\gamma - 1} = 2183.87 \left(\frac{0.0681}{0.4419}\right) = 1033.6 \text{ K}$ **Ans.**

$$p_5 = p_4 \left(\frac{V_4}{V_5}\right)^{\gamma} = 45.92 \left(\frac{0.0681}{0.4419}\right)^{1.4} = 3.349 \text{ bar}$$ **Ans.**

(c) Heat added, $Q_A = mc_v\ (T_3 - T_2) + mc_p\ (T_4 - T_3)$

or $Q_A = 0.5 \times 0.718\ (1285.95 - 803.72) + 0.5 \times 1.005\ (2183.87 - 1285.95)$

$= 624.24$ kJ **Ans.**

(d) Heat rejected, $Q_R = mc_v\ (T_5 - T_1) = 0.5 \times 0.718\ (1033.6 - 308) = 260.49$ kJ **Ans.**

(e) Work done, $Q_A - Q_R = 624.24 - 260.49 = 363.75$ kJ **Ans.**

(f) Thermal efficiency $= \frac{\text{Work done}}{\text{Heat added}} = \frac{363.75}{624.24} = 58.27\%$ **Ans.**

(g) m.e.p. $= \frac{\text{Work done}}{\text{Stroke volume}} = \frac{363.75}{0.4419 - 0.0401} = 905.3 \text{ kN/m}^2 = 9.053$ bar **Ans.**

Example 13.19 During a test on a four-stroke six cylinder engine having 11 cm bore and 16 cm stroke and working on the dual combustion cycle, the fuel consumption had been found to be 0.18 kg/min. The engine runs at 1600 r.p.m. The compression ratio is 10 and calorific value of fuel used is 44000 kJ/kg. At the beginning of compression, the pressure is 0.95 bar and temperature is 45°C. The law of compression and expansion is polytropic $pV^{1.35}$ = constant. Assume that 2/3rds of heat released by the combustion of fuel is at constant volume and remainder 1/3rd at constant pressure. Calculate: (a) heat supplied per cylinder charge, (b) pressure and temperature at each corner of the cycle, (c) work output, (d) cycle efficiency, and (e) mean effective pressure. Assume $c_p = 1.005$ and $c_v = 0.718$ kJ/kg K for air.

Solution Refer to Fig. 13.15.

$$\text{Stroke volume, } V_s = V_1 - V_2 = \frac{\pi}{4} d^2 \times L = \frac{\pi}{4} \times (0.11)^2 \times 0.16 = 1.5205 \times 10^{-3} \text{ m}^3$$

$$\text{Fuel used per cycle per cylinder} = \frac{0.18}{6 \times 1600/2} = \frac{0.18}{6 \times 800} \text{ kg}$$

Since there are $\frac{N}{2}$ cycles in N r.p.m.

$$\text{Heat supplied per cylinder per cycle, } Q_A = 44000 \times \frac{0.18}{6 \times 800} = 1.65 \text{ kJ}$$

(a) Heat supplied per cycle in all six cylinders = $(Q_A)_{\text{total}} = 1.65 \times 6 = 9.9$ kJ **Ans.**

(b) $$V_s = V_1 - V_2 = 1.5205 \times 10^{-3} \text{ m}^3$$

and $$V_2 = \frac{V_s}{r-1}. \quad \text{But } r = \frac{V_1}{V_2} = 10$$

∴ $$V_2 = 1.5205 \times 10^{-3}/9 = 0.16894 \times 10^{-3} \text{ m}^3$$

and $$V_1 = 10 \times V_2 = 10 \times 0.16894 \times 10^{-3} \text{ m}^3$$

$$\text{The mass of air per cylinder} = m = \frac{p_1 V_1}{RT_1} = \frac{0.95 \times 10^2 \times 1.6894 \times 10^{-3}}{0.287 \times 318} = 0.00175 \text{ kg}$$

$$\text{Heat supplied per cycle in each cylinder} = \frac{1.65}{0.00175} = 942.85 \text{ kJ/kg}$$

From polytropic compression process 1–2, we have

$$T_2 = T_1 \left(\frac{V_1}{V_2}\right)^{n-1} = 318 \times 10^{0.35} = 711.9 \text{ K}$$ **Ans.**

$$p_2 = p_1 \left(\frac{V_1}{V_2}\right)^{n} = 0.95 \times (10)^{1.35} = 21.26 \text{ bar}$$ **Ans.**

Heat added per cycle per kg of air in each cylinder during constant volume process 2–3 is

$$q_{2-3} = c_v (T_3 - T_2) = 942.85 \times 2/3 = 628.566 \text{ kJ/kg} = 0.718 (T_3 - 719.91)$$

or $$T_3 = 1595.35 \text{ K}$$ **Ans.**

$$p_3 = \frac{T_3}{T_2} p_2 = \frac{1595.35}{711.91} \times 21.26 = 47.64 \text{ bar}$$ **Ans.**

Heat added per cycle at constant volume per kg of air in each cylinder is

$$q_{3-4} = c_p (T_4 - T_3) = (1/3) \times 942.85 = 314.28 \text{ kJ/kg} = 1.005 (T_4 - 1595.35)$$

or $$T_4 = 1908.06 \text{ K}$$ **Ans.**

Also $$V_4 = V_3 \frac{T_4}{T_3} = 0.16894 \times 10^{-3} \times \frac{1908.06}{1595.35} = 0.202 \times 10^{-3} \text{ m}^3$$ **Ans.**

From expansion process 4–5, we have

$$T_5 = T_4 \left(\frac{V_4}{V_5}\right)^{\gamma - 1} = 1908.06 \left(\frac{0.202 \times 10^{-3}}{1.6894 \times 10^{-3}}\right)^{0.35} = 907.31 \text{ K}$$ **Ans.**

$$p_5 = p_4 \left(\frac{V_4}{V_5}\right)^{\gamma} = 47.6 \times \left(\frac{0.202 \times 10^{-3}}{1.6894 \times 10^{-3}}\right)^{1.35} = 2.706 \text{ bar}$$ **Ans.**

Work done by each cylinder/cycle = Area enclosed by the cycle

$$= p_4 (V_4 - V_3) + \frac{p_4 V_4 - p_5 V_5}{n - 1} - \frac{p_2 V_2 - p_1 V_1}{n - 1} = mR\left[(T_4 - T_3) + \frac{(T_4 - T_5)}{n - 1} - \frac{(T_2 - T_1)}{n - 1}\right]$$

$$= 0.00175 \times 0.287 \left[(1908.06 - 1595.35) + \frac{(1908.06 - 907.31}{0.35} - \frac{711.91 - 318}{0.35}\right]$$

$= 1.0278$ kJ/cylinder **Ans.**

Total work done = 1.0278 × 6 = 6.1668 kJ

(d) $$\text{Thermal efficiency} = \frac{\text{Work done}}{\text{Heat added}} = \frac{1.0278}{1.65} = 62.29\%$$ **Ans.**

(e) $$\text{m.e.p.} = \frac{\text{Work done per cylinder}}{\text{Stroke volume}} = \frac{1.0278}{1.5205 \times 10^{-3}}$$

$$= 67.596 \text{ kN/m}^2 = 6.7596 \text{ bar}$$ **Ans.**

Example 13.20 An engine works on the dual combustion cycle, the compression ratio being 11. The pressure at the commencement of compression is 1 bar and the temperature is 90°C. The maximum pressure in the cycle is 50 bar and the constant pressure heat reception continues for 1/20 of the stroke. Calculate the work done per kg of air and the ideal thermal efficiency. Assume $c_p = 1.005$, $c_v = 0.718$ kJ/kg K.

Solution Refer to Fig. 13.17.

$$\gamma = \frac{c_p}{c_v} = \frac{1.005}{0.718} = 1.4$$

Let $$v_2 = v_3 = 1 \text{ unit}$$

$\therefore \qquad v_1 = r = 11$ unit

and $\qquad$ Stroke volume $= v_s = v_1 = v_2 = 11 - 1 = 10$ units

$$\therefore \qquad v_4 = v_3 + (v_4 - v_3) = 1 + \frac{11-1}{20} = 1.5 \text{ and } \frac{v_4}{v_3} = 1.5$$

From process 1–2, $\qquad T_2 = T_1\, r^{0.4} = 363 \times 1 \times 11^{0.4} = 947.24$ K

$$p_2 = p_1\, r^{\gamma} = 1 \times 11^{1.4} = 28.704 \text{ bar}$$

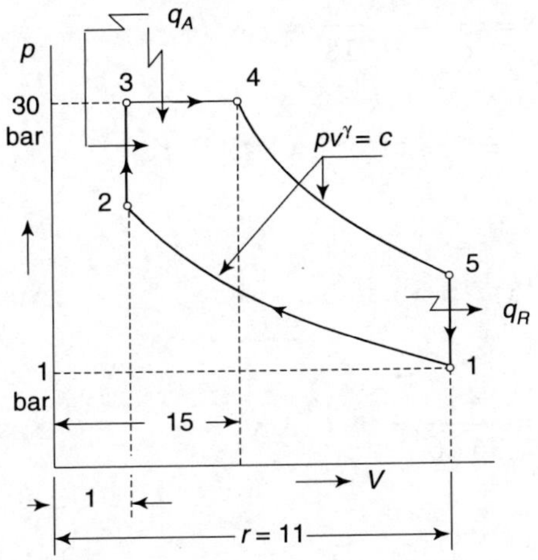

Fig. 13.17 Exmple 13.20

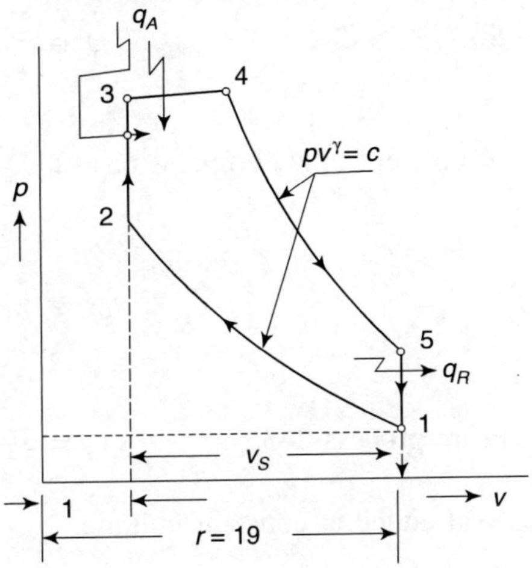

Fig. 13.18 Example 13.21

From process 2–3, $\qquad T_3 = T_2 \dfrac{p_3}{p} = 947.24 \times \dfrac{50}{28.704} = 1650$ K

From process 3–4, $\qquad T_4 = T_3 \dfrac{v_4}{v_3} = 1650 \times 1.5 = 2475$ K

From process 4–5, $\qquad T_5 = T_4 \left(\dfrac{v_4}{v_5}\right)^{\gamma - 1} = 2475 \left(\dfrac{1.5}{11}\right)^{0.4} = 1115.46$ K

Work done = heat added – heat rejected $= c_v\,(T_3 - T_2) + c_p\,(T_4 - T_3) - c_v\,(T_5 - T_1)$

$= 0.718\,(1650 - 947.24) + 1.005\,(2475 - 1650) - 0.718\,(1115.46 - 363)$

$= 973.44$ kJ/kg of air **Ans.**

$$\text{Thermal efficiency} = \frac{\text{Work done}}{\text{Heat added}} = \frac{793.44}{0.718\,(1650 - 947.24) + 1.005\,(2475 - 1650)}$$

$= 59.49\%$ **Ans.**

Example 13.21 A compression ignition engine working on the dual cycle takes in two-thirds of its total heat supply at constant volume and one third at constant pressure. The compression ratio is 13 : 1. The maximum pressure in the cycle is 45 bar. Air intake takes place at 15°C and 1 bar. Calculate (a) the temperature at each point of the cycle, and (b) thermal efficiency of the cycle. Assume $c_p = 1.005$ and $c_v = 0.718$ kJ/kg.

Solution Refer to Fig. 13.18. (a) The clearance volume

$$v_v = v_2 = v_3 = 1 \text{ unit} \quad \text{and} \quad v_1 = 13 \text{ unit}$$

So

$$r = \frac{v_1}{v_2} = 13 \quad \text{and } \gamma = \frac{c_p}{c_v} = \frac{1.005}{0.718} = 1.4$$

From isentropic compression1–2, $T_2 = T_1 \left(\frac{v_1}{v_2}\right)^{\gamma - 1} = 288\ (13)^{0.4} = 803.47\text{ K}$

$$p_2 = p_1 \left(\frac{v_1}{v_2}\right)^{\gamma} = 1\ (13)^{1.4} = 36.26 \text{ bar} \qquad \textbf{Ans.}$$

From process 2–3,

$$T_3 = T_2 \frac{p_3}{p_2} = 803.47 \times \frac{45}{36.26} = 997.13 \text{ K} \qquad \textbf{Ans.}$$

Heat added at constant volume,

$$q_{2-3} = c_v\ (T_3 - T_2) = 0.718\ (997.13 - 803.47) = 139.04 \text{ kJ/kg of air}$$

Heat supplied during process 3–4, $q_{3-4} = \frac{1}{2} \times 139.04 = 69.52$ kJ/kg

Because q_{2-3} is 2/3 of the total heat supplied.

$$\therefore \quad q_{3-4} = 69.52 = c_p\ (T_4 - T_3) = 1.005\ (T_4 - 997.13)$$

or

$$T_4 = 1066.3 \text{ K}$$

$$v_4 = v_3 \frac{T_4}{T_3} = 1 \times \frac{1066.3}{997.13} = 1.069 \text{ m}^3\text{/kg}$$

From process 3–4,

$$T_5 = T_4 \left(\frac{v_4}{v_5}\right)^{\gamma - 1} = 1066.3 \left(\frac{1.069}{13}\right)^{0.4} = 392.51 \text{ K}$$

(b) $$\eta_{th} = 1 - \frac{c_v\ (T_5 - T_1)}{c_v\ (T_3 - T_2) + c_p\ (T_4 - T_3)} = 1 - \frac{0.718\ (392.51 - 288)}{139.04 + 69.52} = 74\% \qquad \textbf{Ans.}$$

13.7 COMPARISON AMONG OTTO, DIESEL AND DUAL CYCLES

In order to know the relative importance of Otto, Diesel and Dual cycles, it is necessary to compare these cycles with regard to thermal efficiency. The comparison can be made in many ways such as equal compression ratio or equal final temperature or equal final pressure.

13.7.1 Equal Compression Ratio and Heat Input

Figure 13.19 shows these three theoretical cycles constructed on *p–V* and *T–S* diagrams for the same compression ratio and heat input.

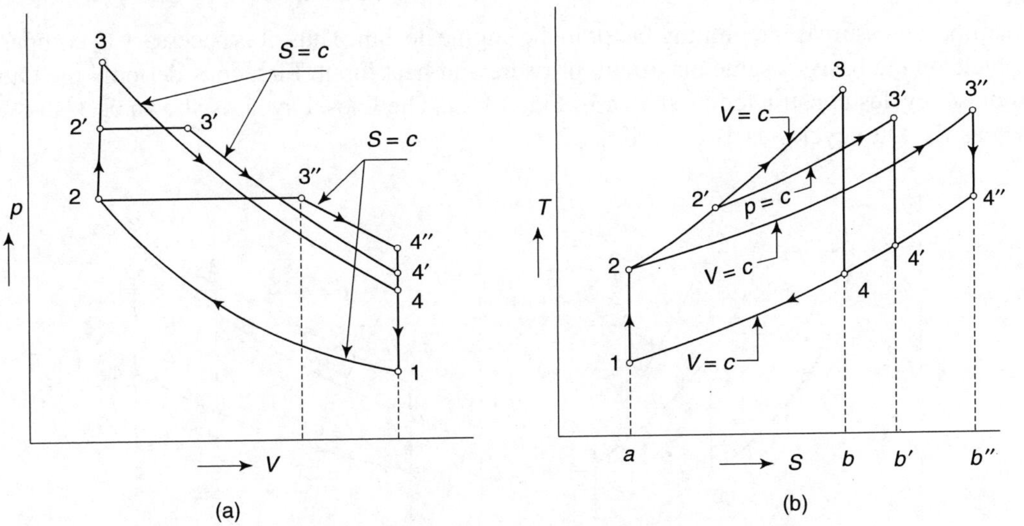

Fig. 13.19 Comparison of Otto, Diesel and Dual cycle for equal compression ratio and heat input

The Otto cycle is shown as 1–2–3–4, the Dual cycle as 1–2–2′–3′–4′ and the Diesel cycle as 1–2–3″–4″. All the cycles start at the same initial temperature, pressure and volume at point 1, with isentropic compression to 2. Heat is added under varying conditions for the different cycles. In order to fulfil the condition of equal heat input, the *T-s* areas *a* – 2–3–*b*, *a*–2–2′–3′–*b*′ and *a*–2–3″–*b*″ must be equal. Since constant volume lines are steeper on the *T–S* diagram than constant pressure line, construction of equal areas for the heat supplied shows point 3″ will lie at the point of greatest entropy. With isentropic expansion to the constant volume line 1–4″, the rejected heat fo each cycle is shown as the approximate area under line 1–4″. As the same amount of heat is supplied to each cycle, the cycle will be more efficient which rejects the least amount of heat after expansion, as obvious from the expression of thermal efficiency.

$$\eta_{th} = 1 - \frac{Q_R}{Q_R} = 1 - \frac{Q_R}{\text{Constant}}$$

We see that is the Otto cycle which rejects least amount of heat shown by area *a*–1–4–*b*. Thus, the Otto cycle has the maximum efficiency. The order of the efficiencies for the three cycles is given as

$$(\eta_{th})_{Otto} > (\eta_{th})_{Dual} > (\eta_{th})_{Diesel}$$

It is to be noted that the condition laid down here for the same compression ratio is purely theoretical. The actual compression ignition engine is operated at a high compression ratio (about double) than spark ignition which is limited by detonation.

13.7.2 Constant Maximum Pressure and Heat Input

Sometimes, pressure is the limiting factor in the engine design. Thus, it is necessary to compare the cycle on the basis of same maximum pressure and heat input. First consider only the Otto and diesel cycles constructed as shown in Fig. 13.20. The Diesel cycle is shown as 1–2–3–4 whereas the Otto cycles as 1–2′–3–4′.

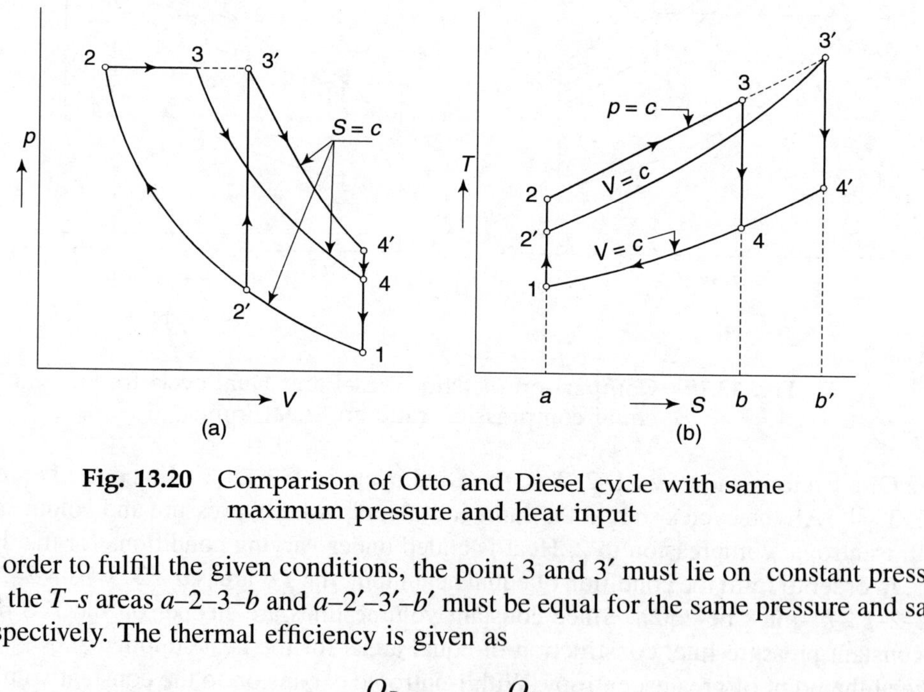

Fig. 13.20 Comparison of Otto and Diesel cycle with same maximum pressure and heat input

Input order to fulfill the given conditions, the point 3 and 3′ must lie on constant pressure lie while the *T–s* areas *a*–2–3–*b* and *a*–2′–3′–*b*′ must be equal for the same pressure and same input respectively. The thermal efficiency is given as

$$\eta_{th} = 1 - \frac{Q_R}{Q_A} = 1 - \frac{Q_R}{\text{Constant}}$$

This means that, that cycle will be more efficient which will reject least amount of heat. It is the Diesel cycle which rejects the least amount of heat as shown by the area *a*–1–4–*b*. A similar argument would show that the performance of the Dual cycle will fall between the two others. Hence, the order of efficiencies are

$$(\eta_{th})_{Diesel} > (\eta_{th})_{Dual} > (\eta_{th})_{Otto}$$

13.7.3 Constant Maximum Pressure and Output

In order to compare the thermal efficiencies of the cycles for constant maximum pressure and output, refer to Fig. 13.20. From the *T–S* diagram, it will be seen that work areas 1–2–3–4 can be only equal to l–2′–3′–4′ when point 3′ lies on the greater entropy. In order to fulfill the constant maximum pressure condition, point 3 must lie on the pressure line passing through 2–3. Hence,

$$\eta_{th} = \frac{W}{Q_A} = \frac{W}{W + Q_R} = \frac{\text{Constant}}{\text{Constant } + Q_R}$$

This equation shows that the cycle will be most efficient which rejects least amount of heat. Hence, the Diesel cycle is more efficient than Otto. Thus,

$$(\eta_{th})_{\text{Diesel}} > (\eta_{th})_{\text{Dual}} > (\eta_{th})_{\text{Otto}}$$

13.7.4 Constant Maximum Pressure and Temperature

In order to compare the cycles with constant maximum pressure and temperature refer to Fig. 13.21. First Otto and Diesel cycles are considered. To fulfill the condition of the same pressure, the Otto cycle must be limited to few compression ratio so that point 3 is at a common state for both cycles. It is clear from construction of Fig. 13.21 that both the cycle will reject the same amount of heat. The thermal efficiency is given as

$$\eta_{th} = 1 - \frac{Q_R}{Q_A} = 1 - \frac{\text{Constant}}{Q_A}$$

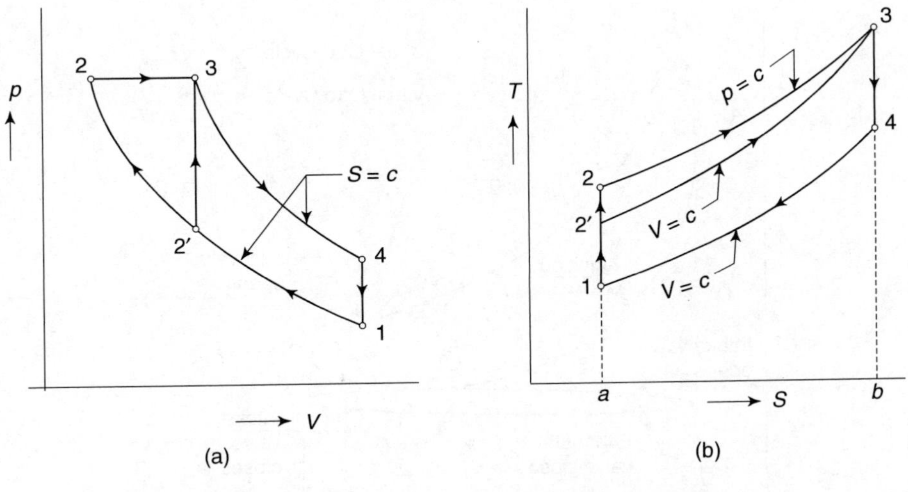

Fig. 13.21 Comparison of Otto and Diesel cycle at the same maximum pressure and temperature

The equation shows with greater heat addition the cycle will be more efficient. We also see that it is the Diesel cycle in which the heat addition is the maximum. Hence, the Diesel cycle is more efficient than Otto. Similarly, it can be proved that Dual cycle will be more efficient that Otto cycle. Hence, the order is

$$(\eta_{th})_{Diesel} > (\eta_{th})_{Dual} > (\eta_{th})_{Otto}$$

13.8 VARIATIONS BETWEEN THE AIR STANDARD OTTO CYCLES AND THE ACTUAL CYCLES

There is great variation between the air standard Otto cycles and the actual cycles. Figure 13.22 shows the *p-v* diagram for an actual Otto engine (spark ignition engine). The air standard Otto cycle has been also incorporated in the diagram by dotted line to mark the deviation.

The main differences between actual and theoretical cycles are as given below:

(i) *Working fluid* The working fluid is not air but fuel-air mixture and the products of combustion. Thus, the properties of the actual working fluid is different than that of air.

(ii) *Variable specific heats* The specific heats are not constant but in actual engine, it is a function of temperature and it increases as temperature increases.

(iii) *Dissociation* At the high temperatures involved, the products of combustion tend to break up or dissociate, into other compounds such as CO, H_2 and O_2 in an endothermic reaction. This reduces the available energy and tends to reduce the thermal efficiency and the net work of the cycle.

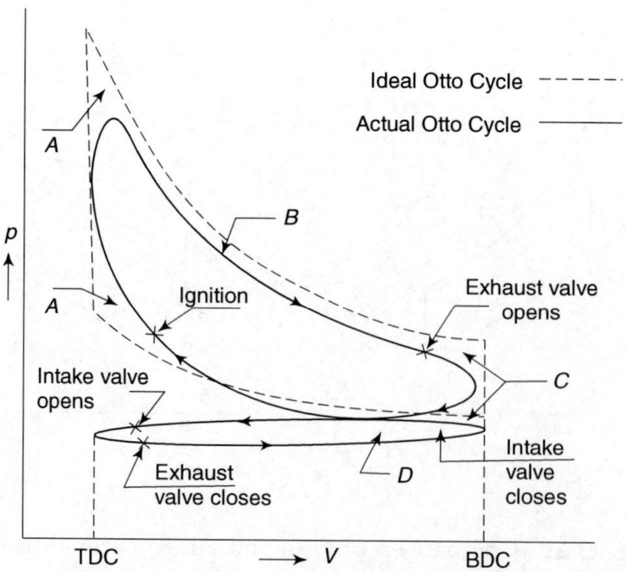

Fig. 13.22 Actual Otto cycle

(iv) *Non-instantaneous burning* In actual engine, the addition of heat is is not instantaneous i.e., the combustion does not occur at constant volume or constant pressure. Combustion takes finite time. This creates a reduction in the net area of the *p-v* diagram (**area A on top**).

(v) *Heat losses* In ideal cycle, there is no heat loss and compression and expansion are isentropic processes. But friction is present in the actual engines and there is a heat transfer to the cylinder wall. This reduces the net work (**area *A***).

(vi) *Valve timing and blow losses* In an ideal cycle, instantaneous rejection of heat is assumed. But in actual engines, it is not possible due to inertia of the exhaust valves and exhaust gases. The exhaust valve opens prior to time at which the piston reaches BDC. This reduces the net work of the actual cycle (**area *C***).

(vii) *Pumping losses* In the actual cycle, the pressure in the cylinder is different during the intake stroke from that during the exhaust stroke (**area *D***).

13.9 VARIATION BETWEEN THE AIR CYCLES AND THE ACTUAL CYCLES OF DIESEL ENGINES

There is much variation in the air cycle and the actual cycle of Diesel engines due to the undermentioned factors not considered in the air cycle.

(i) working fluid, (ii) variable specific heat, (iii) exhaust gas effect, (iv) heat losses, (v) dissociation, (vi) non-constant pressure combustion of the fuel, (vii) blow-down losses, and (viii) pumping losses.

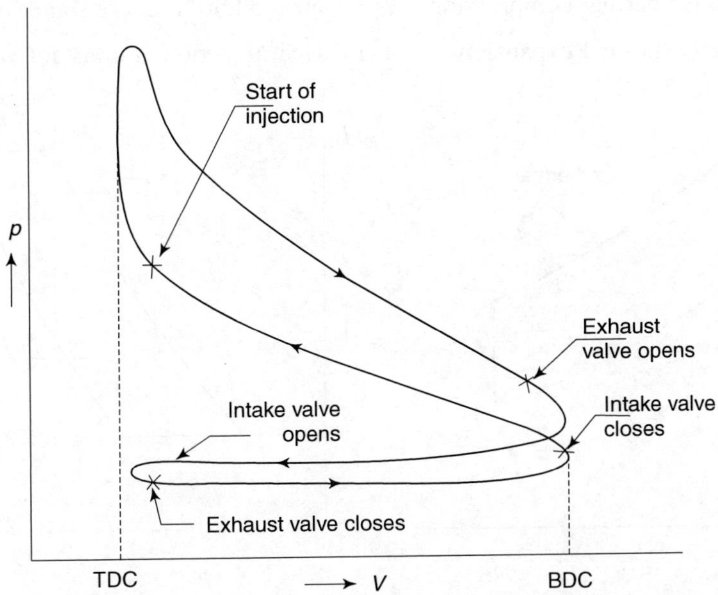

Fig. 13.23 Actual diesel cycle

The effects of first four factors are similar to the actual Otto cycle.

Dissociation Dissociation apparently exists in Diesel engines but it is less pronounced than that in petrol engines. This is due to the heterogeneous mixture existing in the combustion chamber, as well as the excess air present. Both these factors apparently, tend to reduce dissociation by lowering the overall temperature of the combustion.

Not-constant pressure combustion In the actual cycle, combustion does not take place at constant pressure. In fact, in the case of high speed engines, the combustion approaches a constant volume. In some low speed engines, the heat addition is at constant pressure as shown in Fig. 13.23.

Blow-down losses This is similar to spark ignition engines. The two stroke Diesel engine shows greater losses than thc four stroke.

Pumping loss Pumping loss also exists in the Diesel engine but it is rather small compared to petrol engine.

13.10 STIRLING CYCLE (REGENERATIVE CYCLES)

This cycle is named after Dr. Robert Stirling and his brother James Stirling (1816) who devised this cycle to work on the principle of regeneration by using a regenerator within the engine itself which would store the rejected heat energy during heat rejection process and supply the same during heat addition process. Figure 13.24 shows the representation of Stirling cycle on *p–v* and *T–s* diagrams. It consists of two isothermal and two constant volume processes as given below.

1–2 : Reversible isothermal compression 2–3 : Heat addition at constant volume

3–4 : Reversible isothermal expansion 4–1 : Heat rejection at constant volume

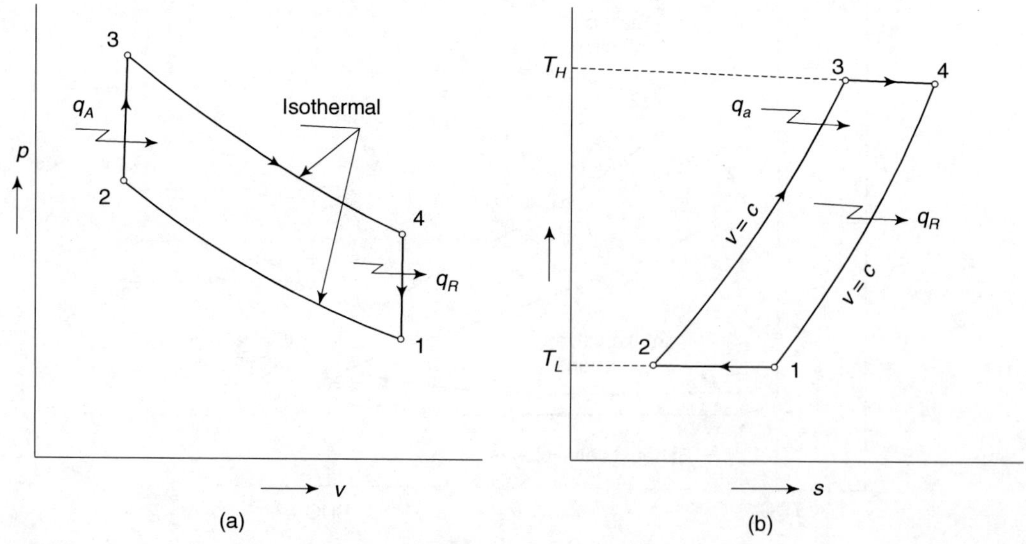

Fig. 13.24 Stirling cycle on *p–v* and *T–s* diagram

The regenerator used in the Stirling cycle consists of a wire mesh stacked in a duct as shown in Fig. 13.25. The air in the space at the lower end is kept at a *high* temperature (T_H) by heating with a furnace below it. In the upper space, a low-temperature (T_L) of air is maintained by a supply of cold water passing through the tubes in contact with it.

Hence, two heat reservoirs, one at high temperature (T_H) and other low temperature (T_L) are obtained. The wire mesh interposed ensures uniform temperature gradient from T_H to T_L. This provides a large heat transfer area and a high heat capacity.

The operation of the cycle is as follows.

Assume the piston at BDC and its position is marked by point 1 on *p-v* and *T-s* diagrams. The air is isothermally compressed from state 1 to state 2 by keeping the cylinder heat in contact with heat sink at temperature T_L. At 2, the air at temperature ($T_1 = T_2 = T_L$) enters the regenerator from top and while flowing through the wire mesh gets heated to temperature ($T_3 = T_H$) at constant volume. Now the hot air is again admitted to the engine cylinder from the bottom portion of the regenerator and undergoes an isothermal expansion in the cylinder by keeping the cylinder head in contact with the heat source at T_H. The expansion of air forces the piston to the position BDC thus producing work. The expansion process is represented as 3–4. At point 4, the air at temperature ($T_4 = T_H$) enters the regenerator from the lower end and while passing upwards transfers its heat to the wire mesh and thus gets cooled at constant volume. Thus, we see, that heat rejected is stored in the regenerator and the same is supplied during heat addition.

It is to be noted that the temperature difference between the working fluid and the wire mesh all along the flow path is infinitesimally small and thus the transfer process may be considered

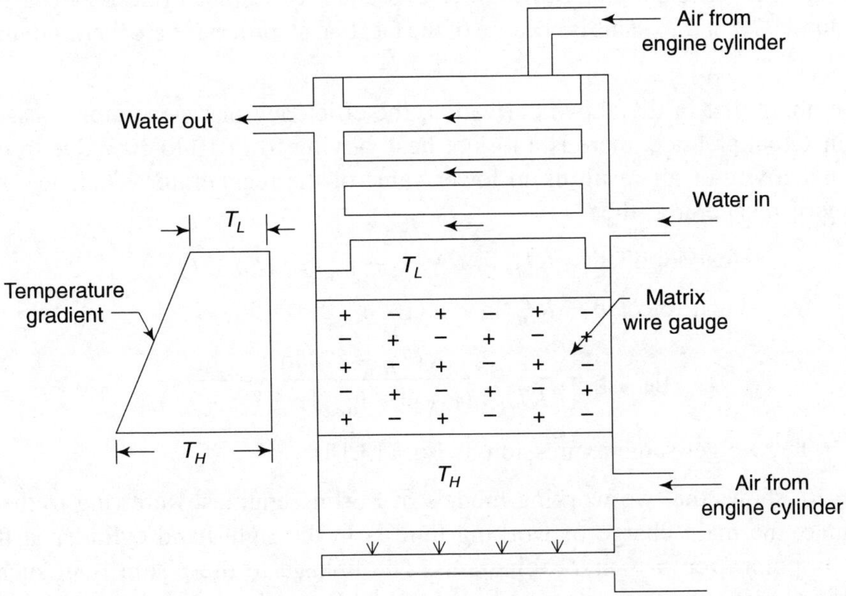

Fig. 13.25 Regenerator

as reversible. In addition, if the regenerator does not lose any heat to the surroundings the cycle heat exchange take place isothermally at T_L and T_H.

Consider 1 kg of working fluid.

In determining the heat added, we note that heat addition process from 2 to 3 and, heat rejected from 4 to 1, cancel each other since the energy taken from the regenerator is returned in the latter. An alternate interpretation is to consider the regenerator as a part of the system; in this way, heat exchange with the surroundings involves only the source at temperature T_H and the sink at T_L. The heat supplied per cycle is therefore during the process from 3 to 4 and heat rejection is during the process 1–2. Note that during the isothermal process, $\Delta u = 0$ since $T = c$.

$$\text{Heat added} = q_A = w_{3-4} = RT_H \ln \frac{V_4}{V_3} = RT_H \ln \frac{v_4}{v_3}$$

$$\text{Heat rejected} = q_R = w_{1-2} = RT_L \ln \frac{V_1}{V_2} = RT_L \ln \frac{v_1}{v_2}$$

$$\therefore \qquad \eta_{\text{thermal}} = \frac{q_A - q_R}{q_A} = \frac{RT_H \ln (v_4 / v_3) - RT_L \ln (v_1 / v_2)}{RT_H \ln (v_4 / v_3)}$$

But $\qquad v_4 = v_1,\ v_3 = v_2$ and $\dfrac{v_1}{v_2} = r$, hence

$$\text{or} \qquad \eta_{\text{thermal}} = \frac{T_H - T_L}{T_H} = 1 - \frac{T_L}{T_H} \qquad \text{...(13.18)}$$

Thus, we see that efficiency of Stirling cycle is equal to Caraot cycle between the same temperature limits, T_H and T_L. This is because of the fact that all processes are thermodynamicaily reversible.

It is to be noted that in the above derivation, the efficiency of regenerator is assumed as 100%. But in actual practice, there is a loss of heat varying from 10 to 20% due to radiation and poor conductivity of air resulting in lower value of the regenerator efficiency. If η_{reg} is the efficiency of regenerator, then

$$\text{Heat supplied} = RT_H \ln r + (1 - \eta_{\text{reg}})\, c_v\, (T_H - T_L)$$

$$\text{Heat rejected} = RT_L \ln r + (1 - \eta_{\text{reg}})\, c_v\, (T_H - T_L)$$

$$\therefore \qquad \eta_{\text{thermal}} = \frac{R (T_H - T_L) \ln r}{RT_H \ln r + (1 - \eta_{\text{reg}})\, c_v\, (T_H - T_L)} \qquad \text{...(13.19)}$$

If $\eta_{\text{reg}} = 100\%$, we get same expression as Eq. (13.18).

Figure 13.26. shows the two working models of Stirling engines. Referring to the double cylinder engine, the main charge of working fluid is in the right-hand cylinder at the state represented by point 1 on *p-v* on *T-s* diagrams. The linkage to the piston is in such a way that the left hand piston does not move when the right hand piston compresses the gas from 1 to 2. Only a small pressure rise is obtained due to heat removal. The left-hand piston now

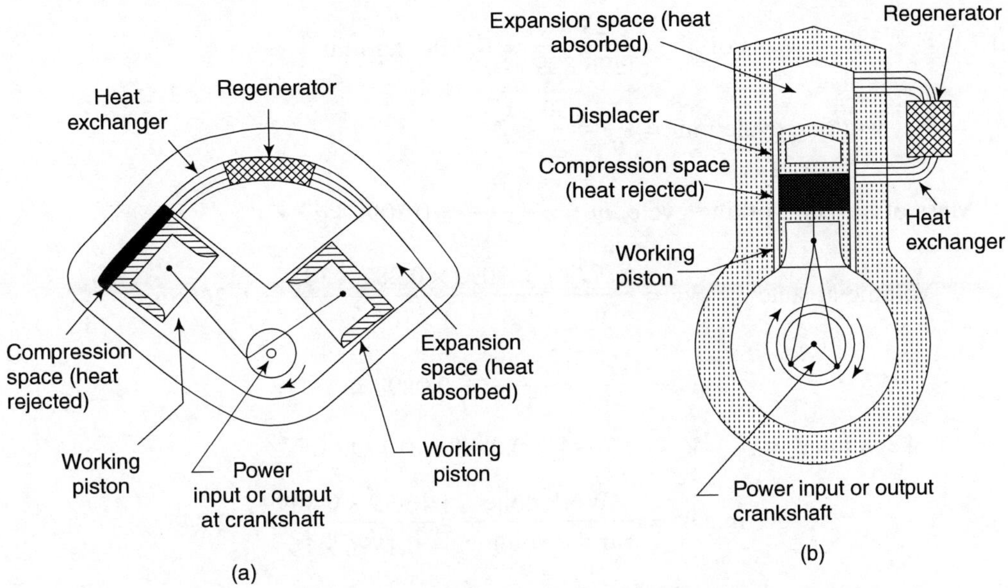

Fig. 13.26 Double cylinder and single cylinder Stirling engines

begins its downward movement while the right-hand piston further compresses the gas and forces it through the regenerator. The net result is a constant volume process in which the gas transferred to the left hand cylinder and changed to state 3. Heat is next added to the cycle at a rate which allows the regenerator into the cooling cylinder.

In single cylinder, the expansion from 3 to 4 to 1 and compression from 1 to 2 to 3 occurs in opposite ends of the same cylinder. A displacer which acts as a second piston is used to alter the proportion of the gas in the expansion and compression spaces. Due to complex drive mechanism, this engine has not been put for commercial use.

Example 13.22 The following particulars relate to an engine working on Stirling cycle.

Compression ratio = 3, Lower temperature = 30°C, Speed = 50 r.p.m., Higher temperature = 550°C, Regenerator efficiency = 90%, Initial pressure = 1 bar, Head added = 2300 kJ/min, Calculate (a) work done/kg, (b) thermal efficiency, (c) mean effective pressure and (d) indicated power.

Solution Refer to Fig. 13.24.

(a) Work done $= R\,(T_H - T_L)\ln r = 0.287\,(823 - 300)\ln 3 = 164.9$ kJ/kg **Ans.**

$$\text{Heat added} = RT_4 \ln r + (1 - \eta_{reg})\, c_v\, (T_H - T_L);\ \text{kJ/kg}$$

$$= 0.287 \times 823 \times \ln 3 + (1 - 0.9)\, 0.718\, (823 - 303) = 296.829;\ \text{kJ/kg}$$

(b) $$\text{Thermal efficiency} = \frac{\text{Work done}}{\text{Heat added}} = \frac{164.9}{296.82} = 55.55\%$$ **Ans.**

(c) $$\text{Mass of air} = \frac{2300}{296.829} = 7.7485 \text{ kg/min}$$

$$\text{Net working cycle/min} = \frac{N}{2} = \frac{50}{2} = 25$$

∴ $$\text{Mass of air per working cycle, } m = \frac{7.7485}{25} = 0.3099 \text{ kg}$$

$$\text{Volume at state 1} = V_1 = \frac{mRT_1}{p_1} = \frac{0.3099 \times 0.287 \times 303}{1 \times 10^2} = 0.2695 \text{ m}^3$$

$$V_2 = \frac{V_1}{r} = \frac{0.2695}{3} = 0.0898 \text{ m}^3$$

∴ $$\text{Stroke volume} = V_1 - V_2 = 0.2694 - 0.0898 = 0.1796 \text{ m}^3$$

$$\text{m.e.p.} = \frac{\text{Work done}}{\text{Stroke volume}} = \frac{164.9 \times 0.3099}{0.1796 \times 10^2} = 2.845$$

(d) $$\text{i.p.} = \frac{100\, p_m\, LAN}{60} = \frac{2.845 \times 0.1796 \times 50 \times 10^2}{60} = 42.58 \text{ kW}$$ **Ans.**

Example 13.23 An engine working on regenerative Stirling cycle takes in air at 1 bar and 20°C. The compression takes place isothermally and then receives heat at constant volume till the pressure rises to 12 bar and temperature 320°C. The cycle is completed by an isothermal expansion and constant volume heat rejection. Calculate the efficiency and mean effective pressure.

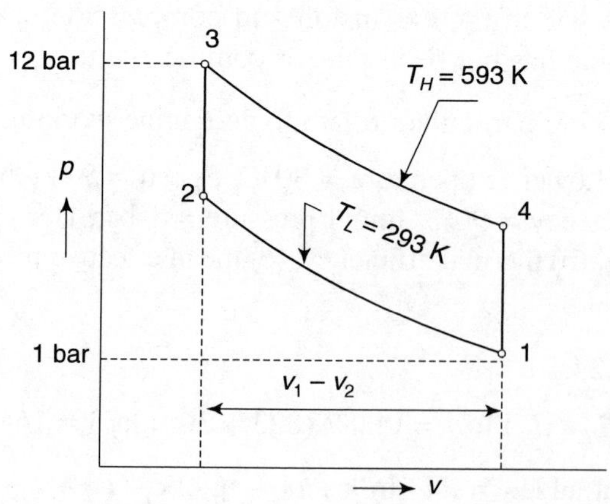

Fig. 13.27 Example 13.23

Solution Refer to Fig. 13.27. Let clearance volume $v_c = v_2 = v_3$, Thus, $v_1 = v_4$

and
$$\frac{v_1}{v_2} = \frac{v_4}{v_3} = \frac{v_4}{v_2} = \frac{v_1}{v_3} = r$$

$$T_1 = T_2 = T_L = 293 \text{ K} \quad \text{and} \quad T_3 = T_4 = T_H = 593 \text{ K}$$

$$v_1 = \frac{RT_1}{p_1} = \frac{RT_L}{p_1} = \frac{0.287 \times 293}{1 \times 10^2} = 0.8409 \text{ m}^3/\text{kg}$$

$$v_3 = \frac{RT}{p_3} = \frac{RT}{p_3} = \frac{0.287 \times 593}{12 \times 10} = 0.1418 \text{ m}^3/\text{kg}$$

$$r = \frac{v_1}{v_2} = \frac{v_1}{v_3} = \frac{0.8409}{0.1319} = 6.375$$

$$\text{m.e.p.} = \frac{[R \ln r][T_H - T_L]}{v_1 - v_2} = \frac{[0.287 \times \ln 5.926][593 - 293]}{(0.8409 - 0.1418)\, 10^2} = 2.281 \text{ bar} \qquad \textbf{Ans.}$$

$$\text{Thermal efficiency, } \eta_{th} = 1 - \frac{T_L}{T_H} = 1 - \frac{293}{593} = 0.5059 \quad \text{or } 50.59\% \qquad \textbf{Ans.}$$

13.11 ERRICSON REGENERATIVE CYCLE

This cycle also works on the principle of regeneration. Figure 13.28 shows the representation of the cycle on *p-v* and *T-s* diagrams. It consists of two isothermal and two constant pressure processes as follows:

1–2 : Reversible isothermal compression of air,

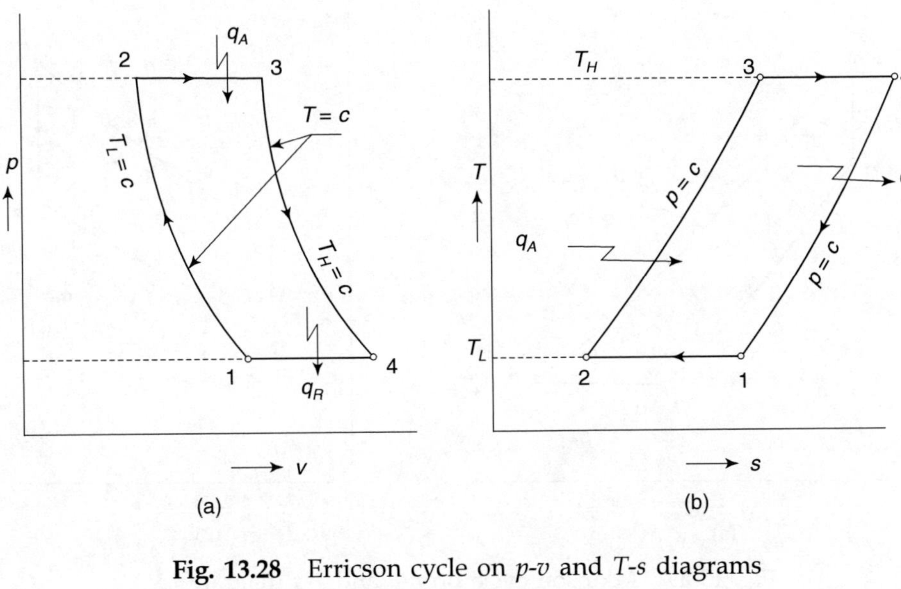

Fig. 13.28 Erricson cycle on *p-v* and *T-s* diagrams

2–3 : Addition of heat at constant pressure

3–4 : Reversible isothermal expansion of air

4–1 : Rejection of heat at constant pressure

The energy during the heat rejection process 4–1 at constant pressure is stored in the regenerator and the same is supplied during heat addition process 2–3. Thus, this cycle is thermally reversible by the use of regenerator. The efficiency of regenerator is assumed as 100%.

$$\text{Heat added} = RT_H \ln \frac{v_4}{v_3} = RT_H \ln r$$

and
$$\text{Heat rejected} = RT_L \ln \frac{v_2}{v_1} = RT_L \ln r$$

$$\therefore \quad \eta_{\text{thermal}} = \frac{RT_H \ln v_4 / v_3 - RT_L \ln v_1 / v_3}{RT_H \ln v_4 / v_3} = \frac{T_H - T_L}{T_H} = 1 - \frac{T_L}{T_H} \quad \text{...(13.20)}$$

The expression of efficiency of Erricson cycle is the same as Carnot or the Striling cycle.

13.12 ATKINSON CYCLE

This cycle is named after its inventor Dr. J. Atkinson, a British engineer. Figure 13.29 shows the representation of the Atkinson cycle on the *p-v* and *T-s* diagrams. It consists of the following four processes:

1–2 : Isentropic compression of air

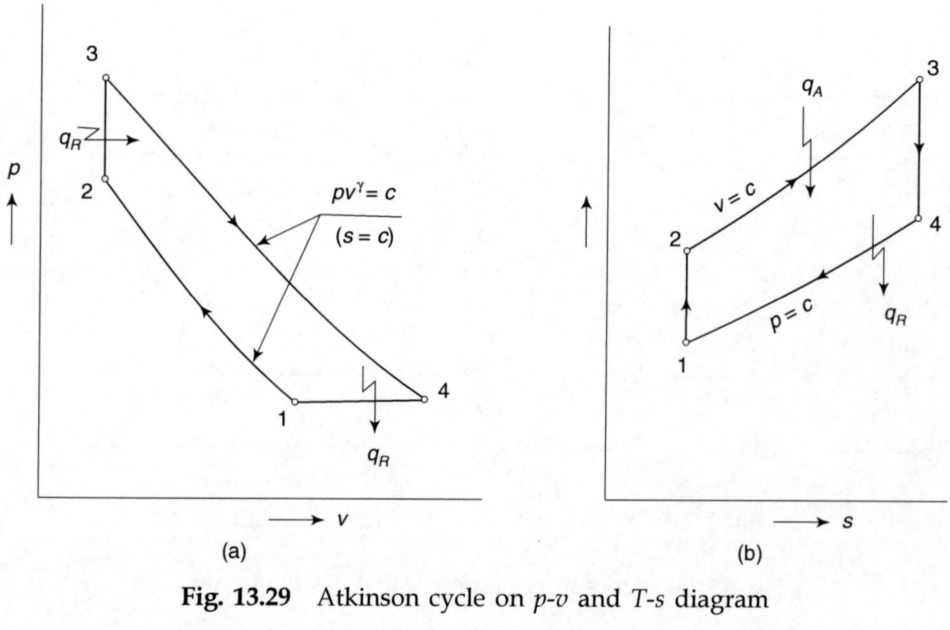

Fig. 13.29 Atkinson cycle on *p-v* and *T-s* diagram

2–3 : Heat supplied at constant volume from an external source. The pressure ratio p_3/p_2 is called the ***explosion ratio*** and denoted by α.

3–4 : Isentropic expansion of air from high temperature and pressure to lowest pressure to obtain work.

4–1 : Rejection of heat at constant pressure with some preliminary compression so that the initial state of the air is restored.

The Atkinson cycle differs from Otto cycle in so far as processes 3–4 and 4–1 are concerned. This cycle is an ideal cycle for an Otto engine exhausting to a gas turbine. Expansion of exhaust gases from Otto engine occurs in the turbine beyond volume v_1.

Consider 1 kg of working fluid.

$$\text{Heat supplied, } q_A = c_v\,(T_3 - T_2)$$

and

$$\text{Heat rejected, } q_R = c_p\,(T_4 - T_1)$$

$$\text{Work done per cycle, } w = c_v\,(T_3 - T_2) - c_p\,(T_4 - T_1)$$

$$\therefore \quad \eta_{\text{thermal}} = \frac{w}{q_A} = \frac{c_v\,(T_3 - T_2) - c_p\,(T_4 - T_1)}{c_v\,(T_3 - T_2)} = 1 - \frac{\gamma\,(T_4 - T_1)}{(T_3 - T_2)} \qquad \text{...(13.21)}$$

Let $\frac{V_1}{V_2} = \frac{v_1}{v_2} = r$ compression ratio and $\frac{p_3}{p_2} = \alpha =$ explosion ratio

From compression process 1–2, $T_2 = T_1\left(\frac{V_1}{V_2}\right)^{\gamma - 1} = T_1\,r^{\gamma - 1}$

From constant volume heat supplied process 2–3, $\frac{p_2}{T_2} = \frac{p_3}{T_3}$

$$\therefore \quad T_3 = T_2\,\frac{p_3}{p_2} = T_2 \cdot \alpha = \alpha\,T_1\,r^{\gamma - 1}$$

From expansion process 3–4.

$$\frac{T_3}{T_4} = \left(\frac{p_3}{p_4}\right)^{(\gamma - 1)/\gamma} = \left(\frac{p_3}{p_1}\right)^{(\gamma - 1)/\gamma} = \left(\frac{p_3}{p_2} \cdot \frac{p_2}{p_1}\right)^{(\gamma - 1)/\gamma} = \left(\alpha \cdot \frac{p_2}{p_1}\right)^{(\gamma - 1)/\gamma}$$

$$= \alpha^{(\gamma - 1)/\gamma}\left(\frac{p_2}{p_1}\right)^{(\gamma - 1)/\gamma}$$

But

$$p_1\,v_1^{\gamma} = p_2\,v_2^{\gamma} \text{ or } \left(\frac{p_2}{p_1}\right) = \left(\frac{V_1}{V_2}\right)^{\gamma} = r^{\gamma}$$

and $$\frac{T_3}{T_4} = \alpha^{(\gamma-1)/\gamma} \cdot r^{(\gamma-1)}$$

or $$T_4 = \frac{T_3}{a^{(\gamma-1)} \cdot r^{(\gamma-1)}} = \frac{\alpha T_1 r^{\gamma-1}}{\alpha^{(\gamma-1)/\gamma} \cdot r^{(\gamma-1)}} = T_1 \alpha^{1/\gamma}$$

Putting the values of T_2, T_3 and T_4 in Eq. (13.21), we get

$$\text{Thermal efficiency} = 1 - \frac{\gamma (T_1 \alpha^{\gamma-1} - T_1)}{(\alpha T_1 r^{\gamma-1} - T_1 r^{\gamma-1})} = 1 - \frac{1}{r^{\gamma-1}} \left[\frac{\gamma (\alpha^{1/\gamma} - 1)}{(\alpha - 1)} \right] \quad \text{...(13.22)}$$

Thus, the required equation for the thermal efficiency of Atkinson cycle.

The operation of Atkinson cycle is achieved by using a special linkage mechanism alternatively providing short and long strokes. The isentropic compression is achieved on a short stroke while the expansion is on long stroke. The main purpose of devising this cycle is to get more work than the Otto cycle shown by area 1–4–5. Note that heat rejection also occurs here at lower to average temperature which gives more efficiency than the Otto cycle. But due to the complex drive mechanism, the engines have not been commercialised yet.

Example 13.24 An engine working on Atkinson cycle takes in air at 1 bar and 25°C. The air is compressed isentropically by a compression ratio 6:1. The heat is added at constant volume resulting in the final pressure to be 20 bar. Now, the air is expanded isentropically to pressure 1 bar. The heat is rejected at constant pressure. Calculate (a) pressure and temperature at various points, (b) work done per kg of air, and (c) cycle efficiency. Compare the efficiency with that of Carnot working with same temperature limits.

Solution Refer to Fig. 13.29. (a) p_1 = 1 bar, T_1 = 298 K

From process 1–2, $T_2 = T_1 (r)^{\gamma-1} = (273 + 25) (6)^{0.4} = 610.206$ **Ans.**

$p_2 = p_1 (r)^{\gamma} = 1 \times 6^{1.4} = 12.286$ bar and $p_3 = 20$ bar **Ans.**

From process 2–3, $T_3 = \frac{p_3}{p_2} T_2 = \frac{20}{12.286} \times 610.206 \; 993.335$ K **Ans.**

$p_4 = p_1 = 1$ bar **Ans.**

From process 3–4, $T_4 = \frac{T_3}{(p_3/p_4)^{(\gamma-1)/\gamma}} = \frac{993.335}{(2)^{0.2857}} = 422.076$ K **Ans.**

Heat added, $q_A = c_v (T_3 - T_2) = 0.718 (993.335 - 610.206) = 275.086$ kJ/kg

Heat rejected, $q_R = c_p (T_4 - T_1) = 1.005 (422.676 - 298) = 125.299$ kJ/kg

(b) Work done = 275.086 – 125.299 = 149.787 kJ/kg **Ans.**

(c) Thermal efficiency, $\dfrac{w}{q_A} = \dfrac{149.787}{275.056} = 54.45\%$

Efficiency of Carnot cycle $= \dfrac{T_H - T_L}{T_H} = \dfrac{993.335 - 298}{993.335} = 70\%$ **Ans.**

13.13 BRAYTON OR JOULE CYCLE*

The air standard Brayton cycle or Joule cycle or constant pressure cycle is an idealised cycle for simple gas turbine power plant shown in Fig. 13.30. Basically, it consists of a compressor, a combustion chamber and a turbine.

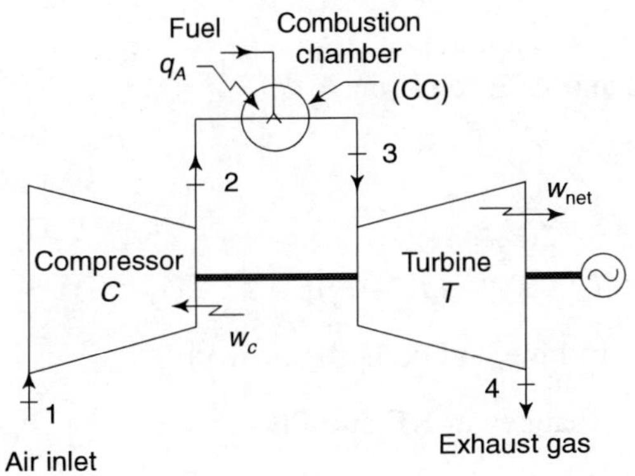

(a) Open cycle

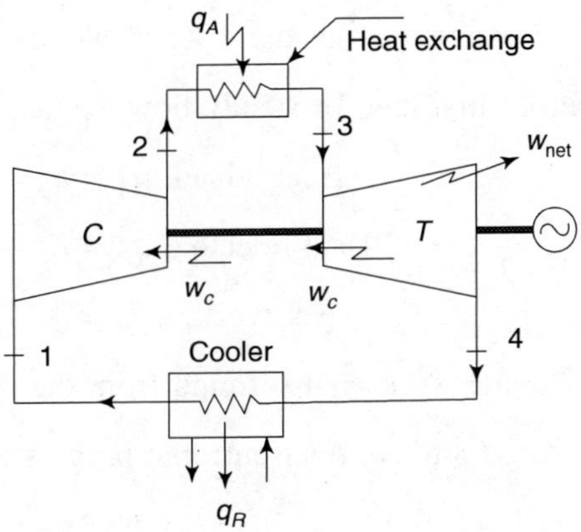

(b) Closed cycle

Fig. 13.30 Simple gas turbine

Atmospheric air is compressed from pressure p_1, to a high pressure p_2 in the compressor and delivered to the combustion chamber where fuel is injected and burnt. The combustion process occurs nearly at constant pressure. Due to combustion, heat is added to the working fluid raising its temperature from T_2 to T_3. The products of combustion from the combustion chamber are expanded in the turbine from pressure p_3 to atmospheric pressure and then discharged into the atmosphere. The turbine and compressor are mechanically coupled, hence the net work is the difference between turbine work and compressor work. In general, gas turbine cycles are open cycles, since new air must be continuously introduced into the compressor. In a closed cycle, heat is added by an indirect method called the heat exchanger, so that, there is no mixing of fuel and air. Similarly, the heat is rejected in another heat exchanger called the ***precooler***.

* For more details, see the Author's book "*Steam and Gas Turbines, Turbomachines and Power Plant Engineering*".

In the ideal Brayton or Joule cycle, the cycle is assumed to be a closed one and heat addition or rejection is only through heat transfer process. Figure 13.36 shows the representation of ideal Brayton cycle on the *p-v* and *T-s* diagrams.

The various processes are:

1–2 : Isentropic compression in the compressor thus raising pressure and temperature from p_1, T_1 to p_2, T_2.

2–3 : Addition of heat at constant pressure raising temperature from T_2 to T_3.

3–4 : Isentropic expansion of air from high pressure and temperature to low pressure and temperature and thus doing work.

4–1 : Rejection of heat at constant pressure to restore the original state of air.

For example, consider 1 kg of working fluid.

From first law for steady flow, neglecting ΔKE and ΔPE, $\delta q - \delta w = dh$

$$\therefore \quad \text{Heat added, } q_A = h_3 - h_2 = c_p\,(T_3 - T_2)$$

$$\text{Heat rejected, } q_R = h_4 - h_1 = c_p\,(T_4 - T_1)$$

$$\text{Net work, } w_{\text{net}} = q_A - q_R = c_p\,[(T_3 - T_2) - (T_4 - T_1)]$$

The net work is also found from the concept of turbine and compressor work.

From first law, for isentropic process neglecting changes in KE and PE,

$$w = -\,\Delta h$$

$$\therefore \quad \text{Work done by turbine, } w_t = h_3 - h_4 = c_p\,(T_3 - T_4)$$

Work consumed by compressor, $w_c = h_2 - h_1 = c_p\,(T_2 - T_1)$

Net work, $w_{\text{net}} = w_t - w_c = c_p\,[(T_3 - T_4) - (T_2 - T_1)] = c_p\,[(T_3 - T_2) - (T_4 - T_1)]$

$$\therefore \quad \eta_{th} = \frac{\text{Net work}}{\text{Heat added}} = \frac{w_t - w_c}{q_A} = \frac{c_p\,\{(T_3 - T_2) - (T_4 - T_1)\}}{c_p\,(T_3 - T_2)} = 1 - \frac{T_4 - T_1}{T_3 - T_2} \quad ...(13.23)$$

From processes 1–2 and 3–4, $\dfrac{T_2}{T_1} = \left(\dfrac{p_2}{p_1}\right)^{(\gamma-1)/\gamma}$ and $\dfrac{T_3}{T_4} = \left(\dfrac{p_3}{p_4}\right)^{(\gamma-1)/\gamma}$

But $p_2 = p_3$ and $p_1 = p_4$, hence $\dfrac{T_2}{T_1} = \dfrac{T_3}{T_4} = \left(\dfrac{p_2}{p_1}\right)^{(\gamma-1)/\gamma}$

or
$$\frac{T_1}{T_2} = \frac{T_4}{T_3} = \frac{T_4 - T_1}{T_3 - T_2}$$

Putting the value of $(T_4 - T_1)/(T_3 - T_2)$ in Eq. (13.23), we get

$$\eta_{th} = 1 - \frac{T_1}{T_2} = 1 - \frac{T_4}{T_3} \quad \text{...(13.24)}$$

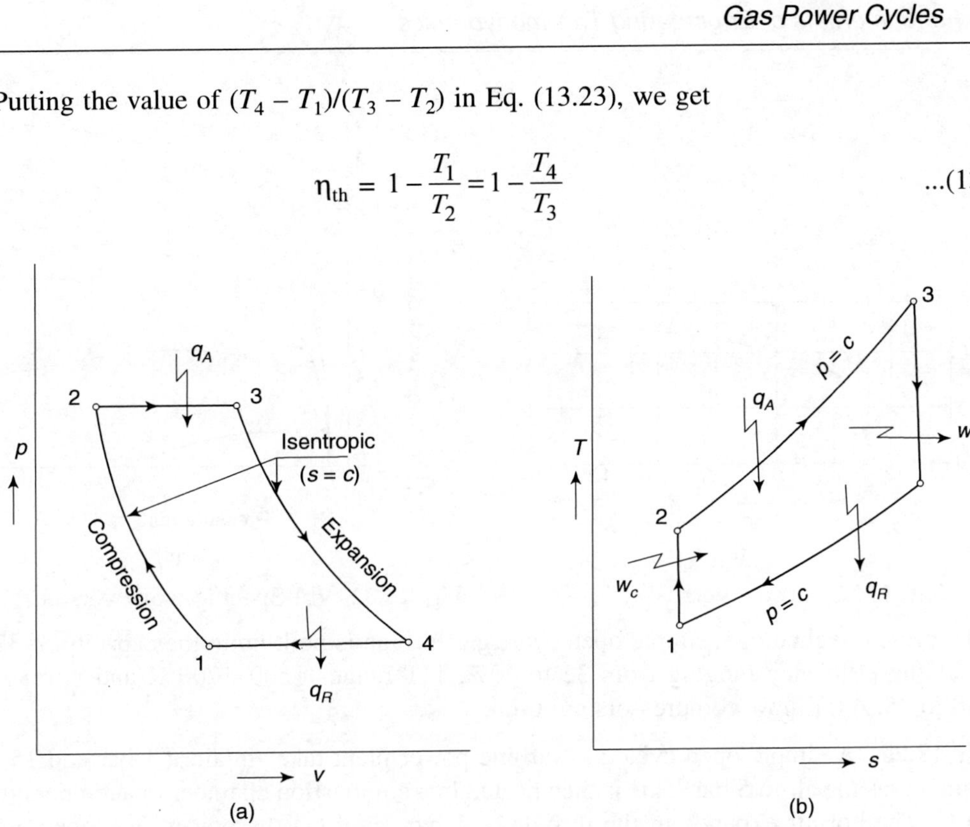

Fig. 13.31 Brayton cycle on *p-v* and *T-s* diagrams

Let r_p = pressure ratio = $\frac{p_2}{p_1}$, so $\frac{T_2}{T_1} = \left(\frac{p_2}{p_1}\right)^{(\gamma-1)/\gamma} = (r_p)^{(\gamma-1)/\gamma}$

$$\therefore \quad \eta_{th} = 1 - \frac{1}{(r_p)^{(\gamma-1)/\gamma}} \quad \text{...(13.25)}$$

From the above equation, we find that the thermal efficiency of Brayton cycle is the same as that of the Otto cycle. **Here, *r* is replaced by r_p.** The variation of thermal efficiency, η_{th} as a function of pressure ratio, r_p is shown in Fig. 13.32(a). It is obvious from Fig. 13.32(a) that the thermal efficiency progressively increases with increasing value of pressure ratio.

The turbine specific work is found to be a function of pressure ratio and turbine inlet temperature as shown in Fig. 13.32(b) in which $T_3/T_1 = T_{max}/T_{min} = t$ and is expressed as

$$\frac{w_{net}}{c_p T_1} = t\left[1 - \frac{1}{(r_p)^{(\gamma-1)/\gamma}}\right] - \left[r_p^{(\gamma-1)/\gamma} - 1\right]$$

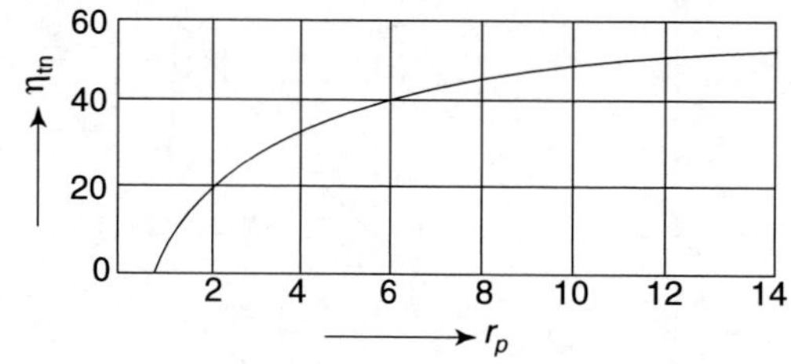

(a)

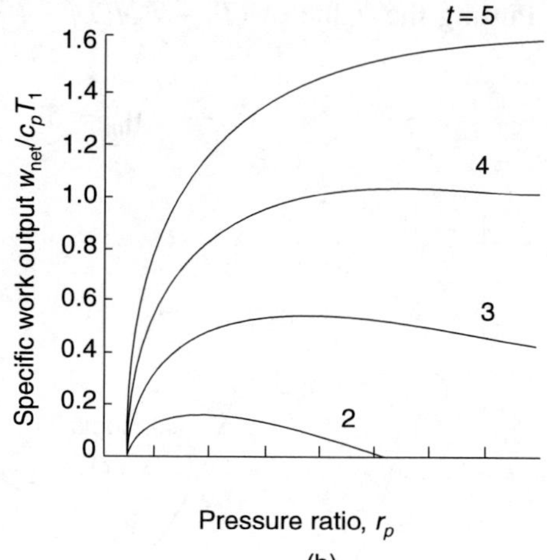

(b)

Fig. 13.32 (a) η_{th} versus r_p **Fig. 13.32** (b) Specific work versus r_p

In the present technology, simple open cycle gas turbine is built up to the capacity of 320 MW yielding efficiency ranging from 32 to 40%, TIT ranges 1500–1700 K and r_p ranges from 10 to 35. Axial flow compressors are used.

Example 13.25 A simple open cycle gas turbine power plant takes in air at 1 bar and 15°C and compresses the air to 5 bar. This is then heated in a combustion chamber to a temperature of 800°C. The hot air expands in the turbine to 1 bar. Find (a) the power developed, and (b) thermal efficiency of the plant. Take $\gamma = 1.4$ and $c_p = 1.005$ kJ/kg K for air.

Solution Refer to Fig. 13.30 (a). The flow is steady flow process

$$T_{2s} = T_1 (r_p)^{(\gamma - 1)} = (273 + 15)(5)^{0.4/1.4} = 456.14 \text{ K}$$

$$\text{Compressor work, } w_c = h_{2s} - h_1 = c_p (T_{2s} - T_1)$$

$$= c_p (456.14 - 288) = c_p \times 168.14 \text{ kJ/kg}$$

$$T_4 = \frac{T_3}{r^{(\gamma-1)/\gamma}} = \frac{800 + 273}{5^{0.4/1.4}} = 677.47 \text{ K}$$

$$\text{Turbine work, } w_t = h_3 - h_{2s} = c_p (T_3 - T_{4s}) = c_p (1073 - 677.47)$$

$$= c_p \times 395.53 \text{ kJ/kg}$$

$$\text{Heat added, } q_A = q_{3-2} = h_3 - h_{2s} = c_p (T_3 - T_{2s}) = c_p (1073 - 456.14)$$

$$= c_p \times 616.86 \text{ kJ/kg}$$

$$\text{Network, } w_t - w_c = c_p (395.53 - 168.14) = c_p \times 227.39 \text{ kJ/kg}$$

$\therefore$ $$\text{Thermal efficienc, } \eta_{th} = \frac{\text{net}}{q_A} \quad \frac{\times 227.39}{c_p \times 616.86} = 0.3686 \text{ or } 36.86\%$$ **Ans.**

$$\text{Power output} = 1.005 \times 227.39 = 228.52 \text{ kJ/kg}$$ **Ans.**

13.14 ACTUAL (IRREVERSIBILITIES) BRAYTON CYCLE

The actual gas turbine cycle differs from the theoretical cycle in the following respects:

1. The fluid velocities are high in turbo-machinery, hence the change in kinetic energy between inlet and outlet of each component cannot necessarily be ignored. This may be taken care by considering stagnation values of properties instead of static values.
2. Due to *frictional losses* in the compressor and turbine, the compression and expansion processes are not frictionless and take place with some increase in entropy (i.e., the processes are irreversible adiabatic). In the ideal case, the compressor and turbine efficiencies were 100 per cent but actual turbine and compressor have efficiency less than 100 per cent. (Generally, $\eta_c = 87\%$ and $\eta_t = 88\%$.)
3. *A small pressure loss (about 2% of inlet pressure) occurs* in the combustion chamber. Combustion is never complete. This loss is so little that it can be neglected for the sake of simplification of the problem wherever necessary. Thus, $p_3 = p_2 - (\Delta p)_{\text{loss, comp}}$.
4. The mass of the gas flowing through the turbine is $(1 + F/A)$ times the mass of air flowing through the compressor where F/A represents the fuel-air ratio.
5. The *specific heat of combustion gas is slightly higher than that of air* ($c_{pg} = 1.20$ kJ/kg K). This increase is so little that the specific heat of combustion gas may be taken as that of air for simplicity wherever necessary.
6. There is a pressure loss in the exhausthood of turbine and as a result the expansion of gas will take place up to a pressure higher than atmospheric so that after pressure loss the exhaust pressure will be equal to the atmospheric pressure. Thus, $p_4 = p_1 + \Delta p_{\text{loss}}$. If $p_4 < p_1$ the exhaust will not go to atmosphere and there will be a back flow.
7. The gas turbine bladings are subjected to high pressure and high temperature and thus they are vulnerable to oxidation and creep. For the safe working of gas turbine blading, it is cooled. Generally, the blade are cooled by air bleeding, from compressor. The amount of the cooling air required varies from 7 to 20 per cent of inlet air flow depending upon turbine inlet temperature. In the present analysis, it is neglected.

h–s diagram for an actual Braytonne cycle is shown in Fig. 13.33. The pressure loss in the combustion chamber is represented by $p_2 - p_3$ and in the exhausthood, by $p_4 - p_1$. In this cycle.

1–2s is isentropic compression 1–2 is actual compression

3 –4s is isentropic expansion 3–4 is actual expansion

The compressor efficiency also known as isentropic compressor efficiency, η_c is

$$\eta_c = \frac{\text{Isentropic compression work}}{\text{Actual compression work}} = \frac{w_c}{w_{ca}}$$

Based on static values, $$\eta_c = \frac{h_{2s} - h_1}{h_2 - h_1} = \frac{c_p (T_{2s} - T_1)}{c_p (T_2 - T_1)} = \frac{T_{2s} - T_1}{T_2 - T_1} \quad ...(13.26)$$

Based on total or stagnation values, $\eta_{ac} = \dfrac{h_{02s} - h_{01}}{h_{02} - h_{01}} = \dfrac{T_{02s} - T_{01}}{T_{02} - T_{01}}$ ...(13.27)

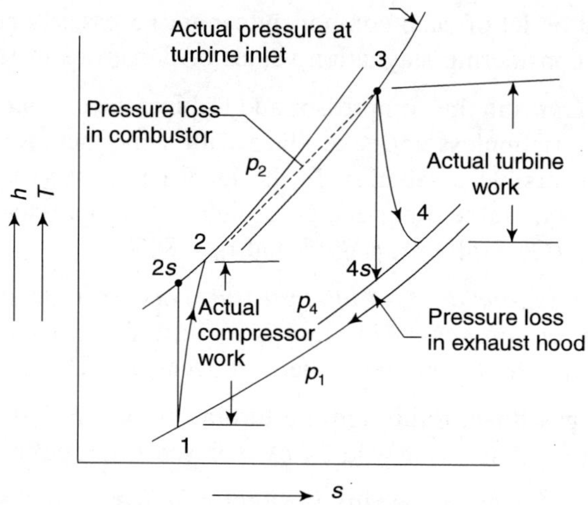

Fig. 13.33 Actual gas turbine cycle representation on *h–s* chart

Here suffix 0 denotes the stagnation values.

The turbine efficiency = $\eta_t = \dfrac{\text{Actual turbine work}}{\text{Isentropic turbine work}} = \dfrac{w_{ta}}{w_t}$

Based on static values, we have

Actual turbine work = $w_{ta} = h_3 - h_4 = c_{pg}\,(T_3 - T_4)$ kJ/kg

If specific heat of combustion gas c_{pg} may be taken as that of air, then

Actual turbine work, $w_{ta} = c_p\,(T_3 - T_4)$ kJ/kg

So, Turbine efficiency, $\eta_t = \dfrac{h_3 - h_4}{h_3 - h_{4s}} = \dfrac{c_p\,(T_3 - T_4)}{c_p\,(T_3 - T_{4s})} = \dfrac{T_3 - T_4}{T_3 - T_{4s}}$ ...(13.28)

Based on total or stagnation values, the turbine isentropic efficiency,

$$\eta_{ot} = \frac{h_{03} - h_{04}}{h_{03} - h_{04s}} = \frac{T_{03} - T_4}{T_{03} - T_{04s}} \quad ...(13.29)$$

The thermal efficiency of the actual cycle is computed as follows for static values:

Actual net work, $w_{neta} = w_{ta} - w_{ca} = [(1 + F/A)\,(h_3 - h_4)] - (h_2 - h_1)$ kJ/kg

where $F/A = \dot{m}_f/\dot{m}_a$ = fuel air ratio

Actual heat supplied, q_{Aa} = [)1 + $F/A)\,h_3 - h_2]$ kJ/kg of air.

So, actual thermal efficiency, $\eta_{tha} = \dfrac{(1+F/A)(h_3-h_4)-(h_2-h_1)}{(1+F/A)\,h_3-h_2}$

$$= \frac{(1+F/A)\cdot c_{pg}\,(T_3-T_4)-c_{pa}\,(T_2-T_1)}{(1+F/A)\cdot c_{pg}\cdot T_3-c_{pa}\cdot T_2} \qquad \text{...(13.30)}$$

If we neglet the mass of fuel in comparison to the mass of air used, then

$$\eta_{tha} = \frac{(h_3-h_4)-(h_2-h_1)}{h_3-h_2} = \frac{c_{pg}(T_3-T_4)-c_{pa}\,(T_2-T_1)}{(c_{pg}\,T_3-c_{pa}\cdot T_2)}$$

If we assume that the specific heat of air and combustion gas are same, then

$$\eta_{tha} = \frac{c_p\,(T_3-T_4)-c_p\,(T_2-T_1)}{c_p\,(T_3-T_2)} = \frac{(T_3-T_4)-(T_2-T_1)}{(T_3-T_2)} \qquad \text{...(13.31)}$$

or
$$\eta_{tha} = \frac{\eta_t\,(T_3-T_{4s})-(T_{2s}-T_1)/\eta_c}{T_3-T_2} = \frac{(T_3-T_{4s})\,\eta_t\cdot\eta_c-(T_{2s}-T_1)}{(T_3-T_2)\,\eta_c}$$

Based on stagnation values, $\eta_{otha} = \dfrac{(T_{03}-T_{04})\,\eta_{0t}\,\eta_{0c}-(T_{02}-T_{01})}{(T_{03}-T_{02})\,\eta_{0c}}$

Based on static values, we have, $\eta_{tha} = \dfrac{(w_t)\,\eta_t\,\eta_c-w_c}{q_{Aa}\cdot\eta_c}$ ...(13.32)

From Eq. (13.22), it is clear that the thermal efficiency is increased by improving either η_t or η_c or both.

It should be brone in mind that the extra work supplied to the compressor because of its inefficiency is not entirely wasteful because as compared to isentropic compression, the heat supplied in the combustion chamber is reduced. It is worth to find an expression for the minimum temperature ratio, $t = t_{min} = T_{03}/T_{01}$ in the case of gas turbine. For stagnation values,

$$w_{ca} = \frac{c_p\,T_{01}}{\eta_{ac}}\left(r_p^{(\gamma-1)/\gamma}-1\right) \text{ and } w_{ta} = \eta_{at}\cdot c_p\,T_{03}\left[\frac{r_p^{(\gamma-1)/\gamma}-1}{r_p^{(\gamma-1)/\gamma}}\right]$$

$$\therefore \qquad w_{neta} = \eta_{ot}\cdot c_p\,T_{03}\left[\frac{r_p^{(\gamma-1)/\gamma}-1}{r_p^{(\gamma-1)/\gamma}}\right]-\frac{c_p\,T_{01}}{\eta_{oc}}\,(r_p^{(\gamma-1)/\gamma}-1)$$

If $w_{ca} = w_{ta}$, the **minimum temperature ratio, $T_{03}/T_{01} = t = t_{min}$** can be claculated for which no net work is possible. This results in

or
$$t = t_{min} = \frac{T_{03}}{T_{01}} = \frac{r_p^{(\gamma-1)/\gamma}}{\eta_{at}\,\eta_{oc}}$$

If $\quad r_p = 5,\ \eta_{oc} = \eta_{at} = 0.85$ and $\gamma = 1.4$, $t = \mathbf{t_{min} = 2.19}$

This means if t_{01} = 15°C, t_{03} = 358°C, then for the net work to be done, t_{03} has to be more than 358°C.

The value of T_3 (TIT), and the ratio *t*, depends upon the maximum temperature which is highly stressed parts of the turbine can withstand for the required working life; it is often called the "**metallurgical limit**". In early gas turbine, *t* was selected between 3.5 and 4 but with the introduction of air cooled turbine blades ***t* is raised to 5 and 6**.

13.15 THE CYCLE AIR RATE, WORK RATIO AND SPECIFIC FUEL CONSUMPTION

13.15.1 The Cycle Air Rate

In addition to the thermal efficiency which is a measure of fuel economy, the size of the plant is equally important in so many applications, particularly in the field of aviation. The size of the plant depends on the rate of flow of air in relationship to the useful power output.

Air rate, (AR), is defined as the air flow required per kWh output.

Thus,
$$\text{AR} = \frac{\dot{m}_a}{\text{kWh output}} = \frac{\text{Mass of air required}}{\text{kWh output}}$$

$$\text{AR} = \frac{\dot{m}_a}{m_a\,(w_{\text{net}}/3600)} = \frac{3600}{w_{\text{net}}\ \text{in kJ/kg}} = \frac{3600}{w_{\text{net}}}\,\frac{\text{kg of air}}{\text{kWh}} \qquad \text{...(13.33)}$$

The reciprocal of air rate is termed as specific power. Air rate is actually the criterion of the size of the plant, i.e., the **lower the air rate the smaller is the plant**.

13.15.2 The Cycle Work Ratio

A good design of gas turbine not only requires a high thermal efficiency and a low air rate but it **must also have a high work ratio**.

The work ratio, (WR), is defined as the ratio of net work to the turbine work. Thus,

$$\text{WR} = \frac{w_{\text{net}}}{w_t} \qquad \text{...(13.34)}$$

From Eq. (13.34), it is also clear that for a good gas turbine, the work ratio should be high. The work ratio may also be used as a guide in the determination of the size of the gas turbine and as an indicator of the sensitivity of the plant to the decrease of the component efficiency.

13.15.3 Specific Fuel Consumption

It is one of the most important parameters and expressed in kg/kWh. Thus,

$$\text{sfc} = \frac{3600\,\dot{m}_f\ \text{(kg/s)}}{W_{\text{net}}\ \text{(kW)}} = \frac{3600\,(\dot{m}_f/\dot{m}_o)}{w_{\text{net}}\ \text{(kJ/kg)}};\ \text{kg/kWh} \qquad \text{...(13.35)}$$

13.16 OPEN CYCLE GAS TURBINE WITH REGENERATION (RECUPERATION)

In this method, the air delivered by the compressor passes through a heat exchanger (regenerator or recuperator) utilizing the gases exhausted from the turbine; the heated air then passes into the combustion chamber and part of it is employed to burn the fuel. Since some heat is added to the air in the heat exchanger itself, so the same turbine gas inlet temperature is attained as that when no heat exchanger is employed; with a lower fuel consumption, hence, the thermal efficiency will accordingly be higher. The compressor, turbine and net work are not affected by the addition of the recuperator or regenerator to the cycle, only the heat required to be supplied in the combustion chamber is decreased which gives the gain in thermal efficiency,

13.16.1 Ideal Cycle

Process 1–2*s*–3–4*s*–6–1 shows the ideal open cycle gas turbine as shown in Fig. 13.34. The high pressure air at state 2*s* is passed through a regenerator (or hear exchanger) where it is heated at constant pressure p_2 from T_2, to T_5 due to which the exhaust gas from the turbine at state 4 is cooled from T_4 to T_6. With perfect heat exchange in the heat exchanger, in the ideal case, the temperature of air leaving the heat exchanger will be equal to the temperature of exhaust gas entering the heat exchanger and the temperature of exhaust gas leaving heat exchanger will be equal to the temperature of air entering the heat exchanger, Thus, $T_5 = T_{4s}$ and $T_{6s} = T_2$.

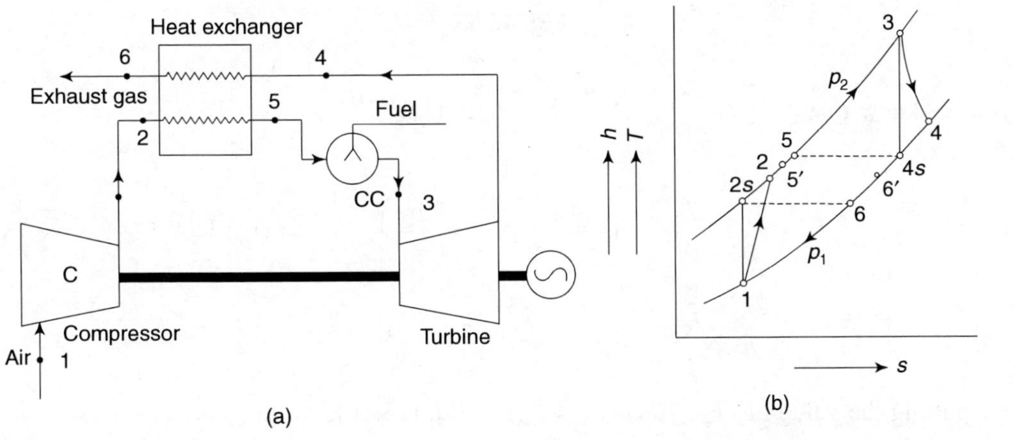

(a) Open cycle with regeneration (b) Process on *T-s* chart with regeneration

Fig. 13.34

The air from the heat exchanger at temperature T_5 then passes through the combustion chamber where it is heated from temperature T_5 to T_3 (i.e., from h_5 to h_3). The advantage of exhaust heat exchange is that the amount of heat supplied in the combustion chamber is reduced from $(h_3 - h_{2s})$ to $(h_3 - h_5)$, i.e., from $c_p\ (T_3 - T_{2s})$ to $c_p\ (T_3 - T_5)$. Thus, the thermal efficiency is increased because the heat supplied is reduced.

Assume 1 kg of air flowing through the compressor.

$$\text{Net work, } w_{\text{net}} = w_t - w_c = c_p\,[(T_3 - T_{4s}) - (T_{2s} - T_1)]$$

$$\text{Heat supplied, } h_3 - h_5 = c_p\,(T_3 - T_5) = c_p\,(T_3 - T_{4s})$$

So,

$$\eta_{\text{th}} = \frac{c_p\,[(T_3 - T_{4s}) - (T_{2s} - T_1)]}{c_p\,(T_3 - T_4)} = 1 - \frac{T_{2s} - T_1}{T_3 - T_{4s}} \qquad \text{...(13.36)}$$

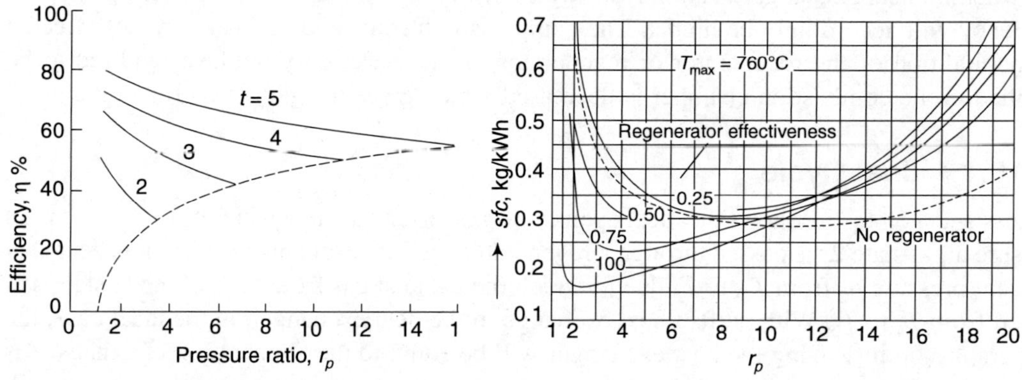

(a) Thermal efficiency of simple cycle with regeneration

(b) Effect of regeneration on η_{th}

Fig. 13.35

But we know that,

$$\frac{T_{2s}}{T_1} = \left(\frac{p_2}{p_1}\right)^{(\gamma-1)/\gamma} = (r_p)^{\gamma - 1/\gamma}$$

and

$$\frac{T_3}{T_{4s}} = \left(\frac{p_3}{p_4}\right)^{(\gamma-1)/\gamma} = \left(\frac{p_2}{p_1}\right)^{(\gamma-1)/\gamma} = (r_p)^{(\gamma-1)/\gamma}$$

So,

$$\frac{T_{2s}}{T_1} = \frac{T_3}{T_{4s}} \quad \text{or} \quad \frac{T_{2s}}{T_3} = \frac{T_1}{T_{4s}} = \frac{T_{2s} - T_1}{T_3 - T_{4s}}$$

By putting the value of $(T_{2s} - T_1)/(T_3 - T_{4s})$ in Eq. (13.36), we get

$$\eta_{\text{th}} = 1 - \frac{T_{2s}}{T_3} = 1 - \frac{T_{2s}}{T_1} \cdot \frac{T_1}{T_3} = 1 - \frac{T_1}{T_3}(r_p)^{(\gamma-1)/\gamma}$$

Let

$$t = \frac{T_3}{T_1} = \frac{T_{\text{max}}}{T_{\text{min}}} \quad \text{and} \quad c = \frac{T_{2s}}{T_1} = \frac{T_3}{T_{4s}} = (r_p)^{(\gamma-1)/\gamma}$$

Then, the thermal efficiency, $\eta_{\text{th}} = 1 - \dfrac{c}{t}$ ...(13.37)

Case 1 When $c = 1$, $\eta_{th} = 1 - \frac{1}{t} = \frac{T_{max} - T_{min}}{T_{max}}$ = Carnot cycle efficiency

Case 2 When $t = c^2$, $\eta = 1 - \frac{1}{c} = 1 - \frac{1}{(r_p)^{(\gamma - 1)/\gamma}}$ = Simple open gas turbine efficiency

Figure 13.35 (a) shows the effect of pressure ratio, r_p, on thermal efficiency for various values of t (i.e., the ratio of turbine inlet temperature to compressor inlet temperature). The thermal efficiency increases with t with a decrease in pressure ratio.

13.16.2 Actual Gas Turbine Cycle with Regeneration

1–2–5–3–4–6–1 shows the actual cycle as shown in Fig. 13.34 (b). In practice, the compression and expansion processes are not isentropic. Assuming 1 kg of air flowing through the compressor and if we neglect the mass of the fuel and assume the same value of c_p for air and exhaust gas, turbine and compressor work may be computed per kg of air as follows:

$$\text{Actual net work, } w_{neta} = w_{ta} - w_{ca} = c_p\,[(T_3 - T_4) - (T_2 - T_1)]$$

$$\text{Actual heat supplied, } q_{Aa} = h_3 - h_5 = c_p\,(T_3 - T_5)$$

$$\text{Actual thermal efficiency, } \eta_{tha} = \frac{w_{neta}}{q_{Aa}} = \frac{(T_3 - T_4) - (T_2 - T_1)}{(T_3 - T_5)} \qquad \text{...(13.38)}$$

or

$$\eta_{tha} = \frac{(T_3 - T_{4s})\,\eta_t - \dfrac{(T_{2s} - T_1)}{\eta_c}}{T_3 - T_5'} = \frac{(T_3 - T_{4s})\,\eta_t \cdot \eta_c - (T_{2s} - T_1)}{\eta_c\,(T_3 - T_5')} \qquad \text{...(13.39)}$$

In practice, perfect heat exchange is never possible in the regenerator. As a result, $h'_5 < h_4$ and $h'_6 > h_2$ (i.e., $T_5' < T_4$ and $T_6' > T_2$) in the actual cycle. The regenerator effectiveness ε_r is defined as

$$\varepsilon_r = \frac{\text{Actual heat exchange}}{\text{Maximum theoretical heat exchange}} \qquad \text{...(13.40)}$$

The actual heat exchange $(h'_3 - h_2)$ or $(h_4 - h'_6)$

The maximum theoretical heat exchange = $(h_4 - h_2)$

Hence,

$$\varepsilon_r = \frac{h_5' - h_2}{h_4 - h_2} = \frac{h_4 - h_6'}{h_5' - h_6'} = \frac{T_5' - T_2}{T_4 - T_2} = \frac{T_4 - T_6'}{T_5' - T_6'} \qquad \text{...(13.41)}$$

If ε_r is known, T_5' can be found out and hence the thermal efficiency can be evaluated.

A drawback of such a heat exchanger, acting as a regenerator or recuperator, is the low overall heat transfer coefficient when both the heating and heated fluids are gases. The maximum effectiveness is about 75 per cent. This, in turn, requires a large heat exchange area. Another drawback is the resistance to flow of the gases, which causes a pressure drop in the compressed air, and increases the back pressure on the turbine. However, by careful design, these losses can be kept low.

Figure 13.35 (b) shows the effect of regeneration on the specific fuel consumption for various values of compression ratio. It is obvious from the figure that up to a certain value of compression ratio, the specific fuel consumption is less except for much lower value of regenerator effectiveness. After certain value of pressure ratio, i.e., when the temperature of compressed air is more than the exhaust temperature of gas. the specific fuel consumption increases. Thus, we can say that the use of regenerator is limited by pressure ratio. Note that the specific fuel consumption is the minimum at lower value of pressure ratio, i.e., a lower value of compressed air temperature.

13.17 OPEN GAS TURBINE CYCLE WITH REHEAT

By reheating or adding heat to the gases after they have passed through a part of the rows of the turbine blading (or a stage), a further increase in work done is obtained. In reheating, the gas temperature, which has dropped due to expansion is brought back to approximately the initial temperature for the expansion in the next stage. Since the working fluid contains about 85 per cent air, additional fuel can be burnt by injecting it into the gases without any additional air supply. GT24 and GT26 models are a reheat gas turbines in operation.

13.17.1 Ideal Cycle

1 –2s–3–4s–5-6s–1 shows the ideal cycle gas turbine with reheat as shown in Fig. 13.36 (a) and 13.36 (b). The combustion gas from the burner (combustor) CC_1 at temperature T_3 is partially expanded in the *h.p.* turbine from pressure p_2 to intermediate pressure p_x. After this, it is then passed through combustion chamber CC_2 (reheater) where it is reheated or additional heat is added at constant pressure (p_x) so that the temperature of the gas is raised from T_{4s} to $T_5 = T_3$. After this, the gas is expanded in the l.p. turbine from pressure p_x to p_1.

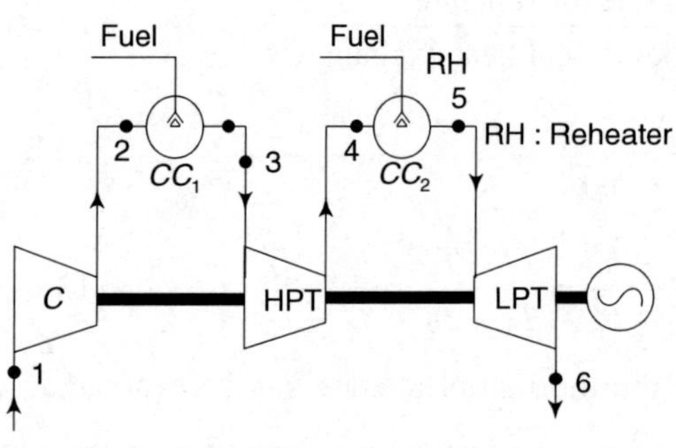

(b) Open cycle gas turbine with reheat

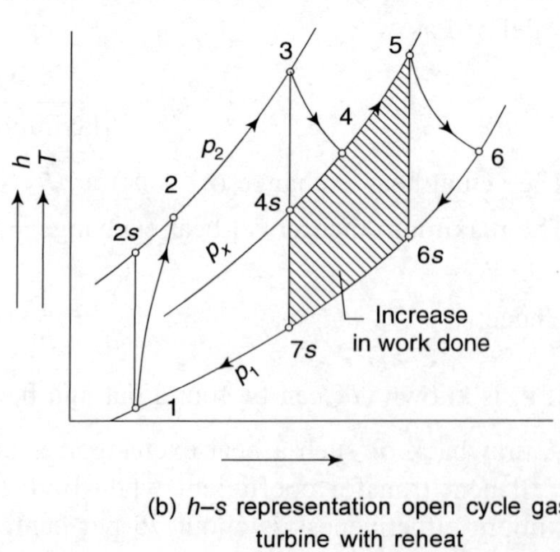

(b) *h–s* representation open cycle gas turbine with reheat

Fig. 13.36

The net work done is more with reheat compared to the ideal cycle (without reheat). The efficiency with which the heat is added to the gas in the combustion chamber (CC_2) at lower pressure (p_x) is lower than the efficiency when this heat is added to the gas in the combustion chamber (CC_1) at higher pressure, p_2. So, efficiency decreases as a result of reheating, only the net work is more by hatched areas 4*s*–5–6*s*–7*s* shown in Fig. 13.37 (b). However, in real cooled cycle, it may increase efficiency at higher compressor pressure ratio at a particular reheat pressure ratio.

Figure 13.37 (a) and (b) shows the variation of thermal efficiency and specific work with pressure ratio and both increase with the increase of t.

$$\text{Net work} = w_{net} = (h_3 - h_{4s}) + (h_5 - h_{6s}) - (h_{2s} - h_1)$$

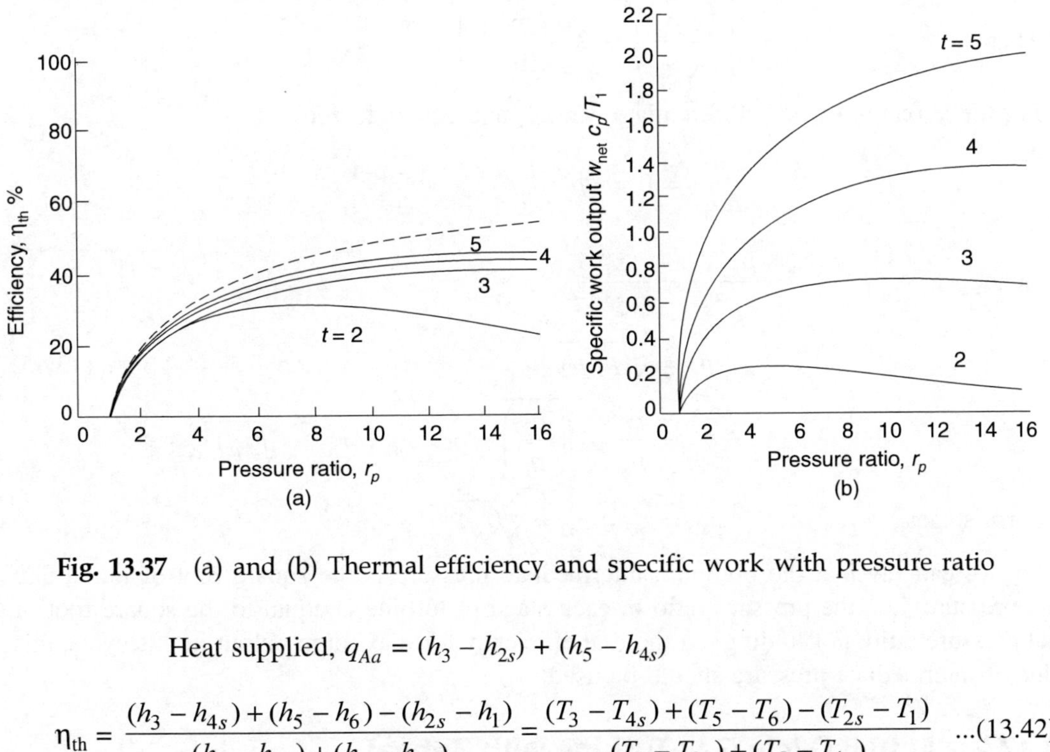

Fig. 13.37 (a) and (b) Thermal efficiency and specific work with pressure ratio

$$\text{Heat supplied, } q_{Aa} = (h_3 - h_{2s}) + (h_5 - h_{4s})$$

$$\eta_{th} = \frac{(h_3 - h_{4s}) + (h_5 - h_6) - (h_{2s} - h_1)}{(h_3 - h_{2s}) + (h_5 - h_{4s})} = \frac{(T_3 - T_{4s}) + (T_5 - T_6) - (T_{2s} - T_1)}{(T_3 - T_{2s}) + (T_5 - T_{4s})} \quad ...(13.42)$$

Let $t = T_3/T_1$ and $c = r_p^{(\gamma-1)/\gamma}$, then

$$\eta_{th} = \frac{2t - c + 1 - 2t/\sqrt{c}}{2t - c - t/\sqrt{c}} \quad ...(13.43)$$

and

$$\frac{w_{net}}{c_p\, T_1} = 2t - c + 1 - \frac{2t}{\sqrt{c}} \quad ...(13.44)$$

For the maximum output, there must be an optimum pressure at which reheating should be done. As we know, the compressor work (w_c) is not affected by reheating, so for the maximum output, we have to find the condition where turbine work (w_t) is the maximum.

$$w_t = c_p\,(T_3 - T_{4s}) + (T_5 - T_6) = c_p \left\{T_3\left(1-\frac{T_{4s}}{T_3}\right)+T_5\left(1-\frac{T_6}{T_5}\right)\right\}$$

$$= c_p\,[T_3\{1 - (p_x/p_2)^{(\gamma -)/\gamma}\} + T_5\,(- (p_1/p_x)^{(\gamma - 1)/\gamma})]$$

Here, $T_3 = T_5$ and c_p, T_3, p_1, p_2 are all constant. The only variable is p_x

Let $c_p\,T_3 = A = \text{constant and } \dfrac{\gamma-1}{\gamma} = a$

Then $$w_t = \left\{A\left(\frac{p_x}{p_2}\right)^a + 2 - \left(\frac{p_1}{p_x}\right)^a\right\}$$

For the maximum work, differentiate w.r.t. p_x and equate to zero.

$$\frac{dw_t}{dp_x} = 0 = ap_x^{a-1}\,p_2^{-a} - ap_x^{-a-1}\,p_1^a$$

Hence, $$\frac{p_x^{a-1}}{p_x^a} = \frac{p_1^a}{p_x^{a+1}}$$

or $$p_x = \sqrt{(p_1\,p_2} \qquad \text{...(13.45)}$$

or $$\frac{p_x}{p_1} = \frac{p_2}{p_x} = \sqrt{\left(\frac{p_2}{p_1}\right)} \quad \text{or} \quad r_{p1} = r_{p2} = \sqrt{(r_p)}$$

For n stages, $$r_{p1} = r_{p2} = r_{p3} = \sqrt[n]{(r_p)}$$

So, we can say that the optimum intermediate pressure is the square root of initial and final pressure, i.e., the pressure ratio in each stage of turbine is equal to the square root of total pressure ratio. For finding out thermal efficiency for maximum output in this cycle, this value of intermediate pressure should be used.

13.17.2 Actual Cycle Gas Turbine with Reheat

1–2–3–4–5–6–1 shows the actual cycle with reheat as shown in Fig. 13.26 (b).

Actual turbine work, $w_{ia} = (h_3 - h_4) + (h_5 - h_6) = c_p\,\{(T_3 - T_4) + (T_5 - T_6)\}$

Actual compressor work, $w_{ca} = h_2 - h_1 = c_p\,(T_2 - T_1)$

Actual heat supplied, $(h_3 - h_2) + (h_5 - h_4) = c_p\,\{(T_3 - T_2) + (T_5 - T_4)\}$

∴ $$\eta_{tha} = \frac{(T_3 - T_4)+(T_5 - T_6)-(T_2 - T_1)}{(T_3 - T_2)+(T_5 - T_4)} \qquad \text{...(13.46)}$$

or $$\eta_{tha} = \frac{(T_3 - T_{4s})\eta_t + (T_5 - T_{6s})\eta_t - \dfrac{(T_{2s} - T_1)}{\eta_c}}{(T_3 - T_2) + (T_5 - T_4)} \quad ...(13.47)$$

In this case also, the condition for maximum output will be the same.

Note With reheat system, two-shaft system may be used; h.p. turbine may drive compressor called compressor turbine and l.p. turbine may produce power by driving generator called power turbine. Thus,

$$w_c = (w_t)_{ct} \quad \text{and power} = (w_t)_{pt}$$

or $$(h_2 - h_1) = (h_3 - h_4)$$

and $$\text{Power} = (h_5 - h_6), \text{ kJ/kg} = \dot{m}\,(h_5 - h_6), \text{ kW}$$

With two-shaft system, the condition of maximum output will not be applicable as it may be possible that the compressor turbine may not be able to drive compressor. Even in simple gas turbine, two shafts may be used.

13.18 OPEN GAS TURBINE WITH INTERCOOLING

The net work of the gas turbine cycle may be increased by saving some compression work. This is done by using several stages of compression with intercooling of air between stages. The air from the low pressure stages compressor (LPC) is cooled in a intercooler approximately to initial temperature before entering the high pressure stage compressor (HPC). The effect of intercooling is to increase the net work and decrease the efficiency as compared to the simple ideal cycle without intercooling.

13.18.1 Ideal Cycle

1–2*s*–3–4*s*–5–6*s*–1 shows the ideal cycle gas turbine with intercooling as shown in Fig. 13.38 (a) and (b). In l.p. stage compressor, atmospheric air is compressed from p_x to p_2, it is then cooled from temperature T_{2s} to $T_3 = T_1$ in the intercooler at constant intermediate pressure p_x and finally compressed from p_x to p_2 in the h.p, stage of compressor. At constant pressure p_2, heat is added in the combustion chamber up to T_5 and after that it is expanded in the turbine from pressure p_2 and temperature T_5 to pressure p_1. It is evident from Fig. 13.38 (b) that the turbine work is the same as for the simple cycle but the work supplied to the compressor is reduced because the constant pressure lines converge towards the left. The work supplied to the compressor is saved due to intercooling which may be shown as area 2*s*–3–4*s*–*a*–2*s*, so this is the amount by which net work of the cycle is increased as compared to ideal cycle without intercooling. But the heat supplied is increased by amount $(h_a - h_{4s})$ because the compressed air delivered by second stage compressor is at lower temperature T_{4s} than in simple cycle. So, the effect of the intercooling is to decrease the thermal efficiency of the cycle. This result is to be expected because the heat supplied in the intercooler is a complete loss, no part of it is converted into useful work.

$$\text{Turbine work, } w_t = h_5 - h_{6s} = c_p\,(T_5 - T_{6s})$$

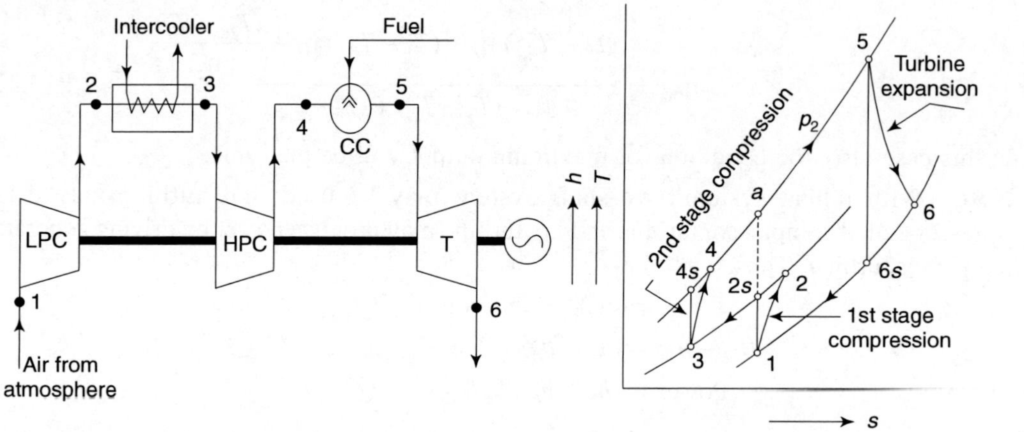

(a) Open cycle gas turbine with intercooling

(b) *h–s* representation of gas turbine with intercooling

Fig. 13.38

Compressor work, $w_c = (h_{2s} - h_1) + (h_{4s} - h_2) = c_p\ \{T_{2s} - T_1) + (T_{4s} - T_3)\}$

Heat supplied = $q_A = h_5 - h_{4s} = c_p\ (T_5 - T_s)$

So, the thermal efficiency, $$\eta_{th} = \frac{T_5 - T_{6s}) - (T_{4s} - T_3) - (T_{2s} - T_1)}{T_5 - T_{4s}} \qquad \text{...(13.48)}$$

For the minimum compression work in the compressor or the maximum work output in this cycle, only the compressor work is responsible. Here $T_3 = T_1$

$$w_c = c_p\ \{(T_{2s} - T_1) + (T_{4s} - T_3)\} = c_p\ T_1 \left\{\frac{T_{2s}}{T_1} - 1 + \frac{T_{4s}}{T_1} - 1\right\}$$

$$= c_p\ T_1\ \{(p_x/p_1)^{(\gamma-1)/\gamma} - 1 + (p_2/p_x)^{(\gamma-1)/\gamma} - 1\}$$

Let $\qquad c_p\ T_1 = A = \text{constant and } (\gamma - 1)/\gamma = a$

$$w_c = A\left[\left(\frac{p_x}{p_1}\right)^a - 1 + \left(\frac{p_2}{p_x}\right)^a - 1\right]$$

For the maximum work output or the minimum compression work, differentiate w_c w.r.t. to p_x and equate to zero.

Thus, $$\frac{dw_c}{dp_x} = 0 = a \cdot p_x^{a-1}\ p_1^{-a} - a \cdot p_2^a \cdot p_x^{-a-1}$$

or $$\frac{p_x^{a-1}}{p_1^a} = \frac{p_2^a}{p_x^{a+1}}$$

or $$p_x = \sqrt{(p_1\ p_2)} \qquad \text{...(13.49)}$$

or $$\frac{p_x}{p_1} = \frac{p_2}{p_x} = \sqrt{\left(\frac{p_2}{p_1}\right)}$$

or $$r_{p1} = r_{p2} = \sqrt{(r_p)} \qquad ...(13.50)$$

For n number of stage, $r_{p1} = r_{p2} = r_{p3} \cdots \sqrt[n]{(r_p)}$

Thus, we see that for the minimum compression work, the optimum intermediate (p_x) is the square root of initial and final pressure. For the maximum work output, this value of intermediate pressure p_x is used.

13.18.2 Actul Cycle Gas Turbine with Intercooling

1–2–3–4–5–6–1 shows the actual cycle gas turbine with intercooling as shown in Fig. 13.38 (b).

$$\text{Actual turbine work, } w_{ta} = h_5 - h_6 = c_p\,(T_5 - T_6)$$

$$\text{Actual compressor work, } w_{ca} = (h_2 - h_2) + (h_4 - h_3) = c_p\,[(T_2 - T_1) + (T_4 - T_3)\}$$

$$\text{Actual net work} = W_{\text{neta}} = c_p\,\{(T_5 - T_6) - (T_4 - T_3) - (T_2 - T_1)\}$$

$$\text{Actual heat supplied, } h_5 - h_4 = c_p\,(T_5 - T_4)$$

$$\text{Intercooler effectiveness, } \varepsilon_1 = \frac{T_2 - T_3}{(T_2 - T_1)}$$

T_3 may be found from known value of intercooler effectiveness

$$\eta_{\text{tha}} = \frac{(T_5 - T_6) - (T_4 - T_3) - (T_2 - T_1)}{(T_5 - T_4)} = \frac{(T_5 - T_{6s})\,\eta_t - \dfrac{(T_4 - T_{3s})}{\eta_c} - \dfrac{(T_2 - T_1)}{\eta_c}}{(T_5 - T_4)} \qquad ...(13.51)$$

In this case also, we can show that for the minimum work input to the compressor $r_{p1} = r_{p2} = \sqrt{r_p}$.

13.19 OPEN GAS TURBINE CYCLE (OGTC) WITH INTERCOOLING, REHEAT AND REGENERATION

The power output as well as the thermal efficiency of this cycle is much greater than that of the simple cycle. It is to be noted that the power output is increased due to reheat and intercooling and thermal efficiency is due to regeneration.

13.19.1 Ideal Cycle

1–2s–3–4s–5–6–7s–8–9s–1 shows such a type of an ideal cycle as shown in Fig. 13.39 (a) and (b). Following are the various processes.

Process 1–2 : Isentropic compression of air in l.p. stage compressor (LPC). For the minimum compression work, the pressure ratio r_{p1} of compressor (LPC) should be equal to $\sqrt{r_p}$.

Procee 2–3 : Cooling of air in the intercooler at constant pressure p_x from temperature T_2 to $T_3 = T_1$.

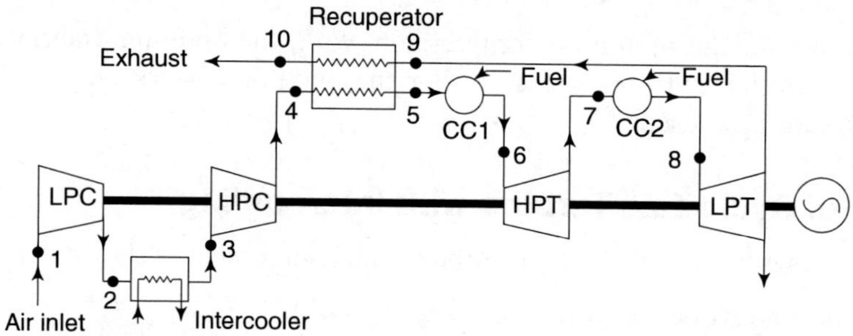

(a) OGTC with intercooling, reheat and regeneration. (complete cycle)

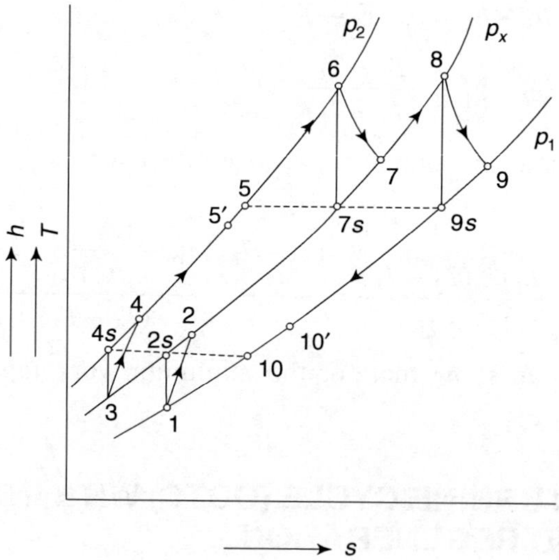

(b) h–s representation of OGTC with intercooling, reheat and regeneration

Fig. 13.39

Procee 3–4 : Isentropic compression of air in h.p. stage compressor (HPC). For the minimum compression work, the pressure ratio r_{p2} of (LPC) should be equal to $\sqrt{r_p}$.

Procee 4–5 : Heating of air in the regenerator at constant pressure p_4 from temperature T_{4s} to T_5. For perfect heat exchange in the regenerator, T_5 should be equal to T_s.

Procee 5–6 : Addition of heat at constant pressure (p_2) in the combustion chamber (CCI) or from temperature T_5 to T_6.

Process 6–7 : Isentropic expansion of gas from pressure p_2 to p_x in the first stage of turbine. For the maximum turbine work, pressure ratio, $p_2/p_x = \sqrt{r_p}$.

Process 7–8 : Reheating of gas at constant pressure (p_x) in the combustion chamber (CC2) from temperature T_{7s} to $T_8 = T_6$.

Process 8–9 : Isentropic expansion of gas from pressure p_x to p_1 in the second stage for turbine, for the maximum turbine work, pressure ratio $p_x/p_1 = \sqrt{r_p}$.

Process 9–10 : Cooling of exhaust gas in the regenerator from temperature T_{9s} to T_{10}. If there is a perfect heat exchange in the regenerator, $T_{10} = T_{2s}$.

Turbine work, $w_t = (h_6 - h_{7s}) + (h_8 - h_{9s})$ and

Compressor work, $w_c = (h_{2s} - h_1) + (h_{4s} - h_3)$

Net work, $w_{net} = (h_6 - h_{7s}) + (h_8 - h_{9s}) - (h_{2s} - h_1) - (h_{4s} - h_3)$

Heat supplied, $q_A = (h_6 - h_5) + (h_8 - h_{7s})$

$$\eta_{th} = \frac{(h_6 - h_{7s}) + (h_8 - h_{9s}) - (h_{2s} - h_1) - (h_{4s} - h_3)}{(h_6 - h_5) + (h_8 - h_{7s})}$$

$$= \frac{(T_6 - T_{7s}) + (T_8 - T_{9s}) - (T_{2s} - T_1) - (T_{4s} - T_3)}{(T_6 - T_5) + (T_8 - T_{7s})} \quad \text{...(13.52)}$$

13.19.2 Actual Gas Turbine Cycle with Regeneration, Reheat and Intercooling

1-2-3-4-5-6-7-8-9-10 shows such an actual cycle as shown in Fig. 13.39 (b).

Actual turbine work, $w_{ta} = (h_6 - h_7) + (h_8 - h_9)$

Actual compressor work, $w_{ca} = (h_2 - h_1) + (h_4 - h_3)$

Net work, $w_{neta} = (h_6 - h_7) + (h_8 - h_9) - (h_2 - h_1) - (h_4 - h_3)$

Heat supplied, $q_{Aa} = (h_6 - h_5) + (h_8 - h_7)$

Actual thermal efficiency $= \eta_{tha}$

$$= \frac{(h_6 - h_7) + (h_8 - h_9) - (h_2 - h_1) - (h_4 - h_3)}{(h_6 - h_5) + (h_8 - h_7)}$$

$$= \frac{(T_6 - T_7) + (T_8 - T_9) - (T_2 - T_1) - (T_4 - T_3)}{(T_6 - T_5) + (T_8 - T_7)}$$

$$= \frac{(T_6 - T_{7s})\,\eta_t + (T_8 - T_{9s})\,\eta_t - (T_{2s} - T_1)/\eta_c - (T_{4s} - T_3)/\eta_c}{(T_6 - T_5) + (T_8 - T_7)}$$

$$\text{Regenerative effectiveness, } \varepsilon_r = \frac{h_5' - h_4}{h_9' - h_4} = \frac{h_9 - h_{10}'}{h_5' - h_{10}'} = \frac{T_5' - T_4}{T_9' - T_4} \quad ...(13.53)$$

If ε_r is known, then T'_5 can be found out. Once T'_5 is known, we can find out the thermal efficiency.

If the pressure loss in intercooler, combustion chambers, regenerator and exhausthood are given the *h–s* representation will be modified accordingly.

It is worth to note that in real cycle for the maximum efficiency, the values of p_x *(intercooling pressure or reheating pressure) will not be the geometric means.*

Example 13.26 An open gas turbine plant works between the fixed absolute temperature limits 300 K and 1500 K, the absolute pressure limits being 1 bar and 14 bar. The isentropic efficiency of compressor is 0.85 and that turbine is 0.86. Estimate the actual thermal efficiency of the plant and the power developed. The calorific value of fuel is 4200 kJ/kg. Assume, $\eta_{comb} = 0.99$, $\eta_m = 0.98$ for whole assembly, $\eta_{gen} = 0.985$ and $\dot{m}_a = 500$ kg/s.

Solution Refer to Fig. 13.39 (a) and 13.40. Given $T_1 = 300$ K, $h_m = 0.98$, $p_1 = 1$ bar, $\eta_{gen} = 0.985$, $p_2 = 14$ bar, $\eta_c = 0.85$ and $\eta_t = 0.86$, $T_3 = 1500$, $\dot{m}_a = 500$ kg/s, $\eta_{comb} = 0.99$, C.V. = 42000 kJ/kg.

$$T_{2s} = T_1 \left(\frac{p_2}{p_1}\right)^{(\gamma-1)/\gamma} = 300\ (14)^{0.4/1.4} = 300 \times 2.125 = 637.5 \text{ K}$$

The compressor efficiency, $\eta_c = 0.85 = \dfrac{T_{2s} - T_1}{T_2 - T_1} = \dfrac{637.5 - 300}{T_2 - 300}$

or
$$T_2 = 300 + \frac{637.5 - 300}{0.85} = 697.05 \text{ K}$$

We have
$$\frac{T_3}{T_{4s}} = \left(\frac{p_1}{p_4}\right)^{(\gamma-1)/\gamma} = (14)^{0.4/1.4} = 2.125$$

$\therefore$
$$T_{4s} = \frac{T_3}{2.125} = \frac{1500}{2.125} = 705.88 \text{ K}$$

or
$$T_3 - T_4 = \eta_t\,(T_3 - T_{4s}) = 0.86\ (1500 - 705.88) = 682.94 \text{ K}$$

or
$$T_4 = T_3 - 682.94 = 1500 - 682.94 = 817.06 \text{ K}$$

Let $\dot{m}_a$ and $\dot{m}_f$ be the mass flow rate of air and fuel, respectively.

$$(\dot{m}_a + \dot{m}_f)\,h_3 - \dot{m}_a\,h_2 = \dot{m}_f \times \text{C.V.} \times \eta_{comb}$$

or
$$(\dot{m}_a + \dot{m}_f)\,c_{pg}\,T_3 - \dot{m}_a\,c_{pa}\,T_2 = \dot{m}_f \times \text{C.V.} \times \eta_{comb}$$

or
$$\frac{\dot{m}_a}{\dot{m}_f} = \frac{\text{C.V.} \times \eta_{comb} - c_{pg}\,T_3}{c_{pg}\,T_3 - c_{pa}\,T_2} = \frac{\text{C.V.} \times \eta_{comb} - c_p\,T_3}{c_p\,(T_3 - T_2)}$$

If $$c_{pg} = c_{pa} = c_p$$

or $$\frac{\dot{m}_a}{\dot{m}_f} = \frac{42000 \times 0.99 - 1.005 \times 1500}{1.005\,(1500 - 697.05)} = 49.65$$

$$\text{Fuel air ratio} = F/A = 1:49.65$$

Actual work consumed by the compressor/kg of air.

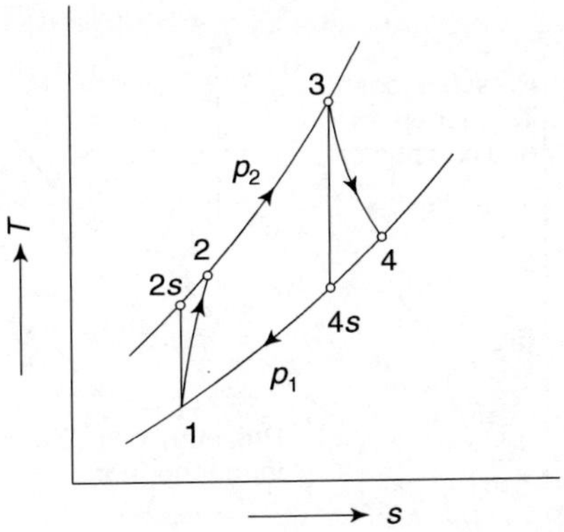

Fig. 13.40 Example 13.26

$$w_{ca} = c_p\,(T_2 - T_1)/\eta_m = 1.005\,(697.05 - 300)/0.98 = 407.17 \text{ kJ/kg}$$

Actual work done by the turbine assuming $c_p = 1.005$ is given by

$$w_{ta} = c_p\left(1 + \frac{\dot{m}_f}{\dot{m}_a}\right)(T_3 - T_4) = 1.005\left(1 + \frac{1}{49.65}\right)(1500 - 817.06) = 700 \text{ kJ/kg}$$

$$\text{Net work} = w_{\text{net}} = 700 - 407.17 = 292.83 \text{ kJ/kg of air}$$

Here, $$\dot{m}_a = 500 \text{ kg/s} \quad \text{and} \quad \dot{m}_f = \frac{500}{49.65} = 10.07 \text{ kg/s}$$

$$\text{Total net power available} = \dot{m}_a \times w_{\text{net}} \times \eta_{\text{gen}}$$

$$= 500 \times 292.83 \times 0.985/10^3 = 144.21 \text{ MW}$$ **Ans.**

$$\text{Thermal efficiency of the plant} = \frac{w_{\text{neta}}}{q_{Aa}} = \frac{w_{\text{neta}}}{\dot{m}_f \times \text{C.V.}} = \frac{144.21 \times 10^3}{10.07 \times 42000} = 34.09\%$$ **Ans.**

Note *In calculation of $\dot{m}_f$, η_{comb} has already been considered.*

Example 13.27 In a gas turbine plant, the air at 10°C and 1 bar is compressed to 12 bar with compression efficiency of 80%. The air is heated in the regenerator and the combustion chamber till its temeprature is raised to 1400°C, and during the process the pressure falls by 0.2 bar. The air then expands in the turbine and passes to regenerator which has 75%

effectiveness, and a pressure drop of 0.2 bar. If the isentropic efficiency of the turbine is 85%, determine the thermal efficiency of the plant.

Solution Refer to Fig. 13.34 (a) and Fig. 13.41. Though there is a fall in pressure in the regenerator and combustion chamber but for simplicity, addition of heat will be assumed at constant pressure, i.e., the proceses 2–3 will be assumed as 2 – *x* – 5′ – 3 and process 4 – 6′ will be assumed as 4 – *z* – 6′. Given : $T_1 = 273 + 10 = 283$ K.

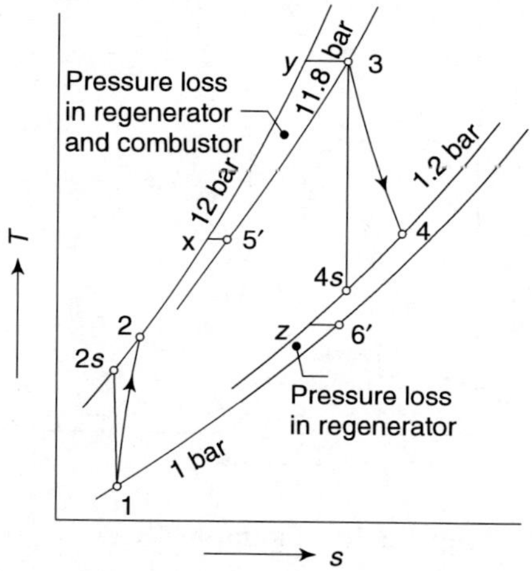

Fig. 13.41 Example 13.27

$$T_{2s} = T_1 \left(\frac{p_2}{p_1}\right)^{(\gamma-1)/\gamma} = 283 \times (12)^{0.286} = 576 \text{ K}$$

Now $$T_2 = T_1 + \frac{(T_{2s} - T_2)}{\eta_c} = 283 + \frac{(576 - 283)}{0.8} = 649.25 \text{ K}$$

$$p_3 = 12 - 0.2 = 11.8 \text{ bar}, \quad p_4 = p_2' = 1 + 0.2 = 1.2 \text{ bar}$$

Again, $$\frac{T_3}{T_{4s}} = \left(\frac{p_3}{p_4}\right)^{0.286} = \left(\frac{11.8}{1.2}\right)^{0.286} = 1.922$$

or $$T_{4s} = \frac{1673}{1.922} = 870.44 \text{ K}$$

and $$T_4 = T_3 - (T_3 - T_{4s})\,\eta_t = 1673 - (1673 - 870.44) \times 0.85 = 990.83 \text{ K}$$

$$w_{ca} = c_p\,(T_2 - T_1) = 1.005\,(649.25 - 283) = 368.08 \text{ kJ/kg}$$

$$w_{ta} = c_p\,(T_3 - T_4) = 1.005\,(1673 - 990.83) = 685.58 \text{ kJ/kg}$$

Net work, $$w_{neta} = w_{ta} = w_{ca} = 685.58 - 368.08 = 317.5 \text{ kJ/kg}$$

The regenerator effectiveness, $\varepsilon_r = \dfrac{T_5' - T_2}{T_4 - T_2} = 0.75$

or $$T'_5 = 649.25 + (990.83 - 649.25) \times 0.75 = 905.43 \text{ K}$$

Heat supplied, $q_{Aa} = c_p\,(T_3 - T'_5) = 1.005\,(1673 - 905.43) = 771.40$ kJ/kg

$$\eta_{th} = \frac{w_{neta}}{q_{Aa}} = \frac{371.5}{771.4} = 41.15\%$$ **Ans.**

Example 13.28 In an open gas turbine plant, the pressure ratio through which air at I5°C is compressed is 14. The same air is then heated to a maximum permissible temperature of 1300°C first in a heat exchanger which is 75% efficient, and then in the combustion chamber. The same air at 1300°C is expanded in two stages such that the expansion work is the maximum. The air is reheated to 1300°C after the high pressure stage. Determine the cycle thermal efficiency, the work ratio and the net shaft work per kg of air. The isentropic efficiencies may be assumed to be 85% and 86% for the compressor and turbine respectively. Calculate the flow rate of air for an output of 240 MW. Assume mechanical for compressor and turbine each and generator efficiencies equal to 99% each.

Solution Refer to Fig. 13.42 and Fig. 13.43. Given : $T_1 = 288$ K, $p_2/p_1 = 14$, $T_4 = T_6 = 1573$ K, $\eta_c = 0.85$, $\eta_t = 0.86$, $\eta_m = 0.99$, $\eta_{gen} = 0.99$.

$$T_{2s} = T_1\left(\frac{p_2}{p_1}\right)^{(\gamma-1)/\gamma} = 288\,(14)^{0.4/1.4} = 612.12 \text{ K}$$

or $$T_2 = T_1 + \frac{T_{2s} - T_1}{\eta_c} = \frac{(612.12 - 288)}{0.85} + 288 = 669.31 \text{ K}$$

For the maximum work output, the reheating should be done at intermediate pressure

$$p_x = \sqrt{(p_1\,p_2} \quad \text{and} \quad \frac{p_x}{p_1} = \frac{p_2}{p_x} = \sqrt{r_p} = \sqrt{14} = 3.74$$

$$\frac{T_4}{T_{4s}} = \left(\frac{p_2}{p_x}\right)^{(\gamma-1)/\gamma} = (3.74)^{0.4/1.4} = 1.457$$

or $$T_{5s} = \frac{1573}{1.457} = 1079.6 \text{ K}$$

Now, $$(T_4 - T_5) = \eta_t\,(T_4 - T_{5s}) = 0.86\,(1573 - 1079.61) = 424.32 \text{ K}$$

$$T_5 = 1573 - 424.32 = 1148.68 \text{ K}$$

Again, $$\frac{T_6}{T_{7s}} = \left(\frac{p_x}{p_1}\right)^{(\gamma-1)/\gamma} = (3.74)^{0.4/1.4} = 1.457$$

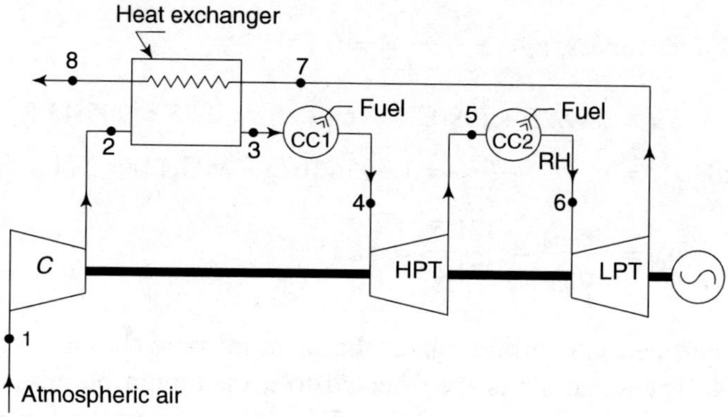

Fig. 13.42 Example 13.28

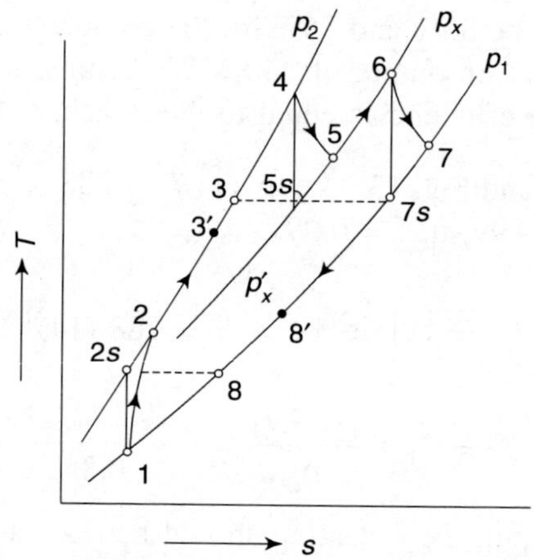

Fig. 13.43 Example 13.28

or $$T_{7s} = \frac{T_6}{1.457} = \frac{1573}{1.457} = 1079.6 \text{ K}$$

and $$(T_6 - T_7) = \eta_t \, (T_6 - T_{7s})$$

or $$T_7 = 1573 - 0.86 \, (1573 - 1079.6) = 1148.68 \text{ K}$$

Alternatively, $$T_7 = T_5 = 1079.61 \text{ K by given data}$$

$$\text{Regenerator effectiveness, } \varepsilon_r = \frac{T_3' - T_2}{T_7 - T_2} = 0.75 = \frac{T_3' - 669.31}{1148.68 - 669.31}$$

or $$T'_3 = 1028.83 \text{ K}$$

$$w_{ca} = c_p \, (T_2 - T_1)/\, \eta_m = 1.005 \, (669.31 - 288)/0.99 = 385.16 \text{ kJ/kg of air}$$

$$w_{ta} = c_p\,[(T_4 - T_5) + (T_6 - T_7)] \times \eta_m$$
$$= 1.005\,[(1573 - 1148.68) + (1573 - 1148.68)] \times 0.99 = 844.34 \text{ kJ/kg}$$

Heat supplied, $q_{Aa} = c_p\,[(T_4 - T'_3) + (T_6 - T_5)]$
$$= 1.005\,[(1573 - 1028.83) + (1573 - 1148.68)] = 973.33 \text{ kJ/kg}$$

Net work, $w_{neta} = (844.34 - 385.16) \times 0.99 = 454.58$ kJ/kg of air **Ans.**

$$\eta_{th} = \frac{w_{neta}}{q_{Aa}} = \frac{\text{Net work}}{\text{Heat supplied}} = \frac{451.58}{973.33} = 46.7\%$$ **Ans.**

$$\text{Work ratio} = \frac{\text{Net work}}{\text{Turbine work}} = \frac{454.58}{844.34} = 0.5283$$ **Ans.**

$$\text{Power output} = \dot{m}_a \times w_{neta} = 240 \times 10^3 = \dot{m}_a \times 454.58$$

or $\dot{m}_a = 527.98$ kg/s

Example 13.29 The following data refer to an open cycle gas turbine using intercooling, regeneration and reheating arrangement.

Pressure ratio = 49, Compressor inlet temperature = 27°C, Turbine inlet temperature = 1300°C, Compressor efficiency = 0.86, Turbine efficiency = 0.87, Regenerator effectiveness = 0.8, Combustion efficiency = 0.99, Mechanical efficiency of whole assembly = 0.97, Generator efficiency = 0.98, Inlet pressure to compressor = 1 bar.

Neglect pressure loses in intercooler, combustion chamber and regenerator.

Determine (a) cycle specific work, (b) cycle thermal efficiency, (c) cycle work ratio, (d) cycle air rate, (e) for 220 MW power output, find the mas flow rate of air.

Solution Refer to Fig. 13.39. Given : $r_p = p_2/p_1 = 49$, $\eta_c = 0.86$, $\eta_t = 0.87$, $\varepsilon_r = 0.8$, $\eta_{comb} = 0.99$, $\eta_{gen} = 0.98$, $\eta_m = 0.97$, $T_1 = 300$ K, $p_1 = 1$ bar, $T_6 = 1573$ K

The optimum pressure for the minimum work of compressor and the maximum work of turbine is given by

$$p_x = \sqrt{p_1\,p_2}\quad \sqrt{49} = 7 \text{ bar}$$

$$T_{2s} = T_1\left(\frac{p_x}{p_1}\right)^{(\gamma-1)/\gamma} = 300\,(7)^{0.4/1.4} = 523.38 \text{ K}$$

Now
$$T_2 = T_1 + \frac{(T_{2s} - T_1)}{\eta_c} = 300 + \frac{523.38 - 300}{0.86} = 300 + 259.74 = 559.74 \text{ K}$$

Work consumed by l.p. compressor = w_{lpca}
$$= c_p\,(T_2 - T_1) = 1.005\,(559.74 - 300) = 261.03 \text{ kJ/kg}$$

For the perfect intercooling and the minimum work to compressor, the work conumed in both stages will be the same. Hence, total work done in two stages of compression including mechanical losses.

$$w_{ca} = (261.03 + 261.03)/0.97 = 538.2 \text{ kJ/kg}$$

$$T_{7s} = \frac{T_6}{(7)^{(\gamma-1)/\gamma}} = \frac{1573}{1.744} = 901.94$$

Again, $$T_7 = T_6 - \eta_t (T_6 - T_{7s}) = 1573 - 0.87 (1573 - 901.94) = 989.18 \text{ K}$$

Work done by the l.p. stage of turbine, $c_p (T_6 - T_7) = 1.005 \times 583.82 = 586.73$ kJ/kg

Assume that the reheating is done at the optimum pressure up to in the initial inlet turbine temperature, i.e., $T_6 = T_8$ and $T_7 = T_9$, hence the work done by both stages will be the same.

Total work done by the turbine, $w_{ta} = 586.73 + 586.73 = 1173.46$ kJ/kg **Ans.**

Since mechanical efficiency as a whole assembly is given so in turbine it will not be considered as it is already considered in compressor work.

(a) Net work, $w_{neta} = (w_{ta} - w_{ca})\, \eta_{gen} = (1173.46 - 538.2) \times 0.98 = 622.55$ kJ/kg **Ans.**

Regenerator effectiveness, $\varepsilon_r = \dfrac{T_5' - T_4}{T_9 - T_4} = \dfrac{T_5' - T_2}{T_9 - T_2}$

or $$T'_5 = 559.74 + 0.8 (989.18 - 559.745) = 343.55 + 559.74 = 903.29 \text{ K}$$

Heat supplied, $q_{Aa} = c_p\, [(T_6 - T_5) + (T_8 - T_7)]$

$$= 1.005\, [(1573 - 903.29) + (1573 - 989.18)]/0.98 = 1285.5 \text{ kJ/kg}$$

(b) Cycle thermal efficiency $= \dfrac{w_{ta} - w_{ca}}{qa} = \dfrac{622.55}{1285.5} = 48.42\%$ **Ans.**

(c) Cycle work ratio $= \dfrac{\text{Net work}}{\text{Turbine work}} = \dfrac{622.55}{1173.46} = 53.05\%$ **Ans.**

(d) Cycle rate $= \dfrac{3600}{w_{neta}} = \dfrac{3600}{622.55} = 5.78$ kJ/kWh **Ans.**

(e) $$\dot{m}_a = \frac{\text{Generator output}}{w_{neta}} = \frac{220 \times 10^3}{622.55} = 353.38 \text{ kg/s}$$

13.20 SUMMARY OF FORMULAE

1. $\eta_{th} = \eta_{as} = \dfrac{Q_A - Q_R}{Q_A}$

2. $\eta_{rel} = \dfrac{\eta_{tha}}{\eta_{as}}$

3. $r = \dfrac{v_1}{v_2}$ and $r = \dfrac{1+c}{c}$

4. $p_m = \dfrac{\text{Work done per cycle}}{\text{Stroke volume}}$

5. i.p. $= \dfrac{100\, p_m\, LAn}{60}$ kW where p_m is in bar

6. $\eta_{\text{carnot}} = \dfrac{T_H - T_L}{T_H}$

7. $(\eta_{\text{th}})_{\text{otto}} = (\eta_{as})_{\text{otto}} = 1 - \dfrac{1}{r^{\gamma-1}}$ where r = compression ratio

8. $(\text{m.e.p.})_{\text{otto}} = p_i \cdot r \left[\dfrac{(\alpha-1)(r^{\gamma-1}-1)}{(r-1)(\gamma-1)}\right]$ bar if p_1 is in bar and $\alpha = p_3/p_2$

9. $(\eta_{\text{th}})_{\text{diesel}} = (\eta_{as})_{\text{Diesel}} = 1 - \dfrac{1}{r^{\gamma-1}}\left[\dfrac{\beta^\gamma - 1}{\gamma(\beta-1)}\right]$ where $\beta = v_4/v_3$

10. $(\text{m.e.p.})_{\text{Diesel}} = p_1\, r\left[\dfrac{\gamma\, r^{\gamma-1}(\beta-1) - (\beta^\gamma - 1)}{(r-1)(\gamma-1)}\right]$

11. $(\eta_{\text{th}})_{\text{Dual}} = (\eta_{as})_{\text{Dual}} = 1 - \dfrac{1}{r^{\gamma-1}}\left[\dfrac{(\alpha\,\beta^\gamma - 1)}{(\alpha-1) + \gamma\,\alpha\,(\beta-1)}\right]$

12. $(\text{m.e.p.})_{\text{Dual}} = \dfrac{p_1 \cdot r}{(\gamma-1)(r-1)}\left[r^{\gamma-1}\{(\alpha-1)+\gamma\,\alpha\,(\beta-1)\} - (\alpha\cdot\beta^\gamma - 1\right]$

13. $(\eta_{\text{th}})_{\text{Stirling}} = \dfrac{R\,(T_H - T_L)\ln r}{R\,T_H \ln r + (1-\eta_{\text{reg}})\, c_v\,(T_H - T_L)} = \dfrac{T_H - T_L}{T_H}$ if $\eta_{\text{reg}} = 100\%$

14. $(\eta_{\text{th}})_{\text{Erricson}} = \dfrac{T_H - T_L}{T_H}$

15. $(\eta_{\text{th}})_{\text{Alkinson}} = 1 - \dfrac{1}{r^{\gamma-1}}\left[\dfrac{\gamma\,(\alpha^{1/\gamma} - 1)}{(\alpha-1)}\right]$ where $\dfrac{p_3}{p_2} = \alpha$

16. $(\eta_{\text{th}})_{\text{Brayton}} = 1 - \dfrac{1}{r_p^{\gamma-1}}$ where r_p compressor pressure ratio

17. $(\eta_{\text{th}})_{\text{act, Brayton}} = \dfrac{w_t \cdot \eta_t, \eta_c - w_c}{q_{Aa} \cdot \eta_c}$

18. $\text{AR} = \dfrac{3600}{w_{\text{net}}}$, kg of air/kWh

19. $\text{WR} = \dfrac{w_{net}}{w_t}$

20. $\text{sfc} = \dfrac{3600\,\dot{m}_f\ (\text{kg/s})}{w_{net}\ (\text{kW})} = \dfrac{3600\,(\dot{m}_f/\dot{m}_a)}{w_{net}\ (\text{kJ/kg})}$; kg/kWh

21. ε_r = regenerator effectiveness = $\dfrac{\text{Actual heat exchange}}{\text{Maximum theoretical heat exchange}}$

PROBLEMS

Things to Think About

13.1. What do you mean by an advertisement of car of 2litre engine?

13.2. List the differences between the events occurring within the cylinder of an actual internal combustion engine and air-standard cycle.

13.3. "A car produces more power with low ambient temperature". Do you agree with this statement? Explain the reasons.

13.4. A car operating at high elevation loses power. Do you agree with this statement? Give reasoning about your answer.

13.5. What are the reasons that external surfaces of a lawn mover engine are covered with fins? Why two stroke engines are not used in cars?

13.6. For the same pressure rise by an air compressor of gas turbine plant, the work input per unit mass flow will be much more than needed by a pump of vapour power plant. Discuss the reasons.

13.7. A lot of development has taken place in gas turbine but still it is not used in passenger cars. Why?

Short and Long Theoretical Questions

13.8. List the various types of thermal power plants in use and research.

13.9. What do you mean by air-standard cycle? What are the assumptions for air standard cycles?

13.10. Prove that the air-standard efficiency for an engine working on the Otto cycle is $\left(1 - \dfrac{1}{r^{\gamma - 1}}\right)$

where r is the ratio of compression and γ is the ratio of sp. heats of air.

13.11. An engine working on the air-standard Otto cycle in which the salient points are 1, 2, 3 and 4 has upper and lower temperature limits, T_3 and T_1. If the maximum work per kg of air is to be done show that the intermediate temperature

$$T_2 = \sqrt{T_1 T_3} = T_4.$$

13.12. Derive an expression for the efficiency of the ideal cycle for the Diesel engine in terms of the compression ratio r, the cut-off ratio β and γ.

13.13. Show that the mean effective pressure for an engine working on an ideal Diesel cycle is given by the following expression.

$$\text{m.e.p.} = p_1 r \left[\frac{\gamma r^{\gamma - 1} (\beta - 1) - (\beta^{\gamma} - 1)}{(r-1)(\gamma - 1)} \right]$$

where p_1 is the pressure at the beginning of compression, r compression ratio, β cut-off ratio and γ is the adiabatic index.

13.14. Derive the expression for the mean effective pressure of Otto cycle.

13.15. Derive an expression for the efficiency of the dual combustion cycle in terms of compression ratio, cut-off ratio and other relevant parameters. Show that this expression reduces to that for Diesel cycle if there is no constant volume rise of pressure and that of the Otto cycle if there is no constant pressure combustion. Also derive the expression for mean effective pressure.

13.16. Compare the efficiency of the Otto, the Diesel and the dual cycle under the conditions of (a) equal compression ratio and heat input, (b) constant maximum pressure and heat input, (c) constant maximum pressure and output, and (d) constant maximum pressure and temperature.

13.17. Discuss the causes of deviation of the actual Otto cycle from that of ideal.

13.18. Show that the efficiency of Stirling and Erricson regenerative cycles is equal to that of the Carnot.

13.19. Derive the air-standard efficiency of the Joule or Brayton cycle and explain why it is the true constant pressure cycle. Explain its use in modern engines. Discuss its irreversibilities.

13.20. In an ideal cycle, the charge is compressed adiabatically, explosion occurs at the end of the stroke and is followed by adiabatic expansion to such a condition that subsequent isothermal compression restores the charge to its original pressure and volume. Show that the air-standard efficiency of the cycle is

$$1 - \frac{\ln x^{\gamma - 1} - \ln r^{\gamma - 1}}{x^{\gamma - 1} - r^{\gamma - 1}}$$

where r = ratio of compression and x = ratio of expansion.

13.21. Derive the expressions for thermal efficiency of simple open cycle gas turbine and also when equipped with regeneration, intercooling, reheat and its combinations.

Numerical Questions

13.22. In a Carnot cycle, the maximum pressure and temperature of the cycle are 17.5 bar and 175°C. The isothermal expansion ratio is 3 while the isentropic expansion ratio is 2. Calculate the thermal efficiency, work done and mean effective pressure.

[*Ans.* 24.25%, 34.3 kJ and 4.642 bar]

13.23 Find the source and sink temperature of an ideal engine working on Carnot cycle in which 1/5 of the heat input is converted into work and the efficiency becomes doubled when the temperature of the sink is reduced by 60°C.

13.24. In an air standard Otto cycle, the compression ratio 10, the compression begins at 37.8°C, 1 bar and the maximum temperature of the cycle is 1060°C. Determine (a) the heat supplied per kg of air, (b) the work done per kg of air, (c) the maximum pressure of the cycle, and (d) the thermal efficiency. [*Ans.* (a) 795.8 kJ/kg, (b) 478.62 kJ/kg, (c) 60.15%, (d) 60.2 bar]

13.25. At the beginning of compression in an air standard Otto cycle, engine cylinder the temperature is 37.8°C, the pressure is 0.99 bar and volume is 0.000707 /m^3. At the end of compression the pressure is 10.55 bar. The heat supplied to the cycle is 1.5 kJ. Calculate (a) the compression ratio, (b) the net work per cycle, and (c) the mean effective pressure.

[*Ans.* (a) 5.42, (b) 0.735 kJ, (c) 12.2 bar]

13.26. The compression ratio of an Otto cycle is 8 : 1. Before the compression stroke of the cycle begins, the pressure is 1 bar and the temperature is 26.7°C. The heat added to the air per cycle is 2079 kJ/kg of air. Determine (a) the pressure and temperature at the end of each process of the cycle, and (b) the thermal efficiency and the mean effective pressure.

[*Ans.* (a) 688.5 K, 18.37 bar 3543 K, 94.52 bar, 1542 K, 5.14 bar (b) 56.47%, 1526 bar]

13.27. An air standard Otto cycle operates with a compression ratio of 8.5. At the beginning of the compression, the air is at 1 bar and 32°C, and during the heat addition process, the pressure is tripled. Calculate (a) the thermal efficiency of the cycle and (b) the efficiency of the Carnot engine operating between the same overall temperature limits. [*Ans.* (a) 57.5%, (b) 85.4%]

13.28. In intake conditions for a theoretical Otto cycle operating with a compression ratio of 8 : 1 are 1 bar and 15°C. If the heat rejection equal 3517 kJ/kg, calculate the air standard efficiency and work ratio. How these quantities would compare with those of a Diesel cycle working under similar conditions? [*Ans.* 60.18%, 0.785; 47.9%, 0.688]

13.29. The intake conditions for a theoretical Otto cycle operating with a compression ratio of 8 : 1 are 1 bar and 15°C. At the beginning of the compression stroke, the cylinder volume is 0.00444 m^3 and 9.45 kJ of heat is added to the gas during the constant volume heating process. Determine the pressure and temperature at the end of each process of the cycle and determine the thermal efficiency and mean effective pressure. [*Ans.* T_2 = 661.65 K, p_2 = 18.37 bar, η_{th} = 56.47%]

13.30. An air standard Diesel cycle has a compression ratio of 16 and a cut-off ratio of 2. At the beginning of compression, the cylinder volume is 0.001415 m^3, the air pressure is 1 bar and the temperature is 4.45°C. Calculate the thermal efficiency. [*Ans.* 61.4%]

13.31. A Diesel engine with cut-off ratio 1.75 and compression ratio 13.5 has a polytropic exponent n for both compression and expansion of 1.35. The pressure at the beginning of compression is 1 bar. Calculate the mean effective pressure and the maximum pressure of the cycle.

[*Ans.* 4.23 bar, 33 bar]

13.32. A Diesel engine with compression ratio of 15 is operating with a fuel having a heating value of 44505 kJ/kg. At half load, this engine requires a 70 to 1 ratio of air to fuel, (a) what value of the cut-off ratio should be used in a ideal cycle if γ is taken as 1.38 during all processes of the cycle and the temperature at the beginning of compression is 93.3°C? (b) what is the ideal cycle efficiency? [*Ans.* (a) 1.592, (b) 60.6%]

13.33. In an air standard Otto cycle, the maximum and minimum temperatures are 1400°C and 15°C. The amount of heal supplied is 800 kJ/kg of air. Find the compression ratio and the thermal efficiency. Find also the ratio of the maximum to the minimum pressures in the cycle.

[*Ans.* 5.26%, 48.8%, 30.5 : 1]

13.34. The swept volume of a four-cylinder petrol engine is 2000 cm^3. The clearance volume of each cylinder is 60 cm^3. Determine the air standard thermal efficiency. If the inlet conditions are 1 bar and 24°C, and the maximum cycle temperature is 1400°C, determine the mean effective pressure. [*Ans.* 59%, 5.27 bar]

13.35. Calculate the thermal efficiency and the mean effective pressure of a dual combustion cycle when the maximum temperature is 2000°C and the maximum pressure is 70 bar. The pressure and temperature at the start of compression are 1 bar and 17°C respectively.

[*Ans.* 63.6%, 10.5 bar]

13.36. An engine operates on the theoretical Diesel cycle with a compression ratio of 12 : 1 and fuel is injected of 10% of the stroke. The pressure of the air entering the cylinder is 0.98 bar and its temperature is 15°C. Calculate (a) cut-off ratio, (b) the temperature at the end of compression process, and (c) the heat input. [*Ans.* (a) 2.1, (b) 753 K, (c) 953.5 kJ/kg]

13.37. A Diesel cycle operating on an air standard cycle has a compression ratio of 15. The pressure and temperature at the beginning of the compression are 1.04 bar and 15°C. If the maximum temperature of the cycle is 2330 K, determine the thermal efficiency and (b) the mean effective pressure. [*Ans.* (a) 56.9%, (b) 16.7 bar]

13.38. In the air standard Diesel cycle, the compression ratio is 15 and the pressure and temperature of the air at the beginning of compression is 1 bar and 288 K. The peak temperature in the cycle is 2700 K. Calculate (a) heat supplied, (b) work done, (c) cycle efficiency, (d) peak pressure of the cycle, (e) cut-off ratio, and (f) m.e.p.

[*Ans.* (a) 1858.62 kJ/kg, (b) 1022.3 kJ/kg, (c) 55.1%, (d) 44.3 bar, (e) 3.17 and (f) 13.25 bar]

13.39. An air-standard Diesel cycle operates with a compression ratio of 14.8 and a cut-off ratio 2. At the beginning of the compression, the air pressure and the temperature are 37.8°C and 1 bar respectively. Calculate, (a) the maximum temperature in the cycle and the heat input per cycle.

13.40. In a dual combustion air standard cycle for a Diesel engine, the maximum pressure 56 bar. At the beginning of compression, the pressure is 0.98 bar and the temperature is 28.6°C. The compression ratio is 14 and the heat supplied is 1265 kJ/kg of air. Find the cut-off ratio and the ratio of expansion. [*Ans.* 1.805, 7.8]

13.41. The compression ratio is 8 for an engine operating on the air-standard dual cycle. The diameter of the cylinder is 25.4 and the stroke is 30 cm. At the start of compression, the air temperature is 21.6°C and the pressure is 1.03 bar. At the end of constant volume process—the pressure is 56 bar. If heat is added at constant pressure during 3% of the stroke, calculate (a) the net work of the cycle, and (b) the thermal efficiency. [*Ans.* (a) 779 kJ/kg (b) 55.8%]

13.42. An engine operating on the ideal air standard dual cycle has a compression ratio of 11 to 1 and the piston displacement is 0.00328 m^3. The initial pressure and temperature at the start of the compression are 1.04 bar and 26.7°C. The volume at the end of constant pressure heat addition process is 1.5 times the clearance volume. The pressure at the end of adiabatic expansion is 4.2 bar. Calculate the heat added during the constant volume process, and (b) the heat added during the constant pressure process and the net work per cycle. [*Ans.* (a) 6.8 kJ, (b) 8.65 kJ]

13.43. An air-standard dual cycle has a compression ratio of 6 and the pressure and temperature at the beginning of compression are 1 bar and 29°C. After constant pressure heat addition, the temperature becomes 1177°C and constant volume heat addition raises the temperature to 16.27°C. Determine (a) cut-off ratio, (b) constant volume pressure ratio, (c) heat added and rejected per kg of air, (d) thermal efficiency.

[*Ans.* (a) 1.31, (b) 2.344, (c) 10.59.29 kJ/kg, (d) 526.kJ/kg (e) 49.93%]

13.44. A four-stroke semi-Diesel engine operates on dual cycle. At the commencement of compression, the pressure and temperature are 1 bar and 37.8°C. The pressure after the end of isentropic compression is 16.4 bar. Heat is added first at constant volume and then at constant pressure of 29.3 bar. The heat of combustion is shared equally between the two phases. If the relative efficiency is 55% estimate (a) the actual specific consumption in kg/kWh and (b) the theoretical power developed per m^3 of displacement at an engine speed of 180 r.p.m. The calorific value of fuel is 43493 kJ/kg of fuel. [*Ans.* (a) 0.293 kJ/kWh, (b) 779 kW/m^3]

13.45. Calculate the index for compression and the mass of the mixture in the cylinder of a gas engine where bore is 18 cm and stroke is 32.5 cm. The ratio of compression is 5. Pressure and temperature at the start of compression is 1 bar and 100°C and the pressure at the end of compression is 6 bar. Determine, also the work of compression and heat interchange through cylinder walls during compression.

[*Ans.* m = 0.00945 kg, n = 1.13, compression work = – 40.5 kJ, Q_R = 27 kJ]

13.46. An internal combustion engine has a bore of 56 cm and a stroke of 76 cm. The compression ratio is 13.5. Pressure and temperature at the end of suction stroke are 1 bar and 43°C respectively. The compression index is 1.37. Calculate (a) pressure and temperature at the end of compression, (b) mass of air in the cylinder, (c) work done, and (d) heat rejected during compression.

[*Ans.* (a) 35.2 bar, 828 K, (b) 0.212 kg, (c) – 82.6 kJ, (d) 7.56 kJ]

13.47. A Stirling regenerative air engine works between temperature of 400°C and 15°C. The ratio of isothermal expansion is 3. Calculate the ideal efficiency of the regenerator is 0.8.

[*Ans.* (a) 57.3%, (b) 45.4%]

13.48. Find the air-standard efficiency of Atkinson cycle if the compression ratio is 5 and pressure at the end of heat reception is 2.5 times that at the beginning. Take γ = 1.41. [*Ans.* 57.5%]

13.49. In an air-standard Brayton cycle, the air enters the compressor at 1 bar and 25°C. The pressure after compression is 3 bar. The temperature at turbine inlet is 650°C. Calculate per kg of air (a) heat supplied, (b) heat rejected, (c) work available at the shaft, (d) temperature of air leaving the turbine, and (e) cycle efficiency.

[*Ans.* (a) 510 kJ/kg, (b) 178 kJ/kg, (c) 138.5 kJ/kg, (d) 875 K, (e) 26.9%]

13.50. A gas turbine power plant operates under the following conditions. Air enters the compressor at 1 bar and 15°C and leaves the compressor at 10 bar. Air leaves the combustion chamber at 1400 K. Calculate the thermal efficiency.

13.51. For an ideal Otto engine operating on air, the temperature at the end of isentropic compression is 452°C and at the end of expansion is 1347°C. The compression ratio is 7.5. Calculate the cycle work and efficiency. Use gas tables. [*Ans.* 1229 kJ/kg 47.6%]

13.52. The fuel-air ratio of an ideal Otto engine with 12% clearance is 0.06 kg fuel/kg of air. The calorific value of the fuel is 41840 kJ/kg. The engine operates on 25 kg/min. of air initially at 1 bar and 37°C. Treating the working substance as air, calculate (a) displacement volume/min, (b) pressure and temperature of all important points, (c) work obtained, (d) thermal efficiency of the cycle, and (e) the m.e.p.

[*Ans.* (a) 20.89 m^3/min, (b) T_2 = 755 K, p_2 = 22 bar, p_3 = 99 bar, T_3 = 3377 K
p_4 = 5.64 bar, T_4 = 1755 K, (c) 1276.6 kJ/kg, (d) 50.85% (e) 15.558 bar]

13.53. An air standard diesel cycle, receives 28.5 kJ/cycle of heat while operating at 3000 r.p.m. At the beginning of compression, the pressure, temperature and volume are 100 kPa, 305 K and 0.0425 m^3 respectively. At the beginning of heat addition, the pressure is 3450 kPa. Determine (a) *p.V.T.* at each point in the cycle, (b) work done (c) power and (d) m.e.p.

13.54. Air enters the compressor of an ideal air standard Brayton cycle at 100 kPa, 300 K with a volumetric flow rate of 5 m^3/s. The compressor pressure ratio is 10. The turbine inlet temperature is 1400 K. Calculate (a) the thermal efficiency of the cycle, (b) the work ratio, (c) the net power developed. [*Ans.* (a) 34.7%, (b) 0.396, (c) 2481 kW]

13.55. In question 13.54, if isentropic efficiency of compressor and turbine are 80% each, calculate the above. [*Ans.* (a) 24.9%, (b) 0.618, (c) 1254 kW]

13.56. Air enters a compressor at 10°C and at atmospheric pressure and is compressed through a pressure ratio of 5. The temperature of gaseous mixture enters the turbine at 850°C and expands to atmospheric pressure. Calculate the thermal efficiency of the cycle under the following conditions.

(a) The system operates on ideal Brayton cycle, (b) the isentropic efficiency of compressor and turbine are 0.85, (c) with condition (b), a regenerator of 75% effectiveness is introduced, (d) with condition (c), an intercooler which cools to 10°C is introduced at optimum pressure ratio. [*Ans.* (a) 36.86% (b) 24.39%, (c) 37.08%, (d) 39.7%]

13.57. The air enters the compressor of gas turbine at 1 bar and 30°C and is compressed to 15 bar with an isentropic efficiency of 85%. The turbine inlet temperature is 1400°C. The isentropic efficiency of turbine is 86%. Calculate the mass flow rate of air, in kg/s for a net electrical power of 200 MW. Assume 2% of pressure loss in combustion chamber, 1% pressure loss in exhausthood, mechanical efficiency of whole assembly 98% and alternator efficiency 98%.

[*Ans.* 142 kg/s]

13.58. A gas turbine plant equipped with regenerative, intercooling and reheat arrangement operating at steady state, takes in air at 100 kPa, 300K at compressor inlet with a mass flow rate of 58.07 kg/s. The pressure ratio across two-stage compressor and turbine is 10. The intercooler and reheat each operate 300 kPa. At the inlet to the turbine stage, the temperature is 1400 K. The temperature at the inlet to the second compressor stage is 300 K. Assume isentropic efficiency each compressor and turbine stage is 80%. The regenerator effectiveness is 80% calculate (a) the thermal efficiency, (b) the work ratio, and (c) the net power developed.

[*Ans.* (a) 44.3%, (b) 45.4%, (c) 20.46 MW]

Software Based Questions

13.59. Develop a software to compare the performance of Otto, Diesel and Dual cycles as a function of compression ratio.

13.60. Develop a software to study the effect of varying compression ratio on performance parameters of engine such as heat addition, network, thermal efficiency, mean effective pressure and sfc of SI and CI engines. Tabulate the results by taking the real data of compression ratio and other independent parameters.

13.61. Develop a software to study the effect of cut-off ratio in a diesel engine on cycle net work, thermal efficiency and mean effect pressure. Plot the results for the following data; p_1 = 95 kPa, T_1 = 300 K, m_a = 12 g, r = 15, cut-off ratios = 1.5 to 2.5.

13.62. Develop a software for the air-standard dual cycle for the following data : p_1 = 1 bar, T_1 = 300 K, Q = 2000 kJ/kg.

Plot the net work per unit of mass in kJ/kg, the mean effective pressure, in bar, and the thermal efficiency versus compression ratio for different fractions of constant volume and constant pressure heat addition by taking compression ratio ranging from 10 to 20.

13.63. Develop a software to compare the performance of simple Brayton gas turbine cycle with regeneration, intercooling and reheating separately and in combinations. Tabulate the results for one realistic data.

13.64. Develop a software to plot the performance curves of simple Brayton gas turbine cycle by varying the compressor pressure ratio from 10 to 30 and turbine inlet temperature from 1400 K to 1800 K for ideal and actual processes.

Design and Open Ended Questions

13.65. There is breakthrough in the development of high strength metal-ceramic materials. With this development, it is possible to build an internal combustion engines with no cylinder wall cooling. These adiabatic engines are capable of working with cylinder wall temperature above 950°C. Prepare a report discussing the important considerations to be taken for adiabatic engine design. Also discuss the possibility of adiabatic engine likely to find widespread application.

13.66. One of the parameters to increase the performance of gas turbine engine is to increase the gas turbine inlet temperature. All the real engines now run at very high inlet temperature which require turbine blade cooling to safeguard the blading from failure. Write a report about the various types of blade cooling with cooling path sketches. Also discuss the current research issue in gas turbine technology.

13.67. In order to meet space vehicles power requirements, closed Brayton power systems, receiving their energy input from radioisotopes, nuclear reactor, or solar collector have been suggested. Write a report discussing the advantages and disadvantages of closed Brayton power cycle for space flight applications and possible design criteria to determine the selection of system and energy source.

13.68. Department of Energy (DOE) of USA has given a project called high efficiency engine turbine (HEET) programme to develop high efficiency combined cycle plant with zero emission.

Prepare a report about the gas turbine engine features and parameters of combined cycle for this programme.

13.69. Use of gas turbine in passenger cars has been suggested long back but still it has not been commercialised. Write a report about the feasibility and advantages and disadvantages of using gas turbines in passenger cars.

13.70. Develop a conceptual gas power cycle different from whatever in common use yielding higher efficiency, higher output with zero emission.

13.71. Inlet refrigeration air cooling in gas turbine plant is now a practice to boost power output and efficiency. Prepare a report on this dealing with latest technology.

Multiple-Choice Answer Type Questions

13.72. Air standard efficiency of Otto cycle is a function

(a) gas turbine inlet temperature

(b) compression ratio

(c) gas turbine exhaust temperature

(d) ratio of maximum to minimum temperature in a cycle

13.73. The mean effective pressure in air standard Otto cycle for a fixed inlet pressure is a function of

(a) compression ratio

(b) compression ratio and exhaust pressure

(c) compression ratio and ratio of maximum to minimum pressure in the cycle

(d) compression ratio and ratio of maximum to minimum pressure in the cycle

13.74. Air standard efficiency of Diesel cycle is a function of

(a) compression ratio and cut-off ratio

(b) compression ratio and ratio of maximum to minimum temperature

(c) compression ratio and ratio of maximum to minimum pressure

(d) compression ratio and ratio of exhaust temperature to inlet temperature

13.75. The air-standard efficiency of Brayton cycle is a function of

(a) compressor pressure ratio

(b) ratio of maximum to minimum temperature

(c) exhaust temperature

(d) ratio of exhaust to inlet temperature

Answers to Multiple-Choice Questions

13.72. (b), 13.73. (c), 13.74. (a), 13.75. (a)

CHAPTER

14

Vapour Power Cycles

OVERVIEW

The steam power plant is one of the most successful systems developed for converting heat energy into mechanical work. Heat energy released by the combustion of fuel or by atomic fission is utilized to vaporize water into steam. The steam thus produced is expanded in a steam engine or turbine to obtain useful work or power. The vapour leaving the turbine is normally condensed and pumped back to its initial state constituting a cycle. Thus, the working fluid changes from liquid to vapour and back to its original state. This succession of processes is designated as vapour cycle to recognise the state of the working substance during the work output process. Here, it may be noted that steam turbine power plant is much superior to the steam engine (obsolete) power plant.

Steam turbines power plants from 1 MW up to 1800 MW units resulting in about 35% to 38% overall thermal efficiency are in current use all over the world. Nowadays, combined (gas and steam turbine) cycle plant is gaining popularity as it yields about 55% to 60% overall thermal efficiency.

Although, the power developed from steam turbine plant is costlier than hydel-one but with the ever growing demand for power, it is necessary to go for this plant because it takes a short time to set up (three to four years) as compared to the hydel which takes nearly ten years and needs a lot of preparatory work in the selection of site. This chapter deals with all types of vapour power cycles.

14.1 THE CARNOT VAPOUR CYCLE

The Carnot vapour cycle consists of four fundamental elements, namely: (i) boiler (steam generator), (ii) turbine, (iii) condenser and (iv) feed pump handling a two-phase mixture- water and its vapour. Figure 14.1 shows a steam power plant operating on the Carnot cycle. The processes of the working fluid (steam, i.e., water) are represented on p-v and T-s diagrams as shown in Fig. 14.2.

Dry and saturated steam at temperature $T_H = T_1$ enters the steam turbine and is expanded isentropically to sink temperature $T_L = T_2$ to do work. The exhaust steam from the turbine is condensed and cooled in the condenser to state 3 at which the working fluid is in a state of two-phase mixture-water and its vapour. The water and its vapour is then pumped by the feed pump to state 4 where the liquid is in a saturated state. This saturated liquid is converted into dry and saturated steam in the boiler at the heat source temperature T_H.

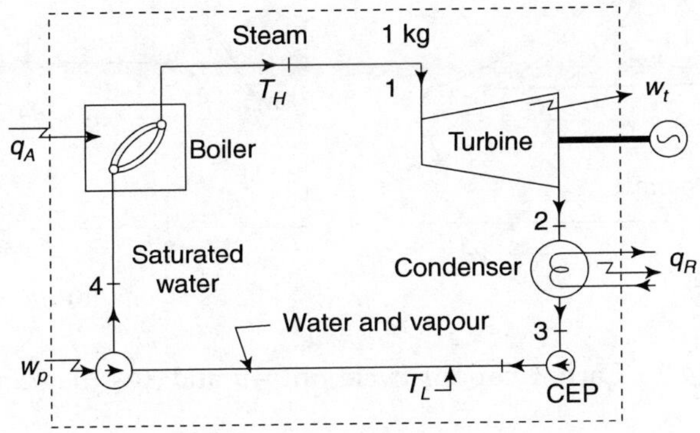

Fig. 14.1. A steam power plant operating on a Carnot cycle (impractical)

Consider 1 kg of working fluid for analysis:

$$\text{For the boiler, } q_A = q_{4-1} = h_1 - h_{f4} \quad \text{since } w = 0$$

$$\text{For condenser, } q_R = q_{2-3} = h_{2s} - h_3 \quad \text{since } w = 0$$

$$\eta_{th} = \frac{q_A - q_R}{q_A} = \frac{(h_1 - h_{f4}) - (h_{2s} - h_3)}{h_1 - h_{f4}} \qquad \text{...(14.1)}$$

The thermal efficiency may also be obtaind from turbine and pump work.

$$\text{For the turbine, } w_1 = w_{1-2s} = h_1 - h_{2s} \quad \text{since } q = 0$$

$$\text{For the pump, } w_p = w_{3-4} = h_{f4} - h_3 \quad \text{since } q = 0$$

$$\therefore \quad w_{net} = w_t - w_p = (h_1 - h_{2s}) - (h_{f4} - h_3) = (h_1 - h_{f4}) - (h_{2s} - h_3)$$

$$\therefore \quad \eta_{th} = \frac{(h_1 - h_{f4}) - (h_{2s} - h_3)}{h_1 - h_{f4}}$$

The thermal efficiency of a Carnot cycle may also be expressed in terms of temperature as discussed earlier in the previous section. From second law,

$$s_1 - s_4 = \frac{q_{4-1}}{T_H} = \frac{q_A}{T_H} \quad \text{or} \quad q_A = T_H (s_1 - s_{f4})$$

$$s_2 - s_3 = \frac{q_{2-3}}{T_L} = \frac{q_R}{T_L} \quad \text{or} \quad q_R = T_L (s_{2s} - s_3)$$

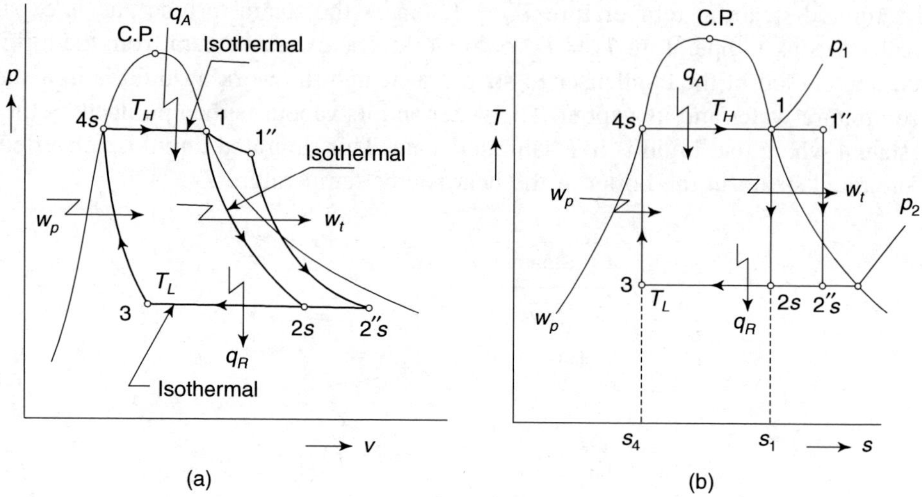

Fig. 14.2 Carnot vapour cycle on p–v and T–s diagrams

But
$$s_1 - s_{f4} = s_{2s} - s_3$$

∴
$$\eta_{th} = \frac{q_A - q_R}{q_A} = \frac{T_H - T_L}{T_H} \qquad ...(14.2)$$

Though, the Carnot cycle is more efficient that any other cycle even though *it is not considered the theoretical cycle for steam power plant for reasons given below.*

(i) Steam is partially condensed in the condenser stopping at state point 3. The condensate at state point 3, in the from of water and its vapour, is pumped to state point 4 where the water is saturated. It is difficult to build a pump which can pump a mixture of water and its vapour at state point 3 and to deliver saturated water at state point 4. In practice, it is easy to completely condense the steam to saturated water and handle it with pump as is done in Rankine cycle.

(ii) In Carnot cycle, superheating of steam is a great problem. Superheating of steam has to be done at constant temperature along path 1–1″, but during this process, pressure drops which is obvious from the T–s and p–v diagrams. This means that heat will be transferred to the vapour in the expansion process. In actual practice, this is very difficult to achieve, i.e., addition of heat at constant temperature. It is easy to superheat the steam at constant pressure as is done in Rankine cycle.

14.2 RANKINE CYCLE

We have seen that Carnot cycle is not the theoretical cycle for steam power plant because it is difficult to build a pump which can suck and pump a mixture of water and vapour and deliver it as saturated water only. This difficulty is eliminated in Rankine cycle by complete condensation of the vapour in the condenser, and then, pumping the water isentropically to boiler pressure. The Rankine cycle is a theoretical cycle for steam turbine power plant. A diagrammatic arrangement of Rankine cycle steam turbine power plant is shown in Fig. 14.3.

The power plant consists of four basic elements:

(i) *Boiler* In the boiler, the working fluid (water), receives heat by the combustion of fuel and is converted into steam.

(ii) *Steam turbine In the steam turbine, the steam from the boiler pressure expands (i.e., steam does work by reducing its pressure, temperature and heat content) and thus performs mechanical work.*

(iii) *Condenser in the condenser, the exhaust steam from the turbine gives up heat to the cooling water which cannot be converted into work and is rejected to restore initial condition of the working fluid. The condenser enables the exhaust steam to be used as a working fluid of the boiler again and again. There is a cooling system for condenser consisting of cooling tower and cooling pump.*

(iv) *Feed pump* The feed pump is used to pump the condensate from the hot-well (in which the condensate is collected) to the boiler at the boiler pressure.

The representation of the various processes of the Rankine cycle on p–v, T–s and h–s diagrams is shown in Fig. 14.4. The various processes are:

Process 3–4 : The water from the hot-well or the surge tank which is at low pressure (p_2) is pumped isentropically into the boiler at high pressure, p_1.

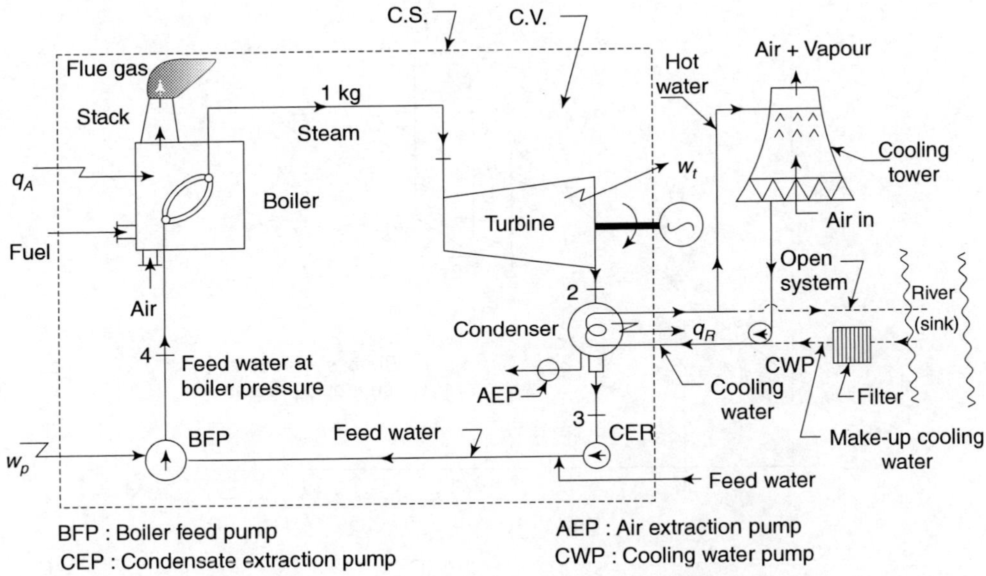

Fig. 14.3 Rankine cycle steam turbine power plant

Process 4–5 : As the water enters the boiler, it is first heated up to the saturation temperature or evaporation temperature (T_1) and during this process, the state point moves along the curve 4*s*–5 called the *sensible heating*. The heat supplied during this process is ($h_{f5} - h_{f4}$) and is represented by the area *L*–3–4*s*–5–*M*, i.e., the sensible heat of water. It may be observed that the point 5 is not shown in power plants because it is the intermediate point is steam generation.

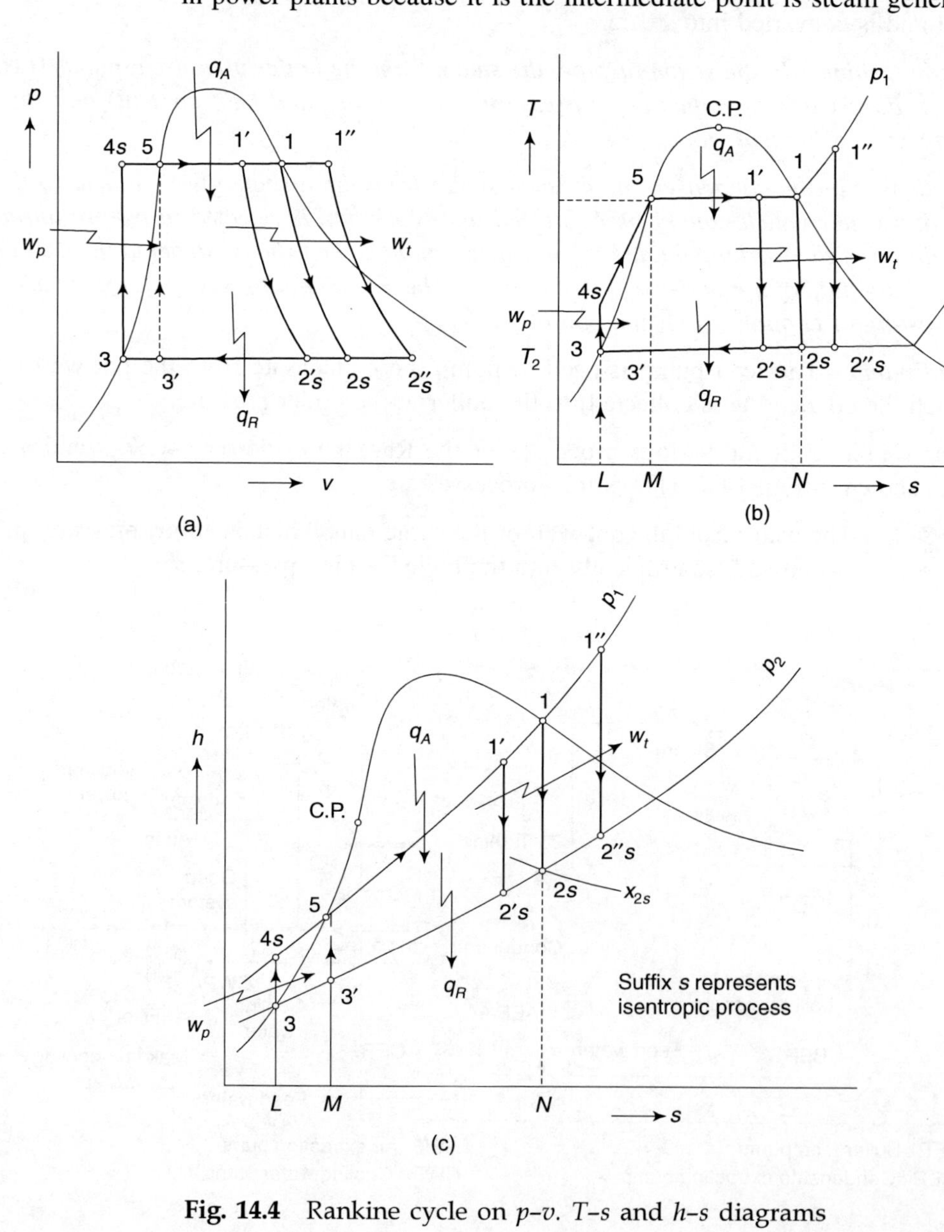

Fig. 14.4 Rankine cycle on *p*–*v*. *T*–*s* and *h*–*s* diagrams

Process 5–1 : At constant pressure p_1 and temperature T_1, water is completely vaporised into steam. The heat added in this process is equal to $(h_1 - h_{f5})$ and is represented by M–5–1–N. i.e., the latent heat of vaporization. The state point 1 shows the dry and saturated condition of steam.

Process 1–2 : This process is the isentropic expansion of steam in the turbine from pressure p_1 to p_2.

Process 2–3 : At constant pressure p_2 and temperature T_2, the exhaust steam is condensed in the condenser giving latent heat to cooling water.

It is possible that steam leaving the boiler may be dry and saturated, wet or superheated and so the corresponding cycle on the T–s and h–s diagrams are 1–2s–3–4s–1, 1′–2′s–3–4′s–5–1′ or 1″–2″s–3–4s–1″ respectively.

Assuming steady flow and negligible kinetic and potential energy changes, the thermal efficieny of the Rankine cycle can be obtained by applying the first law of thermodynamics separately to each of the four components of the Rankine cycle. Let us assume that 1 kg of working fluid flows through the various elements of the cycle. From first law,

$$\delta q - \delta w = dh$$

or

$$q - w = \Delta h$$

For boiler process 4–1, $q_{4-1} = q_A = h_1 - h_{f4}$ since $w = 0$

For condenser process 2–3, $q_{2-3}, = q_{2-3} = q_R = h_{2s} - h_{f3}$ since $w = 0$

For turbine process 1–2, $w_{1-2} = w_t = h_1 - h_{2s}$ since $q = 0$

For pump process 3–4, $w_{3-4}, w_p = h_{f4} - h_{f3}$ since $q = 0$

Pump work, w_p is also given by

$$w_p = h_{f4} - h_{f3} = v_{f3}\,(p_1 - p_3) = v_{f2}\,(p_1 - p_2)$$

where h_f and h_{f4} are the enthaply of water at pressure p_2 and p_1 respectively.

v_{f2} is the specific volume of water in m^3/kg at final pressure p_2 in kN/m^2

The net work = Turbine work – Pump work

or

$$w_{net} = w_t - w_p = (h_1 - h_{2s}) - (h_{f2} - h_{f3})\text{; kJ/kg}$$

or

$$w_{net} = (h_1 - h_{2s}) - v_{f2}\,(p_1 - p_2)\text{; kJ/kg}$$

Note If the pump work is written as $w_p = w_{3-4} = (h_{f3} - h_{f4})$ which is a negative value showing the work input to the pump, then the net work will be the summation of turbine and pump work resulting in the same equation as above. Here, $h_{f3} = h_{f2}$.

$$\eta_{th} = \frac{\text{Net work}}{\text{Heat added}} = \frac{w_{net}}{q_A} = \frac{(h_1 - h_{2s}) - (h_{f4} - h_{f3})}{(h_1 - h_{f4})} \quad \text{...(14.3)}$$

Thermal efficiency may also be calculated from heat supplied q_A and heat rejected q_R.

$$w_{net} = q_A - q_R = (h_1 - h_{f4}) - (h_{2s} - h_{f3}) = (h_1 - h_{2s}) - (h_{f4} - h_{f3})$$

$$\therefore \quad \eta_{th} = \frac{w_{net}}{q_A} = \frac{(h_1 - h_{2s}) - (h_{f4} - h_{f3})}{(h_1 - h_{f4})}$$

or $$\eta_{th} = \frac{(h_1 - h_{2s}) - (h_{f4} - h_{f3})}{(h_1 - h_{f3}) - (h_{f4} - h_{f3})} = \frac{(h_1 - h_{2s}) - w_p}{(h_1 - h_{f2}) - w_p} \quad ...(14.4)$$

Compared to turbine work, pump work is infinitesimally small and it may be neglected because the specific volume of water is very small. Hence, we have

$$w_p = 0 \quad \text{or} \quad h_{f4} = h_{f3} = h_{f2}$$

$\therefore$ $$\eta_{th} = \frac{h_1 - h_{2s}}{h_1 - h_{f3}} = \frac{h_1 - h_{2s}}{h_1 - h_{f2}} \quad ...(14.5)$$

The overall thermal efficiency of a steam power plant varies from 33% to 35%. The thermal efficiency of the Rankine cycle may also be expressed in terms of areas on T–s diagram referring to Fig. 14.4.

$$\eta_{th} = \frac{\text{Area } 2s - 3 - 4s - 5 - 1 - 2s}{\text{Area } L - 3 - 4s - 5 - 1 - N - L}$$

and neglecting pump work, $$\eta_{th} = \frac{\text{Area } 2s - 3 - 5 - 1 - 2s}{\text{Area } L - 3 - 5 - 1 - N - L}$$

Note Turbine may be non-condensing in which the exhaust steam is not condensed in the condenser and thus it is not used as a feed water.

We have so far assumed reversible heat additions and rejection in our analysis. This is, however, not possible in actual power plants. A substantial temperature difference exists between the hot combustion gases and working fluid. But the irreversibility associated with this difference is reduced and the thermal efficiency of the cycle is increased by operating a steam generator at a pressure above the critical. Figure 14.5 shows the super critical pressure Rankine cycle on T–s and h–s diagrams. Expression for efficiency will be the same.

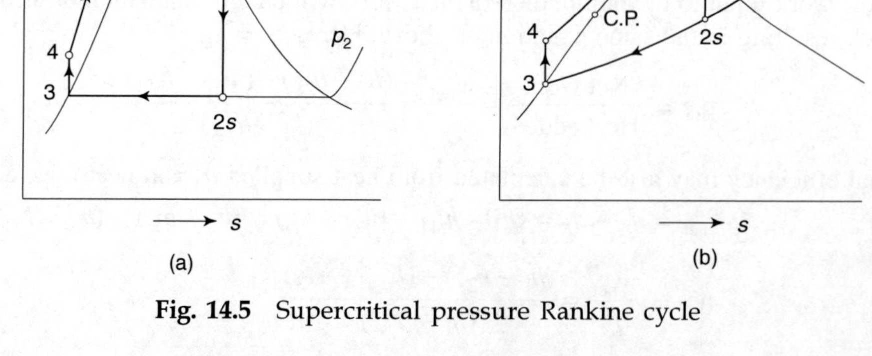

Fig. 14.5 Supercritical pressure Rankine cycle

It may be noted that if the condensate is pumped back to the boiler from the state 3′ instead of 3, then it forms the Carnot cycle. Hence, 1–2*s*–3′–5–1 is the Carnot cycle. It is obvious from Fig. 14.4 that the Rankine cycle which is practical is less efficient than Carnot cycle which is (impractical) working between the same temperature limits. *This is so, because of the fact that the average temperature at which heat is added in the Rankine cycle lies between* T_{4s} *and* T_5 *(or* T_1*) and this is less than constant temperature (*T_1*) at which heat is added in the Carnot cycle. In engineering, our aim is to achieve the efficiency of the modified Rankine cycle equal to Carnot cycle.*

14.3 EFFECT OF OPERATING CONDITIONS ON RANKINE EFFICIENCY

Inlet superheat, inlet pressure and exhaust pressure of steam affect the thermal efficiency of Rankine cycle.

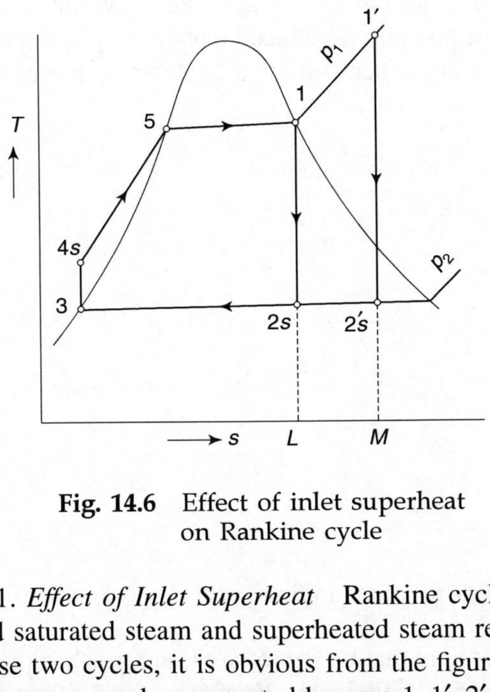

Fig. 14.6 Effect of inlet superheat on Rankine cycle

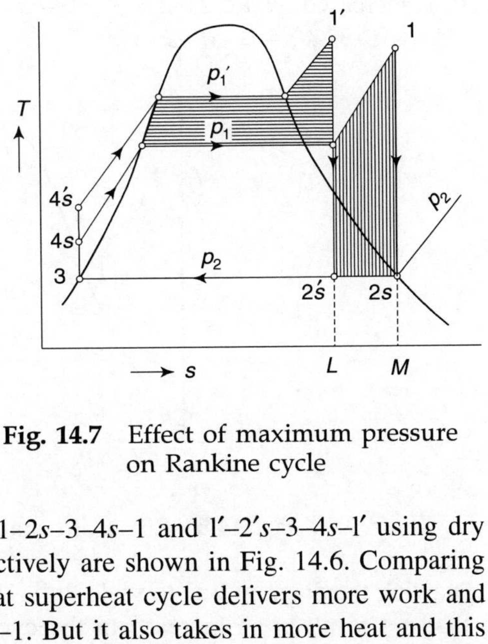

Fig. 14.7 Effect of maximum pressure on Rankine cycle

1. *Effect of Inlet Superheat* Rankine cycles. 1–2*s*–3–4*s*–1 and 1′–2′*s*–3–4*s*–1′ using dry and saturated steam and superheated steam respectively are shown in Fig. 14.6. Comparing these two cycles, it is obvious from the figure that superheat cycle delivers more work and this excess work represented by area 1–1′–2′*s*–2*s*–1. But it also takes in more heat and this is represented by area 1–1′–*M*–*L*–1′. The result is that the efficiency of Rankine cycle using superheated steam is more than that of the dry and saturated steam. From the thermodynamic point of view, this result could have been anticipated. Due to the superheating of steam, the average temperature of heat addition to the cycle increases, so there should be an increase in the thermal efficiency compared to the cycle using dry and saturated steam.

2. *Effect of Inlet Pressure* In Fig. 14.7, two cycles 1–2*s*–3–4*s*–1 and 1′–2′*s*–3–4′*s*–1′ have the same maximum temperature $T'_1 = T_1$ but different maximum pressure p'_1. The condenser

pressure is the same in both cases. It is obvious from the figure that due to increase in the maximum pressure from pressure p_1 to p'_1, the net work increases by an area shown by horizontal lines and decrease by an area shown by vertical lines and these two areas are nearly equal, so the net work is nearly the same but the heat rejected decreases by the area 2*s*–2´*s*–*L*–*M*. Thermal efficiency is given by.

$$\eta_{th} = 1 - \frac{Q_R}{Q_A} = 1 - \frac{\text{Heat rejected}}{\text{Heat added}}$$

Since heat rejection is reduced in the case of increasing the maximum pressure, so thermal efficiency also increases. From the second law, it can be anticipated that the thermal efficiency increases with increase in the maximum pressure because the average temperature of heat addition increases with increase of the maximum pressure.

3. *Effect of Exhaust Pressure* In Fig. 14.8, two cycles 1–2*s*–3–4*s*–1 and 1–2′*s*–3′–4′*s*–1 have the same the maximum pressure and temperature but different exhaust pressure $p'_2 < p_2$. It is obvious from the figure that if the exhaust pressure is reduced from p_2 to p'_2, the net work is increased by are 2*s*–2′*s*–3′–4′*s*–4*s*–3–2*s* and the heat supplied to steam increases by the area a *L*–3′–4′*s* –4*s*–*M*–*L*.

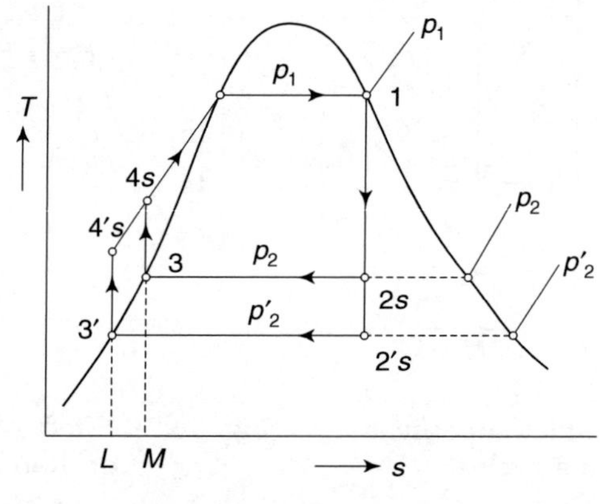

Fig. 14.8 Effect of exhaust pressure on Rankine cycle

Drawing the diagram true to scale, it has been found that these two areas are nearly equal, so that net result is an increase in the cycle thermal efficiency. From *the second law of thermodynamics, it can be anticipated that the thermal efficiency increases with decrease in exhaust pressure because the average temperature of heat rejection decreases with decrease of exhaust pressure.*

Figure 14.9 shows the effect of initial, i.e., throttle temperature and initial pressure on the Rankine thermal efficiency. At a particular value of initial pressure except very high pressure, the thermal efficiency increases approximately linearly as the temperature increases. At a very high pressure in the beginning, the increase in thermal efficiency is quite fast at the start and after that it increases slowly.

The effect of back pressure on thermal efficiency is shown in Fig. 14.10 (a). As the back pressure decreases, the thermal efficiency increases. It is also obvious that as the initial pressure is increased, the thermal efficiency increases.

The effect of boiler (i.e., inlet) pressure on cycle efficiency and specific steam consumption is shown in Fig. 14.10 (b). With the increase in boiler pressure, the latent heat decreases, the less heat is transferred at the maximum cycle temperature. Although, the efficiency increases with boiler pressure over the first part of the range, due to the maximum cycle temperature being raised, it is affected by the lowering of the average temperature at which heat is added to the cycle. The graph of the efficiency rises and reaches a maximum and then falls. The specific steam consumption (s.s.c.) first decreases and then increases with the increase in boiler pressure.

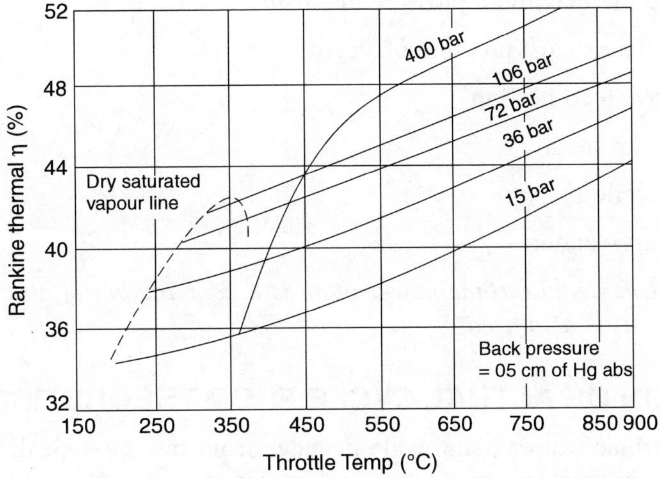

Fig. 14.9 Effect of initial pressure and temperature on thermal efficiency

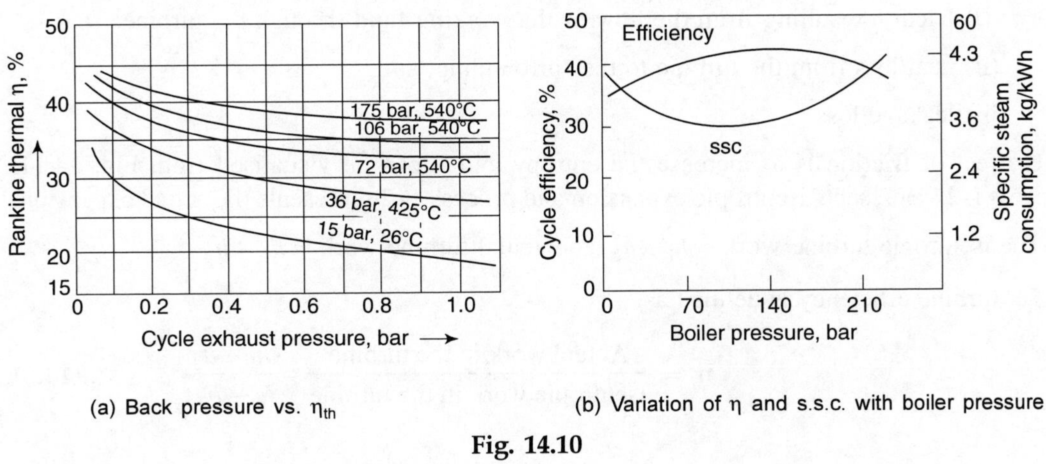

Fig. 14.10

14.4 PRINCIPLE OF INCREASING THE THERMAL EFFICIENCY

From the expression of Carnot cycle efficiency, we conclude that the thermal efficiency of the steam vapour cycle is increased by:

(a) *Increasing the average temperature at which heat is added to the cycle.*

(b) *Decreasing the average temperature at which heat is rejected to the cycle.*

14.5 METHOD OF INCREASING THE THERMAL EFFICIENCY

Based on above two principles, the methods of increasing the thermal efficiency are as follows:

(i) By increasing the superheating temperature of steam

(ii) By increasing the maximum pressure of steam

(iii) By reducing the exhaust pressure of steam

(iv) By regenerative feed heating

(v) By reheating the steam

(vi) By water extraction

(vii) By using binary-vapour

Generally, a modem steam turbine power plant is a combination of all the above except (vii) in some cases (vii) is also used.

14.6 DEVIATION OF ACTUAL CYCLE FROM THEORETICAL CYCLE

The actual steam turbine power plant cycle deviates from the theoretical, i.e., ideal cycle because losses occur in various components like turbine, pump, boiler; and piping. We will now discuss each of them briefly.

(a) *Turbine losses* The losses in turbine arises mainly due to :

(i) friction resulting from the flow of the working fluid through the turbine.

(ii) heat loss from the turbine to the surrounding, and

(iii) leakage loss.

The effect of friction is to increase the entropy and it can be visualised from Fig. 14.11. Process 1–2*s* represents isentropic expansion and process 1–2 represents the actual expansion.

The isentropic turbine work = $h_1 - h_{2s}$ and actual turbine work = $h_1 - h_2$

So, turbine efficiency is defined as

$$\eta_t = \frac{\text{Actual work in the turbine}}{\text{Isentropic work in the turbine}} = \frac{h_1 - h_2}{h_1 - h_{2s}} \quad ...(14.6)$$

η_t *varies from 75% to 86%*

The items (ii) and (iii) are negligible.

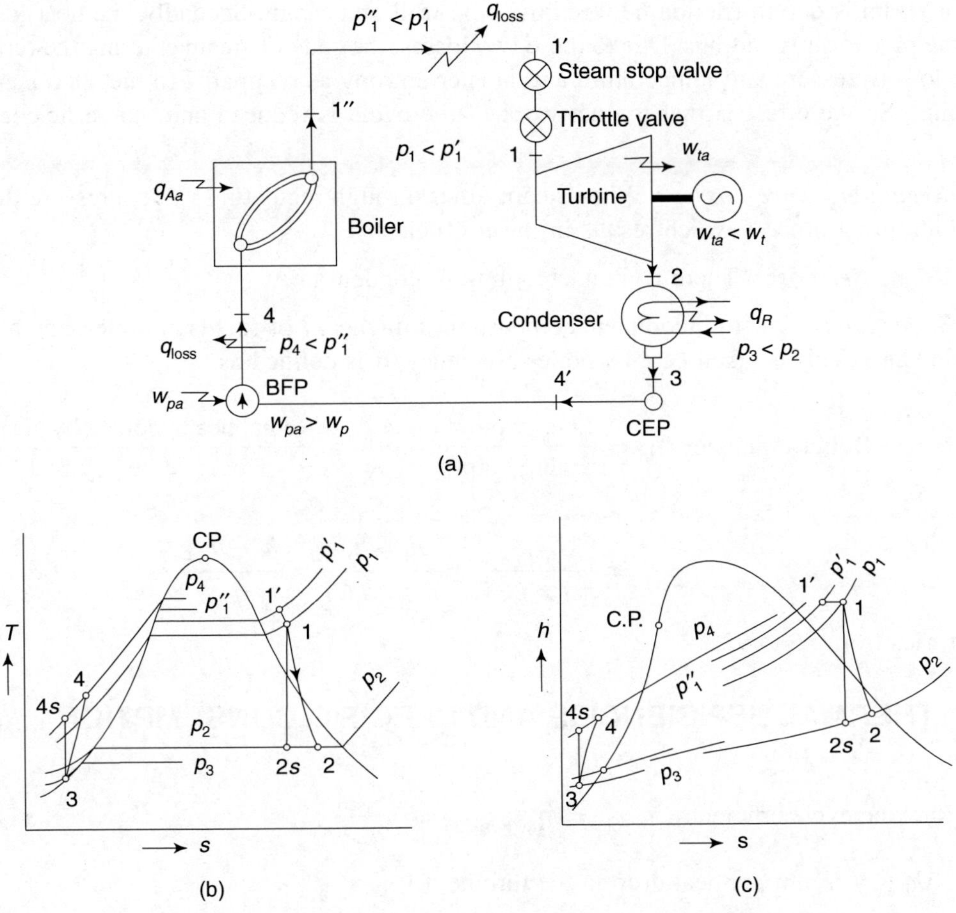

Fig. 14.11 Effect of losses on Rankine cycle

Note In representing isentropic and actual processes, two commonly nomenclatures used are : 1–2*s* is isentropic process and 1–2 is actual process. In other nomenclature, 1–2 is isentropic process and 1–2′ is actual process. Here, the former nomenclature is used.

(b) *Pump Losses* : The losses in pumps are similar to losses in turbine and friction is mainly responsible for this. Heat loss may be neglected because it is very low. The pump efficiency is defined as

$$\eta_{pump} = \frac{\text{Isentropic work supplied}}{\text{Actual work supplied}} = \frac{h_{f4s} - h_{f3}}{h_{f4} - h_{f3}} \quad ...(14.7)$$

Generally, pump work is around 2 to 2.5% of turbine output, so it may be neglected. The pump efficiency varies from 70% to 90%.

(c) *Piping Losses* These are mainly two types of losses in steam pipes. Firstly, the pressure drop of steam is due to friction between the pipe wall and steam. Secondly, the heat loss is from the pipe to surroundings. Due to these two losses, the state of steam entering the turbine is at a lower pressure and temperature and greater entropy as compared to the state leaving the boiler. So the effect is that the net work of the cycle is reduced and thus efficiency is lower.

(d) *Regulating Valve Losses* When steam flows through regulating valve, pressure drops due to throttling process which results in lower efficiency.

(e) *Condenser Loss* There is a undercooling of condensate.

(f) *Boiler Loss* Due to incomplete combustion, full energy of fuel is not released on low calorific value. This is taken care by boiler effieiency. It is defined as

$$\text{Boiler efficiency, } \eta_b = \frac{\text{Energy output}}{\text{Energy input}} = \frac{q_{Aa}}{q_A} = \frac{\text{Actual heat received by steam}}{\text{Heat released by fuel}}$$

$$= \frac{q_{Aa}}{\dot{m}_f \times LCV} = \frac{h_1 - h_{f4}}{\dot{m}_f \times LCV} = \frac{h_1 - h_{f4}}{q_A} \quad \text{...(14.8)}$$

η_b varies from 85 to 92%.

14.7 THERMAL EFFICIENCIES AND SPECIFIC CONSUMPTION

(a) Rankine cycle efficiency, $\eta_{\text{cycle}} = (\eta_{\text{th}})_{\text{Rankine}} = \dfrac{\Delta h_{\text{isen}}}{q_A}$ ...(14.9)

where Δh_{isen} = isentropic heat drop in the turbine, kJ/kg

q_A = heat supplied in the cycle, kJ/kg

$\dot{m}_f$ = mass flow rate of fuel, kg/s

LCV = lower calorific value of fuel, kJ/kg of fuel

(b) Actual thermal efficiency of the turbine plant, $\eta_{\text{tha}} = \dfrac{\Delta h_{\text{cwa}}}{q_A}$ ...(14.10)

where Δh_{cwa} = actual coupling work considering all type of losses

(c) Thermal efficiency ratio, $\eta_r = \dfrac{\eta_{\text{tha}}}{\eta_{\text{cycle}}} = \dfrac{\Delta h_{\text{cwa}}}{q_A} \times \dfrac{q_A}{\Delta h_{\text{isen}}} = \eta_{\text{th}} \times \eta_m$ ...(14.11)

(d) Overall thermal efficiency of power plant is given by actual work done at generator terminal over energy supplied (q_A).

$$\eta_{ath} = \frac{\left[w_{t \cdot isen} \times \eta_t \times \eta_m \times \eta_g - \left(\frac{w_{pump}}{\eta_p}\right)\right]}{q_A} \quad ...(14.12)$$

where q_A = Actual heat supplied $(h_1 - h_{f2}) - w_p/\eta_p$ = Energy output of boiler ...(14.13)

q_A is energy supplied using $\dot{m}_f = \dot{m}_f \times (\text{LCV})_f = q_{Aa} / \eta_{boiler}$

Neglecting pump work, we have

$$\eta_{oth} \approx (\eta_{th})_{Rankine} \times \eta_{turbine} \times \eta_{mech} \times \eta_{boiler} \times \eta_{generator} \quad ...(14.14)$$

Generally, η_{oth} varies from 35% to 38%.

In addition to thermal efficiency and efficiency ratio, the performance criterion of the steam power plant are the work ratio, specific steam consumption and heat ratio.

(e) *Work ratio* (*WR*) It is defined as the ratio of the net work output to the gross turbine output. Thus,

$$\text{Work ratio, } WR = \frac{\text{Net work output}}{\text{Turbine work}} = \frac{w_t - w_p}{w_t} = \frac{w_{net}}{w_t} \quad ...(14.15)$$

(f) *The specific steam consumption* (s.s.c.) also called steam rate is the amount of steam consumed per unit of power developed.

$$\text{s.s.c.} = \frac{3600}{w_{neta}} = \frac{3600}{h_1 - h_2}; \text{ kg/kWh} \quad ...(14.16)$$

where h_1 and h_2 are in kJ/kg. In general s.s.c. varies from 3 to 5 kg/kWh

(g) *Heat rate* (*HR*) This is defined as the heat consumption rate per unit power developed.

$$\text{HR} = \frac{3600}{\eta_{tha}} = \frac{3600\, q_{Aa}}{q_{Aa} - q_R}; \text{ kJ/kWh} \quad ...(14.17)$$

(h) *Specific cooling water consumption in condense* (scwc) is given by

$$\text{scwc} = \frac{\dot{m}_{cw} \text{ (kg/hr)}}{\text{Power output in kW}} \quad ...(14.17a)$$

scwc varies from 120 to 200 kg/kWh

14.8 REQUIREMENTS OF AN IDEAL WORKING FLUID

There are several vapours which have physical properties suitable for working fluid. They are steam, mercury vapour, sulphur dioxide, diphenyl oxide and certain hydrocarbons. The following are the requirements of an ideal working fluid for steam turbine power plant:

(i) Ample amount should be available at low cost.

(ii) Critical temperature should be higher than metallurgical limits.

(iii) Reasonable saturation pressure at the maximum temperature of the cycle from metallurgical point of view of boiler.

(iv) Steep saturated vapour line to the minimise moisture problem in expansion of steam, in the turbine.

(v) It should wet the boiler surface enveloping it and should be chemically stable at the maximum temperature of the boiler.

(vi) Saturation pressure at the minimum temperature of the cycle should be higher than atmospheric, otherwise the maintenance of the condenser will be costly.

(vii) Low liquid specific heat so that most of the heat is added at the maximum temperature.

(viii) Considerable decrease of volume upon condensation to create vacuum.

(ix) Non-toxic and non-corrosive.

(x) Freezing point should be much below the normal atmospheric pressure.

Among all types of working fluids, water satisfies the maximum requirements. Its general, abundance at low cost is of prime importance due to which, it is selected as the **working fluid in steam turbine power plant**.

Example 14.1 Dry and saturated steam at pressure 11 bar is supplied to a turbine and expanded isentropically to pressure 0.07 bar. Calculate the following:

(a) Heat supplied, (b) change of entropy during heat rejections, (c) heat rejected, (d) theoretical thermal efficiency, and (e) overall thermal efficiency if $\eta_t = 80\%$, $\eta_m = 95\%$, and $\eta_g = 96\%$.

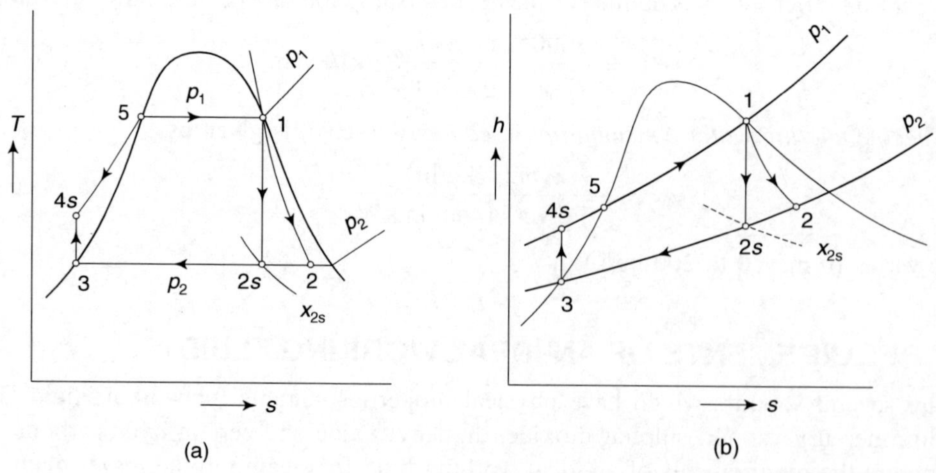

Fig. 14.12 Example 14.1

Solution Refer to Fig. 14.12. From *h–s* chart, processs 1–2*s* is isentropic expansion and 1–2 is actual expansion.

h_1 = 2781 kJ/kg, (at 11 bar), $h_{f3} = h_{f2}$ = 163 kJ/kg (at 0.07 bar), h_{2s} = 2035 kJ/kg

Pump work, $w_p = v_{f2}\,(p_1 - p_2) = 0.001\,(11 - 0.07)\,10^2 = 1.093$ kJ/kg

(a) Heat supplied = $(h_1 - h_{f2})$ – Pump work = (2781 – 163) – 1.093 = 2616.9 kJ/kg **Ans.**

(b) Change in entropy = $s_{f3} - s_2$ = – (6.554 – 0.56) = – 5.994 kJ/kg **Ans.**

(c) Heat rejected = $h_2 - h_{f3}$ = 2035 – 163 = 1782 kJ/kg K **Ans.**

(d) Theoretical thermal efficiency = $\dfrac{\text{Turbine work} - \text{Pump work}}{\text{Heat supplied}}$

$$= \frac{(h_1 - h_{2s}) - 1.093}{2616.9} = \frac{(2781 - 2035 - 1.093)}{2616.9} = 28.465\%$$ **Ans.**

(e) Overall thermal efficiency = $\dfrac{[(2781 - 2035) \times 0.8 \times 0.95 \times 96 - 1.093]}{2616.9} = 20.75\%$ **Ans.**

Note Thermal efficiency low is because inlet pressure and temperature are low.

Exampel 14.2 A steam turbine receives steam at pressure 20 bar and superheated to 88.6°C. The exhaust pressure is 0.07 bar and the expansion of steam takes place isentropically. Using steam table, calculate the following.

(a) Heat supplied, assuming that the feed pump supplies water to the boiler at 20 bar, (b) heat rejected, (e) net work done, (d) work done by turbine, (e) thermal efficiency, (f) theoretical steam consumption, (g) if the actual steam consumption is 5 kg/kWh, what is the efficiency ratio of the turbine.

Solution Refer to Fig. 14.13.

From Steam Tables, t_s at 20 bar = 212.4°C, so t_1 = 212.4 + 88.6 = 300°C

h_1 = 3025 kJ/kg (at 20 bar and 300°C) and s_1 = 6.768 kJ/kg K. At 0.07 bar,

s_{g2} = 8.274 kJ/kg K, s_{f2} = 0.552 kJ/kg K, h_{fg2} = 2409 kJ/kg, h_{f2} = 163 kJ/kg K = h_{f3}.

Now $\quad s_1 = s_2 = x_2\, s_{g2} + (1 - x_2)\, s_{f2}$

$\therefore \quad 6.768 = x_2 \times 8.274 + (1 - x_2) \times 0.552$

or $\quad x_2 = 0.8047$

Again $\quad h_{2s} = h_{f2} + x_2\, h_{f2} = 163 + 0.8074 \times 2409 = 2108$ kJ/kg

Pump work, $w_p = v_{f3}\,(p_1 - p_2)\,10^2 = (20 - 0.07)\,10^2 = 1.993$ kJ/kg

(a) Heat supplied = $h_1 - h_{f4} = h_1 - h_{f2} - w_p$ = 3025 – 163 – 1.993 = 2860 kJ/kg

(b) Heat rejected = $h_{2s} - h_{f2} = 2108 - 163 = 1945$ kJ/kg **Ans.**

(c) Net work done = Turbine work – Pump work

or $\quad w_{\text{net}} = (h_1 - h_{2s}) - w_p = 3025 - 2108 - 1.993 = 915$ kJ/kg **Ans.**

(d) Work done by the turbine = $h_1 - h_{2s} = 3025 - 2108 = 917$ kJ/kg **Ans.**

(e) Net thermal efficiency = $\dfrac{\text{Net work done}}{\text{Heat supplied}} = \dfrac{915}{2860} = 31.99\%$ **Ans.**

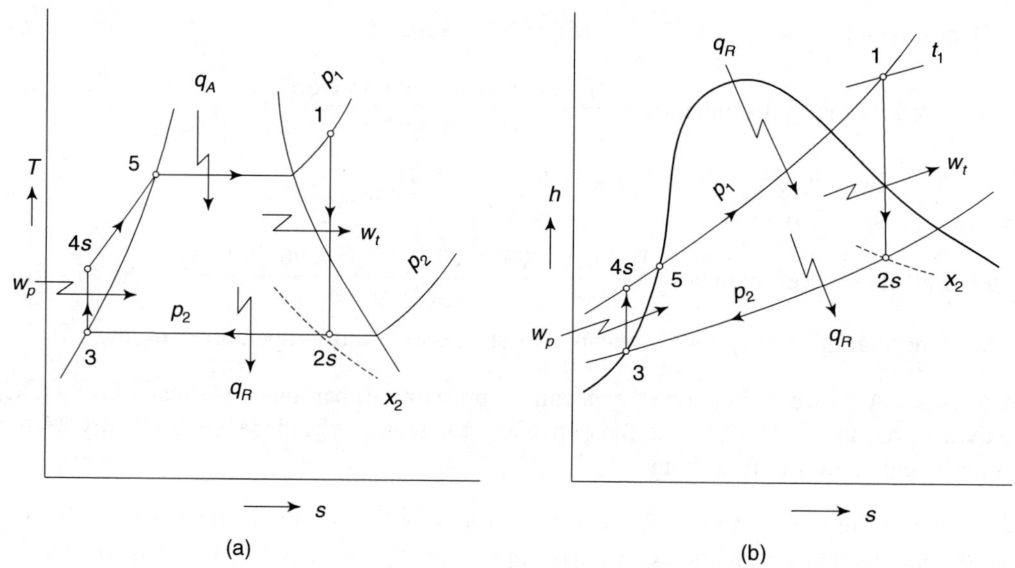

Fig. 14.13 Example 14.2

(f) Theoretical steam consumption = $\dfrac{3600}{\text{Net work done}} = \dfrac{3600}{915} = 3.934$ kg/kWh

(g) Efficiency ratio = $\dfrac{\Delta h_{cw}}{\Delta h_{\text{isen}}} = \dfrac{\text{Theoretical steam consumption in kg/kWh}}{\text{Actual steam consumption in kg/kWh}}$

$$= \frac{3.934}{5} = 78.68\%$$

Example 14.3 A steam turbine plant is supplied with steam at pressure 17 bar and superheated to 100°C. The exhaust pressure is 0.06 bar. The temperature of the condensate in the hot-well is actually 33°C ($v_f = 0.001$ m^3/kg) if the measured steam condensate is 5 kg/kWh, and if generator efficiency is 96%, what is the overall thermal efficiency of the whole plant.

Solution Refer to Fig. 14.13. From Steam Tables t_s at 17 bar = 204.3°C

$\therefore$ Initial temperature of steam, t_1 = 204.3 + 273 = 304.3°C

From h–s chart, h_1 = 3039 at kJ/kg (17 bar and 204.3°C)

h_{2s} = 2110 kJ/kg (at 0.06 bar), h_{f2} = 138.5 kJ/kg (at 33°C)

Pumb work, $v_{f2}(p_1 - p_2)\,10^2 = 0.001\,(17 - 0.06) \times 10^2 = 1.694$ kJ/kg

Heat supplied, $h_1 - h_{f2} - w_p = 3039 - 138.5 - 1.694 = 2898$ kJ/kg

Net work done, $h_1 - h_{2s} - w_p = 3039 - 2110 - 1.694 = 927.3$ kJ/kg

$$\therefore \quad \text{Theoretical steam consumption} = \frac{3600}{927.3} = 3.882 \text{ kg/kWh}$$

$$\text{Efficiency ratio, } \eta_r = \frac{\Delta h_{\text{cwa}}}{\Delta h_{\text{isen}}} = \frac{\text{Theoretical steam consumption in kg/kWh}}{\text{Actual steam consumption in kg/kWh}}$$

$$= \frac{3.882}{5} = 77.64\%$$

$$\text{Thermal efficiency, } (\eta_{\text{th}})_{\text{Rankine}} = \frac{\text{Net work}}{\text{Heat supplied}} = \frac{927.3}{2898.8} = 31.98\% \qquad \textbf{Ans.}$$

$\therefore$ Overall thermal efficiency of the whole power plant

$$\eta_{\text{oth}} = (\eta_{\text{th}})_{\text{Rankine}} \times \eta_r \times \eta_{\text{gen}} = 0.3198 \times 0.7764 \times 0.96 = 23.83\% \qquad \textbf{Ans.}$$

Example 14.4 The following are the mean observations made during a test on a steam turbine :

Duration of test = 2 hours, Power developd (turbine coupling) = 4500 kW, Total mass of steam condensed = 50,000 kg, Steam pressure at stop valve = 14 bar, steam temperatures at stop valve = 300°C, Barometric height = 76 cm of Hg, Vacuum = 72.5 cm of Hg.

Calculate the thermal efficiency of the turbine and also its efficiency ratio. Assume that the condensate is returned to the boiler plant at the exhaust saturation temperature.

Solution Refer to Fig. 14.13.

Absolute exhaust pressure = 76 – 72.5 = 3.5 cm of Hg = 0.04665 bar

From Mollier's chart, h_1 = 3038 kJ/kg (at 14 bar, 300°C),

h_{2s} = 2105 kJ/kg (0.0475) bar).

From Steam Tables, h_{f3} = 133 kJ/kg (at 0.0466 bar)

Neglecting the pump works, Heat supplied = $h_1 - h_{f3}$ = 3038 – 133 = 2905 kJ/kg

Theoretical work done by turbine, $h_1 - h_2$ = 3038 – 2105 = 933 kJ/kg

$$\text{Actual steam consumption} = \frac{50000}{2} = 25000 \text{ kg/h} = \frac{2500}{3600} = 6.944 \text{ kg/s}$$

$$\text{Actual thermal efficiency} = \frac{\text{Actual work done}}{\text{Heat supplied}} = \frac{4500}{2905 \times 6.944} = 22.30\% \qquad \textbf{Ans.}$$

$$\text{Efficiency ratio, } \eta_r = \frac{\Delta h_{\text{cwa}}}{\Delta h_{\text{isen}}} = \frac{4500}{933 \times 6.95} = 69.397\% \qquad \textbf{Ans.}$$

Example 14.5 Two steam turbines *A* and *B* operate with the same initial pressure, namely, 15 bar and the same degree of superheat 102°C. In turbine *A*, the exhaust pressure is 0.1 bar whereas in turbine *B*, the exhaust pressure is 0.03 bar. Calculate the following:

(a) Theoretical steam consumption of turbines *A* and *B* in kg/kWh, (b) percentage reduction in steam consumption due to the lower exhaust pressure in turbine *B*, (c) theoretical thermal efficiency of the turbine *A* and *B*, and (d) percentage increase in thermal efficiency duc to the lower exhaust pressure in turbine *B*.

Solution Refer 14.14. From Steam Tables t_s for 15 bar = I98.3°C

Initial temperature of steam = 198.3 + 102 = 303.3°C

$$h_{f3as} = 192 \text{ kJ/kg (at 0.1 bar)}$$

$$h_{f3bs} = 101 \text{ kJ/kg (at 0.03 bar)}$$

Neglecting the pump work,

Heat supplied to turbine $A = (q_A)_a = h_1 - h_{f3as} = 3039 - 192 = 2847$ kJ/kg

Heat supplied to turbine $B = (q_A)_b = h_1 - h_{f3bs} = 3039 - 101 = 2938$ kJ/kg

Work done by turbine $A = w_{ta} = h_1 - h_{2as} = 3039 - 2177.7 = 861.3$ kJ/kg

Work done by turbine $B = w_{tb} = h_1 - h_{2bs} = 3039 - 2050 = 989$ kg/kg

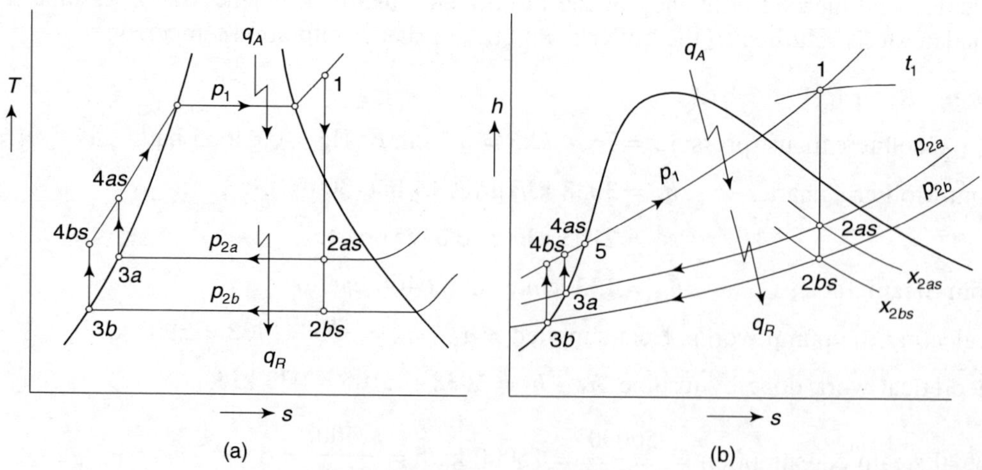

Fig. 14.14 Example 14.5

(a) *Specific steam consumption of turbine A* = $(\text{s.s.c.})_a = \dfrac{3600}{861.3} = 4.179$ kg/kWh **Ans.**

Steam consumption of turbine $B = (\text{s.s.c.})_b = \dfrac{3600}{989} = 3.64$ kg/kWh **Ans.**

(b)% reduction in s.s.c. in turbine $B = \dfrac{4.79 - 3.64}{3.64} = 34.93\%$

(c) Theoretical thermal efficiency of turbine $A = (\eta_{\text{th}})_a = \dfrac{861.3}{2847} = 30.24\%$

Theoretical thermal efficiency of turbine $B = (\eta_{\text{th}}) = \dfrac{989}{2938} = 33.66\%$ **Ans.**

(d) % increase in thermal efficiency of turbine $B = \dfrac{33.66 - 30.24}{30.24} = 11.3\%$ **Ans.**

Example 14.6 In a Rankine cycle, steam enters the turbine at 140 bar and 550°C. If the exhaust pressure is 0.06 bar and all processes are reversible:

(a) Find the cycle thermal efficiency, the work ratio and the steam rate.

(b) If the efficiency of turbine is 80% and the efficiency of pump is 50%, what is the thermal efficiency?

(c) With this value of turbine and pump efficiency if the boiler efficiency is 85%, calculate the fuel flow rate required in kg/kg of steam if LCV of fuel in 42 MJ/kg.

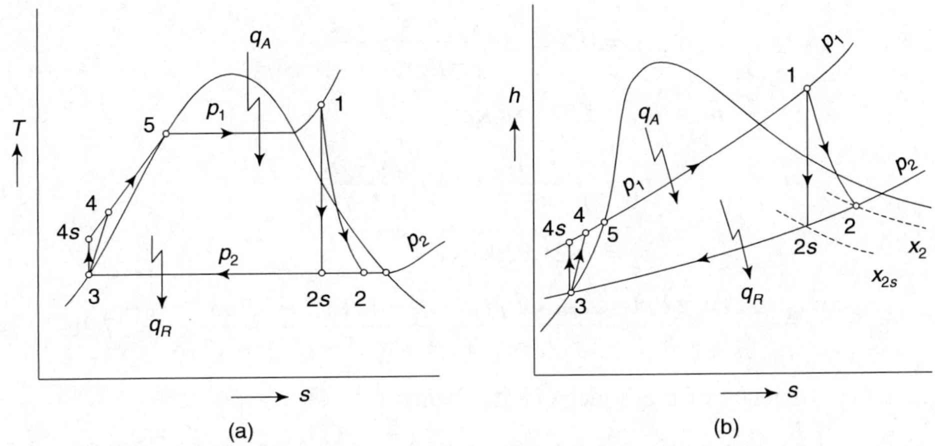

Fig. 14.15 Example 14.6

Solution Refer to Fig. 14.15.

(a) From the Steam Tables at 140 bar and 550°C, h_1 = 3967.5 kJ/kg, s_1 = 7.1166 kJ/kg K. At pressure 0.06 bar, h_{f2} = 151 kJ/kg, h_{fg2} = 2416 kJ/kg, s_{f2} = 0.5209 kJ/kg K.

$$s_{fg2} = 7.8103 \text{ kJ/kg K}, \; v_{f2} = 0.001 \text{ m}^3\text{/kg}$$

The process 1–2 is isentropic, hence

$$s_1 = 7.116 = s_{f2} + x_{2s}\, s_{fg2} = 0.5209 + x_{2s} \times 7.8103$$

or

$$x_{2s} = 0.844$$

∴

$$h_{2s} = h_{f2} + x_{2s}\, h_{fg2} = 151 + 0.844 \times 2416 = 2190.1 \text{ kJ/kg}$$

$$w_t = h_1 - h_{2s} = 3967.5 - 2190.1 = 1777.4 \text{ kJ/kg}$$

$$w_p = v_{f2}\,(p_1 - p_2) \times 10^2 = 0.001\,(14 - 0.06) \times 10^2 = 1.394 \text{ kJ/kg}$$

or

$$w_p = h_{f4} - h_{f3} = h_{f4} - h_{f2} = 1.394 \text{ kJ/kg}$$

(a) Thermal efficiency $= \eta_{th} = \dfrac{w_{net}}{q_A} = \dfrac{w_t - w_p}{q_A} = \dfrac{(h_1 - h_2) - (h_{f4} - h_{f3})}{(h_1 - h_{f3}) - (h_{f4} - h_{f3})}$

$$= \frac{(h_1 - h_{2s}) - w_p}{(h_1 - h_{f3}) - w_p} = \frac{(3967.5 - 2190.1) - 1.394}{(3967.5 - 151) - 1.394} = 46.55\% \quad \textbf{Ans.}$$

Work ratio = WR $= \dfrac{w_{net}}{w_t} = \dfrac{1776}{1777.4} = 99.92\%$ **Ans.**

Specific steam consumption (s.s.c.) $= \dfrac{3600}{w_{net}} = \dfrac{3600}{1776} = 2.027$ kg/kWh **Ans.**

(b) Now,

$$\eta_{pump} = 0.5 = \frac{h_{f4s} - h_{f3}}{h_{f4} - h_{f3}} = \frac{1.394}{h_{f4} - h_{f3}}$$

or

$$h_{f4} - h_{f3} = 2.788 \text{ kJ/kg}$$

$$\eta_t = \frac{h_1 - h_2}{h_1 - h_{2s}} = 0.8 = \frac{h_1 - h_2}{1777.4}$$

or

$$h_1 - h_2 = 1421.92 \text{ kJ/kg}$$

∴

$$\eta_{th} = \frac{w_{reta}}{q_A} = \frac{(h_1 - h_2) - (h_{f4} - h_{f3})}{(h_1 - h_{f3}) - (h_{f4} - h_{f3})} = \frac{1421.92 - 2.788}{(3967.5 - 151) - 2.788} = 37.2\% \quad \textbf{Ans.}$$

(c) Since the efficiency of the boiler is 85%, hence

$$\eta_b = 0.9 = \frac{\text{Energy output}}{\text{Energy input}} = \frac{h_1 - h_{f4s}}{q_A} = \frac{(3967.5 - 152.394) \times 0.85}{\dot{m}_f \times \text{LCV}} = \frac{3815.1 \times 0.85}{\dot{m}_f \times 42000}$$

$$\dot{m}_f = 0.08578 \text{ kg/kg of steam} \quad \textbf{Ans.}$$

The reader should solve this example from Mollier chart.

14.9 REGENERATIVE FEED HEATING CYCLES

We have seen that the efficiency of the Rankine cycle (1–2*s*–4–5*s*–6–1) (Fig. 14.16b) is less than that of the Carnot cycle (1–2*s*–4′–6–1) because the average temperature at which heat is added to the Rankine cycle lies between T_4 and T_1 which is less than average temperature of heat addition at the maximum temperature T_1 in the Carnot cycle. In other words, *the irreversible mixing of the cold condensate with hot water (i.e., sensible heating) reduces thermal efficiency.* Thus, by *eliminating or minimising the irreversibilities* (i.e., *sensible heating*), *the thermal efficiency is increased* or *we may say that the supplying of the feed water to the boiler at a temperature higher than that of Rankine cycle by any arrangement increases its efficiency.* Such an arrangement is possible by regenerative feed heating; the *name regenerative feed heating is given because the expanding steam of the turbine is used to heat the feed water.*

The object of the regenerative feed heating cycle is to supply the feed water to the boiler at a temperature higher than the condenser temperature (i.e., between T_1 and T_2 rather than at T_2) by heating the feed water with the steam expanding in the turbine and by doing so the average temperature of heat addition in the cycle is increased which increases the thermal efficiency of the cycle. It is also called Carnotization of Rankine cycle.

14.9.1 Most Ideal Regenerative Cycle

Figure 14.16a shows the most ideal regenerative feed heating cycle. The condensate from the condenser is pumped to the boiler pressure and then passed through the turbine casing in the direction opposite to the flow of steam in the turbine. The heat from the steam will be transferred to feed water thereby raising its temperature. The transfer of heat from the steam to the feed water would be a reversible process, since at any section, the steam and the feed water would be at the same temperature. Thus, the feed water would get heated along the path 5–6 (Fig. 14.16b) reversibly and the expanding steam will be cooled along 1–2′ resulting in the decrease in entropy. Note that the line 1–2′ must be parallel to 5–6 because the heat exchange is the same for each. Also, since the expansion from 2′ to 3*s* and compressing from 4 to 5*s* are isentropic, the areas under the lines 4–5*s*–6 and 1–2′–3*s* must be the same i.e., area 1–2′–3–4–5*s*–6–1 will be equal to the Carnot cycle are l–2*s*–4′–6–l. This means that heat

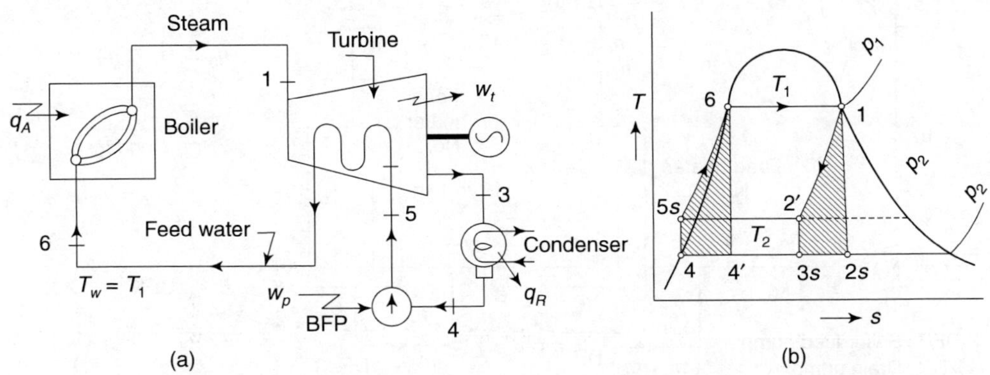

Fig. 14.16 The most ideal regenerative cycle

is added only at T_1 and rejected only at T_2. Hence, the efficiency of this modified cycle will be the same as Carnot cycle operating between T_1 and T_2. Mathematically,

$$\text{Heat added, } q_A = T_1\,(s_1 - s_6)$$

and $$\text{Heat rejected, } q_R = T_3\,(s_{3s} - s_4) = T_2\,(s_{3s} - s_4)$$

$\therefore$ $$w_{\text{net}} = T_1\,(s_1 - s_6) - T_2\,(s_{3s} - s_4)$$

But $$s_1 - s_6 = s_{3s} - s_4 \quad \text{and so } w_{\text{net}} = (T_1 - T_2)\,(s_1 - s_6)$$

$\therefore$ $$(\eta_{th})_{\text{Reg. most ideal}} = \frac{w_{\text{net}}}{q_A} = \frac{(T_1 - T_2)(s_1 - s_6)}{T_1\,(s_1 - s_6)} = \frac{T_1 - T_2}{T_1} \qquad ...(14.18)$$

Thus, efficiency of the most ideal regenerative cycle is equal to that of Carnot cycle. This is only possible **if infinite number of heaters are used**. But the ideal regenerative cycle is not practicable for the following reasons:

(i) It is difficult to pass the feed water through turbine casing or to put an infinite number of heaters.

(ii) Since the feed heating process through the turbine casing changes the condition of the entire steam, it results in excessive moisture at the low pressure stage of the turbine which is not desirable.

14.9.2 Practical Regenerative Feed Heating Cycle (Simplified)

In the practical regenerative feed heating cycle, the steam is extracted from the turbine at one or more points after partial expansion for heating the feed water in heaters. The feed water is heated either in **direct contact feed heaters or surface feed heaters** by taking heat

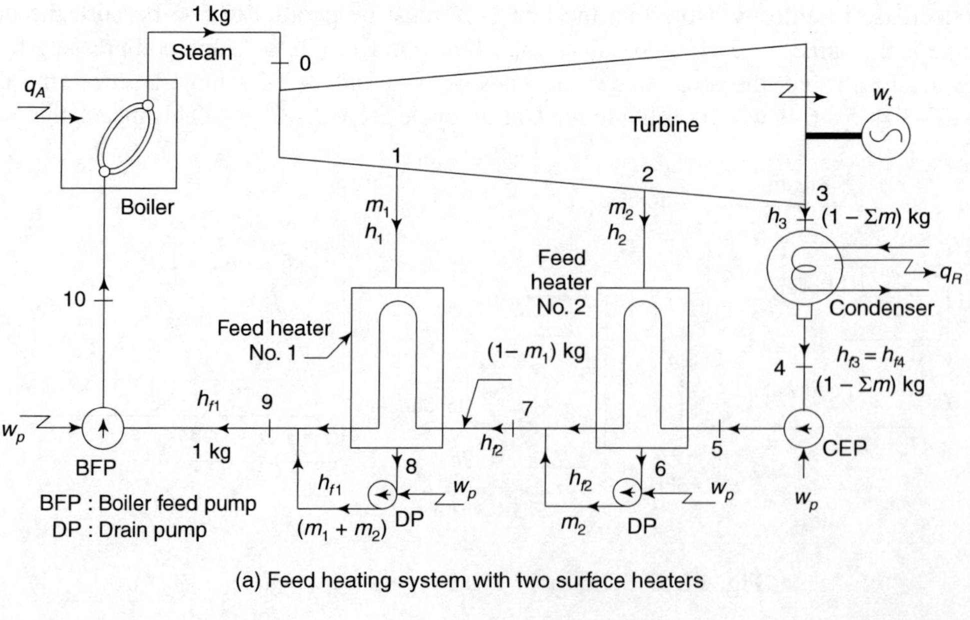

(a) Feed heating system with two surface heaters

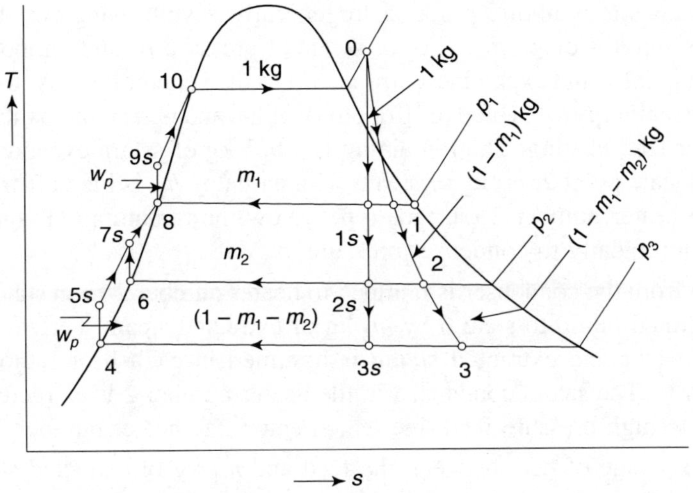

(b) *T–s* Diagram with pump work

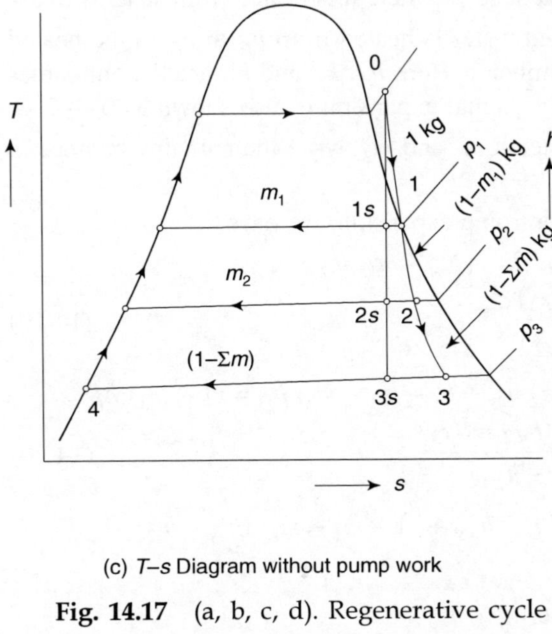

(c) *T–s* Diagram without pump work

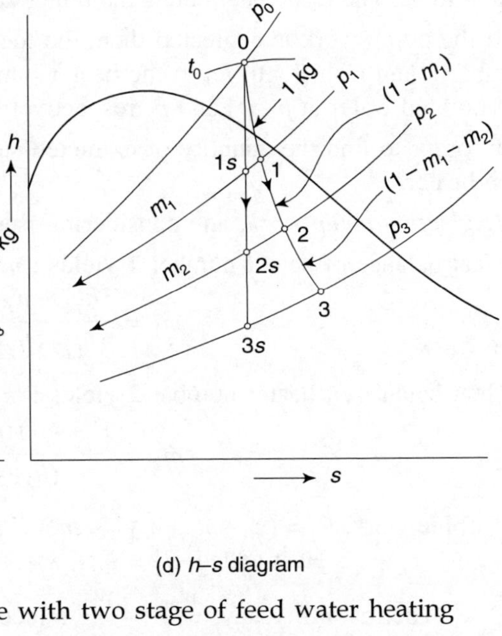

(d) *h–s* diagram

Fig. 14.17 (a, b, c, d). Regenerative cycle with two stage of feed water heating

from steam, thus the steam is condensed. In practice, the high cost of heaters, pipes, valves and accessories put a limit to the numbers of heaters and **so 2 to 10 heaters** are used. It is to be noted that the **bled steam must not be live steam**, otherwise no work would be available from the live steam, nor **must it be completely expanded steam** for then it would useless for feed heating. *The temperature rise per heater* ***varies from 20°C to 35°C***. **Invariably, in steam power plant, a deaerator coupled with storage tank is used ensuring dissolved oxygen free feed water and the feed water pump is located just below it to ensure positive suction head under all conditions of operation.**

Figure 14.17 shows a symbolic practical regenerative cycle using two heaters and its corresponding *T–s* and *h–s* diagrams. 1 kg of steam at pressure p_0 and temperature t_0 enters the turbine at state point 0 and expands isentropically to the state point 1. At the state point 1, m_1 kg of steam with enthalpy h_{1s} is bled off from the turbine and passed on to the heater number 1 (surface type) for feed heating. The remaining $(1 - m_1)$ kg of steam expands isentropically to state point 2. At state point 2, m_2 kg of steam with enthalpy h_{2s} is bled off from the turbine and passed on the heater number 2 (surface type). Now the remaining $(1 - m_1 - m_2)$ kg of steam expands isentropically to condenser pressure, p_3.

The condensate from the condenser is pumped to heater number 2 from state 4 to 5*s* where it is heated to saturated liquid to state 6 by m_2 kg of extracted steam. Perfect heat exchange between the feed water and extracted steam is assumed in each heater though, in actual practice, it is not true. The steam condensed in the heater number 2 is extracted and pumped from state 6 to 7*s* to high pressure feed line which enters the heater number 1.

There will be exchange of heat between the feed and m_1 kg of extracted steam. The feed water will be heated from state 7 to 8 in the heater number 1 and the steam condenses. The condensate from heater number l is extracted and pumped to the high pressure feed pipe from state 8 to 9*s*. The feed water enters the boiler at state 9*s* where it is heated from state 9*s* to 10.

If the pump work is neglected then, the feed water is heated from h_{f3} to h_{f2} in the heated number 2 and from h_{f2} to h_{f1} in the heater number 1. Here h_{f1}, h_{f2} and h_{f3} are the enthalpies of saturated water at p_1, p_2 and p_3 respectively. Actual expansion is also shown as 0–1–2–3.

In order to find the quantity of extracted steam m_1 and m_2, write the enthalpy balance at each heater.

Neglecting pump work, and considering isentropic expansion, we have

Heat balance of heater number 1 yields : $m_1 h_{1s} + (1 - m_1) h_{f2} = h_{f1}$

or
$$m_1 = \frac{(h_{f1} - h_{f2})}{(h_{1s} - h_{f2})} \quad ...(14.19)$$

Heat balance of heater number 2 yields : $m_2 h_{2s} + (1 - m_1 - m_2) h_{f3} = (1 - m_1) h_{f2}$

or
$$m_2 = \frac{(1 - m_1)(h_{f2} - h_{f3})}{(h_{2s} - h_{f3})} \quad ...(14.20)$$

Turbine work, $w_t = (h_0 - h_{1s}) + (1 - m_1)(h_{1s} - h_{2s}) + (1 - m_1 - m_2)(h_{2s} - h_{s3})$

Heat added, $q_A = h_0 - h_{f1}$

$\therefore$ Thermal efficiency, $\dfrac{w_{net}}{q_A} = \dfrac{w_t}{q_A}$ (since $w_p \approx 0$)

$$\eta_{th} = \frac{(h_0 - h_{1s}) + (1 - m_1)(h_{1s} - h_{2s}) + (1 - m_1 - m_2)(h_{2s} - h_{3s})}{h_0 - h_{f1}} \quad ...(14.21)$$

Thermal efficiency can be also expressed as heat added and heat rejected.

$$\eta_{th} = \frac{q_A - q_R}{q_A} = \frac{(h_0 - h_{f1}) - (1 - m_1 - m_2)(h_{3s} - h_{f4})}{(h_0 - h_{f1})} \quad \text{where } h_{f4} = h_{f3} \text{ at } p_3 \text{ bar} \quad ...(14.22)$$

The thermal efficiency of the simple Rankine cycle is given by

$$\eta_{\text{th } R} = \frac{q_A - q_R}{q_A} = \frac{(h_0 - h_{f4}) - (h_{3s} - h_{f4})}{(h_0 - h_{f4})} \quad ...(14.23)$$

Considering pump work, the thermal efficiency of regenerative cycle is

$$w_{\text{net}} = w_1 - \sum w_p; \text{ kJ/kg}$$

and $q_A = (h_0 - h_{f1}) - w_{p3}$; kJ/kg

$$\therefore \quad \eta_{\text{th}} = \frac{w_{\text{net}}}{q_A} = \frac{w_1 - \sum w_p}{(h_0 - h_{f1}) - w_{p3}} \quad ...(14.24)$$

Note Considering pump work, the expressions for m_1 and m_2 will slightly change.

If the flow is accompanied with friction, then the expansion path will be 0–1–2–3. Under this condition for regenerative cycle without pump work,

$$\eta_{\text{th}} = \frac{(h_0 - h_{f1}) - (1 - m_1 - m_2)(h_3 - h_{f4})}{(h_0 - h_{f1})} \quad ...(14.25)$$

Here, h_3 will be calculated using the concept of turbine efficiency.

$$\text{The Rankine cycle efficiency} = \eta_{\text{th } R} = \frac{(h_0 - h_{f4}) - (h_3 - h_{f4})}{(h_0 - h_{f4})} \quad ...(14.26)$$

Note There are many types of feed heating arrangements. Practical feed heating cycles for 120 MW, 250 MW, 350 MW, 500 MW, 600 MW and 1000 MW are out of scope of the book. For these, see the author's book "**Steam Gas Turbines, Turbomachines and Power Plant Engineering**".

Example 14.7 A steam turbine plant equipped with a single regenerative surface feed heater operates under the following conditions.

Initial steam pressure = 16.5 bar, initial surpheat = 93°C

Extraction pressure = 2 bar, exhaust pressure = 0.05 bar,

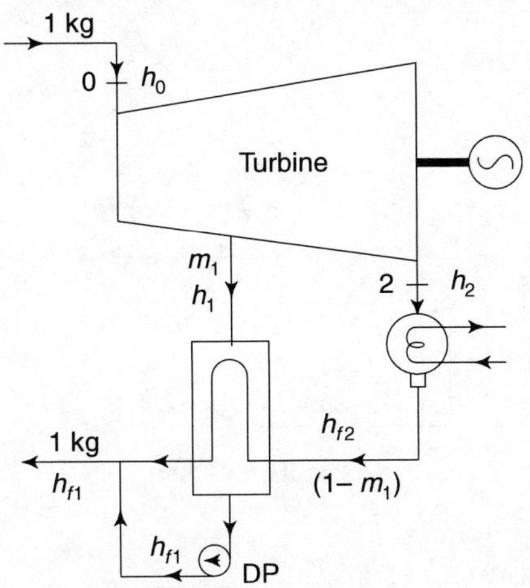

Fig. 14.18 Example 14.7

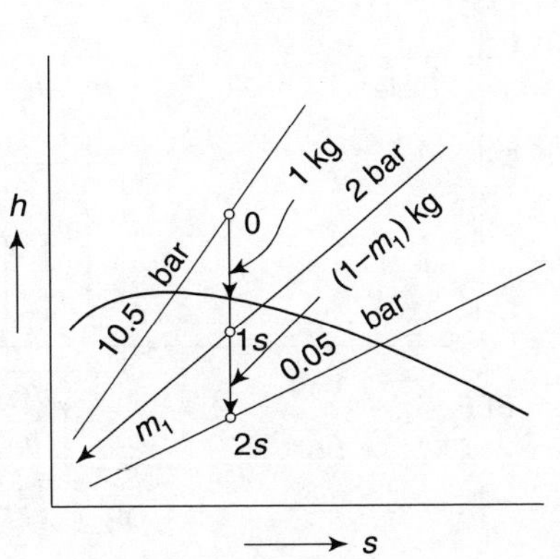

Fig. 14.19 Example 14.7

Compare the regenerative and non-regenerative cycles with respect to the following (a) *steam consumption in kg per kWh, and* (b) *condenser duty steam condensed per kWh). It may be assumed that the expansion is isentropic, but the feed water is heated to the steam saturation temperature in the heater.*

Solution Refer to Figs. 14.18 and 14.19.

From Steam Tables, the initial temperature of steam = 202.85 + 93 = 295.85°C
From chart, h_0 = 3022 kJ/kg (at 16.5 bar and 295.85°C)

(a) Steam consumption per kWh without regeneration = $\dfrac{3600}{907}$ = 3.969 kg/kWh **Ans.**

From heat balance at heater No. 1, m_1 = 0.1497 kg
Steam condensed in the condenser with regeneration is $(1 - m_1)$ kg of entering steam
(b) Hence condenser duty = (1 – 0.1497) × 4.31 = 3.667 kg/kWh **Ans.**
Condenser duty without regeneration = 1 × 3.969 = 3.969 kg/kWh **Ans.**

Example 14.8 A multi-stage steam turbine has to be provided with two stages of regenerative feed heating. The temperature rise of the feed has to be divided equally between the two heaters Initial steam pressure is 27 bar with 300°C and exhaust pressure is 0.07 bar. Temperature of condensate is 35°C. The first tapping is taken at 2 bar. Determine the pressure for the second tapping point and the quantities to be tapped at each point. Calculate the gain in efficiency due to regenerative feed heating and increase in steam consumption rate. Use a stage efficiency of 70%.

Solution Refer to Fig. 14.20 and Fig. 14.21.

As nothing is mentioned about drain arrangement, the system shown in Fig. 14.20 has been assumed with ideal conditions.

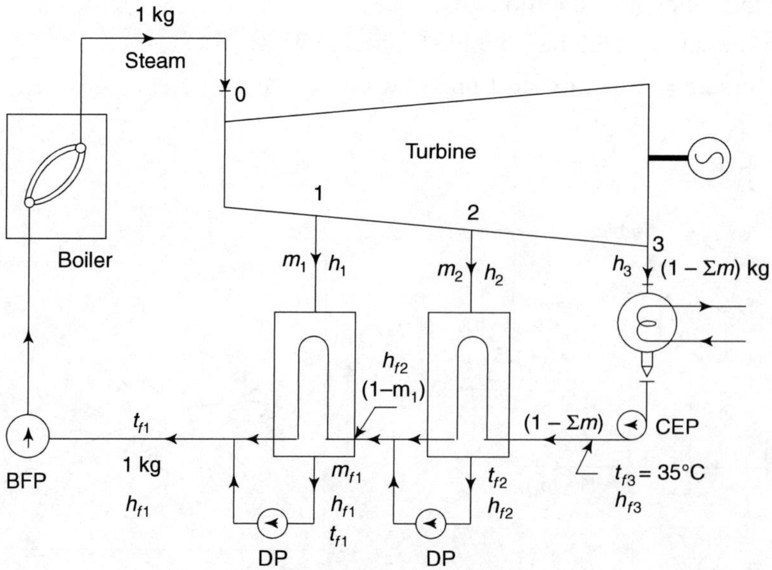

Fig. 14.20 Example 14.8

From Steam Tables, at 27 bar and 3000°C, h_0 – 3000 kJ/kg

At 2 bar, t_{f1} = 120.2°C is saturation temperature corresponding to bar 2

According to question, $t_{f1} - t_{f3} = 2\ (t_{f1} - t_{f2}) = 2\ (t_{f2} - t_{f3})$

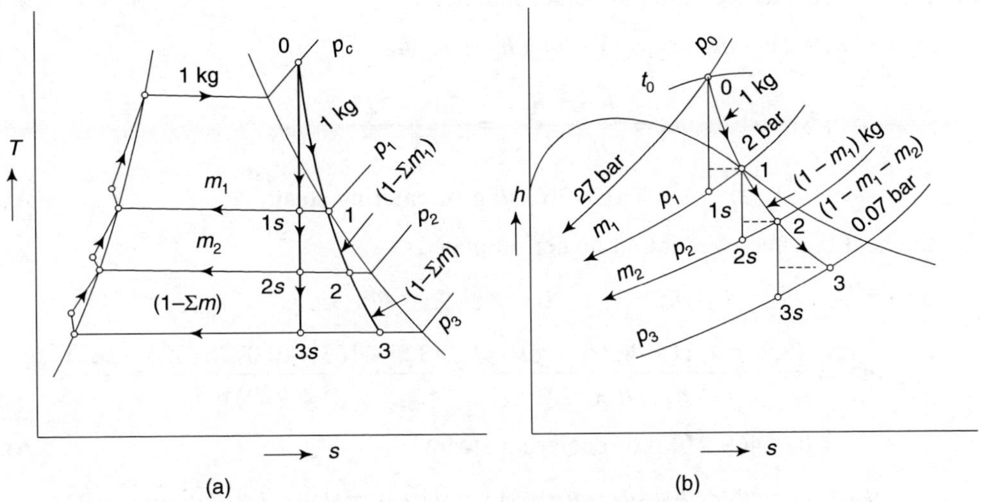

Fig. 14.21 Example 14.8

$$\therefore \qquad t_{f2} = \frac{(t_{f1} - t_{f3})}{2} + t_{f3} = \frac{(120.2 - 35)}{2} + 35 = 77.6°\text{C}$$

Corresponding to 77.6°C, the saturation pressure from Steam Table is 0.43 bar

Now from chart, $h_0 - h_{1s}$ = 3000 – 2490 = 510 kJ/kg

Since stage efficiency is 70%, so

$$h_0 - h_1 = 0.7 \times 510 = 357 \text{ kJ/kg}$$

or

$$h_1 = 3000 - 357 = 2643 \text{ kJ/kg}$$

Thus, with enthaply 2643 kJ/kg and pressure line 2 bar the point 1 is obtained. Again,

$$h_1 - h_{2s} = 2643 - 2400 = 243 \text{ kJ/kg}$$

and

$$h_1 - h_2 = 243 \times 0.7 = 170.1 \text{ kJ/kg}$$

or

$$h_2 = 2643 - 170.1 = 2472.9 \text{ kJ/kg}$$

Hence, the point 2 can be locatted. Again,

$$h_2 - h_1 = 2472.9 - 2230 = 242.9 \text{ kJ/kg}$$

$$h_2 - h_3 = 242.9 \times 0.7 = 170.03 \text{ kJ/kg}$$

or

$$h_3 = 2472.9 - 170.03 = 2302.87 \text{ kJ/kg}$$

Hence, point 3 can be located on chart.

Since temperature at different stages are given, so

$$h_{f1} = 505 \text{ kJ/kg}, h_{f2} = 325 \text{ kJ/kg}, h_{f3} = 163 \text{ kJ/kg}$$

Using the heat balance equation at heater number 1

$$m_1 h_1 + (1 - m_1) h_{f2} = (1 - m_1) h_{f1} + m_1 h_{f1}$$

$$\therefore \quad m_1 = \frac{h_{f1} - h_{f2}}{h_1 - h_{f2}} = \frac{505 - 325}{2463 - 325}$$

$$= 0.0776 \text{ kJ/kg of entering steam} \qquad \textbf{Ans.}$$

Using the heat balance equation at heater number 2

$$m_2 h_2 + (1 - m_1 - m_2) h_{f3} = (1 - m_1 - m_2) h_{f2} + m_2 h_{f2}$$

$$m_2 = \frac{(h_{f2} - h_{f3}) - m_1 (h_{f2} - h_{f3})}{h_2 - h_{f3}} = \frac{(325 - 163) - 0.0776 (325 - 163)}{2472.9 163}$$

$$= 0.06469 \text{ kJ/kg of enetering steam} \qquad \textbf{Ans.}$$

$$q_A = (1 - m_1 - m_2)(h_3 - h_{f3}) = (1 - 0.0776 - 0.0646)(2302.87 - 163)$$

$$= 1835.38 \text{ kJ/kg of steam}$$

$$\text{Heat supplied} = q_A = h_0 - h_{f1} = 3000 - 505 = 2495 \text{ kJ/kg}$$

$$\text{Work done} = w_t = 2495 - 1835.38 = 659.62 \text{ kJ/kg}$$

$$\therefore \quad \text{Thermal efficiency with regeneration} = \frac{659.62}{2495} = 26.43\%$$

Work done with regeneration, $h_0 - h_3 = 3000 - 2302.87 = 697.13$ kJ/kg

Heat supplied without regeneration, $h_0 - h_{f3} = 3000 - 163 = 2837$ kJ/kg

$$\text{Thermal efficiency without regeneration} = \frac{697.13}{2837} = 24.57\%$$

$$\text{Percentage gain in efficiency due to feed heating} = \frac{26.43 - 24.57}{24.57} \times 100 = 7.57\%$$

$$\text{Now, steam consumption with feed heating} = \frac{3600}{w_{net}} = \frac{3600}{2495 - 1835.38}$$

$$= 5.457 \text{ kg/kWh}$$

$$\text{Again steam consumption without feed heating} = \frac{3600}{697.13} = 5.164 \text{ kg/kWh}$$

Hence, increase in consumption rate due to feed heating

$$= 5.457 - 5.164 = 0.293 \text{ kg/kWh} \quad \textbf{Ans.}$$

Example 14.9 Steam enters a turbine at 60 bar and 600°C. Steam is bled off at 7 bar for regenerative feed heating and the remaining steam is condensed in condenser to condenser temperature 30°C. Calculate (a) the amount of bled steam, (b) cycle net work, (c) the ideal thermal efficiency of cycle. For an ideal turbine and with same states, determine (d) ideal turbine work, (e) ideal efficiency, and (f) steam rate in kg/kWh.

Solution Refer to Figs. 14.22 and 14.23. In the absence of specific information, the feed arrangement shown in Fig. 14.22 is assumed. Assuming no undercooling, the condernser pressure corresponding to 30°C = 0.0422 bar. Since cycle network and turbine work are separately required, so pump work has to be considered in cycle work.

For Mollier diagram,

$$h_0 = 3657 \text{ kJ/kg}$$
$$h_{1s} = 3045 \text{ kJ/kg}$$
$$h_{f1} = 697 \text{ kJ/kg}$$
$$h_{2s} = 2170 \text{ kJ/kg}$$
$$h_{f2} = 125.7 \text{ kJ/kg}$$

(a) The total pump work may be approximated to feed pump work, i.e., work required in pumping from p_2 to p_0.

$$\Sigma w_p = v_{f2} (p_0 - p_2) \times 10^2 = 0.0010044 (60 - 0.04242) \times 10^2 = 6.022 \text{ kJ/kg}$$

Writing enthalpy balance equation :

At heater, $m_1 h_{1s} + h_{f3} = h_{f1} + m_1 h_1$ and At hot-well, $(1 - m_1) h_{f2} + m_1 h_{f1} = h_{f3}$

$$\therefore \quad m_1 + m_1 h_{f1} + (1 - m_1) h_{f2} = h_{f1} + m_1 h_{f1}$$

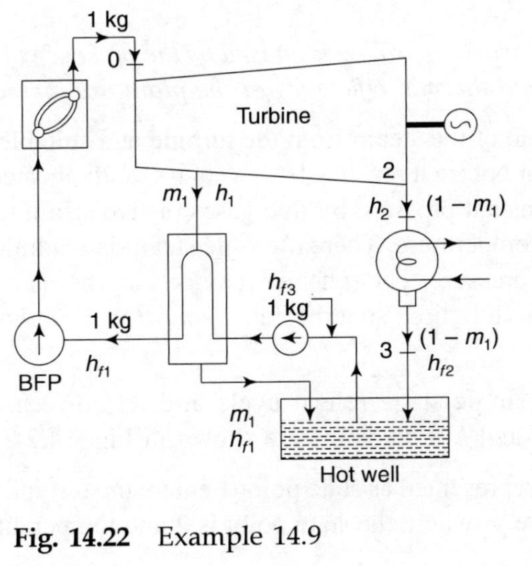

Fig. 14.22 Example 14.9

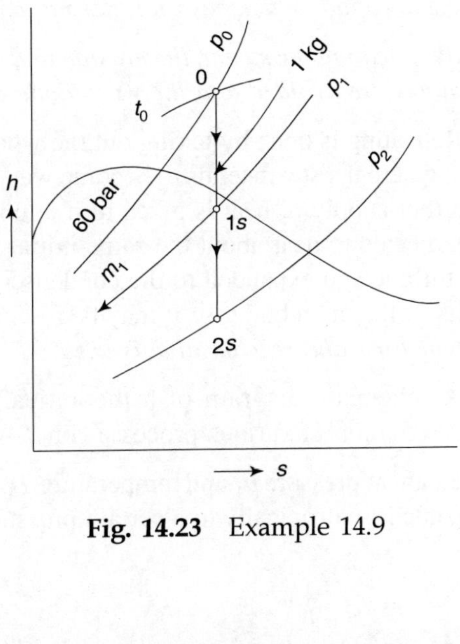

Fig. 14.23 Example 14.9

$$\therefore \quad m_1 = \frac{h_{f1} - h_{f2}}{h_{1s} - h_{f2}} = \frac{697 - 125.7}{3045 - 125.7}$$

$$= 0.1956 \text{ kg/kg of entering steam}$$

(b) Cycle net work, $w_t - \Sigma w_p = (h_0 - h_{1s}) + (1 - m_1)(h_{1s} - h_{2s}) - \Sigma w_p$

$$= (3657 - 3045) + 0.8044 \times (3045 - 2170) - 6.022$$

$$= 1309.28 \text{ kJ/kg}$$ **Ans.**

Heat supplied, $h_0 - h_{f1} - w_p = 3657 - 667 - 6.022 = 2953$ kJ/kg

(c) Cycle thermal efficiency $= \dfrac{w_{net}}{q_A} = \dfrac{1309.28}{2953.97} = 44.32\%$ **Ans.**

(d) Ideal turbine efficiency work $= (h_0 - h_{1s}) + (1 - m_1)(h_{1s} - h_{2s})$

$$= (3657 - 3045) + 0.8044\ (3045 - 2170)$$

$$= 1315.85 \text{ kJ/kg}$$ **Ans.**

(e) Ideal efficiency $= \dfrac{w_t}{q_A} = \dfrac{1315.85}{2960} = 44.45\%$ **Ans.**

(f) Steam rate $= \dfrac{3600}{1315.85} = 2.735$ kg/kWh **Ans.**

14.10 THE REHEAT CYCLE

Due to expansion of steam in the turbine, the steam becomes wet specially in the low pressure stages of the turbine. When the presence of water particles exceeds 10%, it brings about serious erosion and corrosion (eating or grooving) of blades along with a reduction in nozzle efficiency, blade efficiency and thermal efficiency.

By reheating the steam, the erosion and corrosion problems caused by the presence of water particles are avoided and the work done and thermal efficiency of the plant are increased.

Reheating is done by taking out the whole of the steam from the turbine at a suitable point (i.e., where the steam tends to become wet or before it reaches 10% wetness) and reheated, i.e., a further supply of heat is given to it at constant pressure by flue gases or live steam till it is superheated to near about the same initial temperature. Thereafter, the steam is readmitted to the turbine and expanded to the condenser pressure. The reheater may be incorporated in the walls of the main boiler or it may be a separately fired superheater. *Generally, a combination of reheating and regenerative is used.*

A schematic diagram of a theoretical single-stage reheat cycle and its corresponding representation of various processes on *T–s* and *h–s* diagrams are shown in Fig. 14.24.

Steam at pressure p_1 and temperature T_1 represented as state point 1 enters the turbine and is expanded isentropically to a certain pressure p_2 where the state point is 2 and temperature T_2.

At this point, whole of the steam is taken out and is reheated in a reheater to a temperature T_a ($T_a \approx T_1$) and is represented by state point a. After this, the steam is readmitted to the turbine in which it is expanded isentropically to condenser pressure, p_3. In practice, the expansion process before and after reheating will be represented by 1–2 and a–3 respectively. The hatched area shows the increase in work done in the ideal case.

Consider 1 kg of working fluid flowing through the plant.

Ideal case

$$\text{Heat added, } q_A = (h_1 - h_{f3}) + (h_a - h_{s2});\ \text{kJ/kg}$$

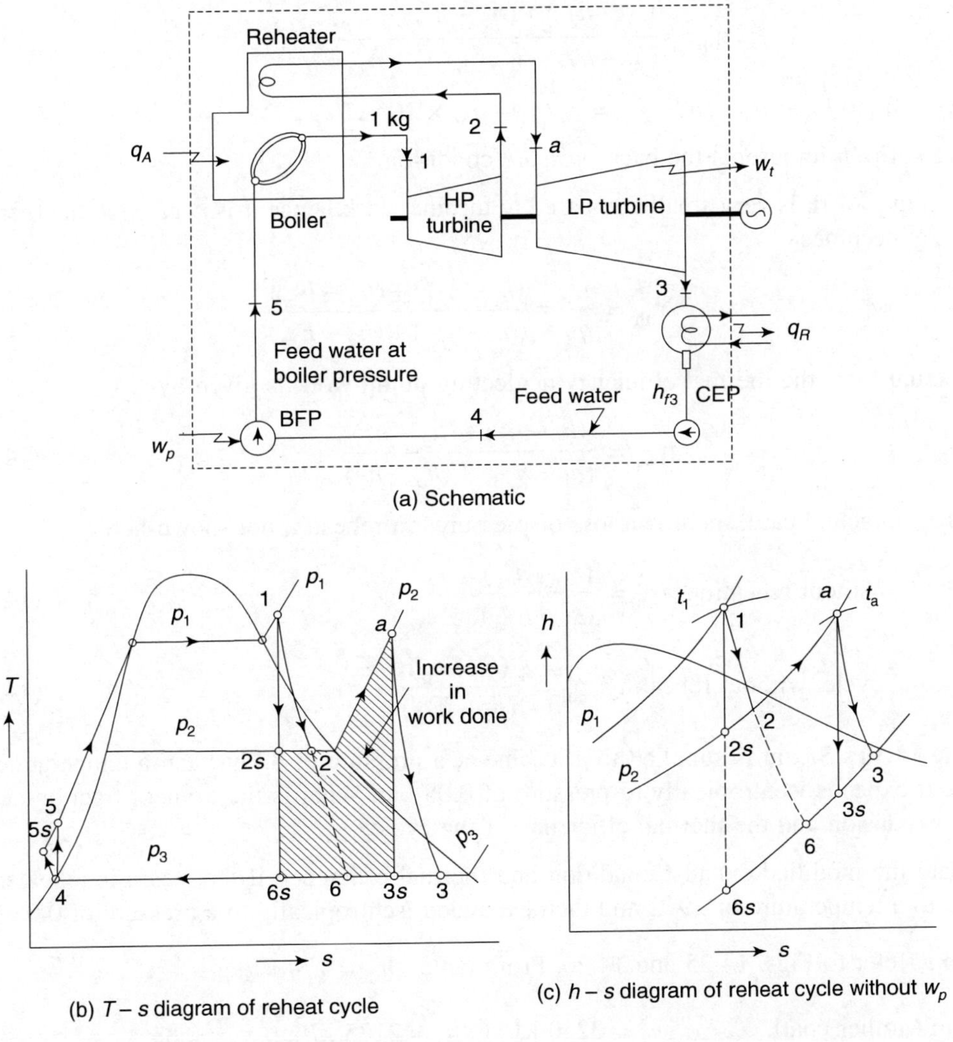

Fig. 14.24 Reheat cycle (Schematic *T-s* and *h-s* representation)

$$\text{Heat rejected, } q_R = (h_{3s} - h_{f3}) = (h_{3s} - h_{f4}) \text{ kJ/kg}$$

$$\text{Turbine work, } w_t = (h_1 - h_{s2}) + (h_a - h_{3s}) \text{ kJ/kg}$$

$$\text{Pump work, } w_p = (h_{f5} - h_{f4}) \text{ kJ/kg}$$

$$\therefore \quad w_{net} = w_t - w_p = (h_1 - h_{2s}) + (h_a - h_{3s}) - (h_{f5} - h_{f3})$$

Also
$$w_{net} = q_A - q_R = (h_1 - h_{2s}) + (h_a - h_{3s}) - (h_{f5} - h_{f4})$$

$$\therefore \quad \eta_{th} = \frac{w_{net}}{q_A} = \frac{w_t - w_p}{q_A} = \frac{(h_1 - h_{2s}) + (h_a - h_{3s}) - (h_{f5} - h_{f4})}{(h_1 - h_{f5}) + (h_a - h_{2s})}$$

or
$$\eta_{th} = \frac{(h_1 - h_{2s}) + (h_a - h_{3s}) - w_p}{(h_1 - h_{f3} - w_p) + (h_a - h_{2s})} \quad \text{...(14.27)}$$

where $h_{f4} = h_{f3}$ at p_3 and $w_p = v_b\,(p_1 - p_b) \times 10^2$ kJ/kg

Here suffix b represents the back pressure condition.

The pump work is very small compared to turbine work hence it is *neglected,* the thermal efficiency becomes

$$\eta_{th} = \frac{w_t}{q_A} = \frac{(h_1 - h_{2s}) + (h_a - h_{3s})}{(h_1 - h_{f3}) + (h_a - h_{2s})}$$

In actual case, the thermal efficiency neglecting pump work is given by

$$\eta_{tha} = \frac{(h_1 - h_2) + (h_a - h_3)}{(h_1 - h_{f3}) + (h_a - h_2)} \quad \text{...(14.28)}$$

Note In actual case, there is a loss of pressures in reheater, not shown here.

$$\text{Without reheating, } \eta_{th} = \frac{(h_1 - h_{6s})}{(h_1 - h_{f3})} \quad \text{...(14.29)}$$

$$\text{In actual cycle, without reheating, } \eta_{th} = \frac{(h_1 - h_6)}{(h_1 - h_{f3})} \quad \text{...(14.30)}$$

Example 14.10 Steam is supplied to a turbine at a pressure of 32 bar and a temperature of 410°C. It expands isentropically to pressure of 0.08 bar. What is the dryness fraction at the end of expansion and the thermal efficiency of the cycle?

Calculate the modified exhaust condition and thermal efficiency if the steam is reheated at 5.5 bar to a temperature of 39°C and then expanded isentropically to a pressure of 0.08 bar.

Solution Refer to Figs. 14.25 and 14.26. From table, $h_{f2} = 174$ kJ/kg

From Mollier chart, $h_1 = 3260$ kJ/kg, $h_2 = 2175$ kJ/kg, $x_2 = 0.83$

$$\text{Work done, } h_1 - h_2 = 3260 - 2175 = 1085 \text{ kJ/kg}$$

Heat supplied, $h_1 - h_{f2} = 3260 - 174 = 3086$ kJ/kg

$$\text{Thermal efficiency} = \frac{\text{Work done}}{\text{Heat supplied}} = \frac{1085}{3086} = 35.15\%$$ **Ans.**

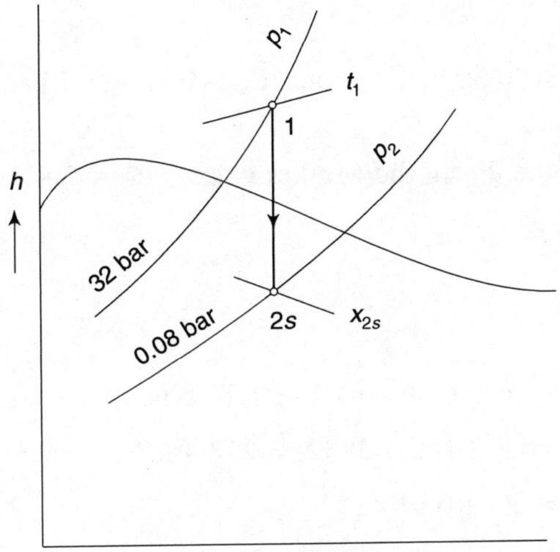

Fig. 14.25 Example 14.10

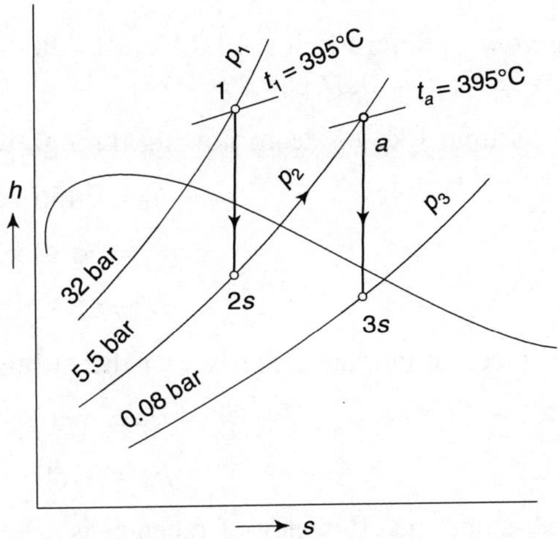

Fig. 14.26 Example 14.10

(b) In the modified case, the reheating of steam takes place at 5.5 bar to a temperature of 395°C. From Mollier chart,

$$h_1 = 3260 \text{ kJ/kg}$$
$$h_{2s} = 2830 \text{ kJ/kg}$$
$$h_a = 3265 \text{ kJ/kg}$$
$$h_{3s} = 2420 \text{ kJ/kg}$$

$$\text{Work done} = (h_1 - h_{2s}) + (h_a - h_{3s}) = (3260 - 2830) + (3265 - 2420)$$
$$= 1275 \text{ kJ/kg}$$

$$\text{Heat supplied} = (h_1 - h_{f3}) + (h_a - h_{2s}) = (3260 - 174) + (3265 - 2830)$$
$$= 3521 \text{ kJ/kg}$$

$$\text{Thermal efficiency} = \frac{1275}{3521} = 36.21\%$$ **Ans.**

Exhaust condition of steam $= x_3 = 0.935$ **Ans.**

Example 14.11 A steam turbine is divided into two sections, HP and LP with a reheater interposed in between the two sections. The steam on its way to the turbine at 30 bar and 500°C passes through a reheater where it gives up heat at constant pressure to heat the steam flowing from the HP turbine to the LP section. The steam then enters the HP turbine at 30 bar and 380°C. The steam leaves the HP turbine at 7 bar and the LP turbine at 0.07 bar. Assuming no loss of pressure between the two sections of the turbine an internal efficiency of 0.8 for both the sections, determine the steam condition at entrance to the LP section and the thermal efficiency of the plant.

Solution Refer to Figs. 14.27 and 14.28. Given : p_1 = 30 bar, t_1 = 500°C, p_2 bar, p_4 = 7 bar, t_2 = 380°, p_3 = 0.07 bar, η_{ti} = 0.8.

Assume 1 kg of steam flowing through the turbine. From the Mollier chart

$$h_1 = 3456 \text{ kJ/kg}$$

$$h_{2s} = 3180 \text{ kJ/kg}$$

$$h_{3s} = 2832 \text{ kJ/kg}$$

Since the internal efficiency of the turbine is 0.8, so actual work in HP turbine

$$h_2 - h_3 = \eta_{ti} (h_2 - h_{3s}) = 0.8 (3180 - 2832) = 278.4 \text{ kJ}$$

$$\therefore \quad h_3 = 3180 - 278.4 = 2901.6 \text{ kJ/kg}$$

Assume the efficiency of reheater is 100% i.e., the heat given by the heating steam must be equal to the heat taken by the system coming from HP turbine.

$$h_1 - h_2 = h_4 - h_3 \quad \text{or} \quad (3456 - 3180) = h_4 - 2901.6$$

or

$$h_4 = 3177.6 \text{ kJ/kg}$$

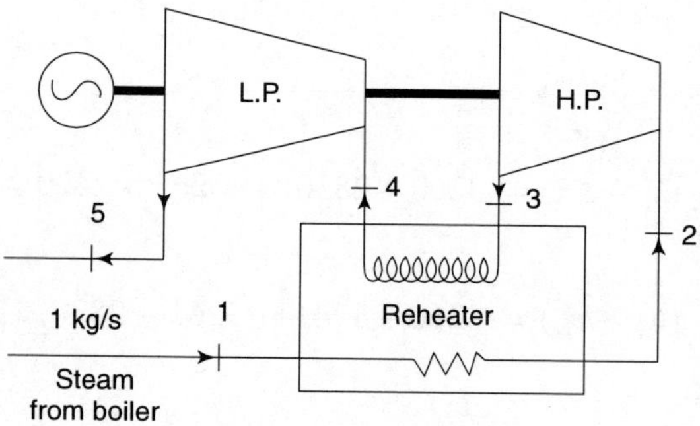

Fig. 14.27 Example 14.11

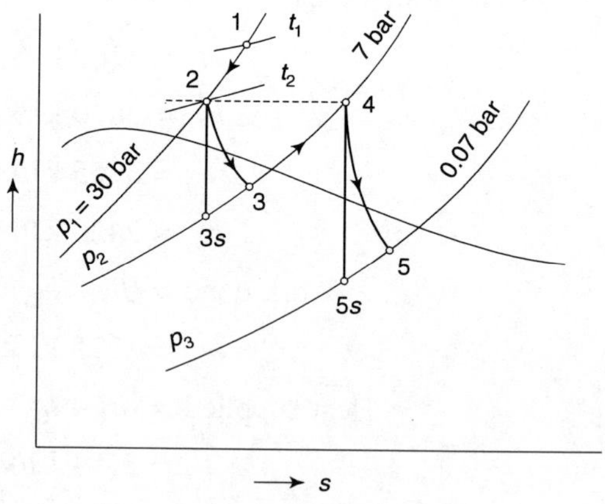

Fig. 14.28 Example 14.11

With the value of h_4 and pressure line 7 bar, the point 4 can be located. Hence, the condition of steam at inlet to LP turbine are 7 bar and 360°C. **Ans.**

Again, from Mollier chart,

$$h_4 - h_{5s} = 3177.6 - 2330 = 847.6 \text{ kJ/kg}$$

$$\therefore \quad h_4 - h_5 = 0.8 \times 847.6 = 678.08 \text{ kJ/kg}$$

Total work done from the plant = useful HP turbine work + useful LP turbine work

$$= (h_2 - h_3) + (h_4 - h_5) = 278.4 + 678.08 = 956.48 \text{ kJ/kg}$$

$$\text{Heat supplied, } h_1 - h_{f5} = 3456 - 163 = 3293 \text{ kJ/kg}$$

$$\text{Thermal efficiency of the plant} = \frac{\text{Work done}}{\text{Heat supplied}} = \frac{956.48}{3293} = 29.04\% \quad \textbf{Ans.}$$

Example 14.12 A steam power plant equipped with regenerative as well as reheat arrangement is supplied with steam to the HP turbine at 80 bar and 470°C. For feed heating, a part of steam is extracted at 7 bar and the remainder of steam is reheated to 350°C in a reheater and then expanded in LP turbine down to 0.035 bar. Determine the following:

(a) Amount of steam bled off for feed heating, (b) amount of steam in LP turbine, (c) heat supplied in boiler and reheater, (d) output of the turbine, and (e) cycle efficiency.

Solution Refer to Figs. 14.29 and 14.30. From Mollier chart.

$$h_0 = 3330 \text{ kJ/kg}, \quad h_{1s} = 2720 \text{ kJ/kg}, \quad h_a = 3170 \text{ kJ/kg}$$

$$h_{2s} = 2236 \text{ kJ/kg}, \quad h_{f1} = 697 \text{ kJ/kg}, \quad h_{f2} = 111.85 \text{ kJ/kg}$$

Let m_1 kg/s be the amount of steam bled off for feed heating. Writing enthalpy balance equation at heater

$$m_1 h_{1s} + (1 - m_1) h_{f2} = h_{f1}$$

$$\text{(a)} \therefore \quad m_1 = \frac{h_{f1} - h_{f2}}{h_{1s} - h_{f2}} = \frac{697 - 111.85}{2720 - 111.85} = 0.2243 \text{ kg/kg of entering steam.} \quad \textbf{Ans.}$$

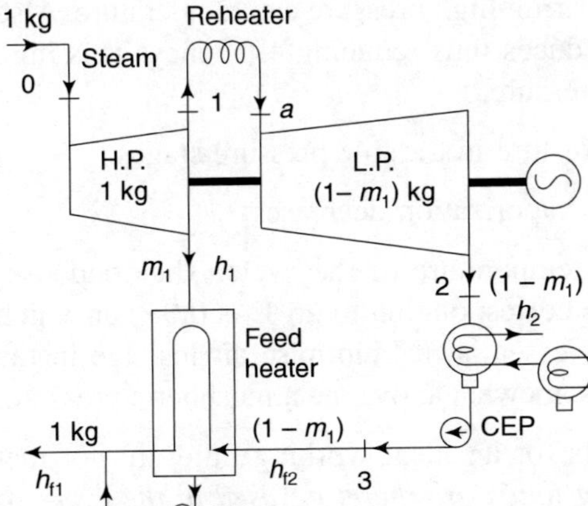

Fig. 14.29 Example 14.12

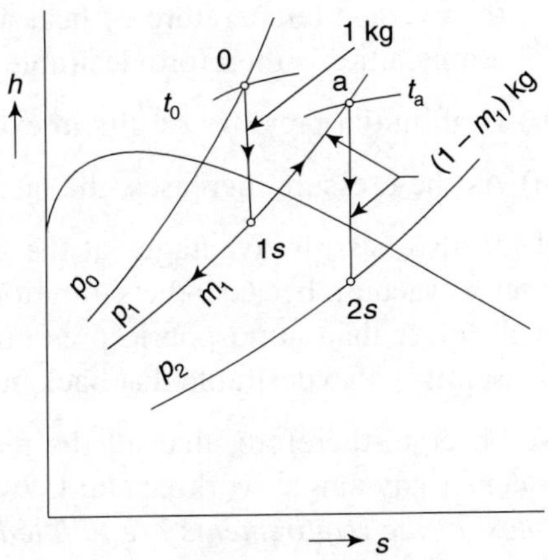

Fig. 14.30 Example 14.30

(b) Amount of steam in LP turbine = $1 - m_1 = 1 - 0.2243 = 0.7757$ kg/kg of steam

(c) Heat supplied in boiler = $h_0 - h_{f1} = 3330 - 697 = 2633$ kJ/kg

Heat supplied in reheater = $(1 - m_1)(h_a - h_{1s}) = (3170 - 2720) \times 0.7757 = 349$ kJ/kg

Total heat supplied = $2633 + 349 = 2982$ kJ/kg

(d)
$$\text{Output} = (h_0 - h_{1s}) + (1 - m_1)(h_0 - h_{2s})$$
$$= (3330 - 2720) + 0.7757\,(3170 - 2236) = 1334.5 \text{ kJ/kg}$$ **Ans.**

(e)
$$\text{Efficiency} = \frac{1334.5}{2982} = 44.75\%$$ **Ans.**

14.11 BINARY-VAPOUR CYCLE

The main aim of a steam turbine designer is to achieve the highest thermal efficiency possible for the plant. We have already seen that the thermal efficiency of the Rankine cycle is increased by (a) increasing the average temperature at which heat is added to the cycle and (b) decreasing the average temperature at which heat is rejected by the cycle. But metallurgical problems of the boiler and the steam turbine limit the maximum temperature of the cycle. A metal can withstand a higher temperature but at a lower pressure. At the present state of technology, this maximum temperature limit is about 650°C. The cooling medium in the condenser is a limit to the minimum temperature of the cycle. This temperature is about 25°C, i.e., the temperature of natural water. If steam is used as the working fluid in Rankine cycle operating over the above temperature range, i.e., from 650°C to 25°C, difficulties arise.

(i) The critical temperature of water is 374°C. To obtain a maximum temperature of 650°C, the steam has to be superheated and this is only possible at a very high pressure. The metallurgical problems which arise, do not permit high pressure and temperature. Also the average temperature of heat addition reduces thus reducing efficiency. This high temperature is therefore desirable at high pressure.

(ii) High initial pressure results in excessive moisture in the low pressure stages.

(iii) As the pressure increases, the latent heat of vaporization decreases.

Water also has disadvantages at the minimum temperature of the cycle. The condenser operates at vacuum because the saturation pressure corresponding to 25°C is 0.03 bar which is much lower than atmospheric pressure. So it is essential to minimise air leakage in the condenser. It is also desirable that back pressure is somewhat above the atmospheric pressure.

We observe, therefore, that all the requirements for an ideal working fluid are not met by water or any single working fluid. *Two working fluids are therefore used in the cycle so that most of the requirements are fulfilled. Such a cycle is a Binary-Vapour Cycle and in this mercury and steam are commonly used as working fluids.* The following are the advantages and disadvantages of mercury as a binary cycle fluid:

Advantages:

(i) Moderate vapour pressure at high temperature. At 540°C, the saturation pressure is less than 14 bar.

(ii) Very stable.

(iii) Liquid mercury has a high density which is advantageous in the separation of the liquid from vapour in boiler and in feeding back to the boiler under hydrostatic head rather than under the pumping head.

(iv) Specific heat of the liquid is only 0.144 so the liquid line is steep, making the cycle closer to Carnot cycle.

(v) Specific enthalpy of mercury vapour is low thus resulting in lower jet velocities in mercury turbines.

(vi) Thermal efficiency is higher than Rankine steam cycle.

Disadvantages:

(i) High cost and limited supply.

(ii) Toxic in nature.

(iii) Pervasive, thus tends to leak through joints, cracks, etc.

(iv) Latent heat is very low, requiring a large amount of mercury for the same heat utilisation.

(v) The whole boiler operates under high temperature and not just the superheater.

(vi) It does not wet surfaces, which causes poor heat transfer, this can, of course, be overcome by chemical treatment.

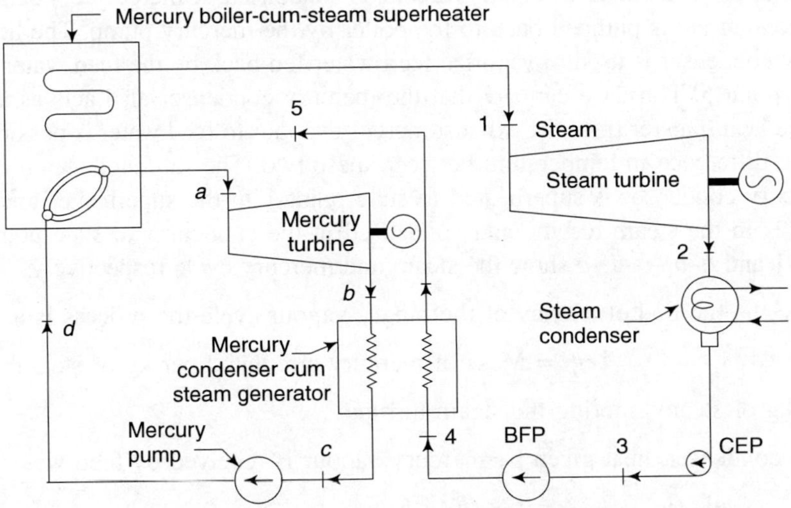

Fig. 14.31 Binary vapour cycle

Hence, it is clear that neither mercury nor any other vapour can replace steam in all the ranges of working. However, mercury is really superior in the high temperature range.

The schematic diagram of a binary vapour (mercury-steam) cycle is shown in Fig. 14.31 and its corresponding *T–s* diagram is shown in Fig. 14.32.

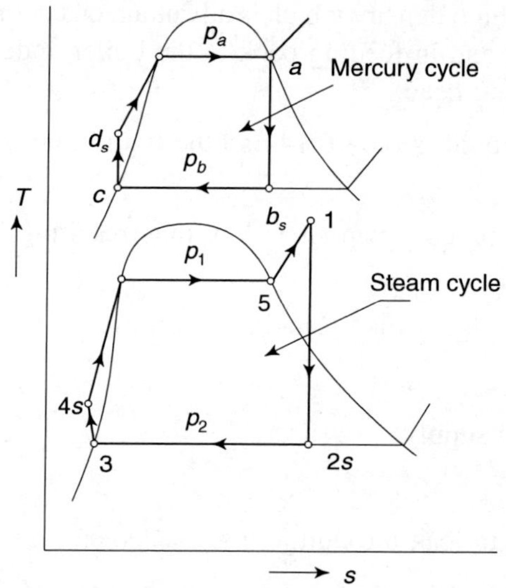

Fig. 14.32 Binary vapour cycle on *T–s* diagram

In the binary vapour (mercury-steam) cycle, there are two different circuits: one for steam and another for mercury. Mercury vapour, dry and saturated, enters the mercury turbine at state point *a* and expands to b_s. The exhaust mercury vapour is condensed in mercury condenser to state point *c*. The condensate is pumped back to the boiler by the mercury pump. The heat rejected in the mercury condenser is used to vaporize feed water fed back by the feed water pump into steam at state point 5. Thus, we can say that the mercury condenser also acts as a boiler for feed water. The heat transfer from the exhaust mercury vapour to feed water is possible because there is a large difference in temperature between these two. The saturated steam coming out from the mercury condenser is superheated to state point 1 in the superheater and expanded to state point 2*s* in the steam turbine and condensed in the condenser to state point 3. Thus, 1–2*s*–3–4*s*–5–1 and a–b_s–c–d_s–a show the steam and mercury cycle respectively.

To calculate the thermal efficiency of the binary-vapour cycle the process is as follows :

Let $\qquad m_m$ = Mass of mercury circulated per kg of steam.

Assume 1 kg of steam entering the steam turbine.

In mercury condenser, heat given by mercury vapour is received by feed water. Thus,

$$m_m (h_{bs} - h_{mc}) = 1 \times (h_5 - h_{f4})$$

or $$m_m = \frac{(h_5 - h_{f4})}{(h_{bs} - h_{mc})} \quad \text{...(14.31)}$$

Net work per kg of steam flow in the cycle

= (mercury turbine work – mercury pump work)

+ (steam turbine work – water pump work)

$= m_m (h_a - h_{bs}) - m_m (h_{mds} - h_{mc}) + (h_1 - h_{2s}) - (h_{f4} - h_{f3})$ kJ/kg of steam

Neglecting pump work, the net work

$w_{net} = m_m (h_a - h_{bs}) + (h_1 - h_{2s})$ kJ/kg of steam

Heat supplied per kg of steam in the cycle $= q_A = m_m (h_a - h_{mds}) + (h_1 - h_5)$

$\approx m_m (h_a - h_{mc}) - (h_1 - h_{2s})$ neglecting pump work

$$\eta_{th} = \frac{m_m (h_a - h_{bs}) + (h_1 - h_{2s})}{m_m (h_a - h_{mc}) + (h_1 - h_5)} \quad \text{...(14.32)}$$

The thermal efficiency of the binary vapour cycle is more than the Rankine steam cycle. The binary vapour cycle may also be called ***combined cycle***.

Note *Now, it is possible to achieve the thermal efficiency about 54% by using potassium-steam binary cycle as compared to 35% to 40% efficiency of conventional system.*

Example 14.13 In a binary-vapour cycle, the steam cycle operates between pressures of 30 bar and 0.07 bar and uses a superheat temperature of 350°C. The mercury cycle works between the pressures of 12.68 bar and 0.07 bar. The mercury vapour, entering the turbine, is in a dry saturated condition.

Compute the efficiency of the combined cycle assuming expansion in both cycle as isentropic. Data for mercury:

p	Saturation	h_{fm}	h_g	s_{fm}	s_g
bar	°C	kJ/kg	kJ/kg	kJ/kg K	kJ/kg K
0.07	236.5	32.395	326.067	0.08548	0.662906
12.68	537.5	71.9796	360.734	0.145798	0.50185

Solution Refer to Fig. 14.33. Mercury cycle.

$s_a = s_{bs} = 0.5018 = s_{fm} + x (s_g - s_{fm}) = 0.08548 + x (0.662906 - 0.08548)$

$$\therefore \quad x = \frac{0.50185 - 0.08548}{0.662906 - 0.08548} = 0.721$$

Enthalpy at b_s $= h_b = h_{fm} + x (h_g - h_{fm}) = 32.395 + 0.721 (326.667 - 32.395)$

$= 244.56$ kJ/kg of mercury

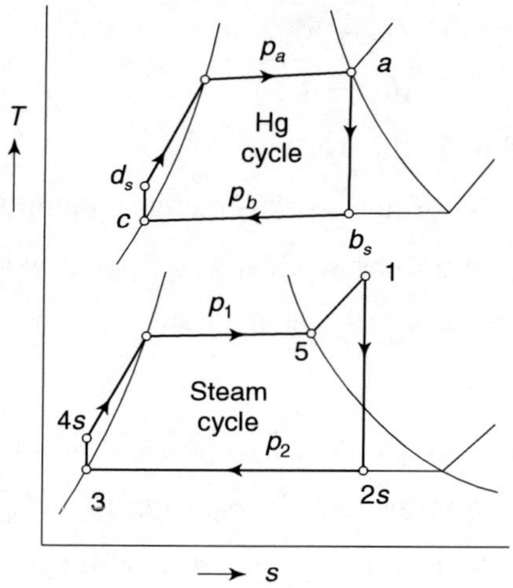

Fig. 14.33 Example 14.13

Isentropic work done by mercury turbine

$$= h_a - h_{bs} = 326.667 - 224.56 = 102.107 \text{ kJ/kg of mercury}$$

Heat rejected by mercury in mercury condenser

$$= (h_{bs} - h_{mc}) = (h_{bs} - h_{fm}) = 244.56 - 32.395$$

$$= 212.165 \text{ kJ/kg of mercury}$$

Heat supplied in mercury boiler

$$= h_a - h_{mc} = h_a - h_{fm} = 326.667 - 32.395 = 294.27 \text{ kJ/kg of mercury}$$

For steam cycle : $h_1 = 3114$ kJ/kg, $h_{2s} = 2090$ kJ/kg, $h_5 = 2794.5$ kJ/kg

Since nothing is mentioned about the heating of feed water, we assume that the feed water comes out from the mercury condenser as dry and saturated steam and superheated to 350°C in superheater.

Insentropic work done by steam turbine, $h_1 - h_{2s} = 3114 - 2090 = 1024$ kJ/kg of steam

Heat to be supplied in steam boiler, i.e., superheater

$$= h_1 - h_5 = 3114 - 2794.5 = 319.5 \text{ kJ/kg of steam}$$

Heat rejected by mercury/kg of steam = heat received by water/kg of steam

$$m_m (h_{bs} - h_{mc}) = (h_5 - h_{f3}) \times 1$$

or

$$m_m = \frac{2794.5 - 161.76}{212.165} = 12.4 \text{ kg of mercury/kg of steam}$$

Total work done in the cycle per kg of steam

= work done by steam turbine/kg of steam + work done by mercury turbine/kg of steam

$$= 1024 + 102.107 \times 12.4 = 2290.126 \text{ kJ/kg of steam}$$

Heat supplied in the cycle

= heat supplied in mercury boiler/kg of steaam + heat supplied in superheat/kg of steam

$$= (h_a - h_{cm})\, m_m + (h_1 - h_5) = 294.27 \times 12.4 + 319.5$$

$$= 3968.44 \text{ kJ/kg}$$

$$\eta_{th} = \frac{\text{Work done}}{\text{Heat supplied}} = \frac{2290.126}{3968.44} = 57.708\% \qquad \textbf{Ans.}$$

Example 14.14 In a binary vapour cycle, mercury vapour dry and saturated enters the mercury turbine at 10 bar and exhausts at 0.1 bar into the mercury condenser steam boiler unit. Steam omes out from this unit at 40 bar dry and saturated and expands in the steam turbine to 0.06 bar. Assuming that the efficiency ratios for mercury and steam turbines are 0.75 and 0.8 respectively, calculate:

(a) Amount of mercury circulating per kg of steam.

(b) Thermal efficiency of the cycle.

(c) Work done by mercury and steam turbine.

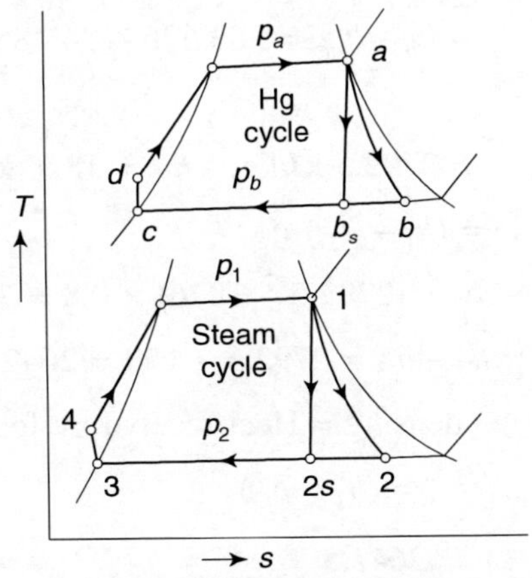

Fig. 14.34 Example 14.14

p (bar)	t_s (°C)	h_{fm} (kJ/kg)	h_g (kg/kg)	s_{fm} (kg/kg K)	s_g (kg/kg K)
10	515.5°C	72.105	362.406	0.14755	0.5158
0.1	249.6°C	34.485	332.975	0.089	0.6604

Solution Refer to Fig. 14.34.

Processes *a–b–c–d* and 1–2–3–1 show the actual mercury and steam cycle respectively. For mercury cycle :

$$s_a = s_{bs} = 0.51587 = 0.089 + x\ (0.6604 - 0.089)$$

$$\therefore \qquad x = \frac{0.5158 - 0.089}{0.6604 - 0.089} = 0.769$$

Enthalpy at $\quad b = h_{bs} = h_{mf} + x\ (h_g - h_{mf})$

$$\therefore \qquad h_{bs} = 34.485 + 0.769\ (332.975 - 34.485) = 257.429 \text{ kJ/kg}$$

Here $\quad h_a = 362.406$ kJ/kg

Hence, isentropic work done = 362.408 – 257.429 = 104.977 kJ/kg of mercury

Actual, work done by mercury turbine

$$= h_a - h_b = 104.977 \times 0.75 = 78.73 \text{ kJ/kg of steam}$$

$$\therefore \qquad h_a = 362.406 - 78.73 = 283.676 \text{ kJ/kg of mercury}$$

Heat supplied in the mercury boiler per kg of mercury

$$= h_a - h_{mc} = 362.406 - 34.485 = 327.92 \text{ kJ/kg}$$

Heat rejected by mercury in mercury condenser

$$= h_b - h_{mc} = 283.676 - 34.485 = 249.191 \text{ kJ/kg}$$

For steam cycle:

From Mollier chart, $\quad h_1 = 2792.5$ kJ/kg, $\quad h_{2s} = 1876$ kJ/kg

Actual work done in the cycle = $(h_1 - h_{2s})\ \eta_r$

$$= h_1 - h_2 = (2792.5 - 1876) \times 0.8 = 733.2 \text{ kJ/kg of steam} \qquad \textbf{Ans.}$$

Heat supplied to feed water = $h_1 - h_{f3}$ = 2792.5 – 150 = 2642.5 kJ/kg of steam

Heat rejected by mercury in condensing = Heat receivd by feed water

(a) $\qquad m_m\ (h_b - h_{mc}) = 1 = (h_1 - h_{f3})$

$$m_m = \frac{2642.5}{249.191} = 10.604 \text{ kJ/kg of steam} \qquad \textbf{Ans.}$$

Total work done in the cycle = 78.73 × 10.604 + 733.2 = 1568.05 kJ/kg of steam

(b) $$\eta_{th} = \frac{\text{Total work done}}{\text{Heat supplied in}} = \frac{1568.05}{327.92 \times 10.604} = 45.09\% \qquad \textbf{Ans.}$$

(c) Work done by mercury turbine = 78.73 × 10.604 = 834.85 kJ/kg of steam **Ans.**

14.12 NUCLEAR POWER PLANT CYCLES*

Figure 14.35 shows a schematic diagram of a nuclear power plant in general. Due to the fissioning or splitting of heavy nuclei such as uranium U^{235}, a chain reaction starts releasing a tremendous amount of heat energy in the reactor. The coolant passing through the reactor takes up the heat and generates steam in the steam generator. The steam thus produced expands in the steam turbine and after expansion, it is condensed in the condenser and again fed back to the steam generator (i.e., heat exchanger). *It is to be noted that the heat evolved by one kg of fissioning of uranium is equivalent to the combustion of 1,13,350 quintals of coal or* 30×10^6 litres of oil. This large release of energy by fissioning results from the conversion of approximately 0.1 per cent of the mass of fissioned uranium into energy based on Einstein theory, $E = mc^2$.

In nuclear power plants, the reactor replaces the familiar boiler. Many types of reactors have been proposed, but only six types have been commissioned so far, namely: (i) *the pressurised* water reactor, (ii) *the boiling water reactor*, (iii) *the heavy water, natural uranium reactor*, (iv) *the sodium-graphite reactor*, (v) *the gas cooled and* (vi) *the fast breader reactor*. The names are generally based on the type of coolant used in the reactor.

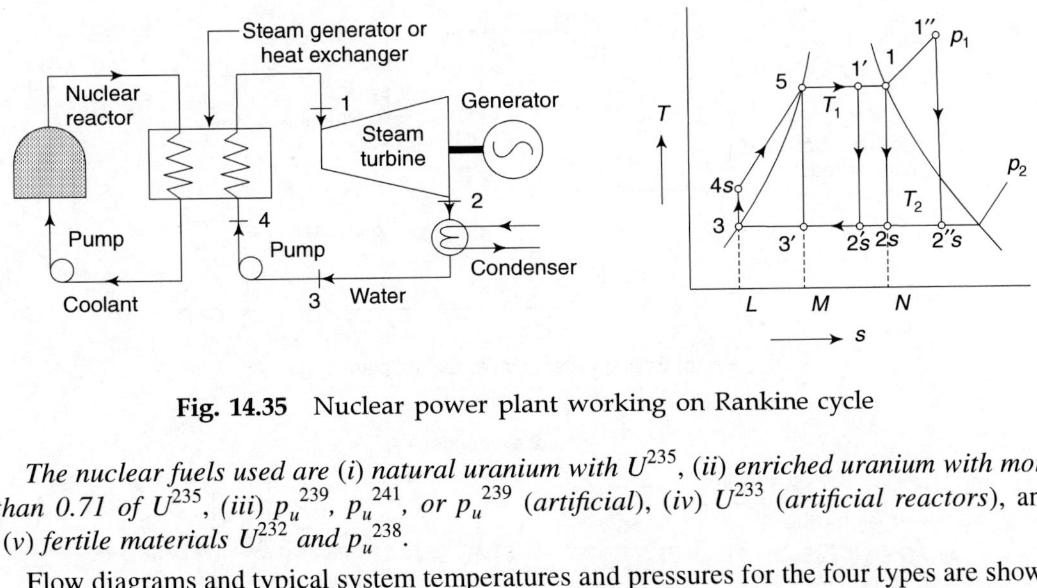

Fig. 14.35 Nuclear power plant working on Rankine cycle

The nuclear fuels used are (i) natural uranium with U^{235}, *(ii) enriched uranium with more than 0.71 of* U^{235}, *(iii)* p_u^{239}, p_u^{241}, *or* p_u^{239} *(artificial), (iv)* U^{233} *(artificial reactors), and (v) fertile materials* U^{232} *and* p_u^{238}.

Flow diagrams and typical system temperatures and pressures for the four types are shown in Fig. 14.36. The cycle of a fast breeder reactor is the same as that for the sodium-cooled thermal reactor. The intermediate coolant loops in every case is the same except in the boiler water reactor system. The coolant loops have been introduced to prevent any radioactivity from reaching the turbine area. The power producing loops are similar to the conventional steam power plants. The steam temperature and pressure, however, are comparable with the

*For more details, see author's book, *Steam and Gas Turbines, Turbomachines and Power Plant Engineering*.

steam generated in the modern boiler for the sodium and gas cooled reactors only; in other cases, it is less than that of boiler. The representation of the process on *T-s* and *h-s* diagrams and the analysis for thermal *efficiency are the same as in the conventional steam power plant system since the power producing loop is the same.*

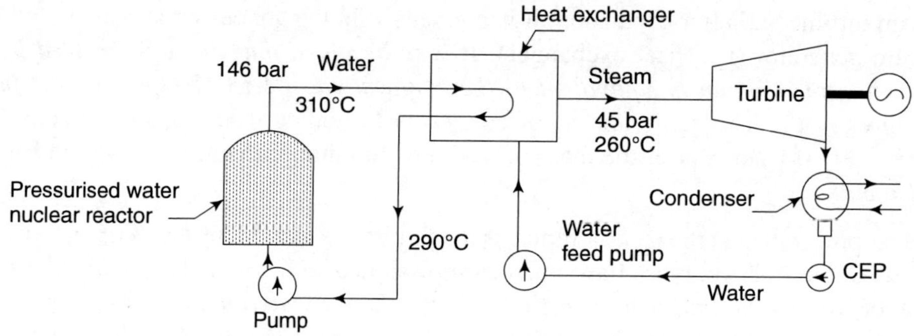

(a) Pressurised water nuclear reactor plant cycle

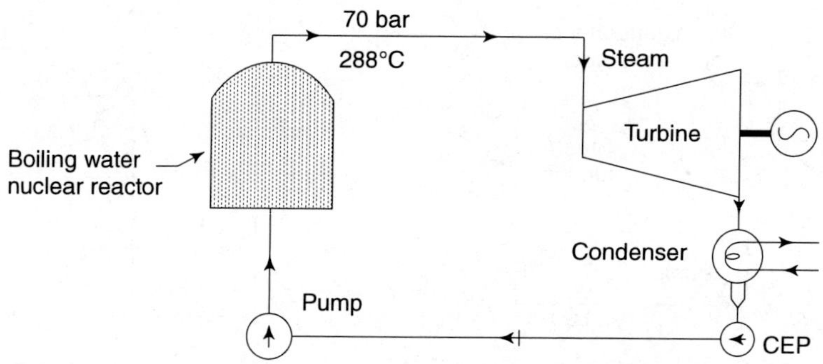

(b) Boiling water nuclear reactor plant cycle

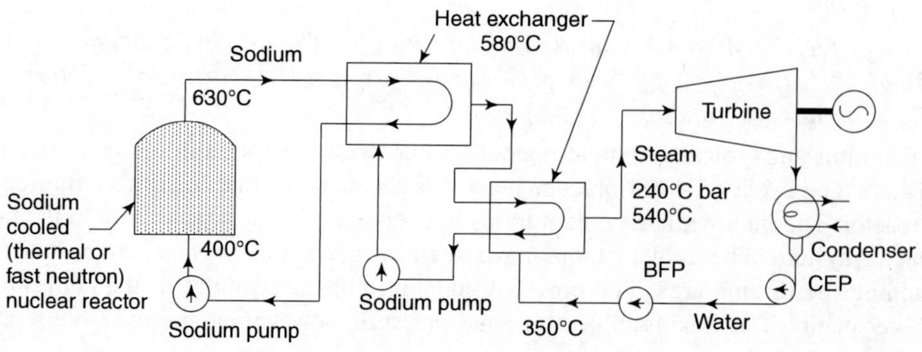

(c) Sodium cooled nuclear reactor plant cycle

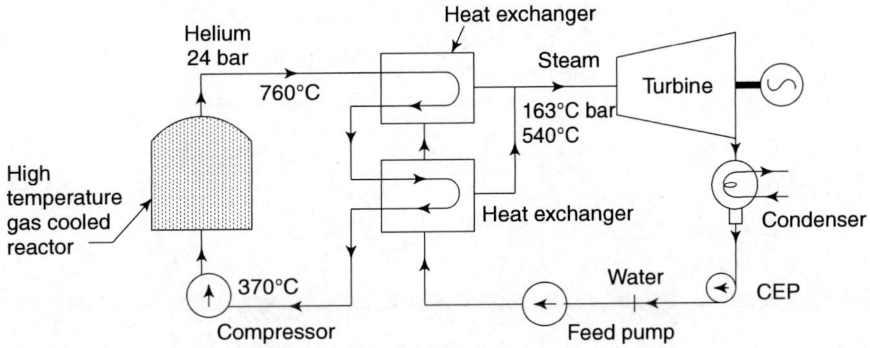

(d) High temperature gas cooled nuclear reactor plant

Fig. 14.36 Four types of nuclear power plant cycles

In space nuclear power plants, the reactor cooling agent is pressurised liquid lithium and the working fluid of the turbine is generally potassium because potassium has a lower specific weight. At Trombay (India) nuclear power plant, the coolant is heavy water and the working fluid of turbine is steam. In India, the electricity produced by the nuclear power plant is more costly than the hydel one. The completion of the nuclear plant takes near about five to six years which is less than the time taken by hydel while the working involves radiation hazards. The radiation emitted from reaction is shielded (by a 2 metre thick concrete layer) but even then if there is a slight leakage, it is very harmful to human life. In spite of this, it has gained ground for electricity generation and propelling ships, anti submarines but due to strong anti-public opinions, their growth is less.

14.13 SOLAR THERMAL ENERGY POWER PLANT

The solar energy incident on a surface oriented normal to the sunrays which is known as solar constant is 1353 W/m^2 at the outer atmosphere of the sky. At the earth's surface, this value gets reduced to about 500 to 1000 W/m^2. There are two methods prevalent to harness solar thermal energy.

In **Central Receiver Type Plant** (Fig. 14.37 a), the sunsrays are collected on reflectors (or concentrators) which are reflected and focussed on a central receiver where intense heat is generated and water is transformed into steam and is then expanded in the turbine to drive generator.

Distributed System (Fig. 14.37 b) using parabolic collector is also common. Steam is generated with the help of sun's rays falling on the collectors and then it is expanded in the turbine. Solar thermal power plant is immensely suitable on the moon where the sun's rays are twice as intense as on the earth. The power producing loop is similar to the conventional steam power plant. *The representation of the various processes on T–s and h–s charts are the same as Rankine cycle*. Though, it is impossible to operate this plant throughout the year due to frosty and cloudy weather even then it is gaining ground. Now, it is possible to store the radiation to be used in frosty weather. 5 to 15 MW units are coming up in USA, USSR and India.

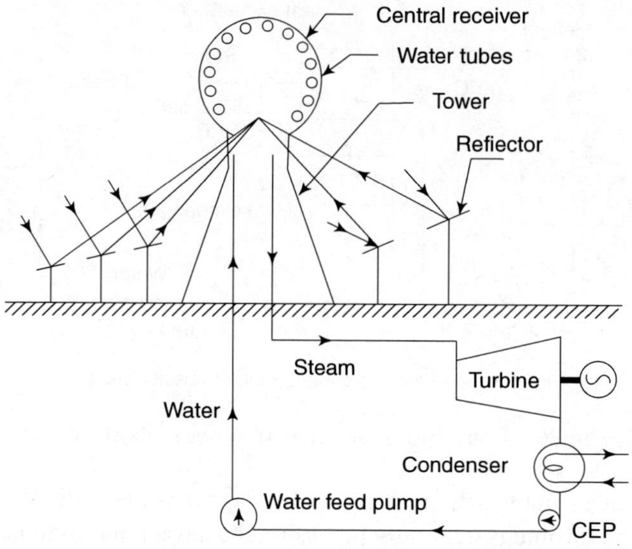

Fig. 14.37 (a) Central receiver type solar thermal energy power plant

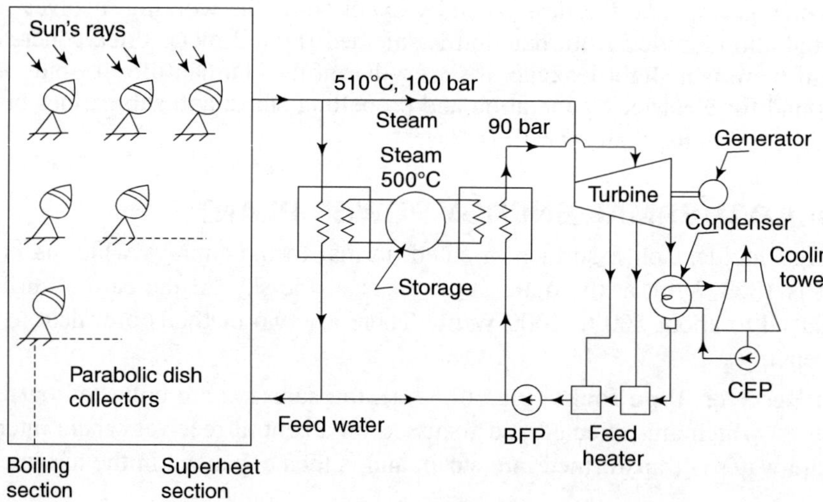

Fig. 14.37 (b) Distributed system parabolic collector solar thermal energy power plant

14.14 GEOTHERMAL POWER PLANT

The heat of the earth's interior is one of the most intensive energy available to the man. Geothermal resources are derived from the distribution of temperature and thermal energy beneath the earth's surface. Depending upon the conditions within the reservoir, the geothermal fluids may be in the form either of *slightly superheated steam* or of pressurised hot water and thus three systems are prevalent for the utilisation of geothermal energy. They are as follows and work on Rankine cycle.

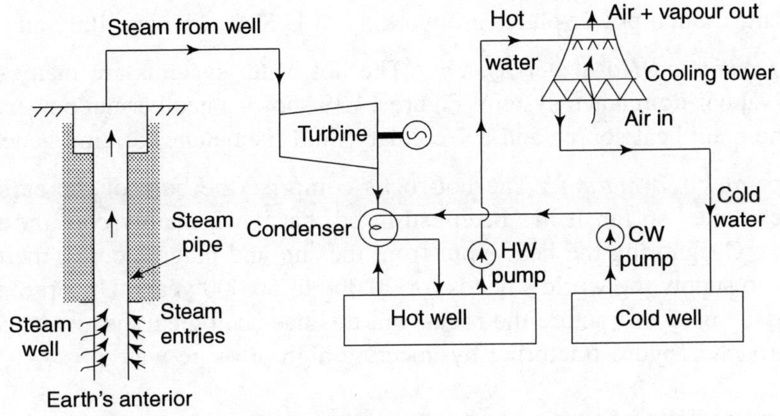

Fig. 14.38 Geothermal power plant using dry steam

(a) Vapour dominated (dry steam) systems

(b) Liquid dominated (hot water) systems

(c) Dry (hot-rock) geothermal formation.

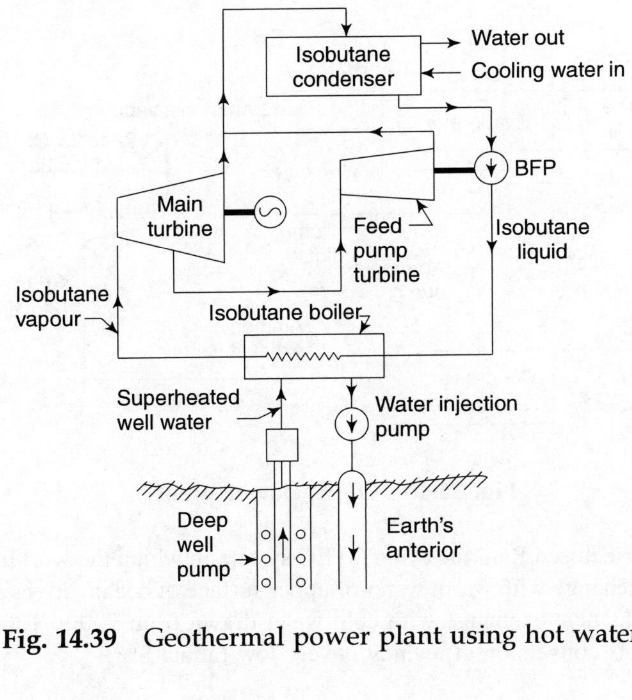

Fig. 14.39 Geothermal power plant using hot water

(a) *Vapour Dominated Systems* Figure 14.38 shows a diagrammatic arrangement of a geothermal power plant using dry steam from well. The inlet pressure is about 7.5 bar and temperature about 180°C. The cycle works on the Rankine cycle. It is the form most sought after and the rarest. Such power plants are working in U.S.A., Mexico, Italy and Japan.

(b) *Liquid Dominated Hot-water System* The hot water systems are many times more abundant than vapour-dominated system. Figure 14.39 shows a geothermal power plant using hot water as the main heat source and a secondary fluid (isobutane) driving generator.

(c) *Dry (Hot-rock) Formation* The hot-rocks comprise the bulk of the earth's surface and are least exploited so far. It has been estimated that if we merely cool the rocky crust under dryland 1°C (ignoring the heat input from the sun and magma core), there would be enough energy to supply the world's needs for 20,000 to 50,000 years. Underground nuclear explosions may be used to fracture the rocks, but this has ecological disadvantages. A more promising approach is hydro-fracturing by injecting high pressure water.

14.15 OCEAN THERMAL ENERGY POWER PLANT

An ocean thermal energy conversion (OTEC) power plant utilises the temperature difference between warm surface water and cold deep water of the ocean to drive a vapour turbine. In semitropical and tropical oceans, the temperature difference is around 20°C and this is sufficient for OTEC operations and is available for 24 hours a day. As per a estimate, 40% of world's total energy requirement in 1983 could be obtained from the ocean in ± 10° latitude. The cycle works on Rankine cycle.

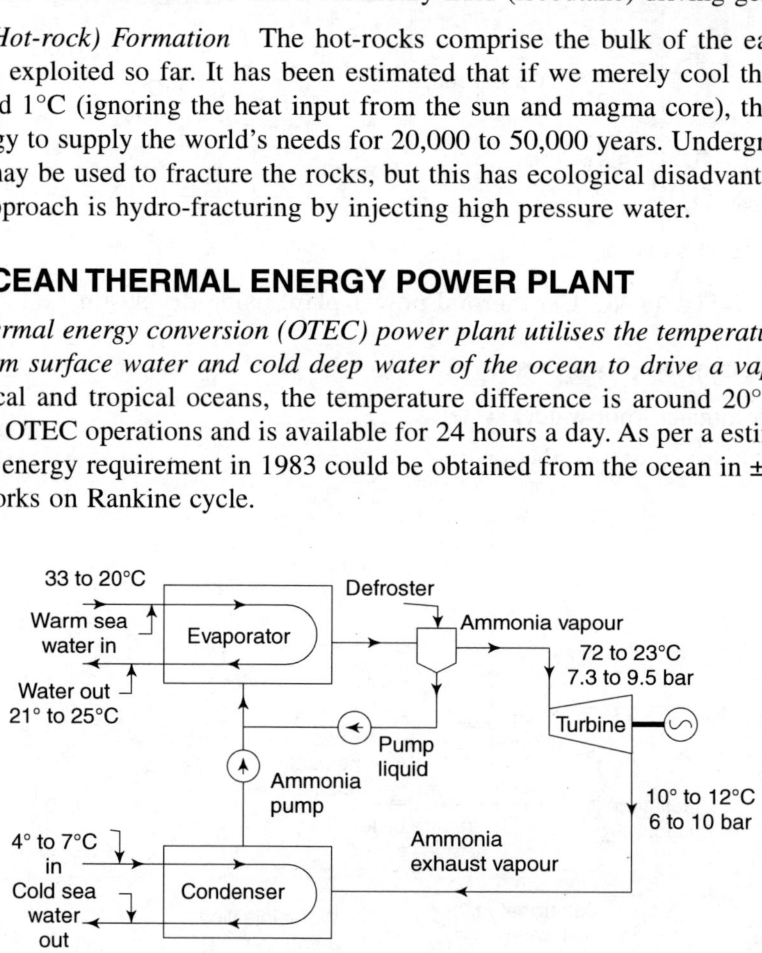

Fig. 14.40 OTEC power plant

Figure 14.40 shows a closed Rankine cycle OTEC system in which the working fluid 'ammonia' is vaporised by heat exchange with warm water of upper surface of ocean, drives a turbine generator and is condensed by the heat exchange with cold water drawn from 700 to 1200 m depth. It is in developing stage and its conversion efficiency is very low (about 3%).

14.16 COMBINED CYCLE AND CO-GENERATION

A considerable amount of heat energy goes as a waste with the exhaust of gas turbine as its temperature is around 500 to 600°C. This energy must be utilised. This is possible by using

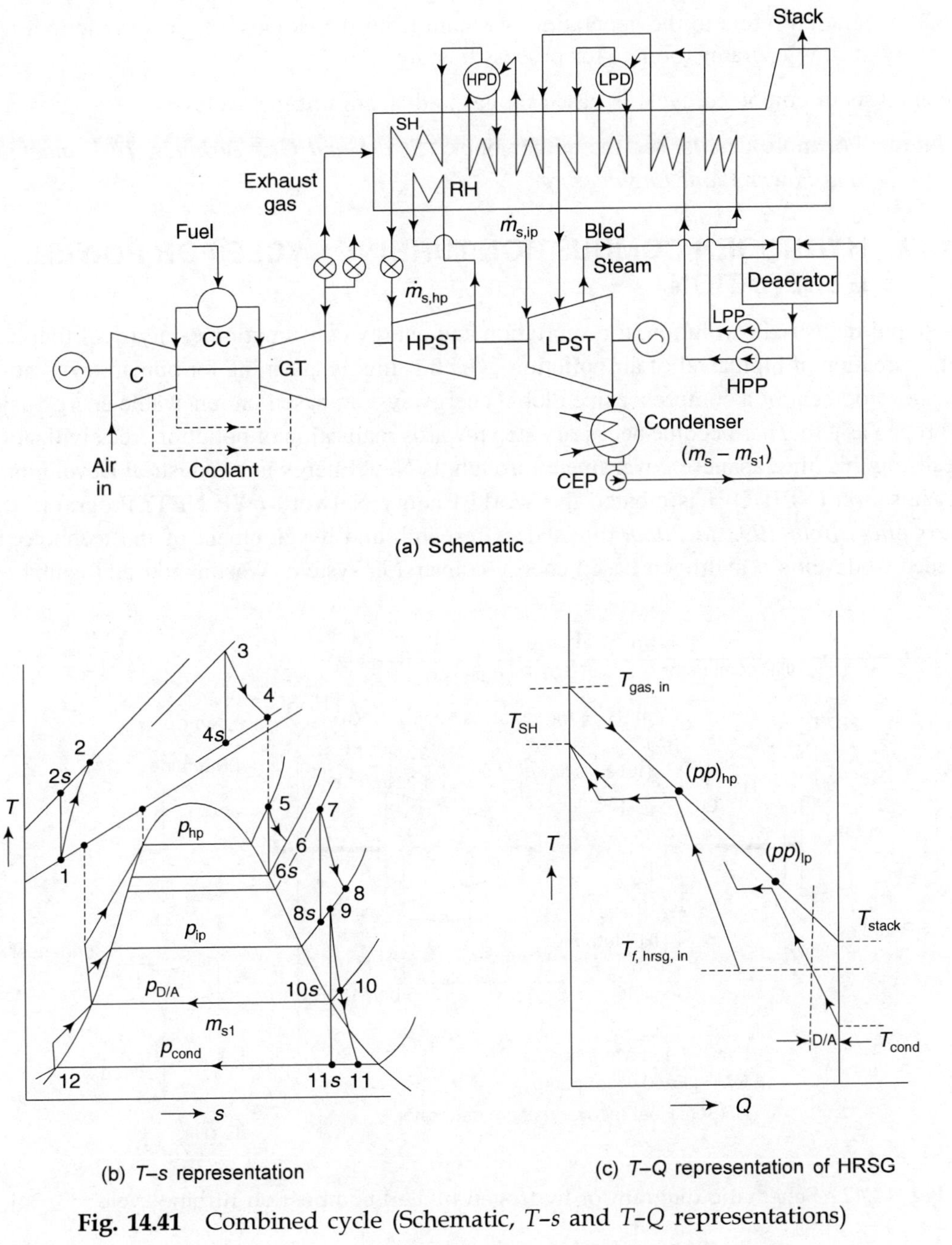

(a) Schematic

(b) T–s representation

(c) T–Q representation of HRSG

Fig. 14.41 Combined cycle (Schematic, T-s and T-Q representations)

gas/steam turbine combined cycle which is shown in Fig. 14.41. The exhaust gas from steam turbine is utilised to generate steam in a heat recovery steam generator (HRSG) and the steam is expanded in a steam turbine. There could be various combinations of combined cycle. The overall thermal efficiency of combined cycle is around 55% to 60%.

Co-generation refers to the generation of steam from the exhaust of gas turbine in a heat recovery steam generator (boiler) for process heating.

An analysis of combined cycle is based on gas and steam turbine cycles.

Note For more details, see the author book "*Steam and Gas Turbines, Turbomachines and Power Plant Engineering*."

14.17 HYDROGEN COMBUSTION TURBINE CYCLE FOR POWER GENERATION

The rapid industrial revolution, transportation and energy consumption using fossil fuel have led to creation of high level of air pollution, which is highly alarming for human civilization. The development of a comprehensive global energy system based on renewable energy using hydrogen as a fuel has become necessary step towards realisation of pollution free civilization. Realising this, the Japanese government through its New Energy and Industrial Development Organisation (NEIDO) has created the World Energy Network (WE-NET) Program, *a 28 years effort from 1993 to 2020*, directed as research and development of the technologies needed to develop a hydrogen-based energy conversion system. Westinghouse, Toshiba and

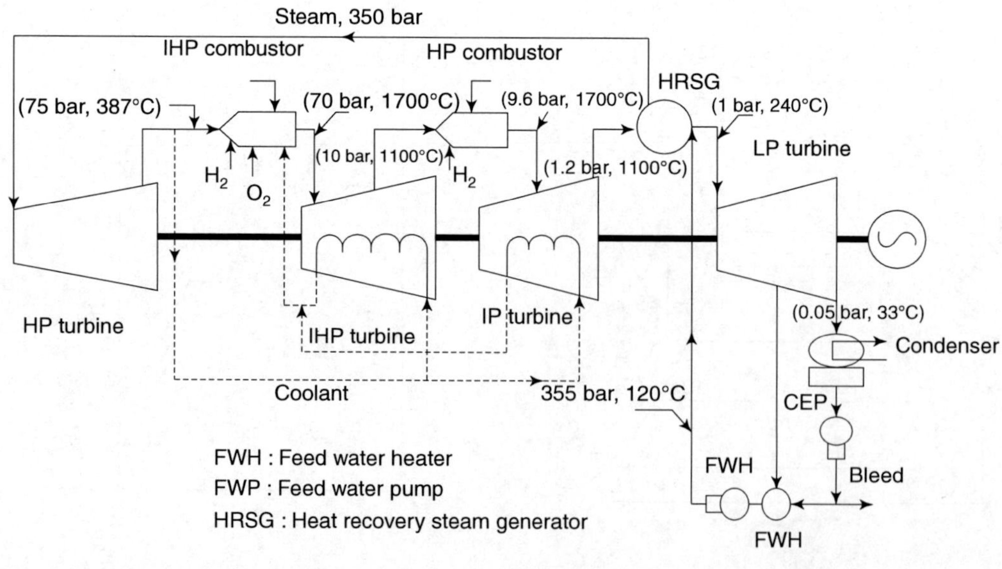

Fig. 14.42 Schematic diagram of hydrogen-fuelled combustion turbine cycle [33, 34]

Mitusubishi heavy industries are working under the time frame to develop a hydrogen fuelled combustion turbine system designed to meet the goals set by WE – NET program which include 500 MW output and greater than 70.9 per cent low heating value thermal efficiency, a reliability-availability-maintainability (RAM) equivalent to current base loaded natural gas fired combined cycle and elimination of CO, CO_2, NO_X and SO_X during power generation.

Figure 14.42 shows the schematic of hydrogen fuelled combustion turbine cycle, a long term plan [33, 34]. *This cycle works on Rankine cycle.* There are four turbines, namely, HP. IHP, IP, and LP and two combustors IHP and IP. The steam with selected maximum pressure (p_{max}) is supplied to HP turbine while IHP and IP turbines receive steam at the selected maximum temperature. In both combustors, H_2 and O_2 are burnt resulting in the increase in steam temperature to a selected value of t_{max}. IPH and IP turbines are cooled by closed loop steam cooling drawn from HP turbine exhaust. The exhaust of IP turbine is used to heat the feed water to generate steam at the maximum cycle pressure in a heat recovery steam generator (HRSG) before expanding to LP turbine while the steam generated is led to the HP turbine for expansion. *The said plant is being designed for the maximum 350 bar and 1700°C steam pressure and temperature respectively.*

Figure 14.43 shows variation of plant efficiency versus plant work for various maximum steam pressure and temperature in the cycle. The plant efficiency and work increase with the maximum steam pressure and temperature. With the success of this plant at comparable cost, there will be a revolution in electricity generation without using fossil fuels and the energy problem of the world may be nearly solved.

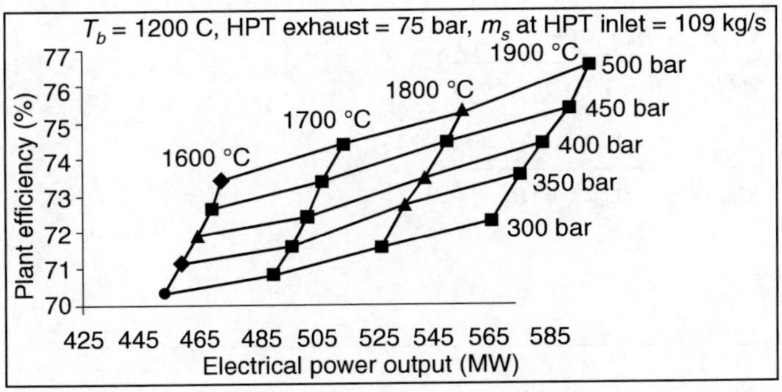

Fig. 14.43 Plant efficiency versus electrical output [33, 34]

14.18 SUMMARY OF FORMULAE

1. $\eta_{carnot} = \dfrac{T_1 - T_2}{T_1}$

2. $(\eta_{th})_{Rankine} = \dfrac{W_{net}}{q_A} = \dfrac{(h_1 - h_2) - w_p}{(h_1 - h_{f2}) - w_p}$

3. $(\eta_{th})_{Rankine} = \dfrac{(h_1 - h_2)}{(h_1 - h_{f2})}$ if $w_p = 0$ 4. $w_p = v_f\,(p_1 - p_2) \times 10^2$, kW

5. $\eta_t = \dfrac{h_1 - h_2}{h_1 - h_{2s}}$ 6. $\eta_p = \dfrac{h_{f4s} - h_{f3}}{h_{f4} - h_{f3}}$

7. $\eta_r = \dfrac{\eta_{tha}}{\eta_{Rankine}} = \eta_{th} \cdot \eta_m$ 8. $\eta_{oth} = \left[w_{t,\,isen} \times \eta_t \times \eta_m \times \eta_g - \dfrac{w_p}{\eta_p} \right] \Big/ q_A$

9. $q_{Aa} = (h_1 - h_2) - w_p/\eta_p = \dot{m}_f \times (\text{LCV})_f \times \eta_{boiler}$ and $q_A = \dot{m}_f \times (\text{LCV})_f = \dfrac{q_{Aa}}{\eta_{boiler}}$

10. $\eta_{oth} = (\eta_{th})_{Rankine} \times \eta_{turbine} \times \eta_{mech} \times \eta_{boiler} \times \eta_{gen}$

11. $\text{WR} = \dfrac{w_{net}}{w_t}$ 12. $s.s.c. = \dfrac{3600}{w_{net}}$

13. $\text{HR} = \dfrac{3600}{\eta_{th}} = \dfrac{3600\, q_A}{q_A - q_R}$ 14. $(\eta_{th})_{Reg.\ most\ ideal} = \dfrac{T_1 - T_2}{T_1}$

15. $(\eta_{th})_{Reg.} = \dfrac{(h_0 - h_{f1}) - (1 - \Sigma m)\,(h_3 - h_{f4})}{(h_0 - h_{f1})}$

16. $(\eta_{th})_{Reheat,\ ideal} = \dfrac{(h_1 - h_{2s}) + (h_a - h_{3s})}{(h_1 - h_{f3}) + (h_a - h_{2s})}$

17. $(\eta_{th})_{Reheat,\ a} = \dfrac{(h_0 - h_2) + (h_a - h_3)}{(h_1 - h_{f3}) + (h_a - h_2)}$

18. $(\eta_{th})_{Binary} = \dfrac{m_m\,(h_a - h_b) + (h_1 - h_2)}{m_m\,(h_a - h_{mc}) + (h_1 - h_s)}$

PROBLEMS

Things to Think About

14.1. List the share of electricity generated in India and USA by different types of power plants.

14.2. Is it true that thermal efficiency of a steam power plant is different in winter than in summer?

14.3 In how many ways solar energy may be used to generate electricity? Discuss the viability and limitations.

14.4. If air is used instead of water to condense steam in condenser, what will be the effect on thermal efficiency?

14.5. About 2/3 of energy input by heat transfer in steam generator is rejected to cooling water in condenser. Is this heat rejection an indicator of the inefficiency of power plant?

14.6. Suggest some ways to use the cooling water exiting the condenser of a large power plant.

14.7. What is the effect on river's ecology if water is used for condenser cooling under closed loop and open loop cooling system?

14.8. What environmental impacts might results from coal based and natural gas based steam power plants?

14.9. What are the future sources of energy for mankind?

14.10. Compare the per capita energy consumption of India with west and USA.

Short and Long Theoretical Questions

14.11. Give reasons why the Carnot cycle cannot be considered as the theoretical cycle for steam power plant even though its efficiency is the maximum.

14.12. List the names of the various power plants which use vapour as a working fluid.

14.13. Derive an expression for thermal efficiency of Rankine cycle.

14.14. List the requirements of an ideal working fluid for steam power plants.

14.15. Show the effects of initial pressure, initial temperature and back pressure of steam on Rankine cycle.

14.16. What are the methods employed to increase the efficiency of a Rankine power cycle?

14.17. Discuss the performance of steam turbine plant.

14.18. What is the thermodynamic principle of regenerative feed heating cycle? Is it possible theoretically to obtain the efficiency of mast ideal regenerative cycle equal to that of Carnot? Explain the method.

14.19. How many heaters are generally employed in actual practice for feed heating?

14.20. What do you mean by reheating of steam?

14.21. What are the causes of erosion and corrosion of turbine blades and how are they removed?

14.22. What do you mean by binary vapour cycle? What are the advantages and disadvantages of mercury as a working fluid.

14.23. Discuss the nuclear power plants.

14.24. Discuss the solar energy power plant and hydrogen fuelled combustion turbine.

14.25. Sketch the Rankine cycle on p–v and T–s diagrams and state in what respects it differs from the Carnot cycle working between the same temperature limits.

Numerical Questions

14.26. Steam is supplied to a turbine at pressure 11 bar with a dryness fraction of 80%. The steam is expanded isentropically to pressure 0.07 bar. Calculate the following : (a) heat supplied (b) total change in entropy (c) heat rejected, and (d) theoretical thermal efficiency. Recalculate the above for turbine efficiency 85%.

[*Ans.* (a) 2213.326 kJ/kg, (b) 5.12 kJ/kg K (c) 1589.779 kJ/kg (d) 28.3%]

14.27. Two steam turbines *A* and *B* operate under the following conditions: Initial pressure, 17 bar; exhaust pressure, 0.046 bar. In turbine A, the steam is initially dry and saturated, whereas in turbine *B*, it is initially superheated by 94.5°C. Calculate:

(a) Theoretical steam consumption, kg/kWh, (b) theoretical thermal efficiency of the two turbines, (c) percentage reduction in steam consumption due to superheating, and (d) percentage increase in thermal efficiency due to superheating.

[*Ans.* (a) 4.28, 3.73. (b) 31.9% (c) 12% (d) 3.18%]

14.28. Two steam turbines operate between the same temperature limits, i.e., 417°C and 26.5°C. In turbine *A*, the initial pressure is 13.5 bar and in *B*, 68 bar. Calculate the theoretical thermal efficiencies of turbine *A* and *B* and the value of dryness fraction after isentropic expansion.

[*Ans.* *A* = 34.56%, 0.86 and *B* = 40.82%, 0.758]

14.29. Steam at 13.5 bar and superheated to a temperature of 417°C expands isentropically to 0.07 bar in steam turbine. Find pressure at which the steam is just dry and saturated and the isentropic heat drop. Calculate the thermal efficiency and specific steam consumption if turbine efficiency is 85%.

14.30. Calculate the overall thermal efficiency and fuel cost per kWh generated in a steam power plant under the following conditions :

Initial pressure = 80 bar, Initial temperature = 500°C, Exhaust pressure = 0.05 bar, Turbine efficiency = 0.82, Boiler efficiency = 0.9, Alternator efficiency = 0.96, Mechanical efficiency = 95%, C.V. of coal = 29260 kJ/kg, Cost to coal = ₹ 800 per tonne.

Assume that the feed water temperature to the boiler is equal to the saturation temperature at 0.05 bar and the 6% of the steam generated is used to cover losses and auxiliary power.

14.31. In a Rankine cycle, steam is supplied at 80 bar and exhausts at a pressure of 0.04 bar. Calculate the overall thermal and specific fuel consumption if boiler, turbine, mechanical and generator efficiencies are 90%, 88%, 95% and 96%, respectively.

14.32. Compare the Rankine and Carnot engine consumptions per kWh if dry saturated steam at 14 bar is supplied and exhaust takes place at 0.07 bar. What will be the difference in efficiency in both cases? Calculate the heat removed in the condenser of Rankine cycle.

14.33. A power plant operating on the Rankine cycle has steam entering the turbine at 37 bar and 425°C. If the turbine output is equivalent to 10,000 kW, determine the efficiency of the cycle and the mass flow rate of steam for condenser pressure of (a) 0.07 bar and (b) 0.035 bar.

[*Ans.* (a) 36.6%. 31800 kg/h; (b) 39.1%, 21,500 kg/h]

14.34. A steam power cycle operates with a turbine inlet temperature of 576°C and an exhaust pressure of 0.07 bar. If the turbine efficiency is 85%, what is the maximum inlet turbine pressure which can be used if the exhaust moisture content is not to exceed 8%? [*Ans.* 70 bar]

14.35. A steam power plant operates between pressure of 42 bar and a condenser pressure of 0.035 bar. Calculate the cycle efficiency, the work ratio and the specific steam consumption.

(a) For a Camot cycle using wet steam.

(b) For a Rankine cycle with dry saturated steam at entry to the turbine.

[*Ans.* (a) 43.2%, 0.739, 4.9 kg/kWh (b) 36.8%, 0.996, 3.64 kg/kWh]

14.36. Steam is supplied, dry saturated at 40 bar to a steam turbine. The condenser pressure is 0.035 bar. Calculate for the Rankine cycle for the following:

(a) The work output neglecting pump work, (b) the feed pump work, (c) the heat rejected by condenser and the amount of cooling water required through the condenser if the temperature rise of water is 5.5°C, (d) the heat supplied, (e) the Rankine efficiency, and (f) the specific steam consumption.

[*Ans.* 982.4 kJ/kg; 1706.6kJ/kg: 74.1 kg; 2685 kJ/kg, 36.6% : 3.66kg/kWh]

14.37. Steam at 56 bar enters a 20,000 kW condensing turbine and exhausts at 0.07 bar. The steam loses 2219 kJ/kg to the condenser cooling water, which increases 8°C in temperature while passing through the condenser. The Rankine cycle efficiency is 32.2%, the turbine adiabatic efficiency is 82% and the actual steam rate is 6.8 kg/kWh. Assuming that the steam condensate leaves the condenser as a liquid subcooled by 2°C, find the temperature of steam entering the turbine.

14.38. A steam generating unit has a capacity of 228 kg of steam per hour. This unit, running on the Rankine cycle is comprised of only of a boiler and a superheater. The boiler receives 1172 × 10^2 kJ/h. Water enters the boiler at 30 bar and 95°C. After passing through the superheater, the steam goes to a turbine which produces net work. The steam is then condensed in the condenser, water 1.3 × 106 kg/h of cooling water is circulated with a temperature rise of 8°C. The steam leaves the condenser at 0.1 bar and 43°C. Calculate (a) the enthalpy of steam leaving the boiler in kJ/kg; (b) enthalpy of steam entering the condenser, (c) from Mollier diagram, the quality of steam entering the condenser, (d) using Mollier diagram enthalpy of steam entering the turbine assuming isentropic expansion from the turbine, and (e) power output of the ideal turbine.

14.39. A steam turbine power cycle operates with turbine inlet condition of 56 bar and 570°C. The condenser pressure is 0.07 bar. Calculate the change in the thermal efficiency if the turbine has an efficiency of only 80% instead of operating isentropically. What is the change in the moisture content leaving the turbine?

14.40. A steam turbine operates under the following conditions :

Initial pressure of steam = 26.5 bar, Initial temperature of steam = 377.5°C, Exhaust pressure = 0.03 bar.

The expansion process is isentropic. At a stage of turbine, where the pressure is 2.7 bar a connection is made for bleeding off the steam to a surface feed water heater in which the feed water is heated by bled steam to a temperature of 127°C. The condensed steam from the feed heater is cooled in a drain cooler before entering the feed heater. The cooled drain water combines with condensate in the well of condenser. Assuming no heat losses in the system, calculate the following : (a) the mass of steam used for feed heating per kg of steam entering the turbine, and (b) thermal efficiency of the cycle. [*Ans.* (a) 0.237 Kg, (b) 36.9%]

14.41. The steam supply to a turbine is at 14 bar with 100°C of superheat. The condenser pressure is 0.07 bar. Steam for feed heating is extracted at 2 bar. Calculate the optimum mass of bled steam per kg of steam supplied to the turbine and cycle efficiency.

14.42. A regenerative steam cycle operates between the limits of 12 bar, 535°C and 2.54 cm of Hg. If only one extraction from, the turbine occurs at 10 bar what percentage of total flow to the turbine is ideally extracted. [*Ans.* 23.6%]

14.43. A regenerative cycle operates between limits of 82 bar 427°C, and 2.54 cm Hg absolute. Assuming the efficiency of turbine and pump are 100%, determine extraction pressure if one heater is used. [*Ans.* 6.4 bar]

14.44. In a regenerative cycle, steam leaves the boiler and enters the turbine at 17.5 bar and 300°C. Steam condition at inlet to the condenser is 0.08 bar and 0.9 dry. There are two stages of steam extraction at 4 bar and 0.8 bar. The low pressure heater releases its drain to condenser and the condensate from die high pressure heater is pumped directly into the boiler feed line. Calculate (a) steam bled off from each heater, (b) thermal efficiency, (c) total power developed by the turbine rotor if steam flow rate is 16000 kg/h, (h) percentage gain in thermal efficiency due to bleed heating, (e) increase in steam consumption rate with regeneration, (f) percentage increase in boiler capacity for given output, and (g) percentage reduction in condenser capacity.

[*Ans.* (a) m_1 = 0.0872 kg, m_2 = 0.09417kg; (b) 26.1%: (c) 282 kW; (d) 6.35% (e) 0.57 kg/kWh; (f) 11%; (g) 9.1%]

14.45. An ideal regenerative cycle operates with throttle conditions at 28 bar 372°C, a single open feed water at 7.6 bar and a condenser pressure of 0.07 bar. Calculate (a) the fraction of steam bled from the turbine for feed water heating, (b) the work output of the turbine in kJ/kg, and (c) the thermal efficiency of the cycle.

14.46. An ideal reheat cycle which produces 125,000 kW of turbine power has the following given conditions; turbine inlet, 70 bar and 600°C, reheat 14 bar to 540°C, condenser pressure 0.07 bar. Compute the required mass flow rate of steam in kg/h and the cycle efficiency.

[*Ans.* 270,000 kg/h, 43%]

14.47. An ideal reheat-regenerative cycle is operated with steam entering the pressure turbine at 21 bar and 600°C and it is expanded to 7 bar. Part of steam is extracted to heat the feed water at 7 bar and the remaining steam is reheated to 600°C and returned to low pressure turbine where it is expanded to 0.035 bar. Determine the kg of steam which must be supplied per hour by the steam generator for an output of 30 MW by the turbine.

14.48. In a turbine, the steam supply is at 16 bar and superheated to 300°C, the exhaust is at 0.035 bar, The friction is 30% of the adiabatic heat drop and the line of condition on the total heat-entropy chart can be assumed as straight line. Find the thermal efficiency if there is one bleeding at 1 bar for feed heating.

14.49. Compare the efficiency of reheat and regenerative cycle which operates with steam supplied at 28 bar and 260°C and condenser pressure of 5 cm. Hg abs. The reheat pressure in the reheat cycle is 2.8 bar and the extraction pressure in the regenerative cycle is 2.8 bar. Assume the efficiency of turbine and pump to be 100% and neglect the condensate pump work.

[*Ans.* Reg 35.8%, Reheat 33.6%]

14.50. In a nuclear power plant, liquid sodium is heated to 370°C in the reheater. The liquid sodium then flows through beat exchanger where it heats and vaporizes water to saturated steam at a pressure of 56 bar. The steam generated flows to a turbine, which has one extraction point at

4.2 bar where steam flows to an open feed water heater. If the turbine efficiency is 75% and the condenser pressure 0.07 bar, calculate heat transfer rate in the reactor for a turbine output of 65,000 kW. [*Ans.* 17.6×10^9 kJ/h]

14.51. A supercritical power plant is being designed to operate at maximum pressure of 280 bar and a temperature of 540°C. The condenser pressure can be maintained at 0.07 bar. Assume turbine efficiency of 85%.

(a) At what point do you recommend reheating? (b) At what pressure should be steam extracted if three direct contact feed water heaters are used? (c) Find the cycle thermal efficiency.

14.52. Liquid sodium leaving nuclear reactor is at 1.4 bar and 370°C. It goes to a heat exchanger where it is cooled to 315°C before returning to the reactor. The sodium flow rate in this circuit (the primary circuit) is 35500 kg/h. In the heat exchanger, heat is transferred to liquid sodium in an intermediate circuit, which in turn, transfers heat to a boiler which produces dry saturated steam at 70 bar. The steam expands adiabatically through a turbine to 0.07 bar. The turbine efficiency is 80%. Sketch a flow diagram showing all three (primary, intermediate and steam) circuits. Assuming (i) negligible pressure drop due to friction, (ii) negligible work required by sodium pumps, (iii) negligible stray heat losses, and (iv) the specific heat of sodium to be constant at 1.3 kJ/kgK, calculate (a) the net amount of power available from the plant, (b) the irreversibility in kJ/h of the process of transferring energy from the primary sodium circuit through the intermediate circuit to the water.

14.53. A simple steam power cycle uses solar energy for the heat input. Water in the cycle enters the pump as saturated liquid at 49°C and it is pumped to 2.1 bar. It then evaporates in the boiler at this pressure and enters the turbine as a saturated vapour. At the turbine exhausts the conditions are 49°C. and 6% moisture. The flow rate is 136 kg/h and the pump is rated at 0.37 kW. Calculate (a) the net work output, (b) the energy conversion efficiency, (c) area of the solar collector needed if collector picks up 5233 kJ/cm^2-h.

14.54. A binary-vapour cycle operates under the following conditions :

Pressure in the mercury boiler = 17.76 bar

Pressure in the mercury condenser = 0.28 bar

The initial steam pressure to steam turbine = 50 bar

Steam condenser pressure = 0.072 bar

Initial condition of mercury and steam is dry and saturated.

Calculate (a) mass of mercury per kg of steam, (b) thermal efficiency of the cycle

p	h_{fm} (kJ/kg)	h_g (kJ/kg)	s_{fm}	s_g
0.28	40	–	0.09914	–
17.76	–	362.406	–	0.49533

[*Ans.* (a) 11.74 kg/kg of steam, (b) 0.554

14.55. Assuming realistic data, calculate thermal efficiency, ssc, sfc of (a) 120 MW steam power plant, (b) 500 MW of combined cycle.

14.56. Calculate the performance of nuclear, OTEC, combined, solar and geothermal power plant, by taking realistic data.

Software Based Questions

14.57. Develop a software to analyse the performance of steam power working on Rankine cycle. The performance parameters may include specific steam consumption, thermal efficiency (electrical), specific fuel consumption, specific power output for various steam inlet conditions, reheating and feed heating parameters. Print the results for 120 MW and 500 MW units.

14.58. Develop a software to analyse the performance of nuclear, ocean thermal energy converter, combined cycle, solar, geothermal and hydrogen fuelled combustion turbine system plants. Print the results for some such standard power units you know.

14.59. Steam enters the turbine of a simple vapour power plant with a pressure of 10 MPa and temperature T, and expands adiabatically to 6 kPa. The isentropic turbine efficiency is 86%. Saturated liquid comes out the condenser at 6 kPa and the isentropic pump efficiency is 84%. Develop a software to calculate the following :

(a) For T = 580°C, calculate the turbine exit quality and the cycle thermal efficiency.

(b) Plot the quantities of part (a) versus T ranging from 580°C to 700°C.

14.60. A steam power plant working on Rankine cycle receives steam at 14 MPa. 600°C. The steam is reheated between the turbine stages to 600°C. For condenser pressure of 6kPa, plot the following by developing software:

(a) cycle thermal efficiency versus reheat pressure for pressure ranging from 2 to 12 MPa, and

(b) cycle thermal efficiency versus number of feed heaters assuming 30°C temperature rise in each heater and taking bleed pressure at 2, 4, 6 and 8 MPa keeping reheat condition same.

Design and Open Ended Questions

14.61. A steam power plant developing 500 MW is to be designed using steam inlet pressure of 300 bar and temperature 650°C. Develop a cycle configuration and list the possible the pressure and temperature at inlet and exit of each element of the cycle. Justify your configuration.

14.62. As compared to earlier steam power plant which was working with inlet conditions of 12 bar and 200°C, the current steam power plant is operating at over 340 bar and 650°C. Prepare a report about the change in design over the years to allow very high pressure and temperature.

14.63. A large amount of coal are reserved in the mines all over the world. Coal based steam power plants pollute the environment. Prepare a report for the best use of coal in power plants with minimum air pollution with schematic diagram of coal processing which you are proposing'.

14.64. Among the options for meeting future power needs are the used of biomass as fuel, nuclear power, geothermal energy and sunlight. Prepare a report discussing advantages and disadvantages of these options including impact on environment and cost.

14.65. Hydrogen is supposed to be the clean energy. Worldwide research is going on to develop power plant using hydrogen as fuel. Prepare a report with layout of hydrogen combustion turbine power of plant starting from production of H_2 and O_2 including listing of various parameters.

14.66. In order to meet the peak demands of electricity, there are some ways to deal with. In one option, the air is compressed during off-peak hours and stored in underground reservoir which develops extra power while expanding through air turbine In other option, ice may be produced during off-peak hours which can be used as a low temperature reservoir for condenser heat rejection during peak demands period. Critically evaluate these options and write a report of your findings.

14.67. Write a report on the advantages of co-generation in terms of cost saving to meet power and heat load as compared to meeting these individually.

14.68. Prepare a conceptual configuration of vapour power plant which you have not heard off.

Multiple-Choice Answer Type Questions

14.69. The thermal efficiency of steam power plant lies in the range of (a) 20 to 30% (b) 30 to 40% (c) 40% to 50% (d) 50 to 60%

14.70. Thermal efficiency of Rankine cycle increases with

(a) increase of inlet steam pressure only

(b) increase of inlet steam temperature only

(c) decrease of condenser pressure only

(d) all of them

14.71. The specific steam consumption expressed in kg/kWh in power plant lies

(a) 0.5 to 2 (b) 2 to 4 (c) 4 to 6 (d) 6 to 8

14.72. The efficiency of superheat Rankine cycle is higher than that of a simple Rankine cycle because

(a) the enthalpy of main steam is higher for superheat cycle

(b) the mean temperature of heat addition is higher for superheat cycle

(c) the temperature in the condenser is high

(d) the quality of steam in the condenser is low

14.73. Which one of the following sequences indicates the correct order of flue gas flow in the steam power plant layout?

(a) Economiser, superheater and air preheater

(b) Air-preheater, economiser and superheater

(c) Economiser air-preheater and superheater

(d) Superheater, economiser and air-preheater

14.74. The temperature rise in feed heater lies in the range of

(a) 20 to 40°C (b) 40 to 60°C (c) 60 to 80°C (d) 80 to 100°C

14.75. The number of feed heater for a 500 MW steam power plant lies in the range of

(a) 2 to 4 (b) 4 to 6 (c) 6 to 8 (d) 8 to 10

14.76. Condenser pressure in steam power plant is limited by

(a) inlet steam pressure (b) inlet steam temperature,

(c) superheater temperature (d) condenser cooling water inlet temperature

14.77. In most ideal regenerative Rankine vapour power cycle steam enters in dry and saturated condition at 100 bar and 311.06°C and exhaust at 1 bar and 99.63°C. The most ideal thermal efficiency is

(a) 47.77% (b) 57.97% (c) 67.97% (d) 77.97%

Answer to Multiple-Choice Questions

14.69. (b)	14.70. (d)	14.71. (b)	14.72. (b)	14.73. (b)	14.74. (a)
14.75. (d)	14.76. (d)	14.77. (c)			

CHAPTER 15

Refrigeration Cycles

OVERVIEW

Refrigeration systems for preservation of foodstuffs and air conditioning play important role in our everyday lives. Heat pumps are used in heating the buildings and for producing industrial process heat. Over and above, there are many other commercial and industrial applications of refrigeration such as air separation to obtain liquid oxygen, liquid nitrogen, liquefaction of natural gas, production of ice, preservation of photographic films, etc.

The chapter deals with the thermodynamic analysis of some refrigeration cycles.

15.1 DEFINITION

The term refrigeration may be defined as the maintenance of a body or a system at a temperature lower than of its surroundings by continuous removal of heat from it. The system maintained

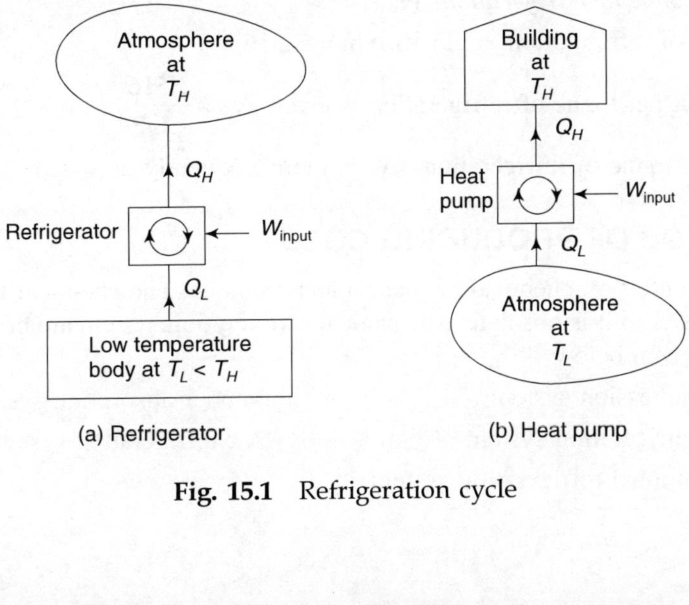

Fig. 15.1 Refrigeration cycle

at lower temperature is called *refrigerated system* while the machine used to maintain lower temperature is called *refrigerator or refrigerating system.*

From the above discussion, it follows that heat flows from lower temperature to higher temperature body but from second law of thermodynamics, it is only possible when mechanical/ electrical energy is supplied to the machine as energy input. This means that the refrigerator and heat pumps work on reverse heat engine cycle. The principles of operation of refrigerator and heat pumps are shown in Fig. 15.1.

15.2 PERFORMANCE PARAMETERS OF REFRIGERATOR AND HEAT PUMP

As discussed earlier, a parameter called ***coefficient of performance*** (***COP***) is used for the performance of a refrigerator. *It is defined as the ratio of net refrigerating effect to work input*. Thus,

$$\text{COP} = \frac{\text{Refrigerating effect}}{\text{Work input}} = \frac{\text{R.E.}}{W} \qquad ...(15.1)$$

The value of COP is greater than unity.

The efficiency of heat pump is called the *reciprocal of thermal efficiency* (*RTE*) or COP of heat pump. *It is defined as the ratio of the heat rejected at higher temperature to work input.*

$$(\text{COP})_{\text{hp}} = RTE = \frac{\text{Heat rejected at higher temperature}}{\text{Work input}} = 1 + (\text{COP})_{\text{Ref.}} \qquad ...(15.2)$$

15.3 UNIT OF REFRIGERATION

The refrigerating machines are usually rated in terms of ***cooling capacity rate***. The unit of refrigeration is commonly expressed in **tonne of refrigeration**. *The heat absorbed by a body or space to be cooled equivalent to the latent heat of fusion of I tonne of ice from 0°C in 24 hours is called 1 tonne of refrigeration*. Thus,

$$1 \text{ tonne of refrigeration} = 211 \text{ kJ/min} = 3.516 \text{ kW} \qquad ...(15.3)$$

$$\text{The power input per tonne of refrigeration is given by} = \frac{3.516}{\text{COP}} \qquad ...(15.4)$$

Now, instead of tonne of refrigeration, kW is more frequently used as refrigeration unit.

15.4 METHODS OF PRODUCING COLD

Cold is produced either by chemical or mechanical methods. The chemical method has not been commercialised so it is obsolete. Mechanical refrigeration system are broadly classified in five groups as given below.

(i) Vapour compression system
(ii) Vapour absorption system
(iii) Steam jet refrigeration system
(iv) Air refrigeration system
(v) Non-conventional refrigeration system.

The working fluid used in the refrigeration system is called ***refrigerant***. The refrigerants used may be NH_3, CO_2, R–11, R–12, R–22, R–110, R–111, R–112, R–114, R–134a, air, etc. The properties of refrigerants are given in tables and calculation of the properties are similar to that of steam.

15.5 CARNOT REFRIGERATION (REVERSE CARNOT) VAPOUR CYCLE

A reversed Carnot heat engine cycle is a refrigeration cycle (Fig. 15.1) in which heat is taken from a low temperature reservoir (T_L) and is rejected to a high temperature reservoir (T_H). Figure 15.2 shows the schematic of Carnot vapour refrigeration cycle and its representation on T–s diagram. The refrigerant is compressed adiabatically and reversibiy from state 1 to 2 in a compressor. The compressor receives work energy input W_c from an external source (electric motor or engine). At state 2, the refrigerant is just dry and saturated vapour at highest temperature, T_H. The refrigerant at state 2, enters a condenser where it is condensed reversibiy by rejecting heat Q_1 at T_H through process 2–3. At state 3, the refrigerant is in saturated liquid state. The refrigerant at state 3, enters an expander (engine or turbine) or expansion valve where it expands adiabatically and reversibly through process 3–4. At state 4, the refrigerant attains the lowest temperature, T_L. From state 4, the refrigerant passes through an evaporator where it absorbs heat Q_2 from the surrounding space at T_L reversibly through process 4–1. By this way, the cycle is completed and it repeats for the next cycle and so on. Assume 1 kg of refrigerant flowing through the system.

Thus, $Q_1 = T_H\,(s_{2s} - s_3)$ and $Q_2 = T_L\,(s_1 - s_{4s})$ but $s_1 - s_{4s} = s_{2s} - s_3$

$\therefore$ $$W_{net} = Q_1 - Q_2 = (T_H - T_L)\,(s_1 - s_{4s})$$

$$(\text{COP})_{\text{ref. Carnot}} = \frac{Q_2}{W_{net}} = \frac{T_L}{T_H - T_L} \qquad ...(15.5)$$

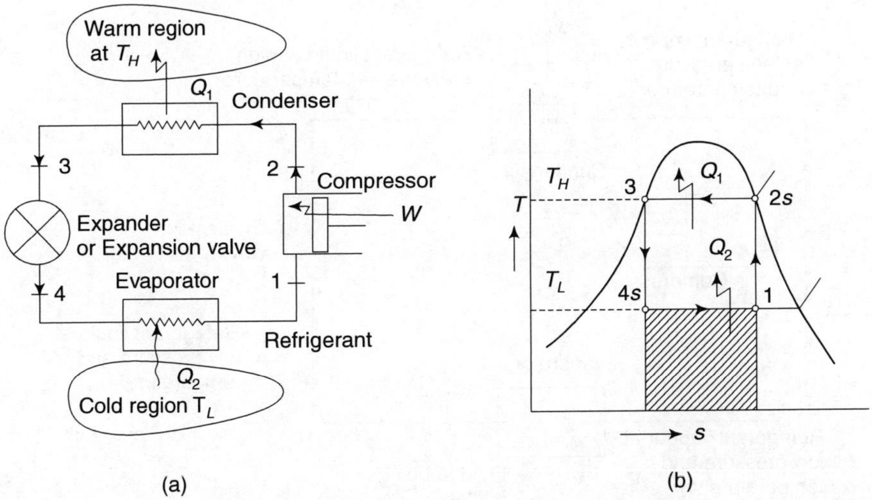

Fig. 15.2 Carnot vapour refrigeration cycle

$$(\text{COP})_{\text{hp, Carnot}} = \frac{Q_1}{W_{\text{net}}} = \frac{T_H}{T_H - T_L} \quad ...(15.6)$$

The COP given by above equations are the maximum values and no other cycle can yield COP more than Carnot cycle. The Carnot refrigeration cycle is an ideal cycle as its all processes are reversible. It is not a practical cycle but ideal cycle. From Eqs. (15.5), and (15.6) for the same T_L or T_H, *COP increases with the decrease in* $(T_H - T_L)$. ***It is therefore concluded that the difference between*** T_H ***and*** T_L ***should be as minimum as possible.***

15.6 VAPOUR COMPRESSION REFRIGERATION CYCLE

Figure 15.3 (a), shows a simple vapour compression cycle and the representations of processes on *p–v*, *T–s* and *h–s* diagrams are shown in Fig. 15.3 b, c and d. It consists of four main parts :

(i) compressor, (ii) condenser, (iii) expansion valve and (iv) evaporator.

The sequence of operation is as follows. Assume isentropic processes.

(i) *Compression process* (1–2*s*) Isentropic compression of refrigerant wet vapour takes place from lower pressure (p_1) and temperature (T_1) represented by state 1 to higher pressure (p_2) and temperature (T_2) represented by state 2. For dry and superheated vapour at the beginning the compression processes are represented by 1′– 2′*s* and 1″– 2″*s*. Generally, reciprocating compressors are used. In big plants, centrifugal compressor may be used.

(ii) *Condensation process* (2*s*–3) Vapour at high pressure and temperature is delivered to the condenser where its heat is rejected to the condenser at constant pressure. This is carried out in two stages. The first stage is through process 2″*s*– 2*s* or 2′*s*– 2*s* where the condenser absorbs the superheat from the high pressure and temperature of vapour through process 2*s*–3 where the vapour finally condensed to a liquid by rejecting latent heat at constant pressure. The cooling medium in the condenser may be air or water.

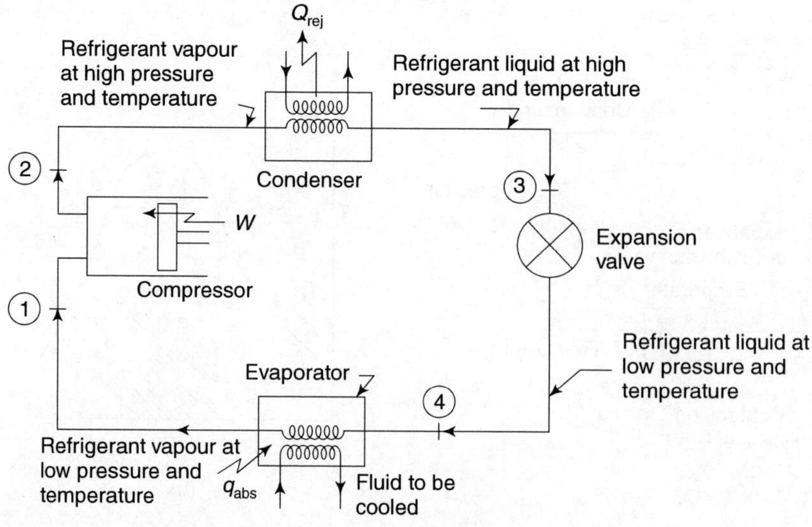

Fig. 15.3 (a) Vapour compression. Refrigeration cycle

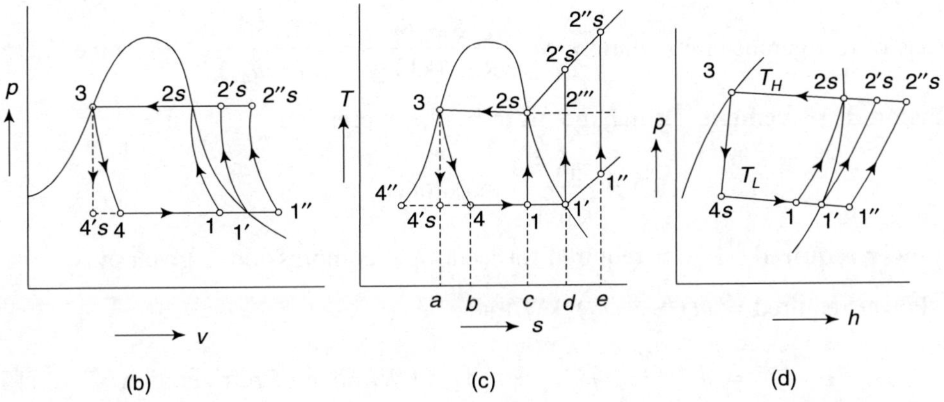

Fig. 15.3 (b), (c), (d), *p–v*, *T–s* and *p–h* representation of vapour compression cycle

(iii) *Throttling process* (3–4*s*) The high pressure saturated liquid refrigerant is expanded irreversibly through an expansion valve or capillary tube to a lower pressure and temperature while enthalpy remains constant.

(iv) *Evaporation process* (4*s*–1) The extremely wet vapour at low pressure and temperature when passes through evaporator coil which may be submerged in brine solution (in ice plant) or may be kept in **closed space** to be cooled (i.e., **refrigerator freezer**) gets evaporated by taking the heat from the liquid or space to be cooled. The final state of the refrigerant depends upon the quantity of heat absorbed and it may be wet, (state 1), dry (state 1′) and superheated (state 1″).

(a) COP The COP and other parameters of ideal vapour compression cycle are calculated as follows :

Assume 1 kg of refrigerant flowing through the cycle. Assume isentropic process.

Wet Compression (1–2s), Referring to *p-h* diagram, work input to the compressor = $h_{2s} - h_1$

Refrigerating effect (R.E.) = $q_{ab} = h_1 - h_{4s}$ ...(15.7)

$$\therefore \qquad \text{COP} = \frac{\text{Refrigerating effect}}{\text{Work input}} = \frac{q_{ab}}{w_{\text{input}}} = \frac{h_1 - h_{4s}}{h_{2s} - h_1} \qquad \text{...(15.8)}$$

Dry Compression. (1′–2′*s* or 1″–2″*s*). The C.O.P. of the dry compression is

$$\text{COP} = \frac{h_1' - h_{4s}}{h_{2s}' - h_1'} \qquad \text{...(15.9)}$$

and

$$\text{COP} = \frac{h_1'' - h_{4s}}{h_{2s}'' - h_1''} \qquad \text{...(15.10)}$$

(b) Mass of refrigerant in circulation

1 ton of refrigeration = 3.516 kW

$\therefore$ Mass of refrigerant in circulation, $\dot{m} = \dfrac{3.516}{\text{R.E. (kJ/kg)}} = \dfrac{3.156}{(h_1 - h_{4s})}$; kg/s tonne ...(15.11)

(c) Piston displacement It may be given by the expression

$$\text{PD} = \frac{v_1 \cdot \dot{m}}{\eta_{vol}} \text{; m}^3\text{/s tonne} \quad ...(15.12)$$

(d) Power required Power required for isentropic compression is given by

$$\text{Power required} = \dot{m}\,(h_{2s} - h_1) \text{ kW/tonne}$$

$$= \dot{m} \cdot \frac{\gamma}{\gamma - 1}(p_2\, v_2 - p_1\, v_1)\text{; kW/tonne of refrigeration} \quad ...(15.13)$$

(e) Heat rejected to condenser The heat rejected to condenser is given by

$$q_{rej} = (h_{2s} - h_{f3})\text{; kJ/kg} \quad ...(15.14)$$

$$\text{Total heat rejected, } Q_{rej} = \dot{m}\,(h_{2s} - h_3)\text{; kW/tonne} \quad ...(15.15)$$

Example 15.1 Calculate the COP of a simple saturated vapour compression refrigeration system utilising Fr–12 for the following data :

Evaporator temperature = – 15°C and Condenser temperature = + 30°C

Also calculate the mass of refrigerant per tonne and power input per tonne of refrigeration.

Take c_p = 0.628 kJ/kg for superheated refrigerant.

Solution Refer to Fig. 15.4. State 1 is taken at saturated vapour line. From refrigeration table,

p_1 = 1.82482 bar, h_1 = 344.927 kJ/kg, v_1 = 0.0911 m^3/kg, s_1 = 1.56323 kJ/kg K

From psychrometric table, corresponding to 30°C,

$$p_2 = 7.44572 \text{ bar}, \quad h'_{2s} = 363.575 \text{ kJ/kg}, \; T'_2 = 303 \text{ K.}$$

$$s'_{2s} = 0.54340 \text{ kJ/kg K}, \quad v'_{2s} = 0.0235 \text{ m}^3\text{/kg}, \; h_{f3} = 228.538 \text{ kJ/kg}$$

Assuming that the compression process is isentropic

$$s_1 = s_{2s} = s'_{2s} + c_p \ln \frac{T_{2s}}{T_1}$$

or

$$T_{2s} = 311.16 \text{ K} \quad \text{or} \quad t_{2s} = 38.16°\text{C}$$

The enthalpy at state point 2*s* is

$$h_{2s} = h'_{2s} + c_p\,(t_{2s} - t'_{2s}) = 363.575 + 0.746 \times (38.16 - 30) = 369.66 \text{ kJ/kg}$$

The work input to the compressor = $h_{2s} - h_1$ = 369.66 – 344.927 = 24.733 kJ/kg

Heat removed in the condenser

$$= q_{2-3} = q_{rej} = (h_{2s} - h'_{2s}) + (h'_{2s} - h_{f3})$$

$$= c_p\,(T_{2s} - T_2') + (h_2' - h_{f3}) = 0.746\,(38.16 - 30) + (363.575 - 228.536)$$
$$= 141.12 \text{ kJ/kg}$$

During throttling enthalpy remains constant,

So, $\quad h_{f3} = h_{4s} = 228.538$ kJ/kg

The refrigerating effect (RE) = $h_1 - h_{4s}$ = 344.927 – 228.538 = 116.389 kJ/kg

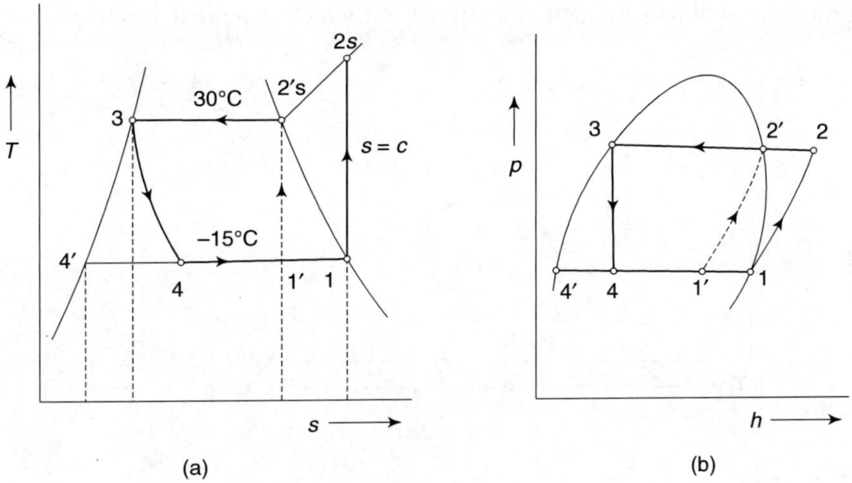

Fig. 15.4 Example 15.1

Let $\dot{m}$ kg/s be the refrigerant to be circulated per tonne of refrigeration, then

$$\dot{m} = \frac{3.516}{\text{Refrigerating effet (kJ/kg)}} = \frac{3.156}{h_1 - h_{4s}} = \frac{3.516}{116.389}$$

$$= 0.0302 \text{ kg/s tonne}$$ **Ans.**

Hence, $$\text{C.O.P} = \frac{\text{Refrigerating capacity}}{\text{Work input}} = \frac{116.389}{24.733} = 4.705$$ **Ans.**

Power input = 0.0302 × 24.733 = 0.746 kW/tonne **Ans.**

15.7 ACTUAL VAPOUR COMPRESSION REFRIGERATION CYCLE

The actual vapour compression refrigeration cycle differs considerably from the theoretical cycle in many respects. Figure 15.5 shows the actual vapour compression cycle on *T–s* diagram.

(i) *Process 1–2–3–4–5–6* This represents the compression process in compressor. In this, we find several departures from the theoretical cycle, namely:

(a) On account of throttling in the suction and discharge valves of the compressor, the actual suction pressure (p_s) in the compressor is lower than the evaporator pressure (p_{eva}) by

Δp_s and the actual discharge pressure p_d will be higher than the condenser pressure (p_{cond}) by Δp_d. Thus, processes 1–2 and 5–6 indicate throttling in the suction and discharge valves, respectively.

(b) At state point 2, the vapour enters the compressor cylinder and heats up on coming into contact with hot cylinder walls, thus its temperature increases. This heating process is represented by 2–3. Likewise, the cooling effect at the discharge is represented by process 4–5.

(c) Process 3–4 shows the actual compression which is neither isentropic nor polytropic.

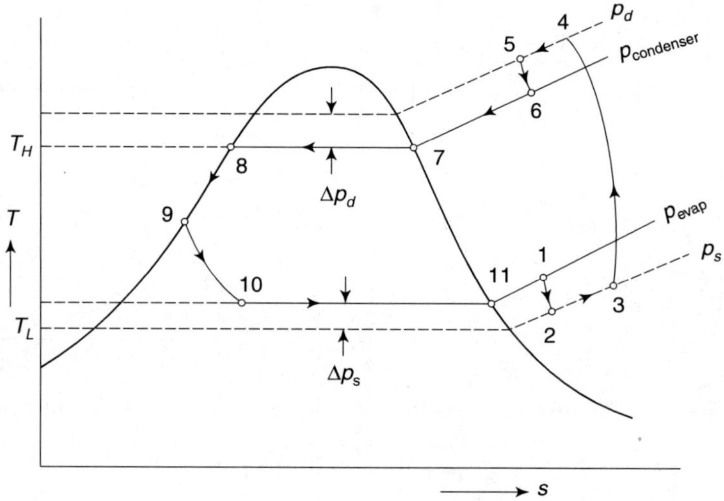

Fig. 15.5 *T–s* diagram of actual vapour compression refrigeration cycle

(ii) *Process 6–7–8–9* During this process, the vapour is condensed in the condenser at pressure p_{cond}. Process 6–7 acts as the remover of superheat 7–8 as the remover of latent heat and 8–9 as the subcooler of vapour.

(iii) *Process 9–10* In this process the sub-cooled vapour is throttled to evaporator pressure.

(iv) *Process 10–11* At the state point 10, the expanded (or throttled) refrigerant enters the evaporator at temperature T_L and pressure p_{evap} and the refrigerant while passing through the evaporator coil absorbs heat, so that the liquid refrigerant becomes superheated at state 1. At state point 1, the refrigerant enters the compressor. The superheating may be due to large scale removal of heat from the evaporator.

15.8 CASCADE AND MULTI-STAGE VAPOUR COMPRESSION REFRIGERATION CYCLES

Various modifications in basic vapour compression refrigeration cycles are used to improve the performance for certain needs. Two variations are presented in this section, namely, cascade vapour compression refrigeration system and two-stage compression with flash intercooling.

15.8.1 Cascade Cycle

Figure 15.6 shows a schematic of cascade vapour compression refrigeration cycle. In this case, two vapour compression refrigeration cycles, designated as *A* and *B*, are arranged in series with a counter flow heat exchanger linking them. In the intermediate heat exchanger, the energy rejected during the condensation of the refrigerant in the lower temperature cycle *A* is used to evaporate the refrigerant in the higher temperature cycle *B*. The desired refrigeration effect is achieved in the low temperature evaporator. The heat rejection from the overall cycle occurs in the high temperature condenser. The coefficient of performance is the ratio of refrigeration to the total work input as given below

$$\text{COP} = \frac{Q_{ab}}{W_A + W_B} \qquad \text{...(15.16)}$$

It is worth to note that the mass flow rates in two cycles (*A* and *B*) normally would be different and they are related by mass and energy balance on the interconnecting counterflow heat exchanger serving as the condenser for cycle *A* and the evaporator for cycle *B*. The cascade system may have more than two cycles.

The main advantage of the cascade system is that the refrigerants in the two or more stages can be selected to have reasonable evaporator and condenser pressure in the two or more temperature ranges.

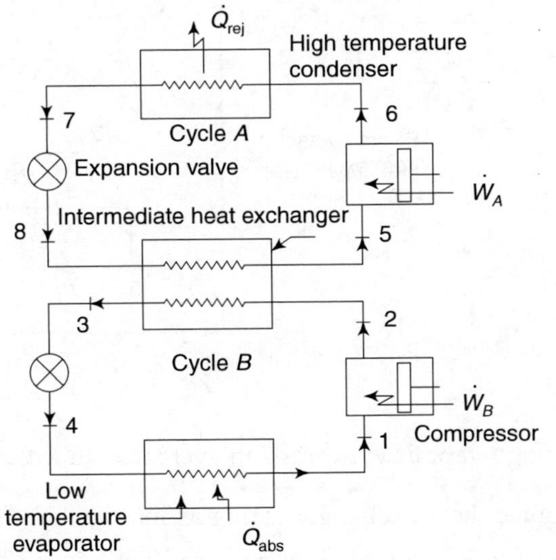

Fig. 15.6 Cascade vapour compression refrigeration cycle

15.8.2 Multi-stage Vapour Compression Cycle with Intercooling

The advantage of multistaging with intercooling lies in saving the work input. Figure 15.7 shows the schematic and representation of process on *p–h* diagram of two-stage vapour compression with flash intercooling. Intercooling is achieved by means of a direct contact heat exchanger. Relatively low temperature saturated vapour enters the heat exchanger at state 9, where it mixes with high temperature refrigerant leaving the first compression stage at state 2. A single mixed stream comes out from the heat exchanger at an intermediate stage at state 3 ($T_3 < T_2$) and is compressed in the second compressor stage to the condenser pressure at state 4. Due to intercooling, the required compressor work is less as single stage compression could have taken the path 1–2–*a* instead of 1–2 and 3–4. Further, since the refrigerant temperature entering the condenser at state 4 is lower than for a single stage compression (process 1-2-a), in which the refrigerant would enter the condenser at state *a*, the external irreversibility associated with heat transfer in the condenser also gets reduced.

The flash chamber used in the system plays a central role. It is a liquid-vapour separator. After the expansion of refrigerant in expansion valve, the state is represented at 6 which is liquid-vapour mixture. Let the quality of vapour at state 6 be x then $(1 - x)$ quantity of refrigerant will be leaving the flash chamber for expansion in second expansion valve while saturated vapour of x quantity will be entering to intercooler (direct contact heat exchanger). The mass and energy balance at flash chamber and direct contact heat exchanger (intercooler) will yield the properties at state 3.

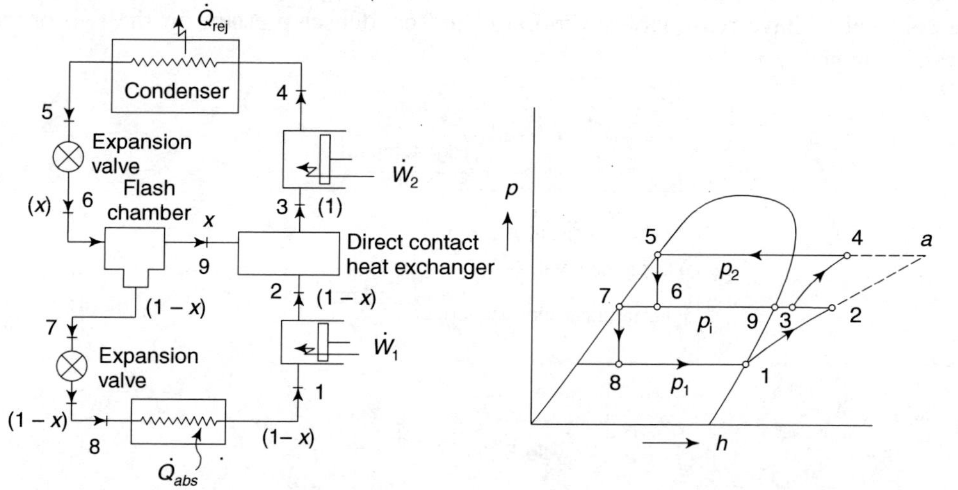

Fig. 15.7 Two-stage vapour compression system with a flash intercooler

In the case of direct contact heat exchanger acting as intercooler, the intercooler pressure, $p_i = \sqrt{p_1 \, p_2}$ and the refrigerant leaving the intercooler is just dry and saturated, (see problem 15.6).

15.9 REFRIGERANTS

The refrigerants used nowadays are a group of halogenated hydrocarbons, marketed under the various proprietary names of freon, genetron, arcton, isotron, frigen, and mafron. These refrigerants are either methane based or ethane based where the hydrogen atoms are replaced by chlorine or fluorine atoms. Up to 1990, the most common class of refrigerant used in vapour compression refrigeration systems was the chlorine containing CFC_5 (chlorofiuorocarbons). R–12 ($CC1_2F_2$) is one of these. Owing to concerns about the effect of chlorine in refrigerants on the earth's protective ozone layer, international agreements have been implemented to phase out the use of HFCs. *Classes of refrigerants containing various amounts of hydrogen in place of chlorine have been developed which has very less effect in depleting atmospheric ozone layer as compared to R–12. One such refrigerant is R–134a* (CF_3CH_2F) *which is supposed to be most environmentally acceptable substitute of R–12.*

Refrigerant 22 ($CHClF_2$) falling under $HCFC_5$ class contains some hydrogen in place of chlorine atoms. Although, R–22 is widely used today but efforts are going on to replace this. NH_3 is widely used in vapour absorption system and previously it was used in vapour compression machine. Since it does not contain chlorine, so it is gaining attention in place of CFC. Hydrocarbons such as CH_4 and C_3H_8 are also under investigation for use as refrigerants.

Properties of refrigerants are prepared in the form of tables and charts and easily available in the market in the form of booklet.

15.10 VAPOUR ABSORPTION CYCLE

In the vapour compression refrigeration system, the purpose of the refrigerant compressor is to withdraw the vapour from the evaporator, then raise its temperature and pressure level so that it can reject heat in the condenser, be ready to expand to the expansion valve again and the heat absorbed in the evaporator. *The major drawback of the vapour compression system is that it requires a large amount of high grade power as input to the compressor.* This drawback is avoided in the case of vapour absorption system which utilises the principle of drawing the refrigerant vapour from the evaporator by absorbing it into a liquid having an high affinity for the refrigerant. Heat (a low grade energy) is supplied to the solution which raises its temperature level and drives off the refrigerant vapour so that it can pass on to the condenser to be liquefied. The liquid refrigerant then flows into the evaporator at a lower pressure and in this way completes the cycle. The absorption process is theoretically governed by the *Raoult's law. This law states that at a given temperature, the ratio of the partial pressure of a volatile component in a solution to the vapour pressure of the pure component at the same temperature is equal to its mole fraction in the solution.*

The absorption system employed in large industrial undertakings or homes consists of a pump to handle the absorbed refrigerant, but domestic refrigerators and small unit for air-conditioning have no such moving parts.

15.11 SIMPLE VAPOUR ABSORPTION CYCLE

Figure 15.8 shows an arrangement of a simple vapour absorption refrigeration system. It consists of an evaporator, an absorber, a generator, a condenser, an expansion valve, a reducing valve and a pump. The pump and expansion valve are eliminated in a modified absorption cycle by using a gaseous third fluid. In this case, the compressor is replaced by an absorber generator, a pressure reducing valve and a pump as shown in Fig. 15.8 enclosed by dotted line. *The most commonly used fluids in the absorption system are water as absorbent and ammonia as refrigerant.* The low ammonia pressure vapour from the evaporator is allowed to be mixed and absorbed in the absorber. The solution in the absorber is called aqua ammonia, and a strong solution contains as much of ammonia as possible, a weak solution contains considerable less ammonia. The heat of absorption generated in the process is rejected from the absorber to the circulating cold water in a heat exchanger employed in the absorber. The strong aqua-ammonia solution from the absorber is pumped up to the condenser pressure and led to the generator. Heat is added to the generator directly from the heat source like fuel burner, steam, etc. And the work input is very small as the fluid handled is liquid with a very low specific volume. The boiling point of the refrigerant NH_3 is lower than that of the absorbing liquid H_2O, hence the vapours leaving the generator are predominantly those of refrigerant. The high pressure high temperature vapour from the generator is fed to the condenser where it is condensed and the liquid refrigerant passing through the liquid receiver and expansion valve enters the evaporator.

The weak ammonia solution left in the generator due to separation of refrigerant vapour, is drained back to the absorber through the pressure reducing valve. The cycle is repeated in this way.

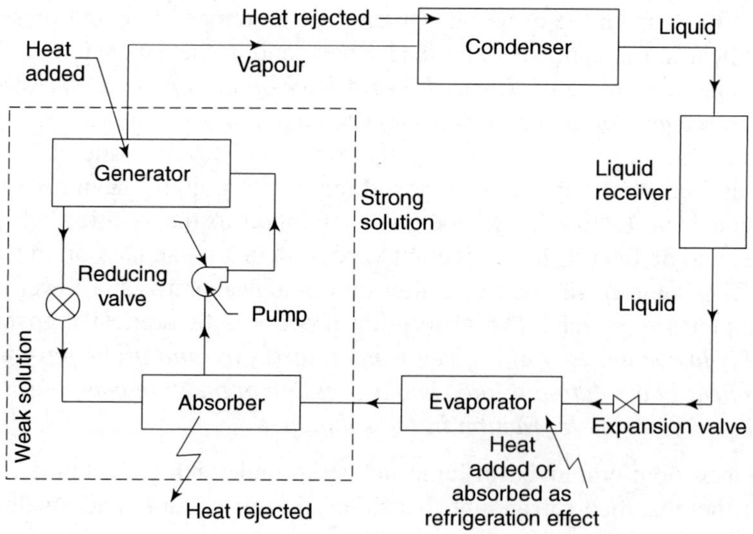

Fig. 15.8 Simple absorption system

15.12 PRACTICAL VAPOUR ABSORPTION CYCLE

The simple vapour absorption system described earlier is not very economical and suffers from many disadvantages. To make the system both practical and economical, many auxiliary equipment are incorporated. In order to obtain the most desirable balance between the initial cost and the operating cost, the number of auxiliaries to be used varies with the application. The basic auxiliaries, however, include an analyser, a rectifier and two heat exchangers. Figure 15.9 shows an arrangement of the practical vapour absorption system. The *basic units* and important auxiliaries are described below.

Generator In the generator, energy is added to the system from an external source such as gas burner, electric heater or steam passing through the coil, etc. The generator unit may consists of a horizontal shell partly filled with strong aqua automatically maintained at a constant level and covering the submerged steam or hot gas tubes. Connections are provided for the strong aqua from the absorber and, the weak aqua to leave.

Analyzer The ammonia vapour leaving the generator may contain moisture and this has to be removed before it is passed on to the condenser and then to the expansion valve otherwise, there is always a chance that the water vapour may freeze in the valve while passing through it and choking it.

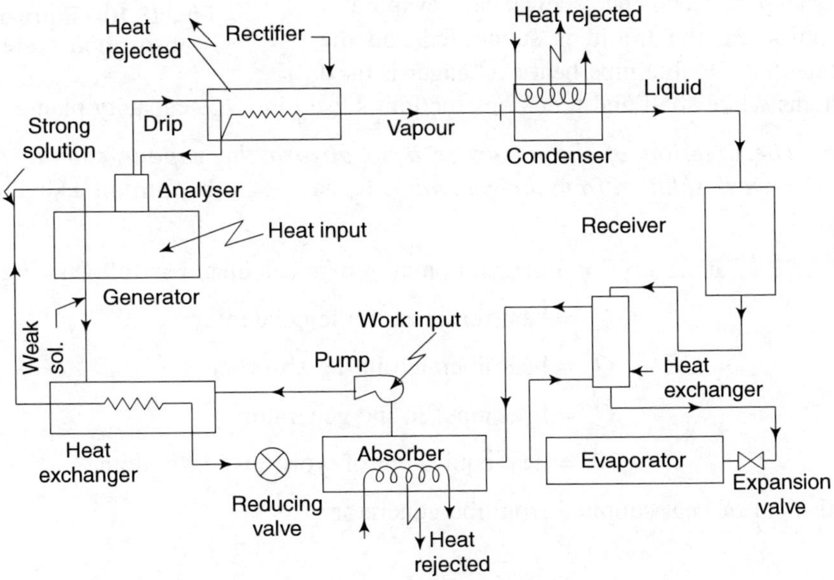

Fig. 15.9 Practical vapour absorption system

To avoid these drawbacks the analyser is used. *The function of the analyser is to remove the moisture as far as possible*. Sometimes, this is also referred to as an open type cooler and may be built as an integral part of the generator or made as a separate piece of equipment. Both the strong aqua from the absorber and the aqua from the rectifier are fed at the top of the analyser and flow downward, usually over trays and into the generator. This exposes a considerable liquid surface area to the vapour rising from the generator and the vapour is thus cooled while most of the water vapour condenses so that only ammonia vapour leaves the top of the analyser. Since the aqua is headed by the vapour less external heat in required in the generator.

Rectifier *The function of rectifier is to further cool the vapour having the analyser so that almost all water vapour is condensed and only dehydrated ammonia gas is allowed to pass to the condenser.* This is a closed type vapour cooler and is generally water cooled and may be of double pipe, shell and coil or shell and tube type. The condensed aqua is drained back (drip) to the analyser and generator.

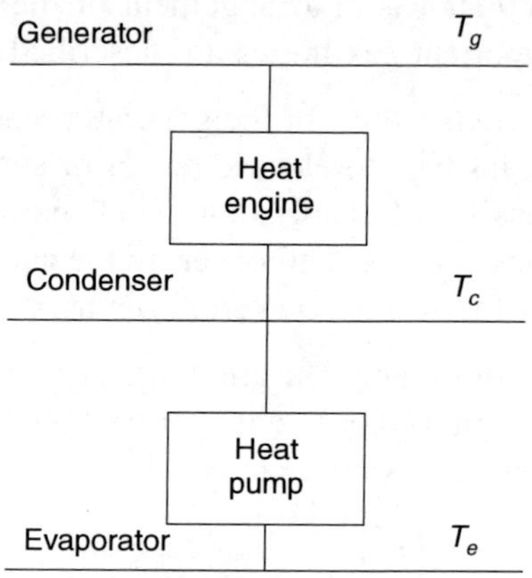

Fig. 15.10 Equivalent absorption system

Aqua Heat Exchanger *The purpose of the heat exchangers is to exchange heat from the higher temperature fluid to the lower temperature fluid so that one is cooled and other is heated.* The heat exchanger located between the absorber and the generator serves to cool the weak aqua and heat the strong aqua. The heat exchanger located between the receiver and evaporator serves two purposes, the liquid is subcooled and the vapour is heated up. Double pipe heat exchanger is used in small systems while shell and tube construction is used in large capacity plants.

Absorber *The function of the absorber is to absorb the vapour coming from the evapourator and weak solution from the generator.* It rnay be in the form of a shell and tube condenser.

The ideal COP of an absorption refrigeration system is calculated as follows (Fig. 15.10).

Let Q_c = heat removed in the condenser

Q_a = heat liberated in the absorber

Q_g = heat input in the generator

W = heat equivalent of work given to pump

Work equivalent of heat supplied from the generator

$$= \frac{T_g - T_c}{T_g} \times Q_g \qquad \text{...(15.17)}$$

Work needed by the heat pump in pumping Q_e units of heat from temperature T_e to T_c.

$$= \frac{T_c - T_e}{T_c} \times Q_e \qquad \text{...(15.18)}$$

If we assume that refrigerator is run by the lower output of the engine, then

$$\frac{T_g - T_c}{T_g} \times Q_g = \frac{T_c - T_e}{T_e} \times Q_e$$

$$\therefore \qquad \text{Ideal C.O.P.} = \frac{Q_e}{Q_g} = \left(\frac{T_g - T_c}{T_c - T_e}\right) \cdot \frac{T_e}{T_g} \qquad ...(15.19)$$

15.13 HEAT PUMP SYSTEMS

The purpose of heat pump is to maintain the temperature within a dwelling or other building above the temperature of surroundings or to provide a heat transfer for certain industrial processes at an elevated temperature. The features of heat pump system have may common similarities with the refrigeration system and it may be either vapour compression or vapour absorption type. Vapour compression heat pump is commonly used for space heating application. The condenser heat rejection is given to the space which is to be heated and the evaporator absorbs heat from the surroundings, i.e., outside air. Absorption heat pump is used for industrial application and also for space heating.

15.14 GAS REFRIGERATION CYCLE

Vapour refrigeration systems involve changes in phase. But in gas refrigeration, the working fluids remains gas throughout. Gas refrigeration systems are used in aircraft cabin cooling and to achieve very low temperature for the liquefaction of air and other gases The Brayton refrigeration gas cycle is one of the important type.

The Brayton refrigeration gas cycle is shown in Fig. 15.11.

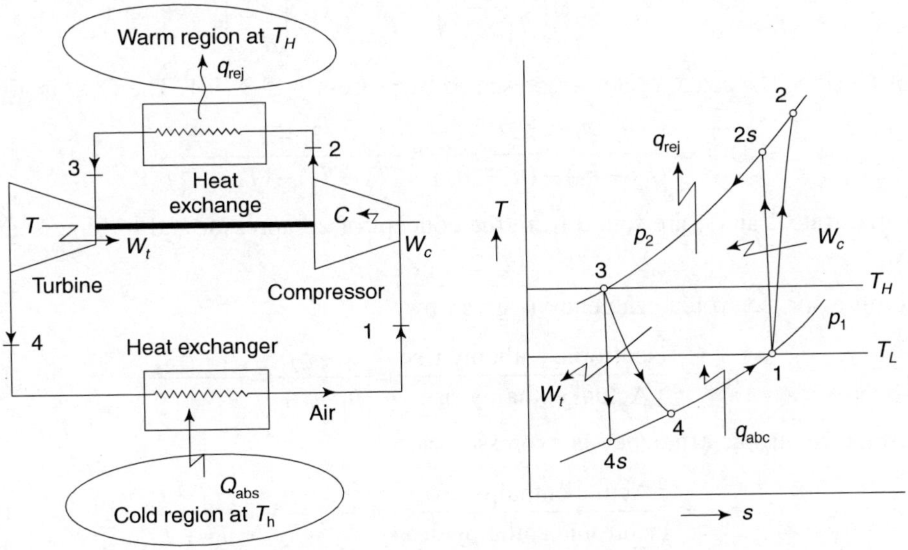

Fig. 15.11 Brayton refrigeration cycle

Ideal Cycle The ideal cycle is represented by 1–2*s*–3–4*s*–1. Process 1–2*s* and 1–2 are the isentropic and actual compression process of air in compressor from pressure p_1 to p_2. Process 2*s*–3 and process 2–3 show the ideal and actual heat rejection process in heat exchanger at pressure p_2. Processes 3–4*s* and 3–4 are the isentropic and actual expansion of air in the turbine. Processes 4*s*–1 and 4–1 show the ideal and actual heat absorption processes in heat exchanger operating at constant pressure p_1. The coefficient of performance is given by

$$\text{COP} = \frac{\text{Refrigerating effect}}{\text{Work input}} = \frac{q_{abs}}{w_{net}}$$

Assume 1 kg of working fluid flowing through the cycle.

$$q_{abs} = h_1 - h_{4s} = c_p (T_1 - T_{4s})$$

$$w_{net} = w_c - w_t = (h_{2s} - h_1) - (h_3 - h_{4s}) = c_p [(T_{2s} - T_1) - (T_3 - T_{4s})]$$

$$\therefore \quad \text{COP} = \frac{T_1 - T_{4s}}{(T_{2s} - T_1) - (T_3 - T_{4s})} = \frac{T_1 - T_{4s}}{T_1\left(\frac{T_{2s}}{T_1} - 1\right) - T_{4s}\left(\frac{T_3}{T_{4s}} - 1\right)} \quad ...(15.20)$$

For isentropic compression and expansion process, we have

$$\frac{T_{2s}}{T_1} = \left(\frac{p_2}{p_1}\right)^{(\gamma-1)/\gamma} = \frac{T_{3s}}{T_4}$$

$$\therefore \quad \text{COP} = \frac{T_1 - T_{4s}}{(T_1 - T_{4s})\left(\frac{T_3}{T_{4s}} - 1\right)} = \frac{1}{\frac{T_3}{T_{4s}} - 1} = \frac{1}{\left(\frac{p_2}{p_1}\right)^{(\gamma-1)/\gamma} - 1} \quad ...(15.21)$$

Actual Cycle The actual cycle is represented by process 1–2–3–4–1. The COP is given by

$$\text{COP} = \frac{h_1 - h_4}{(h_2 - h_1) - (h_3 - h_4)} = \frac{T_1 - T_4}{(T_2 - T_1) - (T_3 - T_4)} \quad ...(15.22)$$

The actual state 2 and 4 are found from the concept of compressor and turbine isentropic efficiency.

The compressor isentropic efficiency is given by

$$\eta_c = \frac{\text{Isentropic enthaply rise}}{\text{Actual enthalpy rise}} = \frac{h_1 - h_{2s}}{h_1 - h_2} = \frac{T_1 - T_{2s}}{T_1 - T_2} \quad ...(15.23)$$

The turbine isentropic efficiency is expressed as

$$\eta_t = \frac{\text{Actual enthalpy drop}}{\text{Isentropic enthalpy drop}} = \frac{h_3 - h_4}{h_3 - h_{4s}} = \frac{T_3 - T_4}{T_3 - T_{4s}} \quad ...(15.24)$$

The work input to the compressor is given by

$$w_{net} = w_c - w_t = (h_2 - h_1) - (h_3 - h_4) = c_p \left[\frac{(T_{2s} - T_1)}{\eta_c} - (T_3 - T_{4s})\,\eta_t\right] \quad ...(15.25)$$

The reverse Brayton and Bell–Coleman cycles were previously used on ship carrying frozen meat because the refrigerant air is non-toxic, non-poisonous and inexpensive. *The system is still used in air conditioning of aircraft cabins. It offers low COP.* In aircraft air conditioning, a required amount of air is bled from compressor exit and is fed to the turbine for expansion. The exhaust air from turbines which is at required, pressure and temperature is supplied to the cabin. There are various configurations of aircraft refrigeration systems.

15.15 LIQUEFACTION OF GASES

As mentioned earlier, gas refrigeration system is used in the liquefaction of gases. A gas may be cooled by two methods. In one method, it may expand in an expander (turbine) up to the desired temperature. *Sometimes it is known as external work method.* In the second method, the compressed gas undergoes *Joule–Kelvin expansion* through a throttle valve called internal work method. The main drawback of Joule–Kelvin method is that it can decrease temperature of fluid only when the temperature before throttling is below the maximum, inversion temperature. The advantage of former method is that it always brings about a temperature decrease. The following are the methods of air liquefaction.

15.15.1 Linde–Hompson System for Liquefaction of Air

This method utilises the technique of Joule–Kelvin effect for cooling and ultimately, liquefying the air. Figure 15.12 shows Linde–Hompson system with its representation of process on *T–s* diagram. The system consists of a two-stage compression with intercooling and aftercooling and then the air is made to pass through the expansion valve and finally to separator where liquefied air is separated from the unliquefied air. Process 1–2 shows the compression process with intercooler and aftercooler and it is nearly an isothermal process. Process 2–3 shows further cooling of air in heat exchanger by the unliquefied air coming out from separator. Process 3–4 shows the expansion in valve. Process 4–5 shows separation of liquefied air and unliquefied air. Processes 4–6 and 6–7 show separation of liquid and air and heating of separated air. Process 7–1 shows the mixing of separated air coming out from heat exchanger and make-up fresh atmospheric air for induction into compressor. *The pressure at state 2 is about 200 atm or above.*

The yield of the system (y) is defined as the ratio of mass of liquid produced to the mass of gas (air) compressed. The energy required per unit mass of liquid produced is known as the ***specific work consumption***

$$w_c = \dot{W}_c / \dot{m}_{La}$$

The mass energy balance for the control volume shown in Fig. 15.12 will give the theoretical yield by assuming perfect insulation. Let $\dot{m}_{La}$ be the rate at which liquid air is produced which compensated to the system by adding make-up water. Let $\dot{m}_a$ be the rate at which air is compressed and then expanded. Thus,

$$y = \frac{\dot{m}_{La}}{\dot{m}_a} \qquad \text{...(15.26)}$$

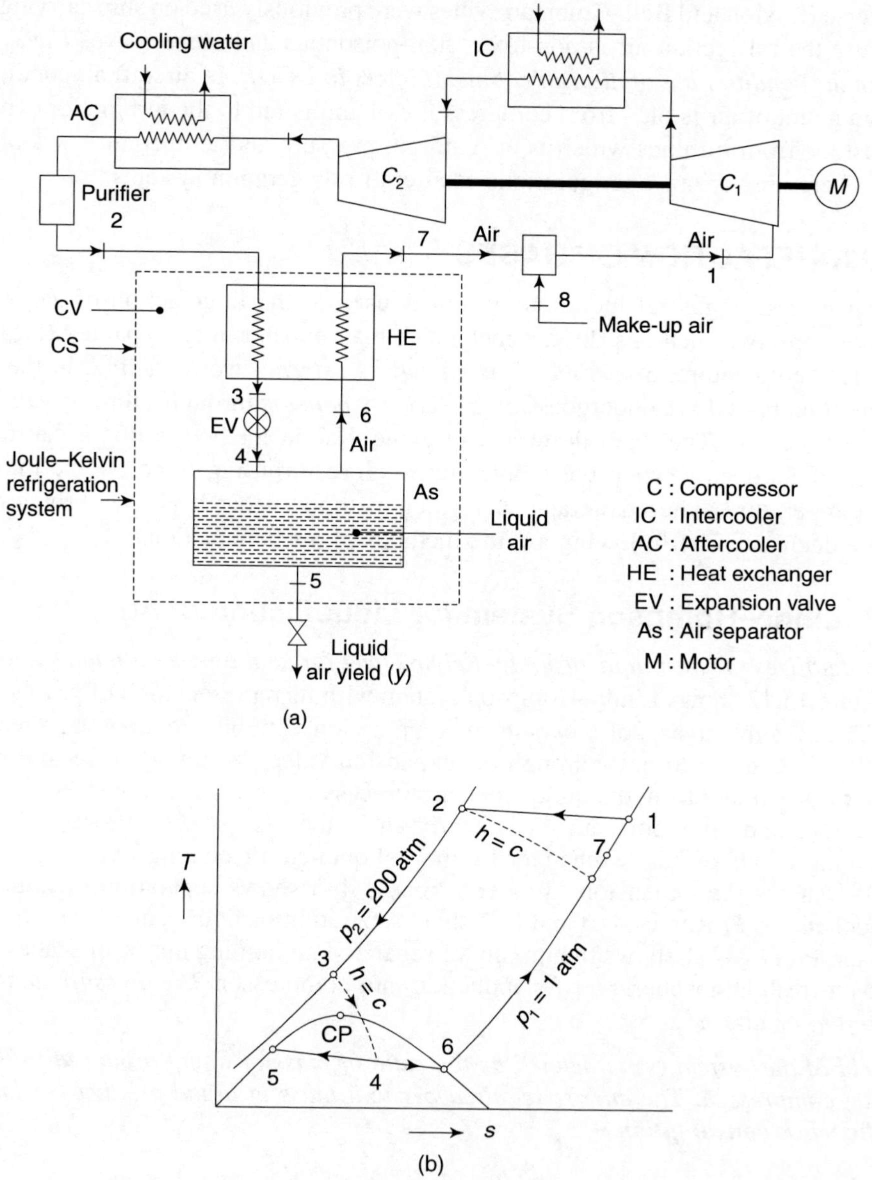

Fig. 15.12 Linde–Hampson cycle for air liquefaction using Joule–Kelvin technique

The energy balance over control volume gives

$$\dot{m}_a h_2 - \dot{m}_{la} h_5 - (\dot{m}_a - \dot{m}_{la}) h_7 = 0$$

or

$$\dot{m}_a (h_2 - h_7) = \dot{m}_{la} (h_5 - h_7)$$

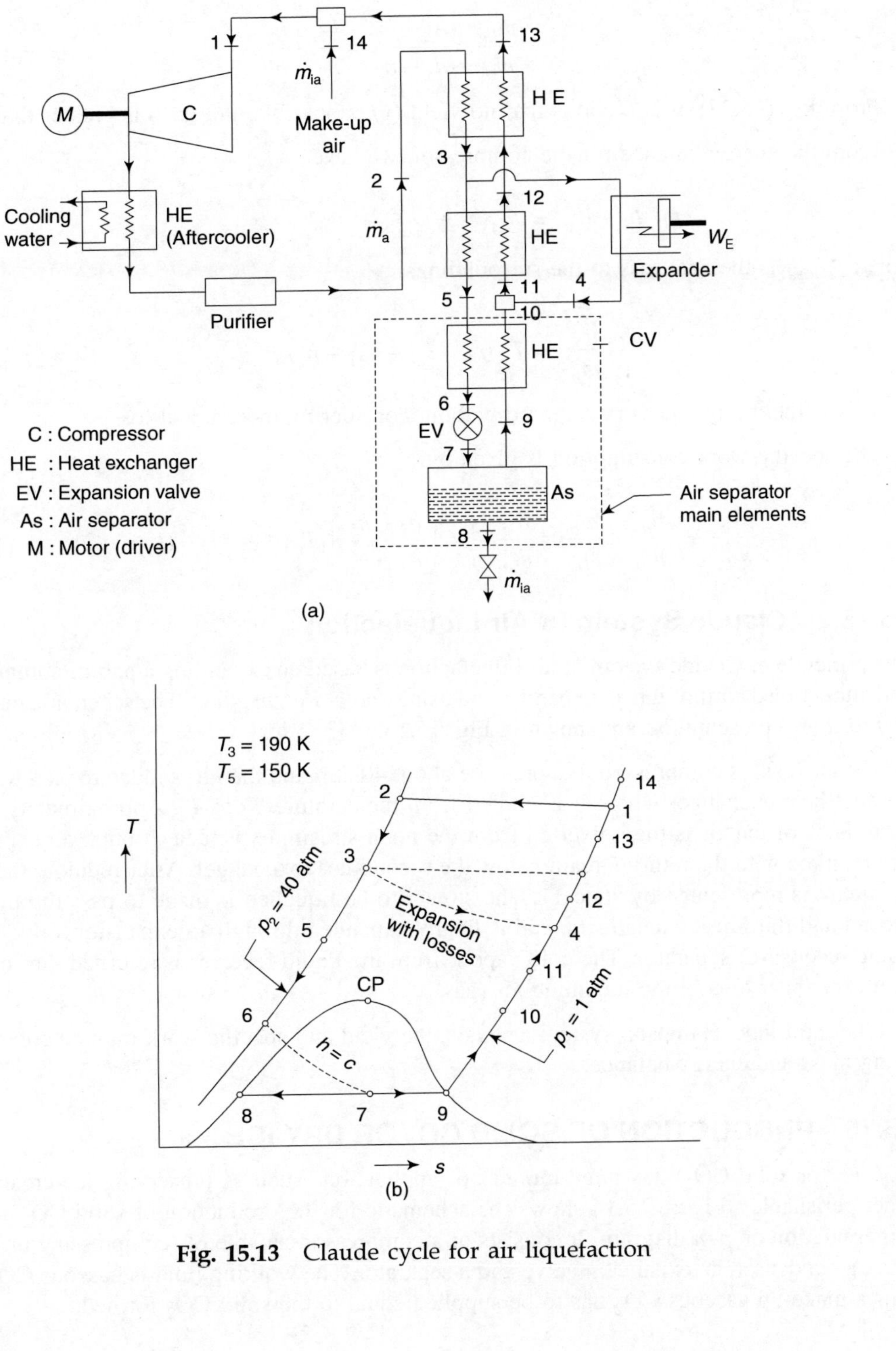

Fig. 15.13 Claude cycle for air liquefaction

or
$$y = \frac{\dot{m}_{la}}{\dot{m}_a} = \frac{h_2 - h_7}{h_5 - h_7} = \frac{h_7 - h_2}{h_7 - h_5} \quad ...(15.27)$$

From Eq. (15.27), it is obvious that no yield (y) is possible unless h_7 is greater than h_2.

From the energy balance for the compressor, we have

$$\dot{m}_a h_1 + \dot{W}_c = \dot{m}_a h_2 + \dot{Q}_{\text{loss}}$$

where $\dot{Q}_{\text{loss}}$ is the heat loss to the surroundings.

or
$$\frac{\dot{W}_c}{\dot{m}_a} = T_1 (s_1 - s_2) - (h_1 - h_2) \quad ...(15.28)$$

This is the minimum work required without considering mechanical losses.

The specific work consumption is given by

$$\frac{\dot{W}_c}{\dot{m}_{la}} = \frac{\dot{W}_c}{\dot{m}_a} \times \frac{\dot{m}_a}{\dot{m}_{la}} = \frac{\dot{W}_c}{\dot{m}_a} \cdot \frac{1}{y} = \left(\frac{h_7 - h_5}{h_7 - h_2}\right)[T_1 (s_1 - s_2) - (h_1 - h_2)] \quad ...(15.29)$$

15.15.2 Claude System of Air Liquefaction

The principle of Claude system of air liquefaction is based on expanding a part of compressed and after cooled air through an expander and using *Joule–Kelvin effect*. The schematic diagram and its T–s representation are shown in Fig. 15.13.

The air (gas) is compressed to a pressure about 40 atm and then it is made to pass through the first heat exchanger where it is cooled from temperature T_2 to T_3. Approximately about 70 to 80% of the air is then diverted from the main stream, expanded through an expander, and reunited with the return stream below the second heat exchanger. After reunion, the state of stream is represented by state 11. The stream to be liquefied is made to pass through the second and third heat exchangers, and is finally expanded through an expansion valve to the liquid receiver or separator. The cold vapour from the liquid receiver is returned through the heat exchanger to cool the incoming air (gas).

Like the Linde–Hampson system analysis, the yield and specific work may be commuted using mass and energy balance.

15.16 PRODUCTION OF SOLID CO_2 OR DRY ICE

Dry ice (or solid CO_2) has large number of applications such as preserving ice-cream and other perishables. Figure 15.14 shows the schematic for the production of solid CO_2 and its representation on p–h diagram. It consists of a compressor capable of (compressing up to 70 atm), a condenser, an expansion valve and a separator. The working fluid is gaseous CO_2 and thus a make-up gaseous CO_2 has to be supplied equal to the solid CO_2 formed.

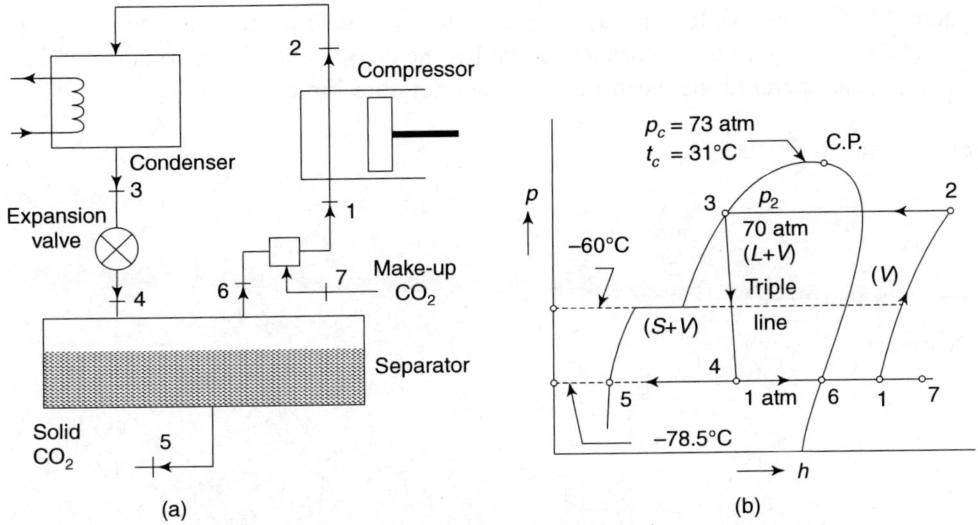

Fig. 15.14 Dry ice (or CO_2) refrigeration cycle

15.17 COMPARISON BETWEEN VAPOUR COMPRESSION AND VAPOUR ABSORPTION SYSTEM

The following points show the difference between vapour compression and vapour absorption system.

Vapour compression system	Vapour absorption system
(i) Energy input is high grade energy (electrical or mechanical)	(i) Energy input is low grade energy (heat)
(ii) More wear, tear and noise due to large number of moving parts in compressor	(ii) Less wear, tear and noise as there is no compressor, only a small pump
(iii) The COP decreases with the decrease in evaporator pressure	(iii) This system can work at lower evaporator pressure without lowering COP
(iv) Poor part load performance	(iv) Better part load performance
(v) The presence of water droplets in the suction line damages the compressor	(v) The presence of water droplets in the suction line offers no danger
(vi) Automatic capacity control is slightly difficult	(vi) Automatic capacity control is easier
(vii) Larger space requirement per unit capacity	(vii) Smaller space requirement per unit capacity
(viii) COP is higher, i.e., energy input is 1/2 to 1/3 of cooling effect	(viii) Poor COP; i.e., energy input is 1.5 times of cooling effect
(ix) In a single compressor system, the capacity is less than 1000 tonnes	(ix) Absorption unit is capable of developing capacity above 1000 tonnes

Example 15.2 An ideal vapour compression refrigeration cycle is working on refrigerant 134 *a*. The cold region of the cycles is at 0°C and the warm region at 26°C. The mass flow rate

of refrigerant is 0.1 kg/s. Calculate (a) the coefficient of performance, (b) the refrigerating capacity, (c) the coefficient of performance, and (d) the coefficient of performance of Carnot refrigerating cycle operating between the same temperature limits.

Solution Refer to Fig. 15.15.

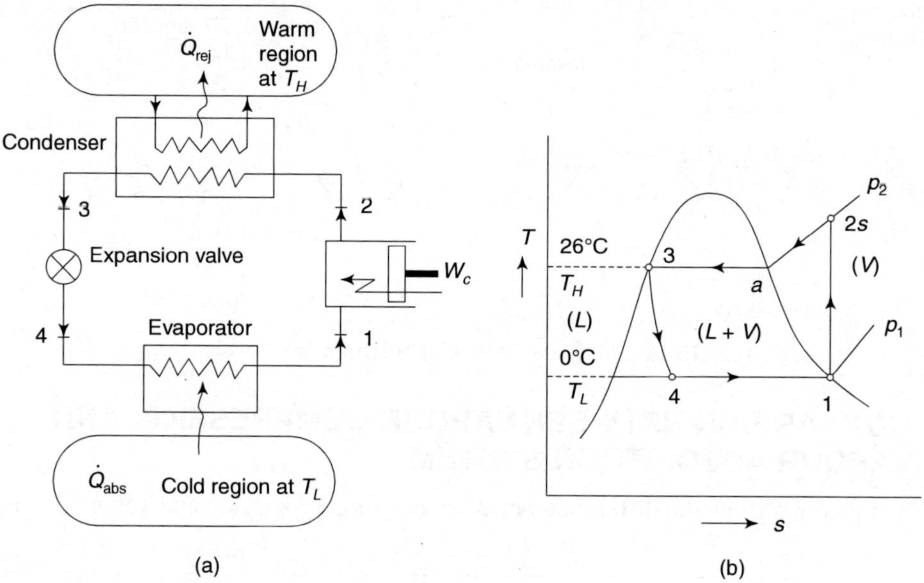

Fig. 15.15 Example 15.2

From refrigeration tables, for refrigerant 134 *a*, we have

at 0°C, $h_1 = 247.22$ kJ/kg, $s_1 = 0.919$ kJ/kg K,

at 26°C, $p_2 = 6.853$ bar

$$h_{2s} = 264.7 \text{ kJ/kg}, \ h_{f3} = 85.75 \text{ kJ/kg} = h_4 = \text{(throttling process)}$$

(a) The compressor work input = $\dot{W}_c = \dot{m}\,(h_{2s} - h_1) = 0.1\,(264.7 - 247.22)$

$= 1.748$ kW **Ans.**

(b) The refrigerating capacity = $\dot{Q}_{abs} = m(h_1 - h_4) = 0.1\,(247.22 - 85.75)$

$= 16.147$ kW **Ans.**

(c) The coefficient of performance = COP = $\dfrac{\dot{Q}_{abs}}{\dot{W}_c} = \dfrac{16.147}{1.748} = 9.237$ **Ans.**

(d) COP of Carnot refrigeration = $\dfrac{T_L}{T_H - T_L} = \dfrac{273}{26} = 10.5$ **Ans.**

Example 15.3 A vapour compression refrigeration system works on refrigerant 134*a*. Saturated vapour enters the compressor at – 10°C and saturated liquid leaves the condenser at a pressure of 9 bar. The temperature of the liquid leaving the condenser is 30°C. The efficiency of compressor is 80%. The mass rate of refrigerant in the cycle is 0.1 kg/s. Calculate (a) the compressor work input, (b) the refrigerating capacity, (c) the coefficient of performance, and (d) the rates of exergy destruction within the compressor and expansion valve. Take $T_0 = 288$ K.

Solution Refer to Fig. 15.16. From refrigerant tables for 134*a*, we have : At 10°C, $h_1 = 241.35$ kJ/kg, $s_1 = 0.9253$ kJ/kg K, $h_{2s} = 272.39$ kJ/kg. The actual enthalpy (h_2) is calculated from the concept of compressor efficiency.

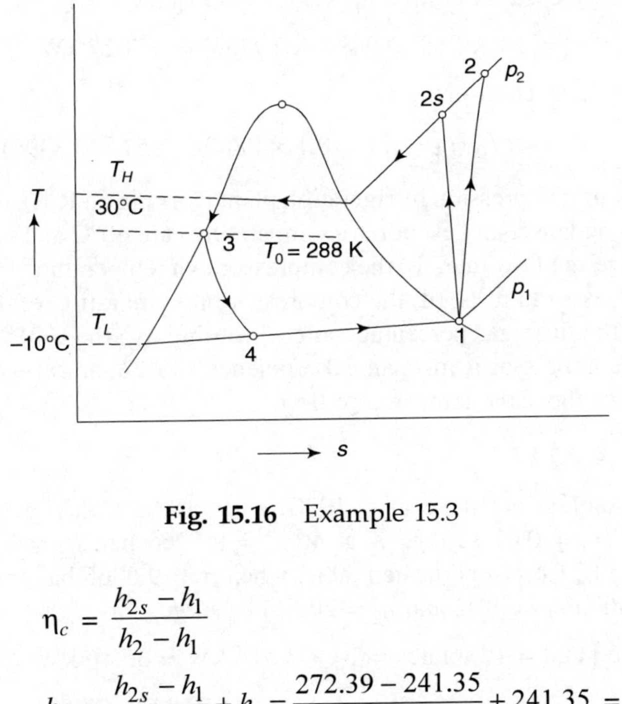

Fig. 15.16 Example 15.3

$$\eta_c = \frac{h_{2s} - h_1}{h_2 - h_1}$$

or

$$h_2 = \frac{h_{2s} - h_1}{\eta_c} + h_1 = \frac{272.39 - 241.35}{0.8} + 241.35 = 280.15 \text{ kJ/kg}$$

From table, at 9 bar and $h_2 = 280.15$, $s_2 = 0.9497$ kJ/kg

$$h_3 = h_{f3} = 91.49 \text{ kJ/kg}, \; s_3 = s_{f3} = 0.3396 \text{ kJ/kg K}$$

$$h_{f4} = 36.96 \text{ kJ/kg}$$

The expansion through valve is throttling process, so $h_{f3} = h_3 = h_4 = h_{f4} + x_4\, h_{fg4}$

or

$$x_4 = \frac{h_{f3} - h_{f4}}{h_{fg4}} = \frac{91.49 - 36.96}{204.38} = 0.2668$$

$$s_4 = s_{f4} + x_4\,(s_{g4} - s_{f4}) = 0.1486 + 0.2668\,(0.9253 - 0.1486)$$

$= 0.3558$ kJ/kg K

(a) The compressor work input = $\dot{W}_c = \dot{m}(h_2 - h_1) = 0.1\,(280.15 - 241.35)$

$= 3.88$ kW **Ans.**

(b) The refrigerating capacity = (RC) = $\dot{m}(h_1 - h_4) = 0.1\,(241.35 - 91.49)$

$= 14.986$ kW $= 4.262$ tonnes **Ans.**

(c) The coefficient of performance = COP = $\dfrac{R.C.}{\dot{W}_c} = \dfrac{14.986}{3.88} = 3.86$ **Ans.**

(d) The rate of energy destruction in compressor = $\dot{m}\,T_0\,(s_2 - s_1)$

$= 0.1 \times 288\,(0.9497 - 0.9253) = 0.7027$ kW **Ans.**

The rate of energy destruction in valve

$= \dot{m}\,T_0\,(s_4 - s_1) = 0.1 \times 288\,(0.3557 - 0.3396) = 0.463$ kW **Ans.**

Example 15.4 A vapour compression refrigeration plant working on R–12 develops 10 tonnes of refrigeration. The condenser and evaporator temperatures are 40°C and –10°C respectively. Calculate (a) the refrigerant flow rate, (b) the compressor exit temperature, (e) the volume flow rate handled by compressor in m^3/s, (d) the compressor pressure ratio, (e) the heat rejected to condenser in kW, (f) the flash gas percentage after throttling, (g) the COP, and (h) the power required to drive the compressor if mechanical efficiency is 90%, and (i) compare COP with Carnot refrigeration for the same temperature limits.

Solution Refer to Fig. 15.17.

From refrigerant tables, we have at – 10°C, h_1 = 183.192 kJ/kg, p_1 = 2.1912 bar, s_1 = 0.7019 kJ/kg K, v_1 = 0.07 m^3/kg, p_2 at 40°C = 9.6066 bar, $h_3 = h_4$ = 74.591 kJ/kg, $s_2 = s_1$ = 0.7019 kJ/kg K, from superheated table, when p_2 = 9.6066 bar and $s_2 = s_1$ = 0.7019 kJ/kg K, by interpolation t_2 = 48°C and h_2 = 209.411 kJ/kg

The capacity of the plant = 10 tonnes = 10 × 3.516 kW = 35.16 kW

Thus, $35.16 = \dot{m}(h_1 - h_4) = \dot{m}\,(183.1921 - 74.591)$

(a) or $\dot{m} = \dfrac{35.16}{183.1921 - 74.592} = 0.3237$ kg/s **Ans.**

(b) The compressor exit temperature = 48°C **Ans.**

(c) Volume flow rate = $\dot{m}\,v_1 = 0.3237 \times 0.077 = 0.02492$ m^3/kg **Ans.**

(d) Pressure ratio = $\dfrac{p_2}{p_1} = \dfrac{9.6066}{2.1912} = 4.39$ **Ans.**

(e) Heat rejected to the condenser = $\dot{m}(h_2 - h_3)$

$= 0.3237\,(209.411 - 74.591) = 43.64$ kW **Ans.**

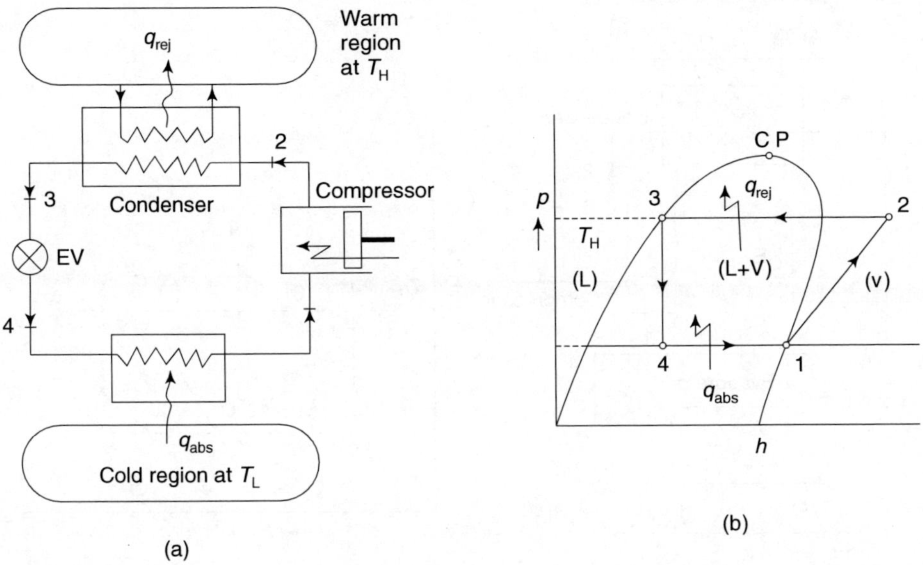

Fig. 15.17 Example 15.4

(f) $$h_4 = 74.59 = h_{f4} + x_4\, h_{fg4} = 26.87 + x_4 \times 156.31$$

or $$x_4 = \frac{74.59 - 26.87}{156.31} = 0.3052$$

∴ Flash gas percentage = 30.52% **Ans.**

(g) $$\text{COP} = \frac{\text{R.E.}}{\dot{W}_c} = \frac{h_1 - h_4}{h_2 - h_1} = \frac{183.191 - 74.59}{209.411 - 183.19} = 4.14$$ **Ans.**

(h) Power required to drive the compressor

$$= \frac{\dot{m}\,(h_2 - h_1)}{\eta_m} = \frac{0.3237\,(209.411 - 183.19)}{0.9} = 9.4308 \text{ kW}$$ **Ans.**

(i) $$\text{Carnot COP} = \frac{T_2}{T_1 - T_2} = \frac{263}{50} = 5.26$$ **Ans.**

Comment *Carnot COP is higher than vapour compression COP.*

Example 15.5 A vapour compression refrigeration plant working on R–12 produces 20 tonnes of refrigeration. The condenser and evaporator operate at 35°C and – 10°C respectively. In order to subcool the saturated liquid leaving the condenser, a suction line heat exchanger is used. The saturated vapour leaving the evaporator is superheated in the suction line heat exchanger to the extent that discharge temperature from the compressor is 60°C under isentropic compression situation. Work out the following using refrigerant R–12 chart.

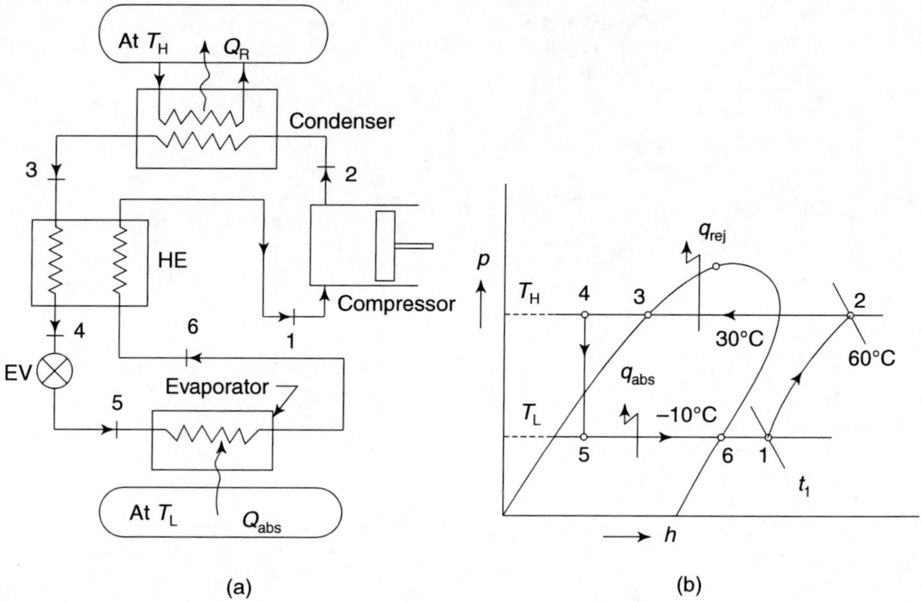

Fig. 15.18 Example 15.5

(a) Draw the system configuration and show the process on *p-h* diagram, (b) the subcooling achieved in the heat exchanger, (c) the refrigerant flow rate, kg/s, (d) the cylinder dimensions of the two cylinder compressor, if the speed of the compressor is 900 rpm and the stroke to bore ratio is 1.0 and volumetric efficiency is 80%, (e) the COP of the plant, and (f) the power required to drive the compressor if mechanical efficiency is 90%.

Solution (a) Refer to Fig. 15.18.

From refrigerant chart, we have, at 35°C, $h_2 = 1034$ kJ/kg, $h_3 = h_{f3} = 882$ kJ/kg, at – 10°C, $h_1 = 1008$ kJ/kg, $h_6 = 998$ kJ/kg, $v_1 = 0.084$ m^3/kg.

Taking energy balance at subcooler (suction line heat exhanger)

$$h_1 - h_6 = h_{f3} - h_{f4}$$

or

$$1008 - 998 = 882 - h_{f4}$$

or

$$h_{f4} = 872 \text{ kJ/kg} = h_5$$

From chart/table, at $h_{f4} = 872$ kJ/kg, $t_4 = 25$°C

(b) The degree of subcooling achieved = 35 – 25 = 10°C **Ans.**

(c) The refrigerant flow rate = $\dot{m} = \dfrac{R.E. \text{ in kW}}{R.E. \text{ in kJ/kg}} = \dfrac{20 \times 3.516}{h_6 - h_5}$

or

$$= \frac{20 \times 3.516}{998 - 872} = 0.558 \text{ kg/s}$$ **Ans.**

(d) The actual sucked volume flow rate in the compressor cylinder having a number of cylinders, $x = 2$, we have

$$\dot{V} = ALN\,\eta_v\,x = \dot{m}\,v_1 = 0.558 \times 0.084 = 0.04687 \text{ m}^3/\text{s}$$

or
$$\frac{\pi}{4} D^2 \times LN\,\eta_v \cdot x = \frac{\pi}{4} \times D^3 \times \frac{900}{60} \times 0.9 \times 2 = 0.04687$$

or
$$D = \left[\frac{0.04687 \times 60 \times 4}{2 \times \pi \times 900 \times 0.9}\right]^{1/3} = 0.1302\ m = 13.02 \text{ cm}$$

or
$$D = L = 13.02 \text{ cm}$$ **Ans.**

(e)
$$\text{COP} = \frac{\text{RE}}{W_c} = \frac{(h_6 - h_5)}{(h_2 - h_1)/\eta_m} = \frac{998 - 872}{(1034 - 1008)/0.9} = 4.36$$ **Ans.**

(f) Power required to drive the compressor

$$\dot{W}_c = \dot{m}\,(h_2 - h_1)/\eta_m = 0.558\ (1034 - 1008)/0.9 = 16.12 \text{ kW}$$ **Ans.**

Example 15.6 A two-stage vapour compression refrigeration system employs a direct contact flash chamber type intercooler between two stages of compressor. The refrigerant used is ammonia and the refrigerating capacity of the plant is 50 tonnes. The evaporator and condensers operate at – 30°C and 40°C, respectively. Calculate (a) the total compressor work input by assuming isentropic compression and mechanical efficiency as 90%, (b) the COP, (c) the percentage of saving in work input over single stage compression, and (d) the percentage increase in COP over single stage compression.

Solution Refer to Fig. 15.19. From NH_3 table, p_1 (at – 30°C) = 1.195 bar

and p_2 (at 40°C) = 15.543 bar

The intermediate pressure is given by

$$p_1 = \sqrt{p_1\,p_2} = \sqrt{1.195 \times 15.543} = 4.309 \text{ bar}$$

From NH_3 tables, we have, at – 30°C, h_1 = 1404.55 kJ/kg, at p_i = 4.309 bar, h_2 = 1574.28 kJ/kg, h_3 = 1443.5 kJ/kg, at 40°C, h_4 = 1628.12 kJ/kg, $s_1 = s_2$ and $s_3 = s_4$, at 40°C, $h_5 = h_{f5}$ = h_6 = 371.7 kJ/kg, at p_i = 4.309 bar, $h_{f7} = h_8$ = 181.5 kJ/kg, h'_4 = 1805.11 kJ/kg, $h'_6 = h_5$ = 371.7 kJ/kg.

The desired refrigerating capacity is achieved in the evaporator. The mass flow rate of refrigerant passing through compressors is given by

$$\dot{m}_1 = \frac{R.E.}{h_1 - h_8} = \frac{50 \times 3.516}{1404.55 - 181.5} = 0.1437 \text{ kg/s}$$

$$\dot{m}_2 = \dot{m}_1 \frac{(h_1 - h_7)}{(h_3 - h_6)} = 0.1437 \times \frac{(1404.55 - 181.5)}{(1443.5 - 371.7)} = 0.1639 \text{ kg/s}$$

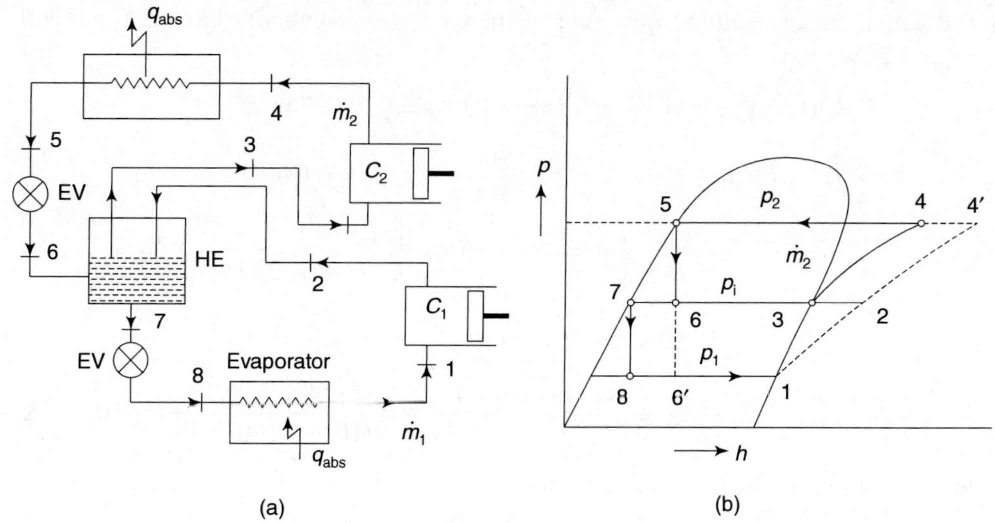

Fig. 15.19 Example 15.6

(a) The total compressor work is given by

$$\dot{W}_c = [\dot{m}_1 (h_2 - h_1) + \dot{m}_2 (h_4 - h_3)]/\eta_m$$

$$= \frac{0.1437\,(1574.28 - 1404.55) + 0.1639\,(1628.12 - 1443.5)}{0.9} = 60.72 \text{ kW} \quad \textbf{Ans.}$$

(b) $$\text{COP} = \frac{R.E.}{\text{Work input}} = \frac{50 \times 3.516}{60.72} = 2.895 \quad \textbf{Ans.}$$

(c) In the case of single stage compression only $\dot{m}_1$ quantity will be flowing through the system and after compression, the state is represented by 4′.

The mass flow rate of refrigerant for single stage compression is

$$(\dot{m}_1)_{ss} = \frac{50 \times 3.516}{(1404.55 - 371.7)} = 0.1702 \text{ kg/s}$$

$$(\dot{W}_c)_{ss} = \frac{(\dot{m}_1)_{ss}\,(h_4' - h_1)}{0.9} = \frac{0.1702\,(1805.11 - 1404.55)}{0.9} = 75.75 \text{ kW}$$

The percentage saving in compressor work input

$$= \frac{75.75 - 60.72}{60.72} \times 100 = 24.75\% \quad \textbf{Ans.}$$

(d) $$(\text{COP})_{ss} = \frac{\text{RE}}{(\dot{W}_c)_{ss}} = \frac{50 \times 3.516}{75.75} = 2.32$$

Percentage increase in COP over single stage compression

$$= \frac{2.895 - 2.32}{2.32} \times 100 = 24.78\% \quad \textbf{Ans.}$$

Example 15.7 A vapour absorption refrigeration system employs aqua-ammonia as refrigerant. The heat is supplied to the generator by condensing steam at 3 bar having 90% quality. The evaporator temperature is –10°C and ambient temperature is 30°C. Calculate (a) the maximum COP of the absorption system.

It the refrigeration load is 40 tonnes and the actual COP is 30% of the maximum COP calculate (b) the mass flow rate of steam required by generator.

Solution Refer to Fig. 15.20.

At 3 bar, $t_{sat} = t_g = 133.55°\text{Cl}\ h_{fg} = 2163.8$ kJ/kg

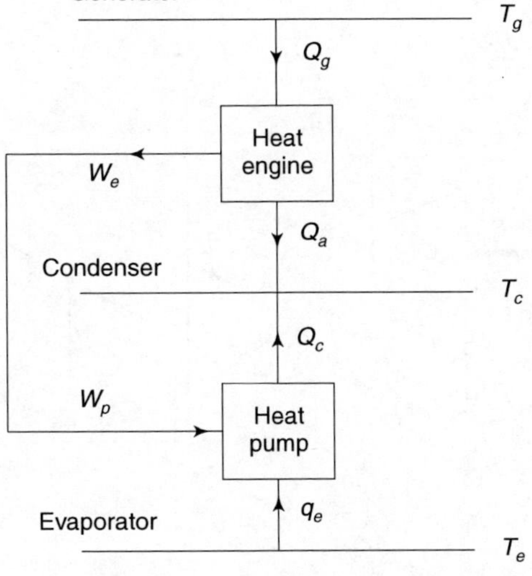

Fig. 15.20 Example 15.7

The maximum COP of the absorption refrigeration system is given by

(a) $$(\text{COP})_{\max} = \frac{(T_g - T_c)\,T_e}{(T_c - T_e)\,T_g} = \frac{(133.35 - 30)\,(273 - 10)}{(30 + 10)\,(273 + 133.55)} = 1.674 \quad \textbf{Ans.}$$

Actual $$\text{COP} = 1.674 \times 0.3 = 0.5022$$

But $$\text{COP} = \frac{\text{RE}}{\text{Energy input}} = \frac{\dot{Q}_e}{\dot{Q}_g} = 0.5022$$

or $$\dot{Q}_g = \frac{\dot{Q}_e}{\text{COP}} = \frac{40 \times 3.516}{0.5022} = 280 \text{ kW}$$

$\therefore$ $$\dot{Q}_g = \dot{m}_s (h_f + xh_{fg}) - h_f = \dot{m}_s h_{fg} = \dot{m}_s \times 2163.8$$

(b) or $$\dot{m}_s = \frac{\dot{Q}_g}{2163.8} = \frac{280}{2163.8} = 0.1294 \text{ kg/s}$$ **Ans.**

Example 15.8 Air refrigeration system used for aircraft cooling of cabin takes in air at 1 bar and 3°C and compresses the air to 5 bar with an isentropic efficiency of 85%. The compressed air is cooled in a heat exchanger up to 60°C at constant pressure and then expands in a turbine up to 1 bar with an isentropic efficiency of 88%. The low temperature absorbs a cooling load of 10 tonnes of refrigeration at constant pressure before re-entering to the compressor which is driven by a turbine. Calculate the air mass flow required.

Solution Refer to Fig. 15.21.

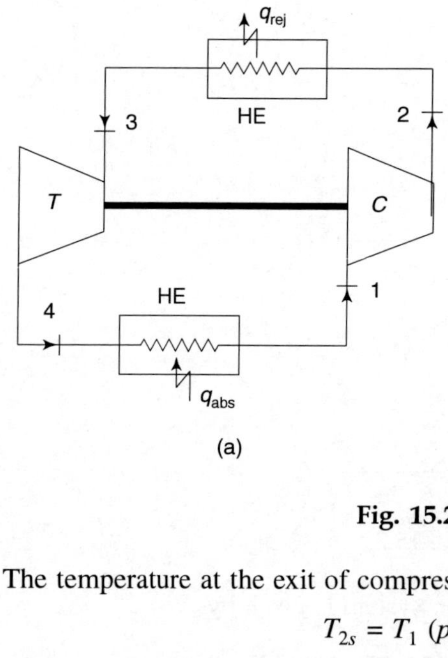

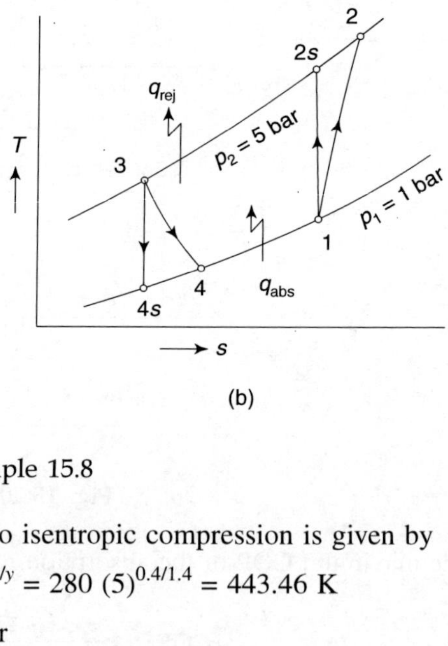

Fig. 15.21 Example 15.8

The temperature at the exit of compressor due to isentropic compression is given by

$$T_{2s} = T_1 (p_2/p_1)^{(\gamma - 1)/\gamma} = 280 (5)^{0.4/1.4} = 443.46 \text{ K}$$

The actual temperature at the exit of compressor

$$T_2 = T_1 + \frac{T_{2s} - T_1}{\eta_c} = 280 + \frac{443.46 - 280}{0.85} = 472.3 \text{ K}$$

Temperature at the exit of turbine due to isentropic expansion is given by

$$T_{4s} = \frac{T_3}{(p_3/p_1)^{(\gamma - 1)/\gamma}} = \frac{333}{(5)^{0.4/1.4}} = 210.25 \text{ K}$$

Actual temperature at the exit of turbine is

$$T_4 = T_3 - \eta_t (T_3 - T_4) = 333 - 0.88 (333 - 210.25) = 224.98 \text{ K}$$

Refrigerating effect (RE) = $c_p (T_1 - T_4) = 1.005 (280 - 224.98) = 55.29$ kJ/kg

Net work input $= c_p [(T_2 - T_1) - (T_3 - T_4)] = 1.005 [(472.3 - 280) - (333 - 224.98)]$

$= 84.70$ kJ/kg

(a) $$\text{COP} = \frac{R.E.}{W_{net}} = \frac{55.29}{84.7} = 0.6527$$ **Ans.**

(b) $$\text{Power input} = \frac{10 \times 3.516}{0.6527} = 53.86 \text{ kW}$$ **Ans.**

(c) Mass flow rate of air = 10 ×3.516/55.29 = 0.6359 kg/s **Ans.**

15.18 SUMMARY OF FORMULAE

1. $(\text{COP})_{HP} = 1 + (\text{COP})_{Ref.}$

2. $(\text{COP})_{Ref.\ Carnot} = \dfrac{R.E.}{W} = \dfrac{T_L}{T_H - T_L}$

3. $(\text{COP})_{Hp,\ Carnot} = \dfrac{T_H}{T_H - T_L}$

4. $PD = \dfrac{v_1 \cdot \dot{m}}{\eta_{vol}}$ m^3/s tonne

5. $\dot{m} = \dfrac{3.516}{h_1 - h_4}$; kg/s tonne

6. $P = \dot{m} (h_2 - h_1)$

7. $(\text{COP})_{gas} = \dfrac{1}{(p_2/p_1)^{(\gamma - 1)/\gamma}}$

8. $w_c = \dfrac{\dot{W}_c}{\dot{m}_{La}}$

9. $(\text{COP})_{abs} = \left(\dfrac{T_g - T_c}{T_c - T_c}\right) \cdot \dfrac{T_e}{T_g}$

10. $\text{COP} = \dfrac{R.E.}{W}$

PROBLEMS

Things to Think About

15.1. List the values of temperature recommended in various compartments of household refrigerator.

15.2. How will you use the household refrigerator to heat the kitchen? If the refrigerator's COP is known, how will you calculate the COP of a heat pump?

15.3. In freezing 1 kg of water 335 kJ of heat is needed. How much ice will be made by an ice maker having 1 tonne refrigeration capacity in 24 hours?

15.4. Is it advisable to replace the capillary tube or expansion valve by a turbine in vapour compression refrigeration system?

15.5. In an automobile air conditioner, the refrigerant is changed time to time but in refrigerator, it is seldom done. What is the reason behind this?

15.6. Which is the most suitable refrigerant at present keeping in view the safeguard of ozone layer?

15.7. List the name of components contained in the outside unit of residential heat pump.

15.8. When you open the door of a refrigerator, you feel a cooling effect on your body. What is the reason behind it?

Short and Long Theoretical Questions

15.9. What is refrigeration? List the applications of refrigeration.

15.10. What are the performance parameters of refrigerator and heat pump?

15.11. What is the unit of refrigeration?

15.12. List the various methods of producing cold.

15.13. Derive the expression for COP in the case of Carnot refrigerator.

15.14. Derive the expression for COP in the case of vapour compression refrigeration system. What is the effect of evaporation and condenser temperature?

15.15. List the main elements of vapour compression refrigeration system.

15.16. How does the actual vapour compression refrigeration cycle differ with ideal one?

15.17. List the types of compressor used in vapour compression refrigeration system.

15.18. What is the purpose of multi-stage compression?

15.19. Discuss the purpose of flash chamber.

15.20. List the refrigerants used.

15.21. With the help of sketch, discuss the working principle of vapour absorption system.

15.22. Discuss the advantages and disadvantages of vapour compression and vapour absorption system.

15.23. Derive the expression for ideal (maximum) COP in the case of vapour absorption system.

15.24. Discuss gas cycle refrigeration system.

15.25. Describe the working principle of Linde–Hampson cycle for air liquefaction with sketch.

15.26. With the neat sketch, discuss the working principle of Claude system of refrigeration.

15.27. With the help of sketch, discuss the working principle of production of solid CO_2 (dry ice).

Numerical Questions

15.28. A Carnot vapour refrigeration cycle uses refrigerant 134*a* as the working fluid. The refrigerant enters the condenser as saturated vapour at 28°C and leaves as saturated liquid. The evaporator operates at a temperature of –10°C. Calculate (a) the work input to the compressor, (b) work developed by the turbine, (c) the refrigerating capacity, and (d) the COP.

15.29. A Carnot vapour refrigeration cycle uses R–12 as working fluid. The evaporator temperature is 0°C. The saturated vapour enters the condenser at 40°C and saturated liquid exists at the same temperature. The mass flow rate of refrigerant is 5 kg/min. Calculate (a) the refrigerating capacity, (b) the power input to the cycle, and (c) the coefficient of performance.

15.30. In an ideal vapour compression refrigeration system, the refrigerant 134*a* enters the compressor as saturated liquid at –16°C with a volumetric flow rate of 1 m^3/min. The refrigerant leaves the condenser at 36°C, 10 bar. Calculate (a) the compressor power, (b) the refrigerating capacity in tonnes, and (c) the coefficient of performance.

15.31. A vapour compression refrigeration system operates at steady state with refrigerant 134*a* as the working fluid. Saturated vapour enters the compressor at –10°C. The isentropic compressor efficiency is 80% and let the temperature of liquid leaving the condenser be 32°C. The mass flow rate of refrigerant is 5 kg/min. Calculate (a) the compressor power, (b) the refrigerating capacity, (c) the coefficient of performance, and (d) the rates of exergy destruction in the compressor and expansion valve. Take T_0 = 28°C.

[*Ans.* (a) 3.73 kW, (b) 3.4 tonnes, (c) 3.21, (d) 0.71 kW, 0.58 kW]

15.32. A vapour compression refrigeration system uses 134*a* as refrigerant. The mass flow rate of refrigerant is 10 kg/min. The refrigerant enters the compressor at – 10°C, 1.4 bar and exits at 7 bar. The isentropic compressor efficiency is 67%. The refrigerant leaves the condenser at 7 bar, 24°C. Calculate (a) the COP, (b) the refrigerating capacity in tonnes, (c) the rates of exergy destruction in the compressor and expansion valve, each is kW, and (d) the changes in specific flow exergy of the refrigerant passing through the evapourator and condenser, respectively, each in kJ/kg. Take T_0 = 15°C, p_0 = 1 bar.

15.33. A vapour compression refrigeration system using ammonia as the working fluid has evaporator and condenser pressure of 2 and 12 bar, respectively. There is no pressure drop in the evapourator and condenser. At the inlet and exit of the compressor, the temperatures are – 10°C and 40°C, respectively. The heat transfer rate from the working fluid passing through the condenser is 20 kW and liquid comes out at 12 bar and 28°C. If the compressor operates adiabatically, calculate (a) the compressor power input, in kW, and (b) the coefficient of performance.

15.34. A vapour compression refrigeration system uses refrigerant 134*a* as working fluid. The system produces 20 tonnes of refrigerating capacity. The refrigerant enters the compressor at 15°C, 4 bar and exists at 12 bar. The compression process can be modeled by $pv^{1.01}$ = constant. At the condenser exit, the pressure is 11.6 bar and the temperature is 44°C. The condenser is water cooled with water is entering at 20°C and leaving at 30°C with negligible change in pressure. Calculate (a) the mass flow rate of refrigerant, in kg/s, (b) the compressor input, in kW, (c) the coefficient of performance, (d) the mass flow rate of cooling water, in kg/s, (e) the rates of exergy destruction in the condenser and expansion valve, each expressed as a percentage of the power input, take T_0 = 15°C.

15.35. A vapour compression refrigeration system using R–12 as refrigerant produces 3 tonnes of refrigeration. The plant maintains a cold store at –15°C. The refrigerant flow rate is 0.133 kg/s. The vapour leaves the evaporator at 5°C superheat. The cooling water is available in plenty at 15°C for the condenser. A suction line heat exchanger subcools the refrigerant before throttling. Calculate (a) the compressor discharge temperature, (b) the COP, (c) the amount of subcooling, (d) the cylinder dimensions of the compressor if the speed is 900 rpm, stroke-to-bore ratio 1.2 and the volumetric efficiency is 95%.

[*Ans.* (a) 66.05°C, (b) 4.1, (c) 125°C, (d) 10.45 cm, 12.5 cm]

15.36. A vapour-compression heat pump with a heating capacity of 1000 kJ/min is driven by a power cycle with a thermal efficiency of 25%. For the heat pump, refrigerant 134*a* is compressed from saturated vapour at –10°C to the condenser pressure of 10 bar. The isentropic compression efficiency is 80%. The refrigerant liquid enters the expansion valve at 9.6 bar, 34°C. For the power cycle, 80% of the heat rejected is transferred to the heated space, calculate (a) the power input to the heat pump compressor, and (b) the ratio of total rate that the heat is delivered to the heated space to the rate of heat input to the power cycle.

15.37. Calculate the ideal COP of a vapour absorption refrigeration plant in which the heating, cooling and refrigeration takes place at 197°C and –3°C respectively. [*Ans.* 5.16]

15.38. A water cooler using R–12 as refrigerant works between 30°C and 2°C respectively. The vapour leaves the evaporator dry and saturated. The output of the cold water is 100 kg/h cooled from 30°C to 8°C, calculate (a) the power of the motor, (b) the volumetric displacement of compressor. Assume volumetric efficiency of compressor 90%, mechanical efficiency of compressor 85%, mechanical efficiency of motor 95% and 20% of useful cooling is lost in the water cooler.

[*Ans.* (a) 0.37kW, (b) 0.0716 m^3/min]

15.39. A vapor-absorption refrigeration system using ammonia as refrigerant has an evaporator temperature of –12°C and a condenser temperature of 150°C. In this cycle, 0.45 kg is transferred to the ammonia in the evaporator for each kg transferred to the ammonia solution in the generator from the high temperature source. Calculate the amount of refrigeration capacity that can be achieved.

15.40. Air enters the compressor of an ideal Brayton refrigeration cycle at 1 bar, 270K. The compressor pressure ratio is 4 and the temperature at the turbine inlet is 310K. Calculate (a) the net work input, in kJ/kg, (b) the refrigeration capacity, (c) the coefficient of performance, and (d) the coefficient of performance of Carnot refrigerator operating between the same temperature limits.

If isentropic efficiency of the compressor and turbine are 80 and 88% respectively, calculate the above and exergy accounting of compressor power input in kJ/kg, Take T_0 = 288K.

Software Based Questions

15.41. Develop a software for the thermodynamic analysis of (a) vapour compression refrigeration system, (b) vapour absorption refrigeration system, and (c) air refrigeration system. Validate the results with realistic data.

15.42. Refrigerant 134*a* enters the compressor of a ideal vapour compression cycle as saturated vapour at –17°C. The condenser pressure is 10 bar. The mass flow rate of refrigerant is 3 kg/min. Develop a software to plot the coefficient of performance and the refrigerating capacity, in tonnes, versus the condenser exit temperature ranging from the saturated temperature of 10 bar to 33°C.

15.43. Develop a software for showing the effect of variables affecting the performance of vapour compression refrigeration system. Take the realistic data, i.e., manufacturer's data of a refrigerator as base input.

15.44. Develop a software for the design of (a) refrigerator, (b) window air-conditioner, (c) ice plant (c) water cooler, and (e) refrigeration system of a big air-conditioning plant. Validate the results with manufacturer's data.

Design and Open Ended Questions

15.45. Write a report on current status of magnetic refrigeration technology for use in the 80 to 300K temperature range. Give a comparative study of this technology with vapour compression refrigeration with respect to economics.

15.46. Refrigerant 22 is widely used as the working fluid in air conditioners and industrial chillers. As a result of ozone depletion, their use in to be phased out and ozone friendly refrigerant is to be used. Prepare a report on the design issues for air conditioners and chillers that would result from changing ozone friendly refrigerants.

15.47. Some of the common *cryogenic refrigeration* applications include air separation to obtain oxygen and nitrogen, large scale production of liquid hydrogen and the liquefaction of natural gas. Prepare a report on the various equipment used to achieve the low temperatures required in these applications with a schematic layout of cryogenic refrigeration system.

15.48. Write a report on economic feasibility of using a waste heat recovery heat pump for domestic water heating that employs ventilation air being discharged from a dwelling as source. Assume a typical hot water use in a family of four living in dwelling of 200 m^2.

15.49. In providing chilled water at 4°C to your engineering college for air conditioning purpose, ammonia is chosen as a refrigerant. Prepare a report listing the major design issues involved and developing a layout of the equipment room and a schematic of the chilled water distribution system including the capacities of each unit.

Multiple-Choice Answer Type Questions

15.50. One tonne of refrigeration is equivalent to

(a) 2.516 kW (b) 3.516 kW (c) 4.516 kW (d) 5.516 kW

15.51. For a definite low temperature to be maintained COP of the refrigerator is related during hot and cold days

(a) $(COP)_{\text{Hot days}} < (COP)_{\text{Cold days}}$ (b) $(COP)_{\text{Hot days}} > (COP)_{\text{Cold days}}$

(c) $(COP)_{\text{Hot days}} = (COP)_{\text{Cold days}}$ (d) none of these

15.52. COP of air refrigeration is related with vapour compression as

(a) $(COP)_{\text{air}} > (COP)_{vc}$ (b) $(COP)_{\text{air}} < (COP)_{vc}$

(c) $(COP)_{\text{air}} = (COP)_{vc}$ (d) none of these

15.53. COP of the vapour compression refrigeration system is increased by

(a) reducing evaporator pressure (b) increasing evaporator pressure

(c) increasing condenser pressure (d) none of these

15.54. COP of the vapour compression refrigeration system is increased by

(a) reducing evaporator temperature (b) increasing evaporator temperature

(c) increasing condenser temperature (d) none of these

Answers to Multiple-Choice Questions

15 50. (b) 15.51. (a) 15.52. (b) 15.53. (b) 15.54. (b)

CHAPTER

16

Psychrometry

OVERVIEW

We have discussed the rules governing the mixtures of perfect gases. But in most of the engineering practices, there are some mixtures where one of the constituents is the vapour. Since the vapour in these mixtures is at a very low partial pressure, it can be treated as an ideal gas. Therefore, the general rules applicable to an ideal gas mixture as described in the preceding sections will also be applicable to a mixture of gas and vapour, but there is a remarkable difference in the behaviour of a gas-vapour mixtures as compared to an ideal gas mixtures, in that the composition of the former may change on heating on cooling while in the latter it does not change normally. For example, in a case where the temperature of the gas-vapour mixture is lowered below a certain limit, it is possible that the vapour may condense or solidify resulting in a change in the composition of the mixture. This we have already seen in the chapter 4 where on cooling the vapour condenses to liquid form if its partial pressure is above the triple point pressure and, solidifies directly if the partial pressure is below the triple point pressure. This example clearly suggests that we have to find out the other governing rules applicable for gas-vapour mixture.

Another apt example of gas-vapour mixture is the normal atmospheric air. The atmospheric air is usually a moist air (i.e., a mixture of N_2, O_2, CO_2, water vapour, etc.) and is regarded as a homogeneous two-component system. *A comprehensive study of the properties of moist air is necessary because of its wide applications, in* **air-conditioning, weather forecasting, evaporative cooling towers, etc**.

The branch of science which deals with the properties of air-water-vapour mixtures is called **psychrometry**.

16.1 PSYCHROMETRIC TERMS

Certain terms commonly used in psychrometry are defined below.

16.1.1 Moist and Dry Air

The normal atmospheric air is the ***moist air****.* In other words, it is the mixture of O_2, N_2, CO_2, Argon, some rare gases and water vapour. *When there is absence of water vapour in the atmosphere, it is called* ***dry air****.* The dry air has the following composition by volume.

O_2 – 20.9%, N_2 – 78.03%, Argon – 0.9%, CO_2 – 0.03% and some traces of rare gases.

Thus, we can say that the moist air is the mixture of dry air and water vapour.

The amount of water vapour present in the atmosphere is very small compared to other constituents. ***Hence, at ordinary temperature, the moisture in the atmosphere remains in superheated state.***

16.1.2 Saturated Air Mixture

It is a mixture of dry air and water vapour in which the partial pressure of vapour is equal to the saturation pressure of water at the temperature of the mixture. The mixture in this condition is said to be saturated with vapour.

16.1.3 Unsaturated Air Mixture

It is a mixture of dry air and superheated water vapour, the partial pressure of the vapour being less than the saturation pressure of water at the temperature of the mixture.

16.1.4 Supersaturated Air Mixture

It is the ratio of dry air and water vapour in which the partial pressure of water vapour is greater than the saturation pressure of water at the temperature of the mixture.

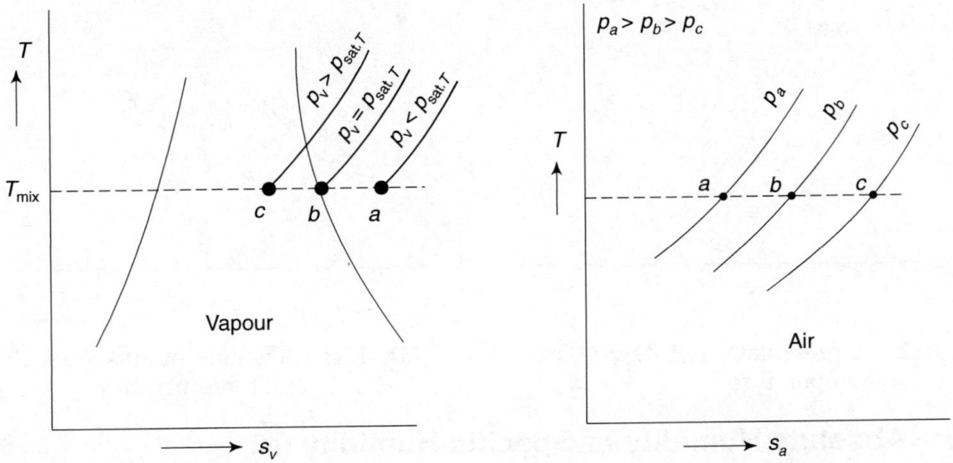

Fig. 16.1 Representation of unsaturated, saturated, and supersaturated air

Figure 16.1 shows that representation for water and dry air on a *T–s* diagram where points *a*, *b* and *c* represent *unsaturated, saturated and supersaturated* mixture respectively at constant temperature. It is to be noted that as the partial pressure of water in a mixture increases, at constant temperature, the partial pressure of air decreases accordingly.

16.1.5 Dry Bulb Temperature (t_{db})

The temperature of a mixture of gases as measured by conventional thermometer is called the dry bulb temperature (d.b.t.). This is denoted as t_{db}.

16.1.6 Dew Point Temperature (t_d)

The temperature at which the mixture becomes saturated or condensation begins when a mixture of dry air and water vapour is cooled at constant pressure from an unsaturated state is called the dew-point temperature (*d.p.t.*). Dew point temperature may be represented on *T-s* diagram of any mixture as shown in Fig. 16.2. The point *A* represents the original state of the water vapour in the mixture which is in superheated region. The pressure at point *A* corresponds to the partial pressure of vapour and the temperature corresponds to the temperature of the mixture. If we cool the mixture at constant pressure till the vapour reaches the saturated state a point *B* is reached after which further cooling will result in the **condensation of vapour or release of moisture. The temperature at B is called the dew point temperature (*d.p.t.*), generally represented as t_{dp}.** Thus, dew point temperature is also the saturation temperature corresponding to initial partial pressure of vapour.

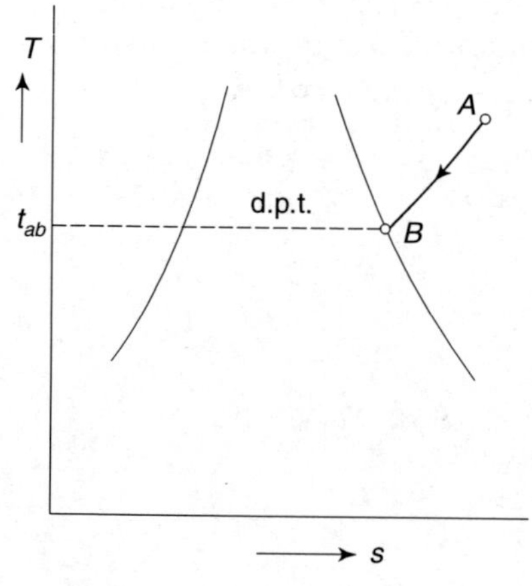

Fig. 16.2 Representation of dew point temperature

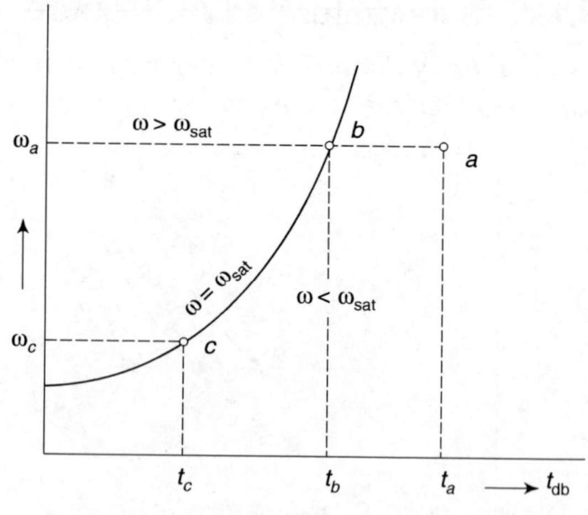

Fig. 16.3 Absolute humidity *vs.* dry bulb temperature

16.1.7 Absolute Humidity or Specific Humidity (ω)

It is defined as the ratio of mass of water vapour, m_v *to the mass of associated dry air,* m_a *in a sample of moist air of mass* ($m_a + m_v$)*. In other words, it is the mass of vapour per unit mass of dry air in any sample of moist air. Thus,*

$$\omega = \frac{m_v}{m_a} \qquad \text{...(16.1)}$$

Applying the perfect gas law, absolute humidity can be expressed as

$$\omega = \frac{m_v}{m_a} = \frac{p_v\, V/R_v\, T}{p_a\, V/R_a\, T} = \frac{p_v\, R_a}{p_a\, R_v}$$

But $$R = R_0/M$$

∴ $$\omega = \frac{p_v\, M_v}{p_a\, M_a} = \frac{M_v}{M_a}\frac{p_v}{p - p_v}$$

or $$\omega = 0.622\,\frac{p_v}{p - p_v} = 0.622\,\frac{p_v}{p_a} \qquad \text{...(16.2)}$$

where p_a and p_v are the partial pressure of air and water vapour respectively in sample of moist air at temperature (T) and volume (V) of the mixture. Here $p = (p_a + p_v)$ is the pressure of the mixture.

16.1.8 Relative Humidity (RH)

It is defined as the ratio of actual partial pressure of water vapour, p_v in moist air at a given temperature to the saturation pressure of the vapour at the temperature. It indicates the degree to which an air-vapour mixture is saturated. In other words, the relative humidity is the ratio of moisture content in the mixture to the moisture content of an equal quantity of saturated air at the same temperature and total pressure. It is expressed on percentage basis. Thus,

$$\phi = \left(\frac{p_v}{p_{v,\,sat}}\right)_T \qquad \text{...(16.3)}$$

If $p_v = p_{v,\,sat}$ then $\phi = 1$, which means that vapour in the moist air is saturated at 100% R. H. Using the perfect gas equation of state, relative humidity is expressed as

$$\phi = \left(\frac{p}{p_{v,\,sat}}\right)_T = \left(\frac{RT/v_v}{RT/v_{v,\,sat}}\right)_T = \left(\frac{v_{v,\,sat}}{v_v}\right)_T = \left(\frac{\rho_v}{\rho_{v,\,sat}}\right)_T \qquad \text{...(16.4)}$$

Equation (16.4) *suggests that the relative humidity is the ratio of density of water vapour at the given temperature to the density of saturated water vapour at the same temperature.*

A relationship between specific humidity, ω and relative humidity, ϕ can be obtained from Eqs. (16.2) and (16.4)

$$\omega = 0.622\,\frac{p_v}{p_a} = 0.622\,\phi\left(\frac{p_{v,\,sat}}{p_a}\right) \qquad \text{...(16.5)}$$

16.1.9 Humidity Ratio or Degree of Saturation (μ)

It is defined, as the ratio of absolute humidity of the moist air to the absolute humidity of saturated air at the same temperature and total pressure. Thus, the degree of saturation (humidity ratio), μ is expressed as

$$\mu = \left(\frac{\omega}{\omega_{sat}}\right)_T \qquad ...(16.6)$$

The saturation ratio, μ can be also expressed in terms of relative humidity, ϕ.

$$\mu = \left[\frac{p_v}{p_a}\cdot\frac{p_{a,\,sat}}{p_{v,\,sat}}\right]_T = \left[\frac{p_v}{p_{v,\,sat}}\cdot\frac{p-p_{v,\,sat}}{p-p_v}\right]_T = \phi\left(\frac{p-p_{v,\,sat}}{p-p_v}\right)$$

or

$$\mu = \phi\left\{\frac{1-(p_{v,\,sat}/p)}{1-(p_v/p)}\right\} \qquad ...(16.7)$$

or

$$\omega = \phi\left\{\frac{1-(p_{v,\,sat}/p)}{1-(p_v/p)}\right\}\omega_{sat} \qquad ...(16.8)$$

For a saturated air mixture, the absolute humidity can be expressed as

$$\omega = 0.622\frac{f(t)}{p-f(t)} \qquad ...(16.9)$$

where saturation pressure of water vapour is expressed as a function to temperature $f(t)$.

At 100% relative humidity, the values of absolute humidity at various temperatures can be calculated by Eq. (16.9) at constant total pressure. Figure 16.3 shows the result of this equation.

In the region below the saturation curve, $\omega < \omega_{sat}$ so that air is unsaturated and the water vapour is superheated. In the region above the saturated curve, $\omega > \omega_{ast}$ and the air is supersaturated with water vapour. The existing pressure of the vapour in this case is larger than the saturation pressure of water at that temperature, so that sudden and irreversible condensation of water, tends to occur, accompanied by a sharp increase in entropy and re-establishment of equilibrium condition. The resulting mixture of air and water vapour is saturated with water vapour.

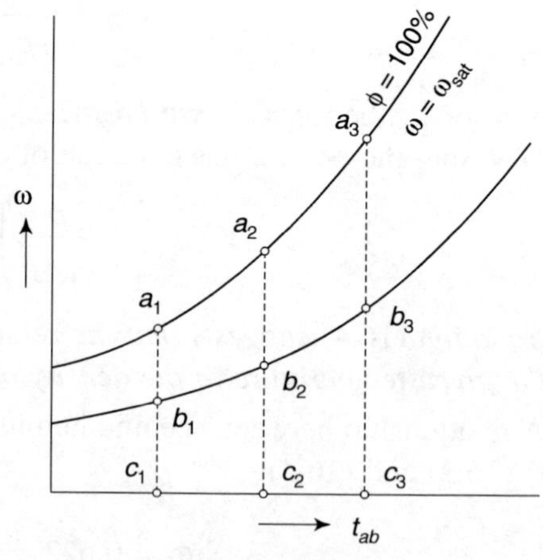

Fig. 16.4 Plot of constant relative humidity lines

It is possible to plot points of constant relative humidity, on a graph of absolute humidity versus temperature shown in Fig. 16.4. For ϕ = 100%, the absolute humidity at various temperature is drawn, using Eq. 16.9. At other selected value of ϕ, absolute humidity is calculated from Eq. 16.5. At room temperature (or lower), the relative humidity can be assumed equal to the saturation ratio and, since the saturation ratio is the ratio of two values of absolute humidity at constant temperature, point of constant (ϕ) can be determined. Hence,

$$\mu_1 = \phi_1 = \frac{b_1\,c_1}{a_1\,c_1} = \frac{b_2\,c_2}{a_2\,c_2} = \frac{b_3\,c_3}{a_3\,c_3}$$

In order to establish the constant relative humidity lines, the ordinates of the saturation curve are first divided into equal parts and points at 10% increment in relative humidity are established from them.

It is to be noted that there is a convenient way to measure ϕ, μ or ω as their expression involves the term p_v and $p_{v,\,sat}$. The value of $p_{v,\,sat}$ may be obtained by two ways.

The first way is to find the dew point temperature and then find the saturation pressure of steam corresponding to that temperature. The second method is to find the wet bulb temperature which will be discussed later on.

Example 16.1 An air-water mixture at 20°C and 1 bar has a relative humidity 80 per cent. Calculate (a) the partial pressures of the vapour and the air, (b) the specific humidity, (c) the humidity ratio, (d) the dew point, (e) the density of the mixture, and (f) if the mixture is cooled at constant pressure to a temperature of 10°C, find the amount of water-vapour condensed per kg of dry air.

Solution From Steam Tables at 20°C, $p_{v\,sat} = 0.02337$ bar

(a) But
$$\phi = \frac{p_v}{p_{v,\,sat}} = 0.8$$

$\therefore$
$$p_v = 0.8 \times 0.02337 = 0.018696 \text{ bar} \qquad \textbf{Ans.}$$
$$p_a = p - p_v = 1 - 0.018696 = 0.9813 \text{ bar} \qquad \textbf{Ans.}$$

(b)
$$\text{Specific humidity, } \omega = 0.622\,\frac{p_v}{p_a}$$
$$= \frac{0.622 \times 0.018696}{0.9813} = 0.01185 \text{ kg/kg of dry air} \qquad \textbf{Ans.}$$

(c)
$$\mu = \frac{\omega}{\omega_{sat}} = \phi\,\frac{p - p_{v,\,sta}}{p - p_v} = 0.8 \times \frac{0.97663}{0.9813} = 0.79319 \qquad \textbf{Ans.}$$

(d) From the Steam Table, at $p_v = 0.01896$ bar
$$t_{dp} = 16.309°\text{C} \qquad \textbf{Ans.}$$

(e)
$$\rho = \rho_a\,(1 + \omega) = \frac{p_a}{R_a\,T_a} \times (1 + \omega)$$
$$= \frac{0.9813 \times 10^2}{0.287 \times 293}\,(1 + 0.01185) = 1.1807 \text{ bar} \qquad \textbf{Ans.}$$

(f) Since 10°C is below the dew point, some of the vapour will condense, leaving a mixture of 100% relative humidity

At
$$10°\text{C},\ p_v = p_{v,\,sat} = 0.01227 \text{ bar}$$

$\therefore \qquad \omega_1 = 0.622 \frac{p_v}{p_a} = 0.622 \times \frac{0.01227}{1 - 0.01227} = 0.007726$ kg/kg of day air

The amount of vapour condensed per kg of dry air is equal to

$$\omega - \omega_1 = 0.01185 - 0.007726 = 0.004124 \text{ kg/kg of dry air} \qquad \textbf{Ans.}$$

Example 16.2 A mixture of air and water vapour possesses a volume of 700 m^3 at 1 bar pressure and temperature 35°C. Its relative humidity is 75%. Find the specific humidity, the dew point, air mass and vapour mass in the mixture.

Solution From Steam Tables at 35°C.

$$p_{v,\,sat} = 0.05622 \text{ bar}$$

But $\qquad \phi = 0.75 = \left(\frac{p_v}{p_{v,\,sat}}\right)_{35°C} = \frac{p_v}{0.05622}$

or $\qquad p_v = 0.75 \times 0.05622 = 0.04216$ bar

(a) $\qquad \omega = 0.622 \frac{p_v}{p_a} = 0.622 \times \frac{0.04216}{1.013 - 0.04216} = 0.0279$ kg/kg of dry air

(b) From Steam Tables, at $p_v = 0.04216$ bar

$$t_{sat} = t_{dp} = 29.8694°C \qquad \textbf{Ans.}$$

(c) $\qquad m_a = \frac{p_a V}{R_a T} = \frac{0.97084 \times 10^2 \times 700}{0.287 \times 308} = 768.76$ kg $\qquad$ **Ans.**

But $\qquad \omega = \frac{m_v}{m_a}$

$\therefore \qquad m_a = 0.0279 \times 768.76 = 20.756$ kg $\qquad$ **Ans.**

16.2 WET BULB TEMPERATURE (t_{wb})

The wet bulb temperature is measured by a thermometer where bulb is covered with a wick saturated with water and placed in moist air (i.e., the air-vapour stream).

Generally, a wet bulb thermometer and a dry bulb (ordinary) thermometer are mounted together and the combined instrument is known as a ***psychrometer***. Figure 16.5 shows a psychrometer known as the *sling psychrometer* which consists of two thermometers side by side on a holder. In order to whirl the whole device in the air, a handle is provided. A vigorous rotation of the psychrometer helps to get a quick and dependable readings on the wet bulb thermometer and it is a must.

If the air surrounding the wet bulb thermometer is unsaturated (i.e., normal atmospheric air), the pressure of the water vapour at the surface of the wick exceeds the partial pressure of water vapour in the mixture. Due to the difference in chemical potential between the liquid and the vapour phases, evaporation of water from the wick and the thermometer which decreases the temperature of the wick, i.e., the wet bulb temperature till equilibrium is attained. Equilibrium is attained when the chemical potentials of the liquid and the vapour phases are identical and at this point, the wick is at the *so called wet bulb temperature. It is to be noted that wet bulb temperature is commonly less than the* dry-bulb temperature.

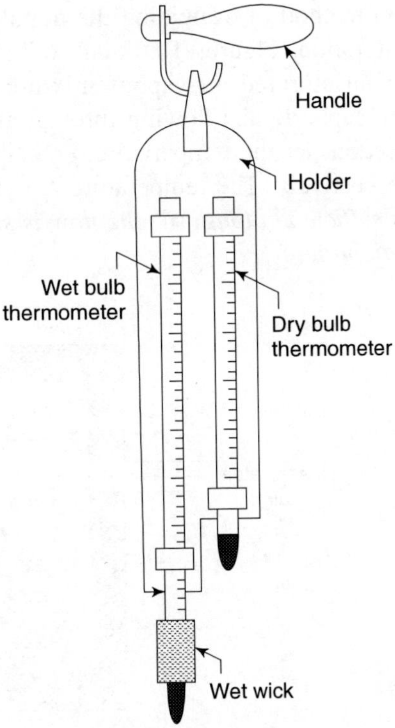

Fig. 16.5 Sling psychrometer

If is evident that if the relative humidity of the ambient air is less, there will be a greater affinity to absorb more moisture, hence evaporation of water from the bulb surface will be greater and greater will be its *wet bulb depression* ($t_{db} - t_{wb}$), i.e., below dry bulb temperature values). If the air surrounding the wet bulb temperature is saturated, there will be no evaporation on the bulb and the same temperature will be read on both dry bulb and wet bulb thermometers. The wet bulb depression (*wbd*) is expressed as

$$wbd = t_{db} - t_{wb}$$

If $\phi = 1, \quad t_{db} = t_{wb},$

Hence, [wet bulb depression]$_{\phi = 1}$

From the above discussion, we can conclude that wet bulb temperature is an indication of indirect moisture content of air. It is difficult to find an analytical expression of the t_{wb} in terms of p_v due to various inaccuracies in measurement.

The following are the empirical relations for the t_{wb} in terms of partial pressure of vapour, p_v:

(a) Modified *Apjohn* Equation

$$p_v = (p_{v,\text{sat}})_{t_{wb}} - \frac{1.8\,(t_{db} - t_{wb})}{2700} \quad \text{...(16.10)}$$

(b) Modified *Ferral* Equation

$$p_v = (p_{v,\text{sat}})_{t_{wb}} - 0.00066\,p\,(t_{db} - t_{wb})\left(1 + \frac{1.8\,t_{db}}{1571}\right) \quad \text{...(16.11)}$$

(c) The Carrier Equation

$$p_v = (p_{v,\text{sat}})_{t_{wb}} - \frac{\{(p - p_{v,\text{sat}})\,t_{wb}\}\,(t_{db} - t_{wb}) \times 1.8}{2854 - 1.325\,(1.8\,t_{db} + 32)} \quad \text{...(16.12)}$$

where p_v = partial pressure of water vapour

$(p_{v,\text{sat}})$ = pressure of saturated vapour at *w.b.t.*

p = barometric pressure

16.3 ADIABATIC SATURATION TEMPERATURE OR THERMODYNAMIC WET BULB TEMPERATURE

On method of evaluating the humidity ratio of a mixture is based on technique of adiabatic saturation. Figures 16.6 and 16.7 show the adiabatic saturation process. Consider a stream of unsaturated air-vapour mixture having relative humidity less than 100% (i.e., normal atmospheric air) flowing through an insulated tank over a surface of water. At the entrance section, let the temperature, enthalpy and relative humidity of the mixture be t_1, h_1 and ω_1 respectively. The temperature t_1 is the dry bulb temperature. It is assumed that (i) *the tank is sufficiently long*, (ii) *the flaw is steady*, (iii) *there is no work done, and* (iv) *the process is adiabatic*.

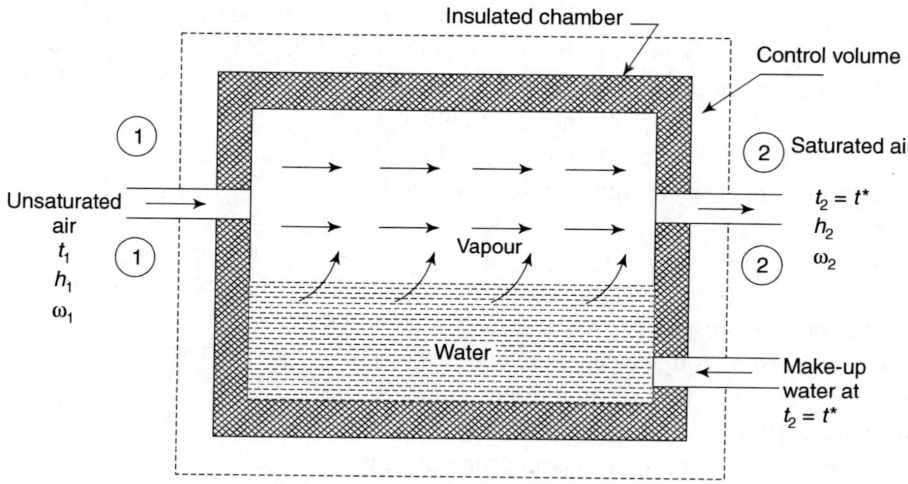

Fig. 16.6 Adiabatic saturation

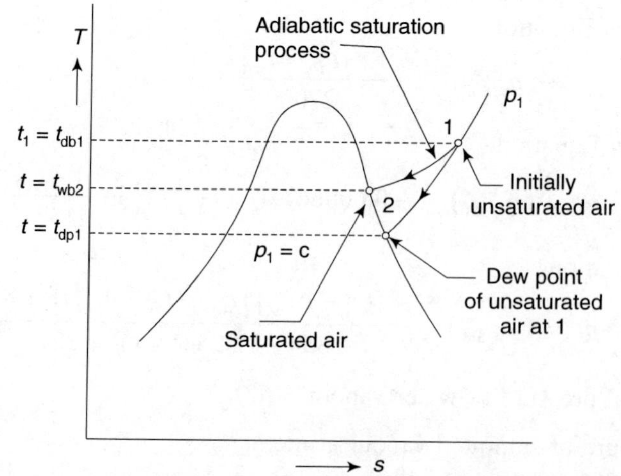

Fig. 16.7 Adiabatic saturation process on *T–s* diagram

Since the moist air entering the tank is unsaturated, when it passes over the water surface, some of water will evaporate and will be carried away by the air resulting in an increase in humidity of air. Heat is required for the evaporation of water, which comes from both the air mixture passing through and the liquid water in the tank. Thus, both air and water cool down due to evaporation till the air becomes saturated, i.e., equilibrium is reached. It is to be noted that the lower the initial relative humidity, the greater is the amount of evaporation and decrease in temperature. This decrease in temperature measures the humidity ratio (μ). It is assumed that by the time the air reaches the exist section 2, it is saturated, and the air at this section attains a temperature called the ***adiabatic saturation temperature*** **(t^*)**. This will also be the temperature of the water in the tank under equilibrium condition. In order to maintain a steady state and thermal equilibrium, make up water must be supplied at the adiabatic saturation temperature. **The adiabatic saturation temperature is also called the *thermodynamic wet bulb temperature*.**

Thus, considering the mixing process quantitatively, an adiabatic saturation temperature is defined as the temperature which results from adding water adiabatically to the air in a steady flow until it becomes saturated, the water being supplied at the final temperature of the mixture.

The adiabatic saturation process may be represented on $T–s$ diagram as shown in Fig. 16.7. The state 1 shows that initial condition of unsaturated air entering the tank.

The process 1-2 shows the adiabatic saturation process. The partial pressure of water vapour increases during the process from p_1 to p_2 while the total pressure p of the mixture remains constant. It is obvious from the diagram that the vapour initially at *d.b.t.*, t_1 is cooled adiabatically to adiabatic saturation temperature, t^*. It is also called thermodynamic *wet bulb temperature*. At the saturation condition, the dry bulb temperature at section 2 will be equal to wet bulb temperature.

It is interesting to note that adiabatic saturation (t^*) indicates indirectly the amount of moisture present in the entering moist air.

Since the system is insulated and no work is done, the first law yields

$$m_a\, h_{a1} + m_{v1}\, h_{v1} + (m_{v2} - m_{v1})\, h_{f2} = m_a\, h_{a2} + m_{v2}\, h_{v2}$$

where m_a = mass of dry air

m_v = mass of vapour

h_a = enthalpy of air per kg dry air, kJ/kg of dry air

h_f = enthalpy of water per kg of dry air, kJ/kg of dry air

h_v = enthalpy of vapour per kg of dry air, kJ/kg of dry air

Suffixes 1 and 2 refer to the inlet and outlet conditions, i.e., *usaturated and saturated conditions,* respectively.

$(m_{v2} - m_{v1})$ = mass of water added to makeup.

After dividing by m_a, the above equation becomes

$$h_{a1} + \omega_1 h_{v1} + (\omega_2 - \omega_t) h_{f2} = h_{a2} + \omega_2 h_{v2} \quad ...(16.13)$$

where, ω = absolute or specific humidity

Solving for ω_1, we have

$$\omega_1 = \frac{h_{a2} - h_{a1} + \omega_2 (h_{v2} - h_{f2})}{h_{v1} - h_{f2}} \quad ...(16.14)$$

At low pressures, the water vapour does not deviate significantly from an ideal gas, even near the saturated vapour state. Consequently, enthalpy is a function of temperature only. Hence, $h_{v1} = h_{g1}$ and $h_{v2} = h_{g2}$ where suffix g stands for gas (dry and saturated steam). Thus, the above equation becomes

$$\omega_1 = \frac{h_{a2} - h_{a1} + \omega_2 (h_{g2} - h_{f2})}{h_{g1} - h_{f2}} \quad ...(16.15)$$

But $$h_{a2} - h_{a1} = c_{pa} (T_2 - T_1)$$

and $$h_{g2} = h_{f2} + h_{fg2}$$

or $$h_{g2} - h_{f2} = h_{fg2}$$

where h_{fg2} is that latent heat of vaporization of water corresponding to adiabatic saturation temperature, $t^* = t_2$

Thus, Eq. (16.14) becomes

$$\omega_1 = \frac{\omega_2 h_{fg2} - c_{pa} (T_1 - T_2)}{h_{g1} - h_{f2}} \quad ...(16.16)$$

The properties at seaction 2, i.e., at adiabatic saturation temperature t^*, may be marked with star for clear understanding, i.e., $h_{fg2} = h^*$

$$\omega_1 = \frac{\omega^* h_{fg}{}^* - c_{pa} (T_1 - T^*)}{h_{v1} - h_f{}^*} \quad ...(16.17)$$

Once the temperature T_1 and $T_2 = T^*$ are measured, every quantity of the right-hand side of the equation is known. The humidity ratio ($\omega^* = \omega_2$) at $T_2 = T^*$ is kown since this state is a saturated state. Thus, the humidity ratio (and hence the relative humidity) of a unsaturated gas vapour mixture could be ascertained by measuring the inlet and outlet temperature and the total pressure of the mixture which undergoes a process of *adiabatic saturation.*

The adiabatic saturation temperature is nearly equal to the wet-bulb saturation temperature. But some difference does exist. The wet bulb temperature depends on both mass transfer and heat transfer from the liquid phase. On the other hand, the adiabatic saturation temperature (also called the thermodynamic wet bulb temperature) involves an equilibrium in which no transfer of heat occurs from the liquid phase to the gas phase. Although, the technique of adiabatic saturation gives the desired results but it is difficult in practice to attain a saturated state by this method without employing an extremely long flow channel which is not possible in practice. Since the difference between the two temperatures is negligible, the *wet bulb temperature is ordinarily used in place of thermodynarric wet bulb temperature* (i.e., adiabatic saturation temperature) *in humidity calculations.*

16.4 ENTHALPY OF MOIST AIR

The enthalpy of moist air is expressed as the sum of the enthalpy of 1 kg of dry air and associated water vapour. Thus,

$$h = h_a + \omega \,.\, h_v = c_p\, t_{db} + \omega \,.\, h_v \quad ...(16.18)$$

where h = enthalpy of moist air, kJ/kg of dry air

c_p = specific heat of air at constant pressure = 1.005 kJ/kg K

t_{db} = d.b.t. of air, °C

ω = absolute humidity of air, kg of vapour per kg of dry air

h_v = enthalpy of vapour, kJ/kg obtained from tables

But
$$h_v = 4.18\, t_{db} + h_{fg\,.\,db} + 1.88\,(t_{db} - t_{dp}) \quad ...(16.19)$$

Here, specifie heat of superheated steam is taken as 1.88 kJ/kg K

∴
$$h = (1.005 + 1.88\,\omega)\, t_{db} + \omega\,(h_{fg.\,db} + 2.3\, t_{dp}) \quad ...(16.20)$$

or
$$h = c_m\, t_{db} + \omega\,(h_{dp} + 2.3\, t_{dp})$$

where c_m = (1.005 + 1.88 ω) is termed as humid heat capacity. In other words, capacity of mixture per kg of dry air or heat capacity of (1 + ω) kg of mixture.

In general case, if the specific humidity and *d.b.t.* of entering moist air are expressed as ω and *t* respectively, Eq. (16.17) becomes

$$\omega = \frac{\omega^* h^*_{fg} - 1.005\,(t - t^*)}{h_v - h^*_f} \quad ...(16.21)$$

or
$$\omega = \frac{\omega^* h^*_{fg} - 1.005\,(t - t^*)}{h^*_{fg} + 1.88\,(t - t^*)}$$

or
$$t^* = t - \frac{h^*_{fg}}{(1.005 + 1.88\,\omega)}(\omega^* - \omega) \quad ...(16.22)$$

or
$$t^* = t - \frac{h_{fg}}{c_m}(\omega^* - \omega) \quad ...(16.23)$$

Equation (16.13) may be rearranged as

$$h_{a1} + \omega_1\, h_{v1} + \omega_2\, h_{f2} = h_{a2} + \omega_2\, h_{v2} + \omega_1\, h_{f2} = \Sigma$$

or $$h_1 + \omega_2\, h_{f2} = h_2 + \omega_1\, h_{f2} = \Sigma$$

Let us define a new function called *sigma function* ($\Sigma = h + \omega\,.\,h_f$). The above equation shows that the sigma function remains constant during an adiabatic saturation process. Thus,

$$\Sigma_1 = \Sigma_2$$

We also know that $\omega \cdot h_f$ is very small and hence it may be neglected, thus,

$$h_1 = h_2 \qquad \text{...(16.24)}$$

Equation (16.24) shows that the enthalpy of a moist air remains constant during an adiabatic saturation process.

Example 16.3 Air at 1 bar has a *d.b.t.* of 25°C and *w.b.t.* of 15°C. Calculate (a) the vpour pressure, (b) the relative humidity, (c) the specific humidity, (d) vapour density in air, (e) the dew point temepratu re, and (f) the enthalpy of mixture.

Solution Given $t_{db} = 25°C,\ t_{wb} = t^* = 15°C,\ p = 1$ bar

(a) From Steam Tables at 25°C, $p_{v.\,sat} = 0.03166$ bar, $v_v = 43.4$ m^3/kg, $h_v = 2547.3$ kJ/kg

At 15°C, $p_{v.\,sat} = 0.01703$ bar, $v_v = 77.98$ m^3/kg, $h_f = 62.94$ kJ/kg.

Using the Carrier equation, the vapour pressure is given by

$$p_v = (p_{v,\,sat})_{t_{wb}} - \frac{\{p - (p_{v,\,sat})_{t_{wb}}\}\,(t_{db} - t_{wb}) \times 1.8}{2854 - 1.325\,(1.8\,t_{db} + 32)}$$

$$= 0.01703 - \frac{(1 - 0.01703)\,(25 - 15) \times 1.8}{2854 - 1.325\,(1.8 \times 25 + 32)} = 0.010606 \text{ bar} \quad \textbf{Ans.}$$

(b) $$\phi = \frac{p_v}{p_{v,\,sat}} = \frac{0.010606}{0.03166} = 0.3349 = 33.49\% \quad \textbf{Ans.}$$

(c) $$\omega = 0.622 \frac{}{p - p} = 0.622 \times \frac{0.010606}{1 - 0.01065}$$

$$= 0.00668 \text{ kg/kg of dry air} \quad \textbf{Ans.}$$

The value of ω may also be obtained from the equation

$$\omega = \frac{\omega^*\, h_{fg} - 1.005\,(t_{db} - t^*)}{h_v - h_f^*} = 0.00668 \text{ kg/kg of dry air}$$

(d) The specific volume of dry and saturated steam at 25°C = 43.40 m^3/kg

$\therefore$ Vapour density $= \dfrac{1}{43.40} = 0.02304$ kg/m^3 of saturated steam

Hence, vapour density at (ϕ = 33.49%) = 0.02304 × 0.3349 = 0.00771 kg/m^3 **Ans.**

(e) For p_v = 0.010606 bar, the saturation temperature is 7.71°C.

$\therefore$ t_{dp} = 7.71°C

(f) The enthalpy of the mixture is

$$h = c_p\, t + \omega \,.\, h_{fg} = (1.005 \times 25) + 0.00668 \times 2547.2 = 42165 \text{ kJ/kg} \quad \textbf{Ans.}$$

16.5 PSYCHROMETRIC CHART

In order to facilitate the computation of process changes for moist air (dry air-water-vapour mixtures), it is convenient to plot parameters on a diagram known as **psychrometric chart**. In other words, the psychrometric chart shows the various properties of moist air. *It is essentially a composition temperature chart. The specific humidity and the partial pressure are the ordinates, the dry bulb temperature is the abscissa.* On the chart shown in Fig. 16.8, lines of constant relative humidity, volume, wet-bulb temperature and enthalpy are indicated. *Most psychrometric charts are constructed from a constant total pressure of 1.013 bar and some charts contain correction curves for pressure other than atmospheric.*

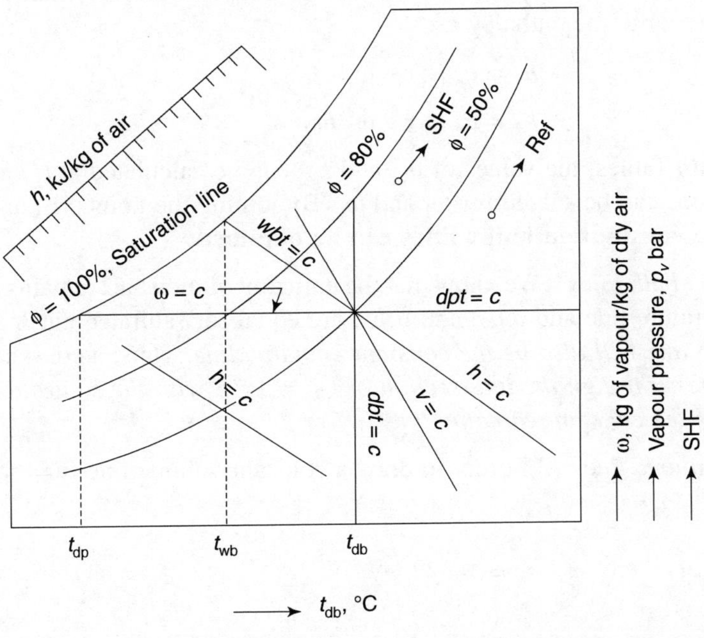

Fig. 16.8 Components of psychrometric chart

The various lines drawn on the chart are discussed below.

(a) *Constant specific humidity lines* are the horizontal lines perpendicular to the vertical specific humidity scale.

(b) *Constant dry bulb temperature lines are vertical lines perpendicular to the horizontal dry-bulb temperature lines.*

(c) *Saturation curve* corresponds to relative humidity of 100%. This curve is shown in the extreme left for different values of dry bulb temperature.

(d) *Constants R.H. lines,* in addition to saturation curve, various constant relative humidity lines say ϕ = 95%, 90%, 85%, etc., are drawn on the chart. Now, suppose we have to draw a constant R.H. line of ϕ = 60%. For this, let us select dry bulb temperatures t_1, t_2, t_3, etc., and find the corresponding vapour saturation pressure $(p_{v,\,sat})_1$; $(p_{v,\,sat})_2$, $(p_{v,\,sat})_3$, etc., *from the steam tables. As* $p_v = \phi\, p_{v,\,sat} = 0.6\, p_{v,\,sat}$, *the value of* p_{v1}, p_{v2}, p_{v3}, *can be calculated. We know that* ω *is expressed as* $\omega = 0.622\,[p_v/(p - p_v)]$, *we can calculate the corresponding values of* ω_1, ω_2, ω_3, etc. for ϕ = *60%. If the points* (t_1, ω_1) (t_2, ω_2) (t_3, ω_3), *etc., are joined by a curve, it will give the curve for* ϕ = *60%.*

(e) *Constant enthalpy lines,* in order to draw the constant enthalpy line, consider the equation

$$h = c_p\,(t) + \omega\, h_v$$

Since $h_v = f\,(t)$, hence along any vertical line, it is a constant. For dry bulb temperature t_1, t_2, t_3, etc., we may write the enthalpy as

$$h_1 = c_p\,(t_1) + \omega_1\, h_{v1}$$

$$h_2 = c_p\,(t_2) + \omega_2\, h_{v2}$$

Using the Steam Tables, the values of h_{v1} and h_{v2} can be calculated at t_1 and t_2. Thus, the above two equations can be solved for ω_1 and ω_2. By joining the points (t_1, ω_1) (t_2, ω_2), etc., for enthalpies, the **constant enthalpy lines can be obtained**.

(f) *Constant wet bulb lines* We know that the enthalpy of moist air remains constant along a wet bulb temperature line and *d.b.t.* and *w.b.t.* are equal for saturated air. *It means that the constant wet bulb line will also be the constant enthalpy line*. Thus, a constant *w.b.t.* line of t_{wb} = 25°C *will pass through the intersection of* t_{db} = 25°C *with the saturation curve and it will coincide with the constant enthalpy line.*

(g) *Constant volume lines* In order to draw a constant volume line ($v = c_1$), consider the equation

$$\omega = 0.622\,\frac{p_v}{p_a}$$

For perfect gas, $$p_a = \frac{R_a \cdot T}{c_1} \quad \text{and} \quad p_v = p - \frac{R_a\, T}{c_1}$$

$$\therefore \qquad \omega = \frac{0.622\,(p - R_a\,T/c_1)}{R_a\,T/c_1} \qquad ...(16.25)$$

The curves drawn through the points (t_1, ω_1) (t_2, ω_2), etc., give the *constant volume line.*

(h) *Sensible heat factor (SHF) scale* The sensible heat factor (S.H.F.) is defined as

$$\text{SHF} = \frac{\text{Sensible heat}}{\text{Sensible heat + Latent heat}} = -\frac{\text{SH}}{\text{SH + LH}} \qquad ...(16.26)$$

In Fig. 16.8, this scale is on the extreme right. This is drawn with reference to the point at a 50% RH line near 27°C *d.b.t.* If we join this point to the appropriate value on the SHF scale, the resulting line represents the inclination of all heating or cooling process of most air which are carried on according to the value of SHF. For example, the slope of line of SHF = 0.75 is shown in Fig. 16.8 represents the cooling process of moist air in a manner that the sensible cooling (temperature drop) is 3 times of the latent heat cooling. Thus,

$$\text{SHF} = 0.75 = \frac{1}{1 + 1/3}$$

Example 16.4 Atmospheric air at 76 cm of Hg has 25°C *d.b.t.* and 15°C *w.b.t.* Using psychrometric chart, calculate (a) relative humidity, (b) humidity ratio, (c) *d.b.t.* (d) enthalpy, (e) partial pressure of vapour, (f) saturation pressure corresponding to *d.b.t.* of 25°C, (g) saturation pressure corresponding to *w.b.t.* of 15°C and (h) specific volume of air.

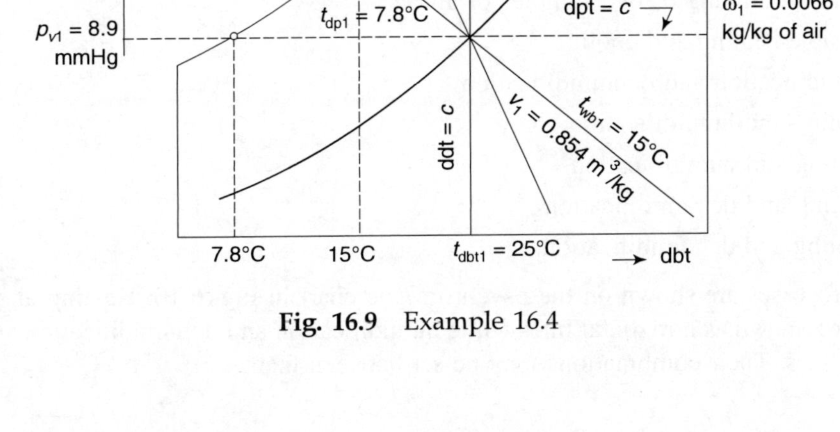

Fig. 16.9 Example 16.4

Solution Refer to Fig. 16.9.

(a) The point 1 on the psychrometric chart is set according to the $t_{db1} = 25°C$ and $t_{wb1} = 15°C$.

From the chart, (a) $\phi_1 = 33.8\%$ **Ans.**

(b) The horizontal line from 1 to the right representing constant humidity ratio line give

$$\omega_1 = 0.0066 \text{ kg/kg of dry air} \qquad \textbf{Ans.}$$

(c) The horizontal line form 1 to the left cuts the saturation line giving

$$t_{dp1} = 7.8°C \qquad \textbf{Ans.}$$

(d) Constant enthalpy line and constant *w.b.t.* line are the same. This gives

$$h_1 = 42.0303 \text{ kJ/kg of dry air}$$

(e) The horizontal line from 1 that cuts the vapour pressure scale gives

$$p_{v1} = 8.9 \text{ mm of Hg} \qquad \textbf{Ans.}$$

(f) The constant *d.b.t.* line (vertical) from 1 cuts the saturation line at point *a*. The horizontal line from *a* to the pressure line gives

$$p_{s,\,dbt} = 23.7 \text{ mm of Hg} \qquad \textbf{Ans.}$$

(g) The *w.b.t* line from 1 cuts the saturation line at *b*. The horizontal line from *b* to the vapour prssure gives

$$p_{s.wbt} = 12.5 \text{ mm kg} \qquad \textbf{Ans.}$$

(h) The specific volume line passing through the point 1 gives

$$v_1 = 0.854 \text{ m}^3\text{/kg of dry air} \qquad \textbf{Ans.}$$

16.6 PSYCHROMETRIC PROCESSES (AIR CONDITIONING PROCESSES)

The most common processes in air conditioning involving air-water mixtures are:

(i) Adiabatic mixing of two samples of air
(ii) Sensible heating and cooling
(iii) Humidification and dehumidification
(iv) Cooling and humidification
(v) Heating and humidification
(vi) Heating and dehumidification
(vii) Cooling and dehumidification.

These processes are shown on the psychrometric chart in Fig. 6.10. Heating and cooling processes are shown as horizontal lines while humidification and dehumidification are shown as vertical lines. Their combinations may be set between them.

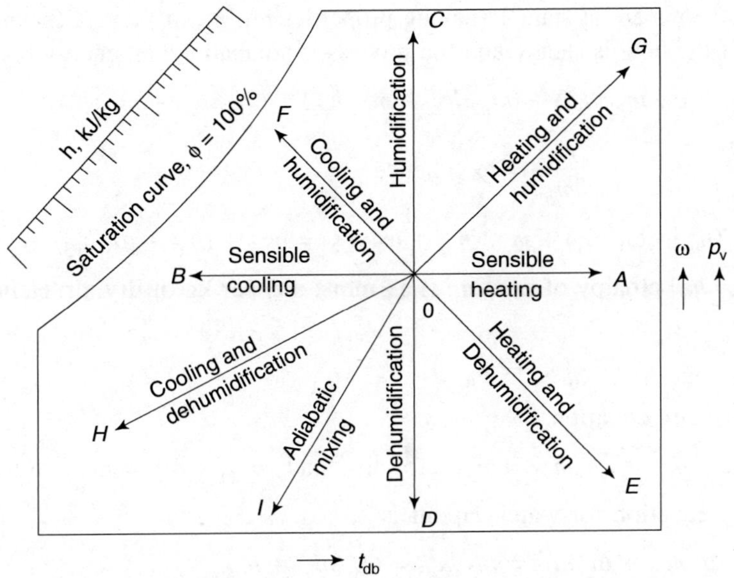

Fig. 16.10 Psychrometric process

16.7 ADIABATIC MIXING OF AIR STREAMS

This process is best illustrated by considering the mixing of two streams of humid air as shown in Fig. 16.11 (a). The proces is represented on the psychrometric chart as shown in Fig. 16.11 (b).

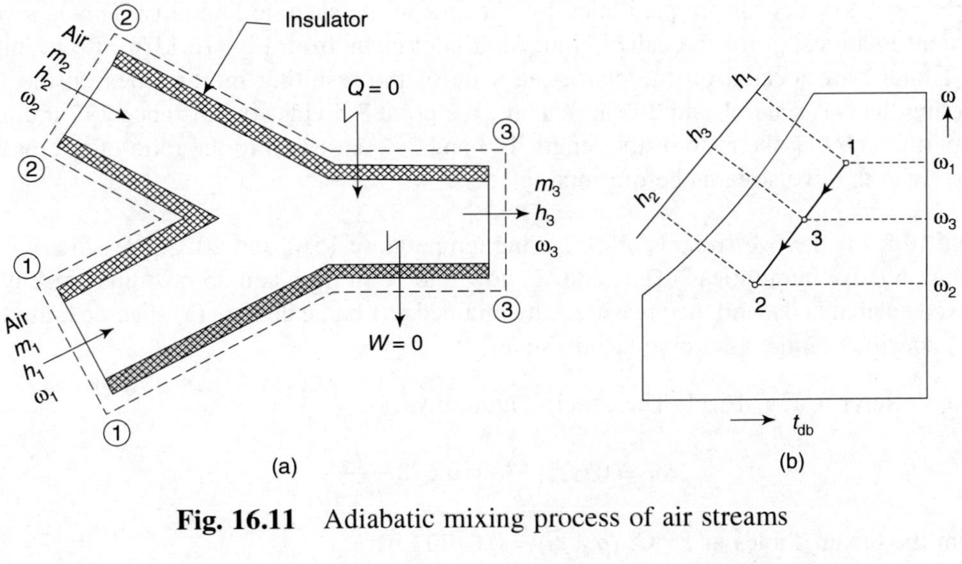

Fig. 16.11 Adiabatic mixing process of air streams

The original streams are at state 1 (having properties m_1, h_1, ω_1) and 2 (having properties m_2, h_2 and ω_2). If the flow is steady and the process is adiabatic, then energy balance leads to

$$(m_{a1}\, h_{a1} + m_{v1}\, h_{v1}) + (m_{a2}\, h_{a2} + m_{v2}\, h_{v2}) = (m_{a3}\, h_{a3} + m_{v3}\, h_{v3})$$

But $$\frac{m_v}{m_a} = \omega, \quad \text{hence}$$

$$m_{a1}\,(h_{a1} + \omega_1\, h_{v1}) + m_{a2}\,(h_{a2} + \omega_2\, h_{v2}) = m_{a3} + (h_{a3} + \omega_3\, h_{v3})$$

But $h_a + \omega h_v = h$ = ethalpy of mixture (i.e., moist air) per kg of dry air. Hence, the above equation becomes

$$m_{a1}\, h_1 + m_{a2}\, h_2 = m_{a3}\, h_3 \qquad \text{...(16.27)}$$

From the continuity equation,

$$m_{a1} + m_{a2} = m_{a3} \quad \text{for air} \quad \text{and} \quad m_{v1} + m_{v2} = m_{v3} \quad \text{for water vapour}$$

The continuity equation for water vapour is

$$\omega_1\, m_{a1} + \omega_2\, m_{a2} = \omega_3\, m_{a3} = \omega_3\,(m_{a1} + m_{a2})$$

The Eq. (16.27) thus becomes $m_{a1}\, h_1 + m_{a2}\, h_2 = (m_{a1} + m_{a2})\, h_3$

or $$\frac{m_{a1}}{m_{a2}} = \frac{h_3 - h_2}{h_1 - h_2} \qquad \text{...(16.28a)}$$

or $$\frac{m_{a1}}{m_{a2}} = \frac{\omega_3 - \omega_2}{\omega_1 - \omega_3} \qquad \text{...(16.28b)}$$

Comparison of Eqs. (16.28 a) and (16.28 b) shows that the changes in enthalpy of the mixtures per kg of dry air are paralleled by changes in the specific humidity. This is a very convenient relationship for the calculation. As it is obvious from Fig, 16.11(b), if the initial states 1 and 2 are located on the chart, the state of the resulting mixture lies on the line connecting the two states 1 and 2 is at 3. Thus, the point 3 divides the distance 1–2 into two equal parts, such that the ratio of the length 1–3 and 3–2 are equal to the ratio of the masses for dry air in the two streams before mixing.

Example 16.5 If the two streams of air having temperature 15°C and 20°C, pressure of both 1.013 bar, relative humidities 20% and 80%, flow rate 18 m^3/min and 25 m^3/min respectively are mixed adiabatically, and the pressure is maintained at 1 bar, calculate (a) specific humidity, and (b) specific volume assuming steady state.

Solution Refer to Fig, 16.11. The specific humidity,

(a) $$\omega_1 = 0.622\,\frac{p_{v1}}{p_a} = 0.622\,\frac{p_{v1}}{p - p_{v1}}$$

From the Steam Tables at 15°C, $(p_{v1.\,sat}) = 0.01703$ bar

$$\phi_1 = 0.2 \quad \frac{p_{v1}}{(p_{v1 \cdot sat})} = \frac{p_{v1}}{0.01703}$$

$$\therefore \quad p_{v1} = 0.01703 \times 0.2 = 0.003406 \text{ bar}$$

$$\therefore \quad \omega_1 = 0.622 \times \frac{0.003406}{1.013 - 0.003406} = 0.002098 \text{ kJ/kg of dry air} \qquad \textbf{Ans.}$$

From Steam Tables at 20°C, $p_{v2.sat} = 0.02337$ bar

$$\phi_1 = 0.8 = \frac{p_{v2}}{p_{2 \cdot sat}} = \frac{p_{v2}}{0.02337}$$

or

$$p_{v2} = 0.018696 \text{ bar}$$

$$\omega_2 = 0.622 \times \frac{0.018696}{1.013 - 0.018696} = 0.01169 \text{ kg/kJ of dry air} \qquad \textbf{Ans.}$$

$$m_{a1} = \frac{p_{a1}\, V_1}{R_a\, T_1} = \frac{(1.013 - 0.003406) \times 10^2 \times 18}{0.287 \times 288} = 21.895 \text{ kg/min}$$

$$m_{a2} = \frac{p_{a2}\, V_2}{R_a\, T_2} = \frac{(1.013 - 0.08696) \times 10^2 \times 25}{0.287 \times 293} = 27.53 \text{ kg/min}$$

The continuity equation for the vapour is

$$\omega_1\, m_{a1} + \omega_2\, m_{a2} = \omega_3\, (m_{a1} + m_{a2})$$

$$\therefore \quad \omega_3 = \frac{0.002098 \times 21.985 + 0.01185 \times 27.53}{(21.985 + 27.53)} = 0.007519 \text{ kg/kg of dry air}$$

The values of ω_1, ω_2 and ω_3 may also be obtained from the Psychrometric chart as

$$\omega_1 = 0.002098, \omega_2 = 0.01185 \quad \text{and} \quad \frac{m_{a1}}{m_{a2}} = \frac{\omega_3 - \omega_2}{\omega_2 - \omega_3}$$

$$\therefore \quad \omega_3 = 0.00768 \text{ kg/kg of dry air}$$

∴ The vapour pressure of the mixture is

$$p_{v3} = \frac{\omega_3\, p_3}{\omega_3 + 0.622} = \frac{0.007519 \times 1}{0.00768 + 0.622} = 0.01194$$

Engery balance gives

$$m_{a1}\, (c_{pa} + \omega_1\, c_{pv})\, t_1 + m_{a2}\, (c_{pa} + \omega_2\, c_{pv})\, t_2 = m_{a3}\, (c_{pa} + \omega_3\, c_{pv})\, t_3$$

$$\therefore \quad 21.985\ (1.005 + 0.002098 \times 1.84) \times 15 + 27.53\ (1.005 + 0.01185 \times 1.84) \times 20$$

$$= 49.425\ (1.005 + 0.00769 \times 1.84)\, t_3$$

or

$$t_3 = 18.18°\text{C}$$

From Steam Tables at $t_3 = 18.18°C$, $p_{v\,.\,sat} = 0.0205$ bar

$$\phi_3 = \left(\frac{p_v}{p_{v\,.\,sat}}\right)_{18.18°C} = \frac{0.01186}{0.0205} = 0.5785 \text{ or } 57.85\%$$

(b) $$v_3 = \frac{v_{a3}}{1+\omega_3} = \frac{R_a\,T_{a3}/\,p_{a3}}{1+\omega_3}$$

$$= \frac{[(0.287 \times 291.18)/(1 - 0.01186)\,10^2]}{1 + 0.00768} = 0.8385 \text{ m}^3\text{/kg} \qquad \textbf{Ans.}$$

From the psychrometric chart also, at $t_3 = 18.18°C$ and $\phi_3 = 57.89\%$ **Ans.**

16.8 SENSIBLE HEATING AND COOLING PROCESS

Heating or cooling of air at the constant moisture content per kg of dry air or at constant humidity is termed as sensible heating or cooling processes respectively. If air is passed over a heating coil, there will be *sensible heating*. The coils may be electric resistance heating coils or steam passed through the coils or hot water passed through the coils. If the air is passed over a cooling coil, there is a *sensible cooling*. In this case, a refrigerant at low temperature evaporating in coils through vapour refrigeration system or cooled water or gas may be flowing through coils.

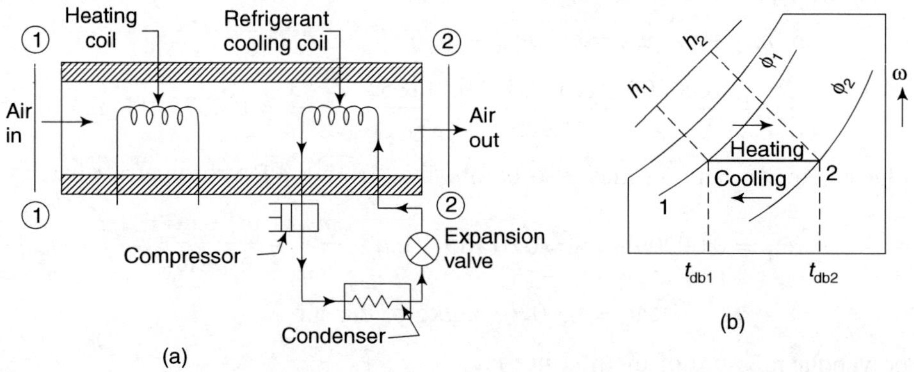

Fig. 16.12 Heating and cooling processes

Figure 16.12 shows the representation of heating and cooling processes on a psychrometric chart. Process 1–2 represents the heating process and 2–1 the cooling process. During heating process 1–2, the temperature of the air increases from t_{dbl} to t_{db2} while humidity ratio remains constant, i.e., $\omega_2 = \omega_1$. It is obvious from Fig. 16.12 that $\phi_2 < \phi_1$, i.e., relative humidity decreases during heating process. The sensible heat added to the air per kg of dry air is

$$h_2 - h_1 = c_m\,(t_{db2} - t_{db1}) \text{ kJ/kg}$$

where c_m is the humid specific heat, (kJ/kg K)

16.9 BY-PASS FACTOR

In the real sense, the by-pass factor is a measure of the efficiency of the heating or cooling coil. Suppose the temperature of the heating coil is t_{db3}. In an ideal case, the air on entering the heating coil at temperature t_{db1} should be able to leave the heating coil at temperature t_{db3} but it leaves at t_{db2}. This difference is due to inefficiency of the heating coil which is termed as **by-pass factor**. Thus, referring to Fig. 16.13.

$$(\text{B.F.})_{\text{Heating}} = \frac{t_{db3} - t_{db2}}{t_{db3} - t_{db1}} \quad \text{...(16.29)}$$

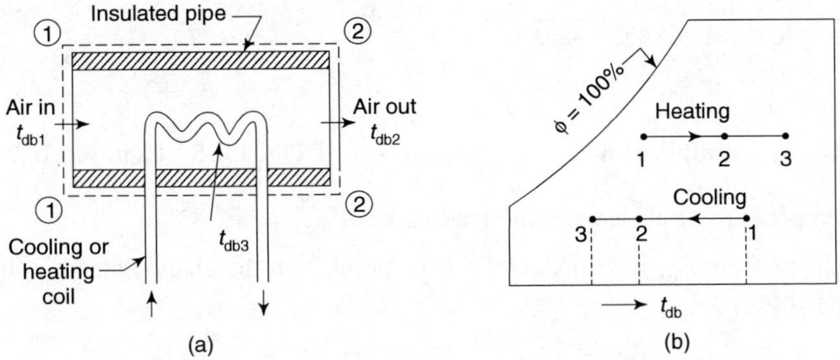

Fig. 16.13 By-pass factor

If t_{db3} be the temperature of cooling coil, then

$$(\text{B.F.})_{\text{cooling}} = \frac{t_{db2} - t_{db3}}{t_{db1} - t_{db3}} \quad \text{...(16.30)}$$

Example 16.6 Atmospheric air at 760 mm of Hg pressure and dry bulb and wet bulb temperature of 16°C and 10°C respectively enters the heating coil whose temperature is 42°C. The by-pass factor of heating coil is 0.4. Calculate the dry bulb temperature, wet bulb temperature, relative humidity of air leaving the heating coil and the sensible heat added to air per kg of dry air.

Solution Refer to Fig. 16.14. Given : $t_{db1} = 16°C$, $t_{wb1} = 10°C$, $t_{db3} = 42°C$

With the help of $t_{db1} = 16°C$ and $t_{wb1} = 10°C$, the point 1 is set on the psychrometric chart. Draw a horizontal line to cut $t_{db3} = 42°C$ at point 3. It is to be noted that point 3 is the ideal condition with by-pass factor equal to zero. We have

$$(\text{B.F.}) = \frac{t_{db3} - t_{db2}}{t_{db3} - t_{db1}} = 0.4 = \frac{42 - t_{db2}}{42 - 16}$$

or

$$t_{db2} = 31.6°C$$

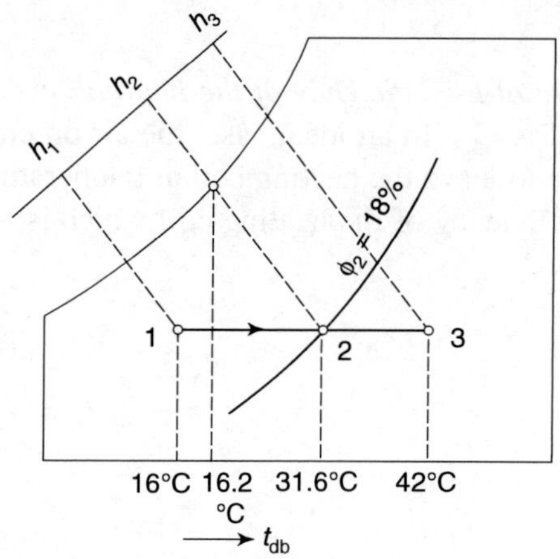

Fig. 16.14 Example 16.6

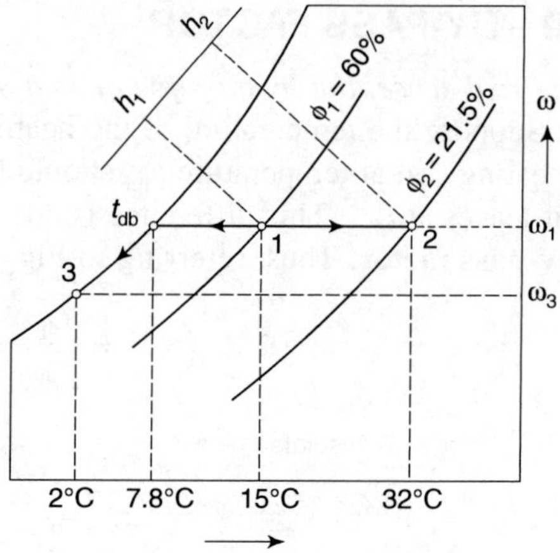

Fig. 16.15 Example 16.8

Dry-bulb temperature of air leaving the heating coil, $t_{db2} = 31.6°C$ **Ans.**

The vertical line from $t_{db2} = 31.6°C$ will set the point 2 on the chart. Thus at point 2, we can see all properties.

$$t_{\omega b2} = 16.2°C \quad \textbf{Ans.}$$

$$\phi_2 = 18\% \quad \textbf{Ans.}$$

and $$h_2 = 45.58 \text{ kJ/kg of dry air} \quad \textbf{Ans.}$$

At point 1, $$h_1 = 29.27 \text{ kJ/kg of dry air}$$

Hence, the sensible heat added to air

$$q = h_2 - h_1 = 45.58 - 29.27 = 16.31 \text{ kJ/kg of dry air} \quad \textbf{Ans.}$$

Example 16.7 A rigid vessel, which has a volume of 0.23 m^3 contains dry air at 25°C. If 0.4 kg of water, also at 25°C, is injected into the vessel so that the pressure becomes 1.03 bar, calulate (a) the relaive humidity of the mixture, (b) the minimum amount of heat that must be transferred to evaporate all the water, also, the resulting temperature and pressure, and (c) the relative humidity if the vessel is heated further to 300°C.

Solution From Steam Tables, at 25°C, $p_{v.\text{sat}} = 0.03166$ bar

If all the injected water were in the vapour state, then its partial pressure would be,

$$p_v = \frac{m_v \cdot R_v \cdot T}{V \times 10^2} = \frac{0.4 \times 0.462 \times 298}{0.23 \times 10^2} = 2.394 \text{ bar}$$

But the partial pressure of water vapour cannot exceed its saturation vapour pressure, hence the vapour contains air at 100% RH and some water is in liquid state

$$p_a = 1.03 - 0.03166 = 0.99834 \text{ bar} \quad \textbf{Ans.}$$

Now, $$m_a = \frac{p_a\, R_v}{R_a\, T_a} = \frac{0.998 \times 0.23}{0.287 \times 298} = 0.2683 \text{ kg}$$

(b) The specific volume of vapour = $v_v = \dfrac{0.23}{0.4} = 0.575 \text{ m}^3/\text{kg}$

From Steam Tables, this specific volume corresponds to $p_v = 3.131$ bar, $t = 134.9°\text{C}$

$$p_a = \frac{m_a\, R_a\, T}{V} = \frac{0.2683 \times 0.287 \times (273 + 134.9)}{0.23 \times 10^2} = 1.365 \text{ bar}$$

The pressure of the mixture, $p_a + p_v = 1.365 + 2.394 = 3.759$ bar

The heat transfer = $Q = \Delta U = m_v\, (\Delta u)_v + m_a\, c_{va}\, (\Delta T)_v$

$= 0.4\,(2742.9 - 104.77) + 0.2683 \times 0.718\,(134.9 - 25)$

$= 1076.42$ kJ **Ans.**

(a) At 300°C and specific volume of 0.575 m^3/kg, the partial pressure of water vapour is

$$p_v = \frac{R_v\, T_v}{v \times 10^4} = \frac{0.462 \times 573}{0.575 \times 10^2} = 4.6039 \text{ bar}$$

This value of p_v may also be obtained from the Steam Tables for superheated steam at 300°C and sp. volume of 0.575 m^3/kg

$$\phi = \left(\frac{p_v}{p_{v,\,\text{sat}}}\right)_{300°\text{C}} = \frac{4.6039}{85.927} = 0.0535 \quad \text{or} \quad 5.35\%$$ **Ans.**

Example 16.8 An air-water vapour mixture at 15°C and 60% R.H. at a pressure of 1.013 bar is heated at constant pressure to a temperature of 32°C. Find (a) the initial and final specific humidities of the mixture, (b) the final relative humidity, the dew point and the amount of heat transferred per kg of dry air, (c) If the initial mixture is cooled isobarically to 2°C, calculate the amount of water vapour condensed and the heat transfer per kg of dry air.

Solution The example may be solved with or without the use of a psychrometric chart as shown in Fig. 16.15.

(a) The point 1 is located on the chart with the help of $t_{db1} = 15°\text{C}$ and $\phi_1 = 60\%$. Horizontal line from 1 and vertical line from $t_{db2} = 32°\text{C}$ set the point 2. Thus, from chart.

$\omega_1 = \omega_2 = 0.00625$ kg/kg of dry air **Ans.**

$h_1 = 31.78$ kJ/kg of dry air, $h_2 = 49.13$ kJ/kg of dry air **Ans.**

(b) $\phi_2 = 21.5\%$ and $t_{db1} = 7.8°\text{C}$ **Ans.**

Heat transferred = $q = h_2 - h_1 = 49.13 - 31.78 = 17.35$ kJ/kg of dry air **Ans.**

(c) Since 2°C is below the dew point of the mixture, condensation will occur. The horizontal line from 1 cuts the saturation line at $t_{dp} = 7.8°\text{C}$. The point 3 is set according to $t_{db} = 2°\text{C}$ on the saturation line. From chart,

$$h_3 = 12.54 \text{ kJ/kg of dry air}$$

$$\omega_3 = 0.0048 \text{ kg/kg of dry air}$$

Mass of vapour condensed $= \omega_1 - \omega_3$

$= 0.00625 - 0.0048 = 0.00145$ kg/kg of dry air **Ans.**

Heat transferred, $h_1 - h_3 = 31.78 - 12.54 = 19.24$ kJ/kg of dry air **Ans.**

Example 16.9 Atmospheric air at 25°C and 60% R.H. is brought to 22°C and 40% R.H. by means of the system comprising of a cooler and a heater. The barometric pressure is 1.013 bar, mass flow rate of atmospheric air including moisture entering the system is 92 kg/h. Calculate the mass flow rate of water vapour removed per hour and heat removed in the cooler per hour. Represent the process on psychrometric chart.

Solution Refer to Fig. 16.16.

From the chart, $\omega_1 = 11.8$ gm/kg of dry air, $\omega_2 = 6.4$ gm/kg of dry air

Heat removed in cooler $= h_1 - h_2 = 55.2 - 23.83 = 31.37$ kJ/kg of dry air

Also, $(1 + \omega_1)\, m_a = 92 = (1 + 0.0118)\, m_a$

$\therefore \quad m_a = 90.9$ kg/h

$\therefore$ Heat removed in the cooler $= 31.37 \times 90.9 = 2851.53$ kJ/kg **Ans.**

Water removed $= (\omega_1 - \omega_2) = (11.8 - 6.4) \times 10^{-3} = 5.4 \times 10^{-3}$ kg/kg of dry air

$= 5.4 \times 10^{-3} \times 90.9 = 0.4908$ kg/h **Ans.**

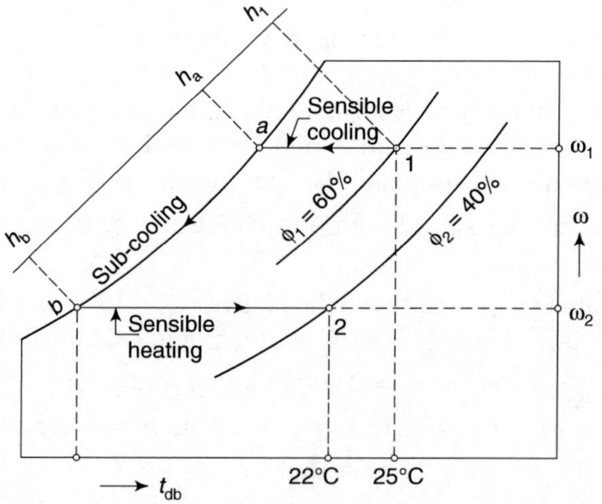

Fig. 16.16 Example 16.9

16.10 HUMIDIFICATION AND DEHUMIDIFICATION PROCESSES

Humidification process is defined as the process in which moisture is added to keeping dry bulb temperature constant. Conversely, dehumidification is defined as process when moisture

is removed keeping dry bulb temperature constant. Thus, these processes will be represented as vertical lines on the psychrometric chart as shown in Fig. 16.17. During humidification process (1–2); relative humidity increases (**i.e.,** $\phi_2 > \phi_1$) while during dehumidification process (2–1), relative humidity decreases. Since dry bulb temperature does not change during these processes, sensible heat remains constant. Thus, the change in enthalpy $(h_2 - h_1)$ per kg of dry air is the latent heat of vaporization of the moisture content (h_{fg}).

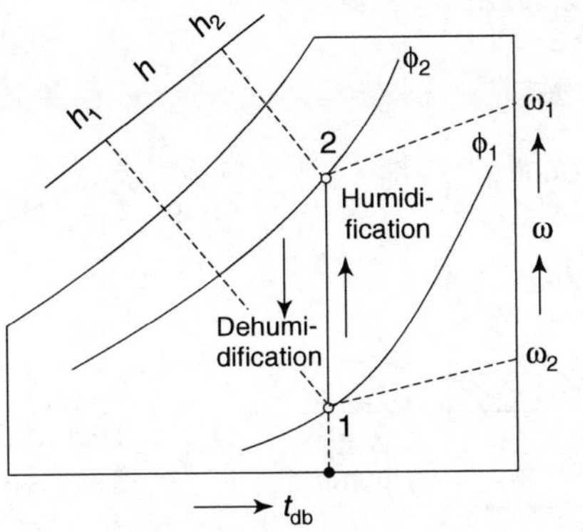

Fig. 16.17 Humidification and dehumidification processes

In practice, pure humidification process is not possible and they may be accompanied by heating or cooling. However, if pure humidification is desired, the spray water through which air passes has to be maintained at dry bulb temperature (t_{db1}) of the entering air.

16.11 COOLING AND HUMIDIFICATION PROCESS

If air is washed through sprays of water maintained at a temperature lower than dry bulb temperature of entering air but higher than the dew point temperature of entering air, the air is cooled and humidified. This process is shown in Fig. 16.18. Three cases arise:

Case (i) If the same water is recirculated again and again, the process is adiabatic, i.e., no heat is supplied or rejected from the spray of water, then it is called adiabatic cooling or **evaporative cooling**.

The air at state 1 is passed through a spray of water. Some of the water evaporates, so that the specific humidity of the air increases while at the same time, the air is cooled.

The temperature of the spray water will reach the thermodynamic wet bulb temperature of the air entering the spray water. This process is represented as 1–2 on the psychrometric chart along the constant wet bulb line. Make-up water = $(\omega_2 - \omega_1)$.

This principle is made use of in desert coolers for cooling buildings in summer season. Since the cooling is not possible below the *w.b.t.*, for the outside air, at which temperature the leaving air will be saturated. This cooling has only limited use, i.e., for summer when the outside wet bulb depression is high. In other words, it will not work during monsoon season.

Case (ii) When the temperature of the spray water is maintained at a temperature *less than the wet bulb temperature* of the entering air (t_{wb1}) by cooling the spray water by coolers before pumping it back to spray nozzles, then this process is shown 1–2′.

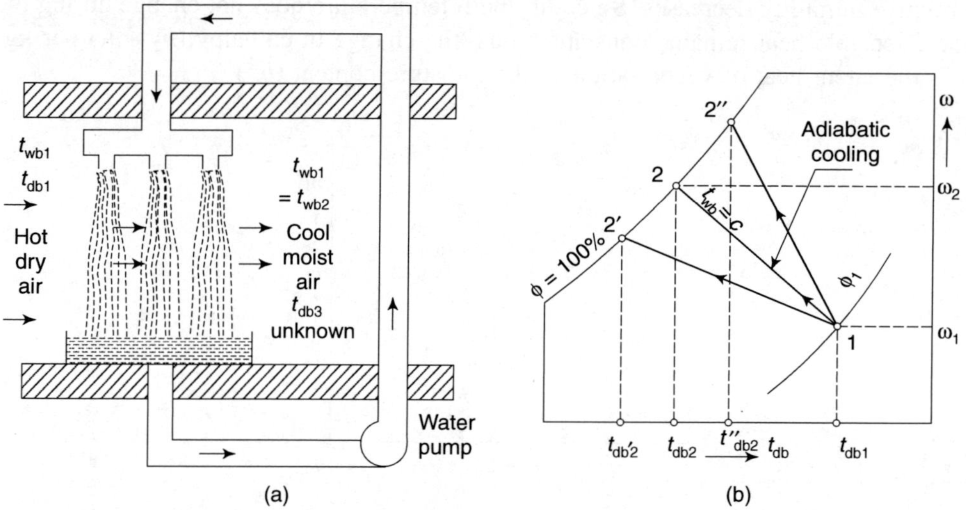

Fig. 16.18 Cooling and humidification process

Case (iii) When the temperature of a spray water is maintained at a constant temperature *higher than the wet bulb temperature* of the entering air but lower than the dry bulb temperature of air by heating the water by heaters before pumping back to the spray nozzles, then this process is shown as 1–2″ psychrometric chart.

The **humidifying efficiency** (η_{humid}) is defined as (for case (i)).

$$\eta_{humid} = \frac{t_{db1} - t_{db3}}{t_{db1} - t_{db2}} \qquad ...(16.31)$$

where $t_{\omega b2} = t_{\omega b1} = t_{db2}$ and t_{db3} **is the actual dry bulb temperature of air at outlet**.

If $\eta_{humid} = 100\%$, $t_{db2} = t_{db3}$

Similarly, for case (ii),

$$\eta_{humid} = \frac{t_{db1} - t_{db3}}{t_{db1} - t_{db2'}} \qquad ...(16.32)$$

and for case (iii),

$$\eta_{humid} = \frac{t_{db1} - t_{db3}}{t_{db1} - t_{db2''}} \qquad ...(16.33)$$

Example 16.10 Moist air at 30°C and relative humidity 20% is brought to saturation by passing through air-water spraying recirculated water. Determine (a) *d.b.t* of the exit air and (b) make-up water.

Solution Refer to Fig. 16.18.

(a) From the chart, *w.b.t.* of entering air = $t_{\omega b2}$ = 15.5°C

Thus, the dry bulb temperature of exit air at saturation condition, $t_{\omega b2}$ = 15.5°C **Ans.**

(b) From the chart, $\omega_1 = 0.005$, $\omega_2 = 0.015$ kg/kg of dry air.

∴ Make-up water required = $\omega_2 - \omega_1 = 0.015 - 0.005 = 0.01$ kg/kg of dry air **Ans.**

16.12 HEATING AND HUMIDIFCATION PROCESS

When the spray water through which air is washed is maintained at a temperature higher than dry bulb temperature of the entring air, the unsaturated air tends to become saturated and the heat of vaporization of water is absorbed from the spray water. In this way, the air is humidifed and heated and the spray water is cooled. The spray water may get cooled to a minimum of the wet bulb temperature of the entering air. *In other words, heating and humidification are achieved when moist air passes through an air washer spraying hot water.* During the process the following points must be kept in mind:

(i) There is an increase of specific humidity, *d.p.t.*, *w.b.t.*, *d.p.t* and enthalpy

(ii) R.H. may increase or decrease.

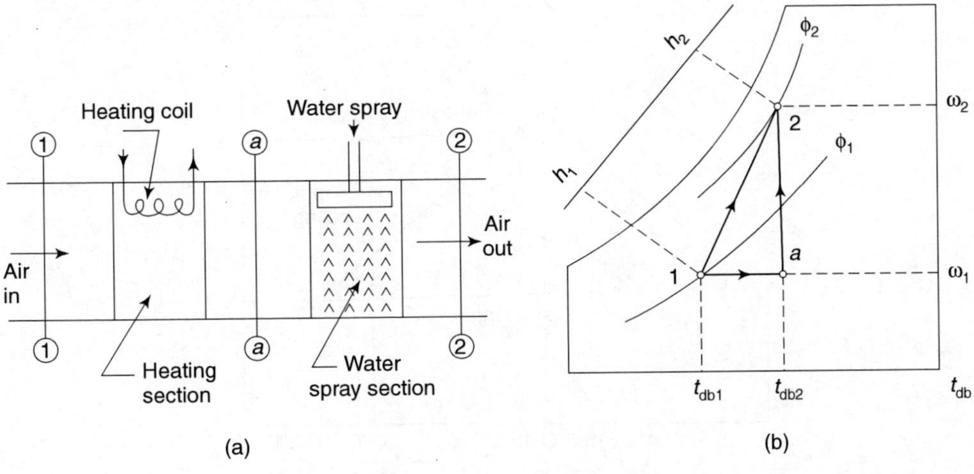

Fig. 16.19 Heating and humidification process

This process is shown on the psychrometric chart (Fig. 16.19) as 1–2 and has practical application in **winter air-conditioning**. This may be broken as process 1–*a* (heating) and process *a*–2 (humidification).

Example 16.11 Air at dry bulb temperature of 4°C and relative humidity of 80% is to be heated and humidified to 30°C and 40% R.H. by the following methods.

(a) By-passing the air through heated water spray air water.

(b) By preheating sensibly and then passing through water spray washer with recirculated water till relative humidity reaches to 90% and then reheating to final state required.

Caclculate (i) total heating required for (a) and (b), (ii) make-up water required in water spray air washer, and (iii) humidifying efficiency of the recirculated spray water.

Solution Refer to Fig. 16.20.

(a) The initial point 1 is set according to $t_{db1} = 4°C$ and $\phi_1 = 80\%$. The final condition 2 is set according to $t_{db2} = 30°C$ and $\phi_2 = 40\%$. The process 1–2 shows the passing of air through heated water spray washer. From the chart

$$h_1 = 14 \text{ kJ/kg of dry air, } \omega_1 = 0.004 \text{ kg/kg of dry air,}$$

$$h_2 = 46.76\ 78 \text{ kJ/kg of dry air, } \omega_2 = 0.0103 \text{ kg/kg of dry air}$$

(i) Heating required = $h_2 - h_1 = 56.87 - 14 = 42.87$ kJ/kg of dry air

(ii) Make-up water required = $\omega_2 - \omega_1 = 0.0103 - 0.004$

$= 0.0063$ kg/kg of dry air **Ans.**

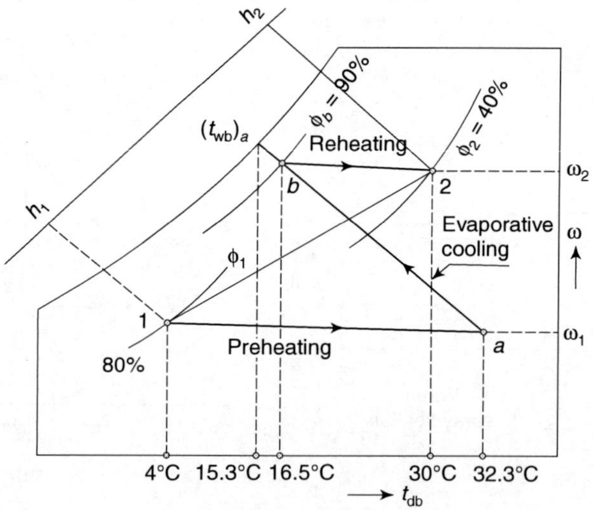

Fig. 16.20 Example 16.11

(b) Horizontal line from point 1 towards right gives sensible heating. Another horizontal line from point 2 is drawn to cut 90% relative humidity line at point *b*. Through point *b*, a constant wet bulb line is drawn to cut horizontal line through 1 at point *a*. Thus, preheating is represented by process 1–*a*, evaporative cooling (i.e., adiabatic cooling) to 90% R.H. by line *a*–*b* and reheating to final condition by line *b*–2.

The amount of preheating required, $h_a - h_1 = 43.07 - 14 = 29.07$ kJ/kg of dry air

The amount of reheating required, $h_2 - h_b = 56.87 - 43.07 = 13.8$ kJ/kg of dry air

∴ Total heat required, $29.07 + 13.8 = 42.87$ kJ/kg of dry air.

Similarly, make-up water = $\omega_b - \omega_a = \omega_2 - \omega_1 = 0.0063$ kg/kg of dry air

Humidifying efficiency of the air washer is

$$\eta_{\text{humidity}} = \frac{(t_{\text{db}})_a - (t_{\text{db}})_b}{(t_{\text{db}})_a - (t_{wb})_a} = \frac{32.3 - 16.5}{32.3 - 15.3} = 0.9294 \text{ or } 92.94\% \qquad \textbf{Ans.}$$

16.13 HEATING AND DEHUMIDIFICATION PROCESS

If air is passed over certain chemicals like silica gel and activated alumina which have an affinity for moisture, the moisture is condensed, and gives up its latent heat resulting in an increase in d.b.t. of air, thus the air is heated and dehumidified. This process is shown by 1–2 on a psychrometric chart in Fig. 16.21. This process may be broken as process 1–*a* (dehumidification) and process *a*–2 (heating). During the process, the specific humidity decreases while the wet bulb temperature of air increases. **This process has an application in industrial air conditioning.**

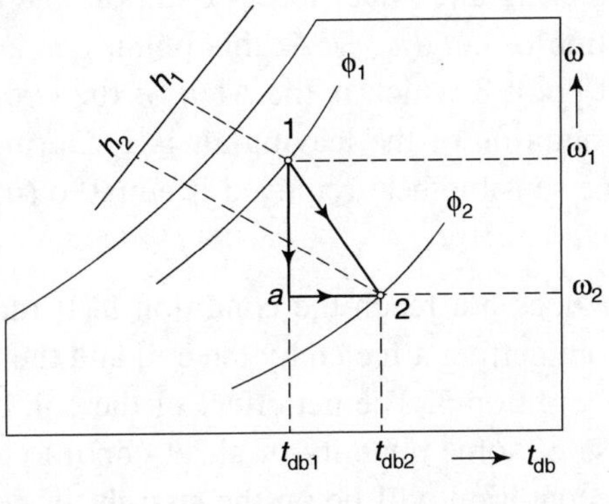

Fig. 16.21 Heating and dehumidification process

16.14 COOLING AND DEHUMIDIFICATION PROCESS

If the moist air is passed through cooling coil whose effective surface temperature is maintained at a temperature below the d.p.t. of the entering air, the air is cooled and condensation of the moisture takes place, i.e., it is dehumidified. This separation of the moisture results in the fall of humidity ratio. The effective surface temperature of the cooling coil which is lower than *d.p.t.* of the entering air is called the **apparatus dew point**, abbreviated as **ADP.** This process is also achieved by passing the air through chilled spray water which is maintained at a temperature lower than *d.p.t.* of the entering air.

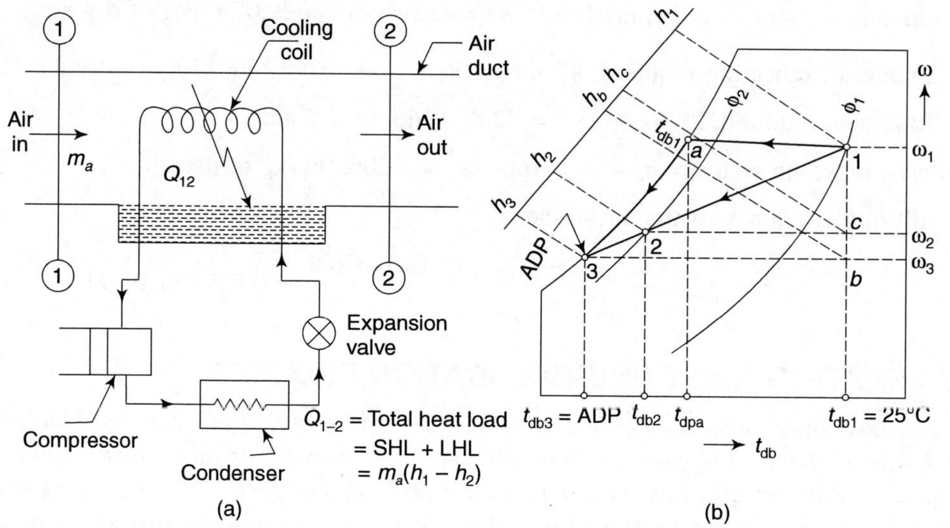

Fig. 16.22 Cooling and dehumidification process

Figure 16.22 shows the representation of the process on a psychrometric chart. The point 1 shows the condition of the entering air. Under ideal condition, the air while passing through the coil will reach the condition *a*, i.e., (t_{dp1}). At this point, condensation will start and the air will be further cooled till point 3 which is the ADP of the cooling coil. In other words, under ideal conditions, the condition of the leaving air is 3. During this process, *d.b.t.* falls from $(t_{db})_1$ to $(t_{db})_3$ = ADP, the sensible heat removed is equal to $(h_1 - h_3)$ and the latent heat removed is equal to $(h_1 - h_b)$.

But in practice, all the air does not reach the condition 3. It may be due to the fact that some air gets by-passed without getting affected by the coil and the remaining air gets ideally cooled and dehumidified to condition 3. The net effect of the coil on the leaving air may be explained as a mixing process of some quantity of air at condition 1 with the remaining air at condition 3. The resulting condition will lie on the straight lie joining 1 and 3 at point 2. Now, we can define the B.F. of the cooling coil as

$$\text{B.E.} = \frac{t_{db2} - t_{db3}}{t_{db1} - t_{db3}} = \frac{h_2 - h_3}{h_1 - h_3} = \frac{t_{db2} - \text{ADP}}{t_{db1} - \text{ADP}} \qquad ...(16.34)$$

or $$\eta_{coil} = 1 - \text{B.F.} \qquad ...(16.35)$$

Thus, during this process, *d.b.t*, and specific humidity of air decrease while relative humidity increases. This process has practical applications in *summer air conditioning*.

Example 16.12 Air at 25°C and 50% relative humidity enters a cooling coil at the rate of 300 m^3/min. Calculate (a) refrigeration in tonnes needed to bring the temperature of the air to the coil temperature of 18°C and also the relative humidity at the condition.

If the effective surface temperature or apparatus dew point (ADP) is 10°C and the by-pass factor is 0.1, calculate (b) refrigeration in tonnes needed and the mass of water condensed out at the cooling coil per minute. Also, calculate the sensible heat factor for the process through the coil.

Solution Refer to Figs. 16.23 and 16.24.

(a) The state point is set on the chart with the help of $t_{db1} = 25°C$ and $\phi_1 = 50\%$. The horizontal line from 1 cuts the saturation line giving $t_{dp} = 14°C$ corresponding to entering air condition. The effective temperature of the cooling coil is above the dew point, hence the process from $t_{db1} = 25°C$ and $\phi_1 = 50\%$ to $t_{db2} = 18°C$ is the sensible cooling process. The state point 2 is set with the aid of horizontal line from 1 and the vertical line from $t_{db2} = 18°C$.

This gives $\phi_2 = 77\%$ **Ans.**

Also from the chart, $h_1 = 50.2$ kJ/kg of dry air

$h_2 = 43.1$ kJ/kg of dry air

$\therefore$ Cooling required $= q = h_1 - h_2 = 50.2 - 43.1 = 7.1$ kJ/kg of dry air

Specific volume of air from the chart, $v_1 = 0.858\ m^3/kg$

$\therefore$ Mass of air flowing per minute $= \dot{m} = \dfrac{300}{0.858} = 349.65$ kg/min

Hence, total refrigeration required

$$= 349.65 \times 7.1 = 2482 \text{ kJ/min.} = \frac{2482.5}{209} = 11.877 \text{ tonnes}$$ **Ans.**

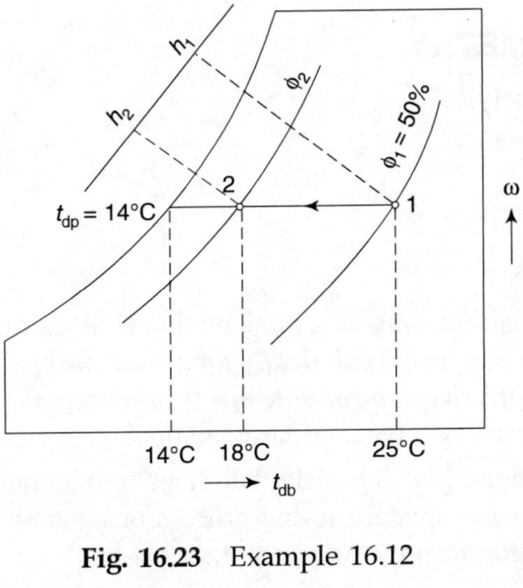

Fig. 16.23 Example 16.12

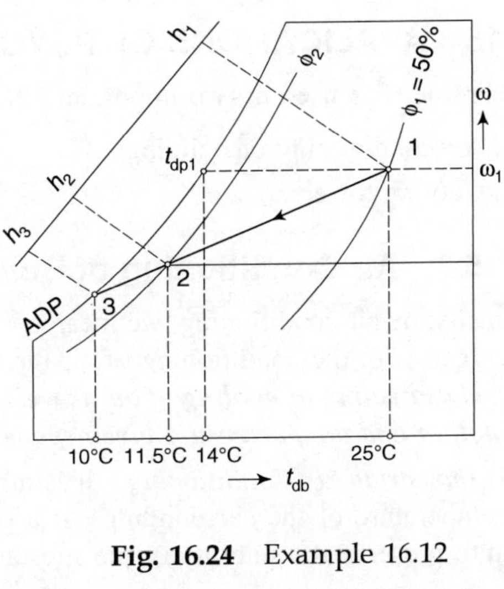

Fig. 16.24 Example 16.12

(b) Refer to Fig. 16.24.

The state point 3 is located with the value of ADP = 10°C. Join points 1 and 3 on the chart. **Since ADP of the cooling coil is less than t_{dp} = 14°C, this is a cooling and dehumidification process**. As the B.F. = 0.1, hence the actual condition of leaving air will be say 2.

$$\text{B.F.} = \frac{t_{db2} - t_{ADP}}{t_{db2} - t_{ADP}} = 0.1 = \frac{h_2 - h_3}{h_1 - h_3} = \frac{t_{db2} - 10}{25 - 10}$$

or $$t_{db2} = 11.5°\text{C}$$

Draw a vertical line from t_{db2} = 11.5°C to cut the line 1–3 at 2. The state point may also be set on the chart if the line 1–3 is divided in such a way that

$$\frac{\text{Intercept } 1-2}{\text{Intercept } 1-3} = \text{B.F.} = 0.1$$

From the chart, h_2 = 31.36 kJ/kg of dry air, ω_2 = 0.00755 kg/kg of dry air

h_1 = 50.2 kJ/kg of dry air, ω_1 = 0.00985 kg/kg of dry air

Refrigeration required = $m_a \times (h_1 - h_2)$ = 349.65 × (50.2 – 31.36) = 6587.4 kJ/min

$$= \frac{6587.4}{309} = 21.31 \text{ tonnes}$$ **Ans.**

Water condensed on the coil = $m_a (\omega_1 - \omega_2)$

= 349.65 (0.00985 – 0.00755) = 0.80 kg/min **Ans.**

Draw the line parallel to 1–2 through the mark gives SHF = 0.72 **Ans.**

16.15 APPLICATIONS OF PSYCHROMETRY

Psychrometry is used in two important fields, namely:

1. Air conditioning of buildings
2. Cooling towers.

16.15.1 Air Conditioning of Buildings

Normally, by air conditioning, we mean cooling of air in summer season, but in practice, it is concerned with the conditioning of air throughout the year. ***Technically air conditioning is defined as heating or cooling of air as well as an effective control of temperature, humidity, air motion and air purity in a given space***. There are two types of air conditioning:

(i) *Industrial Air Conditioning* It is air conditioning of industrial buildings to maintain the temperature of the surroundings of a product, or a manufacturing process or a storage room to preserve the stability of the substance during manufacture or storage.

(ii) *Comfort Air Conditioning* When air conditioning of a space is done to produce conditions most suitable for comfort and health of human beings living and working in that space, it is called comfort air conditioning.

Due to ambient atmospheric conditions, the space to be air-conditioned is subjected to heat load (i.e., heat loss or heat gain). In character, this heat load fluctuates from time to time and day-to-day depending upon the outside weather. This suggests that the amount of conditioned air will have to vary according to the heat load. In order to understand the effect of fluctuating heat load clearly, let us take an example, during the summer season, the supply of air state is at higher temperature and lower specific humidity and its rate are so determined that this air from receiving the heat and moisture gain of the space gives the desired condition of the space.

For more details, see the author's book, "*Applied Thermodynamics, Refrigeration and Air Conditioning*".

16.15.2 Cooling Towers

The function of the cooling tower is to cool water. Though, the psychrometric or thermodynamic principle of cooling tower and air washer is the same but the difference lies in the function. The function of the air-washer is to cool and humidify the air. *Cooling towers are used in cooling the hot water of steam condenser in steam power plant for reuse. Cooling towers are also used in large air-conditioning plant where water is used as secondary fluid to cool air.* There are generally two types of cooling towers:

(i) **Natural draught cooling tower** It utilises, the natural wind velocity for cooling water. This is suitable for small capacities.

(ii) **Artificial draught cooling tower** In this process, a fan forces or induces the air through the tower at a high velocity. This is suitable for large capacities.

Figure 16.25 shows a cooling tower using mechanical induced draught. The water is sprayed down from a header at the top over a larger number of baffles which are placed in the body of tower to give large area for evaporation. Air enters from the bottom at section 1 and rushes up due to mechanical induced draught created by the fan placed at the top. This air is emitted out at section 3 in an approximately saturated condition. The energy balance gives

$$m_{a1}\, h_{a1} + m_{v1}\, h_{v1} + m_{w2}\, h_{f2} = m_{a3}\, h_{a3} + m_{v3}\, h_{v3} + m_{w4}\, h_{f4}$$

or $$h_{a1} + \omega_1\, h_{v1} + \frac{m_{w2}}{m_{a1}}\, h_{f2} = h_{a3} + \omega_3\, h_{v3} + \frac{m_{w4}}{m_{a4}}\, h_{f4}$$

If m_w represents the mass of water per unit mass of dry air, then

$$h_{a1} + \omega_1\, h_{v1} + m_{v2}\, h_{f2} = h_{a3} + \omega_3\, h_{v3} + m_{w4}\, h_{f4}$$

or $$c_p\, T_{a1} + \omega_1\, h_{v1} + m_{w1}\, h_{f2} = c_p\, T_{a3} + \omega_3\, h_{v3} + m_{w4}\, h_{f4}$$

$$m_{w2} - m_{w4} = \omega_3 - \omega_1 \qquad ...(16.36)$$

$$\therefore \quad m_{w4} = m_{w2} - (\omega_3 - \omega_1) \qquad ...(16.37)$$

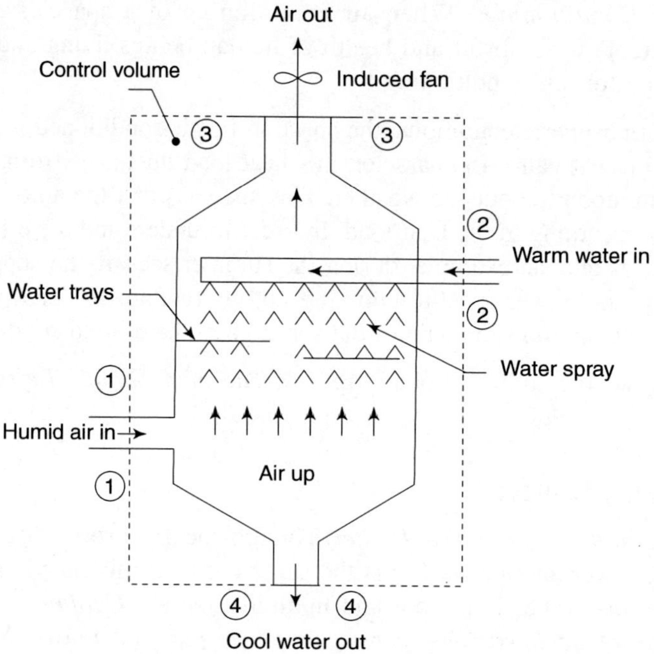

Fig. 16.25 Cooling tower (induced draught)

Example 16.13 It is required to cool 106 kg/h of water from 50°C to 40°C. Air enters the tower at 20°C dry bulb temperature and 15.5°C we bulb temperature. It leaves the tower at 30°C and 90% relative humidity. Calculate (a) fan capacity, and (b) loss of water.

Solution Refer to Fig. 16.25. From psychrometric chart

$$\omega_1 = 0.00925 \text{ kg/kg of dry air (at } t_{db} = 20°\text{C and } t_{wb} = 15°5°\text{C)}$$

$$\omega_3 = 0.024 \text{ kg/kg of dry air (at } t_{db3} = 30°\text{C and } \phi_3 = 90\%)$$

From Steam Tables, $h_{v1} = 2538.2$ kJ/kg at 20°C, $h_{v3} = 2556.4$ kJ/kg at 30°C,

$$h_{f3} = 209.3 \text{ kJ/kg}, \quad h_{f4} = 167.5 \text{ kJ/kg}$$

From the energy balance,

$$(h_{a1} + \omega_1 h_{v1}) + m_{w2} h_{f2} = (h_{a3} + \omega_3 h_{v3}) + m_{w4} h_{f4}$$

or $$c_{pa} (t_1 - t_3) + \omega_1 h_{v1} + m_{w2} h_{f2} = \omega_3 h_{v3} + [m_{w2} - (\omega_3 - \omega_1)] h_{f4}$$

or $$1.005 (20 - 30) + 0.00925 \times 2538.2 + m_{w2} \times 209.3$$

$$= 0.0244 \times 2556.4 + [m_{w2} - (0.0244 - 0.00925)] 167.5$$

or $$m_{w2} = 1.234 \text{ kg of water/kg of dry air}$$

Hence, mass flow of air entering the tower = $m_a = \dfrac{10^6}{1.234} = 810372.7$ kg/h

Specific volume of air at section 3 = $v_{3a} = 0.8895$ m^3/kg

$\therefore$ The induced draft fan capacity = $\dfrac{810372.7 \times 0.8895}{60} = 12013.7$ m^3/min **Ans.**

(b) The loss of water = $(\omega_2 - \omega_1) \times m_{a1}$

$= (0.02444 - 0.00925) \times 810372.7 = 12309.56$ kg/h **Ans.**

Example 16.14 5000 kg of water per hour is cooled from 38°C to 28°C in a cooling tower. The unsaturated air mixture enters the tower at 22°C and 40% R.H. and leaves at 30°C and 90% R.H. Assuming the steady-state process to be isobaric adiabatic, calculate the rate of air flow and the rate at which water is evaporated.

Solution Refer to Fig. 16.25.

From psychrometric chart, at $t_{bd1} = 22°C$ and $\phi_1 = 40\%$, $\omega_1 = 0.0064$ kg/kg of dry air

At $t_{db3} = 30°C$ and $\phi_3 = 90\%$, $\omega_3 = 0.023$ kg/kg of dry air

Mass if water evaporated = $\omega_3 - \omega_1 = 0.023 - 0.0064 = 0.0166$ kg/kg of dry air

The energy balance, from the first law and the continuity equation gives

$$(h_{a1} + \omega_1 h_{v1}) + m_{v2} h_{f2} = (h_{a3} + \omega_3 h_{v3}) + m_{w4} h_{f4}$$

where m_{w2} and m_{w4} are the mass of water per kg of dry air at section 2 and 4. Thus

$$c_{pa}(t_1 - t_3) + \omega_1 h_{v1} + m_{w2} h_{f2} = \omega_3 h_{v3} - [m_{w2} - (\omega_2 - \omega_1)] h_{f4}$$

From Steam Tables, $h_{v1} = 2541.8$ kJ/kg (at 22°C), $h_{v2} = 2556.4$ kJ/kg (at 30°C)

$h_{f2} = 159.04$ kJ/kg (at 28°C), $h_{f4} = 117.31$ kJ/kg (at 28°C)

$$\therefore \quad 1.005\,(22 - 30) + 0.0064 \times 2541.8 + m_{w2} \times 159.04$$

$$= 0.023 \times 2556.4 + [m_{w2} - (0.023 - 0.0064)]\,117.31$$

or $$m_{w2} = 1.1605 \text{ kg/kg of dry air}$$

Mass of air-water vapour mixture introduced

$$m_1 = \frac{5000\,(1 + \omega_1)}{m_{w2}} = \frac{5000\,(1 + 0.0064)}{1.1605} = 4336 \text{ kg/h}$$ **Ans.**

Rate of evaporation of water = $(\omega_3 - \omega_1) \times \dfrac{5000}{1.1605}$

$$= \frac{(0.023 - 0.0064) \times 5000}{1.1605} = 71.5 \text{ kg/h}$$ **Ans.**

16.16 SUMMARY OF FORMULAE

1. $\omega = \frac{m_v}{m_a} = 0.622 \frac{p_v}{p_a} = 0.622 \frac{p_v}{p - p_v}$

2. $\phi = \left(\frac{p_v}{p_{v,\,sat}}\right)_T = \left(\frac{v_{v,\,sat}}{v_v}\right)_T = \left(\frac{\rho_v}{\rho_{v,\,sat}}\right)_T$

3. $\omega = 0.622\ \phi \left(\frac{p_{v,\,sat}}{p_a}\right)$

4. $\mu = \left(\frac{\omega}{\omega_{sat}}\right)_T = \phi \left\{\frac{1-(p_{v,\,sat}/p)}{1-(p_{v,\,sat}/p)}\right\}$

5. $\phi_{\varpi} = (p_{v,\,sat})_{t\omega b} - \frac{1.8\,(t_{bd} - t_{wb})}{2700}$

6. $p_v = (p_{v,\,sat})_{t\omega b} - 0.0006\ p\ (t_{db} - t_{\omega b}) \left(1 + \frac{1.8\ t_{db}}{1571}\right)$

7. $p_v = (p_{v,\,sat})_{t\omega b} - \frac{(p - p_{v,\,sat})_{t\omega b} \times (t_{db} - t_{\omega b}) \times 1.8}{2854 - 1.325\,(1.8\ t_{db} + 32)}$

8. $\omega_1 = \frac{\omega^*, h_{fg}^* - c_{pc}\ (T_1 - T^*)}{h_{v1} - h_f^*}$

9. $h = h_a + \omega h_v$ = enthalpy of moist air

10. $t^* = t - \frac{h_{fg}^*}{(1.005 + 1.88\ \omega)} = t - \frac{h_{fg}}{c_m}(\omega^* - \omega)$

11. $h_1 = h_2$ (adiabatic saturation)

12. $\frac{m_{a_1}}{m_{a_2}} = \frac{h_3 - h_2}{h_1 - h_2} = \frac{\omega_3 - \omega_2}{\omega_1 - \omega_2}$ (Adiabatic mixing of two streams)

13. $(BF)_{heating} = \frac{t_{db3} - t_{db2}}{t_{db3} - t_{db1}}$ (heating coil)

14. $(BF)_{cooling} = \frac{t_{db2} - t_{db3}}{t_{db1} - t_{db3}}$ (cooling coil)

15. $\eta_{humid} = \frac{h_{db1} - t_{db3}}{t_{db1} - t_{db2}}$

16. $BF = \frac{t_{db2} - ADP}{t_{db1} - ADP}$ (cooling and dehumidification)

17. $\eta_{coil} = 1 - BF$

18. $h_{a1} = \omega_1 h_{v1} + m_{w2} h_{f2} = h_{a3} + \omega_3 h_{v3} + m_{w4} h_{f4}$

PROBLEMS

Things to Think About

16.1. Which do you think is most closely related to human comfort, the humidity ratio or the relative humidity?

16.2. How will you explain the different rates of evaporation from a dish of water in winter and summer?

16.3. What is the reason that bathroom mirrors often fog up when showers are taken?

16.4. Is it possible to have an equal value of *d.b.t.* and *w.b.t.*?

16.5. What do you conclude about water dripping from the tailpipe of an automobile on a cold morning?

16.6. Would you recommend to install an evaporative cooling system in Mumbai and Delhi?

16.7. It is well-known fact that water vapour in air is typically a superheated vapour, why can we use the saturated vapour value, $h_g(T)$ to represent its enthalpy?

16.8. What is the mechanisms of maintaining constant temperature in human body throughout the year when there is a lot of variation in weather temperature?

Short and Long Theoretical Questions

16.9. Define the following terms : (a) moist and dry air, saturated, unsaturated and supersaturated air mixture.

16.10. Distinguish clearly between the terms of dry bulb temperature, dew point temperature, wet bulb temperature and adiabatic saturation temperature as used in psychrometry.

16.11. Define the terms specific humidity, relative humidity, and humidity ratio.

16.12. Discuss the methods of preparing a psychrometric chart.

16.13. What are the various psychrometric processes? Show them on a psychrometric chart?

16.14. What do you mean by an adiabatic mixing of air streams? Show the process on a chart.

16.15. What do you mean by by-pass factor of a cooling and heating coil?

16.16. What do you mean by ADP?

16.17. Explain the function of cooling tower.

Numerical Questions

16.18. The moist air is at 30°C *d.b.t.* and 15°C *w.b.t.* Find (a) specific humidity, (b) entropy, (c) relative humidity, (d) specific volume, and (e) *d.p.t.* Check the results with the psychrometric chart.

[*Ans.* (a) 0.000412 kg/kg, (b) 41.5 kJ/kg, (c) 17% (d) 0.8625 m/kg, (e) 2.4°C]

16.19. The barometer reads 750 mm of Hg. The dry bulb temperature is 33°C and wet bulb temperature is 23°C. Calculate (a) the relative humidity, (b) the humidity ratio and (c) *d.p.t.*

[*Ans.* (a) 43.4 per cent, (b) 0.0137 kJ/kg, (c) *d.p.t.* = 18.7°C]

16.20. The humidity ratio of atmospheric air at 27.5°C is 0.016 kg/kg of dry air. Calculate (a) the partial pressure of vapour, (b) relative humidity, and (c) dew point. The barometric pressure is 760 mm of Hg. [*Ans.* (a) 19.05 mm of Hg, (b) 69.3 per cent, (c) 21.33°C]

16.21. Atmospheric air at 72.5 mm of Hg has a 11°C wet bulb depression from 31°C *d.b.t.* during adiabatic saturation process. Find (a) humidity ratio, and (b) vapour pressure and relative humidity at 31°C, (c) *d.p.t.*

Hints: [*w.b.t.* = *d.b.t.* — wet bulb depression.]

16.22. One-stream of air having 5.5 m^3/min at 15°C and 60 per R.H. flows into another stream of air having 35 m^3/min at 25°C and 70 per cent R.H. Find for the mixture (a) *d.b.t.* (b) *w.b.t.* (c) RH, and {d) specific humidity.

[*Ans.* (a) 23.6°C, (b) 19.6°C, (c) 69 per cent, (d) 0.0125 kg/kg]

16.23. A dry-air-water-vapour mixture, at 1.103 bar contains 2 per cent water by volume. If the *d.b.t.* is 22.2°C, calculate (a) *d.p.t.*, (b) the humidity ratio, and (c) the enthalpy.

16.24. The *d.b.t.* of atmospheric air is 26.7°C and the relative humidity is 30 per cent. Find the value of humidity ratio and the partial pressure of water vapour if the total pressure is 1.103 bar. What is the dew point?

16.25. During a steady flow humidification process, water enters the humidifier at 20°C and air enters at 1.1 bar, 20°C and a specific humidity of 0.01 kg/kg of air. The steam leaving the apparatus has a pressure of 1.05 bar, a temperature of 40°C and a relative humidity of 90 per cent. Determine the amount of heat transfer to the humidifier per kg of water evaporated.

16.26. Water at the rate of 25,000 litres per hour enters a cooling tower where it is cooled from 70°C to 34°C. Atmospheric air at 36°C and 25 per cent R.H. is available for the cooling purposes. Assuming that the air leaves the cooling tower saturated with water vapour at 34°C, determine the rate of air leaving the cooling water and the make-up water per hour.

16.27. Air at *d.b.t.* of 50°C, at 2 bar is used in a drying process. Under study flow equilibrium conditions, the air enters the dryer at the rate of 5000 kg/h and leaves at 80 per cent R.H. with no change in total pressure. Calculate (a) *the amount of moisture picked up by the air in kg/h*, (b) *the d.b.t. of the leaving air*.

16.28. Air at 1.103 bar, 40°C *d.b.t.*, 55 per cent R.H. enters a cooling tower and leaves it at 38°C and 95 per cent R.H. Water is cooled from 42°C to 36°C in the tower. The water flow rate to the tower is 7 kg/s. Calculate (a) the necessary air flow rate, and (b) the amount of fraction of water evaporated.

16.29. The *d.b.t.* and *w.b.t.* for atmospheric air are 31.70°C and 23.0°C respectively. Calculate (a) the partial pressure to water vapour, (b) the humidity ratio, (c) the specific volume, and (d) the relative humidity. Use psychrometric chart.

16.30. Atmospheric air at a pressure of 1.103 bar and 32.2°C has a specific humidity of 0.0096 kg/kg of dry air. Calculate from the psychrometric chart (a) the relative humidity, (b) the vapour pressure, (c) total enthalpy, (d) the specific volume, and (e) *d.p.t.*

[*Ans.* (a) 32 per cent, (b) 0.015 bar, (c) 33.45 kJ, (d) 0.87m^3 /kg, (e) 13.3°C]

16.31. One air stream at 13°C and 30 per cent R.H. is mixed adiabatically with another mass flow rate of the hot air stream is twice that of the colder stream, calculate the final temperature and R.H. of the mixture.

16.32. Green lumber is kiln-dried by stacking the lumber and passing heated air through it. Outside air at 26°C and 60 per cent R.H. is first heated to 126.5°C and then circulated over the lumber. The air leaving the kilns has a temperature of 65.5°C and a relative humidity of 30 per cent. What should be the recirculation rate to remove 136.5 kg of moisture per hour from the lumber?

[*Ans.* 36400 kg]

16.33. Atmospheric air at 29°C and 70 per cent R.H. and a rate of 427 m^3/min flows over a set of cooling coils. The liquid which condenses is removed from the system at 10°C. Subsequent heating of the air results in a final temperature of 24°C and a R.H. of 40 per cent. Calculate the heat removed in the cooling section and the heat added in the heating section in kJ/h.

16.34. A 3 m^3 vessel contains a mixture of air and water vapour at 20°C and 1.013 bar. If the R.H. of the mixture is 70 per cent, calculate (a) the specific humidity, (b) the vapour pressure, (c) *d.p.t.* and *w.b.t.* (d) the mass of vapour and air, and (e) the amount of fraction of water evaporates.

Could the desired cooling have been accomplished by simple heat transfer to the air in a heat exchanger? Explain.

16.35. A laboratory requires air at 24°C *d.b.t.* and 50 per cent R.H. to be maintained when the outdoor conditions air 32°C *d.b.t.* and 23°C *w.b.t.* The room sensible and latent heat gains are 123560 kJ/h and 67975 kJ/h respectively and the ventilation is 67.6 m^3/min. Calculate (a) outdoor air load, (b) effective S.H.F., (c) A.D.P. (a) reheat required, (e) supply air quantity, (f) entering and leaving conditions at the apparatus, and (g) supply air condition to the space.

16.36. A room is to be maintained at 24°C *d.b.t.* and 50 per cent R.H. when the outdoor condition are 35°C *d.b.t.* and 24°C *w.b.t.* The room sensible and latent heat gain are 209100 kJ/h and 52275 kJ/h respectively. If the ventilation required is 54 m^3/min, determine (a) outdoor air load, (b) effective sensible hear factor, (c) ADP, and (d) entering and leaving air conditions at the apparatus.

[*Ans.* (a) 945132 kJ/h (b) 0.785, (c) 10°C. (b) entering conditions t_{db}= 29°C, R.H. = 47 per cent; leaving condition: t_{db} = 13.5°C, R.H. = 85 per cent]

16.37. Determine (a) supply air condition, (b) *state of mixture before coil*, (c) *refrigeration load on the coil for an air conditioning with the following data*:

Room conditions : 25°C *d.b.t.*, 20°C *w.b.t.*

Outside conditions : 37.5 °C *d.b.t.* 25°C *w.b.t.*

Room heat load : *Sensible* = 41820 kJ/h, Latent = 24047 kJ/h,

Supply air = 18 m^3/min. 25 per cent fresh and 75 per cent recirculated air.

Software Based Questions

16.38. Develop a software to analyse the thermodynamic performance of various psychrometric processes.

16.39. Develop a software for the design of summer and winter air conditioning systems.

16.40. Develop a software using the relation for specific humidity (ω), to determine the humidity ratio and relative humidity for the following cases:

(a) the dry bulb and wet bulb temperature in a conference room at 1 atm are 24°C and 16°C respectively, and

(b) the dry bulb and wet bulb temperature in a factory space at 1 atm are 24°C and 15°C, respectively.

16.41. Develop a software to calculate the following:

(a) the relative humidity, the humidity ratio, and the specific enthalpy of mixture, in kJ per kg of dry air, corresponding to dry bulb and wet bulb temperature of 30°C and 25°C respectively.

(b) the humidity ratio, mixture specific enthalpy and wet bulb temperature corresponding to dry bulb temperature of 30°C and 60°C relative humidity.

(c) the dew point temperature corresponding to dry bulb and wet bulb tempetature of 30°C and 20°C, respectively.

16.42. At steady state, moist air is to be supplied to a classroom at a specified volumetric flow rate and temperature, *T*. Air is removed from the classroom in a separate stream at a temperature of 27°C and 50% relative humidity. The moisture is added to the room from the occupants at the rate of 5 kg/h. Heat transfer into the occupied space from all sources is estimated to occur at a rate of 35,000 kg/h. The pressure remains uniform at 1 atm. Develop a software to calculate the following:

(a) For a supply air, volumetric flow rate of 42 m^3/min, calculate the supply temperature *T*, in °C and the relative humidity.

(b) Plot the supply air temperature, in °C and relative humidity, each versus the supply air volumetric flow rate ranging from 30 to 100 m^3/min.

16.43. Develop a software to calculate the mixture stream and rate of energy destruction for the adiabatic mixing of two moist air streams at steady state. Show the results by taking some realistic data.

Design and Open Ended Questions

16.44. In general, most of the supermarkets use air-conditioning system designed primarily to control temperature, and these typically produce a relative humidity of about 33%. But at this relative humidity, an appreciable amount of energy must be used to keep frost and condensation from forming refrigerated display cases. Prepare a report about available technologies for reducing the overall humidity levels to 40 to 45% which will reduce the problems associated with frost and condensation and potential cost savings for supermarket.

16.45. As per practice to provide comfort in indoor natatorium air and water temperature should be within 23 and 29°C and the relative humidity in the range of 50 to 60%. But water evaporation from the pool surface provides a major contribution to discomfort and inefficient energy use. At 28°C, an olympic size pool can evaporate upto 100 kg/h. The air-borne water vapour has energy that can be extracted during dehumidification process and may be used to heat the pool water. Prepare a report describing the device for accomplishing this energy-saving measure. The report should also contain the proposed means with sketch for preventing condensation on the natotorium windows for outdoor temperature as a low as –15°C.

16.46. A hospital operating rooms are to be air conditioned. Control of microbial containments is crucial aspect in such rooms. Develop a proposed design of a hospital air-handling system aimed at effective infection control. Whether you will use 100% outside air or recirculated air-filtration arrangement requiring minimum outside air. Whether you will use steam humidification or common water spray method. How many filters you will use at different stages and why? Prepare a comprehensive report discussing all these aspects including the layout of air-handling system.

16.47. The world's largest enclosed ecosystem project entitled "Biosphere 2" is located near Tucscon, Arizona. Based on the philosophy of this project, develop a plan for the air conditioning systems required to maintain a healthy environment for a 10-persons biosphere on Mars and Moon. Also estimate the total power required for this system. It is possible to harness this energy input on Mars and Moon?

16.48. An electricity-generating plant is under expansion of about 30% of generating capacity. Being a consulting engineer, you are supposed to recommend the means for providing condenser cooling. Consider the use of wet cooling towers, dry cooling towers and cooling lakes. Prepare a comprehensive report including relevant environmental issues, the effect of power plant performance and cost.

Multiple-Choice Answer Type Questions

16.49. Humidity ratio can be expressed in terms of partial pressure of dry air (p_a) and water vapour (p_v) as

(a) $0.622\left(\dfrac{p_a}{p_v}\right)$ (b) $0.622\left(\dfrac{p_v}{p_a}\right)$ (c) $0.622\left(\dfrac{p_v}{p_v - p_a}\right)$ (d) $0.622\left(\dfrac{p_v - p_a}{p_v}\right)$

16.50. Wet bulb depression in the case of saturated air is

(a) *d.b.t.* (b) *w.b.t.* (c) *w.b.t./d.b.t.* (d) 0

16.51. In India, for comfort air conditioning requires

(a) 30°C *d.b.t.* and 60% (b) 15°C and 50% R.H.
(c) 26°C *d.b.t.* and 50% R.H. (d) 23°C *d.b.t.* and 60% R.H.

16.52. What processes could be used in summer air conditioning?

(a) cooling (b) humidification (c) dehumidification (d) all of these.

16.53. If air is passed over cooling coil, the process is called

(a) sensible cooling (b) sensible heating
(c) cooling with humidification (d) cooling with dehumidification

16.54. The saturation temperature corresponding to the partial pressure of water vapour present in moist air is given by

(a) *d.b.t.* (b) *w.b.t.* (c) *d.p.t.* (d) subcooled

16.55. Adiabatic cooling means

(a) constant *w.b.t.* (b) constant *d.b.t.* (c) constant R.H. (d) constant *d.p.t.*

Answers to Multiple-Choice Questions

16.49 (b) 16.50 (d) 16.51 (d) 16.52 (d) 16.53 (a) 16.54 (c)
16.55 (a)

CHAPTER

17

Reactive Mixture (Fuels and Combustion)

OVERVIEW

The fossil fuels are responsible for the production of about 70% of the world's total energy today. The remaining is supplied by water power (by erecting dams), nuclear energy by fissioning of nuclear fuel) and other new sources of energy such as solar energy, windmill, and direct energy converters. In India, there are about four nuclear power plants in operation and many new ones are coming up. Nuclear energy was gaining popularity. It was hoped that in near future, nuclear energy will take the place of the fossil fuel whose stock in the world is reducing day by day but this did not happen due to public opinion against this. In this section, we shall deal fossil (conventional) fuels only and its chemical reactions.

A *conventional (fossil) fuel may be defined as a substance which once raised to ignition temperature, will continue to burn without any external support provided sufficient quantity of oxygen is available for the combustion. Most the fuels belong to hydrocarbon family of which the major constituents are carbon and hydrocarbon.* The fuels may either be (a) solid, (b) liquid or (c) gas. These may be natural or artificial.

Before we define the term combustion, let us first understand what is chemical reaction. *A chemical reaction may be defined as a process in which the interatomic bonds of the reactant molecules are broken and a re-arrangement of atoms in a new molecular combination takes place. A chemical reaction which releases heat is called* ***exothermic*** *and the other which absorbs heat is known as* ***endothermic****.*

Combustion means burning and it involves a chemical reaction which releases heat. *Hence,* ***combustion of fuels is an exothermic reaction****.* In order to understand the combustion process fully, we have to turn to chemistry.

17.1 CLASSIFICATION OF FUELS

Fuels are classified in Table 17.1.

(a) Solid Fuels Among the solid fuels, coal is used the most, for example, in boilers to generate steam. Coke is the product of destructive distillation of certain soft coals in closed

Table 17.1 Classification of fuels

Type of Fuels	Natural Fuels	Maufactured Fuels	
	Coal	Coke	Waste fuels
Solid	(i) Bituminos	Pulverised	(i) Bagasse
	(ii) Lignite	Briquettes	(ii) Sawdust
	(iii) Anthracite Peat Wood	Charcoal	(iii) Tanbark
Liquid	Petroleum (crude oil)	Gasoline, kerosene, diesel, fuel oil. alcohol, benzol, shale oil	
Gaseous	Natural gas (mainly consists of CH_4 (80 to 94%), C_2H_6, C_3H_8, C_4H_{10}, N_2 and O_2)	Producer gas Blast furnace gas Carbureted water gas Coal gas Acetylene	

retorts. Briquetted coal is the product of briquetting of dried coal by pressure alone. Pulverised coal is produced by pulverising low grade fuel.

The following methods are used to determine the composition of coal:

(a) *by proximate analysis*, and

(b) *by ultimate analysis*

In proximate analysis, we determine the following:

(i) the moisture content, (ii) the volatile matter, (iii) fixed carbon, and (iv) ash

In *ultimate analysis*, we determine the percentage of the ultimate constituents such as (i) carbon, (ii) hydrogen, (iii) oxygen, (iv) sulphur, (v) nitrogen and (vi) ash. An ultimate analysis of some of the solid fuels is given in Table 17.2.

Proximate analysis is used for commercial purposes as it furnishes data regarding the commercial properties of coal while ultimate analysis is used for calculation and scientific work.

(b) Liquid Fuels When crude petroleum is refined it gives gasoline for the internal combustion engines, such as automobiles and aircraft engines; kerosene for lighting; diesel fuel for diesel engines; lubricating oils for lubrication: fuel oil for domestic heating, steam generation and industrial furnaces; and oil for the manufacture or enrichment of gas. *Petroleum is an intersolution of a mixture of hydrocarbon compounds of the following families.*

$$C_n H_{2n+2},\ C_n H_{2n},\ C_n H_{2n-2},\ C_n H_{2n-4}$$

Table 17.2 Ultimate analysis of solid fuels

Fuel	Carbon C	Hydrogen H_2	Oxygen O_2	Nitrogen N_2	Sulphur S	Incomus-tible	Higher calorific value kJ/kg K	Remarks
Wood (dry)	48.5	6.0	43.5	0.5	–	1.5	17300	
Peat (dry)	58	6.3	30.8	0.9	–	4	19700	Air dried peat contains 20 to 30% moisture. It should be dried before use.
Lignite	66	5.0	20.0	1.0	1.0	3.5	21650	Air dried lignite contains 10 to 20% moisture. It is noncaking. It is a low grade fuel.
Bituminous	81	5.0	8.0	1.5	1.0	3.5	33500	It may be caking and noncaking.
Anthractie	91	3.0					34600	It is noncaking.
Coke	–	–	–	–	–	–	33600	
Charcoal	–	–	–	–	–	–	34300	

The advantages of liquid fuels are:

(i) Higher calorific value, (ii) cleanliness, (iii) economy in space, (iv) easy control of consumption, (v) economy in staff employed for the plant, (vi) ease of lightening up or shutting off operations, and (vii) elimination of wear and tear of grate bars.

Among the important properties of fuel oil are : (i) specific gravity, (ii) heating (i.e., calorific) value, (iii) flash and burning points, (iv) congealing point (pour point), and (v) viscosity.

The flash point of an oil is the temperature at which the oil produces vapours in sufficient quantity to form a momentary flash under certain standardized conditions when a flame is brought near the surface of the oil. Since the flash point is the temperature at which combustible vapours are produced, it is an index of the relative safety with which the fuel oil may be handled and stored. ***A minimum flash point of 71°C is frequently specified.***

The burning point of a liquid fuel is the temperature at which the oil will produce combustible vapours rapidly enough to burn continuously. **It is usually 11°C above the flash point.**

The ***pour point*** *is the lowest temperature at which the oil will flow and is important from the point of view when it has to be used in cold weather.*

The ultimate analysis of some of the liquid fuels is given in Table 17.3.

Table 17.3 Ultimate analysis of liquid fuels

Fuel	Specific gravity	Ultimate analysis or composition by weight			Lower calorific value kJ/kg	Higher calorific value kJ/kg
		C	H_2	S		
Gasoline	0.74	85.4	14.6	–	43700	46820
Paraffin	0.79	86.3	13.6	0.1	43250	46250
Diesel oil	0.87	86.3	12.8	0.9	43250	46000
Heavy oil	0.95	86.1	11.8	2.1	41300	4400
100 octane petrol	–	–	–	–	44000	47300
Propane (LPG)	–	–	–	–	45900	–

(c) The Gaseous Fuels. A *gaseous fuel is supposed to be an ideal fuel.* The main advantages of the gaseous fuels are:

(i) Combustion of gaseous fuels permits exact control in quantity, length of flame, nature of flame (reducing or oxidizing) and temperature.

(ii) Ease of handling and distribution over a vast area even if it is produced at a remote place through pipelines.

(iii) Elimination of smoke and ash.

(iv) Cleanliness.

(v) Greater economy in high temperature work.

(vi) Low grade fuels can be successfully used by gasification and important by-products are recovered.

Many gaseous fuels are extensively used in industrial applications where the product must not be contaminated or the furnace atmosphere must be regulated closely or where uniform heating is required. The following are the gaseous fuels.

(i) *Natural gas* is generally found in petroleum fields. Now, natural gas is extensively used in gas turbines, steam turbines and combined cycle power plants. Its composition varies from well to well. One of the molar compositions are; CH_4, 80.62%; C_2H_6, 5.41%; C_3H_g, 1.87%, C_4H_{10}, 1.6%; and N_2, 10.5%; *natural gas is liquefied and is called LNG.*

(ii) *Coke-oven gas* is obtained as a by-product of the coke oven and is used in furnaces of steel mills. It is also distributed as city gas. It is distilled from coal in the process of making metallurgical coke.

(iii) *Producer gas* is a lean gas with a low heating capacity. It is used in industrial furnaces by passing air with a small quantity of steam through a thick fuel bed where the fuel has been burnt incompletely. The resulting gases have a high content value of hydrogen, nitrogen and carbon monoxide.

(iv) *Blast-furnace gas* is a by-product of the blast furnace which is in effect a large gas producer. It is an important fuel in steel manufacturing.

The volumetric composition of some of the gaseous fuel is shown in Table 17.4.

Table 17.4 Volume composition of gaseous fuels

Fuel	Percentage volumetric composition						Calorific value in MJ/m^3 at S.T.P.	
	H_2	CO	CH_4	C_2H_4	CO_2	N_2	Higher calorific	Lower calorific
Coal gas	27	7	48	13	3	2	20	17.85
Town gas	55	14	23	2.5	2	3.5		
Coke oven gas	40	8	29	4	2	7		
Blast furnace gas	4	28	–	–	8	60	3.41	3.37
Producer gas	6	23	3	0.2	5	62	6.04	6.0
Hydrogen	–	–	–	–	–	–	11.85	10
Natural gas	–	–	–	–	–	–	36.2	32.60
Propane	–	–	–	–	–	–	–	86.42
n-Butane	–	–	–	–	–	–	–	112.41
Methane	–	–	–	–	–	–	–	33.95
Carbon monoxide	–	–	–	–	–	–	–	11.97
Ethane	–	–	–	–	–	–	–	60.43

17.2 COMPLETE COMBUSTION EQUATION (OR STOICHIOMETRIC EQUATION)

In a combustion process, when the fuel is completely oxidized, the combustion is said to be complete. For example, if a fuel contains carbon and hydrogen, complete combustion means, all carbon is burnt to carbon dioxide, and all hydrogen, is reduced to water.

The combustion reaction is written in the form of a chemical reaction expressing the conservation of mass in terms of the conservation of atoms. If we consider the complete combustion of carbon fuel with oxygen it is written as

$$C + O_2 \rightarrow CO_2 \quad \text{...(17.1)}$$

This balanced equation for the complete combustion of fuel such as carbon is called *stoichiometric equation*. It is obvious from Eq. (17.3) that the number of carbon and oxygen atoms are the same before and after the reaction. The following meaning may be drawn from the above equation.

(a) *On molal basis*

$$(1 \text{ mole C}) + (1 \text{ mole } O_2) \rightarrow (\text{I mole } CO_2)$$

(b) *On volume basis*

$$(1 \text{ volume C}) + (1 \text{ volume } O_2) \rightarrow (1 \text{ volume } CO_2)$$

The volume (or mole) of the carbon solid fuel is negligible with the volume of air. Hence, 1 volume of O_2 with carbon gives 1 volume of CO_2. Similarly, 1 mole of O_2 with 1 mole of carbon gives 1 mole of CO_2. Thus, we see that there is no molecular expansion or contraction.

(c) *On mass basis*

$$(12 \text{ kg C}) + (32 \text{ kg } O_2) \rightarrow (44 \text{ kg } CO_2)$$

or

$$1 \text{ kg C} + 8/3 \text{ kg } O_2 \rightarrow 11/3 \text{ kg } CO_2$$

This shows that the total mass of the reactants is equal to the total mass of the products.

Though, we have seen that the number of moles (or volumes) of the reactants is the same as products but this is not always true. For example, consider the combustion of hydrogen with oxygen

$$2H_2 + O_2 \rightarrow 2H_2O$$

In this case, 3 moles of reactants give only 2 moles of the product.

In general, the stoichiometric equation may be written as

$$n_A \cdot A + n_B \cdot B + \dots \rightarrow n_L \cdot L + n_M \cdot M \quad \text{...(17.2)}$$

where the upper case letters *A, B* (not the suffixes), represent the chemical species of the reactants and *L, M,* (not the suffixes), etc., represent the chemical species of the products. The n_i terms are known as the stoichiometric coefficients of the various species.

In the combustion of carbon, $n_A = 1$, $n_B = 1$ and $n_L = 1$, $M = 0$

In an incomplete combustion of carbon in a fuel, the carbon reacts according to the reaction

$$C + \frac{1}{2}O_2 = CO$$

In practice, there is usually CO in the products, even though an excess of oxygen was supplied. This may be attributed either to incomplete mixing during the process or to insufficient time for complete combustion.

In industrial practice, the combustion of fuels is normally accomplished by applying air is the oxidiser rather than pure oxygen. The molecular weight of the air is 28.967. For computation work it is sufficiently accurate to take $M_{air} = 29$.

The composition of air may be taken as follows:

Composition on mass basis : $O_2 = 23.2\%$ and $N_2 = 76.85\%$

∴ For each kg of O_2 supplied by air, we have

$$76.85/23.2 = 3.31 \text{ kg of } N_2 \text{ and } 100/23.2 = 4.31 \text{ kg of air}$$

Composition on volume basis : $O_2 = 21\ \%$, $N_2 = 79\%$

∴ For each mole of O_2, we have

$$79/21 = 3.76 \text{ moles of } N_2 \text{ and } 100/21 = 4.76 \text{ mole of air.}$$

We can write this as

$$1 \text{ kg of } O_2 + 3.31 \text{ kg of } N_2 = 4.31 \text{ kg of air}$$

and $$1 \text{ mole of } O_2 + 4.76 \text{ mole of } N_2 = 3.76 \text{ moles of air}$$

Although, the nitrogen of air does not contribute to the oxidization reactions, it leaves the reacting system at the same temperature from its original energy entrance. This factor is taken into consideration in the energy balance of the combustion process.

The minimum amount of air required for the complete oxidation of a fuel is called the ***theoretical or stoichiometric*** air. For example, the chemical equation for the stoichiometric reaction of propane (C_3H_8) with air is

$$C_3H_8 + 5\ O_2 + 18.8\ N_2 = 3\ CO_2 + 4\ H_2O + 18.8\ N_2 \qquad ...(17.3)$$

This equation indicates that a minimum of 5 moles of O_2 accompanied by 18.8 moles of N_2 are necessary for complete oxidization of 1 mole of C_3H_8 to form 3 moles of CO_2 and 4 moles of H_2O.

The amount of air actually supplied may exceed or may be less than the theoretical amount required. If excess of air is used, oxygen will appear in the products of combustion, but if insufficient air is used, the combustion will be incomplete and the products such as CO, CH, H_2 and unburnt fuel can also appear. In the previous example, when 150 per cent theoretical air or 50% excess air was supplied, the theoretical combustion reaction was:

$$C_3H_8 + 7.5\ O_2 + 28.2\ N_2 \rightarrow 3\ CO_2 + 4\ H_2O + 2.5\ O_2 + 28.2\ N_s$$

In practice, some excess air is generally supplied in order to ensure that no unburnt fuel is contained in the products. But at the same time, any large amount of excess air is also avoided because the additional oxygen and nitrogen absorbs energy from the system, and so lower the temperature of the combustion products resulting in the lowering of the rate of heat transfer from the combustion products.

17.3 AIR-FUEL RATIO AND EQUIVALENCE RATIO

Air-fuel ratio is defined as the ratio of the mass of the air to the mass of fuel. The reciprocal of the above definition is the fuel-air ratio. Hence,

$$\text{A/F} = \text{Air-fuel ratio} = \frac{\text{Mass of air}}{\text{Mass of fuel}} = \frac{m_a}{m_f} \qquad \text{...(17.4)}$$

$$\text{F/A} = \text{Fuel-air ratio} = \frac{\text{Mass of fuel}}{\text{Mass of air}} = \frac{m_f}{m_a} \qquad \text{...(17.5)}$$

For example, in a 100% theoretical combustion of propane, the chemical Eq. (17.3) shows that 23.8 moles of air (5 moles of O_2 + 18.8 moles of N_2) is required per mole of fuel. Therefore, the air-fuel for theoretical combustion of this fuel is

$$\text{A/F} = \frac{\text{Mass of } (5 + 18.8) \text{ mole of air}}{\text{Mass of 1 mole of } C_3\,H_8} = \frac{(5 + 18.8) \times 29}{44} = 15.686$$

Also,

$$\text{A/F} = \frac{\text{Mole of 7.5 mole of } O_2 + \text{Mass of 28.2 mole of } N_2}{\text{Mass of 1 mole of } C_3\,H_8}$$

$$= (5 \times 32 + 18.8 \times 28)/44 = 15.6$$

The fuel-air ratio for the same combustion process will be 0.0641 kg fuel/kg of air. *It is interesting to know that many hydrocarbon fuels obtained from petroleum or natural gas require an air-fuel ratio in the neighbourhood of 15 to 16 for stoichiometric combustion.*

Excess air means the amount of air supplied in excess to that theoretically needed air to burn the fuel completely. Lean mixture means less fuel than stoichiometric and *rich mixture* means more fuel than stoichiometric. Sometimes, a term called **equivalence ratio** is used instead of A/F ratio. *It is the ratio of actual fuel ratio to the stoichiometric (i.e., theoretical) A/F ratio.*

17.4 FLUE GAS ANALYSIS

Among the several devices for measuring the composition of products of combustion (flue gas) are the (i) **orsate analyser**, (ii) **infrared analyser**, (iii) **chromatograph and** (iv) **flame ionization detector**. The first two will be discussed here. *The dew point temperature of exhaust product is very important parameter as it decides the condensation of water vapour in the produced vapour in the product.*

17.4.1 Orsate Apparatus (Volumetric Analysis)

In reality, the combustion is never complete and the resulting products of combustion called flue gases contain CO_2, CO, O_2 and N_2. Orsate appartus is used to analyse the flue gases containing CO_2, CO and O_2 by volume. If the remainder gas is nitrogen, it will also be known by difference. This apparatus gives an accuracy of 0.5% for CO_2 content.

Construction Figure 17.1 show the Orsate apparatus. It consists of three flasks 1, 2 and 3 containing chemicals to absorb the different gases. Flask No. 1 contains a solution of caustic potash KOH (one part KOH to 2 parts of water by weight) for absorbing CO_2. Flask No. 2 contains an alkaline solution of paragollic acid (5 gm paragollic acid in 15 cc of water mixed with 120 gm KOH in 80 cc of water) for absorbing O_2 and flask No. 3 contains cuprous chloride (CuO dissolves in 20 times its weight of concentrated hydrochloric acid till it becomes colourless) for absorbing CO.

In fact, each absorbing flask is a combination of two flasks joined together with a glass tube as shown in Fig. 17.1. Three such flasks are arranged in such a manner that they form two

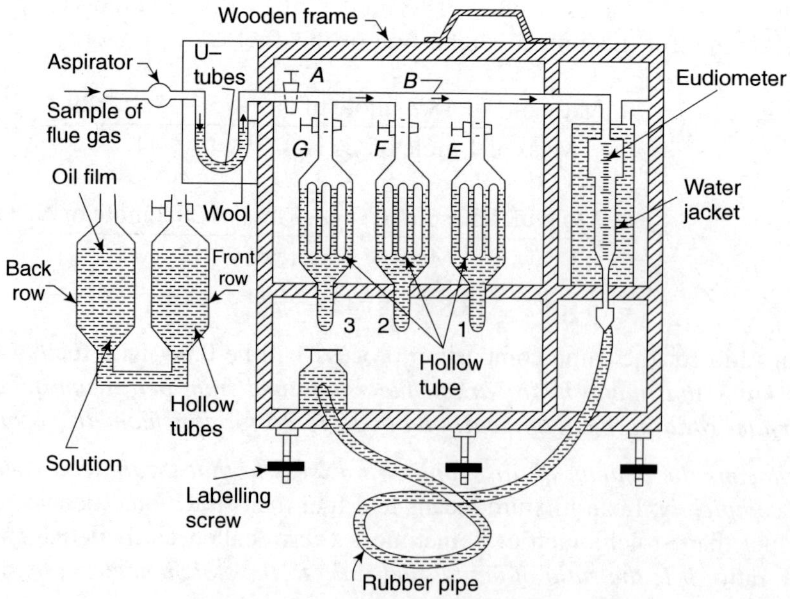

Fig. 17.1 Orsate apparatus

rows, called front rows and the other back row flasks. The top of the front flasks, are connected to a common header *B*. One side of the common header is connected with a three-way valve (*A*) which connects it to flue gas supply. The other side of the common header is connected to an eudiometer (*C*) generally calibrated from 0 cc to 100 cc which is surrounded by a water jacket to keep the temperature constant. A flexible rubber connects the tube eudiometer (*C*) to an aspirator bottle (*D*) filled with salt water which cannot dissolve gas. The position of the aspirator bottle may be raised or lowered according to convenience which changes the level or the salt water in the eudiometer.

In order to admit the controlled gas to the front-row flasks, each flask is provided with one-way glass tap such as *E*, *F* and *G* adjoining the front header. The front row flasks are provided with numerous hollow glass tubes inside them to accelerate the absorption of the gases by increasing the wet surface areas. In order that the solution does not react with the atmosphere, an oil film is provided in the back row flasks.

Test procedure. Before starting the experiment, it is essential that the chemical solutions in the flask are fresh and free from pollution. Then water is filled in the eudiometer jacket and the aspirator bottle (*D*) is raised to fill the eudiometer to the 100 cc mark. The three-way valve (*A*) is then closed and the valve (*E*) is opened. The aspirator bottle is lowered, which in turn, lowers the level of salt water in the eudiometer resulting in the rise in the level of chemical solution in flask No. 1. As soon as the chemical solution touches a mark on the capillary connecting the common header tube, the valve (*E*) is closed. The same procedure is repeated for the chemical solution in flasks No. 2 and 3. By doing so, all the residual gases in the flasks are expelled to the eudiometer and the common header. To expel the residual gases from eudiometer, the three-way valve *A* is opened to the atmosphere and the aspirator bottle is raised so that the salt water reaches up to the mark 100 cc. After carrying out the above procedure, one can be sure that there is no residual gases either in the flasks or in the eudiometer but in the common header tube *B* and the small connecting capillary tube projections there may be some left-over gas which is negligible in quantity.

The apparatus is now ready and three-way valve (*A*) is opened to the flue gases and a sample of 100 cc of flue gas is admitted in the eudiometer through a *U*-tube containing wool to send in only filtered gas by lowering the aspirator bottle.

In order to ensure that there is no leakage, all the valves (*E*, *F* and *G*) are closed and the aspirator bottle is raised. If on raising the aspirator bottle, the water level in the eudiometer rises, it means that there is a leakage. The water jacket in the eudiometer condenses all the moisture present in the sample of gas and hence only a dry sample of 100 cc of flue gas is admitted in the eudiometer. The sample of gas drawn in and expelled a few times ensures a representative unadultered 100 cc of the sample of gas. In order to ensure that the sample collected in the eudiometer is at the constant atmospheric pressure, the water level in the eudiometer and the aspirator bottle is kept the same.

Now, the valve (*E*) is opened and the aspirator bottle is raised, this raises the level of water in eudiometer thus forcing the sample of gas to the flask No. 1 containing KOH. In order to accommodate the gas in the flask, the chemical solution KOH is made to pass to the duplicate flask and the wetted surface of the glass tube CO_2 in the sample of 100 cc of flue gases. The aspirator bottle is again lowered and the solution rises washing the glass tubes. The procedure is repeated a few times ensuring that all CO_2 is absorbed. The valve (*E*) is closed when the solution has risen to the mark on the connecting capillary tube to the header. In order to ensure that the remaining gas in the eudiometer is again at the same atmospheric pressing, the level of the water in the aspirator bottle and the eudiometer is again adjusted to be the same. The reading of the eudiometer thus obtained gives the percentage volume of CO_2 in the flue gas.

The procedure outlined for flask No. 1 is repeated with flask No. 2 to absorb O_2 and flask No. 3 to absorb CO. The percentage of nitrogen in the flue gas is obtained from the difference [i.e., 100 – (CO_2 + CO + O_2)]. All the above procedures are repeated with a fresh sample of the gas to ensure that the readings are correct.

17.4.2 Infra-red Spectra Gas Analyser

The analysis of combustion products carried by the Orsate apparatus is approximate and may not be suitable for research purposes. Analysis by chemical means is a slow process and other methods (such as "Infra-spectra") have been developed to give an accurate analysis, more quickly and continuously.

The basic principle of infra-red spectrum lies in the measurement of the concentration of the constituent gas by recordirg its optical absorption. The novel feature of it is the continuously examining of the mixture by being drawn through a tube, the inside of which is subject to radiation from an infra-red source, each gas absorbing a particular wave-band of the radiation.

Figure 17.2 shows the details of an infra-red spectra gas analyser. The gas being analysed is passed through tube *A* and dry air is passed through tube *B*. Two chambers *C* and *D*, are located just below these tubes and contain pure samples of the constituent to be detected. Radiation is produced from nichrome elements which is passed through tubes *A* and *B* and hence to chambers *C* and *D* where it is absorbed. The effect of radiation is to heat the gases in *C* and *D* which causes an increase in the two chambers. It is to be noted that the two chambers, *C* and *D*, are separated by a thin metal diaphragm which, together with an insulated, perforated plate, forms a capacitor. The capacitor is connected to amplifier, a servo-unit and recording mechanism.

If the constituent being sought is not present in tube *A*, then equal amount of radiation will be absorbed in *C* and *D*. As a result of this, they will be heated equally and there will be no pressure difference across the diaphragm. If the constituent being sought is present in *A*, it will absorb some of the radiation admitted, and the rest will pass on the chamber *C*. This means that the radiation absorbed in *C* will then be less than in *D* so that a greater pressure will be reached in *D* than in *C* resulting in the displacement of the diaphragm. As a result of this displacement, a change in capacitance of the condenser is produced and hence a current is produced which is amplified to give a reading on a micro-ammeter. The micro-ammeter scale is calibrated to give the corresponding concentration of the constituent in the gas being analysed.

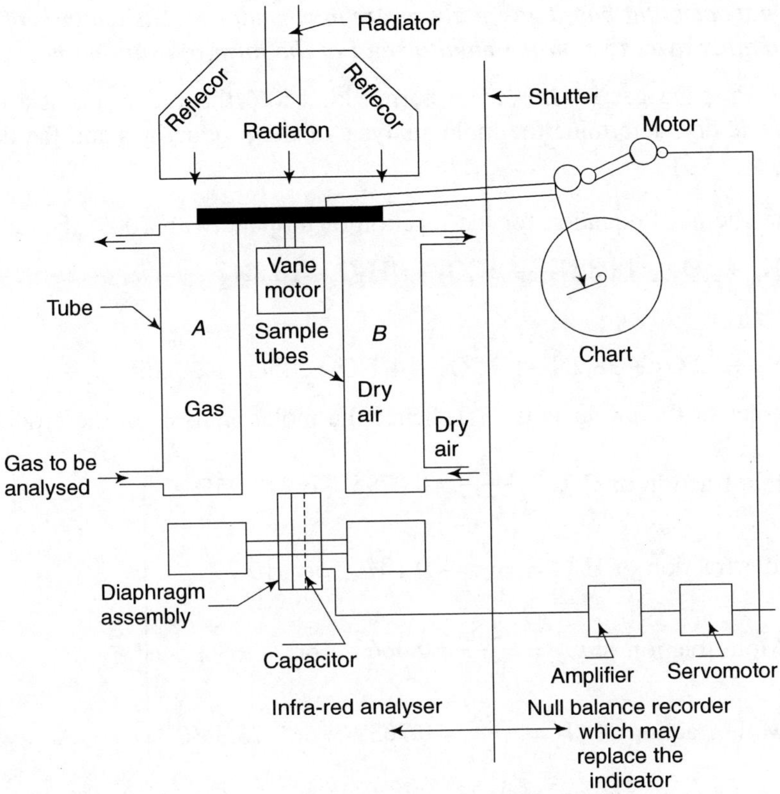

Fig. 17.2 Infra-red spectra gas analyser

In order to avoid zero error, a vane is rotated at low frequency which cuts off radiation from both tubes simultaneously, and allows to fall on them. The pressure changes are then related to the temperature changes produced by the differential absorption in *C* and *D*.

In order to provide a continuously recording instrument, a null balance recorder is employed as shown in the right hand part of Fig. 17.2. The principle employed in the null balance recorder is that the pressure difference created should be nullified by cutting off from the vessel *B* a sufficient amount of radiation to balance that absorbed in *A*. This is achieved by means of a shutter driven by a servo motor which receives a signal from the detector unit. A recording pen linked to the balancing shutter mechanism is used to record the analysed data on a circular chart. The calibration of the instrument is very accurate as it is done by accurately prepared samples of the mixtures.

This type of analyser can analyse the gas mixture consisting of *carbon monoxide, carbon dioxide, sulphur, nitrogen* compounds, methane and other hydrocarbons and organic vapours. *It is to be noted that this analyser cannot analyse the gases which do not absorb infra-red radiation such* as *oxygen, hydrogen, nitrogen, argon, chloride and helium.* O_2 *concentration may be found by difference or another type of instrument may be used.*

Now, personal computer based gas analyser are in current use. The analysis of combustion products are readily available on the monitor and in the form of print output.

Example 17.1 Propane gas is allowed to burn with 150% theoretical air at a pressure of 1 bar. If the air is dry, determine the mole analysis of the product gas and the dew point of the gas mixture.

Solution The chemical equation for the reaction of propane with 100% theoretical air is

$$C_3H_8 + 5O_2 + 18.8\ N_2 \rightarrow 3CO_2 + 4H_2O + 18.8N_2$$

With 150% air, it will be

$$C_3H_8 + 7.5O_2 + 28.2N \rightarrow 3CO_2 + 4H_2O + 2.5O_2 + 28.2N_2$$

The total moles of the products is 37.7, hence the molar analysis of the product is

$$\left.\begin{aligned} \text{Mole fraction of } CO_2 &= \frac{3}{37.7} = 0.0795 \quad \text{or} \quad 7.95\% \\ \text{Mole fraction of } H_2O &= \frac{4}{37.7} = 0.107 \quad \text{or} \quad 10.7\% \\ \text{Mole fraction of } O_2 &= \frac{2.5}{37.7} = 0.0663 \quad \text{or} \quad 6.63\% \\ \text{Mole fraction of } N_2 &= \frac{28.8}{37.7} = 0.7639 \quad \text{or} \quad 76.39\% \end{aligned}\right\} \quad \textbf{Ans.}$$

The partial pressure of water = Mole fraction of H_2O × pressure = 0.107 × 1 = 0.107 bar.

The saturation temperature corresponding to this pressure is 47.3°C. When the gas mixture is cooled to 47.3°C at constant pressure, the dew point of the mixture is reached. If the temperature of the products drops below dew point temperature of 47.3°C, condensation of water will take place.

Example 17.2 An unknown hydrocarbon fuel, C_xH_y reacts with air the Orsate analysis of products of combustion gives :

$$CO_2 = 12.1\%; \quad O_2 = 3.8\%; \quad \text{and} \quad CO = 0.9\%$$

Calculate the (a) chemical eqaution for the actual reaction, (b) composition of the fuel, (c) air-fuel used during the test, and (d) excess or deficiency of air used.

Solution The total percentage of the three gases in the Orsat analysis is 16.8%. Let us assume that the remaining gas in the sample is nitrogen, therefore, the volumetric percentage of nitrogen must be 83.2%. With this knowledge, we may write the general chemical equation for the reaction.

$$C_xH_y + aO_2 + 3.76a\ N_2 \rightarrow 12.1\ CO_2 + 3.8\ O_2 + 0.9\ CO + 83.2\ N_2 + bH_2O$$

From the equation, we find that 83.2 moles of N_2 enter the reaction chamber. However, the composition of air is known to be 1 mole of O_2 per 3.76 moles nitrogen.

Hence, $$a = \frac{83.2}{3.76} = 22.1 \text{ moles of } O_2$$

O_2 balance gives : $$2 \times 22.1 = 2 \times 12.1 + 2 \times 3.8 + 0.9 + b$$

$\therefore$ $$b = 11.5$$

To find the value of x and y, carbon and hydrogen balance should be used.

C balance gives : $$x = 12.1 + 0.9 = 13$$

H_2 balance gives : $$y = 2 \times 11.5 = 23$$

Putting the value of x, y, a and b, we get the correct chemical equation as

$$C_{13}H_{23} + 22.1O_2 + 83.2N_2 \rightarrow 12.1CO_2 + 3.8O_2 + 0.9CO + 83.2N_2 + 11.5H_2O$$ **Ans.**

The air-fuel ratio can be computed from the above balance equation. As 105.3 moles of air used per mole of fuel, so

$$A/F = \frac{\text{Mass of } (22.1 + 83.2) \text{ mole of air}}{\text{Mass of one mole of } C_{13}\,H_{23}}$$

$$= \frac{105.3 \times 29}{179} = 17.059 \text{ kg of air/kg of fuel}$$

The excess or deficiency of air can be found computing the theoretical A/F value of the fuel first.The chemical equation for the combustion of $C_{13}H_{23}$ with the theoretical air requires.

$$C_{13}H_{23} + 18.75O_2 + 3.76 \times 18.75N_2 \rightarrow 13CO_2 + 11.5H_2O + 18.75 \times 3.76N_2$$

Since the actual requirement of O_2 was 18.75 moles but 22.1 moles of O_2 was actually used, the percentage excess air used during the test is

$$\text{Percentage excess} = \frac{22.1 - 18.75}{18.75} = 17.86\%$$ **Ans.**

Example 17.3 Propane gas is allowed to react with air in such a ratio that the Orsate analysis of the products of combustion give CO_2, 11.5% O_2, 2.27% and CO, 0.7%. Calculate the percentage of theoretical air used during the test.

Solution The right way to begin the analysis is to write down the chemical equation for 100 moles of the dry product gases. The total of the product is 14.9%, hence 85.1% is the nitrogen in the product. With this data, the chemical equation may be written as

$$xC_3H_8 + aO_2 + 3.76a\,N_2 \rightarrow 11.5\,CO_2 + 0.7\,CO + 2.7\,O_2 + 85.1\,N_2 + b\,H_2O$$

Since N_2 is inert, it is written down as a side products. On the basis of the composition of air

$$a = \frac{85.1}{3.76} = 22.63 \text{ moles of air}$$

Carbon balance gives: $\quad 3x = 11.5 + 0.7 = 12.2$

or $\quad x = 4.066$

O_2 balance gives: $\quad 2 \times 22.63 = 2 \times 11.5 + 0.7 + 2 \times 2.7 + b$

or $\quad b = 16.16$

Now, employing the hydrogen balance, we check the computation

$$8x = 2b$$

or $\quad 8 \times 4.066 = 2 \times 16.16$

or $\quad 32.28 = 32.32$

This check is fairly accurate and the actual reaction may be written as

$$4.066\ C_3H_8 + 22.63\ O_2 + 85.1\ N_2 \rightarrow 11.5\ CO_2 + 0.7\ CO + 85.1\ N_2 + 16.16\ H_2O$$

Dividing throughout by 4.066, the equation could be reduced to the basis of fuel. The stoichiometric equation for propane is

$$C_3H_8 + 50_2 + 18.8\ N_2 \rightarrow CO_2 + 4H_2O + 18.8\ N_2$$

The moles of oxygen used theoretically per mole of fuel is 5. For the actual combustion, the ratio is 22.63/4.066 = 5.565. Hence, the percentage of theoretical air used in the actual combustion process to percentage of theoretical air

$$= \frac{5.565 \times 100}{5} = 111.3\%$$

The percentage of excess air is 11.3%.

Example 17.4 Calculate the stoichiometric air required for the complete combustion of 1kg of normal heptane C_7H_{16}. Also, calculate the percentage analysis of the products on a mass and a molal basis.

Solution The chemical equation for the combustion of heptane is

$$C_7H_{16} + 11O_2 + 11\ (3.76)N_2 \rightarrow 7\ CO_2 + 8H_2 + 11\ (3.76)\ N_2$$

$$\text{The air-fuel ratio on mass basis} = \frac{11 \times 32 + 11 \times 3.76 \times 28}{7 \times 12 + 16 \times 1} = 15.10 \text{ kg of air/kg of fuel}$$

The analysis of the products on mass and molal basis may be computed as follows:

Gas		By Mass		By Mass
CO_2	7	12.42%	7 × 44 = 308	19.13%
H_2O	8	14.20%	8 × 18 = 144	8.94%
N_2	41.36	73.38%	41.36 × 28 = 1158.08	71.93%
Total	56.37	100%	1610.08	100%

Example 17.5 Five moles of propane C_3 Hg are completely burnt at atmospheric pressure in the theoretical amount of air. Calculate (a) the volume of air used in the combustion process measured at 1 bar and 25°C, (b) the partial pressure of each constituent of the products, and (c) the volumetric analysis of the dry products.

Solution The chemical equation of the complete combustion of 5 moles of propane is

$$5\ C_3H_g + 25\ O_2 + 25\ (3.76)\ N_2 \rightarrow 15\ CO_2 + 20\ H_2O + 25 \times (3.76)\ N_2$$

or
$$5\ C_3H_8 + 25\ O_2 + 94\ N_2 \rightarrow 15\ CO_2 + 20\ H_2O + 94\ N_2$$

From the above equation, we see that 25 moles of O_2 are present in (94 + 25) = 119 moles of air. If air is assumed as perfect gas, then

(a)
$$V = \frac{nR_0\,T}{p} = \frac{119 \times 8.314\,(273 + 25)}{1 \times 10^2} = 2948.31\ \text{m}^3 \quad \textbf{Ans.}$$

(b) Since the partial pressure of the products (constituents) are proportional to the mole fraction, hence

$$p_{CO_2} = x_{CO_2} \cdot p = \frac{15}{15 + 20 + 94} \times p = \frac{15}{129} \times 1 = 0.1162\ \text{bar}$$

$$p_{H_2O} = x_{H_2O} \cdot p = \frac{29}{129} \times p = \frac{20}{129} \times 1 = 0.1550\ \text{bar}$$

$$p_{N_2} = x_{N_2} \cdot p = \frac{94}{129} \times 1 = 0.7286\ \text{bar}$$

Total pressure = 1 bar

(c) The volumetric analysis of the dry products are

$$\frac{V_{CO_2}}{V_{CO_2} + V_{N_2}} = \frac{15}{15 + 94} = 0.1376 \text{ or } 13.76\% \quad \textbf{Ans.}$$

$$\frac{V_{N_2}}{V_{CO_2} + V_{N_2}} = 1 - 0.1376 = 0.8624 \text{ or } 86.24\%$$

Example 17.6 Calculate the air-fuel ratio on both a molar and mass basis for the complete combustion of octane, C_8H_{18}, with (a) the theoretical amount of air, and (b) 160 of theoretical air.

Solution (a) For the complete combustion of octane, C_6H_{18} with the theoretical (Stoichiometric) amount of air, the products of combustion will be carbon dioxide, water and nitrogen only and the chemical equation is given as follows:

$$C_8H_{18} + a\,(O_2 + 3.76\,N_2) \rightarrow b\,CO_2 + c\,H_2O + d\,N_2$$

Applying the principle of conservation of mass to the carbon, hydrogen, oxygen and nitrogen, respectively, we get

$$C : b = 8 \quad H: \quad 2c = 18 \quad O: \quad 2b + c = 2a \quad N: \quad d = 3.76\,a$$

Solving the above equations, we get $a = 12.5, \quad b = 8, \quad c = 9, \quad d = 47$

Thus, the balance chemical equation is expressed as

$$C_8H_{18} + 12.5\,(O_2 + 3.76\,N_2) \rightarrow 8\,CO_2 + 9\,H_2O + 47\,N_2$$

The air-fuel ratio on a molar basis is given by

$$A/F = \frac{12.5 + 12.5\,(3.76)}{1} = 59.5 \text{ kg mole (air)/kg mole (fuel)} \qquad \textbf{Ans.}$$

The air-fuel ratio expressed on mass basis is given by

$$A/F = \left[59.5\,\frac{\text{kg mole (air)}}{\text{kg mole (fuel)}}\right]\left[\frac{28.97\,\dfrac{\text{kg (air)}}{\text{kg mole (air)}}}{114.22\,\dfrac{\text{kg (fuel)}}{\text{kg mole (fuel)}}}\right] = 15.1\,\frac{\text{kg of air}}{\text{kg of fuel}} \qquad \textbf{Ans.}$$

(b) For 160% theoretical air (60% excess air), the chemical equation for complete combustion may be written as follows:

$$C_8H_{18} + 1.6\,(12.5)\,(O_2 + 3.76\,N_2) \rightarrow b\,CO_2 + c\,H_2O + d\,N_2 + e\,O_2$$

Applying conservation of mass to C, H, O and N, we have

$$C : \quad b = 8 \qquad\qquad H: \quad 2c = 18$$

$$O : \quad 2b + c + 2e = 1.6\,(12.5)\,(2) \qquad N: \quad d = 1.6\,(12.5)\,(3.76)$$

Solutions of the above equations yield: $b = 8$, $c = 9$, $d = 75.2$, $e = 7.5$

Thus, the balance chemical equation is expressed as

$$C_8\,H_{18} + 1.6\,(12.5)\,(O_2 + 3.76\,N_2) \rightarrow 8\,CO_2 + 9\,H_2O + 75.2\,N_2 + 7.5\,O_2$$

The air-fuel ratio on molar basis is given by

$$(A/F)_{molar} = \frac{1.6 \times 12.5\,[1 + 3.76]}{1} = 95.2\,\frac{\text{kg mole (air)}}{\text{kg mole (fuel)}} \qquad \textbf{Ans.}$$

On the mass basis, air-fuel ratio is given by

$$(A/F)_{mass} = \left[95.2 \frac{\text{kg mole (air)}}{\text{kg mole (fuel)}}\right] \left[\frac{28.97 \frac{\text{kg (air)}}{\text{kg mole (air)}}}{114.2 \frac{\text{kg (fuel)}}{\text{kg mole (fuel)}}}\right] = 24.14 \frac{\text{kg (air)}}{\text{kg mole (fuel)}} \quad \textbf{Ans.}$$

Example 17.7 In a combustion chamber, methane, CH_4 is burnt with dry air. The molar analysis of the products on a dry basis is CO_2, 9.6%, CO, 0.6%, O_2, 2.95%, and N_2, 86.85% Calculate (a) the air-fuel ratio on both a molar and a mass basis, (b) the per cent theoretical air, (c) the dew point temperature of product if the pressure is 1 atm.

Solution It is assumed that each mole of oxygen in the combustion air is accompanied by 3.76 moles of nitrogen which is inert.

(a) The chemical equation may be written on the basis of 100 kg mole of dry products. In addition to the assumed 100 kg mole of dry products, water must be included as a product,

$$a\ CH_4 + b\ (O_2 + 3.76\ N_2) \rightarrow 9.6\ CO + 2.95\ O_2 + 86.85\ N_2 + c\ H_2O$$

Applying conservation of mass to carbon, hydrogen and oxygen, respectively, we have

$$C: 9.6 + 0.6 = a \qquad H: 2c = 4\ a$$

$$O: (9.6)\ (2) + 0.6 + 2\ (2.95) + c = 26$$

Solution of above equations yields, $a = 10.2$, $b = 23.05$, $c = 20.4$

The balanced chemical equations is written as

$$10.2\ CH_4 + 23.05\ (O_2 + 3.76\ N_2) \rightarrow 9.6\ CO_2 + 0.6\ CO + 2.95\ O_2 + 86.85\ N_2 + 20.4\ H_2O$$

On molar basis, the air-fuel ratio is given by

$$(A/F)_{molar} = \frac{23.05\,(1+3.76)}{10.2} = 10.75 \frac{\text{kg mole (air)}}{\text{kg mole (fuel)}} \quad \textbf{Ans.}$$

On mass basis

$$(A/F)_{mass} = \left[10.75 \frac{\text{kg mole (air)}}{\text{kg mole (fuel)}}\right] \left[\frac{28.97 \frac{\text{kg (air)}}{\text{kg mole air}}}{16.04 \frac{\text{kg (fuel)}}{\text{kg mole (fuel)}}}\right] = 19.41 \frac{\text{kg (air)}}{\text{kg (fuel)}} \quad \textbf{Ans.}$$

(b) The balanced chemical equation for the complete combustion of methane with theoretical amount of air is expressed as

$$CH_4 + 2\ (O_2 + 3.76\ N_2) \rightarrow CO_2 + 2H_2O + 7.5\ N_2$$

The theoretical air-fuel ratio on molar basis is given by

$$(A/F)_{theo} = \frac{2 \times 4.76}{1} = 9.52 \frac{\text{kg mole (air)}}{\text{kg mole (fuel)}}$$

The percentage of theoretical air is given by

$$\text{Percentage of theoretical air} = \frac{\text{A/F}}{(\text{A/F})_{\text{theo}}} \times 100 = \frac{10.75 \times 100}{9.52} = 112.9\%$$

(c) The dew point temperature may be evaluated when partial pressure of water vapour is known. The partial pressure of water vapour is given by

$p_v = y_v\, p$ where, y is the mole fraction of water vapour.

But
$$y_v = \frac{20.4}{100 + 20.4} = 0.169$$

Thus,
$$p_v = 0.169 \times 1 = 0.169 \text{ atm}$$

At 0.169 atm, the saturation temperature from Steam Tables is 56.6°C **Ans.**

Example 17.8 A natural gas has the following molar composition : CH_4, 80.82%, C_2H_6, 5 51; C_3H_g, 1.87%; C_4H_{10}, 1.6%; N_2 , 10.5%. The gas is burnt with dry air. The molar analysis of the combustion products on a dry basis is CO_2, 7.8%; CO, 0.2%; O_2, 7%; N_2, 85%. Calculate (a) the air-fuel ratio on a molar basis, (b) the amount of product in kg mole that would be formed from 100 m^3 of fuel-air mixture of 300 K and 1 bar, assuming ideal gas behaviour as fuel-air mixture, and (c) the percentage of theoretical air.

Solution It is assumed that each mole of oxygen in the combustion air is accompanied by 3.76 moles of nitrogen which is inert and the fuel mixture is modelled as ideal gas mixture.

(a) The solution may be attempted either an assumed amount of fuel mixture or dry products. Let us try the first option, i.e., 1 kg mole of fuel mixture.

$$(0.8082\ CH_4 + 0.0551\ C_2H_6 + 0.0187\ C_3H_8 + 0.016\ C_4H_{10} + 0.105\ N_2) + a\ (O_2 + 3.76\ N_2)$$
$$\rightarrow b\ (0.078\ CO_2 + 0.002\ CO + 0.07\ O_2 + 0.85\ N_2) + c\ H_2O$$

Here, the products consist of b kg mole of dry products and c kg mole of water vapour, each per kg mole of fuel mixture.

The application of conservation of mass to carbon gives

$$b\ (0.078 + 0.002) = 0.8082 + 2\ (0.0551) + 4\ (0.016) + 3\ (0.0187)$$

or
$$b = 12.98$$

The applications of conservation of mass to hydrogen gives

$$2c = 4\ (0.8082) + 6\ (0.0551) + 8\ (0.0187) + 10\ (0.016)$$

or
$$c = 1.9365$$

The application of conservation of mass to oxygen gives

$$2a = 12.98\ [2\ (0.078) + 0.002 + 2\ (0.07)] + 1.9365$$

or
$$a = 2.902$$

Now, the balanced chemical equation may be written as

$0.8082\ CH_4 + 0.0551\ C_2H_6 + 0.0187\ C_3\ H_8 + 0.016\ C_4\ H_{10} + 0.105\ N_2 + 2.902$

$(O_2 + 3.76\ N_2) \rightarrow 12.98\ [0.078\ CO_2 + 0.002\ CO + 0.07\ O_2 + 0.85\ N_2] + 1.9365\ H_2O$

The air-fuel ratio on molar basis is given by

$$(A/F)_{molar} = \frac{2.902\,(1+3.76)}{1} = 13.81\,\frac{\text{kg mole (air)}}{\text{kg mole (fuel)}}$$ **Ans.**

(b) The balanced chemical equation suggests that the total amount of products is $b + c$ = 12.98 + 1.9365 = 14.9165 kg mole of products per kg mole of fuel. From gas law, the amount of fuel in kg mole is given by for 100 m^3 of fuel mixture as

$$n_F = \frac{pv}{RT} = \frac{10^5 \times 100}{8314 \times 300} = 4.01 \text{ kg mole (fuel)}$$

The amount of product mixture $(b + c)\ n_F = 14.9165 \times 4.01$

$= 59.815$ kg mole product gas **Ans.**

(c) The chemical equation for the complete combustion of the fuel mixture with the stoichiometric (theoretical) amount of air is given by

$$(0.8082\ CH_4 + 0.0551\ C_2H_6 + 0.0187\ C_3H_8 + 0.016\ C_4H_{10} + 0.105\ N_2) + a\ (O_2 + 3.76\ N_2) \rightarrow b\ CO_2 + c\ H_2O + d\ N_2$$

Applying conservation of mass

$$C: b = 0,.8082 + 2\ (0.0551) + 3\ (0.0187) + 4\ (0.016) = 1.0387$$

$$H: 2c = 4\ (0.8082) + 6\ (0.0551) + 8\ (0.0187) + 10\ (0.016)$$

$$O: 2a = 2b + c$$

$$N: d = 3.76\ a$$

Solutions of above equations gives, b = 1.0385, c = 1.9365, a = 2, d = 7.62

The balanced chemical equation for complete combustion is

$$(8082\ CH_4 + 0.0551\ C_2\ H_6 + 0.0187\ C_3H_8 + 0.016\ C_4H_{10} + 0.105\ N_2) + 2\ (O_2 + 3.76\ N_2) \rightarrow 1.0385\ CO_2 + 1.9365 + 7.62\ N_2$$

The theoretical air-fuel ratio on molar basis

$$(A/F)_{mole} = \frac{2\,(1+3.76)}{1} = 9.52\,\frac{\text{kg mole (air)}}{\text{kg mole (fuel)}}$$ **Ans.**

The percentage of theoretical air $= \dfrac{13.81}{9.52} \times 100 = 145\%$ **Ans.**

17.5 CALCULATION OF THE MASS OF DRY FLUE GASES PER KG OF FUEL OF KNOWN CARBON CONTENT FROM THE ORSATE TEST

No nitrogen in the fuel:

First Method. We can find the mass of dry flue gases if we write down the dimensionally homogeneous mole equation

$$\frac{\text{kg gas}}{\text{kg fuel}} = \frac{\dfrac{\text{kg gas}}{\text{mole gas}}}{\dfrac{\text{kg C}}{\text{mole C}}} \times \frac{\dfrac{\text{kg C}}{\text{kg fuel}}}{\dfrac{\text{mole C}}{\text{mole gas}}}$$

Let kg C/kg fuel be denoted as C

We have, $\dfrac{\text{kg gas}}{\text{mole gas}} = 44\ CO_2 + 28\ CO + 32\ O_2 + 28\ N_2$

where CO_2, CO, O_2 and N_2 are volumetric analysis or mole fraction

$$\frac{\text{kg gas}}{\text{kg fuel}} = \frac{C\,[44\ CO_2 + 28\ CO + 32\ O_2 + 28\ N_2]}{12\,(CO_2 + CO)}$$

or

$$\frac{\text{kg gas}}{\text{kg fuel}} = \frac{C\,[11\ CO_2 + 8\ O_2 + 7\,(CO_2 + N_2)]}{3\,(CO_2 + CO)} \qquad \text{...(17.6)}$$

17.6 CONVERSION OF GRAVIMETRIC ANALYSIS TO VOLUMETRIC ANALYSIS AND VICE VERSA

The conversion of gravimetric analysis and vice versa are based on the following principles.

(i) Molecular weights of all gases occupy the same volume. Hence, the volume of the constituent gas multiplied by the molecular weight of the constituent gas will give the proportional mass of the constituent gas.

(ii) And the mass of the constituent gas divided, by the molecular weight will give the proportional mass of the constituent gas.

The following examples will show the procedure.

Example 17.9 Volumetric analysis of a flue gas is 12%, CO_2, 4% CO, 4% O_2 and 80% N_2 (by difference). Calculate the gravimetric composition.

Solution Using tabular method (Table 17.5).

Table 17.5

Constituents gas	% Vol. analysis	Molecular weight	Propertional mass	% Gravimetric analysis
	A	B	$C = A \times B$	$D = \frac{C}{\Sigma C} \times 100$
CO_2	12	44	528	$\frac{528}{3008} \times 100 = 17.55$
CO	4	28	112	$\frac{112}{3008} \times 109 = 3.73$
O_2	4	32	128	$\frac{128}{3008} \times 100 = 4.25$
N_2	80	28	2240	$\frac{2240}{3008} \times 100 = 74.47$
			$\Sigma C = 3008$	$\Sigma D = 100$

17.7 MINIMUM AIR REQUIRED IN KG PER KG OF SOLID OR LIQUID FUEL (ANOTHER METHOD)

We have already discussed the method of determination of theoretical air required for complete combustion of a fuel by writing a single complete chemical balanced equation, i.e., stoichiometric equation. *However, the minimum (or theoretical) air required may also be determined by writing the complete combustion equation for every constituent of the fuel.*

For example, a sample of fuel which has composition by mass in 1 kg of fuel as C kg of carbon, H_2 kg of hydrogen, O_2 kg of oxygen and S kg of sulphur.

The combustion of carbon gives

C	+	$O_2 \rightarrow$	CO_2
12		32	44
1		32/12	2

1 kg of carbon requires 32/12 = 8/3 kg of O_2

∴ C kg of carbon requires C × 8/3 kg of O_2

The combustion of H_2 gives

$2H_2$	+	$O_2 \rightarrow$	$2H_2O$
4		32	20
1		32/4	

1 kg of hydrogen requires 32/4 = 8 kg of O_2

∴ H_2 kg of hydrogen requires $H_2 \times 8$ kg of O_2

The combustion of sulphur gives

S+	$O_2 \rightarrow$	SO_2
32	32	

1 kg of sulphur requires 32/32 = 1 kg of O_2

∴ S kg of sulphur requires S × 1 = S kg of O_2

O_2 kg of oxygen is already present in the fuel

∴ The amount of oxygen required for complete combustion of 1 kg of fuel is given by

$$\frac{8}{3}C + 8H_2 + S - O_2 = \frac{8}{3}C + 8\left(H_2 - \frac{O_2}{8}\right) + S \qquad ...(17.7)$$

We know that atmospheric air constant 23% by mass of oxygen. Hence

$$\text{Minimum air/kg of fuel} = \frac{100}{23}\left[\frac{8}{3}C + 8\left(H_2 - \frac{O_2}{8}\right) + S\right] \qquad ...(17.8)$$

$$= 11.6\text{ C} + 34.8\left(H_2 - \frac{O_2}{2}\right) + 4.35\text{ S} \qquad ...(17.9)$$

Equations (17.8) and (17.9) give the minimum air/kg of fuel or air fuel ratio.

This result may also be put in tabular form.

Example 17.10 A sample of coal has the following percentage composition by weight, (i.e., gravimetric).

C = 82%, H_2 = 10%, O_2 = 6%, N_2 = 1% and S = 1%. Calculate the minimum quantity of air requried for the complete combustion of 1 kg of coal, (i.e., A/F ratio). If 20% or excess air is used in the composition, what is the air-fuel ratio?

Solution Here, four methods are attempted.

First method. Consider 1 kg of fuel,

We have

C	+	$O_2 \rightarrow$	CO_2
12		32	44

∴ O_2 required by 1 kg of carbon is fuel = 32/12 kg

∴ O_2 required by 0.82 kg of C in fuel = 0.82 × 31/12 = 2.186 kg

$2H_2$	+	$O_2 \rightarrow$	$2H_2O$
4		32	20

O_2 required by 1 kg of H_2 = 8 kg

∴ O_2 required by 0.1 kg H_2 = 0.1 × 8 = 0.8 kg

$$\begin{array}{lll} S & + \quad O_2 \rightarrow & SO_2 \\ 32 & \quad 32 & \end{array}$$

O_2 required by 1 kg of S 32/32 = 1 kg

∴ O_2 required by 0.01 kg of S = 1 × 0.01 = 0.01 kg/kg

∴ Total O_2 required = 2.186 + 0.8 + 0.01 = 2.996 kg

O_2 available in fuel = 0.06 kg

∴ Net O_2 required = 2.996 – 0.06 = 2.936 kg/kg of fuel

Since atmospheric air contains 23% of O_2 by weight

∴ $$\text{Net air required} = \frac{2.936 \times 100}{23} = 12.765 \text{ kg/kg of fuel}$$ **Ans.**

With 20% of excess air

$$A/F = 12.765 \times 1.2 = 15.318 \text{ kg/kg of fuel}$$ **Ans.**

Second method A single Eq. (17.8) for the minimum air/kg fuel may be obtained based on the above method. The minimum air/kg of fuel or A/F.

$$= \frac{100}{23}\left[\frac{8}{3}C + 8\left(H_2 - \frac{O_2}{u}\right) + S\right]$$

$$= \frac{100}{23}\left[\frac{8}{3} \times 0.82 + \left(0.1 - \frac{0.06}{8}\right) + 0.01\right] = 12.765 \text{ kg/kg fuel}$$ **Ans.**

With 20% excess air **Ans.**

$$A/F = 12.765 \times 1.2 = 15.318 \text{ kg of fuel}$$

Third method The problem may also be solved in tabular form based on the first method.

The minimum air required for the complete combustion

$$\frac{\Sigma C \times 100}{23} = \frac{2.936 \times 100}{23} = 12.765 \text{ kg/kg fuel}$$ **Ans.**

With 20% of excess air, A/F = 12.765 × 1.2 = 15.318 kg/kg of fuel **Ans.**

Constituent	Mass per kg of coal A	O_2 required per kg of constituent B	O_2 required per kg of fuel C = A × B
C	0.82	8/3	2.186
H_2	0.1	8	0.8

Contd.

N_2	0.1	–	–
S	0.01	1	0.01
O_2	0.06	– 1	– 0.06
			ΣC = 2.936

Fourth method The problem may also be solved by considering 100 kg of fuel and assuming an equivalent chemical formula of the form

$$C_a\, H_b\, O_c\, N_d\, S_e$$

Knowing the analysis by weight, (i.e., mass)

$12a = 82$	or	$a = 82/12 = 6.833$
$1b = 10$	or	$b = 10$
$16\,c = 6$	or	$c = 6/16 = 0.375$
$14\,d = 1$	or	$d = 1/14 = 0.071$
$32\,e = 1$	or	$e = 1.32 = 0.031$

Hence, the equivalent formula of coal is $C_{0.833}\, H_{10}\, O_{0.071}\, O_{0.031}$

The unbalanced chemical equation may be written as

$$(C_{6.833}\, H_{10}\, O_{0.375}\, N_{0.071}\, S_{0.031}) + (x\, O_2 + 2.76\, x\, N_2) \rightarrow (p\, CO_2 + q\, H_2O + r\, SO_2 + s\, N_2)$$

The carbon balance gives, $6.833\, C = p\, C$ or $p = 6.833$

The hydrogen balance gives, $10\, H = q\, H_2$ or $q = 5$

The sulphur balance gives, $r = 0.031$

The oxygen balance gives, $\frac{0.375}{2} O_2 + x\, O_2 = \left(p + \frac{q}{2} + r\right) O_2$

$$\therefore \quad x = 6.833 + \frac{5}{2} + 0.031 - \frac{0.375}{2} = 9.1765$$

Nitrogen balance gives, $\frac{0.071}{2} + 3.76 \times 9.1765 = s$

$$\therefore \quad s = 34.539$$

Now the balance chemical equation is given by

$$(C_{6.833}\, H_{10}\, O_{0.375}\, N_{0.031}\, S_{0.031}) + (0.7165\, O_2 + 9.1765 \times 3.76\, N_2)$$

$$\rightarrow 6.833\, CO_2 + 5\, H_2O + 0.031\, SO_2 + 34.539\, N_2$$

Hence, the air-fuel ratio is

$$A/F = \frac{9.1765 \times 32 + 9.1765 \times 3.76 \times 28}{100} = 12.765 \text{ kg/kg fuel.}$$

Example 17.11 A petrol engine uses a fuel C_7H_{16}. Determine the air-fuel ratio which would just suffice of theoretical combustion.

In an actual experiment, an analysis of the exhaust gases showed equal volumes of unused oxygen and carbon dioxide; there was no carbon monoxide. Calculate the ratio of air to fuel, by mass as actually supplied. Assume air contains 23.2% oxygen by mass.

Solution The chemical equation for the complete combustion is

$$C_7\,H_{16} + 11\,O_2 + x\,N_2 \rightarrow 7\,CO_2 + 8\,H_2O + x\,N_2$$

$$100 \text{ kg of fuel} + 11 \times 32 \text{ kg} + 11 \times 32 \times \frac{76.8}{23.2} \text{ kg } N_2 \rightarrow \text{Products}$$

$$\therefore \qquad A/F = \frac{(11 \times 32 + 11 \times 32 \times 76.8 / 23.2)}{100} = 15.172 \text{ kg of air/kg fuel} \qquad \textbf{Ans.}$$

But in actual experiment, we have

$$C_7H_{16} + y\,O_2 + \ldots\, q\,N_2 \rightarrow a\,CO_2 + 8\,H_{20} + p\,O_2 + q\,N_2$$

Let y be the number of kg moles of O_2 supplied/kg mole of fuel. If we compare the moles of carbon, we must have 7 moles of CO_2 if all the carbon is to be burnt.

$$\therefore \qquad y = 7 + 4 + 7 \text{ from } CO_2, H_2O \text{ and } O_2 \text{ respectively} = 18$$

$$\therefore \qquad A/F = \frac{18 \times 32 \times 1}{100 \times 0.232} = 24.8 \text{ kg/kg fuel}$$

Example 17.12 A fuel gas has composition by volume : H_2 = 50%, CO = 40%, CO_2 = 6% and N_2 = 40%. Calculate the volume of air required for complete combustion of 1 m^3 of the gas. If the dry exhaust gas from an engine using this fuel contains 9.2% by volume of carbon dioxide and no CO, calculate the air-to-gas ratio (by volume) which has been used. Also, calculate the percentage of oxygen by volume in the dry exhaust. Assume air contains 21% by volume of oxygen.

Solution (a) For complete combustion of fuel gas, the chemical equation is

$$50H_2 + 40\,CO + 6\,CO_2 + x\,O_2 + \left(\frac{79}{21}x + 4\right) N_2 \rightarrow a\,CO_2 + b\,H_2O + \left(\frac{79}{21}x + 4\right) N_2$$

Here, $x/100$ is the oxygen-fuel ratio by volume.

H_2 balance gives : $b = 50$

C balance gives : $a = 40 + 6 = 46$

O_2 balance gives : $\frac{40}{2} + 6 + x = a + b/2$

or $\qquad x = 45 \text{ m}^3 \text{ of } O_2/100 \text{ m}^3 \text{ of fuel gas}$

$$\therefore \qquad A/F = \frac{45}{100} \times \frac{100}{21} = 2.14 \text{ m}^3 \text{ of air/m}^3 \text{ of gas}$$

(b) In the second case, the combustion equation is given by

$$50\ H_2 + 40\ CO_2 + 6\ CO_2 + y\ O_2 + \left(\frac{29}{21}y + 4\right) N_2$$

$$\rightarrow d\ (CO_2) + 50\ H_2O + e\ (O_2) + \left(\frac{79}{21}y + 4\right) N_2$$

Again, here $y/100$ is the new oxygen-fuel ratio, by volume. It is to be noted that 9.2% of CO_2 is given excluding H_2O.

C balance : $\qquad a = 40 + 6 = 46$

$$CO_2 \text{ balance : } 0.092 = \frac{46}{46 + e + \frac{79}{21}y + 4}$$

$$O_2 \text{ balance : } \frac{40}{2} + 6 + y = 46 + e$$

Solution of the above two equations gives : $y = 104$ and $e = 59$

$$\text{Air-to-gas ratio by volume} = \frac{104}{100} \times \frac{100}{21} = 4.95$$ **Ans.**

% Volume of O_2 in dry gas

$$= \frac{59}{46 + 59 + \frac{79}{21} \times 104 + 4} = \frac{59}{500.23} = 11.79\%$$ **Ans.**

Example 17.13 The composition of a hydrocarbon fuel by mass is 85% carbon, 13% hydrogen and 2% oxygen. Calculate (a) the chemically correct mass of air required for the complete combustion of 1 kg of fuel, (b) the volumetric and gravimetric analysis of the products of combustion and the dew point if the total pressure is 1 bar, (c) if the products of combustion are cooled to 15°C, calculate the mass of water vapour condensed.

Solution Let us consider 1 kg of fuel. We can calculate the air required on mole basis and then convert it to mass basis because that will help in finding molar analysis.

We have

$$C + O_2 \rightarrow CO_2$$

$$1 \text{ mole} + 1 \text{ mole} \rightarrow 1 \text{ mole}$$

and

$$H_2 + \frac{1}{2} O_2 \rightarrow H_2O$$

$$1 \text{ mole} + \frac{1}{2} \text{ mole} \rightarrow 1 \text{ mole}$$

$$\text{We know that mole} = \frac{\text{Mass}}{\text{Molecular weight}}$$

Therefore, the number of moles of oxygen required is

$$= \frac{0.85}{12} + \frac{0.10}{2 \times 2} - \frac{0.02}{32} = 0.102675 \text{ moles of } O_2 \text{ kg of fuel.}$$

But air contains 21% O_2 on mole basis, hence

(a) Mass of air required $= \frac{0.102675 \times 100}{21} \times 28.97 = 14.16$ kg air/kg mole **Ans.**

(b) The analysis of products of combustion by volume and mass basis is given in the tabular form.

Constituent	Moles per kg of fuel A	Molal analysis A/ΣA	kg per kg of fuel B	Gravimetric % analysis (B/ΣB × 100)
CO_2	$\frac{0.85}{12} = 0.07075$	0.136	$\frac{44 \times 0.85}{12} = 3.12$	20.7%
H_2O	$\frac{0.13}{2} = 0.0645$	0.124	$\frac{18 \times 0.13}{2} = 1.16$	7.68%
N_2	3.76 × 0.102375 = 0.385	0.740	0.385 × 28 = 10.82	71.62%
	Σ = 0.5205	1.0	15.1	100.00

The dew point temperature is found from the Steam Tables corresponding to the pressure of the water vapour in the products of combustion. The partial pressure of H_2O is given by

$$p_{H_2O} = \frac{n_{H_2O}}{\Sigma n} \times p = 0.124 \times 1 = 0.24 \text{ bar}$$

The saturation temperature corresponding to this pressure will be the dew point

$$\therefore \quad t_{dp} = 50°C$$ **Ans.**

(c) The product of combustion is cooled to 15°C which is less than the dew point of the water vapour, hence some water vapour will condense. The vapour pressure will then be equal to the saturation vapour pressure corresponding to 15°C.

From Steam Tables 15°, $p_{v\,sat} = 0.0174$ bar and $v_s = 77.99$ m^3/kg.

The volume occupied by the water vapour is equal to the volume occupied by the CO_2 and the N_2 at their partial pressure of 1 – 0.0174 = 0.9826 bar. The volume according to the gas law is given by

$$V = \frac{nR_0T}{p} = \frac{(0.07075 + 0.385) \times 8.314 \times 288}{0.9826 \times 10^2} = 11.1 \text{ m}^3$$

$$\text{Mass of the vapour in products} = \frac{V}{v_s} = \frac{11.1}{77.99} = 0.1423 \text{ kg}$$

$\therefore$ Mass of vapour that is condensed = 1.6 – 0.1465 = 1.0135 kg **Ans.**

17.8 MINIMUM QUANTITY OF AIR REQUIRED FOR THE COMPLETE COMBUSTION OF ONE CUBIC METRE OF GASEOUS FUEL (ANOTHER METHOD)

The main constituents of gaseous fuel are hydrogen, carbon monoxide and methane (commonly known as marsh gas). *The minimum quantity of air required may be found either by writing a complete balanced combustion equation or separate combustion equations for each constituent of the fuel.* For example, consider the separate equations. Let the analysis of 1 m^3 of gaseous fuel be given by H_2 m^3 of hydrogen, CO m^3 of carbon monoxide, CH_4 m^3 of marsh gas, CO_2 m^3 of carbon dioxide and N_2 m^3 of nitrogen. Here CO_2 and N_2 will be incombustible gases.

Combustion of hydrogen

$$H_2 \quad + \quad \frac{1}{2}O_2 \rightarrow \quad H_2O$$

$$1 \text{ vol.} \qquad \frac{1}{2} \text{ vol.} \qquad 1 \text{ vol.}$$

Hence, the quantity of oxygen required for the complete combustion of H_2 m^3 of hydrogen

$$= \frac{H_2}{2} \text{ m}^3 \text{ of air/m}^3 \text{ of fuel}$$

Combustion of carbon monoxide.

$$CO \quad + \quad \frac{1}{2}O_2 \qquad CO_2$$

$$1 \text{ vol.} \qquad \frac{1}{2} \text{ vol.} \qquad 1 \text{ vol.}$$

Hence, the quantity of oxygen required for the complete combustion of CO m^3 of carbon monoxide is

$$\frac{CO}{2} \text{ m}^3 \text{ of air/m}^3 \text{ of fuel}$$

Combustion of marsh gas (CH_4)

$$CH_4 \quad + 2O_2 \rightarrow \quad 2H_2O \quad + \quad CO_2$$

$$1 \text{ vol.} \qquad 2 \text{ vol.} \qquad 2 \text{ vol.} \qquad 1 \text{ vol.}$$

Hence, the quantity of oxygen required for the complete combustion of CH_4 m^3 of marsh gas

$$= 2\ CH_4;\ m^3 \text{ of air/m}^3 \text{ of fuel}$$

∴ Total oxygen required for complete combustion of 1 m^3 of fuel

$$\frac{H_2}{2} + \frac{CO}{2} + 2\ CH_4;\ m^3$$

Since air contains 21% of oxygen by volume, the amount of air required will be

$$= \frac{100}{21}\left[\frac{H_2}{2} + \frac{CO}{2} + 2\,CH_4\right];\ m^3 \text{ of air/m}^3 \text{ of fuel}$$

$$= \frac{100}{21}\ [0.5\ (H_2 + CO) + 2\ CH_4];\ m^3 \text{ of air/m}^3 \text{ of fuel} \quad ...(17.10)$$

17.9 DETERMINATION OF PERCENTAGE OF CARBON IN FUEL BURNING TO CO_2 OR CO FROM ORSATE ANALYSIS OR VOLUMETRIC COMPOSITION

The percentage of carbon C burning to CO_2

$$= \frac{\text{Mass of carbon in } CO_2 \text{ per kg of fuel gas}}{\text{Mass of carbon per kg of fuel gas}} \times 100$$

$$= \frac{CO_2 \times 44 \times 12/44}{(CO_2 \times 44 \times 12/44) + (CO \times 28 \times 12/28)}$$

$$= \frac{CO_2}{CO_2 + CO} \times 100 \quad ...(17.11)$$

In the same way, the percentage of carbon *C* burning to CO is

$$= \frac{CO}{CO_2 + CO} \times 100 \quad ...(17.12)$$

17.10 DETERMINATION OF THE PERCENTAGE OF CARBON IN FUEL BURNING TO CO_2 OR CO FROM GRAVIMETRIC ANALYSIS OF FLUE GASES

The percentage of carbon burning to CO_2 is

$$= \frac{CO_2 \times 12/44}{(CO_2 + 12/44) + (CO \times 12/28)} \times 100 \quad ...(17.13)$$

$$= \frac{7\,CO_2}{7\,CO_2 + 11\,CO} \times 100 \quad ...(17.14)$$

Similarly, the percentage carbon burning to CO

$$= \frac{CO \times 12/28}{(CO_2 \times 12/44) + (CO \times 12/28)} \times 100 \qquad ...(17.15)$$

$$= \frac{11\,CO}{7\,CO_2 + 11\,CO} \qquad ...(17.16)$$

Example 17.14 Calculate the minimum quantity of air required for the complete combustion of a gaseous fuel whose volumetric analysis is 50% H_2, 5% CO, 35% CH_4, 7% N_2 and 3% of water vapour.

Solution The minimum quantity of air required for the complete combustion of 1 m^3 of fuel is

$$= \frac{100}{21} [0.5 (H_2 + CO) + 2\ CH_4] = [0.5 (0.5 + 0.05) + 20.35\ ]$$

$$= 4.64\ m^3/m^3 \text{ of fuel}$$ **Ans.**

Note The problem may also be solved by writing a single chemical balanced equation.

Example 17.15 The volumetric analysis of fuel gas during a boiler trial was :— 100; CO_2; 1.5% CO; 8.5%; O_2 and 80% N_2. The gravimetric analysis of the burnt fuel was : 80%C, 5% H_2, 6% O_2 and the remaining incombustible matter. If the rise in temperature of the flue gases is 350°C, determine (a) mass of air supplied/kg of fuel and excess air, (b) heat carried away by excess air and the products of combustion, (c) heat lost by incomplete combustion, (d) density of dry gases at 1 bar and 15°C, and (e) partial pressure of each gas if the total pressure of dry gases is 1 bar.

Assuming that air contains 23% of O_2 by mass, the calorific value of C to CO_2 is 33,858 kJ/kg and for C to CO is 10,032 kJ/kg, c_p for CO_2, O_2, N_2, CO and air are 1.128, 1.024, 1.086, 1.0199, 0.9927 kJ/kg K respectively.

Solution It will be better first to convert the volumetric analysis of fuel gases into mass analysis.

Constituent	% by volume A	Molecular weight B	Proportional mass C = A × B	Mass analysis $D = \frac{C}{\Sigma C} \times 100$
CO_2	10	44	440	14.7
CO	1.5	28	42	1.41
O_2	8.5	32	272	9.08
N_2	80	28	2240	74.81
			ΣC = 2994	100.00

(a) Mass of carbon per kg of dry flue gases = 2994 kg

(b) Hence, heat carried away by the products of combustion = $c_{p_g}\ m_g\ \Delta t$

$$= 1.0868 \times 17.85 \times 350 = 6789.78 \text{ kJ}$$ **Ans.**

(c) For finding the heat loss due to incomplete combustion, we have to first find the amount of carbon burnt to CO instead of CO_2. The carbon burnt to CO per kg of fuel can be calculated by the equation

$$C + \frac{1}{2} O_2 \rightarrow CO$$

where CO_2 = % of CO_2 by volume in dry flue gases

CO = % of CO by volume in dry flue gases

$$\therefore \quad \text{Carbon burnt to CO} = \frac{1.5 \times 0.8}{10 + 1.5} = 0.1042 \text{ kg}$$

This may also be calculated on mass analysis.

$$\text{Carbon in } CO_2 = \frac{12}{44} \times 0.147 \text{ kg}$$

and

$$\text{Carbon in CO} = \frac{12}{28} \times 0.014 = 0.006 \text{ kg}$$

$$\therefore \quad \text{Total carbon} = 0.04 + 0.006 = 0.046 \text{ kg}$$

$$\therefore \quad \text{Carbon in CO} = \frac{0.006}{0.046} \times 0.8 = 0.1042 \text{ kg}$$

The burning of 1 kg of carbon to CO instead of CO_2 involves a loss of (33858 – 10032) = 2386 kJ in the heat evolved. Therefore, heat loss due to incomplete combustion

$$= 0.1042 \times 23826 = 2482 \text{ kJ}$$ **Ans.**

(d) The molecular weight of flue gases, $M_f = \dfrac{\Sigma n_i \, M_i}{\Sigma n_i}$

where n = number of moles of volume and M = molecular weight

$$\therefore \quad M_f = \frac{10 \times 44 + 1.5 \times 28 + 32 + 80 \times 28}{10 + 1.5 + 8.5 + 40} = 29.94$$

The gas constant for the mixture is

$$R_f = \frac{R_0}{M_f} = \frac{8.314}{29.94} = 0.277 \text{ kJ/kg K}$$

The density of flue gases, $\rho_f = \dfrac{p_f}{R_f T_f} = \dfrac{1 \times 10^2}{0.277 \times 288} = 1.252 \text{ m}^3/\text{kg}$

The partial pressure of the constituent gas, $p_j = (n_j/\Sigma n_i)\, p_{\text{total}}$

$$\therefore \quad p_{CO_2} = \frac{10}{100} \times 1 = 0.1 \text{ bar} \quad \text{and} \quad p_{CO} = \frac{1.5}{100} \times 1 = 0.015 \text{ bar}$$ **Ans.**

$$p_{CO_2} = \frac{8.5}{100} \times 1 = 0.085 \text{ bar} \quad \text{and} \quad p_{N_2} = \frac{80}{100} \times 1 = 0.8 \text{ bar} \qquad \textbf{Ans.}$$

Example 17.16 The gravimetric analysis of a fuel is 65% hexane (C_6H_{14}) having a net calorific value of 45165 kJ/kg and 35% benzene (C_6H_6) having a net calorific value of 40983 kJ/kg. Determine (a) the air-fuel ratio by mass of correct mixture of this blend, and (b) calorific value of one cubic metre of this blend at N.T.P. treating the fuels as gases.

Assume that air contains 21% of oxygen by volume and 1 kg mole of the fuel occupies 22.42 m^3 of volume at N.T.P. Calculate (c) also the percentage by volume if CO in the dry exhaust gases forms a 30% rich mixture assuming complete combustion of hydrogen.

Solution For the complete combustion, the chemically balanced equations are

$$2C_6H_{14} + 19O_2 = 12CO_2 + 14H_2O$$

172 608 528 252

and

$$2C_6H_{16} + 15O_2 = 15CO_2 + 6H_2O$$

156 480 528 108

With the help of the above equations, the oxygen required for complete combustion of hexane is

$$= 0.65 \times \frac{608}{172} = 2.3 \text{ kg}$$

and oxygen required for complete combustion of benzene is

$$= 0.35 \times \frac{408}{156} = 1.08 \text{ kg}$$

∴ 1 kg blend of mixture requires = (2.3 + 1.08) = 3.38 kg of oxygen.

(a) Hence, $$A/F = \frac{3.38 \times 100}{23} = 14.7 \text{ kg/kg fuel} \qquad \textbf{Ans.}$$

(b) The gross calorific value = 0.65 × 45165 + 0.35 × 40983 = 43701 kJ/kg

The constituent of the mixture is expressed on mole basis as follows:

Constituent	Mass *A*	Molecular weight *B*	Number of moles *C = A/B*
C_6H_6	0.35	78	0.00448
C_6H_{14}	0.65	86	0.00756
O_2	3.38	32	0.1058
N_2	14.7 - 3.38 = 11.32	28	0.404
	ΣA = 15.7		ΣC = 0.5218

1 kg mole of mixture gives us a calorific value $= \dfrac{43701}{0.5218} = 83750.5$ kJ/kg mole **Ans.**

At N.T.P., 1 kg mole of the fuel occupies 22.42 m^3, hence

Calorific value of the mixture $= \dfrac{8375.5}{22.42} = 3735.5$ kJ/m^3 **Ans.**

(c) The quantity of air supplied for 30% rich mixture is

$$= \frac{14.7}{12.3} = 11.3 \text{ kg/kg of fuel}$$

The combustion equation on the mole basis is

$$\frac{0.65}{86}(C_6H_{14}) + \frac{0.35}{78}(C_6H_6) + \frac{11.3 \times 0.23}{32}(O_2) + \frac{11.3 \times 0.077}{28}(N_2)$$
$$\rightarrow a\,(CO_2) + b\,(CO) + d\,(H_2O) + eN_2$$

C balance gives : $\dfrac{0.65}{86} \times 6 + \dfrac{0.35}{78} \times 6 = a + b$

$$\therefore \quad a + b = 0.0723 \quad \text{...(i)}$$

H_2 balance gives : $\dfrac{0.65}{86} \times 7 + \dfrac{0.35}{78} \times 3 = d$ or $d = 0.06645$

$$\therefore \quad 2a + b = 0.09644 \quad \text{...(ii)}$$

Solving Eqs. (i) and (ii), we have, $a = 0.02405$ and $b = 0.04825$

N_2 balance gives, $e = \dfrac{11.3 \times 0.77}{82} = 0.31$

For a 30% rich mixture, the percentage of CO by volume is

$$= \frac{b}{a + b - c} = \frac{0.04825}{(0.02405 + 0.04825 + 0.3)} = 12.61\% \quad \textbf{Ans.}$$

17.11 ENTHALPY OF FORMATION

In applying the conservation of energy principle to reacting systems, it is necessary to evaluate their enthalpy, internal energy and entropy.

An enthalpy datum for the study of reacting systems can be established by assigning arbitrarily a value **of zero to the enthalpy** of the stable elements at a state called the *standard reference state and defined by* $\boldsymbol{T_{ref}}$ **= 298.15 K (25°C) and** $\boldsymbol{p_{ref}}$ **= 1 atm**. *Enthalpy of formation is defined as the energy released or absorbed when chemical compound is formed from its elements if the temperature and pressure of the reactants and products are at the reference state.* ***In other words, the enthalpy of a compound at the standard state is the enthalpy of formation, symbolized as*** $\bar{h}_f^0$.

In order to consider the enthalpy of formation concept, refer to a simple combustion chamber as shown in Fig. 17.3. in which carbon and oxygen each enter at T_{ref} (25°C) and p_{ref} (1 atm) and react completely at steady state to form carbon dioxide at the same temperature and pressure. The reaction is as follows

$$C + O_2 \rightarrow CO_2$$

This reaction would be exothermic and there would be heat transfer $\dot{Q}_{cv}$ from the combustion chamber to its surroundings.

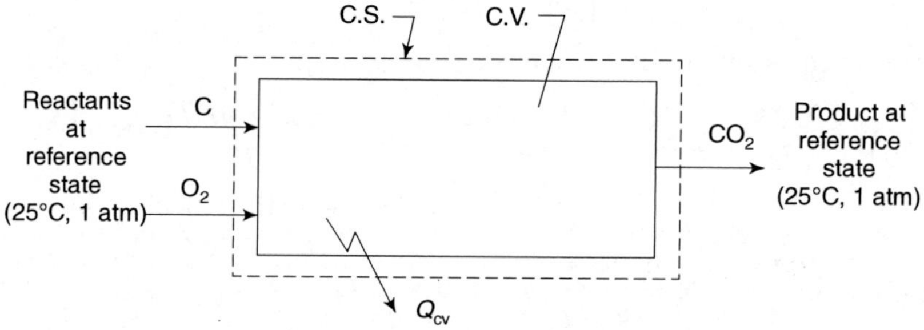

Fig. 17.3 Enthalpy of formation

The measured value of the heat transfer is – 393,522 kJ/kg mole of CO_2 formed. If H_r and H_p refer to the total enthalpy of reactants and products respectively, then applying the first law to the reaction gives

$$H_r + Q_{cv} = H_p \qquad ...(17.17)$$

For all the reactants and products in a reaction, the above equation may be written in *rate form* based on molar flow rate as

$$\sum_r n_i \bar{h}_j + \dot{Q}_{cv} = \sum_p \dot{n}_e \bar{h}_e \qquad ...(17.18)$$

where $\dot{n}$ and $\bar{h}$ represent the molar flow rate and enthalpy per mole, respectively. The suffixes r and p represent reactants and products, respectively. The suffixes i and e denote inlet and exit respectively.

For enthalpies on mass rate basis, Eq. (17.17) is written as

$$\sum_r \dot{m}_i h_i + \dot{Q}_{cv} = \sum_p \dot{m}_e h_e \qquad ...(17.19)$$

where $\dot{m}$ and h denote respectively, mass flow rate and specific enthalpy.

For the above chemical equation, Eq. (17.18) takes the form as

$$\dot{n}_C \bar{h}_C + \dot{n}_{O_2} \bar{h}_{O_2} + \dot{Q}_{cv} = \dot{n}_{CO_2} \bar{h}_{CO_2}$$

or
$$\bar{h}_{(CO_2)} = \frac{\dot{Q}_{cv}}{\dot{n}_{CO_2}} + \frac{\dot{n}_c}{\dot{n}_{CO_2}}\bar{h}_C + \frac{\dot{n}_{O_2}}{\dot{n}_{CO_2}}\bar{h}_{O_2} \quad ...(17.20)$$

Since carbon and oxygen are stable elements at the reference state, $\bar{h}_C = \bar{h}_{O_2} = h_r = 0$, so

$$\bar{h}_{CO_2} = \bar{h}_p = \frac{\dot{Q}_{cv}}{\dot{n}_{CO_2}} = \frac{\dot{Q}_{cv}}{\dot{n}_p} \quad ...(17.21)$$

The above equation suggests that the specific enthalpy of CO_2 (i.e., products) at the standard or reference state equals the heat transfer, per mole of CO_2, between the reactor and the surroundings. Thus,

$$\frac{\dot{Q}_{cv}}{\dot{n}_{CO_2}} = \frac{\dot{Q}_{cv}}{\dot{n}_p} = -\ 393,\ 522 \text{ kJ/kg mole}$$

This is what is known as **enthalpy of formation of CO_2 at reference state 25°C and 1 atm and designated by the symbol $\bar{h}_f^0$. Thus,**

$$\left(\bar{h}_f^0\right)_{CO_2} = \frac{\dot{Q}_{cv}}{\dot{n}_{CO_2}} = \frac{\dot{Q}_{cv}}{\dot{n}_p} = -\ 393,\ 522 \text{ kJ/kg mole}$$

The sign convention is related to the sign convention of heat transfer. If there is heat transfer from the combustion chamber (reactor) in which compound is formed from its element (an exothermic reaction, $\bar{\ }$ has a negative sign. If the heat transfer to the reactor is required (an endothermic reaction), $\bar{h}_f^0$ is positive.

In real situations, the reactants and products are not at standard state. *The specific enthalpy of a compound at a state other than the standard state is found by adding the specific enthalpy change* $(\overline{\Delta h})$ between the standard state and the state of interest to the enthalpy formation. Thus,

$$\bar{h}\,(T, p) = \bar{h}_f^0 + [\bar{h}\,(T, p) - \bar{h}\,(T_{ref,}\ p_{ref})] = \bar{h}_f^0 + \overline{\Delta h} \quad ...(17.22)$$

An arbitrary choice of datum can be used to determine $\overline{\Delta h}$, since it is a difference at constant composition.

Table A8 of Appendices gives the value of enthalpy of formation $(\bar{h}_f^0)$ for a number of substances. The values of $\overline{\Delta h}$ are given in Table A9.

17.12 FIRST LAW APPLIED TO REACTIVE SYSTEM

17.12.1 Open System (Steady Flow Process)

Consider a reaction system for steady flow process as shown in Fig. 17.4. From first law of thermodynamics, we have

$$H_r + Q_{cv} = H_p + W_{cv} \quad ...(17.23)$$

The above equation may be written in the rate form on molar basis.

$$\sum_r \dot{n}_i \bar{h}_i \dot{Q}_{cv} = \sum_p \dot{m}_e \bar{h}_e + \dot{W}_{cv} \quad ...(17.24)$$

On mass rate basis, Eq. (17.23) may be written as

$$\sum_r \dot{m}_i h_i + \dot{Q}_{cv} = \sum_p \dot{m}_e h_e + \dot{W}_{cv} \quad ...(17.25)$$

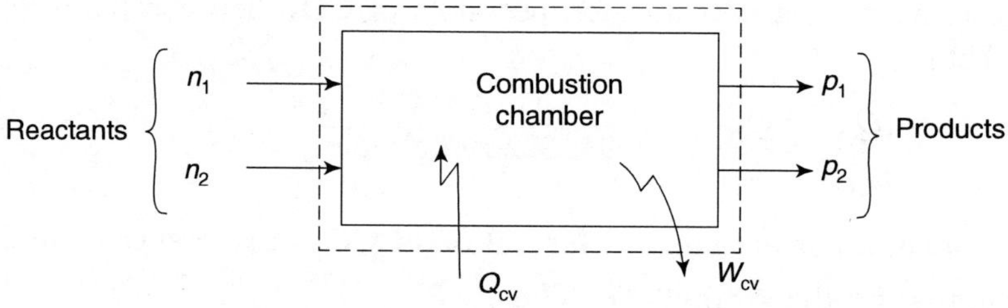

Fig. 17.4 First law for a reactive system

17.12.2 Combustion of Hydrocarbon Under Steady State

Consider a steady state reactor (combustion chamber) shown in Fig. 17.5 in which a hydrocarbon fuel C_aH_b burns completely with the theoretical amount of air. The corresponding reaction is expressed as

$$C_a H_b + \left(a+\frac{b}{4}\right)(O_2 + 3.76\, N_2) \rightarrow a\, CO_2 + \frac{b}{2} H_2O + \left(a+\frac{b}{4}\right) 3.76\, N_2 \quad ...(17.26)$$

Assume that reactants and products modelled as ideal gas and changes in kinetic and potential energies are ignored. The energy equation per mole of fuel basis is given by

$$\frac{\dot{Q}_{cv}}{\dot{n}_F} - \frac{\dot{W}_{cv}}{\dot{n}_F} = \left[a\,\bar{h}_{CO_2} + \frac{b}{2}\bar{h}_{O_2} + \left(a+\frac{b}{4}\right)3.76\,\bar{h}_{N_2}\right]$$

$$-\bar{h}_F - \left[\left(a+\frac{b}{4}\right)\bar{h}_{O_2} + \left(a+\frac{b}{4}\right)3.76\,\bar{h}_{N_2}\right] \quad ...(17.27)$$

where $\dot{n}_F$ represents the molar flow rate of the *fuel*.

The terms inside the first big bracket on the right side is the enthalpy of the exiting gaseous products of combustion per mole of fuel while the terms inside the second bracket is the enthalpy of combustion (reactants) air *per mole of fuel*. The symbol $\bar{h}_F$ denotes the molar enthalpy of fuel. Thus, Eq. (17.27) becomes

$$\frac{\dot{Q}_{cv}}{\dot{n}_F} - \frac{\dot{W}_{cv}}{\dot{n}_F} = \bar{h}_p - \bar{h}_r \quad ...(17.28)$$

where $\bar{h}_p$ and $\bar{h}_r$ represent, respectively, the enthalpies of products and reactants.

Since each component of combustion products is assumed to behave as an ideal gas, its contribution to enthalpy of the products depends solely on the temperature of the products, T_p. Thus,

$$\bar{h} = \bar{h}_f^0 + [\bar{h}(T_p) - \bar{h}(T_{ref})] \qquad ...(17.29)$$

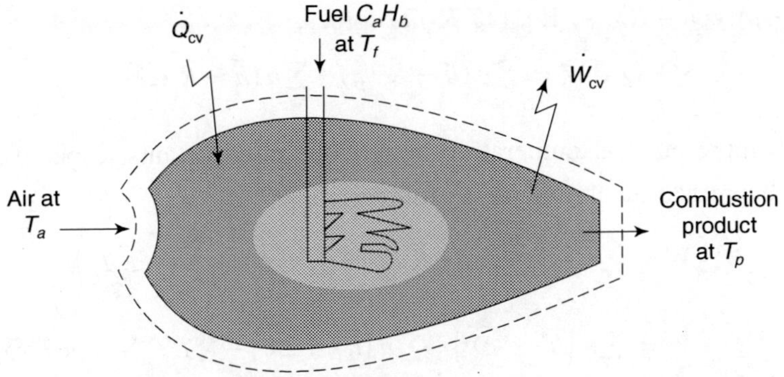

Fig. 17.5 Combustion of hydrocarbon under steady state

In Eq. (17.29), $\bar{h}_f^0$ is the enthalpy of formation. The second term is bracket account for the change in enthalpy from the temperature T_{ref} to the temperature T_p. It is to be noted that $\bar{h}_f^0$ = O for N_2 and O_2. Using Eq. (17.29), the energy Eq. (17.27) for $C_a H_b$ takes the form as

$$\frac{\dot{Q}_{cv}}{\dot{n}_F} - \frac{\dot{W}_{cv}}{\dot{n}_F} = a\left(\bar{h}_f^0 + \overline{\Delta h}\right)_{CO_2} + \frac{b}{2}\left(\bar{h}_f^0 + \Delta h\right)_{H_2O} + \left(a + \frac{b}{4}\right)3.76\,(\overline{\Delta h})_{N_2}$$

$$- \left(\bar{h}_f^0 + \overline{\Delta h}\right)_F - \left(a + \frac{b}{4}\right)(\overline{\Delta h})_{O_2} - \left(a + \frac{b}{4}\right)3.76\,(\overline{\Delta h})_{N_2} \qquad ...(17.30)$$

Equation (17.30) can be written more concisely as

$$\frac{\dot{Q}_{cv}}{\dot{n}_F} - \frac{\dot{W}_{cv}}{\dot{n}_F} = \sum_p n_e\left(\bar{h}_f^0 + \overline{\Delta h}\right)_e - \sum_r n_i\left(\bar{h}_f^0 + \overline{\Delta h}\right)_i \qquad ...(17.31)$$

where i represents incoming fuel and air streams and e the existing combustion products. The coefficients (n_i and n_e) denote the respective coefficients of the reaction equation giving the moles of reactants and products *per mole of fuel*, respectively.

17.12.3 Closed System

Consider a closed system involving a combustion process. In the absence of kinetic and potential energy effects, we have

$$U_p - U_r = Q - W \qquad ...(17.32)$$

where U_r and U_p denote the internal energy of reactants and products, respectively. If the reactants and products form ideal mixture, the energy balance can be expressed as

$$\sum_p n\,\bar{u} - \sum_r n\,\bar{u} = Q - W \qquad ...(17.33)$$

where n stands for moles of each reactant or product

Since $u = h - pv = h - RT$, so Eq. (17.33) becomes,

$$Q - W = \sum_p n\,(\bar{h} - \bar{R}\,T_p) - \sum_r n\,(\bar{h} - \bar{R}\,T_r) \qquad ...(17.34)$$

where T_p and T_r represent the temperature of product and reactants respectively. Equation (17.34) may be expressed involving $\bar{h}_f^0$ as follows :

$$Q - W = \sum_p n\left(\bar{h}_f^0 + \overline{\Delta h} - \bar{R}\,T_p\right) - \sum_r n\left(\bar{h}_f^0 + \overline{\Delta h} - \bar{R}\,T_r\right)$$

or

$$Q - W = \sum_p n\left(\bar{h}_f^0 + \overline{\Delta h}\right) - \sum_r n\left(\bar{h}_f^0 + \overline{\Delta h}\right) - \bar{R}\,T_p \sum_p n + \bar{R}\,T_r \sum_r n \qquad ...(17.35)$$

The above concepts are applicable where gaseous mixture burns in a closed, rigid container.

17.13 ENTHALPY AND INTERNAL ENERGY OF REACTION (COMBUSTION)

The concept of enthalpy of formation underlies the formulations of the energy balances for reactive systems but the enthalpy of formation $(\bar{h}_f^0)$ of fuel is not always known owing to the wide variation in composition. In many cases of practical interest, however, the enthalpy of reaction (or combustion), which is accessible experimentally, can be used to conduct an energy analysis when $\bar{h}_f^0$ data are lacking.

The enthalpy of reaction (combustion) is defined as the difference between enthalpy of products and enthalpy of reactants when complete combustion occurs in a steady flow process at a given temperature and pressure. It is denoted by $\overline{\Delta h}_R$ or $\bar{h}_{rp}$. Thus,

$$\bar{h}_{rp} = \overline{\Delta h}_R = H_p - H_r = \sum_p n_e\,\bar{h}_e - \sum n_i\,\bar{h}_i \qquad ...(17.36)$$

The unit of $\bar{h}_{rp}$ (or $\overline{\Delta h}_R$) is expressed in kJ/kg mole or kJ/kg of the fuel. *Here the suffix rp stands for reactants-products while suffixes R stands for reaction. The term n corresponds to the respective coefficients of the reaction giving the moles of reactants and products per mole of fuel. The suffixes i and e denote inlet and exit conditions.* When the enthalpies of combustion (or reaction) are expressed on a unit mass of fuel basis, it is designated as h_{rp} of Δh_R. The values of enthalpy of reaction (or combustion) of different hydrocarbon fuels at standard state, 25°C and 1 atm designated as $\bar{h}_{rp}^0$ or $\overline{\Delta h}_R$ are given in Table A10. It is represented on $h - T$ diagram as shown in Fig. 17.6.

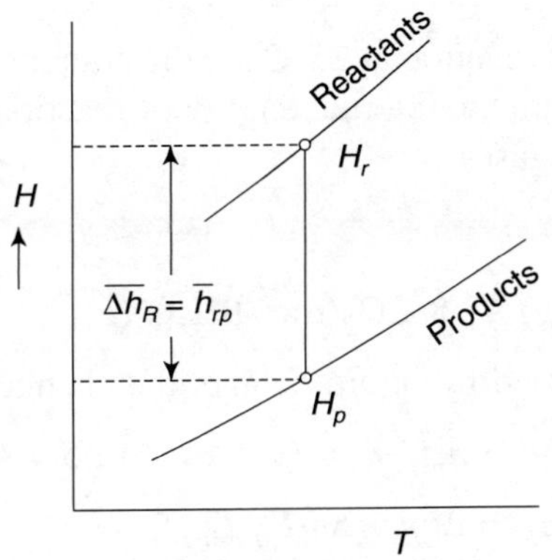

Fig. 17.6 Representation of $\Delta\bar{h}_R = \bar{h}_{rp}$

Equation (17.31) may be remanaged to read

$$\frac{Q_{cv}}{\dot{n}_F} \quad \frac{W_{cv}}{\dot{n}_F} = \left[\sum_p n_e\,(\bar{h}_f^0)_e - \sum_r n_i\,(\bar{h}_f^0)_i\right] + \sum_p n_e\,(\overline{\Delta h})_e - \sum_r n_i\,(\overline{\Delta h})_i$$

The first two terms of right side inside big bracket is just the enthalpy of combustion at T_{ref} and p_{ref}. Thus, the above equation becomes

$$\frac{\dot{Q}_{cv}}{\dot{n}_f} - \frac{\dot{W}_{cv}}{\dot{n}_F} = \bar{h}_{rp}^0 + \sum_p n_e\,(\overline{\Delta h})_e - \sum_r n_i\,(\overline{\Delta h})_i \qquad ...(17.37)$$

The right side of Eq. (17.37) can be evaluated with an experimentally determined value of $\bar{h}_{rp}^0$ and $\overline{\Delta h}$.

The internal energy of reaction $\bar{u}_{rp}$ is defined as the difference between internal energy of products and reactants in a closed system at a given temperature and pressure. Thus,

$$\bar{u}_{rp} = \Delta\bar{u}_R = U_p - U_r \qquad ...(17.38)$$

$$= \sum_p n_e\left[\bar{h}_f^0 + \Delta\bar{h} - p\bar{v}\right]_e - \sum_r n_i\left[\bar{h}_f^0 + \Delta\bar{h} - p\bar{v}\right]_i \qquad ...(17.39)$$

If all the gaseous constituents are considered ideal gases, and the volume of liquid and solid considered is assumed to be negligible compared to gaseous volume, then

$$\bar{u}_{rp} = \bar{h}_{rp} - \bar{R}T\,(n_{\text{gaseous products}} - n_{\text{gaseous reactants}}) \qquad ...(17.40)$$

$\bar{u}_{rp}$ is also generally evaluated at standard state (25°C and 1 atm)

In the case of constant pressure or steady flow process, the negative of enthalpy of combustion is frequently called the heating value at constant pressure. Similarly, the negative value of internal energy of combustion is frequently called heating value at constant volume in the case of combustion.

Example 17.17 The enthalpy of reaction at 25°C for the combustion of *n*-octane with oxygen is – 5113.6 kJ/gm mole. Calculate the internal energy of reaction at the temperature. Assume the products are in the form of gases.

Solution The reaction is

$$C_8H_{18}\ (l) + 12.5\ O_2(g) \rightarrow 8\ CO_2(g) + 9H_2O\ (g)$$

Neglecting the volume occupied by liquid octane, the change in the number of moles is

$$\Delta n = n_p - n_r = (9 + 8) - 12.5 = 4.5 \text{ moles/mole of fuel.}$$

We know that,

$$\Delta u_R = \Delta h_R - \Delta n\ R_0\ T_0$$

$$\Delta u_R = \Delta h_R - \Delta n \times \frac{R_0\ T_0}{M_{C_8\ H_6}}, \text{ kJ/gm mole}$$

$$= \Delta h_R - \frac{\Delta n \times 8.314 \times T_0}{M_{C_8H_{18}}} = -5113.6 - \frac{4.5 \times 8.314 \times 298}{114}$$

$$= -5015.8 \text{ kJ/gm mole}$$ **Ans.**

Example 17.18 Determine the enthalpy of formation of *n*-heptane if the enthalpy combustion is $\Delta \bar{h}_R = 4815.5$ kJ/gm mole.

Solution The combustion equation is $C_7\ H_{16} + 11O_2 \rightarrow 7CO_2 + 8H_2O$

The enthalpy of reaction is $\Delta \bar{h}_R = H_p - H_r = (n\,\bar{h}_f^0)_p - (n\,\bar{h}_f^0)_r$

From the Steam Tables,

$$\bar{h}^0_{f,CO_2} = -393.17 \text{ kJ/gm mole}, \quad \bar{h}^0_{f,H_2O} = -285.7 \text{ kJ/gm mole}$$

$$\therefore \quad \Delta \bar{h}_R = 4815.5 = 7\,\bar{h}^0_{f,CO_2} + 8\,\bar{h}^0_{f,H_2O} - \bar{h}^0_{f,C_7H_{16}}$$

$$= 7\,(-393.32) + 8\,(-285.7 \text{ kJ}) - \bar{h}^0_{f,C_7\ H_{16}}$$

or

$$\bar{h}^0_{f,C_7\ H_{16}} = -187.73 \text{ kJ/gm mole}$$ **Ans.**

17.14 CALORIFIC (OR HEATING) VALUE OF FUEL

The calorific value of fuel is defined as the amount of heat liberated in kJ due to burning of one kg of fuel and it is expressed in kJ/kg of fuel. It is also called *heat of combustion. In general, the calorific value of gaseous fuel is based on 1 m^3 of fuel instead of 1 kg of fuel under standard condition of temperature and pressure.* It is expressed in kJ/m^3 of fuel, however, it is also expressed in kJ/kg. The standard temperature is taken as 15°C (288 K) and the standard pressure is taken a 76 cm of mercury. We therefore, calculate the value of gaseous fuel first under standard condition using the gas equation

$$\frac{p_1 V_1}{T_1} = \frac{p_0 V_0}{T_0}$$

where p_1, V_1 and T_1 refer to given conditions and p_0, V_0 and T_0 refer to standard conditions.

Combustion of a fuel may be at constant volume or constant pressure.

Calorific Value at Constant Pressure. When the fuel is burnt at constant pressure, the heat liberated per kg of fuel is called the ***calorific value at constant pressure***. Thus, calorific value at constant pressure is equal to the enthalpy of reaction (Δh_R) with negative sign showing the liberation of heat

$$(\text{C.V.})_p = \overline{\Delta h_R} = -\bar{h}_{rp} \qquad \text{...(17.41)}$$

Calorific Value at Constant Volume. When a fuel is allowed to burn in a closed vessel, the combustion takes place at constant volume and the heat liberated per kg of fuel which is stored as internal energy is called the ***calorific value at constant volume***. It is equal to the *internal energy of reaction* with negative sign showing the liberation of heat

$$(\text{C.V.})_v = -\Delta\bar{u}_R = -\bar{u}_{rp} \qquad \text{...(17.42)}$$

The relationship between calorific value at constant pressure and constant volume is given as follows :

We have seen that the enthalpy of reaction is related to internal energy of reaction as under.

$$\Delta h_R = \Delta u_R \, \Delta n \, R_0 \, T_0 \qquad \text{...(17.43)}$$

where Δn may be zero, –ve and +ve

But $\quad (\text{C.V.})_p = -\Delta h_R$ and $(\text{C.V.})_v = -\Delta u_R$

or $\quad (\text{C.V.})_p = (\text{C.V.})_v - \Delta n \, R_0 \, T_0 = (\text{C.V.})_v - \Delta n \times 8.314 \, T_0$

or $\quad (\text{C.V.})_p = (\text{C.V.})_v - 8.314 \, \Delta n \, R_0 \, T_0 \qquad \text{...(17.44)}$

Example 17.19 An internal combustion engine uses liquid octane as fuel and operating at steady state consumes 0.002 kg/s. It is mixed with theoretical amount of air for combustion in engine cylinder. The fuel and air enters the engine at 25°C and 1 atm. The mixture burns completely and combustion products leave the engine at 500°C. The engine develops a power output of 45 kW. Calculate the rate of heat transfer from the engine neglecting the changes in kinetic and potential energies.

Solution Refer to Fig. 17.7. Assume each mole of oxygen in the combustion air is accompanied by 3.76 mole of nitrogen.

The chemical equation for complete combustion with theoretical air is given by

$$C_8H_{18} + a\,(O_2 + 3.76\,N_2) \rightarrow b\,CO_2 + c\,H_2O + d\,N_2$$

Applying the conservation of mass to C, H, O and N, we get the balanced chemical equation as

$$C_8H_{18} + 12.5\,O_2 + 47\,N_2 \rightarrow 8\,CO_2 + 9H_2O + 47\,N_2$$

The energy rate balance in the absence of KE and PE effect is given by

$$\frac{\dot{Q}_{cv}}{\dot{n}_F} = \frac{\dot{W}_{cv}}{\dot{n}_F} + \bar{h}_p - \bar{h}_r$$

or

$$\frac{\dot{Q}_{cv}}{\dot{n}_F} = \frac{\dot{W}_{cv}}{\dot{n}_F} + \left\{8\,[\bar{h}_f^0 + \Delta\bar{h}]_{CO_2} + 9\,[\bar{h}_f^0 + \Delta\bar{h}]_{H_2\,o(g)} + 47\,[\bar{h}_f^0 + \Delta\bar{h}]_{N_2}\right\}$$

$$- \left\{[\bar{h}_f^0 + \Delta\bar{h}]_{C_6\,H_{18}(l)} + 12.5\,[\bar{h}_f^0 + \Delta\bar{h}]_{O_2} + 47\,[\bar{h}_f^0 + \Delta\bar{h}]_{N_2}\right\}$$

In the product, $[\bar{h}_f^0]_{N_2} = 0$ and in the reactants

$$[\Delta\bar{h}]_{C_8\,H_{18}(l)}, [\bar{h}_f^0 + \Delta\bar{h}]_{O_2} \quad \text{and} \quad [\bar{h}_f^0 + \Delta\bar{h}]_{N_2} = 0$$

From Table A-8, we have

$$\bar{h}_r = [\bar{h}_f^0]_{C_8\,H_{18}(l)} = -\,249{,}910 \text{ kJ/kg mole}$$

Substituting the values of $\bar{h}_f^0$ and $\overline{\ }$ for various products from Tables A8 and A10 by using $\Delta\bar{h} = \bar{h}\,(T, p) - \bar{h}\,(T_{ref,}\, p_{ref})$, we have

$$\bar{h}_p = 8\,[-\,393{,}520 + (36{,}811 - 9{,}366)]$$

$$+\,9\,[-\,241{,}820 + (31{,}381 - 9{,}902)] + 47\,[26{,}534 - 8{,}673]$$

$$= -\,40{,}75002 \text{ kJ/kg mole (fuel)}$$

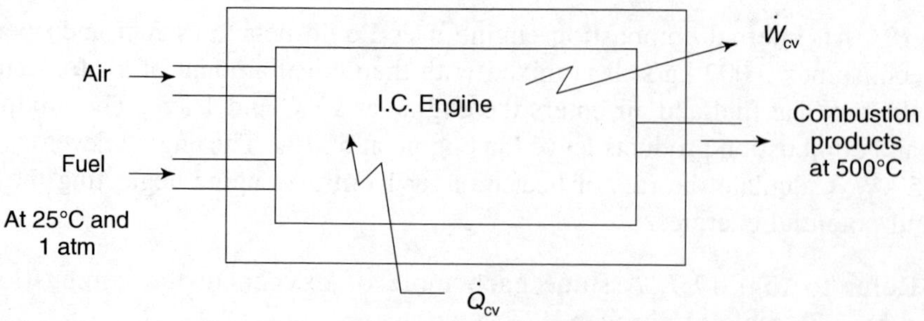

Fig. 17.7 Example 17.19

The molecular weight of fuel is 114.22, thus the molar flow rate of fuel is

$$\dot{n}_F = \frac{0.002}{114.22} = 1.75 \times 10^{-5} \text{ kg mole fuel/s}$$

Substituting the value of $\bar{h}_p, \bar{h}_r, \dot{W}_{cv}$ and $\dot{n}_F$, we have

$$\dot{Q}_{cv} = \dot{W}_{cv} + \dot{n}_F(\bar{h}_p - \bar{h}_r)$$

$$= 45 + 1.75 \times 10^{-5} [40{,}75002 - (-249{,}910)] = -21.939 \text{ kW} \quad \textbf{Ans.}$$

Example 17.20 A combustion chamber receives methane gas at 400 K and 1 atm where it is mixed with air entering at 500 K and 1 atm. The products of combustion exit at 1800 K and 1 atm. The molar analysis of products on a dry basis is CO_2 9.6%; CO, 0.6%; O_2, 2.95%; and N_2, 86.85%. Assuming steady state operation, calculate the rate of heat transfer from the combustion chamber. Assume the average value for the specific heat at constant pressure, $\bar{c}_p$ of methane between 298 K and 400 K as 38 kJ/kg mole. Neglect KE and PE effects.

Solution Refer to Fig. 17.8. Assume that each mole oxygen in the combustion air is accompanied by 3.76 moles of nitrogen, which is inert. The chemical equation is

$$a\,CH_4 + b\,(O_2 + 3.76\,N_2) \rightarrow 0.6\,CO_2 + 0.6\,CO + 2.95\,O_2 + 86.85\,N_2 + c\,H_2O$$

Applying conservation of mass to C, H and O, and dividing by the coefficient of CH_4, we have

$$CH_4 + 2.26\,O_2 + 8.515\,N_2 \rightarrow 0.941\,CO_2 + 0.0588\,CO + 0.289\,O_2 + 8.514\,N_2 + 2\,H_2O$$

Since there is no work done, so the energy rate balance is

$$\frac{\dot{Q}_{cv}}{\dot{n}_F} = \bar{h}_p - \bar{h}_r$$

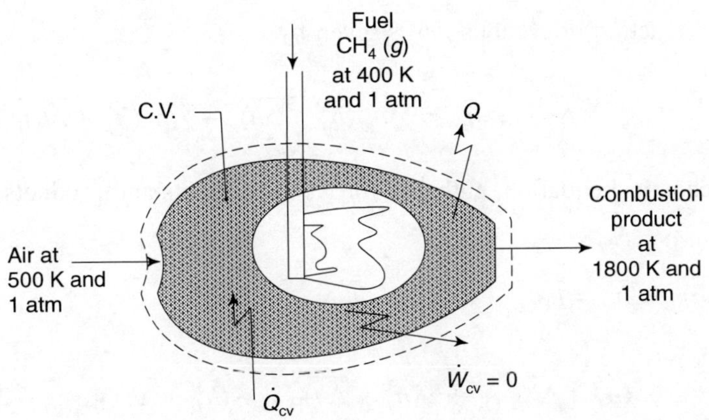

Fig. 17.8 Example 17.20

Here, $\bar{h}_p = [0.941\,(\bar{h}_f^0 + \Delta\bar{h})_{CO_2} + 0.0588\,(\bar{h}_f^0 + \Delta\bar{h})_{CO} + 0.289\,(\bar{h}_f^0 + \Delta\bar{h})_{O_2}$

$$+ 8.514\,(\bar{h}_f^0 + \overline{\Delta h})_{N_2} + 2\,(\bar{h}_f^0 + \overline{\Delta h})_{H_2O}]$$

But $(\bar{h}_f^0)_{O_2}$ and $(\bar{h}_f^0)_{N_2} = 0$ and $\overline{\Delta h} = \bar{h}\,(T, p) - \bar{h}\,(T_{ref}, p_{ref})$,

So taking the value of $\bar{h}_f^0$ and $\Delta\bar{h}$ from table in Appendices A8 and A9, we have

$\bar{h}_p = 0.941\,[-\,393{,}520 + (88{,}806 - 9{,}364)] + 0.0588\,[-\,110{,}530 + (58.19 - 8.669)]$

$+\,0.289\,[60{,}371 - 8{,}682) + 8.514\,[57{,}651 - 8{,}669] + 2\,[-\,241{,}820 + (72{,}513 - 9{,}904)]$

$= -\,228.075$ kJ/kg mole of fuel CH_4.

Here $\bar{h}_r = \left[\bar{h}_f^0 + \overline{\Delta h}\right]_{CH_4} + 2.26\left[\bar{h}_f^0 + \overline{\Delta h}\right]_{O_2} + 8.515\left[\bar{h}_f^0 + \overline{\Delta h}\right]_{N_2}$

But $(\bar{h}_f^0)_{O_2}$ and $(\bar{h}_f^0)_{N_2} = 0$ and $(\overline{\Delta h})_{CH_4} = \bar{c}_p\,\Delta T$. Thus,

$\bar{h}_r = [-\,74{,}850 + 34\,(400 - 298)] + 2.26\,[14{,}770 - 8{,}682] + 8.515\,[14{,}581 - 8{,}669]$

$= -\,6{,}842$ kJ/kg mole of CH_4

Substituting the values of $\bar{h}_p$ and $\bar{h}_r$, we have

$$\frac{\dot{Q}_{cv}}{\dot{n}_f} = \bar{h}_p - \bar{h}_r = -\,228075 - (-\,6842) = -\,2{,}21{,}233 \text{ kJ/kg mole of } CH_4$$

Example 17.21 Determine the enthalpy of reaction (or combustion) of gaseous methane, assuming the reactants and products at 25°C and 1 atm with liquid water in the product and (b) at 25°C and 1 atm with water vapour in the products.

Solution The chemical equation for combustion of $CH_4(g)$ with theoretical air is given by

$$CH_4 + 2O_2 + 7.52\,N_2 \rightarrow CO_2 + 2H_2O + 7.52\,N_2$$

The enthalpy of reaction or combustion is given by

$$\overline{\Delta h}_R = \bar{h}_{rp} = \sum_p n_e\,(\bar{h}_f^0 + \overline{\Delta h})_e - \sum_r n_i\,(\bar{h}_f^0 + \overline{\Delta h})_i$$

For the above chemical equation, enthalpy for N_2 as reactants and products get cancelled, so we have

$$\overline{\Delta h}_R = \bar{h}_{rp} = \bar{h}_{CO_2} + 2\bar{h}_{H_2O} - \bar{h}_{CH_4(9)} - 2\bar{h}_{O_2}$$

$$= (\bar{h}_f^0 + \overline{\Delta h})_{CO_2} + 2\,(\bar{h}_f^0 + \overline{\Delta h})_{H_2O} - (\bar{h}_f^0 + \overline{\Delta h})_{CH_4(g)} - 2\,(\bar{h}_f^0 + \overline{\Delta h})_{O_2}$$

Here $(\bar{h}_f^0)_{O_2} = 0$ by definition. After rearranging, we have

$$\overline{\Delta h}_R = \bar{h}_{rp} = (\bar{h}_f^0)_{CO_2} + 2\,(\bar{h}_f^0)_{H_2O} - (\bar{h}_f^0)_{CH_4(g)} + \left[(\overline{\Delta h})_{CO_2} + 2\,(\overline{\Delta h})_{H_2O} - (\overline{\Delta h})_{CH_4(g)} - 2\,(\overline{\Delta h})_{O_2}\right]$$

Since the reactants and products are of 25°C and 1 atm, so the terms $\overline{\Delta h}$ grouped inside big bracket vanish and $\overline{\Delta h}_R = \bar{h}_{rp}$ becomes $\overline{\Delta h}_R^0 = \bar{h}_{rp}^0$. Thus,

$$\overline{\Delta h}_R^0 = \bar{h}_{rp}^0 = (\bar{h}_f^0)_{CO_2} + 2\,(\bar{h}_f^0) - (\bar{h}_f^0)_{CH_2\,(g)}$$

(a) Now for liquid water in the products, we have

$$\overline{\Delta h}_R^0 = \bar{h}_{rp}^0 = (\bar{h}_f^0)_{CO_2} + 2\,(\bar{h}_f^0)_{H_2O(g)} - (\bar{h}_f^0)_{CH_4\,(g)}$$

Substituting the values from Table A8, we have

$\overline{\Delta h}_R^0 = \bar{h}_{rp}^0 = -393{,}520 + 2\,(-285{,}830) - (-74{,}850) = -890{,}330$ kJ/kg mole of fuel. **Ans.**

On per unit mass basis, the enthalpy of reaction is

$$\Delta h_R^0 = h_{rp}^0 = -\frac{890{,}330}{16.04} = -\,55{,}506 \text{ kJ/kg of fuel} \qquad \textbf{Ans.}$$

The above values agree with the heating value of CH_4 given in Table A8.

(b) If water in the form of vapour, then

$$\overline{\Delta h}_R^0 = \bar{h}_{rp}^0 = (\bar{h}_f^0)_{CO_2} + 2\,(\bar{h}_f^0)_{H_2\,O(g)} - (\bar{h}_f^0)_{CH_4\,(g)}$$

$$= -\,393{,}520 + 2\,(-\,241{,}820) - (-\,74{,}850) = -\,802{,}310 \text{ kJ/kg mole of fuel} \qquad \textbf{Ans.}$$

On unit mass basis,

$$\Delta h_R^0 = h_{rp}^0 = \frac{-\,802{,}310}{16.04} = -\,50019 \text{ kJ/kg of fuel} \qquad \textbf{Ans.}$$

The results agree with heating value given in Table A8.

Example 17.22 In a well-insulated combustion chamber (reactor), liquid octane enters at 25°C and 1 atm. The air also enters at the same pressure and temperature. The combustion takes place and the products comes out of the combustion chamber. For steady-state operation and neglecting the effect of changes in kinetic and potential energies, calculate the temperature of combustion products (or adiabatic temperature) for complete combustion with (a) theoretical amount of air, and (b) 200% theoretical air.

Solution Refer to Fig. 17.9. Since the combustion chamber is insulated and does not exchange any work, so

$\dot{Q}_{cv} = 0$ and $\dot{W}_{cv} = 0$. It is assumed that combustion is accompanied by 3.76 moles of nitrogen, which is inert.

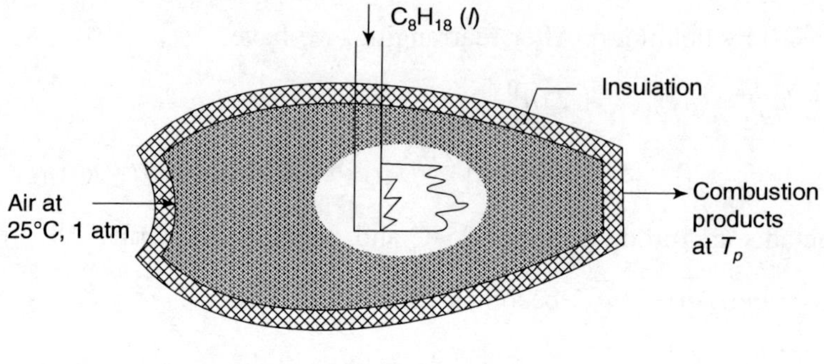

Fig. 17.9 Example 17.22

At steady state, the control volume energy rate balance yields

$$\sum_p n_e \bar{h}_e = \sum_r n_i \bar{h}_i$$

or

$$\sum_p n_e (\bar{h}_f^0 + \overline{\Delta h})_e = \sum_r n_i (\bar{h}_f^0 + \overline{\Delta h})_i$$

After remanaging the above equation, we have

$$\sum_p n_e (\overline{\Delta h})_e = \sum_r n_i (\overline{\Delta h})_i + \sum_r n_i \bar{h}_{fi}^0 - \sum_p n_e \bar{h}_{fe}^0$$

Since the reactants enter at 25°C and 1 bar, so $(\overline{\Delta h})_i = 0$, thus

$$\sum_p n_e (\overline{\Delta h})_e = \sum_r n_i \bar{h}_{fi}^0 - \sum_p n_e \bar{h}_{fe}^0 \qquad \text{...(a)}$$

The chemical equation for the combustion of liquid $C_8 H_{18}$ with the theoretical amount of air is given by

$$C_8H_{18}(l) + 12.5\ O_2 + 47\ N_2 = 8\ CO_2 + 9\ H_2O\ (g) + 47\ N_2$$

Using the chemical equation in Eq. (a), we have

$$8\,(\overline{\Delta h})_{CO_2} + 9\,(\overline{\Delta h})_{H_2\,O(g)} + (\overline{\Delta h})_{N_2}$$

$$= \left[(\bar{h}_f^0)_{C_8H_{18}(l)} + 12.5\,(\bar{h}_f^0)_{0.2} + 47\,(\bar{h}_f^0)_{N_2}\right] - \left[8\,(\bar{h}_f^0)_{CO_2} + 9\,(\bar{h}_f^0)_{H_2\,O(g)} + 47\,(\bar{h}_f^0)_{N_2}\right]$$

Since $(\bar{h}_f^0)_{O_2} = 0, (\bar{h}_f^0)_{N_2} = 0,$ so the above equation becomes

$$8\,(\overline{\Delta h})_{CO_2} + 9\,(\overline{\Delta h})_{H_2O(g)} + 47\,(\overline{\Delta h})_{N_2} = \left[(\bar{h}_f^0)_{C_8\,H_{18}\,(l)}\right] - \left[8\,(\bar{h}_f^0)_{CO_2} + 9\,(\bar{h}_f^0)_{H_2\,O(g)}\right]$$

$= -\ 249{,}910 - [8\ (-\ 393{,}520) + 9\ (-\ 241{,}820)] = 5{,}074{,}630$ kJ/kg mole of fuel

Now, each term of L.H.S. of equation depends on the temperature of the product (T_p) and this value can be obtained only by trial and error through balancing L.H.S. = R.H.S.

Let at 2400 K:

$8\,(\overline{\Delta h})_{CO_2}$	926,304
$9\,(\overline{\Delta h})_{H_2O(g)}$	842,436
$47\,(\overline{\Delta h})_{N_2}$	3,320,597
$\sum\limits_{p} n_e\,(\overline{\Delta h})_e$	5,089,337 kJ/kg mole of fuel

Thus, at $T_p = 2400$ K, L.H.S. is slightly (about 14707) more than R.H.S., so it is estimated that $T_p = 2395$ K will satisfy the equation. Thus, the adiabatic flame temperature or temperature of combustion product will be 2395 K. **Ans.**

17.15 HIGHER AND LOWER CALORIFIC VALUE OF FUEL

Most of the fuels contain hydrogen. During the combustion process, the hydrogen combines with oxygen and forms steam (water vapour). *If this water vapour is condensed at constant temperature, a large amount of additional heat will be released than if it exits complete in the vapour phase*. On account of this, two types of calorific values are defined; the *higher or gross calorific value and the lower or net calorific value*.

(i) Higher calorific value (HCV). *The higher calorific value is defined as the total heat liberated when the water vapor formed by combustion is completely condensed at constant temperature releasing a large amount of additional heat*. The HCV of fuel may be determined accurately by experimental method. The calorific values of elementary fuels at NTP are given below.

Fuel	Higher calorific value (HCV) kJ/kg	Lower calorific value (LCV) kJ/kg
C	33923	–
CO	10174	–
S	92947	–
H_2	144026	121417

If a solid fuel consists of C kg carbon, H kg of hydrogen, S kg of sulphur, O kg of oxygen in 1 kg of fuel, then higher calorific value will be written as

$$\text{H.C.V.} = \left[33913\,C + 144026\left(H_2 - \frac{O_2}{8}\right) + 92947\,S\right] \text{ kJ/kg} \qquad \text{...(17.45)}$$

The quantity $\left(H_2 - \dfrac{O_2}{8}\right)$ in the above expression represents the quantity of hydrogen available for combustion, and that oxygen in the fuel is already wholly in combination with hydrogen. Also, 8 parts of oxygen by weight combine with 1 part of hydrogen by weight to form water. Therefore, hydrogen in combination with oxygen will amount to only $O_2/8$ leaving $(H_2 - O_2/8)$ available for combustion.

Dulong suggested the following empirical formula for calculating the higher calorific value of fuel.

$$\text{H.C.V.} = \left[33829\,\text{C} + 144026\left(\text{H}_2 - \frac{\text{O}_2}{8}\right) + 9294\,\text{S}\right] \text{kJ/kg} \qquad ...(17.46)$$

(ii) *Lower calorific value (LCV). The lower calorific value of fuel may be defined as the net heat liberated when the water formed by combustion exists completely in the vapour phase.* Thus, the lower calorific value is the difference between higher calorific value (HCV) and heat absorbed by hydrogen for formation of steam. In actual practice, this is the net value of heat produced and available for heating purpose.

Now, if we consider the formation of steam by the combination of H_2 and O_2, we find that 5 kg of hydrogen combines with 32 kg of oxygen ($2H_2 + O_2 = 2H_2O$) to form 36 kg of steam. In other words, 1 kg of hydrogen combines with 8 kg of oxygen to form 9 kg value of hydrogen in fuel weight. Thus, the lower calorific value may be expressed as

$$\text{LCV} = \text{HCV} - m_{\text{H20}} \times h_{fg}$$

or

$$\text{LCV} = \text{HCV} - h_{fg}\ \text{(9 times the mass of hydrogen in 1 kg of fuel)}$$

where h_{fg}= Latent heat of steam at saturation temperature of 15°C (NTP) = 2466 I kJ/kg

$$\therefore \quad \text{LCV} = 33913\,\text{C} + 144026\left(\text{H}_2 - \frac{\text{O}_2}{8}\right) + 92947\,\text{S} - 9 \times 2466.1\,\text{H}_2 \qquad ...(17.47)$$

The HCV and LCV at constant pressure and constant volume are governed by the following formula

$$(\text{LCV})_p = (\text{HCV})_p - m_{\text{H}_2\text{O}} \times h_{fg} \qquad ...(17.48)$$

$$(\text{LCV})_y = (\text{HCV})_v - m_{\text{H}_2\text{O}}\,(u_g - u_f) \qquad ...(17.49)$$

Example 17.23 The calorific value of fuel oil ($C_{16}H_{32}$) at constant volume is 52753 kJ/kg at 15°C. Determine the calorific value of fuel at constant pressure at 15°C

Solution The chemical reaction, $C_{16}H_{32} + 24\,O_2 \rightarrow 16\,CO_2 + 16H_2O$

On cooling the products of combustion at STP, the steam will condense so that change in volume is

Δn = 16 moles – 24 moles = – 8 moles per mole of fuel. We know that

$$(\text{CV})_p = (\text{CV})_v - 8.314\,\Delta n\,T_0$$

But $\quad \Delta n = -8$

Molecular weight of $C_{16}H_{32} = 12 \times 16 + 1 \times 32 = 224$

or $$(CV)_p = 52753 - \frac{8.314 \times 8 \times 288}{224} = 52667.4 \text{ kJ/kg}$$ **Ans.**

Example 17.24 The propane reacts with oxygen at 25°C giving the higher calorific value as 2221423 kJ/kg mole at constant pressure. Calculate the lower calorific value at constant volume and at constant pressure.

Solution The combustion reaction for propane is $C_3H_8 + 5\ O_2 \rightarrow 3\ CO_2 + 4\ H_2O$ (gas)

$$(LCV)_p = (HCV)_p - m_{H_2O} \cdot h_{fg}$$

But $$m_{H_2O} = 4 \times 18 = 72 \text{ kg/kg mole of } C_3\ H_8$$

At $$25°C,\ h_{fg} = 2442.5 \text{ kJ/kg of steam}$$

$$(LCV)_p = 2221423 - 72 \times 2442.5 = 2045563 \text{ kJ/kg mole}$$ **Ans.**

Now, $$(LCV)_p = (HCV)_p - m_{H_2O}\ (u_g - u_f)$$

$$(u_g - u_f) = 2307 \text{ kJ/kg at } 25°C$$

But $$(HCV)_v = (HCV)_p + \Delta n\ R_0\ T_0$$

Now, the reaction for higher heating value of is $C_3\ H_8 + 5\ O_2 \rightarrow 3\ CO_2 + 4\ H_2O\ (l)$

$$\Delta n = 3 - (1 + 5) = - 3$$

$$(HCV)_v = 2221423 + 3 \times 8.314 \times 298 = 2228855.7 \text{ kJ}$$

Now, $$(LCV)_v = 2228855.7 - 72 \times 2307 = 2062751.7 \text{ kJ/kg mole}$$ **Ans.**

Example 17.25 A sample of an oil fuel has the following analysis by weight : Carbon 88 per cent and hydrogen 12 per cent. Calculate the higher and lower calorific values of fuel using the fundamental chemical equation and Dulong's formula.

Solution From the equation, 1 kg of oil fuel contains 0.88 kg of carbon and 0.12 kg of H_2.

Using the fundamental chemical equation

$$HCV = 33913\ C + 144026\ (H_2 - O_2/8) + 92947\ S$$

In the absence of sulphur content, $HCV = 33913C + 14406\ (H_2 - O_2/8)$

$$= 33913 \times 0.88 + 144026 \times 0.12 = 47126{,}56 \text{ kJ/kg}$$ **Ans.**

The combustion of 1 kg of oil gives $0.12 \times 9 = 1.08$ kg of steam. Hence, lower calorific value is

$$LCV = H.C.V. - 9 \times 0.12 \times 2466.1$$

$$= 47126.56 - 2663.38 = 44463.18 \text{ kJ/kg}$$ **Ans.**

Using Dulong's Formula

$$HCV = 33829\ C + 144444\ (H_2 - O_2/8) + 92947\ S$$

$$= 33829 \times 0.88 + 144444 \times 0.12 = 47102.8 \text{ kJ/kg}$$

$$\text{LCV} = 47102.8 - 2663.38 = 44439.4 \text{ kJ/kg}$$

17.16 EXPERIMENTAL DETERMINATION OF THE CALORIFIC VALUE OF FUEL BY BOMB CALORIMETER

The bomb calorimeter is normally used to measure the higher calorific value of solid and liquid fuels. Its name is derived from the fact that the fuel is burnt in an enclosed vessel called the bomb calorimeter surrounded by a water bath. Since the combustion of the fuel takes place at constant volume, the calorific value thus determined will be the calorific value at constant volume.

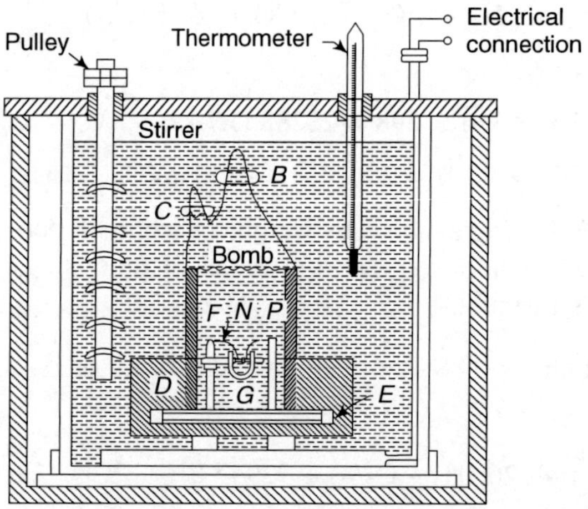

Fig. 17.10 Bomb calorimeter

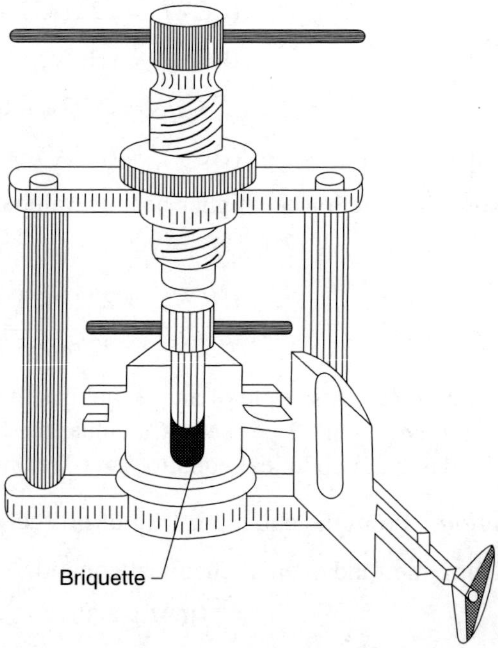

Fig. 17.11 Briquette mould

Construction. Figure 17.10 shows a diagrammatic arrangement of a Professor Schol's bomb calorimeter. It consists of a thick walled, stainless steel vessel (*A*) called the bomb. The inner surface of the bomb is coated with special enamel to prevent corrosion due to the formation of acid, as a result of combustion of fuels containing sulphur and nitrogen. The capacity of bomb is about 650 c.c. and it can withstand a pressure as high as 200 atm. The oxygen required for the burning of the fuel is supplied into the bomb through a non-return valve, *B*. A pressure release valve *C* is also provided in the vessel to discharge the burnt gases. The bottom of the bomb is screwed in a cover (*D*) which forms the base of the bomb. In order to make the arrangement leak proof, a *U*-shaped hydraulic seal (*E*) is provided between the cylinder and bottom. Sometimes, water is also filled in the bomb calorimeter to a specified level to act as a water seal. Undue tightening of base cover is avoided. The bottom carries to guide rods one of which carries a ring to hold the quartz crucible (*G*) they also carry electric connections to fuse wire (*H*) placed in the crucible, *G*.

During test, the bomb is placed in a water bath known as copper calorimeter which contains about 2500 c.c. of water. This calorimeter is itself surrounded by a double walled vessel which acts as an insulator. This vessel is provided with a detachable cover made of insulating material in which arrangement is provided for inserting a stirrer to agitate the water of the calorimeter run by a motor and a thermometer to record the temperature of water. The thermometer used reads up to 1/100 degree centigrade. The apparatus needs other accessories such as (i) a press [also called briquette mould shown in Fig 17.11] to form a standard specimen of fuel in the form of pillet, (ii) a chemical balance and (iii) a high pressure oxygen supply.

Test The specimen of the fuel in standard form is placed in crucible and weighed. If we subtract the known weight, the crucible from the above weight (crucible + fuel) we will get the weight of the fuel. Now, the crucible is placed in the bomb in such a way that the fuse wire touches the fuel. To ensure this, the fine fuse wire is coiled around the fuel. The bottom is then screwed on to the bomb body after pouring a measured quantity of distilled water in it. Oxygen is admitted in the bomb through the oxygen valve to a pressure of about 25 to 30 atm. The two valves (*B* and *C*) are closed tightly. The bomb is then placed in the water bath. The water in the calorimeter is constantly stirred at a moderate rate and temperatures are noted. Due to stirring the temperature of the both will rise slowly. The temperature is noted first after five minutes with an accuracy of 0.001 °C and thereafter every minute for the next five minutes. When the fifth minute is complete, the electric circuit is closed by means of a switch. The fuse wire will ignite the fuel. The firing of the fuse is detected with the help of an ammeter, which for successful firing will indicate the maximum current flow followed by sudden return to zero value. If the ammeter needle stays at its maximum value, it indicates that the fuse has not exploded and the current has to be quickly increased till the fuse goes off. Due to combustion of the fuel, heat will be liberated and the temperature of the water will start rising. The temperatures are first noted every 15 seconds after firing till the maximum temperature is reached. Thereafter, the temperature readings are taken every half minute for about 10 minutes or till the drop in temperature for about five successive readings are uniform. It is to be noted that if the insulation is perfect, there will be no temperature drop. Also, after reaching the top temperature due to heat liberated by the fuel, the temperature may continue to rise steadily and slowly due to mechanical energy input by the stirrer.

After the experiment is over, the bomb is taken out from the bath and the release valve (*C*) is opened to release the products of combustion. The bomb is dried and opened. If there is any unburnt fused wire, it is collected and weighed. With the help of the recorded temperature and time data, the temperature-time curve is plotted.

Calculation In actual practice, perfect insulation is not possible. Therefore, heat transfer from the system to the surroundings by radiation will be clearly indicated by a temperature drop. A correction term for radiation is therefore applied. This is computed from the Newton's law of cooling according to which the rate of heat loss due to radiation is proportional to the temperature difference between the hot body and the surroundings.

Thus, $$\frac{\delta Q}{d\theta} \propto (t_2 - t_1)$$

forms the basis of calculating the cooling rate. The rate of temperature drop is determined from the temperature-time graph for the bomb calorimeter. Let the cooling rate be designated as r°C/min after the maximum temperature is reached and let $(r/2)\ \theta = \alpha$°C be the fall in temperature due to radiation if θ be the time required for the maximum temperature to be reached.

Then True temperature rise = Recorded temperature rise + α

Let $(CV)_f$ = calorific value of fuel, kJ/kg

$(CV)_1$ = calorific value of fuse wire burnt (known, kJ/kg)

M = mass of water contained in the calorimeter, kg

m = water equivalent of the calorimeter, kg

m_f = mass of fuel burnt, kg

m_1 = mass of wire burnt, kg

t_1 = steady temperature before combustion (known, °C)

t_2 = observed maximum temperature after the combustion (known, °C)

θ = time elapsed for the maximum temperature to be reached (known, minutes)

c_{pw} = specific heat of water, kJ/kg K

$t = (t_1 - t_2)$ = observed rise in temperature while the true rise in temperature of water and the container = $t + \alpha$ = temperature rise + cooling correction. Using the energy balance :

Heat liberated by fuel = Heat absorbed by calorimeter

$$\therefore \quad (CV)_f\, m_f + (CV)_1 \,.\, m_1 = (M + m)c_{pw}\,(t + \alpha) \qquad \text{...(17.50)}$$

Equation (17.50) gives the calorific value of fuel.

The cooling correction suggested by Regnault and Pfaunder is given as :

Temperature cooling correction = α

$$= nr + \frac{r_1 - r}{t_1 - t}\left\{\sum_{1}^{n-1}(t) + \frac{1}{2}(t_0 + t_n) - nt\right\} \qquad \text{...(17.51)}$$

where, n = number of minutes between the time of firing and the first reading after the temperature begins to fall from the maximum

r = rate of fall of temperature per minute after the maximum temperature

t and t_1 = average temperature during pre-firing and final periods respectively

$\sum_{1}^{(n-1)} (t)$ = sum of readings during the period between firing and the start of cooling

$\frac{1}{2}(t_0 - t_n)$ = mean of the temperature at the moment of firing and the first temperature after the rate of change

These may be tabulated as follows.

Per-firing	Temp.	Heating time	Temp.	Cooling period	Temp.
0		t_1 6		t_n 10	
1		t_2 7		11	
2		t_3 8		12	
3		t_4 9		13	
4				14	
t_0 5				15	
		$n = 10 - 5 = 5$ min			
		$r = \frac{(t_0 - t_p)}{5}$		$t_1 = \frac{(t_n - t_{15})}{5}$	

Example 17.26 The following data refer to an experimental determination of the calorific value of a sample of coal containing carbon, 88% and hydrogen, 4.2%. Mass of coal = 0.8 gm; mass of fuse wire = 0.02 gm of calorific value 6699 kJ/kg, mass of water in the calorimeter = 1900 gm, water equivalent of calorimeter = 350 gm, the observed temperature rise is 2.9°C. The cooling correction is 0.07°C. Calculate the higher and lower calorific values of fuel.

Solution We know that $(CV)_f\, m_f + (CV)_1\, m_1 = (M + m)\, c_{pw}\, (t + \alpha)$

Here $m_f = 0.8$ gm, $(CV)_1 = 6699$ kJ/kg, $\dot{m}_1 = 0.02$ gm

$M = 1900$ gm, $m = 350$ gm, $\alpha = 0.07°C$, $t = 2.9°C$

$\therefore \quad (CV)_f \times 0.8 + 6699 \times 0.02 = 4.18 \times (1900 + 35)\,(2.97)$

or $\quad (CV)_f = \frac{23888.27}{0.8} = 29{,}860.33$ kJ/kg

$\therefore \quad$ HCV = 29,860.33 kJ/kg **Ans.**

Steam formed per kg of fuel = 9 × 0.042 = 0.378 kg.

$\therefore \quad$ LCV = HCV – 0.378 × 2466.1 = 29,760.33 – 932.18 = 28,928.14 kJ/kg **Ans.**

17.17 EXPERIMENTAL DETERMINATION OF THE CALORIFIC VALUE OF GASEOUS FUEL, JUNKER'S GAS CALORIMETER

Figure 17.12 shows a Junker's gas calorimeter for the determination of the calorific value of gaseous fuel. It consists of a gas meter (*A*) with a thermometer (*B*) attached to it; a gas pressure regulator (*C*) with a manometer (*D*) attached to it, a suitable burner (*E*) which is placed inside a calorimeter (*F*) to maintain the complete combustion of the gas and a set of accurate thermometers attached to the calorimeter. The gas first passes through a gas meter where its volume is measured; and its temperature is measured with the help of thermometer, *B*. Thereafter, the gas passing through the regulator (*C*) is led to the burner *E*. The regulator is so adjusted that a uniform flow rate of gas is maintained. The pressure of the entering gas is measured with the help of a manometer, *D*.

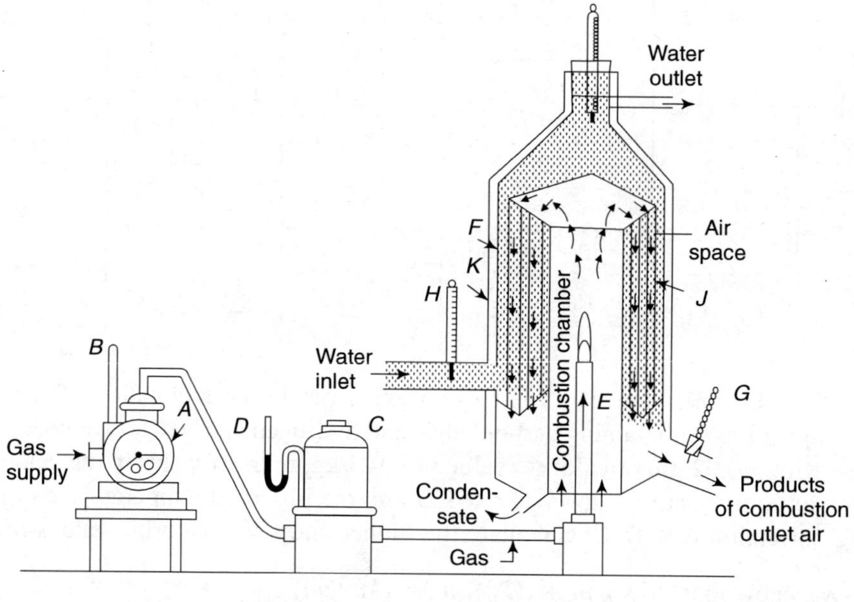

Fig. 17.12 Gas calorimeter

The temperature of the products of combustion is measured by the thermometer (*G*) placed at the exit passage for the products of combustion. The water inlet and outlet temperature are measured by the thermometers (*H* and *J*) respectively. The calorimeter (*F*) is so designed that the gases from the burner flame pass upwards through the centre of the gas calorimeter known as combustion chambers, at the top, the gas is deflected downward and passes through the flues (*J*) which are surrounded by water jackets and finally escape into the atmosphere through the outlet. The water flowing through the jackets absorbs the heat liberated by combustion of the gas and is finally discharged into a measuring tank where the rate of flowing water is measured. In order to prevent radiation losses from the calorimeter, it is covered by an outer cover (*K*) made of insulating material leaving an air gap between the calorimeter and cover.

Every fuel contains some percentage of hydrogen and this forms steam during the burning process. When this steam comes in contact with the cooling water, it condenses and the condensate is collected in a beaker and weighed.

Let V_g = volume of gas used, m^3, h_b = barometric reading, cm of H_g

h_w = manometer reading, cm of water, T_g = gas temperature at inlet, K

p_g = gas pressure at inlet, cm of H_g, T_{w1} = initial temperature of cooling water, K

T_{w2} = final temperature of the cooling water, K

M = total quantity of cooling water, c.c. m = quantity of condensate, c.c.

c_{pw} = specific heat of water, kJ/kg K, HCV = higher calorific value of fuel, kJ/m^3

V_0 = volume of gas reduced to STP, m^3, p_0 = pressure at STP = 76 cm of H_g

T_0 = standard temperature, 288 K

The calorific value of gas is normally expressed in kJ/m^3 at STP. Hence, from energy balance:

Heat given by fuel = Heat taken by water

$$\therefore \quad \text{HCV} = \frac{c_{pw} \cdot M\,(T_{w2} - T_{w1})}{V_0} \text{ at S.T.P.} \qquad ...(17.52)$$

But
$$V_0 = \frac{V_g \cdot p_g}{T_g} \cdot \frac{T_0}{p_0} \qquad ...(17.53)$$

and
$$p_g = \left(\frac{h_w}{13.6} + h_b\right) \text{ cm of Hg} \qquad ...(17.54)$$

$$\therefore \quad \text{HCV} = \frac{c_{pw}\; M\,(T_{w2} - T_{w1})\,T_g\;p_0}{\left(\frac{h_w}{13.6} + h_b\right) V_g\;T_0} \qquad ...(17.55)$$

The lower calorific value of gaseous fuel may be found by subtracting the heat consumed in the formation of steam, which is

$$\frac{m \times 2466.1}{V_0} = \frac{m \times 2466.1 \times T_g \cdot p_0}{V_g \cdot p_g \cdot T_0}$$

$$\therefore \quad \text{L.C.V.} = \frac{c_{pw}\; M\,(T_{w2} - T_{w1})\,T_g\;p_0}{\left(\frac{h_w}{13.6} + h_b\right) V_g\;T_0} - \frac{m \times 2466.1 \times T_g\;p_0}{V_g\;p_g\;T_0} \qquad ...(17.56)$$

Example 17.27 The following data were obtained during a test on Junker's gas calorimeter:

Gas burnt = 0.0025 m^3, Gas temperature at inlet = 20°C, Gas pressure at inlet = 54 mm of water, Barometric reading = 755 mm of H_g, Mass of water passing through the calorimeter

= 1300 cc, Inlet water temperature = 15°C, Oulet water temperature = 26°C, Steam condensed during test = 2.6 cc.

Determine the higher and lower calorific values of fuel.

Solution From the energy balance

$$\text{Heat liberated} = \text{Heat consumed by water}$$

$$\therefore \qquad \text{HCV} = \frac{c_{pv}\, M\,(T_{w2} - T_{w1})}{V_0} \text{ at STP}$$

But
$$V_0 = \frac{V_g \cdot p_g \cdot T_0}{T_g \cdot p_0} \quad \text{and} \quad p_g = \frac{h_w}{13.6} + h_b$$

$$\therefore \qquad \text{HCV} = \frac{c_{pw}\, M\,(T_{w2} - T_{w1})\, T_g\, p_0}{[(h_w/13.6) + h_b]\, V_g\, T_0}$$

Here, $M = 1300$ gm, $c_{pw} = 4.18$ kJ/kg K, $t_{w1} = 15$°C, $t_{w2} = 26$°C

$h_w = 54$ mm of H_2O, $V_g = 0.0025$ m^3, $T_0 = 288$ K, $p_0 = 760$ mm of Hg, $T_g = 273 + 20 = 293$ K, $m = 2.6$ cc $= 2.6$ gm

$$\text{HCV} = \frac{4.18 \times 1300\,(26 - 15) \times 293 \times 760}{[(54/13.5) + 755] \times 0.0025 \times 288 \times 1000} = 24.357 \times 10^3 \text{ kJ/m}^3 \quad \textbf{Ans.}$$

$$\text{Heat in condensate} = \frac{m \times 2466.1}{V_0} = \frac{m \times 2466.1 \times T_g\, p_0}{V_g\, p_g \cdot T_0}$$

$$= \frac{2.6 \times 2466.1 \times 293 \times 760}{0.0025 \times 758.79 \times 288 \times 10^3} = 2616.75 \text{ kJ/m}^3$$

$$\therefore \qquad \text{LCV} = \text{HCV} - 2616.75 = (24357 - 2612.8) \text{ kJ/m}^3 = 21744.2 \text{ kJ/m}^3 \quad \textbf{Ans.}$$

Example 17.28 A gas has the following percentage composition by volume 29% H_2, 2% CH_4, 11% CO, 16% CO_2 and 42% N_2. Determine the volume of air theoretically sufficient for the complete combustion of 1 m^3 of this gas and the volume of the dry products of combustion at the same temperature and pressure as the gas and air. Determine also the lower calorific value of the gas, given that the calorific value of CO is 12707 kJ/m^3 and the lower calorific value of hydrogen and CH_4 are 10885 and 35797 kJ/m^3 respectively.

Solution On the volume basis, for 1 m^3 of fuel

$$0.29\, H_2 + 0.02\, CH_4 + 0.11\, CO + 0.16\, CO_2 + x\, O_2 + (3.76\, x + 0.42)\, N_2$$
$$\rightarrow a\, CO_2 + b\, H_2O + (3.76\, x + 0.42)\, N_2$$

Carbon balance, $0.02 + 0.11 + 0.16 = a$, i.e., $a = 0.29$

H_2 balance, $0.29 + (0.02 + 0.02) = b$, i.e., $b = 0.33$

O_2 balance, $\frac{0.11}{2} + 0.16 + x = 0.29 + \frac{0.33}{2}$ or $x = 0.24$

$$A/F = \frac{0.24 + 3.76 \times 0.24}{1} = 1.1424 \text{ m}^3/\text{m}^3 \text{ of gaseous fuel} \quad \textbf{Ans.}$$

The dry product per m^3 of gas is 0.29 volume of CO_2 and 1.332 volume of N_2. At the same temperature and pressure as the gas and air, we have volume of dry products

$$\frac{0.29 + 1.332}{1} = 1.622 \text{ m}^3 \quad \textbf{Ans.}$$

The lower calorific value of the gas is

$$\text{LCV} = 10885 \times 0.29 + 35797 \times 0.02 + 12707 \times 0.11$$
$$= 5270.36 \text{ kJ/m}^3$$

17.18 ADIABATIC-COMBUSTION (OR FLAME) TEMPERATURE

In some engineering applications of reacting systems, it is necessary to evaluate the highest possible temperature the products of a reaction can attain. In some cases, we sometimes need high temperature products as in the case of rocket propulsion and at other instances, there may be a metallurgical limitation on the temperature that a solid bounding surface can withstand in which case lower temperature has to be managed.

The maximum temperature is possible only when the reaction is an adiahatic process. Under adiabatic conditions, the temperature attained in a reaction is called the adiabatic combustion (or reaction or flame) temperature. In actual practice, the combustion temperature is less than the adiabalic combustion temperature because of (i) heat loss, (ii) incomplete combustion, and (iii) dissociation of the products into other chemical species.

The adiabatic reaction temperature may be determined from the first law (Eq. 17.51) which reduces to

$$\sum_p n_e (\bar{h}_f^0 + \Delta\bar{h}_f^0)_e = \sum_r n_i (\bar{h}_f^0 + \Delta\bar{h}_f^0)_i \qquad ...(17.57)$$

An iteration procedure is to adopted to find flame temperature (see Example 17.31).

17.19 SECOND LAW ANALYSIS OF REACTIVE SYSTEM

In the absence of changes in KE and PE, the reversible work for a steady state flow process is given by

$$\dot{W}_{\text{rev}} = \Sigma \dot{n}_i (h_i - T_0 s_i) - \Sigma \dot{n}_e (h_e - T_0 s_e) \qquad ...(17.58)$$

where suffixes i and e denote, respectively inlet and exit conditions.

For a steady state flow process involving a chemical reaction, we have the rate balance

$$\dot{W}_{rev} = \sum_r n_i \, [\bar{h}_f^0 + \overline{\Delta h} - T_0 \, \bar{s}]_i - \sum_p n_e \, [\bar{h}_f^0 + \overline{\Delta h} - T_0 \, \bar{s}]_e \qquad ...(17.59)$$

The irreversibility for such a process is given by

$$I = \sum_p \dot{n}_e \, T_0 \, \bar{s}_e - \sum_r \dot{n}_i \, T_0 \, \bar{s}_i - \dot{Q}_{cv} \qquad ...(17.60)$$

The availability based on per kg mass is

$$a_{sf} = (h - T_0 \, s) - (h_0 - T_0 \, s_0) \qquad ...(17.61)$$

Equation (17.57) may be written in the form of Gibb's function

$$\dot{W}_{rev} = \sum_r \dot{n}_i \, \bar{g}_i - \sum_p n_e \, \bar{g}_e \qquad ...(17.62)$$

where $\bar{g}$ is the Gibb's function. The Gibb's function of formation, $\bar{g}_f^0$ is defined similar to enthalpy of formation. The value of $\bar{g}_f^0$ is given in Table A8 of Appendices.

17.19.1 Entropy Rate Balance for Combustion of Hydrocarbon at Steady State Flow Process

The entropy rate balance for the combustion of hydrocarbon C_aH_b per mole of fuel is given by

$$O = \sum_i \frac{\dot{W}_f / \dot{n}_F}{T_j} + \bar{s}_F + \left[\left(a + \frac{b}{4}\right)\bar{s}_{O_2} + \left(a + \frac{b}{4}\right) 3.76 \, \bar{s}_{N_2}\right]$$

$$- \left[a \, \bar{s}_{CO_2} + \frac{b}{2} \bar{s}_{H_2O} + \left(a + \frac{b}{4}\right) 3.76 \, \bar{s}_{N_2}\right] + \frac{S_{gen}}{\dot{n}_F} \qquad ...(17.63)$$

where $\dot{n}_F$ is the molar flow of the fuel and coefficients are the same as discussed earlier.

17.20 MAXIMUM EXPLOSION PRESSURE

It is important for the engineer to estimate the maximum possible pressure that could develop during combustion process. This occurs theoretically under adiabatic conditions, with the simultaneous achievement of the adiabatic flame temperature. Under these conditions, the heat release would be almost instantaneous and conditions of internal equilibrium in the combustion process would have to be met. If the ideal gas relation is used for the resulting gas, then the final (maximum) pressure (p_{max}) is given by

$$p_{max} = \frac{n_f \, R_0 \, T_f}{V} \qquad ...(17.64)$$

where T_f = adiabatic flame temperature

n_f = total number of gaseous moles of the products

V = bomb volume

17.21 ENTROPY CHANGES FOR REACTING MIXTURES

The evaluation of entropy changes for reactive mixture requires a means for determining the entropy of a substance, either pure or a mixture, at a pressure other than the reference. We know that

$$ds = c_p \frac{dT}{T} - R\frac{dp}{p} \qquad \text{...(17.65)}$$

where the integration would be carried out from the reference state to the desired state. We also know that, in a gas mixture, an ideal gas exerts a pressure equal to its partial pressure. Hence, the entropy of a gas in a mixture can be found by integrating the above equation from a reference state where the gas is pure and at 1 atm pressure to the mixture state where pressure is its partial pressure p_1. Therefore,

$$(s_i)_{T,p} = (s_i)_{T_{\text{ref.}}\, p_{\text{ref.}}} + \int_{T_{\text{ref}}}^{T} c_p \frac{dT}{T} - R \ln \frac{p_1}{p_{\text{ref.}}}$$

Putting the value of standard state 25°C and 1 atm as the reference state, the above equation becomes

$$(s_i)_{T,p} = s^0_{i,25} + \int_{T_{25}}^{T} c_p \frac{dT}{T} - R \ln \frac{p_i}{p_{\text{ref}}}$$

The integral of $\int_{T_{25}}^{T} c_p \frac{dT}{T} = s_T - s_{25°}$

Thus, $$(s_r)_{T,p} = s_{i,T} - R \ln p_i \qquad \text{...(17.66)}$$

where $s_{i,T}$ is the tabulated value at a given temperature and 1 atm pressure.

For a mixture of ideal gases, the total entropy is given by

$$S_{n,T,p} = \sum_i n_i (s_i)_{T_i, p_1} \qquad \text{...(17.67)}$$

For a chemical reaction, the change in entropy is

$$\Delta S = \sum_i (n_i\, s_i)_{\text{pro}} - \sum_i (n_i s_i)_{\text{reac}} \qquad \text{...(17.68)}$$

where n_i represents the number of moles of the ith component in the mixture and s_i is the value of absolute entropy of a pure substance at the required temperature and partial pressure.

17.22 GIBB'S FUNCTION FOR REACTING SYSTEM

The thermodynamic property known as the Gibb's function plays a role in the exergy analysis. The specific Gibb's function on molar basis is expressed as

$$\bar{g} = \bar{h} - T\bar{s} \qquad ...(17.69)$$

The Gibb's function of formation, $\bar{g}_f^0$ is given as

$$\bar{g}_f^0 = [\bar{h} - T\,\bar{s}]_{T_{ref},\, p_{ref}} \qquad ...(17.70)$$

The value of $\bar{g}_f^0$ is given is Table A8 of Appendices.

The Gibb's function at the state other than the standard state is given by

$$\bar{g}\,(T, p) = \bar{g}_f^0 + [\bar{g}\,(T, p) - \bar{g}\,(T_{ref}, p_{ref})] = \bar{g}_f^0 + \Delta\bar{g} \qquad ...(17.71)$$

where $$\overline{\Delta g} = [\bar{h}\,(T, p) - \bar{h}\,(T_{ref}, p_{ref})] - [T\bar{s}\,(T, p) - T_{ref}\,\bar{s}\,(T_{ref}, p_{ref})] \qquad ...(17.72)$$

17.23 CHEMICAL EXERGY

The objective of this section is to evaluate the work obtainable by allowing the fuel to react with oxygen from the environment to produce the environmental components carbon dioxide and water, each at its respective state in the environment. The *chemical exergy is defined as the maximum theoretical work that could be developed by the combined system. The sum of the thermomechanical and chemical exergies is known as total exergy associated with a given system at a specified state, relative to a specified exergy reference environment* (T_0, p_0).

Consider a combined system formed by an environment and an amount of hydrocarbon fuel, C_aH_b. The reaction is given by

$$C_a\,H_b\left(a + \frac{b}{4}\right) O_2 \rightarrow a\,CO_2 + \frac{b}{2}\,H_2O \qquad ...(17.73)$$

The energy rate balance for a control volume enclosing a fuel cell (for example) by neglecting changes in KE and PE reduces to

$$\frac{\dot{W}_{cv}}{\dot{n}_F} = \frac{\dot{Q}_{cv}}{\dot{n}_F} + H_R - H_P$$

or

$$\frac{\dot{W}_{cv}}{\dot{n}_F} = \frac{\dot{Q}_{cv}}{\dot{n}_F} + \bar{h}_F + \left(a \frac{b}{4}\right)\bar{h}_{O_2} - a\,\bar{h}_{CO_2} - \frac{b}{2}\,\bar{h}_{H_2O} \qquad ...(17.74)$$

The entropy rate balance for the control volume is given by

$$O = \frac{\dot{Q}_{cv}/\dot{n}_F}{T_0} + \bar{s}_F + \left(a + \frac{b}{4}\right)\bar{s}_{O_2} + a\,\bar{s}_{CO_2} - \frac{b}{2}\,\bar{h}_{H_2O} + \frac{\dot{S}_{gen}}{\dot{n}_F} \qquad ...(17.75)$$

From Eqs. (17.74) and (17.75), we have

$$\frac{\dot{W}_{cv}}{\dot{n}_F} = \left[\bar{h}_F + \left(a + \frac{b}{4}\right)\bar{h}_{O_2} - a\,\bar{h}_{CO_2} - \frac{b}{2}\,\bar{h}_{H_2O}\right]$$

$$- T_0\left[\bar{s}_F + \left(a + \frac{b}{4}\right)\bar{s}_{O_2} - a\,\bar{s}_{CO_2} - \frac{b}{2}\,\bar{s}_{H_2O}\right] - T_0\,\frac{\dot{S}_{gen}}{\dot{n}_F} \qquad ...(17.76)$$

The maximum theoretical value for the work developed is obtained when no irreversibilities are present. Thus, setting $T_0\ \dot{S}_{gen} = 0$, the expression for chemical exergy results is

$$\bar{e}_s^{ch} = \left[\bar{h}_F + \left(a + \frac{b}{4}\right)\bar{h}_{O_2} - a\,\bar{h}_{CO_2} - \frac{b}{2}\bar{h}_{H_2O}\right]$$

$$- T_0\left[\bar{s}_F + \left(a + \frac{b}{4}\right)\bar{s}_{O_2} - a\,\bar{s}_{CO_2} - \frac{b}{2}\bar{s}_{H_2O}\right] \qquad ...(17.77)$$

where $\bar{e}_s^{ch}$ is the chemical exergy of fuel on molar basis and the superscript *ch* is used to distinguish it from thermomechanical exergy discussed is Chaper 11.

Since $\bar{s}_i\,(T_0, y_i^e, p_0) = \bar{s}_i\,(T_0, p_0) - \bar{R}\ln y_i^e$

where y_i^e is the mole fraction of component *i* in the environment, so

$$\bar{e}_x^{ch} = \left[\bar{h}_F + \left(a + \frac{b}{4}\right)\bar{h}_{O_2} - a\,\bar{h}_{CO_2} - \frac{h}{2}\bar{h}_{H_2O}\,(9)\right]_{T_0,\,p_0}$$

$$- T_0\left[\bar{s}_F + \left(a + \frac{b}{4}\right)\bar{s}_{O_2} - a\,\bar{s}_{CO_2}\ a\,\bar{s}_{CO_2} - \frac{b}{2}\bar{s}_{H_2O\,(9)}\right]_{T_0,\,p_0} + \bar{R}\,T_0\ln\left[\frac{(y_{O_2}^e)^{a+b/4}}{(y_{CO_2}^e)^a\,(y_{H_2O}^e)^{b/2}}\right] \qquad ...(17.78)$$

In terms of Gibb's function, the above equation is written as

$$\bar{e}_s^{ch} = \left[\bar{g}_F + \left(a + \frac{b}{4}\right)\bar{g}_{O_2} - a\,\bar{g}_{CO_2} - \frac{b}{2}\bar{g}_{H_2O}\,(g)\right] + \bar{R}\,T_0\ln\left[\frac{(y_{O_2}^e)^{a+b/4}}{(y_{CO_2}^e)^a\,(y_{H_2O}^e)^{b/2}}\right] \qquad ...(17.79)$$

Example 17.29 Calculate the chemical exergy of a liquid octance at 25°C, 1 atm having the following composition on a molar basis : N_2, 75.76% O_2, 20.35%, H_2O, 3.12%, CO_2, 0.03% and other 0.83%.

Solution The chemical equation for complete combustion of liquid octane is given by

$$C_8H_{18}\,(l) + 12.5\,O_2 \rightarrow 8\,CO_2 + 9\,H_2O$$

The chemical exergy is given by

$$\bar{e}_x^{ch} = \left[\bar{g}_{C_8H_{18}(l)} + 12.5\,\bar{g}_{O_2} - 8\,\bar{g}_{CO_2} - 9\,\bar{g}_{H_2O\,(g)}\right]_{T_0,\,p_0} + \bar{R}\,T_0\ln\left[\frac{(y_{0.2}^e)^{12.5}}{(y_{CO_2}^e)^8\,(y_{H_2O\,(g)}^e)^9}\right]$$

Since T_0 and p_0 are the reference states, so from Table A8, we have

$\bar{e}_x^{ch}$ = [6610 + 12.5(0) – 8 (– 394,380) – 9 (– 228,590)]

$$+8.314\,(298)\ln\left[\frac{(0.2035)^{12.5}}{(0.0003)^8\,(0.0312)^9}\right] = 5{,}407{,}843 \text{ kJ/kg mole}$$ **Ans.**

Example 17.30 A gasoline engine develops 200 kW by using liquid octane (C_8H_{18}). The fuel enters the engine at 25°C and 200% theoretical air enters at 45°C. The products of the combustion leave the engine at 750 K. The heat transfer from the engine is 260 kW. Assuming complete combustion, calculate the fuel consumption per hour.

Solution Refer to Fig. 17.13.

The stoichiometric chemical equation for $C_8\,H_{18}\,(l)$ is

$C_8H_{18}\,(l) + 12.5\,O_2 + 3.76 \times 12.5\,N_2 \rightarrow 8\,CO_2 + 9\,H_2O + 3.76 \times 12.5\,N_2$

The enthalpy of reactants is expressed as

$$h_r = \sum_r n_i\,[\bar{h}_f^0 + (\bar{h}_f^0 - \bar{h}_{298})]$$

$$= 1(\bar{h}_f)_{C_8H_{18}(l)} + 25\,[\bar{h}_f^0 + (\bar{h}_T - \bar{h}_{298})]_{O_2} + 94\,[\bar{h}_f^0 + (\bar{h}_T - \bar{h}_{298}]_{N_2}$$

Substituting values from Tables A8 and A10,

$$h_r = (-\,249{,}952) + 25 \times 560 + 94 \times 540 = -\,185{,}192 \text{ kJ/kg mole}$$

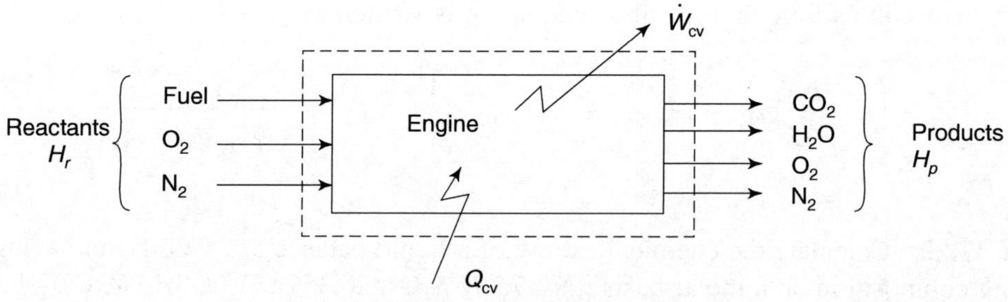

Fig. 17.13 Example 17.30

The enthalpy of products are expressed as

$$h_p = \sum_p n_e\,(\bar{h}_f^0 + \overline{\Delta h})_e$$

= 8 [– 393,522 + 20288] + 9 [– 241,827 + 16087] + 12.5 [14171] + 94 [13491]

= – 8 × 373,234 + 9 (– 225,740) + 177,137.5 + 1268154

= – 2985872 – 2031660 + 177137.5 + 1268154

$= -\ 3572240.5$ kJ/kg mole of fuel

Let n kg mole of fuel per second be consumed. Thus, from first law,

$$\dot{n}\,(h_r - h_p) = \dot{W}_{cv} - \dot{Q}_{cv}$$

or $$\dot{n}\,[(-185{,}192) - (-3572240.5)] = 200 - (-250) = 450 \text{ kW}$$

or $$\dot{n} = \frac{450 \times 3600}{3387048.5} = 0.4782 \text{ kg mole/h.}$$

The fuel consumption rate = $M\dot{n}$ = 114 × 0.4782 = 54.51 = 54.51 kg/h **Ans.**

Example 17.31 A combustion chamber burns propane gas at 25°C and I bar with 300% theoretical air (a) assume that reaction occurs reversibly at 25°C, that the oxygen and nitrogen are separated before the reaction takes place each at 1 bar and 25°C, that the constituents in the products are separated, and that at each at 25°C and 1 bar. Calculate the reversible work for the process, (b) if the reaction occurs adiabatically, and each constituent in the products is at 1 bar pressure and at the adiabatic flame temperature, determine (i) the increase in entropy during combustion, (ii) the irreversibility of the process, and (iii) the availability of the products of combustion.

Solution Refer to Fig. 17.14.

The chemical equation with stoichiometric combustion of propane gas is expressed as

$$C_3\,H_8(g) + 5\,O_2 + 5 \times 3.76\,N_2 \rightarrow 3\,(O_2 + 4H_2O(g) + 3O_2 + 5 \times 3.76\,N_2$$

With 300% excess air, the chemical equation is

$$C_3\,H_8\,(g) + 5 \times (3)\,O_2 + 5(3)\,(3.76)\,N_2 \rightarrow 3CO_2 + 4H_2O(g) + 10O_2 + 56.4\,N_2$$

(a) The reversible work done is expressed by

$$\dot{W}_{rev} = \sum_r n_i\,\bar{g}_i - \sum_p n_e\,\bar{g}_e$$

Since $(\bar{g}_f^0)_{O_2}$ and $(\bar{g}_f^0)_{O_2} = 0$

$$\dot{W}_{rev} = (\bar{g}_f^0)_{C_3H_8(g)} - 3\,(\bar{g}_f^0)_{CO_2} - 4\,(\bar{g}_f^0)_{H_2O(g)}$$

Substituting the values from Table A8, we have

$$\dot{W}_{rev} = -\ 23{,}316 - 3\,(-\ 394{,}374) - 4\,(-\ 228{,}583)$$

$$= 2{,}074{,}138 \text{ kJ/kg mole} = \frac{2{,}074{,}138}{44.097} = 47{,}0358 \text{ kJ/kg}$$ **Ans.**

(b) For adiabatic reaction, $h_r = h_p$

$$(\bar{h}_f^0)_{C_3H_8(g)} = 3\,(\bar{h}_f^0 + \overline{\Delta h})_{CO_2} + 4\,(\bar{h}_f^0 + \overline{\Delta h})_{H_2O(g)} + 10\,(\overline{\Delta h})_{O_2} + 56.4\,(\overline{\Delta h})_{N_2}$$

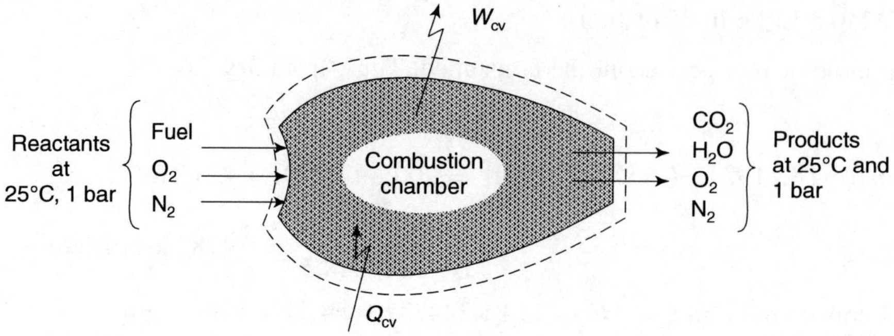

Fig. 17.14 Example 17.31

From Table A8, we have

$$-103{,}847 = 3(-393{,}522 + \overline{\Delta h})_{CO_2} + 4(-241{,}827 + \overline{\Delta h})_{H_2O(g)} + 10(\overline{\Delta h})_{O_2} + 56.4(\overline{\Delta h})_{N_2}$$

Using Table A10 in the Appendix A10, after putting values of $\Delta\bar{h}$, the adiabatic flame temperature is obtained by trial and error and the value comes about 975 K.

The entropy of the reactants is given by

$$s_r = \sum_r (n_i\,\bar{s}_i^{\,0})_{298} = (\bar{s}_{C_3H_8(g)} + 15\,\bar{s}_{0.2}^{\,0} + 56.4\,\bar{s}_{H_2}^{\,0})_{298}$$

Substituting the values from table, we have

$$s_r = 270.019 + 15\,(205.142) + 56.4\,(191.64) = 14154 \text{ kJ/kg mole K}$$

The entropy of the products is expressed as

$$s_p = \sum_p (n_e\,\bar{s}_e^{\,0})_{975} = (3\,\bar{s}_{CO_2}^{\,0} + 4\,\bar{s}_{H_2O(g)}^{\,0} + 56.4\,\bar{s}_{N_2}^{\,0})_{975}$$

$$= 3\,(268.193) + 4\,(231.846) + 10\,(242.845) + 56.4\,(227.47)$$

$$= 16989.72 \text{ kJ/kg mole K}$$

(i) The increase in entropy $= s_p - s_r = 16989.72 - 14154 = 2835.72$ kJ/kg mole K **Ans.**

(ii) The irreversibility in the process is expressed as

$$I = T_0\left[\sum_p n_e\,\bar{s}_e - \sum_r n_i\,\bar{s}_i\right] = T_0\,(s_p - s_r)$$

$$= 288 \times 2835.72 = 816687.36 \text{ kJ/kg mole K} = 18520.24 \text{ kJ/kg mole}$$ **Ans.**

(iii) The availability (exergy) of the products of combustion

$$e_x = w_{rev} - I = 47{,}035.8 - 18520.24 = 28{,}515.56 \text{ kJ/kg}$$ **Ans.**

17.24 SUMMARY OF FORMULAE

1. $\frac{A}{F} = \frac{m_a}{m_f}$

2. $\left(\frac{A}{F}\right)_{\text{theoretical}} = \left(\frac{A}{F}\right)_{st} = 15.686$

3. $\left(\frac{A}{F}\right)_{\min} = \left(\frac{m_a}{m_f}\right)_{\min} = 11.6C + 34.8\left(H_2 + \frac{O_2}{2}\right) + 4.355$; kg/kg of fuel

4. $\left(\frac{A}{F}\right)_{\min} = \frac{100}{21}\left[0.5(H_2 + CO) + 2CH_4\right]$; m^3 of air/m^3 of fuel

5. $\bar{h}(T,p) = \bar{h}_f^0 + \left[\bar{h}(T,p) - \bar{h}(T_{\text{ref}}, p_{\text{ref}})\right] = \bar{h}^0 + \Delta\bar{h}$

6. $\bar{h}_f^0 = \frac{\dot{Q}_{cv}}{\dot{n}_p}$ where $\dot{n}$ = molar flow rate

7. $\sum_r n_i \bar{h}_i + \dot{Q}_{cv} = \sum_p \dot{n}_e \bar{h}_e + \dot{W}_{cv}$ (First law, rate equation, molar)

8. $\sum_r m_i \bar{h}_i + \dot{Q}_{cv} = \sum_r \dot{n}_e \bar{h}_e + \dot{W}_{cv}$ (First law, rate equation, mass basis)

9. $C_a H_b + \left(a + \frac{b}{4}\right)(O_2 + 3.76N_2) \rightarrow a\, CO_2 + \frac{b}{2}H_2O + \left(a + \frac{b}{4}\right)3.76N_2$

10. $\frac{\dot{Q}_{cv}}{\dot{n}_F} - \frac{\dot{W}_{cv}}{n_f} = \bar{h}_p - \bar{h}_r = \sum_p n_e\left(\bar{h}_f^0 + \Delta\bar{h}\right)_e - \sum_r n_i\left(\bar{h}_f^0 + \Delta\bar{h}\right)_i$

11. $Q - W = \sum_p n\left(\bar{h} - \bar{R}T_p\right) - \sum_r n\left(\bar{h} - \bar{R}T_r\right)$ for closed system

12. $Q - W = \sum_p n\left(h_f^0 + \Delta\bar{h}\right) - \sum_r n\left(\bar{h}_f^0 + \Delta\bar{h}\right) - \bar{R}T_p\sum_p n + \bar{R}T_r\sum_r n$ for closed system

13. $\bar{h}_{rp} = \Delta\bar{h}_R = H_p - H_r = \sum_p n_c \bar{h}_e - \sum_r n_i \bar{h}$

14. $\frac{\bar{Q}_{cv}}{\dot{n}_f} - \frac{\dot{W}_{cv}}{\dot{n}_F} = \bar{h}_{rp}^0 + \sum_p n_e\left(\Delta\bar{h}\right)_e - \sum_r n_i\left(\Delta\bar{h}\right)_i$

15. $(C.V.)_p = -\Delta\bar{h}_R = -\bar{h}_{rp} = (C.V.)_v - 8.314\Delta n T_0$

16. $(C.V)_v = -\Delta\bar{u}_R = -\bar{u}_{rp}$

17. H.C.V. $= \left[33829\,C + 144026\left(H_2 \frac{O_2}{8}\right) + 929475\right]$; kJ/kg

18. L.C.V. = H.C.V. – $m_{H_2O} \cdot h_{fg}$

19. $(C.V.)_f \cdot m_f + (C.V.)_1\, m_1 = (M + m)\, c_{pw}\, (t + \alpha)$: Bomb calorimeter

20. $\text{H.C.V.} = \dfrac{c_{p\omega}\, m\,(T_{w2} - T_{w1})\, T_g\, p_0}{\left(\dfrac{h_w}{13.6} + h_b\right) v_g\, T_0}$ {Gas calorimeter

21. $\text{L.C.V.} = \text{H.C.V.} - \dfrac{m \times 2466.1 \times T_g\, p_0}{V_g\, p_g\, T_0}$

22. $\dot{W}_{rev} = \sum\limits_r n_i\, [\bar{h}_f^0 + \Delta\bar{h} - T_0\, \bar{s}]_i - \sum\limits_p n_e\, [\bar{h}_f^0 + \Delta\bar{h} - T_0\, \bar{s}]_e$

23. $\dot{W}_{rev} = \sum\limits_r n_i\, \bar{g}_i - \sum\limits_r n_e\, \bar{g}_e$

24. $I = \sum\limits_p \dot{n}_e\, T_0\, \bar{s}_e - \sum\limits_r n_i\, T_0\, \bar{s}_i - \dot{Q}_{cv}$

25. $p_{max} = \dfrac{n_f\, R_0\, T_f}{V}$

26. $\Delta S = \sum\limits_i (n_i\, s_i)_{prod} - \sum\limits_i (n_i\, s_i)_{reac}$

27. $\bar{g}\,(T, p) = \bar{g}_f^0 + [\bar{g}\,(T, p) - \bar{g}\,(T_{ref.}, p_{ref})] = \bar{g}_f^0 + \Delta\bar{g}$

28. $\bar{e}_s^{ch} = \left[\bar{g}_F + \left(a + \dfrac{b}{4}\right)\bar{g}_{O_2} - a\,\bar{g}_{CO_2} - \dfrac{b}{2}\,\bar{g}_{H_2O(g)}\right] + \bar{R}\, T_0 \ln\left[\dfrac{(y_{O_2}^e)^{a+b/4}}{(y_{CO_2}^e)^a\,(y_{O_2}^e)^{b/2}}\right]$

PROBLEMS

Things to Think About

17.1 Are mass or moles are conserved in a balanced chemical reaction?

17.2. Is it possible to think about incomplete combustion of a hydrocarbon is burnt with theoretical amount of air?

17.3. Why is the combustion of hydrocarbons considered a contributor to global warming?

17.4. Why is combustion of fuel inherently an irreversible process?

17.5. Will the adiabatic flame temperature be higher in complete combustion with theoretical amount of air or oxygen all initially at 298K and 1 atm?

17.6. Why are some enthalpy of formation values positive and others negative?

17.7. Some high efficiency residential natural gas furnaces are equipped with drain tubes. Why is it so?

17.8. Why is the exergy loss maximum in combustor of gas turbine plant?

Short and Long Theoretical Questions

17.9. What are the different types of fuel used in practice?

17.10. What are the advantages of liquid and gaseous fuels over solid fuels?

17.11. Define mole or kilogram mole and fraction.

17.12. What do you mean by stoichiometric equation?

17.13. Discuss methods for finding the volumetric analysis of flue gases.

17.14. Define enthalpy and internal energy of reaction.

17.15. What do you mean by calorific value of fuel : lower and higher calorific value of fuel?

17.16. Describe the construction and working principle of the bomb calorimeter.

17.17. Describe the construction and working principle of the Junker's gas calorimeter.

17.18. Describe the construction and working principle of other types of gas calorimeter.

Numerical Questions

17.19. A gas producer supplies gas to an internal combustion engine and its volumetric composition in percentage is CO_2 = 6.7%, O_2 = 1.3%, CO = 22.5%, H_2 = 2.2%, CH_4 = 0.4%, N_2 = 66.9%. The dry exhaust gas analysis gave (by volume) 0.8% O_2 and no CO. Calculate the ratio of air to gas by volume in the mixture supplied to the engine and estimate the volumetric percentage of CO_2 and N_2 in the dry exhaust gases. Assume the composition of air by volume N_2 = 79% and O_2 = 21%. [*Ans.* A/G. 0.62, CO_2 = 20.2%, N_2 = 79.%]

17.20. The gravimetric analysis of a sample of coal is given as 80% C, 12% H_2, and 8% ash, calculate the stoichiometric A/F ratio and the analysis of the products by volume.

[*Ans.* A/F = 13.3, CO_2 = 13.6%, H_2 = 12.2% and N_2 = 74.2%]

17.21. The analysis of the dry exhaust from an internal combustion engine yielded 12% CO_2; 2% CO; 4% CH_4; 1% H_2; 4.5% O_2; and the remainder N_2. Determine the proportions by mass of carbon to hydrogen in the fuel, assuming it to be a pure hydrocarbon. [*Ans.* C/H_2 = 7.31]

17.22. A sample of bituminous coal resulted in the following ultimate analysis by mass. C 81.9%; H_2 4.9%, O_2 6%; N_2 2.3%; ash 4.9%. Determine:

(a) The stoichiometric AF ratio

(b) The analysis by volume of the wet and dry products of combustion when the air supplied is 25% in excess of that required for complete combustion.

[*Ans.* 10.8 : 1; CO_2 = 14.16%; H_2O = 5.07%; CO = 4.04%,
N_2 = 76.7%; CO_2 = 14.9%; N_2 = 89.9%]

17.23. From a test carried on a four-cylinder petrol engine running on a rich mixture, the following volumetric analysis obtained is: CO_2 = 9.8%, O_2 = 0.2%, CO = 7.6% and N_2 = 82%. Calculate the air-fuel ratio by weight and the probable composition of the fuel on the assumption that this consists solely of hydrogen and carbon. [*Ans.* C = 86.9%; H_2 = 13.1%; A/F = 12.5%]

17.24. In a petrol engine, the dry exhaust analysis gave 1.5% CO, and the O_2 was negligible. If the fuel had an ultimate analysis by weight of C = 0.84 and H_2 = 0.16, calculate the air-fuel ratio supplied. Assume air is 23.1% O_2 by weight and all the carbon in the fuel may be assumed to have been burnt either completely or incompletely. [*Ans.* A/F = 14.75 kg/kg of fuel]

17.25. The gravimetric analysis of coal is C = 0.7, H_2 = 0.03, O_2 = 0.02, N_2 = 0.02, H_2O = 0.1 and ash = 0.13. The volumetric analysis of the dry products CO_2 = 14.6%, CO = 0.2%, O_2 = 5%, N_2 = 79.7%. Calculate (a) the mass of air per kg of coal, (b) mass of dry products per kg coal. If the specific humidity of the air supplied is 0.01 kg/kg of dry air, calculate (c) the amount of water vapour in the products per kg of coal. [*Ans.* (a) 11.54 kg, (b) 12.1 kg, (c) 0.485 kg]

17.26. Propane is burnt with air and the Orsate analysis of the product gives CO_2 = 8.2%, CO = 1.3%, O_2 = 7% and N_2 = 83.5%. Calculate the percentage of theoretical air used for the combustion process. [*Ans.* 140%]

17.27. The dry exhaust gas analysis from an engine using hydrocarbon fuel is CO_2 = 8.7%, O_2 = 03%, CO = 8.9%, CH_4 = 0.3%, H_2 = 3.7%, N_2 = 78.1%. Calculate the air-fuel ratio (a) given that the fuel is C_8 H_{17}, and (b) fuel composition not known. Solve the question by using (i) carbon balance, (ii) hydrogen balance, and (iii) carbon-hydrogen balance.

[*Ans.* (a) 11.31. 11.31, 11.31, (b) 11.31]

17.28. Methane is burnt with 50% excess air. The products of combustion then exist at 1 bar and 250°C. Calculate the dew point temperature of the water vapour in the products of combustion.

[*Ans.* (a) 51°C]

17.29. Liquid benzene in burnt with 200% theoretical air requirements. Calculate (a) dew point of the products, and (b) the air-fuel assuming that the combustion is complete and that the system pressure is 1.03 bar. [*Ans.* (a) 30°C, (b) 26.6%]

17.30. The combustion of methane gas with air leads to the following Orsate analysis during a test : 9.7% CO_2, 0.5% CO and 3% O_2. Calculate the moles of air used per mole of fuel in the test and the per cent theoretical air used. [*Ans.* 10.8, 9.52]

17.31. A fuel gas with the volumetric analysis 60% CO, 20% N_2 is burnt in a steady flow process with dry air to give combustion gases at 260°C. The Orsate analysis of these combustion gases gives 20% CO_2, 5% CO, 2.8% O_2. Calculate the kg of air supplied for each kg of fuel gas burnt.

[*Ans.* 2.47 kg/kg of fuel]

17.32. A gas has the following volumetric analysis by per cent CH_4 = 80.62%, C_2H_6 = 5.41%, C_3H_8 = 1.87%, C_4H_{10} = 1.60% N_2 10.5%. The Orsate analysis of the products of combustion shows 7.8% CO_2, 7% O_2, 0.2% CO. Determine (a) the actual fuel/air ratio, and (b) the dew point temperature of the products. Assume that the combustion air is dry and the pressure of the mixture is 1.03 bar. [*Ans.* (a) 24.6 kg/kg, (b) 51°C]

17.33. A coal had the following analysis by weight, C = 84%, H_2 = 5%, O_2 = 4% and ash = 7%. When burnt in a boiler, the resulting dry flue gas had the following percentage analysis by volume: CO_2 = 13.76%, CO = 1.53%, O_2 = 4.25, N_2 = 80.46. It was found that per kg of coal fired,

0.012 kg of carbon had flown through the fire bars. Calculate (a) mass of air used per kg of coal, (b) the total loss by incomplete combustion. Assume air contains 23.1 of O_2 by weight.

[*Ans.* 13.7 kg of air/kg of coal, loss = 2350 kJ/kg]

17.34. The dry products of combustion of an internal combustion engine has the following volumetric analysis. CO_2 = 11 %, CO = 3%, CH_4 = 4%, H_2 = 1% and O_2 = 5%. The remainder is nitrogen. Assuming the fuel to be pure hydrogen and nothing was left unburnt, calculate the proportion by mass of carbon and hydrogen in the fuel. Assume that air contains 23.3% of O_2 by mass.

[*Ans.* 7.505 :1]

17.35. The gas supplied to an engine has the following percentage volumetric composition : CH_4 = 20%, H_2 = 40%, CO = 15%, CO_2 = 7%, O_2 = 2% and N_2 = 14%. The percentage volumetric analysis of the dry exhaust gases is CO_2 = 8%, O_2 = 9% and N_2 = 83%. Using the mole method, find the volumetric ratio of air to gas in the reactants. Also calculate the mean volumetric heat and the index of adiabatic expansion.

Use the following mean value of c_p, for CO_2 = 50.4, O_2 = 32.86%, N_2 = 30.35 kJ/mole K.

[*Ans.* c_{pav} = 32.19 kJ/mole K, γ = 1.35, *AF* = 5.765 : 1 by volume]

17.36. The percentage composition of a sample of coal was found to be : C = 89.3%, H_2 = 5%, O_2 = 4.2%, N_2 = 1.5% and remainder ash. Calculate the stoichiometric air-fuel ratio by mass. If 30% excess air is supplied, calculate the percentage composition of dry gases by volume. [*Ans.* 11.9, CO_2 = 14.1%, O_2 = 4.9% and N_2 = 81%]

17.37. The higher calorific value of a sample of fuel oil containing 86% of carbon and 12% of hydrogen was found to be 44380 kJ/kg. The room temperature at the time of the experiment was 15.5°C. Using the data, calculate the lower calorific value of fuel. [*Ans.* 41756 kJ/kg]

17.38. A sample of fuel has the following percentage analysis by weight. C = 82%, H_2 = 5%, O_2 = 4% and the remainder being an inert material. Calculate the HCV and LCV of this fuel. The following heating values may be used : C = 33913 kJ/kg and free hydrogen = 144026 kJ/kg. Assume that each kg of steam produced during combustion absorbs 2466 1 kJ/kg.

[*Ans.* 38476.69 kJ/kg, 3625.76 kJ/kg]

17.39. The analysis by weight of the dry coal fed to a boiler is C = 82%. H_2 = 5% and O_2 = 5% and the non-combustible matter 6%. Contained in the coal is 2% of hydroscopic moisture. The ash falling through the fire bars is 6% of the weight of dry coal fired and contains 20% of unburnt carbon, but no hydrogen.

An analysis of the dry flue gas shows 11% by volume of CO_2 and 1% CO. Determine the mass of each of the gases, including the moisture entering the flues per kg of dry coal fired given that air contains 23.1% by weight of oxygen.

[*Ans.* kg CO_2 = 2.783, kg CO = 0.161, kg H_2O = 0.47,
Free O_2 in flue = 1.42 kg, kg N_2 = 12.9]

17.40. A fuel oil containing 84% of carbon and 1.2% of hydrogen by weight is burnt by means of high pressure air burner. The products of combustion are found to contain 12% of carbon dioxide. Determine (a) the proportion of air in excess of that required for theoretical combustion, and (b) approximately the percentage of the potential heat of combustion which would be lost if the waste gases were at a temperature 205°C.

[*Ans.* (a) 27.3% excess, (b) %lost = 14]

17.41. A combustible mixture of air and octane vapour (C_2H_{18}), the air being 15% in excess of that required for chemically correct combustion, is compressed adiabatically through a volume ratio of 6 to I. The temperature at the start of the compression is 49°C. Given that the volumetric heats of octane vapour and air are 92 an 9.6 kJ/kg mole, calculate (a) the value of the adiabatic index for the mixture, and (b) the work required per kg during the compression only. Assume air contains 21% O_2 by volume. [*Ans.* (a) $\gamma = 1.348$, (b) $W = 4312.4$ kJ/kg]

17 42. A mixture of octane C_8H_{18} and ethyl alcohol C_2H_5 OH, in the proportion of 1 molecule of alcohol, is completely burnt in a supply of air 20% in excess of that required for complete combustion. Calculate the volumetric analysis of the dry exhaust products. Assume air contains 20.9% O_2 by volume. [*Ans.* 12% CO_2, 3.7% O_2, 84.3% N_2]

17.43. Calculate the enthalpy of reaction at 25°C when 1 mole of propane C_3H_8 is burnt according to the following equation

$$C_3\,H_8(g) \rightarrow CO_2(g) + 4\,H_2O\,(l)$$

Assume steady flow combustion process.

17.44. If the higher heating value at constant pressure of C_6H_6 at 25°C is 3303720 kJ/kg-mole, calculate:

(a) Lower heating value at constant pressure, (b) Higher heating value at constant volume, (c) Lower heating value at constant volume, and (d) Higher heating value at constant pressure.

17.45. The following data show the enthalpy of formation at 25°C :

$C_2H_2(g)$ = 226899.43 kJ/kg mole $\quad$ $C_6H_6(g)$ = 82982 kJ/k8 mole

$C_6H_{14}(g)$ = – 467304.42 kJ/kg mole $\quad$ $NH_3(g)$ = – 46222.27 kJ/kg mole

$CO_2(n)$ = – 393776.9 kJ/kg mole $\quad$ $H_2O(g)$ = – 241988.58 kJ/kg mole

Calculate the constant pressure lower heating values of C_6H_{14}, and the enthalpy of reaction of

$$4NH_3 + 3O_2 \rightarrow 6H_2O + 2N_2$$

17.46. Ethane gas (C_2H_6) at 25°C and air at 23.2°C enters a combustion chamber in a steady flow at 1 atm. The products of combustion leave at 1100 K and 1 atm. If the per cent excess air is 25%, calculate the heat loss during the process in kJ/mole of fuel assuming complete combustion.

17.47. The combustor of a gas turbine is supplied with dry air at 227°C and liquid *n*-heptane at 25°C. The products of combustion leave at 930°C. Determine the required air-fuel ratio. Assume that the flow is steady, the heat loss to the surroundings is negligible and the combustion is complete.

17.48. Carbon monoxide is burnt with air at an initial state of 25°C. The final temperature of the products is found to be 1100 K and during this process 17793.9 kJ of heat are removed per mole of CO. Calculate the per cent excess air used if combustion is complete. [*Ans.* 260%]

17.49. The combustor of a gas turbine power plant is supplied with air at 227°C and liquid octane at 25°C. The products of combustion leave at 1110°C. Determine the air-fuel ratio if the flow is steady, the combustion is complete and the heat loss is negligible.

17.50. Propane gas (C_3H_8) is burnt at a constant pressure of 1 atm with 20 per cent excess air starting at 25°C. Assuming no dissociation, calculate the maximum adiabatic temperature. [*Ans*. 2120 K]

17.51. Calculate the maximum flame temperature when methane gas is burnt with the stoichiometric amount of air at a pressure of 1 atm both reactants entering at 25°C. Assume that combustion with respect to carbon is only 80% complete : that is, 20% of the carbon in methane is burnt only to CO. There is no dissociation at high temperature. [*Ans*. 2200 K]

17.52. The following data refer to an experimental determination of the calorific value of a coal by using bomb calorimeter.

Mass of coal sample = 1.01 gm, Mass of water = 2500 gm, Heat capacity of apparatus = 744 gm. Temperature rise of water = 2.59°C, Temperature correction for cooling = 0.01°C.

[*Ans*. 35084.38 kJ/kg]

17.53. The following measurements were taken on a gas calorimeter:

Volume of gas used = 10 litres. Temperature of gas used = 11°C, Gas manometer reading = 10 cm of H_2O, Mass of cooling water used during test = 0.005 kg, Cooling water temperature inlet = 14°C, Cooling water temperature outlet = 25°C. Calculate the calorific value of fuel used.

[*Ans*. 17186.8 kJ/m^3]

17.54. Calculate the HCV and LCV of a fuel having the following composition:

C = 80%, H_2 = 10% O_2, = 3%, S = 2%, N_2 = 20% and rest incombustible matter.

[*Ans*. HCV = 41132.84 kJ/kg, LCV = 38924.9 kJ/kg]

17.55. The following data refer for the determination of calorific value of gaseous fuel: barometric pressure = 74 cm Hg, volume of fuel used = 15 liters, pressure of gas = 4 cm of water above atmospheric volume of cooling water collected = 12.5 liters, rise in temperature of cooling water = 5°C, volume of condensate collected = 0.01 liters. Latent heat of steam may be taken as 2466.1 kJ/kg, Sp. gravity of Hg = 13. Calculate HCV and LCV of the fuel.

[*Ans*. 18086.97 kJ/m^3, 16395.5 kJ/m^3]

17.56. A fuel mixture with the molar analysis 70% CH_4, 20% CO, 5% O_2 and 5% N_2 burns completely with 20% excess air. Calculate (a) the balance reaction equation, (b) the air-fuel ratio, both on a molar and mass basis. [*Ans*. (b) 8.25; 12.05]

17.57. Combustion of dodecane ($C_{15}H_{26}$) takes place completely with 150% of theoretical air. Calculate (a) the air-fuel ratio on a molar and mass basis, (b) the dew point temperature of the combustion products in °C when cooled at 1 atm. [*Ans*. (a) 132.09, 22.46, (b) 44.8°C]

17.58. A combustion chamber burns completely a gaseous fuel mixture with a molar analysis of 72% CH_4, 9% H_2, 14% N_2, 2% O_2 and 3% CO_2 with moist air to form gaseous products at 1 atm consisting of CO_2, H_2O and N_2 only. If the dew point temperature of the products is 60°C, calculate the amount of water vapour present in the combustion air, in kg mole per kg mole of mixture. [*Ans*. 0.038]

17.59. Methane (CH_4) enters a furnace and burns completely with 150% of theoretical air entering at 25°C, 94.5 kPa. 75% relative humidity calculate (a) the balanced reaction equation, and (b) the dew point temperature of the combustion products in °C at entry pressure. [*Ans*. (b) 53°C]

17.60. A combustion chamber receives hydrogen (H_2) at the mass flow rate of 2 kg/h and burns with air entering at 30°C, 1 atm with a volumetric flow rate of 120 m^3/h. Calculate the percentage of theoretical air used. [*Ans.* 204.6%]

17.61. An insulated reactor receives carbon monoxide (CO) at 25°C, 1 atm and operates at steady state. The reaction takes place completely with theoretical amount of air entering in a separate stream at 25°C, 1 arm. For the reactor, calculate the rate of entropy production, in kJ/kg mole. K of CO entering. Neglect the changes in KE and PE. [*Ans.* 165.3 kJ/kg mole K]

17.62. A reactor receives a gaseous mixture of ethane (C_2H_5) and theoretical amount of air at 25°C, 1 atm. The reactor operates at steady state and burns completely. Combustion products exit at 627°C, 1 atm. Heat transfer from the reactor takes place at an average surface temperature of T_G. For T_b ranging from 25°C to 600°C. Calculate the rate of exergy destruction within the reactor, in kJ per kg mole of fuel. Takc T_0 = 25° and neglect changes in KE and PE.

17.63. A fuel cell receives stream of methane (CH_4) and oxygen (O_2), each at 259°C, 1 atm and operates at steady state. Streams of carbon dioxide and water exit separately at 25°C, 1 atm. If the fuel cell operates isothermally at 25°C, 1 atm, calculate the maximum theoretical work that it can develop in kJ/kg mole of methane. Neglect the changes in KE and PE.

[*Ans.* 8.18×10^5 kJ/kg mole]

17.64. Carbon monoxide at 25°C, 1 atm enters an insulated reactor operating at steady state and reacts completely with the theoretical amount of air entering in a separate stream at 25°C, 1 atm. Calculate in kJ/kg mole of CO, the following (a) the exergy entering with CO:

(a) the exergy entering with CO, (b) the exergy exiting with products, (c) the rate of exergy destruction. [*Ans.* (a) 275, 364 kJ/kg mole, (b) 226,105 kJ/kg mole (e) 49,259 kJ/kg mole, η_{ex} = 82.1%]

Software Based Questions

17.65. Develop a software to calculate the theoretical air-fuel ratio required for the combustion of any hydrocarbon fuel.

17.66. Develop a software to calculate enthalpy of reaction and adiabatic flame temperature.

17.67. Develop a software to calculate the exergy loss in a combustor. Validate the results with realistic data from a combustor of gas turbine.

17.68. A vessel initially containing 1 kg mole of $H_2O(g)$ and x kg mole of N_2 forms an equilibrium mixture at 1 atm consisting of $H_2O(g)$, H_2, O_2 and N_2 in which 0.5 kg mole of $H_2O(g)$ is present. Develop a software to plot x versus the temperature T for $3000 < T \leq 3000$ K.

17.69. Gaseous propane (C_3H_8) at 25°C, 1 atm enters a reactor operating at steady state and burns with 80% of theoretical air entering separately at 25°C, 1 atm. An equilibrium mixture of CO_2, CO, H_2O(g), H_2 and N_2 exists at 1250°C, 1 atm. Develop a software to calculate the heat transfer between the reactor and its surroundings in kJ per mole of propane entering. Neglect kinetic and potential energy effects.

17.70. Develop a software to calculate the adiabatic flame temperature due to combustion in petrol and diesel engines.

17.71. Deveiop a software for the design of combustor of a gas turbine.

Design and Open Ended Questions

17.72. Spark ignition engines emit pollutants in the form of NO_x, CO, and unburnt hydrocarbon (HC). The pollutants actually present depend on engine design and operating conditions and typically differ significantly from values. Calculate on the basis of chemical equilibrium. Prepare a report discussing both reasons for these discrepancies and possible mechanisms by which such pollutants are formed in actual engines.

17.73. The average production of pollutants from a spark ignition engines upstream of the catalyst, in g per mile of vehicle travel, are nitric oxides, 1.5; hydrocarbons, 2, and CO, 20. For a population of 10 millions of a city, estimate the annual amount, in kg, of each pollutant that would be discharged if automobiles have no emission control device by assuming one car for every four persons. Repeat the estimate by assuming current government emission standards. (EURO-II).

17.74. As per clean air act of any government, NO_x is treated as a significant air pollutant. NO_x is formed via three primary mechanisms : *thermal* NO_x *formation, prompt* NO_x *formation and fuel* NO_x *formation*. Prepare a report discussing these NO_x formation mechanism including the role of NO_x in the formation of ozone and some NO_x reduction strategies.

17.75. A means for converting industrial wastes to CO_2 and water vapour has been described by an U.S. Patent 5,298,233. As per this patent, hydrogen and carbon containing feed, such as organic or inorganic sludge, low grade fuel oil, or municipal garbage, is introduced into a molten bath consisting of two immiscible molten metal phases. The carbon and hydrogen of the feed are converted, respectively, to dissolved carbon and dissolved hydrogen. The dissolved carbon is oxidized in the first molten phase to carbon dioxide, which is released to atmosphere. The dissolved hydrogen migrates to the second molten metal phase, where it is oxidized to form water vapour, which is also released from bath. Prepare a report containing critical evaluation of this technology for waste disposal, its commercial viability. Also discuss alternative waste management practices such as pyrolysis and incineration.

17.76. Prepare a comprehensive report on all modern types of gas analysers with their working merits and demerits.

Multiple-Choice Answer Type Questions

17.77. The method used to determine the composition of coal such as moisture content, volatile matter, fixed carbon, etc. is called

(a) Proximate analysis
(b) Ultimate analysis
(c) Stoichiometric analysis
(d) None of these

17.78. The composition of dry air on mass basis is

(a) 0.232 kg O_2 and 0.768 kg N_2
(b) 0.21 kg O_2 and 0.79 kg N_2
(c) 0.31 kg O_2 and 0.69 kg N_2
(d) None of these

17.79. Enthalpy of combustion of any fuel containing hydrogen in it and H_2O present in liquid form in products is called

(a) Higher heating value (b) Lower heating value

(c) Enthalpy of hydration (d) None of these

17.80. Maximum temperature in the combustion of fuel occurs when the process is

(a) Polytropic (b) Adiabatic (c) Constant volume (d) None of these

17.81. In the combustion of fuel, adiabatic flame temperature is not achieved due to

(a) Heat loss (b) Incomplete combustion

(c) Dissociation (d) All of these

Answers to Multiple-Choice Questions

17.77 (a) 17.78 (a) 17.79 (b) 17.80 (b) 17.81 (d)

CHAPTER

18

Chemical and Phase Equilibrium

OVERVIEW

The objective of this chapter is to consider the concept of equilibrium in greater depth than has been done so far. In this chapter, we will develop the fundamental concept used to study the chemical and phase equilibrium.

18.1 EQUILIBRIUM CRITERIA

A system is said to be in thermodynamic equilibrium if, when it is isolated from the surroundings, there would be no macroscopically observable changes. The thermodynamic equilibrium considers all types of equilibrium such as *mechanical, thermal, electrical and chemical equilibrium.* Mechanical equilibrium requires no unbalance force (i.e., pressure) either inside of the system or between the system and surroundings. If the pressure is not uniform throughout the system, internal changes in the state of system will take place till the mechanical equilibrium is restored. Thermal equilibrium requires uniformity of temperature throughout the system. If this condition is not met spontaneously, heat transfer from one location to another could occur when the system is isolated. Electrical equilibrium requires uniformity of electrical potential throughout the system. These conditions ensure that the system is in mechanical, thermal and electrical equilibrium, but there is still the possibility that complete equilibrium does not exist. A process might occur involving a chemical reaction, a transfer of mass between phases, or both. *For chemical equilibrium, it is essential that their chemical reaction or transfer of matter from one part of the system to the other should not occur.*

The object of this section is to introduce criteria that can be applied to decide whether a system in a particular state is in equilibrium. These criteria will be based on using the *conservation of energy principle and the second law of thermodynamics.*

Let us consider states of a simple compressible system of fixed mass at which temperature and pressure are uniform with position throughout. If there is absence of overall system motion and influence of gravity, the energy balance in differential form is given by

$$\delta Q - \delta W = dU$$

As pressure is uniform with position throughout the system, we have

$$\delta W = pdV$$

Thus,

$$\delta Q = dU + pdV \quad ...(18.1)$$

Since temperature is uniform with position throughout the system, the entropy balance in differential form is

$$dS = \frac{\delta Q}{T} + \delta S_{gen} \quad ...(18.2)$$

From Eq. (18.1) and Eq. (18.2), we have

$$TdS - dU - pdV = T\,\delta S_{gen} \quad ...(18.3)$$

As we know that entropy, is produced in all actual processes and conserved only in the absence of irreversibilities. Thus, Eq. (18.3) provides a constraint on the direction of processes. This suggests that the only processes allowed for those for which

$$TdS - dU - pdV \geq 0 \quad ...(18.4)$$

Equation (18.4) forms one of the basis to study equilibrium under various conditions. For example, a process taking place in an insulated, constant volume vessel, where $dU = 0$ and $dV = 0$ yields

$$[dS]_{U,\,V} \geq 0 \quad ...(18.5)$$

From Eq. (18.5), we conclude that changes of state of a closed system at constant internal energy and volume can occur only in the direction of increasing entropy. Further, this equation also reveals that the entropy approaches a maximum as a state of equilibrium is approached (Fig. 18.1).

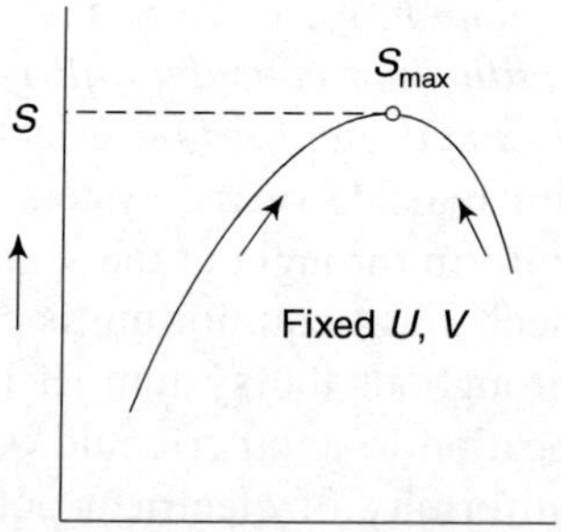

Fig. 18.1 Condition of maximum entropy

One of the most important cases for the study of chemical and phase equilibrium is the one in which temperature and pressure are fixed. For this, it is convenient to employ the Gibbs function in extensive form, i.e.,

$$G = H - TS = U + pV - TS$$

Differentiating of the above gives

$$dG = dU + pdV + Vdp - TdS - SdT$$

or

$$dG - Vdp + SdT = -(TdS - dU - pdV) \quad ...(18.6)$$

The right side of Eq. (18.6) is the same as that in Eq. (18.4) except the minus sign, so Eq. (18.4) takes the form of

$$dG = Vdp + SdT \leq 0 \quad ...(18.7)$$

Here, the inequality reverses direction because of the minus sign in Eq. (18.6).

Equation (18.7) states that any process taking place at a specified temperature and pressure ($dT = 0$ and $dp = 0$) must be such that

$$[dG]_{T,\,p} \le 0 \qquad ...(18.8)$$

The inequality of Eq. (18.8) reveals that the Gibbs function of a system at a fixed T and p decreases during an irreversible process. Each step of such a process results in a decrease in the Gibbs function of the system and brings the system close to the equilibrium. *It is worth to note that the equilibrium state is the one having the minimum value of Gibbs function.* Hence, for equilibrium,

$$[dG]_{T,\,p} = 0 \qquad ...(18.9)$$

Basically, Eq. (18.9) is the criterion for equilibrium (Fig. 18.2). This equation provides a relationship among properties of a system when it is at an equilibrium state. When an equilibrium state is obtained, a system exists at a particular T and p and no spontaneous change can take place.

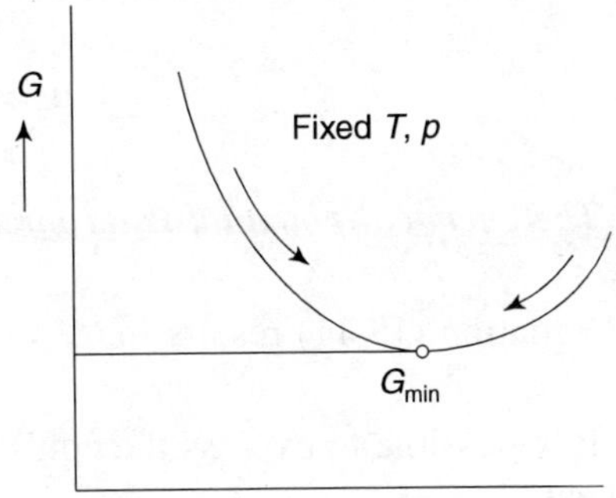

Fig. 18.2 Equilibrium criteria

18.2 CHEMICAL POTENTIAL AND EQUILIBRIUM

In order to find the chemical equilibrium criterion, we will take the help of general equilibrium criterion, $(dG)_{T,\,p} = 0$. We know that any extensive property of a single phase, single component system is a function of two independent intensive properties and the size of the system. Let temperature and pressure be the independent properties and number of moles (n) be the measure of size, the Gibbs function may be expressed in the form of $G = a\ (T, p, n)$. For a single phase, multicomponent system, G may then be considered a function of temperature, pressure and the number of moles of each component, i.e., $G = G\ (T, p, n_1, n_2\ ....\ n_j)$.

Let each mole number be multiplied by α, the size of the system will change by the same factor and so does the value of every extensive property. Thus, for the Gibbs function, it may be written as

$$\alpha\, G(T, p, n_1, n_2\ ....\ n_j) = G\ (T, p, \alpha n_1, \alpha n_2 + ...\ \alpha n_j)$$

Differentiation of above with respect to α while holding temperature, pressure and the mole numbers fixed gives

$$G = \frac{\partial G}{\partial(\alpha\, n_1)} n_1 + \frac{\partial G}{\partial(\alpha\, n_2)} n_2 + ... + \frac{\partial G}{\partial(\alpha\, n_j)} n_j$$

The above equation is true for all values of α. If $\alpha = 1$, the above equation becomes

$$G = \sum_{i=1}^{j} n_i \left(\frac{\partial G}{\partial n_i}\right)_{T,\,p,\,n_i} \qquad ...(18.10)$$

where the subscript n_i denotes that all n'_s are except n_i are hold fixed during differentiation.

In Eq. (18.10), the term $\left(\frac{\partial G}{\partial n_i}\right)_{T, p, n_i}$ is known as the chemical potential of component i introduced in chapter 10 and is denoted as

$$\mu_i = \left(\frac{\partial G}{\partial n_i}\right)_{T, p, n_i} \quad ...(18.11)$$

The chemical potential is an intensive property.

Equation (18.11) results in, $G = \sum_{j=1}^{j} n_i \mu_i$ ...(18.12)

It is possible to express the equilibrium criterion given by Eq. (18.9) in terms of chemical potential.

If $G = G(T, p, n_i, n_2, ... n_i)$ while holding temperature and pressure fixed,

$$[dG]_{T, p} = \sum_{i=1}^{j} \left(\frac{\partial G}{\partial n_i}\right)_{T, p, n_i} dn_i \quad ...(18.13)$$

From Eqs. (18.11) and (18.13), we get, $[dG]_{T, p} = \sum_{i=1}^{j} \mu_i \, dn_i$ ...(18.14)

Thus, the equilibrium criterion may be written as, $\sum_{i=1}^{j} \mu_i \, n_i = 0$ ...(18.15)

Equation (18.15) also forms the basis of equilibrium criterion in terms of chemical potential.

18.3 EVALUATING CHEMICAL POTENTIAL

Chemical potentials for two cases of interest will be discussed here. They are single-phase of a pure substance and an ideal gas mixture.

18.3.1 Single Phase of a Pure Substance

For single phase of pure substance, we have

$$G = n\mu \quad \text{or} \quad \mu = \frac{G}{n} = \overline{g} \quad ...(18.16)$$

Equation (18.16) indicates that the chemical potenital is just the Gibbs function per mole.

18.3.2 Ideal Gas Mixture

The enthalpy and entropy of an ideal gas mixture are expressed by

$$H = \sum_{i=1}^{j} m_i \, \bar{h}_i \,(T) \quad \text{and} \quad S = \sum_{i=1}^{j} n_i \, \bar{s}_i \,(T, p_i)$$

where $p_i = y_i \, p$ is the partial pressure of component i.

Using the above relations, the Gibbs function takes the form

$$G = H - TS = \sum_{i=1}^{j} n_i \, \bar{h}_i \,(T) - T \sum_{i=1}^{j} n_i \, \bar{s}_i \,(T, p_i)$$

$$= \sum_{i=1}^{j} n_i \, [\bar{h}_i \,(T) - T \, \bar{s}_i \,(T, p_i)] \qquad \text{...(18.17)}$$

The molar Gibbs function of component i is given by

$$\bar{g}_i \,(T, p_i) = \bar{h}_i \,(T) - T \, \bar{s}_i \,(T, p_i) \qquad \text{...(18.18)}$$

From Eqs. (18.17) and (18.18), we get $G = \sum_{i=1}^{j} n_i \, \bar{g}_i \,(T, p_i)$ ...(18.19)

From Eqs. (18.19) and (18.20), we get $\mu_i = \bar{g}_i \,(T, p_i)$ for ideal gas mixture ...(18.20)

From Eq. (18.20), we conclude that the chemical potential of component i is an ideal gas mixture and is equal to its Gibbs function per mole of i–evaluated at the mixture temperature and the partial pressure of i in the mixture.

The chemical potenital of component i in an ideal gas mixture may be expressed in another form.

From Eq. (18.18), we have, $\mu_i = \bar{h}_i \,(T) - T \, \bar{s}_i \,(T, p_i)$

$$= \bar{h}_i \,(T) - T \left[\bar{s}_i \,(T) - \bar{R} \ln \frac{y_i \, p}{p_{\text{ref}}} \right]$$

$$= \bar{h} \,(T) - T \, \bar{s}_i \,(T) + \bar{R} \ln \frac{y_p \, p}{p_{\text{ref}}} \qquad \text{...(18.21)}$$

where p_{ref} is 1 atm and y_i is the mole fraction of component i.

The above equation may be written in more compact form as

$$\mu_i = \bar{g}_i + \bar{R} \, T \ln \frac{y_i \, p}{p_{\text{ref}}} \qquad \text{...(18.22)}$$

where $\bar{g}_i$ is the Gibbs function of compressant i evaluated at temperature (T) and pressure of 1 atm.

18.4 TYPES OF EQUILIBRIUM

We have seen that the thermodynamic potentials which control equilibrium in a system depend on the particular constraints. Several types of equilibrium are discussed in this very section as a typical criterion for equilibrium. The Gibbs free energy may increase or decrease or remain constant when a finite change of state occurs in a system at constant temperature and pressure. According to the changes in the Gibbs function, equilibrium is classified as (a) stable, (b) neutral, (c) unstable, and (d) metastable.

(a) *Stable Equilibrium* *A system is said to be in a state of stable equilibrium if for any finite variation of the system of constant temperature and pressure, the appropriate thermodynamic criterion of equilibrium increases.* In other words, at constant temperature and pressure, the stable equilibrium state of a system corresponds to the minimum due to the Gibbs function. Thus,

$$\Delta G_{T,\,p} > 0 \qquad \text{...(18.23)}$$

Analogus criteria of state equilibrium are given by

$$\Delta F_{T,\,v} > 0 \qquad \text{...(18.24)}$$

$$\Delta U_{s,\,v} > 0 \qquad \text{...(18.25)}$$

$$\Delta H_{s,\,p} > 0 \qquad \text{(18.26)}$$

(b) *Neutral Equilibrium* A system is said to be in a state of neutral equilibrium when the appropriate thermodynamic criterion of equilibrium remains at a constant value for all possible variations of finite magnitudes. The criterion of neutral equilibrium for a system at constant temperature and pressure is

$$\Delta G_{p,\,T} = 0 \qquad \text{...(18.27)}$$

The above equation suggests that if the change in Gibbs free energy is zero, the system is in neutral equilibrium. The analogous criteria for neutral equilibrium of other systems are

$$\Delta F_{T,\,v} = 0 \qquad \text{...(18.28)}$$

$$\Delta U_{s,\,v} = 0 \qquad \text{...(18.29)}$$

$$\Delta H_{s,\,p} = 0 \qquad \text{...(18.30)}$$

(c) *Unstable Equilibrium* A system is said to be in a state of unstable equilibrium when the appropriate thermodynamic criterion of equilibrium neither increases nor is zero but reduces. At constant temperature and pressure of the system, the Gibbs free energy for at least one variation reduces, i.e.,

$$\Delta G_{T,\,p} < 0 \qquad \text{...(18.31)}$$

Analogous criteria for neutral equilibrium of other systems are

$$\Delta F_{T,\,v} < 0 \qquad \text{...(18.32)}$$

$$\Delta U_{s,\,v} < 0 \qquad \text{,,,(18.33)}$$

$$\Delta H_{s,\,p} < 0 \qquad \text{...(18.34)}$$

(d) *Metastable Equilibrium* A system is said to be in a state of metastable equilibrium, if it is stable but not to large disturbances. A suitable example of metastable equilibrium is a mixture of oxygen and hydrogen. Since a spark can start a chemical reaction. such a mixture is not in its most stable thermodynamic equilibrium even though in the absence of a spark, it appears to be stable.

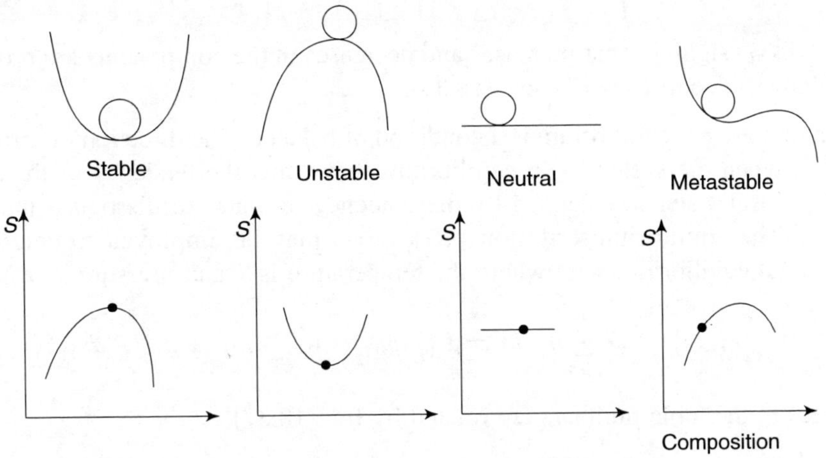

Fig. 18.3 Types of equilibrium

The different types of equilibrium are shown in Fig. 18.3. The entropy has been used as a criterion.

18.5 CHEMICAL EQUILIBRIUM

The equilibrium criterion $[dG]_{T,\,p} = 0$ will be used to study the equilibrium of reacting mixtures. Our objective here is to establish the composition present at equilibrium for a specified temperature and pressure with the help of equation of reaction equilibrium. In determining the equilibrium composition, *equilibrium constant* plays vital role.

18.5.1 Equation of Reaction Equilibrium

Simple Case Consider the reaction of hydrogen and oxygen gases as given below

$$1H_2 + \frac{1}{2}\,O_2 \rightleftarrows 1H_2O \qquad \text{...(18.35)}$$

At equilibrium, the system will consist in general of components, namely, H_2, O_2 and H_2O, for not all of the hydrogen and oxygen initially present need be reacted. Changes in the amounts of these components during each differential step of reaction is given by

$$dn_{H_2O} = -\,dn_{H_2O} \quad \text{and} \quad dn_{O_2} = -\frac{1}{2}\,dn_{H_2O} \qquad ...(18.36)$$

where dn is the differential change in respective component. The minus sign indicates that the amount of hydrogen and oxygen present decrease when the reaction proceeds towards the right. Equation (18.36) may also be expressed as

$$\frac{-\,dn_{H_2}}{1} = \frac{-\,dn_{O_2}}{1/2} = \frac{dn_{H_2O}}{1} \qquad ...(18.37)$$

Equation (18.37) signifies that increases and decreases in the components are proportional to the stoichiometric coefficients of Eq. (18.35).

It is worth to note that equilibrium is a condition of balance. The direction of arrow shown in Eq. (18.35), when the system is in equilibrium shows that the tendency of the hydrogen and oxygen to form water is balanced by the tendency of water to dissociate into oxygen and hydrogen. The equilibrium criterion $[dG]_{T,p} = 0$ may be employed to determine the composition at an equilibrium state where the temperature is T and pressure is p. We have

$$[dG]_{T,p} = \sum_{i=1}^{j} \mu_i\, dn_i = \mu_{H_2}\, dn_{H_2} + \mu_{O_2}\, dn_{O_2} + \mu_{H_2O}\, dn_{H_2O} \qquad ...(18.38)$$

The changes in the mole numbers are related by Eq. (18.37).

Thus, the above equation becomes

$$[dG]_{T,p} = \left(-1\,\mu_{H_2} - \frac{1}{2}\mu_{O_2} + \mu_{H_2O}\right) dn_{H_2O}$$

At equilibrium, $[dG]_{T,p} = 0$, so in the above equation, the term in parenthesis must be zero, Thus,

$$-1\,\mu_{H_2} - \frac{1}{2}\mu_{O_2} + 1\,\mu_{H_2O} = 0$$

or

$$1\,\mu_{H_2} + \frac{1}{2}\mu_{O_2} = 1\,\mu_{H_2O} \qquad ...(18.39)$$

Equation (18.39) is known as the *equation of reaction equilibrium* for the case under study. The chemical potentials are functions of temperature, pressure and composition. The composition will be determined by solving the equation of reaction equilibrium.

General Case Let us consider a closed system containing five components, A, B, C, D and E at a given pressure and temperature, subject to a single chemical reaction of the form given below

$$v_A\, A + v_B\, B \rightleftarrows v_c\, C + v_D\, D \qquad ...(18.40)$$

where v_s are stoichiometric coefficients. Component does not appear in reaction as it is assumed inert. Even though E is not involved in reaction but we will see that it affects coefficients. At equilibrium, the tendency of A and B to form C and D is balanced by the tendency of C and D to form A and B.

There is a definite relationship between stoichiometric coefficients and number of moles as

$$\frac{-dn_A}{\nu_A} = \frac{-dn_B}{\nu_B} = \frac{dn_C}{\nu_C} = \frac{dn_D}{\nu_D} \quad ...(18.41)$$

Here, the minus signs indicate that A and B would be consumed when C and D are produced. E being inert, so $dn_E = 0$.

In Eq. (18.41), a proportionality factor ($d\in$) is introduced, so we get

$$\frac{-dn_A}{\nu_A} = \frac{-dn_B}{\nu_B} = \frac{dn_C}{\nu_C} = \frac{dn_D}{\nu_D} = d\in \quad ...(18.42)$$

or

$$dn_A = -\nu_A\, d\in,\ dn_B = -\nu_B\, d\in,\ dn_C = \nu_C\, d\in \quad \text{and } dn_D = \nu_D\, d\in \quad ...(18.43)$$

The parameter is sometimes called **extent of reaction.**

For the system under consideration, Eq. (18.41) takes the form of

$$[dG]_{T,p} = \mu_A\, dn_A + \mu_B + dn_B + \mu_C\, dn_C + \mu_D\, dn_D + \mu_E\, dn_E \quad ...(18.44)$$

Since $dn_E = 0$, so the above equation becomes

$$[dG]_{T,p} = (-\nu_A\, \mu_A - \nu_B\, \mu_B + \nu_C\, \mu_C + \nu_D\, \mu_D)$$

At equilibrium, $[dG]_{T,p} = 0$, so

$$-\nu_D\, \mu_A - \nu_B\, \mu_B + \nu_C\, \mu_C + \nu_D\, \mu_D = 0$$

or

$$\nu_A\, \mu_A + \nu_B\, \mu_B = \nu_C\, \mu_C + \nu_D\, \mu_D \quad ...(18.45)$$

Equation (18.45) is known as the equation of **reaction equilibrium.**

18.5.2 Determination of Equilibrium Composition

Consider the equation of reaction equilibrium for ideal gas mixture given by Eq. (18.45). Substituting the value of chemical potential from Eq. (18.22), we have

$$\nu_A\left(\bar{g}^\circ{}_A + \bar{R}\,T \ln \frac{y_A \cdot p}{p_{\text{ref}}}\right) + \nu_B\left(\bar{g}^{\circ L}_B + \bar{R}\,T \ln \frac{y_B \cdot p}{p_{\text{ref}}}\right)$$

$$= \nu_C\left(\bar{g}^\circ{}_C + \bar{R}\,T \ln \frac{y_C \cdot p}{p_{\text{ref}}}\right) + \nu_D\left(\bar{g}^\circ{}_D + \bar{R}\,T \ln \frac{y_D \cdot p}{p_{\text{ref}}}\right) \quad ...(18.46)$$

where $\bar{g}^\circ{}_i$ is the Gibbs function of component (i) evaluated at a temperature (T) and pressure $p_{\text{ref}} = 1$ atm.

Collecting like terms of Eq. (18.42) and after rearranging, we have

$$(\nu_C \bar{g}^\circ{}_C + \nu_D \bar{g}^\circ{}_D - \nu_A \bar{g}^\circ{}_A - \nu_B \bar{g}^\circ{}_B)$$

$$= \bar{R}T\left(\nu_C \frac{y_C \cdot p}{p_{\text{ref}}} + \nu_D \ln \frac{y_D \cdot p}{p_{\text{ref}}} - \nu_A \frac{y_A \cdot p}{p_{\text{ref}}} - \nu_B \ln \frac{y_B \cdot p}{p_{\text{ref}}}\right) \quad ...(18.47)$$

The left hand side of the above Eq. (18.47) represents ΔG°, so

$$\Delta G^\circ = \nu_C \bar{g}^\circ{}_C + \nu_D \bar{g}^\circ{}_D - \nu_A \bar{g}^\circ{}_A - \nu_B \bar{g}^\circ{}_B \quad ...(18.48)$$

where $\quad \bar{g}_i = \bar{h}_i - T\bar{s}^\circ{}_i$ so

$$\Delta G^\circ = (\nu_C \bar{h}_C + \nu_D \bar{h}_D - \nu_A \bar{h}_A - \nu_B \bar{h}_B) - T(\nu_C \bar{s}^\circ{}_C + \nu_D \bar{s}^\circ{}_D - \nu_A \bar{s}^\circ{}_A - \nu_B \bar{s}^\circ{}_B) \quad ...(18.49)$$

From Eqs. (18.48) and (18.49), and combining the terms involving logarithms into a single expression, we have

$$\frac{\Delta \ \circ}{RT} = \ln\left[\frac{y^{\nu}_{C^C} \cdot y^{\nu}_{D^D}}{y^{\nu}_{A^A}\, y^{\nu}_{B^B}}\left(\frac{p}{p_{\text{ref}}}\right)^{\nu_C + \nu_D - \nu_A - \nu_B}\right] \quad ...(18.50)$$

Equation (18.50) may be expressed in more concise form in

$$-\frac{\Delta G^\circ}{RT} = \ln K\,(T) \quad ...(18.51)$$

where $\quad K$ = equilibrium constant and is given by

$$K\,(T) = \frac{y^{\nu}_{C^C} \cdot y^{\nu}_{D^D}}{y^{\nu}_{A^A}\, y^{\nu}_{B^B}}\left(\frac{p}{p_{\text{ref}}}\right)^{\nu_C + \nu_D - \nu_A - \nu_B} \quad ...(18.52)$$

For a given value of ν_A, ν_B, ν_C and ν_D and temperature T, the left side of Eq. (18.51) can be evaluated either using Eq. (18.49) together with appropriate data.

It is worth to note that the terms in numerator and denominator of Eq. (18.52) correspond, respectively, to the products and reactants of the reaction given by Eq. (18.40) as it proceeds from left to right as written. For the case of inverse reaction, i.e.,

$\nu_C\, C + \nu_D\, D \rightleftarrows \nu_A\, .\, A + \nu_B\, B$, the equilibrium constant is given by

$$K^* = \frac{y^{\nu}_{A^A}\, y^{\nu}_{B^B}}{y^{\nu}_{C^C} \cdot y^{\nu}_{D^D}}\left(\frac{p}{p_{\text{ref}}}\right)^{\nu_A + \nu_B - \nu_C - \nu_D} \quad ...(18.53)$$

Comparison of Eqs. (18.52) and (18.53) reveals that K^* is just reciprocal of K.

Thus, $\quad \log_{10} K^* = -\log_{10} K \quad ...(18.54)$

Gas Tables can be used to calulate K and K^*.

18.5.3 For Reacting Ideal Gas Mixture

It has been found convenient sometimes to express Eq. (18.52) in terms of number of moles as $y_i = n_i/n$. Hence, Eq. (18.52) may be written as

$$K = \frac{n_C^{v_C}\, n_D^{v_D}}{n_A^{v_A}\, n_B^{v_B}} \left(\frac{p/p_{\text{ref}}}{n}\right)^{v_C + v_D - v_A - v_B} \quad ...(18.55)$$

18.5.4 Equilibrium Constant for Ideal Solution

The procedure adopted to calculate K for ideal gas mixture may be applied to calculate K for general case of reacting mixture. Equation (18.45) may be written for general case.

$$v_A (\bar{g}^\circ{}_A + \bar{R}\,T \ln a_A) + v_B (\bar{g}^\circ{}_B + \bar{R}\,T \ln a_B)$$

$$= v_C (\bar{g}^\circ{}_C + \bar{R}\,T \ln a_C) + v_D (\bar{g}^\circ{}_D + \bar{R}\,T \ln a_D)$$

where a_i is the activity of the component

Using previous procedure, we have $-\dfrac{\Delta g^\circ}{\bar{R}\,T} = \ln\left(\dfrac{a_C^{v_C} \cdot a_D^{v_D}}{a_A^{v_A}\, a_B^{v_B}}\right)$ ...(18.56)

and thus, $K = \dfrac{a_C^{v_C} \cdot a_D^{v_D}}{a_A^{v_A}\, a_B^{v_B}}$ ...(18.57)

Thus, K may also be expressed in terms of fugacity(f)

With the help of equilibrium constant, we can find *equilibrium flame temperature* in a chemical reaction.

18.6 van't HOFF EQUATIONS

The equilibrium constant (K) exhibited in Eq. (18.51) shows its dependency on temperature. An alternative may express this dependency offered by the van't Hoff equation which will be derived in this section. From Eqs. (18.40) and (18.51), we have

$$\bar{R}\,T \ln K = -[v_C\, \bar{h}_C + v_D\, \bar{h}_D - v_A\, \bar{h}_A - v_B\, \bar{h}_B)$$
$$- T\,(v_C\, \bar{s}^\circ{}_C + v_D\, \bar{s}^\circ{}_D - v_A\, \bar{s}^\circ{}_A - v_B\, \bar{s}^\circ{}_B] \quad ...(18.58)$$

Differentiation of above equation results in

$$\bar{R}\,T \frac{d \ln K}{dT} + \bar{R} \ln K = \left[v_C \left(\frac{d\bar{h}_C}{dT} - T\frac{d\,\bar{s}^\circ{}_C}{dT}\right) + v_D\left(\frac{d\bar{h}_0}{dT} - T\frac{d\,\bar{s}^\circ{}_D}{dT}\right)\right]$$

$$- v_A \left(\frac{d\bar{h}_A}{dT} - T\frac{d\,\bar{s}^\circ{}_A}{dT}\right) - v_B\left(\frac{d\bar{h}_B}{dT} - T\frac{d\,\bar{s}^\circ{}_B}{dT}\right)\Bigg] + (v_C\, \bar{s}^\circ{}_C + v_D\, \bar{s}^\circ{}_D - v_A\, \bar{s}^\circ{}_A - v_B\, \bar{s}^\circ{}_B)$$

We have, $\frac{d\bar{s}^\circ}{dT} = \frac{\bar{c}_P}{T}$ and $\frac{d\bar{h}}{dT} = \bar{c}_p$, so the terms in the big brackets get cancelled and finally, we get

$$\bar{R}T\frac{d\ln K}{dT} + \bar{R}\ln K = [\nu_C\,\bar{s}^\circ{}_D + \nu_D\,\bar{s}^\circ{}_D - \nu_A\,\bar{s}^\circ{}_A - \nu_B\,\bar{s}^\circ{}_B] \qquad ...(18.59)$$

From Eqs. (18.58) and (18.59), we get

$$\frac{d\ln K}{dT} = \frac{\nu_C\,\bar{h}_C + \nu_D\,\bar{h}_D - \nu_C\,\bar{h}_B}{\bar{R}T^2} \qquad ...(18.60)$$

or

$$\frac{d\ln K}{dT} = \frac{\Delta h_R}{\bar{R}T^2} \qquad ...(18.61)$$

Equation (18.61) is known as ***van't Hoff equation***. This equation shows that when Δh_R (enthalpy of reaction) is negative (i.e., exothermic reaction), K decreases with temperature, whereas for Δh_R positive (i.e., endothermic reaction), K increases with temperature.

The integration of above equation yields

$$\ln\frac{K_2}{K_1} = -\frac{\Delta h_R}{\bar{R}}\left(\frac{1}{T_2} - \frac{1}{T_1}\right) \qquad ...(18.62)$$

where K_1 and K_2 are equilibrium constants at temperature T_1 and T_2, respectively.

18.7 PHASE EQUILIBRIUM

The equilibrium criterion $(dG]_{T,p} = 0$ will be used to study the equilibrium of multicomponent, multiphase, non-reacting systems.

18.7.1 Equilibrium between Two Phases of a Pure Substance

Let us consider a system consisting of two phases of a pure substance at equilibrium. Being the system at equilibrium, each phase is at the same temperature and pressure. The Gibbs function for the said two-phase system is expressed as

$$G = n'\,\bar{g}'(T, p) + n''\,\bar{g}''(T, p) \qquad ...(18.63)$$

where the primes (′) and (″) represent phases 1 and 2, respectively.

Differentiating the above equation at fixed temperature and pressure, we get

$$[dG]_{T,p} = \bar{g}'\,dn' + \bar{g}''\,dn'' \qquad ...(18.64)$$

As the total amount of the pure substance remains constant, an increase in the amount present in one of the phases must be compensated by an equivalent decrease in the amount present in the other phase, i.e., $dn'' = -\,dn'$ and Eq. (18.64) becomes

$$[dG]_{T,p} = -(\bar{g}' - \bar{g}'')\,dn' \qquad ...(18.65)$$

At equilibrium, $[dG]_{T,p}$ = so the above equation becomes

$$\bar{g}' = \bar{g}'' \qquad \text{...(18.66)}$$

Equation (18.66) signifies that at the equilibrium, the molar Gibbs functions of the phases are equal.

18.7.2 Derivation of the Clapeyron Equation from Phase Equilibrium

It is possible to derive the Clapeyron equation obtained in Chapter 10 using other method. For two phases in equilibrium, variations in pressure are uniquely related to variation in temperature. Thus,

$p = p_{sat}$ (T). The differential of Eq. (18.66) with respect to temperature yields

$$\left(\frac{\partial \bar{g}'}{\partial T}\right)_p + \left(\frac{\partial \bar{g}'}{\partial p}\right)_T \frac{dp_{sat}}{dT} = \left(\frac{\partial \bar{g}''}{\partial T}\right)_p + \left(\frac{\partial \bar{g}''}{\partial p}\right)_T \frac{dp_{sat}}{dT}$$

From general thermodynamic relations (Chapter 10), the above equation becomes

$$-\bar{s}' + \bar{v}' \frac{dp_{sat}}{dT} = \bar{s}'' + \bar{v}'' \frac{dp_{sat}}{dT}$$

After rearrangement, we get

$$\frac{dp_{sat}}{dT} = \frac{\bar{s}'' - \bar{s}'}{\bar{v}'' - \bar{v}'} \qquad \text{...(18.67)}$$

But $\bar{g} = \bar{h} - T\,\bar{s}$, so Eq. (18.66) becomes

$$\bar{h}' - T\,\bar{s}' = \bar{h}'' - T\,\bar{s}''$$

or

$$\bar{s}'' - \bar{s}' = \frac{\bar{h}'' - \bar{h}''}{T} \qquad \text{...(18.68)}$$

From Eqs. (18.62) and (18.63), we have

$$\frac{dp_{sat}}{dT} = \frac{1}{T}\left(\frac{\bar{h}'' - \bar{h}'}{\bar{v}'' - \bar{v}'}\right) \qquad \text{...(18.69)}$$

Equation (18.69) is the Clapeyron equation.

If the specific volume of solid, $\bar{v}'$ is negligible as compared to specific volume of vapour, $\bar{v}'$, and the vapour can be treated as an ideal gas, $\bar{v}'' = \bar{R}p/p_{sat}$, so

Equation (18.69) becomes

$$\frac{dp_{sat}}{dT} = \frac{\bar{h}'' - \bar{h}'}{\bar{R}\,T^2}\,p_{sat}$$

or

$$\frac{d \ln p_{sat}}{dT} = \frac{\bar{h}'' - \bar{h}'}{\bar{R}\,T^2} \qquad \text{...(18.70)}$$

Equation (18.65) is known as **Clausius-Clapeyron equation**. There is a similarity between Clausius–Clapeyron equation and **van't Hoff equation (18.61).**

18.7.3 Equilibrium of Multicomponent, Multiphase Systems

Consider a system consisting of two components A and B in two phases 1 and 2 that are at the same temperature and pressure. Applying Eq. (18.14) to each of the phases of considered system we have

$$\left.\begin{aligned} [dG']_{T,p} &= \mu'_A\, dn'_A + \mu'_B\, dn'_B \\ [dG'']_{T,p} &= \mu''_A\, dn''_A + \mu''_B\, dn''_B \end{aligned}\right| \qquad ...(18.71)$$

where primes signify phases as previous case.

During transfer of matter between two phases, the increase in amount present in one of the phases must be compensated by an equivalent decrease in the amount present in other phase provided there is a absence of chemical reaction. Thus,

$$dn''_A = -\,dn'_A \quad \text{and} \quad dn''_B = -\,dn'_B \qquad ...(18.72)$$

The change in the Gibbs function for the system, we have

$$[dG]_{T,p} = [dG']_{T,p} + [dG'']_{T,p} = (\mu'_A - \mu''_A)\, dn'_A + (\mu'_B - \mu''_B)\, dn'_B \qquad ...(18.73)$$

At equilibrium state, $[dG]_{T,p} = 0$, so

$$(\mu'_A = \mu''_A) \quad \text{and} \quad (\mu'_B = \mu''_B) \qquad ...(18.74)$$

Equation (18.74) signifies that at equilibrium, the chemical potential of each component is the same in each phase.

Let us consider a special case when chemical potential of component B is the same in both phases, i.e., $\mu'_B = \mu''_B$, then

$$[dG]_{T,p} = (\mu'_A - \mu''_A)\, dn'_A \qquad ...(18.75)$$

Here two case arise

- When $\mu'_A > \mu''_A$, it follows that $dn'_A > 0$. This means substance A passes from phase 1 to phase 2.
- When $\mu''_A > \mu'_A$, it follows that $dn'_A < 0$ which means substance A passes from phase 2 to phase 1.
- At equilibrium, $\mu'_A = \mu''_A$, there is no net transfer of A between phases.

18.7.4 Gibbs Phase Rule

At equilibrium, the chemical potential of each component must be the same in all phases. For the case of N components that are present in P phases, there will H $(p - 1)$ equation involving chemical potentials. The number of independent intensive properties is called the ***degree of freedom*** (F) which is given by

$$F = [p\,(N - 1) + 2] - N\,(p - 1) = 2 + N - p \qquad ...(18.76)$$

Equation (18.76) is called **Gibbs phase rule**.

Example 18.1 Calculate the equilibrium constant, ($\log_{10} K$) for the reaction CO + (1/2) O_2 $\rightleftarrows$ CO_2 at (a) 298 K and 1800 K. Compare the result which is available in Gas Tables.

Solution Recalling Eq. (18.9), we have for this case

$$\Delta G^\circ = \left(\bar{h}_{CO_2} - \bar{h}_{CO} - \frac{1}{2}\bar{h}_{O_2}\right) - T\left(\bar{s}_{CO_2} - \bar{s}_{CO} - \frac{1}{2}\bar{s}_{O_2}\right)$$

But we have, $\bar{h}(T, p) = \bar{h}_f^\circ + \Delta\bar{h}$, so the above equation becomes

$$\Delta G^\circ = \left[(\bar{h}^\circ{}_f)_{CO_2} - (\bar{h}^\circ{}_f)_{CO} - \frac{1}{2}(\bar{h}^\circ{}_f)_{O_2}\right] + \left[(\Delta\bar{h})_{CO_2} + (\Delta\bar{h})_{CO} - \frac{1}{2}(\Delta\bar{h})_{O_2}\right]$$
$$- T\left[\bar{s}^\circ{}_{CO_2} - \bar{s}^\circ{}_{CO} - \frac{1}{2}\bar{s}^\circ{}_{O_2}\right]$$

where $\Delta\bar{h}$ is the change in specific enthalpy from T_{ref} = 298 K to the specified temperature T. Here, $(\bar{h}^\circ{}_f)_{O_2} = 0$ by definition.

(a) **At T = 298 K,** $\Delta\bar{h}$ is the above expression becomes zero, so

$$[\Delta G^\circ]_{T = 298\text{ K}} = (\bar{h}^\circ{}_f)_{CO_2} - (\bar{h}^\circ{}_f)_{CO} - \frac{1}{2}(\bar{h}^\circ{}_f)_{O_2} - T\left[\bar{s}^\circ{}_{CO_2} - \bar{s}^\circ{}_{CO} - \frac{1}{2}\bar{s}^\circ{}_{CO_2}\right]$$

Substituting the values from Gas Tables, we have

$$\Delta G^\circ = \left[(-393{,}520) - (-110{,}530) - \frac{1}{2}(0)\right] - 298\left[213.69 - 197.54 - \frac{1}{2} \times 205.03\right]$$
$$= -257{,}253 \text{ kJ/kg mole}$$

From Eq. (18.51), we have

$$\ln K = \frac{-\Delta G^\circ}{\bar{R}T} = \frac{-257{,}253\text{ (kJ/kg mole)}}{(8.314\text{ kJ/kg mole, K})\ (298\text{ K})} = 103.83$$

or $\log_{10}$ K = 45,1 **Ans.**

At 1800 K

$$\Delta G^\circ = \left[(-393{,}520) - (-110{,}530) - \frac{1}{2}(0)\right] + [(88{,}806 - 9364) - (58{,}191 - 8{,}669)]$$
$$- \frac{1}{2}(60{,}371 - 8682)] - 1800\left[302.8 - 255.194 - \frac{1}{2} \times 264.701\right]$$

or $\log_{10} K = 3.667$ **Ans.**

The results are in agreement with values given in Gas Tables separately.

Note Gas Tables are separately available.

Example 18.2 In a reactor, one kg mole of carbon monoxide (CO) reacts with the theoretical amount of air to form an equilibrium mixture of CO_2, O_2, N_2 at 2000 K and 1 atm. Calculate the equilibrium composition in terms of mole fractions.

Solution The chemical equation for a complete reaction of CO with the theoretical amount of air

$$CO + \frac{1}{2} O_2 + 1.88\ N_2 \rightarrow CO_2 + 1.88\ N_2$$

As per question, $CO + \frac{1}{2} O_2 + 1.88\ N_2 \rightarrow z\ CO + \frac{z}{2} O_2 + (1 - z) + 1.88\ N_2$

where z in the amount of CO, in kg mole, present in the equilibrium mixture.

The total number of moles (n) in the equilibrium mixture is given by

$$n = z + \frac{z}{2} + (1 - z) + 1.88 = \frac{5.76 + z}{2}$$

Thus, the composition of equilibrium mixture in terms of mole fraction is given by

$$y_{CO} = \frac{2z}{5.76 + z}, \quad y_{O_2} = \frac{z}{5.76 + z}, \quad y_{CO_2} = \frac{2(1 - z)}{5.76 + z}, \quad y_{N_2} = \frac{3.76}{5.76 + z},$$

At equilibrium, we have $CO_2 \rightleftarrows CO + \frac{1}{2} O_2$

Using Eq. (18.55) for this case

$$K = \frac{z\,(z/2)^{1/2}}{(1 - z)}\left[\frac{p/p_{ref}}{(2 + z)/2}\right]^{1+1/2-1} = \frac{z}{1 - z}\left(\frac{z}{1 - z}\right)^{1/2}\left(\frac{p}{p_{ref}}\right)^{1/2} \quad \text{...(i)}$$

At T = 2000 K, from Gas Tables, $\log_{10} K = -3.693$. Thus, $K = 0.0002027$

Substituting the value of K in Eq. (i), we have, $0.0002027 = \frac{z}{1 - z}\left(\frac{z}{2 + z}\right)^{1/2}$ ...(ii)

Solving the above equation by trial and error, $z = 0.0043$

Thus, the equilibrium composition in terms of mole fraction is given by

$$y_{CO} = \frac{2 \times 0.0043}{5.76 + 0.0043} = 1.491 \times 10^{-3}, \quad y_{O_2} = \frac{0.0043}{5.76 + 0.0043} = 5.2 \times 10^{-4}$$

$$y_{CO_2} = \frac{2(1 - 0.0043)}{5.76 + 0.0043} = 0.3454, \quad y_{N_2} = \frac{3.76}{5.76 + 0.0043} = 0.6522$$

Ans.

Example 18.3 CO_2 at 25°C, 1 atm enters a reactor operating at steady state and dissociates. The equilibrium mixture so obtained is CO_2, CO, and O_2 that comes out at 3250 K and 1 atm. Calculate the heat transfer to the reactor, in kJ per kg mole of CO_2 entering. There is no work done by control volume and the changes in potential and kinetic energy are negligible.

Solution Refer to Fig. 18.3.

By the application of the conservation of mass principle, the overall dissociation reaction is given by

$$CO_2 \rightarrow z\, CO_2 + (1 - z)\, CO + \left(1 - \frac{z}{2}\right) O_2$$

where z is the amount of CO_2, in kg mole, present in the mixture exiting the control volume, per kg mole of CO_2 entering. The total number of mole (n) in the mixture is expressed by

$$n = z + (1 - z) + \left(\frac{1-z}{2}\right) = \frac{3-z}{2}$$

We have already assumed that exiting mixture is an equilibrium mixture. Thus, for the mixture

$$CO_2 \rightleftarrows CO + \frac{1}{2}\, O_2$$

The equilibrium constant for the reaction (Eq. 18.55) takes the form

$$K = \frac{[(1 - z\,(1 - z)/2]^{1/2}}{2}\left[\frac{p/p_{ref}}{(3-z)/2}\right]^{1+1/2-1}$$

After rearranging and setting $p = 1$, we have

$$K = \left(\frac{1-z}{z}\right)\left(\frac{1-z}{3-z}\right)^{1/2} \qquad \text{...(i)}$$

From Gas Tables, At 3000 K, $\log_{10} K = -\,0.054$

so $$K = 1.132$$

Substituting value of K in Eq. (i), we have

$$1.132 = \left(\frac{1-z}{z}\right)\left(\frac{1-z}{3-z}\right)^{1/2}$$

Solving by trial and error, we get $z = 0.31$

The composition of the exiting equilibrium mixture, in kg mole per kg mole of CO_2 entering in

(0.31 CO_2, 0.069 CO and 0.345 O_2) **Ans.**

The energy rate balance when expressed per kg mole of CO_2 entering the control volume is

$$O = \frac{\dot{Q}_{cv}}{\dot{n}_{CO_2}} - \frac{\dot{W}_{cv}}{\dot{n}_{CO_2}} + \bar{h}_{CO_2,\,in} - (0.31\, CO_2 + 0.69\, \bar{h}_{CO} + 0.345\, O_2)$$

$$O = \frac{\dot{Q}_{cv}}{\dot{n}_{CO_2}} - \frac{\dot{W}_{cv}}{\dot{n}_{CO_2}} + \bar{h}_{CO_2,\,in} - (0.31\,\bar{h}_{CO_2} + 0.69\,\bar{h}_{CO} + 0.345\,\bar{h}_{O_2})$$

Since, $\dot{W}_{cv} = 0$, so

$$\frac{\dot{Q}_{cv}}{\dot{n}_{CO_2}} = 0.31\,(\bar{h}^\circ{}_f + \Delta\bar{h})_{CO_2} + 0.69\,(\bar{h}^\circ{}_f + \Delta\bar{h})_{CO} + 0.345\,(\bar{h}^\circ{}_f + \Delta\bar{h})_{O_2}$$

$$- (\bar{h}^\circ{}_f + \Delta h)_{CO_2\ \text{entering}}$$

Here $(\bar{h}^\circ{}_f)_{O_2} = 0$ by definition and $(\Delta\bar{h}^\circ{}_f)_{CO_2,\,in} = 0$ as it is at 25°C. Substituting the values from Gas Table, we have

$$\frac{\dot{Q}_{cv}}{\dot{n}_{CO_2}} = 0.31\ [-\ 293{,}520 + (180{,}800 - 9364)] + 0.69\ [-\ 110,\ 530$$

$$+ (113{,}530 - 8669)] + 0.345\ (118{,}820 - 8662) - (-\ 393{,}520)$$

$$= -\ 68846 - 3911.6 + 38004.5 + 393{,}520$$

$$= 358766.9 \text{ kJ/kg mole of } CO_2 \text{ entering}$$ **Ans.**

18.8 SUMMARY OF FORMULAE

1. $[dS]_{U,\,v} \geq 0$
2. $[dG]_{T,p} \geq 0$
3. $G = \sum_{i=1}^{1} n_i\,\mu_i$
4. $[dG]_{T,p} = 0$
5. $[dG]_{T,p} = \sum_{i=1}^{j} \mu_i\,dn_i = 0$
6. $\mu = \frac{G}{n} = \bar{g}$
7. $\mu_i = \bar{g}_i\,(T, p_i) = \bar{g}^\circ{}_i + \bar{R}\,T \ln \frac{y_i p}{p_{ref}}$
8. $1\,\mu_{H_2} + \frac{1}{2}\mu_{O_2} = 1\,\mu_{H_2O}$
9. $v_A\,\mu_A + v_B\,\mu_B = v_C\,\mu_C + v_D\,\mu_D$
10. $\Delta G^\circ = (v_C\,\bar{h}_C + v_D\,\bar{h}_D - v_A\,\bar{h}_A - v_B\,\bar{h}_B) - T\,(v_C\,\bar{s}^\circ_C + v_D\,\bar{s}^\circ_D - v_A\,\bar{s}^\circ_A - v_B\,\bar{s}^\circ_B)$
11. $\frac{-\Delta G^\circ}{\bar{R}\,T} = \ln\left[\frac{y_C^{v_C}\,y_D^{v_D}}{y_A^{v_A}\,y_B^{v_B}}\left(\frac{p}{p_{ref}}\right)^{v_C + v_D - v_A - v_B}\right]$
12. $\frac{-\Delta G^\circ}{\bar{R}\,T} = K\,(T)$ where $K\,(T) = \frac{y_C^{v_C}\,y_D^{v_D}}{y_A^{v_A}\,y_B^{v_B}}\left(\frac{p}{p_{ref}}\right)^{v_C + v_D - v_A - v_B}$

13. $K^* = \dfrac{1}{K}$ and $\log_{10} K^* = -\log_{10} K$

14. $K = \dfrac{n^{v}_{C}{}^{C}\, n^{v}_{D}{}^{D}}{n^{v}_{A}{}^{A}\, n^{v}_{B}{}^{B}} \left(\dfrac{p_{\text{ref}}}{n}\right)^{v_C + v_D - v_A - v_A}$

15. $\dfrac{-\Delta G^\circ}{RT} = \ln\left(\dfrac{a^{v}_{C}{}^{C}\, a^{v}_{D}{}^{D}}{a^{v}_{A}{}^{A}\, a^{v}_{B}{}^{B}}\right)$

16. $\dfrac{d \ln K}{dT} = \dfrac{\Delta h_R}{\bar{R}\, T^2}$ van't Hoff equation

17. $\mu'_A = \mu''_A$ and $\mu'_B = \mu''_B$ at equilibrium

18. $F = 2 + N - P$

PROBLEMS

Things to Think About

18.1. Justify the statement "Gibbs function is advantageous when studying chemical and phase equilibrium".

18.2. $[dG]_{T,p} = 0$ is a valid equilibrium criterion, explain why p and T should be fixed.

18.3. What is the reason that oxygen content in the iron tank is treated inert in thermodynamics when it rusts the tank?

18.4. In a reaction, how does pressure affect the equilibrium composition?

18.5. List the chemical species that might be present at equilibrium for a system initially containing CO_2 and H_2O held at T, p.

18.6. Why equilibrium is supposed to be one of the important aspects in the study of thermodynamics?

Short and Long Theoretical Questions

18.7. Define thermodynamic equilibrium.

18 8. List the scope of thermodynamic equilibrium and explain it.

18.9. Show that entropy approaches a maximum value as a state of equilibrium is approached.

18.10. Prove that $[dG]_{T,\,p} = 0$ forms the criterion of equilibrium.

18.11. Show that chemical potential forms the basis of equilibrium.

18.12. Show that $\mu_i = \bar{g}^\circ{}_i + \bar{R}\, T \ln \dfrac{y_i \cdot p}{p_{\text{ref}}}$

18.13. Discuss the various types of equilibrium and taking entropy as the basic sketch the various types of equilibrium.

18.14. Derive the expression for equation of reaction equilibrium for general case.

18.15. Derive the expression : $\frac{-\Delta G^\circ}{\bar{R}T} = \ln K(T)$ where K is equilibrium constant.

18.16. Derive the expression for finding equilibrium constant for ideal solution.

18.17. Derive van't Hoff equation, $\frac{d \ln K}{dT} = \frac{\Delta h_R}{\bar{R}T^2}$ and explain the physical significance.

18.18. Show that at equilibrium, the chemical potential of each component is the same as in each phase.

18.19. State Gibbs phase rule.

Numerical Questions

18.20. Calculate the equilibrium constant expressed as $\log_{10} K$ for the reaction $CO_2 \rightleftarrows CO + 1/2\ O_2$ at 500 K. Compare the value from Gas Tables.

18.21. Calculate the equilibrium constant expressed as $\log_{10} K$ for the water-gas reaction for $H_2O \rightleftarrows H_2 + 1/2\ O_2$ at 298 K and 2000 K. Compare the results with Gas Tables.

18.22. Calculate the equilibrium constant for the reaction $CH_4 + H_2O \rightleftarrows 3H_2 + CO$ at 2000 K.

18.23. Carbon dioxide gas at 25°C and 5.5 bar enters a heat exchanger operating at steady state. An equilibrium mixture of CO_2, CO and O_2 exits at 2500 K. Calculate, per kg (a) the composition of the existing mixture, and (b) the heat transfer to the gas stream, in kJ.

18.24. Methane gas at 25°C and 1 bar enters a reactor operating at steady state and burns with 80% of theoretical air entering at 250°C and 1 bar. An equilibrium mixture of CO_2, CO, $H_2O(g)$, H_2 and N_2 exists at 1500 K, 1 bar. Calculate per kg of entering the composition of existing mixture, and (b) the heat transfer between reactor and surroundings, in kJ. Neglect the changes in PE and KE.

18.25. A vessel initially containing 1 kg mole of CO and 4.76 kg mole of dry air forms an equilibrium mixture of CO_2 , CO, O_2, and N_2 at 3000 K and 1 bar. Calculate the equilibrium composition.

18.26. H_2 at 25°C and 1 atm enters an insulated reactor operating at steady state and reacts with 100% of theoretical air entering at 25°C, 1 atm. The products of combustion exist at temperature T and 1 atm. Calculate temperature T in K, if (a) the combustion is complete, and (b) an equilibrium mixture of H_2O, H_2 SO_2 and N_2 exists.

18.27. Methane gas burns with 100% theoretical air to form an equilibrium mixture of CO_2, CO, $H_2O(g)$, H_2 and N_2 at 1500 K, 1 atm. Calculate the composition of the equilibrium mixture, per kg mole of mixture.

18.28. Calculate the enthalpy of reaction at 2200 K, in kJ/kg mole, for $CO_2 \rightleftarrows CO + 1/2\ O_2$ using van't Hoff equation and equilibrium constant data. Compare with the value obtained for the enthalpy of reaction using enthalpy data.

Software Based Questions

18.29. Develop a software for calculation of equilibrium constant and equilibrium composition for various chemical reactions caused by combustion of H_2, CO, CH_4, C_3H_8 and other leading fuels.

18.30. $\log_{10} K$ is nearly linear in 1/7 : $\log_{10} K = C_1 + C_2/T$ where C_1 and C_2 are constant. Develop a software for selected reactions listed in Gas Tables of $\log_{10} K$ and verify this relationship by plotting $\log_{10} K$ versus $1/T$ temperature range from 2000 to 2500 K.

18.31. Two kg mole of CO_2 dissociate to form an equilibrium mixture of CO_2, CO, and O_2 in which 1.8 kg mole of CO_2 is present. By developing a software, plot the temperature of the equilibrium mixture, in K, versus the pressure p for $1 \le p \le = 0$ atm.

18.32. H_2 gas at 25°C and 1 atm enters an insulated reactor operating at steady state and reacts with 250% excess oxygen entering at 230°C and 1 atm. The products of combustion occur at 1 atm. Develop a software to calculate the temperature of products, in K, if

(a) the combustion is complete, and

(b) an equilibrium mixture of H_2O, H_2 and O_2 exits.

Design and Open Ended Questions

18.33. The spark ignition engines produce NO_x, CO and unburnt hydrocarbon (HC) as pollutants in their exhaust. The amounts of pollutants actually depend on engine design and operating conditions and typically differ significantly from values calculated on the basis of chemical equilibrium. Prepare a report discussing the reasons for these discrepancies and possible mechanisms by which such pollutants are formed in an actual engines.

18.34. Prepare a report of formation of air pollutants on daily basis in a city like New Delhi and New York from automobiles taking the real data of vehicles, population and area. Assume that all automobiles do have EURO-VI emission norms. Compare the results with and without catalytic converter in vehicles.

18.35. A gaseous mixture of H_2 and CO enters a catalytic reactor and a gaseous mixture of CH_3OH, H_2 and CO exits. For the preliminary design, calculate the inlet hydrogen mole and the temperature of exiting mixture assuming some realistic constraints.

Multiple-Choice Answer Type Questions

18.36. Thermodynamic equilibrium criterion is governed by

(a) $[dG]_{U,V} = 0$ (b) $[dG]_{T,p} = 0$ (c) $[dG]_{S,T} = 0$ (d) $[dG]_{h,T} = 0$

18.37. Stable equilibrium criterion is governed by

(a) $[\Delta G]_{U,V} = 0$ (b) $[\Delta F]_{T,V} = 0$ (c) $[\Delta G]_{T,p} = 0$ (d) $[\Delta G]_{s,p} < 0$

18.38. Unstable equilibrium criterion is governed by

(a) $[\Delta F]_{T,V} > 0$ (b) $[\Delta U]_{S,V} < 0$ (c) $[\Delta G]_{T,p} < 0$ (d) $[\Delta H]_{s,p} < 0$

18.39. Equilibrium constant is given by

(a) $\dfrac{\Delta G°}{\bar{R}t} = K(T)$ (b) $\dfrac{[\Delta G°]}{RT} = K(T)$ (c) $\dfrac{-\Delta G°}{\bar{R}T} = K(T)$ (d) $\dfrac{-\Delta G°}{\bar{R}T} = \ln K(T)$

Answers to Multiple-Choice Questions

18.36 (b) 18.37 (a) 18.38 (c) 18.39 (d)

CHAPTER

19

Direct Energy Conversion

OVERVIEW

We have already discussed a number of indirect energy converter power plants in the previous chapters. In most of the indirect energy converters, energy is transferred to a fluid; the fluid expands producing mechanical work and finally the mechanical work is converted into electric energy. Several losses are incurred in these conversion steps which reduce the efficiency. Moreover, all the such power cycles are limited by Carnot efficiency since they basically involve the conversion of heat into work.

Direct-energy conversion system eliminates some of the intermediate steps and convert the one form of energy into direct electrical energy. Such devices are of significant practical interest. Direct-energy conversion systems for power generation are currently centred around five different types, namely: (i) *thermoelectric converters*, (ii) *thermionic converters*, (iii) *magneto-hydrodynamic* (*MHD*), (iv) ***Solar cells*** *and* (v) ***fuel cells***. *Thermoelectric, thermionic, solar and fuel cells have been employed in space vehicles, producing from a few watts to several kilo-watts of power.* Now solar cells are commonly used to run pump sets and T.V. sets and fed to grid in huge magnitude through solar cells farm. For large scale power production, the MHD offers a promise of being feasible. Although these systems are currently low in efficiency or difficult to operate and control, experimental work continues to show promise of improved performance. **Fuel cells** *are highly promising*. Now, **hybrid system** (fuel cells and gas turbines) is coming up. A 40 MW hybrid system having efficiency of more than 70% has been installed in USA. It may revolutionize the power sector scenario. *In India, the* ***solar cells installed capacity was 56.951 GW as of 1st June, 2022.*** **Solar cells are the most promising green energy in the world**. In this section of the book, we will discuss the thermodynamic aspects of these systems briefly.

19.1 THERMOELECTRIC EFFECTS

Thermoelectric effect is concerned with the phenomena involving the inter-charge of heat and electric energy. There are three types of thermoelectric reversible effects called the ***Seebeck, Peltier and Thomson effects***. They are defined as follows:

(a) The Seebeck Effect *If the junctions of two dissimilar electrical conductors are maintained at different temperatures, then an electromotive force will exist in open circuit and that will be proportional to the difference in temperature.* Mathematically,

$$\alpha_{AB} = \lim_{\Delta T \to 0} \frac{\Delta V}{\Delta T} = \frac{dV}{dT} \quad ...(19.1)$$

where ΔV = open circuit potential difference.

ΔT = temperature difference between the two junctions.

α_{AB} = a coefficient of proportionality dependent on temperature or *Seebeck coefficient.*

The subscript *AB* denotes that α depends on the materials constituting the junctions. *If the circuit is completed by an external load, the current will flow.* An example of Seebeck effect is the thermocouple. *This Seebeck effect is also utilised for power production.* Heat conduction and electric power loss in the leads are the two main losses that tend to reduce the useful electric power output.

(b) The Peltier Effect *This is reverse of Seebeck effect. An electric current is sent through a junction of two dissimilar materials A and B, which results in either a heating or cooling of the junction. The rate of cooling of heating is proportional to current flow and is given by*

$$Q = \pi_{AB} \cdot I \quad ...(19.2)$$

where π_{AB} = Peltier coefficient which is a function of temperature. The suffix *AB* indicates that the Peltier coefficient also depends on the two materials, *A* and *B*.

The *Peltier effect is utilised in refrigeration.*

(c) The Thomson Effect A temperature gradient along a conducting material may cause heat absorption or heat release when current flows through it. This is known as the Thomson effect. Thus,

$$\tau_A = \frac{\delta Q/dx}{I\,(dT/dx)} \quad ...(19.3)$$

where $\delta Q/dx$ is the rate of heat interaction per unit length of the conductor and dT/dx is the temperature gradient along and length of conductor and τ_A is the Thomson coefficient. Like the Seebeck and Peltier effect, Thomson effect is also reversible. It is to be noted that it only involves a single conductor say *A*. Combining the Seebeck and Thomson effects, we get

$$\frac{d\alpha_{AB}}{dT} = \frac{\tau_B - \tau_A}{T} \quad ...(19.4)$$

These three effects are more pronounced in semiconductors.

19.2 THERMOELECTRIC CONVERTERS

A thermoelectric converter consists essentially of two semiconductors blocks p and n types connected by a conductor as shown in Fig. 19.1. The conductor receives heat from a thermal source at temperature T_h and the open ends of the block reject heat to a low temperature reservoir at temperature T_c. The sides of the semiconductors are insulated so that heat flow takes place unidimensionally through the semiconductors. When heat is transferred to the hot junction, electrons, in *n*-types semiconductors, and holes in *p*-type semiconductors, tend to flow away from the hot junction. This results in the generation of an electric potential and if the circuit is completed at the cold junction, an electric current flows through the load.

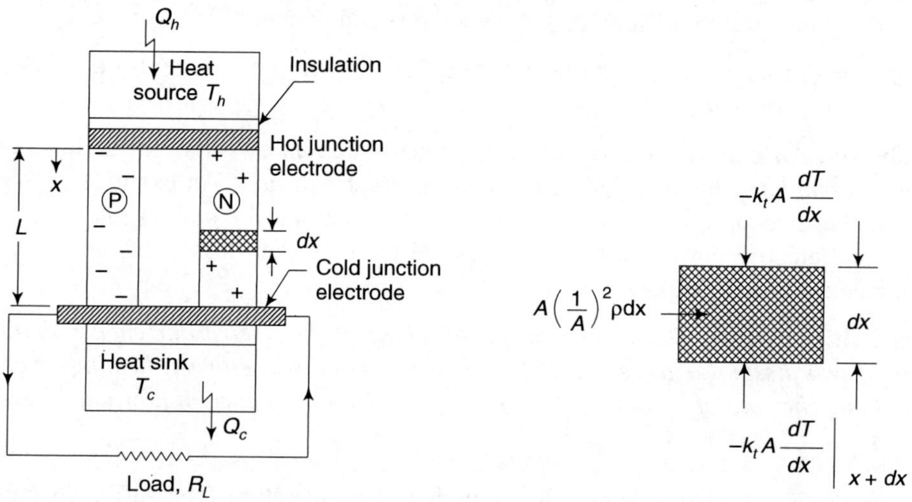

Fig. 19.1 Thermoelectric converter

There are several factors that influence the performance of the thermoelectric generator, namely:

(i) Heat conduction from the heat source to heat sink must pass through the thermoelectric material.

(ii) Joulean heat generation results from current flow in the materials.

(iii) Voltage drops in the material.

(iv) Various losses are caused from contact resistance between materials, like heat losses through insulations, etc.

Using the energy balance, we have

Heat conduction out of the element = Heat conducted into the element + Joulean heat generation

$$\therefore \qquad -k_t\, A \cdot \left.\frac{dT}{dx}\right|_{x+dx} = -k_t\, A \cdot \frac{dT}{dx} + \rho A\, dx \left(\frac{I}{A}\right)^2$$

or $$-k_t\,A\cdot\left(\frac{dT}{dx}+\frac{d^2T}{dx^2}\,dx\right) = -k_t\,A\cdot\frac{dT}{dx}+\rho A\,dx\left(\frac{I}{A}\right)^2$$

or $$k_t\,\frac{d^2T}{dx^2}+\left(\frac{I}{A}\right)^2\rho = 0 \qquad ...(19.5)$$

where p is the electric resistivity of the material and x is measured in the direction shown.

The above equation may be solved with the help of boundary conditions as

$$T = T_h \quad \text{at} \quad x = 0 \quad \text{and} \quad T = T_L \quad \text{at} \quad x = L$$

Thus, the solution of above equation is given by

$$T - T_h = \frac{J^2\,\rho}{2\,k_t}\left[\frac{x}{L}-\left(\frac{x^2}{L}\right)^2\right]-\left(\frac{x}{L}\right)(T_h - T_c) \qquad ...(19.6)$$

where $J = \dfrac{I}{A}$ = current density

The heat flux at the heat source is given by

$$q'' = \frac{Q}{A} = -k_t\left(\frac{dT}{dx}\right)_{x=0} = \frac{k_t\,\Delta T}{L}-\frac{J^2\,\rho L}{2} \qquad ...(19.7)$$

where $\Delta T = T_h - T_c$

The net conduction heat flux into the material is thus the result of conduction heat flux into the material and a Joulean heat flux outward, half of the Joulean heat flows out of each end of the conductor.

For both the p and n legs, the overall **thermal conductance** (K) is

$$K = \left(\frac{k_f\,A}{L}\right)_p + \left(\frac{k_f\,A}{L}\right)_n \qquad ...(19.8)$$

Similarly, the electric resistance is given by

$$R = \left(\frac{\rho L}{A}\right)_p = \left(\frac{\rho L}{A}\right)_n \qquad ...(19.9)$$

In order to get the total heat flow into the hot junction, we must add the Peltier heat to the heat conduction equation given by Eq.(19.7). Thus, we get

$$Q_h = I\cdot\alpha\cdot T_h + K\cdot\Delta T - \frac{I^2 R}{2} \qquad ...(19.10)$$

where α is the Seebeck coefficient, $\alpha = \alpha_n - \alpha_p$ ...(19.11)

The output voltage of the device is the Seebeck voltage minus the losses due to internal resistance. Hence,

$$V_{out} = \alpha \Delta T - IR \quad ...(19.12)$$

The output power or the electric work is given by

$$P = I^2 R_L \quad ...(19.13)$$

or

$$P = W_e = I.V_{out} = I.\alpha.\Delta T - I^2 R \quad ...(19.14)$$

The thermal efficiency of a thermoelectric generator is given by

$$\eta = \frac{\text{Work output}}{\text{Heat input}} = \frac{I \cdot \alpha \cdot \Delta T - I^2 R}{I \cdot \alpha \cdot T_h + K\Delta T - I^2 R/2}$$

$$\eta = \frac{\alpha \cdot \Delta T - I \cdot R}{\alpha T_h + K \cdot \Delta T / I - IR/2} \quad ...(19.15)$$

It is to be noted that the resistance to the external circuit influences the efficiency as it affects the current drawn in the circuit and also results in Joulean heat losses. The Seebeck voltage is, in effect, driving a circuit composed of an internal resistance (R) and an external resistance, R_L. Therefore, the current is expressed as

$$I = \frac{\alpha \cdot \Delta T}{R + R_L} \quad ...(19.16)$$

Putting the value of I into Eq. (19.15) and introducing $m = R_L/R$, we have

$$\eta = \frac{\Delta T}{T_h} \cdot \frac{m}{1 + m - \Delta T/2T_h + (1+m)^2 \, RK/(\alpha^2 \cdot T_h)} \quad ...(19.17)$$

In order to obtain maximum efficiency, it is necessary that the RK product should be the minimum as other terms are constant. With the help of Eqs. (19.8) and 19.9), we get

$$RK = \left[\left(\frac{k_t A}{L}\right)_p + \left(\frac{k_t A}{L}\right)_n\right]\left[\left(\frac{\rho L}{A}\right)_p + \left(\frac{\rho L}{A}\right)_n\right] \quad ...(19.18)$$

For a particular thermoelectric material, the value of ρ and k_t are fixed. Thus, the product of RK will be only minimized with respect to the relative size of the A/L ratios for the two legs.

Thus, $$\frac{d(RK)}{d[(L/A)_n/(L/A)_p]} = 0$$

This gives $$\left.\frac{(L/A)_n}{(L/A)_p}\right|_{opt} = (k_{in} \cdot \rho_n / k_{ip} \cdot \rho_p)^{1/2} \quad ...(19.19)$$

∴ $$(RK)_{opt} = \left[\sqrt{(\rho \cdot k_t)_n} + \sqrt{(\rho \cdot k_t)_p}\right]^2 \quad ...(19.20)$$

It is evident from Eq. (19.17) that for given temperature limits the efficiency depends on the load resistance ralio m and the grouping, or

$$Z = \frac{\alpha^2}{RK} \qquad ...(19.21)$$

where Z is called the **figure of merit** and is the index for rating the thermoelectric converters.

$$m_{\text{opt}} = \sqrt{(1 + ZT)} \qquad ...(19.22)$$

where $T = \dfrac{T_h + T_L}{2}$

Substituting the value of m_{opt} into Eq. (19.17), the optimum efficiency of a thermoelectric generator is

$$\eta_{\text{opt}} = \eta_c \frac{m_{\text{opt}} - 1}{m_{\text{opt}} + T_c/T_h} \qquad ...(19.23)$$

where $\eta_c = \dfrac{T_h - T_c}{T_n}$ is the Carnot efficiency. Thus, the efficiency of the thermoelectric generator is limited by Carnot cycle efficiency. The Carnot cycle efficiency can only be obtained in the case of zero internal resistance (infinite Z).

The figure of merit for p and n-type materials having equal resistivities and thermal conductance is given by

$$Z = \frac{\alpha^2}{\rho\, k_t} \qquad ...(19.24)$$

With the increase in the values of figure of merit, efficiency increases. Hence, this parameter is generally regarded as the most significant variables for selection of materials for thermoelectric power generation. The expression for efficiency may be written as

$$\eta = \eta_c\, \varepsilon \qquad ...(19.25)$$

where ε is called the material efficiency $= \dfrac{m_{\text{opt}} - 1}{m_{\text{opt}} + T_c/T_h}$

For the maximum efficiency and power, the load resistance is equal to the system resistance, i.e., $R = R_L$ or $m = 1$.

The variation of ε with ZT product is shown in Fig. 19.2. Figure 19.2 shows that the ZT product controls the efficiency of the thermoelectric generator. The material properties here are assumed constant, but in practice, they vary hence a proper allowance must be made for this in the analysis of actual power plant. The variation of properties is shown in Fig. 19.3.

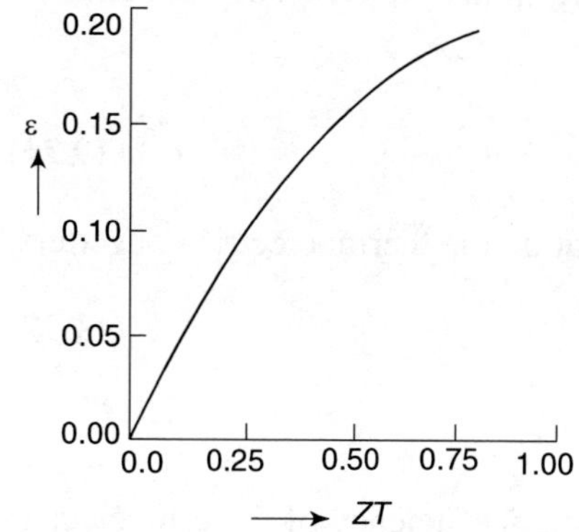

Fig. 19.2 Material efficiency vs ZT

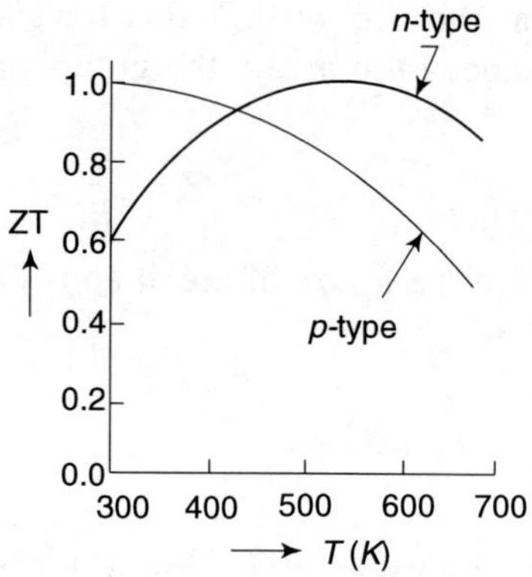

Fig. 19.3 Thermoelectric property vs T

The high performance materials are silicon-germanium alloy, lead-telluride, bismuth-telluride, and germanium-bismuth-telluride.

The output is very low. Its energy conversion efficiency is low, about 10 per cent but it is most reliable.

Example 19.1 Calculate the maximum power output and efficiency of a thermoelectric generator which operates between 1000 K and has the following properties : $\alpha = 0.0025$ volt/K, the resistance of the each element is 0.03 ohm and the thermal conductance of each element is 0.04 W/K.

Solution Electric resistance, $R = 0.03 \times 2 = 0.06$ ohm. Overall thermal conductance, $K = 2 \times 0.04 = 0.08$ W/K. The maximum power output is obtained when the load resistance (R_L) is equal to the system resistance, R. The current is therefore

$$I = \frac{\alpha \cdot \Delta T}{R + R_L} = \frac{\alpha \cdot \Delta T}{2R} = \frac{0.0025 \times (1000 - 500)}{2 \times 0.06} = 10.41 \text{ amp}$$

The maximum power output of the device is

$$P = I\,\alpha\,\Delta T - I^2 R = 10.41 \times 0.0025 \times (1000 - 500) - 10.41^2 \times (0.06) = 6.51 \text{ W} \quad \textbf{Ans.}$$

$$\text{Heat supplied at the hot junction, } Q_h = I\,\alpha . T_h + K\Delta T - \frac{I^2 R}{2}$$

$$= 10.41 \times 0.0025 \times 1000 + 0.08 + (1000 - 500) - \left(\frac{10.41^2 \times 0.06}{2}\right) = 62.8 \text{ W}$$

$$\therefore \text{ Efficiency of generator, } \frac{P}{Q_h} = \frac{6.51}{62.8} = 10.366\% \quad \textbf{Ans.}$$

Example 19.2 A thermo-electric generator uses the following materials with properties at the mean temperature are also listed. It works between the temperature limits of 100°C and 300°C.

The cross-sectional area of the *n*-type element is 1 cm^2 and the length is 1 cm. Using the optimum value of *RK* product, calculate the maximum generator efficiency and the efficiency for maximum power. Also, calculate the maximum power output for both these cases.

n-type	*p*-type
75 per cent B_{t2} T_{e3}	26 per cent B_{12} T_{e3}
25 per cent B_{12} S_{e3}	75 per cent S_{b2} T_{e3}
$\alpha = 190 \times 10^{-6}$ volt/°C	$\alpha = 189.9 \times 10^{-6}$ volt/°C
$\rho = 1.45 \times 10^{-3}$ ohm-cm	$\rho = 1.8 \times 10^{-3}$ odm-cm
$Z = 2.0 \times 10^{-3}$ K^{-1}	$Z = 1.7 \times 10^{-3}$ K^{-1}

Solution The thermal conductivities are given by

$$k_{tn} = \frac{\alpha_n^2}{\rho_n Z_n} = \frac{(189.9 \times 10^{-6})^2}{(1.45 \times 10^{-3})(2.0 \times 10^{-3})} = 0.0124 \text{ W/cm°C}$$

$$k_{tp} = \frac{\alpha_p^2}{\rho_p Z_p} = \frac{(189.9 \times 10^{-6})^2}{(1.8 \times 10^{-3})(1.7 \times 10^{-3})} = 0.0118 \text{ W/cm°C}$$

The figure of merit for the generator will be calculated from the equation.

$$Z = \frac{(\alpha_n - \alpha_p)^2}{(RK)_{\text{opt}}} = \frac{(\alpha_n - \alpha_p)^2}{\left[\sqrt{(\rho k_t)_n} + \sqrt{(\rho k_t)_p}\right]^2} = 1.85 \times 10^{-3} \text{ K}^{-1}$$

In order to get the optimum value of the *RK* product, the area length relations of the *n*- and *p*-type materials are given by

$$\left.\frac{(L/A_n)}{(L/A_{p\,\text{opt}})}\right| = \left(\frac{k_{tn}\,\rho_p}{k_{tp}\,\rho_n}\right)^{1/2}$$

Hence, $\qquad L_n = L_p = 1$ cm and $A_n = 1$ cm^2

$$\therefore \qquad A_p = \left[\frac{(0.0124)(1.8 \times 10^{-3})}{(0.0118)(1.45 \times 10^{-3})}\right]^{1/2} = 1.14 \text{ cm}^2$$

The total thermal and electric conductance will be calculated from

$$K = k_{tn}\frac{A_n}{L_n} + k_{tp}\frac{A_p}{L_p} = (0.024) \times 1 + (0.0118) \times \frac{1.14}{1} = 0.0258 \text{ W/°C}$$

$$R = \rho_n \frac{L_n}{A_n} + \rho_p \frac{L_p}{L_p} = (1.45 \times 10^{-3})\frac{1}{1} + (1.8 \times 10^{-3}) = \frac{1}{1.14} \; 3.02 \times 10^{-3} \text{ ohm}$$

For optimum efficiency, the resistance ratio is given by

$$m_{\text{opt}} = (1 + ZT)^{1/2} \; [(1 + 1.85 \times 10^{-3})\,(200 + 273)^{1/2} = 1.37$$

$\therefore \quad R_L = (1.37)\ (3.03) \times 10^{-3}) = 4.16 \times 10^{-3}$ ohm

The optimum efficiency is given by

$$\eta_{opt} = \frac{\Delta T}{T_h} \cdot \frac{m_{opt} - 1}{m_{opt} + T_c/T_h} = \frac{200}{573} \cdot \frac{1.37 - 1}{(1.37 + 373/573)} = 6.39\% \qquad \textbf{Ans.}$$

This may be compared with the efficiency of the Carnot cycle which is $\eta_c = \Delta T/T_h = 34.9$ per cent.

For maximum power output, $R_L = R$ and $m = 1$.

Hence,
$$\eta = \frac{\Delta T}{T_h} \cdot \frac{m}{1 + m - (\Delta T/2T_h) + (1+m)^2\ RK/(\alpha^2\ T_h)}$$

$$\eta_{max} = \frac{200}{573} \frac{1}{1 + 1 - 200/(2 \times 573) + (1+1)^2\ (4.16 \times 10^{-3} \times 0.0258)/(1.85 \times 10^{-3})\ (573)}$$

$$= 6.24\% \qquad \textbf{Ans.}$$

Here, the optimum efficiency is not very much different from the efficiency at the maximum power output is calculated from both cases as

$$P = I^2\ R_L = \left(\frac{a\Delta T^2}{R + R_L}\right) R_L = \frac{(a \cdot \Delta T)^2}{(L+m)^2\ R_L}$$

For optimum efficiency, the power output is

$$P_{opt} = \frac{(-190 - 190)^2\ (10^{-6})^2\ (200)^2}{(1+1)^2\ (3.03 \times 10^{-3})} = 0.247 \text{ W} \qquad \textbf{Ans.}$$

The maximum power output

$$P_{opt} = \frac{(-190 - 190)^2\ (10^{-6})^2 (200)^2}{(1+1)^2\ (3.03 \times 10^{-3})} = 0.477 \text{ W} \qquad \textbf{Ans.}$$

Thus, open circuit voltage is $\alpha \cdot \Delta T$ or 0.076 volt. **Ans.**

19.3 THERMIONIC CONVERTERS

A thermionic converter is basically an **electric diode** *which utilises the principle of electronic emission from metals at high temperature to convert heat directly into an electric current.* For this reason, *it is potentially used where solar and nuclear energies are tapped.* ***It is a low voltage but high current device***. Figure 19.4 shows a schematic diagram of thermionic converter. It consists of a cathode (also called electron emitter) and an anode (also called electron collector) separated by either an evacuated or a vapour-filled space. The metal cathode plate is maintained at heat source temperature (T_h) while the anode plate is maintained at the heat sink temperature T_c. Due to the high temperature of the cathode (emitter) enough energy is imparted to the electrons to escape from the surface of the cathode and travel across the gap to the anode (collector). An abundance of holes that exists in the cathode, becomes positive whereas the anode becomes negative. If the cathode and anode are connected by an external load, a current will flow, due to the potential difference between the two surfaces and the higher the temperature, the larger number of electrons. Thus, *the thermionic converter works a heat engine operation between temperature limits T_h and T_c and its efficiency is limited by the Carnot principle.*

A major hurdle to the flow of electrons is the space-charge barrier. The space charge barrier is formed by electrons and ends to limit the current density. In order to neutralize this space charge, two techniques are used. In one, the space between the electrodes is reduced to *several microns* of the order of 10 microns (i.e., 0.001 cm) thus limiting the emission current of the space charge. In closed-spaced diode, the spacing is the order of 10 microns and temperature from 1500 K to 2000 K on cathode and 1000 K on the anode are common with a diode output of 2 to 10 W/cm^2.

In the other technique, positive ions are introduced into the inter-electrode space to neutralize the negative space charge. Cesium vapour is commonly employed for the neutralization process because it has a low ionization potential (about 3.9 volts). Ionization results when cesium atoms collide with the emitted electrons or with the hot wall, according to the equation,

$$C_s = C_s^+ + \overline{e}$$

When a vapour is introduced into the spaces, the device is called plasma diode. In *plasma diodes*, the spacing between the electrodes is of the order of 0.1 cm, and the temperature of the cathode varies from 1600 K to 2600 K.

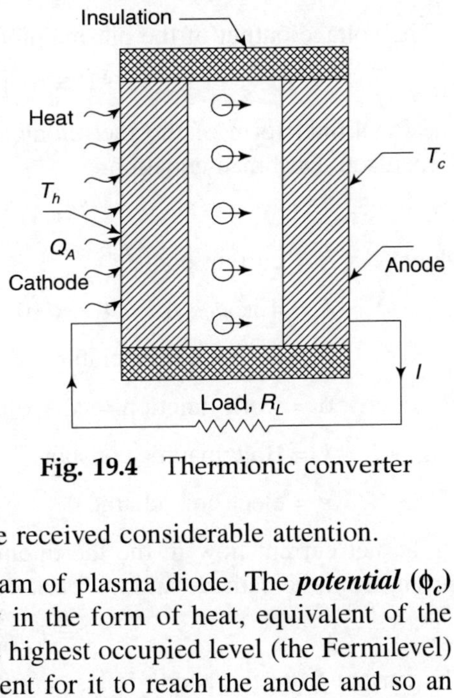

Fig. 19.4 Thermionic converter

Comparing the closed-spaced diode with the plasma diode, we observe that the spacing required in the closed-spaced diode is difficult to achieve and for this reason the plasma diode in which the inter-electrode spacing is of the order of 0.1 cm have received considerable attention.

Figure 19.5 shows an electrostatic potential diagram of plasma diode. The ***potential*** (ϕ_c) ***is called the work function of the cathode***. Energy in the form of heat, equivalent of the work function ϕ_c, must be added to an electron in the highest occupied level (the Fermilevel) in order to leave the cathode. But this in not sufficient for it to reach the anode and so an additional amount of energy is necessary to overcome the space-charge (ϕ_k) set up outside the cathode by the emitted electrons. Thus, for an electron to reach the anode, it must have enough energy to overcome a barrier height of $\phi_c + \phi_k$.

Electrons which cross the space-charge barrier lose energy as they fall through the plasma potential (ϕ_p) to the vacuum level of the anode. This is due to the resistance of the electron gas (Cesium vapour) present in the inter-electrode gap. As the electrons enter the anode, they give up an amount of energy equal to the work function to the anode, $\phi_a - \phi_p$. Here, ϕ_k represents the difference in potential between the cathode (emitter) and anode (collector) surface.

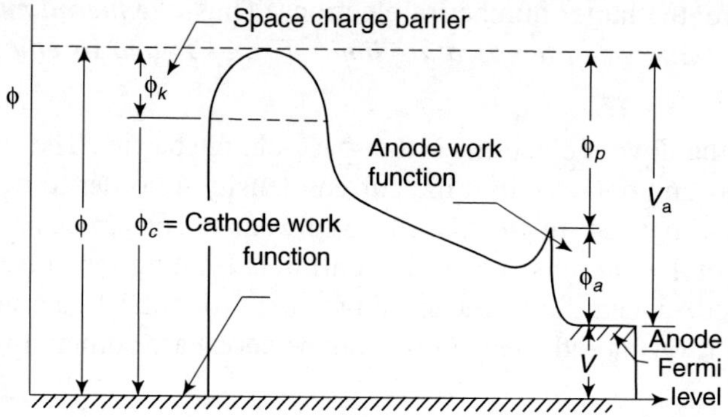

Fig. 19.5 Electrostatic potential diagram of plasma diode

The voltage output of the plasma diode may thus be written as

$$V = \phi_c + \phi_k - \phi_p - \phi_a \quad \text{...(19.26)}$$

The functional form of the thermionic current density (current per unit area) is given by Richardson–Dushman equation

$$j = AT^2 \exp^{-e\phi/kT}$$

where j = current denisty

$A = 4\,\pi\, mek^2/h^3 = 1.2 \times 10^6\ \text{m/m}^2\ k^2$

T = absolute temperature

ϕ = work function = $\phi_c + \phi_k$

k = Boltzmann's constant

e = electronic charge

The net current flow in the thermionic diode is the difference between the emission of current from the cathode and the back emission current from the anode. Thus,

$$j = j_{\text{cathode}} - j_{\text{anode}} \quad \text{...(19.27)}$$

The current from the anode is less compared to what is emitted from the cathode because of the fact that the anode is at a much lower temperature than the cathode. Neglecting the conduction losses, the efficiency of the thermionic converter is given by

$$\eta = \frac{jV}{Q_c/A_c} \quad \text{...(19.28)}$$

where Q_c/A_c = cathode heat flux

The total heat added to the cathode is equal to the sum of the electron energy flux out of the cathode and the thermal radiation heat exchange.

$$\frac{Q_c}{A_c} = j\left(\phi + \frac{2kT_h}{e}\right) + \frac{Q_{\text{rad}}}{A_c} \qquad \text{...(19.29)}$$

where $\phi = \phi_c + \phi_k$ ...(19.30)

$\dfrac{2kT_h}{e}$ = voltage-equivalent to thermal energy of the emitted electrons

$$\frac{Q_{\text{rod}}}{A_c} = \varepsilon\,\sigma\left(T_h^4 - T_c^4\right) \qquad \text{...(19.31)}$$

where σ is the Stefan–Boltzmann constant and has a value of 5.67×10^{-12} joule/s cm^2 K^4 and ε is an appropriate constant (emissivity) which is a function of the radiation properties.

It is to be noted if the collector work function is reduced, a higher thermal efficiency may be obtained. On the other hand, if the work function is very small, the back current is large.

The factors which prevent the thermionic converter from achieving the theoretical Carnot efficiency are:

(i) Impedance to electron flow caused by the space charge between the electrodes

(ii) Thermal radiation between the electrodes

(iii) Heat losses to the surroundings.

The theoretical efficiency of a closely spaced vacuum thermionic converter is less than 10% whereas plasma diode thermionic converters have been built with an efficiency which is as high as 15%, along with power densities of 7.5 W/cm^2.

Example 19.3 A thermionic generator has $\phi = \phi_c + \phi_k = 2.5$ volts, $\phi_a = 2$ volts, $T_h = 2000$ K and $\phi_p = 0.1$ volt. Calculate the efficiency of the thermionic generator and compare with the theoretical Carnot value. Assume $\varepsilon = 0.2$ for the electrode materials and electronic charge $\varepsilon = 1.6 \times 10^{-19}$.

Solution The output voltage, $V = \phi_c + \phi_k - \phi_a - \phi_p = 2.5 - 2 - 0.1 = 0.4$ volt

The cathode electron current, $j_c = AT^2 \exp^{-e\phi/kT}$

$$= (1.2 \times 10^6)\,(2000)^2 \exp\left[\frac{-(1.6\times10^{-19})\,(2.5)}{(1.38\times10^{-23})\,(2000)}\right] = 2.4 \times 10^6 \text{ amp/m}^2$$

$$j_a = (1.2 \times 10^6)(1000)^2 \exp\left[\frac{-(1.6\times10^{-19})\,(2.0)}{(1.38\times10^{-23})\,(1000)}\right] = 1.2 \times 10^6 \text{ amp/m}^2$$

The net current, $j = j_c - j_a = 2.4 \times 10$ amp/m^2

Since the anode current is negligible, the heat added to the cathode is calculated from equation,

$$= \frac{Q_c}{A_c} = j\left(\phi + \frac{2kT_h}{e}\right) + \varepsilon\,\sigma\,(T_h^4 - T_c^4)$$

$$= (2.4 \times 10^6)\left[2.5 + \frac{2 \times 1.38 \times 10^{-23} \times 2000}{(1.6 \times 10^{-19})}\right] + 0.2 \times 5.67$$

$$\times\ 10^{-12} \times 10^{-4}\ (2000^4 - 1000^4) = 7.026 \times 10^6\ \text{W/m}^2$$

The efficiency of the device, $\eta = \dfrac{JV}{Q_c/A_c} = \dfrac{2.4 \times 10^6 \times 0.4}{7.026 \times 10^6} = 13.7\%$ **Ans.**

The Carnot efficiency for the same temperature limits is

$$\eta_c = \frac{T_h - h_c}{T_h} = \frac{1000}{2000} = 50\%$$ **Ans.**

19.4 MAGNETOHYDRODYNAMIC (MHD) GENERATOR

According to the Faraday's law of electromagnetic induction, an electric conductor moving through a magnetic field experiences a retarding force as well as an induced electric field and current. The induced e.m.f. is given by

$$E_{\text{ind}} = u \times B \qquad ...(19.32)$$

where u is the velocity of the conductor and B is the magnetic field intensity. The induced current density is given by

$$J_{\text{ind}} = \partial \cdot E_{\text{ind}} \qquad ...(19.33)$$

where J_{ind} is the induced current and ∂ is the electric conductivity. The retarding force on the conductor is the Lorentz force given by

$$F_{\text{ind}} = J_{\text{ind}} \times B \qquad ...(19.34)$$

According to the law of energy, the movement of the force through a displacement (mechanical work) is converted into electric work (current) flow against potential difference) by means of the electromagnetic induction principle. *The electromagnetic induction principle is utilised in MHD generators.* In the conventional electric generator, the copper conductors cross the lines of magnetic field whereas in the magnetohydrodynamic generator ionised gas, which acts like an electric conductor, flows across the lines of a magnetic field. In either case, a voltage is induced.

Figure 19.6 (a) shows the basic components of a MHD generator. *Hot ionized gas at a high velocity flows through the poles of an electromagnet. This induces a potential difference between a pair of electrodes placed at right angles to the magnetic field. If the electrodes are connected by a resistive load, a current flows through the circuit.*

There are two ways to obtain the ionised gas. In one, ionised gas may be obtained by heating the gas to a temperature between 2200°C and 2500°C. In this, a nuclear fuel serves as an energy source. In the other, the gas may be seeded with substances like cesium or potassium vapour which ionizes the gas at a relatively low temperature (1100°C). This is passed through the MHD generator. After passing through the MHD generator, the ionised gas is usually reclaimed by recirculating it through a regenerator where energy is added to the gas before it is used in the MHD generator again.

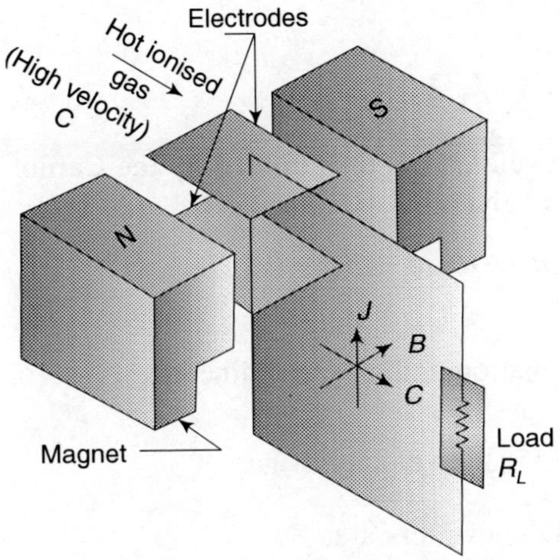

Fig. 19.6 (a) Magnetohydrodynamic generator

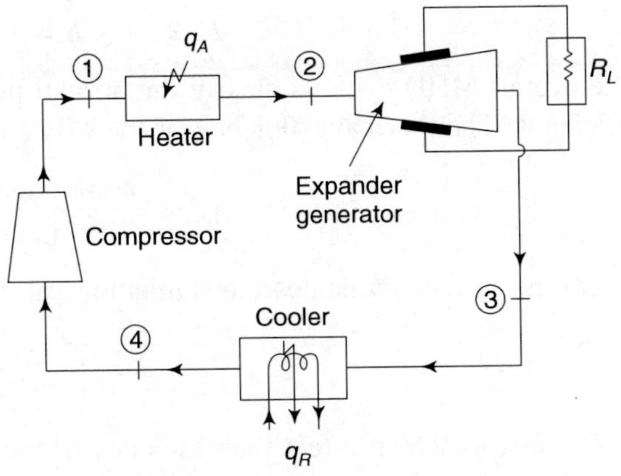

Fig. 19.6 (b) Schematic of MHD power cycle

In the overall power cycle, the MHD converter takes the place of a turbine in a conventional vapour or gas turbine cycle. However, here we must use a compressor to increase the pressure. Heat is added to the heater at high pressure and the velocity of the ionised gas is increased before entering the converter. Figure 19.6 (b.) shows a schematic diagram of a MHD power cycle. The corresponding *T–s* diagram is shown in Fig. 19.7.

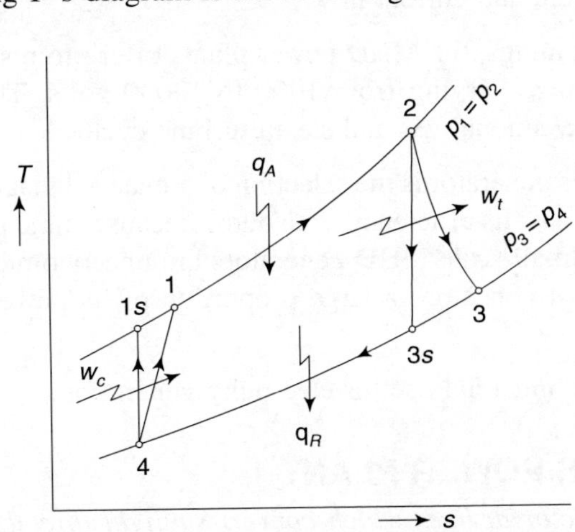

Fig. 19.7 *T–s* diagram for MHD cycle

In addition to the induced electric field $\vec{E}_{ind}$, there is also some applied field $\vec{E}$ (electric field strength) at the electrodes. Therefore, the total current density (J) and the electromagnetic force (F) is given by

$$\vec{J} = \partial(\vec{E} + \vec{C} \times \vec{B})$$

$$\vec{F} = \vec{J} \times \vec{B} = \partial(\vec{E} + \vec{C} \times \vec{B})\,\vec{B} \quad ...(19.35)$$

The overall MHD cycle is clearly the normal power cycle and is thus limited by the Carnot efficiency. The thermal efficiency of the MHD cycle is given by

$$\eta_{th} = \frac{\text{Work output}}{\text{Heat input}} = \frac{(h_{02} - h_{03}) - (h_{01} - h_{04})}{(h_{02} - h_{01})} \quad ...(19.36)$$

where the suffix 0 indicates the stagnation value. Stagnation enthalpy is defined as

$$h_0 = h + \frac{C^2}{2}, \text{ where } C \text{ is the velocity of gas.}$$

The factors which affect the efficiency of the MHD converters are:

(i) The power consumed in magnetizing the magnet.

(ii) Dissipation of energy in the internal resistance of the ionized gas.

(iii) The space charge barrier at the electrode surfaces.

(iv) Heat transfer through the electrode and insulator walls.

(v) Fluid friction losses.

(vi) The Hall effect losses cause a voltage gradient in a direction perpendicular to the applied magnetic field and current flow,

In spite of these shortcomings, the MHD power plants offer promise of large scale power production devices at voltage varying from 1000 to 10000 volts. They are considered as topping generators for conventional gas and steam turbine cycles.

The main problem in this generator is the selection of a material that can withstand the high temperature of the ionised gas. Its efficiency is also low because some power is consumed for magnetising the magnet. Small scale MHD generators are uneconomical because of the low efficiency due to fact that it consumes a large proportion of its power output to magnetise the magnet.

In USSR, MHD is in commercial use for electricity generation.

19.5 SOLAR CELLS POWER PLANT

Solar cells are semiconductor devices which convert sunlight directly into electricity. The selenium solar cell, widely used for a long time (in photographic light meters, for example)

is not suited for space application because its energy conversion efficiency is only 0.3% and cell area required to produce one watt is nearly 0.27 m^2. The other type of solar cell that has been developed for space use is the silicon crystal cell with an efficiency of nearly 20%. Solar cells are ideally suited for such space applications as satellites and space problems where small quantities of power are required and sunlight is available in plenty. The cost of electricity production by this method was very high (₹ 5,000 per watt) in the beginning but now due to cheap production cost of solar cells module it is about ₹ 03 per watt which is at par with the power obtained by conventional power plants. Extensive research is going on in the development of efficient solar cells. A solar photovoltaic cell is made of two types of semiconductors called *p*-type and *n*-type silicon. The *p*-type silicon is produced by adding atoms such as boron or gallium which have one less electron in their outer orbit. The *n*-type silicon is made by including atoms of phosphorus which have one more electron in its outer energy level, not four. When sunlight falls on a solar cell, electrons in the silicon are ejected, which causes the formation of holes due to vacancies left behind by the escaping electrons. If this is happening in the electric field, obviously, the field will move electrons to the *n*-type layer and holes to the *p*-type layer. If *p*- and *n*-type layers are connected with metallic wire, the electrons will travel from the *n*-type layer to the *p*-type layer by crossing the depletion zone and then will travel through external wire back to the *n*-type layer which causes a flow of electricity (Fig. 19.8).

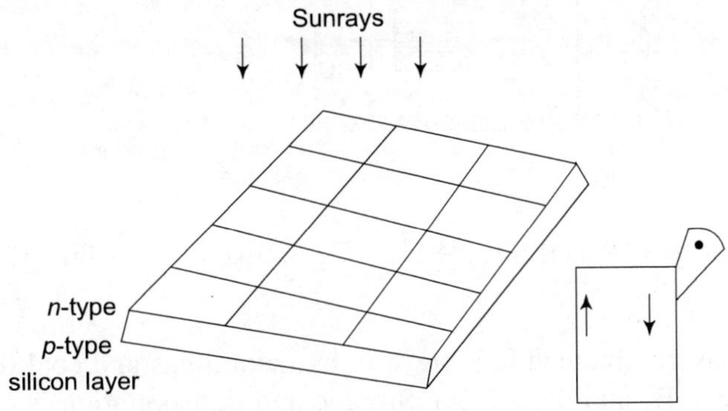

Fig. 19.8 Solar cells

Electricity generation by solar cells using silicon crystal at small scale in Indian villages is common now. Now, India is producing electricity by using solar cell in a highway. In tropical country, solar cell is a boon due to abundance of sunlight. Now solar farms in a big way have come up in the world to generate green energy. The electricity so generated is directly fed to the grid. In India, the solar cells installed capacity was 56.951 GW as of 1st June, 2022. The world is going in a big way for generating green direct energy through solar cell panels which are now very cheap.

19.6 FUEL CELLS

A fuel cell is an electrochemical device which converts the chemical energy of conventional fuel, such as hydrogen, hydrocarbon, etc., directly into electrical energy. Broadly speaking, it consists of a *non-consumable anode, cathode and an electrolyte* (invariant), fuel and oxidant supply system and suitable control. In rural areas, H_2 may be obtained from biomass whereas in urban areas it may be obtained from sewage waste. By this way, fuel cell may be cheaper. H_2 may be obtained by electrolysis process which is costly but gives pure green energy. It

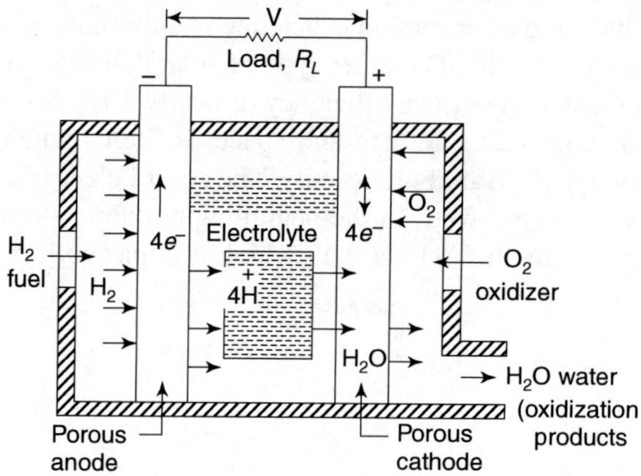

Fig. 19.9 (a) Simple $H_2 - O_2$ fuel cell

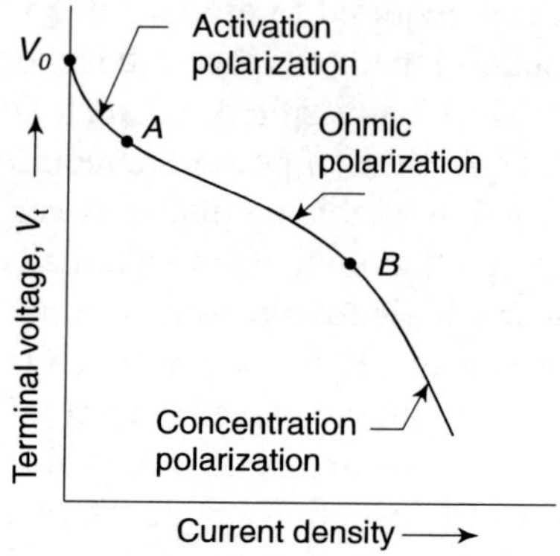

Fig. 19.9 (b) Polarization curve for a fuel cell

may be obtained in a big way by natural gas and coal that may be costly and impure (about 95% H_2 and 05% CO_2). Storage and transportation is costly. Now (H_2-O_2) driven fuel cells are being used in passenger cars and buses more in Japan and next in USA. H_2 is the green future energy. Figure 19.9 (a) shows a schematic diagram of $H_2 - O_2$ fuel cell. The hydrogen gas serves as a fuel while O_2 serves as an oxidiser. Each gas is introduced at high pressure into separate chambers. H_2 is supplied to the anode side and oxygen to the cathode side. *Both electrodes are made of porous material to permit passage of gas. The electrolyte is placed between the electrodes. The electrolyte may be solid or liquid.* Porous ceramic (zirconia oxide) polymers and alkaline solution (potassium hydroxide) have been successfully used as electrolytes. When a solid electrolyte is used, its surface is first impregnated with a catalyst to facilitate the ion migration process. Impregnation allows the two sides of the electrolyte to be used as a electrodes. In the case of liquid electrolyte, separate electrodes are necessary. When the electrodes are connected to a resistive load, current flows. *Now, hydrogen fuel cells are being used to drive passenger cars and buses.* Hydrogen is future green energy.

The principle on which the $H_2 - O_2$ fuel cell works is as follows: At the anode-electrolyte interface, hydrogen, molecules dissociate, forming hydrogen ions and four free electrons per two moles of H_2 molecule. The reaction at the anode surface may be written as

$$2H_2 \rightarrow 4H^+ + 4e^- \qquad ...(19.37)$$

These electrons flow through the external circuit where, they do electrical work and return to the fuel cell at the anode. The hydrogen ions ($4H^+$) migrate through the electrolyte to its cathode interface, where they combine with the returning electrons and with oxygen to form water. The reaction at the cathode is written as

$$4e^- + 4H^+ + O_2 \rightarrow 2H_2O \qquad ...(19.38)$$

The overall reaction for the fuel cell is just the sum of above two equations

$$2H_2 + O_2 \rightarrow 2H_2O$$

In account of the chemical reaction, a quantity of energy representing the enthalpy of combustion of the fuel $\bar{h}°_f$ is released. A part of this energy is available for conversion to electrical energy, the maximum amount is equal to the change in Gibbs function $(w_e = \bar{g}_1 - \bar{g}_2)$. *The remaining part of the energy remains as heat. The electrical energy output per mole of O_2* or per 2 mole of H_2 is expressed as

$$w_e = 4\, N_0 . e . V_r \qquad ...(19.39)$$

where V_r = voltage developed by the reversible cell

N_0 = Avogadro's number

e = electronic charge

The maximum value of electrical work (w_e) under reversible steady-state conditions will be equal to the change in Gibbs free energy,

$$w_e = \bar{g}_1 - \bar{g}_2 \qquad ...(19.40)$$

where $\bar{g}_1$ and $\bar{g}_2$ are the Gibbs free energies of the reactants and products respectively. Hence, the maximum voltage that can be developed in the reversible fuel cell is

$$V_r = \frac{\bar{g}_1 - \bar{g}_2}{n \cdot N_0 \cdot e} \qquad ...(19.41)$$

where n = number of electrons released per molecule of O_2. The maximum ideal efficiency of the fuel cell is

$$\eta_{fc} = \frac{w_e}{(-\bar{h}^{\circ}_f)} = \frac{\bar{g}_1 - \bar{g}_2}{(-\bar{h}°_f)} = \frac{nN_0 \cdot e \cdot V_r}{(-\bar{h}°_f)} \qquad ...(19.42)$$

It is to be noted that in the $H_2 - O_2$ fuel cell, the enthalpy of formation $\bar{h}°_f$, i.e., $\bar{h}°_f$ *is – 286 kJ/gm mole.* **This gives the efficiency of $H_2 - O_2$ fuel cell equal to 83%.**

Fuel cells are not limited by the Carnot efficiency and theoretical efficiency as high as can be expected of it. But high efficiencies are possible only at relatively light load and very low voltage (1 volt). *The main advantages of the fuel cells are its simplicity, its high power-to-weight ratio, emission free and higher efficiency. The disadvantages are its high cost, need for equipment to control the flow of reactants to match the reaction rates, problems of selection of materials and short life, especially, at high temperatures.*

The performance of a fuel cell has been found to depend primarily on the current density (amperes per unit area) and temperature. Figure 19.9 (b) shows a typical voltage current density for a fixed temperature. ***This is known as polarization curve***. Due to the interference at the electrodes the measured no-load voltage (V_0) is less than that for the ideal reversible case, V_r.

It is obvious from Fig. 19.9 (b) that the cell voltage drops rapidly with light loads. This is due to the irreversible effects at the electrodes and is known as *an* ***activation polarization***. The main reason for the linear drop in the operating range *AB* is the ohmic resistance in the electrolyte. Beyond the point *B* electrostatic effects, due to concentration gradients in the electrolyte, begins to limits the current, hence the term concentration polarization for this region of operation.

For an actual fuel cycle, the voltage (V_a) is less than V_r. Hence, the efficiency,

$$\eta_0 = \frac{V_a\, n \cdot N_0 \cdot e}{-\bar{h}^\circ{}_f} \quad \text{...(19.43)}$$

With $V_a = 0.6$ volt, $\eta_0 = 60\%$ at a current density near 1075 amp/m^2. These are the fuel cells which have been used in *Apollo system.*

Now, improved version of H_2–O_2 fuel cells are coming up with efficiency at high load and high voltage. It has now achieved efficiency up to 60%. The main problems are storage and transportation. It is stored at very high pressure even up to 700 bar at 0°C but practically at 20°C as it is highly inflammable.

Example 19.4 A hydrogen-oxygen fuel cell operates at 298 K. Determine the voltage output of the cell, the efficiency and the electrical work output per mole H_2 consumed and per mole of H_2O produced. Also find the heat transfer to the atmosphere.

Solution At 298 K.

$$\bar{g}_1 - \bar{g}_2 = 2{,}37{,}350 \text{ kJ/kg mole } H_2 \text{ and } \bar{h}^\circ{}_f = -2{,}86{,}030 \text{ kJ/kg mole } H_2$$

According to the reaction equation, only two electrons are transferred per mole of H_2. Hence, the voltage developed in the reversible cell,

$$V_r = \frac{\bar{g}_1 - \bar{g}_2}{n\, N_0 \cdot e}$$

Here,

$$N_0 . e = (6.025 \times 10^{23} \text{ mole/g mole}) \; 1.602 \times 10^{-19} \text{ coulomb)}$$

$$= 96{,}500 \text{ coulmbs/g mole} = 96{,}548 \text{ J volt-g-mole}$$

$$\therefore \quad V_r = \frac{2{,}37{,}350}{2 \times 96{,}548} = 1.229 \text{ volt} \qquad \textbf{Ans.}$$

$$\text{Efficiency} = \frac{\bar{g}_1 - \bar{g}_2}{-(\bar{h}^\circ{}_f)} = \frac{237350}{286032} = 82.99\% \qquad \textbf{Ans.}$$

The electrical work output per mole of fuel is calculated by

$$w_e = \bar{g}_1 - \bar{g}_2 = \Delta\bar{g} = 237350 \text{ kJ/kg mole } H_2$$

For each mole of fuel, there is one mole of H_2O produced, therefore, the heat transfer,

$$q = T\Delta s = \bar{h}^\circ{}_f - \Delta\bar{g} = -286030 + 237350 = -48680 \text{ J/mole} \qquad \textbf{Ans.}$$

The negative sign shows the heat is removed from the cell.

19.7 FUEL CELL SYSTEM UNDER DEVELOPMENT

The development of fuel cells are taking place in many directions owing to its usefulness. Figure 19.10 shows a pictorical representation of various types of fuel cells which are under development stage. Out of all fuel cells, hydrogen-oxygen fuel cell is the most developed fuel cell system but now molten carbonate fuel cell is promising one.

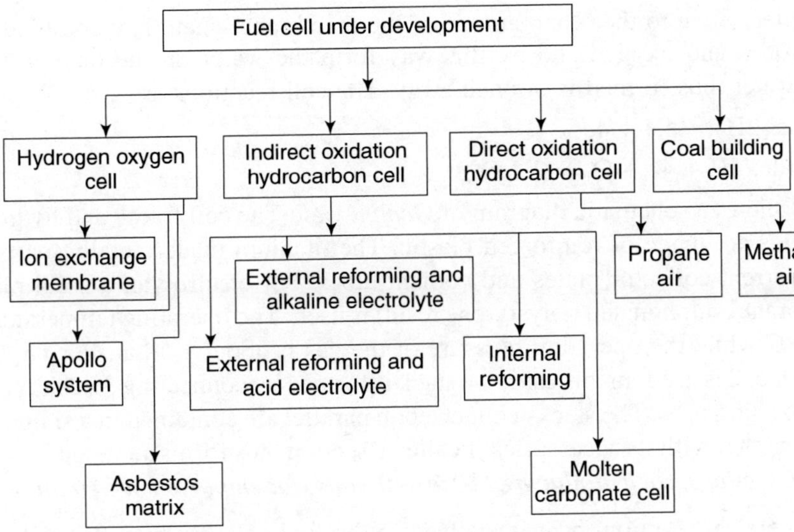

Fig. 19.10 Fuel cells under development

19.8 HYDROGEN–OXYGEN FUEL CELLS

The principle of working of $H_2 - O_2$ fuel cell has been already discussed earlier. There are three types of highly developed fuel cells. They are:

(i) Ion exchange membrane cell of General Electric.

(ii) Porous-nickel-electrode cell of Pratt.

(iii) Asbestos matrix cell of Allis Chalmers.

The main difference between these cells are due to the nature of the electrodes, the method of containing the electrolyte and the method of water removal.

19.8.1 Ion Exchange Membrane Cell

Figure 19.11 shows the working of such a cell. This cell was used in *Gemini V mission in 1965*. Hydrogen enters from one side and oxygen (or air) from the other side the membrane electrolyte and penetrate the thin porous electrodes to react with them near the surface of the membrane. From the hydrogen electrode (i.e., anode), electrons are emitted and are conducted by the external circuit to the cathode. The hydrogen ions travel through

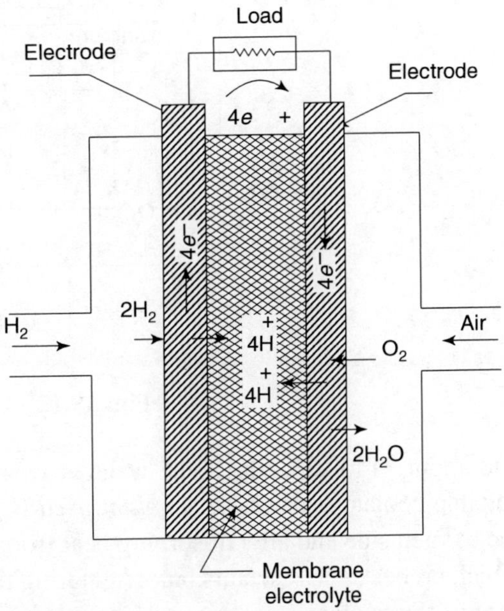

Fig. 19.11 Ion exchange membrane cell

the membrane electrolyte to the other surface of the membrane when they combine with the returning electrodes and oxygen and by this way form the water on the oxygen (cathode) electrode. The water thus formed is drained away. The cell reactions are

At anode: $2H_2 = 4H + 4e$

At cathode: $4H + 4e + O_2 = 2H_2O$

Figure 19.12 shows a schematic diagram of *Gemini Cell*. The cell frame and hydrogen and coolant minefields are made of reinforced plastic. The titanium made metallic parts include the electrodes current collector plates and coolant tubes. For electrolyte, a solid plastic ion exchange membrane (sulphonated polystyrene resin) is used. The operating temperature of the cell is 24 to 32°C while the operating pressure of the gas is about 1.5 bar. For large output, generally stack of cell is used. In this case, the stack is formed by connecting 32 cells (17.5 cm × 20 cm each in size) in series. The stacks connected in parallel are contained in a single canister. Each canister, together with its accessories is called a section thus forms an independent power supply. *Two such sections, each producing 120 to 640 watts was employed in Gemini spacecraft.*

Due to the chemical reaction, heat is evolved in the fuel cell. To keep the cell in proper working condition about 2100 kJ/kWh (if the water formed is in saturated vapour state)

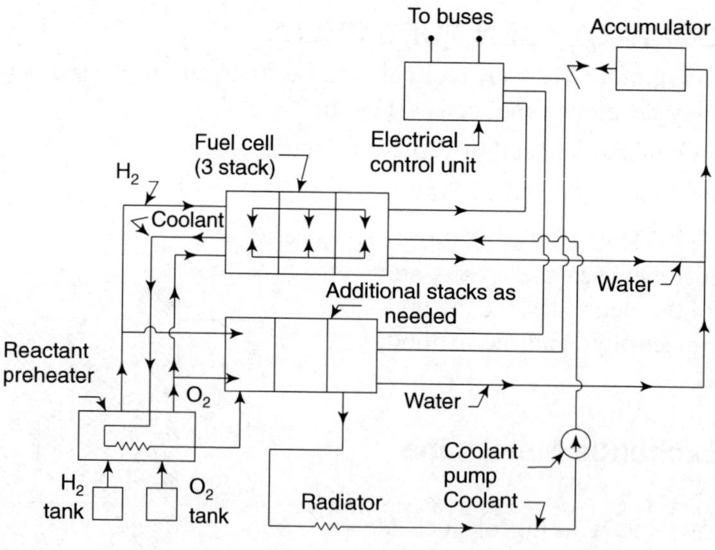

Fig. 19.12 Gemini fuel cell

and about 3140 kJ/kWh (if the water leaves as saturate liquid) has to be removed. Thus, a suitable coolant is used. *The coolant used is glycol-water mixture*. The coolant is pumped to the oxygen side and after absorbing heat from the fuel cell, it passes through a heat exchanger which preheats the reactants before entering the cell. A radiator is used to reject the remaining heat of the coolant.

The other function of the coolant is to remove the water formed in the cell. Nearly 0.21 kg/kWh of heat has to be removed from the cell because the hydrogen and oxygen consumption are 0.0568 and 0.3442 kg/kWh, the method of water removal is shown in Fig. 19.13.

The flow of the coolant through tubes attached to the oxygen side current collector plate, causes a temperature differential between the electrolyte-surface and collector plate. This causes the condensation of the water vapour formed at the oxygen electrode on cloth wicks placed within the channels formed inside the collector plate. By the capillary action of the wicks, this water is collected in a vessel. Note that the auxiliary heat and water removal system consumes 3 to 4% of the gross power output of the system.

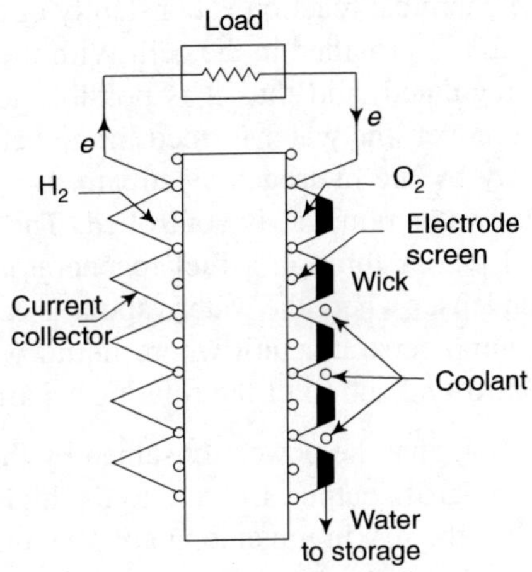

Fig. 19.13 Water removal system of Gemini cell

19.8.2 The Apollo Fuel Cell

This fuel cell has been used in *Apollo Lunar Mission in 1968*. The electrodes are made of porous nickel. Concentrated KOH liquid serves as an electrolyte, The cell operates between 200 to 250°C. The gases are supplied at 4 bar. To increase the power output, 31 cells are connected in series forming a single power unit called a *module*. Such three modules were placed at the Apollo spacecraft, each producing 563 to 1420 watts. Figure 19.14 shows a schematic diagram of Apollo fuel cell. Since the heat and water removal operations are carried

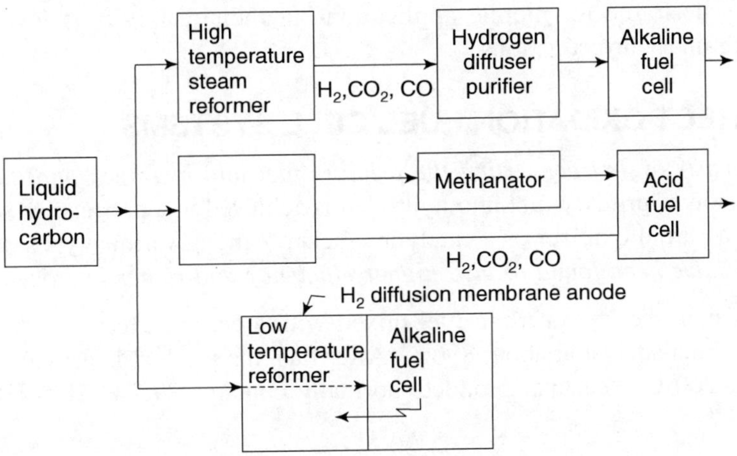

Fig. 19.14 Apollo fuel cell

out at the hydrogen electrode, hence an extra quantity of hydrogen than exactly required for the chemical reaction is constantly circulated past the hydrogen electrode. A separate coolant circuit is provided in the cell. With the help of the coolant, the inlet temperature of hydrogen is regulated, and thus it is possible to remove heat by transferring it the excess hydrogen. Moreover, the water formed in the cell evaporates from the electrodes surface and is carried away by the hydrogen as stream. By varying the humidity of the incoming gas, the rate of this water removal is controlled. The excess hydrogen (moist and hot) leaves the fuel cell and passes through a fuel regenerator where it is partly cooled and then passes through a condenser where the water vapour is condensed. The hydrogen and condensate passes through a pump separator unit where liquid water (condensate) is removed from the gas due to the centrifugal action of the rotating separator. The hydrogen gas is brought back to the fuel cell.

Note that the power consumed by the accessories such as pump and separator is about 10% of the gross output. But due to the high performance of nickel electrodes, the efficiency of the cell is the maximum among all. The total mass of the system more or less same as Gemini cell.

19.8.3 Asbestos Matrix Cell

In this cell, the electrolyte is an asbestos matrix saturated with dilute KOH. The anode is of platinum-catalysed nickel and the cathode is of silver. The operating temperature of the cell varies from 90°C to 100°C. A 2 kW stack consists ot 30–30 pairs of cell connected in series. Heat and water removal systems are quite independent.

19.9 DIRECT OXIDATION HYDROCARBON FUEL CELLS

In this case, there is a direct electrochemical oxidation of hydrocarbon fuels. There are many problems associated with this cell which are yet to be solved. Fuels which are electrochemically more reactive are found to be uneconomical. An example, gasoline and kerosene would be highly desirable for mobile application but their reactivity is low. Many other hydrocarbons are under investigations.

19.10 INDIRECT OXIDATION FUEL CELL SYSTEMS

This method consists of first converting the primary fuel into hydrogen and then fed to the cell. The conversion of primary fuel into hydrogen is achieved in a pre-reactor which may be steam reformer, a partial oxidizer or a catalytic cracker. Among various types of pre-reactor, *steam reformers have been found to give higher efficiency and richer hydrogen yield.*

In steam reformer, the hydrocarbon fuel mixed with a certain excess of steam is passed over a nickel-alumina analyst at about 800°C. After this, the product is passed over an oxide catalyst at about 200°C. The final products normally contains 79.7%, 20% H_2O, CO_2 and 0.3% C.

Figure 19.15 shows four alternative systems currently available for indirect oxidation fuel cells.

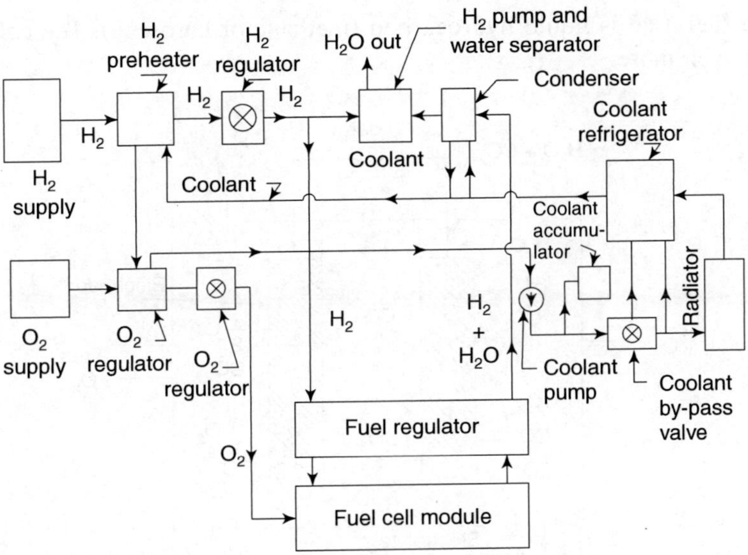

Fig. 19.15 Indirect oxidation fuel cell system

19.10.1 External Reforming and Alkaline Electrolyte

This system is the most advanced at present. In this system, the impure hydrogen from the reformer is purified by passing it through palladium silver diffusers at temperature between 350°C to 600°C. Hydrogen alone diffuses through the alloy membranes and its directly supplied to the fuel cells. Platinum is best suited for the electrocatalyst cells. KOH serves as electrolyte in American Cynamide Cells.

19.10.2 External Reforming and Acid Electrolyte

As the name implies, the impure hydrogen from the reformer is directly fed to the fuel cells. This is possible due to the fact that acid electrolyte (phosphoric or sulphuric) is not affected by CO_2. To protect the platinum anode catalyst from the harmful effect of CO_2, a methanator is placed between the reformer and fuel cell.

19.10.3 Internal Reforming Cell

This process is under development at Pratt and Whitney Aircraft and Leesona Mass Laboratories. In this system, the process of reforming, purification and electrochemical oxidation of the fuel are all carried out at the fuel cell anode.

19.10.4 Molten Carbonate Fuel Cell

Figure 19.16 shows the schematic diagram of molten carbonate fuel cell. In this case, the electrolyte is *molten carbonate* which is generally a mixture of lithium, potassium and sodium carbonates. The fuel used is liquid hydrocarbon (methane or kerosene). The cell operates at about 600°C or even more.

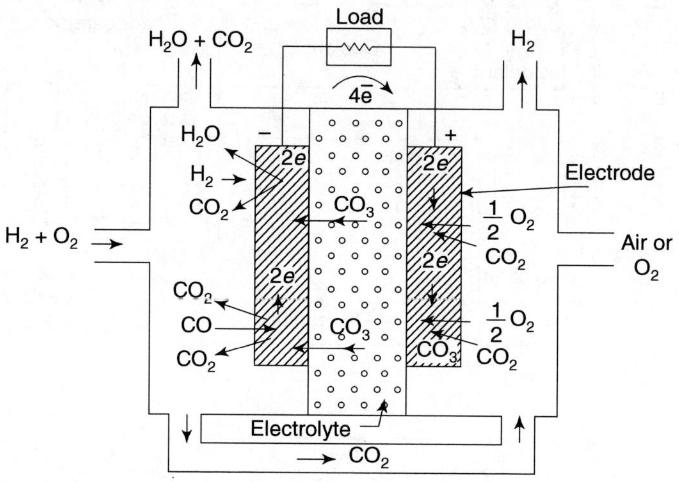

Fig. 19.16 Molten carbonate fuel cell

Electrodes (metallic) are placed in direct contact with this solid electrolyte. The fuel is reacted inside the cell forming H_2 and CO. The reactions are as follows:

At fuel electrodes: $H_2 + CO^-_3 = H_2O + CO_2 + 2e$

$$CO + CO^-_3 = 2CO_2 + 2e$$

At oxygen electrode.

The overall reaction: $H_2 + CO + O_2 = H_2O + CO_2$

The cell has potential to give the maximum efficiency than any other cell but the main drawback is the corrosive effect of molten carbonate electrolytes. However, at present, it is the most promising fuel cell.

19.11 BIOCHEMICAL FUEL CELLS

The biochemical fuel cell is a device which converts the chemical energy of fuel into electrical energy using enzymes as catalysts. The enzymes undergo an oxidation-reduction cycle by sequentially releasing and taking up electrons. The released electrons can be tapped and transferred to an electrode in order to provide electric current in the external circuit. This is under active research.

19.12 HYBRID (FUEL CELL AND GAS TURBINE) SYSTEM

Combination of *fuel cell and gas turbine plant* (hybrid system) came up in 2007 in USA. Through efficient fuel cell using natural gas as fuel, about 80% of total power is produced by fuel cell and about 20% is produced from gas turbine using the gases released at about 1000°C. *A* 40 MW plant is working in USA since 2007.

19.13 SUMMARY OF FORMULAE

For Thermoelectric

1. $P = I \,.\, \alpha \,.\, \Delta T - I^2 R$

2. $\eta = \dfrac{\alpha \cdot \Delta T - IR}{\alpha\, T_H + K \cdot \Delta T / I - IR/2}$

3. $(RK)_{\text{opt}} = \left[(\rho \cdot k_t)_n + (\rho \cdot k_t)_p\right]^2$

4. $Z = \dfrac{\alpha^2}{RK}$

5. $\eta_{\text{opt}} = \eta_c \dfrac{m_{\text{opt}} - 1}{m_{\text{opt}} + T_c / T_h}$ where $m_{\text{opt}} = \sqrt{1 + ZT}$

For Thermionic Converter

6. $\eta = \dfrac{jV}{Q_c / A_c}$

7. $\phi = \phi_c + \phi_k - \phi_p - \phi_a$

8. $j = AT^2 \exp^{-e\phi/kT}$

9. $\dfrac{Q_c}{A_c} = j\left(+\dfrac{2k\,T_h}{e}\right) + \dfrac{Q_{\text{rad}}}{A_c}$

10. $\dfrac{{}_{\text{rad}}}{\quad} = \varepsilon\, \sigma\, (T_h^4 - T_c^4)$

For MHD Converter

11. $\eta_{\text{th}} = \dfrac{(h_{02} - h_{03}) - (h_{01} - h_{04})}{(h_{02} - h_{01})}$

For Fuel Cell

12. $\eta_{fc} = \dfrac{\bar{g}_1 - \bar{g}_2}{(-\bar{h}^\circ{}_f)} = \dfrac{n\, N_0\, e\, V_r}{(-\bar{h}^\circ{}_f)}$

13. $\eta_0 = \dfrac{V_a\, n\, N_0\, e}{-(\bar{h}^\circ{}_f)}$

PROBLEMS

Things to Think About

19.1. Which intermediate steps are eliminated in direct energy converting systems?

19.2. Recently (in 2004) two robot vehicles were sent to Mars and they are roaming and taking pictures and sending back to earth station. With what energy converting devices these two robot vehicles are equipped with?

19.3. What are the main hurdles in commercialisation of MHD power plants?

19.4. What type of energy storage device is used in solar cells power plants?

19.5. Currently hybrid energy converting systems are the talk of the day. What is this? List the maximum capacity plant is under construction with main components.

Short and Long Theoretical Questions

19.6. List the names of various direct energy converters.

19.7. What is the principle of thermoelectric converters and what are its limitations?

19.8. Discuss the thermoelectric power plant in detail by a neat sketch. Derive the expression for the thermal efficiency.

19.9. What is the principle of thermionic generators and what are its limitations?

19.10. Discuss in detail with a neat sketch the thermionic power plant. Derive the expression for its thermal efficiency.

19.11. What is the principle of MHD power plant and what are its limitations?

19.12. Discuss in detail with a neat sketch the MHD power plant. Derive the expression for its thermal efficiency.

19.13. What do you mean by solar cells power plant?

19.14. What is the principle of working of fuel cells power plant? Discuss its limitations.

19.15. Discuss the various types of fuel cells which are under development with their layouts.

Numerical Questions

19.16. A small thermoelectric generator has to be designed to produce a total power output of 20 watts. The generator is constructed of the materials 75% B_{12} T_{c3}^{-} 23% B_{12} S_e^3 (*n*-type) and 25% B_{12} T_{c3} and 75% S_{b2} T_{e2} (*p*-type). It operates between the temperature limits of 20 and 100°C. Determine the number of such couples required to produce this output. Find the heat input and heat rejection for the maximum efficiency and the maximum power output conditions.

19.17. A thermionic generator has $\phi = \phi_c + \phi_k = 4$ volts and $\phi_a = 1.5$ volts and $\phi_b = 0.15$ volts; $T_h = 2500$ K, $T_c = 1200$ K and $\varepsilon = 0.7$ for electrodes. Find the efficiency of the device.

19.18. Show that the maximum efficiency for the methane fuel cell is 93% and ideal cell voltage is 1.15 volts. Calculate the flow rate in kg/h of methane and oxygen would be required to produce a power output of 120 kW? Also, calculate the heat transfer rate.

19.19. A $H_2 - O_2$ fuel cell operates at 600°C and 1 atm. Determine the ideal-cell voltage and efficiency under these condition.

19.20. A constant-velocity MHD generator operates at pressure ratio of 3. The gas used has $\gamma = 1.35$. Find the adiabatic efficiency of this device as a function of the loading factor and plot this.

19.21. Calculate the reversible e.m.f, V_r and the thermal efficiency of carbon fuel cell operating at 25°C, if the reaction is $C + O_2 = 2CO$.

19.22. Calculate the maximum power output and the figure of merit of a thermoelectric convenor which operates between $T_h = 98$°C and $T_c = 32$°C. The average Seebeck coefficient is

2×10^{-2} volt/°C, the thermal conductance is 4.08×10^{-2} watt/°C and the resistance is 0.07 ohm. Calculate the ideal thermal efficiency of the converter.

Software Based Questions

19.23. Develop a software for the performance prediction of all types of direct energy converting devices.

19.24. Develop a software for the design of MHD, solar cells and fuel cells power plant.

19.25. Develop a software for the performance prediction and design of hybrid system.

Design and Open Ended Questions

19.26. As per worldwide energy program, 60,000 MW of electric power will be generated by fuel cells. Prepare a comprehensive report discussing about various programs including the schematic of systems. The report should also include the following:

(a) List the principal features that make fuel cell technology attractive for power generation. Is fuel cell performance limited by Carnot efficiency?

(b) The most promising fuel cells are phosphoric acid, molten carbonate and solid oxide. What are the principal features and reasons of potential applications of these?

(c) Stack life and installed cost are two parameters considered critical for fuel cell development. Discuss these and calculate the cost, in $ per kW for three types of fuel cells.

(d) Discuss the principal issues which have to be resolved for fuel cells to be widely used in power automobiles and trucks.

19.27. Recently DOE of USA has given projects to leading power plant manufacturer's to built 40 MW hybrid system (a combination of fuel all and gas turbines). The fuel used will be natural gas. Prepare a detailed report about this technology with design features and specifications including schematic layout.

Multiple-Choice Answer Type Questions

19.28. The efficiency of solar cells is around

(a) 20% (b) 30% (c) 35% (d) 40%

19.29. Which direct energy converting system is not limited by Carnot efficiency?

(a) Thermoelectric (b) Thermionic (c) MHD (d) Fuel cells

19.30. The ionisation temperature by heating gas in MHD generator lies in the range of

(a) 1500–2000°C (b) 2000°C–2500°C (c) 2500–3000°C (d) 3000–3500°C

19.31. The performance of fuel cells depends primarily on

(a) current density (b) current density and temperature

(c) terminal voltage (d) load

Answers to Multiple-Choice Questions

19.28 (a) 19.29 (d) 19.30 (b) 19.31 (b)

CHAPTER

20

Compressible Fluid Flow

OVERVIEW

In many applications of engineering interest, gases move at a relatively high velocities and appreciable changes in density and the concept of incompressible (no change in density) does not hold good. The flow through nozzles and diffusers of jet engines, wind tunnels, shock tubes, and steam ejectors are some of the important examples. *The flow associated with density change is known as compressible flow*. In this chapter, the principles involved in analysing compressible flow will be discussed.

The following basic equations are used in the analysis of compressible flow.

(i) Conservation of mass (continuity equation)

(ii) Conservation of momentum (Newton's second law of motion)

(iii) Conservation of energy (first law of thermodynamics)

(iv) Second law of thermodynamics (entropy principle)

(v) Equation of state.

All basic equations have been discussed in the preceding sections.

20.1 VELOCITY OF SOUND AND MACH NUMBER

A sound wave is a small pressure disturbance that propagates through a gas, liquid or sound at a velocity that depends on the properties of the medium. It is pertinent to study the expression of sound velocity or some velocity that relates with the other properties.

20.1.1 Modelling Pressure Waves

Consider a pressure wave in a pipe of uniform cross section moving to the right with a sonic velocity of magnitude (a) as shown in Fig. 20.1 (a). The wave is generated by a small

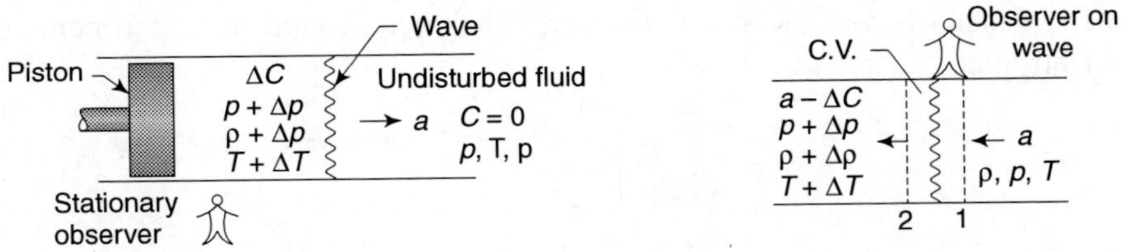

(a) (b)

Fig. 20.1 Propagation of sound waves: (a) stationary observer and (b) observer on waves displacement of piston. The fluid near the piston will have a slightly increased pressure and will be slightly more dense than the fluid away from the piston. The pressure, density, and temperature in the region to the left of the wave, which are designated simply p, ρ and T. After the wave has passed, the fluid to its left is in steady motion with a velocity of magnitude, ΔC.

From the point of view stationary observer, the wave front is shown in Fig. 20.1 (a). It is easier to analyse this situation from the point of view of an observer moving with wave. (i.e., observer at rest relative to the wave) as shown in Fig. 20.1 (b). In order to analyse, consider a control volume enclosing the wave front. To an observer moving with wave, it appears as through the fluid is moving toward the stationary wave from the right with velocity, a, pressure p, density ρ, and temperature (T) and moving away on the left with velocity, a–ΔC, pressure $\rho + \Delta\rho$, density $\rho + \Delta\rho$ and temperature $T + \Delta T$.

Under steady state condition, the conservation of mass for the control volume gives

$$\dot{m}_1 = \dot{m}_2$$

or $$\rho\, A\, a = (\rho + \Delta\rho)\, A\, (a - \Delta C)$$

or $$\rho a = \rho a + \Delta\rho\, .a - \rho\, .\, \Delta C - \Delta\rho\, .\, \Delta C$$

Neglecting the product $\Delta\rho \cdot \Delta C$, we have, $\rho\, \Delta C = a\, d\rho$

or $$\Delta C = \frac{a\, d\rho}{\rho} \qquad ...(20.1)$$

Now, apply the momentum equation to the control volume under consideration. As the thickness of the wave is small, shearing forces at the wall are negligible. The effect of gravity is also ignored. Hence, the only significant forces on the control volume in the direction of flow are forces due to pressure at the inlet and exit. Thus,

$$pA - (p + \Delta p)\, A = \dot{m}\, (a - \Delta C) - \dot{m}\, a = \dot{m}\, (a - \Delta C - a)$$

or $$\Delta p = \rho\, a\, \Delta C \qquad ...(20.2)$$

From Eqs. (20.1) and (20.2), we have $a = \sqrt{\dfrac{\Delta p}{\Delta \rho}}$ ...(20.3)

20.1.2 Sound Waves

In sound waves, the differences in pressure, density, and temperature across the wave are quite small. Hence, the ratio, $\Delta p/\Delta p$, in Eq. (20.3) can be interpreted as the derivative of pressure with respect to density across the wave. Further, it has been found by experiments that the process is isentropic (i.e., $s = c$). Thus,

$$a = \sqrt{\left(\frac{\partial p}{\partial \rho}\right)_s} \quad \text{...(20.4)}$$

In terms of specific volume, $$a = \sqrt{-v^2\left(\frac{\partial p}{\partial v}\right)_s} \quad \text{...(20.5)}$$

The velocity of sound is an intensive property. Its value depends on the state of the medium through which sound propagates.

20.1.3 Velocity of Sound in an Ideal Gas

For an ideal gas in isentropic process, we have,

$$pv^\gamma = \text{constant} \quad \text{or} \quad \frac{p}{\rho^\gamma} = \text{constant}$$

Differentiating the above, we have, $$\frac{dp}{p} - \gamma\frac{d\rho}{\rho} = 0$$

or $$\frac{dp}{d\rho} = \gamma\frac{p}{\rho} \quad \text{or} \quad \frac{\partial p}{\partial \rho} = \gamma\frac{p}{\rho}$$

Since, $$a^2 = \frac{\partial p}{\partial \rho}$$ and $p = \rho RT$ we have,

$$a^2 = \gamma RT \quad \text{or} \quad a = \sqrt{\gamma RT} \quad \text{...(20.6)}$$

20.1.4 Mach Number (*M*)

It is defined as the ratio of velocity (C) at a state in the flowing fluid to the value of sonic velocity (a) at the same state.

Thus, $$M = \frac{C}{a} \quad \text{...(20.7)}$$

When $M > 1$, the flow is called *supersonic*, when $M < 1$, the flow is *subsonic* and when $M = 1$, the flow is *sonic*. The term *hypersonic* is used for flows with Mach numbers much greater than one, and the term *transonic* refers to the flow when the Mach number is close to unity.

20.1.5 Stagnation Properties

It is a common practice to deal with compressible flows with properties evaluated at a reference state known as ***stagnation state***. *Stagnation state is defined as the state a flowing fluid would attain if it were decelerated to zero velocity isentropically.* As mentioned, this is a reference state and is commonly designated with the subscript zero. The stagnation enthalpy (h_0) is related to the enthalpy (h) and velocity of the moving fluid (Fig. 20.2) by

$$h_0 = h + \frac{C^2}{2} \quad ...(20.8)$$

For an ideal gas, $h = h\ (T)$ and c_p is constant, therefore

$$h_0 - h = c_p\ (T_0 - T) \quad ...(20.9)$$

From Eqs. (20.8) and (20.9), we have, $c_p\ (T_0 - T) = \frac{C^2}{2}$

or
$$\frac{T_0}{T} = 1 + \frac{C^2}{2\, c_p\, T} = 1 + \frac{C^2}{2}\frac{(\gamma - 1)}{\gamma RT}$$

or
$$\frac{T_0}{T} = 1 + \frac{\gamma - 1}{2} M^2 \quad ...(20.10)$$

The stagnation pressure (p_0) may be also related to Mach number

$$\frac{p_0}{p} = \left(\frac{T_0}{T}\right)^{\gamma/(\gamma - 1)} = \left[1 + \frac{\gamma - 1}{2} M^2\right]^{\gamma/(\gamma - 1)} \quad ...(20.11)$$

Similarly,
$$\frac{\rho_0}{\rho} = \left[1 + \frac{\gamma - 1}{2} M^2\right]^{1/(\gamma - 1)} \quad ...(20.12)$$

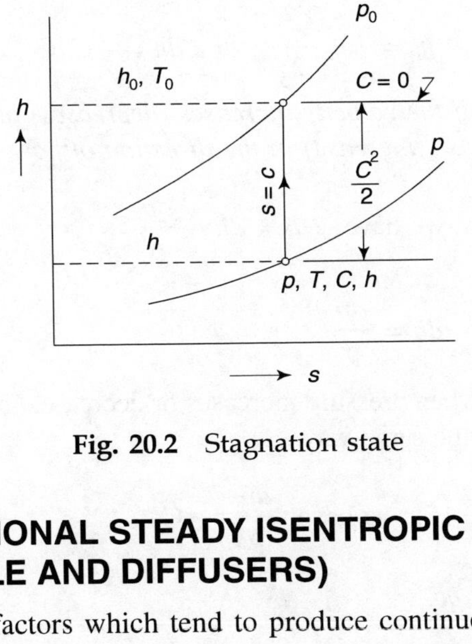

Fig. 20.2 Stagnation state

20.2 ONE-DIMENSIONAL STEADY ISENTROPIC FLOW THROUGH DUCT (NOZZLE AND DIFFUSERS)

The three most common factors which tend to produce continuous changes in the state of the flowing stream are:

(a) Change in cross-sectional areas

(b) Wall friction

(c) Energy effects, such as external heat exchange, combustion, etc.

However, in the present study we shall discuss only the effects of changes in cross-sectional area on fluid, neglecting wall fricition and energy effects.

20.2.1 Effects of Area Change in Subsonic and Supersonic Flows

A nozzle is a smooth duct of varying cross sections in which pressure or head energy is converted into kinetic energy whereas a diffuser converts kinetic energy into pressure or heat energy.

Consider a control volume enclosing a nozzle or diffuser. At steady state, the mass flow rate is constant, so

$$\rho AC = \text{constant}$$

In differential form, $d(\rho AC) = 0$ or $ACdp + \rho AdC + \rho CdA = 0$

On dividing each term by ρAC, we have continuity equation

$$\frac{d\rho}{\rho} + \frac{dC}{C} + \frac{dA}{A} = 0 \quad \text{...(20.13)}$$

Since stagnation enthalpy and stagnation temperature do not change in adiabatic flow, we have

$$h_0 = h + \frac{C^2}{2} \quad \text{or} \quad dh = -\,CdC \quad \text{...(20.14)}$$

This equation shows that if the velocity increases (decreases) in the direction of flow, the specific enthalpy must decrease (increase) in the direction of flow, and conversely.

From the property relation, we have $Tds = dh - \dfrac{dp}{\rho}$

For isentropic flow, $$dh = \frac{dp}{\rho} \quad \text{...(20.15)}$$

This equation shows that when pressure increases or decreases in the direction of flow, the specific enthalpy changes in the same way.

From Eqs. (20.14) and (20.15), we have, $\dfrac{dp}{\rho} = -\,CdC$

or $$\frac{dC}{C} = -\frac{1}{C^2}\frac{dp}{\rho} \quad \text{...(20.16)}$$

Substituting the value of $\frac{dC}{C}$ from Eq. (20.16) in Eq. (20.13), we have

$$\frac{dA}{A} = -\frac{dC}{C} - \frac{d\rho}{\rho} = \frac{1}{C^2}\frac{dp}{\rho} - \frac{dp}{\rho} = \frac{dp}{\rho C^2}\left[1 - \frac{d\rho}{dp}C^2\right)$$

Since $$\frac{dp}{d\rho} = -a^2$$

so therefore, $$\frac{dA}{A} = \frac{dp}{\rho C^2}[1 - M^2] \quad ...(20.17)$$

or $$\frac{dA}{A} = -\frac{dC}{C}(1 - M^2) \quad ...(20.18)$$

Four cases may arise based on the approaching flow and balancing the equation.

Case 1 : Subsonic nozzle : $M < 1, dV > 0 \rightarrow dA < 0$: The duct converges in the direction of flow (Fig. 20.3a)

Case 2 : Subsonic diffuser : $M < 1, dV < 0 \rightarrow dA > 0$. The duct diverges in the direction of flow (Fig. 20.3b)

Case 3 : Supersonic nozzle : $M > 1, dV < 0 \rightarrow dA > 0$: The duct diverges in the direction of flow (Fig. 20.3c)

Case 4 : Supersonic diffuser : $M > 1, dV > 0 \rightarrow dA < 0$. The duct converges in the direction of flow (Fig. 20.3d).

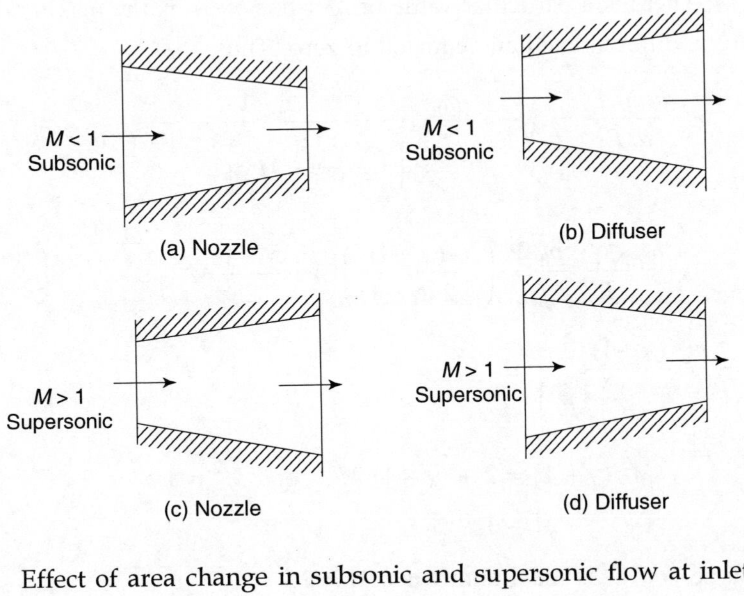

Fig. 20.3 Effect of area change in subsonic and supersonic flow at inlet of duct

The conclusion reached above concerning the nature of flow in subsonic and supersonic cases are summarized in Fig. 20.3.

So from the above results we observed and it is the approaching flow and the required flow at the exit that decides the shape of nozzle and diffuser. So if at the inlet the flow is subsonic in nozzle and if we want supersonic flow at exit, it will be convergent-divergent nozzle.

20.3 CRITICAL PROPERTIES: CHOKING IN ISENTROPIC FLOW

Consider an isentropic flow through a nozzle. Let $\dot{m}$ be the mean flow through the nozzle at any cross-sectional area A with velocity C.

$$\dot{m} = \rho AC$$

or

$$\frac{\dot{m}}{A} = \frac{\rho}{RT}(aM) = \frac{p}{RT}\sqrt{\gamma RT}\cdot M = \frac{p}{p_0}p_0\sqrt{\frac{T_0}{T}}\cdot\sqrt{\frac{1}{T_0}}\sqrt{\frac{\gamma}{R}}\cdot M.$$

or

$$\frac{\dot{m}}{A} = \left(\frac{T_0}{T}\right)^{-\gamma/(\gamma-1)}\cdot\left(\frac{T_0}{T}\right)^{1/2}\cdot\frac{p_0}{\sqrt{T_0}}\frac{\sqrt{\gamma}}{R}\cdot M.$$

Substituting the value of (T_0/T) from Eq. (20.11), we have

$$\frac{\dot{m}}{A} = \sqrt{\frac{\gamma}{A}}\frac{p_0 M}{\sqrt{T_0}}\cdot\frac{1}{\left(1+\frac{\gamma-1}{2}m^2\right)^{(\gamma+1)/2(\gamma-1)}} \qquad ...(20.19)$$

In Eq. (20.19), p_0, T_0 and R are constants, so the discharge per unit area is a function of Mach number, M. There is a particular value of M when $\dot{m}/A$ is the maximum and for this Eq. (20.19) is to be differentiated and equated to zero. Thus,

$$\frac{d(\dot{m}/A)}{dM} = \sqrt{\frac{\gamma}{R}}\cdot\frac{p_0}{\sqrt{T_0}}\cdot\frac{1}{\left(1+\frac{\gamma-1}{2}M^2\right)^{(\gamma+1)/2(\gamma-1)}}$$

$$+\frac{\sqrt{\gamma}}{R}\cdot\frac{p_0 M}{\sqrt{T_0}}\left(-\frac{(\gamma+1)}{2(\gamma-1)}\right)\left(1+\frac{\gamma-1}{2}M^2\right)^{\frac{-\gamma+1}{2(\gamma-1)}-1}\left(\frac{\gamma-1}{2}\cdot 2M\right) = 0$$

or

$$1-\frac{m^2(\gamma+1)}{\left(2\left(1+\frac{\gamma-1}{2}\right)M^2\right)} = 0$$

or

$$M^2(\gamma+1) = 2+(\gamma-1)M^2 \quad \text{or} \quad M^2 = 1$$

or

$$M = 1 \qquad ...(20.20)$$

So, the discharge $(\dot{m}/A)$ is maximum when $M = 1$.

As $C = aM = \sqrt{\gamma RT} \cdot M$, so logarithmic differentiation of above gives

$$\frac{dC}{C} = \frac{dM}{M} + \frac{1}{2}\frac{dT}{T} \qquad ...(20.21)$$

But $$\frac{T}{T_0} = \left(1 + \frac{\gamma - 1}{2} M\right)^{-1}$$, so its logarithmic differentiation yield

$$\frac{dT}{T} = -\frac{(\gamma - 1)M^2}{1 + \frac{\gamma + 1}{2}M^2} \cdot \frac{dM}{M} \qquad ...(20.22)$$

Substituting the value of dT/T into Eq. (20.21), we have

$$\frac{dC}{C} = \frac{1}{1 + \frac{\gamma - 1}{2}M^2} \cdot \frac{dM}{M} \qquad ...(20.23)$$

Substituting the value of dC/C from Eq. (20.18) into Eq. (20.23), we have

$$\frac{dA}{A} = \frac{(M^2 - 1)\,dM}{M\left(1 + \frac{\gamma - 1}{2}M^2\right)} \qquad ...(20.24)$$

If we substitute $M = 1$ either in Eqs. (20.17), (20.18) or in Eq. (20.24), it gives $dA = 0$ or A = constant. This shows that sonic velocity ($M = 1$) occurs only at the throat and nowhere else, and this happens only at the condition of maximum discharge which is also termed as choking flow or critical condition and designated by a superscript asterisk (*).

In convergent nozzle, the velocity increases from subsonic condition to supersonic condition while diffusing flow occurs when inlet is supersonic and exit is subsonic in the same converging-diverging passage.

At the condition of **choking flow**, we have

$$\frac{T_0}{T^*} = 1 + \frac{\gamma - 1}{2}M^2 = 1 + \frac{\gamma - 1}{2} \times 1 = \frac{\gamma + 1}{2}$$

or $$\frac{T^*}{T_0} = \frac{2}{\gamma + 1} \qquad ...(20.25)$$

The critical pressure ratio, p^*/p_0, is expressed as

$$\frac{p^*}{p_0} = \left(\frac{2}{\gamma + 1}\right)^{\gamma/(\gamma - 1)} \qquad ...(20.26)$$

For diatomic gases, like air, $\gamma = 1.4$, and the critical pressure ratio is given by

$$\frac{p^*}{p_0} = \left(\frac{2}{2.4}\right)^{1.4/0.4} = 0.528 \qquad ...(20.27)$$

$$\frac{T^*}{T_0} = \frac{2}{1.4+4} = \frac{2}{2.4} = 0.833 \qquad ...(20.28)$$

and
$$\left(\frac{\rho^*}{\rho_0}\right) = \left(\frac{2}{\gamma+1}\right)^{1/(\gamma-1)} = \left(\frac{2}{2.4}\right)^{1/0.4} = 0.634 \qquad ...(20.29)$$

At the critical condition, the mass flow rate per unit area is given by substituting

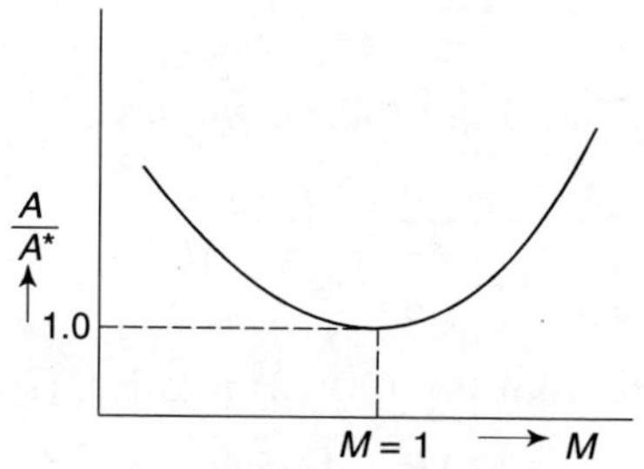

Fig. 20.4 Area ratio at critical condition as a function of Mach number in an isentropic flow through nozzle

$M = 1$ in Eq. (20.19)

$$\frac{m}{A^*} = \sqrt{\frac{\gamma}{R}} \cdot \frac{p_0}{\sqrt{T_0}} \cdot \frac{1}{\left(\frac{\gamma+1}{2}\right)^{(\gamma+1)/2(\gamma-1)}} \qquad ...(20.30)$$

From Eqs. (20.19) and (20.30), we have the area ratio at critical condition and is given by

$$\frac{A}{A^*} = \frac{1}{M}\left[\left(\frac{2}{\gamma+1}\right)\left(1+\frac{\gamma-1}{2}M^2\right)\right]^{(\gamma+1)/2(\gamma-1)} \qquad ...(20.31)$$

The A/A^* is plotted as a function of M as shown in Fig. 20.4.

20.3.1 Dimensional Velcoity, M^*

Sometimes, a new dimensional parameter M^* is used as the value of M increases at high speed. This is defined as

$$M^* = \frac{C}{a^*} = \frac{C}{C^*} \qquad ...(20.32)$$

where $a^* = \sqrt{\gamma RT^*} = C^*$ and $M^{*2} = \frac{C^2}{a^{*2}} + \frac{C^2}{a^2} \cdot \frac{a^2}{a^{*2}} = M^2 \cdot a^2$

But we have for adiabatic flow for ideal gas

$$h_0 = c_p T + \frac{C^2}{2} = = c_p T_0 \quad \text{or} \quad \frac{C^2}{2} + \frac{\gamma RT}{\gamma - 1} = \frac{\gamma RT_0}{\gamma - 1}$$

or
$$\frac{C^2}{2} + \frac{a^2}{\gamma - 1} = \frac{a_0^2}{\gamma - 1} \qquad ...(20.33)$$

$\therefore$
$$\frac{a_0}{a^*} = \frac{\sqrt{\gamma RT_0}}{\sqrt{\gamma RT_0}} = \sqrt{\frac{T_0}{T^*}} = \sqrt{\left(\frac{\gamma + 1}{2}\right)} \qquad ...(20.34)$$

From Eqs. (20.33) and (20.34), we have, $\frac{C^2}{2} + \frac{a^2}{\gamma - 1} = \frac{\gamma + 1}{2} a^{*2} \cdot \frac{1}{\gamma - 1}$

or
$$\frac{C^2}{a^{*2}} + \frac{2}{\gamma - 1} \frac{a^2}{a^{*2}} = \frac{\gamma + 1}{\gamma - 1} \quad \text{or} \quad M^{*2} + \frac{2}{\gamma - 1} \frac{M^{*2}}{M^2} = \frac{\gamma + 1}{\gamma - 1}$$

or
$$M^{*2} = \frac{(\gamma + 1) M^2}{1 + \frac{\gamma - 1}{2} M^2} \qquad ...(20.35)$$

Now from Eq. (20.35), the following relations may be established

When $M < 1, \; M^* < 1$ When $M > 1, \; M^* > 1$

When $M = 1, \; M^* = 1$ When $M = 0, \; M^* = 0$

When
$$M = \infty, \; M^* = \sqrt{\frac{\gamma + 1}{\gamma - 1}} \qquad ...(20.36)$$

Thus, we see that at $M = \infty$, the relation between M and M^* is not identical.

20.4 EFFECT OF BACK PRESSURE ON THE PERFORMANCE OF NOZZLE FLOW

The back pressure is the pressure in the exhaust region outside the nozzle.

20.4.1 Converging Nozzle

Figure 20.5 shows a converging nozzle with stagnation conditions at the iniet, discharging into a region in which the back pressure (p_B) can be varied at will by using the valve located at the discharge size. For the series of cases labelled '*a*' rhrough '*e*', let us consider how the mass flow rate ( ) and nozzle exit pressure (p_E) vary as the back pressure is decreased while maintaining the inlet conditions fixed.

When $p_B = p_E = p_e$, there is no flow, so $\dot{m} = 0$. This corresponds to case of Fig. 20.5(a). With the decrease in back pressure, as in cases *b* and *c*, there will be flow through nozzle. As long as the flow is subsonic at the exit, information about changing conditions in the exhaust region can be transmitted upstream. The results show that the decrease in back pressure thus results in greater mass flow rates and new pressure variations within the nozzle. In each instance, the velocity is subsonic through the nozzle exit. The corresponding pressure is denoted by p^*, called the ***critical pressure*** and this case is represented by point '*d*' on the Fig. 20.5(b).

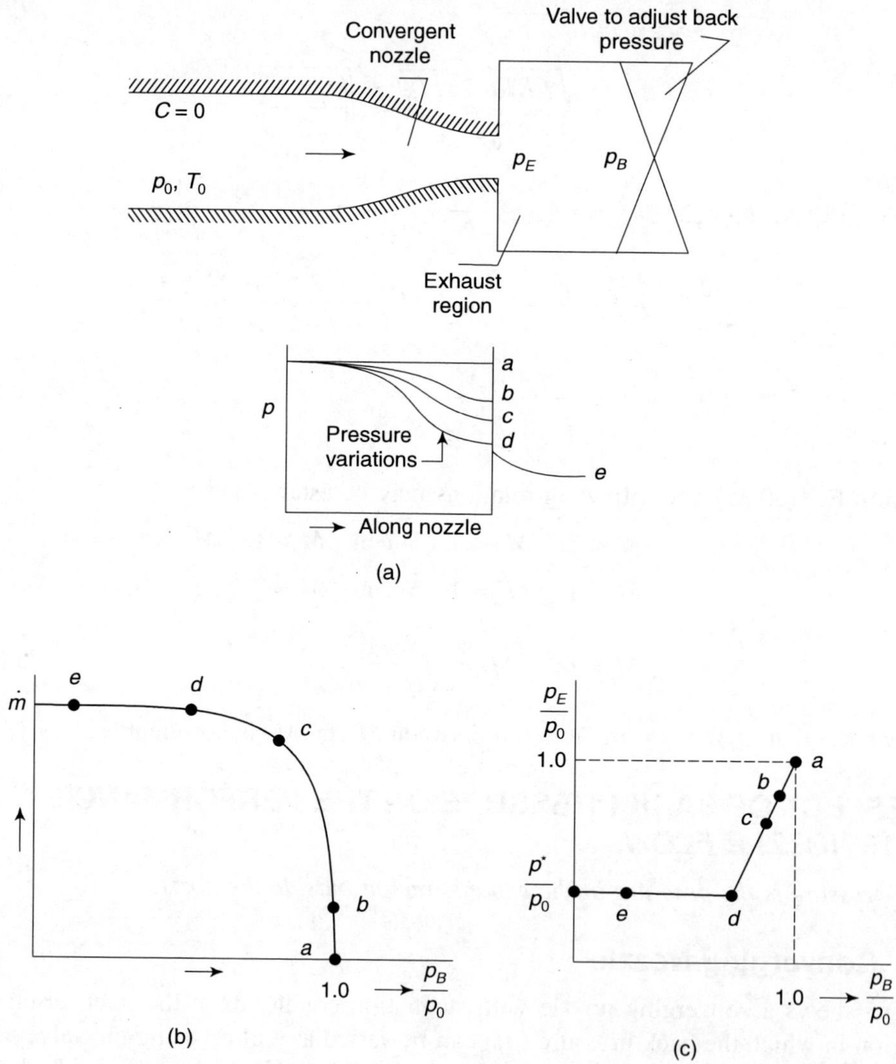

Fig. 20.5 Effect of back pressure on flow through convergent nozzle

In the previous section, it has been established that the Mach number cannot increase beyond unity in a converging nozzle. Now, consider the case when the back pressure is reduced below the value of p^* as represented by case, *e*. It is observed that reduction in p_B below p^* has no effect of flow conditions in the nozzle. Neither the pressure variation within the nozzle nor the mass flow is affected. Under this condition, as mentioned earlier, **the nozzle is said to be choked**. Under the choked condition, the mass flow rate is the maximum possible for the given stagnation conditions. As shown in Fig. 20.5 (a) when p_B less than p^*, the flow expands outside the nozzle to match the lower back pressure as shown by curve, *e*.

20.4.2 Converging-Diverging Nozzles

The effects of back pressure on converging-diverging nozzle are illustrated in Fig. 20.6. The discussions will be centring around three sets of curves.

Curves a, b, c, d Consider the cases designated by curves *a*, *b*, *c* and *d*. Case a corresponds to $p_B = p_E = p_0$ which means there is no flow through nozzle. When the back pressure is slightly less than back p_0 **(case b)**, there is some flow and the flow is subsonic throughout the nozzle. The greatest velocity and lowest pressure occur at the throat and since the approaching flow to the divergent position, is subsonic, so it acts as a diffuser. If the back pressure is reduced further, **(case c)**, the mass flow rate and velocity at the throat are greater than before but still the flow remains subsonic throughout the same as the **case b**. With further reduction in back pressure the Mach number at the throat increases and attains unity **(case d)**. which is called the **choked flow** and it results in the maximum mass flow rate.

Curves e, f and g With the further reduction in back pressure, below that corresponding to **case d**, the flow in the convergent position and at the throat remains unchanged, however, the conditions within the divergent portion can be altered as illustrated by **cases e, f** and **g**. In **case e**, the fluid passing through the throat continued to expand and *becomes supersonic in the diverging portion just downstream of throat but at a certain location an abrupt change in properties occur due to the formation of normal shock wave. The condition of occurrence of shock wave is that when a supersonic flow meets either physical or flow disturbance, it forms a shock wave. Across the shock, there is a rapid and irreversible increase in static pressure accompanied by rapid decrease from supersonic to subsonic flow and decrease in stagnation pressure. Further, downstream of the shock, the diverging portion acts as a subsonic diffuser in which the fluid continues to decelerate and the static pressure increases to match the back pressure imposed at the exit by the valve.* With the further reduction in back pressure **(case f)**, the location of the shock wave moves downstream, but the quality of the flow is the same as in **case e**. As the back pressure further reduces, the shock wave location moves farther downstream of the throat until it extends at the exit **(case g)**. It is worth to note that in this case, the flow throughout the nozzle is isentropic, with subsonic flow at the convergent portion, $M = 1$ at the throat and supersonic flow in the divergent portion. *As the fluid leaving the nozzle passes through a shock, it is subsonic just downstream of the exit plane.* If the valve is removed the **case g** will represent the expansion of fluid in a correctly designed convergent-divergent nozzle.

Curves h, i and j Now consider the final **cases h, i and j** where the back pressure is less than that corresponding to **case g**. In each of these three cases, the flow through the nozzle is not affected and whatever adjustment to changing back pressure occurs, it is outside the nozzle. In **case h**, the pressure decreases continuously as the fluid expands isentropically through the nozzle and then increases to the back pressure outside the nozzle imposed by the valve. *The compression that occurs outside the nozzle involves oblique shock waves.* Whereas in the **case i**, the fluid expands isentropically to the back pressure and no shocks occur within or outside the nozzle. But in **case j**, the fluid expands isentropically through the nozzle and then expands outside the nozzle to the back pressure through ***oblique expansion waves***.

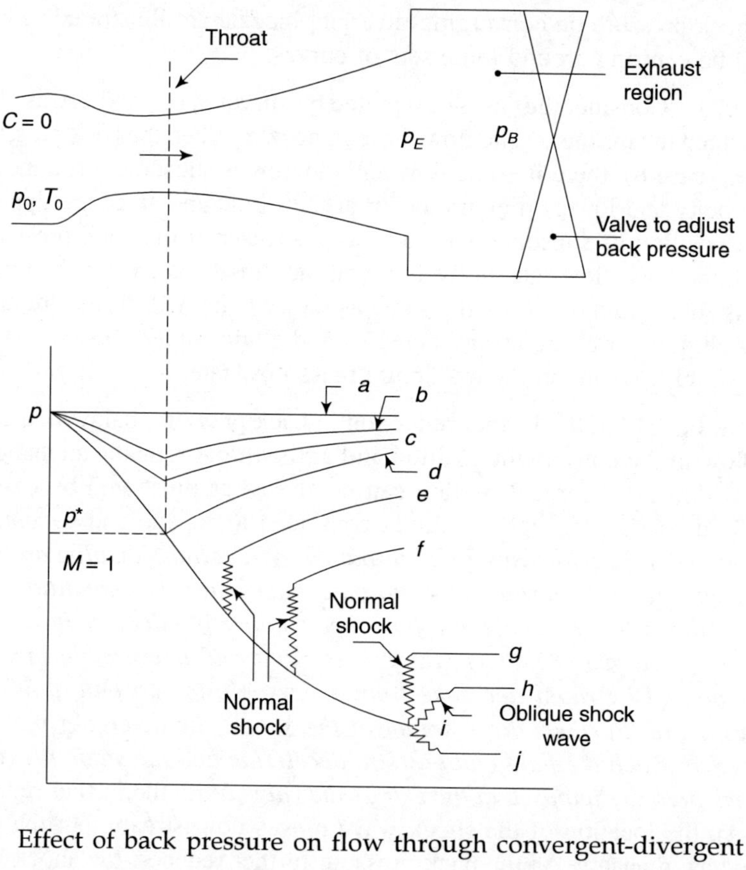

Fig. 20.6 Effect of back pressure on flow through convergent-divergent nozzle

It is to be noted that once M = 1 is achieved at the throat, the mass flow rate is unchanged in all cases starting from d to j.

20.5 FLOW ACROSS A NORMAL SHOCK WAVE

We have observed that under certain conditions, shock waves are formed in the supersonic divergent portion of the nozzle. **A shock wave is a rapid and abrupt change of state. It is a highly localized irreversibilities in the flow**. In a normal shock, this change of state occurs across a plane normal to the direction of flow. *Within the distance of mean free path of molecule, the flow passes from a supersonic state to subsonic state, the velocity decreases* *abruptly, static pressure increases abruptly and total pressure decreases abruptly*. The purpose of this section is to develop means for determining the change of state across a normal shock.

20.5.1 Modelling Normal Shocks

Figure 20.7 shows a control volume enclosing a normal shock. The control volume is assumed as steady state with $\dot{W}_c = 0$ and $\dot{Q}_{cv} = 0$ and negligible effects of potential energy. The thickness of shock wave is very small (on the order of 10^{-5} cm), so it is assumed that there is no significant change in the flow area across the shock. The fluid is assumed to be in thermodynamic equilibrium upstream and downstream of the shock, the properties of which are denoted by the subscripts x and y respectively. Mass, energy, momentum and entropy balance across the control volume may be written to know the change of state.

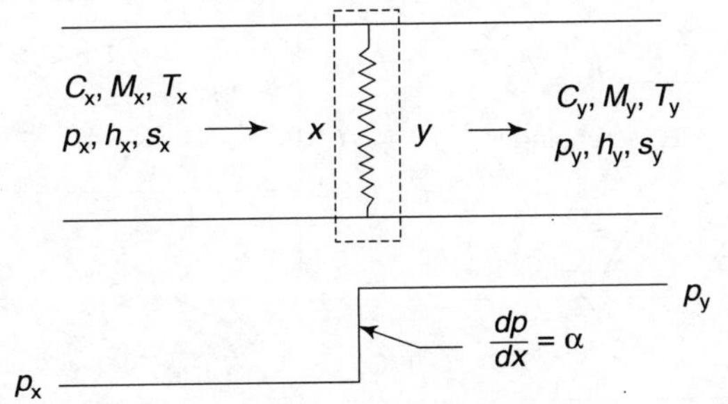

Fig. 20.7 Control volume enclosing a normal shock

Mass balance : $\rho_x C_x = \rho_y C_y = G$ ...(20.37)

where G is the mass velocity (kg/m^{2}s)

Energy balance : $h_x + \dfrac{C_x^2}{2} = h_y + \dfrac{C_y^2}{2}$ ...(20.38)

or $h_{ax} = h_{ay}$ ...(20.38a)

Momentum balance : $p_x - p_y = \rho_y C_y^2 - \rho_x C_x^2$ ...(20.39)

or $p_x + \rho_x C_x^2 = p_y + \rho_y C_y^2$

or $F_x = F_y$ ...(20.39a)

where $F = p/a + \rho A C^2$ is known as **impulse function.**

Entropy balance : $s_y - s_x = (\dot{S}_{gen})_{cv}/\dot{m}$ ...(20.40)

Combining mass and energy balance, we get, $h_0 = h + \dfrac{G^2}{2\rho^2}$ = constant

or $$h = h_0 - \frac{G^2}{2\rho^2}$$ ...(20.41)

20.5.2 Fanno and Rayleigh Lines

The variation between h and $1/\rho$ for various values of G is given in Fig. 20.8. ***The line representing the locus of points with the same mass velocity and stagnation enthalpy is called Fanno line.***

The Fanno line may also be represented on the h–s diagram. For known values of upstream properties h_x, ρ_x, p_x, and C_x, select a particular value of C_y then $\bar{\rho}_y$, h_y and s_y are calculated from mass, energy and entropy balance respectively. By repeating the calculation for various values of C_y, the *Fanno lines* may easily be constructed (Fig. 20.9). It is to be noted that since the momentum equation has not been considered the Fanno lines represent states with the same mass velocity and stagnation enthalpy, but not the same value of the impulse function.

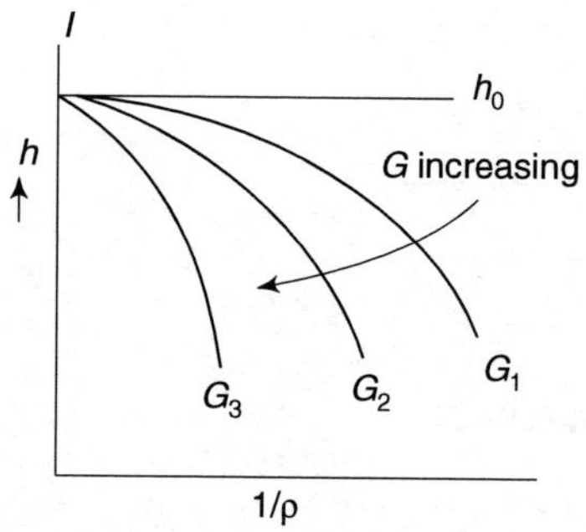

Fig. 20.8 Fanno line on h versus $1/\rho$

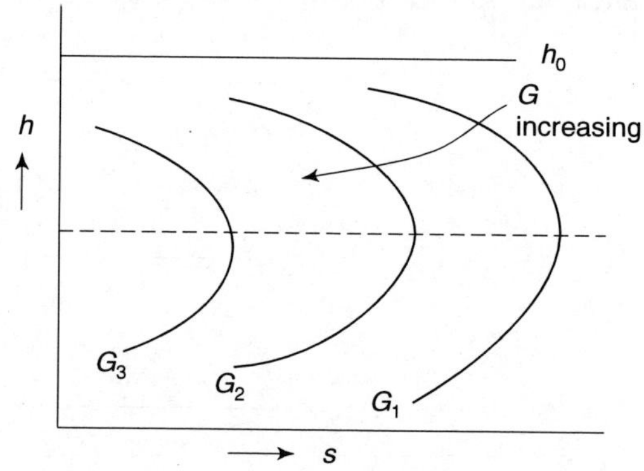

Fig. 20.9 Fanno on h–s co-ordinate

Now, consider the locus of states which are defined by the mass Eq. (20.37), the momentum Eq. (20.39) combined with property relation. The impulse function in this case becomes

$$F = \rho A + \rho A C^2$$

The impulse pressure (I) is given by, $I = \dfrac{F}{A} = p + \rho C^2 = p + \dfrac{G^2}{\rho}$ ...(20.42)

The line representing the locus of states with the same impulse pressure (I) and mass velocity (G) is called the Rayleigh line. The end states of the normal shock must lie on the Rayleigh line, since $I_x = I_y$ and $G_x = G_y$.

The Fanno and Rayleigh lines are generally drawn on h-s diagram. The mass and energy equation combined with property relation for a particular fluid give an equation that when plotted h-s diagram is called **Fanno line**. Similarly, when mass and momentum equation combined with property relation give an equation that when plotted on h-s diagram gives **Rayleigh line**. Figure 20.10 shows the Fanno and Rayleigh lines drawn on h-s diagram. *It is obvious from the diagram that the point of maximum entropy on each line, points a and b corresponds to M = 1. The upper and lower branches of each line correspond to subsonic and supersonic velocity, respectively.*

It is worth to note that the downstream state must satisfy the mass, energy and momentum equations simultaneously, so that y is fixed by the intersection of Fanno and Rayleigh lines passing through state x. *Since $s_y > s_x$, it may be concluded that the flow across the shock can only pass from x to y*. Accordingly, the velocity changes abruptly from supersonic before the shock ($M_x > 1$) to subsonic after the shock ($M_y < 1$). *The stagnation enthalpy does not change across the shock but there is appreciable decrease in stagnation pressure associated with the irreversible process occurring in the normal shock region which causes entropy generation.*

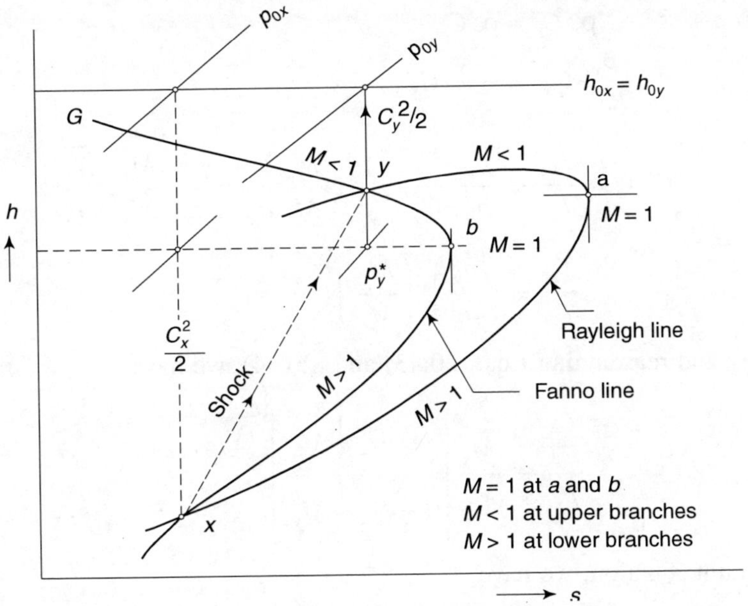

Fig. 20.10 Intersection of Fanno and Rayleigh lines h–s diagram. As a solution to normal shock equation

20.5.3 Normal Shock in an Ideal Gas

In the previous section, the methodology has been discussed to deal with the normal shock behaviour of any fluid. Here, the various governing equations pertaining to an ideal gas are to be established.

The energy for an ideal gas across the shock is given by

Energy balance $$c_p\,T_x + \frac{C_x^2}{2} = c_p\,T_y + \frac{C_y^2}{2} = c_p\,T_0$$

Here, $$h_{0x} = h_{0y} = h_0 \quad \text{and} \quad T_{0x} = T_{0y} = T_0$$

$$c_p = \frac{\gamma R}{\gamma - 1}, \quad a_x = \sqrt{\gamma R T_x} \quad \text{and} \quad a_y = \sqrt{\gamma R T_y}$$

Thus, $\dfrac{T_0}{T_x} = 1+\dfrac{\gamma-1}{2}M_x^2$ and $\dfrac{T_0}{T_{uy}} = 1+\dfrac{\gamma-1}{2}M_y^2$

or
$$\frac{T_y}{T_x} = \frac{1+\dfrac{\gamma-1}{2}M_x^2}{1+\dfrac{\gamma-1}{2}M_y^2} \qquad ...(20.43)$$

From mass (continuity) equation for a constant area duct, we have

Mass balance : $\rho_x C_x = \rho_y C_y$

or
$$\frac{p_x}{RT_x}C_x = \frac{p_y}{RT_y}C_y$$

or
$$\frac{T_y}{T_x} = \frac{p_y}{p_x}\cdot\frac{C_y}{C_x} = \frac{p_y}{p_x}\cdot\frac{M_y\,a_y}{M_x\,a_x} = \frac{p_y}{p_x}\cdot\frac{M_y}{M_x}\sqrt{\frac{T_y}{T_x}}$$

or
$$\frac{T_y}{T_x} = \left(\frac{p_y}{p_x}\right)^2\left(\frac{M_y}{M_x}\right)^2 \qquad ...(20.44)$$

After equating and rearranging Eqs. (20.43) and (20.44), we have

$$\frac{p_y}{p_x} = \frac{M_y}{M_x}\left[\frac{1+\dfrac{\gamma-1}{2}M_x^2}{1+\dfrac{\gamma-1}{2}M_y^2}\right]^{1/2} \qquad ...(20.45)$$

From momentum equation, we have

Momentum balance $p_x+\rho_x C_x^2 = p_y+\rho_y C_y^2$

or
$$p_x+\frac{\gamma\,p_x}{\gamma\,RT_x}C_x^2 = p_y+\frac{\gamma\,p_y}{\gamma RT_y}C_y^2 \quad \text{or } p_x\,(1+\gamma\,M_x^2) = p_y\,(1+\gamma\,M_y^2)$$

or
$$\frac{p_x}{p_y} = \frac{1+\gamma\,M_x^2}{1+\gamma\,M_y^2} \qquad ...(20.46)$$

After equating Eq. (20.45) and Eq. (20.46) and rearranging, we get

$$M_y^2 = \frac{M_x^2+\dfrac{2}{\gamma-1}}{\dfrac{2\gamma}{\gamma-1}M_x^2-1} \qquad ...(20.47)$$

From Eqs. (20.45), (20.46) and (20.47), we have

$$\frac{p_y}{p_x} = \frac{2\gamma}{\gamma+1}M_x^2-\frac{\gamma-1}{\gamma+1} \qquad ...(20.48)$$

Equation (20.48) gives the value of p_y as p_x, M_x and γ are known values.

Substituting the value of M_y^2 from Eq. (20.47) into Eq. (20.44), we have

$$\frac{T_y}{T_x} = \frac{\left(1+\frac{\gamma-1}{2}M_x^2\right)\left(\frac{2\gamma}{\gamma-1}M_x^2-1\right)}{\frac{(\gamma+1)^2}{2(\gamma-1)}M_x^2} \quad ...(20.49)$$

Equation (20.49), gives the value of T_y as M_x, and γ are known.

Further, $$\frac{\rho_y}{\rho_x} = \frac{p_y}{T_y}\cdot\frac{T_x}{p_x} = \frac{\left(\frac{2\gamma}{\gamma+1}M_x^2-\frac{\gamma-1}{\gamma+1}\right)\left(\frac{(\gamma+1)^2}{2(\gamma-1)}M_x^2\right)}{\left(1+\frac{\gamma-1}{2}M_x^2\right)\left(\frac{2\gamma}{\gamma-1}M_x^2-1\right)} \quad ...(20.50)$$

Equation (20.50) gives the value of ρ_y as T_s, M_x and γ are known.

As mentioned earlier, there is a severe total pressure loss across shock. **In fact, the total pressure loss is a measure of degree of irreversibility in the process.** Thus,

$$\frac{p_{0y}}{p_{0x}} = \frac{p_{0y}}{p_y}\cdot\frac{p_y}{p_x}\cdot\frac{p_x}{p_{0x}}$$

But $$\frac{p_{0y}}{p_y} = \left[1+\frac{\gamma-1}{2}M_y^2\right]^{\gamma/(\gamma-1)}$$

and $$\frac{p_{0x}}{p_x} = \left[1+\frac{\gamma-1}{2}M_x^2\right]^{\gamma/(\gamma-1)}$$

Thus, $$\frac{p_{0y}}{p_{0x}} = \left[\frac{\frac{\gamma+1}{2}M_x^2}{1+\frac{\gamma-1}{2}M_x^2}\right]^{\gamma(\gamma-1)} \Bigg/ \left[\frac{2\gamma}{\gamma+1}M_x^2-\frac{\gamma-1}{\gamma+1}\right]^{1/(\gamma-1)} \quad ...(20.51)$$

and $$\frac{p_{0y}}{p_x} = \frac{p_{0y}}{p_y}\cdot\frac{p_y}{p_x} = \left[1+\frac{\gamma-1}{2}M_y^2\right]^{\gamma/(\gamma-1)}\left[\frac{2\gamma}{\gamma+1}M_x^2-\frac{\gamma-1}{\gamma+1}\right] \quad ...(20.52)$$

Equations (20.51) and (20.52) give the value of p_{0y} as p_{0x}, p_x, M_x and γ are known.

The values of M_y, p_y/p_x, p_{0y}/p_{0x}, p_{0y}/p_x, ρ_y/ρ_x, T_y/T_x, p_y/p_x, etc., may be calculated from above equations for different values of M_x and $\gamma = 1.4$ and are given in Appendix.

In order to find the change of entropy across the normal shock for an ideal gas, we have,

$$ds = c_p\frac{dT}{T} - R\frac{dp}{p}$$

$$\therefore \qquad s_y - s_x = c_p \ln \frac{T_y}{T_x} - R \ln \frac{p_x}{p_x} = c_p \left[\ln \frac{T_y}{T_x} - \ln \left(\frac{p_y}{p_x} \right)^{(\gamma-1)/\gamma} \right]$$

As $\quad R = c_p (\gamma - 1)/\gamma$, so

$$s_y - s_x = c_p \ln \frac{T_y/T_x}{(p_y/p_x)^{(\gamma-1)/\gamma}} = - R \ln \frac{T_{0y}/T_{0x}}{(p_{0y}/p_{0x})^{(\gamma-1)/\gamma}}$$

or

$$s_y - s_x = - R \ln \frac{p_{0y}}{p_{0x}} \qquad \text{...(20.53)}$$

Sometimes, a parameter called ***strength of shock wave (P) is used to describe the shock wave. It is defined as the ratio of static increase in pressure to the initial static pressure***. Thus

$$P = \frac{p_y - p_x}{p_x} = \frac{p_y}{p_x} - 1$$

Substituting the value of p_y/p_x from Eq. (20.48), we have

$$P = \frac{2\gamma}{\gamma + 1}(M_x^2 - 1) \qquad \text{...(20.54)}$$

From the above discussions, it follows that **normal shock wave can only arise in a supersonic flow, and that is produces a sudden reduction in velocity to a subsonic value**. Thus, the normal shock is a compression process and $p_y/p_x > 1$ for all $M_x > 1$. **Owing to the irreversibility introduced by the shock wave which causes stagnation pressure loss, i.e., $p_{0y}/p_{0x} < 1$.**

Figure 20.11 shows the changes in final Mach number M with inlet Mach number M_1. The variation of pressure cross shock wave is shown if Fig. 20.12.

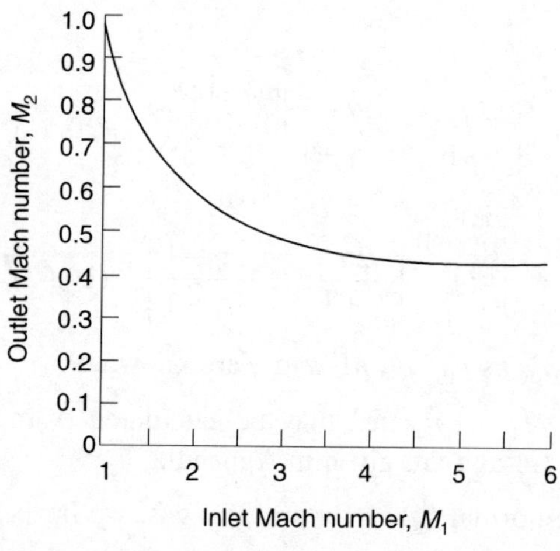

Fig. 20.11 Changes in Mach number across normal shock

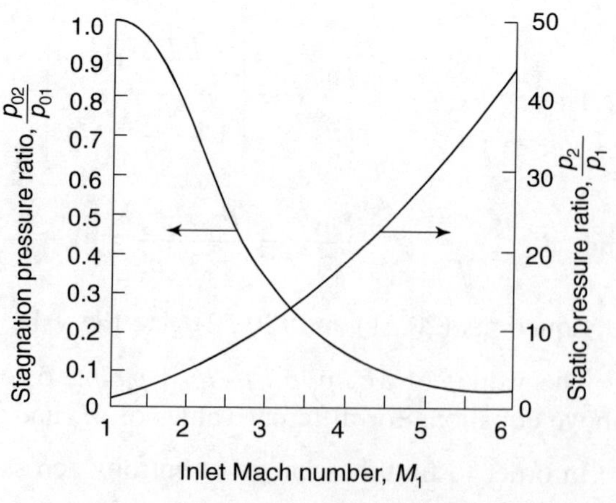

Fig. 20.12 Changes in pressure across a plane normal shock wave

It is clearly of interest to see how efficient the normal shock is as a compressor, because it can be made use of at the front of intakes of turbojet and ramjet engines when these are operating at supersonic flight speeds. The isentropic efficiency of compression is

$$\eta = \frac{T_{ys} - T_x}{T_y - T_x} = \frac{[(p_y/p_x)^{(\gamma-1)/\gamma} - 1]}{T_y - T_x}$$

Figure 20.13 shows the isentropic efficiency of plane shock waves. Efficiency falls of rapidly as M_1 is increased, Nevertheless, when it is realised that $p_y/p_x = 4.5$ for $M_2 = 2$ (Fig 20.11) and that this pressure ratio has been achieved in a negligible length of intake, an efficiency of 78% (Fig. 20.13) is not too low to be useful.

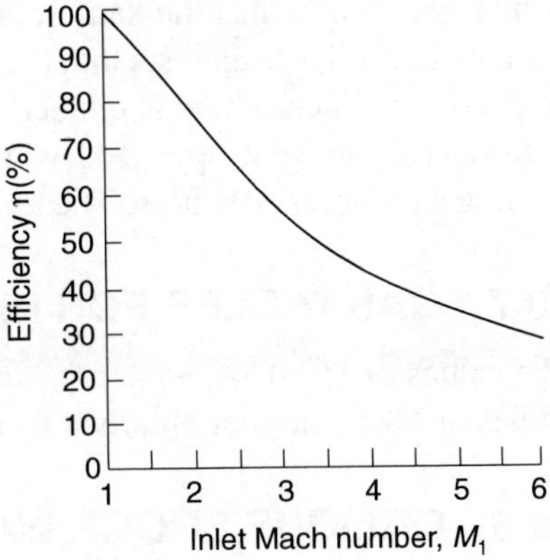

Fig. 20.13 Isentropic efficiency of plane shock wave

20.6 ADIABATIC FLOW WITH FRICTION AND DIABATIC FLOW WITHOUT FRICTION

The Fanno line representing the states of constant mass velocity and constant stagnation enthalpy also holds good for adiabatic flow in a constant area duct with friction. For adiabatic flow with friction the entropy must increase in the flow direction. Therefore, a Fanno process must follow its Fanno line to the right (Fig. 20.14), and the Mach number of subsonic flows increase in the downstream section. Similarly, in supersonic flows, friction acts to decrease the Mach number and thus drives the flow to the sonic point ($M = 1$).

Diabatic flow is a flow with heating or cooling. Diabatic flow without friction can be represented by Rayleigh line (Fig. 20.15). The process is reversible, and the direction of

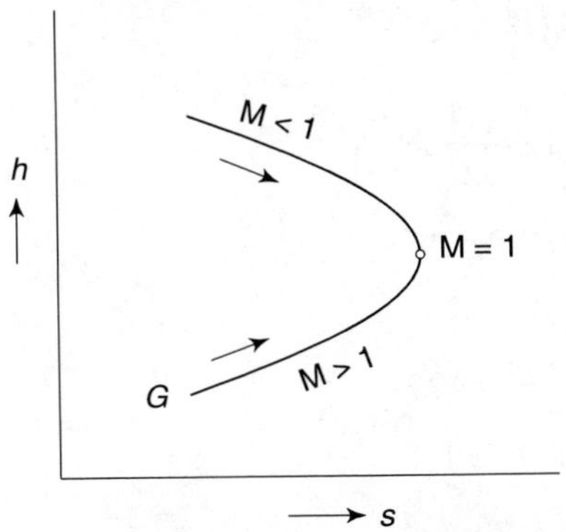

Fig. 20.14 A Fanno line in adiabatic flow with friction

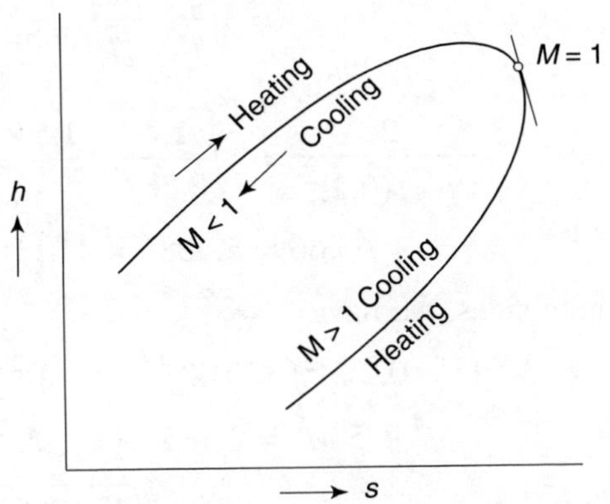

Fig. 20.15 A Rayleigh line in diabatic flow without friction

entropy change is determined by heating or cooling, i.e., sign of heat transfer. Heating a compressible flow has the same effect as friction, and Mach number for subsonic flow goes towards unity. This suggests that there is a maximum heat input for a given flow rate which can be passed by duct, which is choked. The cooling of the flow increases the stagnation pressure with decrease of entropy. It is worth to note that non-mechanical pump is not feasible by cooling a compressible flow. The reason is attributed to the strong effect of friction in the flow.

20.7 GAS TABLES FOR ISENTROPIC FLOW

The values of M^*, A^*/A, p/p_0, ρ/ρ_0 and T/T_0 computed for ideal gas having $\gamma = 1.4$ for different values of Mach number obtained from the section 20.3 are appended in Appendix (Table A13).

20.8 OBLIQUE SHOCK WAVES

In a certain type of supersonic flow, the shock wave is formed at an inclination to the direction of flow. Strictly speaking, the analysis of such phenomena involves a study of two-dimensional flow, although, some equations obtained for plane normal shock waves can be used with a little modification. The analysis of oblique shock wave is out of the scope of the book.

Example 20.1 In a duct of 20 cm diameter, a stream of air flows at the rate of 2 kg/s. The stagnation temperature is 40°C. At any section of the duct, the static pressure is 1.5 bar. Calculate the Mach number, velocity, and stagnation pressure at this section.

Solution Data given : $T_0 = 40 + 273 = 313$ K, $p = 1.5$ bar, $\gamma = 1.4$

The mass flow rate per unit area is expressed as

$$\frac{\dot{m}}{A} = \rho C = \rho a M = \frac{p}{RT}\sqrt{\gamma RT} \cdot M$$

$$= \sqrt{\frac{\gamma}{R}} \cdot \frac{pM}{\sqrt{T_0}}\sqrt{\frac{T_0}{T}} = \sqrt{\frac{\gamma}{R}} \cdot \frac{pM}{\sqrt{T_0}}\left[1 + \frac{\gamma - 1}{2}M^2\right]^{1/2}$$

or $$\frac{2}{(\gamma/4)(0.2)^2} = \sqrt{\frac{1.4}{0.287}} \times \frac{1.5 \times 10^2\, M}{\sqrt{313}}\left[1 + \frac{0.4}{2}M^2\right]^{1/2}$$

or $$63.66 = 2.2086 \times 44.11\, M\, [1 + 0.2\, M^2]^{1/2}$$

Squaring both sides, we have

$$M^2\,[1 + 0.2\, M^2] = 0.427 \quad \text{or} \quad M^2 + 0.2\, M^4 = 0.427$$

or $$M^4 + 5\, M^2 = 2.135 \quad \text{or} \quad (M^2)^2 + 5\, M^2 - 2.135 = 0$$

or $$M^2 = \frac{-5 + \sqrt{25 + 8.54}}{2} = \frac{-5 + 5.79}{2} = 0.395$$

or $$M = 0.628$$ **Ans.**

We have, $$\frac{T_0}{T} = 1 + \frac{\gamma - 1}{2} M^2 = 1 + \frac{0.4}{2} \times 0.682^2 = 1.079$$

or $$T = \frac{313}{1.079} = 290 \text{ K}$$

Sound velocity, $a = \sqrt{\gamma RT} = \sqrt{1.4 \times 0.287 \times 10^3 \times 290} = 341.35$ m/s

$$C = aM = 341.35 \times 0.628 = 214.36 \text{ m/s}$$ **Ans.**

$$\frac{p_0}{p} = \left(\frac{T_0}{T}\right)^{\gamma/(\gamma-1)} = \left(\frac{313}{290}\right)^{1.4/0.4} = 1.306$$

or $$p_0 = 1.5 \times 1.306 = 1.959 \text{ bar}$$ **Ans.**

Example 20.2 A conical diffuser used in a certain installation receives air at 1.6 bar, static temperature 315K and velocity 280 m/s. The intake area is 0.12 m^2 while exit area is 0.45 m^2. Calculate (a) the mass flow rate of air passing through diffuser, (b) the Mach number, static pressure and static temperature leaving the diffuser and (c) the net thrust acting upon the diffuser if the atmospheric air surrounding the diffuser is at 1 bar.

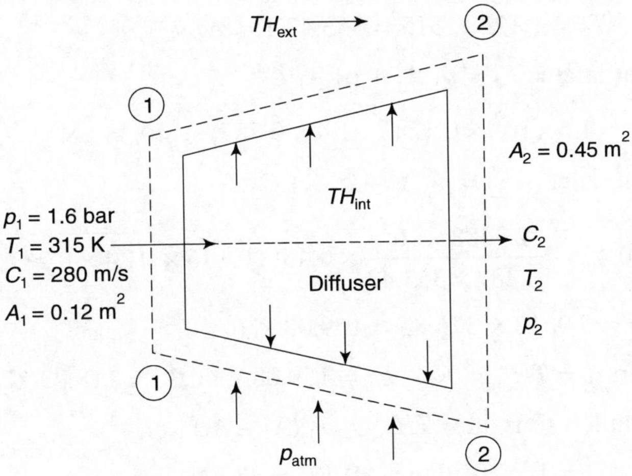

Fig. 20.16 Example 20.2

Solution Refer to Fig. 20.16.

The mass flow rate of air passing through the diffuser is given by

(a) $$\dot{m} = \rho_1 A_1 C_1 = \left(\frac{p_1}{RT_1}\right) A_1 C_1 = \frac{1.6 \times 10^2}{0.287 \times 315} \times 0.12 \times 280 = 59.46 \text{ kg/s}$$ **Ans.**

Sonic velocity, $a_1 = \sqrt{\gamma RT_1} = \sqrt{1.4 \times 0.287 \times 315 \times 10^3} = 355.76$ m/s

(b) $$M_1 = \frac{C_1}{a_1} = \frac{280}{355.76} = 0.787$$ **Ans.**

From gas tables given for isentropic flow in Appendix, for $M_1 = 0.787$

$$\frac{A_1}{A^*} = 1.04, \quad \frac{p_1}{p_{01}} = 0.657 \quad \text{and} \quad \frac{T_1}{T_{01}} = 0.887$$

$$\frac{A_2}{A_1} = \frac{0.45}{0.12} = 3.75 = \frac{A_2/A^*}{A_1/A^*}$$

or $$\frac{A_2}{A^*} = 3.75 \times 1.04 = 3.9$$

From isentropic table at $\frac{A_2}{A^*} = 3.9$, we have, $M_2 = 0.136$, $\frac{p_2}{p_{02}} = 0.965$, $\frac{T_2}{T_{02}} = 0.996$

$$\frac{p_2}{p_1} = \frac{p_2/p_{02}}{p_1/p_{01}} = \frac{0.965}{0.657} = 1.468 \quad \text{and} \quad \frac{T_2}{T_1} = \frac{T_2/T_{02}}{T_1/T_{01}} = \frac{0.996}{0.887} = 1.122$$

or $$p_2 = 1.468 \times 1.6 = 2.3488 \text{ bar}$$ **Ans.**

$$T_2 = 1.122 \times 315 = 353.43 \text{ K}$$ **Ans.**

Impulse function at inlet $= F_1 = p_1 A_1 + \rho_1 A_1 C_1^2$

$$= p_1 A_1 (1 + \gamma M_1^2) = 1.6 \times 10^2 \times 0.12\,(1 + 1.4 \times 0.787) = 40.35 \text{ kN}$$

Impulse function at oulet $= F_2 = p_2 A_2 + \rho_2 A_2 C_2^2$

$$= 2.3488 \times 10^2 \times 0.45 + \frac{2.3488 \times 10^2}{0.287 \times 353.43} \times 0.45 \times 0.136 \times \sqrt{1.4 \times 0.287 \times 10^3 \times 353.43}$$

$$= 105.696 + 2.315 \times 0.0612 \times 376.83 = 159.08 \text{ kN}$$

$$\text{Internal thrust} = TH_{int} = F_2 - F_1 = 159.08 - 40.35 = 118.73 \text{ kN}$$

External thrust from left to right $= TH_{ext} = p_0 (A_2 - A_1)$

$$= 1 \times 10^2 (0.45 - 0.12) = 33 \text{ kN}$$

$$\text{Net trust} = TH_{net} = TH_{int} - TH_{ext} = 118.73 - 33 = 85.73 \text{ kN}$$ **Ans.**

Example 20.3 A converging nozzle has an exit area of 0.002 m^2. Air enters the nozzle with negligible velocity of 15 bar and a temperature of 4000 K. For isentropic flow of an ideal gas ($\gamma = 1.4$), calculate the mass flow rate, the exit Mach number for back pressure of (a) 7.5 bar, and (b) 8.5 bar.

Solution Refer to Fig. 20.17.

The first step is to check whether the flow is closed or not.

For choked flow, $$\frac{p^*}{p_0} = 0.528$$

$$p^* = p_0 \times 0.528 = 15 \times 0.528 = 7.92 \text{ bar.}$$

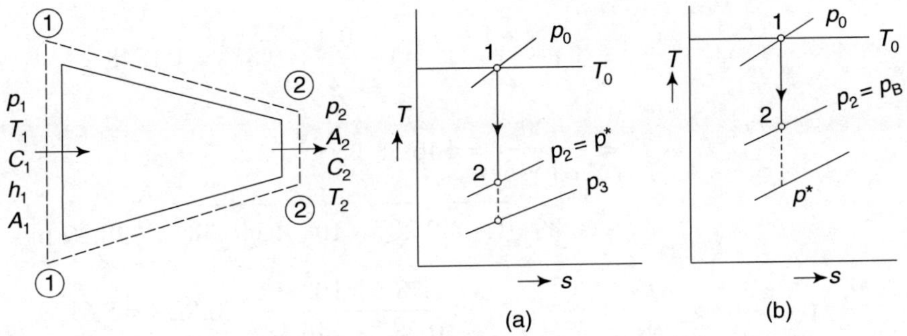

Fig. 20.17 Example 20.3

The critical pressure is 7.92 bar, hence the flow is choked. The mass flow rate is the maximum value at this condition. For a back pressure of 7.5 bar, the Mach number is unity at the exit and the nozzle is choked.

(a) From the above discussion, it follows that for a back pressure of 7.5 bar, the nozzle is choked. At the exit, $M_2 = 1.0$ and the exit pressure is equal to the critical pressure 7.92 bar. The mass flow rate is given by

$$\dot{m} = \rho_2 A_2 C_2 = \frac{p_2}{RT_2} A_2 C_2$$

Also, $$\frac{T_0}{T_2} = 1 + \frac{\gamma - 1}{2} M_2^2 = 1 + \frac{\gamma - 1}{2} M^{*2}$$

or $$\frac{400}{T_2} = 1 + \frac{0.4}{2} \times 1^2 = 1.2$$

or $$T_2 = \frac{400}{1.2} = 333.33 \text{ K}$$

Since $M_2 = 1$, so $C_2 = \sqrt{\gamma RT_2} = \sqrt{1.4 \times 0.287 \times 10^3 \times 333.33} = 365.96$ m/s **Ans.**

Thus, $$\dot{m} = \frac{7.92 \times 10^2}{0.287 \times 333.33} \times 0.002 \times 365.96 = 6.059 \text{ kg/s}$$ **Ans.**

(b) Since the back pressure is 8.5 bar which is greater than critical pressure of 7.92 bar, so the flow throughout the nozzle is subsonic and the exit pressure is equal to the back pressure, $p_2 = 8.5$ bar. We have,

$$M_2 = \left\{\frac{2}{\gamma - 1}\left[\left(\frac{p_0}{p_2}\right)^{(\gamma-1)/\gamma} - 1\right]\right\}^{1/2} = \left\{\frac{2}{0.4}\left[\left(\frac{15}{8.5}\right)^{0.4/1.4} - 1\right]\right\}^{1/2} = 0.88 \quad \textbf{Ans.}$$

$$C_2 = M_2\, a_2 = M_2\, \sqrt{\gamma\, RT_2}$$

$$— = 1 + \frac{\gamma - 1}{2} M^2 = 1 + \frac{0.4}{2} \times 0.88^2 = 1.1548$$

or $$T_2 = \frac{400}{1.1548} = 346.38 \text{ K}$$

∴ $$C_2 = 0.88\sqrt{1.4 \times 0.287 \times 10^3 \times 346.38} = 328.29 \text{ m/s}$$

Thus, $$\dot{m} = A_2\, \rho_2\, C_2 = A_2 \frac{p_2}{RT_2} C_2 = 0.002 \times \frac{8.5 \times 10^2}{0.287 \times 346.38} \times 328.29 = 5.6139 \quad \textbf{Ans.}$$

We observe that the mass flow rate is less than the maximum discharge as per critical condition.

Example 20.4 A convergent-divergent nozzle is employed to conduct air from a high pressure tank to a large container. While working at off-design condition, a normal shock occurs in the divergent part of the nozzle at a section where the cross-sectional area is 20 cm². The stagnation pressure and temperature at the inlet of nozzle are 3 bar and 40°C respectively. The throat area is 10 cm² and the exit area is 30 cm². Calculate (a) the exit Mach number, (b) the exit pressure, (e) the loss in stagnation pressure, and (d) the increase in entropy during the flow from high pressure tank to large container.

Solution Refer to Fig. 20.18. Let x and y represent the location before and after the control volume which encloses the shock.

At shock location, $$\frac{A_x}{A^*} = \frac{20}{10} = 2.0$$

Up to the shock, the flow is isentropic. From the isentropic flow table, we have

$$M_x = 2.2 \quad \text{and} \quad \frac{p_x}{p_{0x}} = 0.094$$

The static pressure stream of the shock, $p_x = 0.094\, p_{0x} = 3 \times 0.09 = 0.282$ bar

From the gas tables on normal shocks,

(a) at $M_x = 2.2$, M_y 0.54706, $\frac{p_y}{p_x} = 5.48, \frac{p_{0y}}{p_x} = 6.7163, \frac{p_{0y}}{p_{0x}} = 0.62812$ **Ans.**

∴ $p_y = 5.48 \times 0.282 = 1.545$ bar and $p_{0y} = 0.62812 \times 3 = 1.8843$ bar

The loss of stagnation pressure occurs only across the shock.

(c) $$p_{0x} - p_{0y} = 3 - 1.885 = 1.115 \text{ bar}$$ **Ans.**

From the shock section to the exit of nozzle, the flow is again isentropic.

At $M_y = 0.54706$, from isentropic table, we have $\frac{A_y}{A^*} = 1.264$

$$\frac{A_2}{A^*} = \frac{A_2 A_y}{A_y A^*} = \frac{30}{20} \times 1.264 = 1.8976$$

At $\frac{A_2}{A^*} = 1.896$, from isentropic table, we have $M_2 = 0.33$ **Ans.**

$$\frac{p_2}{p_{0y}} = 0.93 \quad \text{so} \quad p_2 = 0.93 \times p_y = 0.93 \times 1.545 = 1.4368 \text{ bar}$$ **Ans.**

Increase in entropy, $s_y - s_x = -R \ln \frac{p_{0y}}{p_{0x}} = -0.287 \ln \frac{1.885}{3} = -0.133$ kJ/kg K **Ans.**

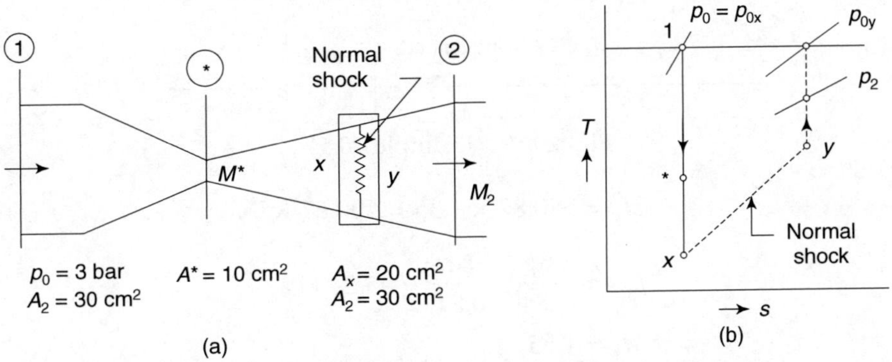

Fig. 20.18 Example 20.4

Example 20.5 In a supersonic wind tunnel, a curved shock wave but treated as normal shock wave forms due to a certain condition. A pitot tube travelling in this wind tunnel gives values of 20 kPa and 80 kPa for the static pressure upstream of the shock and stagnation pressure at the mouth of the tube respectively. Calculate the Mach number of the tunnel. The stagnation temperature is 300°C. Calculate (a) the static temperature, (b) the Mach number after the shock, and (c) stagnation pressure loss across the shock.

Solution Refer to 20.19. $p_x = 20$ kPa, $p_{0y} = 80$ kPa, $T_{0x} = T_{0y} = 573$ K.

$$\frac{p_{0y}}{p_x} = \frac{80}{20} = 4$$

From the gas tubes for normal shock,

At $$\frac{p_{0y}}{p_x} = 4, \quad M_x = 1.65, \quad \frac{p_y}{p_x} = 3.05, \quad \frac{\rho_y}{\rho_x} = 1.895$$

$$\frac{T_y}{T_x} = 1.35, \quad \frac{p_{0y}}{p_{0x}} = 0.875 \quad \text{and} \quad M_y = 0.654$$

$$T_{0x} = T_{0y} = 573 \text{ K}$$

$$T_{0x} = \left[1 + \frac{\gamma - 1}{2} M_x^2\right] T_x = [1 + 0.2 \times 1.65^2]\, T_x = 1.5445\, T_x$$

or $$T_x = \frac{573}{1.545} = 370.87 \text{ K}$$

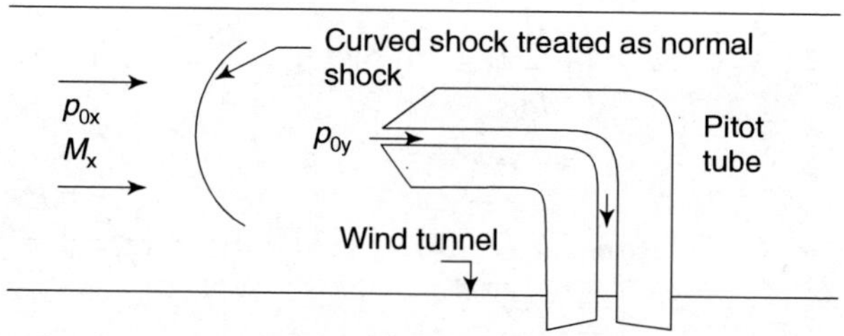

Fig. 20.19 Example 20.5

(a) $$T_y = 370.82 \times 1.35 = 500.67 \text{ K}$$ **Ans.**

$$p_{0x} = \frac{p_{0y}}{0.84} = \frac{80}{0.84} = 95.23 \text{ kPa}$$

(b) $$M_x = 1.65$$ **Ans.**

(c) $$p_{0x} - p_{0y} = 95.23 - 80 = 15.23 \text{ kPa}$$ **Ans.**

20.9 SUMMARY OF FORMULAE

1. $a = \sqrt{\gamma RT}$

2. $M = C/a$

3. $h_0 = h + \dfrac{C^2}{2}$

4. $\dfrac{T_0}{T} = 1 + \dfrac{\gamma - 1}{2} M^2$

5. $\dfrac{p_0}{p} = \left[1 + \dfrac{\gamma - 1}{2} M^2\right]^{\gamma/(\gamma - 1)}$

6. $\dfrac{\rho_0}{\rho} = \left[1 + \dfrac{\gamma - 1}{2} M^2\right]^{1/(\gamma - 1)}$

7. $\dfrac{dA}{A} = -\dfrac{dC}{C}(1 - M^2)$

8. $\dfrac{dA}{A} = \dfrac{(M^2 - 1)\, dM}{M\left(1 + \dfrac{\gamma - 1}{2}\right) M^2}$

9. $\dfrac{p^*}{p_0} = \left(\dfrac{2}{\gamma - 1}\right)^{\gamma/(\gamma - 1)}$

10. $\dfrac{T^*}{T_0} = \dfrac{2}{\gamma - 1}$

11. $\dfrac{A}{A^*} = \dfrac{1}{M}\left[\left(\dfrac{2}{\gamma - 1}\right)\left(1 + \dfrac{\gamma - 1}{2}\right)\right]^{(\gamma+1)/2(\gamma-1)}$

12. $M^{*2} = \dfrac{\left(\dfrac{\gamma + 1}{2}\right)M^2}{1 + \dfrac{\gamma - 1}{2}M^2}$

13. $\dfrac{p_y}{p_x} = \dfrac{2\gamma}{\gamma + 1} M_x^2 - \dfrac{\gamma - 1}{\gamma + 1}$

14. $\dfrac{p_{0y}}{p_x} = \left[1 + \dfrac{\gamma - 1}{2} M_x^2\right]^{\gamma/(\gamma-1)} \left[\dfrac{2\gamma}{\gamma + 1} M_x^2 - \dfrac{\gamma - 1}{\gamma + 1}\right]$

15. $P = \dfrac{2\gamma}{\gamma + 1}(M_x^2 - 1)$

PROBLEMS

Things to Think About

20.1. In which of the following media is the sonic velocity the greatest – air, steel or water? Does sound propagate in vacuum?

20.2. Can a shock stand upstream of the throat of a converging-diverging nozzle?

20.3. In a supersonic fighter plane, what is done to the nose of jet of plane to avoid shock?

20.4. Is shock wave desirable in a flow stream?

20.5. Is it true that there is a serious total pressure loss after shock wave?

20.6. Does shock wave act as a compressor?

Short and Long Theoretical Questions

20.7. Define compressible fluid and mention its basic laws.

20.8. Derive the expression of sonic velocity.

20.9. Show that the sonic velocity in an ideal gas depends on the temperature and nature of the gas.

20.10. What is Mach number? What do you mean by stagnation state and properties. Show on *T-s* diagram.

20.11. Define nozzle and diffuser.

20.12. Discuss the effect of area change in subsonic and supersonic flow by deriving the governing equations.

20.13. Discuss the effect of back pressure in converging and converging-diverging nozzles.

20.14. What is a shock wave? What are its characteristics?

20.15. Discuss the modelling of flow across a normal shock.

20.16. Discuss the Fanno and Rayleigh lines on h-s diagram as a solution to normal shock equation.

20.17. Derive the various governing equations dealing with shock wave for an ideal gas.

20.18. Discuss the isentropic efficiency of a plane shock wave.

20.19. Diccuss the nature of adiabatic flow with friction and diabatic flow without friction.

20.20. Derive the various expressions for critical properties-choking in isentropic flow.

Numerical Questions

20.21. A supersonic wind tunnel is to be designed for $M = 2.5$ with a throat section, 0.15 m^2 in area. The inlet pressure and temperature at the nozzle inlet are 80 kPa and 40°C respectively. Calculate the mass flow rate, the exit area and the fluid properties at the throat and exit. Assume $\gamma = 1.4$ and negligible velocity at nozzle inlet.

20.22. Air at $p_0 = 2$ bar, $T_0 = 300$ K expands isentropically through a converging nozzle and discharges to the atmosphere at 1 bar. The exit plane area is 0002 m^2. Calculate the mass flow rate.

20.23. Air enters a nozzle operating at steady state at 4 bar, 450 K, with a velocity of 500 m/s. Calculate (a) the exit pressure, (b) the ratio of exit area to inlet area, and (c) whether the nozzle is divergent only or convergent only or convergent-divergent.

20.24. Air enters a diffuser operating at steady state at 5 bar, 300 K with a velocity of 550 m/s. Plot the velocity, Mach number, and the area ratio A/A^* for locations in the flow corresponding to pressure ranging from 5 to 15 bar assuming isentropic flow.

20.25. A converging-diverging nozzle operates at state. It has a throat area of 4 cm^2 and exit area of 8 cm^2. Air as an ideal gas enters the nozzle at 10 bar and 500 K and a Mach number of 0.2 and flows isentropically throughout. If the nozzle is choked and diverging portion acts as a supersonic normal, calculate the mass flow rate Mach number, pressure and temperature at the exit.

20.26. In an air duct, normal shock is formed. The upstream condition of the shock $p_x = 0.1$ bar and 290 K and $M_x = 2.0$. Calculate (a) the downstream pressure (p_y) of the shock, (b) the stagnation pressure p_{0x}, (c) the stagnation temperature T_{0x}, and (d) the change in specific entropy across the shock.

20.27. A converging-diverging nozzle operating at steady state receives air at 4 bar, 550 K and a Mach number of 0.4. A normal shock stands in the diverging section at a location where the Mach number is $M_x = 1.8$. The flow is isentropic before and after the shock. For an ideal gas calculate (a) the stagnation temperature upstream of shock, T_{0x}, (b) the stagnation pressure upstream of shock, p_{0x}, (c) the pressure p_x, (d) the pressure p_y, (e) the stagnation pressure, p_{0y}, (f) the stagnation temperature T_{0y}. If the throat area is 8×10^{-4} m^2 and the exit plane pressure is 2.5 bar, calculate (g) the mass flow rate, and (h) exit area.

20.28. Air while flowing through a nozzle encounters a shock. The Mach number upstream of the shock is 1.8 and the static temperature downstream of the shock is 450 K. Calculate the velocity changed across the shock.
[*Ans.* 378 m/s]

20.29. Diffuser is employed at the inlet of jet engine. The jet engine travels through the air with a speed of 1000 km/h at an attitude of 6000 m where the pressure is 40 kPa and the temperature is –12°C. The air exits at diffuser with a velocity of 100 m/s. Assuming isentropic flow, calculate (a) the pressure and temperature leaving the diffuser and (b) the ratio of inlet to exit area.

[*Ans.* (a) 61 kPa, 259.2 K and (b) 0.4875]

20.30. A converging-diverging nozzle operating at steady state has a throat area of 8 cm^2 and exit area of 14 cm^2. Air enters the nozzle with a negligible velocity at a pressure of 7 bar and a temperature of 550°C. Calculate, the mass flow rate, the exit pressure, and exit Mach number for each of the following five cases, (a) Isentropic flow with $M = 0.65$ at the throat, (b) Isentropic flow with $M = 1$ at the throat, (c) Isentropic flow with $M = 1$ at the throat and diverging portion acting as nozzle, (d) Isentropic flow through the nozzle with a normal shock standing at the exit. (e) A normal shock forms in the diverging section at a location where the area is 12 cm^2, and elsewhere the flow is isentropic.

Software Based Questions

20.31. Develop a data based software for the isentropic flow and shock wave.

20.32. Develop a software for the analysis of diffuser and nozzle using data based software.

20.33. Develop a software for the analysis of normal shock wave.

Design and Open Ended Questions

20.34. Using manufacturer's data list of the all available sonic and supersonic fighter planes, prepare a report on their main specifications concerning aerodynamics. Discuss the main design features.

20.35. A three-spool aircraft with turbofan engine travels at supersonic speed. Develop a design procedure for its nozzle.

Multiple-Choice Answer Type Questions

20.36. The sound velocity (a) is expressed as

(a) $\sqrt{-v\left(\frac{\partial p}{\partial u}\right)_s}$ (b) $\sqrt{\left(\frac{\partial}{\partial}\right)}$ (c) $\sqrt{T\left(\frac{\partial p}{\partial v}\right)_s}$ (d) $\sqrt{-T^2\left(\frac{\partial p}{\partial v}\right)_s}$

20.37. The density ratio (ρ_0/ρ) is expressed as

(a) $\left[1-\frac{\gamma-1}{2}m\right]^{1/(\gamma-1)}$ (b) $\left[1+\frac{\gamma-1}{2}m\right]^{1/(\gamma-1)}$

(c) $\left[1+\frac{\gamma-1}{2}M^2\right]^{1/(\gamma-1)}$ (d) $\left[1+\frac{\gamma-1}{2}M^2\right]^{\gamma-1}$

20.38. The temperature ratio (T_y/T_x) across a normal shock is given by

(a) $\dfrac{1+\frac{\gamma-1}{2}M_x^2}{1+\frac{\gamma-1}{2}M_y^2}$ (b) $\dfrac{1-\frac{\gamma-1}{2}M_x^2}{1-\frac{\gamma-1}{2}M_y^2}$ (c) $\dfrac{1-\frac{\gamma-1}{2}M_x^2}{1+\frac{\gamma-1}{2}M_y^2}$ (d) $\dfrac{1-\frac{\gamma-1}{2}M_x}{1+\frac{\gamma-1}{2}M_y}$

20.39. The static pressure ratio (p_y/p_x) across a normal shock wave is given by

(a) $\dfrac{M_y}{M_x} \cdot \dfrac{1+\dfrac{\gamma-1}{2} M_x}{1+\dfrac{\gamma-1}{2} M_y}$

(b) $\dfrac{M_y}{M_x} \cdot \dfrac{1+\dfrac{\gamma-1}{2} M_x^2}{1+\dfrac{\gamma-1}{2} M_y^2}$

(c) $\dfrac{M_y}{M_x} \cdot \left[\dfrac{1+\dfrac{\gamma-1}{2} M_x^2}{1+\dfrac{\gamma-1}{2} M_y^2}\right]^{1/2}$

(d) $\dfrac{M_y^2}{M_x^2} \cdot \dfrac{1+\dfrac{\gamma-1}{2} M_x^2}{1+\dfrac{\gamma-1}{2} M_y^2}$

Answers to Multiple-Choice Questions

20.36 (b) 20.37 (c) 20.38 (a) 20.39 (c)

CHAPTER

21

Heat Transfer

OVERVIEW

Heat transfer is that science which predicts the rate of energy transfer which may take place between material bodies as a result of temperature difference between them. According to thermodynamics, this energy transfer is defined as heat. The science of heat transfer not only explains how heat energy may be transferred but also **predicts the rate** at which the exchange will take place under certain specified conditions. *There is a difference between thermodynamics and heat transfer study and it is the* ***transfer rate*** *which forms the basis for difference. Thermodynamics deals with the system in equilibrium and predicts the amounts of energy required to change a system from one equilibrium state to another. It does not consider the mechanism of heat transfer and time required to transfer the heat.*

The importance of heat transfer study can be understood from the following examples.

In order to *design the thermal equipment*, engineers will have to make a detailed heat transfer analysis. The *design of boilers, heaters, refrigerators and heat exchangers* depend not only on the amount of heat to be transmitted but also on the rate at which the heat is to get transferred under given conditions. In *chemical engineering problems* such as evaporation, condensation heating and cooling of fluids, heat transfer analysis is essential. *In civil engineering problems such as construction of dams, structures and building design, a heat transfer study must be made. In some machines, heat is generated and it may be detrimental to the equipment.* For example, *walls of I.C. engines, combustion chamber, gas turbine blades, nuclear reactor and outer surface of space vehicle all depend for durability on continuous rapid heat removal from their surfaces. In the design of electrical machines, transformers, insulation provided on electric wires and bearing, a heat transfer analysis must be made to avoid conditions which will cause overheating and damage to the equipment.*

From the above examples, it follows that in almost every branch of engineering, heat transfer problems are encountered which are not capable of solution by thermodynamic reasoning alone, but requires heat transfer analysis.

21.1 MODES OF HEAT TRANSFER

There are three modes of heat transfer namely (i) **conduction,** (ii) **convection,** and (iii) **radiation**. *They have in common that temperature difterence that must exist so that heat may be transferred in the direction of decreasing temperature. But they differ entirely in the physical mechanism and laws by which they are governed.*

The following examples will distinguish clearly between the three modes of heat transfer.

A class teacher wants to distribute apples to a row of students. He may cause the apple to reach the farthest most student by any one of the following methods:

(i) He could deliver the apple to the student nearest to him who delivers the same to the next and so on till it finally reaches the last student.

(ii) He could go to the last student and deliver the apple to him.

(iii) He could throw the apple so that it reaches the last student.

Here, we assume that the apple corresponds to heat energy while the students correspond to the molecules. *Conduction, convection and radiation correspond to first, second and third methods of delivering apple to the last boy.*

21.1.1 Conduction

In conduction heat flow, the energy is transmitted by direct molecular communication without appreciable displacement of molecules. In other words, the phenomenon of heat conduction is a result of transfer of internal energy from one molecule to other. *From the concept of electron drift, heat conduction is primarily due to the mobility of free or valence electrons.*

Conduction is the only mechanism by which heat can flow in opaque solids. Conduction is also important in fluids and gases. The best example of conduction is heating a metallic rod from one end.

The heat transfer by conduction is given by Fourier's law

$$q = kA\,(T_1 - T_2)/L = -\,kA\Delta T/L \text{ ; W}$$

where k = thermal conductivity of material, W/m K

A = area normal to the direction of heat transfer, m^2

ΔT = temperature difference, K

The negative sign is provided to make q positive.

21.1.2 Convection

Convection is a process involving mass movement of fluids. Thus, the mechanism of convection is simply the transfer of energy by actual physical movement from one molecule to another of a fluid in which the energy is stored. *When a temperature difference produces a density difference which results in the mass movement, the process is called* **free or natural convection.** When an external source such as *pump or fan causes mass motion to take place*, the process is called **forced convection**. The best example of convection is the heating of water kept in a metallic vessel.

The heat exchange rate is given by Newton's law of cooling

$$q = h\,A_s\,(T_s - T_\infty)\ ;\ \text{W}$$

where h = the average heat transfer coefficient, W/m^{2}K

A = heat transfer surface area, m^2

T_s = metal surface temperature, K

T_∞ = free stream fluid temperature, K

21.1.3 Radiation

Radiation heat transfer is due to the property of matter to emit and absorb thermal energy (*or different kinds of rays*) *in the form of electromagnetic waves and to the fact that no material medium is required for the transfer of heat*. The best example of heat transfer is the heat received by the earth from the sun. The radiant energy waves coming from the sun towards earth pass through almost perfect vacuum (called ether) and also through atmospheric layer surrounding the earth.

The heat emitted by radiation is given by Stefan–Boltzmann law (see section 21.16).

$$q = \sigma\,AT^4$$

where σ = Stefan–Boltzmann constant = 5.67×10^{-8} W/m^2 K^4

A = heat transfer area, m^2

T = absolute temperature of the radiating body, K

21.2 STEADY AND UNSTEADY HEAT TRANSFER

Heat transfer through any material medium may be steady or unsteady. *Steady state means that there is no variation of temperature with time, i.e.,* $dT/d\tau = 0$. *Unsteady state means that the temperature of the body varies with time, i.e.,* $dT/d\tau \neq 0$.

21.3 FOURIER'S LAW OF HEAT CONDUCTION

The fundamental law of heat conductance was proposed by the French scientist, J.B.J. Fourier in 1822. With reference to Fig. 21.1, one-dimensional form of Fourier's law states *that the rate of heat flow by conduction in a material is equal to the product of three quantities, namely,*

(a) *k, a property of the material known as thermal conductivity* (W/m K)

(b) *Area A, normal to the direction of heat flow* (m^2)

(c) $\partial T/\partial x$, *the temperature gradient at the section, i.e., the rate of change of temperature* (*T*) *with respect to the distance in the direction of heat flow. Mathematically,*

$$\frac{\delta Q}{\partial \tau} = -kA\frac{\partial T}{\partial x}; W \qquad ...(21.1)$$

where δQ = quantity of heat conducted in time $\partial \tau$

In the above equation, negative sign has been arbitrarily affixed to make Q positive because of the fact that flow always occurs in the direction of decreasing temperature, i.e., the temperature gradient $(\partial T/\partial x)$ will always be negative.

In the case of steady state heat conduction, we have simply

$$\frac{\delta Q}{\partial \tau} = \frac{Q}{\tau} = q, \quad \text{hence } q = -kA\frac{dT}{dx}; W \qquad ...(21.1a)$$

$$q'' = -k\frac{dT}{dx} \quad \text{where } q'' = q/A = \text{heat flux, W/m}^2 \qquad ...(21.2)$$

To find the steady state linear heat conducted through a uniform slab in which area A does not vary with x as shown in Fig. 21.1, Eq. (21.2) may be integrated with the help of separation of variable. Thus,

$$q\int_{x_1}^{x_2} dx = -kA\int_{T_1}^{T_2} dT$$

or
$$q(x_2 - x_1) = kA(T_2 - T_1)$$

or
$$q = \frac{kA(T_1 - T_2)}{(x_2 - x_1)}$$

If $x_2 - x_1 = L$ = the thickness of the slab, then

$$q = kA\frac{T_1 - T_2}{L} \qquad ...(21.3)$$

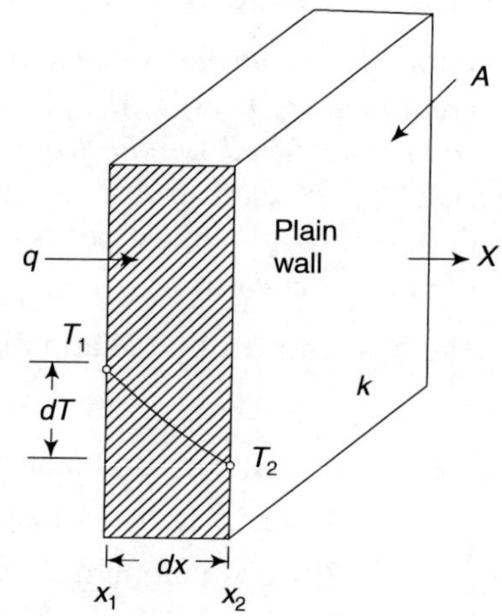

Fig. 21.1 Linear heat conduction

If q is in W (watt), L is in m, area A is in m^2 and $(T_1 - T_2)$ is in K, then the unit of thermal conductivity (k) is given by

$$k = \frac{qL}{A(T_1 - T_2)} = \frac{W \cdot m}{m^2 \cdot K} = \frac{W}{m \cdot K} \quad \text{or W/m°C}$$

Note: $(T_1 - T_2)$ may be expressed either °C or K, as it is the difference of temperature.

Note that k is not necessarily uniform for any one material but may be non-uniform say, as a function of local temperature T.

Equation (21.3) may be written as, $q = kA(T_1 - T_2)/L = \dfrac{T_1 - T_2}{L/kA}$ ...(21.4)

Here, *kA/L is known as* ***internal thermal conductance****. The unit of internal thermal conductance is W/K. The term k/L is known as unit thermal conductance. The reciprocal of internal thermal conductance (L/kA) is known as* ***internal thermal resistance***, i.e.,

Internal thermal resistance, $R = \frac{L}{kA}$; K/W

The thermal resistance R, is similar to the resistance against the flow of electric current. From Ohm's law:

$$I = \frac{V_1 - V_2}{R} = \frac{\text{Potential difference}}{\text{Resistance}} \qquad ...(21.5)$$

This means that the heat conduction problems may be transformed to electrical problems. Thus, $q = (T_1 - T_2)/R$ where R is **internal thermal resistance.**

21.4 CONDUCTION THROUGH A PLANE WALL

Consider steady state heat flow through a one dimensional plane wall as shown in Fig. 21.2. From Fourier's equation

$$q = -kA\frac{dT}{dx} \quad \text{or} \quad dx = -\frac{kA}{q}dT$$

Integrating, we get $$x = -\frac{kAT}{q} + C$$

Using the boundary conditions, at $x = 0$, $T = T_1$ and at $x = L$, $T = T_2$

This gives, $$q = \frac{kA}{L}(T_1 - T_2) = \frac{-kA}{L}(T_2 - T_1) = \frac{kA(T_1 - T_2)}{L} \qquad ...(21.6)$$

If the thermal conductivity (k) is not uniform, then the above equation will not hold good.

Let $$k = k_o(1 + \alpha T)$$

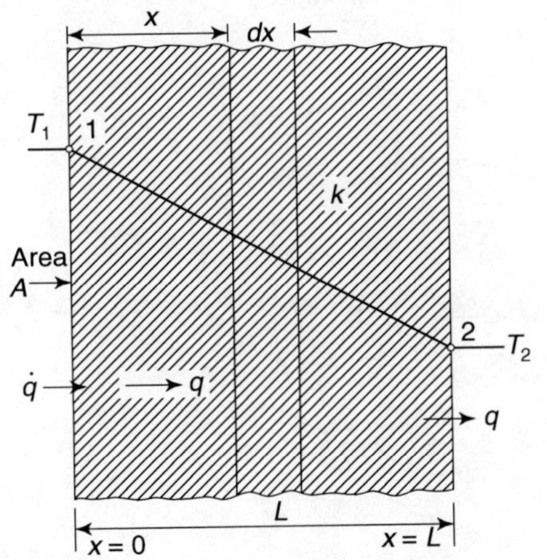

Fig. 21.2 Heat flow through plane wall

where k_o = thermal conductivity at 0°C and α = coefficient

The resulting equation for the heat transfer is given by

$$q = \frac{-k_o A}{L}[(T_2 - T_1) + (\alpha/2)(T_2^2 - T_1^2)] \qquad ...(21.7)$$

The general three-dimensional heat condition equation in Cartesian co-ordinates is out of scope of the book and is given as

$$\frac{\partial^2 T}{\partial x^2} + \frac{\partial^2 T}{\partial y^2} + \frac{\partial^2 T}{\partial y^2} + \frac{q_g'''}{k} = \frac{1}{\alpha}\frac{\partial T}{\partial \tau} \qquad ...(21.8)$$

where α = thermal diffusivity = $k/\rho c$ expressed in m^2/s

q_g''' = heat generation per unit volume

For one-dimensional steady state heat conduction without heat generation is $\partial^2 T/\partial x^2 = 0$. From this equation also, we can find Eq. (21.6).

21.5 CONDUCTION THROUGH CYCLINDER

Consider steady state radial heat flow through a hollow cylinder of radius r_1 and r_2 and of length L as shown in Fig. 21.3. From Fourier's law

$$q = -kA\frac{dT}{dr}$$

But $A = 2\pi rL$, hence $q = -2\pi krL \dfrac{dT}{dr}$

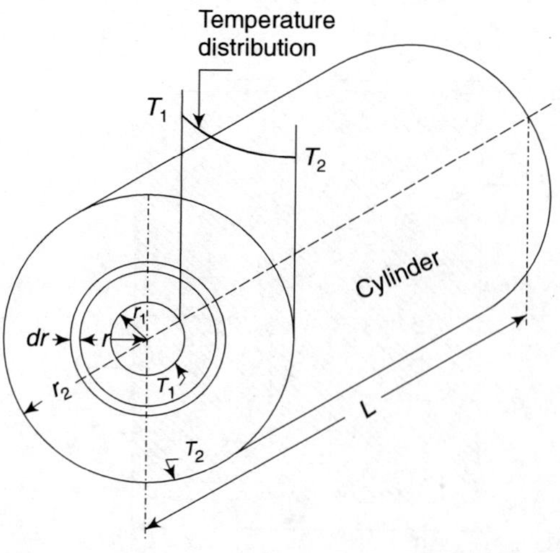

Fig. 21.3 Heat flow through cylinder

or $$dT = \frac{-q\,dr}{k\,2\pi\,Lr}$$

Integrating, we get $$T = \frac{-qdr}{k2\,\pi L}\ln r + C$$

Using the boundary conditions, $T = T_1$ at $r = r_1$ and $T = T_2$ at $r = r_2$

This gives $$q = \frac{2\,\pi\,kL\,(T_1 - T_2)}{\ln\,(r_2/r_1)} \qquad ...(21.9)$$

The temperature distribution, $$T = T_1 = -\frac{q\ln\,(r/r_1)}{2\,\pi\,kL} \qquad ...(21.10)$$

And the thermal resistance, $$R = \frac{\ln\,(r_2/r_1)}{2\,\pi\,kL} \qquad ...(21.11)$$

Example 21.1 A wire whose resistance per cm length is 0.1 ohm is embedded along the axis of a cylindrical cement tube of radii 0.06 cm and 1.2 cm. A current of 6 amp flows in the wire to keep a steady temperature difference of 130°C between the inner and outer surfaces. Calculate the conductivity of cement and the amount of heat supplied per cm length.

Solution Heat generated = $I^2 R$; W

Resistance per cm length = 0.1 ohm

Resistance per m length = 0.1 × 100 = 10 ohm

Heat generated per m length = $q/L = 6^2 \times 10 = 360$ W/m **Ans.**

We know $$T_1 - T_2 = \frac{q}{2\,\pi L}\left[\frac{\ln\,(r_2/r_2)}{k}\right] \quad \text{or} \quad 130 = \frac{360}{2\,\pi k}\ln\,(1.2/0.06)$$

or $$k = 1.32 \text{ W/m K}$$ **Ans.**

21.6 CONDUCTION THROUGH A SPHERE

Consider heat flow through a hollow sphere of inner radius r_1 as shown in Fig. 21.4. Let the inner surface be at temperature T_1 and outer at T_2. Within the sphere, consider a small element of dr at a radius r.

From Fourier's law,

$$q = -kA\frac{dT}{dr} = -k\,(4\,\pi\,r^2)\frac{dT}{dr}$$

or $$dT = \frac{q}{4\,\pi\,k}\frac{dr}{r^2}$$

Integrating the above equation,

we get, $$T = \frac{q}{4\,\pi\,k}\frac{1}{r} + C$$

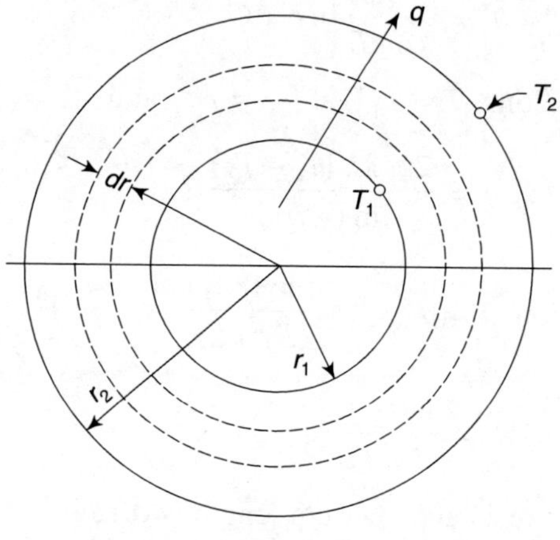

Fig. 21.4 Heat conduction through a sphere

The boundary conditions are

at $r = r_1$, $T = T_1$, and

at $r = r_2$, $T = T_2$

This gives $$T_1 - T_2 = \frac{q}{4\,\pi\,k}\frac{r_2 - r_1}{r_1\,r_2}$$

The temperature distribution is given by $$T = T_1 - \frac{q}{4\,\pi k}\frac{r_2 - r_1}{r_1} \qquad ...(21.12)$$

or $$q = \frac{4\,\pi k\,(T_1 - T_2)\,r_1\,r_2}{r_2 - r_1} = \frac{T_1 - T_2}{(r_2 - r_2)/4\pi k\,r_1\,r_2} = \frac{T_1 - T_2}{R} \qquad ...(21.13)$$

Here, R = thermal resistance = $\dfrac{r_2 - r_1}{4\,\pi k\,r_1\,r_2}$

21.7 CONDUCTION THROUGH A COMPOSITE STRUCTURE

Composite structures made of two or more layers of different materials have practical significance. For examples, steam pipes have the cylindrical steel wall of the pipe, a layer of insulating material and then a layer of protecting plaster. Similarly, cold storage walls have a layer of bricks, a layer of thick insulation and plaster on both sides. The other examples are the atomic reactor wall, boiler furnace wall and pressure vessel used in chemical and fertilizer factories.

In these cases, the systems are exposed to an external heat source and sink, so that free face surface temperatures are unknown. This means that to find heat transfer rate, the surface resistances must be accounted for as well as the individual resistance within the structure itself. Let us consider two cases of composite structures.

21.7.1 Series of Composite Wall

Consider a series of composite wall made of three different materials as shown in Fig. 21.5. The inner layer which is exposed to hot gas is made of fire brick. The intermediate layer consists of an insulation brick which is surrounded by atmospheric layer. It is assumed that gas and air form a thin film over the inner and outer surfaces respectively.

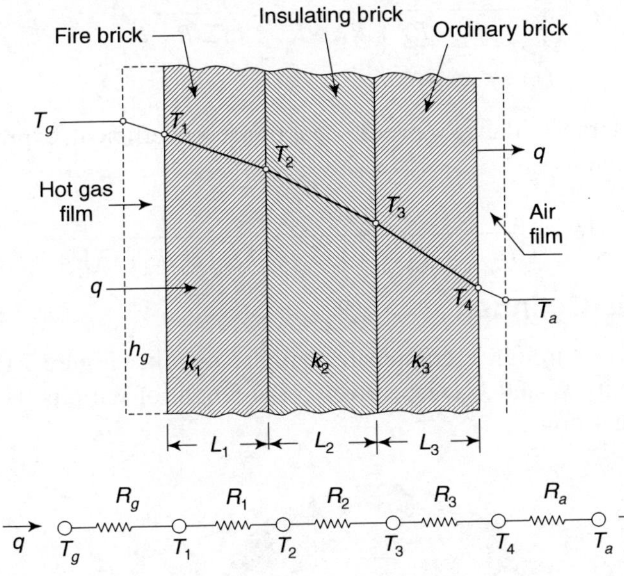

Fig. 21.5 Steady one-dimensional heat transfer through a series of composite walls

Let suffix g and a denote gas and air. Let k_1, k_2 and k_3 be the thermal conductivities of three materials respectively.

The heat flow through the composite wall is given by

$$q = h_g A (T_g - T_1) = \frac{k_1 A (T_1 - T_2)}{L_1} = \frac{k_2 A(T_2 - T_3)}{L_2}$$

$$= \frac{k_3 A(T_3 - T_1)}{L_3} = h_a A (T_4 - T_a) \qquad \text{...(21.14)}$$

The above equations may be also written in terms of thermal resistance as

$$q = \frac{T_g - T_1}{1/h_g A} = \frac{T_1 - T_2}{L_1/k_1 A} = \frac{T_2 - T_3}{L_2/k_2 A} = \frac{T_3 - T_4}{L_3/k_3 A} = \frac{T_4 - T_a}{1/h_a A}$$

$$T_g - T_1 = qR_g \quad \text{where} \quad R_g = 1/h_g\, A$$

$$T_1 - T_2 = q\, R_1 \quad \text{where} \quad R_1 = L_1/k_1\, A$$

$$T_2 - T_3 = q\, R_2 \quad \text{where} \quad R_2 = L_2/k_2\, A$$

$$T_3 - T_4 = q\, R_3 \quad \text{where} \quad R_3 = L_3/k_3\, A$$

$$T_4 - T_a = q\, R_a \quad \text{where} \quad R_a = 1/h_a\, A$$

Adding both sides of the above equation, we get

$$q = \frac{T_g - T_a}{R_g + R_1 + R_2 + R_3 + R_a} = \frac{T_g - T_a}{\Sigma R} \qquad ...(21.15)$$

or

$$q = UA\,(T_g - T_a)$$

where U = thermal transmittance = overall heat transfer coefficient, expressed in W/m^2 K and given as

$$U = \frac{1}{(1/h_g) + (L_1/k_1) + (L_2/k_2) + (L_3/k_3) + (1/h_a)} \qquad ...(21.16)$$

21.7.2 Composite Cylinder

Consider radial heat flow through a composite cylinder or tube. Figure 21.6 made of two layers having conductivity k_1 and k_2, respectively. The length of the pipe (L) is sufficiently long to ensure radial heat flow.

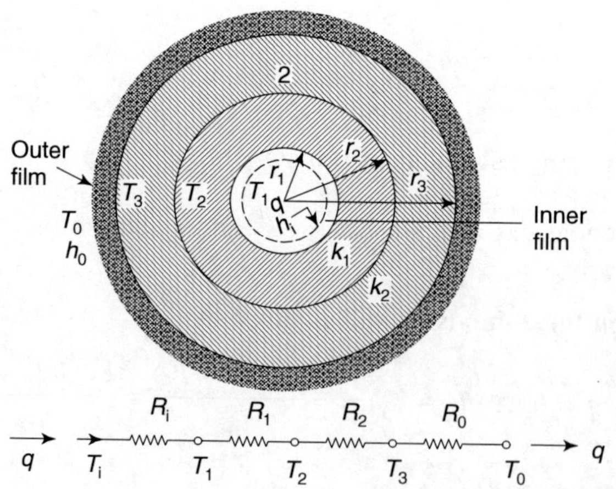

Fig. 21.6 Composite cylinder

Let suffixes i and o denote the inside and outside conditions of composite tube.

In the steady state, the rate of heat flow through each section is the same and is represented by

$$q = 2\,\pi\, r_1\, Lh_1\,(T_i - T_1) = \frac{T_i - T_1}{R_i} \quad \text{for the inner film}$$

$$q = \frac{2\,\pi\, k_1\, L\,(T_1 - T_2)}{\ln (r_2/r_1)} = \frac{T_2 - T_3}{R_1} \quad \text{for the inner surface}$$

$$q = \frac{2\,\pi\, k_2\, L\,(T_2 - T_3)}{\ln (r_3 - r_2)} = \frac{T_2 - T_3}{R_3} \quad \text{for outer surface}$$

$$q = 2\,\pi r_3 L h_0\,(T_3 - T_0) = \frac{T_3 - T_0}{R_0} \quad \text{for the outer film}$$

The above equation may be written as

$$T_i - T_1 = \frac{q}{1/2\,\pi\, r_1\, Lh_1} = \frac{q}{R_i}$$

$$T_1 - T_2 = \frac{q}{\ln (r_2/r_1)/2\,\pi\, k_1\, L} = \frac{q}{R_1}$$

$$T_2 - T_3 = \frac{q}{\ln (r_3/r_2)/2\,\pi\, k_2 L} = \frac{q}{R_2}$$

$$T_3 - T_0 = \frac{q}{1/2\pi\, r_3\, h_0\, L} = \frac{q}{R_0}$$

Adding the above equations, we gat

$$q = \frac{T_i - T_0}{\dfrac{1}{2\,\pi\, r_1\, L\, h_i} + \dfrac{\ln (r_2/r_1)}{2\,\pi\, k_1\, L} + \dfrac{\ln (r_3/r_2)}{2\,\pi\, k_2\, L} + \dfrac{1}{2\,\pi\, r_3\, h_0\, L}} = \frac{T_i - T_0}{\Sigma R} \quad \text{...(21.17)}$$

Also, $$q = U_0\, A_0\,(T_i - T_0) = U_0(2\pi\, r_3\, L)\,(T_i - T_0)$$

where U_o = overall heat transfer coefficient based on outer surface.

$\therefore$ $$U_o = \frac{1}{\dfrac{r_3}{r_1\, h_1} + \dfrac{r_3 \ln (r_2/r_1)}{k_1} + \dfrac{r_3 \ln (r_3/r_2)}{k_2} + \dfrac{1}{h_0}} \quad \text{...(21.8)}$$

Note A similar expression can be developed for **composite sphere.**

21.8 CRITICAL THICKNESS OF INSULATION

By insulation, we generally mean that it reduces the heat transfer. But it is not always true, such as, in cylinder and tubes. In electric wire carrying current, insulation is provided to dissipate more amount of heat. There exists a critical thickness beyond which the heat transfer rate decreases. Therefore, in the case of wire carrying current, the insulation is provided such that the outer radius should be less than critical radius, r_c or equal to r_c to dissipate the maximum amount of heat (Fig. 21.7). The maximum heat transfer is at point B.

To calculate the value of critical radius r_c, recall Eq. (21.17). The heat transfer rate from an insulated pipe having inner radius, r_1 and outer radius r_2 is given by

$$q = \frac{2\pi L (T_i - T_0)}{(1/h_i\, r_1) + (1/k)\ln(r_2/r_1) + (1/r_2\, h_0)}$$

If we consider r_1 (the radius of bare pipe) fixed and thereby $q = f(r_2)$, then q will be the maximum at the particular value of r_2 for which

$$\frac{dq}{dr_2} = \frac{-2\pi L (T_i - T_0)[1/k\, r_2 - \left(1/h_0\, r_2^2\right)]}{[(1/h_i\, r_1) + (1/k)\ln(r_2/r_1) + (1/h_0\, r_2)} = 0$$

$$\therefore \quad r_2 = k/h_0$$

The value of r_2 obtained is called the **critical radius r_c, hence**

$$r_c = k/h_0 \qquad \text{...(21.19)}$$

where k = thermal conductivity of insultating material (W/m K)

h_0 = heat transfer coefficient from outer surface to atmosphere (W/m^2 K)

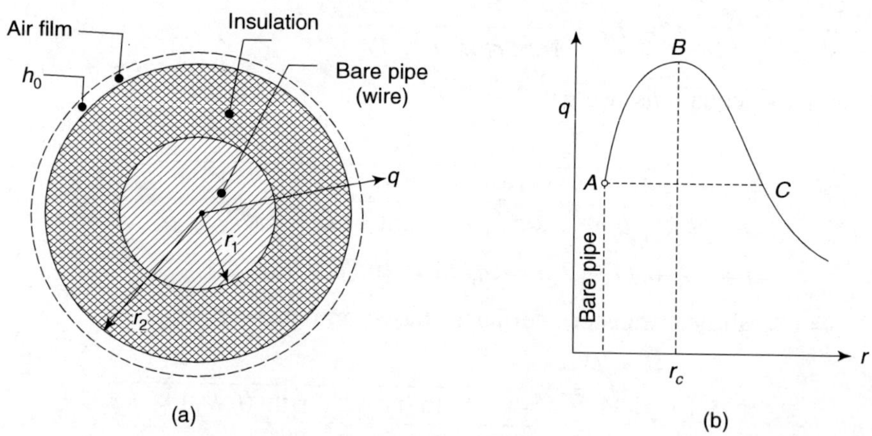

Fig. 21.7 Critical radius

Example 21.2 A flat plate consists of two layers of materials A and B. A is 6 cm thick having thermal conductivity equal to 209 W/m K. B is 3 cm thick having thermal conductivity equal to 46.5 W/m K. The plate is in contact with a hot gas at a temperature of 150°C and having heat transfer coefficient equal to 14 W/m^2 K. The plate B is in contact with a cold fluid at temperature 30°C having heat transfer coefficient equal to 23 W/m^2 K.

Find the overall coefficient of heat transfer and the rate of heat transfer through 1 m^2 surface.

Solution The total thermal resistance is

$$\Sigma R = \frac{1}{h_g A} + \frac{L_1}{k_1 A} + \frac{L_2}{k_2 A} + \frac{1}{h_0 A}$$

or

$$\Sigma R = \frac{1}{14 \times 1} + \frac{0.06}{209 \times 1} + \frac{0.03}{46.5 \times 5} + \frac{1}{23 \times 1} = 0.11573 \text{ K/W}$$

But

$$q = \frac{\Delta T}{\Sigma R} = \frac{150 - 30}{0.11573} = 1036.89 \text{ W} \qquad \textbf{Ans.}$$

Also

$$q = UA\Delta T = U \times 1 \times (150 - 30)$$

or

$$U = \frac{1036.89}{120} = 8.64 \text{ W/m}^2 \text{ K} \qquad \textbf{Ans.}$$

Example 21.3 A furnace wall is made of 10 cm thick of refractory brick (k = 1.5 W/m K), 15 cm thick insulating fire brick (k = 0.29 W/m K) and 2 cm thick plaster coating (k = 0.15 W/m K). The refractory brick and the insulating fire brick are separated from each other by an air gap having a thermal resistance of 0.43 K/W. The temperature of wall is 1000°C and the outer surface of the plaster coating is exposed to air having temperature 18°C and h_a = 16 W/ m^2 K. During steady state condition, compute the heat flow rate through the wall and the temperature at the interfaces.

Solution Refer to Fig. 21.8

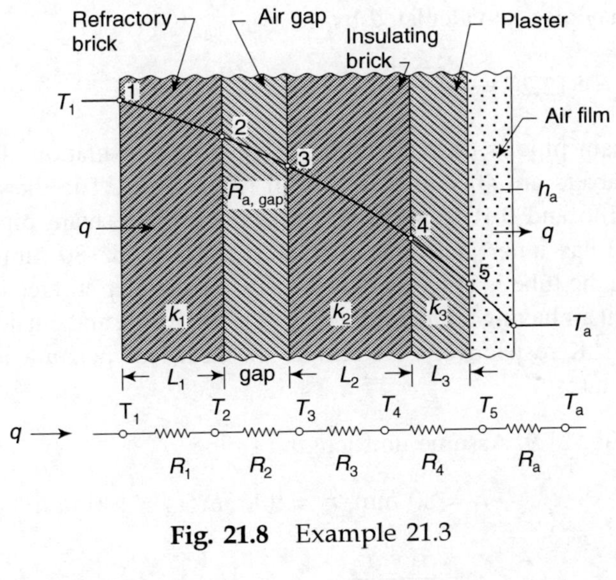

Fig. 21.8 Example 21.3

The total resistance is given by $= \Sigma R = R_1 + R_{gap} + R_2 + R_3 + R_a$

$$= \frac{L_1}{k_1 A} + R_{a,\,gap} + \frac{L_2}{k_2 A} + \frac{L_3}{k_3 A} + \frac{1}{h_0 A}$$

Let $$A = 1\text{ m}^2$$

$$\therefore \quad \Sigma R = \frac{0.10}{1.5 \times 1} + 0.43 + \frac{0.15}{0.29 \times 1} + \frac{0.02}{0.15 \times 1} + \frac{1}{16 \times 1}$$

$$= 0.0666 + 0.43 + 0.517 + 0.1333 + 0.0625 = 1.209 \text{ K/W}$$

$$q = \frac{\Delta T}{\Sigma R} = \frac{1000 - 18}{1.209} = 812.24 \text{ W}$$ **Ans.**

During steady state conditions, the heat flowing through each section of the composite wall is the same, hence

$$q = \frac{T_1 - T_2}{R_1} \quad \text{or} \quad 812.24 = \frac{1000 - T_2}{0.0666} \quad \text{or} \quad T_2 = 945.9°\text{C}$$ **Ans.**

$$q = \frac{T_2 - T_3}{R_{a,\,gap}} \quad \text{or} \quad 812.24 = \frac{945.9 - T_3}{0.43} \quad \text{or} \quad T_3 = 596.63°\text{C}$$ **Ans.**

$$q = \frac{T_3 - T_4}{R_2} \quad \text{or} \quad 812.24 = \frac{596.63 - T_4}{0.517} \quad \text{or} \quad T_4 = 176.7°\text{C}$$ **Ans.**

$$q = \frac{T_4 - T_5}{R_3} \quad \text{or} \quad 812.24 = \frac{176.7 - T_5}{0.1333} \quad \text{or} \quad T_5 = 68.42°\text{C}$$ **Ans.**

The value of T_5 may also be calculated by

$$q = 812.24 = \frac{T_5 - 18}{0.0625} \quad \text{or} \quad T_5 = 68.76°\text{C}$$

Example 21.4 A steam pipe is covered with two layers of insulation. The thicknesses of the inner and outer layers are 40 mm and 60 mm respectively. The thermal conductivities of insulations are 0.186 and 0.1046 W/m K, respectively. The steam pipe itself is of steel (k = 58 W/m K) and has inner and outer diameter of 160 and 180 mm respectively. The steam passing though the tube is at temperature 350°C. The outer surface of the insulation is surrounded by ambient air having temperature of 30°C. The inside and outside film coefficients are 30 and 6.98 W/m^2 K respectively. Determine the heat loss per unit length of pipe and layer contact temperatures.

Solution Refer to Fig. 21.9. Assume unit length of pipe.

We have, $$r_1 = 80 \text{ mm},\ r_2 = 90 \text{ mm},\ r_3 = 130 \text{ mm},\ r_4 = 190 \text{ mm}$$

We have, $$q = \frac{T_1 - T_0}{\Sigma R}$$

But $$\Sigma R = R_i + R_1 + R_2 + R_3 + R_o$$

$$= \frac{1}{2\,\pi L}\left\{\frac{1}{h_i\,r_1} + \frac{\ln(r_2/r_1)}{k_1} + \frac{\ln(r_3/r_2)}{k_2} + \frac{\ln(r_4/r_3)}{k_3} + \frac{1}{h_0\,r_4}\right\}$$

$\therefore$ $$\Sigma R = \frac{1}{2\,\pi\times 1}\left\{\frac{1}{30\times 0.08} + \frac{\ln(90/80)}{58} + \frac{\ln(130/90)}{0.186} + \frac{\ln(190/130)}{0.1046} + \frac{1}{6.98\times 0.19}\right\}$$

$$= \frac{1}{2\,\pi}[0.41666 + 0.00203 + 1.977 + 3.628 + 0.754] = \frac{1}{2\,\pi}\times 6.777$$

$\therefore$ $$q = \frac{(350-30)\times 2\,\pi}{6.777} = 296.68 \text{ W/m}$$ **Ans.**

Under steady state condition, the heat loss from each layer is the same, hence

$$q = \frac{T_i - T_1}{R_i} = \frac{(350 - T_1)\,2\,\pi}{0.41666} = 296.68 \quad \text{or} \quad T_1 = 330.32°\text{C}$$ **Ans.**

$$296.88 = \frac{T_1 - T_2}{R_1} = \frac{(330.32 - T_2)\,2\,\pi}{0.00203} \quad \text{or} \quad T_2 = 320.22°\text{C}$$ **Ans.**

$$296.68 = \frac{T_2 - T_3}{R_2} = \frac{(330.22 - T_3)\,2\,\pi}{1.977} \quad \text{or} \quad T_3 = 236.86°\text{C}$$ **Ans.**

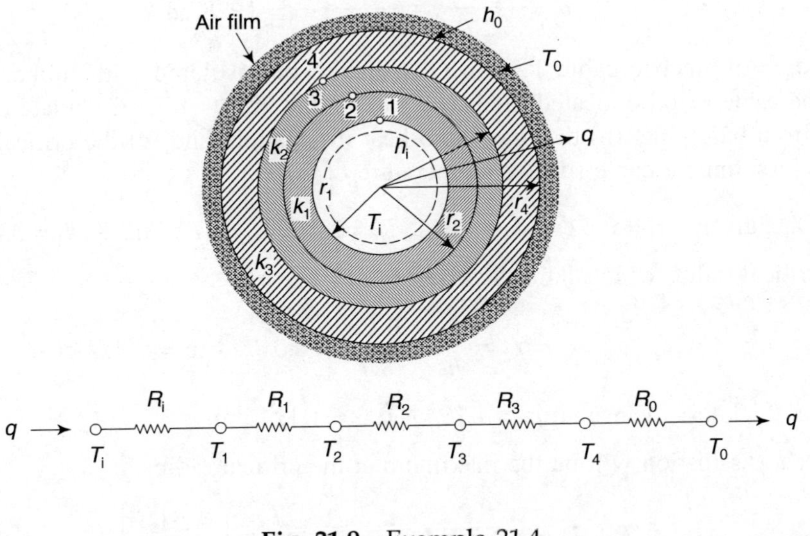

Fig. 21.9 Example 21.4

$$296.68 = \frac{T_3 - T_4}{R_3} = \frac{(236.86 - T_4)\,2\,\pi}{3.628} \quad \text{or} \quad T_4 = 65.55°\text{C} \quad \textbf{Ans.}$$

Example 21.5 A hemispherical oven is built of an inner layer of insulating firebrick 125 mm thick (k = 0.31 W/m^2 K) and an outer covering of 85 % magnesia 40 mm thick (k = 0.05 W/m K). The inner surface of the oven is at 800°C and the heat transfer coefficient for the outside surface is 10 W/m^2 K. The room temperature is 20°C. Calculate the heat loss through the hemisphere if the inside radius is 60 cm.

Solution For the insulating fire brick, the mean area is

$$A_{m_1} = \frac{4\,\pi\,r_1\,r_2}{2} = \frac{4\,\pi \times 0.6 \times 0.725}{2} = 2.73 \text{ m}^2$$

$$\text{The resistance to fire brick, } R_1 = \frac{x_1}{k_1\,A_{mi}} = \frac{0.125}{0.31 \times 2.73} = 0.1477 \text{ K/W}$$

$$\text{For the 85\% magnesia, the mean area, } A_{m_2} = \frac{4\,\pi \times 0.725 \times 0.765}{2} = 3.485 \text{ m}^2$$

$$\text{The resistance of \% magnesia, } R_2 = \frac{x_2}{k_2\,A_{m_1}} = \frac{0.04}{0.05 \times 3.485} = 0.2295 \text{ K/W}$$

$$\text{The resistance of outside air film, } R_0 = \frac{1}{h_0\,A_0} = \frac{2}{10 \times 4\,\pi \times 0.765^2} = 0.0272 \text{ K/W}$$

The total resistance, $\Sigma R = R_1 + R_2 + R_0 = 0.1477 + 0.2295 + 0.0272 = 0.4044$ K/W

$$q = \frac{T_i - T_0}{\Sigma R} = \frac{800 - 20}{0.404} = 1928.78 \text{ W} \qquad \textbf{Ans.}$$

Example 21.6 An electric cable 1.2 cm O.D. has to be insulated with rubber (k = 0.15 W/m K). The cable is to be located in air (h_0 = 8.7 W/m^2 K) at 18°C. Calculate (a) the heat dissipation from bare pipe, (b) the maximum heat dissipation, and (c) the critical thickness of insulation assuming a cable surface temperature of 70°C.

Solution Data given $r_1 = 0.6$ cm, $T_1 = 70$°C, $T_0 = 18$°C, $h_0 = 87$ W/m^2 K, $k_1 = 0.15$ W/m K.

(a) The critical radius of insulation is given by

$$r_c = r_2 = \frac{k_1}{h_0} = \frac{0.15}{8.7} = 0.01724 \text{ m} = 1.724 \text{ cm}$$

Hence, the thickness of insultaion = 1.724 – 0.6 = 1.124 cm **Ans.**

(b) The heat dissipation will be the maximum at the critical radius.

Hence,
$$T_i - T_o = q\,\Sigma R = \frac{q}{2\,\pi\,L}\left[\frac{1}{k_1}\ln\left(\frac{r_2}{r_1}\right) + \frac{1}{h_0\,r_2}\right]$$

or $$(70-18)\,2\,\pi = \frac{q}{L}\left[\frac{1}{0.15}\ln\left(\frac{1.724}{0.6}\right)+\frac{1}{8.7\times 1.724\times 10^{2}}\right]$$

or $$\left(\frac{q}{L}\right)_{max} = \frac{(70-18)\,2\pi}{13.703} = 23.84 \text{ W/m}$$ **Ans.**

(c) For bare pipe, the heat dissipation rate per unit length is

$$\left(\frac{q}{L}\right)_{bare} = 2\,\pi\,h_0,\ r_1\,(T_1 - T_0) = (70-18)\,2\,\pi \times 8.7 \times 0.6 \times 10^{-2} = 17.05 \text{ W/m}$$ **Ans.**

Example 21.7 An electric cable of 12 mm diameter is insulated with rubber which is exposed to atmosphere at 30°C. Calculate the most economical thickness of the rubber insulation when the cable surface temperature with and without insulation is at 70°C. The environmental temperature is 30°C. Assume that the heat transfer coefficient from the bare or insulated cable is 14 W/m^2 K. Calculate the conductivity of the insulating material.

Solution The heat dissipation rate per metre length from the bare wire is

$$q_1 = 2\,\pi\,r_1\,L\,h_0\,(T_i - T_0) = 2\,\pi\,r_1\,h_0\,(T_i - T_0) = 2\,\pi \times 6 \times 10^{-3} \times 14\,(70-30) = 21.11 \text{ W/m}$$

The heat dissipation with insulation will be the maximum when

$$r_2 = r_c = k/h_0$$

Hence, the maximum heat dissipation rate from the insulated cable is

$$q_2 = \frac{T_i - T_0}{\Sigma R} = \frac{T_i - T_0}{\dfrac{1}{2\,\pi\,L}\left[\dfrac{1}{k}\ln\left(\dfrac{r_2}{r_1}\right)+\dfrac{1}{q_2\cdot h_0}\right]} = \frac{2\,\pi\,(T_i - T_0)}{\dfrac{1}{k}\left[\ln\dfrac{k\times 1000}{14\times 6}+1\right]}$$

$$= \frac{2\,\pi\times(70-30)}{\dfrac{1}{k}\left[\ln\left(\dfrac{k\times 1000}{14\times 6}+1\right)\right]} = \frac{80\,\pi}{\dfrac{1}{k}[\ln 11.904\,k+1]}$$

Now $$\frac{q_2}{q_1} = \frac{I_2^2\,R_{el}}{I_1^2\,R_{el}} = \frac{80\,\pi\,k}{[\ln(11.904\,k)+1]\times 21.11}$$

Since, $$I_2 = 1.15\,I_1,\ \text{hence } \frac{q_2}{q_1} = (1.15)^2 = 1.3225 = \frac{11.905}{[\ln(11.904\,k)+1]}$$

or $$1 + \ln 11.904\,k = 9\,k$$

By trail and error : Let $k = 0.1$, then

$$1.174 = 0.9$$

Let $k = 0.05$, then $0.481 = 0.45$

Thus, $k \approx 0.052$ W/m K **Ans.**

21.9 CONVECTION

The heat exchange between a solid surface and a fluid in contact with it is one of the very important problems in engineering practice. This mechanism of heat exchange is involved in all types of heat exchangers such as intercoolers, condensers, evaporators, radiators, nuclear reactors, boilers, etc. and is known as convective heat transfer.

On the basis of method of fluid motion obtained, the convective heat transfer is classified as

(i) *Forced convection* in which the fluid is forced to flow along the solid surface by means of a pump or fan.

(ii) *Free or natural convection* in which the fluid motion is caused due to density difference produced by temperature gradients.

The heat exchange rate in the case of forced convection is more than free because the fluid velocity in forced convection is much higher than that of free convection.

The rate of heat transfer by convection between a solid boundary and a fluid is given by Newton's law of cooling

$$q = h\, A_s\, (T_s - T_\infty);\ \text{W} \qquad \text{...(21.20)}$$

where h = average heat transfer coefficient, W/ m^2 K

A_s = heat transfer area, m^2

T_s = solid surface temperature, °C or K

T_∞ = temperature of the fluid free steam, °C or K

The main problem in convective heat transfer is to find out the average heat transfer coefficient. **The heat transfer coefficient depends on the following factors**:

(i) surface conditions – roughness and cleanliness,

(ii) geometry and orientation of the surface-plate, tube, cylinder, sphere placed vertically or horizontally,

(iii) nature of fluid flow – laminar of turbulent,

(iv) boundary layer configuration,

(v) prevailing thermal conditions, and

(vi) thermo-physical properties of the fluid density, viscosity, specific heat, thermal conductivity and coefficient of expansion.

Table 21.1 shows typical average values of convective heat transfer coefficient.

Table 21.1 Heat transfer coefficient

	Forced convection (h)	Free convection (h)
Water	300 – 10^4 W/m^2 K	liquids 30 – 300 W/m^2 K
Oil	60 – 3810 W/m^2 K	gases 2 – 20 W/m^2 K
Air and superheated steam	300 – 10^4 W/m^2 K	air 3 – 7 W/m^2 K

21.10 DIMENSIONLESS NUMBERS AND ANALYSIS

In the study of convective heat transfer, a large number of dimensionless numbers are used. Some of them are given below.

(a) Reynols number, Re *It is defined as the ratio of inertial forces to viscous forces.*

$$\text{Re} = \frac{\text{Inertia force}}{\text{Viscous force}} = \frac{\rho\, u_\infty^2\, L^2}{\mu\, u_\infty\, L} = \frac{\rho\, u_\infty\, L}{\mu}$$

or

$$\text{Re} = \frac{L u_\infty}{v} \quad \text{...(21.21)}$$

(b) Prandtle number, Pr *It is defined as the ratio of kinematic viscosity to thermal diffusivity.*

$$\text{Pr} = \frac{\mu\, c_p}{k} = \frac{v}{\alpha} \quad \text{...(21.22)}$$

It is indicative of relative ability of the fluid to diffuse momentum and internal energy by molecular mechanism.

(c) Nusselt number, Nu *It may be defined as the ratio of heat flow by convection process under a unit temperature gradient through a stationary thickness of L metre.* Thus,

$$\text{Nu} = \frac{q_{\text{conv}}}{q_{\text{cond}}} = \frac{h}{k/L} = \frac{hL}{k} \quad \text{...(21.23)}$$

(d) Stantan number, St *It is defined as the ratio of Nu to the product of* **Re** *and* **Pr**. Thus,

$$\text{St} = \frac{\text{Nu}}{\text{Re Pr}} = \frac{hL}{\rho\, u_\infty\, c_p} \quad \text{...(21.24)}$$

(e) Grashoff number, Gr *It is defined as the ratio of product of inertia force and buoyancy force to the square of viscous force.*

$$\text{Gr} = \frac{(\text{Inertia force}) \times (\text{Buoyancy force})}{(\text{Viscous forces})^2} = \frac{(\rho\, u_\infty^2\, L^2)(\rho\, \beta\, g\, \Delta T\, L^3)}{(\mu\, u_\infty\, L)^2} = \frac{\rho^2\, \beta\, g\, L^3\, \Delta T}{\mu^2} \quad \text{...(21.25)}$$

where β = thermal expansion coefficient, ΔT = temperature difference, g = acceleration due to gravity, ρ = density, μ = dynamic viscosity

It is indicative of the relative strength of buoyant to viscous force.

21.11 LAMINAR AND TURBULENT FLOW

Laminar flow *In this type of flow, the fluid particles move in flat or curved unmixing layers or streams; follow a smooth continuous path and parallel to each other. There is*

no intermixing of fluid; the paths of fluid movement are well defined and the fluid particles retain their relative position throughout cross section of the fluid passage. A crude analogy to laminar flow is the parade by soldiers.

Turbulent flow *In this type of flow, the motion of fluid particles is irregular and erratic. There is too much intermixing of fluid layers.* Commuters at the railway junction during the rush hour provide a somewhat crude analogy to turbulent flow.

If Re > 2500 in the case of tube, the flow is turbulent otherwise laminar. If Re > 5 × 10⁵ in the case of flow over flat plate, the flow is turbulent otherwise laminar.

21.12 DIMENSIONAL ANALYSIS APPLIED TO FORCED CONVECTION

In forced convection, the heat transfer coefficient will be the function of following factors.

$$h = f(\rho, L, u_\infty, \mu, c_p, k)$$

or $$f(h, \rho, L, u_\infty, \mu, c_p, k) = 0 \qquad ...(21.26)$$

where ρ is density, u_∞ is velocity, μ is viscosity; c_p is specific heat and k is thermal conductivity.

In the above equation, there ate 7 physical quantities containing 4 fundamental dimensions. So according to Buckingham π theorem, this problem can be controlled by (7 – 4) = 3 non-dimensional functions which are known as π terms.

$$\therefore \qquad f(\pi, \pi_2, \pi_3) = 0$$

According to Buckingham π theorem, terms marked in Eq. (21.26) which contain all four fundamental dimensions must form a non-dimensional group with each remaining term

$$\therefore \qquad \pi_1 = \mu\, h^a \rho^b L^c u_\infty^d$$

Writing the dimensions of each term, we get

$$M^\circ L^\circ T^\circ \theta^\circ = (ML^{-1} T^{-1})(MT^{-3} \theta^{-1})^a (ML^{-3})^b (L)^c (LT^{-1})^d$$

where M = mass, L = length, T = time, θ = temperature,

Equating the powers of MLT and θ of both sides, we get the following equations

$$1 + a + b = 0,\ -1 - 3b + c + d = 0,\ -1 - 3a - d = 0 - a = 0$$

Solution of the above equation gives $a = 0$, $b = -1$, $c = -1$, $d = -1$

$$\therefore \qquad \pi = \mu/\rho\, L\, u_\infty \qquad ...(21.27)$$

Similarly, $$\pi_2 = c_p (h)^{a_1} (\rho)^{b_1} (L)^{c_1} (u_\infty)^{d_1}$$

or $$M^\circ L^\circ T^\circ \theta^\circ = (L^2 T^2 \theta^{-1})(MT^{-3} \theta^{-1})^{a_1} (ML^{-3})^{b_1} (L)^{c_1} (LT^{-1})^{d_1}$$

Equating the powers of MLT and θ of both sides, we get the following equations

$$a_1 + b_1 = 0,\quad 2 - 3b_1 + c_1 + a_1 = 0,\quad -2 - d_1 - 3a_1 = 0 \quad \text{and} \quad -1 - a_1 - = 0$$

Solution of the above equation gives: $a_1 = -1, b_1 = 1, c_1 = 0$ and $d = 1$

$$\therefore \quad \pi_2 = c_{p_p} u_\infty / h$$

As dimensions of h and k/L are same, hence

$$\pi_2 = c_p \rho u_\infty L/k \qquad \text{...(21.28)}$$

Similarly, $$\pi_3 = k (h)^{a_3} (\rho)^{b_3} (L)^{c_3} (u_\infty)^{d_3}$$

or $$M^\circ L^\circ T^\circ \theta^\circ = (MLT^{-3}\theta^{-1})(ML^{-3}\theta^{-1})^{a_2}(ML^{-3})^{b_2}(L)^{c_2}(LT^{-3})^{d2}$$

Equating the powers of ML T θ of both sides, we get

$$1 + a_2 + b_2 = 0, \quad 1 - 3b_2 + c_2 + d_2 = 0$$

$$-3 - 3a_2 - d_2 = 0, \quad -1 - a_2 = 0$$

Solution of the equations gives : $a_2 = -1, b_2 = 0, c_2 = 1, d_2 = 0$

$$\pi_3 = k/hL$$

According to Buckingham π theorem,

$$\pi_3 = f(\pi_1, \pi_2)$$

$$\therefore \quad \left(\frac{k}{hL}\right) = f\left(\frac{\mu}{\rho L u_\infty}\right)^{n'} \left(\frac{c_p \rho L u_\infty}{k}\right)^{m'}$$

where n' and m' are constants

Assuming $n' > m'$, we can write

$$\frac{hL}{K} = f\left(\frac{\mu}{\rho L u_\infty}\right)^{m'} \left(\frac{\mu}{\rho L u_\infty}\right)^{n'-m'} \left(\frac{c_p \rho L u_\infty}{k}\right)^{m'} \qquad \text{...(21.29)}$$

$$= f\left(\frac{\mu}{\rho L u_\infty}\right)^{n'-m'} \left(\frac{\mu}{\rho L u_\infty} \frac{c_p \rho L u_\infty}{k}\right)^{m'} = f\left(\frac{\mu}{pL}\right)^{n'-m'} \left(\frac{\mu u_\infty c_p}{k}\right)^{m'}$$

$$= f\left(\frac{\mu}{\rho} L u_\infty\right)^{n} \left(\frac{\mu u_\infty c_p}{k}\right)^{m}$$

Here, n is substituted for $(n' - m')$ and m for m'.

But **hL/k is called Nusselt number, Nu**

$\frac{u_\infty L}{v}$ is called Reynolds number, R_e (Re = $\frac{u_\infty D}{v}$ for pipe)

$\mu c_p/k$ is called Prandtle number, Pr

$$\therefore \quad \text{Nu} = f(\text{Re})^n (\text{Pr})^m \qquad \text{...(21.30)}$$

21.13 FREE CONVECTION

In free convection, the heat transfer coefficient (h) depends upon the same variables as forced convection. Thus, h would be a function of μ_{∞}, ρ k, c_p and L. But the fluid motion in free convection is not due to external agency but due to difference in density between various fluid layers caused by temperature gradient. Therefore, the term velocity (u_{∞}) itself will depend upon the following factors.

(i) Temperature difference between the surface and bulk fluid, ΔT

(ii) Coefficient of volume expansion of the fluid, β

(iii) Acceleration due to gravity

The heat transfer coefficient (h) is of the following terms

$$h = f(\rho, L, \mu, k, c_p, \beta, g, \Delta T) \quad ...(21.31)$$

In this particular problem, (β g ΔT) can be considered as one physical factor which is causing fluid notion. Thus, the above equation may be written as

$$f[\rho, k, \mu, k, c_p (\beta g \Delta t)] \quad ...(21.31a)$$

In the above equation, there are 7 physical quantities containing 4 fundamental dimensions, so according to Buckingham π theorem, this problem can be controlled by (7 – 4) = 3 non-dimensional groups.

According to Buckingham π theorem, four terms marked in equation containing all four fundamental dimensions must form a non-dimensional group with each of the remaining term

$$\therefore \quad \pi_1 = h(\rho)^a (L)^b (\mu)^c (k)^d$$

$$M^\circ L^\circ T^\circ \theta^\circ = (ML^{-3}\theta^{-1})(ML^{-3})^a (L)^b (ML^{-1}T^{-1})^c (MLT^{-3}\theta^{-1})^d$$

Equating the powers of MLT and θ from both side, we get

$$a + b + c + d + 1 = 0, \quad -3a + b - c + d = 0$$

$$-c - 3d - 3 = 0, \quad -d - 1 = 0$$

Solution of the above equation gives : $a = 0$, $b = 1$, $c = 0$, $d = -1$

$$\therefore \quad \pi_1 = \frac{hL}{k} \quad ...(21.32)$$

Similarly,

$$\pi_2 = c_p (\rho)^{a_1} (L)^{b_1} (\mu)^{c_1} (k)^{d_1}$$

$$M^\circ L^\circ T^\circ \theta^\circ = (L^2 T^{-2} \theta^{-1})(ML^{-3})^{a_1} (L)^{b_1} (ML^{-1}T^{-1})^{c_1} (MLT^{-3}\theta^{-1})^{d_1}$$

Equating the powers of M, L, T and θ of both sides, we get the following equations.

$$a_1 + c_1 + d_1 = 0, \quad -3a_1 + b_1 - c_1 + d_1 + 2 = 0$$

$$-c_1 - 3d_1 - 2 = 0, \quad -d_1 - 1 = 0$$

Solution of the above equation gives

$$a_1 = 0,\ b_1 = 0,\ c_1 = 0,\ d_1 = -1$$

$$\therefore \quad \pi_2 = \mu\, c_p/k$$

Similarly,

$$\pi_2 = (\beta\, g\, \Delta\, T),\ (\rho)^{a_2}\ (L)^{b_2}\ (\mu)^{c_2}\ (k)^{d_2}$$

$$M^\circ\, L^\circ\, T^\circ\, \theta^\circ = (L^{-1}\, T^{-2})\ (ML^{-3})^{a_2}\ (L)^{b_2}\ (ML^{-1}\, T^{-1})^{c_2}\ (ML^{-1}\, T^{-3}\, \theta^{-1})^{d_2}$$

Equating the power of M, L, T an θ of both sides, we get the following equations

$$a_2 + c_2 + d_2 = 0, \quad -3\, a_2 + b_2 - c_2 + d_2 + 1 = 0$$

$$-a_2 - 3\, d_2 - 2 = 0, \quad -d_2 = 0$$

Solution of the above equation gives : $a_2 = 2$, $b_2 = 3$, $c_2 = -2$ and $d_2 = 0$

$$\therefore \quad \pi_2 = \frac{(\beta\, g\, \Delta\, T)\, \rho^2\, L^3}{\mu^2} = \frac{(\beta\, g\, \Delta\, T)\, L^3}{\nu^2}$$

This non-dimesnional number is known as **Grashoff number, Gr**

Thus,

$$\pi_1 = f\,(\pi_2,\, \pi_3)$$

$$\frac{h\,L}{k} = f\left(\frac{\mu\, c_p}{k}\right)^m \left(\frac{\beta\, g\, \Delta\, T\, L^3}{\nu^2}\right)^n$$

$$\therefore \quad Nu = f\,(Pr)^m\,(Gr)^n \qquad ...(21.33)$$

where m and n are constants.

21.14 EMPIRICAL EQUATIONS FOR CONVECTIVE HEAT TRANSFER

The following are the empirical equations for convective heat transfer.

(A) Forced Convection

(i) *Laminar flow over flat plates* (from exact solution)

$$Nu = \frac{hL}{k} = 0.664\ (Pr)^{1/3}\ (Re)^{1/2} \qquad ...(21.34)$$

Properties are evaluated at mean bulk temperature $[T_b = (T_1 + T_2)/2]$

(ii) *Laminar flow inside tubes*

$$Nu = 1.86\left(Re_b\ Pr_b\ \frac{D}{L}\right)^{1/3} \left(\frac{\mu_b}{\mu_w}\right)^{0.14} \qquad ...(21.35)$$

Here, all the properties are evaluated at the mean bulk temperature, except μ_w which is evaluated at wall temperature.

(iii) *Turbulent flow over flat plates*

$$\text{Nu} = 0.037\,(\text{Re})^{0.8}\,(\text{Pr})^{1/3} \quad \text{...(21.36)}$$

$$\text{Nu}_x = 0.0296\,(\text{Re}_x)^{0.8}\,(\text{Pr})^{1/3} \quad \text{...(21.37)}$$

(iv) *Turbulent flow inside tubes*

McAdams has proposed a general equation

$$\frac{hD}{k_b} = 0.023\left(\frac{u_\infty D \rho}{\mu_b}\right)^{0.8}\left(\frac{\mu c_p}{k}\right)^{0.4}$$

or
$$\text{Nu} = 0.023\,(\text{Re})^{0.8}\,(\text{Pr})^{0.4} \text{ for heating} \quad \text{...(21.38)}$$

and
$$\text{Nu} = 0.023\,(\text{Re})^{0.8}\,(\text{Pr})^{0.3} \text{ for cooling} \quad \text{...(21.39)}$$

Here, all the properties are evaluated at the bulk temperature for the fluid

(v) *Reynolds analogy*

$$\text{St} = \frac{f}{2}$$

where St = Stanton number = $\dfrac{\text{Nu}}{\text{Re Pr}}$ and f = friction factor = $\dfrac{\tau_w}{(\rho u_\infty^2/2)}$

τ_w = shear stress at the well

(B) Free Convection

Laminar flow over vertical planes and cylinder

$$\text{Nu}_x = 0.508\left(\frac{\text{Pr}}{0.592 + \text{Pr}}\right)^{1/4}(\text{Gr}_x\ \text{Pr})^{1/4} \quad \text{...(21.40)}$$

or
$$\text{Nu} = 0.667\left(\frac{\text{Pr}}{0.592 + \text{Pr}}\right)^{1/4}(\text{Gr}\cdot\text{Pr})^{1/4} \quad \text{...(21.41)}$$

Turbulent flow over vertical planes.

$$(\text{Gr}\,.\,\text{Pr}) = 10^9 \text{ to } 10^{12} \quad \text{...(21.42)}$$

(Note that suffix b indicates bulk temperature [$t_b = (T_1 + T_2)/2$] and f indicates film temperature $(T_s + T_\infty)/2 = T_f$ at which properties should be taken.

Example 21.8 Water flows through a 6 cm diameter tube at a velocity 70 cm/s. The tube length is 4 m. Calculate that heat transfer coefficient if wall surface temperature is 70°C and bulk temperature of water is 30°C.

Solution The properties of water is taken at $T_f = (30 + 70)/2 = 50$°C. They are from table.

	30°C	50°C
ρ (kg/m^3)	1000	988.1
c_p (kJ/kg K)	4.18	4.18
k (W/m K)	0.628	0.647
v (m^2/s)	9.2×10^{-7}	5.56×10^{-7}
Pr	6.13	3.54

First of all, it is necessary to see whether the flow is laminar or turbulent. For this, let us find Re

$$\text{Re} = \frac{u_\infty \cdot D}{v} = \frac{0.7 \times 0.06}{9.2 \times 10^{-7}} = 4.565 \times 10^4$$

Hence, the flow is turbulent. Using McAdam's equation, we have

$$\frac{hD}{k_b} = 0.023 \left(\frac{D u_\infty \rho}{\mu_b}\right)^{0.8} \left(\frac{\mu c_p}{k}\right)^{0.4}$$

$$= 0.023\ (4.565 \times 10^4)^{0.8}\ (6.13)^{0.4} = 0.023 \times 5340 \times 2.065 = 253.62$$

$$h = \frac{253.62 \times 0.628}{0.06} = 2654.55 \text{ W/m}^2 \text{ K}$$

Example 21.9 Air at atmospheric pressure and 30°C flows over a flat plate at a velocity of 1.5 m/s. The plate is maintained at 90°C. The length of the plate is 2 m. Calculate the heat transfer rate per unit width using exact solution.

Solution The mean bulk temperature of the fluid is (30 + 90)/2 = 60°C. The properties of air at the mean bulk temperature of 60°C are:

$$\rho = 1.06 \text{ mg/m}^3, \quad \mu = 7.221 \text{ kg/hm}, \quad c_p = 1.005 \text{ kJ/kg K}$$

$$v = 18.97 \times 10^{-6} \text{ m}^2\text{/s}, \quad k = 2.894 \times 10^2 \text{ W/m K}$$

$$\text{Reynolds number} = \text{Re} = \frac{u_\infty \cdot L}{v} = \frac{1.5 \times 2}{18.97 \times 10^{-6}} = 1.58 \times 10^5$$

and

$$\text{Pr} = \frac{\mu c_p}{k} = \frac{7.221 \times 1.005}{2.824 \times 10^{-2} \times 3600} = 0.071$$

Since, Re < 5 × 10^5, the flow is laminar. Using exact solution

$$\text{Nu} = \frac{hL}{k} = 0.664\ (\text{Pr})^{1/3}\ (\text{Re})^{1/2}$$

or

$$\frac{hL}{k} = 0.664 \times (0.071)^{1/3}\ (1.58 \times 10^5)^{1/2} = 109.29$$

or $$h = \frac{109.29 \times 2.894 \times 10^{-2}}{2} = 1.581 \text{ W/m}^2 \text{ K}$$ **Ans.**

21.15 HEAT EXCHANGERS

A heat exchanger is a device in which heat is transferred from a hot fluid to a colder fluid. The examples of heat exchangers are intercoolers used in compressors, oil coolers, boilers, condensers, regenerators, precooler, etc. Commonly, there are three types or flow path in heat exchanger.

(1) *Parallel flow,* (2) *Counter flow,* (3) *Cross flow.*

Figure 21.10 shows the flow path and temperatures distribution in the above three types of heat exchangers.

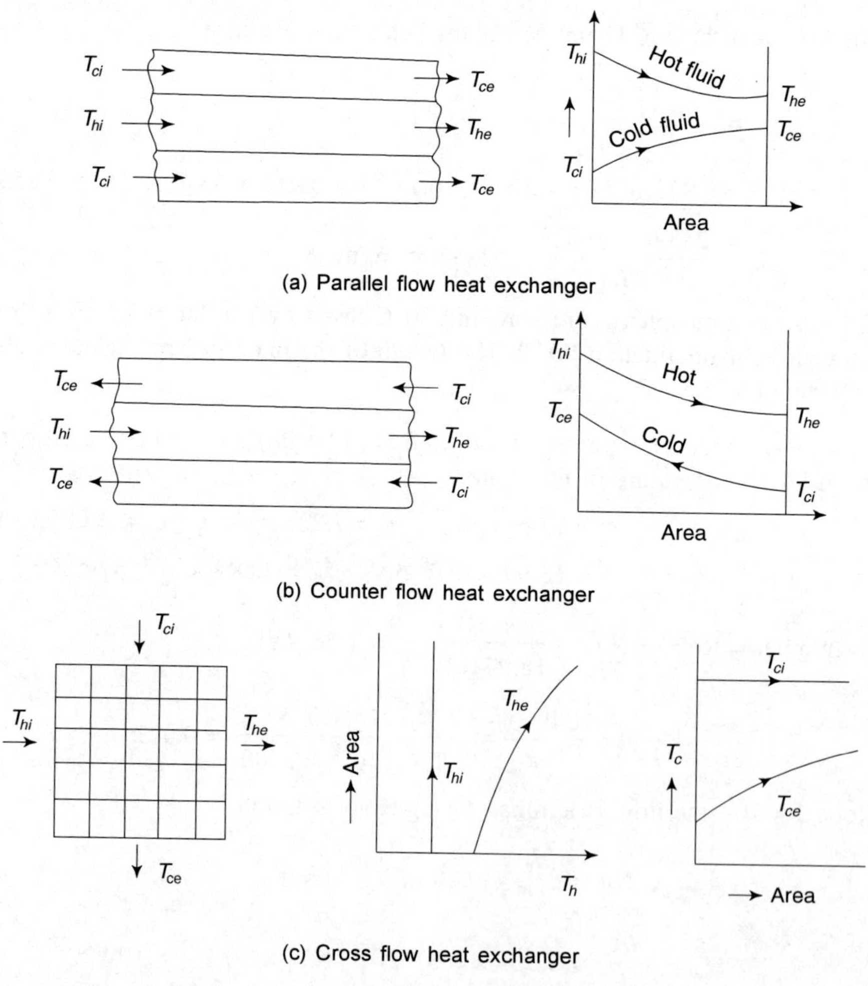

Fig. 21.10 Heat exchangers

21.16 LONG MEAN TEMPERATURE DIFFERENCE

We have seen that the temperatures of the fluids in the heat exchangers are generally not constant but vary from point to point because of the fact that the heat flows from the hotter to the colder fluid. However, the rate of heat flow varies along the path of heat exchanger. Therefore, a concept of **long-mean mean temperature difference** is devised to study heat transfer rate. Thus,

$$q = A\ U\ \Delta\ T_m \qquad \text{...(21.43)}$$

where ΔT_m **is the log-mean temperature difference.**

To find the expression for LMTD, consider the temperature distribution in a single pass parallel flow heat exchanger as shown in Fig. 21.11.

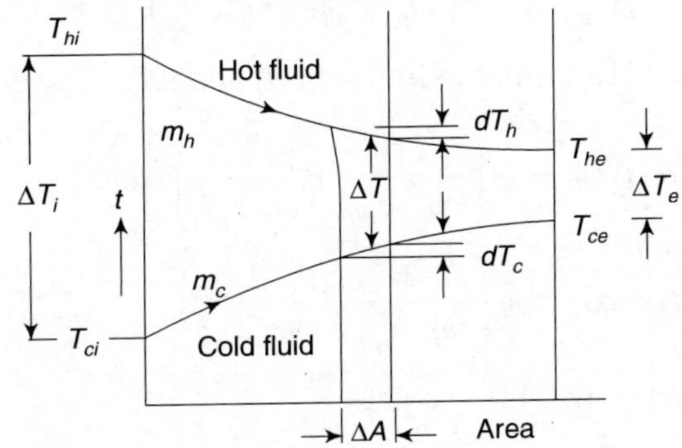

Fig. 21.11 Calculation of LMTD for parallel flow

Suffixes i and e represent the inlet and exit condition of the heat exchanger.

Suffixes h and c represent the hot and cold fluids respectively.

Let us consider an elementary area (dA) of the heat exchanger. The heat flow rate through this elementary area is

$$dq = U\ dA\ \Delta T \qquad \text{...(21.44)}$$

Due to heat transfer (dq) through the area dA, the hotter fluid cools down by dT_h whereas the colder fluid heats up by dT_c. Thus,

$$dq = -\ m_h\ .\ c_{ph}\ dT_h = m_c\ .\ c_{pc}\ .\ dT_c = U\ dA\ (T_h - T_c) \qquad \text{...(21.45)}$$

From the above equation, $dT_h = -\ dq/m_h\ .\ c_{ph}$ and $dT_c = q/m_c\ .\ c_{pc}$

$$\therefore \qquad d\ (\Delta T) = dT_h - dT_c = -\ dq\left(\frac{1}{m_h \cdot c_{ph}} + \frac{1}{m_c \cdot c_{pc}}\right) \qquad \text{...(21.46)}$$

Generally, $m_h\ .\ c_{ph}$ and $m_c\ .\ c_{pc}$ are written as capacity C_h and C_c respectively. Putting the value of dq from Eq. (21.45), the above equation becomes

$$d\,(\Delta T) = -\,UdA\,\Delta T\left(\frac{1}{m_h \cdot c_{ph}} + \frac{1}{m_c \cdot c_{pc}}\right)$$

Integrating the above equation in the limit of ΔT_i and ΔT_c, we get

$$\int_{\Delta T_i}^{\Delta T_e} \frac{d\,(\Delta T)}{\Delta T} = -\left(\frac{1}{m_h \cdot c_{ph}} + \frac{1}{m_c \cdot c_{pc}}\right)\int_0^A U \cdot dA$$

or $$\ln\frac{\Delta T_e}{\Delta T_i} = -\,UA\left(\frac{1}{m_h \cdot c_{ph}} + \frac{1}{m_c \cdot c_{pc}}\right)$$

or $$\ln\frac{\Delta T_i}{\Delta T_e} = UA\left(\frac{1}{m_h \cdot c_{ph}} + \frac{1}{m_c\, c_{pc}}\right) \qquad ...(21.47)$$

Now integrating the Eq. (21.46), in the limit of ΔT_i and ΔT_c, we get

$$\int_{\Delta T_i}^{\Delta T_e} d\,(\Delta T) = -\left(\frac{1}{m_h \cdot c_{ph}} + \frac{1}{m_c\, c_{pc}}\right)\int dq$$

$$\Delta T_i - \Delta T_e = q\left(\frac{1}{m_h \cdot c_{ph}} + \frac{1}{m_c\, c_{pc}}\right) \qquad ...(21.48)$$

Dividing Eq. (21.47) by Eq. (21.48), we get

$$\ln\frac{(\Delta T_i / \Delta T_e)}{\Delta T_i - \Delta T_c} = \frac{U \cdot A}{q}$$

$$q = U \cdot A\,\frac{\Delta T_i - \Delta T_e}{\ln\,(\Delta T_i / \Delta T_e)} = U \cdot A \cdot \Delta T_m \qquad ...(21.49)$$

where $\Delta T_m = \text{LMTD} = \dfrac{\Delta T_i - \Delta T_e}{\ln\,(\Delta T_i / \Delta T_e)}$

Similar expressions result in for counter flow, condensers and evaporators.

Example 21.10 A tubular heat exchanger cools 25 kg/min of exhaust gas passing through it from 400°C to 130°C by water initially at 10°C. The specific heat of gases may be taken as 1.13 kJ/kg K and overall heat transfer coefficient as 140 W/m^2 K. Find the surface area required if the water flows at the rate of 30 kg/min for (a) parallel, and (b) counter flow.

Solution Heat given by gases = Heat taken by water

or $25 \times 1.13\,(400 - 130) = 30\,(T_{ce} - 10) \times 4.18$

$\therefore$ Cooling water outlet temperature $= T_{ce} = \dfrac{25 \times 1.13 \times 270}{30 \times 4.18} + 10 = 70.8°\text{C}$

(a) Parallel flow

$$\text{LMTD} = \Delta T_m = \frac{\Delta T_i - \Delta T_e}{\ln(\Delta T_i/\Delta T_e)} = \frac{(400-10)-(130-70.8)}{\ln\dfrac{(400-10)}{(130-70.8)}}$$

$$= \frac{330.8}{\ln\left(\dfrac{390}{59.2}\right)} = 175.466°\text{C} = 175.46 \text{ K}$$

Heat exchange, $q = AU\,\Delta T_m$

$$\therefore \quad A = \frac{15 \times 1.13 \times 1000 \times (400-130)}{140 \times 175.46 \times 60} = 5.175 \text{ m}^2 \qquad \textbf{Ans.}$$

(b) Counter flow

$$\text{LMTD} = \frac{(400-70.8)-(130-10)}{\ln\dfrac{(400-70.8)}{130-10}} = \frac{209.2}{\ln(329.2/120)} = 207.2°\text{C}$$

$$\therefore \quad \text{Surface area} = A = \frac{25 \times 1.13 \times 1000\,(400-130)}{60 \times 140 \times 207.29} = 4.38 \text{ m}^2 \qquad \textbf{Ans.}$$

Example 21.11 Air is heated while flowing through a 2.5 cm bore copper tube maintained at 280°C. The air enters at 15°C and leaves at 270°C at a mean velocity of 30 m/s. Calculate the length of the tube. Assume $f = 0.0791\,(\text{Re})^{-1/4}$ for the turbulent flow and all properties are at the mean film temperature.

Solution The mean film temperature can be found as

$$T_f = \frac{T_b + T_w}{2} = \frac{[(15+270)/2]+280}{2} = 211.25°\text{C}$$

From tables at 211.25°C, the properties of air, $v = 3.591 \times 10^{-5}$ m^2/s

$$\text{Re} = \frac{u_\infty \cdot D}{v} = \frac{30 \times 0.025}{3.591 \times 10^{-5}} = 2088.5$$

Since Re > 2100, hence the flow is turbulent, thus $f = \dfrac{0.0791}{(20885.5)^{1/4}} = 0.00658$

From Reynolds analogy, $\quad \text{St} = f/2$

or

$$\text{St} = \frac{\text{Nu}}{\text{Re Pr}} = \frac{f}{2} = \frac{0.00658}{2} = 0.00329$$

or

$$\text{Nu} = \text{Re Pr} \times 0.00329 = 20885.5 \times 0.681 \times 0.00329 = 46.8 = hD/k$$

or

$$h = \frac{46.8 \times 3.938 \times 10^{-5}}{0.025} = 0.0737 \text{ W/m}^2 \text{ K}$$

$$\dot{m} = (\pi/4) \times 0.025^2 \times 30 \times 0.73 = 0.01075 \text{ kg/s}$$

$$q = \dot{m}\, c_p\, \Delta T = 0.01075 \times 1.027\ (270 - 15) = 2.815 \text{ kW}$$

$$q = UA\, \Delta T_m = hA\, \Delta T_m$$

But $$\Delta t_m = \frac{\Delta T_i - \Delta T_e}{\ln (\Delta T_i / \Delta T_e)} = \frac{(280 - 15) - (280 - 270)}{\ln [(280 - 15)/(280 - 270)]} = 77.8°\text{C}$$

$$q = 2.825 = 0.07377 \times 77.8\ A$$

or $$A = \frac{2.825}{0.0737 \times 77.8} = 0.49 \text{ m}^2$$

The tube length $= L = \dfrac{A}{\pi D} = \dfrac{0.49}{\pi \times 0.025} = 6.228$ m **Ans.**

21.17 RADIATION

As discussed earlier, *thermal radiation is the electromagnetic radiation emitted by a body as a result of its temperature*. There are many types of electromagnetic radiations and thermal radiation is only one. It travels at the speed of light, 3×10^{10} cm/s. This speed is equal to the product of the wavelength and frequency of radiation

$$c = \lambda v \qquad \text{...(21.50)}$$

where c = speed of light, λ = wavelength in micron (μ) and v = frequency.

Thermal radiation lies in the range from about 0.1 to 100 μm. The propagation of thermal radiation takes place in the form of discrete quanta, each quantum having an energy of

$$E = hv \qquad \text{...(21.51)}$$

where h is Planck's constant and has the value $h = 6.6256 \times 10^{-34}$ J.s.

A very rough physical picture of the radiation propagation may be obtained by considering each quantum as a particle having energy, mass and momentum just as we considered the molecules of gas flowing from one place to another. Thus,

$$E = mc^2 = hv$$

or $$m = hv/c^2$$

By assuming the radiation on a gas, it can be shown from thermodynamics point of view that the energy density of the radiation is proportional to absolute temperature to fourth power i.e.,

$$E_b = \sigma T^4, \text{ W/m}^2 \qquad \text{...(21.52)}$$

where suffix b stands for a black body.

σ = Stefan–Boltzmann constant = 5.67×10^{-8} W/m^2 K^4. The above equation is known as **Stefan–Boltzmann law**.

21.18 RADIATION PROPERTIES

When radiation energy impinges on a material surface, part of the radiation is absorbed, part is transmitted and part is reflected as shown in Fig. 21.12. *Absorptivity* (α) *is defined as the fraction of incident radiation absorbed. Similarly, reflectivity* (ρ) *is defined as the fraction of incident radiation reflected and the transmissivity* (τ) *is defined as the fraction of incident energy transmitted.* Let G be the total incident radiation per unit time and per unit surface and G_α, G_ρ and G_τ be the amount of radiation absorbed, reflected and transmitted.

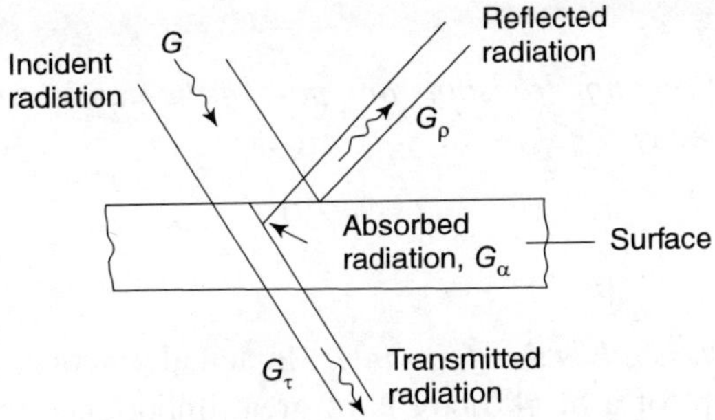

Fig. 21.12 Types of radiation

Thus, $\alpha = G_\alpha/G, \quad \rho = G_\rho/G \quad \text{and } \tau = G_\tau/G$

But $G_\alpha + G_\rho + G_\tau = G$

$$\therefore \quad \alpha + \rho + \tau = 1 \qquad \text{...(21.53)}$$

Similarly, the monochromatic absorpitivity (α_λ), reflectivity, (ρ_λ) and transmissivity (τ_λ) may be defined as

$$\alpha_\lambda = \frac{(G_\alpha)_\lambda}{G_\lambda}, \quad \rho_\lambda = \frac{(G_\rho)_\lambda}{G_\lambda} \text{ and } \tau_\lambda = \frac{(G_\tau)_\lambda}{G_\lambda}$$

$$\therefore \quad \alpha = \frac{G_\alpha}{G} = \frac{1}{G}\int_0^\infty \alpha_\lambda G_\lambda \, d\lambda \qquad \text{...(21.54)}$$

$$\rho = \frac{G_\rho}{G} = \frac{1}{G}\int_0^\infty \rho_\lambda G_\lambda \, d\lambda \qquad \text{...(21.55)}$$

$$\tau = \frac{G_\tau}{G} = \frac{1}{G}\int_0^\infty \tau_\lambda G_\lambda \, d\lambda \qquad \text{...(21.56)}$$

On the basis of the three properties α, ρ and τ, the bodies are classified as follows.

21.18.1 Transparent Body

It transmits part of the incident radiation falling on a surface. In this category comes glass.

For an opaque body, $\tau = 0$

Therefore, $\alpha + \rho = 1$ ...(21.57)

This means that when the radiation falls on a surface of an opaque body, part of it is reflected and the remainder is absorbed by a very thin layer of materials *approximately 1 μ thick. Almost all engineering materials may be considered as opaque bodies because they have so much thickness that transmissivity is zero.*

21.18.2 Black Body

It is one which neither reflects nor transmits any part of the incident radiation, but absorbs all of it. Thus, for a back body

$$\rho = 0 \quad \text{and} \quad \tau = 0$$

Therefore, $\alpha = 1$ ...(21.58)

Due to this a black cloth is advisable for winter. In actual practice, a black body does not exist. However, the concept of a black body is of great importance in the study of thermal radiation, similar to the concept of ideal gas in thermodynamics.

21.18.3 White Body

It is one which reflects all the incident radiation and does not absorb or transmit any part of it

For a white body, $\alpha = 0, \tau = 0$

∴ $\rho = 1$ ...(21.59)

Due to this reason, it is advisable to put on white cloth in summer.

21.18.4 Grey Body

It is one in which the absorptivity of the surface does not vary with temperature and wavelength of the incident radiation. For a grey body

$$\alpha = \alpha_\lambda \text{ constant} \quad \text{...(21.60)}$$

21.19 PLANCK'S DISTRIBUTION LAW

As discussed earlier, the amount of radiation emitted by a surface varies with the temperature so with a black body. At a given wavelength, the radiation energy emitted by a black body increases with temperature. Similarly, at a given temperature the radiation emitted varies with wavelength. The energy emitted by a surface in all directions at a given wavelength is called the ***monochromatic emissive power of surface***. Thus, monochromatic emissive power of a black body may be denoted as $(E_\lambda)_b$.

According to Max Planck, the monochromatic emissive power of a body is expressed as

$$(E_\lambda)_b = 2\,\pi\,c^2\,h\,\lambda\,\frac{\lambda^{-5}}{\exp\,(ch/\lambda\,kT) - 1} \qquad \ldots(21.61)$$

where c = velocity of light = 2.998×10^8 m/s

h = Planck's constant = 6.6256×10^{-34} J.s

k = Boltzmann constant = 1.3805×10^{-23} J/K

λ = wavelength, µm

T = absolute temperature, K

Thus, the unit of $(E_\lambda)_b$ is W/m^2 · µm

Figure 21.13 shows Planck's distribution law for several temperatures. Thus, law has been found to be in close agreement with the experiments. The following conclusions may be drawn from this law.

(i) As temperature increases, $(E_\lambda)_b$ increases for a particular value of wave length.

(ii) At a particular value of temperature, $(E_\lambda)_b$ first increases as λ increases but reaches a maximum value and then decreases.

(iii) The peak value of $(E_\lambda)_b$ shifts towards the lower wave length range as temperature increases.

(iv) E_λ of a real surface is much less than $(E_\lambda)_b$ at the same temperature and wavelength and the curve is very irregular in nature.

21.20 DERIVATION OF STEFAN–BOLTZMANN LAW FROM PLANCK'S DISTRIBUTION LAW

From the definition of total emissive power at any temperature, the value of $(E_\lambda)_b$ is represented by the total area under the curve of $(E_\lambda)_b$ (Fig. 21.13) pertaining to that temeprature.

Thus,
$$E_b = \int_0^\infty (E_\lambda)_b\,d\lambda$$

Putting the value of $(E_\lambda)_b$ from Planck's law, we get

$$E_b = C_1 \int_0^\infty \frac{\lambda^{-5}\,d\lambda}{\exp\,(C_2/\lambda T) - 1}$$

where $C_1 = 2\,\pi c^2\,h$ and $C_2 = ch/K$

Let,
$$C_2/\lambda T = x, \text{ hence } \lambda = C_2/x\,T \text{ and } d\lambda = \frac{-C_2}{Tx^2}$$

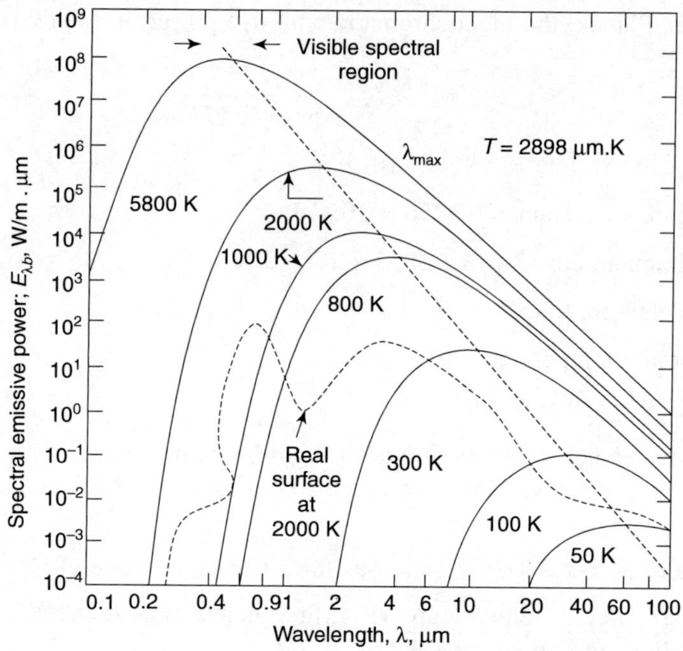

Fig. 21.13 Planck's distribution law

Putting these values, we get

$$E_b = -C_1 \int_0^\infty \frac{x^5\,T^6}{C_2^5(e^x-1)\,x^2\,T}\,dx = \frac{C_1\,T^4}{C_2^4}\int_0^\infty \frac{x^3}{e^x-1} = \frac{C_1}{C_2^4}T^4\,\frac{\pi^{-4}}{15}$$

$$= 5.67 \times 10^{-8}\,T^4;\ \text{W/m}^2\,.\,k^4$$

$$\therefore \qquad E_b = \sigma\,T^4$$

21.21 EMISSIVITY (ε)

It is defined as the ratio of the emissive power of the surface to the emissive power of a hypothetical black body at the same temperature. Emissivity depends upon temperature as well as wavelength. Commonly, emissivity is classified as:

(i) Monochromatic emissivity, (ii) total emissivity, and (iii) normal total emissivity.

(i) Monochromatic emissivity, ε_λ *It is the ratio of monochromatic emissive power of a surface to monochromatic emissive power of a black body.*

$$\varepsilon_\lambda = E_\lambda/(E_\lambda)_b \qquad ...(21.62)$$

(ii) Total emissivity, ε *It is the ratio of total emissive power of a surface to the total emissive power of a black body.*

$$\varepsilon = E/(E_\lambda) \qquad ...(21.63)$$

(iii) Normal total emissivity, ε_n *It is the ratio of normal component of total emissive power to the normal component of total emissive power of a black body.*

$$\varepsilon_n = E_n/(E_b)_n \qquad ...(21.64)$$

Note that emissivity is the property of a surface and its value depends upon the type of surface and independent of wavelength.

21.22 KIRCHHOFF'S LAW

It states that at any temperature, the ratio of total emissive power (E) to the absorptivity (α) is a constant for all substances which are in thermal equilibrium with each other.

Thus,
$$\frac{E_1}{\alpha_1} = \frac{E_2}{\alpha_2} = \frac{E_3}{\alpha_3} = \text{constant}$$

If the third body is a black body, then

$$\frac{E_1}{\alpha_1} = \frac{E_2}{\alpha_2} = \frac{E_b}{\alpha_b} = E_b \quad \text{since } \alpha_b = 1$$

$$\therefore \quad \frac{E_1}{E_b} = \alpha_1, \quad \text{and} \quad \frac{E_2}{E_b} = \alpha_2$$

But
$$\frac{E_1}{E_b} = \varepsilon_1, \quad \frac{E_2}{E_b} = \varepsilon_2$$

$$\therefore \quad \alpha_1 = \varepsilon_1 \quad \text{and} \quad \alpha_2 = \varepsilon_2$$

Therefore,
$$\alpha = \varepsilon \qquad ...(21.65)$$

It means that if a body is in thermal equilibrium with its environment, its absoptivity (α) is equal to its emissivity, ε.

Kirchhoff's law is also true for monochromatic radiation.

21.23 INTENSITY OF RADIATION

It is defined as the rate of emission of radiation of all wavelengths in a given direction from a surface per unit solid angle and per unit projected area of the surface on a plane perpendicular to the direction of radiation. The unit of intensity of radiation is W/m^2 Steradian.

To find a relation between the total emissive power of a surface to the intensity of radiation, consider the radiation emitted by a small area (*dA*), of a surface located at the centre of a hemisphere of radius (*R*) as shown in Fig. 21.14. The amount of this radiation falling on a portion of the hemispherical surface (dA_2) per unit time is given by

$$dq_{1-2} = I_1\ (dA_1 \cos\theta)\ \frac{dA_2}{R^2} \qquad \text{...(21.66)}$$

where dq_{1-2} = rate of radiation emitted by dA_1, falling on the surface, dA_2

I_1 = intensity of radiation of surface dA_1, W/m^2 Steradian.

$dA_1 \cos\theta$ = projected area of dA_1 on a plane perpendicular to the line.

If q_{1-2} denotes the total energy emitted by the surface dA_1 in all directions per time, then

$$q_{1-2} = dA_1\ E_1 \qquad \text{...(21.67)}$$

Also from Eq. (21.67),

$$q_{1-2} = \int I_1\ (dA_1 \cos\theta)\ dA_2 / R^2$$

But $\quad dA_2 = (r\ d\theta)\ (R d\beta) \quad \text{and} \quad r = R \sin\theta$

$\therefore \quad dA_2 = R^2 \sin\theta\ d\theta\ d\beta$

Putting the value of dA_2, we get

$$q_{1-2} = I_1\ dA_1 \int_0^{\pi/2} \int_{\beta}^{\beta=2\pi} (\cos\theta \sin\theta)\ d\theta\ d\beta = \pi\ I_1\ dA_1 \qquad \text{...(21.68)}$$

Comparing Eq. (21.67) and Eq. (21.68), we get

$$E = \pi I \qquad \text{...(21.69)}$$

This means that the total emissive power of a surface is π times the intensity of radiation.

21.24 RADIATION HEAT EXCHANGE BETWEEN BLACK SURFACES

It is not only the emission, absorption and reflection characteristics of the surface but also the geometrical arrangement which govern the radiation heat exchange between two surfaces. Consider two black surfaces of area A_1 and A_2 respectively as shown in Fig. 21.15.

The projected area of ΔA_1 normal to the line joining ΔA_1 and $\Delta A_2 = \Delta A_1 \cos\theta_1$

The solid angle subtended by ΔA_2 with centre on $\Delta A_1 = \dfrac{\Delta A_1 \cos\theta_2}{S^2}$

$\therefore$ The rate at which energy emitted by ΔA_1 impinges on ΔA_2 is given by

$$= (I_{b1})(\Delta A_1 \cos\theta_1)\frac{(\Delta A_2 \cos\theta_2)}{S^2}$$

As A_2 is black, hence all the above energy will be absorbed. Thus, the rate at which energy emitted by ΔA_1 is absorbed by A_2 is given by

$$\Delta q_{12} = \frac{I_{b1} \cos\theta_1 \cdot \cos\theta_2\ \Delta A_1 \cdot \Delta A_2}{S^2} \qquad \text{...(21.70)}$$

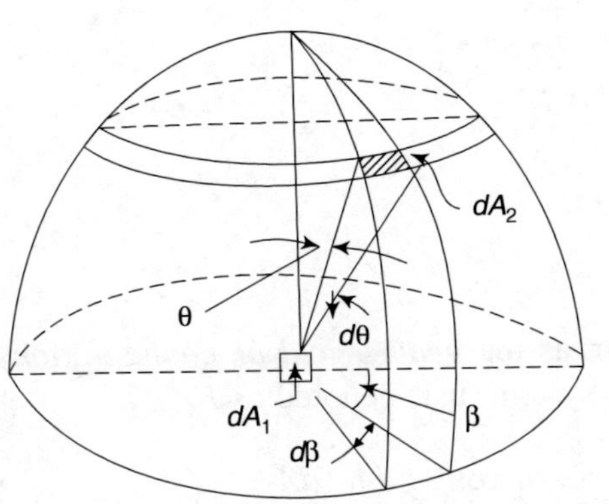

Fig. 21.14 Determination of intensity of radiation

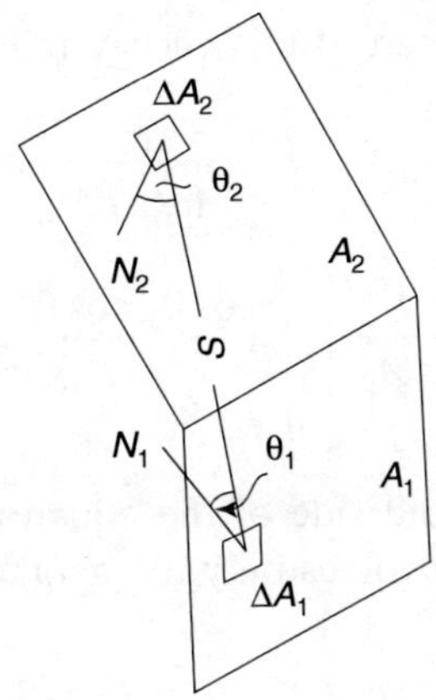

Fig. 21.15 Thermal radiation on between two black bodies

Similarly, the heat energy emitted by ΔA_2 and absorbed by ΔA_1 is given by

$$\Delta q_{21} = \frac{I_{b2} \cos\theta_1 \cos\theta_2 \Delta A_1 \Delta A_2}{S^2} \quad ...(21.71)$$

$\therefore$ Net heat exchange, $\Delta q_{12} - \Delta q_{12} = (I_{b1} - I_{b2}) \dfrac{\cos\theta_1 \cos\theta_2 \, \Delta A_1 \cdot \Delta A_2}{S^2}$ ...(21.72)

But

$$I_{b1} = \frac{E_{b1}}{\pi} \text{ and } I_{b2} = \frac{E_{b2}}{\pi}$$

$\therefore$

$$(\Delta q_{12})_{net} = \frac{(E_{b1} - E_{b2}) \cos\theta_1 \cos\theta_2 \, \Delta A_1 \cdot \Delta A_2}{\pi S^2} \quad ...(21.73)$$

or

$$(\Delta q_{12})_{net} = \frac{\sigma (T_1^2 - T_2^4) \cos\theta_1 \cos\theta_2 \, \Delta A_1 \cdot \Delta A_2}{\pi S^2} \quad ...(21.74)$$

In the limit,

$$(dq_{12})_{net} = \frac{\sigma (T_1^4 - T_2^4)}{\pi} \frac{\cos\theta_1 \cos\theta_2 \, dA_1 \cdot dA_2}{S^2} \quad ...(21.75)$$

The net heat transfer from the entire surface area A_1 and A_2 can be found by integrating

$$(dq_{12})_{net} = \frac{\sigma (T_1^4 - T_2^4)}{\pi} \int_{A_1} \int_{A_2} \frac{\cos\theta_1 \cos\theta_2 \, dA_1 \cdot dA_2}{S^2} \quad ...(21.76)$$

Moreover, integration of Eq. (21.75) gives

$$(q_{12}) = \frac{\sigma T_1^4}{\pi} \int_{A_1} \int_{A_2} \frac{\cos\theta_1 \cos\theta_2 \, dA_1 \cdot dA_2}{S^2} \quad ...(21.77)$$

As only a part of this energy falls on A_2,

Hence, $$\frac{q_{12}}{A_1 \sigma T_1^4} < 1 \quad \text{and} \quad \frac{q_{12}}{\sigma T_1^4} < A_1$$

or $$\frac{1}{\pi A_1} \int_{A_1}\int_{A_2} \frac{\cos\theta_1 \cos\theta_2 \, dA_1 \cdot dA_2}{S^2} < 1 \qquad \text{...(21.78)}$$

The left hand side of the equation is known as the *geometrical or configuration factor* because this involves only the geometry of the system. It is denoted as F_{12}.

Hence, $$F_{12} = \frac{1}{A_1 \pi} \int_{A_1}\int_{A_2} \frac{\cos\theta_1 \cos\theta_2 \, dA_1 \cdot dA_2}{S^2} < 1 \qquad \text{...(21.79)}$$

Note that F_{12} is the configuration factor of surface A_1 with respect to surface A_2 and it signifies the fraction of the total energy emitted by A_1 and intercepted by A_2.

Hence, $$q_{12} = A_1 F_{12} \sigma T_1^4. \quad \text{Similarly, } q_{21} = A_{21} F_{21} \sigma T_2^4$$

Net heat exchange $$(q_{12})_{net} = A_1 F_{12} \sigma T_1^4 - A_2 F_{21} \sigma T_2^4 \qquad \text{...(21.80)}$$

If $$T_1 = T_2, \text{ then } (q_{12})_{net} = 0, \text{ i.e., heat exchange ceases.}$$

$$\therefore \quad (q_{12})_{net} = 0 = \sigma T_1^4 (A_1 F_{12} - A_2 F_{21}) \qquad \text{...(21.80a)}$$

As σ and T_1 cannot be zero, hence $A_1 F_{12} = A_2 F_{21}$ ...(21.81)

The above equation is called ***reciprocity theorem***. In general.

$$(q_{12})_{net} = A F \sigma (T_1^4 - T_2^4)$$

The value of F is calculated by graphical methods.

The heat exchange between two real bodies is given by

$$(q_{12})_{net} = A F \sigma \varepsilon (T_1^4 - T_2^4) \text{ where } \varepsilon \text{ is emissivity.} \qquad \text{...(21.82)}$$

Example 21.12 A black body of 300 cm^2 area is at temperature of 900 K. Determine the total rate of energy emission. Compare this value with a real body having emissivity equal to 0.9.

Solution From Stefan–Boltzmann law

$$q = \sigma A T^4 = 5.67 \times 10^{-8} \times \left(\frac{300}{10^4}\right) \times (900)^4 = 11160.26 \text{ W} \quad \textbf{Ans.}$$

For real body, $$q = \varepsilon \sigma A T^4 = 0.9 \times 11160.26 = 10044.2 \text{ W} \quad \textbf{Ans.}$$

Example 21.13 A polished metal rod of 6 cm diameter has its outer surface at temperature of 100°C. Find the heat exchange by radiation to the surroundings and by natural convection per metre length of the rod if the ambient temperature is 20°C and emissivity of the surface is 0.2 and heat transfer coefficient is 11.6 W/m^2. Also calculate the overall heat transfer coefficient by convection and radiation.

Solution Heat transfer by radiation = $q_r = \sigma\, \varepsilon\, A\, (T_1^4 - T_2^4)$

$$= (5.67 \times 10^{-8}) \times 0.2 \times \left(\pi \times \frac{6}{100} \times 1\right) (373^4 - 293^4) = 25.62 \text{ W}$$

Heat transfer by convention,

$$q_c = h\, A_s\, \Delta T = 11.6 \times \left(\frac{(\pi \times 6 \times 1)}{100}\right) \times (100 - 20) = 174.92 \text{ W}$$

$\therefore$ Total rate of heat transfer, $q = 25.62 + 174.9 = 200.54$ W **Ans.**

Also $q = UA\Delta T$

or $200.54 = U \times (\pi \times 0.06 \times 1)\,(100 - 20)$ **Ans.**

or $U = 13.298$ W/m^2 K

Example 21.14 Two black discs of diameter 0.6 m are placed directly opposite at a distance of 1 m. The temperatures are maintained at 1000 K and 500 K respectively. Determine the heat flow between the discs when no other surface are present.

Solution The radiation heat transfer = $(q_{12})_{\text{net}} = A_1\, F_{12}\, \sigma\, (T_1^4 - T_2^4)$

$$A_1 = \frac{\pi}{4}(0.6)^2 = \frac{0.36\,\pi}{4}$$

From the graph, $F_{12} = 0.8$

$$(q_{12})_{\text{net}} = \frac{0.36 \times \pi}{4} \times 0.8 \times (5.67 \times 10^{-8})\,(1000^4 - 500^4) = 12023.66 \text{ W} \quad \textbf{Ans.}$$

Example 21.15 An insulated steam pipe of outside diameter 75 mm is laid down in a room in which the air and the walls are at a temperature of 20°C. The surface temperature and emissivity areas 220°C and 0.85 respectively. Calculate the rate of heat loss from the surface per unit length of the pipe if $h = 16$ W/m^2 K.

Solution Two modes of heat transfer will occur in this case, one is by convection and other by radiation

$$q = hA\,(T_s - T_\infty) + \varepsilon\, A\, \sigma\, (T_s^4 - T_{\text{wall}}^4) = h\,(\pi\, DL)\,(T_s - T_\infty) + \varepsilon\,(\pi\, DL)\,\sigma\,(T_s^4 - T_{\text{wall}}^4)$$

or $$\frac{q}{L} = q' = 16\,\pi \times 0.075 \times (220 - 20)$$

$$+\, 0.85 \times \pi \times 0.0745 \times 5.67 \times 10^{-8}\,(493^4 - 293^4) = 1341.1 \text{ W/m} \quad \textbf{Ans.}$$

Example 21.16 The hot combustion gases of a furnace are separated from ambient air and its surroundings which are at 30°C by a brick wall (k = 1.2 W/m K) of 0.15 cm thick having surface emissivity of 0.8. Under steady state condition, the temperature of the outer surface is 100°C, and h_0 = 20 W/m^2 K. Calculate the brick inner surface temperature. Take F_{12} = 1.

Solution Surface energy balance gives $E_{in} = E_{out}$

or
$$q''_{cond} = q''_{conv} + q''_{rad}$$

or
$$k\frac{(T_1 - T_2)}{L} = h(T_2 - T_\infty) + \varepsilon\,\sigma\,(T_2^4 - T_\infty^4)$$

or
$$1.2\frac{(T_1 - 100)}{0.15} = 20\,(100 - 30) + 0.8 \times 5.67 \times 10^{-8}\,(373^4 - 303^4)$$

or
$$T_1 = 336.92°\text{C}$$ **Ans.**

21.25 SUMMARY OF FORMULAE

1. $q = -kA\dfrac{dT}{dx} = \dfrac{-kA}{L}(T_2 - T_1)$

2. $k = k_0\,(1 + \alpha T)$

3. $\dfrac{\partial^2 T}{\partial x^2} + \dfrac{\partial^2 T}{\partial y^2} + \dfrac{\partial^2 T}{\partial z^2} + \dfrac{q_g'''}{k} = \dfrac{1}{\alpha}\dfrac{\partial T}{\partial \tau}$ General heat conduction equation

4. $q = -kA\dfrac{dT}{dr} = \dfrac{2\pi k L (T_1 - T_2)}{\ln (r_2/r_1)}$ for cylinder

5. $q = -kA\dfrac{dT}{dr} = \dfrac{4\pi k (T_1 - T_2) r_1 r_2}{r_2 - r_1}$ for sphere

6. $q = \dfrac{T_g - T_a}{\Sigma R} = UA\,(T_g - T_a)$ for composite wall

7. $U = \dfrac{1}{(1/h_g) + (L_1/k_1) + (L_2/k_2) + (L_3/k_3) + (1/h_a)}$ for composite wall

8. $q = \dfrac{T_i - T_0}{\Sigma R} = \dfrac{(T_i - T_0)}{\dfrac{1}{2\pi r_1 L h_1} + \dfrac{\ln (r_2/r_1)}{2\pi k_1 L} + \dfrac{\ln (r_3/r_2)}{2\pi k_2 L} + \dfrac{1}{2\pi r_3 h_0 L}}$

9. $r_c = \dfrac{k}{h_0}$

10. $q = h\,A_s\,(T_s - T_\infty)$

11. $\text{Re} = \dfrac{L \cdot u_\infty}{v}$

12. $\text{Pr} = \dfrac{\mu \cdot c_p}{k} = \dfrac{v}{\alpha}$

13. $\text{Nu} = \dfrac{hL}{k}$

14. $\text{St} = \dfrac{\text{Nu}}{\text{Re} \cdot \text{Pr}} = \dfrac{h\,L}{\rho\, u_\infty\, c_p}$

15. $\text{Gr} = \dfrac{\rho^2\, \beta\, g\, L^3\, \Delta T}{\mu^2}$

16. $\text{Nu} = f\,(\text{Re})^n\,(\text{Pr})^m$ for force convection

17. $\text{Nu} = f\,(\text{Pr})^m\,(\text{Gr})^n$ for free convection

18. $\text{Nu} = \dfrac{h\,L}{k} = 0.664\,(\text{Pr})^{1/3}\,(\text{Re})^{1/2}$ for flate plate (laminar flow)

19. $\text{Nu} = 0.037\,(\text{Re})^{0.8}\,(\text{Pr})^{1/2}$ for flate plate (Turbulent flow)

20. $\text{Nu} = 0.023\,(\text{Re})^{0.8}\,(\text{Pr})^{0.4}$ for heating

21. $\text{Nu} = 0.023\,(\text{Re})^{0.8}\,(\text{Pr})^{0.3}$ for cooling

(20 and 21: for tubes, turbulent flow)

22. $\text{St} = \dfrac{f}{2} = \tau_w/(1/2)\,\rho\, u_\infty^2$

23. $\text{Nu} = 0.667\left(\dfrac{\text{Pr}}{0.592 + \text{Pr}}\right)^{1/4}(\text{Gr}.\text{Pr})^{1/4}$ for free convection, laminar

24. $\Delta T_m = \dfrac{\Delta T_i - \Delta T_e}{\ln\,(\Delta T_i/\Delta T_e)}$

25. $q = UA\,\Delta T_m$ for heat exchangers

26. $E_b = \sigma\, AT^4$

27. $E = \varepsilon\,\sigma\,A_1\,F_{12}\,(T_1^4 - T_2^4)$

28. $\alpha + \rho + \tau = 1$

29. $(E_\lambda)_b = 2\,\pi\,c^2\,h\lambda\,\dfrac{\lambda^{-5}}{\exp\,(ch/\lambda\, kT) - 1}$

30. $\alpha = \varepsilon$

31. $E = \pi I$

32. $F_{12} = \dfrac{1}{A_1\,\pi}\displaystyle\int_{A_1}\int_{A_2}\dfrac{\cos\theta_1 \cdot \cos\theta_2\; dA_1 \cdot dA_2}{S^2} < 1$

PROBLEMS

Things to Think About

21.1. What is the temperature of outer surface of the sun?

21.2. Do gas and liquid have thermal conductivity?

21.3. What is electron drift theory of heat conduction?

21.4. What is the reason that moon and stars do not shine in day time?

21.5. Does radiation need media to travel?

Short and Long Theoretical Questions

21.6. Discuss the modes of heat transfer.

21.7. Derive the expression for heat transfer rate through cylinder, sphere, and composite cylinder.

21.8. Show the non-dimensional relation in forced and free convection.

21.9. Derive expressions for LMTD in cross and counter flow heat exchangers.

21.10. Define thermal radiation.

21.11. Give a concept of black body.

21.12. Discuss the Planck's distribution law of radiation.

21.13. Define Kirchhoff's law and intensity of radiation.

21.14. What do you mean by geometrical factor?

Numerical Questions

21.15. A furnace wall is made of 2.5 cm thick earth insulating (k = 0.11 W/mK), 6.25 cm insulating brick (k =1.1 W/m K), 6 cm fire brick (k =7 W/mk) and 0.3 cm steel casting (k = 71 W/m K). The coefficient of heat transfer is 17 W/m^2 K at the inner surface and 12.5 W/m^2 K at the outer surface. The temperatures are 1150°C inside and 35°C outside the furnace. Determine (a) loss of heat flux, and (b) overall coefficient of heat transfer. [*Ans.* 11426 W, 2.28 W/m K]

21.16. Steam at 935 bar flows through a pipe. The unit surface conductance of the steam side film is 5640 W/m^2 K and that of outside film at pipe is 30 W/m^2 K. If the pipe is placed in a room at a temperature of 25°C, find the temperature at the outer surface of the pipe. [*Ans.* 174.57°C]

21.17. A steam pipe (ID = 100 mm and O.D. = 170 mm) is covered with two layers of insulation to reduce heat transfer, the layers are 30 mm and 50 mm thick respectively. The thermal conductivities of the insulating materials are 0.175 and 0.093 W/m K respectively while that of steel is 50. The temperature of inner surface of the pipe is 300°C while the outer surface is 50° C. Calculate the heat loss from the pipe and the layer contact temperatures.

[*Ans.* 279 W/m, 300°C, 223°C]

21.18. Water at an average temperature of 27°C is flowing through a smooth 5 cm ID pipe at a velocity of 1 m/s. If the temperature of the inner surface of the pipe is 50°C, calculate (a) the unit surface conductance, and (b) heat flux. [*Ans.* (1) 4233 W/m^2 K, (b) 15,293 W]

21.19. Mercury at mean bulk temperature of 65°C flows through a copper tube 5 ID at a velocity of 7.2 m/s. The temperature at the inside surface of the tube is 80°C. Calculate the rate of heat transfer if the tube is 9 m long. [*Ans.* 9.07×10^5 W]

21.20. Determine the radiant energy loss from a 1 cm diameter opening in a thin-walled furnace located in a large enclosure, if the temperature within the furnace is 900°C and the surroundings are at 20°C. [*Ans.* 7.55 W]

21.21. What is the net radiant exchange between the floor of a furnace 5 m × 6 m and a side wall 5 m × 2.5 m, if the temperatures of the floor and wall are 600°C and 310°C respectively? Assume both surfaces are black.

21.22. A steam condenser is transferring 25 kW of thermal energy at a condensing temperature of 65°C. The cooling water enters the condenser at 20°C with a flow rate of 7500 kg/h. Calculate the log mean temperature difference. If overall heat transfer for condenser surface is 1250 W/m^2 K, calculate the surface area required to handle this load. [*Ans.* 28.25°C, 7.08 m^2]

21.23. A tubular heat exchanger is to be designed for cooling oil from a temperature of 80°C to 30°C by a large body of stagnant water which may be assumed to remain constant at a temperature of 20°C. The heat transfer surface consists of 30 m long straight tube of 2 cm inside diameter, c_p of oil = 2.5 kJ/kg K and specific gravity 0.8, the velocity of oil is 50 cm/s. Calculate the overall heat transfer coefficient.

21 24. Air at 80 kg/h flows through a 10 cm internal diameter tube. At a particular point, the pressure and temperature of air are 1.1 bar and 333 K respectively whilst the surface temperature of the tube is maintained at 380 K. Find heat transfer coefficient and rate from 1 m length of the tube in this region. [*Ans.* 12 W/m^2, 177.3 W]

21.25. Steam at 300°C flows through a pipe of 30 cm in diameter. The pipe is placed horizontally and exposed to atmospheric air at 20°C. Calculate the heat transfer per metre of length. Assume pipe surface temperature to be the same as that of steam. Use free convection.

21.26. A horizontal heated plates at 200°C and facing upwards has been placed in still air at 20°C. If the plate measures 1.5 m × 1 m, calculate the heat loss by natural convection. Take : $h = 3.02\ (\theta)^{1/4}$ W/m^2 K.

when θ = mean film temperature. [*Ans.* 3000 W]

Software Based Questions

21.27. Develop a software for the design of composite wall made of different materials used in (a) cold storage, and (b) furnace wall. All possible materials may be considered in database.

21.28. Develop a software for the design of composite cylinder pipe which carries steam from boiler to steam turbine. Consider all possible materials in database.

21.29. Develop a software for the design of the following type of heat exchangers:

(a) Liquid to liquid, (b) liquid to gas, (c) gas to gas, and (d) surface condenser of power plant.

21.30. Develop a software for the design of radiator of a passenger car.

21.31. Develop a software for the design of a condenser used in power plant.

Design and Open Ended Questions

21.32. In space vehicle launched from earth station must be adequately protected by insulating material against aerodynamic heating and radiation from the sun. Prepare a report dealing with heat transfer mechanism, material of insulating material, its thickness and its fixation on the surface.

21.33. Gas turbine blades are cooled as the temperature of gas is around 1500 to 1700 K; otherwise it will fail. They are cooled by internal convection, film and transpiration cooling using bled air from compressor, steam generated by turbine exhaust or water. Prepare a comprehensive report about heat transfer mechanism and a method to calculate coolant requirement.

21.34. Prepare a report on the cooling of fuel element of nuclear reactor including the analysis of heat transfer so that there is no hot spot in fuel element.

Multiple-Choice Answer Type Questions

21.35. The critical radius of insulation is given by

(a) $\dfrac{k}{c_p}$ (b) $\dfrac{k}{h_0}$ (c) $\dfrac{h_0}{k}$ (d) $\dfrac{c_p}{k}$

21.36. Grashoff number (Gr) is given by

(a) $\dfrac{\sigma g \beta L \Delta T}{v}$ (b) $\dfrac{\rho^2 \beta g L^3 \Delta T}{\mu^2}$ (c) $\dfrac{\rho^2 \beta^2 g L^2 \Delta T}{\mu^2}$ (d) $\dfrac{\rho^2 \beta g L^3 dT}{\mu^3}$

21.37. Stanton number St is given by

(a) $\dfrac{hL}{\rho c_p}$ (b) $\dfrac{hL}{\rho u_\infty^2 c_p}$ (c) $\dfrac{hL}{\rho u_\infty c_p}$ (d) $\dfrac{h \cdot L}{u_\infty c_p}$

21.38. For white body

(a) $\alpha + \rho + \tau = 1$ (b) $\alpha = 1$ (c) $\tau = 1$ (d) $\rho = 1$

Answers to Multiple-Choice Questions

21.35 (b) 21.36 (b) 21.37 (c) 21.38 (d)

Appendix

APPENDIX A

SI Units

In I960, the CGPM (Conference Generals des Poids et Measures) recommended the universal adoption of units based on metre (m), kilogram (kg), second (s), degree kelvin (K) and candela (cd) to cover the entire system of science and engineering and gave it the name "**System Internal d' Units**", **abbreviated SI**. Most of the countries including India have switched over to this unit. The fundamental and supplementary units of the SI system are depicted in Table A1. The derived units from the base and supplementary units are presented in Table A2. Table A3 shows multiples and submultiples.

Table A 1 Fundamental and supplementary SI units

Quantity	Unit	Symbol
length	metre	m
mass	kilogram	kg
time	second	s
temperature	degree kelvin	K
current	ampere	amp
luminous intensity	candela	cd
amount of substance	mole	mole
	Supplementary Unit	
plane angle	radian	rad
solid angle	steradian	sr

Note In addition to thermodynamic temperature scale degree Kelvin (K), the degree Celsius (°C) is also in use.

Table A2 Derived units

Quantity	Unit	Symbol
area	square metre	m^2
volume	cubic metre	m^3
velocity	metre per second	m/s
acceleration	metre per second square	m/s^2
density	kilogram per cubic metre	kg/m^3
force	newton	N; kg m/s^2
pressure, stress	pascal, or newton per square metre	$P_a = N/m^2$, 1 P_a = 1 N/m^2 (1 J = 1 Nm)
work, heat, energy	joule	J; Nm
power	watt	W; J/s (1 W = 1 J/s)
specific heat	joule per kilogram per degree kelvin	J/kg K
enthalpy	joule per kilogram	J/kg
entropy	joule per degree kelvin	J/K
rate of heat flow	watt	W
heat flux rate	watt per square metre	W/m^2
thermal conductivity	watt per metre per degree kelvin	W/mK
heat transfer coefficient	watt per square metre per degree kelvin	W/m^2K
radiation factor	watt per square metre per degree kelvin in the fourth power	W/m^2K^4
Frequency	hertz	Hz = c/s
Dynamic viscosity	pascal-second or newton per second per square metre	P_a s; Ns/m^2
Kinematic viscosity	square metre per second	m^2/s

Note **Pressure is expressed in pascal, N/m^2 or bar. 1 bar = 10^5 N/m^2, 1 atm. pressure = 101.3 kN/m^2 = 1.013 bar.**

Table A 3 Multiples and sub-multiples

tetra	T	10^{12}	milli	m	10^{-3}
giga	G	10^9	micro	μ	10^{-6}
mega	M	10^6	nano	n	10^{-9}
kilo	k	10^3	pico	p	10^{-12}
deci	d	10^{-1}	femto	f	10^{-15}
centi	c	10^{-2}	atio	a	10^{-18}

Conversion Factors

A4 Unit Conversion Factor, g_c

Table A4 Table shows the magnitude and unit of g_c (which is unit conversion factor)

System	Mass (*M*)	Length (*L*)	Time (θ)	Force (*F*)	g_c
C.G.S.	gm	cm		dyne	1.0 gm cm/dyne s^2
S.I.	kg	m		newton	1.0 kg . m/newton s^2
M.K.S.	kg	m		kgf	9.806 kg m/kgf s^2
F.P.S.	1 b_m	ft		lbf	32.174 1 bm.ft/lbf s^2
	slug	ft		lbf	1.0 slug ft/lbf s^2
	1 b_m	ft		poundal	1.0 1 bm ft/poundal s^2

Table A5 M.K.S. and S.I. units

Quantity to be measured	M.K.S.	S.I.	Conversion
Length L.	m	m	
Mass m	kg	kg	
Time θ	s	s	
Temperature	°C or K	K	
Plane angle	–	–	
Area *A*	m^2	m^2	
Volume *V*	m^3	m^3	
Density ρ	kg/m^2	kg/m^3	
Velocity *C*	m/s	m/s	
Acceleration *a*	m/s^2	m/s^2	
Angular acceleration	rad/s^2	rad/s^2	
Force *F*	kgf	newton (N)	1 kgf = 9.806 N
Pressure *p*	kgf/cm^2	Pa or N/m^2	1 kgf/cm^2 = 9.806 $\times 10^4$ $N/m^2 \approx 10^5$ N/m^2
Specific volume *v*	m^3/kg	m^3/kg	
Work *W*	kgf. m	Joules (Nm)	
Energy *E*	kgf. m	Joules {Nm)	1 kcal = 4.1868 kJ
Heat *Q*	kcal	Joules (Nm)	1 kgf.m/s = 9.806 W
Power *P*	kgf. m/s or HP = 75 kgf.m/s	Watt (w) 1 W = 1 J/s watt (W) 1 W = 1 J/s	= 9.806 J/s 1 H.P. = 735.3 W

Contd.

Table A4 Contd.

Dynamic viscosity μ	kg/m.s	Ns/m^2	
Kinematic viscosity *v*	m^2/s	m^2/s	
Weight *W*	kgm = kgf on the earth	Newton (N) In fact in SI Unit weight term is not used but mass is used	
Specific work	kgf m/kg	J/kg	1 kgf.m/kg = 9.806 J/kg
Specific heat *c*	kcal/kg K	J/kg K	1 kcal/kgK = 4186.8 J/kg K = 4.1868 kJ/kg K
Entropy	kcal/K	J/K	1 kcal/K = 4.1865 kJ/K.

Table A 6 Conversion factors

		LENGTH		
1 cm	= 03937 in		1 in	= 2.54 cm
1 m	= 3.28 ft		1 ft	= 0.3048 m
1 km	= 0.621 mile		1 mile	= 1.609 km
		AREA		
1 sq. cm	= 0 155 sq. in		1 sq. in	= 6.452 sq. cm
1 sq. m	= 10.764 sq. ft.		1 sq. ft	= 0.0929 sq m.
		VOLUME		
1 cc	= 0.063 cu in		1 cu in	= 16.39 cc
1 cc m	= 35.31 cu ft.		1 cu. ft.	= 0.0283 cu. m
1 litre	= 0.22 Imperial gallon		1 Imperial gallon	= 4.546 litres
1 litre	= 0.2642 US gallon		1 US gallon	= 3.785 liters
		MASS		
1 kg	= 2.204 lbm		1 lbm	= 0.4536 kg
1 tonne	= 0.984 tonne		1 tonne	= 1.01605 tonne
		FORCE		
1 newton	= 10^5 N/m^2		1 dyne	= 10^{-5} newton
1 newton	= 0.1019 kgf		1 kgf	= 9.8066 newton
1 newton	= 0.2248 lbf		1 lbf	= 4.4482 newton
		PRESSURE		
1 bar	= 10^5 N/m^2		1 Pa	= 1 N/m^2
1 bar	= 0.1 MPa		1 $dyne/cm^2$	= 0.1 N/m^2
1 bar	= 1.0197 kgf/cm^2		1 kgf/cm^2	= 0.1 N/m^2
1 bar	= 0.9869 atm		1 atm	= 1.01325 bar

Contd.

Table A6 Contd.

1 bar	= 14.504 Ibf/m^2		1 Ibf/m^2	= 6894.6 N/m^2
1 bar	= 750.06 mm Hg		1 mm Hg	= 133.32 N/m^2
1 bar	= 10.197 m H_2O		1 mm H_2O	= 9.8066 N/m^2
		ENERGY		
1 joule	= 1 W.s. = 1 Nm = 10^7 ergs		1 dyne-cm	= 10^7 ergs
1 kJ	= 0.2388 kcal		1 kcal	= 4.1868 kJ
1 kJ	= 0.2777 Wh		1 kWh	= 3600 kJ
1 kJ	= 101.97 m kgf		1 m kgf	= 9.8066 joules
1 kg	= 0.9478 Btu		1 Btu	= 1.055 kJ
		SPECIFIC ENERGY		
1 kJ/kg	= 0.2388 kcal/kg		1 kcal/kg	= 4.1868 kJ/kg
1 kJ/kg	= 0.4299 But.lbm		1 But/Ibm	= 2.326 kJ/kg
1 kJ/kg	= 101.97 m kgf/kg		1 m kgf/kg	= 9.8066 kJ/kg
		POWER		
1 kW	= 1 kJ/s = 0.2388 kcal/s		1 kcal/s	= 101.97 m kgf/s
1 kW	= 1.36 metric hp		1 metric hp	= 75 m kgf/s
1 kW	= 1.34 British hp		1 British hp	= 550 ft lbf/s
1 metric hp	= 736 watts		1 British hp	= 746 watts
1 watt	= 0.8598 kcal/h		1 watt	= 3.4121 Btu/h
		SPECIFIC VOLUME		
1 m^3/kg	= 1000 liters/kg		1 litre/kg	= 1 cc/gm
1 m^3/kg	= 16.018 ft^3/lbm		1 ft^3/lb	= 0.0624 m^3/kg
		SPECIFIC HEAT		
1 kJ/kg K	= 0.2388 kcal/kg K = 0.2388 Btu/lbm R			
1 Btu/lbm R	= 1 kcal/kg K = 4.1868 kJ/kg K			
		DYNAMIC VISCOSITY		
1 kg/ms	= 10 gm/cm s		1 poise	= 1 gm/cm s
1 kg/ms	= 10 poise		1 Ns/m^2	= 1 Pa.s
1 kg/ms	= 1 Ns/m^2		1 kgf /ms	= 9.8066 kg/ms
1 kg/ms	= 0.10197 kg/s/m^2		1 lbm/ft s	= 1.4881 kg/ms
1 kg/ms	= 0.6719 Ibf/ft s		1 lbm/s/ft^2	= 47.8812 kg/ms
		KINEMATIC VISCOSITY		
1 m^2/s	= 10^4 stokes		1 stoke	= 1 cm^2/s

Contd.

Table A6 Contd.

1 m^2/s	= 10.7639 ft^2/s		1 ft^2/s	= 0.0929 m^2s
THERMAL CONDUCIVITY				
1 W/mK	= 0.5898 kcal/hr mK		1 kcal/hr mK	= 1.1629 W/mK
1 W/mK	0.5778 But/hr ft *R*		1 But/hr ft R	= 1.4881 kcal/mK
HEAT FLUX PER UNIT AREA				
1 W/m^2	= 0.8598 $kcal/m^2$ h K		1 But/ft^2 h R	= 4.882 6 $kcal/m^2$ h K
	HEAT TRANSFER COEFFICIENT			
1 W/m^2 K	= 0.8598 $kcal/m^2$ h K		1 But/ft^2 h R	= 4.882 6 $kcal/m^2$ h K
REFRIGERATOR EFFECT				
1 tonne of refrigeration = 50 kcal/min = 3.489 kW				
1 kW = 1 kJ/s = 0.2866 tonnes of refrigeration				
ENTROPY				
1 kcal/K = 4.1868 kJ/K 1 kcal/kg K = 4.1868 kJ/kg K				
PHYSICAL CONSTANTS				
		Universal Gas Constant R_0		
R_0 = 8.314 kJ/kg mol K		= 1.986 But/lbmol *R* = 1.986 kcal/kg mol K		
		= 1545 ft. lbf/lbmol *P* = 848 m kg/kg mol K		

Quantity	Value and Unit
Avogadro's number, N	$= 6.02252 \times 10^{-26}$ kg mol^{-1}
Planck's constant, h	$= 6.625 \times 10^{-31}$ Js
Stefan-Boltzmann constant, σ	$= 5.668 \times 10^{-8}$ watt $metre^{-2}$ K^{-4} $= 1.355 \times 10^{-12}$ cal sec^{-1} cm^2 K^{-4} $= 0.1712 \times 10^{-8}$ But hr^{-1} ft^{-2} R^{-4}
Wein displacement constant	$= 2.8978 \times 10^{-3}$ mK
Boltzmann's constant, K	$= 1.38054 \times 10^{-23}$ joule K^{-1}
Velocity of light, C	$= 2.99793 \times 10^{10}$ cm s^{-1}
Electronic charge, |c|	$= 4.80286 \times 10^{-10}$ esu (State Coulomb) = 1.60206 $\times 10^{-19}$ Coulomb
Electron rest mass, m_e	$= 9.1083 \times 10^{-28}$ gm
Ratio of proton mass to electron mass, m_n/m_e	= 1836.12
Joule's constant, J	$= 4.1840 \times 10^{-7}$ erg cal^{-1} = 4.1840 joule cal^{-1}

Thermodynamic Properties

Table A7 Thermodynamic properties of saturated moist air (76 cm Hg)

Temperature °C	Specific volume (v) m^3/kg	Enthalpy (h) kJ/kg of dry air	Specific humidity (ω) kg/kg of dry air
- 10	0.7466	- 6.0959	1.597
- 5	0.7622	1.138	2.474
0	0.7778	9.445	3.771
5	0.7947	18.585	5.400
6	0.7980	20.594	5.793
7	0.8014	22.654	6.210
8	0.8048	24.777	6.654
9	0.8082	27.00	7.127
10	0.8115	29.292	7.629
11	0.8189	31.657	9.33
12	0.8189	31.097	8.729
13	0.8223	36.657	9.333
14	0.8262	39.284	9.071
15	0.8296	42.035	10.65
16	0.8353	44.84	11.37
17	0.8376	47.813	12.13
18	0.8416	49.911	12.94
19	0.8456	54.093	13.79
20	0.8496	57.442	14.70
21	0.8541	60.917	15.66
22	0.8580	64.518	16.68
23	0.8626	68.328	17.76
24	0.8670	72.222	18.89
25	0.8715	76.325	20.09
26	0.8759	80.637	21.37
27	0.8811	85.117	22.72
28	0.8858	89.806	24.14
29	0.8908	94.705	25.65
30	0.8958	99.771	27.23

Table A8 Thermo-chemical properties of selected substances at 298 K and 1 atm

Substance	Formula	Molar mass M (kg/kg mol)	Ethalpy of formation $\bar{h}_f^0$ (kJ/kg mol)	Gibbs function of formation $\bar{g}°_f$ (kJ/kg mol)	Absolute entropy $\bar{s}°$ (kJ/kg mol K)	Heating values higher HHV (kJ/kg)	lower LHV (kJ/kg)
Carbon	C(s)	12.01	0	0	5.74	32,770	32,770
Hydrogen	H_2(g)	2.016	0	0	130.57	141,780	119,950
Nitrogen	N_2(g)	28.01	0	0	191.50	–	–
Oxygen	O_2(g)	32.00	0	0	205.03	–	–
Carbon monoxide	CO(g)	28.01	- 110,530	- 137, 150	197.74	–	–
Carbon dioxide	CO_2(g)	44.01	- 393,520	- 394,380	213.69	–	–
Water	H_2O(g)	18.02	- 241,820	- 228,590	188.72	–	–
Water	H_2O(l)	18.02	- 235,830	- 237,180	69.95	–	–
Hyd. peroxide	H_2O_2(g)	34.02	- 136,310	- 105,600	232.63	–	
Ammonia	NH_3(g)	17.03	- 46,190	- 16,590	192.33	–	–
Oxygen	O(g)	16.00	249,170	231,770	160.95	–	–
Hydrogen	H(g)	1.008	218,000	203,290	114.61	–	–
Nitrogen	N(g)	14.01	472,680	455,510	153 19	–	–
Hydroxyl	OH(g)	17.01	39,460	34,280	183.75	–	–
Methane	CH_4(g)	16.04	- 74,850	- 50,790	186.16	55,510	50,020
Acetylene	C_2H_2(g)	26.04	226,730	209.170	200.85	49.910	48,220
Ethylene	C_2H_2(g)	28.05	52,280	68,120	219.83	50,300	47,160
Ethane	C_2H_6(g)	30.07	{ 84,680	- 32,890	229.49	51,870	47,480
Propylene	C_3H_6(g)	42.08	20.410	62,720	266.94	48,920	45.780
Propane	C_3H_8(g)	44.09	- 103,850	- 23,490	269.91	50,350	46,360
Butane	C_4H_{10}(g)	58.12	- 126,150	- 15,710	310.03	49.500	45,720
Pentane	C_5H_{12}(g)	72.15	- 146,440	- 8,200	348.40	49,010	45.350
Octane	C_8H_{18}(g)	114.22	- 208,450	17,320	463,67	48,260	44,790
Octane	C_8H_{18}(l)	114.22	- 249,910	6,610	360.79	47.900	44,430
Benzene	C_6H_6(g)	78.11	82,930	129,660	269.20	42.270	40.580
Methyl alcohol	CH_3OH(g)	32.04	- 200,890	- 162,140	23170	23,850	21,110

Contd.

Table A8 Contd.

Methyl alcohol		32.04	- 238,810	- 166,290	126.80	22.670	19,920
Ethyl alcohol		46.07	- 235,310	- 168,570	282.59	30,590	27,720
Ethyl alcohol		46.07	- 277,690	174,890	160.70	29,670	26,800

Source Based on JANAF Thermo-chemical Tables, NSRDS-NBS-37, 1971; Selected Values of Chemical Thermodynamic Properties, NBS Tech. Note 270-3, 1968; and API Research Project 44, Carnegie Press, 1953.

Note : Table A9 is given on page 1046 (next page)

Table A10 Enthaply of combustion ($-\bar{h}_{rp} = -\Delta h_R$ at 25°C and 1 atm)

Substance	Symbol	H_2O and $CO_2(g)$ kJ/mg mole	$H_2O(g)$ and $CO_2(g)$ kJ/mg mole
Acetylene	C_2H_2 (g)	1300.5	1256.4
Benzene	C_6H_6(g)	3303.5	3171.5
n-Butane	C_4H_{10}(g)	2870.6	2658.8
Isobutane	C_4H_{10}(g)	3870.6	2650.4
I-Butene	C_4H_8(g)	2719	2543
Carbon	C (graphite)	393.77	–
Carbon monoxide	CO (g)	283.1	–
n-Decane	$C_{10}H_{22}$ (g)	6834.2	6350.6
n-Dodecane	$C_{12}H_{26}$ (g)	8152.6	7580
Ethane	C_2H_6(g)	1560.9	1428.7
Ethylene	C_2H_4(g)	1411.9	1323.8
n-Heptane	C_7H_{16}(g)	4856.7	4505.3
n-Hexane	C_6H_{14}(g)	4197.56	3889.2
Hydrogen	H_2(g)	286	240.9
Methane	CH_4 (g)	890.9	802.8
n-Nohane	C_9H_{22} (g)	6175	5734.9
n-Octane	C_8H_{18}(g)	5515.8	5119.4
n-Pentane	C_5H_{12}(g)	3538.5	3274.2
Isopentane	C_5H_{12}(g)	3530.4	3266.2
Propane	C_3H_8(g)	2221.5	2045.37
Propylane	C_3H_6(g)	2059.8	1927.7

Table A 9 Ideal gas properties of selected gases

T (K), $\bar{h}$ and $\bar{u}$ (kJ/kg mole), $\bar{s}°$ (kJ/k mol K)																
T	Carbon dioxide CO_2 ($\bar{h}_f^0$ = - 393,520 kJ/kg mole)			Carbon monoxide, CO ($\bar{h}_f^0$ = - 110,530 kJ/kg mole)			Water vapour, H_2O ($\bar{h}_f^0$ = - 241,820 kJ/kg mole)			Oxygen, O_2 ($\bar{h}_f^0$ = 0 kJ/kg mole)			Nitrogen, N_2 ($\bar{h}_f^0$ = 0 kJ/kg mole)			T
	$\bar{h}$	$\bar{u}$	$\bar{s}°$	$\bar{h}$	$\bar{u}$	$\bar{s}°$	$\bar{h}$	$\bar{u}$	$\bar{s}°$	$\bar{h}$	$\bar{u}$	$\bar{s}°$	$\bar{h}$	$\bar{u}$	$\bar{s}°$	
0	0	0	0	0	0	0	0	0	0	0	0	0	0	0	0	0
250	7,627	5,548	207.337	7,266	5,188	192.411	8.294	6.215	182.831	7.275	5,197	199.885	7,266	5,188	186.370	250
298	9,364	6,8851	213.685	8,669	6,190	197.543	9,904	7,425	188.720	8,682	6,203	205.033	8,669	6,190	191.502	298
300	9,431	6,939	213.915	8,723	6,229	197.723	9,966	7,472	188.928	8,736	6,242	205.213	8,723	6,229	191.682	300
400	13,372	10,046	225.225	11,644	8,319	206.125	13,356	10,030	198.673	11,711	8,384	213.765	11,640	8,314	200.071	400
500	17,678	13,521	234.814	14,600	10,443	212.719	16,828	12,671	206.413	14,770	10,614	220.539	14,581	10,4.23	206 630	500
600	22,280	17,291	243.199	17,611	12,622	218.204	20,402	15,413	212.920	17,929	12,940	226.346	17,563	12,574	212.066	600
700	27,125	21,305	250.663	20,690	14,870	222.953	24,088	18.268	218.1610	21,184	15,364	231.358	20,604	14,784	216.756	700
800	32,179	25,527	257.408	23,844	17.193	227.162	27,896	21,245	223.693	24,523	17,872	235.810	23,714	17,061	220.907	800
900	37,405	29,922	263.559	27,066	19,583	230.957	31,828	24,345	223.321	27,928	20,445	239.823	26,890	19,407	224.647	900
1000	42,769	31,455	269.215	30,355	22,41	234.121	35,882	27.568	232.597	31,389	23,075	243.471	31,129	21,815	228.127	1000
1100	48,258	39,112	274.445	33.702	24,557	237.609	40,071	30.925	236.584	34,899	25,753	246.818	33,426	24,280	231 199	1100
1200	53,848	43,871	279.307	37,095	27,118	240.663	44,380	34,403	240.333	38,447	28,469	249.906	36,777	26,799	134.115	1200
1300	59,522	48,713	283.847	40,534	29,725	243.316	48,807	38,00	243.877	42,033	31,224	252.776	40,170	29,361	236.831	1300
1400	65,271	53,631	288.106	44,007	32,367	245.889	53,351	41,711	247.241	45,648	34,008	255.454	43,605	31,964	239.375	1400
1500	71,078	58,606	292.114	47,5 17	35.046	248.312	57,999	45.528	250.450	49,024	36,821	258.965	47,073	34,601	241.768	1500
1600	76,944	63,741	295.901	51,53	37,750	250.592	62,748	49,445	253.512	52,961	39,658	260.333	50,571	37,268	244.028	1600
1700	82,856	68,721	299.482	54,609	40,474	252.751	67,589	53,455	256.450	56,652	42,517	262.571	54,099	39,965	246.165	1700

Contd.

Table A9 Contd.

1800	88,806	73,840	302.884	58,191	43,225	254.797	72,513	57,547	259.262	60,371	45,405	264.701	57,651	42,685	248.195	1800
1900	94,793	78,996	306.112	61,794	45,997	256.746	77,517	61,720	261.969	64,116	48,319	266.722	61,220	45,423	250.128	1900
2000	100,804	84,185	309.210	65,408	45,780	258.600	82,593	65,965	264.571	67,881	51,253	268.655	64,810	48,181	251.969	2000
2100	106,864	89,404	312.160	69,044	51,584	260.370	87,735	70,275	267.081	71,668	54,208	270.504	68,417	50,957	253.726	2100
2200	112,939	94,648	314.988	72,688	54,396	262.065	92,940	74,649	269.500	75,484	57,192	272.278	72,040	53,49	255.412	2200
2300	119,035	99,912	31.695	76,345	57,222	263.692	98,199	79,076	271.839	79,316	60,193	273.981	75,676	56,553	257.027	2300
2400	125,152	105,197	320.302	80,015	60,060	265.253	103,508	83,553	274.098	83,174	63,219	275.625	79,320	59,366	258.580	2400
2500	131,290	110,504	322.808	83,692	62,906	266.755	108,868	88,082	276.286	87,057	66,271	277.207	82.981	62,195	260.073	2500
2600	137,449	115,832	325.222	87,383	65,766	268.202	114,273	92,656	278.407	90,956	69,339	278.738	86,650	65,033	261:512	2600
2700	143,620	121,172	327.549	91,077	68,628	269.596	119,717	97,269	280.462	94.881	72,433	280.219	90,328	67,880	262.902	2700

Table A11 Variation of $\bar{c}_p$ with temperature for selected ideal gases

$\frac{c_p}{R} = \alpha + \beta T + \gamma T^2 + \delta T^3 + \varepsilon T^4$ T is in K, equations valid from 300 to 1000 K					
Gas	α	$\beta \times 10^3$	$\gamma \times 10^6$	$\gamma \times 10^9$	$\varepsilon \times 10^{12}$
CO	3.710	- 1.619	3.692	- 2.032	0.240
CO_2	2.401	8.735	- 6.607	2.002	0
H_2	3.057	2.677	- 5.810	5.521	- 1.812
H_2O	4.070	- 1.108	4.152	- 2.964	0.807
O_2	3.626	- 1.878	7.055	- 6.764	2.156
N_2	3.675	- 1.208	2.324	- 0.632	- 0.226
Air	3.653	- 1.337	3.294	- 1.913	0.2763
SO_2	3.267	5.324	0.684	- 5.281	2.559
CH_4	3.826	- 3.979	24.558	- 22 733	6.963
C_2H_2	1.410	19.057	- 24.501	16.391	- 4.135
C_2H_4	1.426	11.383	7.989	- 16.254	6.749

Table A12 Ref : p_0 = 1 atm T_0 = 0 K thermodynamic properties of air at low pressure

T_r K	h_r kJ/kg	pr	u_r kJ/kg	v_r	$s°$ kJ/kg K
100	99.76	0.02990	71.06	2230	1.4143
130	129.81	0.07474	92.51	1159.8	1.6773
160	159.87	0.154.31	113.95	691.4	1.8853
190	189.94	0.28114	135.40	450.6	2.0575
220	219.99	0.4690	156.84	312.8	2.2043
250	250.05	0.7329	178.29	227.45	2.3325
280	280.14	1.889	199.78	171.45	2.4461
310	310.24	1.5546	221.27	132.96	2.5483
340	340.43	2.149	242.86	105.51	2.6412
370	370.67	2.892	264.47	85.31	2.7264
400	400.98	3.806	286.19	70.07	0.8052
430	431.43	4.915	308.03	58.34	2.8786
460	462.01	6.245	329.99	49.11	2.9473
490	492.74	7.824	352.11	41.76	3.0120
520	523.63	9.684	374.39	35.80	3.0733
550	554.75	11.858	396.89	30.92	3.1314

Contd.

Table A12 Contd.

580	586.04	14.380	419.56	26.89	3.1869
610	617.53	17.297	442.43	23.51	3.2397
640	649.21	20.64	465.55	20.674	3.2905
670	681.15	24.46	488.88	18.266	3.3392
700	713.27	28.80	512.37	16.205	3.3861
730	745.62	33.72	536.12	14.434	3.4314
770	789.10	41.27	568.10	12.440	3.4894
800	821.94	47.75	592.34	11.172	3.5312
830	855.01	55.00	616.79	10.062	3.5718
860	888.28	63.09	641.44	9.090	3.6111
890	924.75	72.08	666.31	8.233	3.6493
920	955.38	80.05	691.33	7.476	3.6865
950	989.22	93.08	716.57	6.805	3.7227
980	1023.25	105.24	741.99	6.209	3.7580
1040	1091.85	133.34	793.35	5.201	3.8259
1200	1277.79	238.0	933.40	3.362	3.9922
1220	1301.33	254.7	951.19	3.194	4.0117
1260	1348.55	290.8	986.92	2.889	4.0497
1280	1372.25	310.4	1004.88	2.750	4.0684
1320	1419.77	352.5	1040.93	2.497	4.1049
1360	1467.50	399.1	1077.17	2.272	4.1406
1400	1515.41	450.5	1113.62	2.072	4.1753
1440	1563.49	506.9	1150.23	1.8942	4.2092
1480	1611.80	568.8	1187.03	1.7350	4.2422
1500	1635.99	601.9	1205.47	1.6617	4.2585

* Adapted from Table 1 in Gas Tables, by Joseph H. Keenan and Joseph Kaye. Copyright 1948, by Joseph H. Keenan and Joseph Kaye. Published by John Wiley & Sons, Inc., New York.

Table A13 Isentropic flow

M	M^*	$\frac{A}{A^*}$	$\frac{p}{p_0}$	$\frac{\rho}{\rho_0}$	$\frac{T}{T_0}$
0	0	∞	1.00000	1.00000	0.10
0.10	0.10943	5.8218	0.99303	0.99502	0.99800
0.20	0.21822	2.9635	0.97250	0.98027	0.99206
0.30	0.32572	2.0351	0.93947	0.95638	0.98232
0.40	0.43133	1.5901	0.89562	0.92428	0.96899

Contd.

Table A13 Contd.

0.50	0.53452	1.3398	0.84302	0.88517	0.95238
0.60	0.63480	1.1882	0.78400	0.84045	0.93284
0.70	0.73179	1.09437	0.72092	0.79158	0.91075
0.80	0.82514	1.03823	0.65602	0.7400	0.88652
0.90	0.91460	1.00886	0.59126	0.68704	0.86058
1.00	1.00000	1.00000	0.52828	0.63394	0.83333
1.10	1.08124	1.00793	0.46835	0.58169	0.80515
1.20	1.1583	1.03044	0.41238	0.53114	0.77640
1.30	1.2311	1.06631	0.36092	0.48291	0.74738
1.40	1.2999	1.1149	0.3424	0.43742	0.74738
1.50	1.3646	1.1762	0.27240	0.39498	0.68965
1.60	1.4254	1.2502	0.23527	0.35573	0.66138
1.70	1.4825	1.3376	0.20259	0.31969	0.63372
1.80	1.5360	1.4390	0.17404	0.28682	0.60680
1.90	1.5861	1.5552	0.14924	0.25699	0.58072
2.00	1.6330	1.6875	0.12780	0.23005	0.55556
2.10	1.6769	1.8369	0.10935	0.20580	0.53135
2.20	1.7179	2.0050	0.09352	0.18405	0.50813
2.30	1.7563	2.1931	0.07997	0.16458	0.48591
2.40	1.7922	2.4031	0.06840	0.14720	0.46468
2.50	1.8258	2.6367	0.05853	0.13169	0.44444
2.60	1.8572	2.8960	0.05012	0.11787	0.42517
2.70	1.8865	3.1830	0.04295	0.10557	0.40684
2.80	1.9140	3.5001	0.03685	0.09462	0.38941
2.90	1.9398	3.8498	0.03165	0.08489	0.37286
3.00	1.9640	4.2346	0.02722	0.07623	0.35714
3.50	2.0642	6.7896	0.01311	0.04523	0.28986
4.00	2.1381	10.719	0.00658	0.02766	0.23810
4.50	2.1936	16.562	0.00346	0.01745	0.19802
5.00	2.2361	25.000	$189(10)^{-5}$	0.01134	0.16667
6.00	2.2953	53.180	$633(10)^{-6}$	0.00519	0.12195
7.00	2.3333	104.143	$242(10)^{-6}$	0.00261	0.09259
9.00	2.3772	327.189	$474(10)^{-7}$	0.000815	0.04762
10.00	2.3904	535.938	$236(10)^{-7}$	0.000495	0.04762
∞	2.4495	∞	0	0	0

Table A14 Atomic or molecular weight and critical properties of selected elements and compounds

Substance	Chemical formula	M (kg/kg mole)	T_c (K)	p_c (bar)	$Z_c = \frac{p_v v_c}{RT_c}$
Acetylene	C_2H_2	26.04	309	62.8	0.274
Air (equivalent)	–	28.97	133	37.7	0.284
Ammonia	NH_3	17.03	406	112.8	0.242
Argon	Ar	39.94	151	48.6	0.290
Benzene	C_6H_6	78.11	563	49.3	0.274
Butane	C_4H_{10}	58.12	425	38.0	0.274
Carbon	C	12.01	–	–	–
Carbon dioxide	CO_2	44.01	304	73.9	0.276
Carbon monoxide	CO	28.01	133	35.0	0.294
Copper	Cu	63.54	–	–	–
Ethane	C_2H_6	30.07	305	48.8	0.285
Ethyl alcohol	C_2H_5OH	46.07	516	63.8	0.249
Ethylene	C_2H_4	28.05	283	51.2	0.270
Helium	He	4.003	5.2	2.3	0.300
Hydrogen	H_2	2.016	33.2	13.0	0.304
Methane	CH_4	15.04	191	46.4	0.290
Methyl alcohol	CH_3OH	32.04	513	79.5	0.219
Nitrogen	N_2	28.01	126	33.9	0.291
Octane	C_8H_{18}	114.22	569	24.9	0.258
Oxygen	O_2	32.00	154	50.5	0.290
Propane	C_3H_8	44.09	370	42.7	0.276
Propylene	C_3H_6	42.08	365	46.2	0.276
Refrigerant 12	CCl_2F_2	120.92	385	41.2	0.278
Refrigerant 22	$CHClF_2$	86.48	369	49.8	0.267
Refrigerant 134a	CF_3CH_2F	102.03	374	40.7	0.260
Sulphur dioxide	SO_2	64.06	431	78.7	0.268
Water	H_2O	18.02	647.3	220.9	0.233

Sources Adapted from International Critical Tables and L.C Nelson and E.F. Obert, Generalized Compressibility Charts, Chem. Engg., 61 : 203 (1954).

Table A15 Properties of selected solids and liquids : c_p, ρ, and k

Substance	Specific heat, c_p (kJ/kg K)	Density, ρ (kg/m^2)	Thermal conductivity, k (W/m . K)
Selected Solids, 300K			
Aluminium	0.903	2700	237
Coal, anthracite	1.260	1350	0.26
Copper	0.385	8930	401
Granite	0.775	2630	2.79
Iron	0.447	7870	80.2
Lead	0.129	11300	35.3
Sand	0.800	1520	0.27
Silver	0.235	10500	429
Soil	1.840	2050	0.52
Steel (AISI 302)	0.180	8060	15.1
Tin	0.227	7310	66.6
Building Materials, 300K			
Brick, common	0.835	1920	0.72
Concrete (stone mix)	0.880	2300	1.4
Glass, plate	0.750	2500	1.4
Hardboard, siding	1.70	640	0.094
Limestone	0.810	2320	2.15
Plywood	1.220	545	0.12
Softwoods (fir, pine)	1.380	510	0.12
Insulating Materials, 300K			
Blanket (glass fibre)	–	16	0.046
Cork	1.800	120	0.039
Duct liner (glass fibre, coated)	0.835	32	0.038
Polystyrene (extruded)	1.210	55	0.027
Vermiculite fill (flakes)	0.835	80	0.068
Saturated Liquids			
Ammonia, 300K	4.818	599.8	0.465
Mercury, 300K	0.139	13529	8.540
Mercury, 300K	1.267	1183.1	0.085
Refrigerant 22, 300K	1.434	11997	0.081
Refrigerant 134a, 300K	1.909	884.1	0.145

Contd.

Table A15 Contd.

Water, 275K	4.211	999.9	0.574
300K	4.179	996.5	0.613
325K	4.182	987.1	0.645
350K	4.195	973.5	0.668
375K	4.220	956.8	0.681
400K	4.256	937.4	0.688

Source Drawn from several sources, these data are only representative. Values can vary depending on temperature, purity, moisture content and other factors.

Note: Properties of Steam and Refrigerants are not given in Appendix as they are easily available in the form of Booklets and for study and appearing in tests, students procure these.

References

1. **Angrist, S.W.,** "*Direct Energy Conversion*" Allyn and Bacon, Inc." Boston, 1965.
2. **Bruges, E.A.,** "*Available Energy and Second Law Analysis*", Academic Press, 1959.
3. Calibration of Temperature Measuring Instruments, NPL, Notes Applied Science, No. 12 (HMSO, 1964).
4. **Callen, H.B.,**"*Thermodynamics*" John Wiley and Sons, Inc., New York, 1960.
5. **Chang, S.L.**, "*Energy Conversion*", Prentice Hall, Inc., Englewood, Cliffs. N.J. 1963.
6. **Doolittle, J.S.**, "*Thermodynamics for Engineers*", 2nd ed., McGraw-Hill Book Co., New York 1961.
7. **Eastop, T.D. and A. McConky**, "*Applied Thermodynamics for Engineering Technologists*", 3rd ed., Longman, 1978.
8. **Faires, V.M.**, "*Thermodynamics*", 4th ed., The Macmillan Company, New York, 1962.
9. **Fast J.D.,** "*Entropy*", McGraw Hill Book Company, New York, 1963.
10. **Giedt Warren**, "*Thermophysics*", VNR, 1971.
11. **Harry, A. Sorensen**, "*Principles of Thermodynamics*", Amerind Publishing Co., 1972.
12. **Hatsopoulos, G.N., and J.H. Kennen**, "*Principles of General Thermodynamics*", John Wiley & Sons, New York, 1965.
13. **Hodge, J**.,"*Cycles and Performance Estimations.*", Butterworths, 1955.
14. **Holman, J.P.,** "*Thermodynamics*" 3rd ed., McGraw-Hill Book Co., New York, 1988.
15. **Horlock, J.H.**, "Combind Cycle Plants, Pergaman Press. 1992.
16. **Jones E. Ahern,** "*The Exergy Method of Energy Systems Analysis*", John Wiley & Sons, 1980.
17. **Jones J.B. and G.A. Hawkins**, "*Engineering Thermodynamics*", John Wiley & Sons, Inc., New York. I960.
18. **Knuth, E.S.**, "*Introduction to Statistical Thermodynamics*", McGraw-Hill Book Co., New York, 1966.

19. **Lewis, G.N. and M. Randall**, "*Thermodynamics*". McGraw-Hill Book Co., New York, 1966.
20. **Mayhew, Y.R., and Rogers G.F.C**. "*Thermodynamics and Transport Properties of Fluids*". S.I. Units, Basil Blackwell, 1967.
21. **Moran, M.J. and Shapiro, H.N.**, "*Fundamentals of Engineering Thermodynamics*", 4th ed. John Wiley & Sons. 2000.
22. **Obcrt, F.E.,**"*Concepts of Thermodynamics*". McGraw-Hill Book Co., New York, 1960.
23. **Paolino, M.A. and Burghardt, M.D.**, "*Energy Conservation and Second Law Efficiency*", Journal of Engg.; for Power. Trans. ASME, Jan. 1982, Vol. 104, p. 241.
24. **Present, R.D.**, "*Kinetic Theory of Gases*", McGraw-Hill Book Co., New York, 1958.
25. **Reynolds. W.C.**, "*Thermodynamics*" McGraw-Hill Book Co., New York, 1969.
26. **Rogers, G.F.C., and Y.R. Mayhew**, "*Engineering Thermodynamics, Work and Heat Transfer*", 2nd ed., Longman, 1967.
27. **Sad**, "*Engineering Thermodynamics*", McGraw-Hill Book, Co. New York.
28. **Sears, F.W.**, "*Thermodynamics*", 2nd ed., McGraw-Hill Book Co., New York. 1971.
29. **Sonntag, R.E., Borgnakke, C., and Van Wylen G.J.**, "*Fundamental of Thermodynamics*", John Wiley & Sons, Inc., New York. 5th ed.; 2000.
30. **Spalding, D.B. and Cole, E.H.**, "*Engineering Thermodynamics*". Edward Arnold, 1958.
31. **Sutton, G.W.**, "*Direct Energy Conversion*", McGraw-Hill Book Co., New York, 1966.
32. **Wark, Wenneth**, "*Thermodynamics*". 2d ed., McGraw-Hill Book Co., New York. 1971.
33. **Bannister, R.L., Newby, R.A. and Yang, W.C.**, "Final Report on the Development of Hydrogen Fueled Combustion Turbine Cycle for Power Generation, Journal of Engg. for Gas Turbine and Power, Trans. of ASME, Vol. 121; 1999, pp. 38–45.
34. **Yadav, R.; Yadav, N.P. and Saraswati Samir**; "*Thermodynamic Evaluation of Hydrogen Fueled Combustion Turbine Cycle for Power Generation*", Proceedings of International Conference on Energy and Environment, Dec 11–13, 2003, Shanghai, China.
35. **Zemansky, M.W. and H.C.**, "*Basic Engineering Thermodynamics*", McGraw-Hill Book Co.. New York, 1966.
36. **C.P. Arora,** "*Thermodynamics*", Tata McGraw-Hill, New Delhi, 1998.
37. **P.K. Nag,** "*Engineering Thermodynamics*", 3rd edition, 2005, The McGraw-Hill companies, New Delhi.
38. **Yunus A. Cengel and Michael A. Boles**, "*Thermodynamics – An Engineering Approach*", McGraw Hill Education (India) Pvt. Ltd., 2017, Chennai, TN, (India).

Index

E

I

J

K

L

M

U

V

W

Z